Limnology

SECOND EDITION

Robert G. Wetzel

Michigan State University

SAUNDERS COLLEGE PUBLISHING

Harcourt Brace College Publishers
Fort Worth   Philadelphia   San Diego
New York   Orlando   Austin   San Antonio
Toronto   Montreal   London   Sydney   Tokyo

Text Typeface: Melior
Compositor: University Graphics, Inc.
Acquisitions Editor: Michael Brown
Project Editors: Carol Field and Maryanne Miller
Copy Editor: Sarah Fitz-Hugh
Managing Editor & Art Director: Richard L. Moore
Art/Design Assistant: Virginia A. Bollard
Text Design: Phoenix Studio
Cover Design: Lawrence R. Didona
Production Manager: Tim Frelick
Assistant Production Manager: Maureen Iannuzzi

Cover credit: Volvox colonies with developing colonies living in fresh-water. © James Bell

Library of Congress Cataloging in Publication Data

Wetzel, Robert G.
    Limnology.

    Bibliography: p.
    Includes index.

    1. Limnology.   I. Title.
QH96.W47 1983   551.48'2   81-53073
ISBN 0-03-057913-9

LIMNOLOGY, Second Edition   ISBN 0-03-057913-9

Printed in the United States of America.

6 7 8 9 0 1 2 016 18 17 16 15 14

To my parents, William and Eugenia (née Wagner) Wetzel, and to Theodore Jones, who took the time to inspire a fledgling at an early, critical time.

PREFACE TO THE SECOND EDITION

Use of the first edition of *Limnology* has been more widespread than had been anticipated. The text has been widely adopted in limnology courses as an updated résumé of the field; this use was my primary intent in initially undertaking the arduous task of writing it. The book has been of value as an interpretative reference work, and significant portions of *Limnology* have now been included in other limnology texts.

Numerous persons have encouraged preparation of a second edition in this rapidly advancing field of study. I have received a great number of useful suggestions for modifications, additions, and corrections for this revision. The response has been most gratifying, and I greatly appreciate the interest of so many persons.

A number of significant changes have been made in this edition. Over one-third of the original text has been completely revised, updated with improved information, or conceptually altered. Throughout the text I have attempted to simplify the often cumbersome prose.

Another major change is organizational. The initial chapters cover relatively distinct and concise topics. Although modified internally, the introductory comments and discussions of water as a substance, geomorphology, water economy relations, light, thermal properties, and hydrodynamics remain in the general format of the earlier edition. I believe that knowledge of these topics is critical for effective discussions of biogeochemical cycling, population dynamics, and metabolism. A brief discussion chapter is included on productivity and ecosystem trophic relationships, the central theme of the entire book and, indeed, of limnology in general.

Some instructors prefer to integrate biogeochemical dynamics ("ecological chemistry") after they have treated the biota and population dynamics. This approach has certain advantages and disadvantages. I have elected again in this edition to introduce the fundamental chemical dynamics with a continual integration of the biota, for microflora and macroflora predominate as metabolic effectors of biogeochemical cycling, and the subtle, complex roles of animals in these processes are still very incompletely understood.

As before, the biota dominate the book. I have, however, completely reorganized presentation of this material. The fascinatingly complex biotic interrelationships are presented in the commonly accepted, albeit artificial, divisions—planktonic, littoral, and benthic. Each of these reasonably distinct community regions has been divided into essentially producer, consumer, and decomposer groups while the primary integration of each has been maintained. The result is more but shorter topic divisions, perhaps more readily assimilable by beginning students.

I do not deviate in my belief in the fundamental importance of dissolved and particulate detritus as the overall stabilizer in the metabolism of ecosystems. Although I am still in a minority in maintaining this viewpoint, substantiating evidence is rapidly accumulating from freshwater as well as other ecosystems. The treatment of the organic carbon cycle in fresh waters remains as the integrating chapter; it is hoped that the essence of the messages is more clear than it had been previously.

The importance of paleolimnological studies in an understanding of the ontogeny of lake ecosystems is a subject that deserves and has been given separate emphasis. This treatment leads into generalizations, admittedly simplistic, of the general evolution and maturation of lakes.

In two of my original perceptions of the first edition I remain steadfast in this revision. First and foremost, I believe strongly in an order and unity in complex biological systems. Limnologists, like most ecologists, are not alone in being plagued with the common difficulty of becoming so engrossed with the detailed differences among biological systems that unifying relationships do not emerge easily. One of the most satisfying aspects of science is the aim of discerning that unity. As Bronowski (1956:27) eloquently said, "Science is nothing else than the search to discover unity in the wild variety of nature—or more exactly, in the variety of our experience."

Generalization, however, reaps damnation from many. Genetic adaptation and diversity are so great that unity is not often readily apparent at our current stage of knowledge. Simplification is valuable in an attempt to generalize the operation of freshwater ecosystems. Such attempts are especially useful at the early stages of beginning students' conceptual understanding of a subject.

Second, I remain steadfast in my opinion that limnology is not an easy subject. The beauty and fascination of the subject should not be diluted to the point where early students are given only superficial exposure. Detail must be provided to represent adequately the state of our understanding, the foundation upon which general statements are based, and to provide interested students with inroads into the plethora of literature on any given subject.

I recognize, as has been called to my attention, that integral concepts and general statements can be difficult for beginning students to separate from the many examples that I detail in the first edition of the text. Therefore, in this edition I have attempted to clarify major points by separation of exemplary detail in a different type font. Furthermore, summaries of topics are given in some detail at the end of each chapter.

Many persons have asked what the symbols at each chapter heading mean. The symbols originate from alchemy and were somewhat modernized by the artists of the publisher. My original use of them was purely for fun. Upon further inquiry, I became so thoroughly engrossed in this interesting subject that it was detracting from tasks at hand. In the end, symbols were chosen that had some relationship to the content of the chapters, occasionally stretching the relationship with tongue in cheek. A listing of the symbols, their original meaning, and my interpretation can be found at the end of the text.

The preface of a text permits one to thank the many persons who have been so generous and helpful with suggestions for improving this edition. In addition to those persons cited in the preface to the first edition, I wish to acknowledge in particular the arduous and unselfish labors of Arthur J. Stewart. Dr. Stewart patiently worked through my entire manuscript, pointing out endless awkward areas in my prose and suggesting many additional scientific studies that strengthened or altered points being made. His wife, Dr. Joyce A. Dickerman, also read through the entire manuscript and made many helpful suggestions. Dr. Robert E. Moeller read most of the manuscript, and offered many improvements as well as saving me from several erroneous or misleading statements. Dr. W. T. Edmondson gave many generous and continuing insights into reorganization of this edition, based upon his lengthy, effective teaching of limnology. I appreciate particularly the assistance of these friends and colleagues, as well as their steadfast belief in the value of my efforts to upgrade a detailed summary of limnology to serve both beginning as well as advanced students and workers in this field.

Numerous other friends and colleagues assisted with parts of the revisions by casting a critically constructive eye on statements made in a number of parts of chapters: John E. Hobbie, Dan A. Livingstone, Amelia K. Ward, David A. Culver, David A. Brakke, and David A. Francko offered widespread counsel. Others assisted in smaller but important ways: Peter Kilham, Clifford H. Mortimer, Gene E. Likens, Max Tilzer, William G. Crumpton, Robert E. Keen, Gordon L. Godshalk, Donald J. Hall, G. R. Marzolf, Robert Stallard, Kenneth G. Wood, Thomas Dunne, R. Anton Hough, John L. Confer, P. M. Jónasson, M. Søndergaard, N. A. Andresen, Kenneth A. Stewart, Alan J. Faller, W. J. Widmer, Michael F. Coveney, H. H. Hannan, David Hart, Earl E. Werner, Richard F. Losee, Michael J. Klug, and Peter A. Tyler. Other assistance was received from R. W. Pennak, P. H. Rich, G. Woloschak, J. F. Kitchell, R. C. Bubeck, P. Weiler, S. H. Smith, B. B. Jana, H. Taube, F. E. Trainor, L. R. Zúniga, D. M. Long, C. B. Reif, K. Thomasson, T. T. Macan, Charles Holt, Bruce Cowell, and L. Joseph Hendricks.

The encouragement of George H. Lauff in my research and syntheses is appreciated. Anita J. Johnson continued to patiently type the manuscript and to prepare the many figures; her uncompromising attention to detail was most helpful in bringing all to fruition. C. Hammarskjold and M. Jacobs assisted in location of my many requests for literature. My wife Carol and children continued to bear the brunt that my work purloins from family responsibilities; their understanding is appreciated more than is realized. Portions of the unpublished results cited in the text were supported by subventions by the Department of Energy and the National Science Foundation. Despite all of the help from others, the responsibility rests with me.

Robert G. Wetzel
Michigan State University
June 1982

CONTENTS

A fundamental feature of the earth is an abundance of water, which covers 71 per cent of its surface to an average depth of 3800 meters. Thus we find an immense quantity of water constituting the hydrosphere, over 99 per cent of which is deposited in ocean depressions (Table 1-1). The relatively small amounts of water that occur in freshwater lakes and rivers belie their basic importance in the maintenance of terrestrial life.

OUR FRESHWATER RESOURCES

Human growth and utilization of fresh water on a sustained exponential basis are preeminent components of any analysis of inland water resources. Man must be recognized for what he is: an animal whose population growth is in an exponential phase. In spite of its absurdity, a belief prevails that the earth's supply of finite water resources can be increased constantly to meet exponential demands. Fresh waters are a finite resource that can be increased only slightly, e.g., via desalinization, and at the present time at a tremendous energy cost. Society as a whole, and many freshwater ecologists, have tended to ignore man, and his use and misuse of freshwaters, as an influential factor in the maintenance of lake ecosystems. The issue immediately concerns resource utilization, governed by the spiraling relationships of an expanding situation in which supply is constantly made to respond to demand. The unfortunate effect of essentially uncontrolled growth is that consumption increases in response to rising supply. Every increase in supply is met by a corresponding increase in consumption, because in contemporary society, voluntary control over consumption is ineffective.

DEMOPHORIC GROWTH

As discussed in a particularly penetrating analysis by J. R. Vallentyne, the impending environmental crisis is not only the result of population growth. It is also a result of technological growth, both directly in the sense of increased per capita production and consumption, and indirectly in that technology has furthered the growth of population and urbanization. This concept of growth, encompassing the combined effects of population in a biological sense, and of production-consumption in a technological sense, has been termed *demophora*. The importance of the concept of demophoric growth lies in its emphasis on both production and consumption. It describes the cycle in which biospheric degradation occurs as a result of production utilization of the environment

1

TABLE 1-1   Water in the Biosphere*

	VOLUME (THOUSANDS OF KM³)	PER CENT OF TOTAL	RENEWAL TIME
Oceans	1,370,000	97.61	3100 years†
Polar ice, glaciers	29,000	2.08	16,000 years
Groundwater (actively exchanged)‡	4000	0.29	300 years
Freshwater lakes	125	0.009	1–100 years§
Saline lakes	104	0.008	10–1000 years§
Soil and subsoil moisture	67	0.005	280 days
Rivers	1.2	0.00009	12–20 days¶
Atmospheric water vapor	14	0.0009	9 days

*Modified from Vallentyne, J. R., after Kalinin and Bykov. *In:* The Environmental Future. London, Macmillan Publishers, Ltd. Reprinted by permission of Macmillan London and Basingstoke.

†Based on net evaporation from the oceans.

‡Kalinin and Bykov (1969) estimated that the total groundwater to a depth of 5 km in the earth's crust amounts to $60 \times 10^6 km^3$. This is much greater than the estimate by the U.S. Geological Survey of $8.3 \times 10^6 km^3$ to a depth of 4 km. Only the volume of the upper, actively exchanged groundwater is included here.

§Renewal times for lakes vary directly with volume and mean depth, and inversely with rate of discharge. The absolute range for saline lakes is from days to thousands of years.

¶Twelve days for rivers with relatively small catchment areas of less than 100,000 km²; 20 days for major rivers that drain directly to the sea.

and consumption of technological products, both leading to pollution of water resources. In this way, attention is properly focused on all of the aspects of technological growth, i.e., the technological metabolism of man.

The demands that demophoric growth have imposed upon freshwater resources are monumental. Although about 105,000 km³ of precipitation, the ultimate source of the freshwater supply, fall on the land surface per year, only about one-third of it (approximately 37,500 km³ year⁻¹) reaches the oceans as river discharge (Vallentyne, 1972). About two-thirds of the annual water supply is returned to the atmosphere by evaporation and plant transpiration. If the potential water supply of 37,500 km³ year⁻¹ were divided evenly among the 3.7 billion humans now existing on earth (1971; cf. Clark, 1967), each person would have potentially 10,000 m³ year⁻¹ or 27,000 l day⁻¹. These values would be halved at the projected population level of 7.0 billion humans in the year 2000. Even though these quantities seem large in comparison to the human physiological requirement of 2 l caput⁻¹ day⁻¹, they are insufficient in view of modern technological demands. Domestic consumption averages 250 l caput⁻¹ day⁻¹, the average industrial consumption is 1500 l caput⁻¹ day⁻¹ in developed countries, and agriculture uses up to several thousand l caput⁻¹ day⁻¹ in countries with hot, dry climates (Vallentyne, 1972).

Historical patterns of environmental resource utilization have resulted in a series of crises, at least to populations of developed countries (Fig. 1–1). As population levels increase, the severity of the crises increases. Particular resource-related problems are confronted with various actions by man. Most of these actions are therapeutic adaptations and adjustments rather than true corrective measures. As a result, stress levels increase. The margin between coping successfully with environmental resource limitations and disastrous situations of overexploitation decreases continually. Stability becomes increasingly tenuous.

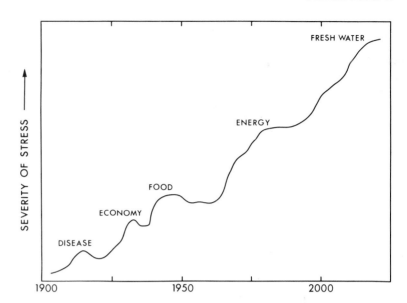

Figure 1-1. Stresses of increasing severity imposed upon the populations of developed countries resulting largely from excessive demophoric growth. (Modified from Wetzel, 1978.)

Vagaries in climate (e.g. Bryson and Murray, 1977) are superimposed upon the immediate demophoric pressures of resource utilization. Fluctuations in climate have resulted in catastrophic crises both in the availability of resources and effective resource utilization. Rapid climatic changes affect not only the availability of adequate food resources, but availability of fresh water as well; the two are inexorably coupled. Although major climatic changes are difficult to predict, such changes are certain and, on the basis of long-term trends, significant change is imminent. In order to cope effectively with these natural climatic changes, we must maintain sufficient flexibility in our use of freshwater resources and an adequate margin below the maximum carrying capacity.

HUMAN IMPACT ON FRESHWATER ECOSYSTEMS

The potential freshwater supply is in reality much less because of many factors. First, rainfall is not evenly distributed over land surfaces, nor has man distributed himself proportionally in relation to water availability. This disparity results in a great expense of energy for distribution systems. Second, total consumption has increased exponentially with demophoric growth. Expansion of distribution systems to areas of low precipitation, such as irrigation of semiarid regions, results in disproportionately high use of water because of very high losses by evapotranspiration. Third, potentially the most serious factor stemming from demophoric growth is the severe degradation of the water quality. The effect is a severe reduction of the water supply available for other purposes.

We can still look to fresh waters for purposes other than water supply, e.g., for recreation, transportation systems, and the like. However, it is clear that the demands of exponential demophoric growth have been given total precedence over the use of fresh waters for other purposes. The most fundamental laws of resource utilization may be recognized by most agencies and industries, but they are not being implemented to a significant degree. The fuel 'crisis' of 1973-74, one of the consequences of demophoric growth, demonstrated the prevailing human behavioral pattern of virtually ignoring all diagnoses or prognostications of impending crisis. When the crisis finally occurs, the public response is not to seek corrective measures that strike at the root of the problem, such as stabilization of population and technological growth, but rather to expand the distribution system to more remote sources of energy. Only when disaster is imminent, when risk of survival of a large population group is glaringly obvious to the majority, does a unified global response occur to save man from himself.

While, unfortunately, the above remarks are pessimistic, they are an accurate assessment of existing patterns of utilization of our water resources. It is clear that demophoric growth will continue to impose increasing demands upon freshwater supplies until either inefficient utilization creates a disastrous situation threatening the survival of a major segment of the human race, or until the expenditures of energy needed to obtain water exceed tolerable operational levels. Looking back at the repetitious history of responses to impending environmental disasters, we can be optimistic about the future only until such time as our understanding of the operation of the biosphere, and our knowledge of freshwater ecosystems in particular, is adequate to allow us to recognize the point of irreversibility. Reflecting on the progress that has been made in limnology since its inception nearly a century ago, it becomes apparent that our time is disconcertingly limited. We need more time to learn to understand freshwater ecosystems sufficiently to judge their resiliency and capacity for change in response to exponential demophoric loading. There is a need to extend the existing knowledge of fresh waters to a greater percentage of the population being educated at the college level so that this information can be effectively infused into the population at all levels.

THE STUDY OF LIMNOLOGY

It is of major importance, therefore, that we understand the structure and function of fresh waters. Man is a component of these ecosystems, and his effects on them will increase markedly until demophoric growth is stabilized. Emotionalism and alarmist reactions to the momentum of exploitation of the finite biosphere by the technological system accomplish little, and, as has been repeatedly demonstrated, are often antagonistic to improvement. Strict conservation and isolation of resource parcels, in the belief that such areas are exempt from technological alterations of the atmosphere and water supply, are naive and likewise contribute little to solution of the overall problem.

Understanding of metabolic responses of aquatic ecosystems is necessary in order to confront and offset the effects of these alterations, and in order to achieve maximum, meaningful management of freshwater resources. All waters, of course, cannot be managed directly. Rather, an integration of man's demophora with the metabolism of fresh waters is required to minimize detrimental changes. A well-documented effect of human impact upon aquatic ecosystems is eutrophication, a multifaceted term generally associated with increased productivity, structural simplification of biotic components, and a reduction in the ability of the metabolism of the organisms to adapt to imposed changes (reduced stability). In this condition of eutrophication, excessive inputs commonly seem to exceed the capacity of the ecosystem to be balanced. In reality, however, the systems are out of equilibrium only with respect to the freshwater chemical and biotic characteristics desired by man for specific purposes. In order to have any hope of effectively integrating man as a component of lake ecosystems, and of monitoring his utilization of these resources, it is mandatory that we comprehend in some detail the functional properties of fresh waters. Only then can we evaluate, with reasonable certainty, the influence that man's activities will have on the metabolic characteristics of these systems.

In broad terms, limnology is the study of the functional relationships and productivity of freshwater communities as they are affected by their physical, chemical, and biotic environment. The analyses of topics that follow focus primarily on standing (lentic) waters. The rationale for this approach is several-fold. The lake ecosystem is a system intimately coupled with the land surrounding it in its drainage area and its running (lotic) waters that transport, and metabolize en route, components of the land to the lake. The limnology of running waters has been given a masterful, eloquent, and succinct treatment in a text by H. B. N. Hynes (1970). No attempt will be made to encapsulate his review here, except to emphasize the significant roles that inflowing waters have in lake dynamics.

In many sectors of the earth, a majority of the lakes are of glacial origins. Because most limnological research has been concentrated in northern temperate regions, a strong bias occurs in current instruction by the disproportionate knowledge of natural temperate lakes. Greater understanding of tropical and warm water lakes of other regions is emerging. Furthermore, the number of man-made reservoirs has increased to the point where they form major lake systems over large areas of the world. Although reservoirs can possess many characteristics fundamentally different from lakes, a firm grasp of the dynamics of lakes permits a relatively easy transition to an understanding of the more variable and individual characteristics of reservoirs. Underlying all of these systems are basic metabolic similarities. In this treatment of lake ecosystems, I will attempt to introduce fundamental, functional similarities without becoming mired in a plethora of individual detail.

Selecting material to include in such a study is difficult not only because of individual biases, but because there is a lack of complete understanding of many subjects. Some background detail is needed to appreciate even the rudiments of the field. It is hoped that, in the end, the examples presented here are balanced and provide a basic overview of contemporary comparative limnology, and a basic minimal understanding of freshwater ecology at the undergraduate level. A serious major in limnology will realize that he or she needs much greater depth of understanding to comprehend the subject thoroughly. Many of the reference works cited, such as G. E. Hutchinson's classic, perceptive treatise (1957, 1967, 1975, and forthcoming volumes), are only initial introductory summaries of specialized subjects. I hope that the forefronts of contemporary limnology, as well as the gaps in need of more intensive investigation, will be evident from the ensuing discussions.

Hegel (1807) stated, *"Das Wahre ist das Ganze"* (The truth is the whole), yet holism alone is inadequate for a comprehensive understanding of limnology. Integration of our knowledge about the individual operational components and environmental factors that regulate productivity of freshwater ecosystems is fundamental. Previous treatments have been deficient in this respect and almost totally directed toward open water pelagic communities, to the neglect of other major components of lake systems (i.e., microbial decomposition of dissolved and particulate detritus; the intensive metabolism of the littoral zone and adjacent wetlands; metabolism in the sediments; influxes of nutrients and organic matter from outside the lake; the rapidity of nutrient cycling; and regulatory feedback mechanisms of fauna on production and decomposition processes). Integrated study of the dynamics of all components of lake ecosystems requires alterations in former perspectives on freshwater ecology.

SUMMARY

1. Over 99 per cent of the immense amount of water of the biosphere occurs in the oceans and polar ice deposits. The turnover time of this water is very long (i.e., the rate of renewal is very slow).
2. Most of the remaining water occurs in freshwater lakes; its renewal time is much shorter than in marine systems.
3. Finite freshwater resources are being exploited and degraded at an alarming rate by the activities of humankind.
    a. Demands upon surface and groundwater supplies result from both population growth and expanding utilization and consumption in technological growth. This growth in utilization of freshwater and other finite environmental resources is termed *demophora*. Biospheric degradation results from both biological population expansion and growth of technological production and consumption.
    b. Unless the demands of exponential demophoric growth upon fresh waters are rapidly controlled, a freshwater crisis is imminent. The severity of the freshwater crisis will exceed that of past and contemporary resource-utilization problems. The crisis is already apparent and will become acute early in the 21st century.
4. *Limnology* is the study of the functional relationships and productivity of freshwater communities as they are regulated by the dynamics of their physical, chemical, and biotic environments.

THE CHARACTERISTICS OF WATER

Water is the essence of life on earth, and totally dominates the chemical composition of all organisms. The ubiquity of water in biota as the fulcrum of biochemical metabolism rests on its unique physical and chemical properties.

The characteristics of water regulate lake metabolism. The unique thermal-density properties, high specific heat, and liquid-solid characteristics of water allow the formation of a stratified environment that controls the chemical and biotic properties of lakes to a marked degree. Water provides a tempered milieu in which extreme fluctuations in water availability and temperatures are ameliorated relative to conditions facing aerial life. Coupled with a relatively high degree of viscosity, these characteristics have enabled the biota to develop a large number of adaptations that improve sustained productivity.

MOLECULAR STRUCTURE AND PROPERTIES

The unique properties of water center upon its atomic structure and bonding, and the association of water molecules in solid, liquid, and gaseous phases. In a state of equilibrium, the nuclei of a water molecule form an isosceles triangle, with a slightly obtuse bond angle of 104.5° at the oxygen nucleus (Eisenberg and Kauzmann, 1969). The bond length from the center of the oxygen atom to that of each hydrogen atom is 0.96×10^{-8} cm. The nuclei of the molecules are in a continual state of vibration. As the molecule resonates, an electrical dipole moment occurs such that a complex wave function results in a valence electron density pattern in which electron density is highest near the atoms and along the bonds (Fig. 2-1). The electronic charge of the molecule is not restricted to the planar configuration depicted, but is distributed in multiple directions.

Although a majority (approximately 56 per cent) of water molecules have their valances balanced, excitation and ionization do occur. The most common ionized state is characterized by removal of one of the hydrogen atoms, which is charged positively, and is essentially functioning as a proton. Higher ionization potentials, that is, the energy required to remove electrons from the molecule, are needed to reach further ionization states of water, and occur less frequently. The weak Coulombic characteristics of the bonding of hydrogen atoms to the weakly electronegative oxygen atom

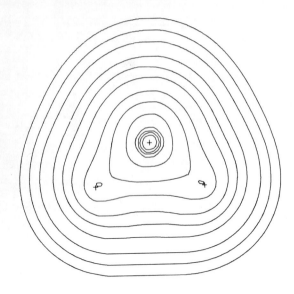

Figure 2-1 Contour map of the electronic charge distribution of the water molecule in the plane of the nuclei (positions indicated by crosses). Contours increase in value of atomic units (au) (1 au = $6.7e^-$ per $Å^3$) from the outermost contour in steps of 2×10^n, 4×10^n, 8×10^n. The smallest contour value is 0.002 with n increasing in steps of unity to yield a maximum value of 20. (Unpublished figure courtesy of Dr. R. F. W. Bader, Department of Chemistry, McMaster University.)

result in both ionized and covalent states that simultaneously maintain the integrity of water. Water is nearly the only known compound that possesses these characteristics.

The density properties of water that are so germane to limnology and life in fresh waters result from the aggregation and bonding characteristics of water molecules. It is instructive to look first at the structure of ice, whose physical properties are understood better than those of liquid water. Every oxygen atom is at the center of a tetrahedron formed by four oxygen atoms, each about 2.76×10^{-8} cm distant (Eisenberg and Kauzmann, 1969; Horne, 1972). Every water molecule is hydrogen-bonded to its four nearest neighbors; its O-H bonds are directed towards lone pairs of electrons on two of these adjacent molecules, forming two O-H—O hydrogen bonds. In turn, each of its lone pairs is directed towards an O-H bond on one of the other adjacent molecules, forming two O—H-O hydrogen bonds. This arrangement leads to an open lattice in which intermolecular cohesion is great (Fig. 2-2). This structure results in both parallel and perpendicular voids between the molecules, creating an open tetrahedral structure that permits ice to float upon liquid water.

Molecules in liquid water near 0°C experience about 10^{11} to 10^{12} reorientational and translational movements per second, whereas ice molecules near 0°C experience only about 10^5 to 10^6 movements per second. Increasing the temperature of water increases the rate of reorientations and molecular displacements and results in decreased viscosity, decreased molecular relaxation times, and increased rates of self-diffusion. By virtue of their dipolar nature, water molecules interact to form quasi-stable polymers.

Although we have viewed water molecules as diffusionally averaged structures both in vibrational time and in space between molecules, variations in temperature change the intermolecular distances. Thermal increases result in increased agitation of molecules, which distorts or breaks down the hydrogen-bonded networks. As the ice melts, increased bond dislocation and rupture occur, which disrupt and fill in the open

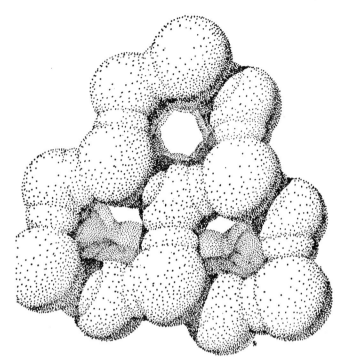

Figure 2-2 A diagrammatic representation of an ice crystal showing the van der Waals radii of the atoms and open voids between the aggregated molecules. (From Pimental, G. C., and McClellan, A. L.: The Hydrogen Bond. San Francisco, W. H. Freeman and Co., copyright © 1960.)

spaces of the ice lattice structure. The result is an increase in density, and this effect predominates and reaches a maximum at 4°C. As water is heated above 4°C, intermolecular vibrations increase in amplitude and increase interatomic distances. The result is expansion of the liquid and decreased density. It is a characteristic of water that minimum volume, hence maximum density of water occurs at 3.98°C, a point at which competition between the negative configurational contribution of hydrogen bonding and its positive vibrational contribution is maximized.

ISOTOPIC CONTENT

Known isotopes of hydrogen are ^1H, ^2H or D (deuterium), and ^3H (tritium). Tritium is radioactive with a half-life of 12.5 years, sufficiently short that after its decay to ^3He, most of which is lost from the atmosphere into space, concentrations of tritium in uncontaminated natural water are very low (approximately 1 ^3H atom per 10^{18} ^1H atoms) and do not accumulate. Six isotopes of oxygen are known: ^{14}O, ^{15}O, ^{16}O, ^{17}O, ^{18}O, and ^{19}O. The isotopes ^{14}O, ^{15}O, and ^{19}O are radioactive but are short-lived, and do not occur significantly in natural water. The precise isotopic content of natural water depends on the origin of the sample but, within the limits of variation, the abundances of $H_2^{18}O$, $H_2^{17}O$, and $HD^{16}O$ are 0.20, 0.04, and 0.03 mole per cent, respectively (Eisenberg and Kauzmann, 1969). Other isotopic combinations are exceedingly rare in natural water (Hutchinson, 1957).

Changes in the ratios of these isotopes in compounds and remains of organisms formed within a lake over geological time permit an estimation of paleotemperatures. The technique is based on the temperature-dependence of fractionation of the oxygen isotopes in the carbon dioxide-water-carbonate system (Stuiver, 1968). During slow precipitation of the carbonate, the temperature of the solution is reflected in the ratio of ^{18}O to ^{16}O in the carbonates of sediments and mollusks. The ratio of oxygen isotopes in the precipitated carbonates depends not only on the temperature, but also on the ratio of ^{18}O to ^{16}O isotopes in the water in which carbonate is formed. The measured difference in the ratio of ^{18}O to ^{16}O isotopes between a carbonate fossil and a contemporaneous sample is the result of changes in both temperature and composition of water isotopes. Before paleotemperature determinations can be made, an estimate of the change in ratio of oxygen isotopes of the water is needed; this is more difficult to do in lakes where variations over time are greater than in the oceans. However, estimates can be done with reasonable precision.

SPECIFIC HEAT

The specific heat (that amount of heat in calories that is required to raise the temperature 1°C of a unit mass of a substance) of liquid water is very high (1.0), exceeded only by a few substances such as liquid ammonia (1.23), liquid hydrogen (3.4), and lithium at high temperatures. Other substances of the biosphere, such as many rocks, have much lower specific heats (approximately 0.2). The high specific heat of water, as well as a high latent heat of evaporation, is a function of the relatively large amounts of heat energy required to disrupt the hydrogen bonding of liquid water molecules.

These heat-requiring and heat-retaining properties of water provide a much more stable environment than is found in terrestrial situations. Fluctuations in water temperature occur very gradually, and seasonal and diurnal extremes are small in comparison to those of aerial habitats. The high specific heat of water can have profound effects on the climatic conditions of adjacent air and land masses. This thermal inertia of the hydrosphere (Hutchinson, 1957) occurs on either a large or a small scale, in relation to the volume of water body. Examples are many, and include the warm Gulf Stream currents of the Atlantic Ocean, which improve the climate of Western Europe, and the prevailing air movements across the Great Lakes, which moderate the climate of Michigan and other states adjacent to and east of the lakes (Leighly, 1942). Mild winters with higher precipitation rates are found in these areas than in the continental interiors, since the water masses cool and yield heat to the air more slowly than land. Similarly, areas adjacent to large water masses experience moist, cool, summer periods. Fall fogs are common over and adjacent to lakes, since water vapor from evaporation of the warm lake water condenses and stagnates in the cold overlying air. This phenomenon is described in common parlance as "the lake is steaming."

The specific heat (thermal capacity) of ice below 0°C is about half (0.5 cal g^{-1} °C^{-1}) that of water and decreases progressively at temperatures below 0°C. However, the amount of heat required to change ice to liquid water is large (latent heat of melting = 79.72 cal g^{-1}) and very much larger for disruption of hydrogen bonding in evaporation of water (540 cal g^{-1}) or for direct sublimnation of ice to water vapor (latent heat of sublimation = 679 cal g^{-1}). Similarly, a large amount of heat must be lost for the fusion of molecules of 0°C water to ice (latent heat of fusion = 79.72 cal g^{-1}). Because of these properties, large energy inputs are required to melt ice in the spring and large energy losses are required to form ice cover in the winter.

DENSITY RELATIONSHIPS

Without question, the regulation of the entire physical and chemical dynamics of lakes and the resultant metabolism is governed to a very great extent by differences in density. Throughout the remaining chapters of this book, we will refer repeatedly to the fundamental importance of this unique property of water.

The density or specific gravity of pure ice at 0°C is 0.9168, about 8.5 per cent lighter than liquid water at 0°C (0.99987). The density of water increases to a maximum of 1.0000 at 3.98°C, beyond which molecular expansion and decreasing density occur at a progressively increasing rate (Fig. 2–3). The density differences are small but highly significant. It is important to examine the difference in density between water at a given temperature and water at a temperature 1°C lower, referred to as the *density difference per degree lowering* (Vallentyne, 1957). The magnitudes of this density difference for water of different temperatures are shown on the right side of Figure 2–3. The density difference per degree lowering increases markedly as the temperature goes above or below 4°C. Physical work is required to mix fluids of differing density, as, for example, when mixing cream into milk, and the amount of energy input required is proportional to the difference in density. The amount of work required to mix layered water masses between 29 and 30°C is 40 times, and between 24° and 25°C 30 times that required for the same masses between 4° and 5°C.

Density also increases with increasing concentrations of dissolved salts in an approximately linear fashion (Table 2–1). The salinity of a majority of inland waters is within a range of 0.01 and 1.0 g l^{-1}, usually between 0.1 and 0.5 g l^{-1}. Inorganic salinity of highly saline lakes can commonly exceed 60 g l^{-1} (Rawson and Moore, 1944; International Association of Limnology, 1959; Wetzel, 1964). However, in a majority of lakes,

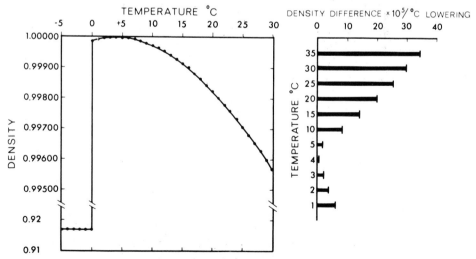

Figure 2-3 Density (g ml^{-1}) as a function of temperature for distilled water at 1 atm. The density difference per °C lowering is shown in the right-hand portion of the figure at various temperatures. (Modified from Vallentyne, 1957.)

TABLE 2-1 Approximate Changes in the Density (g ml⁻¹) of Water with Salt Content

SALINITY (PARTS PER THOUSAND, ‰)	DENSITY (AT 4°C)
0	1.00000
1	1.00085
2	1.00169
3	1.00251
10	1.00818
35 (mean, sea water)	1.02822

After Ruttner, 1963.

the salinity is very low and varies less than 0.1 g l^{-1} spatially and seasonally. While salinity-induced variations in density may be small, they cannot be ignored, for under certain conditions of lake stratification, inorganic salts can accumulate temporarily or permanently (cf. Chapter 6).

Salinity also decreases the temperature of maximum density of water at a rate of approximately 0.2°C per g l^{-1} (‰) increase. The temperature of maximum density of sea water (mean 35 g l^{-1}) is −3.52°C, which does not occur at surface pressures and is below the freezing point (−1.91°C). In most lakes, however, the change in point of maximum density due to salinity is very small.

Hydrostatic pressure can be great enough to compress water sufficiently to lower the temperature of maximum density. Pressure increases one atmosphere per 10 m depth. The temperature of maximum density decreases about 0.1°C per 100 m of depth (Strøm, 1945). In very deep lakes, temperatures below 4°C are found to be partially, but not totally, related to the relatively high pressures encountered at great depth (Eklund, 1963, 1965; Johnson, 1964, 1966; Bjerke, et al., 1979).

VISCOSITY-DENSITY RELATIONSHIPS

The density of water is 775 times greater than that of air at standard temperature and pressure (0°C, 760 mm Hg). The greater density of water makes aquatic organisms relatively buoyant against gravitational pull, and consequently reduces the amount of energy that an organism must expend in order to support or maintain its position. Although a marked reduction in supporting tissue is seen in many freshwater animals, especially among the lower invertebrates, these adaptations are particularly conspicuous among aquatic vascular plants. The truly submersed angiosperms are good examples; these plants are almost totally limited to fresh waters and have immigrated and adapted to aquatic conditions relatively recently. Many modifications have been observed in these plants, particularly with respect to the reduction of their vascular tissue. Submersed organs exhibit a reduction in the extent of lignification of the xylem or water-conducting and supporting elements, and most of the vascular strands are condensed into a weakly developed central cylinder. The supportive tissue of these hydrophytes develops strongly only in those plants or parts that are adapted to aerial existence as surface-floating or emergent forms.

Water viscosity is influenced to a minor extent by the salinity of the water, but to a considerable extent by temperature. Viscosity decreases as temperature increases; water viscosity doubles as the temperature is lowered from 25°C to 0°C. The viscosity of water offers roughly 100 times the frictional resistance to a moving organism or particle that air does, depending upon surface exposure area, speed, and the temperature and chemical composition of the fluid. As will be discussed further on, organisms with locomotion must expend considerable energy to overcome changes in viscosity. The sinking rates and distribution of passive organisms, such as planktonic algae or sedimenting particles, are influenced by density-related changes in viscosity.

SURFACE TENSION

The quasi-polymeric bonding properties of liquid water molecules discussed above are disturbed at the interface with air. At the interface plane, the molecular attractions are unbalanced and exert an inward adhesion to the liquid phase. The result is an interface surface or film under tension. The surface tension at the air–water interface of pure water is higher than that for any other liquid except mercury. Surface tension decreases with increasing temperature, and increases slightly with dissolved salts (Table 2-2).

The surface tension of water is reduced markedly by the addition of organic compounds, a phenomenon reflected in lakes or portions of fresh waters where dissolved organic matter concentrates (Table 2-2). For example, the surface tensions of relatively pristine waters containing few algae differ little from those of distilled water. Bog lakes, which are heavily stained with dissolved organic compounds, exhibit depressions in surface tension of 6 to 7 dynes cm^{-1} (range 0–20). Where growth of floating algae or of higher aquatic plants is particularly great, the surface tension is depressed by about 20 dynes cm^{-1}. It will be shown further on that the concentrations of dissolved organic

TABLE 2-2 Surface Tension of Water with Changes in Temperature and Its Depression in Natural Waters under Various Conditions

°C	WATER DYNES cm^{-1}	CONDITION	DEPRESSION OF SURFACE TENSION, RANGE IN DYNES cm^{-1}
Pure Water			
0	75.6	Oligotrophic lakes	0–2
5	74.9	Eutrophic lakes	0–20
10	74.4	Bog lakes	0–20
15	73.5	Lake water with foam	2–9
20	72.7	Near floating-leaved angiosperms	5–20
25	72.0	Near submersed angiosperms	1–2
30	71.2	During a blue-green algal bloom	0–20
35	70.4	Open sea	<1
40	69.6	Plymouth Sound, near muddy beach	6–20
		Harbor, heavy boat traffic	15–>20
Sea Water			
ca. 5	75.0		

From data of Adam, 1937, and Hardman, 1941.

matter in the more productive waters are high. Additionally, natural populations of algae and submersed angiosperms secrete large quantities of organic compounds during active photosynthesis, as well as during senescence and lysis. The effects of artificial introduction of organic pollutants into fresh waters on surface tension are readily apparent (cf. Jarvis, 1967; Jarvis, et al., 1967).

The air–water interface forms a special habitat for organisms adapted to living in surface film. This community is collectively referred to as neuston, and will be discussed in some detail further on (Chapter 8). The surface tension is sufficient to serve as a supporting surface for many organisms of considerable size, for example for gyrinid and other beetles, and can destroy others, such as cladoceran microcrustacea, that happen to become caught in the interface through wave action and cannot reenter their normal submersed habitat.

SUMMARY

1. Water is a unique substance. Its thermal-density properties, high specific heat, and liquid-solid characteristics influence extensively the physical, chemical, and metabolic behavior of freshwater ecosystems.
2. The molecular aggregation and bonding characteristics of water result in unique density properties that permit water molecules to form quasi-stable polymers.
    a. The maximum density of water ($1.00000 \text{ g ml}^{-1}$) occurs at $3.98°C$. Molecular expansion and decreasing density occur at progressively increasing rates both above and below $4°C$. The density of ice at $0°C$ (0.9168) is significantly less than that of liquid water at $0°C$, and ice consequently floats.
    b. The density difference between water at a given temperature and $1°C$ lower increases markedly at temperatures above and below $4°C$. The amount of energy required to mix water of differing densities increases proportionally to the density differences.
3. The specific heat of liquid water is very high. As a result, thermal conditions within standing waters change slowly, are more stable, and permit a longer growing season than is commonly found in aerial habitats. The heat-absorbing and heat-retaining properties of bodies of water also modify the climate of adjacent land masses.
4. Water viscosity is far greater than the viscosity of air and is influenced inversely by temperature. Aquatic organisms have adapted to the more buoyant aquatic environment and to the resistance to locomotion; their distribution is often influenced by thermally induced stratification of water masses of differing density and viscosity.
5. A high surface tension occurs at the air–water interface. Surface tension decreases with increasing temperature, salinity, and concentrations of organic compounds.

CHAPTER THREE

LAKES — THEIR DISTRIBUTION, ORIGINS, FORMS

The amount of fresh water on earth is very small in comparison to the water of the oceans, but has much more rapid renewal times (cf. Table 1-1). On a volumetric basis, fresh water is concentrated in the large, deep basins of several great lakes and about 20 per cent is contained in Lake Baikal, USSR. The number of individual depressions of smaller lakes and reservoirs is extremely large, however, and most of these lakes are located in temperate and subarctic regions of the Northern Hemisphere.

Catastrophic events of glacial, volcanic, and tectonic activity have aggregated many fresh waters into lake districts. The morphometry and geological substrates of the lake basins are of major importance in determining sediment–water interactions, the resultant productivity, and the significance of littoral productivity to that of the entire lake. Shallow lakes, which have more sediment area per unit of water volume, are generally more productive than deep lakes, and a greater proportion of their total productivity is due to attached littoral communities.

DISTRIBUTION OF FRESH WATERS

Inland waters cover less than 2 per cent of the earth's surface, approximately 2.5 $\times$ 10^6 km^2. About 20 lakes are extremely deep (in excess of 400 m), and a significant portion of the world's fresh water lies within these basins. For example, approximately 20 per cent of the world's fresh water is contained in Lake Baikal, Siberian USSR (A = 31,500 km^2; z_m = 1620 m; $\bar{z}$ = 740 m), the deepest lake of enormous volume (23,000 km^3). Like Lake Baikal, nearly all other extremely deep lakes are tectonic or volcanic in origin or have formed from fjords that have subsequently become fresh. Although a few lakes of glacial origin, such as the Laurentian Great Lakes and Great Slave Lake of Canada, have basins of great depth, few exceed a depth of 300 m. Most of the very deep lakes are found in mountainous regions along the western portions of North and South America, of Europe, and in the mountainous areas of central Africa and Asia. Lakes Baikal in Asia and Tanganyika in Africa are the only lakes known to have maximum depths in excess of 1000 m and mean depths over 500 m.

On an areal basis, only a few lakes are of very large surface area; most of these are compared in Figure 3-1. If we exclude the Black Sea, which is an inland water really connected to the ocean, the Caspian Sea, a large (436,400 km²) salt lake, is the largest inland basin separated from the ocean. The Laurentian Great Lakes of North America—lakes Superior, Huron, Michigan, Ontario, and Erie—constitute the greatest continuous mass of fresh water on earth, with a collective area of 245,240 km² and a volume of 24,620 km³. Lake Superior has the greatest area (83,300 km²) of any purely freshwater lake.

While these very large lakes constitute a major freshwater resource, most lakes are much smaller. Most lakes are of catastrophic origin, formed by glacial, volcanic, or tectonic processes. As a result, many lakes are localized into lake districts in which large numbers of lake basins are concentrated. Glacial activity during the most recent period of major ice advance and retreat created literally millions of small depressions that subsequently were filled with water and modified. Thus large numbers of lakes are found in the Northern Hemisphere, where large land masses of North America and Europe-Asia interfaced with glacial movements. Consequently, numerous small, shallow basins are found throughout the arctic, subarctic, and northern temperate zones.

More recently, man has created large numbers of reservoirs and ponds. A majority of both natural and man-made lakes are small and relatively shallow, usually <20 m in depth. Because of these morphometric characteristics, an increasing proportion of the lake volume is exposed to and interacts with the chemical and metabolic processes of soil and sediments. A great proportion of shallow lakes possess properties necessary for the development of sessile littoral flora, which generally results in markedly increased overall productivity.

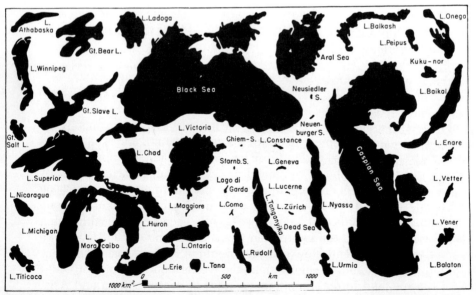

Figure 3-1 An approximate comparison of the surface areas of many of the larger inland waters of the world, all drawn to the same scale. (After Ruttner, F.: Fundamentals of Limnology. Toronto, University of Toronto Press, 1963.)

GEOMORPHOLOGY OF LAKE BASINS

The origins of lake basins and their morphometry are of much more than casual interest. The geomorphology of lakes is intimately reflected in physical, chemical, and biological events within the basins and plays a major role in the control of a lake's metabolism, within the climatological constraints of its location. The geomorphology of a lake controls the nature of its drainage, inputs of nutrients to the lake, and the volume of influx in relation to flushing-renewal time. Thermal and stratification patterns are, as discussed in the following chapters, markedly influenced by basin morphometry and volume of inflow. These patterns in turn govern the distribution of dissolved gases, nutrients, and organisms, so that the entire metabolism of freshwater systems is influenced to varying degrees by the geomorphology of the basin and how it has been modified throughout its subsequent history.

The shape of a lake basin often dictates its productivity. Steepsided U- or V-shaped basins, often formed by tectonic forces, are usually deep and relatively unproductive. In such lakes a proportionally smaller volume of water is contiguous with sediments. Shallow depressions with greater percentage contact of water with the sediments generally exhibit intermediate to high productivity.

The following brief résumé on the origin of lakes is based on Hutchinson's (1957) summary of the subject which was drawn from an array of sources. Hutchinson differentiates 76 types of lakes on the basis of geomorphological inception. Despite this detailed classification, numerous minor exceptions have been cited since it appeared (e.g., Horie, 1962, among others). Discussion here will be limited to nine distinct groups of lakes, each formed by different processes.

TECTONIC BASINS

Tectonic basins are depressions formed by movements of deeper portions of the earth's crust, and are differentiated from lake basins resulting from volcanic activity. The major type of tectonic basin forms as the result of faulting in which depressions occur between the masses of a single fault displacement or in downfaulted troughs (Fig. 3-2). The latter type of basin is referred to as a *graben*, and is the mode of origin of a large number of the most spectacular relict lakes of the world. Foremost among these is Lake Baikal of eastern Siberia, the deepest lake in the world, which has a continuous lacustrine history from at least the early Tertiary period. Lake Baikal, and many other relict lakes, most of which lie in tectonic grabens, are of particular interest because they contain a large number of relict endemic species. For example, of 1200 animal species and at least half that number of plant species known from Lake Baikal, over 80 percent of those occurring in the open water are endemic to this lake (Kozhov, 1963). Lake Tanganyika of equatorial Africa is a similar deep (z_m = 1435 m) graben lake formed in rift valley displacements along crustal fractures, and contains a large number of relict endemic species of plants and animals. Pyramid Lake of Nevada and Lake Tahoe of California are familiar examples of American graben lakes.

Tectonic movements causing moderate uplifting of the marine sea bed have isolated several very large lake basins. The relict marine basins of Eastern Europe, which include the Caspian Sea and the Sea of Aral, were separated by the formation of

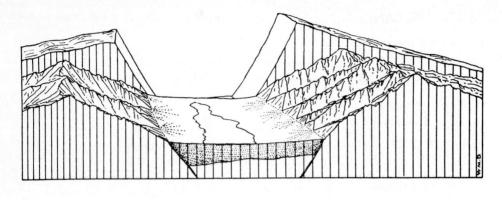

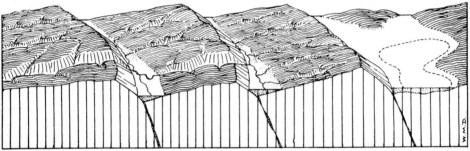

Figure 3-2 Tectonic lake basins. *Upper:* in the background, a depressed fault block between two upheaved fault blocks; in foreground, the same after a considerable period of erosion and deposition. *Lower:* Diagram of the great fault blocks of the northern Sierra Nevada Mountains with the plain of Honey Lake to the east. (From Davis, W. M.: Calif. J. Mines Geol. 29:175, 1933.)

uplifted mountain ranges in the Miocene period. Upwarping of the earth's crust in lesser degrees also has resulted in the formation of many large lake systems. Lake Okeechobee in Florida resulted from minor depressions (z_m = ca 4 m) in the sea floor as it uplifted in the Pliocene epoch to form the Floridian peninsula, and it forms one of the largest lakes (approximately 1880 km^2) completely within the United States, exceeded in area only by Lake Michigan. Similarly, Lake Victoria in central Africa resulted from upwarping of the margins of the plateau in which the lake basin lies. This uptilting also created a natural impoundment of the major river system flowing into the valley. Drainage outlets of the Great Salt Lake basin of Utah were eliminated by uplifted deformations, forming a closed lake basin. Upwarping of the earth's crust also contributed, along with the primary glacial scouring activity, to the formation of the Great Lakes of America. As the massive glacial sheet receded, the release of pressure resulted in a significant crustal rebound.

Lake basins are occasionally formed in areas of localized subsidence that result from earthquake activity. Many of these depressions are dry or temporarily contain water, depending on the porosity of the basin material, while others become permanent lakes, usually open basins with outlet drainage.

LAKES FORMED BY VOLCANIC ACTIVITY

Catastrophic events associated with volcanic activity can generate lake basins in several different ways. As volcanic materials are ejected upward and create a void, or as released magma cools and is distorted in various ways, depressions and cavities are created. If these depressions are undrained, they may contain a lake (Fig. 3-3). Because of the basaltic nature of these lake basins and their drainage areas, which are often very restricted, many lakes associated with volcanic activity contain low concentrations of nutrients and are relatively unproductive.

Small crater lakes are occasionally found occupying cinder cones of quiescent volcanic peaks. However, crater lakes in depressions formed by the violent ejection of magma (Fig. 3-4), or by collapse of overlying materials where underlying magma has been ejected to create a cavity, are more common in all areas of recent volcanic activity. Craters of explosive origin, termed *maars*, are generally very small depressions with diameters less than 2 km, and result from lava coming into contact with ground water or from degassing of magma. Maars are usually nearly circular in shape and can be extremely deep (>100 m) in relation to their small surface area. Basins formed by the subsidence of the roof of a partially emptied magmatic chamber are termed *calderas*, and can be somewhat larger than maars (minimum diameter about 5 km). Among the most spectacular of lakes formed by the collapse of the center of a volcanic cone is Crater Lake, Oregon, with an area of 64 km^2 and a depth of 608 m (seventh deepest lake in the world). Caldera lakes can be modified in various ways by secondary peaks partially filling the original caldera depressions, or by the occurrence of faulting over an emptied magma chamber.

Some volcanic lakes originated by a combination of large-scale volcanic and tectonic processes. In some situations caldera collapse occurred on such a large scale that extensive portions of the surrounding land subsided, in addition to the central portion of the volcano. Such subsidence usually takes place along prexisting fault fractures. Some of the largest lakes associated with volcanic activity were formed in this manner.

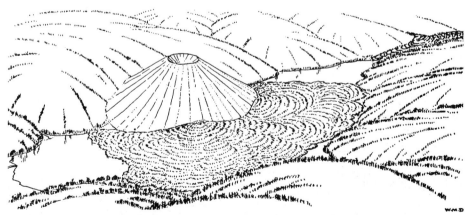

Figure 3-3 Volcanic lakes. A caldera lake within the volcanic cone and several lakes within the valleys dammed by lava flows. (From Davis, W. M.: Calif. J. Mines Geol. 29:175, 1933.)

Figure 3-4 Lake Okama on Mt. Zao, northern Japan, a caldera lake. (Photo courtesy of I. Saijo)

Many examples of these lakes exist in equatorial Asia and New Zealand (cf. Bayly and Williams, 1973).

Lava flows from volcanic activity can form lakes in several ways. As lava streams flow, cool, and solidify, surface lava frequently collapses into voids created by the continued flow of the underlying molten lava. A lake basin may be formed in this manner when the unsupported overstory crust collapses, and may be filled when the depression extends below groundwater level. Lava streams also commonly flow into a preexisting river valley and form a dam, behind which a lake can collect (Fig. 3-3). If the dam is of sufficient magnitude, the entire hydrology of the region can be changed by the resultant reversal of the river system. In some cases, the river flow is forced underground in order to pass the lava obstruction.

LAKES FORMED BY LANDSLIDES

Sudden movements of large quantities of unconsolidated material in the form of landslides into the floors of stream valleys can cause dams and create lakes, often of very large size (Fig. 3-5). Such landslide dams may result from rockfalls, mudflows, iceslides, and even flows of large amounts of peat, but they are usually found in glaciated mountains. The landslides usually are brought about by abnormal meteorological events, such as excessive rains acting on unstable slopes. More spectacular slides are occasionally initiated by earthquake activity. Lakes formed behind landslide dams are often transitory, existing only for a few weeks to several months. This is because unless the slide is very massive, the unconsolidated dam is susceptible to rapid erosion

Figure 3-5 Lakes formed by a large landslide into a steep-sided stream-eroded canyon *(upper)* and in a hollow behind a recent slide with tilted trees *(lower)*. Four mountain landslides are shown in the background. (From Davis, W. M.: Calif. J. Mines Geol. 29:175, 1933.)

by the effluent of the newly formed lake. Many disastrous floods have resulted from the rapid erosion of the dam material by the effluent which, once started, can quickly empty the accumulated basin. As with damming by lava flows, if the landslide dam is sufficiently large, the lake can become permanent and reverse the direction of drainage flow from the river valley.

LAKES FORMED BY GLACIAL ACTIVITY

By far the most important agents in the formation of lakes are the gradual but none-theless catastrophic erosional and depositional effects of glacial ice movements. Land surfaces that are now glaciated include Greenland and several smaller arctic islands, Antarctica, and numerous small areas in high mountains throughout the world. These contemporary glacial activities are small, however, in comparison to the massive Pleis-tocene glaciation that advanced and receded in four major episodes of activity in the Northern Hemisphere. With the retreat of the last Pleistocene glaciers, an immense number of small lakes were created. Lakes of glacial origin are far more numerous than lakes formed by other processes. The action of glaciers in mountainous regions of high relief usually produces lake basins that are quite different from those that result from the movements of large ice sheets on regions of more mature and gentle relief.

A number of lakes, often temporary in nature, occur on the surfaces of, within, or beneath existing glacial ice masses in areas of transitory thaw. For example, numerous meltwater subice lakes several kilometers in diameter occur below 3000 to 4000 m of ice of the Antarctic ice sheet (Oswald and Robin, 1973). In high mountain regions the fronts and lateral arms of glaciers, as well as terminal morainal deposits supported by the ice, often will function as effective dams in river valleys that originate almost totally from glacial meltwater.

Glacial ice-scour lakes refer to a vast number of small lakes formed by ice moving over relatively flat mature rock surfaces that are jointed and contain fractures. The ice-scour lakes are particularly common in mountainous regions where glacial movements have removed loosened rock material along fractures. When the glaciers retreat, the rock basins thus formed fill with meltwater. Such glacial scour lakes may be found on the upland peneplains of Scandinavia, in the United Kingdom, and in the great Canadian shield region.

A frequently occurring type of ice-scour lake forms in the upper portion of glaciated valleys of mountainous areas, where the valleys are shaped into structures resembling amphitheaters by freezing and thawing ice action (Fig. 3–6). The amphitheater-like formation is referred to as a *cirque*, and lakes within such depressions are *cirque lakes*.* Water is held in the cirque depression either by a rock lip above the depression, or by morainal deposits. Cirque lakes, which are generally small and relatively shallow (<50 m), are often found in tandem arrangement within a glaciated trough of a mountain valley, with the higher lake "hanging" above the lower succeeding lake in a stairway-like fashion. True cirque lakes, which are common to all major mountainous regions, are formed at approximately the snow line in glaciated valleys. When the glaciers extend well below the snow line of constant freezing and thawing, the corrosive action of the ice can form rock basins within the glacial valley. When such valley rock basins form a chain of small lakes in a glaciated valley resembling a string of beads, they are referred to as *paternoster lakes*. Where there are mountains adjacent to the sea, as in many areas of Norway and western Canada, *fjord lakes* can be formed within narrow, deep basins in glacially deepened valleys.

The English Lake District, a small group of lakes in a roughly circular area of northwestern England, is worthy of special notice because of the long history of limnological investigations that have centered in this region (reviewed in detail by Macan, 1970). The geological formations of the Lake District consist of an elevated dome of Ordovician and Silurian slates and volcanic intrusions. Rock formations overlying the early uplifted region were later eroded into nine major valleys. In the process of widening and deepening the troughs, glacial erosive activity scoured piedmont basins in each of these valleys. The result was the formation of a series of valleys radiating from the dome, most of which contain one large lake depression and several smaller basins situated in deposits of morainal till.

---

*Sometimes referred to as *tarns*.

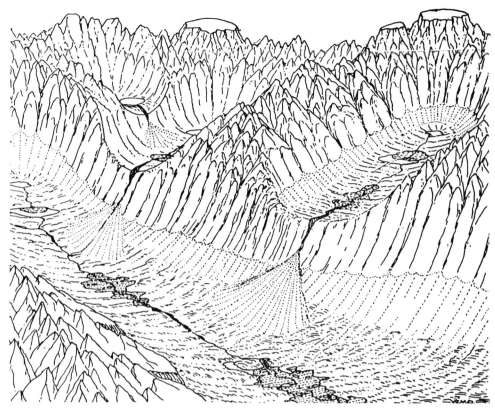

Figure 3-6 Several small cirque lakes within a mountain group from which several converging, cirque-headed branch troughs all join the same trunk trough. (From Davis, W. M.: Calif. J. Mines Geol. 29:175, 1933.)

Where continental ice sheets encountered weak areas in primary basal rock formations, glacial scouring of lake basins occurred on a massive scale in nonmountainous piedmont areas. Great Slave Lake and Great Bear Lake in the central Canadian subarctic region are examples of scouring of preexisting valleys to very great depths by the ice sheet movements. The most impressive example, however, of large rock basins produced by glacial continental ice erosion are the Great Lakes of the St. Lawrence drainage, which collectively form the largest continuous volume of liquid fresh water in the world.

The entire Great Lakes region was covered by continental ice during the glacial maxima. Three important events led to the formation of the contemporary Great Lakes during the retreat of the Wisconsin ice sheet (Fig. 3–7), approximately 15,000 years B. P.* (Hough, 1958). In the early phases of retreat, lakes formed in the Mich-

*Before Present

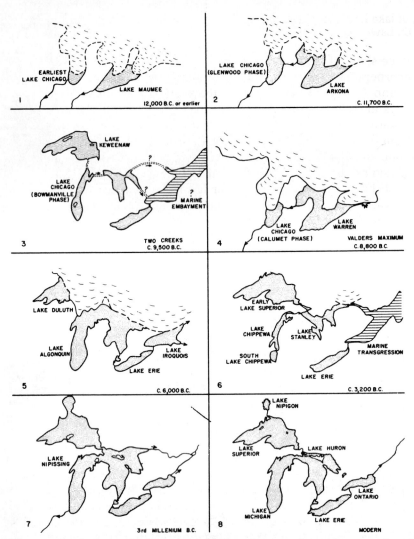

Figure 3-7 The history of the Laurentian Great Lakes. *1*, Cary substage; *2*, Late Cary; *3*, Cary-Valders interstadial, low water in eastern basins, marine transgression in Ontario Basin; *4*, Valders maximum; *5*, Post-Mankato retreat; *6*, Postglacial thermal maximum; *7*, Lake Nipissing with triple drainage; *8*, Modern lakes. (From Hutchinson, G. E.: A Treatise on Limnology. Vol. 1, New York, John Wiley & Sons, Inc., 1957, after various sources; see Hough, 1958, for details.)

igan and Huron–Erie basins against the retreating ice lobes and drained southwestward into the Mississippi valley. As the ice sheet repeatedly retreated and advanced over a period of nearly 8000 years, discharge channels from previous scouring were uncovered and altered, resulting in drainage changes to an easterly direction. Another factor that contributed to changes in drainage was the crustal rebound uplifting that took place as the weight of the overlying ice was removed. The flow patterns and configurations of the modern Great Lakes were fixed about 5000 years B. P., except for a lowering of lake levels and diversion of all discharge through the Erie–Ontario basins to the St. Lawrence drainage.

As continental glacial ice sheets retreated in the late Pleistocene, vast amounts of rock debris, moved and incorporated into the ice during former advances, were deposited in terminal moraines and laterally to the lobes of the retreating glacier. These deposits dammed up valleys and depressions in an irregular way and formed lake basins. Some depressions were below later groundwater levels; others were filled by meltwater and drainage from the surrounding topography. Morainal damming of preglacial valleys, usually at one end but occasionally at both ends, created many of the numerous lakes of the glaciated northern United States. Lake Mendota and many other Wisconsin lakes were formed in this way, and the Finger Lakes of New York underwent complex glacial modification in which morainic damming occurred at both ends of their deep, narrow valley basins (von Engeln, 1961).

In regions glaciated by continental ice sheets, a very common process of formation of lakes was associated with deposition of meltwater outwash left at the border of the retreating ice mass, or with blocks of ice buried in this debris. The major features of basins formed in this manner, which accounts for the origin of thousands of small lakes of northern North America, are shown in Figure 3–8. Glacial drift material was depos-

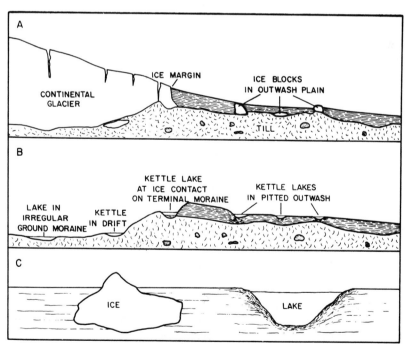

Figure 3-8 The formation of various types of kettle lakes, *A*, An outwash plain of retreating continental ice containing ice blocks; *B*, Lakes formed in the outwash plain and morainal till; *C*, Irregular slopes and shelves of a kettle lake formed by deposition of overburden till on an irregular block of ice. (From Hutchinson, G. E.: A Treatise on Limnology, Vol. 1, New York, John Wiley & Sons, Inc., 1957, after Zumberge, 1952.)

ited terminally and beneath the ice washed out into plains and often contained large segments of ice broken from the decaying glacier. Based on paleolimnological evidence, we know that these blocks of ice, deposited in glacial drift or in the outwash plain, often took several hundred years to melt completely (Florin and Wright, 1969), and resulted in the formation of *kettle lakes*. As the ice blocks melted, the morphometry of the resulting depressions was modified variously by the overburden of rock debris (Fig. 3–8C). The ensuing lake basins are highly irregular in shape, size, and slope, corresponding to the irregularities of the original ice blocks. Kettle lakes often exhibit variable underwater relief, with multiple depressions separated by irregular ridges and mounds. These lakes rarely exceed 50 m in depth, their shallow basin being related to the limiting depth of crevasse formation in the fracturing of the terminal glacier portions.

Most arctic lake basins result from glacial activity, as discussed above, or are *cryogenic lakes* formed from the effects of permafrost (Hobbie, 1973). The most abundant type of water body found in the Arctic is the shallow pond formed inside of an ice-wedge polygon which grows in the permafrost from water seepage through cracks in the surface of the ground. Eventually, polygonal networks of ridges are formed that often contain ponds from 10 to 50 m in diameter. Millions of these ponds exist in coastal northern Alaska, Canada, and flat regions of Siberia. The shallow ponds may coalesce into large ponds, or large amounts of ice deeper in the permafrost may melt, especially if the plant cover is disturbed or destroyed, and thereby form a shallow *thermokarst lake*.

A very large number of Arctic thaw lakes are elliptical, with the long axis oriented in a northeast-southwest direction across the prevailing winds. It is apparent, at least among the smaller elliptical lakes of this region of nearly constant wind, that the prevailing system of currents would tend to erode and thaw permafrost at the ends of the long axis of the ellipse lying across the wind (Livingstone, 1954, 1963, Livingstone, et al., 1958).

SOLUTION LAKES

Lake depressions can be created in any area from deposits of soluble rock that are slowly dissolved by percolating water. While many rock formations are readily soluble (salts such as sodium chloride [NaCl], calcium sulfate [$CaSO_4$], and ferric and aluminum hydroxides), most solution lakes are formed in depressions resulting from the solution of limestone (calcium carbonate, $CaCO_3$) by slightly acidic water containing carbon dioxide (CO_2). Solution lakes are very common in limestone regions of the World, notably the karst regions of the Adriatic, especially in Yugoslavia, the Balkan Peninsula, the Alps of Central Europe, and in Michigan, Indiana, Kentucky, Tennessee, and particularly Florida in North America.

Solution basins are usually very circular and conically shaped sinks, termed *dolines*, which develop from the solution and gradual erosion of the soluble rock stratum. Percolating surface or ground water dissolves limestone most readily at the joints and points of fault fracture through which it drains. Adjacent dolines may eventually fuse to form compound depressions that are more irregular in conformity. Alternatively, dissolution of limestone frequently occurs in large subterranean caves. Continued erosion of the superstructure by ground water results in its weakening to the point where the roof collapses, forming a reasonably regular conical doline.

The level of water in the solution basin is often highly variable. Usually, the depressions are sufficiently deep to extend well into the groundwater table and permanently contain water. Other basins that just reach the water table can undergo fluctuations in water level in response to seasonal and long-term variations in groundwater supply.

LAKE BASINS FORMED BY RIVER ACTIVITY

The running waters of rivers possess considerable erosive power that may create lake basins along their courses from elevated land to large lakes or the sea. In the upper reaches of rivers where gradients are steep, excavation by water can produce rock basins that may persist as lakes after the course of the river has been diverted. *Plunge-pool lakes*, excavated at the foot of waterfalls, provide a rare but spectacular example of such destructive fluvial action. Several large lakes of the Grand Coulee system of the State of Washington were surely plunge pools shaped by the former interglacial course of the Columbia River system.

A combination of destructive erosional and obstructive depositional processes occurs as rivers flow down more gentle slopes to form many lakes in the river flood plains. Many lakes were formed along large rivers when sediments of the main stream were deposited as levees across the mouths of tributary streams. In this way, the obstruction of the tributary flow continued until the side valley was flooded and a lateral lake was formed. Lateral lakes are frequently found in tributary valleys along major river systems in all continents, especially in the upper portion of the drainage. The reverse situation of fluviatile dams holding lakes in the main river channel occasionally occurs as a result of deposition by a lateral tributary. If more sediment than the main stream can remove is deposited, the resulting lake may be permanent.

Where rivers enter the relatively quiescent waters of a lake or the sea, sedimentation results in the formation of deltas, often of very large size. Occasionally, a delta is sufficiently large when entering a long, narrow lake on one side to form a barrier that divides the original lake in half. Much more common, however, are lakes that are formed in the deltas of all major rivers of the world. As the river velocity becomes reduced and sediments are deposited, the water tends to flow around the sediments in a U-shaped pattern where the open end extends seaward. As alluvial deposition extends further seaward, the inward depressions eventually are isolated as shallow lakes, often of large size. Subject to the influence of tides and other water movements of the sea, these *deltaic lakes* often receive salt water and frequently are brackish.

As a river meanders within the irregularities of the topography, greater turbulence and erosion occur on the outer, concave side of the river bend, while deposition occurs on the inner, convex side where currents and turbulence are reduced. With time, continued erosion and concavity takes place until the U-shaped meander of the river closes in upon itself (Fig. 3–9). The main course of the river cuts a channel through the initial portion of the meander, and levee deposits eventually isolate the loop, referred to as

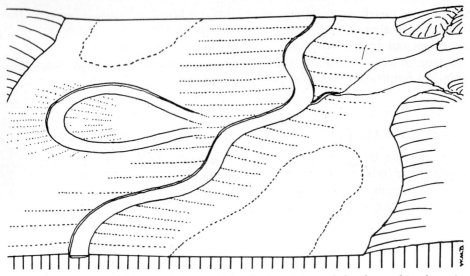

Figure 3-9 Diagram of lakes, including an isolated oxbow lake, and sloughs resulting from river activity. (From Davis, W. M.: Calif. J. Mines Geol. 29:175, 1933.)

an *oxbow lake*, from the river channel. The outer erosional side of oxbow lakes is usually deeper than the inner concave side.

WIND-FORMED LAKE BASINS

Wind action operates in arid regions to create lake basins by deflation or erosion of broken rock, or by redistribution of sand, resulting in the formation of *dune lakes*. Such lake depressions may be solely or partially the result of wind action, and the water they contain is often temporary and dependent upon fluctuations in climate.

Bayly and Williams (1973) differentiate several types of dune lakes based on those found in Australia and New Zealand, although they occur in many other parts of the world as well. Dune barrage lakes form behind sand dunes that are moving inland to block a river valley draining toward a coast. These lakes are typically triangular in shape, with the deepest part close to the sand dune, and are found inland in desert regions as well as in coastal areas. If a large amount of deflation occurs so that little or no sand is left on the floor between the trailing arms of parabolic dunes and an impervious rock floor is exposed, a lake may develop. Organic additions from vegetation assist in creating impervious, organically bonded sand-rock in dune depressions. Deflation depressions that are permeable to water can form lakes when they extend below an extensive water table. Numerous small, shallow lakes of this type occur along the eastern shore of Lake Michigan, among wind-blown depressions of the sand dunes (e.g. Barko, et al., 1977). All of these lakes are extremely transitory.

Deflation basins are also common where material is moved and eroded from horizontal strata of rock or clay. The wind-eroded material may be deposited in crescent-shaped mounds downwind or removed completely from the area, permitting the formation of large pans on nearly level areas, often called *playas.** Deflation basins commonly fill with water during rainy seasons or wet periods, and become increasingly saline with evaporation, and dry during opposing seasons or dry periods. These ephemeral lakes are common in large portions of Australia, South Africa, endorheic (see p. 40) regions of Asia and South America, and the plains and arid regions of the United States.

BASINS FORMED BY SHORELINE ACTIVITY

When the coast line of a large body of water, such as the sea or a large lake, possesses some irregularity or indentation, the potential exists for the formation of a bar across the depression to form a coastal lake. A longshore current flowing along the shore line and carrying sediment will, on encountering a bay, deposit the material in the form of a bar or spit across the mouth of the indentation (Fig. 3–10). Often the spit can eventually separate the bay from the sea or large lake to form a coastal lake.

Marine coastal lakes commonly result from bar formation across the mouths of old estuaries that have been inundated by rising water levels or slight subsidence of the land. Often river discharge and tidal currents are sufficient to prevent complete separation of the lake from the sea. The result is an alternation between fresh and brackish water in the lake, the salinity depending upon the ratio of freshwater inputs and salt water intrusions. Other coastal lakes are completely separated from the sea.

Numerous coastal lakes are found inland, adjacent to large lakes. Formation of these water bodies occurs in analogous fashion, by the deposition of bars across bays and river valleys. Spits are known to have formed on two sides of a lake, as a result of current patterns in relation to specific morphometry. The spits may then join in the middle, dividing the lake into two.

LAKES OF ORGANIC ORIGIN

The array of lakes created by the damming action of plant growth and associated detritus is incompletely known. It is clear that plant growth can be sufficiently profuse to dam the outlet of shallow depressions, and alter the drainage pattern of small lakes. The effectiveness of this method of water retention in flat regions, such as the arctic tundra, is unknown. The photosynthetic activities of attached blue-green algae and other plants can induce massive precipitation of $CaCO_3$ from calcareous waters. In time, these deposits can form barriers and isolate small bodies of water from stream systems (cf. Golubić, 1975).

Two mammals, the beaver and man, are particularly effective in constructing dams across river valleys to impound water into lakes. The American beaver created numerous large, long-lived lakes, many of which became permanent by means of sediment deposited against the dams. Man has created artificial lakes by damming streams for at least 4000 years. Only in the last two centuries, however, has this activity become

---

*Playa is a geological term for flat and generally barren lower portions of arid basins that periodically flood (a *playa lake*) and accumulate sediment. Most basins of playas have multiple origins (see detailed review of Neal, 1975).

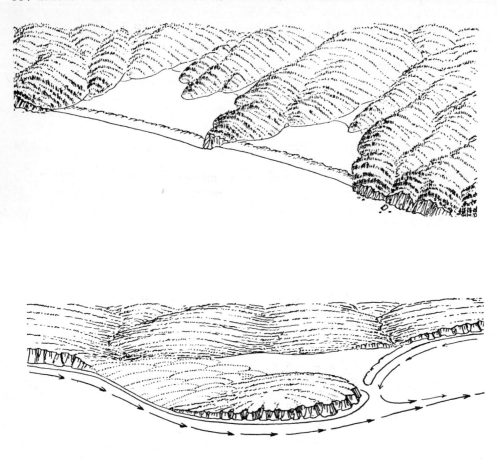

Figure 3-10 *Upper:* Coastal lakes formed by enclosure of lagoons by wave and current-built bars on a shore line embayed by slight subsidence. *Lower:* A land-bound island with a bay behind it, closed by a sand-reef beach formed by reverse eddy currents. (After Davis, W. M.: Calif. J. Mines Geol. 29:175, 1933.)

highly significant for the purposes of flood control, and the provision of power and water supplies for urban populations.

Reservoirs are being constructed on an unprecedented scale in response to the exponential demophoric demands of man. Plans for the construction of enormous reservoirs, approaching the size of the Laurentian Great Lakes, are underway in Canada and the Soviet Union. Such massive alterations of large drainage systems will result in major modifications in topography and regional climate that are not yet fully recognized or even partially understood. A great need exists for careful planning in the construction of reservoirs, because deleterious effects commonly exceed expected benefits (e.g. Dussart, et al., 1972).

Small shallow reservoirs, in the form of farm ponds and moderate-sized inundations, have created literally millions of small lakes. Particularly characteristic of the morphometry of these lakes is that they are generally shallow and possess large areas where macrophytic vegetation can grow. Such plant life radically alters the productivity of the lake system. Moreover, small reservoirs generally receive high nutrient inputs in relation to their volume, which further increases their productive capacity. Most reservoirs are relatively short-lived because of the extensive sediment load delivered by the influents.

MORPHOLOGY OF LAKE BASINS

The morphology of a lake basin has important effects on nearly all of the physical, chemical, and biological parameters of lakes, as will be shown repeatedly in the ensuing chapters. The forms of lake basins are extremely varied, and reflect their modes of origin, subsequent modifying events of water movements within the basin, and the degree of loading from the drainage basin.

The morphology of a lake is best described by a detailed bathymetric map which is required for evaluation of all major morphometric parameters. The preparation of such a map requires a survey of the shore line by standard methods, often in combination with aerial photography. From the map of the shore line a detailed bathymetric map of depth contours must be constructed using a series of accurate soundings along intersecting transects. Methods commonly employed for both lake and stream mapping are discussed by Welch (1948), Wetzel and Likens (1979), and Håkanson (1981). Accurate sonar transects, which permit acquisition of much greater detail than was previously possible by manual sounding methods, are now used frequently.

MORPHOMETRIC PARAMETERS

The many morphometric parameters that can be determined from a detailed bathymetric map are discussed at length by Hutchinson (1957). The most commonly used parameters are defined here.

Maximum Length (l).   The distance on the lake surface between the two most distant points on the lake shore. This length is the maximum effective length or fetch for wind to interact on the lake without land interruption.

Maximum Width or Breadth (b).   The maximum distance on the lake surface at a right angle to the line of maximum length between the shores. The mean width $(\bar{b})$ is equal to the area divided by the maximum length; $\bar{b} = A/l$.

Area (A).   The area of the surface and each contour at depth z is best determined by planimetry (cf. Welch, 1948; Wetzel and Likens, 1979) or, less precisely, by a grid enumeration analysis (Olson, 1960).

Volume (V).   The volume of the basin is the integral of the areas of each stratum at successive depths from the surface to the point of maximum depth. The volume is closely approximated by plotting the areas of contours, as closely spaced as possible, against depth and the area of this curve, integrated by planimetry, corresponds to basin volume. Alternatively, the volume can be estimated by summation of the frusta of a series of truncated cones of the strata:

$$V = \frac{h}{3} (A_1 + A_2 + \sqrt{A_1 A_2})$$

where h is the vertical depth of the stratum, A_1 the area of the upper surface, and A_2 the area of the lower surface of the stratum whose volume is to be determined.

Maximum Depth (z_m).   The greatest depth of the lake.

Mean Depth $(\bar{z})$.   The volume divided by its surface area; $\bar{z} = V/A_0$.

Relative Depth (z_r).   The ratio of the maximum depth as a percentage of the mean diameter of the lake at the surface, expressed as a percentage.

$$z_r = \frac{50\ z_m\ \sqrt{\pi}}{\sqrt{A_0}}.$$

Most lakes have a z_r of less than 2 per cent, whereas deep lakes with a small surface area usually have $z_r > 4$ per cent.

Shore Line (L).   The intersection of the land with permanent water is nearly constant in most natural lakes. The shore line, however, can fluctuate widely in ephemeral lakes and especially in reservoirs in response to variations in precipitation and discharge. The length of the shore line can be determined directly or from maps with a map measurer (chartometer; rotometer; cf. Welch, 1948; Wetzel and Likens, 1979).

Shoreline Development (D_L).   The ratio of the length of the shore line (L) to the circumference of a circle of area equal to that of the lake

$$D_L = \frac{L}{2\ \sqrt{\pi A_0}}.$$

Very circular lakes, such as crater lakes and some kettle lakes, approach the minimum shoreline development value of unity. The conformation of most lakes, however, deviates strongly from the circular. Many are subcircular or elliptical in form, with D_L values of about 2. A more elongated morphometry increases the value of D_L markedly, as for example is found in the dendritic outlines of lakes occupying flooded river valleys. Shoreline development is of considerable interest because it reflects the potential for greater development of littoral communities in proportion to the volume of the lake.

HYPSOGRAPHIC AND VOLUME CURVES

The hypsographic curve, or depth-area curve, is a graphic representation of the relationship between the surface area of a lake and its depth. This relationship may be expressed in terms of per cent of the lake area overlying a given depth contour, or in absolute units such as m^2, hectares, or km^2 (Fig 3–11). The depth-volume curve is closely related to the hypsographic curve, and represents the relationship of lake volume to depth. Similarly, the units of expression can be either in per cent of total lake volume above a specific depth, or in actual volume units.

The importance of these area and volume curves in limnological investigations stems from the relationship between lake morphology and biological productivity. The hypsographic curve represents the relative proportion of the bottom area of the lake, which is included between the strata under consideration. However, it is only an approximation of the area of the exposed lake bottom, since areal measurements are related to the plane of the lake surface, whereas the actual area of the sediments is

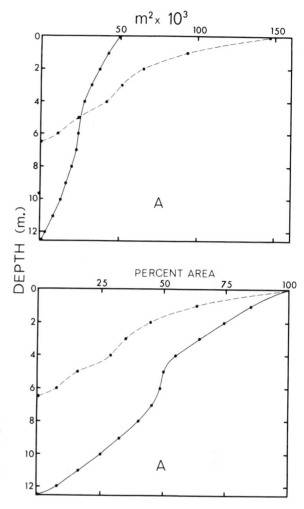

Figure 3-11 Hypsographic (depth-area) curves (A) and depth-volume curves (B) of oligotrophic Lawrence Lake (————) and eutrophic Winter-green Lake (— — — —), southwestern Michigan. (From Wetzel, unpublished data.)

greater. In lakes with otherwise comparable conditions, biological productivity is generally greater in those with more interfacing zones of photosynthetic production and of decomposition (Thienemann, 1927; Strøm, 1933; Rawson, 1955, 1956). The extent of shallow water in a lake is a determining factor in the interrelationship of these zones (cf. Chapter 6), and determines the area available for growth of rooted aquatic plants and associated littoral communities (cf. Chapters 18 and 19).

The ratio of mean to maximum depth ($\bar{z}:z_m$) is an expression similar to the ratio of the volume of the lake to that of a cone of basal area A and height z_m $\left[A\bar{z}/ \left(\frac{1}{3} z_m A \right) = 3 \; \bar{z}/z_m \right]$. The ratio $\bar{z}:z_m$ thus gives a comparative value of the form of the basin in terms of volume development. For most lakes the value $\bar{z}:z_m$ is >0.33, which is the value that would be given by a perfect conical depression. The ratio exceeds 0.5

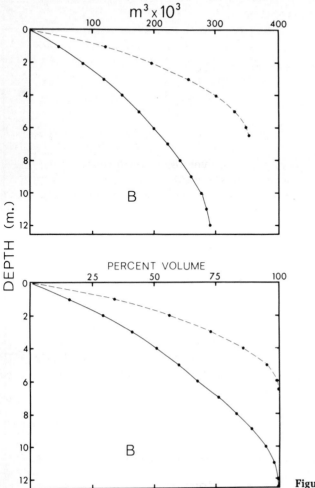

Figure 3-11 *Continued*

in many caldera, graben, and fjord lakes, whereas most lakes in easily eroded rock usually have ratios between 0.33 and 0.5. Very low values of $\bar{z}:z_m$ occur only in lakes with deep holes, such as solution or kettle lakes.

In his examination of the morphometry of a large number of lakes, Neumann (1959) has shown that the average shape of lake basins approximates an elliptic sinusoid (Fig. 3-12). The elliptic sinusoid is a geometric body whose base is an ellipse, so that planes perpendicular to the base of an ellipse passing through the center of the latter intersect the surface of the body along troughs of sine curves. The volume of such an elliptic sinusoid is:

$$V = 4 \left(1 - \frac{2}{\pi} \right) abz_m = 1.456 \ abz_m,$$

where a and b are the half-axes of the lake surface ellipse. Since the area of the lake surface ellipse concerned is πab, $\bar{z} = V/A = 0.464 z_m$, the ratio $\bar{z}:z_m = 0.464$. This value

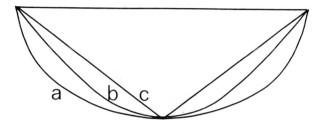

Figure 3-12 Vertical cross-sections through three forms of lake basins: *a*, half an ellipsoid of revolution; *b*, elliptic sinusoid; and *c*, right elliptic cone. (Modified from Neumann, 1959.)

is very close to the average value (0.467) of the ratio for over 100 lakes that have been evaluated (cf. also Hayes, 1957; Anderson, 1961; Koshinsky, 1970). Therefore the elliptic sinusoid serves as a good model for an average lake in which irregularities and sub-merged depression individuality are not severe. The mean depth of an average lake is slightly less than one-half (0.46) of its maximum depth.

Since the pioneering work of Thienemann, much attention has been focused on the importance of lake morphometry, especially mean depth, to lake productivity in relation to the effects of climatic and edaphic factors (Hutchinson, 1938; Rawson, 1952, 1955, 1956; Edmondson, 1961; Patalas, 1961; Hayes and Anthony, 1964). Mean depth is regarded as the best single index of morphometric conditions and clearly shows a general inverse correlation to productivity at all trophic levels among large lakes. This relationship deteriorates among small lakes and indicates, as will become apparent in later discussions, that regulation of the dynamics of metabolism and productivity in aquatic ecosystems is multifaceted. Morphometry is only one, although important, interacting parameter.

SUMMARY

1. Some 40 per cent of the total volume of fresh water is contained in the great lake basins. Most lakes and reservoirs are much smaller, however, and are concentrated in the subarctic and temperate regions of the Northern Hemisphere. Most of the millions of lakes are small and relatively shallow, usually < 20 m in depth.

2. Most natural lakes were formed by catastrophic events.

    a. Tectonic lake basins are depressions formed by displacements of the earth's crust. Many of the deepest relict lakes of the world, such as *graben* lake basins, were formed by faulting movements. Uplifting of the earth has created a number of lake basins; these basins were often modified by glacial scouring activity.

    b. Lakes can result from volcanic activity. Small, deep *maar* lakes can form in volcanic cones; collapse of the roof of partially emptied magmatic chambers can result in large, deep *caldera* lakes within a volcano. Lava flows often dam preexisting river valleys and create isolated lakes.

    c. Temporary or permanent lakes can result from *landslides* into stream valleys. Lakes can also form in depressions created behind the landslide material.

d. The erosional and depositional activity of glaciers was the most important agent of lake formation. Many lakes were formed from the outwash morainal deposits at the retreating edges of continental glaciers or from melted blocks of ice buried in this morainal debris *(kettle lakes)*. In mountainous regions terminal and lateral morainal deposits can effectively dam river valleys and form lakes. Glaciers often scour lake basins in glaciated valleys of mountainous areas and form amphitheater-shaped depressions termed *cirque lakes*. Several cirque lakes can form in the trough of a mountain valley in a series of connected lakes of successively lower elevation *(paternoster lakes)*. Many arctic and subarctic lakes also formed from glacial activity. Some arctic lakes *(cryogenic lakes)* are very shallow and form by water seepage into the permafrost which, on freezing, forms a polygonic network of ridges that contains subsequent melt water.

3. Other natural lakes form by gradual events.
   a. Solution lakes result from sinks, termed *dolines,* formed by the gradual dissolution of rock, such as limestone, along fissures and fractures. Eventually the superstructure is weakened to the point of collapse into the depression.
   b. The erosional and depositional action of river water can isolate depressions to form lakes (e.g. *plunge-pool lakes* below former waterfalls, *oxbow lakes* of former river channels).
   c. Wind erosion can form shallow depressions which often contain water temporarily or seasonally (e.g. *dune lakes* in sandy areas, *playa lakes* of flat, arid or semi-arid regions).
   d. *Coastal lakes* often form along irregularities in the shore line of the sea or large lakes. Longshore currents deposit sediments in bars or spits that eventually isolate a fresh or brackish-water lake.
   e. *Reservoirs* are impoundments created largely by man by damming of river valleys. Because of high rates of sedimentation, reservoirs, smaller inundation lakes, and farm ponds are very short-lived.

4. Variation in basin morphology is great; most lake basins approximate an elliptic sinusoid shape. The mean depth of lakes is about half (0.46) of the maximum depth.

5. The morphology of a lake basin has profound effects on nearly all physical, chemical, and biological properties of lakes. Morphometry of lake basins and geological substrates of the drainage basin influence sediment–water interactions and lake productivity, especially extremely productive littoral communities. Greater productivity of large lakes is usually correlated with higher water-sediment interface area per water volume (i.e. lower mean depth).

CHAPTER FOUR

WATER ECONOMY

The balance of water in lakes is expressed by the basic hydrological relationship in which change in water storage is governed by inputs from all sources less water losses. Water income from precipitation, surface influents, and groundwater sources is balanced by outflow from surface effluents, seepage to ground water, and evapotranspiration. Each of these incomes and losses varies seasonally and geographically, and is governed by the characteristics of the lake basin, its drainage basin, and the climate.

The distribution of water over continental land masses depends on the global hydrological cycle, in which excessive oceanic evaporation is counterbalanced by greater precipitation over land. The hydrological cycle, which can be altered by extensive man-induced changes of surface water systems, determines the distribution of lakes regionally in relation to the distribution of suitable catchment lake basins.

THE HYDROLOGICAL CYCLE

The hydrological cycle of the earth basically evaluates the cyclical budgetary processes of water movement. These processes include movement from the atmosphere, inflow, and temporary storage on land, and outflow to the primary reservoir, the oceans. The cycle consists of three principal phases: precipitation, evaporation, and surface and groundwater runoff. Each phase involves transport, temporary storage, and a change in the physical state of the water.

EVAPORATION AND PRECIPITATION

Evaporation of water into the atmosphere occurs from land, the oceans, and other water surfaces. Major sites include evaporation from precipitation, from precipitation intercepted by vegetation, from the oceans, lakes, and streams, from soils, and from transpiration of plants. The evaporation rates of each of these and other sources of water are determined by an array of dynamic environmental parameters (cf. Meinzer, 1942; Grey, 1970; and others). The mass of atmospheric water vapor, although it is smallest in relation to the amount of total global water, is stored as vapor for the least amount of time (average renewal time of 8.9 days) before returning to the earth as rain, snow, sleet, hail, or condensates (dew and frost), either on land or on the oceans.

The precipitated water may be intercepted or transpired by plants, may run off over the land surface to streams (surface runoff), or may infiltrate the ground (Fig. 4-1).

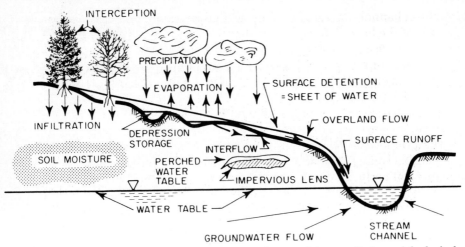

Figure 4-1 Simplified representation of the major pathways of the runoff phase of the hydrological cycle. (From Grey, D. M., ed.: Handbook on the Principles of Hydrology. Ottawa, National Research Council of Canada, 1970. Reprinted by permission of Water Information Center, Inc., Manhasset Isle, Port Washington, New York, after Davis and DeWeist, 1966.)

Much of the intercepted water and surface runoff (up to 80 per cent) is returned to the atmosphere by evaporation. Infiltrated water may be temporarily stored (average renewal time approximately 280 days) as soil moisture prior to being evapotranspired. Some of the water percolates to deeper zones to be stored as groundwater (average renewal time of 300 years). Groundwater is actively exchanged and may be used by plants, flow out as springs, or seep to streams as runoff. The runoff phase is exceedingly complex and variable because of the extensive involvement with biota, the extreme heterogeneity of soil structure and composition, and variations in climate.

RUNOFF FLOW PROCESSES

The soil and rock substrate of the lake drainage basins is of major importance in regulating the pathways and rates of hillslope runoff of water received as rain and melt-water (cf. review of Dunne, 1978). These pathways of runoff influence many characteristics of the landscape, the uses to which land can be put, and the requirements for effective land management.

When the rate of rainfall or meltwater influx exceeds the absorptive capacity of the soil, the excéss water flows over the surface as overland flow. Overland flow is most common in arid and semiarid regions; it also is found in humid areas where the original vegetation and soil structure have been disturbed, or in areas where normally porous soil contains a thin layer of frost and consequently cannot absorb meltwater (Dunne and Black, 1970a, 1970b, 1971).

When precipitation is first absorbed by the soil, the water may be stored or it may move by gravity toward stream channels along several pathways (Dunne, 1978). If the soil or rock is deep and of relatively uniform permeability, the subsurface water moves vertically to the zone of saturation, and then follows a generally curving path to the nearest stream drainage channel. This simple pattern of groundwater flow often is disrupted by irregularities of the base geological structure (Davis and DeWiest, 1966; Leopold, Wolman, and Miller, 1964). Rates of groundwater flow are generally low and the pathways long; hence much of this groundwater contributes to the sustained baseflow of streams between periods of precipitation. Drainage of water from storms is added to this more uniform groundwater discharge. Although the long route of groundwater flow generally dominates the baseflow of streams, the rate of flow can increase in very permeable rock formations such as limestones and jointed basalts, and add significantly to runoff from stormflows to recipient drainage streams.

When shallow surface soil and weathered rock is permeable and underlaid with relatively impermeable soil horizons, percolating water will be diverted horizontally as subsurface stormflow. The shallow subsurface pathway of drainage to a stream channel or lake basin is much shorter than the pathway of groundwater flow, and generally occurs in soil of high permeability. Therefore, subsurface stormflow can be volumetrically greater than runoff, particularly along steep gradients in narrow drainage valleys.

As subsurface water flows down the gradient of a hillslope, vertical and horizontal percolation can saturate the soil. Shallow subsurface flow encountering these saturated areas emerges from the soil surface as return flow, and continues to the recipient channel or basin as overland flow. Similarly, direct precipitation onto saturated areas flows over the surface of the soil. These processes assume greater significance in gently sloping, wide valleys with thin soils.

Each of these processes of rainfall or meltwater runoff responds differently to variations in topography, soil, and characteristics of precipitation, and indirectly to variations in climate, vegetation, and land use. Therefore runoff flow processes control the volume, periodicity, and chemical characteristics of contributions to receiving streams and lake basins.

GLOBAL WATER BALANCE

The global water balance reflects the fact that more water evaporates from the oceans than is returned via precipitation (Fig. 4–2), whereas on the land more water is received via precipitation than is lost by evaporation. A majority of continental water income originates from evaporation of the oceans. The water of land masses is not uniformly distributed over the major continents (Table 4–1). The total and groundwater runoff are greatest in South America, nearly twice that per area of other continents.

Until recent times, the global water balance fluctuated very little. Man has introduced regional fluctuations by extensive environmental modifications for irrigation, industrial, and domestic uses, such as land clearing to obtain more arable land, changes in drainage patterns, and exploitation of groundwater reserves (Flohn, 1973). Water demand for agricultural and industrial purposes is projected to increase the man-made fraction of continental evaporation from its present nearly 3 per cent to 10 per cent in approximately 30 years, and to 50 per cent in about 70 years. The result will be an accelerated rate of continental freshwater turnover. Since freshwater supplies are inadequate where demand is highest, large-scale desalinization of seawater, at great

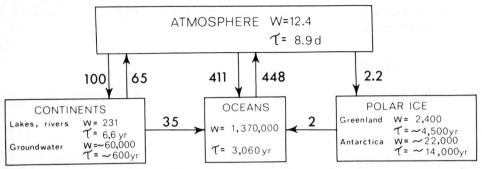

Figure 4-2 Global water balance. W = water content in 10³ km³, values on arrows = transport in 10³ km³ yr⁻¹, and τ = retention time. Estimate of ground water is to a depth of 5 km in the earth's crust; much of this water is not actively exchanged. (Modified from Flohn, 1973, after Lvovitch.)

energy expense, is the only practical contemporary alternative for continued expansion of fresh waters at sites of high demand. Manipulation of surface waters on a massive scale will inevitably lead to irreversible modifications of climate. Modifications of regional climatic conditions have already occurred as a result of extensive alterations of large river systems in which surface waters and evaporation are greatly increased.

Among the continental land masses, three hydrological regions have been recognized (Hutchinson, 1957). The distribution of lakes is related partly to distribution of lake basins and partly to that of water. *Exorheic regions*, within which rivers originate and from which they flow to the sea, contain the major lake districts of the world and most of the lakes. *Endorheic regions*, within which rivers arise but never reach the sea, occur between subtropical deserts and the tropical and temperate humid regions. *Arheic regions*, within which no rivers arise, are desert areas that occur in the latitudes of the trade winds, and between which lies the zone of equatorial rains. Endorheic

TABLE 4-1 Water Resources and Annual Water Balance of the Continents of the World*

	EUROPE†	ASIA	AFRICA	N. AMERICA‡	S. AMERICA	AUSTRALIA§	TOTAL
Area (10⁶ km²)	9.8	45.0	30.3	20.7	17.8	8.7	132.3
Precipitation (km³)	7165	32690	20780	13910	29355	6405	110305
River runoff (km³)							
Total	3110	13190	4225	5960	10380	1965	38830
Underground	1065	3410	1465	1740	3740	465	11885
Surface	2045	9780	2760	4220	6640	1500	26945
Total soil moistening (infiltration and renewal of soil moisture)	5120	22910	18020	9690	22715	4905	83360
Evaporation	4055	19500	16555	7950	18975	4440	71475
Underground runoff (% of total)	34	26	35	32	36	24	31

*After data from Lvovitch, 1973.
†Includes Iceland.
‡Includes Central America but not the Canadian archipelago.
§Includes New Zealand, Tasmania, and Papua New Guinea.

regions, transitional in nature between the other two, can shift to exorheic or arheic characteristics with relatively small changes in climate.

WATER BALANCE IN LAKE BASINS

The water balance of a lake is evaluated by the basic hydrological equation in which the change in storage of the volume of water in or on the given area per unit time is equal to the rate of inflow from all sources less the rate of water loss. Water income to a lake includes several sources:

(a) *Precipitation directly on the lake surface.* Although most lakes, largely in exorheic regions, receive a relatively small proportion of their total water income from direct precipitation, this percentage increases in very large lakes. Extreme examples include equatorial Lake Victoria, which receives most (>70 per cent) of its water from precipitation on its surface, and the endorheic Dead Sea, which receives practically no water from direct precipitation.

(b) *Water from surface influents of the drainage basin.** The amount of the total water income to a lake from surface influents is highly variable. Lakes of endorheic regions receive nearly all their water from surface runoff. The rate of runoff from the drainage basin and corresponding changes in lake level are strongly influenced by the nature of the soil and vegetation cover of the drainage basin. One of the best examples of this effect resulted from the experimental forest cutting and use of herbicides to prevent vegetation regrowth in the Hubbard Brook drainage in New Hampshire (Likens, et al., 1967, 1970, 1977; Bormann and Likens, 1979). Annual stream-flow increased 39 per cent the first year and 28 per cent the second year above the values prior to selective deforestation.

(c) *Groundwater seepage below the surface of the lake.* Seepage of groundwater is commonly a major source of water for lakes in rock basins and lake basins in glacial till that extend well below the water table. Sublacustrine groundwater seepage forms the major source of water flow into, and out of, karst and doline lakes of limestone substrata.

(d) *Groundwater entering lakes as discrete springs.* Sublacustrine springs from groundwater occur frequently in hard-water lakes of calcareous drift regions, where the basin is effectively sealed from groundwater seepage by deposits within the basin. For example, groundwater largely from springs contributed nearly 40 per cent of the annual water income to calcareous Lawrence Lake, Michigan, in comparison to 10 per cent from precipitation, and the remainder from surface runoff in two streams (Table 4-2). Groundwater inputs were closely related to rates of precipitation and evapotranspiration each month.

In *drainage lakes,* loss of water occurs by *flow from an outlet* (Birge and Juday, 1934), and in *seepage lakes* losses occur by seepage *into the groundwater* through the

---

*Reference is to the drainage or catchment area *(Einzugsgebiet; bassin versant)* which is, in American usage, equivalent to watershed, the region or area drained by a river. Watershed, as used in the United Kingdom, refers to the ridge or crest line dividing two drainage areas, and is defined similarly in German *(Wasserscheide)* and French *(ligne de partage des eaux).*

TABLE 4-2   Annual Water Budget for
Lawrence Lake, Michigan, 1971*

SOURCE/LOSS	10^3 m^3	PER CENT
Inputs		
Inlet 1	146.6	32.1
Inlet 2	87.5	19.1
Ground water	178.1	39.0
Precipitation	44.6	9.8
Total inputs	456.8	100.
Outputs		
Outlet	436.5	90.4
Evapotranspiration	46.2	9.6
Seepage losses†	0.	0.
Total outputs	482.7	100.

*After Wetzel and Otsuki, 1974.
†Indirect evidence indicated that seepage losses were negligible. Subsequent direct measurements have shown that seepage losses are negligible in Lawrence Lake (Wetzel, unpublished data).

basin walls. Deposition of clays and silts commonly forms a very effective seal in drainage lakes, from which most or all of the outflow leaves by the outlet (e.g., Table 4-2). In seepage lakes the sediments over much of the deeper portions of the basin also often form an effective seal; losses to groundwater usually occur from the upper portions of the basin in these lakes.

Further losses of water come about *directly by evaporation, or by evapotranspiration* from emergent and floating-leaved aquatic macrophytes. The extent and rates of evaporative losses are of course highly variable according to season and latitude, and are greatest in endorheic regions. Lakes of semiarid regions commonly have no outflow and lose water only by evaporation. Such lakes are termed *closed* in contrast to *open lakes* that have outflow by an outlet or seepage.

Evaporative losses are greatly modified by the transpiration of emergent and floating-leaved aquatic plants, a subject treated in detail by Gessner (1959). Rates of transpiration and evaporative losses to the atmosphere vary greatly with an array of physical (e.g., wind velocity, humidity, temperature) and metabolic parameters, as well as with species (Brezny, et al., 1973). Further, plant growth and evapotranspiration are predominantly seasonal in lakes of exorheic regions; in tropical lakes, many of the large hydrophytes are perennials and grow more or less continually. In most situations, transport of water from the lake to the air is greatly increased by a dense stand of actively growing littoral vegetation, as compared to evaporation rates from open water (Table 4-3). Since a majority of lakes are small and often possess well-developed littoral flora, these communities contribute significantly to the water balance of many lakes.

Analytical techniques for a detailed evaluation of the water balance of a lake and its drainage basin are complex and require a great deal of work and effort. Because of numerous local fluctuations in climate from year to year, analyses should extend over several years. Methodology varies greatly in relation to the objectives, and is beyond

TABLE 4-3 Comparison of Water Loss from a Stand of the Emergent Aquatic Macrophyte *Phragmites communis* to that of Open Water, Berlin, Germany, 1950*

DATE	EVAPOTRANSPIRATION (kg m⁻² day⁻¹)	EVAPORATION (kg m⁻² day⁻¹)	RATIO OF TRANSPIRATION TO EVAPORATION
11 May 1950	3.20	3.24	1.0
25 May 1950	2.50	1.44	1.6
27 July 1950	9.82	2.24	4.4
22 August 1950	16.01	2.29	7.0
17 October 1950	2.79	0.79	3.9

*Modified from Gessner, 1959, after Kiendl.

the scope of this work. Those interested in such analyses are referred to specific hydrological and isotopic compilations of methods such as those of Chow (1964), Stout (1967), Gray (1970), Rodda, et al. (1976) and Kirkby (1978).

SUMMARY

1. The hydrological cycle consists of the global processes affecting the distribution and movement of water.
   a. Greater evaporation from the oceans is counterbalanced by greater precipitation onto land masses.
   b. Although the amount of water in the atmosphere is small, its retention time is low, and it cycles on the average every 9 days.
   c. Once moved from sites of evaporation, water is returned by precipitation. Much of this water is returned to the atmosphere by evaporation and plant transpiration.
   d. On land, water is absorbed by soil, stored within ground water, and moves by gravity to stream channels and lake depressions. Retention times within groundwater reservoirs are variable, and depend on the composition of the soil and rock, slope gradients, vegetation cover, and climate. Groundwater flow rates are generally slow and the pathways long.
   e. Retention times in lakes are generally short (6-7 years on the average).
   f. Man's modification of the environment can result in alterations in the global water balance and the climate.
2. Changes in water storage and retention in lakes result from alterations in the balance between input rates from all sources and rates of water losses.
   a. Water income results from:
      i. Precipitation directly on the lake surface.
      ii. Water from surface influents of the drainage basin.
      iii. Groundwater seepage below the surface of the lake through the sediments or as discrete subsurface springs.

b. Losses of water from lakes occur by:
    i. Flow from an outlet in the most common *drainage lakes* or by seepage through the basin walls into the ground water in *seepage lakes*.
    ii. Direct evaporation from the lake surface.
    iii. Evapotranspiration from emergent and floating-leaved aquatic plants.

Solar radiation is of fundamental importance to the entire dynamics of freshwater ecosystems. Almost all energy that controls the metabolism of lakes is derived directly from the solar energy utilized in photosynthesis, either autochthonously (within the lake) or allochthonously (within the catchment basin and brought to the lake in various forms of organic matter). The utilization of this energy received by the lake and from its drainage basin, and factors that influence the lake's efficiency of conversion of solar energy into potential chemical energy, are basic to lake productivity.

In addition to these effects, the absorption of solar energy and its dissipation as heat have profound effects on the thermal structure, water mass stratification, and circulation patterns of lakes. The array of attendant effects on nutrient cycling, distribution of dissolved gases and biota, and behavioral adaptations of organisms exert major controls on the environmental milieu. Therefore, the optical properties of lakes and reservoirs are important regulatory parameters in the physiology and behavior of aquatic organisms. They deserve detailed scrutiny.

LIGHT AS AN ENTITY

The term *light* is often confusing, in part because of different usages in absolute physical terms, reactions of visual receptors to light, and responses of plants to light energy. For the purposes of this discussion, it is essential that it be viewed physically, as part of the radiant energy of the electromagnetic spectrum. Light is energy, that is, something that is capable of doing work and of being transformed from one form into another, but it can neither be created nor destroyed. Radiant energy is transformed into potential energy by biochemical reactions, such as photosynthesis, or to heat. Energy transformations are far from 100 per cent efficient in a system such as a lake, and most of the radiant energy is lost as heat.

ELECTROMAGNETIC SPECTRUM

The electromagnetic spectrum is expressed as units of frequency and wavelength. At one extreme are cosmic rays of very high frequency (10^{24} cps) and short wavelength (10^{-14} cm), and at the other are radio and power transmission waves of low frequency (1 cps) and long wavelength (to 10^{10} cm). For all practical purposes, solar radiation constitutes all of the significant energy input to aquatic systems. This solar flux of energy

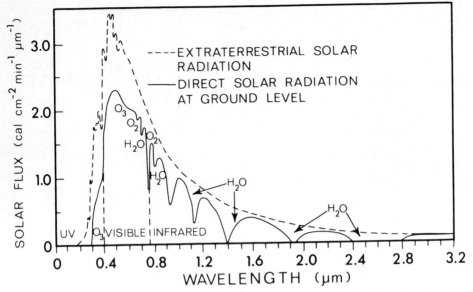

Figure 5-1 Extraterrestrial solar flux and that at the surface of the earth showing major absorption bands from atmospheric O_2, O_3, and water vapor. (Modified from Gates, 1962.)

consists of wavelengths of 100 to >3,000 nm (1000 to >30,000 Å), from the ultraviolet (UV) to infrared radiation, as it is received extraterrestrially (Fig. 5-1). As solar radiation penetrates and diffuses in the atmosphere of the earth, the energy of certain wavelengths is strongly absorbed and attenuated by scattering. The visible portion of the spectrum, with maximum energy flux in the blue and green portions (480 nm) of the visible range, is only a small amount of the total energy radiated by the sun (Fig. 5-1). Ultraviolet energy is strongly absorbed by ozone and oxygen, and infrared wavelengths are absorbed by water vapor, ozone, and carbon dioxide.

With respect to the mechanisms by which life receives light energy, it is important to view light as the radiation of packets of energy termed *quanta* or *photons*. A photon is a pulse of electromagnetic energy; as this energy propagates it has an electric (E) and magnetic (H) field, with respect to direction of flux and wave characteristics of wavelengths (λ) and amplitude (A) (Fig. 5-2). Hence, light is effectively a transverse wave of energy that behaves as a movement of particles with defined mass. The photon carries energy in a wave conformation.

ABSORPTION OF LIGHT

Absorption of light by atoms and molecules can occur when the electrons of the atoms and molecules resonate at frequencies that correspond to a photon's energy state. In the collision of an electron and a photon, the electron gains the quantum of energy lost by the photon. It is important to keep this basic photochemical relationship in mind, since the quantum energy imparted by the photon functions in relation to frequency, and each molecular or atomic species has a unique set of absorption characteristics or bands. Life responds to quantum energy of photons at specific frequencies.

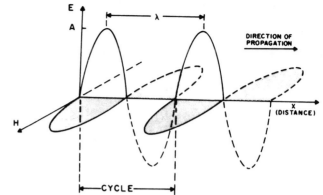

Figure 5-2 Instantaneous electric (*E*) and magnetic (*H*) field strength vectors of a light wave as a function of position along the axis of propagation (*x*), showing the amplitude (*A*), wavelength (λ), cycle, and direction of propagation. (From Bickford, E. D., and Dunn, S.: Lighting for Plant Growth. Kent, Ohio, Kent State University Press, 1972.)

If the energy distribution of solar flux is plotted against wavelength, as was done in Figure 5-1, the maximum monochromatic intensity of sunlight appears to occur in the blue-green portion of the visible spectrum, an illusion caused by the manner of presenting the data. It is more meaningful to express energy against frequency, where the area under any portion of the curve is directly proportional to energy (Fig. 5-3). The

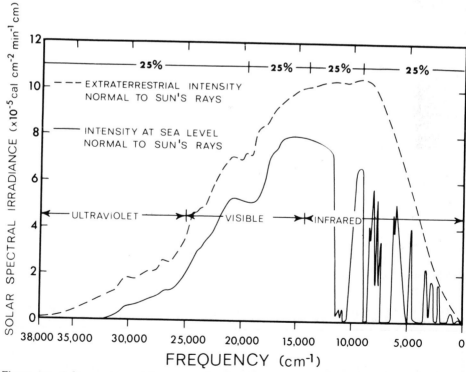

Figure 5-3 Solar spectrum at the mean solar distance from the earth as received outside of the earth's atmosphere and at sea level on a surface perpendicular to the solar rays (solar constant of 2.00 cal cm^{-2} min^{-1}). (Modified from Gates, 1962.) Frequency (cm^{-1}) of the abscissa refers to the wavenumber (v^1), or number of wavelengths per cm, and is therefore equal to the reciprocal of the wavelength.

TABLE 5-1   Speed of Light (589 nm, Sodium
                  D-lines)

MEDIUM	SPEED (cm sec^{-1})
Vacuum	2.9979×10^{10}
Air (760 nm, 0°C)	2.9972×10^{10}
Water	2.2492×10^{10}
Glass	1.9822×10^{10}

energy of the photon in the electromagnetic spectrum is proportional to frequency and inversely proportional to wavelength in accordance with Planck's Law:

$$\epsilon = h\nu$$

where:

ϵ = energy of the photon, ergs
h = Planck's constant, $6.63 \cdot 10^{-27}$ erg-seconds
ν = frequency of the radiation in cycles per second.

When energy is expressed against frequency (Fig. 5–3), the true maximum energy of solar irradiance is found in the infrared at wavelengths somewhat greater than 1000 nm or 1μm. The median value of irradiance occurs in the near infrared at a frequency of 14,085 cm^{-1} (= a wavelength of 710 nm), slightly above the visible range. A major portion (29 per cent) of the incoming radiation occurs at wavelengths greater than 1,000 nm (frequency <10,000 cm^{-1}) and 50 per cent beyond the red portion of the visible range. This distribution of energy shifts somewhat as the solar radiation passes through the earth's atmosphere (Fig. 5–3). The point to be made, however, is that *a large portion of irradiance impinging on the surface of a lake is in the infrared portion of the solar spectrum and has major thermal effects on the aquatic system.*

Wavelength (λ) is a quantitative parameter of any periodic wave motion, not only of light but also of water movements, as will be discussed later on. It is defined simply as the linear distance between adjacent crests of waves, and is equal in cm to the speed of light (c = 2.998×10^{10} cm sec^{-1}) divided by the frequency (ν) in cycles per second (cps):

$$\lambda = c/\nu$$

Wavelength is also often expressed as the wave number (ν^1) [or k], which is the number of wavelengths per cm, or the reciprocal of the wavelength:

$$\nu^1 = 1/\lambda$$

The speed of light is reduced in a roughly linear fashion as it passes through transparent materials of increasing density (Table 5–1). (A conversion table for commonly used units of length and irradiance is given in the Appendix.)

Light Impinging on Lakes

The amount of solar energy that reaches the surface of a lake is dependent upon an array of dynamic factors. The amount of direct solar energy per unit of time from

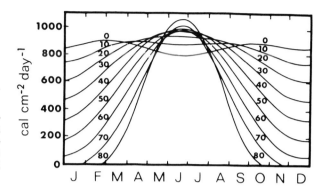

Figure 5-4 Daily totals of the undepleted solar radiation received on a horizontal surface for different geographical latitudes as a function of the time of year (solar constant 1.94 cal cm^{-2} min^{-1}). (After Gates, 1962.)

the sun, incident upon a surface outside the atmosphere perpendicular (normal) to the rays of the sun at an average distance of the earth from the sun, is referred to as the *solar constant*.* The amount of energy received is a function of the angular height of the sun incident to the earth and is greatly influenced by latitude and season (Fig. 5–4). The angle of light rays impinging on the water has a marked effect upon the productivity of lakes, as will be illustrated repeatedly in subsequent chapters. In equatorial regions, sunlight impinges vertically and leads to relatively constant energy inputs. This contrasts strongly with temperate and polar areas, where the sun's angle changes with the sequence of the seasons. The time of day is another factor that strongly influences the solar flux reaching the surface of a lake, for time of day influences the position of the sun and the distance of the path the light must travel through the absorbing atmosphere. In the polar extremes, for example, direct solar energy decreases to zero for over one-third of the year, and polar waters then receive thermal radiation only from indirect sources.

The absorptive capacities of the atmosphere for solar radiation are governed largely by oxygen, ozone, carbon dioxide, and water vapor, as discussed previously. Additionally, atmospheric transparency can be strongly modified in some regions by both industrial- and urban-derived contaminants. Scattering and absorption also increase in moist air, which is more common on the downwind side of large water bodies. The elevation of a lake and the angular height of the sun both determine the quantity of atmosphere through which the radiation must travel. In sum, the amount and spectral composition of *direct* solar radiation reaching the surface of a water body vary markedly with latitude, season, time of day, altitude, and meteorological conditions.

Indirect solar radiation from the sky is largely the result of scattering of light as it passes through the atmosphere. The extent of scattering is a function of the fourth power of the frequency, and therefore the UV and shorter wavelength radiation of high frequency is reduced by about one-fourth as a result of scattering. The result is the blue sky we see directly overhead on clear days. The factors influencing direct solar radia-

*The solar constant is difficult to measure but recent evidence from satellite instrumentation indicates a value of 1.94 cal cm^{-2} min^{-1} (Drummond, 1971; Gates, personal communication); Hickey, et al. (1980) found 1.97 cal cm^{-2} min^{-1} (1376.0 W m^{-2}).

tion also influence scattering, but of particular importance are solar height and the atmospheric distance through which the light must pass. The percentage of indirect radiation increases significantly as deviations of the rays from the perpendicular increase, e.g., 20 to 40 per cent at a sun elevation from the horizontal of 10°, in contrast to 8 to 20 per cent at a sun elevation of 40°.

DISTRIBUTION OF RADIATION IMPINGING ON LAKES

Solar radiation that impinges upon the surface of an inland water body does not all penetrate the water. A significant portion is *reflected* from the surface, and is lost from the system unless it returns to the lake after being backscattered from the atmosphere or surrounding topography.

The extent of reflectivity of solar radiation varies greatly with the angle of incidence, the surface characteristics of the water, the surrounding topography, and the meteorological conditions. The reflection (R) of unpolarized direct sunlight as a fraction of the incident light is a function of Fresnel's law:

$$R = \frac{1}{2}\left[\frac{\sin^2(i-r)}{\sin^2(i+r)} + \frac{\tan^2(i-r)}{\tan^2(i+r)}\right]$$

where i = angle of incidence, and r = angle of refraction. This function states simply that the reflectivity is dependent upon the solar height from the zenith (Fig. 5–5A), i.e., the greater the departure of the angle of the sun from the perpendicular, the greater the reflection will be. Indirect radiation from the sky is also reflected from a water surface, but is less affected by solar height (Fig. 5–5B). Under an overcast sky the

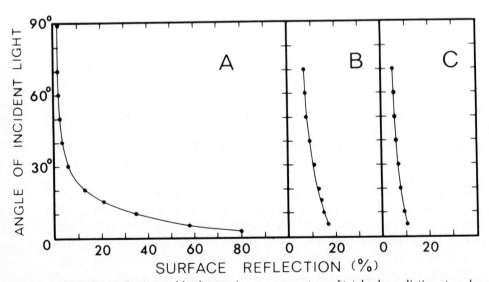

Figure 5-5 Surface reflection and backscattering as a percentage of total solar radiation at angles of incidence varying from the horizontal. A, clear, cloudless conditions; B, reflection of diffuse light under moderate cloud cover; C, heavily overcast conditions. (Generated from data in Steleanu, 1961, and from Sauberer, 1962.)

amount of indirect light that is reflected decreases (Fig. 5–5C). An average reflectance loss value of 6.5 per cent is common, although reflective losses can be reduced further if surrounding topography, mountains for example, moderates low-angle radiation (Sauberer, 1962).

When the surface of the water is disturbed by wave action, reflection increases by about 20 per cent at low angles of incident light (approximately 5°) to approximately a 10 per cent increase at higher angles (5 to 15°). The difference is small at angles of incidence at greater than 15° from the horizontal. Reflection may decrease slightly when the waves are very large and the light is exposed to the water surfaces at angles more closely approaching the perpendicular. Ice and especially snow cover markedly affect the reflectivity of light from the surfaces of lakes. Although data are very meager, clear, smooth ice acts with reflection characteristics similar to those of the undisturbed liquid phase. Changes in texture of the ice generally result in increases in reflectivity. Reflection increases markedly with the greatly increased angles and quantity of surface planes of granular ice in the form of snow cover. On the average, about 75 per cent of incident light striking snow is reflected, and under some conditions, the amount reflected can be as high as 95 per cent.

Of the total incident light impinging upon the surface of a lake, a reasonable average amount that is reflected on a clear, summer day is 5 to 6 per cent. This mean value increases to about 10 per cent during winter. Qualitatively, light in the red portion of the spectrum is reflected to a slightly greater extent than light of higher frequencies, particularly at low angles of incidence. About one-half of the total quantity of light leaving the lake is by reflection, and half by scattering of light.

SCATTERING

Scattering of light from the water results in the loss of large amounts of light energy from the lake. This phenomenon is apparent to anyone who has looked down into relatively clear waters where the surface reflection is eliminated. Of the total light energy entering the water, portions are absorbed by the water and its suspensoids, as will be discussed in detail further on, and a significant portion is scattered. The scattering of light is the result of deflection of quanta by the molecular components of the water and its solutes but also, to a large extent, by particulate materials suspended in the water.

The scattering of light energy can be viewed in a simple way as a composite of reflection at a massive array of angles internally within the lake. The energy scattered in all directions within a volume of water varies greatly with the quantity of suspended particulate matter and its optical properties. Volcanic siliceous materials, for example, scatter light much less than suspended particulate matter of lower transparency.

Scattering of light can change significantly with depth, season, and location in the lake and in response to variations in the distribution of particulate matter. When particulate matter is concentrated in the middle zone of great density change (metalimnion) of a thermally stratified lake, either as a result of reduced rates of sinking as the particles encounter increased water densities, or as a result of the development of large populations of plankton in certain strata, scattering of light can increase. Scattering can also increase markedly in areas of the lake where wind-induced currents and wave action agitate and temporarily suspend littoral and shore deposits of particulate matter

(Tyler, 1961b). When dimictic or amictic lakes (see next chapter) undergo complete circulation, a significant portion of the recent sediments of the lake basin are brought into resuspension (Davis, 1973; Wetzel, et al., 1972; White and Wetzel, 1975) and can affect the scattering properties of the lake for an extensive period (weeks). Similarly, the variable influxes of suspended inorganic and organic matter from stream inflows to reservoirs can radically increase the scattering of light nonuniformly within the lake basin. These alterations can be short- or long-lived depending upon the composition and density of the material and the density characteristics of the recipient water at that particular time.

A significant amount of light can be reflected and scattered from the sediments both in the littoral zone and in the shallow areas of the lake, as well as from the bottom of moderately deep, clear lakes (Fig. 5-6). The amount of light returned to the water is dependent upon the composition of the sediments; sand sediments or those sediments rich in $CaCO_3$ (calcium carbonate, marl) reflect considerably more light than dark-colored sediments of high organic content.

Differential scattering of light also depends on scattering coefficients for different wavelengths, as well as on absorption characteristics for different wavelengths. In very clear water, scattering occurs predominantly in the blue portion of the visible spectrum. As the quantity and size of suspensoids increase, radiation of longer wavelengths is scattered preferentially and greater absorption of the light of high frequencies occurs. Hard-water lakes with large amounts of suspended $CaCO_3$ particles characteristically backscatter light that is predominantly blue-green; lakes rich in suspended organic materials appear more green or yellow.

The diffuse light from scattering and reflecting sources is of obvious importance to organisms that utilize it directly in photosynthesis or indirectly in behavioral responses. In lake systems where light enters the environment unidirectionally from above, diffuse light can form a major supplementary source of energy. The amount of light scat-

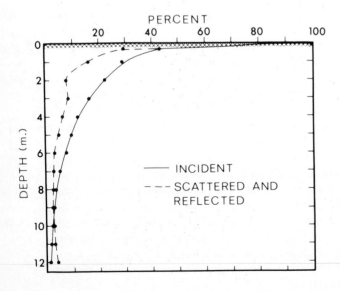

Figure 5-6 Comparison of incident light penetration (per cent of total surface light) with depth to that backscattered from concentrations of plankton, especially at 3 m, and from calcareous sediments. Lawrence Lake, Michigan; 17 March 1972; 23 cm cloudy ice.

tering can easily be one-fourth of that light absorbed by the water. The values are likely to be even higher, since most of the existing data on scattering are based on unidirectional (2π—flat) instruments. When values using this type of sensor are compared with those obtained with instruments approaching 4π (spherical) geometry (e.g., Rich and Wetzel, 1969), higher values are commonly found. A portion of the scattered light is returned to the surface of the lake, and much (80 to 90 per cent) of that which returns to the surface is lost to the atmosphere. Because of differences in light-wave refraction properties at different frequencies, somewhat less of the red scattered light reaches the surface than the blue.

Several terms are in common usage in relation to light radiation used by photosynthetic organisms. *Photosynthetically active radiation* (PAR) is defined as radiation in the 400–700-nm waveband. The photosynthetically active range may extend below 400 nm to wavelengths as low as 290 nm (Halldal, 1967; Klein, 1978), but the amount of subsurface irradiance available at the shortest wavelengths is very small, especially if the water contains significant amounts of dissolved organic compounds (as most lakes do; see following discussion). PAR is a general radiation term that is applicable to both energy terms and the preferred photon (quantum) terms.

Photosynthetic irradiance is the radiant energy (400 to 700 nm) incident per unit time on a unit surface (i.e., radiant energy flux density of PAR in units such as watts m^{-2}). *Photosynthetic photon flux density* is the number of *photons* (quanta) in the 400–700-nm waveband incident per unit time on a unit surface (i.e., photon flux density of PAR in units in μeinsteins second^{-1} m^{-2}).*

Total underwater light received by a receptor system, such as that received by an algal cell from all angles, is the optimum measure of radiant energy available for photosynthesis. This realistic value is the *photosynthetic photon flux fluence rate* or PPFFR (that is, photon scalar irradiance or scalar quantum irradiance, Smith and Wilson, 1972), which is defined as the integral of photon flux radiance at a point over all directions about the point. In other words, PPFFR is a measure of the total number of photons of PAR per unit time and area arriving at a point from all directions about the point when all directions are weighted evenly (Smith and Wilson, 1972). Such spherical, 4π quantum sensors are now available commercially and respond equally to all photons in the 400–700 nm range (units in μE sec^{-1} m^{-2}).

Photosynthetic and phototropic responses of organisms are related to the number of quanta of light of specific frequencies (wavelengths) impinging upon biochemical receptor systems. The attenuation of illuminance under water between a spectral range of 350 to 700 nm is not exactly the same for energy units as it is for quanta (Steemann Nielsen and Willemoës, 1971; Lewis, 1975). The rate of attenuation of quanta is approximately equal to that of energy in moderately clear water bodies containing intermediate concentrations of dissolved organic carbon (approximately 5 mg l^{-1}). In more transparent waters, energy is absorbed at a slower rate than quanta; the opposite is true in more deeply stained waters containing greater amounts of dissolved organic matter. Divergence in penetration of quanta and energy can increase in certain specific portions of the spectral range and thus affect the utilization by action spectra of organisms. For example, small differences in the utilization of illuminance during photosynthesis between diatoms, green algae, and blue-green algae with differing action spectra of photosynthetic pigments occur when the rate of illu-

---

*The einstein has been used to represent both the quantity of energy in Avogadro's number of photons and also Avogadro's number of photons (Incoll, et al., 1977). When used as the quantity of photons, 1 einstein (E) = 1 mole = 6.02×10^{23} photons.

mination is measured in energy units. The ratio of total quanta to total irradiance energy within the spectral region of photosynthetic activity, however, varies by no more than $\pm$ 10 per cent and is $2.5 \pm 0.25 \times 10^{18}$ quanta sec^{-1} watt^{-1} within a number of waters differing in optical characteristics (Morel and Smith, 1974). The ratio can be used to determine accurately the total quanta available for photosynthesis from measurements of total energy; the converse is also true.

Thermal Radiation in Lake Water

Lakewater behaves like a blackbody* to radiation of low frequency (wavenumber <14,000 cm^{-1}) and long wavelengths (750 to >12,500 nm). Because of surface reflection, only about 97 per cent of thermal radiation is emitted into the atmosphere. The atmosphere does not function as a blackbody, but thermal radiation from it is influenced by water vapor pressure and the degree of cloud cover. At night, the net emission of thermal radiation from the surface of a lake, in cal cm^{-2} day^{-1}, approximates [11 ·(°K of water − °K of air)] or simply [11 · (temperature of the water − that of the air)] (cf. discussion in Hutchinson, 1957).

NET RADIATION

The net amount of solar radiation affecting a lake can be referred to as the *net radiation surplus* (Q_B):

$$Q_B = Q_S + Q_H + Q_A - Q_R - Q_U - Q_W,$$

where:

Q_S = direct solar radiation
Q_H = indirect scattered and reflected radiation from sky and clouds
Q_A = long-wave thermal radiation from the atmosphere and from surrounding topography (the latter is usually insignificant)
Q_R = radiation reflected from the lake
Q_U = radiation scattered upward and lost
Q_W = emission of long-wave radiation.

At night, most components are negligible, and the net radiation surplus becomes equal to the long-wave thermal radiation of the atmosphere minus that emitted from the water

$$Q_B = Q_A - Q_W$$

or, approximately, $Q_B = -11$ (temperature of the water − the temperature of the air) in cal cm^{-2} day^{-1}. We will return to these relationships when discussing heat budgets in the following chapter. It should be emphasized, however, that even though the mean value of Q_B may be positive, the lake can be losing heat through evaporation and convective heating of the air (Hutchinson, 1957).

The collective value for the inputs ($Q_S + Q_H + Q_A$) can be measured directly by

---

*A blackbody absorbs all of the radiant energy incident upon it. The thermal radiation (Q) from a blackbody is proportional to the fourth power of the absolute temperature (K). $Q = 8.26 \times 10^{-9}(K)^4$

sensitive Moll thermopile pyranometers. The subject of radiation measurement is treated excellently by Latimer (1972), Bickford and Dunn (1972), Šesták, Čatský, and Jarvis (1971), and Coulsen (1975).

Transmission and Absorption of Light by Water

The quantity of light energy penetrating the water is dispersed by the mechanisms discussed above, and absorbed. The diminution of radiant energy with depth, by both scattering and absorption mechanisms, is referred to as light *attenuation*, whereas *absorption* is defined as diminution of light energy with depth by transformation to heat (cf. Westlake, 1965). It is important to understand the selective absorptive properties of water, first in pure water and then in lake waters of differing optical characteristics.

The transmission and absorption of light in water can be approached in several ways. Perhaps the most direct way is to look first at the percentage of transmission or absorption of monochromatic light through given depths of pure water. This percentile absorption, or Birgean percentile absorption (after E. A. Birge who used the relationship extensively), is based on the expression

$$\frac{100 \ (I_0 - I_z)}{I_0}$$

where

I_0 = irradiance at the lake surface
I_z = irradiance at depth z, in this case taken as 1 m.

In distilled water, the percentile absorption is very high in the infrared region of the spectrum, decreases rapidly in the lower wavelengths to a minimum absorption in the blue, and then increases again in the violet and especially UV wavelengths (Table 5-2). These absorption relationships usually are expressed graphically in linear or, even better, in logarithmic form (Fig. 5-7). Generally about 53 per cent of the total light energy is transformed into heat in the first meter.

The light intensity or irradiance, I_z, at depth z is a function of intensity at the surface (I_0) and the log of the negative extinction coefficient (η) times the depth distance, z, in meters:

$$I_z = I_0 e^{-\eta z}$$
$$\text{or } \ln I_0 - \ln I_z = \eta z.$$

The extinction coefficient (η) is a constant for a given wavelength; approximate values for pure water are given in Table 5-2. This relationship is imperfect in nature because sunlight is not monochromatic, but is instead a composite of many wavelengths.

Direct sunlight rarely enters the water at right angles to the surface, and indirect irradiance is not perpendicular to the surface at any time. Moreover, the natural total extinction coefficient (η_t) is influenced not only by that of the water itself (η_w), but also by absorption of particles suspended in the water (η_p), and particularly by dissolved, colored compounds (η_c). Thus the in situ extinction coefficient (η_t) is a composite of these components (Åberg and Rodhe, 1942):

$$\eta_t = \eta_w + \eta_p + \eta_c.$$

TABLE 5-2   Absorption Coefficients (Percentile Absorption) and Extinction Coefficients of Monochromatic Light of Liquid Water at 21.5°C by Laser Optoacoustic Spectroscopy*

WAVELENGTH (nm)	WAVE NUMBER (v^1) (cm^{-1})	ABSORPTION COEFFICIENT (10^{-4} cm^{-1}) ("PERCENTILE ABSORPTION")	EXTINCTION COEFFICIENT (η m^{-1})
820 (infrared)	12,200	(91.1)	2.42
800	12,500	(89.4)	2.24
780	12,820	(90.1)	2.31
760	13,160	(91.4	2.45
740	13,515	(88.5)	2.16
720	13,890	(64.5)	1.04
700	14,285	59.0	0.89
689.2	14,500	47.6	0.646
680.1 (red)	14,700	42.6	0.555
666.5	15,000	38.7	0.489
645.0	15,500	31.4	0.377
624.8 (orange)	16,000	29.6	0.351
605.9	16,500	24.8	0.285
588.0	17,000	12.3	0.131
574.5 (yellow)	17,400	8.1	0.084
546.3	18,300	5.3	0.054
526.2 (green)	19,000	4.0	0.041
512.7	19,500	3.48	0.0354
499.9	20,000	2.33	0.0236
487.7	20,500	1.86	0.0188
473.1	21,000	1.79	0.0181
465.0 (blue)	21,500	2.06	0.0208
454.4	22,000	2.21	0.0224
446.3	22,400	2.38	0.0241

*For data from 700 to 446 nm (Tam and Patel, 1979). Values in parentheses are older data determined by 1-meter path transmission measurements from James with Birge (1938).

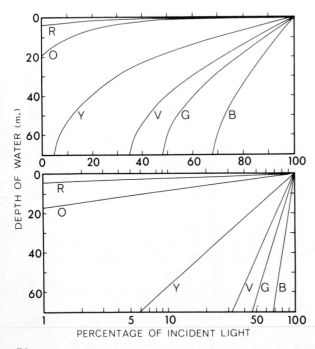

Figure 5-7   Transmission of light by distilled water at six wavelengths (R-720, O-620, Y-560, G-510, B-460, V-390 nm). Percentage of incident light that would remain after passing through the indicated depths of water expressed on a linear (upper) and a logarithmic (lower) scale. (After Clark, 1939.)

At low concentrations, the particulate suspensoids have relatively little effect on absorption. With high turbidity, however, the effect is quite significant, particularly at lower wavelengths of the visible spectrum. In detailed analyses of the absorption of lake water and its dissolved components, the particulate fraction is commonly removed by filtration or centrifugation.

EFFECTS OF ORGANIC COMPOUNDS

The effects of dissolved organic compounds on the absorption of light energy are very marked, and are best introduced by examples taken from the extensive work of James with Birge (1938). In comparison to distilled water, lake water with increasing concentrations of dissolved organic compounds, particularly humic acids, not only drastically reduces the transmission of light, but shifts the absorption selectively (Table 5-3). Common to all waters is a very high absorption of infrared and red wavelengths, which results in significant heating effects in the first meter of water. At the other extreme, while distilled water absorbs relatively little UV light, even very low concentrations of dissolved organic compounds increase UV absorption greatly. In lakes highly stained with humic compounds, such as Helmet Lake, absorption of UV, blue,

TABLE 5-3 Percentile Absorption of Light of Different Wavelengths by One Meter of Lake Water, Settled of Particulate Matter, of Several Wisconsin Lakes of Progressively Greater Concentrations of Organic Color*

WAVELENGTH (nm)	DISTILLED WATER	CRYSTAL LAKE	LAKE MENDOTA	ALELAIDE LAKE	MARY LAKE	HELMET LAKE
800	88.9	89.9	90.5	92.4	91.7	93.2
780	90.2	91.3	91.9	93.5	93.0	94.5
760	91.4	93.5	92.6	94.5	94.8	96.0
740	88.5	89.3	91.5	92.7	93.0	96.2
720	64.5	67.6	71.0	78.0	78.0	86.9
700	45.0	50.4	49.7	66.3	70.7	82.5
685	38.0	45.2	42.2	65.7	71.7	86.6
668	33.0	40.3	36.8	65.0	72.3	88.0
648	28.0	37.0	31.9	64.5	75.2	91.2
630	25.0	34.4	28.9	65.8	77.8	94.0
612.5	22.4	32.1	26.3	66.8	80.3	96.0
597	17.8	27.5	22.5	67.0	83.2	97.6
584	9.8	22.0	17.6	67.1	85.7	98.2
568.5	6.0	19.3	14.0	67.6	88.5	98.6
546	4.0	19.2	13.5	70.9	91.6	99.3
525	3.0	19.8	14.1	74.5	94.8	–
504	1.1	20.7	15.2	81.0	97.4	–
473	1.5	21.7	21.7	88.6	99.4	–
448	1.7	23.8	27.8	92.2	–	–
435.9	1.7	24.4	31.0	95.2	–	–
407.8	2.1	28.1	44.3	99.0	–	–
365	3.6	40.0	80.0	–	–	–
Color Scale (Pt units)	0	0	6	28	101	264

*Selected data from James and Birge, 1938.

and green wavelengths is essentially complete in much less than a depth of 1 meter. This relationship of intense absorption of UV light by dissolved organic compounds has been used extensively as a relative assay of their concentrations, as will be discussed in a subsequent chapter on organic matter.

The major effects of dissolved organic matter and particulate suspensoids on absorption of light at varying wavelengths are illustrated graphically in Figure 5-8. The

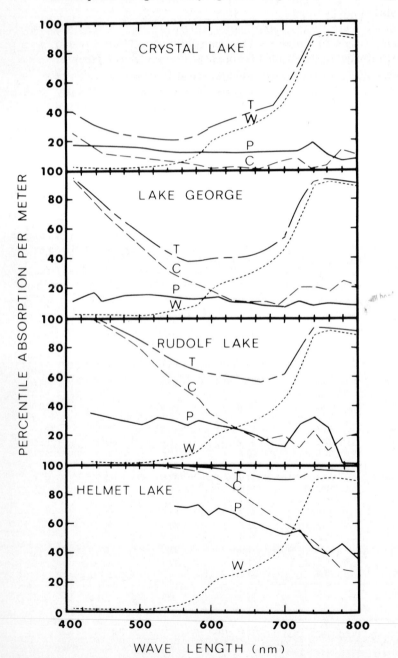

Figure 5-8 Percentile absorption of light at different wavelengths passing through 1 meter of water of 4 lakes of northern Wisconsin of increasing concentrations of dissolved organic matter. T = total absorption; C = absorption by dissolved organic color; P = absorption by suspended particulate matter; and W = absorption by pure water. (Modified from James and Birge, 1938.)

total absorption characteristics in the spectrum (T) are compared to the percentile absorption values through 1 meter of distilled water (W), absorption attributable to particulate suspensoids (P), and absorption by lake water that has been filtered to remove particles >1 μm (C), termed *color absorption*. Absorption by dissolved organic compounds ("dissolved color") is selective and greatest in the UV, blue, and green wavelengths. The absorption by dissolved color at the red end of the spectrum is less selective, and is probably unrelated to the organic compounds absorbing at lower wavelengths. The extinction coeffcients of dissolved color (η_c) increase directly with the color units of the water, which are measured by the relative visual comparisons of the color of the filtered lake water under standard conditions to the color of a specific mixture of platinum-cobalt compounds in serial dilution (discussed further on). In the examples given in Figure 5-8, the water of Crystal Lake indicated zero platinum (Pt) units, Lake George, 24, Rudolf Lake, 50, and that of Helmet Lake, 236 Pt units, which is about the color of weak tea.

Also apparent from this classic work is the relationship that absorption resulting from particulate suspensoids (P) is relatively unselective at different wavelengths, particularly at lower concentrations. The η_p functions essentially independently of the η_c but, along with the absorption of water (η_w), η_p values are additive for a particular lake at depth at a given time of year (cf. Åberg and Rodhe, 1942).

ANALYSIS OF LIGHT TRANSMISSION

The transmission or absorption within a lake of the total white light from direct insolation and indirectly from the sky has been analyzed in many ways. The vertical extinction coefficient is most commonly determined from the percentile absorption of surface light through depth (Fig. 5-9). The isopleths* of these examples of depth–time distribution of light indicate some of the marked fluctuations that are found in natural waters, seasonally and vertically. The composite mean η_t of all depths in Lawrence Lake, an unproductive hard-water lake with rather high concentrations of particulate and colloidal $CaCO_3$ suspensoids, was 0.39 m^{-1} (n = 1,746), within an annual range of 0.05 to 1.02. The same value for extremely productive Wintergreen Lake was 1.00 m^{-1} (range 0.46 to 1.68). In the former case, the η_t is largely constant over an annual period, whereas in the latter situation the marked fluctuations in particulate suspensoids of algae are reflected in the mean η_t m^{-1} and mean percentage transmission m^{-1} of the water column (Fig. 5-10).

Calculations of the vertical extinction coefficient are not very reliable in the first meter below the surface because of surface agitation. Average calculations often exclude this region. Direct calculations are made using the formula given earlier, or changed to

$$\eta z = \ln I_0 - \ln I_z.$$

A nomogram for estimating the extinction coefficient directly from light transmission data is given in the appendices.

---

*An isopleth is a line on a graph of a specified constant value showing the occurrence of a parameter as a function of two variables (depth and time in this case). See Wetzel and Likens (1979:26) for details of preparation of these diagrams.

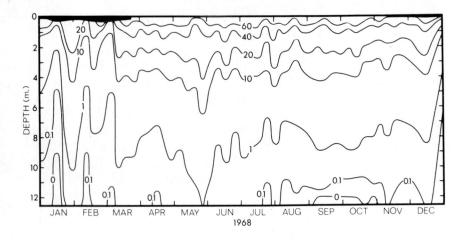

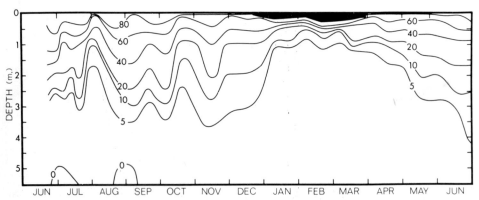

Figure 5-9 Isopleths of the percentage transmission of light at the surface with depth and time in unproductive Lawrence Lake (*upper*), and extremely productive Wintergreen Lake (*lower*), southwestern Michigan. (From Wetzel, unpublished observations.)

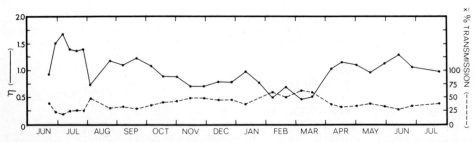

Figure 5-10 Average extinction coefficient (η_t) per meter and percentage transmission of light per meter of the water column, Wintergreen Lake, Michigan, 1971–72. (From Wetzel, unpublished data.)

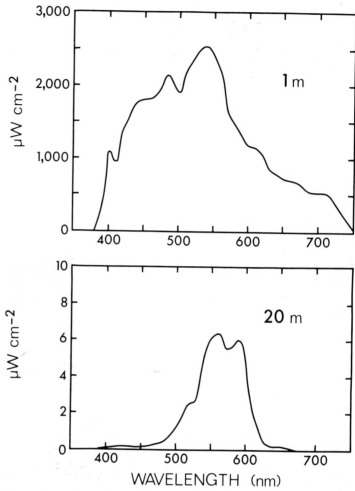

Figure 5-11 Comparison of the spectral distributions of energy using a scanning spectroradiometer, Gull Lake, Kalamazoo–Barry counties, Michigan, 16 November 1975 at 1 m (*upper*) and 20 m (*lower*). From Wetzel and Likens (1979).

The values cited for the vertical extinction coefficient (η_t) obviously represent a composite for all of the wavelengths, each of which is variously influenced by the absorption characteristics of water, particulate matter, and dissolved organic matter. The η_t values for natural lake waters vary from approximately 0.2 (about 80 per cent transmission m^{-1}) per meter in very clear lakes, such as Crystal Lake, Wisconsin, Lake Tahoe, California, and Crater Lake, Oregon, to about 4.0 m^{-1} in highly stained lake waters or lakes with very high biogenic turbidity. Where turbidity is extremely high, such as in reservoirs near major river inflows under flood conditions (e.g. Roemer and Hoagland, 1979), or in lakes receiving fine materials, such as volcanic ash, that remains in colloidal suspension, extinction coefficients in excess of 10 m^{-1} are not unusual.

As the composite distribution of light transmission with depth is dissected, the spectral selectivity of light reaching deep water should be apparent from the foregoing discussion. It is rather common for the green portion of the spectrum to penetrate most deeply (Fig. 5–11). In very clear waters, the deepest penetration of light is in the blue portion of the spectrum; in Crater Lake, maximum penetration occurred at approximately 469 nm, and in Lake Tahoe at approximately 475 nm (Tyler and Smith, 1970; Smith, et al., 1973). In lakes that are darkly stained with dissolved organic matter, practically no light of wavelengths below 600 nm penetrates below 1 meter.

The rapid attenuation of light by dense populations of algae or bacteria at certain depths in stratified lakes is common, particularly in the lower depths where certain microorganisms are adapted to low light intensities (Fig. 5–12). Similarly, abiogenic suspensoids may be brought into a water basin through an inlet at temperatures colder than the surface strata. When this occurs, the suspensoid-laden water will penetrate to deep strata of comparable densities, and the stratified particles can markedly influence the vertical extinction of light.

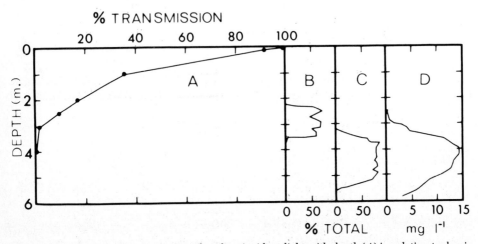

Figure 5-12 Percentage transmission of surface incident light with depth (A) in relation to dominating bacterial plate of *Thiopedia* (B) at 2.5 to 3.5 meters and *Clathrochloris* (C) at 3.5 to 5.5 meters (as a percentage of total plankton) and bacteriochlorophyll d (D) in Wintergreen Lake, Michigan, 4 July 1971. (From Wetzel, unpublished data, and Caldwell, et al., 1975.)

TRANSMISSION THROUGH ICE AND SNOW

The percentage of light transmission through clear, colorless ice is not greatly different from that through water (Fig. 5-13). Attenuation of light increases greatly, however, if the ice is stained with organic matter or is cloudy, i.e., contains air bubbles or forms irregular crystals upon freezing (Adams, 1978). With the addition of snow over ice (Table 5-4), reduction of light is great. It is not unusual for light to be attenuated to essentially zero at a depth of as little as a meter below the underside of ice and wet, heavy snow. Prolonged periods of heavy snow and ice cover, with attendant severe reduction or elimination of light from the water and from photosynthetic organisms, can have profound effects on the entire metabolism of the lake. In very productive lakes, consumption of dissolved oxygen by catabolic processes can exceed augmentation by photosynthesis, which leads to severe reductions in dissolved oxygen or even to total anoxia. Oxygen reduction or anoxia resulting from excessive snow or ice cover

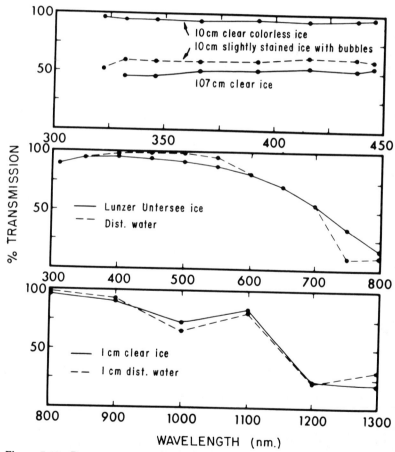

Figure 5-13   Percentage spectral transmission of light through ice in comparison to water. (From data of various sources cited in Sauberer, 1950.) See also data of Adams (1978).

TABLE 5-4 Light Penetration through Ice and Snow Cover of Lakes under Different Conditions*

ICE-SNOW CONDITIONS	THICKNESS (cm)	PERCENTAGE TRANSMISSION OF SURFACE INSOLATION
Clear ice	43	72
Clear ice with vestige snow	39	53
Milky ice with bubbles	29	54
Wet ice with bubbles	39	41
Translucent ice ("snow ice")	25	11–18
Ice with irregular surface	29	58
New snow	0.5	34
	5.0	20
	10.	9
	17–20	8.8–6.7
Compacted old snow	17–20	5–1

*Selected data from Albrecht, 1964.

commonly occurs in shallow, productive lakes and ponds in temperate latitudes (cf. Chapter 9), and often leads to the death of many organisms. This phenomenon is termed *winterkill*.

Ice and particularly snow are not uniformly distributed over lake surfaces (Adams, 1978; Adams and Lasenby, 1978; Adams and Prowse, 1978). Thickness of ice and snow can vary as much as a factor of two within short distances (<25 m) and lead to considerable patchiness in both reflectance of light and attenuation of light through ice cover. This patchiness in light distribution beneath ice–snow cover can, in turn, influence horizontal distribution of algal photosynthesis and behavioral responses of planktonic animals.

The characteristics and amount of light penetrating the littoral zone of lakes are greatly altered by the type and extent of development of floating-leaved and emergent higher aquatic plants, as well as by reflection and scattering from the substratum. Since the structure of the littoral macroflora changes seasonally, dense vegetative stands commonly reduce incident light by 50 to over 90 per cent (Szumiec, 1961; Gessner 1955; Ondok, 1973a, 1973b, 1978).

Color of Natural Waters

The observed color of lake water is the result of light being scattered upward from the lake after it has passed through the water to various depths and undergone selective absorption en route. Since molecular scattering of light in the water is a function of the fourth power of the frequency, observed light and therefore color is greater for shorter than for longer wavelengths, and blue dominates. Scattering of light from particulate suspensoids, however, is increasingly less selective with increasing particle size (Hutchinson, 1957). Colloidal $CaCO_3$, common to hardwater lakes, scatters light in the greens and blues and gives these waters a very characteristic blue-green color. Most of the color of lake waters results from dissolved organic matter and its rapid, selective absorption of the shorter wavelengths of the visible spectrum. The result is a dominance of emitted scattered light in the green portion of the spectrum and, with increasing concentrations of organic matter, especially humic compounds, an increase in yellows and reds.

When the density of particulate matter suspended in the water becomes great, a seston* color can be imparted to the water in spite of the relatively nonselective scattering properties of the particles. Suspension of large amounts of inorganic materials, such as clays and volcanic ash, can yield a yellow to brownish-red coloration. Seston color, however, is usually associated wtih large concentrations of suspended algae or, less frequently, with pigmented bacteria or microcrustaceans. Where blue-green algae or diatoms occur in great profusion and often accumulate in the surface waters, they may produce blue-green or yellowish-brown colors, respectively. Blood-red color is not an infrequent occurrence in lakes where conditions are temporarily ideal for massive development of populations of algae, such as *Glenodinium*. At low concentrations of suspended algae, as in the open ocean, the chlorophyll bands of algal pigments have no appreciable influence on water color (Yentsch, 1960).

COLOR SCALES

Any color (shade or tint) always has two decisive characteristics: color intensity (brightness) and light intensity (lightness) (Albers, 1963). This duality in color intervals results in an extremely subjective ability to discriminate colors. Moreover, visual memory is very poor in comparison with auditory memory. Therefore, the psychophysical nature of reactions of visual organs to light and color has led to several attempts to standardize observations by means of various color scales.

Several color scales have been devised to empirically compare the true color of lake water, after filtration to remove suspensoids, to various combinations of inorganic compounds in serial dilutions. Platinum units† is the most widely used comparative scale in the United States. Very clear lake water would yield a value of zero Pt units, and heavily stained bog water about 300 (cf. Table 5-3). In Europe the Forel-Ule color scale, involving comparisons to alkaline solutions of cupric sulfate ($CuSO_4$), potassium chromate (K_2CrO_4), and cobaltus sulfate ($CoSO_4$), is commonly used. A strong correlation exists between the brown organic color, which is derived chiefly from peat and marsh detritus, and the amount of dissolved organic carbon in the surface waters (Juday and Birge, 1933). Frequently color units increase with depth in strongly stratified lakes; this is most likely related to increased concentrations of dissolved organic matter and ferric compounds near the sediments. Subjectivity of color evaluations can be reduced greatly by rather elaborate optical analyses and comparisons with standardized chromaticity coordinates (e.g., Smith, et al., 1973).

---

*Seston is a collective term for all particulate material present in the free water. Seston includes both *bioseston* (plankton and nekton; see Chapter 8) and *abioseston* or *tripton* (nonliving particulate material).

†1000 Pt units = the color from 2.492 g potassium hexachloroplatinate (K_2PtCl_6), 2 g cobaltic chloride hexahydrate ($CoCl_2 \cdot 6H_2O$), 200 ml concentrated hydrochloric acid (HCl), and 800 ml water.

Transparency of Water to Light

The in situ evaluation of the vertical extinction and spectral characteristics of light in lakes is now commonly accomplished using modern instrumentation. An approximate evaluation of the transparency of water to light was devised by an Italian scientist, Secchi, who observed the point at which a white disk lowered into the water was no longer visible. This method continues to be widely used owing to its simplicity. The Secchi disk transparency is the mean depth of the point where a weighted white disk, 20 cm in diameter, disappears when viewed from the shaded side of a vessel, and that point where it reappears upon raising it after it has been lowered beyond visibility.

The Secchi disk transparency is essentially a function of the reflection of light from its surface, and is therefore influenced by the absorption characteristics both of the water and of its dissolved and particulate matter. Within limits, a parabolic relationship exists between dissolved organic matter and Secchi disk transparency (Fig. 5-14). However, both theoretical analyses and a large number of empirical observations have shown that the reduction in light transmission in relation to Secchi transparency measurements is associated to a greater extent with increased scattering by particulate suspensoids (Štepánek, 1959; Szczepanski, 1968). This is particularly true in very productive lakes and in a very generalized way has been used to estimate the approximate density of phytoplankton populations. Although the transparency depth is independent of surface light intensity to a significant extent, results become erratic near dawn and dusk (Fig. 5-15), and the Secchi disk preferably should be used at midday.

Observed Secchi disk transparencies range from a few centimeters to over 40 m in a few rare clear lakes. The Secchi disk transparency correlates closely with percentage transmission. At the extremes, Secchi disk transparency depths can represent from 1 to 15 per cent transmission (Štepánek, 1959; Beeton, 1958; Tyler, 1968; and others). The variation is related to differences in the sensitivity of underwater photometers and to the size of suspensoids. In general, the Secchi disk transparency depth corresponds to the depth of approximately 10 per cent of surface light. The relationship between Secchi disk transparency and the extinction coefficient is very good during ice-free periods. The relationship, based on empirical data, of $\eta \ m^{-1} = 1.7/z_{sd}$, where z_{sd} is the Secchi

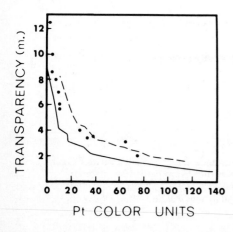

Figure 5-14 Relationship between Secchi disk transparency and dissolved color from Swedish lakes (points and dashed line) and 470 Wisconsin lakes (solid line). (From data of Åberg and Rodhe, 1942.)

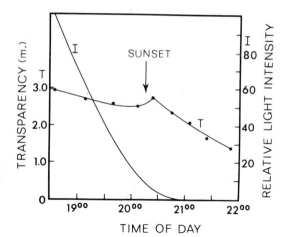

Figure 5-15 Changes in the Secchi disk transparency (T) in relation to light intensity (I) at the surface. I_0 at 2130 hours was 1.1×10^{-6} that at 1845 hours. (Slightly modified from Åberg and Rodhe, 1942.)

disk transparency depth in meters (Poole and Atkins, 1929), has been shown to be approximately correct in a variety of inland waters (Idso and Gilbert, 1974), although Walker (1980) indicates that $\eta\ m^{-1} = 1.45/z_{sd}$ is more in agreement with results for the sea. Seasonal variations, such as depicted in Figure 5–16, are common for lakes of the temperate zone. The examples cited include an unproductive hard-water lake and a highly productive lake with dense algal populations throughout much of the year, especially in the period of late winter through early summer. Even in the clearest lakes, such as alpine lakes, similar annual trends are found.

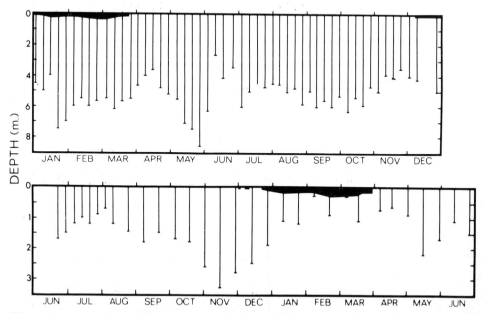

Figure 5-16 Annual variations in the Secchi disk transparencies of hardwater Lawrence Lake, 1968 (upper), and very productive Wintergreen Lake, 1971-72 (lower), southwestern Michigan. (From Wetzel, unpublished data.)

Colored Secchi disks have been employed in estimations of the spectral distribution of light with depth (Štěpánek, 1959; Elster and Štěpánek, 1967). Within general limits, comparison of transparencies with a series of colored disks to that of white provides an approximate evaluation of the spectral characteristics of lake waters.

Utilization of Solar Radiation

Solar radiation is the major energy source driving the productivity of aquatic ecosystems, whether it is incorporated into potential energy by biochemical conversions within the lake directly by its flora, or by terrestrial components within the watershed and imported as organic matter to the lake. Algae and macrophytes use between 4 and 9 quanta of light energy per molecule of carbon dioxide (CO_2) reduced. The photosynthetic receptor is the chloroplast, which consists of a lamellar structure from 1 to 10 μm in diameter and approximately 0.025 μm in thickness. Absorption of light energy is specific for different chlorophyllous pigments: The absorption peaks of the primary pigment chlorophyll a are at 640 nm and 405 nm, and those of chlorophyll b at 620 nm and 440 nm. Certain plants, such as wheat, have broad responses to quanta of energy with relatively high efficiency of utilization in the green wavelengths, in addition to those of chlorophyll a. Other potential adaptations of plants to certain photic environments exist. For example, in the blue-green alga *Microcystis*, the action spectra indicate that the accessory blue-green algal pigment phycocyanin gains importance when this alga is grown in red light, and the cells lose chlorophyll a. Other auxiliary pigments supplement the excitation of chlorophyll a and improve the efficiency of photosynthesis at long wavelengths.

In addition to the adaptations of photosynthetic organisms to the aquatic photic environment, utilization of light by aquatic animals is also specific. The visual receptors of animals have a quantum efficiency of 1, the maximum possible. Among the invertebrates, for example, the opossum shrimp *Mysis* possesses visual receptors with a peak at 515 nm, and is found only in deep portions of clear lakes, that is, in an optically blue environment. The water flea *Daphnia* migrates diurnally from deep to surface waters, and is adapted for life in changing photic environments with four visual receptor peaks at 370, 435, 570, and 685 nm (Chapter 16). The dragonfly, which spends much of its life cycle as an immature aquatic nymph, has visual receptors with a peak only in the green (530 nm) and possibly the blue (420 nm) portion of the spectrum (Ruck, 1965). Upon its shifting to the aerial adult stage, an increase in the number of receptor peaks (380, 420, 518, 530, and 550 nm) occurs. Similar situations are found among the fishes; those living only deep in a blue photic environment have maximum reception at approximately 485 nm, whereas shallow-living fishes possess receptors sensitive to longer wavelengths.

Behavioral adaptations to the utilization of light are also common. For example, *Daphnia* uses light as a cue in body orientation while swimming. The long axis of the body is oriented with the vector of maximum light energy, which affects directional swimming of these crustaceans during vertical migrations; hence the distribution of *Daphnia* populations within the lake is light-dependent.

It is now essential that we consider, in addition to these direct physiological effects and utilization responses, the effects of absorption of solar energy and its dissipation as heat on the physical, chemical, and biotic structure of lake systems.

1. Solar radiation is the major energy source to aquatic ecosystems. The productivity and internal metabolism of fresh waters are driven and controlled by energy derived directly from the solar energy utilized in photosynthesis. Photosynthetic biochemical conversions utilizing solar energy occur both within fresh waters directly by aquatic flora as well as by terrestrial and wetland plants within the drainage basin and imported to the streams and lakes as organic matter. In addition, absorption of solar energy and its dissipation as heat affects the thermal structure, water mass stratification, and hydrodynamics of lakes and reservoirs. These characteristics have marked attendant effects on all chemical cycles, metabolic rates, and population dynamics.

2. Solar radiation can be viewed as pulses of electromagnetic energy, *quanta* or *photons,* that move transversely in a wave conformation of characteristic wavelengths and amplitudes. Absorption of light by atoms and molecules occurs from resonation of electrons of atoms and molecules at frequencies that correspond to the quantum energy state of impinging light. Each molecular or atomic species, whether photochemical organic compounds in biochemical reactions or inorganic, has a unique set of absorption frequencies at which quantum energy lost by the photon is gained.

3. Much of the incoming irradiance (over 50 per cent) impinging on the surface of water bodies occurs at wavelengths in the infrared ($>1,000$ nm) portion of the solar spectrum and has major thermal effects on aquatic systems. Within the visible portion of the spectrum, most of the energy is in the longer wavelengths (reds) of lower frequencies.

4. The amount and spectral composition of solar energy impinging on water bodies is influenced by many dynamic factors.

    a. The angle of light incidence is more perpendicular in equatorial regions, during summer, and at midday than in regions of higher latitude, winter, or at times of the day closer to dawn or dusk. As a result light must pass through greater distances of atmosphere as the angle of incidence increases from the perpendicular. Selective absorption by oxygen, ozone, carbon dioxide, and water vapor occurs as light passes through the atmosphere, and shorter wavelengths of higher frequencies are rapidly removed (hence the red sky at dawn and dusk).

    b. Indirect solar radiation from the sky results from atmospheric molecular scattering. Short wavelength radiation of higher frequency is scattered more than longer wavelength light (hence the blue appearance of clear sky at midday).

    c. A portion of light reaching water is reflected away from the surface. The greater the departure of the angle of the sun from the perpendicular, the greater the reflection. When the surface of the water is disturbed by wave action, reflection increases somewhat (10 to 20 per cent). Cloudy ice and snow greatly increase the amount of reflected light.

5. Of the total light energy entering the water, a portion is scattered and the remainder is absorbed by the water itself, dissolved compounds, and sus-

pended particulate matter. The total diminution of radiant energy by both processes is called *light attenuation*.

a. Scattering of light results from deflection of light energy by molecules of water, its solutes, and suspended inorganic and organic particulate materials. Therefore the extent of this internal reflection varies greatly with depth, season, and the dynamics of inorganic and organic loading in relation to water stratification, productivity and distribution of organisms, and other factors.

b. *Light absorption* is the diminution of light energy with depth by transformation to heat.

c. The molecular structure of water results in selective absorption of light. Percentile absorption of light by water alone is very high in the infrared and red, lowest in the blue, and increases again somewhat in the violet-ultraviolet portions of the visible spectrum.

d. Dissolved organic compounds, especially humic compounds, drastically increase absorption of light. Selective absorption is greatest in the infrared and red wavelengths; coupled to the high infrared absorption of water itself, the combined effect is rapid absorption and heating of surface water. Even very low concentrations of dissolved organic compounds absorb greatly the UV, blue, and green wavelengths. Hence, in stained waters, the dominant wavelengths penetrating to significant depths are in the yellow and red portions of the spectrum.

e. Light absorption by particulate suspensoids, when not in extremely high concentrations, is relatively unselective at different wavelengths.

f. The vertical extinction coefficient (η per meter) is an expression of the exponential attenuation of irradiance at depth in relation to that at the surface. The total extinction (η_t) of natural waters is the sum of absorption by the water itself (η_w), dissolved compounds (η_c), and particles suspended in the water (η_p). Although the mean η_t varies considerably with depth and over a year, the annual mean is fairly representative of the transparency of a lake.

g. Although clear, colorless ice absorbs little more light than water, cloudy ice (with air bubbles, irregular crystals) and snow attenuate light severely.

CHAPTER SIX

FATE OF HEAT

The absorption of solar energy by lake water is, as we have seen, influenced by an array of physical, chemical, and, under certain conditions, biotic properties of the water. These characteristics are dynamic, and change seasonally and over geological time for individual lake systems.

The amount of light energy absorbed by a solution increases exponentially with the distance of the light path through the solution. For light of a wavelength of 750 nm, 90 per cent is absorbed in one meter with only 1 per cent transmitted through 2 meters of pure water. Absorption is increased markedly by dissolved organic matter. Since much of solar energy is of low frequency in the infrared portion of the spectrum at wavelengths greater than 750 nm, the upper two meters of lake water absorb over one-half of the sun's incoming radiation.

The high specific heat of water permits the dissipation of light energy and accumulation as heat. The retention of heat is coupled with factors that influence its distribution within the lake system: physical work of wind energy, currents and other water movements, morphometry of the basin, and water losses. The resulting patterns of thermal stratification influence in fundamental ways physical and chemical cycles of lakes, which in turn govern lake production and decomposition processes.

DISTRIBUTION OF HEAT

It is evident from previous discussion that the greatest source of heat to lakes is solar radiation, and most is absorbed directly by the water. Some transfer of heat from the air and from the sediments does occur, but in lakes of moderate depth, this input is small compared to direct absorption. In shallow waters, sediments can absorb significant quantities of solar radiation and this heat may be transferred in part to the water. However, terrestrial heat is generally very small in comparison to direct absorption of solar radiation by the water. Exceptions include lakes that receive significant percentages of their volume from surface runoff in short periods of time, such as small or shallow reservoirs that have short retention times. Indirect heating also can be significant in lakes that have high volume inputs from groundwater sources and springs, especially in the case of hot springs and certain volcanic lakes. Condensation of water vapor at the water surface also provides some heat input to the lake from the air.

Some heat is lost from lakes by thermal radiation (cf. Chapter 5). However, because the thermal conductivity of water is very low, heat loss by thermal radiation is predom-

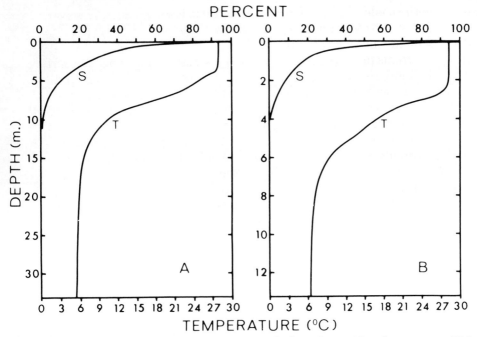

Figure 6-1 Vertical depth profiles of the penetration of solar radiation (*S*) and temperature (*T*) in Crooked Lake (*A*) and interconnected, adjacent Little Crooked Lake (*B*), Noble-Whitley counties, Indiana, July 18, 1964. (From Wetzel, unpublished data.) Note differences in depth scales; Crooked Lake has a much larger surface area (79 ha) than Little Crooked Lake (5.3 ha).

inantly a surface phenomenon, restricted to the first few centimeters of water. Measurable amounts of heat are lost nonetheless by specific conduction both to the air and, to a much lesser degree, to the sediments. Heat is also lost through evaporation. The rate of evaporation increases with higher temperatures, reduced vapor pressure, lower barometric pressure, increased air movement over the water or ice surface, and decreases with increasing salinity.

For water at 0°C to freeze, 79.7* cal g^{-1} must be released, and conversely when ice at 0°C melts to water at 0°C, an equal amount of heat must be absorbed. This latent heat of fusion is among the highest of known compounds. In addition, as ice sublimes to water vapor, 679 cal g^{-1} must be absorbed (see full discussion in Chapter 2).

A significant portion of the heat increments to lakes or especially reservoirs can be lost through outflow. Outflow usually consists of surface waters, and surface waters are often much warmer than underlying water strata.

The inputs and outputs of heat are therefore largely surface phenomena. Solar influx would be expected to dominate in the warmer seasons of the year, and the vertical thermal structure would then approximate the attenuation profile of solar radiation (Fig. 6–1). We might expect that warmer, less dense and very stable heated water

---

*The correct value is 79.72. The often cited value of Weast (1970) and earlier editions is in error and is corrected in newer editions of this handbook (personal communication).

would successively overlie cooler and more dense water. Such is not the case, however, and what is observed (Fig. 6-1) is a relatively uniformly mixed upper portion of the lake that is isothermal often well below the photic zone.

Some convection currents do occur at night when the surface waters cool, become more dense, and sink. Similarly, surface waters can cool during brief shifts in local meteorological conditions, such as under cloudy conditions, increased evaporation rates, cold rain, etc., or after seasonal decreases in air temperatures. Surface influents also may cool surface waters and induce convection currents.

Under most conditions of typical lakes of temperate latitudes, mixing of the surface waters by convection currents is weak (up to 3 m) and insufficient to produce the thermal profiles observed during the heating period of thermal stratification.

Direct absorption of solar radiation accounts for only about 10 per cent of the observed distribution of heat (Birge, 1916). Most of the heat distribution profile results from the action of the wind. As air currents move across the water interface, a frictional wind stress on the surface water generates mixing and currents proportional to the wind velocity (cf. Chapter 7).

Thermal Stratification

It is instructive to begin a discussion of thermal stratification in lakes by examining typical conditions of lakes of the temperate zone that experience strong contrasts in seasonal conditions. A majority of lakes are concentrated in the temperate and especially northern latitudes, and this permits some degree of generalization. Many exceptions to this general pattern can be found, and will be indicated when applicable.

Ice cover on the lake commonly deteriorates in the spring in a relatively slow, progressive way until it is permeated with air columns and/or saturated with water. Warm rains often accelerate this process of ice erosion. Loss of ice cover is usually rapid, often taking place in a few hours, especially if associated with a strong wind. At this time, the water at all depths is near the temperature of maximum density. Slight changes from the temperature of the maximum density, either cooling below or warming above 4°C due to vagaries in weather conditions, result in very small changes in density difference per unit change in temperature (cf. Chapter 2). As a result, there is relatively little thermal resistance to mixing, and only small amounts of wind energy are required to mix the water column. In most lakes, the amount of wind energy impinging on the surface is adequate to circulate the entire water column. Circulation, aided to some extent by convection currents induced by cooling at night and by evaporation, can continue for varying periods of time. The duration of the spring turnover is governed by many factors. Lakes of small surface area, especially if protected from the wind by surrounding topography or vegetation, may circulate only briefly in the spring, often for only a few days. Large lakes, in contrast, often circulate for a period of weeks, and, weather conditions permitting, the temperature of the entire water mass can increase to well above that of maximum density. During this period of mixing in spring turnover, the extent of heating is also a function of the volume of water that must be heated relative to net solar income. In very deep, large lakes such as the Great Lakes and Lake Baikal, the heat income can be insufficient to increase significantly the temperatures of the deep water, even if the lake circulates completely. The spring period of circulation in shallow lakes, on the other hand, may allow water temperature to increase to well above 10°C.

As spring progresses, the surface waters of lakes of sufficient depth are heated more rapidly than the heat is distributed by mixing. Usually, the rapid surface heating occurs during a warm, calm period of several days. As the surface waters are warmed and become less dense, the relative thermal resistance of mixing increases markedly. A difference of only a few degrees is then sufficient to prevent circulation (Fig. 6-2; an identical analysis is given for Lake Mendota by Birge, 1916). From that time onward, the water column is thermally divided into three regions, which are exceedingly resistant to mixing with each other. The initial temperature of the lowest stratum (the *hypolimnion*) is thus determined by the final water temperature during the spring turnover. The temperature of the hypolimnetic water changes little throughout the period of summer stratification, especially in deep lakes.

The period of summer stratification is characterized by an upper stratum of more or less uniformly warm, circulating, and fairly turbulent water, the *epilimnion* (Fig. 6-3). The epilimnion essentially floats upon a cold and relatively undisturbed region, the hypolimnion. The stratum between the epilimnion and the hypolimnion exhibits a marked thermal discontinuity, which is termed the *metalimnion*. The metalimnion is defined as the water stratum of steep thermal gradient demarcated by the intersections of the nearby homoiothermal epilimnion and the hypolimnion (graphically depicted in Fig. 6-3). The term *thermocline* has been defined variously, but correctly refers to the *plane* of maximum rate of decrease of temperature with respect to depth. An extensive discussion of these terms and their conceptual basis is given by Hutchinson (1957). Terms in wide use that are functionally synonymous with the above definition of the metalimnion include the German *Sprungschicht*, and discontinuity layer as used in the United Kingdom.

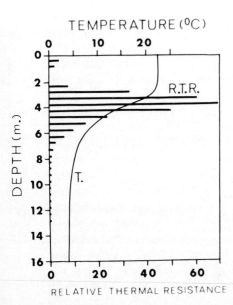

Figure 6-2 A summer temperature profile (single line) and relative thermal resistance to mixing (bars) for Little Round Lake, Ontario. The relative thermal resistance (R.T.R.) to mixing is given for columns of water 0.5 m deep. One unit of R.T.R. = 8×10^{-6}, i.e. the density difference between water at 5° and at 4°C. The R.T.R. of the lake water columns is expressed as the ratio of the density difference between water at the top and bottom of each column to the density difference between water at 5° and 4°C. (Modified from Vallentyne, 1957.)

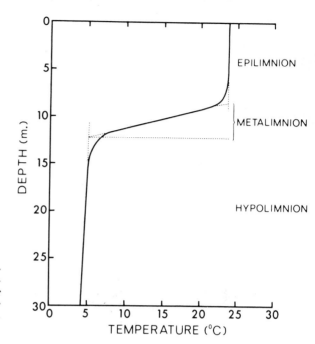

Figure 6-3 Typical thermal stratification of a lake into the epilimnetic, metalimnetic, and hypolimnetic water strata. Dashed lines indicate planes for determining the approximate boundaries of the metalimnion (see text).

During the transition from spring turnover to summer stratification, the depth of the stratum of greatest thermal discontinuity (usually accepted as a change of $>1°C$ per meter) varies greatly among lakes and from year to year in relation to weather conditions. If the water is being heated rapidly and wind-induced mixing of water is moderately intense but insufficient to circulate the water from top to bottom, the stratum of greatest thermal change will occur deep in the water column (Fig. 6-4). This thermal

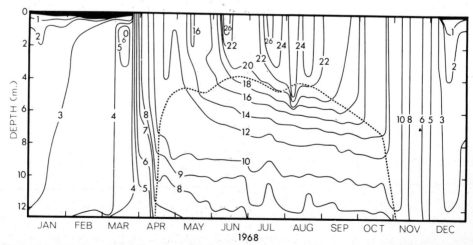

Figure 6-4 Depth–time diagram of isotherms (°C) in Lawrence Lake, Michigan, 1968. Dashed line indicates the upper metalimnetic–lower epilimnetic boundary. Ice-cover drawn to scale. (Modified from Wetzel, et al., 1972.)

discontinuity rises rapidly as further heating occurs and density differences increase (see also Fig. 6-9). In contrast, if the weather at this time is relatively calm and hot, strong thermal discontinuities will occur at the surface—in this case, the metalimnion essentially begins at the surface and moves deeper as wind-induced mixing establishes the epilimnion. From that time on, the epilimnion undergoes periods of temporary heating during hot, calm weather and alternates with periods of strong mixing by wind (Fig. 6-4).

The depth of the stratum of steep thermal change is gradually depressed as the upper portion of the hypolimnion slowly gains small amounts of heat throughout the period of thermal stratification (Fig. 6-4 and also particularly Fig. 6-8). It is clear from the early analyses of Ricker (1937) and of Hutchinson (1941) that the heating of a stratified lake proceeds by a combination of direct solar radiation, turbulent conduction, and density currents. Turbulence of the epilimnion is carried into the metalimnion and hypolimnion. Conduction of heat by turbulence decreases as the stability of stratification increases throughout the heating season, but within the hypolimnion, heat conduction varies little with increasing depth. Chemical density currents follow the contour of the basin along the sediments, and as they move to the deepest portions of the basin they can be effective in heat transport. Heat generated by biological oxidations during decomposition is entirely inadequate to account for the observed heat gain by the hypolimnion.

Solar heating of the hypolimnion can occur in lakes that are particularly transparent, such as in alpine situations. In a small clear mountain lake that is protected from much wind action, an estimated 65 to 85 per cent of the heating of the upper hypolimnion can be accounted for by direct solar heating (Bachmann and Goldman, 1965; these estimates are likely too high, however; cf. Jassby and Powell, 1975). As might be anticipated, the relative importance of direct solar heating and the transport of heat by turbulence from the upper water strata varies greatly with conditions of individual lakes. While wind-induced transport of heat to the hypolimnion is usually of greater importance than direct solar heating in most lakes, particularly where light is rapidly attenuated with increasing depth, a spectrum of conditions can exist in a lake that will cause solar heating of the hypolimnion to dominate.

STABILITY OF A STRATIFIED LAKE

Stability (S) per unit area of a lake is the quantity of work or mechanical energy in ergs required to mix the entire volume of water to a uniform temperature without addition or subtraction of heat (Birge, 1915; Schmidt, 1915, 1928; Idso, 1973). Therefore, during summer stratification, stability expresses the amount of work needed to prevent the lake from developing thermal stratification. More importantly, stability quantifies the resistance of that stratification to disruption by the wind, and therefore the extent to which the hypolimnion is isolated from epilimnetic and surface movements.

The stability of a lake is very strongly influenced by its size and morphometry (Table 6-1). The amount of work (B) contributed by the wind in distributing heat throughout the lake increases proportionally as the size and volume of the lakes decrease, whereas the stability (S) or amount of work per unit area to mix a lake to a uniform vertical density is greater in larger lakes. The sum of B and S is the amount of

TABLE 6-1 Work Required to Distribute Heat by Wind (B), to Render the Lake
Homoiothermal (S), and to Maintain Hypothetical Homoiothermal Conditions Throughout the
Heating Season (G) for Several Lakes*

LAKE	AREA (km²)	MAXIMUM DEPTH (m)	B (g — cm cm⁻²)	S (g — cm cm⁻²)	G (g — cm cm⁻²)
Atitlan (Guatemala)	136.9	341	3741	21500	25241
Pyramid (Nevada)	532	104	4055	10255	14310
Mendota (Wisconsin)	39.2	25.6	1209	514	1723
Amatitlan (Guatemala)	8.2	33.6	752	415	1167
Güija (El Salvador)	44.3	26.0	958	175	1133
Lunzer Untersee (Austria)	0.68	33.7	535	390	925
Lawrence (Michigan)	0.050	12.6	301	208	509
Mirror (New Hampshire)	0.150	10.9	281	135	416
Findley (Washington)	0.114	24.0	175	133	308
Wingra (Wisconsin)	1.370	6.1	110	7	117
Marion (British Columbia)	0.133	4.5	56	11	67

*Modified from Hutchinson (1957) and Johnson, et al. (1978).

work per unit area (G) needed by the wind to distribute the heat in a lake uniformly at the minimum winter temperature. It should be apparent that the metalimnetic thermal discontinuity represents an effective mixing barrier. While not a total barrier, it is nearly as effective as the sides of the lake basin. An immense amount of work is required to disrupt it.

An excellent example of the usefulness of thermal energy and stability parameters is seen in the comparative analyses of five North American lakes (Fig. 6–5 and Table 6–1). The five lakes are all small, occur at similar latitudes, and, with the exception of alpine Findley Lake, are at similar elevations (Loucks and Odum, 1978). The maximum and mean depths of the lakes (Table 6–1) and water renewal times* differ considerably (Fig. 6–5), both seasonally and among the lakes (Johnson, et al., 1978). Marion Lake of this series has a renewal rate of about 76 times per year.

Lakes Wingra and Marion are distinctively unstable, and exhibit an order of magnitude less stability (10^1 g-cm cm⁻²) than lakes Lawrence, Mirror, and Findley (10^2 g-cm cm⁻²). Lawrence and Mirror lakes are virtually identical with respect to their overall thermal and convective regimes, although Lawrence Lake is more stable than Mirror because of its smaller surface area and greater depth. Latent heat of fusion is the dominant factor in the thermal regime of Lake Findley, which is the coldest lake in this comparison. Winter stability is uniformly low (S < 10 g-cm cm⁻²) for all five lakes, because the presence of ice cover prevents wind work from being exerted on the water column (B → 0). Advection (horizontal movement of water) dominates the circulation of Lake Marion (approximately 76 renewals per year) compared with Mirror, Lawrence, and Wingra lakes (1–2 renewals per year). Advection is seasonally dominant in Findley Lake (> 100% renewal per month during some months). Coincidence of the ice-free period with relaxation of advection (August–November) allows considerable stability to become established.

Density differences that regulate stratification in lakes are not always thermally induced. Differences in salinity can produce a similar stratification, often in combi-

---

*Water renewal is the amount of water required to replace the lake volume during a given time interval (reciprocal water residence time).

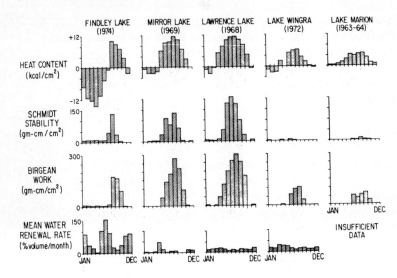

Figure 6-5. Monthly distribution of heat content, stability, Birgean work, and mean water renewal rate parameters in five North American lakes. (From Johnson, et al. 1978. Verh. Int. Ver. Limnol. *20*:562–567.)

nation with temperature differences. The stability so produced can exceed that resulting from thermal density differences, and may lead to conditions where disruption of stratification is intermittent or never occurs (see discussion of meromictic lakes below).

LOSS OF STRATIFICATION

In late summer and fall, declining air temperatures result in a negative heat income to the lake, and loss of heat exceeds inputs from solar radiation. Surface waters cool and become more dense than underlying warmer epilimnetic water. As the surface water sinks, it is mixed by a combination of convection currents and wind-induced epilimnetic circulation. The penetration of surface waters into the metalimnion continues as the lake continues to cool. A progressive erosion of the metalimnion from above can be observed as the stratum of thermal discontinuity is reduced and the homoiothermous epilimnion increases in thickness. On an annual basis, the result is an apparent lowering of the metalimnetic thermocline. The isopleth diagram of Figure 6–4 illustrates lines of equal temperature (isotherms), over time against depth. Depth-time diagrams of quantities or intensities of a given parameter are particularly useful in limnology because they aggregate hundreds and often thousands of data points taken at different depths and times into an annual picture that indicates the seasonal dynamics of physical, chemical, or biological properties of the lake. In the example from Lawrence Lake, it can be seen that stratification was maintained down to a depth of 11 m until late October. Finally, the entire volume of lake water was included in the circulation, and *fall turnover* was initiated. The transition from the final stages of weak summer stratification to autumnal circulation usually is dramatic, and can occur in a few hours, especially if associated with the high wind velocities of a storm.

Circulation continues with gradual cooling of the water column. The rate and duration of cooling are highly variable from lake to lake and are dependent upon basin morphometry, the volume of water that must be cooled, and prevailing meteorological conditions. For example, in Figure 6–4, Lawrence Lake, with a surface area of 5 hectares (ha) and a volume of 0.292×10^6 m³, cooled down to and below the temperature of maximum density by early December. Waters of nearby Gull Lake, with an area of 827 ha and a volume of 102.1×10^6 m³, usually circulate well into January before reaching comparable temperatures. Very large lakes, such as the Great Lakes, which are of similar latitudes to those just mentioned, circulate all winter and rarely reach conditions that permit the formation of an ice cover. In large, very deep lakes, for example Great Bear Lake of the Canadian Northwest Territories, late summer circulation often does not extend to the lower depths, e.g., below 200 m, because stability at lower levels is sufficiently strong to prevent mixing (Johnson, 1966). The temperature of the maximum density of the water below 200 m is 3.53°C, because of hydrostatic pressure. In cold years with slow spring heating, when the upper 200 m of water are cooled to 3.53°C in the autumn, the upper waters reach the same temperature of maximum density as the water strata below 200 m. Circulation then extends to the bottom, which is 450 m in Great Bear Lake.

WINTER STRATIFICATION

As the temperature of the water reaches the point of maximum density—4°C—surface ice can form rapidly with loss of heat as could occur on a calm, cold night. The difference in water density between 4°C and 0°C is, as discussed earlier, very small, and results in only a minor density gradient just below the surface. Hence, *inverse stratification*, in which colder water lies over warmer water, is easily disrupted by a small amount of wind energy. Under stormy weather conditions common at this time of year, it is not unusual for a lake to continue to circulate and cool to well below the termperature of maximum density. Water temperatures of between 3 and 4°C are very common before conditions are such that ice cover can form, as for example is seen in Figure 6–4. Isothermy to less than 1°C has been observed frequently before the formation of ice cover, especially in large lakes that are subject to much wind action.

After the lake has frozen, the ice cover effectively seals it off from the effects of the wind. Immediately below the ice, water of increasing density at the circulation temperature of the water prior to ice formation underlies water of 0°C. The steep thermal gradient near the undersurface of the ice is referred to as *inverse thermal stratification*. The term inverse stagnation should not be used for this condition, because it implies that the water beneath the ice is stagnant and not subject to water movement. Most lakes exhibit considerable water movement under ice (cf. Chapter 7).

Heating of the water under ice may occur throughout winter to considerable depths (cf. Fig. 6–4). Although clear ice is, as we have seen, quite transparent to solar radiation, the type and amount of snow cover on the ice affect solar radiation transmission. From the few detailed studies on winter heating, it is clear that most (>75 per cent) is from solar radiation (Birge, et al., 1927). The relative contribution of heat stored in the sediments during the summer varies from lake to lake in relation to morphometry, summer heating conditions, and other factors (Hutchinson, 1957). Density currents, generated by heating of water through the ice in the shallow littoral areas, can flow toward the central portion of the basin along the sediments. Thermal-density instability is prevented by diffusion of solutes from the sediments, which raises the density slightly, especially as winter progresses, and accounts for the commonly observed temperatures of above 4°C near the sediments.

It is not unusual to observe apparent thermal-density anomalies under the ice. For example, in Figure 6-4, a cell of water in excess of 6°C was found centered around 1 m in late March in this extremely hardwater lake. The ice at this time was weak and porous from heavy, warm rains. Water very low in dissolved solutes, which had entered through the ice or possibly from the margins of the lake, and was running under the ice, was less dense than that of the normal water high in dissolved ions. By this means, solar heating through the ice was able to increase the temperatures in this dilute layer without producing instability. Most of these situations are quite transitory and variable (Koźmiński and Wisniewski, 1935), and may be related in part to the early onset of certain phytoplankton populations, which increase light absorption beneath the ice (cf. Chapter 15).

A detailed résumé of the types and characteristics of lake ice is given in Hutchinson (1957), based primarily on the extensive work of Wilson, Zumberge, and Marshall. Two books by Pivovarov (1973) and Ficke and Ficke (1977) summarize in detail the diverse literature on the properties of ice and the thermal characteristics of lakes and rivers under differing conditions of ice cover.

MODIFICATIONS IN STRATIFICATION

The stratification picture just described, in which a typical temperate lake of moderate size undergoes two periods of mixing, one in the spring and one in the fall, is called *dimictic*, and is not always so consistent. Many variations are found in relation to local or regional differences in climate, individual characteristics of the lake morphometry, and movement of the water masses.

A common divergence from the typical summer stratification pattern is the formation of secondary or multiple discontinuity layers (Fig. 6-6). In these situations, which usually last only for a few days or weeks, secondary metalimnia are formed in the epilimnion of the initial stratification during the heating period. These conditions are common when intense heating alternates with periods of extensive mixing. The initial spring metalimnion forms at various depths as discussed earlier and stabilizes at a depth characteristic of the lake morphometry and prevailing meteorological conditions. This can be seen in the general limits of the epilimnetic boundary approximately indicated by the dashed line in the example given in Figure 6-4. The mixing of the entire epilimnion can cease during a calm, hot period, during which the surface waters can be intensely heated and perhaps mixed by light breezes for only a few meters of depth. In this way, a secondary metalimnion can be formed, overlying the first. The stability of these multiple metalimnia is usually not as great as the primary deeper one, and they are susceptible to disruption during periods of cooling and strong mixing of the epilimnion; on a small scale this process is analogous to normal autumnal erosion of the metalimnion and ensuing circulation.

Thermal stratification can be modified by inflow-outflow relationships, particularly if the volume of influent is large in relation to the volume of the epilimnion. For exam-

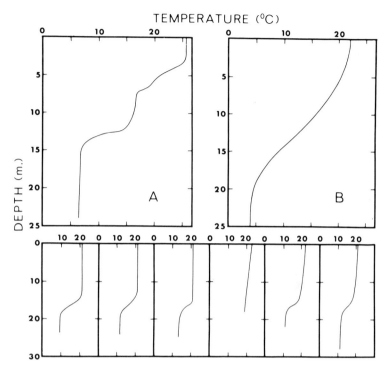

Figure 6-6 Variations in summer thermal stratification. *A*, Example of multiple thermal discontinuities (generalized from several sources); *B*, Example of thermal discontinuity greatly reduced by high through-flow volumes (generalized). *Lower series*, Profiles of thermal stratification in submersed depressions of Douglas Lake, Michigan, August 10–12, 1929. (From data of Welch and Eggleton, 1932.)

ple, in reservoirs, high inflow from stream discharge, often cooler than the water of the epilimnion, can cause much turbulence and reduce the thermal gradient appreciably (Fig. 6-6B). A similar phenomenon is observed frequently in alpine and northern lakes that receive large flows of glacial or snow meltwater during the later portions of the summer stratification.

Many other variations occur in relation to individual characteristics of lake heating and mixing. Individuality is a prominent characteristic of the observed patterns of thermal structure, and is governed strongly by climatic variations, volume of inflow and outflow in relation to the volume of the basin, basin configuration, surface area of the lake, position of the basin in relation to wind action, and other factors. Where the morphometry of the basin is complex, different stratification patterns can be observed within the same basin (e.g., Wetzel, 1966a). Lakes with many submersed depressions characteristically exhibit variations in both the position and stability of the metalimnion (Fig. 6-6), which can strongly influence the productivity of these different regions. In the shallow water, depth is insufficient to maintain a typical epilimnion and hypolimnion, but these layers exhibit a reduction in temperature with increased depth. Such diminutive stratification is common in protected shallow areas of large lakes, or in entire shallow lakes.

Marked horizontal differences in thermal profiles are also observed in steep-sided lake basins. For example, thermal microstructures increase as shallow water is approached along a steep lake-bottom slope (Caldwell, et al., 1978). These stepped thermal structures result from the mixing activity of internal waves, which are accentuated near the sides of the basin (cf. Chapter 7).

TYPES OF STRATIFICATION

It is useful to scrutinize closely the types of stratification found in lakes of the world as they deviate from that described for a typical dimictic situation. The effects of the interactions of climate, morphometry, and chemistry on stratification then become increasingly evident, as does the importance of climatic influences on phyto- and zoogeography and lake productivity.

F. A. Forel, who is often referred to as the father of limnology, published extensive monographs on Lake Geneva, Switzerland, in 1892, 1895, and 1904, which served as a classical foundation in the field for many years. Forel, who proposed a classification of lakes based on their thermal conditions, recognized three types: (1) temperate lakes, which undergo a regular annual alternation of summer and inverse winter stratification between two circulation periods at the temperature of maximum density; (2) tropical lakes, in which the water is never cooled below 4°C and is stratified, except for a single period of winter circulation; and (3) polar lakes, in which temperatures never rise above 4°C, and the water is inversely stratified except for a single period of summer circulation. However, these terms suggest differences in latitude that are unfortunate, because many exceptions to the above classification can be found. An ideal "tropical lake," for example, occurs in the northern temperate region.

Forel's terminology was modified in a number of ways, particularly by Whipple (1898, 1927), Birge (1915), and Yoshimura (1936). Whipple's widely adopted modification of this system subdivided each of the categories into orders on the basis of surface- or deep-water temperatures. According to his scheme, temperatures of bottom waters in *first-order* lakes remain at 4°C throughout the year. Circulation periods are often absent. In lakes of the *second order*, water at greatest depth is near or above 4°C in the summer, and one or two complete circulation periods occur annually between the periods of thermal stratification. *Third-order* lakes are those in which thermal stratification never occurs, and circulation is more or less continuous, except when the lake is frozen.

As useful as these early categorizations were, it became apparent that alternative groupings were necessary as more information on polar, high-altitude, and tropical lakes became available. The following classification, introduced by Hutchinson and Löffler (1956) and Hutchinson (1957), is now generally accepted as most useful and as having the least ambiguity. The definitions center on circulation patterns, as the roots of the names indicate, and refer to lakes that are of sufficient depth to form a hypolimnion.

Amictic Lakes.   Lakes defined as amictic are sealed off perennially by ice from most of the annual variations in temperature. Amictic, perennially ice-covered lakes are rare and largely limited to Antarctica (Goldman, 1972), or, more rarely, to very high mountains. Only a very few cases of such lakes have been recorded in the Arctic, mostly from Greenland, wherein general conditions necessary for the formation of a permanent ice cover are rare (Hobbie, 1973). It is clear that most of the heating of these lakes is by means of light transmission through the ice and conduction of heat from the surrounding land through the sediments (Ragotzkie and Likens, 1964).

Cold Monomictic Lakes.   Lakes with water temperatures never greater than 4°C, and with only one period of circulation in the summer at or below 4°C are called cold monomictic lakes. The category comprises, for the most part, Arctic and mountain lakes, which, although they may be ice-free for brief periods in the summer, are in frequent contact with glaciers or permafrost. Thermal cycles of Arctic lakes exhibit large variations in relation to their location and summer climate. Shallow lakes are extremely numerous in the subarctic, but usually lack sufficient depth to stratify. Water temperatures often rise above 10°C, but never exceed 15°C (Hobbie, 1973). Very deep lakes that become ice-free for brief periods in the summer are truly cold monomictic lakes in which temperatures never reach 4°C. Others, such as Char Lake, Cornwallis Island, Canada (latitude 76°), warm to 4–5°C, and are therefore technically dimictic (see below), but do not always stratify. Lake Schrader in the Brooks Range, Alaska, was dimictic and stratified in 1958, with the epilimnion at 10°C and the hypolimnion at 4°C. The following year, summer ice breakup occurred a month later, and the lake did not warm above 4°C, i.e., it was cold monomictic (Hobbie, 1961).

Dimictic Lakes. Lakes are called dimictic if they circulate freely twice a year in the spring and fall and are directly stratified in summer and inversely stratified in winter. Dimictic lakes represent the most common type of thermal stratification observed in most lakes of the cool temperate regions of the world. Their characteristics were described in detail earlier in this chapter. Such lakes also are found commonly at high elevations in subtropical latitudes.

Warm Monomictic Lakes. In warm monomictic lakes, temperatures do not drop below 4°C; they circulate freely in the winter at or above 4°C, and they stratify directly in the summer. Warm monomictic lakes are common to warm regions of the temperate zones, particularly those influenced by oceanic climates, and to mountainous areas of subtropical latitudes. Most lakes of the central and eastern portions of North America and the interior of Europe exhibit a distinct continental type of dimictic stratification, whereas a warm monomictic stratification is prevalent in many coastal regions of North America and Northern Europe.

Oligomictic Lakes. Generally tropical, oligomictic lakes have rare circulation periods at irregular intervals, and have temperatures always well above 4°C. Lakes of small to moderate area, or lakes of very great depth in the tropics, often maintain stable stratification, even though only a small temperature difference may exist between the surface and bottom strata. These lakes are common in equatorial regions of high humidity, and may circulate only at irregular intervals during periods of abnormally cold weather. Circulation can be quite rare and occurs between periods of several years of continuous feeble stratification (Vollenweider, 1964; Talling, 1969).

Polymictic Lakes. Lakes with frequent or continuous circulation are called polymictic. The category has been further divided (Ruttner, 1963). *Cold polymictic lakes* circulate continually at temperatures near or slightly above 4°C. These lakes commonly are of large area and moderate depth, and are found in equatorial regions of high wind and low humidity, where little seasonal change in air temperatures occurs. At very high altitudes in equatorial regions, cold polymictic lakes gain a significant amount of heat during the day, but nocturnal losses are sufficient to permit complete mixing during the night.

Warm polymictic lakes are usually tropical lakes that exhibit frequent periods of circulation at temperatures well above 4°C. Annual temperature variations are small in the equatorial tropics and result in repeated periods of circulation between short

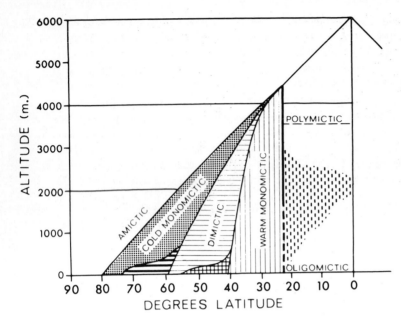

Figure 6-7 Schematic arrangement of thermal lake types with latitude and altitude. *Black dots:* cold monomictic; *black and white horizontal bars:* transitional regions; *horizontal lines:* dimictic; *crossed lines:* transitional regions; *vertical lines:* warm monomictic. The two equatorial types occupy the unshaded areas labelled oligomictic and polymictic, separated by a region of mixed types, mainly variants of the warm monomictic type *(broken vertical lines).* (Modified from Hutchinson and Löffler, 1956.)

intervals of heating and weak stratification, followed by periods of rapid cooling. Under these circumstances, convectional circulation is sufficient, in combination with wind, to disrupt stratification.

A useful schematic arrangement of these six lake types based on thermal and circulation characteristics is presented in Figure 6-7, in which generalizations are drawn on altitudinal and latitudinal distribution. Such generalizations are difficult to maintain; many exceptions exist, particularly at low altitudes, where there is a strong influence of oceanic climate. The diagram does, however, demonstrate the general observed geographical distribution of lake types.

The discussion of thermal lake types above refers to lakes with sufficient depth to form a hypolimnion. Numerically speaking, when considering the millions of small, relatively shallow lakes that virtually cover the tundra and northern temperate region, it becomes evident that most lakes do not become stratified or become stratified only briefly during ice-free periods. The depth required to become thermally stratified varies so greatly with surface area, basin orientation in relation to prevailing wind, depth–volume relations, protection by surrounding topography and vegetation, and other factors, that generalizations in this regard would be misleading.

Meromixis

The types of stratification discussed thus far describe circulation that occurs throughout the entire water column, as is the case with *holomictic* lakes. A number of lakes do not undergo complete circulation, and the primary water mass does not mix with a lower portion. Such lakes are termed *meromictic* (Findenegg, 1935; Hutchinson,

1937). In meromictic lakes, the deeper stratum of water that is perennially isolated is the *monimolimnion*; this underlies the upper *mixolimnion*, which periodically circulates. These two strata are separated by a steep salinity gradient which is called the *chemocline*.

Although the causes of chemically induced stratification have been treated and named variously, the divisions of Hutchinson (1937) and Walker and Likens (1975; cf. also Dickman and Hartman, 1979) are particularly appropriate. As pointed out in Chapter 2, a salt concentration of 1 g l^{-1} increases the density of water by approximately 0.0008. This change in specific gravity is very large in relation to density changes associated with temperature. For example, the density difference between 4° and 5°C is 0.000008, and it would only require 10 mg l^{-1} of salt to give the same effect in resistance to mixing. The salinity gradient with depth during stratification is almost always much greater than 10 mg l^{-1}. In a majority of lakes, salinity gradients are insufficient to increase stability to a point where wind energy does not cause holomixis. A large number of lakes, however, do exhibit temporary or permanent meromixis as a result of salinity gradients. In such cases, the term "concentration stability," in contrast to the thermal stability of holomictic lakes discussed earlier, is more appropriate (Berger, 1955).

Ectogenic Meromixis. The condition that results when some external event brings salt water into a freshwater lake or fresh water into a saline lake is called ectogenic meromixis. The result in either case is a superficial layer of less dense, less saline water overlying a monimolimnion of saline, more dense water. As might be expected, such situations are found often along marine coastal regions where catastrophic intrusions of salt water from storms associated with unusual tidal activity are fairly common events. In estuarine lakes, such events also occur routinely. A more strict definition of ectogenic meromictic lakes includes only those that are isolated from routine marine influxes of water. Such isolation may be recent or long-standing. An example of the latter is southern Norwegian Lake Tokke, which most likely has been permanently meromictic for about 6000 years; the isolation of the lake probably took place during a period when the former fjord depression and the surrounding land were elevated some 60 m above sea level (Strøm, 1955).

A most interesting variant of ectogenic meromixis is found in many saline lakes of arid regions or in small depressions that occasionally receive salt water intrusions (Hudec and Sonnenfeld, 1974). Often these saline lakes overlie large deposits of such soluble salts as magnesium sulfate, and are only a few meters in depth. Infusion of large amounts of fresh water, either naturally during wet periods or artificially as a result of increased irrigation of nearby land, creates a strong meromictic stratification that may persist for many years. A striking example is Hot Lake, a shallow (3.5 m) saline body of water occupying a former epsom salt excavation in north central Washington (Anderson, 1958). Energy of solar radiation passing through the overlying mixolimnion of fresh water can accumulate as heat in the chemolimnion and monimolimnion, where circulation is absent. The resulting temperatures can be in excess of 50°C, and even when the surface of the lake is frozen, a temperature of nearly 30°C can be found at a depth of 2 m (Fig. 6–8). These *dichothermic* conditions, in which the monimolimnion is considerably warmer than the overlying water of the chemolimnion, are frequently observed in these meromictic lakes (Fig. 6–8).

Man's activities have been known to form meromictic lakes. Creating a connecting channel between a freshwater lake and the sea for purposes of navigation

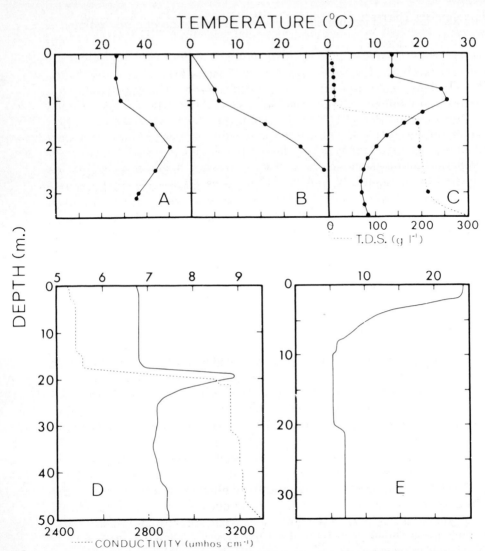

Figure 6-8 Thermal discontinuities in meromictic lakes. *Upper series:* Hot Lake, Washington (after Anderson, 1958): *A,* 23 July 1955; *B,* Heating under thin ice cover, 7 December 1954; *C,* Heating in a deeply colored layer, 0.5–1.0 m, 1 May 1956; *T.D.S.* = Total dissolved solids on evaporation. *Lower series: D,* Depth distribution of temperature and conductivity, Fayetteville Green Lake, New York, 16 December 1966 (from Brunskill and Ludlam, 1969); *E,* Thermal stratification in Ulmener Maar, Germany, 8 August 1911. (Modified from Ruttner, 1963, after Thienemann). The thermal gradient observed earlier in Ulmener Maar has subsequently been modified (cf. Stewart and Hollan, 1975). Note variations in depth and temperature scales.

can result in the intrusion of saline water into the lake and cause permanent mero-mixis. Another event that can form a large number of small meromictic lakes is the introduction of salt caused by runoff from street de-icing (Judd, 1970). The excellent analyses of the paleolimnological record of Längsee, Austria, by Frey (1955) clearly show how the clearing of forests surrounding the lake in prehistoric times introduced ectogenic meromixis. The resulting silt-laden runoff flowed into the hypolimnion and initiated a monimolimnion, which was augmented subsequently by salinity inputs of biogenic origin (see further on).

Crenogenic Meromixis. Crenogenic meromixis results from submerged saline springs that deliver dense water to deep portions of lake basins. The saline water displaces the water of the mixolimnion. The chemolimnion will stabilize at a depth related to the rate of influx, density differences, and the degree of wind mixing of the mixolimnion. Crenogenic meromixis is really a subclassification of the general group of externally mediated (ectogenic) meromixis (Walker and Likens, 1975).

Examples of crenogenic meromictic lakes are numerous. In Ulmener Maar (Fig. 6–8E), the temperature of the monimolimnion stabilized at 7°C. Meromictic lakes of this type recently have been found in interior Alaska, when subsurface springs introduce saline water into small, deep lakes originating in the thawed craters of *pingos* (Likens and Johnson, 1966). Pingos are formed when water, rising by hydraulic pressure through gaps in the permafrost, freezes and uplifts a mound of ice covered by a layer of alluvium. The overburden usually ruptures and the ice of the depression can thaw to form small lakes.

Biogenic Meromixis. Biogenic or endogenic meromixis results from an accumulation of salts in the monimolimnion, which is usually liberated from decomposition in the sediments and sedimenting organic matter. Often biogenic meromixis is initiated when abnormal meteorological conditions prevent circulation of a normally dimictic lake. Hypolimnetic accumulation in biochemically derived salinity may be sufficient to permit meromixis to persist either temporarily or permanently. Temperature inversions are usually found in the monimolimnion, such as is seen in Figure 6–8D. The high temperatures observed in the chemolimnion are the result of absorption of solar radiation by the dense bacterial plate that occurs at this level. The processes associated with the increasing temperatures in the lower monimolimnion, which are consistently observed, are less obvious. It is clear that the heat of metabolism in decomposition is inadequate to account for such increases (Hutchinson, 1941). When direct solar heating is not possible because of the depth, such heating is likely to be the result of warmer salt-laden water near the sediments that form density currents from overlying strata and descend into the basin along its contours. Heat from the surrounding land and conduction through the sediments is a minor heat source.

Biogenic meromixis increases in frequency in lakes that are very deep. For example, nearly all of the extremely deep lakes of the equatorial tropics are meromictic. But biogenic meromixis is common among lakes of small surface area, moderate depth, and that are sheltered, especially in continental regions that experience long, severe winters. *Partial* or *temporary meromixis*, when a normally dimictic lake skips a circulatory period (usually the spring period), is a result of dynamic processes of decomposition in the lower strata of the lake under unusual weather conditions (Findenegg, 1937). For example, after an unusually long and severe winter followed by a rapid thaw and warm conditions, Martin Lake did not undergo spring circulation (Fig. 6–9); the chemically stabilized water of the temporary monimolimnion was carried over from winter accumulation and was augmented throughout the subsequent summer (Wetzel, 1973). Particularly notable is the carryover of low temperatures from the circulation during the preceding fall, which persisted throughout the summer and fall, prior to the ice-cover

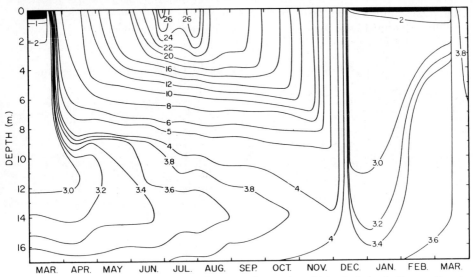

Figure 6-9 Isothermal variations (°C) in the major depression of Martin Lake, LaGrange County, Indiana, 1963–64. (From Wetzel, R.: Hydrobiological Studies, 3:91–143, 1973.)

formation. Autumnal circulation in this protected lake did not occur until December, and then only briefly before ice formation. The following spring, a normal complete circulation was observed. This type of temporary, partial meromixis was also found in two other lakes in northeastern Indiana, of the 13 investigated during the same period (Wetzel, 1966b). In his classical work on phosphorus cycles and enrichment effects in the small eutrophic Lake Schleinsee, Einsele (1941) found that complete spring turn-over occurred, on the average, only every other year. In Lawrence Lake, a small (4.9 ha), deep (12.6 m) hard-water lake of southern Michigan, temporary meromixis occurs about every third year (Wetzel, et al., 1972, and unpublished data). Many other examples exist, but these are adequate to emphasize that this phenomenon of intermittent biogenic meromixis is quite common. It is not difficult to envision a combination of circumstances that can shift a lake from temporary to permanent meromixis. Both conditions have profound effects on biota and productivity, as will be discussed later.

Another quite different cause of endogenic meromixis apparently occurs from precipitation of salts of upper water strata by "freezing out" from the surface ice layer. The precipitating salts may accumulate in deeper waters in sufficient concentrations to cause meromixis. In Algal Lake, Antarctica, this condition has been termed *cryogenic meromixis* (Goldman, et al., 1972; cf. also Priddle and Heywood, 1980).

Thermal Energy Content: Heat Budgets of Lakes

Temperature and density stratification in lakes are dominant regulators of nearly all physicochemical cycles, and consequently of lake metabolism and productivity. It is important, therefore, to understand a lake's relative energy content (budget) or thermal capacity* both in the amount of heat energy required to develop stratification and the amount of heat a lake must lose in order to overcome density differences.

Heating capacity and energy demand to reach a given state are governed, in the case of a lake, by heat inputs, the volume and accessibility of the water at depth to heat incomes, and heat losses. Although many means of determining the heat content of lakes were developed in the pioneer days of limnology, particularly by Forel, the extensive works of Birge (1915, 1916) on this subject provide the basis for much of the ensuing discussion (cf. also Ragotzkie, 1978).

The heat content of a body of water is the amount of heat in calories that would be released on cooling from its maximum temperature to 0°C or the amount of heat that has been transferred to the lake to heat it from 0°C to its maximum temperatures at various depths. Of course, the lowest temperatures of many lakes are not near 0°C, or even its point of maximum density at 4°C. Hence, the minimum and maximum temperatures of the individual strata are used to determine heat content of a lake. Although the specific heat of water is unity, which simplifies computations, calculations of the heat budget of a lake require consideration of changes in volume with depth, since the lower strata contain smaller water volumes and make a smaller contribution to the total heat content than do the upper water layers.

A Birgean heat budget for a typical dimictic lake can be defined as follows (Birge and Juday, 1914; Birge 1915; Hutchinson, 1957):

Summer Heat Income (θ_s) is the amount of heat necessary to raise the temperature of the lake from an isothermal condition at 4°C to the maximum observed summer heat content. This heat income is primarily a result of the heating of the surface layers by direct insolation, and the distribution of this heat by wind. It therefore has been called (Birge, 1915) the wind-distributed heat, even though not only wind energy is involved in its distribution.

Winter Heat Income (θ_w) is the amount of heat necessary to raise the temperature from its minimum heat content to 4°C. The winter heat income is usually not considered as wind-distributed, even though wind is involved in loss of heat prior to the lake's reaching the minimum temperature during autumnal–early winter cooling. Accurate winter heat incomes must be corrected for the latent heat of fusion of ice. This correction can be sizable when the ice cover is considerable. One-cm-thick ice is equivalent to about 66.7 cal cm^{-2} (79.72 cal g^{-1}, where $\rho_{ice} = 0.9168$) or a correction of about 835 cal cm^{-2} for 12.5 cm of ice (see Adams and Lasenby, 1978, and Wetzel and Likens, 1979).

Annual Heat Budget (θ_a) is the total amount of heat necessary to raise the water from the minimum temperature of winter to the maximum summer temperature. From the standpoint of the whole lake, this refers to the total amount of heat needed between the period of its lowest and its highest heat content.

The heat budget may be calculated by plotting the product $A_z (\theta_{sz} - \theta_{wz})$ against depth z, where θ_{sz} and θ_{wz} are the respective maximum summer and minimum winter temperatures, and A_z is the area of the water stratum in cm^2 at depth z. The area of the resulting curve is integrated by planimetry† and divided by the surface area of the lake

---

*The use of the word "budget" here is incorrect in terms of its normal definition. Reference actually is to the heat storage capacity of the lake, but the term "budget" is firmly entrenched in limnology.

†The proper use of these ingeniously devised instruments is described in detail in Welch (1948) and Wetzel and Likens (1979).

(A_o) because lake heat gains and losses are primarily surface phenomena. To separate the heat budget into summer or winter heat incomes, 4°C is substituted for the respective summer and winter temperatures at depth. A sample computation is given in Figure 6-10 for a small dimictic lake of moderate depth in southwestern Michigan. Note particularly the decreasing contribution of lower water strata to the overall heat budget of a rather typical lake.

A large number of annual heat budgets have been computed; about 100 are listed in Hutchinson (1957), and many have appeared in the literature since then. A few exemplary values are given in Table 6-2 to emphasize several points in regard to factors influencing the observed heat budgets.

Although a majority of the heat budgets have been derived for large, deep lakes, a few generalizations are possible. In temperate lakes that are sufficiently deep to maintain hypolimnetic temperatures at or near 4°C, annual heat budgets are usually between 30,000 and 40,000 cal cm^{-2} year^{-1}, most of which (60 to 80 percent) is gained during the spring–summer transition. Very few lakes have an annual heat budget in excess of 50,000 cal cm^{-2} year^{-1}. Notable exceptions include Lake Baikal and Lake Michigan. Annual heat incomes of polar region lakes are small (Θ_s values are in the range of 10,000 cal cm^{-2} for large lakes). However, annual values are likely to be very much larger than this when latent heat of fusion for melting ice is included. The same is true for the apparently low heat budget values of lakes at high altitudes. In the temperate zone, annual heat budgets are strongly correlated with mean depth, area, and lake volumes, and rise continuously with lake dimensions, though at a decreasing rate (Gorham, 1964). For lakes of a given volume, deeper ones of lesser area take up slightly more heat than shallower lakes of greater area. Dimictic lakes have slightly higher heat budgets than warm monomictic lakes of similar size.

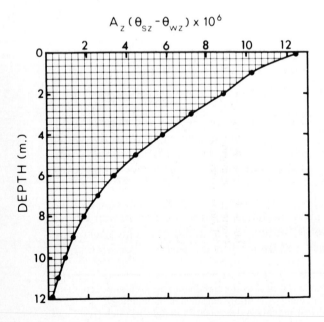

Figure 6-10 Curve generated for calculation of the annual heat budget of Lawrence Lake, Michigan, 1968. Calculations resulted in a budget, including a correction for ice, of 10,918 cal cm^{-2} yr^{-1} (From Wetzel, unpublished data.)

TABLE 6–2 Heat Budgets of Various Lakes (cal cm⁻² yr)*

LAKE	SURFACE AREA (A_o) [km²]	MAXIMUM DEPTH (z_m) [m]	MEAN DEPTH ($\bar{z}$) [m]	SUMMER HEAT INCOME θ_s (cal cm⁻² yr⁻¹)	WINTER HEAT INCOME θ_w (cal cm²yr⁻¹)	ANNUAL HEAT BUDGET θ_a (cal cm⁻² yr⁻¹)
Baikal, USSR	31500	1741	730	42300	22800	65500
Michigan, USA	57850	282	77	40800	11600	52400
Washington, Washington	128	65	18	43000	[−2600]	43000
Geneva, Switzerland	581.5	310	154.4	36000	[−23200]	36000
Tahoe, California–Nevada	499	501	313	34800	0	34800
Green, Wisconsin	29.7	72.3	33.1	26200	7800	34000
Pyramid, Nevada	532	104	57	33600	[−7000]	33600
Tiberias, Israel	167	50	24	33500	[−26400]	33500
Ladoga, USSR	18150	223	56	18000	15300	33300
Mendota, Wisconsin	39.2	25.6	12.4	approx. 18250	approx. 5800	24073
Atitlan, Guatemala	136.9	341	183	22110	[−288300]	22110
Windermere, England						
North Basin	8.16	67	26.0	17500	[−2900]	17500
South Basin	10.5	44	17.7	15680	[−2450]	15680
Fureso, Denmark	9.9	36	12.3	14400	2700	17100
Lunzer Untersee, Austria	0.68	33.7	19.8	12300	1400	13700
Lawrence, Michigan	0.049	12.6	5.9	9168	1750	10918
Schrader, Alaska						
1958	13.2	57	33	9050	–	–
1961				8900	–	–
Chandler, Alaska	15	21	13.5	5760	–	–
Lanao, Philippines	357	112	60.3			
1970				7250	[−121300]	7250
1971				4500	[−121300]	4500
Ranu Klindungan, Java	2	134	90	approx. 3410	[approx. −189000]	approx. 3410
Hula, Israel	14	4	1.7	approx. 2240	[−1600]	2240

*From numerous sources.

TABLE 6-3   Variations in the Annual Heat Budgets of Several Lakes*

LAKE	NUMBER OF ANNUAL OBSERVATIONS	MEAN DEPTH $(\bar{z})$	MEAN θ_a (cal cm^{-2} yr^{-1})	RANGE (cal cm^{-2} yr^{-1})
Green, Wisconsin	5	33.1	34200	32300–36400
Geneva, Switzerland	4	154.4	32325	22000–40200
Mendota, Wisconsin	5	12.4	24073	22308–25953
Menona, Wisconsin	3	7.7	17559	17256–18041
Waubesa, Wisconsin	3	4.6	11362	10948–11739
Mize, Florida	3	4.0	4720	3767–6003

*From data of Birge, 1915, Nordlie, 1972, and Stewart, 1973.

Heat content of a lake can vary significantly (10 per cent) within 2 or 3 days; accurate determinations of annual heat budgets therefore require frequent measurements over long periods (Stewart, 1973). Examples of this variation are given in Table 6-3.

The annual income and loss of heat in tropical lakes is very small compared with that of lakes at higher latitudes or altitudes. Also, the mixing patterns are complex, often variable or oligomictic in nature. Lewis (1973) demonstrated the weak, variable stratification with the repeated formation of multiple, temporary metalimnia common to many tropical lakes of large size. The minimum yearly heat content, given as negative values in brackets in Table 6-2, represents the heat content above 4°C in lakes that never cool to that temperature. In such lakes at northern latitudes, as in England, these values are very small; they are of course high in equatorial regions.

The stratification patterns and thermal structure of reservoirs with high through-flow volumes are enormously complicated (Wunderlich, 1971). Where withdrawal of major volumes for generators occurs daily and intermittently, the patterns can be further disrupted and altered. Although draw-down usually occurs from the epilimnion, sometimes water is drawn from the hypolimnion. Influent waters also vary greatly in relation to surface runoff, heat content, and retention time in series of reservoirs in tandem on major river systems. Normal Birgean heat dynamics must be modified considerably in order to be applicable to complex reservoir systems. Among the most detailed analyses are those of the Slapy Reservoir in Czechoslovakia, where seasonal thermal dynamics have been determined for over a decade (Hrbáček and Straškraba, 1966; Straškraba, Hrbáček, and Javornický, 1973), and of Lake Powell, Utah–Arizona (Johnson and Merritt, 1979).

Heat of Lake Sediments

The sediments of lakes absorb an appreciable amount of heat from the water during the warmer periods of the year, and transmit heat to the water during the winter period. Although few detailed studies are available, data from Lake Mendota illustrate this characteristic (Birge, Juday, and March, 1927). It will be noted from Figure 6-11 that the temperatures of the surface sediments at a water depth of 8 m follow the annual cycle of the epilimnetic water temperatures rather closely. This thermal variation is attenuated with depth in the sediments, so that at a depth of 5 m into the sediments,

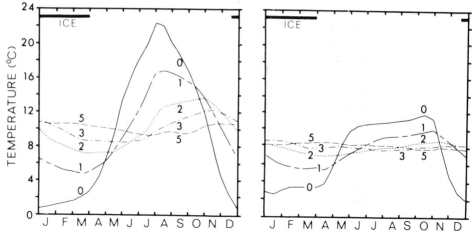

Figure 6-11 Annual temperature curves at 0, 1, 2, 3, and 5 m within the sediments of Lake Mendota, Wisconsin (means for 1918, 1919, and 1920). *Left*, Sediments at a water depth of 8 m; *Right*, sediments at 23.5 m. (Modified from data of Birge, Juday, and March, 1927.)

little seasonal oscillation is observed. A similar relationship has been observed in the sediments of the hypolimnion at a water depth of 23.5 m, but the amplitude and heat energy exchange are much less (Fig. 6–11). Heat content of the sediments lags behind hypolimnetic water temperature, and the lag duration increases about one month per meter depth of sediment.

Spatial variations in temperatures within the sediments at similar water depths have been found that are apparently due to the type of adjacent land forms of the basin (Matveev, 1964). Shoreline sediments of deep bog lakes were found to be considerably warmer than sediments beneath the open water, but annual variations were largely damped at a sediment depth of 4 m (Likens and Johnson, 1969). Heat flux from the relatively warm shoreline sediments occurred both lakeward and landward. In a shallower bog lake, deep sediments beneath the open water were found to be warmer than those of comparable depth beneath shoreline sediments (Reif, 1969) indicative of a heat flux away from the lake. Differences between the thermal characteristics of these lakes may be related to their maximum depths, for in the shallower lake, absorption of solar energy by the sediments is greater (cf. also Dale and Gillespie, 1977).

Determinations of the heat budget of sediments based on the differences between their minimum and maximum mean temperatures showed that in sediments at a lake depth of 8 m, this value was 2950 cal cm^{-2} yr^{-1}, at 12 m, 2200, and at the maximum of 23.5 m, 1100 cal cm^{-2} yr^{-1}. An estimated average heat budget of all of the lake sediments was ca. 2000 cal cm^{-2} yr^{-1} for Lake Mendota. As lakes become more shallow, the heat budget of the sediments becomes a greater portion of the whole (Table 6–4). All of the examples of annual heat budgets of lakes cited above (Tables 6–2 and 6–3) are based on water temperatures, and ignore sediment temperatures. While in large deep lakes such a simplifying assumption may be justified, it should be noted that the sediments of shallow lakes are significant to their heat budgets.

TABLE 6-4 Heat Budgets of Lakes in Which the Heat Budget of the Sediments Is Compared to that of the Water*

LAKE	MEAN DEPTH (m)	HEAT BUDGET OF WATER θ_s (cal cm^{-2} yr^{-1})	HEAT BUDGET OF SEDIMENTS θ_m (cal cm^{-2} yr^{-1})	HEAT BUDGET OF LAKE (cal cm^{-2} yr^{-1})	% SEDIMENTS OF TOTAL HEAT BUDGET
Mendota, Wisconsin	12.1	23500	2000	25500	7.8
Stewart's Dark, Wisconsin	4.3	7000	730	7730	9.4
Beloie, USSR	4.15	8000	2500	10500	23.8
Tub, Wisconsin	3.6	8000	970	8970	10.8
Hula, Israel	1.7	2290	1400	3690	37.9

*Modified from Hutchinson, 1957, with data of Likens and Johnson, 1969.

SUMMARY

1. The high specific heat of water results in dissipation of much of the incoming light energy and accumulation as heat. In lake systems, the way retained heat is distributed and altered is governed by a number of coupled factors: the physical work of wind energy, currents and other water movements, basin morphometry, and water losses. Resulting patterns of density-induced stratification influence physical and chemical properties and cycles of lakes and, in turn, govern their productivity and decomposition.

2. Temperature is a measure of the *intensity* (not the amount) of heat stored in a volume of water. Heat is measured in calories, and is the product of the weight of the substance (in grams), temperature (°C), and the specific heat (cal g $-$ °C^{-1}).

3. Heat income to lakes results from several processes:
    a. Direct absorption of solar radiation; usually the dominant source
    b. Transfer of heat from the air
    c. Condensation of water vapor at the water surface
    d. Transfer of heat from sediments to the water
    e. Heat transfer from terrestrial sources via precipitation, surface runoff, and groundwater inputs

4. Heat losses from lakes occur by:
    a. Specific conduction of heat to the air, and, to a lesser extent, to the sediments
    b. Evaporation
    c. Outflow, especially of surface water

5. Although differential heating and cooling of surface waters, as well as influents, can cause convection currents, mixing of surface waters by convection is weak and insufficient to produce the long-term stratification patterns commonly observed in most lakes. Wind energy mechanically distributes most of the heat in a lake.

6. In temperate and other regions, many lakes of moderate depth (> approximately 10 m) exhibit *thermal stratification*.

a. During warmer periods of the year, the surface waters are heated, largely by solar radiation, more rapidly than the heat is distributed by mixing. As the surface waters are warmed and become less dense, the *relative thermal resistance* to mixing increases.

b. The lake becomes stratified into three zones (Fig. 6-3):

   i. *Epilimnion*: an upper stratum of less dense, more or less uniformly warm, circulating, and fairly turbulent water.

   ii. *Hypolimnion*: the lower stratum of more dense, cooler, and relatively quiescent water lying below the epilimnion.

   iii. *Metalimnion*: the transitional stratum of marked thermal change between the epilimnion and hypolimnion.

c. The *thermocline* is the plane or surface of maximum rate of decrease of temperature in the metalimnion.

d. The resistance of thermal and chemical-density stratification to mixing is a measure of the lake's mechanical *stability*.

   i. *Stability* per unit area of a lake is the quantity of work or mechanical energy in ergs required to mix the entire volume of water to a uniform temperature by the wind without the addition or loss of heat. Stability quantifies the resistance of the stratification to disruption by wind.

   ii. The amount of work by the wind per unit area needed to mix a lake to a uniform vertical density increases as the size and volume of the lake increases.

7. At the end of thermal stratification when loss of heat exceeds heat inputs, surface waters of the epilimnion cool, become more dense, and are mixed with deeper strata of similar density by wind and convection currents. The metalimnion is progressively eroded as the relative thermal resistance to mixing is reduced. Eventually, the entire volume of water is included in the circulation and *fall turnover* is initiated. Fall turnover continues with progressive cooling, often to the temperatures of maximum density of 4°C or less.

8. Ice cover forms on the surface under calm, cold conditions. *Inverse stratification* of water temperatures occurs under the ice, in which colder, less dense water overlies warmer, more dense water near the temperature of maximum density at 4°C. Some gradual heating of the water occurs during the winter under ice cover.

9. When the ice cover melts in the spring, the water column is nearly isothermal. If the lake receives sufficient wind energy, as is commonly the case, the lake circulates completely and undergoes *spring turnover*.

10. Lakes undergoing complete circulation in spring and fall separated by thermal summer stratification and winter inverse stratification are called *dimictic lakes*. Dimictic lake types are very common among temperate lakes of moderate size.

11. Many other types of density-related stratification patterns occur among lakes in relation to the interacting effects of climate, morphometry, and chemistry. The types of lakes are classified on the basis of circulation patterns and refer to lakes of sufficient depth to form a hypolimnion.

TYPE OF LAKE	ANNUAL VERTICAL MIXING	DOMINANT FACTORS CONTRIBUTING TO OR PREVENTING VERTICAL MIXING
1. Amictic	No appreciable amount	Permanent ice cover; slow internal mixing
2. Holomictic	Complete, at least once	Predominantly wind energy; convection currents
a. Monomictic	Once	
b. Dimictic	Twice	
c. Oligomictic	Irregular, infrequent	
d. Polymictic	More than twice	
3. Meromictic	Permanently stratified or interruption of stratification patterns at irregular intervals	Chemically enhanced density stratification
a. Ectogenic	Permanently stratified	(1) Surface inflow of (a) fresh water overlying a pre-existing saline stratum, (b) saline water underlying a pre-existing fresh stratum, or (c) turbidity currents of particulate-laden water. (2) Subsurface inflow of fresh or saline water (= crenogenic).
b. Endogenic	Permanently or temporarily stratified	
i. Biogenic	Permanently stratified	Chemically enhanced density stratification contributed by biological processes with accumulations of bicarbonate or iron/manganese ions in lower stratum and shelter afforded by morphometry of lake basin and surrounding topography.
ii. Temporary biogenic	Temporary elimination of complete circulation (spring or fall)	Same factors as biogenic meromixis but borderline conditions where unusual climatic conditions lead to omission of spring or fall circulation and accumulation of hypolimnetic ions.
iii. Cryogenic	Permanently stratified	Deep-water accumulation of salts precipitated by freeze concentration from a surface ice layer.

12. Water strata of meromictic lakes are termed differently but analogously to those of holomictic lakes:
   a. *Mixolimnion*: the circulating upper stratum
   b. *Monimolimnion*: the deeper stratum of water that is perennially or periodically isolated
   c. *Chemolimnion*: the interfacing stratum of steep salinity (density) gradient between the mixolimnion and the monimolimnion.

13. The heat budget of a lake is a measure of the heat storage capacity under existing conditions of morphometry and the climate to which it is exposed.

The annual heat budget is the total amount of heat accumulated between the period of its lowest and its highest heat content. The annual income and loss of heat in tropical lakes is small in comparison to that of lakes at higher latitudes and altitudes. Annual heat budgets of lakes of the temperate zone generally increase with mean depth, area, and lake volume. The sediments of lakes absorb considerable heat from the water during warmer periods of the year and transmit heat to the water during the winter period.

CHAPTER SEVEN
WATER MOVEMENTS

THE HYDRODYNAMICS OF WATER MOVEMENTS IN LAKE BASINS

Lakes behave like large mechanical oscillators. They respond in numerous, complex ways to applications of force, and are frictionally damped by viscous forces associated with turbulence. Water movements generated by the transfer of wind energy to the water give rise to a spectrum of rhythmic motions (oscillations), both at the water surface and internally deep within the basin. The oscillations and their attendant currents may be in phase or in opposition; their final fate is to degrade into arrhythmic turbulent motions. Each basin has its own set of free modes of oscillation, both surface and internal, which depend on gravity, on basin shape and size, and on the internal distribution of water density. Which of these modes is brought into play during any particular disturbance or storm depends on the duration, the periodicity, and the spatial distribution of the applied force. In other words, the morphometry of the lake basin, the lake's stratification structure, and exposure to wind are important factors that determine water movement.

The turbulence resulting from water movements is of major significance to the biota and productivity of the lake. Limnological thought is still pervaded by the idea that summer and winter density stratification results in stagnation of lake strata. There is little basis for this view. Water movements are integral components of the lake system, and consideration must be given to their effects on changes in distribution of temperature, dissolved gases and nutrients, and other chemical parameters. Water movements influence not only the aggregation and distribution of nutrients and food, but also the distribution of microorganisms and plankton.

Flow of Water

TURBULENT AND LAMINAR FLOW

At sufficiently slow speeds, flow of water in a smooth tube moves along the interface in an apparent orderly manner. If the velocity of movement is increased at the interface, a critical speed is reached above which the smooth and unidirectional laminar flow becomes disordered or turbulent. The transition from laminar to turbulent flow also occurs at the interface of two miscible fluids moving in opposing directions. An example would be opposing horizontal movements between two water strata of differing densities, such as the epilimnion and metalimnion of a lake.

Figure 7-1 Stages in vortex formation during shear instability on the interface of a stratified two-layer system of differing densities. (From Mortimer, C. H.: Mitteilungen Int. Ver. Limnol., 20:131, 1974.)

Appreciable experimental work on turbulent shearing stresses (Reynolds stresses) has demonstrated that only a low velocity is needed to shift from laminar flow to turbulent current. The critical velocity is a function of the fluid viscosities and densities, and decreases in proportion to increase in size of the channel or tube. In lakes, velocities of only a few mm sec^{-1} can induce turbulent flow, at which point disturbances arising on the interface between the layers will no longer be suppressed by buoyancy (gravity) forces, which tend to keep layers of different densities separated. As a result, laminar flow is almost never found in aquatic systems. It is important that the basic properties of turbulent flow and diffusion be understood in order to appreciate the magnitude of the resultant alterations in distribution of physicochemical parameters and lake biota.

If a critical velocity difference along a given density interface is exceeded, disturbances grow in amplitude and break into vortices (Fig. 7-1) (Mortimer, 1961, 1974; Smith, 1975). Such vortex formation increases mixing of the two layers by generating a transitional layer, across which there is both a velocity gradient (shear) and a density gradient. This model represents in simplified form what occurs when layers of nearly uniform velocity and density stream parallel to one another, as in the movements in the epi-, meta-, and hypolimnion discussed further on.

If the rate (*a*) of supply of energy to turbulent eddies* from the shear stresses exceeds the rate (*b*) at which they have to do work against gravity in disrupting the

---

*The rotating region of fluid, an *eddy*, refers to an assemblage of shear waves of a spectrum of many lengths or "eddy diameters" (Mortimer, 1974).

density stratification, turbulence increases (Richardson, 1925; Mortimer, 1961). Rate a is proportional to the square of the shear, while rate b is proportional to the density gradient. The magnitude of the ratio b/a determines whether turbulence increases or decreases. In nondimensional form, that ratio is the Richardson number (R_i) expressed as:

$$R_i = \frac{g(d\rho/dz)}{\rho(du/dz)^2}$$

where
  g = acceleration of gravity
  ρ = density
  u = velocity
  z = depth.

When R_i of a stratified fluid subjected to shearing flow falls below about one-quarter, flow becomes unstable. Or, put in another way, when R_i falls below 0.25, there is a sudden shift in the eddy spectrum from the molecular microturbulence of stable flow toward macroturbulence of the large vortices associated with unstable flow. The shift is accompanied by increases in friction and in mixing *perpendicular* to the direction to the current.

Thus in a lake with stratified shear layers, little mixing and friction between the layers occurs as long as the flow remains stable ($R_i > 0.25$). When flow increases and becomes unstable ($R_i < 0.25$), the stirring action of vortex formation (macroturbulence) increases the interfacial area many times. Mixing is then rapidly completed by microturbulence and molecular diffusion. Soluble components (nutrients) and suspensoids (microalgae, microfauna) are carried along with this dispersion of water into the newly formed, thicker layer in which density and velocity gradients are less than those that were initially present.

The shear instability that results when R_i falls below 0.25 causes a sudden shift in turbulent energy and the layers dissipate and collapse into a complex array of turbulent patches of nearly uniform density (Fig. 7-2). These patches are exceedingly transient in nature, and only the composite final product, after stability is restored, is what is most often observed. The thermal microstratification, commonly observed in detailed studies of the metalimnion (Whitney, 1937; Simpson and Woods, 1970), is possibly a result of dissipation of shear instabilities in the thermocline region.

EDDY DIFFUSION AND CONDUCTIVITY

The distribution of turbulent motion must be viewed in a statistical sense, for the distribution of fluid properties such as heat or solute content is a composite of many small random movements. The diffusion models for fluids, developed by Taylor and by Schmidt (Mortimer, 1974), are analogous to those that measure random movement of molecules in molecular diffusion, but are obviously on much larger scales.

Heat transfer across a thermal gradient within a layer in a lake may be used to estimate the extent of turbulent transport, because turbulent mixing and heat conduction are quite similar. In a hypothetical layer of water devoid of motion but with a

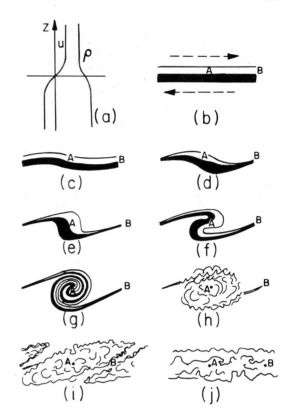

Figure 7-2 Growth of shear instability leading to turbulent mixing in a stratified fluid with the velocity and density (ρ) distribution shown in a. A and B are fixed points, the arrows indicate direction of flow, and the lines represent surfaces of equal density. (From Mortimer, C. H.: Mitteilungen Int. Ver. Limnol., 20:134, 1974, after Thorpe.)

temperature gradient, heat equilibrium is brought about only by conduction. If flow is applied across that gradient and it is turbulent ($R_i < 0.25$), the equalization of temperature is greatly accelerated. Thus, heat (relatively warm water) is transported through a plane perpendicular to the gradient by eddy turbulence. The rate of transport (diffusivity or conductivity) across a fluid plane is the product of the gradient perpendicular to the plane and an exchange coefficient (K_z). If expressed as a substance dissolved in the water across the gradient, it may be written as the gradient concentration per unit volume of water in terms of passage across unit area, as $g \; cm^{-2} \; sec^{-1}$.

Thus, the *coefficient of eddy diffusion* (K_z) is a measure of the rate of exchange or intensity of mixing across the plane. It is not a constant, but varies with average density, velocity of vertical motion, and mixing length. The horizontal dimensions of lakes are very much larger than the vertical dimensions of stratification; therefore, large differences exist between the magnitudes of horizontal and vertical eddy diffusion coefficients. The concept of an average K_z is subject to many variations related to the distance and is therefore an oversimplification. In reality, dispersion and mixing occur as a result of eddy movements over a wide range of spatial scales in which smaller eddies occur within larger ones. It is necessary to view the turbulent movements and consequent diffusion as taking place across a spectrum of turbulence (the eddy spectrum of Mortimer, 1974).

Turbulent transport causes particles to move apart at right angles to the direction of flow. When particles are close together, their rate of separation is governed by the small eddies of turbulent motion. But as the distance of separation increases, the dispersion of the particles is influenced by eddies of increasing size. Over significant ranges of dispersion, the mean rate at which particles diffuse is proportional to the ⅔ power of the distance between them (Stommel, 1949), a law that has been found to hold true for separation of particles over a range of 10 to 10^8 cm (Olson and Ichige, 1959; Verduin, 1961). The most probable value of turbulent diffusion velocity is in the range of 0.3 to 1.0 cm sec^{-1}, with a single diffusion velocity of approximately 1.0 cm sec^{-1} being valid for all scales of diffusion processes (Noble, 1961).

The coefficient of eddy diffusion is, for practical purposes, identical to the coefficient of eddy conductivity, which permits, in a general way, estimations of turbulence from changes in temperature. Estimations of eddy conductivity coefficients are useful as they permit insight into the magnitude of current-induced turbulence in the hypolimnion. The magnitude of K_z and changes observed vertically with stratification, however, depend very much upon the methods of analysis and assumptions about the magnitude of direct solar heating of the upper hypolimnion, convective currents, lateral heating from the sediments, and accuracy of measurements of small temperature differences (Hutchinson, 1957; Dutton and Bryson, 1962; Lerman and Stiller, 1969; Hesslein and Quay, 1973; Powell and Jassby, 1974; Jassby and Powell, 1975; Di Cola, et al., 1977; Imboden and Emerson, 1978).

Coefficients of eddy diffusion in the hypolimnion decrease with increasing stability. Therefore, in lakes of large area and depth that have longer fetch for wind exposure, eddy diffusion coefficients of the hypolimnion are greater than in smaller shallow stratified lakes (Mortimer, 1941, 1942). In addition, the K_z of a turbulent epilimnion is

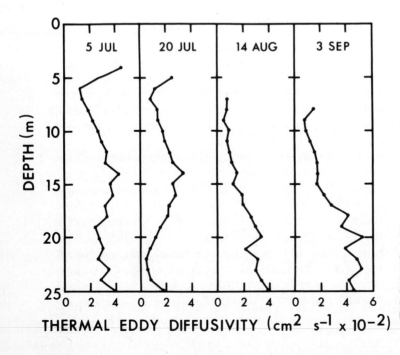

Figure 7-3 Selected depth profiles of the vertical thermal eddy diffusivity in Castle Lake, a small cirque lake in northern California, during the summer of 1972. The first value on each date marks the average top of the metalimnion at that date. (Drawn from data of Jassby and Powell, 1975.)

markedly greater than the K_z in the more stable metalimnion of a stratified lake (Fig. 7-3). This relationship holds seasonally as well: As the gradient of thermal density stratification increases from spring to summer, the coefficient of eddy diffusion decreases, i.e., K_z is inversely proportional to stability. In the deep hypolimnion, K_z tends to increase slightly, especially near the bottom where the effects of other water movements further complicate analyses of turbulent transport.

Surface Water Movements

SURFACE WAVES (PROGRESSIVE WAVES)

The frictional movement of wind blowing over water sets the water surface into motion, producing a wind drift. Wind also sets the surface into oscillation, producing *traveling surface waves*. If these waves become large enough to break, their momentum is transferred to the water. Although these surface waves are the most conspicuous periodic responses to an observer looking at a lake, their limnological importance is largely confined to shore areas. Traveling surface waves are confined to surface layers and cause little displacement of the deeper water masses.

Short surface waves cause the surface water particles to move in a circular path or orbit; in cross section, the path is *cycloid* (Fig. 7-4), with very little significant motion other than horizontal translocation (Smith and Sinclair, 1972). In a given wave, water is displaced vertically and returned by gravity to an equilibrium state. Similarly, neighboring short surface waves are oscillating periodically in cycloid fashion. In synchrony, these movements result in a traveling wave whose path may be described as analogous to a point on the rim of a wheel moving along a plane where the hub is at the mean surface of the water (Fig. 7-4). Except in shallow beach areas, the wavelengths (λ) of short surface waves are less than the water depth and, consequently, they are dispersive, i.e., they travel at speeds proportional to $\lambda^{1/2}$ of deepwater gravity waves (discussed later).

Of greater interest than the small horizontal movement of short surface waves is the influence such periodic oscillations have vertically, although the height (h) of this vertical oscillation is attenuated rapidly with depth (Fig. 7-4). The decrease of vertical

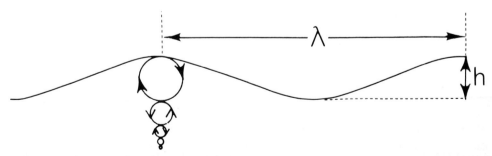

Figure 7-4 Surface wave indicating wavelength (λ), height (h), and attenuation of height ($h/2$, the displacement positive or negative from the equilibrium in a sinusoidal wave) of the cycloid movement with depth.

motion with increasing depth can be approximately described as a halving of the cycloid diameter for every depth increase of $\lambda/9$. As we will see shortly, the amplitude or height of surface waves is not directly proportional to wavelength, although a ratio of about 1:20 of h:λ is a common average. If, for example, short surface waves had a $\lambda = 18$ m and h $= 1$ m, the vertical oscillation at a depth of 4 m $(2/9\lambda)$ would be 25 cm, at 8 m or $4/9\lambda$, 6.25 cm, and at 18-m depth, 1.95 mm. In other words, the magnitude of vertical displacement at a depth equal to the wavelength will be $\frac{1}{512}$ (i.e. $\frac{1}{2^9}$) of that at the surface.

Short surface waves with a wavelength greater than 6.28 cm $(2\pi$ cm) are referred to as *gravity waves*. Waves of a length less than this minimum are called *ripples* or capillary waves. The ratio of wave height (h) to wavelength (λ) is highly variable within the range of 1:100 to about 1:10. At great height, the crests of the waves sharpen and become less stable. When the angle of the wave height exceeds a h:λ ratio of about 1:10, the peak collapses and some apical water may be blown off, forming a *whitecap*.

For a given wind speed, wave height appears to be nearly independent of depth in small lakes, but in lakes of great area, wave height and length increase with increasing water depth (Hutchinson, 1957). The height of the highest waves observed on a lake appears, without good theoretical explanation, to be proportional to the square root of the *fetch*, or distance over which the wind has blown uninterrupted by land. Thus the maximum height (cm, crest to trough), where x = the fetch distance in centimeters, is expressed as

$$h = 0.105 \sqrt{x}$$

For example, the maximum wave height observed in Lake Superior was 6.9 m with a fetch of 482 km $(4.82 \times 10^7$ cm), which agrees well with a maximum of 7.3 m predicted by this relationship.

The waves described in the foregoing paragraph are "deep water" or "short" waves, in which wavelength is much less than water depth. Where this condition no longer holds, and the wavelength becomes more than 20 times the water depth, the wave is transformed into a "shallow water" or "long" wave, and the cycloid motions are transformed into a to-and-fro sloshing, which extends to the bottom of the water column. Shallow water waves are no longer dispersive, because their speed is proportional to the square root of the water depth, rather than being determined by wavelength. Therefore, as deep water waves enter shallow water areas, their velocity decreases as the square root of depth decreases, and there is a simultaneous reduction in wavelength (Iverson, 1952; Mason, 1952; Hutchinson, 1957). Wave height first decreases slightly, then increases markedly and becomes asymmetrical and unstable. The collapse of water over the front of this now asymmetrical wave is termed a *breaker*, and occurs in a spectrum of forms between two extreme types (Fig. 7–5). In a *plunging breaker*, the forward face of the wave becomes convex and the crest curls over, but collapses with insufficient depth to complete a vortex. The crest of a *spilling breaker* collapses forward, spilling downward over the front of the wave.

Because in shallow water the wave action extends to the bottom, sedimentation can be prevented in sufficiently shallow water, and movement may be too severe for most aquatic plants to grow. In lakes with high rates of sedimentation in the littoral

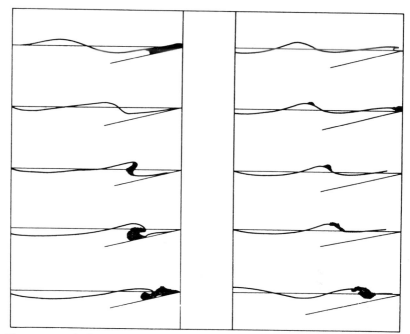

Figure 7-5 The breaking of waves on beaches. *Left*, Plunging breaker. *Right*, Spilling breaker. (From Hutchinson, G. E.: A Treatise on Limnology. Vol. 1, New York, John Wiley & Sons, Inc. 1957; after Iversen, 1952, and Mason, 1952.)

zone, such as in calcareous hardwater lakes where marl banks or benches extend out from the shoreline for many meters, recently sedimented material is constantly resuspended and removed to deeper areas. As a result, in suitably shallow water of about a meter, marl benches on the steep sides of these lakes do not increase in height. In a hardwater lake of northern Indiana, for example, the surface sediments of a submersed lakemount island were found to be very old (Wetzel, 1970). The lakemount ceased accretion of sediments abruptly 1 m from the lake surface about 2740 years ago, for more recent sediments were continually moved into deeper parts of the lake.

SURFACE CURRENTS

Currents are nonperiodic water movements generated by external forces. Forces responsible for producing currents include the wind, changes in atmospheric pressure, horizontal density gradients caused by differential heating or by diffusion of dissolved materials from sediments, and the influx of water into a lake. The latter subjects have been treated in other chapters, and the following remarks are therefore concerned primarily with wind-derived surface currents of open water masses.

Geostrophic effects of the deflecting (Coriolis) force due to the earth's rotation are found in all currents in moderate- to large-sized lakes. In the Northern Hemisphere, surface water currents are deflected to the right relative to the direction of the wind. In the Southern Hemisphere, the Coriolis force causes surface waters to deflect to the left relative to the wind direction.

The combined effects of wind and geostrophic deflection cause surface water to move downwind and to the right in the Northern Hemisphere. Under stratified conditions, there is also a tendency for denser water to collect on the left side of the current and less dense water on the right side of the current.

The resulting wind-drift current is deflected 45° from the direction of the wind in a spiral manner (the Ekman spiral) in open waters of large, deep lakes. As the size, and especially the depth, of the lakes decrease, the magnitude of the deflection angle decreases until, in a lake with a depth of less than about 20 m, the angle of declination becomes insignificant. In Lake Mendota, which is of moderate size (39.2 km^2) and depth (25.6 m), a rotation of wind-driven surface currents had an average angle of deflection of 20.6° to the right (Shulman and Bryson, 1961). The depth of frictional influence in Lake Mendota was found to be between 2 and 3.5 m. Stress exerted by the wind on the water surface was found to be a linear function of wind speed.

The wind factor, defined as the ratio of surface water current velocity to wind velocity, is quite variable. In general, the velocity of wind-driven currents is about 2 per cent of the speed of the wind generating them, and is largely independent of the height of the surface waves. An average wind factor of 1.3 per cent for the upper half-meter of water was found in Lake Mendota at low wind velocities (Haines and Bryson, 1961). Water velocity in the surface layers increases with wind velocity until a critical wind speed is reached. Above the critical speed, surface water velocity decreases, and the wind factor becomes nonlinear. In Lake Mendota, the critical wind speed was 5.7 to 6.1 m sec^{-1}.

The surface currents of very large lakes tend to circulate in large swirls or *gyrals*. Lake Michigan, for example, is typically exposed to predominately westerly winds during summer, and its large, conspicuous gyrals are clearly influenced by geostrophic rotational forces (Noble, 1967). In this and other examples, gyral movements are complicated by numerous other water movements related to standing water motions (see later discussion). Inertial currents on a smaller scale also occur in all of the Laurentian Great Lakes, not only at the surface, but at all depths and in all seasons (Verber, 1964, 1966). The geostrophic right-hand acceleration of currents also occurs under ice-cover. Many complex flow patterns have been observed: straight-line flow, sinusoidal or oscillatory, repeated crescent, circular or spiral, and rotary or screw flows. Gyrals in lakes the size of Lake Michigan are not simply the result of geostrophic forces; their direction is strongly modified by large, long waves and especially by shifts in duration of strong prevailing winds.

LANGMUIR CIRCULATION AND STREAKS

Turbulence generated by surface water movement and wave action is often sporadic and independent of wave direction. Sporadic turbulence occurs, for example, in the dispersion of wave energy in the mixing of the epilimnion when the lake is stratified, or during the mixing of the general water mass. Langmuir (1938) demonstrated that under some circumstances the motions associated with turbulent transport are organized into vertical helical currents in the upper layers of lakes (Fig. 7–6). Convection from this vertical motion generates *streaks*, which are oriented approximately parallel to the direction of the wind. Streaks coincide with lines of surface convergence and

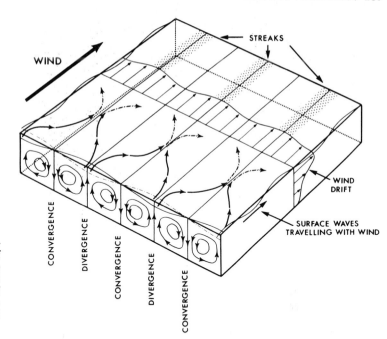

WIND

STREAKS

WIND
DRIFT

SURFACE WAVES
TRAVELLING WITH WIND

CONVERGENCE

DIVERGENCE

CONVERGENCE

DIVERGENCE

CONVERGENCE

Figure 7-6 Diagrammatic representation of the helical flow of Langmuir circulations in surface waters with streaks of aggregated organic matter occurring at lines of convergence. (Based on the experimental work of Faller, 1978, and others.)

downward movement, and are marked by windrows of aggregated particulate and surface-active materials. Between the streaks are zones of upwelling. Such *Langmuir circulation* occurs in all waveswept lakes of any significant size at wind speeds above 2 to 3 m sec^{-1}. The streaks are seldom observed at wind speeds above about 7 m sec^{-1}, even though the Langmuir circulations are still occurring (Scott, et al., 1969; Ottesen Hansen, 1978), for at high wind speeds, surface turbulence is apparently great enough to disrupt the surface aggregations.

The mechanisms responsible for the generation of Langmuir circulations in lakes have been the subject of much debate (Stewart and Schmitt, 1968; Myer, 1969; Scott, et al., 1969; Faller, 1971; Harris and Lott, 1973). It is evident from Faller's (1969, 1978) and Faller and Caponi's (1978) laboratory work that the primary mechanism generating Langmuir circulations probably is the wave and shear-flow interaction mechanisms described in a series of theoretical analyses by Craik and Leibovich (Craik, 1977; Craik and Leibovich, 1976; Leibovich, 1977). In these experiments, the helical circulations were present only when both the wind and the wave patterns existed at the same time, in agreement with the theoretical predictions that both a shear flow due to wind and the Stokes drift of progressive waves (see below) are essential. From this complex interaction, wind energy is converted into organized helical circulations as well as into waves, random turbulence, and the mean shear flow. The resulting Langmuir circulation is one of the primary ways in which turbulence is transported downward and the upper layers of water are mixed.

The Langmuir convectional helices form a series of parallel clockwise and counterclockwise rotations that result in linear alternations of divergences and convergences (Fig. 7-6). In Lake George, New York, Langmuir found the velocity of downward convergence currents to be 1.6 cm sec^{-1} with a wind speed of 6 m sec^{-1}. At this time, there was a horizontal surface current of 15 cm sec^{-1} and a current of 10 cm sec^{-1} at a depth of 3 m. Downward movement velocities at convergences were about three times the upward current velocities at divergences. Such currents are sufficient to markedly influence the distribution of microflora and fauna suspended in the surface waters of lakes. Algae and zooplankton, with limited or no powers of locomotion, tend to be aggregated in streaks in areas of water divergence (George and Edwards, 1973). The result is a patchiness in horizontal distribution that is important not only in sampling of the microorganisms for estimates of population size and distribution, and in metabolic measurements, but also in determining the distribution of predators, which congregate in these zones of higher prey density.

Internal Water Movements

METALIMNETIC ENTRAINMENT IN STRATIFIED LAKES

Surface waves and Langmuir circulations cause turbulence, but the barrier formed by the density gradient in the metalimnion suppresses the transmission of turbulence into the metalimnion and hypolimnion. However, as the wind blows over the surface for a reasonable period of time, wind drift causes water to pile up with a rise in surface level at the lee end of the lake. The accumulated mass of water is pulled downward by gravity and, when it encounters the more dense water of the metalimnion, it flows back in a direction opposite that of the prevailing wind (Fig. 7-7). The thermocline, or plane separating the two layers of different density (termed pycnocline by Hutchinson, 1957),

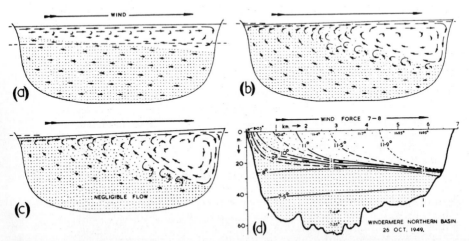

Figure 7-7 Stages of wind drift and circulation that led, after about 12 hours of strong wind, to the thermal situation depicted in (d), Lake Windermere, England, in late fall. Broken lines show equilibrium levels of water surface in (a), (b), and (c), and of the thermocline in (a). The initial layer below the thermocline is stippled; speed and direction of flow are roughly indicated by arrows. (From Mortimer, C. H.: Verhandlungen Int. Ver. Limnol. 14:81, 1961.)

is tilted in the process. We will return to the vertical movement of this entire water mass in the following section, after discussing the circulation pattern.

Progressive erosion of the metalimnion occurs by *entrainment,* with a corresponding lowering of the thermocline level (Mortimer, 1961). This erosion, which is characteristic of the autumnal cooling phase, begins when the net heat flux across the water surface changes from downward to upward, and ends in the fall overturn. During the early stages of entrainment (Fig. 7–7A), when the return current above the thermocline and the entrained layers below it both are moving upwind, shear turbulence in the metalimnion is low. As long as R_i does not fall below the critical value, the thermocline remains a slippery surface with little mixing across it. If the shear at the thermocline increases ($R_i < 0.25$), flow becomes unstable and large vortices appear in the return current at the downwind end (Fig. 7–7B,C). These vortices are carried upwind, where they become mixed into the surface drift. This process increases the density gradient at the downwind end until it offsets the local shear to produce neutral turbulence stability, with R_i near 0.25. The resultant whole-basin disposition of isotherms becomes fan-shaped (Fig. 7–7D), and the shape of the metalimnetic density profile varies in different regions of the basin, correspondingly.

As the wind velocity decreases in a subsequent calm, the displaced layers slide over each other to redistribute into a new equilibrium. The oscillations so created may or may not be sufficient to create internal instability and waves. The resulting thermal profile retains much of its original form, but the mean metalimnion depth has been pushed to a lower level, depending upon the amount of eddy conductivity. It is important to emphasize that this type of internal water motion is not continuous. Entrainment and metalimnetic erosion are progressive and forced by storms. Erosion can be counteracted by calm periods of heating, but in the fall, when cooling exceeds heating, the erosion continues in step-wise progression with storms. A strong inverse relationship occurs between the magnitude of entrainment and the mean density gradient across the metalimnion, and a strong positive correlation exists between entrainment and basin dimensions (Blanton, 1973).

Entrainment rates are obviously of major importance in the control of the extent of heat intrusion into the metalimnion and nutrient return from the hypolimnion to the epilimnion. A certain degree of compensatory motion occurs in the hypolimnion in a direction opposite that in the metalimnion. Turbulence in the hypolimnion is very much weaker than in the overlying layers, but increases in large lakes where space is sufficient for development of large-scale entrainment.

Surface and epilimnetic water movements and the effects of internal metalimnetic and hypolimnetic currents can be demonstrated very clearly in simple laboratory models of lakes (cf. Mortimer, 1951, 1954; Vallentyne, 1967; Wetzel and Likens, 1979). Such models, stratified either thermally or with colored solutions of different chemical densities, are so effective in demonstrating entrainment, seiches, and internal waves (discussion below) that their use is strongly urged in all introductory courses. For example, the upwelling areas of the epilimnion (Fig. 7–7) are seen clearly with tracer dyes in these models. Even more strikingly shown are the vortices of internal turbulence (curved arrows of Fig. 7–7B,C) which, with strong wind, become unstable and break as waves.

Water Movements Affecting the Whole Lake

LONG STANDING WAVES

Displacement of the water mass of the whole lake, as illustrated in Figure 7-7, gives rise to rhythmic motions, including both oscillations of the water surface and internal oscillations of isotherm depth. These motions take the form of very long waves, which have wavelengths of the same order as basin dimensions. These long waves are reflected at the basin boundaries and combine into standing wave patterns. The water surface or the thermocline oscillates up and down like a seesaw about a line of no vertical motion (a node, Fig. 7-8), which coincides with regions of maximum to-and-fro horizontal motion of the water masses. Such standing waves, surface or internal, are referred to as *seiches*, a term originally referring to the periodic "drying" exposure of shallow littoral zones. Figure 7-8 illustrates a constant depth model with two nodes, beneath which the horizontal to-and-fro sloshing of the whole water mass is at a maximum. Seiche motion, particularly the internal variety, has far-reaching effects on the vertical excursion of water masses, much larger than observed at the surface, and on their horizontal motions in the form of alternating currents.

The most common cause of seiches is the wind-induced tilting of the water surface and of the thermocline, as described in the previous section. When the wind stress is removed, the tilted water surface and the tilted thermocline flow back towards equilibrium. Momentum, however, is great and equilibrium is overshot (Fig. 7-8), resulting in a rocking motion about one or more *nodal points*. No vertical movement occurs at the

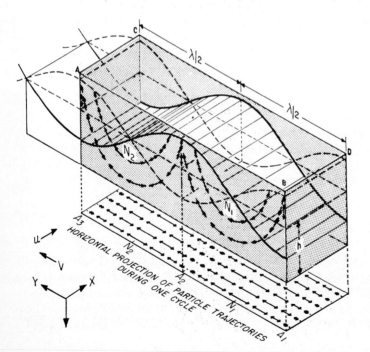

Figure 7-8 A long standing wave, without rotation, in a uniform depth (*h*) model. (From Mortimer, C. H.: Mitteilungen Int. Ver. Limnol., 20:157, 1974.)

nodal point, whereas maximum vertical movement takes place at the ends or *antinodes*. The nodal point is in reality a nodal line in this nonrotating rectangular situation running the width of the basin.

Surface Seiches. The most conspicuous example of long standing waves is the *surface seiche*, which conforms to one or more resonant oscillatory frequencies (free modes) of a particular basin. The free mode of the surface seiche is a barotropic or surface wave, which affects the motion of the entire water mass of the lake, whether stratified or not, and attains its maximum amplitude at the surface. The surface seiche contrasts with the baroclinic or internal seiche, discussed further on, which is associated with the density gradient in stratified basins and attains its maximum amplitude at or near the thermocline.

The periodicity of vertical movement at the antinodes is a function of the length and depth of the basin. Particles at the node move back and forth horizontally with the same oscillatory rhythm as the vertical motion at the antinodes. The horizontal oscillation at the nodal line is large when the basin is very long in relation to depth, even though the vertical amplitude of the seiche is small at the antinodes. In a simple rectangular situation, analogous to many lakes in which the length greatly exceeds the mean depth, the period of the uninodal surface oscillation is approximated by:

$$t = \frac{2l}{\sqrt{g\bar{z}}}$$

where:

t = period (seconds)
l = length of the basin at the surface (cm)
$\bar{z}$ = mean depth of the basin (cm)
g = acceleration of gravity (980.7 cm sec^{-2}).

Once the surface seiches are set into motion, frictional and gravity damping of the oscillations begins and the water mass gradually returns to equilibrium. The magnitude of damping varies with water depth and complexity of basin shape. In deep lakes of uncomplicated shapes, for example, weakly damped oscillations may persist with slowly diminishing amplitudes long after the impact of the storm that set them into motion has passed. A few examples in Table 7–1 illustrate the periods of the surface seiche modes, which correspond to values calculated from hydrodynamic theory. In most lakes, the period of the second mode (t_2) is more than half of that of the first mode (t_1).

In long, narrow lakes, transverse seiches are found with periodicities and amplitudes much less than those of longitudinal seiches, which move the length of the longest axis—for example, the conspicuous transverse seiche of 132 min period in Lake Michigan (Mortimer, 1965). Where lake length differs little from breadth, complicated seiches can result when wind direction changes.

Uninodal seiches, as described earlier, are common even in very large lakes. If pressure is exerted on the surface in the center of a basin and released, or periodically exerted and released, binodal surface seiches are generated. Multinodal seiches, with up to 17 nodes, have been observed.

TABLE 7-1 Periodicity of Longitudinal Seiches

		OBSERVED PERIOD (MIN)			OBSERVED PERIOD (MIN)
Loch Earn, Scotland	t_1	14.52	Lake Vetter, Sweden t_1		179.0
	t_2	8.09		t_2	97.5
	t_3	6.01		t_3	80.7
	t_4	3.99		t_4	57.9
	t_5	3.54		t_5	48.1
	t_6	2.88		t_6	42.6

LAKE	LENGTH (KM)	$\bar{z}$ (M)	FIRST MODE PERIOD (t_1) (MIN)
Altausseer, Austria	2.8	34.6	5.3
Garry, Scotland	6	15.3	10.5
Mendota, Wisconsin			
NS axis	6.8	12.0	25.6
E axis	9.1	12.8	25.8
Ness, Scotland	38.6	132	31.5
Constance, Germany	66	90	55.8
Huron, USA—Canada	444	76	400
Erie, USA—Canada	400	21	786

Compiled from data cited in Hutchinson, 1957, from various sources, and from Mortimer, personal communication.

The amplitudes of surface seiches are generally rather small in comparison to internal seiches. For example, in Lake Mendota, the amplitude of a surface seiche was 1 to 2 mm with a periodicity of 25.8 minutes. In larger Lake Geneva, Switzerland, the maximum observed surface seiche amplitude was 1.87 m (t = 73 min), although the amplitude damped gradually in two weeks to less than 10 cm. In still larger lakes, such as Lake Erie, where amplitudes of surface seiches can exceed 2 m (t = 14 hours), the effects can have both destructive and beneficial results, e.g., flushing of river delta and harbor areas. Surface seiches can occur under ice-cover in similar ways.

True lunar tides, the result of gravitational attraction of water masses by the moon and sun, have been poorly investigated in lakes. However, it is certain that true tidal movement, even in the largest lakes, is small (a few millimeters). The maximum observed amplitudes in Lake Baikal, USSR, are less than 15 mm and in Lake Superior, 20 mm (Hutchinson, 1957). All observations are about one-half or less of the theoretical value. The elastic yielding of the earth to the tide-generating forces presumably accounts for the difference (Mortimer, 1965).

Internal Seiches.   We have seen that the amplitude of surface seiches is relatively small, even in large lakes, and that their periodicity increases with basin dimensions (Table 7-1). When the lake is stratified, water layers of differing density oscillate relative to one another. Most conspicuous is the successive oscillation of the metalimnion (Fig. 7-9), in which both the period and amplitude of this internal standing wave or *internal seiche* are much greater than those of the surface seiche (Fig. 7-10).

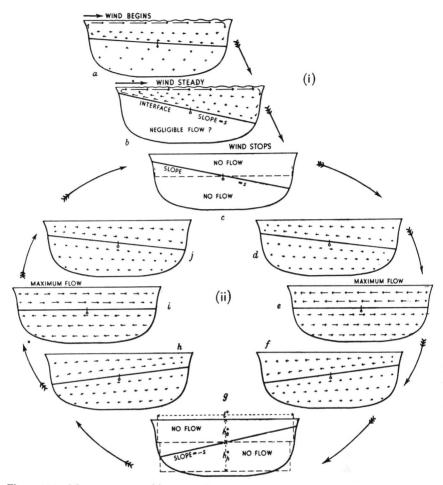

Figure 7-9 Movement caused by (*i*) wind stress and (*ii*) a subsequent internal seiche in a hypothetical two-layered lake, neglecting friction. Direction and velocity of flow are approximately indicated by arrows. ○ = nodal section. (From Mortimer, C. H.: Proc. Royal Soc. London, Series B, 236:355, 1952.)

In a simple, rectangular, trough-like lake or lake model, without considering rotation, one observes the resultant horizontal flow depicted in Figure 7-9. Horizontal flow in this oversimplified stratified situation, which contains only an epilimnion and hypolimnion of differing densities, is maximal at the equilibrium point (node) of the oscillation; it ceases at the point of maximum vertical deflection. Because of the much greater water movement associated with internal seiches, resulting currents that rhythmically flow back and forth in opposing directions are the major deepwater movements of lakes. They lead to vertical and horizontal transport of heat and dissolved substances and significantly alter the distribution and productivity of phytoplankton and zooplankton, either directly or indirectly, by means of changes in thermal and chemical stratification (cf. Thomas, 1951).

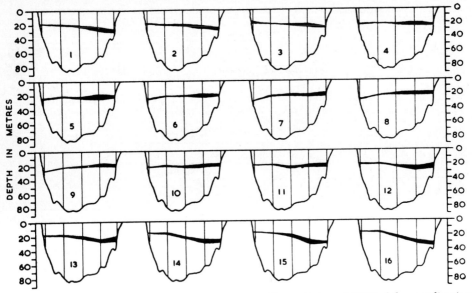

Figure 7–10 Successive hourly positions (0100 to 1600 hours, 9 August 1911) of the metalimnion bounded by the 9° and 11°C isotherms on a longitudinal section of Loch Earn, Scotland. (From Mortimer, C. H.: Schweiz. Zeitschrift f. Hydrologie, 15:94, 1953, after Wedderburn.)

In the simplest model, consisting of a rectangular basin with a homogeneous "epilimnion" of thickness z_e and density d_e and a "hypolimnion" of thickness z_h and density d_h, the period (t) of the first internal seiche mode is given by

$$t = 2l \left/ \sqrt{\dfrac{g(d_h - d_e)}{\dfrac{d_h}{z_h} + \dfrac{d_e}{z_e}}} \right.$$

where:

l = length of the basin (cm)

g = acceleration due to gravity (980.7 cm sec^{-2}).

It should be noted that this equation is simply an extension of that for the periodicity of a surface seiche in which the density of the upper medium, air, was neglected as very small in comparison to that of water. Although more elaborate formulae have been developed for continuous density gradients in stratified lakes, the above formula yields reasonable estimates of internal seiche periodicity.

The detailed analyses of internal seiches by Mortimer (1952, 1953) permit a few generalizations (Table 7–2). Of all internal waves that are theoretically possible in stratified lakes, the one most commonly set in motion is a uninodal seiche on the metalimnetic boundary (thermocline). In basins of fairly regular shape ranging in length (Table 7–2) from 1.5 km (Lunzer Untersee) to 74 km (Lake Geneva), the uninodal internal seiche along the medial axis always appears as the main resonance. In basins possessing topographic features that may impede a uninodal internal seiche, such as constric-

TABLE 7-2   Periodicity of Uninodal Metalimnetic Seiches of Several Lakes

LAKE	DATES	LENGTH (KM)	CALCULATED PERIOD (HOURS)	OBSERVED PERIOD (HOURS)
Lunzer Untersee, Austria	Aug. 1927	1.51	4.0	3.7
Windermere, England				
Northern Basin	Aug–Sept. 1951	6.5	14.4	12–14
Southern Basin	June–July 1951	8.9	23.1	23–25
Loch Earn, Scotland	Aug. 1911	9.6	17.2	16
Madüsee, Germany	July–Aug. 1910	13.8	27.3	25
Loch Ness, Scotland	Sept. 1904	37.0	62.4	60
Geneva, Switzerland	July–Aug. 1942	73	96	72–108
Baikal, USSR	Sept. 1914	675	1848	912

From data cited in Mortimer, 1953, from various sources.

tions, or shallows as in the southern end of Lake Windermere, the observed period is appreciably longer than the theoretical period. Multinodal internal seiches form a dominant type of resonance in very large lakes, especially when forced and damped by wind and other short-period disturbances.

GEOSTROPHIC EFFECTS ON SEICHES

Coordinates that appear fixed to an earthbound observer are in fact rotating relative to coordinates in space (Mortimer, 1955). Inertia causes a moving water mass to attempt to follow a straight line in space, and the resultant track appears curved to the earthbound observer. The observer can account for the curvature by postulating a small force, the Coriolis force, directed at right angles to the line of motion. This force, which produces a deflection to the right in the Northern Hemisphere and to the left in the Southern Hemisphere, is zero at the Equator and maximum at the poles; it is a function of latitude and is proportional to the speed of the current. In sufficiently large lake basins in which frictional forces are small and constraints of the basin sides are negligible, a water mass, once set in motion, follows a circular track, owing to the effects of the Coriolis force. The time taken to complete an "inertial circle" of this type depends only on latitude. If the basin is stratified, this motion may be associated with internal waves, the period of which is referred to as the period of the inertia oscillation. For example, in Loch Ness, Scotland, at a latitude of 57° 16′ when the velocity of a water mass was 10 cm sec^{-1}, the radius of the circle would be 810 m with a period of the inertia oscillation of 14.2 hours (Mortimer, 1955).

Motion in the inertia circle will be encountered only in basins whose width is many times the radius of the inertial circle. In lakes whose width is of the same order or less than that radius, side constraints may be expected to modify the movement. If a lake is stratified and conforms to the simplified two-layered situation in Figure 7–9, then the flow depicted in that figure will be deflected by the Coriolis force onto the righthand shore (Northern Hemisphere) in the manner indicated in Figure 7–11. This rotating counterpart of the nonrotating conditions of Figure 7–9 has been demonstrated very clearly in Loch Ness (Mortimer, 1955), and subsequently in Lake Michigan and other large lakes. In addition to the morphometric parameters of the basin, important

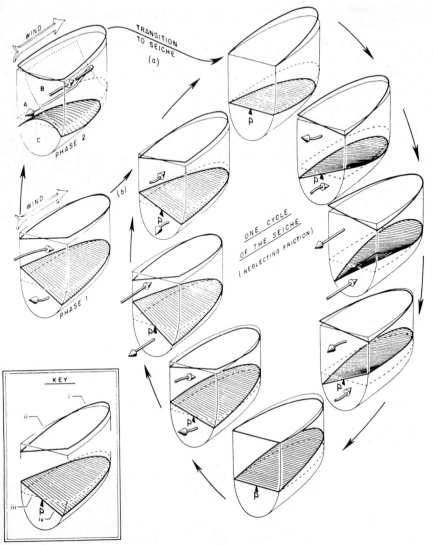

Figure 7-11 An internal seiche in a rotating two-layered lake model. In the inset key diagram, *i*, the oscillating lake surface is shown by a heavy line (this is the surface signature of the internal seiche mode, i.e., is *not* a surface seiche mode); *ii*, the equilibrium lake surface position is shown by a thin line; *iii*, the equilibrium interface position is shown by a broken line; and *iv*, the oscillating interface is shown by a shaded surface.

The two diagrams at upper left illustrate a hypothetical distribution of the layers during the application of the wind stress, which set the seiche in motion. B and A respectively indicate the wind-driven surface and return currents in the upper layer, both deflected to the right by the Coriolis force. C indicates the lower layer, the greater part of which has become displaced out of the half-basin shown.

Eight phases of one oscillation cycle of the first mode internal seiche are shown. Directions of flow in the upper and lower layers are shown by heavy arrows. The nodal point, *P*, the only point of zero elevation change, takes the place of the nodal line in the nonrotating model. Around this point, the internal seiche and its surface (out-of-phase) counterpart rotate counterclockwise. (From Mortimer C. H.: Mitteilungen Int. Ver. Limnol., 20:169, 1974.)

control factors include the timing of wind stresses, the time required to tilt the thermocline plane, and the period of the internal seiche.

The amplitude of internal seiches and the currents generated by them are of major importance. For example, in Lunzer Untersee with a fetch of 1.6 km, an internal seiche with a period of 4 hours had an antinodal amplitude of 1 m and generated maximal horizontal currents of 1 cm sec^{-1} at the node. Similarly, currents of 1 cm sec^{-1} were found in Lake Mendota. In larger lakes, for example at the ends of the long basin of Loch Ness and in Lake Michigan, amplitudes of > 10 m are common, with horizontal currents of > 10 cm sec^{-1} near the nodes.

INTERNAL PROGRESSIVE WAVES

The horizontal water movements associated with shearing flow at the metalimnetic interfaces can generate large *internal progressive* waves if the Richardson number falls below the 0.25 stability criterion (see Fig. 7–2). Internal waves on the thermocline are roughly an order of magnitude or more larger than waves found on the surface of large lakes. Examples include:

	Period	Mean Amplitude
Lake Mendota (Bryson and Ragotzkie, 1960)	1.5–7.9 min	0.15 m (max 1.0 m)
Lake Michigan (Mortimer, McNaught, and Stewart, 1968)	3–5 min	1.03 m (max 6 m)
	7 min	0.6 m
	10 min	0.3 m

Internal progressive waves propagate and break much the same as surface waves do (Bryson and Ragotzkie, 1960). Again, these wave movements can be demonstrated dramatically in simple lake models containing dyed strata of differing densities (thermally or by salinity) (cf. for example the photographs of Mortimer, 1952). The turbulence associated with internal waves is analogous to that at the surface, but occurs on a much larger scale and is influential in the transfer of heat and other properties through the metalimnion. Since they break at the sides of the basin, their effects, coupled with the vertical movement of the internal seiche on which they move, are particularly significant.

In very large lakes, large circulatory cells of water movement can develop over the troughs of internal progressive waves and extend through the epilimnion to the surface. The result is a series of large, widely spaced convergences and divergences that operate analogously to Langmuir currents in aggregation and dispersion of suspensoids at the surface. Such aggregations are referred to as *slicks,* and are of much larger size and more dispersed than the streaks of Langmuir currents. It is important to note, in addition, that the origins of these circulation patterns are quite different.

Other Water Movements

LONG SURFACE AND INTERNAL WAVES

Up to this point, discussion has centered on long standing waves and short progressive waves. In the latter case, the wavelengths (λ), except at the beach, are less than water depth. Consequently, short waves are dispersive, i.e., they travel at speeds proportional to $\lambda^{1/2}$ (Mortimer, 1963, 1974). In contrast, *long waves*, with a λ long in comparison to and much greater than the water depth, are nondispersive and travel at a speed independent of wavelength. A nonrotating model of such a wave, traveling in water of uniform depth, is illustrated in Figure 7–12. Its speed of progression is gz (in which g is the acceleration of gravity and z is the water depth), and the wave crests are horizontal. A combination of two such progressive waves, of equal amplitude and traveling in opposite directions, yields a model of the long standing waves (seiches) described earlier (Fig. 7–8).

The nonrotating models represented by Figures 7–8 and 7–12 provide simplified descriptions of events in small lakes. When basin dimensions exceed about 15 km, geostrophic effects of the earth's rotation must be introduced, and two separate model components emerge to provide the simplest interpretations of long surface and internal waves (Mortimer, 1963, 1974, 1975). Both model components, applicable to channels and basins of constant depth, were developed many years ago in connection with tidal theory and are named after the mathematicians concerned. The first component, associated with a shoreline, is illustrated in Figure 7–13. In the Kelvin wave model, the rightward deflection by the earth's rotation (Coriolis force) is everywhere exactly balanced by the components of gravity along a wave crest which slopes downward from right to left in the direction of progress (Northern Hemisphere), so

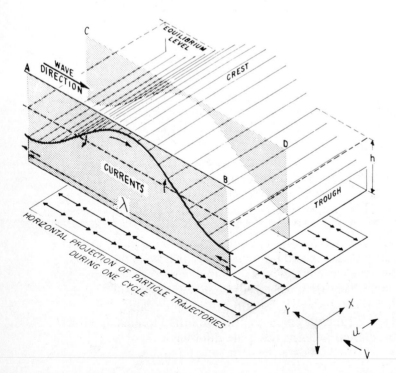

Figure 7–12 A long progressive wave, without rotation, in a uniform depth (h) model basin. (From Mortimer, C. H.: Mitteilungen Int. Ver. Limnol., 20:155, 1974.)

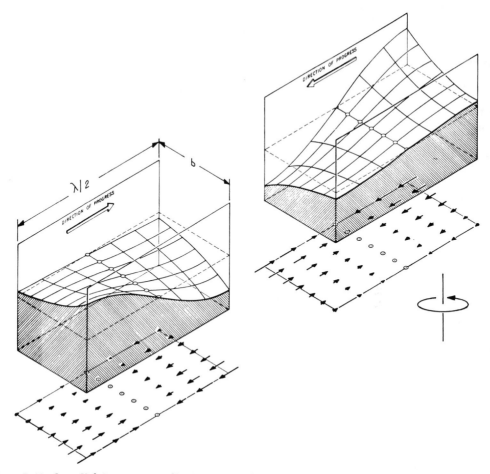

Figure 7-13   Long Kelvin waves traveling in opposing directions in straight, rotating (counterclockwise) channels. Half-wavelength portions are shown. Horizontal components of the wave currents are projected as vectors onto the plane below the channels. (From Mortimer, C. H.: Verhandlungen Int. Ver. Limnol., 19:65, 1975.)

that the righthand shoreline coincides with maximum wave amplitude. The condition of exact cancellation of the Coriolis force can only be met by an exponential decay of wave amplitude and wave-associated currents along a line normal to the shore. The result is that the currents are parallel to the shore, which satisfies the boundary condition of zero onshore current component along the shore.

As progressive Kelvin waves travel in opposite directions along a lake basin channel, the long undulations induce currents along the shores (Fig. 7–14). These wave-induced currents are parallel to the direction of the wave progression and the shores of the basin sides. As portions of the water mass rise, others are depressed in a geostrophic balance (Mortimer, 1975). The balance keeps the currents shore-parallel with no cross-channel (basin) current at the basin sides. It should be noted that when rotation from geostrophic effects becomes significant in lakes of medium to large width, the nodal line in the nonrotating models (comparable to those of small lakes) is replaced by a single midchannel nodal point (P in Fig. 7–14).

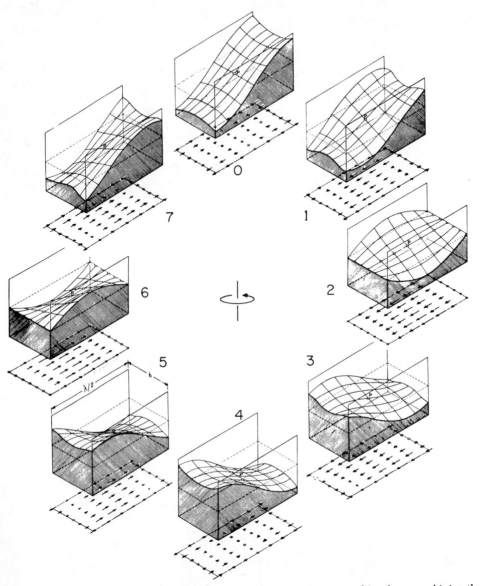

Figure 7-14 Successive phases (⅛ cycle) of the long standing wave resulting from combining the two Kelvin waves of Figure 7-13 traveling in opposite directions in a straight, rotating lake basin channel (width of 7 km) of rectangular cross section. The wave topography is at the metalimnion interface, with the upper layer omitted. Horizontal wave current components are projected as vectors onto the plane below the channel. The point of zero elevation change is shown at P, around which the wave rotates counterclockwise. (From Mortimer, C. H.: Verhandlungen Int. Ver. Limnol., 19:67, 1975.)

In very large lakes, such as the Laurentian Great Lakes, the basin width is sufficiently large to permit the long waves to travel without the interference of the shore basin boundaries. These *Poincaré waves* are also influenced by Coriolis geostrophic effects, but in this case the wave amplitude does not decrease exponentially away from the shore as in Kelvin waves. That is, Poincaré waves occur in the open water of large lakes, and are transformed into Kelvin waves when encountering shore boundaries. Poincaré waves undulate in a standing wave pattern both across and along the model channel or basin, yielding a cellular pattern of alternating rising and falling hills and valleys with a corresponding cellular pattern of wave-associated currents (Fig. 7–15). Their direction rotates clockwise once every wave cycle. In very large basins, for example Lake Michigan, the period of observed internal Poincaré waves, equivalent to the Figure 7–15 model, is a little less than the period of motion

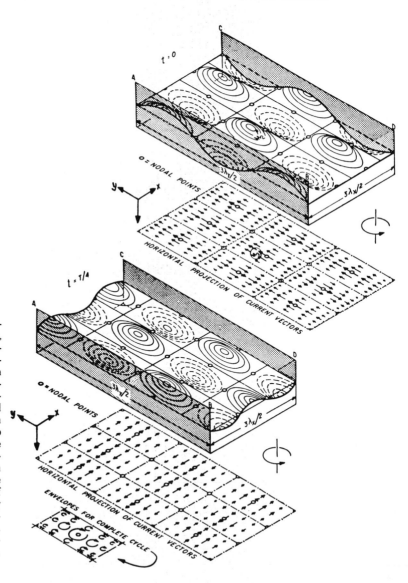

Figure 7–15 A standing Poincaré wave in which two phases, separated by ¼ cycle, of the oscillation are shown for the cross-channel trinodal case, with a ratio of long-channel to cross-channel wavelengths of 2:1. Current vectors rotate clockwise and attain their maximum values in the center of the cells. The two phases of the figure show the oscillations separated by a quarter period in which, in addition to the clockwise rotation of the vectors, the current directions of one cell remain approximately parallel and opposite in direction to those in neighboring cells. (From Mortimer, C. H.: Mitteilungen Int. Ver. Limnol., 20:179, 1974.)

of the inertial circle. This period depends on latitude, and is 17.5 hours for the central part of Lake Michigan.

The important difference between the Kelvin wave and Poincaré wave components is that the Poincaré wave extends with undiminished amplitude right across the channel or basin, whereas the Kelvin wave decreases in amplitude away from the shore and is thereby "trapped" or constrained to travel along the shore. In actual basins we must speak of Kelvin-type and Poincaré-type waves, although the mathematical models illustrated in Figures 7–13, 7–14, and 7–15 (in which the wave surfaces can be visualized either as a water surface or a metalimnetic interface surface) are valid only for constant-depth conditions, not met in natural lakes. Nevertheless, even though they are oversimplified, the models do provide useful interpretations of what is observed in large basins (Mortimer, 1963, 1974, 1975).

When the Kelvin wave model is applied to typical conditions of summer stratification in a lake of "medium" or "large" width, for example Lac Léman and Lake Michigan, with respective widths of the order of 10 and 100 inertial circle radii, Table 7–3 predicts that an internal Kelvin wave traveling along one shore will have decreased to about one-sixth of its initial amplitude on the other shore in Lac Léman and to a negligible amplitude on the opposite shore of Lake Michigan. In the open water of the latter lake, therefore, the Poincaré component should theoretically dominate the wave pattern, with upwellings and Kelvin-type wave responses restricted to nearshore bands of some 20 km width. This pattern in fact has been observed in Lake Michigan (Mortimer, 1974). One must visualize a transition in the wave-induced current patterns from the cellular, clockwise-rotating near-inertial periodicity of the offshore Poincaré-type waves in very large lakes, and the shore-parallel Kelvin-type currents which increase in amplitude as the shore is approached. This transition is sketched in Figure 7–16. In smaller basins of Lac Léman size, on the other hand, one can predict that the internal Kelvin-type long standing wave response will dominate. Metalimnetic oscillations with a counterclockwise rotation of the internal wave motion in Lac Léman can be interpreted as an internal Kelvin-type wave progressing counterclockwise around the basin in the manner illustrated in Figure 7–14, with a periodicity of a little over three days (Mortimer, 1963, 1974). As this usually coincides with the periodicity of storm passages, the oscillatory response of Lac Léman is

TABLE 7-3   Parameters of Single Kelvin Waves in Lac Léman and Lake Michigan Representative of Nearshore Conditions in Late Summer

LAKE CONDITIONS	LAC LÉMAN		LAKE MICHIGAN	
	SURFACE	INTERNAL	SURFACE	INTERNAL
Inertial period (t_p) in hr	16.6		17.4	
Thickness of upper layer (m)		15		15
Thickness of lower layer (m)		85		60
Mean °C of upper layer		19°		20°
Mean °C of lower layer		6°		6°
Density difference × 10⁻³		1.54		1.74
Mean width (km)	7.9	7.4	102	100
Wave velocity (km hr⁻¹)	113	1.59	104	1.60
Offshore distance intervals over which wave amplitude is successively halved	(207)	2.9	(190)	3.1
Per cent of wave amplitude still remaining at opposite shore	98%	17%	70%	$< 1 \times 10^{-7}\%$

After data cited in Mortimer, 1963.

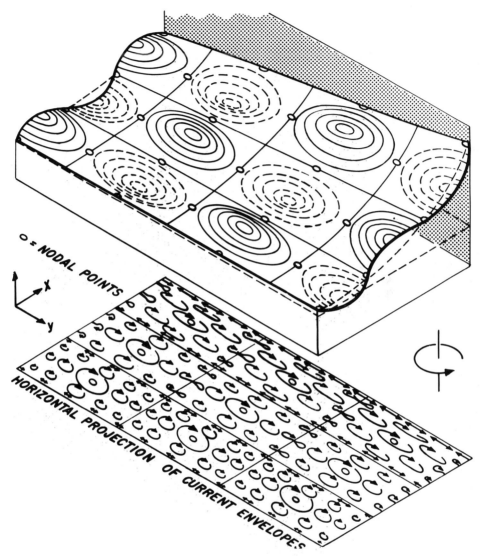

Figure 7–16 Combination of a multinodal standing Poincaré wave in a very wide channel model, of which only one side is shown, and a portion of a Kelvin wave traveling along the side. The elevation and current amplitudes associated with the Kelvin wave are at a maximum at the channel side (elevation amplitude, a), and decrease exponentially away from the side in the x-direction, falling to 1/e of the onshore value at a distance d_e from the side. The figure illustrates the transition, in current trajectories, from a nearshore pattern dominated by the Kelvin wave to a pattern dominated by the Poincaré wave at distances $> 2d_e$ offshore. (From Mortimer, C. H.: Large-scale oscillatory motions and seasonal temperature changes in Lake Michigan and Lake Ontario. Special Report no. 12, Center for Great Lakes Studies, University of Wisconsin-Milwaukee, 1971.)

strong. In Lake Michigan, by contrast, the internal Kelvin wave requires about four weeks to complete the circuit of the shores. This cycle is rarely completed before a new storm disturbs its progress.

CIRCULATION CAUSED BY THERMAL BARS

Another feature of large lakes is a steady circulation set in motion along the shore as a result of density gradients arising when shallow, nearshore waters heat more rapidly than the open water mass (Fig. 7–17). In the illustrated Lake Ontario example, the shallow waters develop stratification, while the main water mass remains in its isothermal (usually below 4°C) mixed winter condition. A narrow transition zone, called a *thermal bar*, consisting of a nearly vertical 4°C isotherm, develops between the open water mass and the littoral stratified water. Thermal density differences drive downward flowing currents along the vertical thermal bar both into the lower portions of the inshore and offshore water masses, and the earth's rotation combines with the density gradient to induce a counterclockwise coastal current inside the bar.

The thermal bar moves progressively further from shore as the heat influx continues to warm the larger open water mass. Little mixing occurs between inshore and offshore waters, and a significant portion of inshore water originates from runoff (Spain, et al., 1976). Finally, thermal differences between the inshore and offshore regions lessen to the point where stratification of the whole basin occurs (Rodgers, 1966). This mechanism results in temporary isolation of inshore waters where, when augmented by surface runoff and river discharge, chemical enrichment can occur. As a consequence, increased productivity can occur much earlier in the inshore regions than in the open water.

Thermal bars commonly occur to some extent in all lakes. In small lakes, the phenomenon is transitory, often lasting only a few days. In large lakes, however, the transition to stratification of the whole basin may take weeks, as seen in the example from Lake Ontario (see Fig. 7–17).

CURRENTS GENERATED BY RIVER INFLUENTS

When a river enters a lake or reservoir, the incoming water will flow into a density layer in the lake which is most similar to its own density; this process is governed by temperature, dissolved material, and suspensoids. Depending on the density differences between the inflowing water and the lake water, three basic types of inflow water movements can result (Fig. 7–18). *Overflow* occurs when the inflow water density is less than the lake water density ($\rho_{in} < \rho$), and *underflow* results when the inflow water density is greater than the lake water density ($\rho_{in} > \rho$). If the density of inflow is greater than that of the epilimnion but less than that of the metalimnion or hypolimnion, flow enters in a plume at an intermediate depth, and *interflow* occurs ($\rho_{in} > \rho_1$, $\rho_{in} < \rho_2$).

A certain amount of turbulent entrainment and mixing nearly always accompanies the flow of water into a lake, given adequate influx velocities. In many lakes, and especially in reservoirs, inflows enter through elongated bays which tend to inhibit lateral mixing. As the water enters the basin, the depth of inflow increases and the veloc-

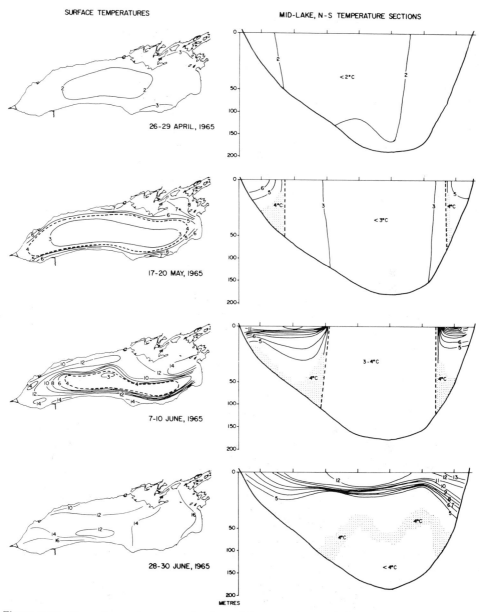

SURFACE TEMPERATURES

MID-LAKE, N-S TEMPERATURE SECTIONS

26-29 APRIL, 1965

17-20 MAY, 1965

7-10 JUNE, 1965

28-30 JUNE, 1965

Figure 7-17 Formation and progress of a thermal bar in Lake Ontario from winter to full summer stratification. (From Rodgers, G. K.: Publications Gt. Lakes Res. Div. Univ. Mich., 15:372, 1966.)

ity is gradually reduced until a critical section depth (d_0) is reached, characterized by the densimetric Froude number (F_0) (Wunderlich, 1971):

$$F_0 = \frac{(v/A_0)}{\sqrt{g\,(\Delta\rho/\rho)\,d_0}}$$

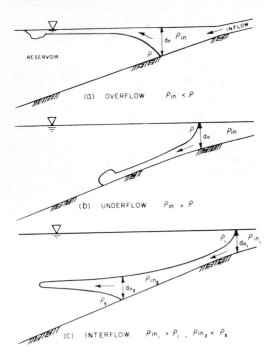

Figure 7-18 Types of inflow into lakes and reservoirs. (From Wunderlich, W. O.: The dynamics of density-stratified reservoirs. *In* Hall, G. E., ed., Reservoir Fisheries and Limnology. Washington, D.C., American Fisheries Society, 1971.)

where:

v = flow rate

A_0 = cross sectional area at the critical section

ρ = density of inflow

$\Delta\rho$ = density difference between the incoming and receiving water.

Overflow or underflow initiates at this critical section, with a reduction of flow velocity that may be accompanied by deposition of suspended materials. A critical analysis of density differences and distribution of inflowing water in Lake Biwa, Japan, showed that the inflowing river waters do not mix readily if density differences are reasonably large, and flow occurs instead in complex overflow and interflow patterns (Morikawa, et al., 1959).

Variations in density differences between incoming water and those of the recipient water body are so great that generalizations are very difficult to make. Discharge also varies widely seasonally, not only in volume but in the accompanying load of dissolved and suspended materials. For example, in alpine situations, the density of river water is greater (cold, high dissolved and particulate load) than the recipient water and underflow is common, particularly if glacial erosion at the head of the river contributes to the suspended load. During the summer, although alpine river inflow water is still cold, its density is reduced by high-volume dilution with snow and ice melt runoff. In the summer, then, alpine rivers interflow into the metalimnion. This situation has been documented frequently, but a particularly striking case is seen in the Bodensee, Germany, where light penetration is abruptly attenuated in the metalimnion over large areas of the lake by the intrusion of river water with a high suspended load (Lehn, 1965).

The extent of intrusion and current generation in the receiving lake is also a function of discharge volume of the river in relation to the volume of the lake. The theoretical retention time of a lake or reservoir, based on the relation of the total influent-outflow to the total volume of the lake, is realized only approximately in most lakes. Retention time varies with the dimensions and shape of the basin, seasonal rates of inflow, and stratification characteristics (Kajosaari, 1966). At high discharge rates, overflow and interflow of rivers may channel across or through the water mass of the stratified basin, whereas at lower discharge rates, river water may penetrate more into the main water mass and may be mixed through more normal circulatory mechanisms. In larger lakes, geostrophic deflection of the incoming water intrustions, surface or interflows, is consistently observed.

CURRENTS UNDER ICE-COVER

It is clear that the water of a lake under ice-cover, even in lakes of closed basins with no appreciable inflow and outflow, exhibits water movement by currents. Much of the evidence for the existence of weak under-ice currents stems from careful direct measurements in which radioactive sodium (^{24}Na) was released at various points beneath the ice, and its short-term distribution was measured for several days. Slow horizontal currents, with velocities too low to be measured by mechanical current meters, have been observed in a small ice-covered lake in Wisconsin (Likens and Hasler, 1962) and in a large, permanently ice-covered lake in Antarctica (Ragotzkie and Likens, 1964). Similarly, radioactive tracers have been used to measure slow horizontal movements in a small lake of Nova Scotia (McCarter, et al., 1952), in a meromictic lake in Washington (Hutchinson, 1957), and in the monimolimnion of a small meromictic lake in Wisconsin (Likens and Hasler, 1960).

Horizontal displacements of radiosodium were found to be generally asymmetrical in a small ice-covered, closed basin bog lake (Fig. 7–19). Horizontal velocities of at least 10 m day^{-1} (42 cm hr^{-1}) were observed near the bottom at the deepest point of the basin, and maximum velocities of 15 to 20 m day^{-1}. The vertical component of this motion was between one and three orders of magnitude less than the horizontal component (Table 7–4). In further experiments, Likens and Ragotzkie (1966) also found a definite rotary movement of horizontal currents, which was cyclonic near the center of the lake and anticyclonic near the perimeter (Fig. 7–20).

It is apparent from these detailed studies, as well as from studies of a larger lake in Swedish Lapland (Mortimer and Mackereth, 1958) that under-ice water movements are the result of convective cells of density currents generated primarily by relatively large quantities of heat flowing from the sediments. If the temperature of the water into which the heat flows is greater than 4°C, that flow will generate ascending buoyant currents. If, however, the lake has cooled to well below 4°C before the ice-cover is established, the heat flux will increase the density of the water in contact with the sediments, causing it to flow as a density current down the sloping sides of the basin into the nearest depression. Chemical exchanges also occur between sediment and water; for example, oxygen is taken up by the sediments. If the lake basin contains more than one topographically separated depression, the density flows may produce significantly different temperatures and chemical compositions in the bottom waters of these depressions (Mortimer and Mackereth, 1958).

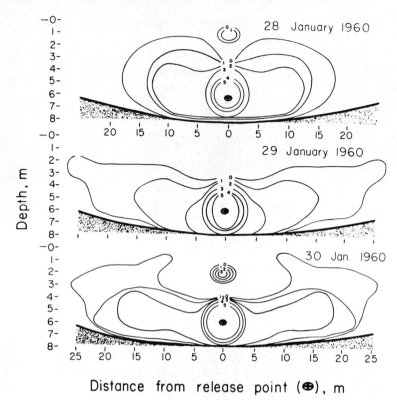

Depth, m

Distance from release point (⊕), m

Figure 7-19 Diagram of the horizontal and vertical dispersal of ^{24}Na in Tub Lake, Wisconsin, under ice-cover 24, 48, and 72 hours after release. Integers indicate the logarithmic value of the radioactivity concentration in counts min^{-1}. (From Likens, G. E., and Ragotzkie, R. A.: Journal of Geophysical Research, vol. 70, p. 2333, 1965. Copyright by American Geophysical Union.)

Most of the heat of the sediments is gained during the previous summer period and fall turnover (cf. Chapter 5) and is dissipated slowly. Birge et al. (1927) found that about 650 cal cm^{-2} were conducted from the sediment to the water of Lake Mendota during the period of ice-cover (110 days), a heat gain of 5.9 cal cm^{-2} day^{-1}. Thermal gains of 3 to 4 cal cm^{-2} day^{-1} are common in small temperate lakes during the winter. Tub Lake

TABLE 7-4 Vertical Motion in Ice-Covered Tub Lake, Wisconsin, Calculated from Measurements of ^{24}Na Dispersal*

DEPTH (m)	VELOCITY (w) IN CM SEC^{-1} × 10^{-2}*		
	28-30 JAN 1960	26-27 JAN 1961	27-28 JAN 1961
0.5	0	0	0
1.5	0	+0.0132	+0.0378
2.5	+0.0033	−0.0057	+0.0047
3.5	−0.0041	+0.0007	−0.0158
4.5	−0.0233	−0.0498	−0.0508
5.5	−0.0066	−0.0497	−0.0219
6.5	+0.0381	−0.0583	−0.1180
7.5	+0.0661	−0.6990	−1.55

From data given in Likens, G. E., and Ragotzkie, R. A.: Journal of Geophysical Research, 70:2333-2344,1965. Copyright by American Geophysical Union.

*Negative values of w indicate downward motion and positive values indicate upward movement.

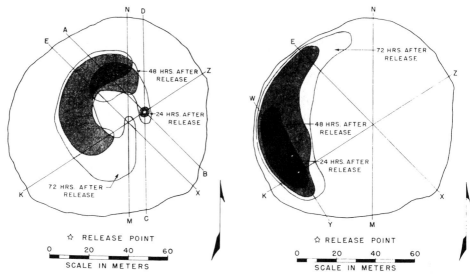

Figure 7-20 Maximum horizontal displacement of [24]Na following its release at a depth of 5 m *(right)*, and of ^{131}I following its release at a depth of 3 m *(left)* in Tub Lake, Wisconsin, January 1962. (From Likens, G. E., and Ragotzkie, R. A.: Verhandlungen Int. Ver. Limnol., 16:126, 1966.)

gained about 3 cal cm^{-2} day^{-1} and larger Lake Torneträsk, Swedish Lapland, about 1.0 cal cm^{-2} day^{-1}. The sediment heat source is adequate to explain the observed currents. Biological oxidation during respiration in the sediments represents a caloric equivalent of about 0.04° cm^{-3}, which is totally inadequate as a heat source.

Similarly, excess of solar radiation at the lake surface over heat losses is appreciable only at the surface and in late spring; it is absorbed primarily in melting of ice and snow. Geothermal heat flow (approximately 1.23 × 10^{-6} cal cm^{-2} sec^{-1}) from the interior of the earth represents only about 0.1 cal cm^{-2} day^{-1} through the sediments of the lake. Ground water that is relatively warm and circulating near the basin probably transmits some heat through the sediments to the lake water. In Wisconsin, this potential heat source was estimated to be about 1 cal cm^{-2} day^{-1}, a significant source of thermal conductivity in addition to that from heat storage from the previous summer. Thus the major operator of thermally induced convection currents under winter conditions appears to be sediment heat that accumulated during summer.

SUMMARY

1. Water movements occur in response to forces, especially wind, that transfer energy to the water. Rhythmic motions (oscillations) result both at the surface of the water and internally deep within the basin. These motions and their attendant currents may be in phase or in opposition. The ultimate fate of these movements is to degrade into arrhythmic turbulent motions, which disperse the water and chemicals and organisms within it.

2. As water of a given density moves along an interface, such as the wall of a tube or water of a different density, frictional shearing stresses occur between them. Below a certain speed of flow, water movement is smooth and undisturbed at the interface. This ordered movement is called *laminar flow*. As the velocity of movement is increased, the damping effects of gravity are exceeded, and vortices occur at the interface with a decrease in stability. Water movement then becomes disordered in *turbulent flow*, in which the frictional increases result in mixing of the two fluids of different density perpendicular to the direction of the current movements. Since the velocities needed to induce turbulent flow are very low (only a few mm s^{-1}), laminar flow is rarely found in aquatic systems.

3. The coefficient of eddy diffusion (K_z) measures the rate or intensity of mixing (turbulent exchange) across the plane of a density gradient, most often a thermal gradient. K_z decreases markedly in the transition from the turbulent epilimnion into the more stable metalimnion and as the thermal density gradient increases seasonally during summer stratification. K_z also decreases, i.e., there is less turbulent mixing, in the metalimnion and hypolimnion of smaller lakes of greater stability than in lakes of large area with longer fetch for wind exposure.

4. Movement of air over water sets the water surface into an oscillating motion, called progressive or *traveling surface waves.*
    a. The water of surface waves moves in a *cycloid path* (Fig. 7–4). Water is displaced upward and returned to equilibrium by gravity along a circular path (therefore sometimes called gravity waves).
    b. The height of the vertical oscillation is attenuated rapidly with depth, and decreases by half of the cycloid diameter for every depth increase of ⅑ the wavelength (distance from crest to crest).
    c. While water entrained in surface waves oscillates considerably up and down, horizontal movement is small. The wavelength is much less than water depth.
    d. When the wavelength of these short surface waves is less than 6.3 cm (2π), the surface waves are called *ripples* (capillary waves).
    e. The height of the highest surface waves on a lake is proportional to the square root of the fetch, that distance over which the wind has blown uninterrupted by land.

5. When surface waves occur near the shore in shallow water, their cycloid motions are transformed into a to-and-fro sloshing which extends to the bottom of the water column. The wavelength becomes more than 20 times the water depth. As deep water surface waves enter shallower water, their velocity decreases proportional to the square root of depth. A reduction in wavelength occurs with a marked increase in wave height. With increased height, the waves become asymmetrical and unstable. Collapse or breaking of water over the front of the wave results in a *breaker* (Fig. 7–5).
    a. In a *plunging breaker* the forward face of the wave becomes convex and the crest curls over and collapses.
    b. The crest of a *spilling breaker* collapses forward, spilling water downward over the front of the wave.

    c.   Shallow water wave energy is effective in moving littoral sediments to deeper water and inhibiting the growth of many organisms not adapted to such turbulence.

6.   Currents are nonperiodic water movements. Although many external forces contribute to the generation of currents, wind is dominant in deriving *surface currents* of open water.

    a.   The velocity of wind-driven water currents is about two percent of wind speed. Water velocity of surface layers increases in a linear way with wind velocity until a critical wind speed (approximately 6 m s^{-1}) is reached, beyond which the ratio becomes nonlinear.

    b.   The Coriolis force of the Earth's rotation creates *geostrophic effects* by deflecting the direction of surface currents from that of the wind. Deflection is to the right of wind direction in the Northern Hemisphere and to the left in the Southern Hemisphere. The resulting geostrophic wind-drift deflection is 45° from the wind direction in very large lakes. The angle of declination decreases as the size and especially depth of lakes decrease.

7.   *Langmuir circulations* are large currents of turbulent transport of the upper strata of lakes that are organized into vertical helices. Langmuir circulations move water and entrained particles in circular cells along cylindrical patterns both vertically and in a helical manner parallel to the wind direction and perpendicular to the lines of surface waves (Fig. 7–6).

    a.   Downward movement velocities at cell convergences are about three times greater than upward current velocities at divergences.

    b.   Aggregations of surface particles which occur at lines of convergence are called *streaks*.

    c.   Langmuir circulations are generated by surface waves at wind velocities above 2 to 3 m sec^{-1}. Wind energy is converted through surface wave energy into turbulence; surface instability is then dispersed downward and drives the convection pattern.

    d.   Langmuir circulation is a primary way in which turbulent motion is transported downward and the upper layers of water are mixed.

8.   Water movements in surface strata cause water to pile up at the downwind end of the lake basin. The epilimnetic water there flows downward by gravity until it encounters more dense water of the metalimnion, and then flows back (upwind) along the epilimnetic–metalimnetic interface.

    a.   A portion of the metalimnetic water is entrained (incorporated by mixing) into that of the epilimnion. As a result, the epilimnion can become deeper than was previously the case as the upper portion of the metalimnion is eroded.

    b.   If the shear at the thermocline is sufficiently great, resulting in low Richardson numbers ($R_i < 0.25$), turbulent flow and vortices occur in the upwind direction. Internal waves can form at the shear interface.

    c.   As entrained water of the metalimnion reaches the upwind end of the basin, especially as the storm event producing the movement ends, some flow of the metalimnion is transferred downward to the hypolimnion. In a similar manner, although on a much smaller scale, movement occurs along the metalimnetic–hypolimnetic density-gradient interface in the windward direction.

9. Wind-induced tilting of the water surface and of the metalimnion (or barometric pressure on one part of a large lake) results in a displacement of more water at one end of the lake than at the other. When the wind stops as the storm passes, the tilted water strata flow back towards equilibrium. The equilibrium point is overshot, however, by momentum. The resulting rocking motion of the entire water mass of the upper lake strata results in *long standing waves* termed *seiches*.

   a. Seiches have the form of very long wave patterns with wavelengths approximating the dimensions of the basin of many lakes.

   b. The surface of the metalimnion oscillates up and down at the lake ends like a seesaw about a line, the *node*, of no vertical motion. Maximum vertical motion occurs at the ends, the *antinodes*, often at the ends of the lake basin. Horizontal movement is maximum at the nodal line and nil at the antinodes. The horizontal movements reverse directions with each long oscillation and can create currents of sufficient size to develop instability, vortices, and waves. The seiches return to equilibrium as a result of gravity and frictional damping with a progressive reduction in the duration of the oscillation periodicity.

   c. *Surface seiches* affect the motion of the entire water mass of the lake, whether stratified or not, and have maximum amplitude at the surface.

      i. Horizontal oscillation at the nodal line is large when the basin is very long in relation to depth.

      ii. The amplitude of surface seiches is relatively small (to several meters, maximum) but is sufficient to cause significant flooding and erosion effects of shoreline areas in large lakes.

      iii. The periodicity of surface seiches increases with increasing basin dimensions (see Table 7–1).

   d. *Internal seiches* occur when the lakes are stratified and the metalimnion is set into oscillation as an internal standing wave.

      i. Both the periodicity and the amplitude of internal seiches are much greater (nearly an order of magnitude) than those of surface seiches.

      ii. Most internal seiches are uninodal, although multinodal seiches occur in very large lakes.

      iii. Since the nodal horizontal oscillations of internal seiches are much larger than those of surface seiches, shearing flow at the metalimnetic interfaces can generate large *internal progressive waves*. These internal waves on the metalimnion are about 10 times larger than surface waves, but they propagate and break internally much as surface waves do.

      iv. The geostrophic effects of Coriolis force increase north and south of the Equator. The geostrophic effects deflect the oscillation patterns and flows associated with internal seiches and cause large-scale twisting undulations of the metalimnion (Fig. 7–11).

    v.   The internal currents and waves cause considerable mixing of water, solutes, and organisms across the epilimnetic–metalimnetic interface.

  e. *Long progressive waves* of both the surface and metalimnion waters have wavelengths that are much greater than the water depth. Long progressive waves are nondispersive.

    i.   As the long waves oscillate in lakes sufficiently large for Coriolis geostrophic effects to influence motion and intersect the shore line, the rotational effects and decay of wave energy by gravity generate currents that are parallel to the shore and to the progression of the waves (Fig. 7–13). The wave amplitude of these *Kelvin waves* decreases exponentially away from the shore.

    ii. In very large lakes, the long waves can travel in open water without the interference of shore boundaries. These *Poincaré waves* are influenced by Coriolis geostrophic forces and form a wave pattern of alternating hills and valleys with corresponding cellular patterns of currents (Fig. 7–15).

    iii. The amplitude of Poincaré waves increases as the shore boundary is reached and these waves are transformed into the Kelvin wave configuration (Fig. 7–16).

10. A narrow zone of warm water, termed a *thermal bar,* develops between the shore line and the open water mass in the spring. The shallow nearshore waters heat more rapidly than the open waters. A steady circulation is set into motion along the shore as a result of density gradients and little mixing occurs between inshore and offshore waters until the whole lake is thermally stratified.

11. Incoming river water to a lake or reservoir flows into strata of density similar to its own. Inflow water movements are dependent on density differences from temperature, dissolved materials, and particulate suspensoids, and vary greatly seasonally and among lake ecosystems.

12. Slow water currents occur in lakes even though the water is sealed from wind by ice cover and there is no appreciable inflow and outflow. Horizontal current velocities tend to be greater than vertical velocities. The currents are thermally induced by convection from heat flowing from the sediments. Most of the heat released from the sediments during winter was accumulated during the previous summer.

CHAPTER EIGHT

STRUCTURE AND PRODUCTIVITY OF AQUATIC ECOSYSTEMS

Discussion so far has been directed to basic physical aspects of water itself, how water is distributed to lake basins, and to the origins of lake basins and how basin morphology is modified with time. We then considered both the penetration of solar radiation and the distribution of heat in various lake systems, and explained how absorbed heat interacts with wind energy to influence water movements.

Now, the basic nutrient chemistry of aquatic ecosystems will be evaluated. The biota are inseparably coupled with the dynamics of many chemical elements, in particular the nutrient distribution and their biogeochemical regulation. Before discussing the major inorganic constituents in detail, however, a brief overview of the ecosystem components and how they interrelate is useful.

LAKE ECOSYSTEM CONCEPT

Several early investigators emphasized the functional relationships of organisms within lakes (e.g., Forbes, 1887; Lindeman, 1942). The innovative perceptions of these studies led to intensive evaluation of interrelationships among organisms and changes in organism populations in response to alterations in physical, chemical, and biotic properties of the environment. Lakes were viewed as microcosms, functionally isolated from the rest of the landscape and biosphere.

In their time, these analyses were perceptive and they led to extensive investigations of the interrelationships between biotic and physical components within lakes. These studies, which continue to the present time, are of fundamental importance; they have resulted in a comprehensive understanding that is unrivaled by any other area of ecology. Still, within a broad context, these in-lake analyses encompass but a small part of the lake ecosystem.

A number of early workers called attention to the importance of nutrients from the drainage basin in regulating metabolism in lakes. This insight recognized what is commonly accepted now—that the lake ecosystem consists of the lake and its entire drainage basin. Recognition of the importance of the drainage basin, however, is still largely limited to loading of inorganic nutrients. Comprehensive analyses of the manner in which the terrestrial biota regulate loading to streams and recipient lakes are much more rare (e.g., Likens, et al., 1977; Bormann and Likens, 1979).

The importance of the catchment area to inputs of both dissolved and particulate organic matter to lakes, and of the regulatory influence of organic matter once it reaches the lake, is still largely unappreciated. Extremely heterogeneous and productive wetland-littoral areas often lie at the interface between the terrestrial drainage basin and the open-water zone of the lake (Fig. 8-1). These complex wetland-littoral areas are exceedingly important in regulating lake metabolism (Wetzel, 1979). Since a majority of lakes of the biosphere are small and relatively shallow, the metabolically active wetland and littoral components regulate the productivity of most lakes of the world. These complex interface regions are the least understood lake ecosystem component; major aspects will be discussed later (cf. Chapters 18, 19, 20, and 22).

The effects that terrestrial, wetland, and littoral biota have on the quality and quantity of inorganic and organic loading to a lake can be profound. Water laden with inorganic and organic substances flows from higher elevations to the recipient lake

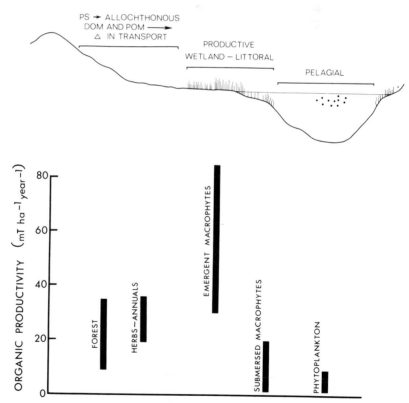

Figure 8-1. The lake ecosystem, showing the drainage basin with terrestrial photosynthesis (*PS*) of organic matter, movement of nutrients and dissolved (*DOM*) and particulate (*POM*) organic matter in surface and groundwater flows towards the lake basin, and chemical and biotic alteration of these materials en route, especially as they pass through the highly productive and metabolically active wetland-littoral zone of the lake per se (net organic productivity in metric tons per hectare per year).

basin both in groundwater and in surface streams. Chemical and biological reactions occur en route that selectively modify the quality and quantity of nutrients entering the lake. Surface flows often pass through the wetland-littoral complex, and can further selectively lose or gain inorganic and organic compounds before reaching the open water.

Finally, appreciable loading of inorganic and organic substances to the atmosphere is common from the industrial and agricultural activities of man. These compounds reach the drainage basin and lake itself by dryfall and with precipitation. This source of loading is often a significant portion of the total nutrient loading to a lake.

LACUSTRINE ZONATION AND TERMINOLOGY

The bottom of a lake basin is separable from the free open water, the *pelagial* zone, and is further divisible into a number of rather distinct transitional zones from the shore to the deepest point (Fig. 8-2). The *epilittoral* zone lies entirely above the water level and is uninfluenced by spray; the *supralittoral* zone also lies entirely above the water level, but is subject to spraying by waves. The *eulittoral* zone encompasses that shoreline region between the highest and lowest seasonal water levels, and is often influenced by the disturbances of breaking waves. The eulittoral zone and the *infralittoral* zone collectively constitute the *littoral* zone. The infralittoral zone is subdivided into three zones in relation to the commonly observed distribution of macrophytic vegetation: *upper infralittoral* or zone of emergent rooted vegetation; *middle infralittoral* or zone of floating-leaved rooted vegetation; and *lower infralittoral* or zone of submersed rooted or adnate macrophytes.

Below the littoral is a transitional zone, the *littoriprofundal*, occupied by scattered photosynthetic algae and bacteria. The littoriprofundal zone is often adjacent to the metalimnion of stratified lakes. The upper boundary of the littoriprofundal zone at the

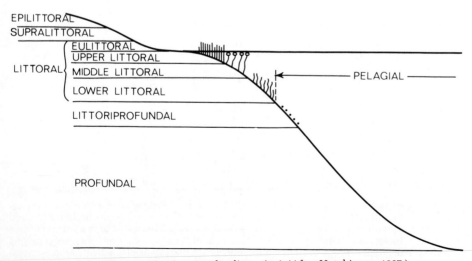

Figure 8-2. Lacustrine zonation (see text for discussion). (After Hutchinson, 1967.)

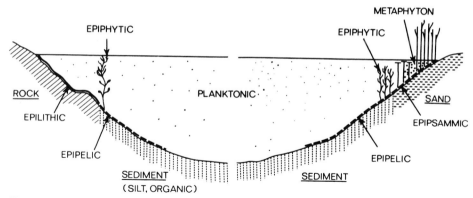

Figure 8-3. Major terms of microfloral communities associated with different substrata in fresh waters.

lower edge of macrovegetation of the lower infralittoral is usually quite distinct. The lower boundary of the littoriprofundal consists of a gradient of benthic algae, especially blue-green algae, and photosynthetic bacteria, and is less sharply demarcated. The remainder of the sediments, which consists of exposed fine sediment free of vegetation, is referred to as the *profundal zone*.

The philology associated with bacteria and algae variously attached to substrata in aquatic systems is involved (cf. reviews by Naumann, 1931; Cooke, 1956; Sládečková, 1962; Wetzel, 1964; Round, 1964a, 1965a; and Hutchinson, 1967). The number of terms is large and many are unnecessarily complex and quite confusing, especially when removed from their original definitions. The subject is rather unrewarding because of the great variation in microhabitats and association of organisms with substrata. A few widely used terms in limnology need to be discussed.

The term *benthos*, originating from the Greek word for "bottom," was initially defined broadly to include the assemblage of organisms associated with the bottom or, better, any solid-liquid interface in aquatic systems. *Benthos* is now nearly uniformly applied to animals associated with substrata.* The terms *Aufwuchs* (German: "growth upon"), and to a lesser extent *haptobenthos*, generally connote all organisms adnate to, but not penetrating, a solid surface. Since the term *Aufwuchs* is so broad and ambiguous, it is of little meaning and should be abandoned.

Periphyton, although variously used, usually refers to microfloral growth upon substrata. Reference to more specific subdivisions by complex phrases (e.g., epiphytic periphyton) results in involved and often redundant nomina. A much more explicit manner of expression, and one championed by Round, is to refer to the organisms with appropriate modifiers descriptive of the substrata upon which they grow in natural habitats. Hence, among the algal communities, one can readily differentiate the following (Fig. 8-3): (a) *epipelic* algae as the flora growing on sediments (fine, organic); (b) *epilithic*

Herpobenthos as used by Hutchinson (1967) is synonymous with *benthos* when the term *benthos* is unqualified, and means "growing on" or "moving through sediment."

algae growing on rock or stone surfaces; (c) *epiphytic* algae growing on macrophytic surfaces; (d) *epizooic* algae growing on surfaces of animals; and (e) *epipsammic* algae as the rather specific organisms growing on or moving through sand. The general word *psammon* refers to all organisms growing or moving through sand (cf. Cummins, 1962, for a detailed discussion of particle size differentiation among sediments).

A group of algae found aggregated in the littoral zone is the *metaphyton,** which is neither strictly attached to substrata nor truly suspended. The metaphyton commonly originates from true floating algal populations that aggregate among macrophytes and debris of the littoral zone as a result of wind-induced water movements. In other situations, the metaphytonic algae can originate from fragmentation of dense epipelic and epiphytic algal populations. A surprisingly large number of descriptions exist of clustering of metaphytonic algae and macrophytes into "lake balls," which are densely packed aggregations of algae, plant parts, or both. These balls are formed by the alternating rolling movements of wave action in the littoral zone (Nakazawa, 1973).

The *phytoplankton* consist of the assemblage of small plants having no or very limited powers of locomotion; they are therefore more or less subject to distribution by water movements. Certain planktonic† algae move by means of flagella, or possess various mechanisms that alter their buoyancy. Most, however, are "freely floating," a term commonly used when referring to *plankton* in general, even though it is obviously somewhat of a misnomer, in that most algae do not float. Most algae are slightly denser than water, and sink, or sediment from, the water. Phytoplankton are largely restricted to *lentic* ("standing") waters and large rivers with relatively low current velocities (cf. Wetzel, 1975b). Phytoplankton in streams of moderate flow, which either break loose from the algal communities attached to substrata or enter a stream from the outflow of lakes, are usually rapidly fragmented or killed by the abrasive action of turbulence and substrata (see, e.g., Chandler, 1937). Distinctly macroscopic algae with long filamentous forms usually inhabit parts of the littoral zone. The common groups of phytoplankton, their population and community dynamics, and their productivity have been studied extensively (cf. Chapter 15).

Animals of fresh waters are extremely diverse, and include representatives of nearly all phyla. The *zooplankton* include animals suspended in water with limited powers of locomotion; they are subject to dispersal by turbulence and other water movements. Like phytoplankton, zooplankton are usually denser than water, and constantly sink by gravity to lower depths. The planktonic protozoa have limited locomotion, but the rotifers, cladoceran and copepod microcrustaceans, and certain immature insect larvae often move extensively in quiescent lake water. These important zooplankton are treated in detail in subsequent chapters (Chapters 16 and 19).

The distinction between suspended zooplankton having limited powers of locomotion, and animals capable of swimming independently of turbulence—the latter

---

*Metaphyton (Behre, 1956) is essentially synonymous with the terms *tychoplankton* and *pseudoplankton* used much earlier by Naumann (1931), and the term *pseudoperiphyton* used by Sládečková (1960).

†The word *planktonic*, while etymologically incorrect (see discussion by Rodhe, 1974, and Hutchinson, 1974), is so ingrained in aquatic ecology that change is not desirable.

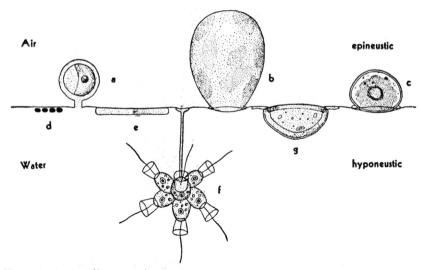

Figure 8-4. Exemplary organisms of the neuston at the surface film. Epineustonic: *A, Chromatophyton* (Chrysophyceae); *B, Botrydiopsis* (Xanthophyceae); *C, Nautococcus* (Chlorococcaceae). Hyponeustonic: *D, Lampropedia* (Coccaceae); *E, Navicula* (Bacillariophyceae); *F, Codonosiga* (Craspedomonadaceae); *G, Arcella* (Rhizopoda). (From Ruttner, F.: Fundamentals of Limnology. Translated by Frey, D. G., and Fry, F. E. J., 3rd ed., Toronto, University of Toronto Press, 1963.)

referred to as *nekton*—is often diffuse. Certain zooplankton and early larval stages of fish are initially planktonic but distinctly nektonic in later life stages (cf. Chapter 16).

A number of specialized organisms, the *pleuston*, are adapted to the interface habitat between air and water. There are a number of macroscopic organisms such as many of the duckweeds (Lemnaceae) and adult insects (e.g., gerrid hemipterans) that are morphologically adapted to the interface. Numerous species of bacteria and algae, including some diatoms, chrysophytes, and xanthophytes, are specialized to live at the interface. The microscopic components of the pleuston are collectively termed *neuston*, and are separated into those organisms adapted to living on the upper surface of the interface film (the *epineuston*) and those living on the underside of the surface film (*hyponeuston*). A majority of the neustonic organisms live on the upper surface of the interface. A few examples of algal and protozoan neuston are illustrated in Figure 8-4.

Although the development of neuston is most extensive in sheltered, quiescent waters, neuston are by no means restricted to these conditions, as attested to by their development in marine systems (cf. treatise by Zaitsev, 1970). Conspicuous development of neuston, particularly of the epineuston, is common in small, quiet waters in the littoral or along the margins and backwater areas of streams. Often populations of epineustonic organisms become so large that light is reflected from the chromatophores, and the water appears to be covered with a dry film of varying coloration. A few cladoceran zooplankton and insect larvae are adapted to feed on the neuston from the underside of the interface film.

Very little is known of the physiology and metabolism of the neuston of fresh waters in comparison with that of marine waters (cf. Zaitsev, 1970; Hardy, 1973). Although the neustonic productivity and contributions to the total metabolism of fresh waters probably are small in a majority of situations, their roles in shallow lake and marsh systems deserve much further study.

DEAD ORGANIC MATTER

The living biota of lakes constitute only a very small portion of the total organic matter of ecosystems, even in the most productive of freshwater habitats. Most of the organic matter in fresh waters is nonliving, and is collectively referred to as *detritus*.

Important aspects of the functioning of aquatic ecosystems center on the cycling of organic carbon between living and nonliving components. Because of the pivotal relationships between the productivity of the living components and the massive amounts of nonliving organic matter, the subject will be discussed in detail later (Chapter 22). Two aspects, however, should be emphasized at this point.

Detritus consists of all nonliving organic matter, in both dissolved and particulate forms. Dissolved organic matter is approximately ten times more abundant than particulate organic matter in natural waters, and the living biota constitute only a small fraction of the total particulate organic component. The living components, however, are important users and transformers of the detrital components, which they metabolize. The large amount of detritus in aquatic ecosystems, most of which exists in a dissolved form, provides a large organic reserve of materials and energy that is utilized in the system at variable rates.

Much of the organic matter newly synthesized by photosynthesis is not consumed by animals, but instead enters nonpredatory pools of dissolved and particulate detritus (Wetzel, et al., 1972; Rich and Wetzel, 1978). Metabolism by living biota of detrital dissolved and particulate organic matter from allochthonous, wetland-littoral, and pelagial components provides a fundamental stability to ecosystems. While populations of individual species commonly undergo rapid oscillations in response to an array of dynamic factors governing their growth, the detrital system comprises many species of organisms and spans all trophic levels; consequently, detrital metabolism is slower and more evenly sustained on a much larger organic reserve (cf. Chapter 22).

POPULATION GROWTH AND REGULATION

A *population* of organisms is a spatially defined assemblage of individuals of one species. A population of a species could broadly include all individuals of that species inhabiting the biosphere, but ecologists generally deal with individuals of a species that occupy, or populate, a more localized area, such as a particular lake or section of flowing water. If movement into and out of the defined area is small, relative to the number of births and deaths occurring within the area, population changes within the area can be studied with reasonable accuracy. The number of individuals per unit area or volume of water, the population *density*, is an index of population size at any given time.

Growth among populations and their competitive interactions for available resources are fundamental to nearly all aspects of biological limnology. Analyses of population growth have been the subject of intensive quantitative and theoretical study.* While a detailed analysis is inappropriate here, several fundamental relationships and concepts are discussed below, since reference is made to these basic parameters and characteristics in subsequent evaluations of freshwater population and community dynamics.

The growth of a population can be very rapid among smaller organisms. The rate of growth (g) is a geometric expression of increase of the numbers (N) of the population. The time (t) that it takes a population number to double under ideal conditions is:

$$t = \frac{\log_e 2}{\log_e g} = \frac{0.69}{\log_e g} \tag{8-1}$$

Environmental and competitive restrictions quickly limit the size of the populations either by increasing death rate or by decreasing birth rate.

The geometric constant (g, Eq. 8-1) for rate of population change is difficult to use in comparative growth analyses among populations because of differences in population sizes. This problem is mathematically avoided by expressing population growth using an exponential constant, r, where $r = \log_e g$, or $g = e^r$. Calculation of r from population growth data is done by obtaining the difference between the logarithms of the population sizes at the beginning and the end of a period of growth and dividing by the length of the time period over which the observations are conducted:

$$r = \frac{\log_e N_t - \log_e N_o}{t} \tag{8-2}$$

Factor r is an instantaneous, relative measure of population growth rate, in which the percentage rate of increase in a population is determined at every time interval.

The growth rate of a population can be characterized by the logistic equation, which incorporates a term for the intrinsic capacity of a population to grow and a term corresponding to density-dependent limitations upon that growth potential imposed by competition of others in the population for available environmental resources. Among small, rapidly growing, asexual or parthenogenetic organisms, such as bacteria, many algae, and zooplankton, the differential form of the logistic:

$$\frac{dN}{dt} = rN \left(\frac{K - N}{K} \right) \tag{8-3}$$

or

$$\frac{dN}{dt} = rN \left(1 - \frac{N}{K} \right) \tag{8-4}$$

can adequately model a population (N) growing from a small number of individuals to an upper limiting population K, the latter being governed by nutrient supply, space, inhibitory metabolic products, and other density-dependent factors. As the population increases towards the carrying capacity (K), fewer resources are available for reproduction, and reproductive rates decline. As the population density approaches the car-

---

*The basic discussion of population analyses is presented in many texts on population biology (e.g., Wilson and Bossert, 1971; Hutchinson, 1978; Ricklefs, 1979; cf. also Kingsland, 1982).

rying capacity, $(1 - N/K)$ approaches zero and population growth ceases. The resulting logistic growth curve has a sigmoid shape (Fig. 8–5), and is commonly observed among natural and experimental populations. The basic logistic model discussed here has been variously modified (cf. Hutchinson, 1978), but is commonly found among populations.

A *life table* and its resulting *survivorship curve* provide the parameters needed for determining the survival and reproductive performance of a population. When a large group (e.g., 1000) of individuals of the same age or nearly the same age, referred to as a *cohort*, are counted at time intervals, one can obtain an estimate of the number of survivors in each time interval. The fraction (l_x) of the original cohort surviving to time x is plotted logarithmically against x to yield a survivorship curve (Fig. 8–6). Such a survivorship curve, in which the population is followed as it changes with time, is age-specific, since all surviving individuals have similar ages at any time (Hutchinson, 1978). The *specific death rate* (q_x) over any time interval x to x + 1 is the number of individuals dying during that interval $(l_x - l_{x+1})$ divided by l_x. The slope of the survivorship curve at any point corresponds to an instantaneous value of q_x.

Population size tends to oscillate about the equilibrium population level (K) that can be supported by the environment. The rate of increase per individual will decline steadily throughout time as the population grows following the logistic growth equation. The density-dependent decline can result either from a decrease in birth rate (b) or an increase in mortality (death rate, d) as N increases $(r = b - d)$. In contrast, density-independent changes in population size can be caused by environmental factors that affect all individuals in a population with equal probability. Density-dependent changes in b or d are particularly important in the regulation of population sizes under uniform environmental conditions. When environmental conditions fluctuate, density-

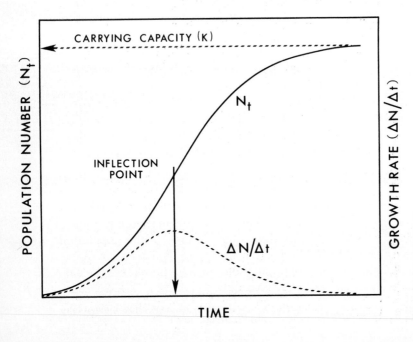

Figure 8–5. Common sigmoid growth curve of a population with time (N_t) as it approaches the environmental carrying capacity (K). The growth rate $(\Delta N/\Delta t)$ denotes the difference in number of organisms over each time interval. (Modified from Ricklefs, 1979.)

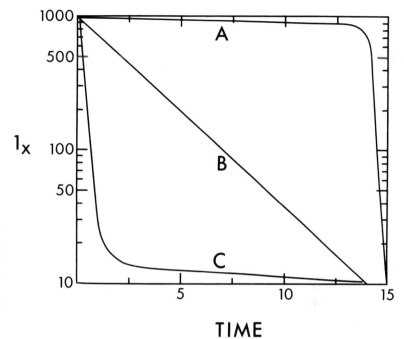

Figure 8-6. Idealized survivorship curves of a population *A*, with a determined physiological life expectancy (negatively skewed rectangular curve), *B*, with a constant death rate (diagonal), and *C*, with high juvenile mortality (positively skewed rectangular). (Modified from Hutchinson, 1978.)

independent constraints to population growth can assume greater importance. Examples of these population changes are discussed in detail in subsequent chapters.

Community Structure and Interrelationships

Although variously defined, a *community* usually refers to a group of interacting populations. Some practical spatial boundaries are most often applied to studies of communities (e.g., planktonic communities) because habitats are diverse and often organisms move from one habitat to another.

Throughout subsequent discussions of freshwater biota, the effects of density-dependent and dynamic environmental factors on population and community changes will be evaluated. It is useful to summarize the basic concepts of these relationships and competition among populations of a community in a general way before evaluating specific cases.

ECOLOGICAL NICHE

The concept of the ecological *niche* has had a long history of development and interpretation in ecology. A given habitat consists of a number of interacting environmental gradients. The gradients include both physical and biotic factors, many of which are dynamic, changing constantly at differing rates.

In seminal analyses, Hutchinson (1944a, 1957, 1978) formally defined the niche concept as a certain biological activity space in which an organism exists in a particular habitat. This space is influenced by the physiological and behavioral limits of a species and by the effects of environmental parameters (physical and biotic, such as temperature and predation) acting upon it. Each of these parameters can be ordinated on an axis, and can be thought of as a dimension in space. The fundamental niche, then, can be viewed as an n-dimensional space or hypervolume, with each of its n axes or dimensions corresponding to the range of an environmental parameter over which the organism can exist. Since many physical and biotic factors interact, each species occupies only a portion of its fundamental niche; this portion can be referred to as the species' realized niche.

Exact quantitative definition of a species's niche, realized or fundamental, is impossible because of its infinitely large number of axes or dimensions. In addition, while it is mathematically and statistically possible to describe multidimensional spaces, we can readily understand physical or graphical explanations in only three or fewer dimensions. In many cases, however, only a few dimensions are needed to adequately describe specific niche interactions (cf. Hutchinson, 1978; Morowitz, 1980).

COMPETITION

For each species and each resource requirement, a range of possible existence is bounded by extremes of tolerance and bracketing an optimal growth zone. Species differ in resource requirements for optimal growth and reproductive success, but rarely does a species occupy only optimal zones within its requirements. Certain variables (e.g., temperature) are continuous, whereas others (e.g., prey availability) are discrete or discontinuous. Some species have similar requirements along one or more resource axes. When two or more species overlap in terms of resource utilization, competition occurs.

The logistic growth equation can be extended to gain insight into interspecific competition and coexistence of populations in a given habitat over a period of time. In the logistic growth equation (Eq. 8-4), rN is the growth potential of the population in the absence of interspecific competition. The term $1 - N/K$ indicates the change in growth, usually a decrease, caused by competition among individuals of the population (N) with respect to the carrying capacity (K) of the environment. As population growth approaches the capacity (K) of the environment to support it, N/K approaches 1, and the resulting growth potential ($1 - N/K$) approaches zero.

When two species occur in the same habitat, competition for limiting resources may occur that reduces growth potential and influences the population levels attained. The Volterra equations describe competition between two species.

$$\frac{dN_1}{dt} = r_1 N_1 \left(1 - \frac{N_1}{K_1} - \frac{\alpha_{12} N_2}{K_1} \right)$$

$$\frac{dN_2}{dt} = r_2 N_2 \left(1 - \frac{N_2}{K_2} - \frac{\alpha_{21} N_1}{K_2} \right)$$

(8-5)

The *competition coefficients* (α_{12} and α_{21}) indicate the intensity of the effect that each species has on the population growth of the other. The growth rate of each population consequently becomes dependent upon the number of individuals in the populations of both species.

An increase in population size of one species will result in displacement of the other. Numerous combinations of competitive effects have been found. Often, individuals of one species have a greater effect on individuals of a second species than on members of their own (the first) species, and the competitive effect of individuals of the second species on the first is less than the effects on its own (the second) species ($\alpha_{12} > K_1/K_2$ and $\alpha_{21} < K_2/K_1$). When each species has a greater effect on the other species than it does upon its own growth ($\alpha_{12} > K_1/K_2$ and $\alpha_{21} > K_2/K_1$), the surviving species will depend upon the initial proportion of the two species in the growing mixed populations (cf. detailed treatment in Hutchinson, 1978).

If each species has a greater effect on its own members than it does on individuals of the other species (i.e., $\alpha_{12} < K_1/K_2$ and $\alpha_{21} < K_2/K_1$), coexistence of the two species can occur. For competitive exclusion of one or more species by a dominant competing species to occur, the environmental medium must be homogeneous, and sufficient time in relation to reproduction must be available. Often, particularly in planktonic environments, conditions change too rapidly for competitive effects to drive a species to extinction. As a result, a number of apparently competing species with similar environmental requirements coexist. Such interactions will be discussed repeatedly in subsequent chapters.

Ecosystem Interrelationships

In most cases, energy required for growth and maintenance of organisms enters ecosystems as light and is converted to chemical energy by plant photosynthesis (autotrophy). Biological communities are based on the photosynthetic production of organic matter, produced either within a community (autochthonous production) or derived from an external source (allochthonous production) and transported in dead or decomposing state to a community for utilization (heterotrophy).

TROPHIC STRUCTURE OF FOOD CYCLES

In self-sustaining biological communities, functionally similar organisms can be grouped into a series of operational levels. Each level, which usually consists of many species competing with each other for available resources, forms a *trophic level*. The *trophic structure* of a community refers to the pathways by which energy is transferred and nutrients are cycled through the community trophic levels.

Photosynthetic organisms are primary producers, and they represent the first level (Λ_1) of the trophic structure.* The Λ_1 organisms are eaten by primary consumers or herbivores (Λ_2), which in turn are successively consumed by secondary (Λ_3), tertiary (Λ_4), etc., consumers (carnivores). Since energy is lost in each transfer, more than six trophic levels can rarely be sustained.

Organisms exist in complex *food webs*, in which nutrients and energy of one trophic level are utilized by organisms from several different trophic levels. Although

---

*The notations, which were initially introduced by Hutchinson (unpublished), were later formalized by Lindeman (1942).

greatly oversimplified, Lindeman's (1942) conceptualization is useful for understanding the general patterns of ecosystem structure and energy flow.

The productivity of each trophic level (Λ_n) was considered to be the rate of energy content entering (λ_n) the trophic level from the lower one (Λ_{n-1}):

$$\frac{d\Lambda_n}{dt} = \lambda_n + \lambda'_n \tag{8-6}$$

where λ'_n is the rate of energy leaving the trophic level by metabolic dissipation (respiration with dispersion as heat), returned to the environment as excrement or dead organisms, or being passed on to the next trophic level (Λ_{n+1}) as food.

The amount of energy available for metabolism decreases with each increase in trophic level because energy is lost with each transformation. Since organisms expend a considerable amount of energy for maintenance, only a portion of the energy of one trophic level is available for transfer and use by higher trophic levels. Therefore, at equilibrium (Hutchinson, 1978):

$$\lambda_n = -\lambda'_n \tag{8-7}$$

and

$$\ldots \lambda_{n-1} > \lambda_n > \lambda_{n+1} \ldots \tag{8-8}$$

Total biomass and numbers of individuals at each successively higher trophic level also usually decrease, although some exceptions will be discussed later.

Efficiency of energy or material transfer through a trophic level is simply the ratio, expressed as a percentage, of energy or material leaving a level or system to the quantity of energy or material entering the level or system. In the case just discussed, efficiency would equal the ratio of these productivities, λ_n/λ_{n-1} (critically discussed by Kozlovsky, 1968). The *ecological efficiency* is this percentage of transfer of energy from one trophic level to the next. The rate of conversion of solar energy (Λ_0) into chemical energy by photosynthesis (λ_1) by the first trophic level (Λ_1) of primary producers is small, usually less than 1 per cent, because of the great losses of light energy by nonbiological scattering and absorption and dissipation as heat (Chapter 5). The expenditures of energy for maintenance and losses to the detrital dynamic structure (see Chapter 22) increase with higher trophic levels. As a result, the efficiencies of energy transfer are low from one trophic level to the next (5 to 15 per cent) and usually decrease as Λ_n increases.

DIVERSITY

The number of species in a community (species richness, S) increases with the complexity of food webs and with the extent of niche overlap or species packing (i.e., the number of species-niche hypervolumes that a given habitat can contain), especially in the lower trophic levels of the food chain. Evaluations of species diversity within a community are complicated by dominant species and an abundance of rare species (Pielou, 1977; Sugihara, 1980). Various indices of diversity have been used, the most common of which account for both relative abundance as well as the number of species. The Shannon diversity index (H') is based on information theory (Shannon and Weaver, 1949):

$$H' = - \sum_{i=1}^{s} p_i \log_2 p_i \qquad (8\text{-}9)$$

in units per individual per unit volume or area, where p_i is estimated from n_i/N as the proportion of the total population of individuals (N) belonging to the ith species (n_i) and using logarithms to the base 2. In an ecological context, H' measures the diversity in a many-species community.

Productivity

In any analysis of productivity, it is important that the values obtained be comparable among different ecosystems, among different community components of the same system, or among different responses of the components to environmental dynamics. The conceptual framework behind productivity historically has been characterized by numerous definitions largely based on agriculture and economics. Perhaps more so than in any other field, in aquatic ecology the questions of productivity have been addressed in detail, and an array of often incongruous terms and definitions has been proposed. These definitions, many of which are unnecessarily complex, have caused much confusion that has led to ambiguous thinking and expression.

The basic terminology and definitions of the concepts of productivity have been discussed in detail by MacFadyen (1948, 1950), Elster (1954a), Balogh (1958), Davis (1963), and Westlake (1963, 1965a). Much of the confusion emanates from early concepts that considered productivity as the maximum growth and development of organisms under optimal conditions (Thienemann, 1931), i.e., the potential production of organisms or organic matter per unit volume or surface area per unit time (Dussart, 1966). While the potential of organisms to produce and increase towards infinity may be a useful conceptual framework, in the real world environmental constraints regulate these increases. Optimal conditions for an organism, population, community, or ecosystem can, at best, only be approximated after extensive investigation. Even when the detailed physiological optima for an organism are determined, their applicability to theoretical maximum growth under the dynamics of a nearly infinite number of competitive interactions in natural systems is relegated to the abstract. Such information is much more useful in the interpretation of abiotic and biotic environmental growth controls under in situ natural or disturbed conditions.

It is much more meaningful to define the terms *production* and *productivity* in relation to realized or actual production of organisms, a functional group of organisms, or an ecosystem. Changes in production are related to time and dynamics of environmental regulatory parameters. However, to generate separate terms for productivity under arbitrarily defined "natural conditions" and under perturbed conditions, as has been variously done, has introduced anthropocentricity and unnecessary complexity into the problems being addressed. The following definitions of terms represent common usage, and all are in basic agreement with the long history of theoretical discussions (reviewed in detail by Westlake, 1963; 1965a).

TERMINOLOGY AND DEFINITIONS

Standing crop (also referred to as *standing stock*) is the weight of organic material that can be sampled or harvested at any one time from a given area. *Standing crop* does not necessarily include the whole population, because certain species or inaccessible parts of the sampled species may be omitted by the sampling procedure. The term derives from agricultural usage of the word *crop*, which is the total weight of organic material removed from a given area over a period of time in the course of normal harvesting practice. The crop is a much less variable measurement than the standing crop, which depends on the time of measurement. For example, a wheat crop is the annual maximum standing crop of aboveground foliage, an alfalfa crop is often the annual total of several standing crops of plant tops, and a sugar cane crop may be a standing crop of tops after nearly two years. Underground organs can constitute a large portion of the total plant mass, and there may be great differences in the standing crop according to the time of sampling.

Complications have resulted from the wide usage of standing crop in limnology. When applied to plankton, *standing crop* (or stock) is synonymous with *biomass*. *Biomass* is the weight of all living material in a unit area at a given instant in time. Ambiguities arise when *standing crop*, which refers only to the upper aboveground portions, is applied to aquatic macrophytes; *biomass* includes the entire plant. Biomass evaluations are essential in any analyses of an aquatic plant population or of productivity dynamics. Thus, because of these inherent differences between the terms, the measurement of standing crop and use of the term should be abandoned in limnology and only biomass should be used. If, for some reason, e.g., in herbivory, the foliage above ground is specifically of interest, it should be labeled as such: *aboveground biomass*.

Yield is the crop expressed as a rate.

Production is the weight of new organic material formed over a period of time, including any losses that occurred during that period. *Production* thus refers to the increase in biomass observed over a period adjusted for losses attributable to respiration, excretion, secretion, injury, death, or grazing. Thus primary production is the quantity of new organic matter created by photosynthesis (or chemosynthesis), or the stored energy this material represents. If a photosynthetic organism also uses organic substrates (i.e., is mixotrophic), this additional energy flow is secondary production, even though new organic material may be produced by transformation.

Productivity is the rate of formation of organic matter averaged over some defined period of time, such as a day or year. Natural systems have so many factors causing rapid, frequent, and irregular changes in the instantaneous rates that only average rates can be determined in normal study. It is imperative, as we will stress repeatedly, that the evaluations be done with consideration of the generation times of the organisms under consideration. Weekly evaluations of the productivity of bacteria, for example, are relatively meaningless because much of their population dynamics will be excluded; a weekly sampling regime may be adequate for certain animal components.

Losses of production at any ecosystem level occur as a result of nonpredatory losses (respiratory generation of CO_2 and heat, excretion and secretion of dissolved organic materials, and death or injury) and predatory losses (grazing by herbivores or by carnivores). *Gross productivity* (sometimes termed *real productivity*) refers to the observed change in biomass, plus all predatory and nonpredatory losses, divided by the time interval. Thus, *gross primary productivity* is the rate of production of new organic matter, or fixation of energy, including that which is subsequently used and lost during that time interval. *Net productivity* (occasionally termed *apparent productivity*) is the gross accumulation or production of new organic matter, or stored energy, less losses, divided by the time interval. Sometimes only respiratory losses are subtracted, particularly among evaluations of plant productivity in which losses from processes other than respiration are small. However, all losses should be considered in a true evaluation of net productivity.

Biomass and Productivity

Numerous criteria have been used in evaluations of the biomass and productivity of aquatic organisms, including enumeration, volume, wet (fresh) weight, dry weight, organic weight, content of carbon, pigments, energy as heat on combustion, and ATP, and the rates of exchange of oxygen and carbon dioxide. Often these criteria have been used uncritically, which makes comparability and interpretation difficult (cf. reviews of Lund and Talling, 1957; Strickland, 1960; Westlake, 1965b; Vollenweider, 1969b; Edmondson and Winberg, 1971; Stein, 1973).

Enumeration and Volume. Enumeration of organisms per unit volume is a commonly used method to evaluate microorganisms. While numbers afford great advantages in permitting qualitative differentiation between species and differentiation of organisms from detrital particles, numbers do not give a true evaluation of biomass because organisms differ greatly in size (Table 8–1). The abundance of each species instead can be expressed as the total volume of all individuals (numbers times average cell volume determined from mean dimensions of the cells) per unit volume of water or area of sediment. Among larger organisms, such as zooplankton, benthic fauna, and fish, biomass can be determined using allometric relationships specific to the organisms under study. Volume is a poor measure of freshwater macrophytes because of the large number of internal air spaces; thus the relationship between volume and biomass can be extremely variable.

Weight. Fresh weight, the weight of the organism without any adherent water is, with appropriate precautions, essentially equivalent to wet weight. Owing to the highly variable water content of nearly all organisms, however, wet weight should be avoided. If elaborate precautions are taken, wet-weight analyses can be converted to dry weight on a species basis from a particular environment. Dry weight is variable at temperatures below 105°C; hence, dry weights at 105°C are recommended. If the small percentage of volatile organic constituents lost at 105°C is germane to the analyses, lyophilization (freeze-drying) is the preferred method.

Although dry weight is employed widely in production analyses, organic (ash-free) dry weight, reasonably determined in most situations by the loss in weight after ignition at 550°C, is the preferred measure of biomass for larger organisms. The difficulty of separating bacteria, algae, and other small microorganisms from detrital particulate organic matter restricts the application of ash-free dry-weight measurements to larger organisms.

Cellular Constituents. Perhaps the most satisfactory means of measuring the biomass of photosynthetic organisms is to oxidize the organic plant material back to carbon

TABLE 8-1 Calculated Mean Volumes of Representative Species of Freshwater Plankton Organisms (in μm^3)

CLASSIFICATION	VOLUME (μm^3)	CLASSIFICATION	VOLUME (μm^3)
Cyanophyta		*Bacillariophyceae*	
Anabaena flos-aquae (col.)	80,000	Amphiphora ornata	17,650
Aphanocapsa delicatissima	4	Asterionella formosa (Michigan)	350
Aphanothece clathrata	10	(Europe)	700
Aphanothece nidulans	5	Cyclotella bodanica	10,000
Chroococcus limneticus (col.)	400	Cyclotella comensis	400
Chroococcus turgidus (col.)	1000	Cymatopleura solea	80,000
Coelosphaerium naegelianum (col.)	15,000	Diatoma vulgare	4350
Dactylococcopsis smithii (col.)	1500	Fragilaria crotonensis (1 mm)	200,000
Gloeocapsa rupestris	18	Nitzschia gracilis	240
Gomphosphaeria lacustris (col.)	2000	Melosira granulata (1 mm)	60,000
Merismopedia tenuissima	8	Melosira islandica (1 mm)	80,000
Microcystis flos-aquae	50	Stephanodiscus astraea	2000
Microcystis aeruginosa (col.)	100,000	Stephanodiscus hantzschii	
Oscillatoria limnetica (1 mm)	17,500	var. pusillus	200
Oscillatoria rubescens (1 mm)	30,000	Stephanodiscus niagarae	5000
Synechococcus aeruginosus	350	Synedra acus	250
		Synedra acus angustissima	1000
Chlorophyta		Synedra capitata	950
Ankistrodesmus falcatus	250	Synedra delicatissima	300
Chlamydomonas subcompleta	250	Synedra ulna	50
Botryococcus braunii (col.)	10,000	Tabellaria fenestrata (Michigan)	3000
Chlorella vulgaris	200	(Europe)	4000
Closterium aciculare	4000		
Cosmarium phaseolus	3000		
Cosmarium reniforme	30,000	*Pyrrophyta*	
Gloeococcus shroeteri (col.)	5000	Ceratium hirundinella	4000
Pandorina morum (col.)	4000	Gymnodinium fuscum	10,000
Oocystis solitaria	400	Gymnodinium helveticum	20,000
Scenedesmus quadricauda	1000	Gymnodinium ordinatum	400
Staurastrum paradoxum	20,000	Peridinium cinctum	40,000
Tetraedron minimum	40	Peridinium willei	40,000
Ulothrix zonata (1 mm)	6000		
		Euglenophyta	
Cryptophyta		Trachelomonas hispida	4200
Chroomonas nordstedtii	35	Trachelomonas volvocina	1800
Cryptomonas erosa (Michigan)	1000		
(Europe)	2500		
Cryptomonas ovata	2500		
Rhodomonas lacustris (Michigan)	175		
(Europe)	200		
Rhodomonas minuta	200		
Chrysophyta			
Chromulina pyriformis	50		
Dinobryon borgei	1500		
Dinobryon divergens	800		
Dinobryon sociale	800		
Mallomonas caudata	12,000		
Mallomonas urnaformis	1200		
Rhizochrysis limnetica	1200		
Uroglena americana (col.)	90,000		

After Nauwerck, 1963, Findenegg, unpublished in Vollenweider, et al., 1969, and Wetzel and Allen, unpublished. Numerous values for tropical species are given in Lewis (1977).

dioxide, from which it originated in photosynthetic reduction. The organic carbon content of plants is one of the least variable constituents, and falls nearly without exception within the range of 40 to 60 per cent of ash-free dry weight. The average carbon content among algae is 53 $\pm$ 5 per cent, and among aquatic macrophytes 47 per cent of ash-free dry weight. Exclusion of extraneous organic detritus is not practical at present, and analyses of particulate organic carbon in pelagial samples (cf. Chapter 22) include a very high percentage of nonliving organic material. Algal carbon content usually is estimated from average species content, and is extrapolated to natural populations by volume measurements of the populations, since the carbon-to-volume relationships are allometric (Mullin, et al., 1966).

A number of other cellular constituents have been variously employed to estimate changes in population biomass. Except for carbon, however, measurement of biomass by other elements is so complicated, owing to the extreme variability of cellular composition in response to environmental changes, that their use is limited to specific physiological analyses. While pigment content also varies appreciably with environmental parameters, being able to correct accurately for pigment degradation products in order to measure only the functional pigment content of plants (separate from particulate detritus) permits effective analyses of composite population dynamics among algae.

Several conversion factors have appeared in the literature that allow biomass estimates of one cellular component to be made from another. Even under the most favorable conditions, such factors must be used with utmost caution; under most conditions their use cannot be justified, and is best not attempted.

Productivity. Rates of production have been estimated by numerous techniques, most of which are discussed in ensuing treatments of specific groups of organisms. Among aquatic macrophytes, and to a certain extent attached algae, primary productivity can be estimated from changes in biomass over time (Chapter 18). Estimates of production rates by planktonic microflora from changes in biomass are much more difficult (Vollenweider, 1969b). A temporal set of biomass measurements results in minimal estimates or underestimations of net productivity because of losses attributable to grazing, current transport, sedimentation, death, and decomposition. Rarely are these parameters sufficiently evaluated to permit analyses of production rates from changes in biomass; exceptions include the investigations by Lund (1949, 1950, 1954), Grim (1952), and Crumpton and Wetzel (1982).

The primary productivity of phytoplankton has received an extraordinary amount of attention in limnology, and has been measured in great detail in a number of aquatic systems. The reasons for the abundance of information on rates of primary production are manifold.

First, phytoplanktonic productivity represents a major synthesis of organic matter of aquatic systems and can be summarized in the universal equation:

$$6 \, CO_2 + 12 \, H_2O \xrightarrow[\text{pigment receptor}]{\text{light}} C_6H_{12}O_6 + 6 \, H_2O + 6 \, O_2.$$

This equation is a gross oversimplification of the complex Calvin-Benson metabolic pathway of photosynthesis, which is basically a redox reaction:

$$CO_2 + 2 \, AH_2 \xrightarrow{h \cdot v} (HCOH) + 2 \, A + H_2O,$$

in which AH_2 represents a hydrogen donor. Normally, the hydrogen donor is water but, as discussed later, types of donors can include an array of reduced sulfur (e.g., H_2S) or organic carbon compounds among autotrophic and certain other bacteria. In large lakes, phytoplanktonic productivity often represents the dominant input of new organic matter and potential energy that drives the system.

Second, as is often the case in ecology, the development of reasonably accurate techniques for the measurement of in situ rates of primary production has led to their wide application, unfortunately often without any sound rationale as to why the measurements were being made. Only now, after many years of experimentation, is there an adequate understanding of the methodology. The appeal of these techniques lies in being able to directly measure rates of metabolism in situ; most other evaluations of productivity are forced to use indirect methods.

In practice, changes in oxygen production or rates of carbon uptake are usually measured on isolated samples of the natural communities that are incubated for brief periods at the points of collection or under simulated natural conditions aboard ship. Certain environmental factors, e.g., temperature and light, are simulated closely; other factors such as turbulence, nutrient replenishment, and grazing, can differ to varying degrees in the isolated samples. Alternatively, productivity estimates can be made from measurements of changes in oxygen, pH, carbon, or conductivity over short intervals in the natural environment on nonisolated communities. This approach, in addition to possessing numerous complications, yields estimates of community metabolism.

The light- and dark-bottle techniques for estimating primary productivity have received wide application. In the oxygen method, samples of phytoplankton populations are incubated at the depths from which they were collected in clear and opaqued bottles. The initial concentration of dissolved oxygen (c_1) can be expected to decline to a lower value (c_2) by respiration in the darkened bottles, and increase to a higher concentration (c_3) in the clear bottles, according to the difference between photosynthetic production and respiratory consumption. The difference $(c_1 - c_2)$ represents the respiratory activity per unit volume over the time interval of incubation. The difference $(c_3 - c_1)$ is equal to the net photosynthetic activity, and the sum $(c_3 - c_1) + (c_1 - c_2) = (c_3 - c_2)$ corresponds to the gross photosynthetic activity. Numerous assumptions are made in the method that can alter the photosynthetic measurements appreciably, e.g., respiration rates are not necessarily the same in light and dark, since photorespiration clearly occurs in algae, other processes such as photo-oxidative consumption utilize oxygen separately from apparent respiratory uptake, nonphotolysis of water by bacterial photosynthesis, etc. Under many circumstances, these errors are small, but the technique can be considered only as a reasonable estimate. It is probable, however, that these and analytical errors in determination of oxygen concentrations are appreciably less than sampling errors of heterogeneous plankton populations in the lake for the analyses. Large portions of recent methodological works are devoted to detailed discussions of this and the following ^{14}C techniques, and should be read critically prior to using the methods (cf. especially Strickland, 1960; Vinberg, 1960; Vollenweider, et al., 1969b; Strickland and Parsons, 1972; Wetzel and Likens, 1979; Westlake, et al., 1980).

The incorporation of ^{14}C tracer into the organic matter of phytoplankton during photosynthesis has been used as an extremely sensitive measure of the rate of primary production. If the content of total CO_2 of the experimental water is known, and if a known amount of $^{14}CO_2$ is added to the water, then determination of the content of labeled carbon in the phytoplankton after incubation permits calculation of the total amount of carbon assimilated. Numerous methodological and physiological problems confront application of the ^{14}C light and dark technique; however, with care, most technical problems can be overcome and errors evaluated, e.g., respiratory losses of CO_2, and secretion rates of soluble organic products of photosynthesis. Rates of respiration are difficult to evaluate directly using this technique. In many situations, the ^{14}C method yields a measure close to net photosynthetic rates. Comparison of the oxygen and ^{14}C methods under optimal conditions shows close agreement, with a photosynthetic quotient ($PQ = \Delta O_2 / -\Delta CO_2$, by volume) somewhat greater than unity (Fogg, 1963). The photosynthetic quotient varies from near unity when carbohydrates are the principal photosynthetic products, to as high as 3.0 during synthesis of lipids.

Assuming a photosynthetic quotient of 1.2 and a statistical significance level of $p < 0.05$, the smallest amount of photosynthesis that the oxygen light and dark technique can measure under ideal conditions is about 20 mg C m^{-3} hour^{-1}, with a potential error range $\pm$ 20 mg C m^{-3} (Strickland, 1960). The limit of sensitivity of the ^{14}C method is some 50 to 100 times greater, on the order of 0.1 to 1 mg C m^{-3} hour^{-1}.

Secondary productivity in fresh waters by invertebrates and vertebrates is much more difficult to estimate accurately than is primary productivity. Trophic relationships are complex and often change during the life cycle of a species or from one ecosystem to another. The size of animals varies greatly from immature to adult stages and the diet can also vary greatly throughout the life history. Because most animals are mobile, they actively distribute themselves in response to environmental stimuli, which results in heterogeneous distributions and makes accurate sampling more difficult.

Productivity measurements of animals are based on estimation of numbers, biomass, and growth rates. Sampling accuracy of changes in population size decreases from zooplankton to benthic organisms to fish. In contrast, growth rates are easier to evaluate in many temperate fishes and in long-lived invertebrates than they are in small, rapidly reproducing zooplankton and other invertebrates. The basic methods of estimating productivity of each of these main animal groups will be discussed in subsequent chapters.

When evaluating the population dynamics and productivity of any animal, estimating food incorporation or ingestion and the utilization of ingested food is imperative. Assimilation means the absorption of food from the digestive system, and the efficiency of assimilation refers to the percentage fraction of ingested food that is digested and absorbed into the body. Assimilation efficiency is not constant, and varies greatly with the food quality and rates of food ingestion.

Measurements of assimilation are based on the simple relation:

$$\text{Assimilation} = \text{ingestion} - \text{egestion, or}$$
$$\text{Assimilation} = \text{growth} + \text{respiration.}$$

While these relationships are simple, their accurate measurement among natural populations of zooplankton and larger organisms is extremely problematic. Methods employed are discussed critically in Edmondson and Winberg (1971), and Winberg (1971). Although ingestion can be measured in situ with reasonable sophistication, accurate measurements of egestion rates are often beyond the capacity of contemporary

TABLE 8-2 Mean and Range of P/B Ratios among Trophic Groups
of Freshwater Ecosystems

	MEAN	RANGE
Bacteria	141.0	73–237
Phytoplankton	113.0	9–359
Herbivorous Zooplankton	15.9	0.5–44.0
Carnivorous Zooplankton	11.6	1.5–30.4
Herbivorous Benthic Invertebrates	3.7	0.6–12.8
Carnivorous Benthic Invertebrates	4.8	1.0–25.0

After data of Saunders, et al., 1980 and Brylinsky, 1980.

technology. Growth is used in a general sense in that it includes the production of eggs or young, much in the same way that respiration includes excretion and other losses as well as normal biochemical respiration.

Assimilation and its efficiency can be approximated using the first of the preceding equations by measuring changes in radioactivity of the animal after allowing it to feed first on radioactive food, and then on nonradioactive food (cf. Sorokin, 1968; Saunders, 1969; Edmondson and Winberg, 1971). Growth and respiration can be estimated by the second equation by measuring the feeding rate (e.g., in calories per time), the growth rate, and respiration. Respiration measured as oxygen consumption can be converted to energy units if the respiratory quotient (CO_2 produced/O_2 consumed) is known.

PRODUCTIVITY/BIOMASS (P/B) RATIOS

The productivity (P) and biomass (B) of an organism or group of organisms are often related in P/B ratios. The P/B ratios can be used to estimate the turnover rate of organisms, and for an entire trophic level give a general index of the rate of energy flow relative to the biomass. Determination of P/B values is useful in making comparisons among different trophic levels, as well as among similar trophic levels under different environmental conditions.

Average annual P/B ratios generally decrease with increasing trophic level (Table 8-2), and range over two orders of magnitude (Brylinsky, 1980). As would be expected, smaller organisms tend to have greater P/B values, and P/B ratios decrease with increasing organism size. Production per unit biomass generally increases with decreasing latitude of the aquatic ecosystem since the active growing season is longer, allowing increased numbers of generations per growing season at lower latitudes. Organisms of oligotrophic lakes tend to have lower P/B values than those in eutrophic lakes (Saunders, et al., 1980; Brylinsky, 1980).

SUMMARY

1. Aquatic ecosystems consist of entire drainage basins: the nutrient and organic-matter content of drainage from the catchment area is modified in each of the terrestrial, stream, and wetland-littoral components, as well as

in the lake or reservoir per se. Productivity is generally low to intermediate in the terrestrial components, highest in the wetland interface region between land and water, and lowest in the open-water portion of the lake.

2. The lake is separated into the open-water *pelagial zone* and the *littoral zone,* the latter consisting of the bottom of the lake basin colonized by macrovegetation. The sediments free of vegetation that lie below the pelagial zone are referred to as the *profundal zone* (Fig. 8–2). The *littoriprofundal zone* is the transitional area of the sediments occupied by scattered benthic algae.

3. Various terms have been applied to groups of organisms living within these zones. *Plankton* are small organisms with no or limited powers of locomotion that are suspended in the water; they are subject to dispersal by turbulence and other water movements. Both the small plant plankton, the *phytoplankton,* and animal plankton, the *zooplankton,* are usually denser than water, and sink by gravity to lower depths. Organisms with relatively good swimming powers of locomotion are termed *nekton.*

   a. Bacteria and algae growing attached to substrata are collectively called *periphyton* and have been further named in relation to the type of substrata (sediments, rock, plant, animal, sand) upon which they grow (Fig. 8–3).

   b. *Benthos* refers to nonplanktonic animals associated with freshwater substrata at the sediment–water interface.

   c. Specialized organisms adapted to the air–water interface are called *pleuston.* The pleuston is dominated by microflora collectively termed *neuston.*

4. Living organisms constitute only a very small portion of the total organic matter of ecosystems. Most organic matter is nonliving and is collectively called *detritus.*

   a. Detritus consists of all dead particulate and dissolved organic matter. Dissolved organic matter is about 10 times more abundant than particulate organic matter.

   b. Much of the newly synthesized organic matter of photosynthesis is not consumed by animals, but instead enters the detrital pool and is decomposed.

5. A *population* is a defined assemblage of individuals of one species. The growth rate of a population is characterized by both its intrinsic capacity to grow and reproduce, and the limitations imposed upon that growth potential by the environment and competition of others in the population.

   a. The population growth of many small organisms can be accurately described with the logistic growth equation. As a growing population approaches its environmental carrying capacity, reproduction usually declines because resources become limiting. Under idealized conditions, when a population attains its carrying capacity, reproductive rates decline to zero.

   b. Survivorship curves permit an analysis of the survival and reproductive performance of a population at different time intervals.

6. A group of interacting populations forms a *community.*

7. The success of organisms is governed by individual physiologic and behavioral limits and by physical and biotic environmental parameters acting upon them. Each of the regulating parameters forms a dimension; the *niche* refers to an *n*-dimensional resource space, or hypervolume. Each species occupies only a portion of its niche hypervolume.

   a. Species niches can overlap, and overlap in resource utilization can lead to *competition.* The logistic growth equation can be extended to include intra- and interspecific competition and predation, allowing more realistic mathematical descriptions of population growth.

   b. In multispecies systems, better differential growth of one species will result in exclusion or displacement of another. Coexistence of competing species can occur where intraspecific competitive effects are greater than interspecific effects, through selective predation, or in cases where the environment changes faster than the time required for competitive exclusion.

8. In biological communities, functionally similar organisms can be grouped into *trophic levels* based on similarities in patterns of food production and consumption. Energy is transferred and nutrients are cycled within an overall ecosystem trophic structure.

   a. The productivity of each trophic level (Λ_n) is the rate at which energy enters the trophic level (λ_n) from the next lower level (Λ_{n-1}).

   b. Since organisms expend considerable energy for maintenance, and because death of an organism routes energy and nutrients into the detrital pool, only a portion of the energy of one trophic level is available for transfer and use by higher trophic levels. Available energy decreases progressively at higher trophic levels, so that rarely can more than five or six trophic levels be supported.

   c. The efficiency of energy transfer from one level to the next is low (5 to 15 per cent), and often decreases as Λ_n increases.

9. *Production* is the amount of new organic biomass formed over a period of time, and includes any losses from respiration, excretion, secretion, injury, death, and grazing. *Productivity* usually refers to an average rate of production over a distinct period of time (e.g., day, year).

   a. Estimates of primary productivity by photosynthesis can be obtained directly by following changes in oxygen production or rates of inorganic carbon assimilation.

   b. Secondary productivity by invertebrates and vertebrates is difficult to estimate because of complex trophic relationships, changes in diet during the organism's life history, and animal mobility. Productivity measurements of animals are based on changes in numbers, biomass, and growth rates.

   c. The ratio of productivity to biomass (P/B ratios) estimates the turnover rates of energy flow relative to biomass; P/B values generally decrease with increasing trophic level.

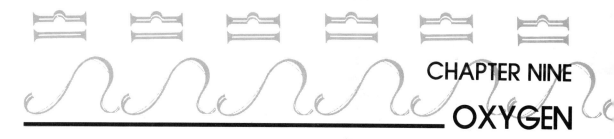

CHAPTER NINE

OXYGEN

THE OXYGEN CONTENT OF FRESH WATERS

Oxygen is the most fundamental parameter of lakes, aside from water itself. Dissolved oxygen is obviously essential to the metabolism of all aerobic aquatic organisms. Hence, the solubility and especially the dynamics of oxygen distribution in lakes are basic to the understanding of the distribution, behavior, and growth of aquatic organisms.

The mechanisms that control dissolved oxygen concentrations are important in determining oxygen availability to the lake biota. The dynamics of oxygen metabolism are, however, only one facet of the oxygen content of fresh waters. The rates of supply of dissolved oxygen from the atmosphere and from photosynthetic inputs, and hydromechanical distribution of oxygen are counterbalanced by consumptive metabolism. The rate of oxygen utilization in relation to synthesis permits an approximate evaluation of the metabolism of the lake as a whole.

The resulting distribution of oxygen strongly affects the solubility of many inorganic nutrients. The changes in nutrient availability are governed by seasonal shifts from aerobic to anaerobic environments in regions of lakes. The changes in distribution of nutrients result in rapid growth of many organisms capable of taking advantage of this altered nutrient availability. Population responses may be temporary and transient. If, however, long-term changes in oxygen-regulated nutrient availability are sustained, the productivity of the entire lake can be radically altered. Although oxygen content is discussed here largely as a separate entity, its integration with all facets of biotic metabolism will become more apparent in subsequent discussions.

Solubility of Oxygen in Fresh Water

Air contains about 20.95 per cent oxygen by volume, and the remainder is nitrogen, except for a very small percentage of other gases. Because oxygen is more soluble in water than nitrogen, the amount of oxygen dissolved in water equilibrated with air is greater than nitrogen.

Solubility of oxygen is affected nonlinearly by temperature, and increases considerably in cold water. The general relationship of oxygen solubility as a function of temperature is given in Table 9-1, although these values are subject to minor modifications

TABLE 9-1 Solubility of Oxygen in Pure Water in Relation to Temperature in Equilibrium with Air at Standard Pressure Saturated with Water Vapor

EQUILIBRIUM TEMPERATURE (t) (°C)	OXYGEN (C*) (mg l^{-1})	OXYGEN (C†) (μgat kg^{-1})
0	14.621	913.9
1	14.216	888.5
2	13.829	864.4
3	13.460	841.3
4	13.107	819.2
5	12.770	798.2
6	12.447	778.0
7	12.139	758.7
8	11.843	740.2
9	11.559	722.5
10	11.288	705.5
11	11.027	689.2
12	10.777	673.6
13	10.537	658.6
14	10.306	644.1
15	10.084	630.3
16	9.870	616.9
17	9.665	604.1
18	9.467	591.7
19	9.276	579.8
20	9.092	568.3
21	8.915	557.2
22	8.743	546.5
23	8.578	536.1
24	8.418	526.1
25	8.263	516.4
26	8.113	507.1
27	7.968	498.0
28	7.827	489.2
29	7.691	480.7
30	7.558	472.4
31	7.430	464.4
32	7.305	456.6
33	7.183	449.0
34	7.065	441.6
35	6.949	434.4
36	6.837	427.3
37	6.727	420.5
38	6.620	413.7
39	6.515	407.2
40	6.412	400.8

From Mortimer (1981) after data of Benson and Krause (1980). Unit standard concentrations (C* by volume in mg l^{-1} or mg dm^{-3} and C† by weight in μgat kg^{-1}) of oxygen (1 mole = 31.9988 g) in pure water in equilibrium with air at standard pressure (1 atm = 760 mm Hg = 101.325 Pa) saturated with water vapor and with the dry air mole fraction for oxygen taken as 0.20946. To convert mg to cm^3 (ideal gas S.T.P.) or to μgat, multiply mg by 0.70046 or 62.502, respectively. To convert μgat to μmol, divide by 2. To convert C* to C†, at temperature (t)° and with the accuracy of this table, multiply C* by the density of pure water at t°.

as analytical assays improve (extensively discussed in Ohle, 1952; Mortimer, 1956, 1975, 1981; Hutchinson, 1957; Weiss, 1970; and Benson and Krause, 1980). More detailed data are given in the references cited.

The solubility of gases in water is affected by pressure as well as by temperature. Therefore, equilibrium of oxygen of the atmosphere with the oxygen concentration in the water depends on the atmospheric partial pressure of oxygen. Hence the elevation of the lake also influences the concentrations of oxygen. Usually, saturation is considered in relation to the pressure at the surface of the lake. The percentage saturation of oxygen in water has been extensively treated by Ricker (1934) and Mortimer (1956, 1975, 1981), both of whom have prepared nomograms to aid in such computations. The improved nomogram of Mortimer is included in this work (Appendix), although in routine analyses it is strongly recommended that one use the enlarged version, which is printed on material that does not change with humidity.

The amount of a gas that will remain in a dissolved phase is influenced by the atmospheric pressure to which the lake is exposed, by meteorological conditions, and also by hydrostatic pressure exerted by the stratum of water overlying a particular depth. The amount of gas that can be held in water by the combined atmospheric and hydrostatic pressures at a particular depth is called the absolute saturation (Ricker, 1934). The actual pressure, P_z, in atmospheres at a given depth is equal to that at the surface, P_o, plus 0.0967 times the depth z in meters, or

$$P_z = P_o + 0.0967_z.$$

The importance of this relationship lies in the amount of pressure required to prevent the formation of bubbles, which rise and escape to the surface. Such bubbles are a composite of many gases, the most common of which are oxygen derived from photosynthesis, and methane resulting from anaerobic decomposition under certain conditions. The degree of supersaturation of oxygen necessary for bubble growth increases with depth. In still water, bubble growth may occur as a result of oxygen tension at depths greater than about one meter (Ramsey, 1962a). At greater depth, an unusually high degree of supersaturation is required, as well as an absence of turbulence or internal waves that may induce a depth displacement and change in pressure. Very small bubbles can be stabilized by an adsorbed organic film. Such bubbles, on rising to the surface, either rupture or accumulate on the surface as foam (Ramsey, 1962b). Depending upon the turbulence patterns, oxygen produced at depths greater than 1 to 4 meters can remain dissolved because of the hydrostatic pressure. Oxygen can accumulate to several hundred per cent supersaturated relative to the pressure at the surface of the lake, and still be below its absolute saturation at depth.

Salinity also reduces the solubility of oxygen in water, and must be considered in analyses of oxygen in inland saline and brackish waters. Oxygen solubility declines exponentially with increases in salt content, and is reduced by about 20 per cent in normal sea water as compared to the amount in fresh water. Revised tables and a nomogram for oxygen saturation in saline water are given in Green and Carritt (1967) and Weiss (1970).

Although the methodology involved in the physical, chemical, and biological aspects of aquatic ecology is very complex and a full discussion would involve a work several times the length of this book, the importance of dissolved oxygen warrants a few comments. In spite of recent advances in electrode technology (Ohle, 1953; Golterman, 1969), the mainstay of oxygen analysis is the Winkler method, based on chemical fixation of the oxygen and colorimetric titration against reagents of known reaction with changing concentration. Experimenters should be aware of the numerous compounds that interfere with precise measurements, such as iron, nitrates, and organic matter, and become significant, particularly in polluted waters. Modifications of the basic method are discussed in great detail in Welch (1948), *Standard Methods* (APHA, 1976), Golterman (1969), and Wetzel and Likens (1979). In inland saline waters, other methods must be used (Walker, et al., 1970; Ellis and Kanamori, 1973).

Distribution of Dissolved Oxygen in Lakes

The diffusion of gases into water is a very slow process. To establish an equilibrium between the atmospheric and dissolved oxygen within a reasonable period, the water must circulate, as it does during periods of turnover or in the epilimnion of stratified lakes. If the dissolved oxygen concentrations at depth are near saturation, equilibrium at prevailing temperatures and altitudinal pressure is established relatively quickly, usually in a matter of a few days. Very deep lakes, however, require longer periods for complete equilibrium saturation, and saturation may or may not be achieved before thermal stratification effectively terminates circulation for a seasonal interval.

In an idealized lake, the oxygen concentration at spring circulation is at or near 100 per cent saturation, which would be between 12 to 13 mg O_2 l^{-1} if occurring at about 4°C and at an altitude near sea level (Fig. 9–1). This ideal concentration at 100 per cent saturation is based on physical control of diffusion, mixing, and saturation. Deviations are commonly found, most frequently in the form of slight supersaturation resulting

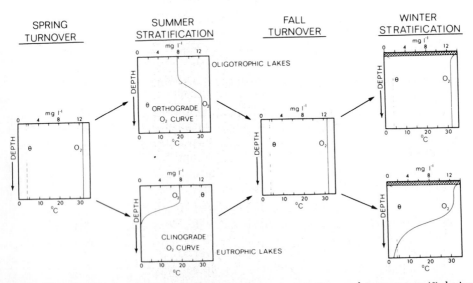

Figure 9–1 Idealized vertical distribution of oxygen concentrations and temperature (θ) during the four main seasonal phases of an oligotrophic and an eutrophic dimictic lake.

from photosynthetic activity in excess of losses to the atmosphere. Alternately, bio-chemical oxidations can result in slight undersaturation during conditions where consumption slightly exceeds the circulation equilibrium mechanisms.

ORTHOGRADE OXYGEN PROFILE

Where the lake is *oligotrophic* (low in nutrient inputs with low organic production), the oxygen concentration with depth is regulated largely by physical means as summer stratification occurs. The oxygen concentration in the circulating epilimnion decreases as the temperature increases (Fig. 9–1). With decreasing temperature in the metalimnion and hypolimnion, the oxygen concentrations increase, that is, they have remained in the hypolimnion essentially at saturation from the period of spring turnover just before summer stratification began. This oxygen profile has been termed (Åberg and Rodhe, 1942) *orthograde*. The important characteristic to note is that the percentage of saturation is more or less 100 per cent with increasing depth.

CLINOGRADE OXYGEN PROFILE

An idealized orthograde curve is found only in a few extremely unproductive lakes or in moderately oligotrophic lakes during the very early stages of summer stratification. Oxidative processes occur constantly in the hypolimnion, and their intensity is proportional to the amount of organic matter reaching the hypolimnion from the upper zones of the lake. As a result, the oxygen concentration of the hypolimnion becomes progressively more reduced and undersaturated. The oxygen content of the hypolimnion of *eutrophic* (high in nutrients, with high organic production) lakes is depleted rapidly by oxidative processes. The resulting curve, in which the hypolimnion is anaerobic, is termed *clinograde* (Fig. 9–1). The hypolimnetic oxygen content of highly eutrophic lakes is often depleted after only a few weeks of summer stratification. The hypolimnion remains anaerobic throughout this stratification period.

The loss of oxygen from the hypolimnion results primarily from the biological oxidation of organic matter, both in the water and especially at the sediment–water interface, where bacterial decomposition is greater. Although plant and animal respiration can consume large, often catastrophic, amounts of dissolved oxygen, major consumption from the lake is associated with bacterial decomposition of sedimenting organic matter. Oxygen consumption in the free water by bacterial respiration is intensive at all depths, but in the hypolimnion is rarely offset by renewal mechanisms of circulation and photosynthesis that occur in the epilimnion and metalimnion. Oxygen consumption is most intense at the sediment–water interface, where accumulating organic matter and bacterial metabolism are greatest. The interface region very rapidly becomes anaerobic during summer stratification, and the hypolimnetic depletion is usually observed first at the lowermost stratum of the hypolimnion. Diffusion of oxygen into this depleted zone from overlying layers occurs slowly. Horizontal distribution of the oxygen profile in the hypolimnion can be modified by vertical turbulence, horizontal translocations, and density currents that move along the basin sediments.

In addition to oxygen consumption by animals, plants, and especially by bacterial respiration in the open water, purely chemical oxidation of dissolved organic matter occurs. Lakes highly stained with humic organic compounds are frequently undersaturated even in epilimnetic strata. Although the mechanisms are not entirely clear, it is apparent that at least some of the oxygen uptake is the result of purely chemical oxidation or photochemical oxidation induced by ultraviolet light (Gjessing and Gjerdahl, 1970). This chemical oxidation most likely is masked in very productive lakes by the intense bacterial biochemical demands.

The relative importance of different mechanisms of hypolimnetic oxygen depletion varies from lake to lake. In large, deep lakes, bacterial respiration of organic matter of phytoplanktonic origin may dominate, and benthic decomposition then will play a minor role. In shallow lakes with relatively large inputs of organic matter from terrestrial, stream, wetland, and littoral sources, benthic decomposition dominates. And in highly stained bog lakes that receive large inputs of dissolved humic compounds, chemical oxidation may assume greater significance. With the shift from an aerobic to anaerobic hypolimnion, a large, often major volume of the lake is excluded from habitation by most animals and many plants. Another major change is the shift from aerobic to anaerobic bacterial metabolism, which markedly reduces overall efficiency of decomposition of organic matter.

Fall overturn begins with the complete loss of summer stratification. During the terminal stages of stratification, with progressive deepening of the epilimnion, circulation of approximately oxygen-saturated water extends deeper into the hypolimnion. When circulation is complete, oxygen concentration remains at saturation in accordance with solubility at existing temperatures (Fig. 9–1).

With the advent of ice formation, exchange of oxygen with the atmosphere ceases for all practical purposes. Based on solubility relationships, one would expect an oxygen concentration profile essentially at saturation in relation to temperature at any given depth. Such would be the case in an ultra-oligotrophic lake where biotic influences are minor. These profiles are very rarely found in dimictic lakes. Much more frequently observed is a reduction in oxygen concentration with depth, which becomes particularly acute near the sediment.

In eutrophic lakes, photosynthetic production of organic matter, although less than in other seasons, continues throughout the winter, and is often vigorous at the later stages of winter ice-cover. Light penetration is variable with changing conditions of ice- and snow-cover, but the photic zone generally is confined to the upper layers. Respiratory utilization and chemical oxidations increase with depth as during summer stratification, although at a lower rate because of the lower temperatures. During winter, the water of the main water mass under ice is commonly colder than 4°C. Water in the littoral areas can be heated slightly through the ice by solar radiation. This warmer, slightly denser water will sink and flow in profile-bound density currents along the sediments to the deeper portions of the basin. Generally, water movement is sufficiently slow so that en route, the oxygen of this water is reduced or depleted as it passes over the sediment–water interface. At the bottom of depressions it displaces water upward and results in a conspicuous depletion at the sediment–water interface, accentuating the decompositional utilization already occurring there.

The stages of the annual cycle discussed above can be illustrated by isopleths of oxygen concentration and percentage saturation for a typical dimictic lake of moderate productivity (Fig. 9–2). Noteworthy are the high concentrations (near 100 per cent sat-

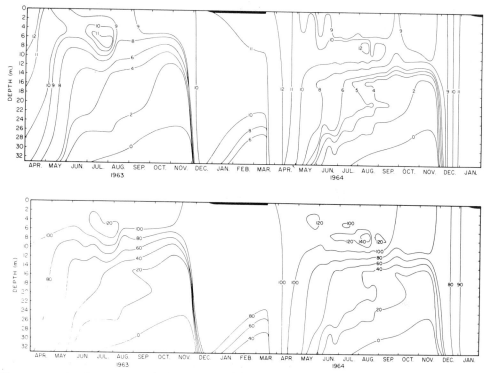

Figure 9-2 Depth–time diagram of isopleths of dissolved oxygen concentrations in mg l^{-1} (*upper*) and percentage oxygen saturation (*lower*), Crooked Lake, Noble-Whitley counties, northeastern Indiana. Ice-cover drawn to scale. (From Wetzel, unpublished data.)

uration) during the colder periods of spring and fall circulation. Epilimnetic concentrations of oxygen are reduced, but, because of reduced solubility in the warmer water, are still at saturation. The high concentrations in the metalimnion are the result of photosynthesis, and will be discussed in detail below. The slow, progressive reduction and final depletion of oxygen in the lower hypolimnion is seen in mid- and late summer. This same reduction, but not depletion, is seen under ice cover at the greater depths in Crooked Lake. More extreme changes are seen in the oxygen concentrations of smaller eutrophic Little Crooked Lake (Fig. 9-3), in which the changes are similar to those of Crooked Lake but greater and more accelerated.

Variations in Oxygen Distributions

METALIMNETIC OXYGEN MAXIMA

Variations in the vertical and horizontal distributions of dissolved oxygen from the general conditions just discussed are often observed. The most common variation is an increase in oxygen in the metalimnion during stratification (Figs. 9-2, 9-3, 9-4, 9-6); the metalimnetic maximum is referred to as a *positive heterograde curve* (Åberg and

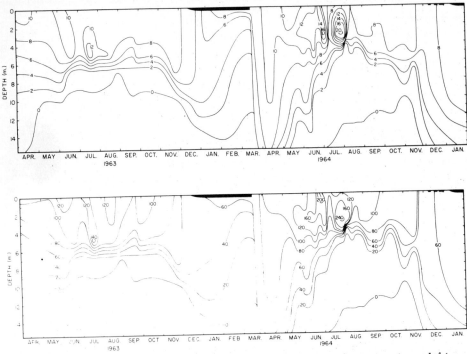

Figure 9-3 Depth-time diagram of isopleths of dissolved oxygen concentrations in mg l^{-1} (upper) and percentage oxygen saturation (lower), Little Crooked Lake, Whitley County, northeastern Indiana. Ice-cover drawn to scale. (From Wetzel, unpublished data.) Isopleth lines appear to merge where gradients are extremely steep; in reality, the lines are separated by small depth-time intervals.

Rodhe, 1942). As the solubility of the epilimnion decreases with increasing summer temperatures, and oxygen consumption in the hypolimnion results in the typical clinograde reduction with depth, the result is an absolute oxygen maximum in the metalimnion which may be at or above saturation.

Metalimnetic oxygen maxima often are extremely pronounced, with supersaturated values above 200 per cent (Fig. 9-3). Concentrations of nearly 36 mg O_2 l^{-1} (400 per cent saturation) have been recorded (Birge and Juday, 1911). It is clear that the maxima are nearly always the result of oxygen produced by algal populations that develop more rapidly than they sink out into lower water strata (Fig. 9-4A). The algae are commonly stenothermal, adapted to growing well at low temperatures and low light intensities, but have access to nutrient concentrations that are usually higher in the lower metalimnion than in the epilimnion. Blue-green algae, especially *Oscillatoria*, are often major contributors to this phenomenon (Eberly, 1959, 1963, 1964; Wetzel, 1966a, 1966b). The depth at which metalimnetic oxygen maxima occur is correlated directly with the transparency of the water (Thienemann, 1928; Yoshimura, 1935). In a majority of lakes metalimnetic maxima are found between 3 to 10 m, but in very clear lakes at depths as great as 50 m.

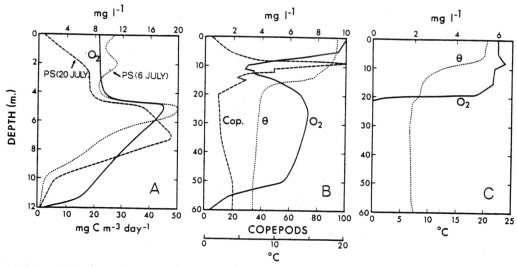

Figure 9-4 *A*, Metalimnetic oxygen maximum, showing a positive heterograde curve, in relation to rates of phytoplanktonic photosynthesis (*PS*), Lawrence Lake, Michigan, 20 July 1971. (From Wetzel, et al., 1972.) *B*, Metalimnetic oxygen minimum, showing a negative heterograde curve, in relation to abundance of copepod microcrustaceans (*Cop.*) and temperature (*θ*), Lake Washington, Washington, 18 August 1958. (Drawn from data of Shapiro, 1960.) *C*, Oxygen concentrations in permanently meromictic Fayetteville Green Lake, New York, 3 September 1935. (Data of Eggleton, 1956.)

Extreme metalimnetic oxygen maxima are observed much more frequently in lakes that are strongly stratified. Such lakes commonly have high relative depths,* i.e., their surface area is small relative to their maximum depth, and they are protected from wind action by surrounding topography and vegetation. The average peak in oxygen concentration and relative depth of over 50 lakes that exhibited extreme metalimnetic maxima was 17.2 mg l^{-1} and 4.30 per cent, respectively (Eberly, 1964). Most lakes have a relative depth of less than 2 per cent.

Metalimnetic oxygen maxima occur in many lakes. When characteristics of morphometry and biota are suitable for their development, the maxima occur consistently from year to year at approximately the same time and depth. For example, in Lawrence Lake, Michigan, the metalimnetic maximum has occurred between 4 and 6 m, July through August for fifteen consecutive years of observation. The maxima are superimposed exactly over maxima of photosynthesis by phytoplankton and summer growth maximum of dense beds of rooted aquatic plants on the steep slopes of the littoral zone. It is likely that water in the littoral zone, enriched with oxygen, dissipates into the metalimnion layer of equal density and contributes to the observed oxygen peak. Ruttner (1963) and Dubay and Simmons (1979) reported a similar situation in small clear

---

*Relative depth or z_r is an expression of the maximum depth (z_m) as a percentage of the mean diameter, and is determined by $z_r = \dfrac{50\, z_m\, \sqrt{\pi}}{\sqrt{A_o}}$ where A_o is the surface area.

lakes with precipitous slopes, in which the metalimnetic oxygen maximum coincided seasonally with zones of maximum production of submersed macrophytes.

METALIMNETIC OXYGEN MINIMA

The converse condition, a *metalimnetic oxygen minimum* exhibiting a *negative heterograde curve* (Fig. 9-4B), is much less frequently observed. Numerous causal mechanisms have been associated with this metalimnetic reduction, and most or all are likely to be in effect simultaneously in situations when such minima are observed. To single out one mechanism as operational and exclude others (as, for example, Czeczuga did in 1959) can be misleading.

Oxidizable material produced in the epilimnion or brought into the lake from outside the basin continuously sinks. Its sinking rate will be slowed down when it encounters the denser metalimnetic water, and here there will be more time for it to decompose (Birge and Juday, 1911; Thienemann, 1928). Moreover, decomposition rates are usually higher in the metalimnion, where temperatures are greater than in the colder hypolimnion. As a result, more readily oxidizable organic matter is decomposed at this level, with concomitant consumption of oxygen by bacterial respiration; more resistant organic matter settles slowly into the hypolimnion (Kuznetsov and Karsinken, 1931; Vinberg, 1934; Åberg and Rodhe, 1942). In a review of the subject, Czeczuga (1959) presented some evidence from a lake in which, conversely, decomposition rates were greater in the epilimnion. However, renewal of epilimnetic oxygen and time of residence of sedimenting organic matter in the metalimnion were not clearly delineated. It is clear that a balance between the transparency of the trophogenic zone and the depth of the metalimnion is important in relation to the depth at which photosynthetic maxima occur, and in relation to whether oxygen inputs are sufficient to offset decompositional consumption (Ruttner, 1933). It is easy to conceive of a situation in which a pronounced metalimnetic maximum could shift to a minimum within a summer season (Fig. 9-5).

In certain situations, concentrations of massive numbers of zooplankton microcrustacea in the metalimnion can contribute to a severe reduction of oxygen. Most

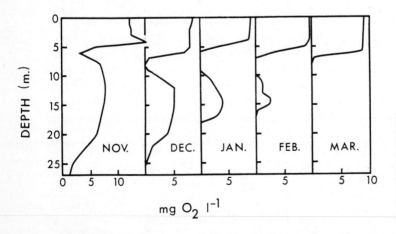

Figure 9-5 Depth profiles of dissolved oxygen concentrations during the summer (Southern Hemisphere) in hypereutrophic Lake Johnson, New Zealand, 1970-71. Similar distributions were observed in the summers of 1969-70, 1971-72, and 1974. (Modified from Mitchell and Burns, 1979.)

likely, such respiratory consumption was the major cause of the metalimnetic minima observed in Zürichsee, Switzerland (Minder, 1923), and the same condition surely prevails in Lake Washington (Fig. 9-4B), where nonmigrating copepods develop in profusion in late summer (Shapiro, 1960). In the extreme case shown in Fig. 9-5, less than 10% of the metalimnetic oxygen maximum was accounted for by zooplankton respiration (Mitchell and Burns, 1979). The metalimnetic anoxia was apparently caused mainly by intensive bacterial decomposition.

The basin morphometry of a lake can also contribute to metalimnetic oxygen minima. In cases where the slope of the basin is gentle where it coincides with the prevailing metalimnion, a greater area of the sediments with high bacterial utilization of oxygen will be in contact with metalimnetic water than is the case adjacent to the lower strata. Horizontal mixing and streaming of water from internal water movements is greatest in the metalimnion, where vertical density stability is greatest. As a result, reduction in oxygen content at the sediment-water interface is greater where the slope of the sediments is less. The reduction in the metalimnion may be sufficient to extend laterally over the entire lake or, as in the case of Skärshultsjön, Sweden (Alsterberg, 1927), it may occur more strongly in the metalimnion closer to the sediment-metalimnion interface.

Biogenic oxidation of methane by methane-oxidizing bacteria can result in metalimnetic oxygen minima in certain productive lakes (Ohle, 1958; Kuznetsov, 1935, 1970). Methane from anaerobic fermentation in sediments rises from the sediments and is carried to water strata such as the metalimnion where, with warmer temperatures and greater oxygen, methane oxidation can occur rapidly. In such cases, severe oxygen reduction can occur in a few days if adequate dissolved inorganic nitrogen is available (Rudd, et al., 1975, 1976).

DISTRIBUTION IN MEROMICTIC LAKES

Extreme clinograde oxygen profiles occur in permanently meromictic lakes (Fig. 9-4C). The permanent monimolimnion receives constant inputs of organic matter or sulfates from saline water, which rapidly deplete any oxygen intrusions.

Temporary (partial) meromixis, discussed in Chapter 6, can lead to highly variable, anomalous oxygen distributions. If a normally dimictic lake is quite productive, a strongly clinograde oxygen distribution both during summer and winter stratification would be anticipated, often with anaerobic deeper layers in both seasons. If spring or fall circulation and renewal of oxygen is incomplete, the reduced or anaerobic conditions of the lower strata will be carried over to the next stratified period and be further reinforced. Such was clearly the case in Little Crooked Lake in the fall of 1963 (Fig. 9-3). Mixing was incomplete in Lawrence Lake in the spring of 1968 (Fig. 9-6), which resulted in an accentuated reduction of the hypolimnetic oxygen the following summer, and anoxia below 11 m for three months. In normal years with complete spring circulation, as in the following four years (cf. Wetzel, et al., 1972), the hypolimnion did not become anaerobic or only very briefly just before fall turnover in a half-meter stratum immediately overlying the sediments. In 1973, incomplete mixing occurred again in the spring, resulting in an oxygen distribution pattern similar to that of 1968. The chemical nutrient and biotic implications of aerobic and anoxic hypolimnia are very important for the productivity of lakes, and will be pursued in detail further on.

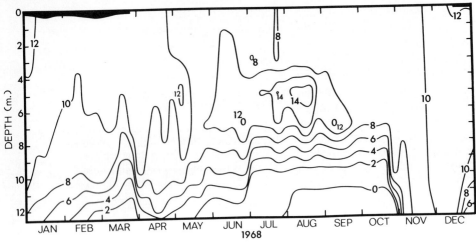

Figure 9-6 Depth-time diagram of isopleths of dissolved oxygen concentrations in mg l^{-1} of Lawrence Lake, Michigan, 1968. Ice-cover drawn to scale. (From Wetzel, R. G., et al: Memorie Istituto Italiano Idrobiol., 29, Suppl., 185–243.)

HORIZONTAL VARIATIONS AND DIEL CYCLES

Horizontal variations in the distribution of dissolved oxygen of the open-water pelagic zone are generally small during the periods of spring and fall circulation. This is not always the case during the productive summer period and under ice-cover. Littoral aquatic plants often cover large areas of more productive lakes, to a depth of 3 to 10 m. Photosynthesis by these macrophytes and of sessile algae growing attached to them and to the sediments generates large amounts of oxygen. Similarly, large quantities of oxygen are utilized in respiration at night and by littoral bacteria in a zone rich in dissolved and particulate organic matter. The oxygen regime of the littoral zone is usually totally different from that of the open water and undergoes marked diurnal cycles (Fig. 9–7). Moreover, the vertical distribution within the vascular plant beds can be highly stratified in accordance with the macrophyte portions that are actively photosynthesizing or those that are predominately respiring (Fig. 9–7).

The oxygen content of the littoral zone is periodically severely reduced. Such is the case when large populations of macrophytes (vascular hydrophytes and macroalgae) senesce and die at the end of their growing season. Severe oxygen deficits caused by intense decomposition persist for several months and may extend from the littoral zone into the open water for some distance (Thomas, 1960). In shallow lakes where submersed hydrophytes grow over the entire basin, macrophyte decomposition at the end of the growing season, in late summer when temperatures are still high, can be so intense that the oxygen content of the entire lake may be severely reduced to near anoxia. Such a catastrophic event can result in massive die-offs of many animals; often entire fish populations are lost in such a *summerkill*.

Where lakes are extremely eutrophic, phytoplankton algae often occur in such profusion that they severely attenuate the light, and submersed macrophytes are limited to very shallow areas. Under such conditions, diurnal fluctuations in oxygen concen-

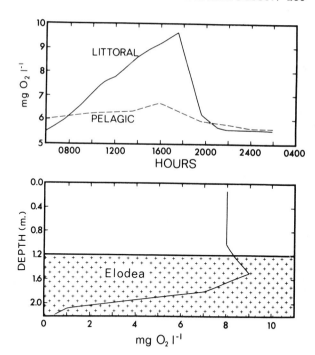

Figure 9-7 *Upper:* Changes in dissolved oxygen in the littoral and open water areas over a diurnal period in eutrophic Winona Lake, Indiana, 9 August 1922. (From data of Scott, 1924.) *Lower:* Vertical stratification of oxygen within the littoral zone of Parvin Lake, Colorado, 9 July 1955, in a luxuriant stand of the submersed macrophyte *Elodea*. (Generated from data of Buscemi, 1958.)

trations of the epilimnion are often just as marked as those of the littoral zone, as shown in Figure 9–7. Changes of 4 to as much as 6 mg 0_2 l^{-1} are seen commonly between midafternoon and early morning darkness (e.g. Lingeman, et al., 1975). Under clear ice-cover in such productive waters, oxygen can accumulate, commonly resulting in diurnal supersaturation. Oxygen analyses in surface waters of productive lakes undergoing rapid diurnal changes, when taken only at one time of day, show only one stage of a much more complex situation.

Large horizontal variations in oxygen content are also found in lakes of complex basin morphometry. When a lake consists of many bays, the oxygen concentrations are very different in these areas and each embayment may operate essentially as an individual lake (Welch and Eggleton, 1932; Wetzel, 1966a).

In reservoirs, the structure of oxygen distribution becomes highly variable horizontally, vertically, and seasonally. The complex hydrodynamics are unique for each reservoir in relation to variations in its influents, its morphometry, characteristics of draw-down and discharge, and other factors (see, for example, Straškraba, et al., 1973). Seasonal variations in influxes of organic matter from inflowing streams can place varying oxidative demands on the oxygen content of reservoirs.

Horizontal variations in oxygen are particularly conspicuous in the winter under ice-cover, especially in the surface strata. Fluctuations were pointed out earlier in response to seasonal changes in photosynthesis (Figs. 9–2, and 9–3). Photosynthesis by algae is strongly suppressed by snow-cover on ice, but often accumulated snow-cover is very spotty on the surface of windswept-ice. The result is variegated photosynthesis and oxygen production below the ice. Heavy snow-cover can plunge the lake into virtual darkness; if sustained for several weeks, heavy respiratory demands by decomposition can reduce the oxygen content to very low levels, or to anoxia (Greenbank, 1945). The resulting conditions are intolerable to many animals. For example, most fish cannot survive, even at low temperatures, at less than 2 mg O_2 l^{-1}. Low oxygen concentrations can lead to selective fish mortality with profound effects on intraspecific population structure (Casselman and Harvey, 1975) and community relationships of subsequent years. Such *winterkill* conditions of low winter oxygen concentrations are common in shallow eutrophic lakes (e.g. Barica and Mathias, 1979).

An example of the rapid seasonal changes in oxygen concentrations that can occur can be seen in Green Lake, a small reservoir lake in southern Michigan (Fig. 9-8). In early February, heavy ice and snow-cover had severely reduced the oxygen content of the lake, except in the area immediately adjacent to the small inlet. Three days later, following late winter rains on deteriorating ice, conditions had improved, although horizontal variations in oxygen concentrations still were conspicuous. A week later, oxygen brought in primarily from increased flow of the inlet had virtually obliterated the former distribution.

Methane and hydrogen can be formed in anaerobic sediments of productive lakes in sufficient quantities to bubble up and rise to the surface. Rossolimo (1935) and Kuznetsov (1935) have shown that the methane (CH_4) and hydrogen (H_2) are effectively oxidized by bacteria in the water. Under ice-cover, this oxidation can cause severe reductions in oxygen content (Fig. 9-9). Generation of these two gases, and oxygen utilization, is particularly great over the deeper depressions of the lake basin. The central oxygen depletion is probably accelerated by profile-bound density currents that slowly

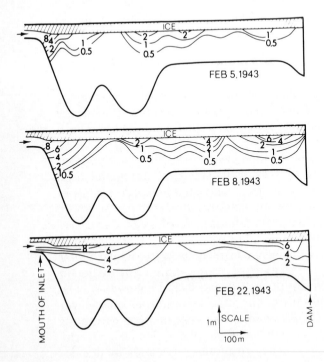

Figure 9-8 Horizontal isopleths of oxygen concentrations (mg l^{-1}) during a two-week period in Green Lake, Michigan, winter 1943. (Modified from Greenbank, 1945.)

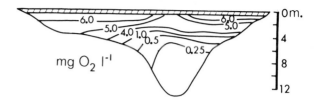

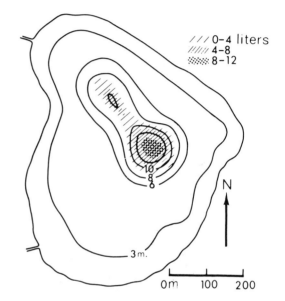

Figure 9-9 *Upper:* Isopleths of oxygen concentration in Lake Beloie, USSR, 15 March 1928; *lower:* quantity (liters m^{-2}) of gas, containing 20-24% CH$_4$, permeating the underside of the thawing ice in late winter. (Modified from Rossolimo, 1935.)

accumulate in the depression and force overlying water upward. Oxygen concentration in these currents is reduced as they flow along the sediments to the deepest point.

Thus the oxygen regime of most dimictic lakes of the temperate zone is governed to a large degree by the quantities of oxidizable matter received during periods of stratification. In lakes that tend toward permanent stratification (either meromictic variants or oligomictic lakes of low elevation in equatorial regions), the oxygen distribution is clinograde and is determined by the duration of the stratification in relation to the inputs of oxidizable organic matter. Lakes containing large quantities of dissolved organic matter, such as bogs and bog lakes, have oxygen concentrations that are appreciably below the saturation point.

From the numerous examples of oxygen distribution and control in natural water bodies, the effects of artificial loading of lakes with pollutional sources of organic matter and other oxidizable materials should be readily apparent. Organic effluents from agricultural activities, sewage, and industry can effectively and rapidly exceed the processing capacity of lakes. The resulting effects on aerobic organisms are direct and immediate. Indirect effects on the whole biogeochemical cycling and productivity are more gradual but, as we will see, are more effective in producing generally undesirable conditions.

Oxygen Deficits

The volumetric rate of oxygen loss from the hypolimnion during summer stratification increases not only with depth, but with time during the stratification period. The difference in amount of oxygen present at the beginning and at the end of stratification below a given depth is referred to as the *oxygen deficit*. In its simplest meaning, the oxygen deficit quantifies a metabolic relationship of the *trophogenic zone*, the superficial stratum of a lake in which photosynthetic production predominates, to that in the underlying *tropholytic zone*, the aphotic deep stratum where decomposition of organic matter predominates. Hence the amount of organic matter synthesized in the trophogenic zone that rains into and decomposes in the tropholytic zone is reflected in the rate of utilization of hypolimnetic oxygen. Although the dynamics of oxygen utilization are not that simple, the amount of oxygen consumption in the hypolimnion during stratification (oxygen deficit) provides an approximate estimate of the productivity of the lake.

The recognition and development of the theory of oxygen deficits as an approximation of lake productivity has a long history since its foundations in the fundamental work of Birge and Juday (1911) and Thienemann (1928). Since several of the older means of estimating oxygen deficit are still in use in spite of serious shortcomings, they should be briefly defined. All must be used cautiously, and only in a general way.

ACTUAL, ABSOLUTE, AND RELATIVE DEFICITS

The *actual oxygen deficit* of the water refers to the difference between the observed oxygen content and the saturation value of that same quantity of water at its observed temperature and atmospheric pressure at the lake surface. A shortcoming of this measurement is that it assumes that the water was saturated at the observed temperature during spring turnover, and that oxygen intrusion from the trophogenic zone during stratification by currents was negligible. The same criticism can be leveled at the *absolute oxygen deficit*, the difference between the observed oxygen concentration and the saturation value at 4°C at the pressure of the lake surface (Alsterberg, 1929). The assumption that the oxygen content during spring circulation is at saturation at 4°C is not always borne out; indeed, more often this is not the case. Therefore, Strøm (1931) proposed and used the *relative oxygen deficit*, which is the difference between the oxygen content of the hypolimnion and that empirically determined at the end of spring turnover.

Use of relative oxygen deficit greatly improved the estimations of rates of consumption, because it took account of different oxygen deficits in layers below the hypolimnetic surface. Thus, consideration was given to the most basic property of changing volumes of the strata with depth, and the importance of different hypolimnetic volumes among lakes. Thienemann (1926, 1928) emphasized this point clearly (Fig. 9–10). In comparing two lakes having identical volumes in the trophogenic zones, it would be assumed that the production and influx of organic matter in the trophogenic zone of each would be the same, and that similar amounts of oxidizable organic matter would sediment from the trophogenic layer into the tropholytic zones. However the volume

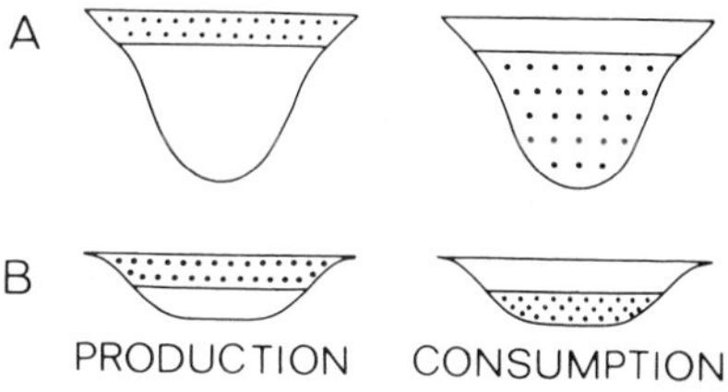

Figure 9-10 Diagrammatic representation of two lakes of equal production in the trophogenic zone but of differing volumes of their tropholytic zones, in which oxidative consumption occurs. (Modified from Thienemann, 1926, 1928.)

of the first lake (Fig. 9-10A) is much larger than that of the second lake (B), and the first lake contains a much larger amount of oxygen to be utilized in decomposition. As a result, during the same interval of time, the oxygen content of the hypolimnion of the shallow lake (B) would be reduced more rapidly than the oxygen in the hypolimnion of the deep lake (A). One might anticipate a clinograde oxygen curve in the former case, and a more or less orthograde curve in the latter, deep lake. Much of the oxygen utilization in the shallow lake may occur after the organic material has settled into the sediments. Turbulence and horizontal currents in the hypolimnion, although small, are usually sufficient to distribute much of the lower deoxygenated water near the sediments into much of the hypolimnetic water. Thus, the relative oxygen deficit, which compensates for differences in hypolimnetic volume with depth, incorporates the intensity of trophogenic production into the consumptive capacity of the hypolimnion or tropholytic zone.

This *relative areal deficit*, as introduced by Strøm (1931) and modified by Hutchinson (1938, 1957), is the mean oxygen deficit below one cm^2 of hypolimnetic surface and consists of the summation of the individual deficits of a series of layers of decreasing volume with increasing depth. A computation from Gull Lake, southwestern Michigan, illustrates the general approach (Table 9-2). The data consist of measurements of oxygen concentration at several depths; a detailed hydrographic map is essential for accurate estimates of both volume and area. Since the oxygen concentrations did not differ greatly from one depth to the next in Gull Lake at this time, it is possible to find the average concentration in any layer by taking the mean of the concentration at the top of the layer and that at the bottom of the layer. When concentration gradients are steep, simple arithmetic means are not sufficiently accurate; a planimetric integration should then be used (Wetzel and Likens, 1979). This concentration is then multiplied by the volume of the layer to give the total amount of dissolved oxygen in the stratum, and the amounts for each layer of the hypolimnion are added. Thermal and morphometric data show that the surface of the hypolimnion of Gull Lake is 34.98 ha or 349.8 × 10^8 cm^2. By division the difference in oxygen is 2.67 mg cm^{-2} for the 55-day period between 6 June and 1 August, 0.0486 mg cm^{-2} day^{-1} or 1.46 mg cm^{-2} month^{-1}.

Many of the calculations of areal oxygen deficits are from older investigations that make direct comparisons of deficits with various indices of biological productivity difficult. Nonetheless, as would be expected, there is a strong tendency in lakes of moderate depth for the relative areal deficit to increase with increasing productivity. Based on limited data, Hutchinson (1938, 1957) set the ranges among the rather ambiguous general categories of unproductive to productive lakes as: oligotrophic, < 0.017 mg O$_2$ lost cm^{-2} day^{-1} and eutrophic, > 0.033 mg cm^{-2} day^{-1}. Mortimer (1941; Hutchinson,

TABLE 9-2 Sample Computation of the Relative Areal Oxygen Deficit of Gull Lake, Summer 1972

CALCULATIONS FOR 6 JUNE 1972:

STRATA (m)	VOLUME (m³)	MEAN O₂ (mg l⁻¹)	TOTAL O₂ in STRATA (metric t)*
15–20	12.910×10^6	9.42	121.612
20–25	5.221×10^6	8.90	46.464
25–30	0.203×10^6	8.68	1.762
30–33.5	0.0428×10^6	8.35	0.358
			170.196

Calculations for
  1 August 1972:
Difference:

metric t in hypolimnion
76.719
93.477 t or 93.477×10^9 mg

The area of the surface of the hypolimnion at 15 m is 3.4980×10^6 m², or 34.98×10^9 cm².

By division, the difference in oxygen or areal oxygen deficit was:
  2.67 mg cm⁻² 55 days⁻¹
or 0.0486 mg cm⁻² day⁻¹
or 1.46 mg cm⁻² month⁻¹

Unpublished data of Tague, Lauff, and Wetzel.
*1 mg l⁻¹ = 1000 metric tons km⁻³ = 1 metric t $(10^6$ m³$)^{-1}$

1957) believes that the limits of 0.025 mg cm⁻² day⁻¹ for the upper limit of oligotrophy and 0.055 mg cm⁻² day⁻¹ for the lower limit of eutrophy are more realistic. These general limits are supported by other criteria for evaluating productivity (e.g. Lasenby, 1975).

Oxygen deficits are subject to numerous errors that detract from true representation of an index of trophogenic productivity. Significant inputs of allochthonous particulate and dissolved organic matter can cause large overestimates, while pelagial productivity of moderately to highly stained waters (> 10 Pt units) may be underestimated, owing to ultraviolet-light-mediated oxygen consumption by dissolved organic matter. Although some pelagial and littoral photosynthesis commonly occurs in the upper portion of the hypolimnion in both shallow and in deep, clear lakes, this source of error is minimized in larger dimictic lakes with a maximum depth > 20 m.

It is further assumed that the organic matter decomposing in the water column and sediments of the hypolimnion includes most of that produced in the trophogenic zone and that a relatively constant amount sinks into the hypolimnion. Gliwicz (1979) has shown, in particularly insightful analyses, that the steeper the vertical gradient of water temperature in the metalimnion (greater thermal resistance to mixing), the less the sinking speed of sedimenting organic particles in the metalimnion. As a result, not only is a larger part of the sedimenting seston prevented from leaving the more turbulent epilimnion and metalimnion, but the residence time for nutrients is prolonged in the epilimnion. Consequently, the hypolimnetic relative oxygen deficits are lower in lakes with steeper metalimnetic thermal gradients. Epilimnetic productivity is greater, and respiratory oxidative consumption is largely complete before the organic matter ever reaches the hypolimnion.

Further critical analyses of areal hypolimnetic oxygen deficits have demonstrated a number of additional relationships to lake parameters (Ohle, 1956; Edmondson, 1966; Lasenby, 1975; Stewart, 1975; Cornett and Rigler, 1979, 1980; Charlton, 1980). Areal hypolimnetic oxygen deficits of a number of lakes indicate that (1) the deficit is positively correlated with primary productivity of phytoplankton algae; (2) the deficit is inversely proportional to epilimnetic transparency (Secchi disc depth); (3) lakes with higher phosphorus concentrations have higher oxygen deficits; (4) the deficit is greater in lakes with higher mean summer hypolimnetic temperatures; and (5) oxygen deficits are larger in lakes of greater hypolimnetic thickness. That a lake with a thicker hypolimnion has an oxygen deficit that is greater than a lake with a shallow hypolimnion is still congruous with earlier theoretical evaluations, for lakes with shallow hypolimnia have higher rates of oxygen depletion per unit volume of the hypolimnion but lower rates of oxygen depletion per unit area than do lakes with thicker hypolimnia.

A number of methods exist for evaluating lake productivity that are better than hypolimnetic oxygen deficits. Autochthonous productivity can be measured directly, and fluxes of particulate and dissolved organic matter can be evaluated in detail by chemical and biotic assays. Further, the hypolimnetic oxygen deficit method cannot be used in lakes in which the hypolimnion becomes anaerobic during the period of stratification. When detailed data on productivity are lacking, however, the oxygen deficit can be informative about the general trophic status of a lake.

Changes in the hypolimnetic oxygen deficit rates over long periods of time can be indicative of overall changes in the productivity of the lake. An example of this is Douglas Lake, in the northern portion of the southern peninsula of Michigan. Although much

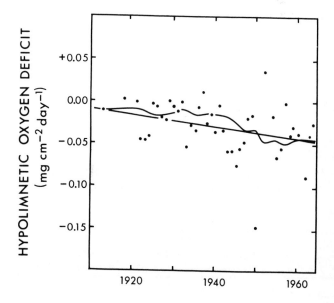

Figure 9-11 Change in estimated annual hypolimnetic oxygen deficits over a 53-year period in Douglas Lake, Michigan (negative values in all except two cases indicating a progressively greater rate of hypolimnetic oxygen reduction). The straight line represents the line of best fit by linear regression; the smooth curve represents an exponentially smoothed one-year forecast at each year. (Modified from data of Bazin and Saunders, 1971.)

is known about the characteristics and biota of this lake, which has been studied for many years, no good estimates of annual productivity exist (Lind, 1978). Analyses of the annual amount of oxygen below the metalimnion showed a progressive, slow decrease between 1911 and 1964 during each year considered except two (Fig. 9–11). Although considerable scatter in the annual rate of oxygen depletion was found, the rate of change appears to be increasing in recent years (Bazin and Saunders, 1971; Lind, 1978). The trend reflects an accelerated nutrient input and eutrophication associated with human activity first as a result of deforestation, and second as a result of the development of the area for recreational purposes. This pattern has been repeated many times in other lakes, but long-term data are available only rarely. The well-known rapid eutrophication of a much larger lake, Lake Erie, has been followed in a similar way, largely on the basis of losses of hypolimnetic oxygen and benthic fauna (Carr, 1962). Other indices of eutrophication will be discussed in other chapters. Hypolimnetic oxygen deficits are but one index and must be used with great caution with due consideration of differences in hypolimnetic temperatures and thickness.

SUMMARY

1. Dissolved oxygen is essential to the respiratory metabolism of most aquatic organisms. The dynamics of oxygen distribution in lakes are governed by a balance between inputs from the atmosphere and photosynthesis, and losses due to chemical and biotic oxidations. Oxygen distribution is important for the direct needs of many organisms, and affects the solubility and availability of many nutrients, and therefore the productivity of aquatic ecosystems.

2. Solubility of oxygen in water decreases as temperature increases. The solubility of oxygen decreases somewhat with lower atmospheric pressures at higher altitudes, and increases with greater hydrostatic pressures at depth within the lake. Oxygen solubility decreases exponentially with increases in salt content.

3. Since diffusion of oxygen from the atmosphere into and within water is a relatively slow process, turbulent mixing of water is required for dissolved oxygen to be distributed in equilibrium with that of the atmosphere. The subsequent distribution of oxygen in the water of thermally stratified lakes is controlled by a combination of solubility conditions, hydrodynamics, inputs from photosynthesis, and losses to chemical and metabolic oxidations.

   a. In unproductive oligotrophic lakes that stratify thermally in the summer, the oxygen content of the epilimnion decreases as the water temperature increases. The oxygen content of the hypolimnion is higher than that of the epilimnion because the saturated colder water from spring turnover is exposed to limited oxidative consumption. Such a vertical distribution is called an *orthograde oxygen profile* (Fig. 9–1).

   b. The loading of organic matter to the hypolimnion and sediments of productive eutrophic lakes increases the consumption of dissolved oxygen.

As a result, the oxygen content of the hypolimnion is reduced progressively during the period of summer stratification, usually most rapidly at the deepest portion of the basin where strata of less volume are exposed to the more intensive decomposition in surficial sediments. The resulting vertical distribution is termed a *clinograde oxygen profile* (Fig. 9–1).

c. Oxygen saturation at existing water temperatures returns to all water strata during fall circulation.

d. Exchange of oxygen with the atmosphere ceases for all practical purposes with the advent of ice formation. The oxygen content and saturation levels are reduced at lower depths in productive lakes, but not to the extent observed during summer stratification, because of prevailing colder water temperatures (greater solubility; reduced respiration).

e. *Metalimnetic oxygen maxima* are often observed, in which the oxygen content in the metalimnion is much greater than, and often supersaturated in relation to, levels in the epilimnion and hypolimnion. The resulting *positive heterograde oxygen curve* is usually caused by photosynthetic oxygen production by algae in excess of oxidative consumption in the metalimnion.

f. *Metalimnetic oxygen minima,* resulting in *negative heterograde oxygen curves* (Fig. 9–5), occur less commonly than metalimnetic maxima. Metalimnetic oxygen minima result from a combination of mechanisms. Most often, oxidation during decomposition exceeds oxygen inputs as the rate of organic matter sedimentation from the epilimnion decreases upon encountering the colder, denser metalimnetic water. Respiration by dense populations of microcrustacea in the metalimnion can also contribute to oxygen loss in this stratum.

g. Extreme clinograde oxygen distributions occur in meromictic lakes, in which the monimolimnion may be permanently anoxic.

h. Horizontal variations in oxygen content can be great, especially in lakes where the photosynthetic production of oxygen of littoral flora exceeds that of the open-water algae. Rapid decay of massive littoral flora or phytoplankton can result in marked reductions in oxygen content in small, shallow lakes to the demise of many animals (called *summerkill*). Oxygen distribution in reservoirs can vary greatly on an annual basis in relation to differences in organic loading from influents, open-water productivity, and discharge from the basin.

i. Under ice and snow cover where light is severely attentuated, photosynthetic augmentation of oxygen content can be eliminated. Respiratory oxidative consumption can reduce oxygen to levels insufficient to support organisms, and result in *winterkill.* Winterkill conditions are common in shallow, productive temperate lakes.

j. In productive lakes, epilimnetic oxygen concentrations can vary markedly on a diel basis. Daily photosynthetic augmentation and nighttime respiratory consumption can exceed turbulent exchange with the atmosphere, and can result in rapid (hours) fluctuations between super-saturation and undersaturation.

4. The volumetric rate of dissolved oxygen loss from the hypolimnion during summer stratification, the *relative areal hypolimnetic oxygen deficit,* has been used as an approximate index of the productivity of dimictic lakes. This method assumes that the organic matter production of the trophogenic zone is reflected in the oxygen consumption that occurs in the hypolimnion. Because of this simplicity of determination, oxygen deficits have often been used uncritically. Recent analyses indicate that differences in hypolimnetic temperatures and volumes (thicknesses) must be considered in comparisons among lakes. Empirical analyses using limited data indicate positive correlations between hypolimnetic oxygen deficits and primary productivity of phytoplankton algae, phosphorus concentrations of the water, mean summer hypolimnetic temperatures, and hypolimnetic thickness, and a negative correlation with epilimnetic water transparency (Secchi disc depth). Used properly and with caution, areal hypolimnetic oxygen deficits can serve to compare productivity approximately for large, deep dimictic lakes of low to moderate productivity if little allochthonous loading of organic matter occurs.

SALINITY OF SURFACE WATERS

IONIC COMPOSITION OF SURFACE WATERS

The ionic composition of fresh waters is dominated by dilute solutions of alkalies and alkaline earth compounds, particularly bicarbonates, carbonates, sulfates, and chlorides. The amounts of silicic acid, which occur largely in undissociated form, are usually small but occasionally are significant in hardwater lakes. The concentrations of four major cations, Ca^{++}, Mg^{++}, Na^+, K^+, and four major anions, HCO_3^-, $CO_3^=$, $SO_4^=$, and Cl^-, usually constitute the total ionic *salinity* of the water for all practical purposes. Concentrations of ionized components of other elements such as nitrogen (N), phosphorus (P), and iron (Fe), and numerous minor elements are of immense biological importance, but are usually minor contributors to total salinity.

Salinity is the correct chemical term for the ionic composition of fresh waters. Salinity is best expressed in mg l^{-1} or meq l^{-1} which are essentially equivalent as mass or volume in dilute solutions. In open lakes with an outlet, chemical composition of the water is governed largely by the composition of influents from the drainage basin and the atmosphere. Water salinity in closed basins is increased by evaporation and modified by precipitation of salts. *Soft waters* refers to waters of low salinity, which are usually derived from drainage of acidic igneous rocks (Hutchinson, 1957). *Hard waters* contain large concentrations of alkaline earths, usually derived from drainage of calcareous deposits.

The salinity of fresh waters is best expressed as the sum of the ionic composition of the eight major cations and anions in mass or milliequivalents per liter. The quantity *total solids*, an estimation of inorganic materials dissolved in water by evaporation to dryness (105°C), is less satisfactory. Combustion of the residue at 550°C yields the *nonvolatile solids* per unit mass or volume, and results in the loss of CO_2 from organic carbon. However, $MgCO_3$, some alkalis and chlorides, which are true nonvolatile solids at normal temperatures, will also release CO_2 at 550°C, thereby causing a considerable underestimate of the nonvolatile solids.

SALINITY DISTRIBUTION IN WORLD SURFACE WATERS AND CONTROL MECHANISMS

The salinity of the surface waters of the world is highly variable and depends upon ionic influences of drainage and exchange from the surrounding land, atmospheric sources derived from the land, ocean, and human activity, and equilibrium and exchange with sediments within the water body. The global mean composition given in Table 10-1 is based on a more recent evaluation than Clarke's 1924 monographic treatment of the subject. The global mean salinity of river water of 120 mg l^{-1} is somewhat lower than Clarke's average but based on better data, which compensate for the large amounts of dilute water draining tropical regions in major river systems. Principal differences among the continents are in the amounts of calcium and carbonate ions (all bicarbonate expressed as carbonate equivalents). Low values of these normally dominating ions are found especially in South America and Australia.

The three major mechanisms controlling the salinity of world surface water are rock dominance, atmospheric precipitation, and the evaporation-precipitation process (Gibbs, 1970; Feth, 1971; Kilham, 1975; Stallard, 1980; Stallard and Edmond, 1981). Waters of the rock-dominated end of a spectrum are rich in calcium and bicarbonate ions (Fig. 10-1), and are more or less in equilibrium with materials of their drainage basins. Positions within this grouping depend on the climate, basin relief, and the composition of rock material in the basin. The chemical composition of low-salinity waters, dominated by sodium and chloride, is influenced by dissolved salts derived from atmospheric precipitation derived from the ocean. These fresh waters are limited to immediate maritime coastal regions. Inland tropical humid areas of South America and Africa generally have high rainfall of an ionic composition that is either influenced primarily by terrestrial biological emissions and particulate dust or, once fallen, is then altered by rapid, intense rock weathering (Kilham, 1975; Stallard and Edmond, 1982). The third major process that influences salinity of surface waters is evaporation and fractional crystallization with subsequent sedimentation of mineral salts. Fresh waters extend in a series from calcium- and bicarbonate-rich, to low salinity waters of rock

TABLE 10-1   Mean Composition of River Waters of the World (mg l^{-1})

	Ca^{++}	Mg^{++}	Na$^+$	K$^+$	CO$_3^-$ (HCO$_3^-$)	SO$_4^-$	Cl$^-$	NO$_3^-$	Fe (as Fe$_2$O$_3$)	SiO$_2$	SUM
North America	21.0	5.	9.	1.4	68.	20.	8.	1.	0.16	9.	142
South America	7.2	1.5	4.	2.	31.	4.8	4.9	0.7	1.4	11.9	69
Europe	31.1	5.6	5.4	1.7	95.	24.	6.9	3.7	0.8	7.5	182
Asia	18.4	5.6	5.5	3.8	79.	8.4	8.7	0.7	0.01	11.7	142
Africa	12.5	3.8	11.	—	43.	13.5	12.1	0.8	1.3	23.2	121
Australia	3.9*	2.7	2.9	1.4	31.6	2.6	10.	0.05	0.3	3.9	59
World	15.	4.1	6.3	2.3	58.4	11.2	7.8	1.	0.67	13.1	120
Cations (meq)	0.750	0.342	0.274	0.059	–	–	–	–	–	–	1.425
Anions	–	–	–	–	0.958	0.233	0.220	0.017	–	–	1.428

Selected data from Livingstone, 1963, and Benoit, 1969.
*Values of calcium are likely less, on the average, than Na and Mg in Australian surface waters (Williams and Wan, 1972).

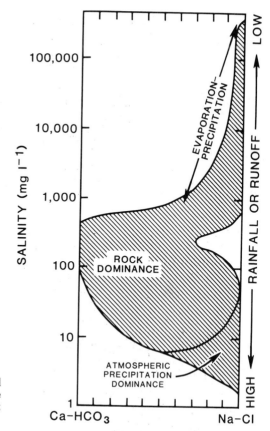

Figure 10-1. Diagrammatic representation of the general processes controlling the salinity of surface waters of the world. (Composite relationships based on data of Gibbs, 1970; Kilham, 1975; and Stallard, 1980.)

dominance or to sodium-dominated, high-salinity waters (Fig. 10-1). Rivers and lakes at the extreme saline end of this series are generally located in hot and arid regions.

The data discussed here represent average concentrations and general conditions of distribution on a global basis. Within regions, the distribution of salinity of inland waters is highly variable and localized. Hence, some areas or lake districts can be found in which, because of uniform rock formations, the ionic proportions are relatively consistent (Table 10-2). By contrast, in southwestern Michigan, extremely softwater seepage lakes may be found immediately adjacent to very hardwater calcareous lakes, because of complex, varying patterns in the deposition of glacial till.

Over large regions of the temperate zone, dominance by calcium and bicarbonate ions prevails in open lake systems:

$$\text{Cations: } Ca > Mg \geq Na > K$$
$$\text{Anions: } CO_3 > SO_4 > Cl$$

The relationships are very consistent with those of the average world river water (Table 10-1), and there is a distinct tendency for the composition of lake waters to approach this average concentration (Rodhe, 1949; Gorham, 1955). These salinity relationships led Rodhe (1949) to propose a general standard of lake water concentration, given in Table 10-3 in abbreviated form. Deviations from these proportions are common, par-

TABLE 10-2 Mean Ionic Salinity in Equivalent Percentages of Several Natural Waters

NATURAL WATERS	Ca^{++}	Mg^{++}	Na$^+$	K$^+$	CO$_3^-$	SO$_4^-$	Cl$^-$
Wisconsin soft waters (Juday, et al., 1938; Lohuis, et al., 1938)	46.9	37.7	10.9	4.8	69.9	20.5	9.9
N. German soft waters (Ohle 1955a)	36.0	14.3	43.0	6.7	42.4	14.1	43.5
Water from igneous rock (Conway, 1942)	48.3	14.2	30.6	6.9	73.3	14.1	12.6
Mean sedimentary source material (Hutchinson, 1957, after Clarke, 1924)	53.2	34.0	8.0	4.8	93.8	6.2	–
Swedish rivers, average entire Sweden (Ahl, 1980)	54.9	18.4	22.6	4.1	42.7	39.1	16.5
Ugandan river and lake waters (Viner, 1975)	32.5	26.9	32.1	8.7	62.9	28.0	9.1
Upland Swedish sources (Rodhe, 1948, after Lohammar)	67.3	16.9	13.6	2.2	74.3	16.2	9.5
Hardwater lakes of N.E. Indiana (Wetzel, 1966b)	79.3	14.4	3.3	3.0	60.0	37.9	2.1
Hypereutrophic Sylvan Lake, Indiana* (Wetzel, 1966a)	57.3	11.8	26.9	3.4	48.7	34.7	16.6
Average river content of world (Livingstone, 1963)	52.6	24.0	19.3	4.1	67.1	16.3	15.4

*A hardwater lake that received large amounts of domestic effluents that had high sodium content, most likely from detergents. Nutrient loading to this lake has subsequently been reduced.

ticularly in the continents of South America and Australia, and in drainage basins of igneous source materials and soft waters (see Table 10-2). Drainage from igneous rock commonly has a salinity of less than 50 mg l^{-1} and cationic proportions of:

$$Ca > Na > Mg > K$$

The proportions of anions in softwater systems shift toward a decrease in carbonates and an increase in halides:

$$Cl \geq SO_4 > CO_3$$

TABLE 10-3 Composition of Swedish Bicarbonate Fresh Waters (in mg l^{-1}) in Relation to Specific Conductance

SALINITY	Ca	Mg	Na	K	CO$_3$	SO$_4$	Cl	μmhos cm^{-1} (20°C)
10.5	2.5	0.4	0.7	0.3	4.4	1.5	0.7	20
32.9	7.9	1.3	2.2	0.8	13.8	4.7	2.2	60
56.5	13.5	2.3	3.8	1.4	23.7	8.0	3.8	100
92.1	22.0	3.7	6.2	2.3	38.6	13.1	6.2	160
117.3	28.1	4.7	7.9	2.9	49.2	16.6	7.9	200
155.1	37.1	6.2	10.4	3.8	65.2	22.0	10.4	260
180.8	43.3	7.3	12.2	4.4	75.8	25.6	12.2	300
219.7	52.6	8.8	14.8	5.4	92.1	31.2	14.8	360
246.2	59.0	9.9	16.6	6.0	103.2	34.9	16.6	400

From data of Rodhe, 1949, and Hutchinson, 1957.

Saline lakes form and persist when (a) outflow of water is restricted, as in hydro-logically closed basins, (b) evaporation exceeds inflow, and (c) the inflow is sufficient to sustain a standing body of water (Eugster and Hardie, 1978). The relative proportions of the major solutes vary greatly from one saline lake to the next. Most saline waters are dominated by Na; very few saline lakes are found in which Ca or Mg is the major cation. Dominating major anions are highly variable, although Cl very commonly pre-dominates. Certain elements, normally present in only trace amounts in surface waters, can occur in very high concentrations in saline lakes, e.g. bromine, strontium, phos-phate, or boron (Wetzel, 1964; Eugster and Hardie, 1978; Hammer, 1978).

The specific conductance of lake water is a measure of the resistance of a solution to electrical flow. The resistance of an aqueous solution to electrical current or electron flow declines with increasing ion content. Hence, the purer the water is, i.e., the lower its salinity, the greater its resistance to electrical flow will be. By definition, specific conductance of an electrolyte is the reciprocal of the specific resistance of a solution measured between two electrodes 1 cm^2 in area and 1 cm apart. The conductance is expressed in μmhos cm^{-1} (reciprocal of ohms).* The temperature of the electrolyte affects the ionic velocities; conductance increases about 2 per cent per °C. The inter-national chemical standard reference of 25°C is recommended in all cases, even though 18°C has been used widely in past limnological studies.

The specific conductance of the common bicarbonate-type of lake water is closely proportional (see Table 10–3) to concentrations of the major ions (Juday and Birge, 1933; Rodhe, 1949). Once the concentrations of the major ions are known, changes in specific conductance reflect proportional changes in ionic concentrations (Otsuki and Wetzel, 1974). This relationship is not true, however, for minor constituents of lake water (e.g. N, Fe, Mn [manganese], Sr [strontium], or especially P) (Rodhe, 1951). As would be anticipated, there is a positive correlation between conductance and pH in the inter-mediate pH range of bicarbonate fresh waters, but this relationship deteriorates among lakes of low salinity and high dissolved organic matter content (Strøm, 1947).

The relative concentrations of the eight major ions of salinity can be shown graphically using ionic polygonic diagrams (Maucha, 1932). A 16-sided regular poly-gon of standard area is divided into eight equal sectors, the four to the right of the vertical diameter representing the cations and the four to the left the anions (Fig. 10–2). Lines are drawn from the center to bisect each sector. The length of each radius bisecting a sector is proportional to the concentration of a particular ion in equivalent percentage of the total cations or anions.† Lines from the radii of the ionic sections are joined to the polygon and shaded.

---

*Specific conductance (μmhos cm^{-1} = $\dfrac{\text{R of 0.00702 N KCl}}{\text{R of sample}}$ × 1000) where R is the resistance in ohms (cf. Rainwater and Thatcher, 1960; Golterman, 1969; Wetzel and Likens, 1979).

†The radius (r) within the sectors of a 16-sided polygon is

$$\frac{r^2 (\sin 22.5)}{2} = A/16$$

where A represents the area of the polygon, for example 200 mm^2. Thus r = 25 mm/0.38268, or r = 8.082 mm. The total equivalent per cent for each of the four cations or four anions is equal to 100 per cent, or in this example: 100 mm^2 (1 mm^2 = 1 equivalent %). Details of calculation are given in Gessner, 1959, and in Broch and Yake, 1969.

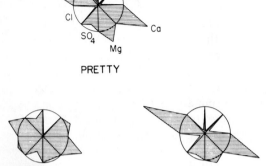

PRETTY

SYLVAN   CROOKED

Figure 10-2. Ionic polygonic diagrams in which the relative concentrations in equivalent percentage of the major cations and anions are indicated for average concentrations of Pretty, Sylvan, and Crooked lakes, northeastern Indiana. The length of each radius bisecting a sector is proportional to the concentration of a particular ion in equivalent percentage of the total cations or total anions. (After Wetzel, 1966b.)

SOURCES OF SALINITY

WEATHERING OF SOIL AND ROCK

The composition of soil and rock and their ion-exchange capacities influence both rates of weathering and ion supply to runoff and percolating water. The adsorption and release of ions is dependent upon (a) the cationic availability, (b) ionic concentrations and proportions in a soil solution or leaching water, (c) the nature and number of exchange sites on the exchange complex of soils, and (d) the volume of water in contact with the exchange complex. Four general processes of weathering control ion supply: solution, oxidation-reduction, the action of hydrogen ions, and the formation of complexes (reviewed in detail by Gorham, 1961, and Carroll, 1962).

Ordinary solution not involving hydrolysis or acid weathering is important, primarily in sedimentary deposits rich in soluble salts. Leaching of marine salt deposits results in an enrichment of sodium, potassium, and chlorides relative to other ions in recipient lake waters. Relative leaching rates of different ions are time-dependent, and thus marked fluctuations in ionic proportions of runoff water occur during periods of high (flooding) or low rates of runoff because of differences in solubility.

Oxidation and reduction processes primarily affect iron, manganese, sulfur, nitrogen, phosphorus, and carbon compounds in soils. Iron sulfides are common constituents of rocks and water-saturated soils. Oxidation of these sulfides can be a major source of sulfate for natural waters, usually in the form of dilute sulfuric acid, which solubilizes other rock and soil constituents as well. Microbial decomposition of sulfur-containing organic compounds, especially in woodland soils and peaty bog waters, adds sulfate to natural waters usually as sulfuric acid. Within the lake, oxidation-reduction processes play dominant roles in the sulfur cycle (see Chapter 14).

The action of hydrogen ions from the dissociation of carbonic acid, discussed at length in the following chapter, is of major importance in the weathering of soils and rocks. Concentrations of CO_2 and H^+ ions increase in soils in which intensive microbial decomposition of organic matter occurs. The importance of carbonic acid in weathering is illustrated by the high proportion of bicarbonate ions in most river waters. The proportion is high even in waters of low salinity (<50 mg l^{-1}) draining igneous rocks. Strong acids (H_2SO_4, HNO_3) in rainfall originating from air pollution accelerate the rate of weathering (Likens, et al., 1972, 1979). Colloidal acids of humic compounds and acid clays can also supply large amounts of H^+ ions for weathering. The 'black' waters of tropical rivers draining acidic podzols are colored brown by humic substances originating from incomplete decomposition of forest and swamp plants. The low pH (3.5 to 5) of these waters results largely from dissociation of humic acids, and the fact that the soils of lowland tropics contain very low levels of bases and of calcium.

Certain soluble organic molecules can chelate or complex ions by bonding, thereby preventing these ions from reacting with others, and maintaining solubility of the complex. The magnitude of this process in weathering and transport of ions probably is quite significant, but as yet it has not been studied adequately.

The exchange of ions between soils and the soil solution is governed by exchange equilibria between hydrogen ions and ions attached to the soil particles. Soil colloids have a finite absorptive capacity, and the exchange rate is highly variable. In humid, well-drained regions, water selectively removes cations from weathering rocks and soils. In areas of limited rainfall and stream activity, an accumulation of sodium and chloride deposited by rain in the soil water can result. The proportion of divalent to monovalent cations in the exchange positions on soil particles is a function of cation concentrations and their relative proportions in the soil water. Sodium ion adsorption to clay particles decreases with increasing concentrations of Ca^{++} and Mg^{++}. Calcium and magnesium are the dominant exchangeable cations in most neutral or alkaline soils. In well-drained situations, the cations added by rainfall will have little effect on exchange leaching from soils or weathering rock. In contrast, when the soils are saturated, cations in exchange positions are in equilibrium with those in the soil water and groundwater. Cations added by rainwater to the water surrounding clay minerals will establish equilibrium with those in exchange positions, and little weathering or removal of cations will occur. Under arid conditions, salts in soil solutions accumulate, for there is no removal of cations derived from rainfall. Exchange sites contain predominantly Ca^{++} and Mg^{++}; as the soil water becomes saturated with Na^+, the Ca^{++} and Mg^{++} are gradually replaced by Na^+ until all sites are filled. This Na^+ can then be released again during periods of rainfall, resulting in a much higher concentration of Na^+ than of divalent cations. Clay minerals are affected very little by these exchanges and do not weather appreciably.

The salinity of natural waters is influenced further by the depth and mode of water percolation. Seepage lakes receiving relatively large amounts of surface runoff are usually very dilute in comparison to open lakes that drain deeper soil horizons, for salinity of groundwater is generally much higher than that of surface runoff.

ATMOSPHERIC PRECIPITATION AND FALLOUT

The atmosphere is a significant source of salinity for many dilute fresh waters and for some saline lakes of arid regions. Rainfall carries much of the atmospheric salt to lake and river waters (cf. reviews of Gorham, 1961, and Carroll, 1962). In less humid regions, dry fallout of salt particles can occur in significant quantities, often exceeding those washed down with rainfall. Snow is less efficient in removing atmospheric salts than rain, but at high latitudes, contaminants form an appreciable supply of salinity (e.g. Barica and Armstrong, 1971). All of the major anions in natural waters are cycled in part through the atmosphere as gases as well as in dissolved and particulate form.

The sea is the major source of atmospheric sodium, chloride, magnesium, and sulfate. Sea spray carries large amounts of sea water into the atmosphere which, on evaporation, forms salt particles that are capable of acting as nuclei for cloud and raindrop condensation. Atmospheric salinity can be carried for great distances. Although a majority of it is precipitated with rainfall in the coastal regions, decreasing amounts are carried inland before deposition occurs. Continental rain generally contains more sulfate than chloride; chloride ion concentrations usually increase with proximity to the sea. In a similar fashion, sodium is the dominant cation of rain, and its concentration decreases in relation to amounts of magnesium and calcium with increasing distance from the sea. The effects of atmospheric transport of ions can be seen in lakes enriched with Na^+ and Cl^- in coastal maritime regions. This enrichment is particularly evident in igneous coastal region mountains, in which many of the maritime lakes of Japan can be found (Sugawara, 1961).

Windblown dust from the soil often contributes salts, especially of calcium and potassium, to rain and snow. This is the case, for example, in calcareous regions in Sweden. Wind-transported salts from salt pans of lakes in endorheic regions, such as in western Australia or the United States, can be moved large distances to drainage basins of other lake systems.

An additional major source of atmospheric salinity is industrial and domestic air pollution. Although numerous ions and particles are emitted into the air, chlorides, calcium, and especially sulfates and nitrates are major contaminants. The magnitude of the consequences of air pollution on the chemistry of surface waters has now reached such major proportions that global action is required to curtail its effects. Although much of the atmospheric pollution returns to the ground in the areas immediately adjacent to the industrial sites of production, sufficient sulfuric, nitric, and hydrochloric acids enter the atmosphere to influence the water of precipitation and surface waters of entire countries or portions of continents. The most infamous example is in Scandinavia, which receives air currents from southwestern Europe for much of the year. Contaminants from heavily industrialized regions of the United Kingdom, the Ruhr Valley, and elsewhere have caused a 200-fold increase in acidity of rain in a decade (Likens, et al., 1972). Values of pH less than 3 have been recorded, and over large areas of northern Europe, pH of the rain is less than 4.

An analogous situation has developed in eastern New York and in the smaller eastern states. Acid rain falling in these areas can be traced to industrial origins in the central states and Canada. In addition to direct effects on the leaching rates of nutrients from plant foliage and soil, plant metabolism is altered and major nutrient inputs to and acidification of surface waters can result in changes in entire drainage basins. In the Hubbard Brook drainage basin of the White Mountains of New Hampshire, sulfate was found to be the principal ion in precipitation (Fisher, et al., 1968), and supplied most of the sulfate that was discharged by the streams (Table 10–4). The input of ammonium and nitrate exceeded the discharge of these constituents. The annual deposition of hydrogen ion exceeded that of sulfate, and is a major determinant of pH of the waters in that region.

TABLE 10-4   Allocation of Elements (in Percent) within the
Hubbard Brook Experimental Forest Drainage Basin

	Ca	K	Mg	Na	N	S
Source						
Precipitation input	9	11	15	22	31	65
Net gas or aerosol input	–	–	–	–	69	31
Weathering release	91	89	85	78	–	4
Storage or Loss						
Biomass accumulation						
Vegetation	34	68	17	2	43	6
Forest floor	6	4	5	<1	37	4
Streamflow						
Dissolved substances	59	22	74	95	19	90
Particulate matter	<1	6	5	3	<1	<1

After Likens, et al. (1977).

ENVIRONMENTAL INFLUENCES ON SALINITY

Climate has a marked effect on the balance between precipitation and evaporation, and on the salinity concentrations of surface waters. This effect already has been discussed in global terms. From a more localized standpoint, climatic effects are manifested in a general increase in salinity with decreasing elevation of lakes. This correlation is related to the fact that most of the salinity of rain and particulate fallout is deposited at lower elevations.

The salinity of lake waters of closed drainage basins is governed not only by inputs of dissolved ions from runoff, but by the fate of these materials upon evaporation (Hutchinson, 1957). A majority of closed lakes occur in regions with fluctuating, long-term (a period of several years) climate, and are often exposed to periods of severe aridity. Commonly very shallow, these lakes may evaporate completely, or enough to expose large expanses of sediments. Loss of salts then can occur by wind deflation.

Saline lakes are generally categorized on the basis of dominating anionic concentrations into carbonate, chloride, or sulfate waters. A spectrum of intermediate combinations may occur. The range of salinity in these lakes is extraordinary, from several hundred to over 200,000 mg l^{-1} in the Great Salt Lake of Utah and the Dead Sea. Borax Lake of northern California, one of the few shallow saline lakes that has been studied in detail over an annual period, exhibited a nearly fourfold decrease in volume and over a twofold increase in salinity, from less than 28,000 mg l^{-1} to nearly 60,000 mg l^{-1} (Wetzel, 1964). An analogous condition was recorded in Lake Chilwa, Malawi, which is typical of many shallow lakes of endorheic regions (Moss and Moss, 1969; Kalk, et al., 1979).

Other significant climatic factors influencing salinity are temperature and wind. Temperature influences the rate of rock weathering. Tropical waters, for example, which drain strongly weathered soils, are usually poor in electrolytes, and a large part of their total composition consists of silica. Wind direction and speed may strongly affect the chemical composition of atmospheric precipitation by altering the amount of incorporated sea salinity and sites of inland deposition. Losses of atmospheric salinity are greater in low-elevation, turbulent air masses. The type of vegetation growing on the drainage basin and its requirements for major ions are also influenced by climate. Mineral cycling in tropical perennial forests of dense vegetative cover, for example, differs greatly in rates of utilization and leaching of soil nutrients, as compared to nutrient cycling in deciduous vegetation of temperate regions.

DISTRIBUTION OF MAJOR IONS IN FRESH WATERS

The spatial and temporal distribution of the major cations and anions of salinity are separable into (1) conservative ions, whose concentrations within a lake undergo relatively minor changes from biotic utilization or biotically mediated changes in the environment, and (2) dynamic ions, whose concentrations can be influenced strongly by metabolism. Of the major cations, magnesium, sodium, and potassium ions are relatively conservative both in their chemical reactivity under typical freshwater conditions and their small biotic requirements. Calcium is more reactive and can exhibit marked seasonal and spatial dynamics. Of the major anions, inorganic carbon is so basic to the metabolism of fresh waters that it is treated in a separate chapter (Chapter 11), and later coupled with organic carbon cycling (Chapter 22). Similarly, sulfate is greatly influenced by microbial cycling and the chemical milieu, and is treated separately (Chapter 14), along with iron and silica. Chloride is relatively conservative. The total ionic salinity, composed almost entirely of the eight major ions, is of major importance in osmotic regulation of metabolism and in the distribution of biota, and is briefly discussed in the concluding section of this chapter.

CALCIUM

Calcium has been implicated in numerous ways in the growth and population dynamics of freshwater flora and fauna. Calcium is a required nutrient of normal metabolism of higher plants. A universal requirement for calcium has not been demonstrated for algae, but most likely it is required by the green algae, and is considered an essential inorganic element of algae. Where essential, calcium is usually needed as a micronutrient. Substitution of calcium by a closely related element, strontium, is readily acceptable to some algae (e.g., *Chlorella, Scenedesmus*) that require calcium and are indifferent to strontium; in other algae, calcium utilization is strongly inhibited by strontium.

The two principal membranes of plant cells, the outer cytoplasmic plasmalemma and the inner cytoplasmic tonoplast, operate in ion transport, one of the basic biological functions of cell membranes (see reviews of Eppley, 1962, and Epstein, 1965). The central vacuole is the main repository of accumulated ions, and both cations and anions may be absorbed from dilute solutions until the vacuolar concentrations vastly exceed external concentrations. Membranes of higher plants have a very low permeability to free ions; ionic exchange is more common among algae. Active transport mechanisms presumably involve high-energy phosphate compounds of phosphorylation. The ion transporting mechanisms are ion-selective and are often species-specific; the external concentrations of ions and ratios of divalent to monovalent cations can affect transport mechanisms markedly. Calcium is essential for maintenance of the structural and functional integrity of cell membranes.

The distribution of certain algae has been correlated to differing concentrations of calcium. While causality cannot be established definitely for such relationships by singling out one affecting chemical species from an array of simultaneously interacting parameters that regulate growth and distribution, it is clear that calcium is involved indirectly in the metabolism of certain organisms. The desmids are a large algal group that is found in large part in acidic (pH 5–6) waters of low salinity, especially waters of low calcium content. Of the few species of desmids studied, many have a narrow tolerance to calcium concentrations. Their restriction to soft waters is by no means universal. Species of the larger genera are divisible into those adapted to acidic (pH < 6), calcium-deficient (<10 mg Ca l⁻¹) waters, through a series of those adapted to increasingly alkaline, calcium-rich waters. The detailed studies of Höll on this subject have been verified and are reviewed by Hutchinson (1967). Again, causality between calcium concentration and the metabolism of these algae has not been established through experimental studies, even though evidence for such an interaction is very strong.

A ubiquitous relationship has been found between calcium, calcium-specific regulatory proteins, and cyclic adenosine 3′:5′-monophosphate (cAMP) in eukaryotic and prokaryotic organisms. These molecules function synergistically to regulate a broad spectrum of important metabolic functions, including ion transport and energy transduction reactions (Kretsinger, 1979; Marx, 1980). The relationships between these regulatory molecules in aquatic systems are only beginning to be appreciated. Cyclic AMP is produced and released by aquatic photoautotrophs and bacteria (Francko and Wetzel, 1980, 1981, 1982) and is found in highest concentrations in lake water and plant tissues in midsummer during periods of high photosynthetic demand for carbon and resultant epilimnetic and littoral decalcification (Francko and Wetzel, 1982).

Hardly a group of freshwater animals exists in which the distribution of some species has not been related to calcium concentration (cf. Macan, 1961). Like algae and certain macrophytes, in some groups most of the species are found in calcareous waters, and their numbers decrease as the concentration of calcium declines. Other species, which are frequently closely related to those in hard waters, are characteristic of waters poor in calcium. In particular, mollusks, leeches, and tricladian flatworms are divisible into hard-water species (≥20 mg Ca l⁻¹), and those that can tolerate less than this concentration. Analogous but less clear correlations occur among the crustaceans. Although the requirements for calcium in invertebrates are known to some extent (cf. Robertson, 1941), little is known about how small differences in calcium concentrations affect distribution. Attempts to explain the metabolism and distribution of animal species on the basis of single chemical differences have been numerous but largely unrewarding, except where conditions are extreme. Calcium concentrations have been implicated in the aging of rotifers, a process that affects longevity and morphology by accumulation of calcium (Edmondson, 1948). These laboratory findings fit remarkably well into the observed distribution and population dynamics of rotifers in lakes of varying calcium concentrations, in which longevity and population success increase in waters of lesser calcium content.

The calcium content of soft-water lakes remains well below saturation levels, and these concentrations exhibit minor seasonal variations with depth. Usually, the amount of calcium utilized by the biota is so small in comparison to existing levels that depletion by biota cannot be seen in normal analyses. Decomposition processes can lead to some calcium accumulation in the hypolimnion of productive soft-water lakes during stratification.

The calcium concentrations of hard-water lakes, however, undergo marked seasonal dynamics. The changes depicted in Figure 10–3 for a hard-water lake in southern

Michigan are quite typical. Between rather uniform concentrations during spring and fall periods of circulation, one can see a conspicuous stratification that is repeated annually with minor variations. Both the levels of calcium (Fig. 10-3) and bicarbonate (Fig. 11-3) decreased markedly as a result of precipitation of $CaCO_3$ during the summer months from May through September. Similar losses were observed during the period of winter stratification. In late winter, the decrease just beneath the ice is associated in part with dilution by rains permeating the ice, as well as with increases in photosynthesis just before ice loss. Decreases of concentrations of calcium and of inorganic carbon in the epilimnion and metalimnion are directly associated with rapid increases in the rates of photosynthesis by phytoplankton and littoral flora (Otsuki and Wetzel, 1974), indicating the major role of photosynthetic induction in epilimnetic decalcification.

Detailed analyses of the calcium and bicarbonate budgets of this hard-water lake were determined for the influxes, dynamics within the lake, and outflow over an annual period. Calcium inputs of two spring-fed inlet streams were uniform and contributed from 93 to 98 percent of the total monthly inputs to the basin during the summer months. After compensation for potential evapotranspiration and precipitation, inputs of calcium from groundwater influxes varied from 2 to 58 per cent of total monthly inputs. Inputs of bicarbonate of the two inlet streams were also uniform annually, and constituted from 82 to 97 per cent of the total monthly influx during the summer period. Contributions from the streams decreased anywhere from 35 to 67 per cent of the total inorganic carbon input during other times of the year because of the increase in groundwater influxes. The $CaCO_3$ precipitation rate was calculated to be 446 g m^{-2} year^{-1}. This loss of $CaCO_3$ influences the metabolism of hardwater lakes by coprecipitation of inorganic nutrients, such as phosphorus, and selective removal of humic, particularly yellow organic acids, and other organic compounds by adsorption (Stewart and Wetzel, 1981).

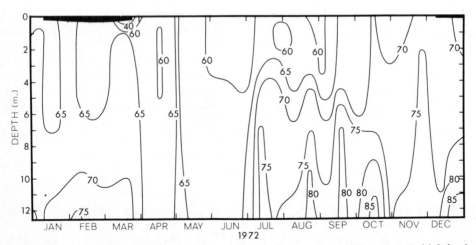

Figure 10-3. Depth-time distribution of isopleths of calcium concentrations (mg Ca^{++} l^{-1}) of oligotrophic, hardwater Lawrence Lake, Michigan, 1972. Opaque areas = ice-cover to scale. (From Wetzel, unpublished data.)

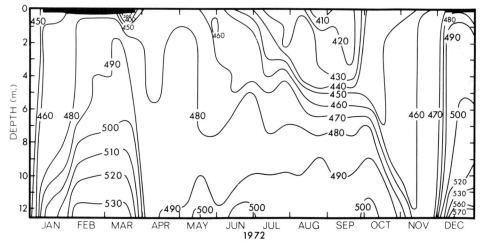

Figure 10-4. Depth-time distibution of isopleths of specific conductance (μmhos cm^{-1} at 25°C) of hardwater Lawrence Lake, Michigan, 1972. Opaque areas = ice-cover to scale. (From Wetzel, unpublished data.)

The epilimnetic decalcification is reflected simultaneously in the distribution of specific conductance of hard-water lakes (Fig. 10-4). Since concentrations of Mg, Na, K, and Cl are relatively conservative, as discussed further on, the specific conductance follows the changes in Ca^{++} and HCO_3^- concentrations in nearly a 1:1 relationship (r = 0.997; Otsuki and Wetzel, 1974). A portion of the precipitating $CaCO_3$ is resolubilized in the hypolimnion, which is reflected in both the increased concentrations of Ca^{++} and the specific conductance in hypolimnetic waters. Some of the $CaCO_3$ is entrained permanently in the sediments, and commonly constitutes > 30 per cent of the sediments by weight in moderately hard-water lakes (Wetzel, 1970). Adsorption of organic compounds to $CaCO_3$ lowers the rates of dissolution in hypolimnetic strata of reduced pH. Further, calcium complexes with humic acids, especially in the sediments, alter exchange equilibria in the hypolimnion (Ohle, 1955a; Stewart and Wetzel, 1981a).

Biogenically induced decalcification of the epilimnion reaches an extreme in very productive hard waters. For example, the calcium concentrations of the epilimnion of Wintergreen Lake, Michigan, were reduced from nearly 50 mg l^{-1} to below analytical detectability in a few weeks (Fig. 10-5). The magnitude of this loss from the trophogenic zone and the resulting increase in monovalent:divalent cation ratios should be recalled in subsequent discussion of the effects of the calcium content and of cation ratios on species distribution.

MAGNESIUM

Magnesium is required universally by chlorophyllous plants for the magnesium porphyrin component of chlorophyll molecules, and as a micronutrient in enzymatic transformations, especially in transphosphorylations by algae, fungi, and bacteria. The demands for magnesium in metabolism are minor in comparison to quantities generally

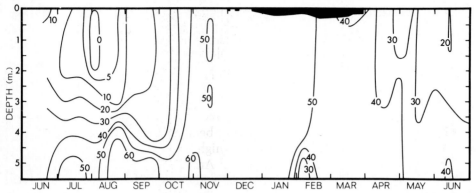

Figure 10-5. Depth-time distribution of isopleths of calcium concentrations (mg Ca^{++} l^{-1}) of hypereutrophic Wintergreen Lake, Kalamazoo County, Michigan, 1971–72. Opaque area = ice-cover to scale. (From Wetzel, et al., unpublished data.)

available in fresh waters. Magnesium compounds, moreover, are much more soluble than their calcium counterparts. As a result, significant amounts of magnesium rarely precipitate. The monocarbonates of hard waters are usually > 95 per cent CaCO$_3$ under ordinary CO$_2$ pressures. MgCO$_3$ and magnesium hydroxide precipitate significantly only at very high pH values (>10) under most natural conditions. The concentrations of magnesium are extremely high in certain closed saline lakes.

Because of magnesium's solubility characteristics and its minor biotic demand, the concentrations of magnesium are relatively conservative and fluctuate little (Fig. 10-6). This attribute has been used to advantage by employing the magnesium budget of inputs and outflows to determine the groundwater influxes to a lake by magnesium mass balance (Wetzel and Otsuki, 1974).

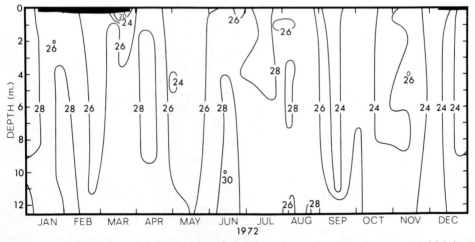

Figure 10-6. Depth-time distribution of isopleths of magnesium concentrations (mg Mg^{++} l^{-1}) of hard-water Lawrence Lake, Michigan, 1972. Opaque areas = ice-cover to scale. (From Wetzel, unpublished data.)

Low available magnesium has been implicated as one of several factors influencing phytoplanktonic productivity in an extremely oligotrophic Alaskan lake (Goldman, 1960). However, such conditions are exceedingly rare in comparison to limitations imposed by restricted availability of other major nutrients.

SODIUM, POTASSIUM, AND MINOR CATIONS

The monovalent cations sodium and potassium are involved primarily in ion transport and exchange. An absolute sodium requirement has been demonstrated in only a few plants. The sodium requirements are particularly high in some species of blue-green algae (Allen, 1952; Gerloff, et al., 1952; Allen and Arnon, 1955). Potassium and other elements of this series—lithium (Li), rubidium (Rb), and cesium (Cs)—cannot substitute for Na. A threshold level of 4 mg Na l^{-1} is required for near optimal growth of several species (Kratz and Myers, 1954), a concentration near mean for numerous hardwater lakes. Presumably sodium, along with many other factors, can influence the development of large populations of this algal group. Maximal growth of several blue-green algae was found at 40 mg Na l^{-1}. The enrichment of waters with high levels of sodium and phosphorus, as is the case, for example, in domestic effluents with very high concentrations from synthetic detergents, was indicated as a potential contributor to effective competition among the blue-green algae under bloom conditions (Provasoli, 1958; Wetzel, 1966a, Ward and Wetzel, 1975). Both sodium and potassium occur in such abundance as highly soluble cations of numerous salts that alteration of their concentrations in content of natural waters is not common.

The spatial and temporal distribution of sodium and potassium in lakes is uniform and demonstrates very little seasonal variation, indicating the conservative nature of these ions (Fig. 10-7; see also Stangenberg-Oporowska, 1967). Moderate epilimnetic reduction in potassium concentration, which has been observed in extremely productive lakes, is presumably related to potassium utilization by the massive algal populations and by submersed macrophytes (Mickle and Wetzel, 1978a), and to biotic sedimentation to the hypolimnion. An analogous situation has been found in fertilized, productive farm ponds (Barrett, 1957).

The concentrations of rarer alkaline earth and alkali cations of natural waters vary considerably in relation to the lithology of the drainage basins. Their distribution is discussed by Durum and Haffty (1961), Livingstone (1963), and Cowgill (1976, 1977a, 1977b) in general, and in closed-basin lakes in particular by Whitehead and Feth (1961). The general ratios to major cations are given in Table 10-5. Nutritional requirements

TABLE 10-5 Approximate Ratios of Minor Alkaline Earths and Alkalies to the Major Cations

Na/Li 1500	Ca/Ba 400
Na/Rb 3600	Ca/Sr 3000-4500
Na/Cs 31900	Ca/Be ca 40000
	Ca/Ra 5 × 10^{10}

From data of Livingstone, 1963.

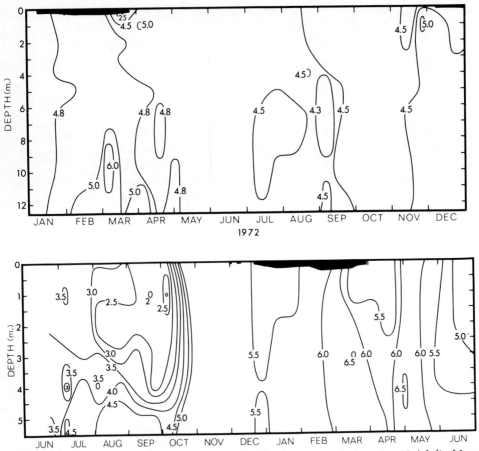

Figure 10-7. Depth-time distribution of isopleths of sodium concentrations (mg Na$^+$ l^{-1}) of Lawrence Lake, 1972 (upper) and potassium concentrations (mg K$^+$ l^{-1}) of hypereutrophic Wintergreen Lake, Michigan, 1971–72 (lower). Opaque areas = ice-cover scale.

for these relatively rare elements have not been demonstrated, although they can frequently substitute for the major cations in metabolic pathways.

CHLORIDE AND OTHER ANIONS

In the prior discussion of the general distribution of chloride ions in lakes, it was emphasized that this anion is usually not dominant in open lake systems. Lakes of maritime regions often receive significant inputs of chlorides from atmospheric transport from the sea. Pollutional sources of chlorides can modify natural concentrations greatly. For example, the concentration of chloride in Lake Erie increased threefold in 50 years, largely (approximately 70 percent) from industrial sources, road salting, and municipal wastewaters (Ownbey and Kee, 1967). Chloride is influential in general osmotic salinity balance and ion exchange, but metabolic utilization does not cause

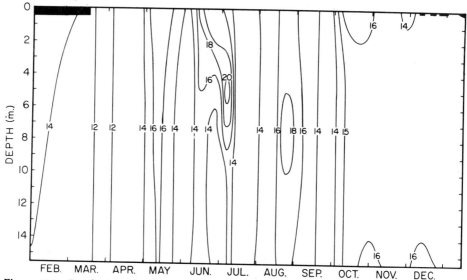

Figure 10-8 Depth–time distribution of isopleths of chloride concentrations (mg Cl⁻ l⁻¹) of eutrophic Little Crooked Lake, Whitley County, Indiana, 1964. Opaque areas = ice-cover to scale. (From Wetzel, unpublished data.)

significant variations in the spatial and seasonal distribution within a lake (Fig. 10–8). Variations observed in Lake Erie were associated with the hydrology of the basin.

The concentrations of minor halides (Table 10–6) vary somewhat with the drainage lithology of the basin, and higher concentrations often occur in lakes in proximity to marine regions or in those that possess marine rock formations within their drainage basins. Boron is of greater limnological interest because it is a micronutrient required by many algae and other organisms. Concentrations of boron in natural waters are relatively high in comparison to other minor elements. The average concentration in rain and snow is also high (4.7 μg B l⁻¹) and apparently is of terrestrial origin rather than from the sea (Nishimura, et al., 1973). Concentrations reach exceedingly high levels in certain closed, saline lakes, and can approach one gram

TABLE 10-6 Approximate Average Concentrations of Halides and Boron in Natural Fresh Waters

ELEMENT	AVERAGE CONCENTRATION (mg l⁻¹)	CHLORIDE RATIO
Chloride	8.3	
Fluoride	0.26	Cl/F 32
Bromine	0.006	Cl/Br 1400
Iodine	0.0018	Cl/I 4600
Boron	0.01 (approx.)	

From data of Livingstone, 1963.

per liter. As is the case with most micronutrients, high concentrations are generally toxic to most organisms, as was exemplified by the well-known excessive use of the micronutrient copper as an herbicide. Organisms adapted to high concentrations of boron, to > 800 mg l^{-1}, were affected little by additions of further borate (Wetzel, 1964). Aquatic angiosperms normally living in waters of very low boron content show a high resistance to large concentrations of boron. Boron additions were found to be stimulatory to photosynthesis to levels of 100 mg l^{-1}, beyond which inhibitory responses occurred (Baumeister, 1943).

CATION RATIOS

The question of the importance of the ratio of monovalent to divalent cations (M:D) is particularly interesting in regard to the distribution and dynamics of algae and larger aquatic plants in fresh waters. Three major genera of diatoms common to oligotrophic waters, *Fragilaria*, *Asterionella*, and *Tabellaria*, are stimulated by high levels of calcium (Chu, 1942; Vollenweider, 1950). Increasing levels of potassium permit increased tolerance of these species to high concentrations of Ca and Mg. Numerous studies, particularly by Provasoli and coworkers (Provasoli, 1958), indicate that the M:D ratio is significant to the observed growth of these algae. The concentrations of Ca and Mg can be manipulated over a wide range as long as the M:D ratio is maintained within reasonably narrow limits for different species. Calcium and Mg are widely interchangeable, and many species are quite tolerant to different Ca:Mg ratios. These experimental results provide excellent confirmation of the much earlier perceptive ecological observations of Pearsall (1922; 1932), who considered a M:D ratio below 1.5 favorable to diatoms and much higher ratios favorable to desmid algae. Diatoms dominate the algal flora of very hard-water lakes with M:D ratios much less than 1.5. As the epilimnion undergoes biogenically induced decalcification in early summer in calcareous waters, calcium concentration following the spring diatom maximum is often halved. It is unknown whether, along with other factors, the resulting shift in M:D ratio, which is concomitant with the shift to mixed populations of predominantly green algae with diatoms, influences phytoplankton succession.

The effects of cations on the photosynthesis and release of dissolved organic matter were studied on a submersed angiosperm, *Najas flexilis*, a macrophyte that grows poorly in extremely calcareous hard-water lakes (Wetzel, 1969; Wetzel and McGregor, 1968). Concentrations of Ca over 30 mg l^{-1} and Mg over 10 mg l^{-1}, both greatly exceeded in hard-water lakes, suppressed the rates of carbon fixation and altered the secretion rates of dissolved organic matter. Increasing the concentrations of Na above levels commonly found in hard-water lakes increased both the rates of fixation and secretion. The overall effects observed were decreased rates of photosynthetic carbon fixation with decreasing M:D ratios.

The potential of dissolved organic compounds of fresh waters to regulate M:D ratios and indirectly influence rates of photosynthesis was emphasized by Wetzel (1968). Sequestering of divalent cations by chelation by amino substances and peptides is well known. Complex formation is also possible by pyrophosphates, binding with macromolecules such as proteins, and the formation of peptized metal hydroxides of humic acids. The mechanisms have not been adequately investigated in relation to their control over algal succession and dynamics, but it is clear that changes in dissolved organic matter can exert a major influence on productivity. This subject is discussed in greater detail in Chapter 15.

SALINITY, OSMOREGULATION, AND DISTRIBUTION OF BIOTA

ORIGINS AND DISTRIBUTION OF FRESHWATER BIOTA

The salinity of inland waters is generally very low in comparison to that of the sea, even though in semiarid regions, salinity of closed-basin lakes occasionally exceeds that of sea water by several times. The distribution of biota in fresh waters has been influenced by a long evolutionary history of physiological adaptations to a wide range of salinities. These changes have developed against a background of large differences between salinity of the environment and that of the cytoplasm or body fluids. The general distribution of the freshwater biota and their tentative origins in terrestrial, freshwater, or marine sources are summarized in the detailed review of Hutchinson (1967). That summary, coupled with several major reviews of the physiological mechanisms and adaptation of osmotic regulation (Krogh, 1939; Beadle, 1943, 1957, 1959; Gessner, 1959; Robertson, 1960; and Potts and Parry, 1964), provides an introduction to the adaptations to freshwater life. The distribution of biota in brackish water interface regions between marine and fresh waters is summarized in great detail in the symposium on brackish waters (1959), and especially by Remane and Schlieper (1971).

The number of species living in brackish water is very much smaller than the number living in marine regions with similar habitats, and much smaller than the number of species in fresh water (Fig. 10-9). However, it should be pointed out immediately

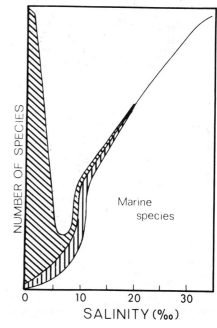

Figure 10-9   Number of species in relation to salinity. Diagonal hatching: proportion of freshwater species; vertical hatching: proportion of brackish-water species; lower open area: marine species. (After Remane and Schlieper, 1971.)

that, although there is a paucity of species in this region, this in no way implies a low productivity; productivity by adapted organisms can be exceedingly high. The lowest number of species occurs in salinities of about 5 to 7‰. The salinity gradient seen here, with its associated osmotic and ionic properties, is the predominant factor influencing the distribution of biota in brackish waters. In these transitional waters, as in the spectrum of salinities of inland waters, the salinity range occupied by a species depends on the efficiency of the physiological mechanisms by which it is adapted to changes in salinity in the environment.

OSMOTIC ADAPTATIONS OF AQUATIC PLANTS AND ANIMALS

Osmotic regulation functions primarily in the maintenance of a difference in concentrations of ions inside and outside of the cells at appropriate operational physiological levels. The aquatic bacteria and blue-green algae (Monera: lacking mitochondria or chromoplastids) and the more primitive Protista (algae, fungi, and protozoa with mitochondria and, if they are photosynthetic, with chromoplastids) demonstrate high evolutionary adaptability to changes in salinity (cf. Hutchinson's 1967 description of evolutionary euryhalinity) through relatively small genetic changes. Most freshwater bacteria and blue-green algae are relatively homoiosmotic, tolerating only a narrow range of salinity, but adapt to increasing salinity relatively rapidly by means of genetic change. Extensive adaptive radiation is seen among these groups. Members of the Protista, which are largely single-celled, retain considerable evolutionary euryhalinity and are widely distributed with respect to salinity, although some groups, especially among the green algae, are restricted to fresh water. The contractile vacuole is the primary osmoregulatory organelle among the Protista.

Higher aquatic plants, which are of terrestrial origin, have developed adaptations to fresh water secondarily. Only a few major groups of angiosperms have developed extensively in fresh waters, and very few groups extend into saline waters of brackish and marine areas or hypersaline closed-basin lakes.

Nearly all of the higher freshwater animals originated from the sea, although most aquatic insects are of terrestrial origin. Both in terms of evolutionary and contemporary life in fresh water, osmoregulation is a major problem, for which diverse mechanisms have developed to regulate salt and water content. Adaptation to low salinities by some marine animals has been achieved without osmoregulation. Such *poikilosmotic* animals adjust the osmotic pressure of their body fluids to become more or less isotonic with the salinity of the medium. In contrast, a *homoiosmotic* animal will tend to retain its initial internal osmotic concentration upon being exposed to modest changes of salinity of the medium. The general relations between the osmotic pressure among different types of animals, expressed as salinity of the blood versus salinity of the external water, can be visualized in Figure 10–10. The range depicted by area A extends over a wide variation in osmotic pressure of body fluids that is found in brackish-water animals, which tend to be more poikilosmotic at high concentrations and more homoiosmotic at lower salinities. The range of the osmotic pressure curve extends from the most homoiosmotic (a_1) to the most poikilosmotic species (a_2). The lower limits are very variable, represented by the undefined left-hand edge of area A, but all of these species have failed to colonize in fresh water.

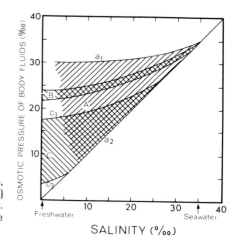

Figure 10-10. Relationship between the osmotic pressure, expressed as salinity, of body fluids of brackish-water organisms (A) and freshwater animals (B, C) and the salinity of external water. (After Beadle, 1959.) Relationships of the curves are detailed in the text.

A few species have succeeded in penetrating fresh water without a renal osmoregulatory mechanism. These brackish-water animals are partially homoiosmotic (area B, Fig. 10-10), and maintain a very high osmotic pressure in hypertonic body fluids by the active uptake of ions from the water. Excretory organs are not involved, since the urine produced is isotonic with the blood.

In the majority of freshwater animals, however, the osmotic pressures of the body fluids have decreased to levels equivalent to 5 to 15‰ salinity to reduce osmotic gradients. These organisms have developed excretory organs that effectively recover ions and produce urine hypotonic to the body fluids. The majority of freshwater animals therefore effect osmoregulation by active uptake of ions and by a renal mechanism of ion retention. Extremes in osmotic pressures of blood delineate area C of Figure 10-10 by curves c_1 and c_2. The isotonic line along the right-hand edge of area C indicates the upper salinity tolerance limit of most freshwater animals, and that they are incapable of hypotonic regulation. While most freshwater animals are capable of living in water of low salinity, the adaptation of body fluids of low osmotic pressure is apparently irreversible, and with few exceptions freshwater organisms are restricted to salinities of < 10‰.

Because of the slow diffusion of oxygen in water relative to that in air, movement of water over permeable membranes or tissue surfaces for respiratory needs is almost universal among aquatic animals. The pumping process places high energetic demands on the animals, and additionally exposes cellular surfaces to osmotic gradients. Mechanisms for taking up salts against a concentration gradient vary greatly among freshwater animals. In a few organisms, incorporation of salts with the food may be adequate, but more often organs have developed for this purpose. Aside from resorption mechanisms of the excretory organs, which are advantageous energetically, active uptake mechanisms for ions, especially sodium and chloride, are often associated with the respiratory organs (commonly gills) of many invertebrates and vertebrates.

The fauna of extremely saline inland waters are relatively insensitive to the high salinity of these lakes and to large fluctuations in the chemical composition of the water (Beadle, 1969). Most of the animals of saline lakes are of freshwater origin, and include particularly representatives of the aquatic insects, phyllopods, copepod and cladoceran crustacea, and rotifers, all of which belong to predominantly freshwater groups. The blood of these saline-inhabiting animals maintains osmotic pressures at levels characteristic of those living in fresh waters. The body surface of these animals exhibits very low permeability, and they possess effective excretory mechanisms for maintaining the body fluids strongly hypotonic to the external medium. Further, the water balance of saline inland waters frequently fluctuates widely; freshwater animals with resting stages capable of withstanding desiccation have a distinct advantage over marine animals, which lack strong development of this characteristic.

SUMMARY

1. The total *salinity* of inland waters is usually dominated by four major cations [calcium (Ca^{++}), magnesium (Mg^{++}), sodium (Na^+), and potassium (K^+)] and the major anions [bicarbonate (HCO_3^-), carbonate ($CO_3^=$), sulfate ($SO_4^=$), and chloride (Cl^-)].

2. The salinity of surface waters has a world average concentration of about 120 mg l^{-1}, but varies among continents and with the lithology of the land masses.
    a. Salinity is governed by ionic contributions via leaching from rock and soil and runoff of the drainage basin, atmospheric precipitation and deposition, and the balance between evaporation and precipitation.
    b. The concentrations of major ions of many surface waters of the world tend to exist in the proportions of Ca > Mg $\geq$ Na > K and $CO_3 - HCO_3$ > SO_4 > Cl. In soft waters and in surface waters of coastal regions, Na and Cl often occur in greater equivalent concentrations.

3. The release of ions as soil and rock undergo weathering is controlled by the processes of solution, oxidation-reduction, the action of hydrogen ions, and the formation of organic complexes. A major source of salinity to many dilute fresh waters and certain saline lakes of arid regions is the ionic content of atmospheric precipitation and particulate deposition.

4. Concentrations of the cations magnesium, sodium, and potassium and the major anion chloride are relatively conservative, and undergo only minor spatial and temporal fluctuations within a lake from biotic utilization or biotically mediated environmental changes. Calcium, inorganic carbon, and sulfate are dynamic, and concentrations of these ions are strongly influenced by microbial metabolism.
    a. The proportional concentrations of major cations and the ratios of monovalent:divalent cations can influence the metabolism of many organisms, especially certain algae and macrophytes, as much as absolute concentrations can.
    b. Factors that influence the availability of some cations disproportionately to others (e.g., organic complexing of calcium or biologically induced

decalcification of the epilimnion) can indirectly affect seasonal population succession and productivity.

5. The relatively low salinity of fresh waters has greatly influenced the distribution of biota and their long evolutionary history of physiological adaptations for osmotic and ionic regulation in an extremely hypotonic environment.

   a. Although some groups of bacteria and algae are relatively homoiosmotic and can tolerate only a narrow range of salinity, most of the lower flora and fauna are euryhaline, i.e., adaptable to a wide range of salinity.

   b. Most higher freshwater animals originated from the sea or from land, and adapted to fresh water secondarily. In comparison to marine forms, nearly all of these organisms have reduced osmotic pressures of body fluids and have developed efficient mechanisms for active uptake of ions and renal mechanisms for ion retention.

CHAPTER ELEVEN

INORGANIC CARBON

THE OCCURRENCE OF INORGANIC CARBON IN FRESHWATER SYSTEMS

Carbon Dioxide and Its Solution in Water

The carbon dioxide (CO_2) content of the atmosphere varies with locality and potential enrichment from industrial pollution. The global average is approximately 0.033 per cent by volume as of 1975, and is increasing (Machta, 1973; Broecker, et al., 1979). Carbon dioxide is very soluble in water, some 200 times greater than oxygen, and obeys normal solubility laws within the conditions of temperature and pressure encountered in lakes. The amount of CO_2 dissolved in water from atmospheric concentrations is about 1.1 mg l^{-1} at 0°C, 0.6 at 15°C, and 0.4 mg l^{-1} at 30°C.

As CO_2 dissolves in water, the solution contains unhydrated CO_2 at about the same concentration by volume (approximately 10 μM) as in the atmosphere (reviewed extensively by Hutchinson, 1957; Kern, 1960; Stumm and Morgan, 1970; J. C. Goldman, et al., 1972):

$$CO_2 \text{ (air)} \rightleftharpoons CO_2 \text{ (dissolved)} + H_2O$$

Dissolved CO_2 hydrates by a slow reaction (a half-time of approximately 15 seconds) to yield carbonic acid:

$$CO_2 + H_2O \rightleftharpoons H_2CO_3$$

This reaction predominates at a pH of less than 8, with the equilibrium concentration of H_2CO_3 only about 1/400 that of the unhydrated CO_2. Above a pH of 10, $CO_2 + OH^- \rightleftharpoons HCO_3^-$ is the dominant reaction.

H_2CO_3 is a fairly weak acid that dissociates rapidly relative to the hydration reaction:

$$H_2CO_3 \rightleftharpoons H^+ + HCO_3^-$$
$$HCO_3^- \rightleftharpoons H^+ + CO_3^=$$

The pK_1 dissociation value of the first reaction, including both hydrated and unhydrated CO_2 as the undissociated molecule, is 6.43 at 15°C. The pK_2 of the second reaction is 10.43 (15°C).

The bicarbonate and carbonate ions also dissociate to establish an equilibrium:

$$HCO_3^- + H_2O \rightleftharpoons H_2CO_3 + OH^-$$
$$CO_3^= + H_2O \rightleftharpoons HCO_3^- + OH^-$$
$$H_2CO_3 \rightleftharpoons H_2O + CO_2.$$

The hydroxyl ions generated in the first two of the above reactions result in alkaline waters (above a pH of 7) in lakes and streams that have a naturally high content of carbonates derived from surface and groundwater of the drainage basin. As water percolates through soil of the drainage basin, it becomes enriched with CO_2 from plant and microbial respiration. The carbonic acid that forms solubilizes limestone of calcium-enriched rock formations, and produces calcium bicarbonate $(Ca(HCO_3)_2)$, which is relatively soluble in water, and increases the amount of ionized Ca^{++} and HCO_3^- of the water. As HCO_3^- and $CO_3^=$ increase in hard waters of calcareous regions, common to much of the glaciated temperate region, the pH also increases from the release of hydroxyl ions. In normal hard waters of the midwestern United States, for example, HCO_3^- is the dominant anion (ca. approximately 60 per cent), so that the pH of the water is > 8 and $[HCO_3^-] > 100$ mg l^{-1}. Saline lakes of endorheic regions often have a carbonate dominance $(CO_3^= > 10,000$ mg l^{-1} and $HCO_3^- > 1000$ mg l^{-1} with pH > 9.5). Such highly saline carbonate brines, in which total inorganic carbon exceeds several moles per liter, are rare.

Photosynthesis and respiration are two major factors that influence the amounts of CO_2 in water. However, the equilibria of the reactions given above result in the buffering action of alkaline waters, which contain appreciable amounts of bicarbonate. Water tends to resist change in pH as long as these equilibria are operational. An addition of hydrogen ions neutralizes hydroxyl ions formed by the dissociation of HCO_3^- and $CO_3^=$, but more hydroxyl ions are formed immediately by reaction of the carbonate with water. Consequently, the pH remains essentially unaltered, unless the supply of carbonate or bicarbonate ions is exhausted. Similarly, when hydroxyl ions are added they react with bicarbonate ion.

$$HCO_3^- + OH^- \rightleftharpoons CO_3^= + H_2O.$$

If the pH of a solution is held constant with buffer reactions, and it is permitted to equilibrate with gaseous carbon dioxide, the total hydrated and unhydrated CO_2 in solution is independent of pH, while the bicarbonate and carbonate concentrations increase with pH in accordance with the pK values. These equilibria are influenced by temperature and by salt concentration (ionic strength). At the salinity of seawater, the pK_1 for HCO_3^- is about 0.5 pH unit, and the pK_2 for $CO_3^=$ is about 1 unit lower than in fresh water. Both oceanic and fresh waters are close to equilibrium with atmospheric CO_2. In the marine habitat, the inorganic carbon pool contains about 2 millimoles C l^{-1}, largely as HCO_3^-, a reservoir some 50 times that of the atmosphere. In fresh waters, the total inorganic carbon (ΣCO_2) is much more variable, within a typical range of 50 micromoles to 10 millimoles l^{-1}, and more pH-dependent. From these dissociation relationships, the proportions of CO_2, HCO_3^-, and $CO_3^=$ at various pH values can be evaluated (Fig. 11-1). Free CO_2 dominates in water at pH 5 and below, while above pH

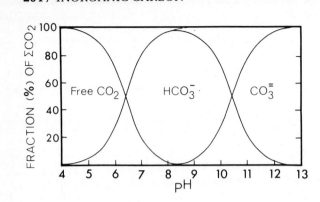

Figure 11-1 Relation between pH and the relative proportions of inorganic carbon species of CO_2 ($+H_2CO_3$), HCO_3^-, and $CO_3^=$ in solution. (Slightly modified from Golterman, H. L. (ed.), Methods for Chemical Analysis of Fresh Waters. IBP Handbook No. 8. Oxford England, Blackwell Scientific Publications, 1969.)

TABLE 11-1 Proportions of CO_2, HCO_3^-, and $CO_3^=$ in Water at Various pH Values

pH	TOTAL FREE CO_2	HCO_3^-	$CO_3^=$
4	0.996	0.004	1.25×10^{-9}
5	0.962	0.038	1.20×10^{-7}
6	0.725	0.275	0.91×10^{-5}
7	0.208	0.792	2.6×10^{-4}
8	0.025	0.972	3.2×10^{-3}
9	0.003	0.966	0.031
10	0.0002	0.757	0.243

From Hutchinson, G. E.: A Treatise on Limnology. New York, John Wiley & Sons, Inc., 1957, p. 657.

9.5 $CO_3^=$ is quantitatively significant (Table 11-1). Between pH 7 and 9, HCO_3^- predominates.

CO_2 Exchange Between the Atmosphere and Water

The diffusion of CO_2 from the atmosphere and the dissociation kinetics of dissolved carbonates are obviously of major importance to photosynthetic organisms dependent on the availability of inorganic carbon. The magnitude of CO_2 exchange between the atmosphere and water cannot be determined by partial pressure differences alone. Many lakes with surface waters near neutrality are slightly supersaturated with CO_2 relative to the atmospheric pressure of CO_2. Other waters, especially in alkaline bicarbonate lakes containing large amounts of carbonate, are apparently not in equilibrium with the CO_2 of the atmosphere, although they may be with other gases such as oxygen. Diffusion of atmospheric CO_2 has been studied by Broecker and coworkers (1965, 1968, 1971, 1973), who utilized techniques of radium-226 and flux of radon-222 to determine gas transfer between the atmosphere and water. When these methods were applied to a soft-water lake of the Canadian Shield of very low ΣCO_2, atmospheric invasion of CO_2 was adequate (0.12 ± 0.06 g C m^{-2} day^{-1}) to account for 30 to 90 per cent of the carbon fixed by phytoplankton (Schindler, et al., 1972). In this relatively sheltered lake, an invasion rate of CO_2 from the air of 17 ± 8 mol CO_2 m^{-2} day^{-1} was estimated through a hypothetical "stagnant boundary layer" at the surface of about 300 μm in thickness (Emerson, et al., 1973). This layer decreases in thickness with increased exchange as wind velocities increase, especially above 1.5 m sec^{-1}. Organic substances dissolved in or present on the water surface can also decrease exchange rates of gas and evaporation. In more productive waters, consumption of CO_2 by epilimnetic photosynthesis enhances the flux of atmospheric CO_2 to the water (Weiler, 1974; Emerson, 1975).

The flux of CO_2 into poorly buffered soft waters can be enhanced as CO_2 (dissolved) and H_2CO_3 concentrations decline and hydroxide concentrations increase (Wood, 1974, 1977). At high pH values, the realized rates of CO_2 flux across the air–water boundary can exceed theoretical rates.

At the opposite extreme, in a hard-water lake in Michigan that receives high inputs of inorganic carbon, Otsuki and Wetzel (1974) determined that the estimated partial pressure in two inlet waters and groundwater was about 30 times higher than that in the atmosphere. In the outlet, the partial pressure of CO_2 (dissolved) in water was seven times higher during the spring period, and three times higher even during the summer period of maximum utilization and losses through precipitation of $CaCO_3$. These estimates indicate that CO_2 in this lake was being released into the atmosphere throughout the year.

Proportions of Carbonates in Fresh Waters

The total inorganic carbon concentration in fresh water depends on the pH, which is governed largely by the buffering reactions of carbonic acid and the amount of bicarbonate and carbonate derived from the weathering of rocks. Carbonates exist as a number of polymorphic and hydrated forms. The most important carbonate of aquatic systems is $CaCO_3$, which occurs in natural waters principally as calcite and the metastable polymorphic aragonite.

The solubility of CO_2 increases markedly in water that contains carbonate. A definite amount of free CO_2 will remain in solution after equilibrium is reached between calcium, bicarbonate, carbonate, and undissociated calcium carbonate. The amount of excess CO_2 required to maintain stability of $Ca(HCO_3)_2$ in solution increases very rapidly with increasing content of bicarbonate in the water derived from carbonates. If the amount of free CO_2 is increased above that required to maintain a given amount of $CaCO_3$ in solution at equilibrium as $Ca(HCO_3)_2$, this aggressive CO_2, as it is termed, will dissolve additional $CaCO_3$.

If a solution of calcium bicarbonate in equilibrium with CO_2, H_2CO_3, and $CO_3^=$ loses a portion of the CO_2 required to maintain the equilibrium (e.g., CO_2 assimilated by photosynthetic organisms), $CaCO_3$ will precipitate until the equilibrium is reestablished by the formation of CO_2:

$$Ca(HCO_3)_2 \rightleftharpoons CaCO_3 \downarrow + H_2O + CO_2.$$

The excess CO_2 that is required to maintain large amounts of HCO_3^- in solution at equilibrium can be lost in several ways, resulting in massive precipitation of $CaCO_3$.

Groundwater of limestone regions, heavily enriched with CO_2 from terrestrial decomposition, can release much CO_2 into the atmosphere when it flows to the surface, with resulting precipitation of $CaCO_3$. When such spring water, rich in bicarbonate, surfaces in streams or lakes, all substrata become covered with a dense encrustation of $CaCO_3$. A major cause of the loss of aggressive CO_2 in lakes and certain streams is photosynthetic utilization of CO_2 by littoral flora and phytoplankton. Hard-water lakes rich in bicarbonate commonly undergo massive epilimnetic decalcification during the summer stratification period due to photosynthetic removal of CO_2 (a subject discussed later). In the littoral zone of hard-water lakes, the submersed macrophytic vegetation is densely encrusted with $CaCO_3$ precipitated by photosynthetic utilization of CO_2 by the macrophytes and epiphytic algae. The marl encrustations are frequently massive, and often exceed the weight of the plant (Wetzel, 1960). Blue-green algae growing attached to substrata in the littoral of lakes and in streams also produce large deposits of carbonates (Golubić, 1973).

In addition to highly dynamic demands for CO_2 from and inputs of CO_2 to fresh water, complex shifts in precipitation and dissolution reactions of carbonate occur spatially and temporally. In alkaline hard-water lakes, often twice the concentrations of calcium and bicarbonate are found than would be expected on the basis of equilibrium with atmospheric pressures of CO_2 (Ohle, 1934, 1952; Wetzel, 1966b, 1972; Otsuki and Wetzel, 1974). The solubility product of $CaCO_3$ is low (0.48×10^{-8}), and $CaCO_3$ can start precipitating from calcareous waters when the pH is sufficiently high in a uniformly buffered system, or where equilibria are shifted in microzones associated with sites of active photosynthesis. However, large amounts of inorganic carbon can exist as carbonate and $CaCO_3$ in metastable conditions. The rate of $CaCO_3$ precipitation is slow, unless induced metabolically, as by photosynthesis. The result is a supersaturation with respect to both Ca^{++} and HCO_3^- at concentrations often two to three times that predicted on the basis of equilibria. There is strong evidence that appreciable $CaCO_3$ occurs in stable colloidal form in hard-water lakes (White and Wetzel, 1975). The trophogenic zone in moderately hard Lake Mendota was found to be supersaturated with respect to Ca^{++} and HCO_3^- in all seasons except winter (Morton and Lee, 1968). Extremely hard Lawrence Lake was found to be supersaturated continually (Otsuki and Wetzel, 1974), a situation common where dissolved CO_2 is not in equilibrium with atmospheric CO_2.

The importance of colloidal $CaCO_3$, in addition to larger particulate $CaCO_3$, is just beginning to be appreciated in relation to indirect effects upon metabolism and flux rates of organic carbon (Wetzel and Rich, 1972). Organic compounds (amino acids, fatty acids, humic acids) adsorb to particulate and colloidal $CaCO_3$ (Chave, 1965; Chave and Suess, 1970; Suess, 1968, 1970; Meyers and Quinn, 1971a; Wetzel and Allen, 1970; Otsuki and Wetzel, 1973; Orlov, et al., 1973; Stewart and Wetzel, 1981). Although such adsorption could be viewed as a scavenging and concentrating process whereby labile dissolved organic carbon from dilute solution is rendered more concentrated for utilization by bacteria, empirical evidence indicates, rather, a chemical competition with the bacteria for the organic substrates. During photosynthetic removal of CO_2, a large fraction of $CaCO_3$ is precipitated by algae and macrophytic vegetation. Frequently, the plant cells serve as a nucleus for particulate $CaCO_3$ formation, which is where organic compounds are being secreted. This association of dissolved organic detrital carbon with $CaCO_3$ is a component of certain freshwater systems that affects the chemical milieu without clearly defined energetic transformations (Wetzel, et al., 1972). The organic coatings also reduce the rate of dissolution of sedimenting $CaCO_3$ in lakes, and form a major sink for inorganic and organic detrital carbon (Wetzel, 1970, 1972).

Alkalinity and Acidity of Natural Waters

Natural waters exhibit wide variations in relative acidity and alkalinity, not only in actual pH values, but also in the amount of dissolved material producing the acidity or alkalinity. The concentration of these compounds and the ratio of one to another determine the actual pH and the buffering capacity of a given water. Since lethal effects of most acids begin to appear near pH 4.5 and of most alkalis near pH 9.5, that buffering can be of major importance in the maintenance of life.

Alkalinity of waters, as usually interpreted, refers to the quantity and kinds of compounds present which collectively shift the pH to the alkaline side of neutrality. The property of alkalinity is usually imparted by the presence of bicarbonates, carbonates, and hydroxides, and less frequently in inland waters by borate, silicate, and phosphates. The CO_2–HCO_3^-–$CO_3^=$ equilibrium system is the major buffering mechanism in fresh waters.

The terms *alkalinity, carbonate alkalinity, alkaline reserve, titratable base,* or *acid-binding capacity* are frequently used to express the total quantity of base (usually in equilibrium with carbonate or bicarbonate) that can be determined by titration with a strong acid (Hutchinson, 1957). The milliequivalents of acid necessary to neutralize the hydroxyl, carbonate, and bicarbonate ions in a liter of water are known as the *total alkalinity*. Alkalinity is numerically the equivalent concentration of titratable base, and is determined by titration with a standard solution of a strong acid to equivalency points dictated by pH values at which the alkaline contributions of hydroxide, carbonate, and bicarbonate are neutralized.

The least ambiguous usage of alkalinity is to express values as mass per unit volume; i.e., milliequivalents per liter (meq l^{-1}). In effect, alkalinity measures the proton deficiency with respect to the reference proton level CO_2–H_2O (this topic is reviewed at length by Stumm and Morgan, 1981). Alkalinity is often expressed in milligrams per liter (or parts per million) of $CaCO_3$. This expression assumes that alkalinity results only from calcium carbonate and bicarbonate, which in some lakes, e.g., closed alkaline lakes, implies much greater calcium than actually is present. In moderately hard waters, nearly all of the base is present as bicarbonate, and the term *bicarbonate alkalinity*, as mg HCO_3^- l^{-1}, is sometimes used. The clearest expression, however, is milliequivalents per liter ($=50$ mg l^{-1} as $CaCO_3$).

The term *hardness* is frequently used as an assessment of the quality of water supplies. The hardness of a water is governed by the content of calcium and magnesium salts, largely combined with bicarbonate and carbonate (temporary hardness) and with sulfates, chlorides, and other anions of mineral acids (permanent hardness). The carbonate hardness can be removed by boiling, which causes precipitation of $CaCO_3$. The fraction of calcium and magnesium remaining in solution as sulfates, chlorides, and nitrates after boiling constitutes the residual noncarbonate hardness. The extent of

TABLE 11-2 **Various Scales of Hardness of Water**

1 German degree of hardness, dH°	= 10 mg CaO l^{-1}
	= 7.14 mg Ca l^{-1}
	= 17.9 mg Ca(HCO$_3$)$_2$ l^{-1}
1 French degree of hardness, French H°	= 10 mg CaCO$_3$ l^{-1}
1 English degree of hardness, English H°	= 10 mg CaCO$_3$ 0.7 l^{-1}
	= 0.8° dh
1° dH	= 1.25° English H°
	= 1.79 French H°
1 French H°	= 0.56 German dH°
	= 0.7 English H°
1 American degree of hardness	= 1 mg CaCO$_3$ l^{-1}
	= 0.056° German dH°
International degree of hardness, mval	= 1 meq l^{-1}
	$= \dfrac{\text{German dH°}}{2.8}$

After Höll, K.: Water: Examination, Assessment, Conditioning, Chemistry, Bacteriology, Biology. Berlin, Walter de Gruyter, 1972. See also Hochmüller and Simoneth (1980a, 1980b).

hardness has been expressed numerically in a remarkably heterogeneous system of scales among different countries (equivalents are given in Table 11-2). For example, one degree of hardness (H°) in the United States equals 1 mg CaCO$_3$ l^{-1}; one German H° corresponds to a concentration of 10 mg lime (CaO) l^{-1}. Although Höll (1972) has proposed that the international unit be expressed in mval (1 mval = 1 milliequivalent per liter of the material concerned), this terminology has been rescinded by the Système International d'Unités (SI) (see the critical evaluation of the unit Val (= gram equivalent) by Hochmüller and Simoneth, 1980a, 1980b).

Acidity is used infrequently as a parameter in limnological investigations. Uncombined carbon dioxide, organic acids such as tannic, humic, and uronic acids, mineral acids, and salts of strong acids and weak bases are usually responsible for the acidity of natural waters. Free CO$_2$ of most waters is seldom present in large quantities, because of its reactions in the carbonate equilibria and exchange with the atmosphere, discussed earlier.

In practice, acidity is a measure of the quantity of strong base per liter required to attain a pH equal to that of a solution of sodium carbonate (Na$_2$CO$_3$) equivalent to the total inorganic carbon; i.e., it is a measure of the active or free CO$_2$ expressed as meq or mg l^{-1} CaCO$_3$ rather than as a concentration of CO$_2$. Mineral acid acidity measures those materials present, other than CO$_2$, which result in the pH value of water below 4.5. Details on methodology for evaluation of alkalinity and acidity are given in the American Public Health Association publication (1976), Golterman and Clymo (1969), and Wetzel and Likens (1979).

Hydrogen Ion Activity

Pure water dissociates weakly to H$^+$ and OH$^-$ ions. The dissociation constant is very small (10^{-14}), however, and the amounts of H$^+$ and OH$^-$ present are 10^{-7} g-ions per liter. Natural waters are, of course, not pure, and salts, acids, and bases contribute to the H^+ and OH^- ions in varying ways, depending on the individual circumstances. Since the dissociation constant of water is fixed, addition of one ion will result in a decrease of the other. The pH is usually defined as the logarithm of the reciprocal of the concentration of free hydrogen ions.* The "p" of pH refers to the power (puissance) of the hydrogen ion activity. Therefore, more H^+ activity in an acid reaction increases the power from neutrality (10^{-7} or pH 7) to say 10^{-4} (pH 4). In more alkaline reactions, H^+ ion activity is decreased from neutrality to, for example, 10^{-10} (pH 10). By definition, pH values cannot be averaged arithmetically, but the average must be estimated from the logarithm of the reciprocals.

The pH of natural waters is governed to a large extent by the interaction of H^+ ions arising from the dissociation of H_2CO_3 and from OH^- ions produced during the hydrolysis of bicarbonate. The pH of natural waters ranges between the extremes of < 2 to 12. Nearly all waters with pH values less than 4 occur in volcanic regions that receive strong mineral acids, particularly sulfuric acid. The oxidation of pyrite of rocks and clays in drainage basins can result in sulfuric acid drainage to lakes.

Low pH values are found in natural waters rich in dissolved organic matter, especially in bogs and bog lakes that are dominated in the littoral mat by the moss *Sphagnum*. The pH of *Sphagnum* bogs is usually in the range of 3.3 to 4.5 (0.5 to 0.03 meq H^+ l^{-1}). The sources of the H^+ ion activity are several (Clymo, 1963, 1964). Although acidic precipitation resulting from industrial pollution can influence the pH of poorly buffered waters significantly (Likens, et al., 1972), in bog areas with no great pollution, the supply of H^+ in rain is unlikely to exceed 30 per cent of the total annual input of H^+. The metabolism of proteins and the reduction of $SO_4^=$ by sulfur-metabolizing bacteria may contribute some H^+ ions to these waters, but their contributions are likely to be small. Although live *Sphagnum* plants secrete organic acids, concentrations of these acids are usually insufficient to account for the observed acidity. Most of the H^+ ion concentrations appear to result from the active cation exchange by the cell walls of *Sphagnum*, during which H^+ is released.

Very high pH values of lakes are usually found in endorheic regions, where lake water contains exceedingly high concentrations of soda (e.g., Na_2CO_3).

The range of pH of a majority of open lakes is between 6 and 9. Most of these lakes are the "bicarbonate type," i.e., they contain varying amounts of carbonate and are regulated by the CO_2–HCO_3^-–$CO_3^=$ buffering system. Calcareous hard-water lakes commonly are buffered strongly at pH values above 8. Seepage lakes and lakes of igneous rock catchment areas are less well buffered and more acidic, with pH values usually somewhat less than 7.

Spatial and Temporal Distribution of Total Inorganic Carbon and pH in Lakes

Total inorganic carbon (ΣCO_2) is distributed uniformly with depth during periods of circulation in dimictic lakes, and in shallow lakes of insufficient depth for thermal

---

*It should be noted that measurements of pH involve not the concentration but the activity of the hydrogen ion. One measures the differences of hydrogen ion activity, rapidly being released and incorporated from proton-donor molecules, between unknown solutions and standard buffers of known pH values.

stratification. The ΣCO_2 content of the water is dependent upon the equilibria established between atmospheric CO_2, the bicarbonate-carbonate system, contributions from metabolic respiration, and utilization in photosynthesis. Metabolism is markedly influenced by numerous parameters; in the spring, increasing light and temperature exert major controls on photosynthetic uptake of CO_2, and generation of CO_2 from microbial decomposition of organic matter is temperature- and oxygen-dependent.

During the period of thermal stratification, several conspicuous changes occur in the vertical distribution of ΣCO_2. Oligotrophic lakes which exhibit an orthograde oxygen curve and do not possess high concentrations of bicarbonate and carbonate usually also have an orthograde ΣCO_2 curve (Fig. 11–2). Where the dissolved CO_2–H_2CO_3 component constitutes an appreciable part of the total CO_2, as in soft-water lakes, the warmer epilimnetic waters contain less CO_2 because of decreased solubility. Photosynthetic utilization may exceed rates of replacement if mixing of the epilimnion is relatively incomplete, but this phenomenon is uncommon in the open water. A slight increase in the ΣCO_2 is often observed in the hypolimnetic water overlying the sediments. The vertical pH distribution exhibits a pattern approximately the inverse to that of ΣCO_2.

The vertical distributions of ΣCO_2 and pH are strongly influenced by various biologically mediated reactions (each of which is treated separately elsewhere). Most conspicuous is the photosynthetic utilization of CO_2 in the trophogenic zone, which tends

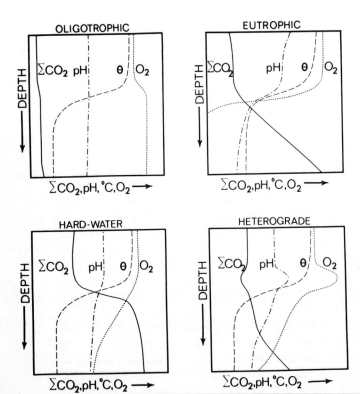

Figure 11-2 Generalized vertical distributions of ΣCO_2 and pH in stratified lakes of very low and high productivity, hard water calcareous lakes exhibiting pronounced epilimnetic decalcification, and a lake with a distinct positive heterograde oxygen curve (θ = temperature).

to reduce CO_2 content and to increase pH, and the respiratory generation of CO_2 throughout the water column and sediments, which tends to decrease pH. In addition to heterotrophic degradation of organic matter, generation of CO_2 and reduction of pH are augmented by microbial methane fermentation, nitrification of ammonia, and sulfide oxidation. Further, denitrification of nitrate to molecular nitrogen and reduction of sulfate to sulfide result in a slight net increase in pH. The combination of decompositional processes results in an increase in ΣCO_2 of the hypolimnetic waters and a decrease in pH.

As the intensity of decomposition increases in the tropholytic zone, the amount of CO_2 and especially of HCO_3^- increases markedly. The accumulation of ΣCO_2, both free and combined, exceeds the rate of oxygen consumption, and decomposition shifts from aerobic to anaerobic as the hypolimnion becomes anoxic. The origin of increasing concentrations of HCO_3^- in the hypolimnion, especially near the sediments, stems in part from bacterial production of ammonium bicarbonate in the sediments (Ohle, 1952). Ferrous and manganous ions are released as bicarbonates from the sediments when the redox potential is reduced sufficiently under anoxic conditions. In hard-water lakes, part of the $CaCO_3$ sedimenting from the epilimnion undergoes dissolution in the colder, more acidic hypolimnion, and results in increases in HCO_3^- concentrations. It is clear, however, as discussed previously, that sorbed coatings of dissolved organic matter reduce the rates of $CaCO_3$ dissolution. The hypolimnetic increases of cations, especially calcium, usually are delayed somewhat from proportional increases in hypolimnetic HCO_3^-. Under oxidized conditions, Ca^{++} is complexed with the sediments. Under the reducing conditions of anoxia, Ca^{++} is released nearly in proportion to bicarbonate, although some Ca^{++} may be complexed with humic acids (Ohle, 1955; Stewart and Wetzel, 1981). Hutchinson (1941) demonstrated conclusively that during stratification, the extent of sediment contact per volume of water (and therefore the morphology of the basin) in the metalimnion and hypolimnion is related directly to the development of increasing HCO_3^- as the period of stratification progresses.

Therefore, in eutrophic lakes possessing a clinograde oxygen curve, a marked inverse clinograde ΣCO_2 curve can be observed (Fig. 11–2) and pH decreases markedly in the hypolimnion.

The vertical distribution of ΣCO_2 of very hard calcareous lakes shifts seasonally, because photosynthetic utilization of CO_2 in the trophogenic zone occurs rapidly, and induces the precipitation of $CaCO_3$ (Fig. 11–3). This epilimnetic biogenic decalcification was described long ago (e.g., Minder, 1922; Pia, 1933), and has been observed frequently. While precipitation of $CaCO_3$ can be induced by many physical and biotic agents (increasing temperature, bacterial metabolism), photosynthetic utilization of CO_2 by algae and submersed macrophytes is by far the dominant mechanism (cf. Otsuki and Wetzel, 1974). The result is a marked decrease in the ΣCO_2 of the epilimnion by the end of summer stratification (see Fig. 11–2), and a slow progressive increase in the hypolimnion (Fig. 11–3). In the example shown, the hypolimnion became anoxic below a depth of 11 m in September and October, 1968, as reflected in the increasing HCO_3^- in this layer. The water of the trophogenic zone of unproductive calcareous lakes fluctuates very little seasonally in pH (Fig. 11–4). The bicarbonate buffering capacity is adequate to compensate for metabolic changes in ΣCO_2. The pH of the hypolimnion progressively decreases throughout the period of stratification. Similar changes also are

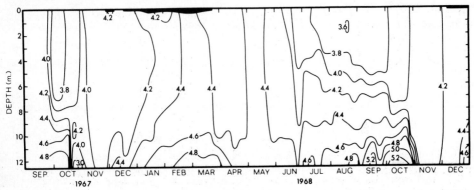

Figure 11-3 Depth–time diagram of isopleths of alkalinity, predominately bicarbonate, in meq l⁻¹, Lawrence Lake, a calcareous hard water lake of southwestern Michigan. Opaque area = ice-cover to scale. (From Wetzel, unpublished data.)

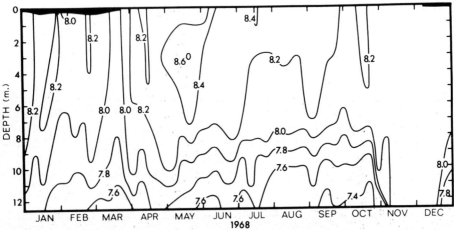

Figure 11-4 Depth–time diagram of isopleths of pH in hardwater Lawrence Lake, Michigan. Opaque area = ice-cover to scale. (From Wetzel, unpublished data.)

observed under ice-cover, but they are less marked than during the summer period, when rates of production and decomposition are lower.

Vertical changes in ΣCO_2 and pH occur much more rapidly in eutrophic lakes (Figs. 11–5 and 11–6). In the example given in the figures, the hypolimnion below 3 m became anoxic within a month after the onset of thermal stratification, and became nearly anoxic under ice-cover. The pH values of the epilimnion represent midmorning measurements. Under the conditions of intensive photosynthesis by phytoplankton, the pH can undergo appreciable diurnal fluctuations, often exceeding 10 in the late afternoon and decreasing below 8 during darkness. Biogenic decalcification of the epilimnion also is apparent in this moderately hard hypereutrophic lake (Fig. 11–5).

Where photosynthetic activity in a layer is particularly intense, as might occur in the metalimnion, a positive heterograde oxygen curve is frequently found (see Chapter

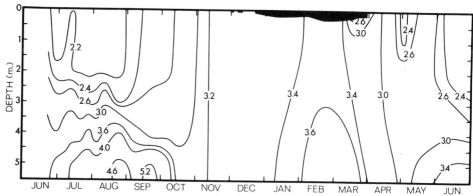

Figure 11-5 Depth–time diagram of isopleths of alkalinity, predominately bicarbonate, in meq l^{-1}, of moderately calcareous, hypereutrophic Wintergreen Lake, southwestern Michigan, 1971–72. Opaque area = ice-cover to scale. (From Wetzel, et al., unpublished data.)

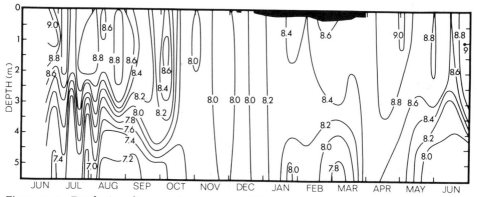

Figure 11-6 Depth–time diagram of isopleths of pH in hypereutrophic Wintergreen Lake, Michigan, 1971–72. Opaque area = ice-cover to scale. (From Wetzel, et al., unpublished.)

9). Commonly associated with this metabolism is a corresponding positive heterograde pH curve and a negative heterograde ΣCO_2 curve (Fig. 11-2, and at 3 m in May of Fig. 11-4). The opposite situation of a negative heterograde oxygen curve, with concomitant ΣCO_2 and pH curves, occasionally is found in strata of intensive respiration, e.g., containing plates of bacteria or aggregations of zooplankton (see Nagasawa, 1959).

Hypolimnetic CO$_2$ Accumulation in Relation to Lake Metabolism

As in the case of hypolimnetic oxygen deficits discussed earlier, changes in the concentrations of ΣCO_2 in the hypolimnion have been used to indirectly estimate the organic production of the trophogenic zone. The concept is that the accumulation of CO_2 in the hypolimnetic tropholytic zone from decomposition is proportional to the production of organic matter in the trophogenic zone that settles into the hypolimnion. This approach to an estimate of lake metabolism is complicated by inputs of organic matter that do not originate in the trophogenic zone, and by losses of carbon from the system that are not represented in CO_2 changes in the tropholytic zone. For many lake systems the measurement of hypolimnetic CO_2 accumulation can provide a reasonable estimate of the intensity of lake metabolism.

The idea of estimating production from CO_2 accumulation was first suggested by Ruttner in his treatise on tropical lakes (1931). Einsele (1941) compared the CO_2 accumulation in the hypolimnion with several other estimates of production in his critical evaluation of the effects of phosphorus fertilization on productivity of algae, and found it to be in agreement. By far the most comprehensive application of the hypolimnetic CO_2 accumulation principle to lakes, and its comparison with other measures of production, especially the oxygen deficit, was done by Ohle (1934, 1952, 1956). The CO_2-accumulation approach has the advantage of being able to follow metabolically mediated changes under both aerobic and anaerobic conditions, whereas the oxygen deficit method is applicable only when the hypolimnion is oxic. More fundamentally, carbon, which is the initial and end product of organic metabolism, is one of the best parameters by which to evaluate productivity.

There are several assumptions underlying calculations of hypolimnetic CO_2 accumulation. First, it is important that the water under consideration be relatively well isolated from the atmosphere and from CO_2 exchange; these conditions may be found in the hypolimnia of small, sheltered lakes with steep thermal stratification gradients or when the entire lake is under ice-cover. Second, it is assumed that the allochthonous inputs of organic matter from outside the basin are very small in comparison to autochthonous production within the basin, a situation that probably is not realized very often (cf. Chapter 22). Loss of organic production from the trophogenic zone via the outlet is assumed to be small, a condition which, again, is highly variable in open lakes. A major assumption is that most of the synthesized organic matter of the trophogenic zone decomposes only after it has sedimented to the tropholytic zone. This assumption certainly causes underestimations, because much decomposition of algae, littoral macrophytes, and sessile algae occurs in the epilimnetic and metalimnetic layers before reaching the relatively quiescent hypolimnion. The rates of decomposition in the trophogenic zone depend upon many physical and biological factors. Important factors include the proportion of production by the littoral flora vs. phytoplankton, seasonal shifts in the proportion of phytoplankton algae reaching the hypolimnion, e.g., silicious diatoms sediment more rapidly than small algae, which are more neutrally buoyant, and water temperature. A significant amount of organic production will be incorporated permanently into the sediments without complete decomposition; the proportion permanently lost in this way generally increases in lakes of greater productivity. Perhaps the greatest problem rests in the assumption of a respiratory quotient (RQ = CO_2/O_2) of 0.85, an average figure for the *aerobic* decomposition of a mixture of carbohydrate, lipid, and protein organic compounds. As will be discussed later (Chapter 22), anaerobic decomposition produces CO_2 utilizing alternate electron acceptors after oxygen has been depleted, and often leads to RQ values much greater than 1 (Rich, 1975, 1980; Rich and Wetzel, 1978). Despite these numerous limitations, many estimates of hypolimnetic CO_2 accumulation correlate with the general lake productivity and are directly proportional to mean depth of the basin.

Analysis of hypolimnetic CO_2 accumulation requires evaluation of the various components of total CO_2 and their origins. In bicarbonate-poor waters, the difference in CO_2 content at the beginning and after a period of stratification is determined. In bicarbonate-rich waters, the amount of free CO_2 is very small at the onset of hypolimnetic isolation. As stratification is maintained, respiration during aerobic and anaerobic decomposition allows the accumulation of free CO_2 and bicarbonate. In addition, a portion of the bicarbonate is present as "volatile" ammonium carbonate of metabolic origin; additional carbon exists as volatile bicarbonates, half of which are of metabolic origin. Half of the nonvolatile bicarbonate is bound CO_2 of $CaCO_3$, which largely becomes incorporated into the sediment. The sum of these CO_2 inputs is assumed to be an estimate of lake metabolism. A sample calculation is given in Table 11–3 (see Wetzel and Likens, 1979). If CO_2 accumulation is calculated for the entire hypolimnion, and this is divided by the volume of the epilimnion, the quantity termed the *relative assimilation intensity* results. In the example given (Table 11–3):

$$(10.10 \text{ mg } CO_2 \text{ l}^{-1} \text{ month}^{-1}) \left[\frac{4.620 \times 10^6 \text{ m}^3}{7.442 \times 10^6 \text{ m}^3} \right] = 6.3 \text{ mg } CO_2 \text{ l}^{-1} \text{ month}^{-1}.$$

This estimate implies that, in the epilimnion of this lake, approximately 6.3 mg CO_2 were photosynthetically combined into carbohydrate per liter of water per month

TABLE 11–3 Summary of Method of Evaluating the Hypolimnetic CO₂ Accumulation of a Lake with a Sample Calculation

CALCULATIONS	LAKE EXAMPLE
(1) $NH_4:HCO_3 = 18.04:61.02$	(1) 1.85 mg NH_4 l^{-1} present as NH_4HCO_3
(2) $HCO_3 = \dfrac{(NH_4)(61.02)}{18.04} = (3.38)(NH_4) = \beta$	(2) $\beta = (1.85)(3.38) = 6.25$
(3) $HCO_3:CO_2 = 61.02:44.01$	(3) $b = (6.25)(0.721) = 4.51$
$\quad CO_2 = \dfrac{(HCO_3)(44.01)}{61.02} = (0.721)(HCO_3)$	
$\quad b = \beta (0.721)$	
(4) $\alpha = HCO_3;\ \alpha - \beta = x$	(4) $\alpha = 32.4$ mg HCO_3 l^{-1}
	$\quad x = 32.4 - 6.25 = 26.15$
(5) $a = \dfrac{x(0.721)}{2} = x(0.361)$	(5) $a = (26.15)(0.361) = 9.43$
(6) $c = CO_2$	(6) $c = 13.10$ mg CO_2 l^{-1}
(7) $\Sigma CO_2 = a + b + c$	(7) $\Sigma CO_2 = 9.43 + 4.51 + 13.10 = 27.04$
(8) $\delta O_2 = $ actual O_2 deficit	(8) mg O_2 l$^{-1} = 0.56$
$\quad \dfrac{CO_2}{O_2} = \dfrac{44}{32} = 1.375$	
$\quad \delta O_2 (1.375) = \delta CO_2$	$\quad °C = 6.7,\ O_2$ at turnover $= 12.27$ mg l^{-1}
$\quad$ R.Q. $= 0.85*$	$\quad \delta O_2 = 12.27 - 0.56 = 11.71$
$\quad \gamma = \delta CO_2 (0.85)$	$\quad \gamma = (11.71)(1.375)(0.85) = 13.69$
(9) $\Sigma CO_2 - \gamma = \Delta^1 CO_2$	(9) $\Delta^1 CO_2 = 27.04 - 13.69 = 13.35$
(10) $z = \Delta^1 CO_2 (2)$	(10) $z = (13.35)(2) = 26.70$
(11) $\xi = z + \gamma$	(11) $\xi = 26.70 + 13.69 = 40.39$
$\quad \dfrac{\xi (30)}{\text{No. of days}} = \xi$ per month	$\quad \xi$ per month $= \dfrac{(40.39)(30)}{120} = 10.1$ mg CO_2 l^{-1}

Note: Example based on: 1.85 mg NH_4 l^{-1}; 13.1 mg CO_2 l^{-1}; 32.4 mg HCO_3 l^{-1}; 0.56 mg O_2 l^{-1}; $°C = 6.7$; period of stratification of 4 months. After Ohle, 1952.

*An RQ value (CO_2/O_2) of 0.85 is a mean value based on a number of analytical analyses of plant and animal respiration, and agrees approximately with some whole closed lake evaluations (but see discussion of RQ variations in Chapter 22).

during the period of observation. The CO_2-accumulation method of calculating production usually leads to underestimation because of the confounding factors mentioned earlier. Further, some CO_2 loss from the upper hypolimnion occurs by turbulence, and some bicarbonate is released from the reduction of ferric hydroxide as redox potential decreases during stratification.

The hypolimnetic CO_2 accumulations in several lakes are compared in Table 11–4, along with several other indices of increasing productivity (Ohle, 1955b). The general correlations among various criteria hold. Einsele (1941) applied the CO_2 accumulation evaluation of productivity, along with several other measures, to the trophogenic zone of Schleinsee, Germany, before and after extensive fertilization of the lake with phosphorus (Table 11–5). Schleinsee lends itself well to such analyses because of its limited inflow and outflow. Although Einsele's estimations were quite approximate, again the comparison was good. When used in a relative manner, this method is of value.

TABLE 11–4 Comparison of the Relative Assimilation Intensity to Other Indices of Productivity in Lakes of Northern Germany

	RELATIVE ASSIMILATION INTENSITY (mg CO_2 l^{-1} EPILIMNION MONTH^{-1})	CHLOROPHYLL OF EPILIMNION (μg l^{-1})	ESTIMATE OF PHOTOSYNTHETIC EFFICIENCY OF SOLAR RADIATION (%)	HYPOLIMNETIC O_2 DEFICIT (mg O_2 l^{-1})
Schaalsee	1.91	1.3	0.76	3.30
Schöhsee	2.13	1.4	0.56	6.96
Zanzen	2.33	1.6	0.96	–
Schmaler Lucin	2.70	1.8	0.80	3.16
Breiter Lucin	4.64	3.1	1.90	2.16
Techiner Binnensee	5.45	3.6	1.65	7.86
Tollensesee	7.04	4.1	2.09	5.49
Edebergsee	14.6	9.7	2.27	10.0

After data of Ohle, 1952, 1955b.

TABLE 11–5 Estimations of Total Production in the Trophogenic Zone of Schleinsee, Germany, during Summer Stratification (May–September) in the Years Before and in the Year After Application of Phosphorus Fertilization

BASIS FOR CALCULATED PRODUCTION	Kg DRY WEIGHT		Ratio
	1935–1937	1939	
Biogenic decalcification	1900	5000	2.6
Ammonia N accumulation	2100	5900	2.8
Hypolimnetic CO_2 accumulation	3800	9600	2.5
Apparent total production	5600	17200	3.1
Related to surface area	43 g m^{-2}	117 g m^{-2}	2.7

After data of Einsele, 1941.

Utilization of Carbon by Algae and Macrophytes

Algae and submersed aquatic macrophytes require an abundant and readily available source of carbon for high sustained growth. Abundant physiological evidence indicates that free CO_2 is most readily utilized by nearly all algae and larger aquatic plants. A number of algae and submersed macrophytes, particularly the mosses, can utilize only free CO_2. Many algae and aquatic vascular plants are capable of utilizing CO_2 from bicarbonate ions when free CO_2 is in very low supply and HCO_3^- is abundant; a few species of algae require HCO_3^- and cannot grow with free CO_2 alone. There is no clear evidence that algae or higher aquatic plants assimilate $CO_3^=$ directly as a carbon source, although it has been implicated in growth at very high pH values (see Felföldy, 1960). The most comprehensive review of the subject is given by Raven (1970; cf. Allen and Spence, 1981).

HETEROTROPHIC AUGMENTATION

A large number of algae are heterotrophic, i.e., they can remain viable in the absence of light in bacterial-free culture by chemo-organotrophic uptake of dissolved organic compounds (cf. Danforth, 1962; Provasoli, 1963; Stanier, 1973; Droop, 1974; Neilson and Lewin, 1974). However, the rates of growth under the experimental conditions used are very low in comparison with normal light-mediated autotrophic growth, and the levels of substrate concentrations employed are usually exceedingly high, often several orders of magnitude in excess of what would be found in natural waters. The range of organic compounds that support growth in the dark in blue-green algae is very narrow, because the pentose phosphate pathway is the energy-yielding dissimilatory pathway. Consequently, only exogenous substrates that are readily convertible to glucose-6-phosphate can support growth in the dark. Among obligately photoautotrophic algae, suitable enzymes, e.g., glucose permease, are absent. Investigations conducted at naturally occurring substrate concentrations have consistently shown that bacteria possessing active permease enzyme systems effectively out-compete algae for organic substrates under natural conditions (Wright and Hobbie, 1966; Wetzel, 1967; Hobbie and Wright, 1965; and others). Where active heterotrophic uptake of simple organic substrates is found under natural conditions, carbon assimilation is still low in comparison to the amount assimilated in photoautotrophy, even under very low light conditions. It might be anticipated that any significant augmentation of autotrophy by chemo-organotrophy would be confined largely to the blue-green algae, which have numerous morphological and physiological similarities to bacteria. This potential heterotrophic augmentation is seen in the blue-green alga *Oscillatoria*, which grows in massive densities in the microaerobic zone at the metalimnetic-hypolimnetic interface of some productive lakes (Saunders, 1972). Photoheterotrophic growth in the light in the absence of exogenous CO_2 has been shown in two species of blue-green algae (Ingram, 1973a, 1973b). Studies indicate that respired CO_2 from the substrate oxidation is assimilated by the photosynthetic reactions within the cells. Photoheterotrophy in natural waters was first critically evaluated by McKinley (1977; McKinley and Wetzel, 1979; see also Ellis and Stanford, 1982). Photoheterotrophic contributions to overall carbon cycling were greatest in the morning periods at depths of low light and during spring turnover and late summer stratification. Photoheterotrophic and chemoheterotrophic (not light-mediated) utilization of organic substrates was small in this moderately productive lake in comparison with normal photosynthetic utilization of inorganic carbon.

All of the existing evidence for heterotrophic growth by algae indicates that under natural conditions, photoautotrophy by inorganic carbon uptake overwhelmingly dominates in freshwater systems. Supplementation of photosynthesis by algal heterotrophy is limited to a few specialized conditions. Although probably insignificant for overall plant synthesis, these supplementary processes are potentially important to the survival of some species and in selective competitive interactions.

In most freshwater systems, dissociation rates of ionic CO_2 species and the maintenance of near-equilibrium conditions between atmospheric CO_2 and that of the water are sufficiently rapid that severe inorganic carbon limitation to algal photosynthesis is unlikely, even under conditions of low ΣCO_2. Below a pH of 8.5, the theoretical dissociation kinetics of the CO_2-HCO_3^--$CO_3^=$ complex, derived from pure solutions, agree very well with the concentrations of the ionic species found in natural waters (Talling, 1973). Above pH 8.5, however, the total inorganic carbon dioxide concentrations fall considerably below the values calculated on the basis of the apparent dissociation constants K_1' and K_2' of carbonic acid, as derived from measurements of alkalinity (Fig. 11-7). Although the explanation for this phenomenon is not completely clear, it appears to be associated with the presence of a noncarbonate, nonhydroxide alkalinity component resulting from ionized silicate, and can be corrected for (Talling, 1973).

Experimental evidence on freshwater photosynthetic utilization of carbon indicates a strong relationship between physiological availability and the forms of ΣCO_2 (Raven, 1970; Wetzel and Hough, 1973; Hough and Wetzel, 1977; Smith and Walker,

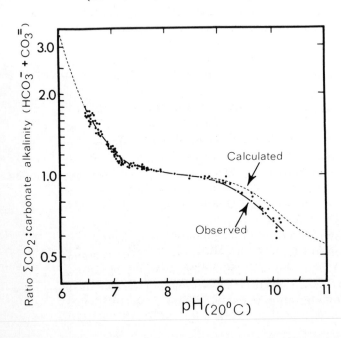

Figure 11-7 Variation in the ratio of ΣCO_2 to carbonate alkalinity resulting from HCO_3^- and $CO_3^=$ with pH for samples of water from the surface and bottom (14 m) layers of Esthwaite Water, England. The broken line indicates the relationship calculated from appropriate values of pK_1' (6.38) and pK_2' (10.32). (Slightly modified from Talling, 1973.)

1980). In a majority of plants, the first stable major product of fixation of inorganic carbon is 3-phosphoglycerate. Up to 5 per cent of the inorganic carbon fixed is by β-carboxylation, a pathway involved in the formation of carbon skeletons or amino acid and porphyrin synthesis. In some tropical plants, all inorganic carbon fixed in photosynthesis occurs via the β-carboxylation pathway, but this mode of metabolism has not been found to occur to any significant extent in aquatic plants. The enzyme carboxydismutase is involved in the aquatic carboxylation reaction, which produces 3-phosphoglycerate as the first product of carbon fixation, and uses free, unhydrated CO_2 as its immediate substrate. Carbonic anhydrase, an enzyme that catalyzes the reversible hydration of carbon dioxide, is found universally in photosynthetic cells of plants. Where CO_2 is the carbon species entering the cell, there is no obvious biochemical role for carbonic anhydrase found in the plants, although some evidence indicates that this enzyme accelerates the diffusion of CO_2. Bicarbonate ions have been implicated as a critical factor in the evolution of oxygen during photosynthesis (Stemler and Govindjee, 1973). At high pH values and HCO_3^- concentrations, bicarbonate moves less effectively from binding sites of the chloroplast.

UTILIZATION OF BICARBONATE IONS

The ability to assimilate bicarbonate ions is variable among planktonic algae, macroalgae, and submersed angiosperms (Raven, 1970; Allen and Spence, 1981). When this ability does occur, an additional reaction is needed for HCO_3^- assimilation, which is not required for CO_2 assimilation. Active bicarbonate transport followed by HCO_3^- dehydration within the cytoplasm is apparently required, and is coupled with a similarly active stoichiometric excretion of hydroxyl ions from the cells. Carbonic anhydrase activity is found in plants that cannot use bicarbonate as well as in those that can, although such activity generally increases among the latter in algae and certain macrophytes (Weaver and Wetzel, 1980).

When aquatic plants have similar affinities for CO_2 and the HCO_3^- ion, utilization of bicarbonate generally occurs when the bicarbonate concentration exceeds that of CO_2 by more than 10 times. Free CO_2 concentrations (about 10 micromolar) of most fresh waters and of the sea are in approximate equilibrium with the atmosphere, although many fresh waters contain bicarbonate concentrations far in excess of 10 times that quantity. Equilibrium concentrations of dissolved CO_2, particularly in alkaline hard-water lakes with a pH > 8.5, are inadequate to saturate photosynthesis in plants adapted to utilize bicarbonate. As these waters become more productive, and in densely populated littoral zones of less productive lakes, pH is rapidly altered by metabolism on a diurnal basis (pH ranges from a low of 6 to a high of 10 or more per 24 hours), and can be associated with reduced carbon fixation and bicarbonate assimilation. Under stagnant conditions, common to heavily colonized littoral zones of lakes, the shift to bicarbonate metabolism, as well as the increased pH, is often associated with severe reduction in CO_2.

Stagnant layers around plant cells (algae) and tissues (macrophytes) clearly have major effects on limiting the assimilation of both CO_2 and HCO_3^- by actively photosynthesizing plants (Smith and Walker, 1980). Even when the plants are exposed to water movements by turbulence or sinking, a residual layer (approximately 10 μm thick) parallel to the surface remains. The rate of solute movement across this residual layer approximates that of molecular diffusion. Diffusion of CO_2 and HCO_3^- through these stagnant layers is an important rate-limiting process to both availability of CO_2 and to membrane transport of HCO_3^- ions. Mucilage sheaths surrounding algal cells, especially among blue-green algae, can further reduce the uptake rates of inorganic carbon (Chang, 1980).

Bicarbonate assimilation in media of high pH assumes greater significance in large aquatic macrophytes that have morphologically long diffusion paths. Active transport of HCO_3^- originating from the dissociation of calcium bicarbonate and secretion of OH^- ions results in the precipitation of $CaCO_3$. Many submersed angiosperms and algae can utilize bicarbonate. Aquatic mosses are able to utilize only free CO_2, and are restricted to waters of relatively low pH and abundant CO_2, e.g., springs, mountain streams, bogs, or shallow waters or the lower trophogenic zone where free CO_2 is higher than elsewhere in the lake. Moreover, many angiosperms with large intercellular gas lacunae refix CO_2 of respiration and photorespiration rather efficiently (Hough and Wetzel, 1972; Søndergaard and Wetzel, 1980). The efficiency of refixation and photosynthetic efficiency of carbon fixation must be highly plastic, related in part to induced shifts to bicarbonate assimilation, and can affect rates of net primary production significantly.

In summary, evidence is available that, among the diversity of concentrations and states of inorganic carbon in fresh waters, there exists a large number of situations where free CO_2, even in equilibrium with the atmosphere, may be inadequate for metabolism at high sustained levels (see especially Talling, 1976). In other cases, free CO_2 may be inadequate as a result of slow diffusion rates and chemical losses from the system. Although it is highly doubtful that inorganic carbon per se is seriously limiting to photosynthetic metabolism under most natural situations, assimilation of bicarbonate at metabolic expense is known and can be induced among plants that have an affinity for both CO_2 and HCO_3^- ions under conditions of low free CO_2 and high bicarbonate concentrations (e.g., Goldman and Graham, 1981; Beer and Wetzel, 1981). Possession of an affinity for bicarbonate is an adaptive advantage in fresh waters.

SUMMARY

1. Most carbon of fresh waters occurs as equilibrium products of carbonic acid (H_2CO_3). A smaller amount of carbon occurs in organic compounds as dissolved and particulate detrital carbon, and a small fraction occurs as carbon of living biota.
2. The complex equilibrium reactions of inorganic carbon and the distribution of the chemical species of total CO_2 [$\Sigma CO_2 = CO_2 + HCO_3^- + CO_3^=$] are understood in considerable detail.
    a. As atmospheric CO_2 dissolves in water containing bicarbonate (primarily associated with the calcium cation), dissociation kinetics of the inorganic carbon of fresh waters below a pH of 8.5 closely follow those predicted by pure dilute solution chemistry.
    b. Free CO_2 is very soluble, and in water some becomes hydrated to form carbonic acid:

$$CO_2 \text{ (air)} \rightleftarrows CO_2 \text{ (dissolved)} + H_2O \rightleftarrows H_2CO_3$$

c. H_2CO_3 is a weak acid and rapidly dissociates to bicarbonate and carbonate:

$$H_2CO_3 \rightleftharpoons H^+ + HCO_3^- \rightleftharpoons H^+ + CO_3^=$$

d. At equilibrium, bicarbonate and carbonate ions dissociate and hydroxyl ions (OH^-) are formed by the hydrolysis of carbonic acid:

$$HCO_3^- + H_2O \rightleftharpoons H_2CO_3 + OH^-$$
$$CO_3^= + H_2O \rightleftharpoons HCO_3^- + OH^-$$
$$H_2CO_3 \rightleftharpoons H_2O + CO_2$$

e. As carbonic acid percolates over rock and through soils, carbonates are solubilized; ionized Ca^{++} and HCO_3^- are released to the water. The dilute solution of bicarbonate of many fresh waters is weakly alkaline because slightly greater concentrations of OH^- ions than H^+ ions result from the dissociation of HCO_3^-, $CO_3^=$, and H_2CO_3.

f. Losses of CO_2 by photosynthetic utilization or additions of CO_2 from biotic respiration tend to change the pH (H^+ ion activity) of the water. The *buffering action* of the water, however, tends to resist changes in pH as long as the equilibria of the ΣCO_2 complex are operational. Addition of H^+ ions is neutralized by OH^- ions formed by the dissociation of HCO_3^- and $CO_3^=$. The pH remains essentially the same as before, unless the supply of HCO_3^- and $CO_3^=$ is exhausted. Similarly, added OH^- ions react with HCO_3^- ions:

$$HCO_3^- + OH^- \rightleftharpoons CO_3^= + H_2O.$$

g. If the solution of calcium bicarbonate in equilibrium with CO_2, H_2CO_3, and $CO_3^=$ loses a portion of the CO_2 required to maintain the equilibrium, such as by photosynthetic uptake exceeding replacement of CO_2, the relatively insoluble calcium carbonate (marl) will precipitate:

$$Ca(HCO_3)_2 \rightleftharpoons CaCO_3\downarrow + H_2O + CO_2$$

until the equilibrium is reestablished by the formation of sufficient CO_2. As $CaCO_3$ forms and precipitates, inorganic (e.g., $PO_4^=$ ions) and organic compounds can adsorb to or coprecipitate with the $CaCO_3$ and be carried out of the trophogenic zone of lakes. Metabolic activity can be reduced as a result of this adsorption and/or coprecipitation.

3. The distribution of ΣCO_2 and pH in surface waters varies both seasonally and vertically in lakes in relation to loading from allochthonous sources, physical conditions, and with biotic inputs and consumption.

a. Although the exchange of CO_2 of the water with that of the atmosphere is rapid and relatively complete in aerated open water of streams and lakes, the spatial and temporal distribution of ΣCO_2 and pH is altered by microbial respiration metabolism and by photosynthetic consumption, especially in quiescent strata in the pelagial zone of stratified lakes and in productive littoral areas.

b. Relatively unproductive lakes exhibiting an orthograde oxygen curve generally have an orthograde ΣCO_2 curve.

c. In productive waters exhibiting a clinograde oxygen curve, an inverse clinograde ΣCO_2 curve is generally found. The pH decreases in the hypolimnion in relation to the increased hypolimnetic concentrations of CO_2 and bicarbonate.

d. The ΣCO_2 distribution of hard-water calcareous lakes is greatly modified by biologically induced decalcification of the epilimnion.

4. In thermally stratified lakes of low to moderate productivity with minimal inputs of organic matter from outside the basin, the accumulation of CO_2 in the hypolimnion is positively correlated to rates of organic production in the trophogenic zone.

5. Inorganic carbon is a major nutrient of photosynthetic metabolism. However, phosphorus and nitrogen limit photosynthesis more frequently than does inorganic carbon, which occurs in much greater abundance. Because assimilation of CO_2 and HCO_3^- occurs more rapidly than resupply, and conditions of reduced availability result from diffusion resistance at the uptake sites, potential photosynthetic productivity by both algae and submersed macrophytes is not always fully realized. Many algae and vascular aquatic plants have developed compensatory mechanisms to enhance utilization and recycling of respired CO_2.

SOURCES AND TRANSFORMATIONS OF NITROGEN IN WATER

Nitrogen occurs in fresh waters in numerous forms: dissolved molecular N_2, a large number of organic compounds from amino acids, amines to proteins and refractory humic compounds of low nitrogen content, ammonia (NH_4^+), nitrite (NO_2^-), and nitrate (NO_3^-). Sources of nitrogen include: (a) precipitation falling directly onto the lake surface, (b) nitrogen fixation both in the water and the sediments, and (c) inputs from surface and groundwater drainage. Losses of nitrogen occur by (a) effluent outflow from the basin, (b) reduction of NO_3^- to N_2 by bacterial denitrification with subsequent return of N_2 to the atmosphere, and (c) permanent sedimentation loss of inorganic and organic nitrogen-containing compounds to the sediments.

Nitrogen in Precipitation and Fallout

The amount of influent nitrogen to lakes and their drainage areas from atmospheric sources generally has been considered to be minor in comparison with that from direct terrestrial runoff. However, on closer inspection, and in view of exponentially increasing inputs of nitrogen from atmospheric sources, the amount of nitrogen reaching lakes in this way is often significant to the nitrogen cycle and for productivity. For example, in relatively oligotrophic mountainous regions of granitic bedrock, precipitation is a major source of nitrogen (Likens and Bormann, 1972; Likens, et al., 1977). Inorganic nitrogen input by precipitation similarly was found to be a major source of loading to the drainage basin of Lake Tahoe, California–Nevada (Coats, et al., 1976).

Nitrogen from precipitation and direct bulk (particulate) fallout is extremely variable depending on local meteorological conditions, wind patterns, and the location of the lake with respect to industrial and agricultural outputs.

Nitrogen may enter a lake in many forms: as dissolved N_2, nitric acid, NH_4^+, and NO_3^-, as NH_4^+ adsorbed to inorganic particulate matter, and as organic compounds, which can occur in either dissolved or particulate phases. No direct relationship exists between the volume of rainfall or snowfall and the quantity of nitrogen influx per area of land or water (Chapin and Uttormark, 1973). Dry fallout can contain as much as ten times the quantity of nutrients commonly found in rain. The nitrogen content of snow is often much higher than that of rain, and can contribute up to half of the total annual nitrogen influx to a lake, even though snow generally constitutes a small proportion of total precipitation. N_2-fixing bacteria occur in rainwater in low numbers (Visser, 1964a), but their contribution to the total input of nitrogen is most likely small in comparison to other sources. In nonpolluted areas, most of the combined nitrogen in the atmosphere is ammonia, much of which originates from the decomposition of terrestrial organic matter (Hutchinson, 1944). Atmospheric ammonia associated with dust particles can be oxidized to nitrate so that precipitation contains both ammonia and nitrate (Hutchinson, 1944, 1975).

The inputs of NO_3^- and NH_4^+ from atmospheric sources average about 0.1 g N m^{-2} year^{-1} over the continental United States, but are highly irregular in distribution. The north central and eastern regions, especially bordering the southern Great Lakes, receive the largest input of nitrogen from precipitation, on the order of 0.3 to 0.35 g N m^{-2} yr^{-1} (Chapin and Uttormark, 1973). Dry fallout sources increase these values by a factor of 3 to 4, so that in the Great Lakes region, atmospheric contributions of nitrogen occur at a rate of approximately 1 g N m^{-2} yr^{-1}. Neglecting other nutrients for the time being, this influx alone corresponds to the approximate amounts of nitrogen that are generally required to shift shallow lakes of a mean depth of <5 m from moderate to high productivity (Vollenweider, 1968). In this region of the country, which also receives high inorganic inputs from runoff of nitrogen-containing sedimentary rock formations, it would be very difficult to control productivity of lakes by limiting nitrogen inputs because of large atmospheric contributions. We will return to this subject in later discussions.

Molecular Nitrogen and N₂ Fixation

Although N_2 is not particularly soluble in water, N_2 is usually saturated with respect to surface temperature and pressure in streams and during periods of circulation of lakes (Birge and Juday, 1911). Maximum concentrations are found in winter owing to increased solubility at colder temperatures (approximately 15 to 20 ml l^{-1}). During summer thermal stratification, the heating of the epilimnion decreases N_2 solubility, while N_2 concentrations in the metalimnion and much of the hypolimnion may be slightly supersaturated as a result of hydrostatic pressure maintaining a higher gas concentration despite the heating in these water strata (Fig. 12-1). A decrease in the N_2

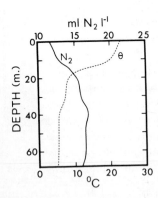
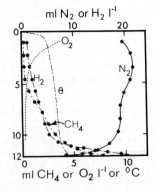

Figure 12-1 Vertical distribution of dissolved gases *(left)* in Green Lake Wisconsin, summer, 1906 (Birge and Juday, 1911) and *(right)* in Beloie Lake, USSR, March 1938. θ = °C. (After Kuznetsov and Khartulari, 1941.)

content in the lower hypolimnion above the sediments has been observed (Fig. 12-1), which presumably is related to bacterial fixation of N_2 in productive lakes (Kuznetsov and Khartulari, 1941). Conversely, late in the summer stratification of Lake Mendota, N_2 of the hypolimnion increased somewhat above what would be expected on the basis of solubility (Brezonik and Lee, 1971); this may be the result of denitrification of nitrate in this lake (see following discussion).

NITROGEN FIXATION: BLUE-GREEN ALGAE

The importance of in situ nitrogen fixation to the productivity of lakes has been shown only within the last decade, even though blue-green algae were implicated in this process many years earlier (Burris, et al., 1943). An excellent comparative review is given by Hardy, et al. (1973). Thus far, the occurrence of nitrogen fixation in the open waters of lakes has been strictly correlated with the presence of blue-green algae that possess heterocysts (Fogg, 1971a, 1974). Heterocysts are specialized cells that occur singly in most filamentous blue-green algae, except for the Oscillatoriaceae, and are the sole site of nitrogen fixation in aerobically grown, heterocyst-forming blue-green algae (Wolk, 1973). Nitrogen fixation has been found to occur in some unicellular forms which do not produce heterocysts (Fogg, 1974). However, among many species of blue-green algae, such as *Anabaena* spp., numbers of heterocysts correspond approximately to observed nitrogen-fixing capacity (Horne and Goldman, 1972; Horne, et al., 1972, 1979).

In plankton of open water, nitrogen fixation is primarily light-dependent in that it requires reducing power and adenosine triphosphate (ATP), both of which are generated in photosynthesis. Nitrogen-fixing algae and some photosynthetic bacteria can fix limited quantities of N_2 in the dark (Horne, 1979). Determinations in situ show a relationship of nitrogen fixation with depth similar to that of photosynthesis (Ward and Wetzel, 1980a). In full sunlight N_2 fixation is often inhibited at the surface, reaches a maximum some depth below the surface, and shows rapid, nearly exponential decrease with greater depth (Fig. 12-2; cf. Chapter 5) (Dugdale and Dugdale, 1962; Goering and Neess, 1964; Horne and Fogg, 1970; Horne, 1979). Comparison of N_2-N, NO_3-N, and

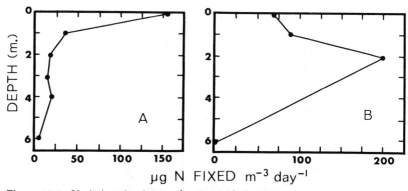

Figure 12-2 Variations in nitrogen fixation with depth (*A*) in Lake Windermere, and (*B*) in Esthwaite Water, England, 30 August 1966. (Modified from Horne and Fogg, 1970.)

NH_4-N as nitrogen sources for *Aphanizomenon flos-aquae* (heterocystous) and *Microcystis aeruginosa* (nonheterocystous) at high, low, and variable light intensities revealed that highest growth rates always occurred with NH_4-N, followed by NO_3-N, and then N_2-N (Ward and Wetzel, 1980a). These findings were consistent with those expected from energy expenditures required to assimilate these nitrogen sources (i.e., N_2-N > NO_3-N > NH_4-N). *Aphanizomenon* would not grow at low light intensities with N_2-N as the sole nitrogen source, although growth could be maintained with either NO_3-N or NH_4-N.

Readily utilizable sources of combined nitrogen suppress synthesis of the nitrogenase complex rather than the activity of any existing enzyme (Fogg, 1971, 1974; Wolk, 1973). Suppression of heterocyst formation by nitrate, even at very high concentrations, is often only partial (Ogawa and Carr, 1969; Ohmori and Hattori, 1972). Similarly, ammonia at low concentrations represses the formation of additional nitrogenase, but does not affect its activity. In general, then, one would expect an inverse relationship between the rate of nitrogen fixation by blue-green algae and the concentration of combined nitrogen in lake waters; this is often the case. But because of the carryover of residual nitrogenase activity, the relationship is not always consistent. Thus N_2 fixation by blue-green algae sometimes may occur at reduced levels in the presence of appreciable quantitites of combined nitrogen in the water. The importance of the microenvironment surrounding the cell also should be pointed out. Molecular N_2 is in higher concentrations and diffuses more readily than either ammonium or nitrate ions. Within the massive mucilage sheaths surrounding many blue-green algae, a steep gradient could easily develop in which other inorganic nitrogen sources are greatly reduced in comparison to N_2 availability, regardless of the inorganic concentrations of the water (cf. Chang, 1980).

Nitrogen fixation has also been positively correlated with concentrations of dissolved organic nitrogen occurring in the water (Horne and Fogg, 1970; Horne, et al., 1972). Algae secrete many simple and complex organic carbon and nitrogen compounds. Paerl (1978) found that *Anabaena*, a common nitrogen-fixing blue-green alga, was colonized by bacteria, especially during N_2-fixing algal blooms. The bacteria generally do not parasitize the algae but apparently increase the algal N_2-fixing capabilities. The bacteria showed a preference for typical algal excretion products and were enhanced by such organic products. Algal nitrogenase activity increases when the bacteria create oxygen-consuming microzones around the nitrogenase-bearing heterocysts.

Diurnal rates of nitrogen fixation in open lake water are typically low in the early morning, reach a maximum midday at maximum insolation, and then decline to low rates during the afternoon and evening (Rusness and Burris, 1970; Horne, 1979; Ward and Wetzel, 1980b). The most commonly observed seasonal pattern in temperate lakes is for both the percentage and absolute rates of N_2 fixation to increase to maximum levels as heterocyst-bearing blue-green algal populations develop and sources of combined nitrogen are reduced or depleted. High concentrations of total phosphorus are necessary (Lean, et al., 1978; Lundgren, 1978; Flett, et al., 1980). Rates of N_2 fixation decline abruptly as the blue-green populations decrease (cf. Chapter 15 on algal successions). In winter, the N_2 fixation is nonexistent or greatly reduced (Billaud, 1968; Horne and Fogg, 1970; Toetz, 1973). In productive tropical lakes where the periodicity of physicochemical factors and algae is less marked, nitrogen fixation rates probably are more uniform throughout the year.

To determine the nitrogen cycle of a lake, estimates of the total nitrogen fixed per annum are required. Extensive measurements are required because of the spatial and temporal heterogeneity of blue-green algal populations. For lack of data to the contrary, it had been assumed previously that the magnitude of nitrogen income to lakes as N_2 fixation by the algal phytoplankton was very small or insignificant in comparison to other sources. It is now evident that a spectrum of conditions exists, ranging from lakes where N_2 fixation is insignificant, to eutrophic lakes in which extensive N_2 fixation permits higher rates of production to occur than would be possible otherwise. Direct estimates of N_2 fixation obtained by measuring $^{15}N_2$ uptake and acetylene reduction in oligotrophic lakes, or in lakes of moderate productivity but which contain large pools of combined inorganic nitrogen, are consistently undetectable (Ward and Wetzel, 1980a). In moderately productive Lake Windermere, the nitrogen fixation by benthic blue-green algae growing in abundance on littoral rocks and stones to a depth of 3.5 m was estimated conservatively to be 1 g N m^{-2} year^{-1} (Horne and Fogg, 1970). For a minimal rocky area of 2.86 km^2, the estimated total annual fixation by the benthic blue-green algae in Windermere is 0.8 megagrams. In this example, the littoral benthic contribution is considerably larger than the nitrogen fixed by the phytoplankton (Table 12–1). Even though these values are large, the total N_2 fixation by blue-green algae contributes at most a small proportion (less than 1 per cent) to the total combined nitrogen income of this lake. An estimate (probably high) for small, eutrophic Chernoye Lake, USSR, indicated that the nitrogen fixation constituted about 13 per cent of the total nitrogen income (Kuznetsov, 1959). In eutrophic Clear Lake, California, biological N_2 fixation, largely associated with blue-green algal blooms, contributed at least 43 per cent of the nitrogen income, almost as much as NO_3^- from river inflows.

TABLE 12–1 Estimated Annual Rates of Nitrogen Fixation by Phytoplankton and Benthic Algae

LAKE	AREA (km²)	N₂ FIXATION (g N m⁻²)	TOTAL N FIXED (megagrams = metric t)
Phytoplankton			
Windermere, England			
North Basin 1966	8.2	0.037	0.30
South Basin 1965	6.7	0.287	1.92
1966	6.7	0.107	0.72
Esthwaite, England			
1965	1.01	0.127	0.13
1966	1.01	0.061	0.06
Clear Lake, California, 1970			
Upper Arm	127.0	0.352	361
Lower Arm	37.2	0.759	49
Oaks Arm	12.5	0.250	50
Total for Lake	176.7	0.384	460
Benthic Algae			
Windermere, England	2.9	1.0	0.8
Taharoa, New Zealand			
Exposed Beach	–	0.62	–
Submersed Beach	–	0.15	–
California Stream	–	0.04–0.36	–

From data of Horne and Fogg, 1970, and Horne and Goldman, 1972, Horne and Carmiggelt, 1975, and Lam, et al., 1979.

A detailed study of the nitrogen fixation in a small subarctic lake of interior Alaska demonstrated that ammonia was more important than N_2 fixation (Billaud, 1968). Of two main algal production periods, the first consisted largely of microflagellates under the ice, and depended on ammonia as a nitrogen source. Immediately after the ice melted from the lake, an algal population composed almost exclusively of the blue-green *Anabaena flos-aquae* developed. During the peak of the *Anabaena* bloom, molecular nitrogen constituted over 25 per cent of the nitrogen assimilated by the plankton. Nitrogen fixed by the plankton in the south basin of Lake Windermere in 1965 amounted to 72 per cent of the amount available as nitrate, although in 1966 and in other water bodies this proportion was less (<0.2 to 48 per cent) (Horne and Fogg, 1970; Tison, et al., 1977; Ward and Wetzel, 1980a). Rates of N_2 fixation are greatly enhanced when the productivity of a lake is increased by phosphorus fertilization (e.g., Lean, et al., 1978; Lundgren, 1978).

NITROGEN FIXATION: BACTERIA

Quantitative information on bacterial nitrogen fixation in lakes is sparse. Heterotrophic nitrogen fixation is commonly disregarded, on the premise that nitrogen-fixing bacteria are limited by low availability of exogenous carbohydrate. One to 25 mg of nitrogen can be fixed per gram of carbohydrate utilized (Stewart, 1969). Sufficiently large quantities of soluble carbohydrate rarely are available in natural waters, and there is an intense competition for these substrates by heterotrophic bacteria incapable of N_2 fixation.

The most common heterotrophic N_2 fixing bacteria comprise several species of *Azotobacter* and *Clostridium pasteurianum*, which are found in fair abundance living free in the water, epiphytically on submersed aquatic plants, and in the sediments (Kuznetsov, 1959, 1970). The numbers of these bacteria are generally lowest in the open water, where soluble organic concentrations are low, and tend to increase, with *Azotobacter* being dominant, in bog lakes which contain high concentrations of dissolved humic organic matter (Table 12–2). Based on scant data from eutrophic Chernoye Lake, USSR, numbers of open-water *Clostridium* increase in the spring just before and after

TABLE 12–2 Occurrence of Nitrogen-fixing Bacteria in Water and Sediments of Lakes

LAKE TYPE	NUMBER LAKES STUDIED	IN WATER (nos ml⁻¹)		IN SEDIMENT (nos g⁻¹ wet wt)	
		AZOTOBACTER	CLOSTRIDIUM PASTEURIANUM	AZOTOBACTER	CLOSTRIDIUM PASTEURIANUM
Oligotrophic	5	0	0	0–10	0–10
Mesotrophic	5	0–10	0–10	0–10	0–1000
Eutrophic	10	0–10	1-20	0–10	100–10000
Eutrophic with high humic content	2	10	1	–	–
Dystrophic with high humic content	3	1–10	0–1	–	–
Eutrophic reservoir	1	5–10	1–5	10–1000	1000–10000

Modified from Kuznetsov, 1970.

ice-loss, decrease to a low in summer, and increase in autumn. This pattern is somewhat analogous to the seasonality of other, non-N$_2$-fixing bacteria populations that lag slightly behind the productivity pulses of phytoplankton (cf. Chapter 17). In contrast, Niewolak (1972) observed a maximum of *Azotobacter* in spring and summer in Polish lakes.

      Azotobacter is found in particular abundance as an epiphyte on submersed aquatic angiosperms and submersed portions of emergent macrophytes. A symbiotic relationship between the *Azotobacter* and the macrophytes is possible since the plants secrete many dissolved organic compounds (Wetzel, 1969) that can serve as substrates for nitrogen-fixing bacteria. The nitrogen released by the bacteria may be utilized by the macrophytes. In the littoral water, populations of the planktonic bacterium *Azotobacter* are higher than those of the open water (J. Overbeck, personal communication). Although few quantitative data are available, the littoral zone may serve as a major site of nitrogen fixation by both heterotrophic bacteria and sessile blue-green algae. Further investigation is needed.

      Large numbers of N$_2$-fixing bacteria occur in the sediments of lakes; most are concentrated in the upper 2 cm. Bacterial numbers are much higher in productive lakes than in oligotrophic lakes, particularly in the case of *Clostridium* (Table 12–2). Seasonal changes in the numbers of *Clostridium* in the sediments of Chernoye Lake were similar to those observed in the open water: maximum numbers occurred in the fall, followed by low winter populations which increased to another peak in the spring. Lowest populations in the sediments were found in midsummer. Similar results were observed for *Azotobacter* populations of sediments in Polish lakes (Niewolak, 1970, 1972).

      Several methane-oxidizing bacteria are capable of N$_2$ fixation (reviewed by Zeikus, 1977; Rudd and Taylor, 1980). The N$_2$ fixation of these bacteria increases when sources of combined inorganic nitrogen are depleted either within water or at the sediment–water interface where methane is abundant. Although N$_2$ fixation is physiologically important to the methane-oxidizing bacteria, these nitrogen fixers have not yet been demonstrated to constitute a major source of fixed nitrogen to a lake (Rudd and Taylor, 1980).

      The green photosynthetic sulfur bacterium *Chlorobium* (see Chapter 14) also fixes N$_2$ even in the presence of NH$_4$-N (Bergstein, et al., 1981). In the case studied, however, these nitrogen fixers also failed to contribute significantly to the organic nitrogen load of the lake.

In two lakes (Mary and Mize; Table 12–3), N$_2$ fixation by planktonic heterotrophic bacteria *(Azotobacter)* occurred at a rate several orders of magnitude lower than rates in lakes dominated by the blue-green alga *Anabaena*. In waters rich in dissolved humic organic compounds (such as in lakes Mary and Mize), blue-green algae are rare and nitrogen-fixation rates are very low.

Unlike heterotrophic bacteria, in which nitrogen fixation is limited to a few groups, nearly all photosynthetic bacteria are capable of fixing N$_2$. The photosynthetic bacteria include facultative aerobes and strict anaerobes. As will be discussed in greater detail in Chapter 14, photosynthetic bacteria can develop in great densities in highly structured strata at the interface regions between the aerobic epilimnion or metalimnion and the anaerobic hypolimnion of eutrophic or meromictic lakes, given sufficient light to permit photosynthesis. N$_2$ fixation occurs only in the light, and intensive rates occur only under anaerobic conditions among the green and purple photosynthetic bacteria (Kondrat'eva, 1965). N$_2$ fixation by photosynthetic bacteria occurs simultaneously with

TABLE 12-3 Estimates of Fixation of Molecular Nitrogen by Heterotrophic Bacteria and Blue-Green Algae in Several Lakes

LAKE	ORGANISMS	N₂ FIXATION	N₂ FIXATION IN g, WHOLE LAKE, SUMMER STRATIFICATION PERIOD
Chernoye, USSR 1937	*Azotobacter*	1.6×10^{-11} mg cell^{-1} day^{-1}	0.14
	Anabaena scheremetievi	1.12×10^{-9}	13,100
Mendota, Wisconsin 1960, 1967	Blue-green algal dominance	0.07–43 μg N l^{-1} hr^{-1}	
Sanctuary, Pennsylvania 1959	Blue-green algal dominance	0–6	
Wingra, Wisconsin 1961	Blue-green algal dominance	0.005–1	
Smith, Alaska 1963	*Anabaena* dominance	0–1	
Mary, Wisconsin June 1968	Heterotrophic bacteria	0.003–0.047	
Mize, Florida July 1968 August 1968	Heterotrophic bacteria	0.083–0.308 0.000–0.083	

After many sources cited in the text; also Brezonik and Harper, 1969.

the release of molecular H₂ by a noncyclic electron flux resulting from photophosphorylation. The source of these electrons is an exogenous H-donor, such as thiosulfate.

Very little quantitative information is available on the in situ rates of N₂ fixation by photosynthetic bacteria. In the example given in Figure 12-3, nitrogen fixation is clearly associated with green photosynthetic bacteria, mainly *Pelodictyon*, of the chem-

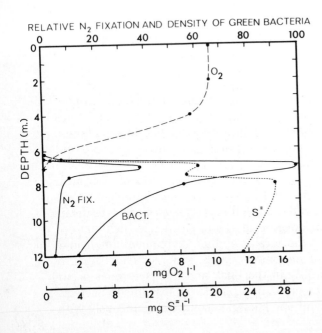

Figure 12-3 Nitrogen fixation in a Norwegian meromictic lake, June 1966, in relation to a dense layer of green photosynthetic bacteria, primarily *Pelodictyon*, on the surface of the monimolimnion. (Drawn from data of Stewart, 1968.)

olimnion in this meromictic lake. These data indicate the potential importance of bacteria to the nitrogen income of lakes under certain circumstances.

NITROGEN FIXATION: WETLAND SOURCES

In addition to the inputs of nitrogen from drainage runoff from terrestrial sources, lakes and streams often are bordered by dense stands of shrublike trees that fix nitrogen from the atmosphere. Common species of the genera *Alnus* and *Myrica* are non-leguminous angiosperms that form large nodules containing an actinomycetal fungal endophyte at or just below the soil surface. Nitrogen fixation by dense stands of *Alnus* can be as high as 22.5 g N m^{-2} year^{-1}, most of which enters the plant. Subsequently, leachate from direct leaf-fall or release during decomposition of foliage can reach nearby streams and lakes.

Alnus trees have been implicated as a significant nitrogen source for streams and lakes in glaciated regions of Alaska that are particularly nitrogen-deficient (Goldman, 1960; Dugdale and Dugdale, 1961). In Castle Lake, an alpine cirque lake of northern California, alder *(Alnus)* trees were abundant on only one side of the lake (Goldman, 1961). *Alnus* leaves contained over four times as much nitrogen as those of other deciduous species. Soils, lake sediments, and spring waters draining to the lake from the alder side all contained higher nitrogen levels than those from the nonalder side of the lake. Bioassay of in situ rates of photosynthesis by the phytoplankton demonstrated that these nitrogen sources were stimulatory, and also that the primary productivity of the lake was significantly higher along the side on which the alders grew.

Heterotrophic bacterial N$_2$-fixation rates can add significant amounts of nitrogen to wetlands. For example, the nitrogen fixed annually by heterotrophic bacteria on decaying plant debris of calcareous wetlands ranged from 530 to 2,100 mg N m^{-2} and was considerably less (70 mg N m^{-2}) among acidic bog vegetation (Waughman and Bellamy, 1980).

Inorganic and Organic Nitrogen

In addition to atmospheric nitrogen in the form of precipitation, dry fallout, and fixation of N$_2$, discussed above, a major source of nitrogen income to lakes is from influents, both from surface land drainage and from groundwater sources. Inputs of nitrogen by ground water can be a major part of the annual nitrogen loading in many lakes, especially of regions rich in limestone. In detailing the nitrogen cycle, the forms of nitrogen must be discussed individually, and then integrated with rates of biogeochemical utilization and transformations as well as nitrogen losses from the lake ecosystem. An obvious loss is via inorganic dissolved and organic (both dissolved and particulate) compounds in effluents flowing out of lake basins. Further losses occur as a result of permanent interment of partially decomposed biota and inorganic and organic nitrogen compounds adsorbed to inorganic particulate matter in the sediments. Nitrogen can also be lost by volatilization of compounds from the water surface, such as ammonia at high pH, and as N$_2$ formed in microbial denitrification of nitrate.

Combined nitrogen occurs as ammonia (NH$_4^+$), hydroxylamine (NH$_2$OH), nitrite (NO$_2^-$), nitrate (NO$_3^-$), and dissolved and particulate organic nitrogen. NH$_4$-N can range from 0 to 5 mg l^{-1} in unpolluted surface waters, although concentrations are usually low, to well above 10 mg l^{-1} in anaerobic hypolimnetic waters of eutrophic lakes. The intermediate product NH_2OH is rapidly oxidized and occurs only in very low concentrations. Similarly, NO_2-N levels of natural lake waters are generally very low, in the range of 0 to 0.01 mg l^{-1}, although concentrations of up to 1 mg l^{-1} have been found in the interstitial waters of deep (>90 cm depth) sediments of Lake Mendota (Konrad, et al., 1970). Concentrations of NO_2-N increase in the anaerobic hypolimnion of lakes under reducing conditions (e.g. Overbeck, 1968; Brezonik and Lee, 1968) and in streams and lakes receiving heavy organic-matter pollution. Concentrations of NO_2-N are usually low under oxygenated conditions, but significant deep-water maxima (to 10 μg l^{-1}) have been found in the upper portions of the hypolimnion (Mortonson and Brooks, 1980). Concentrations of NO_3-N range from undetectable levels to nearly 10 mg l^{-1} in unpolluted fresh waters, but are highly variable seasonally and spatially. Organic nitrogen, much of which occurs in forms resistant to rapid bacterial degradation, commonly accounts for more than one-half of the total dissolved nitrogen.

Recently, much attention has been devoted to the importance of nitrogen concentrations of fresh waters in the regulation of algal productivity (for example, cf. Vollenweider, 1968). Although, as we will see in Chapter 13, phosphorus-cycling rates frequently regulate high sustained productivity, the loading rates of nitrogen are of major importance to the maintenance of high flux rates within the nitrogen cycle. Though there are a number of exceptions, a positive correlation has been found between high sustained productivity of algal populations and average concentrations of inorganic and organic nitrogen (Table 12-4). Mean chemical mass, however, must be used with caution when little consideration is given to the rates of mineralization, microbial transformation, and recycling.

In comparing the distribution of organic and inorganic forms of nitrogen in oligotrophic and eutrophic lake sediments, little correlation was found between concentrations of total or organic forms of nitrogen in sediments and general productivity of the lake (Table 12-5) (Keeney, et al., 1970; Konrad, et al., 1970). The percentage of organic nitrogen as hexosamine-N decreased while amino acid-N increased with increasing lake fertility. NH_3-N of interstitial water was somewhat higher in sediments of eutrophic lakes than in those of oligotrophic lakes.

TABLE 12-4 General Relationship of Lake Productivity to Average Concentrations of Epilimnetic Nitrogen

GENERAL LEVEL OF LAKE PRODUCTIVITY	CHANGE IN ALKALINITY IN EPILIMNION IN SUMMER (meq l^{-1})	INORGANIC N (mg m^{-3})	APPROXIMATE AVERAGE ORGANIC N (mg m^{-3})
Ultra-oligotrophic	<0.2	<200	<200
Oligo-mesotrophic	0.6	200–400	200–400
Meso-eutrophic	0.6–1.0	300–650	400–700
Eutrophic		500–1500	700–1200
Hypereutrophic	>1.0	>1500	>1200

Modified from Vollenweider, R. A.: Scientific Fundamentals of the Eutrophication of Lakes and Flowing Waters, with Particular Reference to Nitrogen and Phosphorus as Factors in Eutrophication. OECD Report No. DAS/CSI 68.27, Paris, OECD, 1968, after data of E. A. Thomas and Lueschow, et al., 1970.

TABLE 12-5 Average Distribution of Nitrogen in Some Wisconsin Lake Sediments

| LAKE TYPE | SEDIMENT-N (mg kg^{-1}) | | INTERSTITIAL WATER-N (mg l^{-1}) | | | TOTAL ORGANIC N (%) | ACID HYDROLYZABLE (% of Total N) |
	Fixed NH$_4$	Exchangeable NH$_4$	Organic	NH$_4$	NO$_3$		
Softwater, oligotrophic	69	167	2.1	3.8	0.2	1.53	83.1
Softwater, eutrophic	44	415	2.2	9.6	0.3	3.26	80.4
Hardwater, oligotrophic	66	66	–	–	–	0.52	84.4
Hardwater, eutrophic	134	120	2.0	11.4	0.3	0.80	82.0

Modified from Keeney, 1973.

AMMONIA

Ammonia is generated by heterotrophic bacteria as a primary end product of decomposition of organic matter, either directly from proteins or from other nitrogenous organic compounds. Intermediate nitrogen compounds are formed in the progressive degradation of organic material, but rarely accumulate, because deamination by bacteria proceeds rapidly. Although ammonia is a major excretory product of aquatic animals, this nitrogen source is quantitatively minor in comparison to that generated by bacterial decomposition.

Ammonia in water is present primarily as NH_4^+ and as undissociated NH_4OH, the latter being highly toxic to many organisms, especially fish (Trussell, 1972). The proportions of NH_4^+ to NH_4OH are dependent on the dissociation dynamics which are governed by pH and temperature. The approximate ratios of NH_4 to NH_4OH are as follows (Hutchinson, 1957):

pH 6	3000:1
pH 7	300:1
pH 8	30:1
pH 9.5	1:1

Detailed dissociation relationships with pH and temperature are given by Trussell (1972) and Emerson, et al. (1975). Ammonia is strongly sorbed to particulate and colloidal particles, especially in alkaline lakes containing high concentrations of humic dissolved organic matter.

Since NO_3^- must be reduced to NH_4^+ before it can be assimilated by plants, ammonia is an energy-efficient source of nitrogen for plants. The energy necessary to assimilate nitrogen is lowest for NH_4-N and increases for NO_3-N (and N_2-N for N_2-fixing blue-green algae). Among the blue-green algae, highest growth rates always occurred with NH_4-H as the nitrogen source at many different light intensities (Ward and Wetzel, 1980a, 1980b). Reports of better growth of some algae with NO_3-N as the nitrogen source rather than with NH_4-N may be, in part, the result of toxicity of NH_4OH at high pH values that can prevail both in culture and in situ during periods of high daily photosynthesis in very eutrophic lakes (cf. Rodhe, 1948). Inhibition of NH_4-N assimilation by natural populations of phytoplankton was observed at high concentrations of NH_4-N,

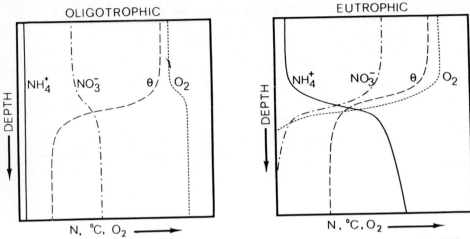

Figure 12-4 Generalized vertical distribution of ammonia and nitrate nitrogen in stratified lakes of very low and high productivity.

but at low concentrations NH₄-N uptake rates exceeded those of NO₃-N (Toetz, et al., 1977; Toetz and Cole, 1980).

The distribution of ammonia in fresh waters is highly variable regionally, seasonally, and spatially within lakes and depends upon the level of productivity of the lake, and the extent of pollution from organic matter. While generalizations are difficult to make, the concentration of NH₄-N in well-oxygenated waters is usually low. In the trophogenic zone, NH₄-N is rapidly assimilated by algae and represents the most significant source of nitrogen for the plankton in many lakes (Liao and Lean, 1978). Thus, concentrations of NH₄-N are commonly low in unproductive oligotrophic waters, in the trophogenic zones of most lakes, and in most lakes after periods of circulation (Fig. 12-4). When appreciable amounts of sedimenting organic matter reach the hypolimnion of stratified lakes, NH₄-N can accumulate. The accumulation of NH₃-N greatly accelerates as the hypolimnion becomes anoxic. Under anaerobic conditions, bacterial nitrification of NH₄⁺ to NO₂⁻ and NO₃⁻ ceases as the redox potential is reduced to below about +0.4 V. Moreover, with the loss of the oxidized microzone at the sediment–water interface under anoxic hypolimnetic conditions, the adsorptive capacity of the sediments is greatly reduced (Kamiyama, et al., 1977). A marked release of NH₄⁺ from the sediments then occurs. If light reaches the sediments in amounts sufficient to support benthic algae, blue-green algal growth can assimilate NH₄-N from the interstitial water and totally prevent the flux of NH₄-N from the sediment to the water (Jansson, 1980). The algae release large amounts of dissolved organic nitrogen to the water above the sediments.

NITRIFICATION

Nitrification may be broadly defined as the biological conversion of organic and inorganic nitrogenous compounds from a reduced state to a more oxidized state (Alex-

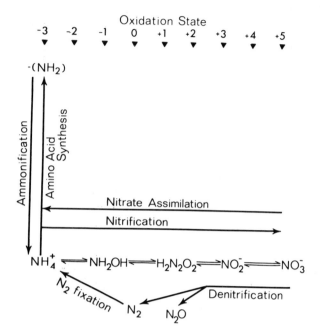

Figure 12-5 Biochemical reactions that influence the distribution of nitrogen compounds in water. (After Stadelmann, 1971, and Kuznetsov, 1970.)

ander, 1965). Of the numerous oxidation and reduction stages outlined in Figure 12–5, initial nitrification by bacteria, fungi, and autotrophic organisms involves (Kuznetsov, 1970):

$$NH_4^+ + 1\tfrac{1}{2} O_2 \rightleftharpoons 2H^+ + NO_2^- + H_2O \; [\Delta G_0' = -66.0 \text{ kcal}],^*$$

which proceeds by a series of oxidation stages through hydroxylamine and pyruvic oxime to nitrous acid:

$$NH_4^+ \rightarrow NH_2OH \rightarrow H_2N_2O_2 \rightarrow HNO_2$$

The intermediate products are labile to physical and heterotrophic oxidation, and are found only rarely in significant quantities relative to other forms of combined nitrogen (cf. Baxter, et al., 1973). Much of the energy (total exothermic energy -84.0 kcal mol^{-1}) released by the oxidation series is used to reduce CO_2 in the formation of organic matter; detailed reactions are discussed by Alexander (1965a), Kuznetsov (1970), and Fenchel and Blackburn (1979).

The nitrifying bacteria capable of the oxidation of $NH_4^+ \rightarrow NO_2^-$ are largely confined to *Nitrosomonas* (Nitrobacteriaceae, order Pseudomonadales), although several other taxa, including methane-oxidizing bacteria, are known to be capable of this process (Alexander, 1965; O'Neill and Wilkinson, 1977). These bacteria are mesophilic, with a wide temperature tolerance range (1 to 37°C), and grow optimally at a pH near neutrality.

---

*Gibbs energy of formation, kcal mol^{-1}.

Oxidation of nitrite proceeds further to nitrate by:

$$NO_2^- + \tfrac{1}{2}O_2 \rightleftharpoons NO_3^- \ [\Delta G_0' = -18.0 \text{ kcal}]$$

Nitrobacter is the primary bacterial genus involved in this oxidation. *Nitrobacter* is somewhat less tolerant of low temperatures and high pH, conditions that can lead to a slight accumulation of NO_2-N. The release of energy for synthesis of organic matter by the oxidation of nitrite is much lower at -18.0 kcal per mole than that of NH_4^+ to NO_2^-.

The overall nitrification reactions,

$$NH_4^+ + 2O_2 \rightarrow NO_3^- + H_2O + 2H^+$$

require two moles of oxygen for the oxidation of each mole of NH_4^+. Although conditions must be aerobic in order for nitrification to occur, these processes will continue until concentrations of dissolved oxygen decline to about 0.3 mg l^{-1}. Below this concentration, diffusion rates of oxygen to the bacteria become critical. Nitrification is greatly reduced in undisturbed sediments because oxygen is very low or absent (Chen, et al., 1972). Some nitrate may diffuse to the water following nitrification in the well-oxygenated surficial sediments of the littoral zone, or during periods of circulation of the water and disturbance of sediments (cf. Laurent and Badia, 1973; Landner and Larsson, 1972). Oxidation of ammonia by autotrophic bacteria was found to be of relatively minor importance during summer stratification of eutrophic Pluss See, Northern Germany (Gode and Overbeck, 1972). Several genera of heterotrophic nitrifying bacteria were found in much higher concentrations in both aerobic and anaerobic water strata. Experimental results suggested that these bacteria were most likely responsible for a majority of the nitrification that occurred in this lake.

Nitrification is severely inhibited by certain dissolved organic compounds, especially by tannins and their decompositional derivatives (Rice and Pancholy, 1972, 1973). Although this mechanism has been demonstrated only in soil systems, there is a strong possibility that an analogous situation exists in freshwater systems that contain relatively high concentrations of humic materials, for ammonia concentrations are often greater in highly stained lakes than in aerobic waters of other lakes. Therefore, it is likely that rates of nitrification are lower in neutral or alkaline waters containing high concentrations of dissolved humic organic matter. Such was apparently the case in several European waters (Nygaard, 1938; Karcher, 1939). Moreover, nitrification proceeds slowly in acidic waters, such as in acid bogs and acidic bog lakes, where the pH is 5 or less. Nitrate produced in such lakes is probably utilized as rapidly as it is formed, so that most of the time only very low or undetectable quantities are found.

NITRATE REDUCTION AND DENITRIFICATION

As nitrate is assimilated by algae and larger hydrophytes, it is reduced to ammonia. Molybdenum is required in the enzyme systems associated with this reduction. In a few lake regions, such as granitic mountainous regions, concentrations of molybdenum are extremely low. Rates of carbon fixation by phytoplanktonic algae in lakes of these regions can be increased by additions of molybdenum (Goldman, 1960, 1972). These circumstances presumably are related to the rates of nitrate reduction.

The assimilation of nitrate and its reduction by green plants are clearly dominant processes in the trophogenic zone of lakes. As much as 60% of the photoassimilated NO_3-N can be excreted as dissolved organic nitrogen compounds, some of which are simple amino acids readily utilized by bacteria (Chan and Campbell, 1978). Nitrate assimilation in oligotrophic lakes may be adequate to reduce the observed concentrations in the trophogenic zone, as depicted in Figure 12-4. Alternatively, NO_3-N assimilative losses may be counterbalanced by nitrification and inflow sources of NO_3-N. In eutrophic lakes, denitrification (see below) is a major process influencing the vertical distribution of NO_3-N. Nitrate assimilation can, however, greatly exceed sources of income and generation, in some cases to the point of reducing NO_3-N to below detectable concentrations.

The ratio of NO_3-N to NH_4-N in fresh waters is variable in relation to natural and pollutional sources of both forms of combined nitrogen. In regions draining calcareous sedimentary landforms, unpolluted lakes can have NO_3-N:NH_4-N ratios of 25:1. In other areas where natural sources of NO_3-N are low, the ratio may approach 1:1; where slight to moderate sewage contamination or agricultural applications of nitrogen fertilizers influence the lake, ratios in the range of 1:10 are common.

Denitrification by bacteria is the biochemical reduction of oxidized nitrogen anions, NO_3-N and NO_2-N, with concomitant oxidation of organic matter. The general sequence of this process is:

$$NO_3^- \rightarrow NO_2^- \rightarrow N_2O \rightarrow N_2,$$

which results in a significant reduction of combined nitrogen that can, in part, be lost from the system if it is not refixed.

Many facultative anaerobic bacteria, particularly of the genera *Pseudomonas, Achromobacter, Escherichia, Bacillus,* and *Micrococcus,* can utilize nitrate as an exogenous terminal H acceptor in the oxidation of organic substrates (Alexander, 1961). The denitrification reactions are associated with the enzyme nitrogen reductase and require cofactors of iron and molybdenum. Denitrification operates similarly under both aerobic and anaerobic conditions (Bandurski, 1965).

An exemplary reaction of the oxidation of glucose and concomitant reduction of nitrate is (Hutchinson, 1957):

$$C_6H_{12}O_6 + 12\ NO_3^- = 12\ NO_2^- + 6\ CO_2 + 6\ H_2O\ [\Delta G_0' = -460\ \text{kcal mol}^{-1}];$$

and for the reduction of nitrite to molecular nitrogen:

$$C_6H_{12}O_6 + 8\ NO_2^- = 4\ N_2 + 2\ CO_2 + 4\ CO_3^= + 6\ H_2O\ [\Delta G_0' = -720\ \text{kcal mol}^{-1}].$$

Approximately as much free energy results as in the aerobic oxidation of glucose by dissolved O_2 ($\Delta G_0' = -699$ kcal mol^{-1}). The denitrification reactions occur intensely in anaerobic environments, such as in the hypolimnia of eutrophic lakes (Fig. 12-4) and in anoxic sediments, where oxidizable organic substrates are relatively abundant.

A specialized case, of much less general quantitative significance than the heterotrophic denitrification discussed above, is denitrification of nitrate concurrently with the oxidation of sulfur. The process is accomplished by denitrifying sulfur bacteria, particularly *Thiobacillus denitrificans*, that utilize S° or reduced sulfur compounds such as thiosulfate (Kuznetsov, 1970):

$$5S + 6KNO_3 + 2H_2O \rightarrow 3N_2 + K_2SO_4 + 4KHSO_4$$
$$5Na_2S_2O_3 + 8\,KNO_3 + 2NaHCO_3 \rightarrow 6Na_2SO_4 + 4K_2SO_4 + 2CO_2 + H_2O + 4N_2$$

Both processes occur chemosynthetically under dark, anaerobic conditions, and yield relatively small changes in free energy.

The rate of denitrification, as of nitrification, decreases in acidic waters (Keeney, 1973) and is very slow at low temperatures (around 2°C). Optimum rates of denitrification occur well above the temperatures of most natural fresh waters. At high temperatures the primary product is N_2, while at lower temperatures nitrous oxide (N_2O) predominates. However, N_2O is rapidly reduced to N_2, and has not been found in most lakes in appreciable quantities (Goering and Dugdale, 1966; Kuznetsov, 1970; Macgregor and Keeney, 1973).

Rates of denitrification by nitrate-reducing bacteria in eutrophic Pluss See, Germany, showed marked seasonal and depth variations (Tan and Overbeck, 1973). Rates of nitrate reduction were particularly high, and were correlated with cell numbers during the early portion of summer stratification, before hypolimnetic nitrate concentrations were greatly reduced. High concentrations of oxygen and low levels of nitrate depressed rates of nitrate reduction, but had relatively little effect on cell numbers, especially in winter.

TABLE 12-6   Approximate Rates of Bacterial Denitrification in Lake Water and Sediments by Direct Measurements

LAKE	RATE OF DENITRIFICATION		SOURCE
	Water (μg N l^{-1} day^{-1})	Sediment (mg N m^{-2} day^{-1})	
Lake 227, Ontario	0–30	15	Chan and Campbell, 1980
Norrviken, Sweden	0.53	100	Tirén, et al., 1976
Ramsjön, Sweden	0.65	120	
Smith Lake, Alaska	15	(90)	Goering and Dugdale, 1966
Lake Mendota, Wisconsin	8–26	–	Brezonik and Lee, 1968; Keeney, et al., 1971
Enriched drainage ditch, Netherlands	–	160	van Kessel, 1978
Lake Kasumiga-ura, Japan	–	3–74*	Yoshida, et al., 1979
Boyrup Langsø, Denmark	–	57.5	Andersen, 1977
Kvind sø, Denmark	–	34.2	
Stream sediments, Ontario			
Without worms	–	50	Chatarpaul, et al., 1980.
With tubificid worms	–	90	
Littoral sediments,			Chan and Knowles, 1979
Lake St. George, Ontario			
Without submerged macrophytes	–	2.6	
With submerged macrophytes	–	2.2	

Note: Measurements often use $^{15}NO_3$; other approaches use mass balance techniques, but one can only calculate total lake denitrification by that approach (see text).

*Approximate extrapolations from values expressed as rates per weight of sediment and the area sampled by the coring device.

Nitrification and denitrification can occur simultaneously. In lake sediments, denitrification of added $^{15}NO_3$ was rapid; within two hours up to 90 per cent of added NO_3-N was reduced, much to $^{15}N_2$ (Chen, et al., 1972). Much of the NO_3-N of lake sediments is incorporated into bacterial organic matter. Keeney, et al. (1971) found that up to 37 per cent of experimentally added NO_3-N (at the level of 2 mg NO_3-N l^{-1}) became incorporated into the organic fraction. The remainder was denitrified. Denitrification rates of sediments are three to four orders of magnitude greater than those of the overlying water (Table 12–6). Denitrification activity within the sediments is related to the reducing conditions (Jones, 1979). In sediments of the littoral zone or those in contact with oxygenated water, denitrifying activity is depressed at the sediment surface and increases at sediment depths (10 to 15 mm) where reducing conditions increase (electrode potential, E_h, of 210 mV; see Chapter 14). As the overlying water becomes anoxic and reducing, increased denitrifying activity occurs at the sediment–water interface.

DISSOLVED AND PARTICULATE ORGANIC NITROGEN

Much of the basic understanding of major nitrogenous fractions and their distribution in lakes has evolved from the classical studies of Wisconsin lakes, especially Lake Mendota (Peterson, et al., 1925; Domogalla, et al., 1925; Birge and Juday, 1926). Although analytical methodology has improved greatly since these early investigations, their results are still generally valid for many lakes of the temperate region.

The dissolved organic nitrogen (DON) of fresh waters often constitutes over 50 per cent of the total soluble nitrogen. Geographic variation is great, however, in relation to inputs of inorganic nitrogen from natural and artificial sources. Over one-half of the DON is in the form of amino nitrogen compounds, of which about two-thirds is in the form of polypeptides and complex organic compounds, and less than one-third occurs as free amino nitrogen (Table 12–7). The qualitative nature of the numerous nitrogen compounds is incompletely known (see reviews of Vallentyne, 1957, and Schnitzer and Khan, 1972). The simple amino acids are substrates that are readily utilized by bacteria; rates of decomposition are high, and result in low instantaneous concentrations of free amino acids in fresh waters (Chapters 17 and 22).

As with organic carbon, the DON of lakes and streams is from 5 to 10 times greater than particulate organic nitrogen (PON) contained in the plankton and seston (Table 12–7) (Peterson, et al., 1925; Stadelmann, 1971; Barica, 1970; Manny, 1972a; Manny and Wetzel, 1973, 1974: Serruya, et al., 1975). The ratios of DON to PON decrease as the lakes become more eutrophic, are closer to 1:1 in the trophogenic zone, and increase in the tropholytic zones. More organic nitrogen is apparently synthesized by small phytoplanktonic algae (<10 μm) per unit cell volume than by larger forms (Manny, 1972b). A significant portion of algal intracellular nitrogen (10 to 20 per cent in the blue-green alga *Oscillatoria*) is released extracellularly, mainly as protein and ammonia nitrogen with smaller amounts of nitrite and amino acid nitrogen (Meffert and Zimmerman-Telschow, 1979). Bacterial utilization of these organic compounds, particularly the amino acids, is extremely rapid. Algae, especially blue-green algae, also excrete polypeptides and other organic compounds which are capable of forming complexes with metals such as iron and copper, and with phosphates, and of altering their solubility and physiological availability (Fogg and Westlake, 1955; Murphy, et al., 1976; Tuschall and Brezonik, 1980). Similar nitrogenous compounds

TABLE 12-7   Particulate Organic Nitrogen of the Seston and
Dissolved Organic Nitrogen of Several Lakes

LAKE	PARTICULATE ORGANIC N		DISSOLVED ORGANIC N	
	(μg N l^{-1})			
	0 m	20 m	0 m	20 m
Mendota, Wisconsin (means)	103	86		
Total soluble N			593	842
Amino N less free acids			170	177
Free amino N			88	88
Peptide N			181	173
Organic nonamino N			187	181
Furesø, Denmark	30–190		440–640	
Ysel, Netherlands	250–1400		590–1840	
Smith, Alaska	80–750			
Bodensee, Germany	10–160		50–150	
Lucerne, Switzerland (surface)	70–390		80–180	
Rotsee, Switzerland (surface)	180–1200		270–660	
Wintergreen, Michigan (surface)	50–2350		500–1320	
Augusta Creek, Michigan	ca 50–300		150–870	
Lawrence Lake, Michigan				
Pelagic	20–110		80–240	
Surface inlet streams			60–1550	
Groundwater			50–650	
Lake Kinneret, Israel				
Epilimnion (3 m)	50–283		89–545	
Hypolimnion (30 m)	55–398		40–343	

After several sources cited in the text.

have been found to be secreted by larger aquatic plants (Wetzel and Manny, 1972); in some situations where the littoral zone is extensively developed, release of DON by macrophytes can form a major source of organic nitrogen to the lake. Furthermore, as aquatic vascular vegetation decomposes, large quantities of organic nitrogen are released (Nichols and Keeney, 1973). Much of this organic nitrogen is absorbed by the sediments, where decomposition can rapidly become limited by inorganic nitrogen, especially under anaerobic conditions (cf. Chapter 22).

Seasonal Distribution of Nitrogen

The seasonal changes in nitrogen vary greatly from lake to lake. Several general trends, however, do emerge from the seasonal patterns in oligotrophic and productive lakes. Although many seasonal cycles of inorganic nitrogen have been studied, the cycles of the dissolved and the biota-bound particulate organic nitrogen are often overlooked. The ensuing discussion concerns five lakes: (a) Lawrence Lake, an oligotrophic, hard-water lake of southern Michigan that receives high natural inputs of inorganic nitrogen, and where N_2 fixation is very low; (b) mesotrophic Vierwaldstattersee (Lake of Lucerne), Switzerland, a lake in which the hypolimnion remains aerobic throughout summer stratification; (c) moderately eutrophic Lake Mendota, Wisconsin; (d) highly eutrophic Wintergreen Lake, southern Michigan, a lake which undergoes normal dimictic circulation; and (e) Rotsee, Switzerland, an extremely eutrophic lake that circulates incompletely in spring and autumn.

The nitrogen inputs to Lawrence Lake, which is located in calcareous glacial till high in nitrates, occur largely as nitrate (Manny and Wetzel, 1982). Concentrations of NO_3-N of two small spring-fed inlet streams varied from 1 to over 20 mg NO_3-N l^{-1}; groundwater inflow directly into the basin contained lower concentrations (mean 3 mg l^{-1}), but contributed about 25 per cent to the annual nitrogen income. Ammonia in the influents was much lower (between 0 to 190 μg NH_4-N l^{-1}), and represented a relatively minor source of nitrogen to this lake. Nitrite concentrations were very small. The total dissolved organic nitrogen (TDON) of the surface and groundwater influents, however, was a significant source and reached levels as high as 1.5 mg l^{-1}. This TDON was strongly correlated with the metabolic activity of the surrounding marsh vegetation through which the streams flow.

Figure 12–6 contrasts the concentrations of NO_3-N + NO_2-N in mg l^{-1} (upper) and NH_4-N in μg l^{-1} (lower) in seasonal depth–time diagrams. NH_4-N levels of the epilim-

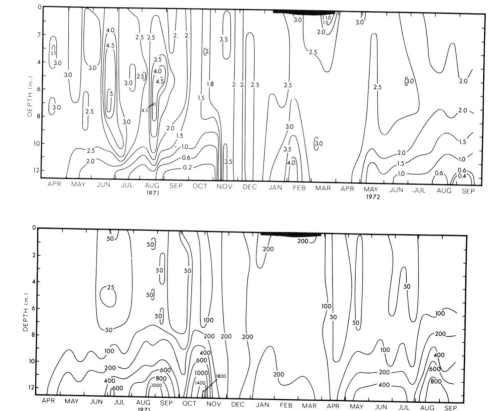

Figure 12-6   Depth-time diagrams of seasonal concentrations of NO_3-N + NO_2-N in mg l^{-1} (upper) and NH_4-N in μg l^{-1} (lower), Lawrence Lake, Michigan, 1971–72. Opaque areas = ice-cover to scale. (From Manny and Wetzel, unpublished data.)

nion and metalimnion were very low during stratification, and increased conspicuously only below 11 m, where anoxic conditions existed in late summer and autumn. This increase was accompanied by a simultaneous denitrification of NO_3 in the lower hypolimnion. NO_2-N concentrations were always very low (<20 μg l^{-1}). During turnover in November, and throughout much of the winter period of ice-cover, NH_4 concentrations were distributed uniformly with depth at levels over 200 μg l^{-1}. Nitrification was seen in the greater depths during the winter, and marked decreases in NO_3-N concentrations occurred immediately below the ice, consonant with intensive growth of algal populations. After the spring period of uniformity during circulation, the pattern was repeated.

The dissolved organic nitrogen (DON) and particulate organic nitrogen (PON) of the open water of Lawrence Lake generally were found to be about equal in concentration (50 to 1000 μg l^{-1}). Concentrations of PON closely followed the dynamics of the biomass of the plankton; those of DON were maximal in the summer in the epilimnion and minimal during the winter in the hypolimnion.

The nitrogen budget of a large bay of Vierwaldstattersee was studied in detail by Stadelmann (1971). The nitrate concentrations (Fig. 12-7, *left*) of the epilimnion decreased sharply in the summer because of algal utilization, and this decrease was reflected in the amounts of planktonic particulate organic nitrogen (Fig. 12-7, *right*). NO_2-N concentrations followed a similar seasonal cycle, but never exceeded 11 μg NO_2-N l^{-1}. Ammonia concentrations were always low, usually only in trace quantities; maximum levels (70 μg NH_4-N l^{-1}) occurred in late summer near the sediments. During this summer period when epilimnetic concentrations of inorganic nitrogen approached

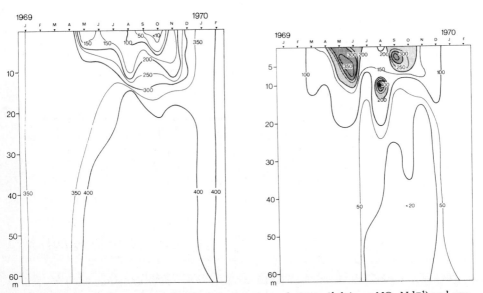

Figure 12-7 Depth–time diagrams of the concentrations of nitrate (*left* in μg NO_3-N l^{-1}) and particulate organic nitrogen (*right*, in μg N l^{-1}) in Vierwaldstattersee, Switzerland. (After Stadelmann, P: Schweiz. Zeitschrift f. Hydrologie, 33:1–65, 1971.)

zero, blue-green algal and heterocyst numbers reached their maximum, indicative of a heavy reliance on molecular N_2 fixation. Aerobic hypolimnetic waters of this lake did not undergo significant denitrification.

As lakes become more eutrophic, the processes of assimilation of nitrogen in the trophogenic zone and denitrification in the tropholytic zone intensify. Vertical gradients become extreme. As is exemplified in Wintergreen Lake (Fig. 12–8), nitrate and nitrite are reduced by algal assimilation and denitrification extremely rapidly, from over 1000 μg l^{-1} to below detectability in slightly over a month (Wetzel, et al., 1982). Fixation of molecular N_2 was limited to the 0 to 3 m depths in Wintergreen Lake, and very high rates of fixation by *Anabaena* and *Aphanizomenon* were found only during periods when levels of combined inorganic nitrogen were very low (Duong, 1972; Ward and Wetzel, 1980a). Ammonification and denitrification were particularly intensive in the anaerobic tropholytic zone below a depth of 3 m. As the name of the lake (Wintergreen) implies, and as is common in hypereutrophic lakes (Wetzel, 1966a), production continues vigorously beneath the ice and the denitrification process continues, although more slowly, during winter stratification. An analogous situation occurred in the extremely hypereutrophic Rotsee, Switzerland (Stadelmann, 1971). However, in this lake, which is protected from major wind-induced water movements by the surrounding topography, circulation in the spring and autumn was weak and incomplete. The hypolimnion was anaerobic all year long, except for a slight intrusion of oxygen for a week in the spring; this condition was reflected in the high concentrations of ammonia throughout much of the lake most of the year (Fig. 12–9). Hypolimnetic concentrations of nearly 10 mg NH_4-N l^{-1} are not uncommon near the sediments.

The seasonal distribution of the various forms of nitrogen in eutrophic Lake Mendota (Fig. 12–10) is similar to that commonly found in numerous lakes of temperate regions (Domogalla, et al., 1926; Domogalla and Fred, 1926; Barica, 1970). In many cases, autumnal nitrogen minima gradually increase in late fall to maxima in late winter and early summer. Looking at the seasonal processes involved in the surface and hypolimnetic strata over an annual period (Fig. 12–11), the net effects of nitrification and nitrate reduction-assimilation can be seen. In the oxygenated upper waters, nitrate decreases during the spring maximum due to increased rates of assimilation by the plankton and nitrate reduction by bacteria. Nitrification decreases rapidly in the tropholytic zone after thermal stratification develops, because the hypolimnion becomes anaerobic. Additional decreases in rates of nitrification in the hypolimnion can occur when nitrifying bacteria suffer losses from grazing of zooplankton (Cavari, 1977). Nitrification ceases and nitrate reduction increases markedly until stratification is disrupted in the fall.

Carbon-to-Nitrogen Ratios

The carbon content of organic matter is on the average at least an order of magnitude greater than that of nitrogen. The complex mixtures of organic compounds in particulate and dissolved phases are decomposed and mineralized to inorganic carbon primarily as CO_2, and to inorganic nitrogen. The proteolytic metabolism of fungi and

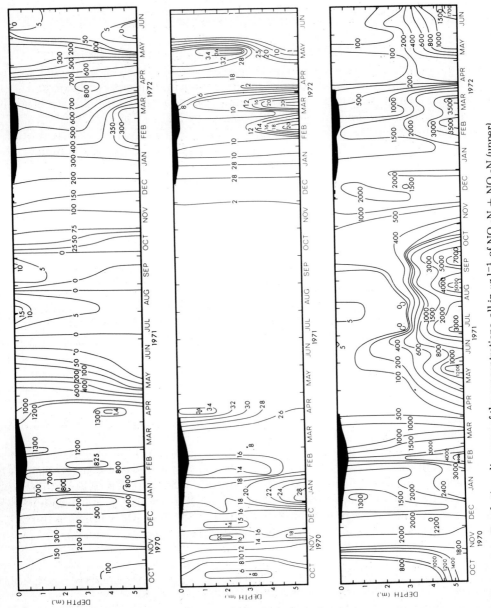

Figure 12-8 Depth–time diagrams of the concentrations, all in $\mu g\,l^{-1}$, of NO_3-N + NO_2-N (upper), NO_2-N (middle), and NH_4-N (lower), Wintergreen Lake, Michigan, 1970–72. Opaque areas = ice-cover to scale. (From Wetzel, et al., unpublished data.)

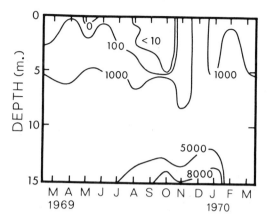

Figure 12-9 Depth–time diagram of the concentrations of ammonia (μg NH_4-N l^{-1}), Rotsee, Switzerland, 1969–70. (Redrawn from Stadelmann, 1971.)

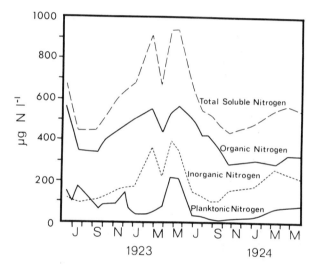

Figure 12-10 Average seasonal distribution of forms of nitrogen in Lake Mendota, Wisconsin, June 1922, through May 1924. (Redrawn from Hutchinson, 1957, after Domogalla, et al., 1926.)

bacteria removes proportionately more nitrogen than carbon. The rates of decomposition become slower with the greater resistance of the residual organic compounds, and the selective removal of nitrogen by microbes results in a net increase in the C:N ratios.

If we look at the general distribution of organic carbon and nitrogen in many lakes, as Birge and Juday (1934) did in several hundred Wisconsin lakes, certain trends become apparent (Table 12–8). There are large seasonal and spatial variations in the particulate organic matter in lakes and streams, but most of the organic matter is in the dissolved fraction, and concentrations of dissolved organic matter are very constant (cf. Wetzel, et al., 1972). Lakes can be grouped into categories of increasing organic carbon and dissolved organic matter. There is a clear decrease in organic nitrogen with increasing organic carbon content, which results in increasing C:N ratios. This relationship suggests an increase in the refractory nature of the organic compounds in lakes that contain greater amounts of dissolved organic matter. Such is indeed the case when waters containing high concentrations of dissolved organic matter receive increasingly

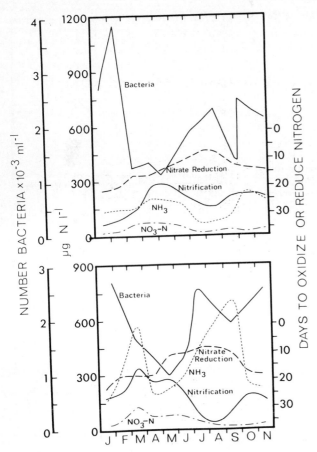

Figure 12-11 Rates of nitrification and denitrification, determined by the time required to oxidize or reduce added nitrogen, bacterial numbers, and the concentrations of ammonia and nitrate nitrogen in Lake Mendota, Wisconsin, 1925. (Redrawn from Domogalla, et. al., 1926.) *Upper:* surface water of trophogenic zone; *lower:* near bottom water of tropholytic zone.

TABLE 12-8 Approximate Composition of Organic Matter of Water from Numerous Wisconsin Lakes

TOTAL PARTICULATE AND DISSOLVED ORGANIC CARBON (mg l^{-1})	PARTICULATE ORGANIC MATTER (mg dry wt l^{-1})	DISSOLVED ORGANIC MATTER*				
		Dry Weight (mg l^{-1})	Crude Protein† (%)	Lipid Material‡ (%)	Carbohydrate (%)	C:N Ratio
1.0–1.9	0.62	3.1	24.3	2.3	73.4	12.2
5.0–5.9	1.27	10.3	19.4	1.3	79.0	15.1
10.0–10.9	1.89	20.5	14.4	0.4	85.2	20.1
15.0–15.9	2.32	31.3	12.9	0.2	86.9	22.4
20.0–25.8	2.22	48.1	9.9	0.2	89.9	29.0

After data of Birge and Juday, 1934.
*Includes colloidal material; particulate matter removed by centrifugation.
†Total nitrogen content of organic fraction × 6.25.
‡Ether extract.

greater proportions of their organic content from organic plant material produced in wetland and littoral marsh areas surrounding the water (Wetzel, 1979). Material of wetland and littoral origin dominates the dissolved organic matter pool of bogs and bog lakes.

Organic matter of terrestrial and marsh areas undergoes varying degrees of decomposition prior to and during transport to the lake or stream and during which much of the organic nitrogen can be utilized. Therefore, in general (cf. Hutchinson, 1957), allochthonous organic matter contains about 6 per cent crude protein and has a C:N ratio from 45:1 to 50:1. The dissolved organic matter contains a high percentage of humic acid compounds that are low in nitrogen content (Shapiro, 1957; Schnitzer and Khan, 1972), and which impart a stained brown color to the water. In contrast, autochthonous organic matter produced by decomposition of plankton within the lake contains about 24 per cent crude protein and has a C:N ratio of about 12:1.

Summary of the Nitrogen Cycle

A diagrammatic representation of the major nitrogen inputs, transformation pathways, and outputs to a general lake system is given in Figure 12–12. Although numerous pathways are presented, it is obviously a simplification of the complex mechanisms and

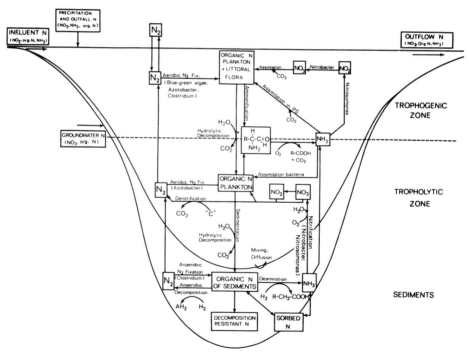

Figure 12-12 Generalized nitrogen cycle for fresh waters. *PS* = photosynthesis. (Greatly modified from Kuznetsov, 1970.)

processes. Numerous analogies in the nitrogen cycle exist between lake and river systems.

The components of the cycle have already been discussed; only a few points need to be emphasized. The processes indicated in the tropholytic zone represent a composite situation under aerobic conditions. Clearly, the obligately aerobic decomposition processes would cease in anaerobic hypolimnia of productive lakes. The metabolism of the littoral flora, including the vascular macrophytes and their epiphytic associations of bacteria and algae, is not only a major or dominant source of organic nitrogen synthesis in many lakes, but can influence significantly the flux of nitrogen from the sediments to the water as well.

Many components and processes are variable seasonally and spatially. For example, inputs of nitrogen from guano of migratory waterfowl that briefly reside on a lake in extraordinary densities (one per m^2) can represent a major input of nitrogen and phosphorus to certain lakes (Manny, et al., 1975; cf. also the extremely detailed review of Hutchinson, 1950). Sewage inputs of organic N, NO_3-N, and NH_4-N to rivers and recipient lakes are often highly pulsed. Similarly, agricultural applications of fertilizers to drainage basin areas are seasonal, and can alter nitrogen inputs to lakes and rivers. Nitrogen fixation, normally a minor component of the total nitrogen income, can become a significant driving source of the system at certain times of the year.

The nitrogen dynamics of the sediments are poorly understood. Lake sediments typically contain on the order of 50 to 200 kg of nitrogen per hectare to 10 cm sediment depth (Table 12–9), much of which is immobilized and sorbed to inorganic particles (Keeney, 1973). The interstitial water in sediments usually has a much higher concentration of soluble nitrogen compounds, mainly as NH_4-N and organic N, than that of the overlying water. Diffusion rates are exceedingly slow, but mixing of surficial sediments occurs to some extent, even when the lake is strongly stratified, as a result of deep-water water movements, activities of benthic organisms, and loss of gas bubbles from the sediments. At periods of turnover, 1 to 10 cm of the surface sediment can be mixed, and much of it resuspended in the water column (e.g., Wetzel, et al., 1972; Davis, 1973). Surface sediments are usually oxidized for only a few millimeters in depth, but an oxidized microzone is critical to the solubility and sorption properties of the sediments for ammonia, and can fundamentally alter rates of microbial transformations. Nitrogen exchange between sediments and water also varies greatly with sediment composition. For example, release of nitrogen from sediments of the Rybinsk Reservoir was greatest in silts high in organic matter (Table 12–10). In this reservoir, the anaerobic

TABLE 12–9 Approximate Nitrogen Content of Various Components of Bantam Lake, Connecticut

COMPONENTS	kg N PER LAKE	PER CENT
Lake water	10700	13
Algae	29200	33
Vascular aquatic plants	2500	3
Sediment (surface 1 cm)	44000	51
	86400	100

Modified from estimates of Frink, 1967.

TABLE 12–10   Average Daily Exchange of Nitrogen Between the Sediments and Benthic Water Layers in Rybinsk Reservoir, USSR, in mg N m^{-2} day^{-1}

SEDIMENT	NO$_3$-N	NH$_4$-N	ALBUMINOID N	ORGANIC N	TOTAL N
Sand	+0.14	+1.14	+0.52	−6.00	−4.72
Unflooded soil	−0.44	+0.36	+0.41	+2.13	+2.03
Grey silt (fluvial areas)	−1.20	+21.43	+3.19	+10.98	+31.28
Redeposited peat	−4.3	+1.2	+1.3	+7.1	+4.0

From Kuznetsov, S. I.: Limnol. Oceanogr., 13:211–224, 1968.

decomposition of sediment organic matter resulted in considerable loss of nitrogen in the form of N$_2$.

Because of the functional complexity of the many components of the nitrogen cycle, no direct measurements exist for all processes simultaneously in any lake or stream over an annual period. Where detailed experimentation has dissected individual process rates using ^{15}N isotopic techniques, results from certain times of the year emphasize the major importance of the sediments not only as a large reservoir of nitrogen but as the site of much of the nitrogen metabolism.

Lake Wingra in southern Wisconsin is a shallow, eutrophic lake in which submersed macrophytes constitute a major part of the total lake productivity. An overall average nitrogen content of the various compartments in Lake Wingra was estimated for midsummer conditions (Table 12–11). Much (97 per cent) of the total nitrogen

TABLE 12–11   Estimates of the Nitrogen in Various Compartments in Lake Wingra, Wisconsin, Midsummer

COMPARTMENT	NITROGEN (METRIC TONS)	PER CENT OF TOTAL N	
		A*	B†
Soluble and particulate N in water	3.2	0.5	50.6
Macrophyte N	1.2	0.2	19.3
Sediment, upper 10 cm			
Organic N	138	23.0	–
Interstitial water			
NH$_4$-N	0.74	0.1	11.5
Organic N	0.19	0.03	3.0
Exchangeable NH$_4$-N	1.0	0.2	15.6
Sediment, 10 to 30 cm			
Organic N	460	74.3	–
Interstitial water			
NH$_4$-N	4.6	0.8	–
Organic N	0.51	0.08	–
Exchangeable NH$_4$-N	5.4	0.9	–
Total			
Total N in system	615		
Total N excluding sediment organic N and all N below 10 cm in sediment	6.3		

Modified from Isirimah, et al., 1976.
*Total N in system.
†Total N excluding sediment organic N and all N below 10 cm in sediment.

occurred in the sediments as organic nitrogen (Column A) but was not readily available for metabolism in the lake system (Isirimah, et al., 1976). Neglecting these relatively unreactive nitrogen reservoirs, about 50% of the "available" nitrogen existed in the water column, mostly in dissolved form, 20% in the macrophytes, and 30% in the interstitial water of surficial sediments (Column B). Rapid turnover rates of NH_4-N occurred in the water but not in the sediments. NO_3-N turnover was slower in the water than in the sediments where about 80% of the NO_3-N was rapidly denitrified to N_2.

NITROGEN BUDGETS

Detailed evaluations of the nitrogen cycle that include close interval quantitative measurements of inputs, metabolic dynamics, and outputs are not available for freshwater systems. The cycle is obviously complex, and would require accurate analyses of the dynamics of all components for several years. Only a few approximate analyses of the nitrogen budgets of lakes are available. In these analyses the rates of bacterial metabolism are obtained indirectly, by assumptions of microbial processing or sedimentation by differences. Nonetheless, the proximate calculated budgets are instructive.

The nitrogen budget of Lake Mendota, Wisconsin, demonstrates the roughly equivalent contributions from runoff, groundwater, and precipitation (Table 12–12). Major losses of nitrogen occur via sedimentation, denitrification, and outflow. Loss by seepage out of the basin was probably very small, since most lake basins are well sealed (cf. Wetzel and Otsuki, 1974). In Table 12–13, the nitrogen budget for a large reservoir of complex morphometry accounts for inflows from all river influents and losses by sedimentation and outflow. Although the total nitrogen content of this reservoir decreased in 1962 because of low water levels, inputs of atmospherically derived nitrogen amounted to roughly 10,000 tons per year. About 1300 metric tons entered with rain and snow; the remainder may be attributed largely to biological fixation.

TABLE 12–12 Estimated Nitrogen Budget of Lake Mendota, Wisconsin

SOURCES	NITROGEN INCOME		LOSSES	NITROGEN LOSSES	
	kg N year^{-1}	%		kg N year^{-1}	%
Municipal and industrial wastewater	21200	10.4	Outflow	41300	20.4
			Denitrification	28100	13.9
Urban runoff	13700	6.8	Fish catch	11300	5.6
Rural runoff	23500	11.6	Weed removal	3250	1.6
Precipitation on lake surface	43900	21.6	Loss to groundwater	*	*
Groundwater			Sedimentation and other†	118850	58.6
Streams	35900	17.7			
Seepage	28500	14.1		202800	100
Nitrogen fixation	36100	17.8			
Marsh drainage	‡	‡			
	202800	100			

Modified from Brezonik and Lee, 1968, with data improvements by Keeney, 1972.
*Unknown; likely very small in this lake.
†By difference between total of other estimated losses and sum of income sources.
‡Considered significant, but data unavailable.

TABLE 12–13   Estimated Nitrogen Budget of the Rybinsk Reservoir, USSR, in Metric Tons N per Year

	FACTORS MEASURED	TIME		
		1 June 60– 1 June 61	1 April 61– 1 April 62	1 June 62– 1 June 63
A	Measured nitrogen balance of water mass	+12519	+5399	+12170
B	Difference between inflow and outflow	−180	−7329	+1565
A − B = C	Increase in water mass	+12699	+12728	+10605
D	Precipitated in sediments (20-year average)	+12500	+12500	+12500
C − D = E	Theoretical balance in water mass	+199	+228	−1895
A − E	Nitrogen input from atmosphere and fixation	+12320	+5171	+14065

From Kuznetsov, S. I.: Limnol. Oceanogr., 13:211–224, 1968.

General budgets of this type leave much to be desired because they provide little insight into the dynamics of internal processing or metabolic control mechanisms. From an applied point of view, however, such budgetary information is useful in relation to nitrogen loading of aquatic systems, and effects of this loading on increased productivity.

NITROGEN LOADING: EFFECTS OF MAN'S ACTIVITIES

Enrichment of fresh waters with nutrients needed for plant growth occurs commonly as a result of losses from agricultural fertilization, loading from sewage and industrial wastes, and enrichment via atmospheric pollutants (especially nitrate). Sufficient information exists about the general responses of many lakes to loading of major nutrients, especially phosphorus and nitrogen, to predict the potential changes in their productivity. The phytoplankton productivity of infertile, oligotrophic lakes is often limited by the availability of phosphorus. As phosphorus loading to fresh waters increases and lakes become more productive, nitrogen often becomes the nutrient limiting to plant growth. Excessive loading of these nutrients permits increased plant growth until other nutrients or light availability become limiting.

The nitrogen loading concepts in relation to increased lake productivity are best discussed simultaneously with phosphorus loading and limitations. Both will be treated in the following chapter.

SUMMARY

1. Nitrogen, along with carbon, hydrogen, and phosphorus, is one of the major constituents of cellular protoplasm of organisms. Nitrogen is a major nutrient that affects the productivity of fresh waters.

2. A major source of nitrogen of the biosphere originates from fixation of atmospheric molecular nitrogen (N_2). The nitrogen cycle is a complex biochemical process in which nitrogen in various forms is altered by nitrogen fixation, assimilation, and reduction of nitrate to N_2 by denitrification. For all practical purposes, the nitrogen cycle of lakes is microbial in nature: Bacterial oxidation and reduction of nitrogen compounds are coupled with photosynthetic assimilation and utilization by algae and larger aquatic plants. The direct role of animals in the nitrogen cycle is certainly very small; under certain conditions, however, their grazing activities can influence microbial populations and nitrogen transformation rates, as well as nitrogen utilization rates by photosynthetic organisms. Although the nitrogen cycle of fresh waters is understood qualitatively in appreciable detail, the complex dynamics of quantitative transformation rates have not been clearly delineated, especially at the sediment–water interface where intensive bacterial metabolism occurs.

3. Dominant forms of nitrogen in fresh waters include: (a) Dissolved molecular N_2; (b) ammonia nitrogen (NH_4^+); (c) nitrite (NO_2^-); (d) nitrate (NO_3^-); and (e) a large number of organic compounds (e.g., amino acids, amines, nucleotides, proteins, and refractory humic compounds of low nitrogen content).

4. The nitrogen cycle consists of a balance between nitrogen inputs to and losses of nitrogen from an aquatic ecosystem.

   a. Sources of nitrogen include: (i) nitrogen contained in particulate "dry fallout" and precipitation falling directly on the lake surface; (ii) nitrogen fixation both in the water and the sediments; and (iii) inputs of nitrogen from surface and groundwater drainage.

   b. Losses of nitrogen occur by: (i) outflow from the basin; (ii) reduction of NO_3^- to N_2 by bacterial denitrification with loss of N_2 to the atmosphere; and (iii) sedimentation of inorganic and organic nitrogen-containing compounds to the sediments.

5. Microbial fixation of molecular N_2 in soils by bacteria is a major source of nitrogen. In lakes and streams, N_2 fixation by bacteria and certain blue-green algae is quantitatively less significant, except under certain conditions of severe depletion of combined inorganic nitrogen compounds.

   a. N_2 content of water is usually in equilibrium with N_2 of the atmosphere during periods of turbulent mixing. In stratified, productive lakes, concentrations of N_2 may decline in the epilimnion because of reduced solubility as temperatures rise, and increase in the hypolimnion from denitrification of NO_3-N.

   b. N_2 fixation by blue-green algae is usually much greater than fixation by bacteria. In blue-green algae, N_2 fixation is light dependent and usually coincides with the spatial and temporal distribution of these algae. NH_4-N assimilation requires less energy expenditure than NO_3-N, and NO_3-N less than N_2-N. N_2 fixation by blue-green algae increases when NH_4-N and NO_3-N concentrations decrease in the trophogenic zone.

   c. Bacterial N_2 fixation in wetlands surrounding lakes or adjacent to streams can add significant amounts of combined nitrogen to lake ecosystems.

6. Ammonia is generated by heterotrophic bacteria as the primary nitrogenous end product of decomposition of proteins and other nitrogenous organic

compounds. Ammonia is present primarily as NH_4^+ ions and is readily assimilated by plants in the trophogenic zone.

   a. NH_4-N concentrations are usually low in aerobic waters because of utilization by plants in the photic zone. Additionally, bacterial nitrification occurs, in which NH_4^+ is oxidized through several intermediate compounds to NO_2^- and NO_3^-.

   b. When the hypolimnion of a eutrophic lake becomes anaerobic, bacterial nitrification of ammonia ceases. The oxidized microzone at the sediment–water interface is also lost, which reduces the adsorptive capacity of the sediments. A marked increase in the release of NH_4^+ from the sediments then occurs. As a result, the NH_4-N concentrations of the hypolimnion increase (Fig. 12-4).

   c. Bacterial nitrification proceeds in two stages: (i) The oxidation of $NH_4^+ \rightarrow NO_2^-$, largely by *Nitrosomonas* but also by other bacteria, including methane oxidizers; and (ii) the oxidation $NO_2^- \rightarrow NO_3^-$, in which *Nitrobacter* is the dominant bacterial genus involved.

   d. Nitrite (NO_2^-) is readily oxidized and rarely accumulates except in the metalimnion, upper hypolimnion, or interstitial water of sediments of eutrophic lakes. Concentrations are usually very low ($<100\ \mu g\ l^{-1}$) unless organic pollution is high.

7. Nitrate is assimilated and aminated into organic nitrogenous compounds within organisms. This organic nitrogen is bound and cycled in photosynthetic and microbial organisms. During normal metabolism of these organisms, and at death, their nitrogen is liberated as ammonia. Additionally, organisms release a variety of organic nitrogenous compounds which are resistant to proteolytic deamination and ammonification by heterotrophic bacteria to varying degrees.

   a. Nitrate (NO_3^-) is the common form of inorganic nitrogen entering fresh waters from the drainage basin in surface waters, groundwater, and precipitation. In certain oligotrophic waters in basaltic rock formations, nitrate loading from atmospheric sources, especially if contaminated by man-produced combustion emission products, can dominate nitrogen loading.

   b. Bacterial denitrification is the biochemical reduction of oxidized nitrogen anions (NO_3^- and NO_2^-), concomitant with the oxidation of organic matter: $NO_3^- \rightarrow NO_2^- \rightarrow N_2O \rightarrow N_2$. Nitrous oxide ($N_2O$) is rapidly reduced to N_2, and has never been found in lakes in appreciable quantities. Denitrification is accomplished by many genera of facultative anaerobic bacteria, which utilize nitrate as an exogenous terminal hydrogen acceptor in the oxidation of organic substrates. Denitrification occurs in anaerobic environments, such as in the hypolimnia of eutrophic lakes (Fig. 12-4) or in anoxic sediments, where oxidizable substrates are relatively abundant.

8. Dissolved organic nitrogen (DON) often constitutes over 50 per cent of the total soluble nitrogen in fresh waters.

   a. Over half of the DON occurs as amino nitrogen compounds, mostly as polypeptides and complex nitrogen compounds.

   b. The ratios of DON to particulate organic nitrogen (PON) of streams and lakes are usually from 5:1 to 10:1. As lakes become more eutrophic, DON:PON ratios decrease.

9. The distribution of nitrogen in a lake can change rapidly. Examples of the depth–time distributions of the different forms of nitrogen demonstrate that as lakes become more productive from nutrient loading, concentrations of NO_3-N and NH_4-N in the trophogenic zone can be severely reduced and depleted by photosynthetic assimilation. Blue-green algae with the capability of nitrogen fixation may then come to dominate. In anaerobic hypolimnia, NO_3-N is rapidly denitrified to N_2, which is either fixed or lost to the atmosphere. NH_4-N concentrations accumulate from decomposition of organic matter and release of NH_4-N from sediments under anaerobic conditions.

10. Organic carbon-to-nitrogen ratios (C:N) indicate an approximate state of resistance of complex mixtures of organic compounds to decomposition, because proteolytic metabolism by fungi and bacteria removes proportionally more nitrogen than carbon. Higher C:N ratios commonly occur in residual organic compounds which are more resistant to decomposition.
    a. Organic materials from allochthonous and wetland sources commonly have C:N ratios from 45:1 to 50:1, and contain many humic compounds of low nitrogen content.
    b. Autochthonous organic matter produced by the decomposition of plankton tends to have higher protein content and C:N ratios of about 12:1.

11. Much of the total nitrogen occurs in the sediments in forms that are relatively unavailable for biotic utilization. Of the readily available nitrogen, a majority occurs in soluble form in the water and in the interstitial water of surficial sediments (and in littoral vegetation in shallow, productive lakes). Turnover rates of NH_4-N are rapid in water but slower in the sediments. In contrast NO_3-N turnover is slower in the water than in sediments, where, under anoxic conditions in eutrophic lakes, NO_3-N is rapidly denitrified to N_2.

12. Increased loading of inorganic nitrogen to rivers and lakes frequently results from agricultural activities, sewage, and atmospheric pollution by man. In unproductive oligotrophic lakes, phosphorus availability is often the principal limiting nutrient for plant growth. As phosphorus loading to fresh waters increases and lakes become more productive, nitrogen becomes more important as a growth-limiting nutrient.

CHAPTER THIRTEEN

THE PHOSPHORUS CYCLE

PHOSPHORUS IN FRESH WATERS

No other element in fresh waters has been studied as intensively as phosphorus. A great number of quantitative data exist on the seasonal distribution of phosphorus in lakes and loading rates of phosphorus from the surrounding drainage basins. Ecological interest in phosphorus stems from its major role in biological metabolism, and the relatively small amounts of phosphorus available in the hydrosphere. In comparison to the rich natural supply of other major nutritional and structural components of the biota (carbon, hydrogen, nitrogen, oxygen, sulfur), phosphorus is least abundant, and most commonly limits biological productivity.

The Distribution of Organic and Inorganic Phosphorus in Lakes

In contrast to the numerous forms of nitrogen in lake systems, the most significant form of inorganic phosphorus is orthophosphate (PO_4^{---}). A very large proportion, greater than 90 per cent, of the phosphorus in fresh water occurs as organic phosphates and cellular constituents in the biota adsorbed to inorganic and dead particulate organic materials. It is instructive to discuss first the general aspects of the forms and distribution of phosphorus that occurs in fresh waters, before analyzing dynamics of exchange between the compartments.

Total inorganic and organic phosphorus has been separated in various ways for chemical analyses; often, these fractions relate poorly to the metabolism of phosphorus. Perhaps the most important measure is the total phosphorus content of unfiltered water, which consists of the phosphorus in the particulate and in 'dissolved' phases (Juday, 1927; Ohle, 1938). Both compartments consist of several components. Particulate phosphorus includes: (1) Phosphorus in organisms as (a) relatively stable nucleic acids DNA, RNA, and phosphoproteins, which are not involved in rapid cycling of phosphorus, (b) low-molecular-weight esters of enzymes, vitamins, etc., and (c) nucleotide phosphates, such as adenosine diphosphate (ADP) and adenosine 5-triphosphate (ATP) used in biochemical pathways of respiration and CO_2 assimilation. (2) Mineral phases of rock and soil, such as hydroxyapatite, in which phosphorus is adsorbed onto inorganic complexes such as clays, carbonates, and ferric hydroxides. (3) Phosphorus adsorbed onto dead particulate organic matter or in macroorganic aggregations. In contrast to the phosphorus of particulate matter, dissolved phosphorus is composed of: (1) orthophosphate (PO_4^{---}), (2) polyphosphates, often originating from synthetic detergents, (3) organic colloids or phosphorus combined with adsorptive colloids, and (4) low-molecular-weight phosphate esters.

Because of the fundamental importance of phosphorus as a nutrient and major cellular constituent, much emphasis has been placed on its analytical evaluation. Chemical analyses of phosphorus center around the reactivity of phosphorus with molybdate, and changes in reactivity during enzymatic and acidic hydrolysis of complex forms of phosphorus compounds as they are converted to orthophosphate. Detailed analyses recognize eight forms of phosphorus, which are differentiated on the basis of its reactivity with molybdate, ease of hydrolysis, and particle size (Strickland and Parsons, 1972). Four operational categories result: (a) soluble reactive P, (b) soluble unreactive P, (c) particulate reactive P, and (d) particulate unreactive P. However, these operational methods do not necessarily correspond to either the chemical species of phosphorus or to their role in biotic cycling of phosphorus.

Most of the phosphorus data for fresh waters refer to total phosphorus and inorganic soluble phosphorus (orthophosphate), although in more detailed studies, four general fractions have been identified (Hutchinson, 1957). These four fractions are similar to the four operational groups already cited: (a) soluble phosphate phosphorus, (b) acid-soluble suspended (sestonic) phosphorus, mainly ferric phosphate and calcium phosphate, (c) organic soluble and colloidal phosphorus, and (d) organic suspended (sestonic) phosphorus.

Total phosphate concentrations in nonpolluted natural waters extend over a very wide range from less than 1 μg l^{-1} to more than 200 mg l^{-1} in some closed saline lakes. The total phosphorus concentrations of most uncontaminated surface waters are between 10 to 50 μg P l^{-1}. Variation is high, however, and can be related to characteristics of regional geology. Phosphorus levels of fresh waters are generally lowest in mountainous regions of crystalline bedrock geomorphology, and increase in lowland waters derived from sedimentary rock deposits. Lakes rich in organic matter, such as bogs and bog lakes, tend to exhibit high total phosphorus concentrations. A few sedimentary coastal areas, such as in the southeastern United States, are rich in phosphatic rock. Lakes with drainage from these deposits have abnormally high phosphorus levels.

TABLE 13-1   **General Relationship of Lake Productivity to Average Concentrations of Epilimnetic Total Phosphorus***

GENERAL LEVEL OF LAKE PRODUCTIVITY	CHANGE (REDUCTION) IN ALKALINITY IN EPILIMNION DURING SUMMER (meq l^{-1})	TOTAL PHOSPHORUS (μg l^{-1})
Ultra-oligotrophic	<0.2	<5
Oligo-mesotrophic	0.6	5–10
Meso-eutrophic	0.6–1.0	10–30
Eutrophic		30–100
Hypereutrophic	>1.0	>100

*Modified from Vollenweider, R. A.: Scientific Fundamentals of the Eutrophication of Lakes and Flowing Waters, with Particular Reference to Nitrogen and Phosphorus as Factors in Eutrophication. OECD Report No. DAS/CSI/68.27, Paris, OECD, 1968, after numerous sources.

TABLE 13-2 Fractionation of Total Phosphorus in Lakes Analyzed by Different Techniques of Separation

LAKES	LAKES	SOLUBLE INORGANIC P		SOLUBLE ORGANIC P		SESTONIC P		TOTAL ORGANIC P		TOTAL P
		(μg l^{-1})	(%)	(μg l^{-1})	(%)	(μg l^{-1})	(%)	(μg l^{-1})	(%)	(μg l^{-1})
Northern Wisconsin* (Juday and Birge, 1931)		3	13.0	14	60.9	6	26.1	20	87.0	23
Michigan Lakes† (Tucker, 1957)		1.5	12.0	5.7	46.9	5.0	41.1	10.7	88.0	12.2
Linsley Pond, Connecticut‡ (Hutchinson, 1957)		2	9.5	6	28.6	13	61.9	19	90.5	21
Ontario Lakes‡ (Rigler, 1964)		—	5.9	—	28.7	—	65.4	—	94.1	—

*Centrifugation techniques.
†Paper (No. 44 Whatman) filtration.
‡Membrane filtration (0.5 μm).

In an extremely detailed treatment relating phosphorus and nitrogen to lake productivity, Vollenweider (1968) demonstrated by several criteria that the amount of total phosphorus generally increases with lake productivity (Table 13-1). Although there are a number of exceptions to this relationship, it demonstrates a general principle that is useful when dealing with applied questions of eutrophication. The relationships of total phosphorus content to loading rates of phosphorus from the drainage basin and from lake sediments are discussed in the concluding sections of this chapter.

Separation of the total phosphorus into inorganic and organic fractions in an appreciable number of lakes indicates that a large majority of the total phosphorus is in an organic phase (Table 13-2). Of the total organic phosphorus, about 70 per cent or more is within the particulate (sestonic) organic material, and the remainder is present as dissolved or colloidal organic phosphorus. Rigler (1964) and Lean (1973b) demonstrated conclusively that data of former researchers who employed centrifugation and paper filtration methods of fractionation underestimated the importance of the sestonic organic phosphorus. Soluble organic phosphorus includes a significant quantity of phosphorus in a colloidal state. Inorganic soluble phosphorus is consistently very low, constitutes only a few percent of total phosphorus, and as will be seen further on, is cycled very rapidly in the zones of utilization. The ratio of inorganic soluble phosphorus to other forms of phosphorus of approximately 1:20 or <5 per cent as inorganic phosphate phosphorus is remarkably constant in a large variety of lakes within the temperate zone. The percentage of total phosphorus occurring as truly ionic orthophosphate is probably considerably less than 5 per cent in most natural waters (e.g., Tarapchak, et al., 1982; Prepas and Rigler, 1982).

The phosphorus distribution within the fractions just discussed is the picture generally observed in the trophogenic zones of lakes. Phosphate, pyrophosphate, triphosphate, and higher polyphosphate anions additionally form complexes, chelates, and insoluble salts with a number of metal ions (Stumm and Morgan, 1981). The extent of complexing and chelation between various phosphates and metal ions in natural waters depends upon the relative concentrations of the phosphates and the metal ions, the pH, and the presence of other ligands (sulfate, carbonate, fluoride, and organic species).

Because phosphate concentrations are generally low, complex formations involving these major cations and various phosphate anions will have little effect on the distribution of metal ions, but may have marked effects on the phosphate distribution (cf. Golachowska, 1971). Metal ions, such as those of ferric iron, manganous manganese, zinc, copper, etc., are present in concentrations comparable to or lower than those of phosphates. For these ions, complex formation can significantly affect the distribution of the metal ion, the phosphates, or both. For example, the solubility of aluminum phosphate ($AlPO_4$) is minimal at pH 6 and increases at both higher and lower pH values. Ferric phosphate ($FePO_4$) behaves similarly, although it is more soluble than $AlPO_4$. Calcium concentration influences the formation of hydroxylapatite [$Ca_5(OH)(PO_4)_3$]. In an aqueous solution lacking other compounds, a calcium concentration of 40 mg l^{-1} at a pH of 7 limits the solubility of phosphate to approximately 10 μg l^{-1}. A calcium level of 100 mg l^{-1} lowers the maximum equilibrium of phosphate to 1 μg l^{-1}. Elevation of the pH of waters containing typical concentrations of calcium should lead to apatite formation. Moreover, increasing pH leads to formation of calcium carbonate, which coprecipitates phosphate with carbonates (Otsuki and Wetzel, 1972). Sorption of phosphates and polyphosphates on surfaces is well known, particularly onto clay minerals (cf. Stumm and Morgan, 1981), by chemical bonding of the anions to positively charged edges of the clays and by substitution of phosphates for silicate in the clay structure. In general, high phosphate adsorption by clays is favored by low pH levels (approximately 5 to 6).

Following from these interactions, and the distribution of phosphorus in inorganic and organic fractions, the general tendency is for unproductive lakes with orthograde oxygen curves to show little variation in phosphorus content with depth (Fig. 13-1). Similarly, during periods of fall and spring circulation, the vertical distribution of phosphorus is more or less uniform. Oxidized metals, such as iron, and major cations, particularly calcium, can induce precipitation of phosphorus.

Lakes exhibiting clinograde oxygen curves during the periods of stratification, however, possess much more variable vertical distributions of phosphorus. Commonly, there is a marked increase in phosphorus content in the lower hypolimnion, especially during the later phases of thermal stratification (Fig. 13-1). Much of the hypolimnetic increase is in soluble phosphorus near the sediments. The sestonic phosphorus fraction is highly variable with depth. Sestonic phosphorus in the epilimnion fluctuates widely

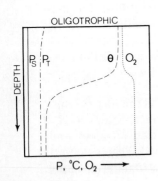 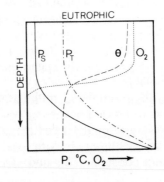

Figure 13-1. Generalized vertical distribution of soluble (P_s) and total (P_T) phosphorus in stratified lakes of very low and high productivity.

with oscillations in plankton populations. Sestonic phosphorus in the metalimnion and hypolimnion varies with sedimentation of plankton, depth-dependent rates of decomposition, and the development of deep-living populations of bacterial and other plankton (e.g., euglenophyceans).

Phosphorus and the Sediments

The exchange of phosphorus between sediments and the overlying water is a major component of the phosphorus cycle in natural waters. There is an apparent net movement of phosphorus into the sediments in most lakes. The effectiveness of the net phosphorus sink to the sediments and the rapidity of processes regenerating the phosphorus to the water depend upon an array of physical, chemical, and metabolic factors. There is little correlation between the amount of phosphorus in the sediments and the productivity of the overlying water, and the phosphorus content of the sediments can be several orders of magnitude greater than that of the water. The important factors are (1) the ability of the sediments to retain phosphorus, (2) the conditions of the overlying water, and (3) the biota within the sediments that alter exchange equilibria and effect phosphorus transport back to the water.

EXCHANGES ACROSS THE SEDIMENT INTERFACE

Exchanges across the sediment interface are regulated by mechanisms associated with mineral-water equilibria, sorption processes (notably ion exchange), oxygen-dependent redox interactions, and the activities of bacteria, fungi, plankton, and invertebrates. The exchange rates depend on local diffusion coefficients and on environmental control of inorganic and organic, i.e., enzymatic, reactions. The sediment–water interface separates two very different domains. In all but the upper few millimeters of sediment, exchange is controlled by motions on molecular scales with correspondingly low diffusion rates (Duursma, 1967). In the water, exchange is regulated by much higher and more variable rates of turbulent diffusion (Mortimer, 1971).

OXYGEN CONTENT OF THE MICROZONE

The most conspicuous regulatory features of the sediment boundary are the mud–water interface and the oxygen content at this interface. The oxygen content at this microzone is influenced primarily by metabolism of bacteria, fungi, planktonic invertebrates that migrate to the interface, and sessile benthic invertebrates. Microbial degradation of dead particulate organic matter that settles into the hypolimnion and onto the sediments is the primary consumptive process of oxygen in deepwater areas of lakes. The rate of oxygen depletion is governed by the rates of organic loading to the hypolimnion and lake or reservoir morphology (cf. Chapter 9). For example, it has been estimated that 88 per cent of the hypolimnetic oxygen consumption in the central basin of Lake Erie resulted from bacterial degradation of algae sedimenting from the trophogenic zone (Burns and Ross, 1971). Decomposition of more labile organic fractions, largely of plant origin, occurs en route to the sediments and, depending on rates of input and sedimentation, the sediments often receive organic residues that are relatively resistant to further decomposition.

Sediment demand for oxygen is high and is governed by the intensity of microbial and respiratory metabolism, slow rates of diffusion, and from the fact that inorganic elements, such as Fe^{++}, accumulate in reduced form when released into the sediment from decomposing biota. Diffusion regulates transport and is essentially molecular in the sediments, unless the superficial sediments are disturbed by overlying water turbulence. Oxygen from well-aerated overlying water, as in oligotrophic lakes or in more productive lakes at periods of complete circulation, will penetrate only a few centimeters into the sediments by diffusion. Oxygen penetration into the sediments is governed by the rate of oxygen supply to the sediments, turbulent mixing of superficial sediments, if any, and by the oxygen demand per unit volume of the sediment. The superb experimental and observational work of Mortimer (1941, 1942, 1971) has demonstrated the importance of an oxidized microzone to chemical exchanges, especially of phosphorus, from the sediments. At the sediment surface, a difference of a few millimeters in oxygen penetration is the critical factor regulating exchange between sediment and water. These relationships are exemplified by two lakes with organic sediments: The first lake maintained oxygen concentrations at the sediment interface greater than 8 mg l^{-1} throughout summer stratification, while in the second lake, oxygen levels at the interface decreased to <1 mg l^{-1}.

In the first situation, illustrated by Mortimer's studies of Lake Windermere, England, oxygen concentration at the sediment surface did not fall below 1 or 2 mg l^{-1}. Electrode potentials, which approximate composite redox potentials (cf. Chapter 14), of the oxygenated overlying water and surficial sediments to a depth of approximately 5 mm, were uniformly high (+200 to 300 mv). Below 40 to 50 mm in the sediments, the potentials were uniformly low (approximately −200 mv), indicative of extreme reducing conditions and total anoxia (Fig. 13-2). The sediment remained oxidized to a depth of about 5 mm throughout the period of summer stratification. Seasonal differences were observed in sediment depth at which the transition from high to low potential occurred, but the region of low potential never extended into the water. After five months of stratification, the point of zero mv moved towards the surface of the sediments to −5 mm from approximately −12 mm at the time of spring turnover, and

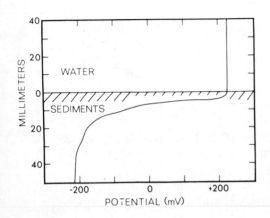

Figure 13-2. Diagrammatic profile of composite electrode potentials, not corrected for pH variations, across the sediment–water interface in undisturbed cores from the deepest portion of Lake Windermere before, during, and after stratification. (Based on data of Mortimer, 1971.)

moved downwards to -10 mm during fall circulation. The integrity of the oxidized microzone was maintained in a thin but operationally very significant layer during stratification periods. The oxidized microzone was further maintained by diffusion and by turbulent displacement of the uppermost sediments to the overlying water during turnover periods (cf. Gorham, 1958). The effectiveness of the oxidized microzone in preventing significant release of soluble components from the interstitial waters of the sediments to the overlying water was demonstrated in experimental chambers for over five months (Fig. 13-3, *left*). Phosphorus, in particular, was prevented from migrating upward.

The ability of sediments to retain phosphorus beneath an oxidized microzone at the interface is related to several interacting factors. Much of the organic phosphorus reaching the sediments by sedimentation is decomposed and hydrolyzed (Sommers, et al., 1970). Most of the sediment phosphorus is inorganic, for example apatite, derived from the watershed, and phosphate adsorbed onto clays and ferric hydroxides (Frevert, 1979a, 1979b). Additionally, phosphate coprecipitates with iron, manganese, and carbonates (Mackereth, 1966; Harter, 1968; Wentz and Lee, 1969; Otsuki and Wetzel, 1972). Work on Wisconsin lake sediments and the Great Lakes indicated that phosphorus was present in the sediments predominantly as apatites, organic phosphorus, and orthophosphate ions covalently bonded to hydrated iron oxides (Shukla, 1971; Williams, et al., 1970, 1971a, 1971b, 1971c; Williams and Mayer, 1972). In calcareous sediments of hard-water lakes containing 30 to 60 per cent $CaCO_3$ by weight, $CaCO_3$ levels were not directly related to inorganic and total phosphorus. These sediments had a lower capacity to adsorb inorganic phosphorus than noncalcareous sediments. The observations imply that $CaCO_3$ sorption is less important than iron-phosphate complexes in controlling the concentrations of phosphorus in sediments (Andersen, 1975; Frevert, 1980). Although phosphorus exchange by adsorption and desorption within the sediments between sediment particles and interstitial water can be as rapid as a few minutes (Hayes and Phillips, 1958; Li, et al., 1972), the rate of transfer across the sediment–water interface depends on the state of the microzone. The oxidized layer forms an efficient trap for iron and manganese (cf. Chapter 14), as well as for phosphate, thereby greatly reducing transport of materials into the water and scavenging materials such as phosphate from the water.

As the oxygen content of water near the sediment interface declines, the oxidized microzone barrier weakens. As seen from Mortimer's experiments (Fig. 13-3, *righthand series*), the release of phosphorus, iron, and manganese increased markedly as the redox potential decreased. With the reduction of ferric hydroxides and complexes, ferrous iron and adsorbed phosphate were mobilized and appeared in the water. The same general reactions were observed in the hypolimnetic water just overlying the sediments in eutrophic Esthwaite Water (Fig. 13-4), a pattern that has been observed repeatedly in productive dimictic lakes since its initial detailed description by Einsele (1936). A sudden release of ferrous iron and phosphate into the water occurs at the time when the $+0.20$ isovolt ($E_7 = +200$ mv) emerges above the interface surface. This event is preceded by nitrate reduction and the slow release of bases (alkalinity), CO_2, and ammonia. Manganese is reduced and mobilized at a higher redox potential than iron.

The introduction of oxygen during autumnal circulation causes ferrous iron to be oxidized, and produces a simultaneous reduction of phosphate, part as ferric phos-

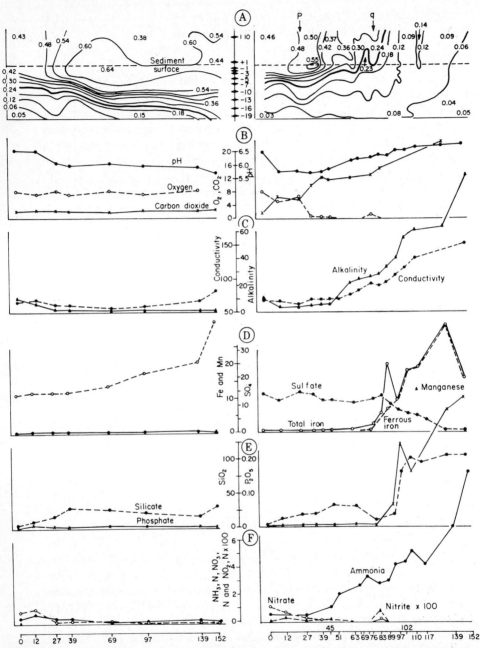

Figure 13–3. Variation in chemical composition of water overlying deepwater Lake Windermere sediments over 152 days in experimental sediment–water tanks. *Lefthand series:* aerated chamber; *Righthand series:* anoxic chamber. A, Distribution of redox potential (E_7; Eh adjusted to pH 7) across the sediment–water interface in mm; B, pH, concentrations of O_2 and CO_2 in mg l^{-1}; C, alkalinity expressed as mg $CaCO_3$ l^{-1} and conductivity in μmhos cm^{-1} at 18°C; D, iron (total and ferrous as Fe) and SO_4 in mg l^{-1}; E, phosphate as P_2O_5 and SiO_2 in mg l^{-1}; F, nitrate, nitrite × 100, and ammonia, all as N, in mg l^{-1}. (From Mortimer, C. H.: Limnol, Oceanogr., 16:396, 1971, and J. Ecol., 29:280, 1941.)

phate, which is less soluble than ferric hydroxide, and part by adsorption onto ferric hydroxide and $CaCO_3$. Although manganese is oxidized more slowly than iron, it nonetheless is effectively precipitated at the time of overturn. Ferrous iron released from the sediments is always in excess of phosphate, and when oxidized, it precipitates much of the phosphate. Some of the ferric phosphate in particulate form may slowly hydrolyze and restore some phosphate to the upper waters and littoral areas (Hutchinson, 1957). However, most phosphate is returned eventually to the sediments.

In very productive lakes where hypolimnetic decomposition of sedimenting organic matter produces anoxic conditions and hydrogen sulfide, some ferrous sulfide (FeS) is precipitated. Ferrous sulfide, like many other metal sulfides, is exceedingly insoluble and forms at a redox potential of about +100 mv. If large quantities of FeS precipitate, sufficient iron can be removed to permit some of the phosphate accumulated in the hypolimnion to remain in solution during autumnal circulation. The addition of sulfate to a lake in order to increase the bacterial production of hydrogen sulfide

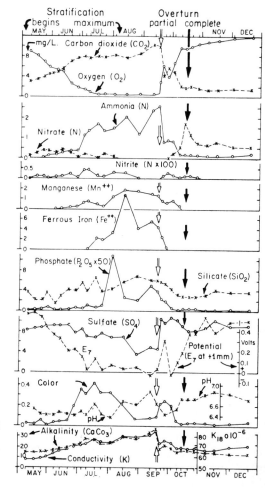

Figure 13-4. Seasonal distribution in composition (mg l^{-1}) and properties of water within 30 cm of the sediments at 14 m in Esthwaite Water, England. Components as in Figure 13-3; color in arbitrary units. (From Mortimer, C. H.: Limnol. Oceanogr., 16:387, 1971.)

(H_2S) and to accelerate the loss of iron has been suggested as a method of fertilizing lakes by regenerating phosphate from the sediments (Hasler and Einsele, 1948).

PHOSPHORUS RELEASE FROM THE SEDIMENTS

Because of the obvious importance of phosphorus as a nutrient that often accelerates the productivity of fresh waters, much interest has been devoted to the phosphorus content of sediments and its movement into the overlying water. Lake sediments contain much higher concentrations of phosphorus than the water (Olsen, 1958, 1964; Holden, 1961; Hepher, 1966; Holdren, et al., 1977; and many others). Under aerobic conditions, the exchange equilibria are largely unidirectional toward the sediments. Under anaerobic conditions, however, inorganic exchange at the sediment–water interface is strongly influenced by redox conditions (Frevert, 1979b). The depth of the sediment involved in active migration of phosphorus to the water is considerable. In undisturbed anoxic sediments, given sufficient time (2 to 3 months), phosphorus moved upward readily from a depth of at least −10 cm to the overlying water, regardless of whether the sediments were calcareous eutrophic muds or acidic and peaty in nature (Hynes and Greib, 1970). Comparison of movement in sterile sediments and sediments with anaerobic bacteria showed no significant difference; thus, diffusion predominated.

Phosphorus-mobilizing bacteria, especially of the genera *Pseudomonas*, *Bacterium*, and *Chromobacterium*, were abundant to at least 15 cm in reservoir sediments (Gak, 1959, 1963). Their abundance and vertical distribution varied with the type of sediments. Low numbers occurred in sandy sediments with small amounts of silt, and the bacteria were concentrated near the interface. Their numbers increased more uniformly with depth in sandy sediments with moderate amounts of organic matter and silts. The greatest numbers of bacteria were found in silts high in organic matter. Hayes and Anthony (1958, 1959, and Anthony and Hayes, 1964) examined the relationship of bacterial densities and organic content of sediments in detail, and found only a weak correlation in a large number of lakes. Bacterial numbers in the sediments increased proportionally, especially at the interface, with several indices of increasing lake productivity in lakes of neutral or alkaline pH and with low concentrations of humic compounds. Bacterial biomass in sediments of organically stained acidic bog lakes was also high.

While bacteria are of major importance in the dynamics of phosphorus cycling in the water, as will be discussed in the following section, their role in expediting phosphorus exchange across the sediment interface is relatively minor in comparison to chemical equilibria processes (Hayes, 1964). Bacterial decomposition is proportional to bacterial densities at the interface and directly related to the general productivity of the lake (Hayes and MacAuley, 1959; Hargrave, 1972). The sediment microflora is important in increasing the concentrations of phosphorus dissolved in interstitial water of the sediments (Fleischer, 1978). Bacterial metabolism at the interface, however, has relatively little effect on biogenic fixation and removal of phosphate from the overlying water. Using sterilized and natural sediments, it was found that microbial fixation and transport of phosphorus from the water to the sediments amounted to less than 5 per cent of the total movement under anaerobic reducing conditions (Hayes, 1955; Olsen, 1958; Macpherson, et al., 1958; Pomeroy, et al., 1965). Under aerobic conditions, bacteria of the interface did significantly increase the microbial transport of phosphorus to the sediments (Hayes, 1955; Hayes and Phillips, 1958), and this loss was related to the amount of microbial phosphorus sedimenting to the interface (cf. also Frevert, 1979a).

The rate of phosphorus release from lake sediments increases (about doubles) markedly if the sediments are disturbed by agitation from turbulence (Zicker, et al., 1965). Covering anaerobic sediments with sand or polyethylene sheeting greatly impedes the loss of oxygen in the overlying water, and decreases sediment release of phosphorus, iron, and ammonium (Hynes and Greib, 1970).

> Algae growing on sediments are able to effectively utilize phosphorus from the sediments (Golterman, et al., 1969). Moreover, algae suspended in water with various particulate inorganic compounds of extremely low solubility were capable of extracting sufficient phosphorus for growth; without sediment phosphorus sources, the phosphorus content of the water limited algal growth under experimental conditions. The presence or absence of bacteria had little effect on algal utilization of phosphorus. These results stress the importance of extractable phosphates in the sediments if they are agitated into the water column, as in shallow lakes, even though their solubilities may be extremely low.

PHOSPHORUS CYCLING MEDIATED BY AQUATIC ANGIOSPERMS AND BENTHIC ORGANISMS

The importance of submersed, floating-leaved, and emergent angiosperms to the dynamics of the phosphorus cycle has been implicated by several studies. Although phosphorus uptake by roots of plants and cycling return of phosphorus (primarily by decomposition of organic matter) are well-known phenomena of terrestrial plants, for many years this cycle was believed not to exist or not to be of importance in aquatic habitats. The structural and vascular systems of aquatic angiosperms are greatly reduced. Furthermore, the leaf morphology of these plants is simplified with reduction or complete loss of cuticle on epidermal cell walls. Observing these relationships, in addition to numerous physiological studies that indicated active uptake of nutrients by submersed leaves, led to the generalization that nutrients were absorbed primarily through the leaves, and that the roots of these angiosperms functioned only in anchorage. Nutrient uptake by foliage absorption or by the root–rhizome system in aquatic plants varies from species to species. Evidence for shifts in functional dominance of uptake by leaves to that by roots during the life cycle is discussed in more detail in Chapter 18 and by Sculthorpe (1967).

The importance of the littoral vegetation has been emphasized repeatedly in early attempts to analyze the dynamics of phosphorus cycling in lakes through the introduction of radioactive phosphorus to the surface waters (e.g., Hutchinson and Bowen, 1947, 1950; Coffin, et al., 1949; Hayes, et al., 1951, 1952; Hayes and Phillips, 1958). These studies clearly indicated rapid uptake of phosphorus by the littoral vegetation, as well as by the phytoplankton, when phosphorus was added to the water. Moreover, it was observed that phosphorus was released slowly from the plants, and more rapidly from epiphytic algae on the plants. With the decay of annual macrophytes at the end of summer, releases of phosphorus were postulated.

A detailed investigation by Solski (1962) on the release of phosphorus from the leaves and roots of emergent, floating, and submersed macrophytes after death indicated the importance of the macrovegetation as a source of phosphorus to many aquatic systems. Leaching of phosphorus from dead macrophytes under sterile conditions was rapid, and resulted in a loss of from 20 to 50 per cent of total phosphorus content in a few hours, and 65 to 85 per cent over longer periods. Leaching of phosphorus was more rapid from dried than from wet plants, but rates were affected very little by fragmentation of plant material or by different temperatures within the normal limnological range. Rates of leaching were greater from roots than from leaves. Because the content of organically bound phosphorus of the plant tissue changes during the vegetative period, generally decreasing with increasing age, the quantity of phosphorus leached to the water varies seasonally, and depends upon the total phosphorus content of the plants. Phosphorus released rapidly (within days) from decaying macrophytes is mineralized quickly, and can be utilized either by bacterial and algal metabolism, or lost to the sediment-oxidized microzone of the sediments (Solski, 1962; Nichols and Keeney, 1973; Landers, 1982).

The presence of the macrophytes *Eriocaulon* or *Utricularia* in experimental sediment-water systems greatly increased the movement of radiophosphorus from the water to the sediments (Hayes, 1955). The absorption and translocation of phosphorus were studied in bacteria-free cultures of the macroalga *Chara*, a common littoral macrophyte of hard-water lakes (Littlefield and Forsberg, 1965). All parts of the plants could absorb ^{32}P about equally, and approximately the same proportion of absorbed phosphorus was translocated from the apices or from rhizoids to other parts of the plants. Similar results were obtained for other freshwater plants (e.g., Schwoerbel and Tillmanns, 1964; DeMarte and Hartman, 1974).

Much insight into the role of aquatic macrophytes in phosphorus cycling has been provided by the studies of McRoy, et al. (1972). Rates of phosphorus uptake and excretion by both roots and leaves of eelgrass, *Zostera marina* L., were found to be dependent on the orthophosphate concentration of the medium. The interstitial inorganic phosphorus concentrations were nearly two orders of magnitude greater than those of the water. The plants absorbed 108 mg P m^{-2} day^{-1} from the sediments, assimilated 45 mg in the production of fresh plants, and excreted 62 mg into the water (Fig. 13–5). An amount equivalent to 41 percent of the inorganic reactive phosphorus excreted was exported from the plant beds to the open water. Although this pathway of pumping reactive phosphorus from the sediment to water by aquatic vascular plants was demonstrated in a marine littoral area, there is no reason to suspect that it is quantitatively less important in freshwater systems. Indeed, much of the information on the quantitative cycling of the open water of lakes can be explained by the continual export of phosphorus from the sediments, coupled with the rapid cycling of phosphorus by the phytoplankton.

LITTORAL FLORA AND THE SITES OF PHOSPHORUS FLUX

One way of distinguishing the major sites of phosphorus flux is by dividing them into three compartments: (a) the open water and organisms of the epilimnion, (b) the littoral organisms, and (c) the hypolimnion and sediments (Fig. 13–6). The lake has contact with the drainage basin via the epilimnion, and phosphorus enters with inflowing water and leaves with outflowing water and by sedimentation. In this division, the littoral flora, which is a major component in many if not a majority of lakes, is included

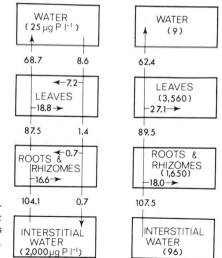

Figure 13-5. Phosphorus flux in a stand of the eelgrass *Zostera marina* L. *Left:* Phosphorus gradient, units in μg P (g plant)$^{-1}$ day^{-1} *Right:* Net daily movement of phosphorus (mg m^{-2}). Amount of phosphorus (mg m^{-2}) in each compartment is in parentheses. (After McRoy, et al., 1972.)

as part of the epilimnion and the trophogenic zone, both of which extend virtually to the same depth in many small lakes.

The application of compartmental analysis to the results of tracer experiments has demonstrated that phosphorus in the epilimnion is extremely mobile (Rigler, 1964), a subject treated at length in the following section. The turnover time of phosphorus in the epilimnion (that is, the time in which an amount of phosphorus equivalent to the total amount in a compartment leaves that compartment and a similar amount enters it) was found to be 3.6 days (Rigler, 1956). Within 20 minutes of the time it entered, over 95 per cent of the added phosphorus was taken up by the plankton; it had a turnover time of less than 20 minutes. In Toussaint Lake, a small acidic lake with a well-developed littoral zone of rooted aquatic plants, the littoral region was the most important contributor to the turnover of phosphorus in the epilimnion; phosphorus was lost to this compartment ten times more rapidly than to the hypolimnion and sediments, and 50 times more rapidly than its loss through the outlet. Return of phosphorus from the lit-

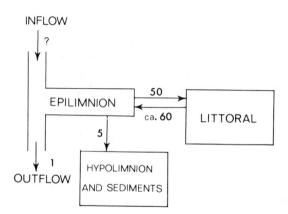

Figure 13-6. The three major compartments of phosphorus flux in a lake. The numbers indicate the relative fluxes of phosphorus between compartments in Toussaint Lake, Ontario, in midsummer. (After Rigler, 1964.)

toral zone during the summer was about 20 per cent higher than loss. In comparable experiments in an acidic bog lake of Nova Scotia with extensive developments of littoral sphagnum moss, nearly all of the tracer phosphorus was taken up by the plankton and sphagnum (Coffin, et al., 1949; Hayes and Coffin, 1951). Essentially no phosphorus reached the hypolimnion. When these studies are added to those on the effectiveness of the larger plants in assimilation and transport of phosphorus from the sediments, it becomes clear that the littoral flora can play a major role in the dynamics of the phosphorus cycle.

Although the rate constants of phosphorus loss from the epilimnion cannot be derived from existing data, Rigler (1973) reanalyzed the results of several studies. He found that turnover time of phosphorus in the epilimnion, which ranged from 20 to 45 days (Table 13–3), was inversely correlated with areas of the lakes and the estimates of development of the littoral vegetation.

Certain apparently conflicting conclusions on the importance of the littoral flora to the phosphorus circulation of lakes, based on summertime additions of tracer radioactive phosphorus, can be resolved by analyses of the differences among the communities of the lakes studied (critically reviewed by Confer, 1972). In experimental pond systems with regulated steady-state inputs of phosphorus to and outputs from the epilimnion, filamentous algae of the simulated littoral zone were a major site of phosphorus uptake before it reached the epilimnion. As the filamentous algae decayed, phosphorus was released to the open water. In this relationship, the equilibrium exchange system between phosphorus input to or release from the littoral flora and from or to the epilimnetic water varies with the amounts of phosphorus input from the drainage basin (Chamberlain, 1968). Also, phosphorus uptake by and release from littoral macrophytes and attached algae will vary with the physical constraints of littoral development determined by basin morphology. Moreover, this relationship will shift in proportion to seasonal changes in the active growth of the littoral flora and subsequent decay of annual plants in late summer. With perennial submersed vascular plants, or in tropical lakes where littoral growth is more or less continuous, a steady exchange of phosphorus between the littoral zone and the epilimnion must be generally the rule.

BENTHIC INVERTEBRATES AND THE TRANSPORT OF PHOSPHORUS

The effect of benthic invertebrates on the dynamics of phosphorus cycling between the sediments and the water is not completely understood. It is apparent that the burrowing activities of largely benthic invertebrates can significantly alter the surficial oxidized zone by physical penetration. The effects of such disruption by dense populations of benthic fauna or bottom-feeding fishes (such as carp or perch) on phosphorus exchange as a result of physical manipulation of the interface are unclear, but are most likely quite small in comparison to overriding chemical-microbial regulation. In the development of populations of benthic invertebrates, phosphorus is incorporated into the fauna from the organic material fed upon in the sediments. Adsorption or direct assimilation of inorganic phosphorus is quantitatively insignificant, at least among the microcrustacea (Rigler, 1964). When the benthic invertebrates emerge as adults, they may emigrate from the sediments, thereby transporting phosphorus to other compartments of the system. For example, in streams a significant upstream migration of phosphorus by fish and invertebrates has been found (Ball, et al., 1963a, 1963b). However,

TABLE 13-3 Calculated Rate Constants of Phosphorus Transport in Three Lakes of Differing Littoral Development*

LAKE	AREA (ha)	LITTORAL VEGETATION (rank)	P TURNOVER TIME (days)	RATE CONSTANTS	
				k (Out of Epilimnion)	k (Sedimentation from Epilimnion)
Toussaint	4.7	1	20	0.05	0.01
Upper Bass	5.8	2	27	0.04	—
Linsley Pond	9.4	3	45	0.02	0.02

*Modified from Rigler, 1973; lakes are ranked in order of decreasing amount of littoral vegetation, subjectively estimated.

this displacement of phosphorus by invertebrates plays only a small part in the overall quantitative cycling of phosphorus in the lake system.

The role of microinvertebrate activity at the sediment interface in relation to transport of phosphorus to the water is also unclear. Ciliates associated with the sediments are capable of hydrolyzing dissolved organic acids and releasing inorganic phosphate to the water (Hooper and Elliot, 1953). Low oxygen concentrations, however, not only produce an unfavorable environment for the ciliates, but also inhibit the release of phosphate by the cells. Negatively phototactic cladoceran zooplankton, which migrate to the sediment interface region during daylight, presumably feed actively on the rich microflora of that region. The extent of phosphorus transport to the epilimnion during their subsequent nighttime migration is unknown, but worthy of investigation. Although it was found that the experimental addition of snails and ostracod microcrustacea to the sediments altered plankton abundance in ponds, these benthic organisms produced no corresponding changes in uptake rates of phosphorus in open water (Confer, 1972).

Studies of benthic macroinvertebrates have demonstrated the potential for these organisms to disrupt the sediment–water interface by burrowing and influence the exchange of phosphorus. Davis, et al. (1975) showed that tubificid worms only slightly influenced phosphorus release and may actually enhance phosphorus deposition at the sediment–water interface. Chironomid midge larvae, however, can cause increases in the phosphorus content of the overlying water (Gallepp, et al., 1978; Gallepp, 1979). Larger species caused a greater effect. Release of total phosphorus increased approximately linearly (0.3 to 9.4 mg P m^{-2} d^{-1}) over a range of 0 to 6585 larvae m^{-2}. Most of the phosphorous released to the overlying water was as orthophosphate, which is readily assimilated by bacteria (Johannes, 1964a, 1964b) or algae if light is available. It is likely that the phosphorus was excreted directly by the invertebrates, as has been found for zooplankton (see following discussion). When larval invertebrates reach extreme densities (e.g., over 100,000 m^{-2}; Lindegaard and Jónasson, 1979), their activities can influence the transport of phosphorus across the sediment–water interface.

Phosphorus Cycling within the Epilimnion

The classical studies of Einsele (1941) and many subsequent theoretical and applied analyses of the circulation and fate of phosphorus in the open water of the epilimnion have shown that phosphorus is very rapidly incorporated into phytoplanktonic algae and bacteria. Much recent study has focused on rates of movement of phosphorus among biologically important forms in lake water. Two broad classes of investigations have resulted: (1) biotically mediated phosphorus-transfer mechanisms, and (2) abiotic complexation reactions. The former includes studies on the transfer of phosphorus among seston and various forms of dissolved phosphorus, including the biotically mediated formation of colloidal phosphorus and the enzyme-mediated utilization of dissolved phosphate esters. The latter category includes studies on the sorption and desorption of phosphorus to dissolved humic compounds, colloidal calcium carbonate, and other particles.

The analyses of Rigler and his coworkers have generated a framework that has stimulated considerable further study of phosphorus dynamics (Lean, 1973a, 1973b; Lean and Rigler, 1973; Rigler, 1973). Lean proposed a quantitative, steady-state model of phosphorus exchanges in epilimnetic water in the summer, with the following composition (Fig. 13–7): (a) *particulate phosphorus*, the fraction removed when the water is filtered through a 0.5 μm pore-size filter, which contains the bulk of the phosphorus; (b) reactive inorganic *soluble orthophosphate* (PO_4^{-3}), which has an extremely short turnover time; (c) a *low molecular weight* (approximately 250) *organic phosphorus compound* (XP); and (d) a soluble macromolecular *colloidal phosphorus* of a molecular weight >5,000,000. An exchange mechanism predominates between the inorganic phosphate and the particulate fraction, but some phosphorus is excreted by the microorganisms in the form of the low-molecular-weight compound (XP). Polycondensation of the low molecular weight compound (XP) produces the high molecular weight colloidal compound. Both fractions, but primarily the latter, release phosphate in the soluble inorganic fraction, which then becomes available to the plankton.

The rate constants shown in Figure 13–7 demonstrate the rapidity of phosphorus dynamics in this algal-bacterial compartment for an exemplary lake in midsummer:

(a) k_1, the uptake of PO_4 = 0.9 relative mass units min^{-1} or 0.002 μg liter^{-1} min^{-1} of total biologically active phosphorus.

(b) k_2, the release of XP, and k_3, the binding of XP by condensation to colloidal P = 0.022 min^{-1}. The excretion of extracellular organic phosphorus compounds is well known among marine and freshwater algae during periods of rapid growth and during senescence (Kuenzler, 1970; Fogg, 1971; Lean, 1973a).

(c) k_4, hydrolysis of colloidal P to PO_4 = 0.0017 min^{-1}. After the rapid formation of colloidal P via XP, in less than two minutes, subsequent additions of XP displace P from the colloid, again making the soluble inorganic PO_4 form available for uptake.

(d) k_5, the release of inorganic PO_4 by the particulate seston, was found to be very much greater than excretion of soluble organic phosphorus (XP). Mass balance calculations showed that k_5 is about 70 times k_2.

(e) k_6, the direct hydrolysis of XP to PO_4, was insignificant in comparison to release via the colloidal form.

(f) k_7, the rate constant for the loss of colloidal phosphorus, is small relative to the other rate constants. However, further aggregation of the colloidal fraction to particulate form or sorption to particulate matter followed by sedimentation represents a phosphorus sink from the epilimnion. Similarly, sedimentation of particulate phosphorus represents a slow but continuous loss from the epilimnion. Thus, during active growth of algae and heterotrophic bacteria, phosphorus must be continuously replaced through input from influents or from the littoral zone, or recycled by zooplankton ingestion and subsequent excretion (see following discussion). In experi-

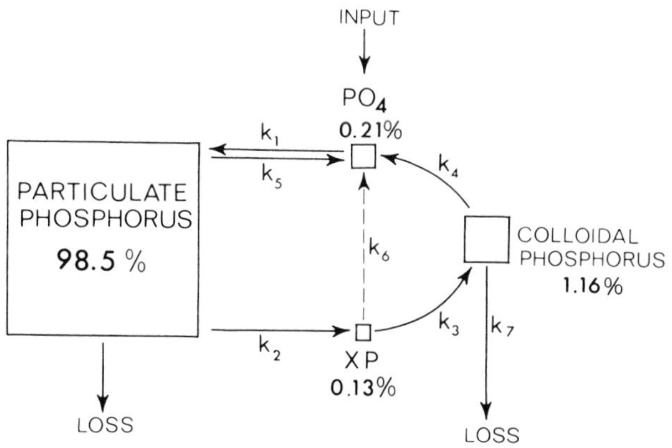

Figure 13-7. Phosphorus movement within the epilimnetic open water zone of Heart Lake, Ontario, elucidating the exchange mechanism between the phosphate and the particulate fractions. (Slightly modified from Lean, D. R. S.: Science, 179:678–680, 1973a, copyright 1973 by the American Association for the Advancement of Science; and J. Fish. Res. Bd. Canada, 30:1525–1536, 1973b.)

mental systems, much of the colloidal fraction settles out and becomes biologically unavailable after one to five days. Other workers, however, have shown that high-molecular-weight colloidal phosphorus compounds can be used directly as a phosphorus source by algae possessing alkaline phosphatase enzymatic activity (Paerl and Downes, 1978). These rates of particulate loss are congruous with those found in natural waters to which phosphorus additions have been made.

From these initial insights and subsequent studies into the cycling of phosphorus in the open water, it is apparent that sestonic organic phosphorus must be separated into at least two fractions: (a) A rapidly cycled fraction, which is exchanged with soluble forms. In this fraction, phosphate is transferred rapidly through the particulate phase to low-molecular-weight compounds. (b) A fraction of the sestonic phosphorus that is released more slowly. The transfer of colloidal phosphorus material from the phytoplankton to the water appears within minutes after uptake of soluble phosphate by bacteria and algae (Paerl and Lean, 1976; Lean and Nalewajko, 1976). In addition to the formation of high-molecular-weight fibrillar and amorphous particles of colloidal size (0.05 to 0.1 μm), bacteria and algae excrete significant amounts of dissolved organic phosphorus compounds. As discussed below, the uptake of phosphate is greater by bacteria than algae, but since algal biomass is greater than bacterial biomass, algal phosphate incorporation usually dominates uptake and release pathways in the epilimnion.

The uptake and turnover kinetics of phosphorus have been studied in a number of lakes of low to high productivity and at different seasons of the year (Table 13-4). Phosphorus turnover rates are extremely rapid in the summer periods of high demand and relatively low loading inputs, but become as much as two orders of magnitude slower during the winter periods. Phosphorus turnover rates are faster under more oligotrophic conditions of greater phosphorus deficiency (e.g., Peters, 1979).

Phytoplankton contain many complex phosphorus esters (e.g., sugar phosphates, nucleotide phosphates, polyphosphates). These low-molecular-weight phosphorylated compounds can be released to fresh water upon cell death or by active release by microorganisms (Berman, 1970; Kuenzler, 1970). In the Lean model (Fig. 13-7), direct enzymatic hydrolysis of low-molecular-weight XP fractions is presumed to be negligible. In reality, it is probable that more complex phosphorus regenerative mechanisms exist.

TABLE 13-4   Turnover Times of Phosphate in Fresh Waters*

LAKE	TURNOVER TIMES (min)	SOURCE
Lago Maggiore, Italy		
Epilimnion, summer	10–200	Peters, 1975
Epilimnion, winter	200–10,000	
Hypolimnion	1000–100,000	
15 European Lakes (epilimnia)	4–74,400	Peters, 1975
Southern Ontario Lakes (epilimnia)		Rigler, 1964
Summer	2–8	
Winter	10–10,000	
Lake Kinneret, Israel		Halmann and Stiller, 1974;
Summer	11–280	Halmann and Elgavish, 1975
Winter and spring	40–545	
East African Lakes	1–1,000	Peters and MacIntyre, 1976
Lake 227, Ontario		Lavine, 1975
Summer	0.4–15	
Winter	120–11,700	
Lake 302, Ontario (Summer)	5–26	Levine, 1975
Lake Memphremagog, Quebec		Peters, 1979
Summer	9–26	
Winter	83–770	
Jordan River, Israel (Winter)	630	Halmann and Stiller, 1974

*The turnover time is the reciprocal of turnover rate, which is the fraction of phosphate incorporated by the seston per unit time.

Enzyme-mediated hydrolysis of naturally occurring phosphate esters is one of the most plausible mechanisms for phosphorus regeneration in the epilimnion. Direct enzymatic breakdown of dissolved inorganic phosphorus compounds and polyphosphates to dissolved inorganic phosphorus by relatively nonspecific phosphatase (phosphomonoesterase) activity has been the subject of intensive study. Numerous investigations demonstrate that membrane-bound algal phosphatase activity increases as phosphorus deficiency become acute. Increases in alkaline phosphatase activity (APA) under conditions of phosphate limitation, either from increased rates of enzyme synthesis or from derepression of pre-existing enzyme, can enhance algal competitive ability by permitting algae to utilize organophosphate or inorganic polyphosphate substrates as alternative phosphorus sources (Berman, 1970; Jones, 1972; Heath and Cooke, 1975; Jansson, 1976; Stevens and Parr, 1977; Pettersson, 1980; Wetzel, 1981). Although free, dissolved alkaline phosphatase is short-lived (Reichardt, et al., 1967; Pettersson, 1980), substantial quantities of APA (often in excess of 50 per cent of the total) can be found in dissolved phase (Pettersson, 1980; Wetzel, 1981; Stewart and Wetzel, 1981b). The importance of phosphatase activity to phosphorus cycling in lakes is further confounded because bacterial production of alkaline phosphatase occurs in pelagial environments (Jones, 1972), but need not be induced strictly by phosphate limitation (e.g., Wilkins, 1972).

Soluble APA and APA associated with algae and nonalgal particulate matter are highly variable seasonally and spatially (Wetzel, 1981; Stewart and Wetzel, 1981b). Epilimnetic APA is commonly high during the spring and summer periods of maximal phosphorus demand. During years when spring circulation is incomplete or a lake experiences temporary meromixis, extremely high APA and phosphorus limitations were found in the phytoplankton during spring and summer months (Wetzel, 1981).

Phosphomonoesters (PME) were found in lake water in substantial quantities (up to 55 μg PME-P l^{-1} in eutrophic Twin Lakes, Ohio; Heath and Cooke, 1975). An inverse relationship was found between PME concentrations and APA during summer stratification. Potential hydrolysis rates of PME of over 0.05 μmoles per hour by APA were found. These values are similar to those found in marine systems using algal alkaline phosphatase as the hydrolytic enzyme (Rivkin and Swift, 1980). These results, in addition to widely differing chemical analyses of the complex phosphorus pool in lake water (e.g., Lean, 1973a, 1973b; Lean and Nalewajko, 1979; Paerl and Downes, 1978; Downes and Paerl, 1979; Francko and Heath, 1979, 1981, 1982), suggest that the composition of the soluble phosphorus pool may vary greatly among different lake types. While some lakes may approach the model of phosphorus cycling presented in Figure 13-7, other lakes may possess a phosphorus cycle dominated by the production and biotically mediated hydrolysis of PME.

Certain physicochemical mechanisms can influence the cycling of phosphorus and, as a result, affect primary productivity. Certain complex humic materials of both low and high molecular weights contain phosphorus (e.g., Koenings and Hooper, 1976). High-molecular-weight humic compounds can be photoreduced by low-intensity ultraviolet light such as occurs in sunlight (Francko and Heath, 1979; Stewart and Wetzel, 1981a). Orthophosphate adsorbed to ferric-humic compounds can be released by UV-induced photoreduction of ferric iron (Francko and Heath, 1982). Low-molecular-weight humic compounds can release orthophosphate upon exposure to alkaline phosphatase.

The importance of phosphorus-humic complexes to phytoplankton is not fully appreciated at this time. For example, rates of carbon assimilation by phytoplankton were reduced when the algae were exposed to low concentrations of dissolved humic materials of littoral and wetland origin (Stewart and Wetzel, 1982). The effects were species-selective and greatest among smaller (1 to 5 μm) phytoplankton growing under low light conditions. Alkaline phosphatase activity was greatly stimulated by low concentrations of dissolved humic compounds of low molecular weight. Evidence indicates that these humic materials may act as sequestering agents for organophosphorus compounds, thereby reducing phosphorus availability to phytoplankton. These relationships can be important to both phosphorus cycling and phytoplankton productivity, for major rain events can simultaneously introduce substantial quantities of phosphorus and dissolved humic materials into the pelagial areas of small and moderate-sized lakes.

PHOSPHORUS AND NITROGEN RECYCLING BY ZOOPLANKTON

The uptake of nutrients, especially phosphorus and nitrogen, by ingestion of food particles by herbivorous zooplankton can be large at certain times of the year (cf. Chapter 16). During egestion, nutrient release occurs in the form of soluble phosphate ions, some organic phosphorus compounds, and ammonia (Lehman, 1980a). Some phosphorus and nitrogen remains in particulate fecal materials which settle out of the trophogenic zone.

Nutrients released by zooplankton are rapidly reassimilated by phytoplankton. Under certain circumstances, when the concentrations of phosphorus in the trophogenic zone are low, phosphorus and nitrogen regenerated by the herbivorous zooplankton can constitute a significant portion of the phosphorus and nitrogen requirements of the algae (reviewed by Lehman, 1980a, 1980b). For example, based on nutrient budgets and in situ measurements, epilimnetic zooplankton in Lake Washington supply ten times more phosphorus and three times more nitrogen to the surface-mixed layer during the summer months (June to September) than enters from all external sources combined. Examples of the rates of regeneration of phosphate and ammonia by zooplankton are given in Table 13-5.

The magnitude of nutrient uptake and recycling depends on the nutritional status of the algal cells (Lehman, 1980a). Compared with nutrient-sufficient cells, nutrient-deficient algal cells assimilated phosphate and ammonia more rapidly and exhibited lower rates of remineralization.

Zooplankton grazing accelerates rates of phosphorus and nitrogen regeneration. It is also probable that planktonic algae and bacteria immediately sequester released phosphorus. Even though average phosphorus (and often combined nitrogen) concentrations in the epilimnion can be low, the release of nutrients from zooplankton can cause a micro-heterogeneous patchiness of nutrient concentrations, since the zooplankton distribution is far more clumped than the phytoplankton distribution.

Less clear is the importance of nutrient recycling to the overall rates of algal productivity on an annual basis. Removal of algae and bacteria by grazing zooplankton fluctuates widely on a seasonal and spatial basis (Chapter 16). Moreover, grazing of algae is often size-selective. This selectivity may enhance the availability of nutrients to those algae that are not as effectively grazed. The interactions are furthermore complicated by seasonal patterns of selective predation on zooplankton by planktivorous fishes. Such predation reduces the effectiveness of nutrient regeneration by zooplankton. For example, Bartell and Kitchell (1978) estimated an order of magnitude of reduction in rates of phosphorus release from pelagial zooplankton during June and July when fish predation reached its seasonal maximum and allochthonous supplies were at a seasonal low.

Finally, it must be recalled that most of the phosphorus is bound in particulate matter: algae, bacteria, and detritus. Rates of nutrient regeneration from these sources as the phytoplankton senesce and die are poorly understood. Certainly, these organisms with short generation times rapidly recycle nutrients in the trophogenic zone before they settle to lower depths. The relative magnitude of their recycling to that of the zooplankton is not clear, but must be variable both spatially and seasonally. At this time, one can only conclude that zooplankton grazing and concomitant nutrient recycling can enhance rates of primary productivity and algal growth during periods when heavy nutrient demands exceed loading inputs to the trophogenic zone from external sources.

ALGAL REQUIREMENTS FOR PHOSPHORUS

Compounds containing phosphorus play major roles in nearly all phases of metabolism, particularly in the energy transformation of phosphorylation reactions during photosynthesis. Phosphorus is required in the synthesis of nucleotides, phospholipids,

TABLE 13-5   Regeneration Rates of NH₄-N and PO₄-P by Some Freshwater Zooplankton

SPECIES*	TEMPERATURE (°C)	µg DRY WT ANIMAL^{-1}	µg (mg dry wt)$^{-1}$ day^{-1} N	P	SOURCE
Daphnia magna (F)	20–22	250.0	–	0.8	Rigler, 1961
D. rosea (F)	20–22	13.4	–	2.0†	Peters and Lean, 1973
D. pulex (F)	15, 20, 25	26.0	5.1	–	Jacobsen and Comita, 1966
D. pulex (F)	20	–	27.9	5.6	Lehman, 1980a
Mostly Cladocera (U)	5	1.0	–	2.0	Barlow and Bishop, 1965
	20	0.9	–	4.6	
Mostly Copepoda (U)	5	3.6	–	1.9	
	20	2.9	–	6.0	
Mixed Cladocera, Copepoda, Rotifers (U)					
Epilimnion	3–24				Ferrante, 1976
>0.3 mm			–	58.4‡	
<0.3 mm			–	132.8	
Hypolimnion	4–9				
>0.3 mm			–	120.1	
<0.3 mm			–	170.1	
Mixed epilimnetic zooplankton (U)					
27–28 Sep '77	approx. 20		11.9	1.7	Lehman, 1980b
29–30 Sep '77			12.4	2.0	
30 Sep–1 Oct '77			16.7	4.3	
15–16 Sep '77	approx. 20		19.2	2.0§	
				3.1¶	

From the sources cited, especially Lehman, 1980b; cf also Taylor and Lean, 1982.
*Water for the incubations was either filtered (F) or unfiltered (U) to remove phytoplankton.
†Total P, including dissolved organic P.
‡Annual mean excretion rates; values fluctuated two orders of magnitude over an annual period.
§Soluble reactive P.
¶Dissolved P, including organic P.

sugar phosphates, and other phosphorylated intermediate compounds. Further, phosphate is bonded, usually as an ester, in a number of low-molecular-weight enzymes and vitamins essential to algal metabolism.

The importance of phosphorus in algal physiology is of special interest to the limnologist, because phosphorus is the least abundant element of the major nutrients required for algal growth in a large majority of fresh waters. Furthermore, the most important form of phosphorus for plant nutrition is ionized inorganic PO_4. Although phytoplanktonic algae can utilize organic phosphate esters, such as glycerophosphates, this ability is variable among species, as is their ability to obtain phosphate groups enzymatically or by release of exoenzymes to the water for catalytic dissociation (see reviews of Krauss, 1958, Provasoli, 1958, and Overbeck, 1962a, 1963). A majority of the phosphorus released to the water during active growth of algae is inorganic soluble phosphate and organic esters, which are, in turn, very rapidly recycled. During cell lysis and decomposition, most of the algal phosphorus released is organic and undergoes bacterial degradation (Kraus, 1964).

PHOSPHATE UPTAKE AND LIGHT

The uptake of phosphate from the water is influenced by numerous external factors, the physiological importance of which has not been clearly established. Among many algae studied in culture, the initial absorption of phosphate and subsequent uptake are greater in the light, especially under CO_2 limitation. The result is a reversible accumulation of cellular phosphate, much of which can be released to the medium. Natural algal populations are more or less synchronized in their productive and growth cycles, and protein synthesis predominates in the initial portion of the daily light cycle (Soeder, 1965). In the green alga *Scenedesmus*, release of phosphate from the algae has been found to be greatest during the latter portion of the light period, both prior to and during cell division in the dark (Overbeck, 1962).

Absorption of phosphate per cell in the light is generally dependent on phosphorus concentration in the medium, within a range that is rather specific for a given algal species. Uptake by phosphorus-deficient cells in the dark is independent of phosphate concentrations (Kuhl, 1962, 1974). When phosphate is supplied in excess, the phosphorus content per cell remains nearly constant; the amount in excess of physiological needs is often incorporated and stored as polyphosphates. Phosphorus absorption rates are quite specific to groups of algae and often to species within a genus. Whereas nitrate absorption is independent of phosphate concentrations, optimal growth of many algae occurs at higher concentrations of phosphate when nitrate, rather than ammonia, is the nitrogen source.

PHOSPHATE UPTAKE AND pH

Many species of algae exhibit optimal growth and uptake of phosphorus within a distinct range of pH of the medium. The pH may alter rates of phosphate absorption by direct effects on the activity of enzymes, on the permeability of the cell membrane, or by changing the degree of phosphate ionization. For example, uptake of phosphorus by the diatom *Asterionella* is greatest at pH values between 6 and 7 (Mackereth, 1953). Uptake rates of phosphorus have been correlated directly with the presence of numerous ions and compounds of the water, such as potassium, availability of micronutrients, and organic compounds. The mechanisms involved are not clearly understood.

PHOSPHORUS CONCENTRATIONS REQUIRED FOR GROWTH

From an ecological standpoint, the growth of algae in both natural and laboratory cultures exhibits dependency on the amount of available phosphorus and the rate at which it is cycled. Extensive investigations of minimal and maximal phosphorus concentrations, especially by Chu (1943) and Rodhe (1948), grouped freshwater algae into categories according to whether their tolerance ranges fell below, around, or above 20 μg PO_4-P l^{-1}:

(a) Species whose optimum growth and upper tolerance limit is below 20 μg PO_4-P l^{-1}, e.g., *Uroglena*, and some species of the macroalga *Chara*.

(b) Species whose optimum growth is below 20 μg PO_4-P l^{-1}, but whose tolerance limit is well above that level, e.g., *Asterionella* and other diatoms.

(c) Species whose optimal growth and upper tolerance limit is above 20 μg PO$_4$-P l^{-1}, e.g., green algae such as *Scenedesmus*, *Ankistrodesmus*, and many others.

In these studies, phosphorus concentrations of the culture media required for optimal growth were almost always higher than those of the water in natural habitats where the algae were growing. Many explanations were offered for this difference, such as the presence of unknown organic growth factors in the lake but not in the artificial media of the bacteria-free cultures. It is now apparent that the chemical mass of inorganic phosphorus of the water has relatively little relation to growth kinetics. Of overriding importance is the rapidity with which phosphorus is cycled and exchanged between the particulate phosphorus and soluble inorganic and organic phases, as discussed in detail earlier. Secondly, it is of great ecological importance that many algae, when provided with a sufficient supply of phosphorus, can absorb phosphorus in quantities far in excess of their actual needs (Kuhl, 1974). A portion of this phosphorus is lost in normal active growth both as inorganic and organic compounds, and is recycled rapidly, at least in part. But a large majority (>95 per cent) of the phosphorus is in the particulate phase of algae and dead organic seston. Assimilation of surplus amounts of phosphorus by algae, often referred to as "luxury" consumption, can provide sufficient phosphorus to maintain algal growth in the epilimnion, even though the external concentration of phosphorus may be very low or depleted.

Of course, this recycling and utilization of stored phosphorus reserves cannot persist for very long. Losses continually occur from the colloidal phosphorus component, as well as from sedimentation from the particulate phosphorus component. Inputs of phosphorus to the system are needed continually, either from the littoral zone, turbulent transfer from lower depth strata, or externally, in order to sustain growth for extended periods. The amount of phosphorus input or loading, then, is important to sustenance of algal growth, and is one of many factors influencing the types of algae that are growing in a particular lake at a particular time of year.

It is therefore more relevant to the question of increasing algal productivity to view

TABLE 13-6   The Minimal Phosphorus Requirements Per Unit Cell Volume of Several Algae Common to Lakes of Progressively Increasing Productivity

ALGAE	MINIMUM P REQUIREMENT, IN μg mm^{-3} CELL VOLUME
Asterionella	<0.2
Fragilaria	0.2–0.35
Tabellaria	0.45–0.6
Scenedesmus	>0.5
Oscillatoria	>0.5
Microcystis	>0.5

From data of Vollenweider, R. A.: Scientific Fundamentals of the Eutrophication of Lakes and Flowing Waters, with Particular Reference to Nitrogen and Phosphorus as Factors in Eutrophication. OECD Report No. DAS/CSI/68.27, Paris, OECD, 1968, after numerous sources.

phosphorus concentrations in terms of total phosphorus, since most of the phosphorus is bound in the particulate component at any given time. From the few studies available in which common algal species of lakes of differing productivity were studied in relation to phosphorus requirements, the minimum phosphorus required per cell volume can be evaluated (Table 13-6). The colonial diatom *Asterionella formosa* is often found in oligotrophic waters, has a low phosphorus requirement, and can reach maximum cell densities at very low phosphorus concentrations. The quantity of phytoplankton, expressed as cell volume, which may be produced by 1 μg P l^{-1}, is 2 to 5 mm^3 l^{-1} (Vollenweider, 1968). This quantity of algae is common in lakes of low productivity. *Tabellaria* and *Fragilaria* are diatoms that reach maximum population densities at concentrations of approximately 45 μg PO$_4$-P l^{-1}, while *Scenedesmus* needs higher concentrations of around 500 μg l^{-1}. The blue-green alga *Oscillatoria rubescens* does not reach its maximum phosphorus content until initial phosphate concentration of the media used is about 3000 μg PO$_4$-P l^{-1}. It is of ecological interest to note that cations stimulate phosphate uptake by the blue-green alga *Synechococcus* (= *Anacystis*). Calcium ions in particular cause a pronounced reduction in the half-saturation concentration for orthophosphate uptake by increasing active transport of phosphorus into the cells (Rigby, et al., 1980;cf. also Lehman,1976).

Studies on the kinetics of phosphate uptake and growth of *Scenedesmus* in continuous culture have shown that phosphorus uptake velocity is a function of both the concentrations of internal cellular phosphorus compounds and the external substrate concentrations (Rhee, 1973). The apparent half-saturation constant (K_m) of phosphorus uptake was 0.6 μM (approximately 18 μg P l^{-1}), whereas the apparent half-saturation constant for growth was less than K_m by an order of magnitude. Growth was a function of cellular phosphorus concentrations. Internal polyphosphate content appeared to regulate growth rate, particularly in the initiation of cell division. The activity of alkaline phosphatase, the primary enzyme involved in the release of phosphates from polyphosphates, exhibited a relationship that was inversely related to growth rate and was correlated to both polyphosphates and internally stored surplus phosphorus. During phosphate limitation, polyphosphates are mobilized for the synthesis of cellular macromolecules (protein, RNA, and DNA). Similar results were found for species of diatoms (Fuhs, et al., 1972) and for green and blue-green algae (Healey and Hendzel, 1975; Senft, 1978).

PHOSPHORUS UPTAKE OF ALGAE VS. BACTERIA

Minimum phosphorus content, growth rates when phosphorus was nonlimiting, the Michaelis half-saturation uptake constant, and the maximum rate for uptake of phosphorus were compared for two diatoms and three bacteria (Table 13-7). These particularly illuminating findings by Fuhs and collaborators have far-reaching implications. The range of concentrations for limitation of growth rate by inorganic phosphate is less than 3μg P l^{-1}, provided that depletion at the cell surface does not occur. Maximum growth rates were found to vary widely, but the phosphorus content depended on protoplasm volume. Maximum uptake rates of orthophosphate phosphorus per unit area of cell surface were similar among the organisms. Bacteria showed lower affinity for

TABLE 13-7 Phosphorus-Dependent Growth and Phosphorus Uptake Kinetics of Two Diatoms and Three Bacterial Species

ALGAE AND BACTERIA	MEAN CELL VOLUME (μm^3)	MEAN CELL SURFACE (μm^2)	a_o (10^{-15} g-atom P)	μ_m (doublings per day)	K_m (10^{-6} g-atom P liter^{-1})	V_m (10^{-15} g-atom P μm^{-2} day^{-1})
Diatoms						
Cyclotella nana	77.5	103	0.95	1.6	0.6	2.0
Thalassiosira fluviatilis	1570	826	12.5	1.6	1.7	7.3
Bacteria						
Corynebacterium bovis	0.71	6	0.19	4.8	6.7	7.7
Pseudomonas aeruginosa	0.41	4.2	0.10	48	12.2	17.9
Bacillus subtilis	0.39	3.9	0.15	12	11.3	12.5

Modified from Fuhs, et al., Characterization of phosphorus-limited plankton algae. *In* Likens, G. E., ed., Nutrients and Eutrophication. Milwaukee, American Society of Limnology and Oceanography, 1972.

a_o = minimum P content per cell.
μ_m = growth rate during unrestricted growth at 22°C, other conditions at or near optimum.
K_m = Michaelis half-saturation constant of uptake mechanism.
V_m = maximum uptake rate for orthophosphate P per unit area of cell surface.

orthophosphate than did the algae, but potentially could outgrow the algae because of a more favorable surface-area-to-volume ratio.

Studies on the competition for phosphate between the alga *Scenedesmus* and the bacterium *Pseudomonas* have demonstrated that algal growth is severely limited in the presence of the bacteria, but that the growth of the bacteria is little affected by the algae (Rhee, 1972). Competition, seen as the cessation of exponential algal growth, was observed some time after the external phosphate concentrations were exhausted. The concentrations of stored internal polyphosphates within the algae decreased to a near-zero critical value at the time of suppressed algal and much faster bacterial growth.

MAN AND THE PHOSPHORUS CYCLE IN LAKES

Sources of Phosphorus

PRECIPITATION

The phosphorus content of precipitation and of particulate material fallout is highly variable, and results therefore in variable phosphorus loading to lakes (Chapin and Uttormark, 1973). In general, the contribution of phosphorus from precipitation is less than that of nitrogen. In heavily fertilized agricultural regions, the phosphorus content of precipitation is much higher during the active growing season than in winter. Prevailing winds and water of storms traveling from oceans onto land are usually low in phosphorus content but can be higher in coastal regions (Graham, et al., 1979). The major source of phosphorus in precipitation is from dust generated over the land from soil erosion and from urban and industrial contamination of the atmosphere.

The phosphorus content of precipitation is generally low, at <30 μg P l^{-1} in non-populated regions over land. Around urban-industrial aggregations, this content increases considerably to well over 100 μg l^{-1}. Based on relatively few data, atmospheric contributions of phosphorus are approximately 0.01 to 0.1 g m^{-2} year^{-1} (0.1–1.0 kg ha^{-1} year^{-1}), with a majority of contemporary values in the lower portion of that range. This input from the atmosphere, however, is more significant than is generally believed when compared with the 0.07 g m^{-2} year $^{-1}$ value that Vollenweider (1968) estimated as a general permissible phosphorus loading rate for lakes. Values of 0.13 g m^{-2} year^{-1} or higher are considered dangerous from the standpoint of eutrophicational control in lakes having a mean depth of less than 5m.

A great number of chemical analyses of wet and dry precipitation from atmospheric sources have been and are being determined on an international basis because of the high potential loadings from atmospheric pollution. The literature contains many reports of atmospheric loadings for individual drainage basins. Loadings from atmospheric sources are highly variable, but in most cases the amounts of phosphorus and nitrogen are a significant part of the total loading and cannot be ignored. It should be noted also that nutrients accumulate in snowpacks and on the ice during winter, and can be released rapidly in large amounts during the spring thaw (e.g., English, 1978; Adams, et al., 1979).

GROUNDWATER

The phosphorus content of groundwater is generally low; average concentrations are about 20 μg P l^{-1}, even in areas where soils contain relatively large quantities of phosphorus. The low phosphorus content is a result of the relatively insoluble nature of phosphate-containing minerals, and the scavenging of surface phosphate by biota and soil.

LAND RUNOFF AND FLOWING WATERS

In general, regional chemical characteristics of surface waters are closely related to the soil characteristics of their drainage basins (Keup, 1968; Vollenweider, 1968). Soils reflect the regional geological and climatic characteristics. Surface drainage is often a major contributor of phosphorus to streams and lakes. The quantities of phosphorus entering surface drainage vary with the amount of phosphorus in soils, topography, vegetative cover, quantity and duration of runoff flow, land use, and pollution.

The parent rock material from which soil evolves by weathering is extremely variable in phosphorus content, and this variability increases with the thickness and heterogeneity of the stratified soil layers. Basic igneous rock contains relatively little phosphorus as apatite; phosphorus percentages of other rocks are as follows: sandstone is approximately 0.02, gneiss 0.04, unweathered loess 0.07, andesite 0.16, and diabase 0.03. Limestone containing approximately 1.3 per cent P and rare deposits of rock phosphate (10 to 15 per cent P), in which biotic accumulations either from the organisms themselves or from guano were concentrated, are largely of sedimentary origin (cf. Hutchinson, 1950). Surface layers of soil are relatively rich in organic phosphorus from plant detritus in various stages of decomposition by soil fungi and bacteria. The exchange capacity of soils for phosphorus depends on the composition of the soil, and increases with greater quantities of organic and inorganic colloids. Phosphorus is most available and easily leached from soils having a pH of 6 to 7. At lower pH values, phosphorus combines readily with aluminum, iron, and manganese, while at pH 6 and above, progressively greater amounts of phosphate are associated with calcium as apatites and calcium phosphates.

The topography of the catchment basin influences the extent of erosion and subsequent export of nutrients. Flat lands with little runoff and relatively high infiltration rates contribute less nutrient load to runoff than similar lands with steeper gradients (Table 13–8). Further, the relative erosion is influenced markedly by the type of vegetation and use to which the land is put (Table 13–9). The literature on this subject is large, and the reader is referred to the reviews examining phosphorus inputs from runoff by Vollenweider (1968), Keup (1969), Biggar and Corey (1969), Cooper (1969), Griffith, et al. (1973), Dillon and Kirchner (1975), Meyer and Likens (1979), and Loehr, et al. (1980). An attempt has been made to classify natural and agricultural areas on the basis of the quantity of nitrogen and phosphorus in runoff (Table 13–10). Applications of fertilizers and land management practices, both in agriculture and forestry, will modify and generally increase these values considerably. Urbanization results in increases in the amount of phosphorus discharged to surface waters in approximately direct proportion to population densities (Weibel, 1969). Phosphorus originating from heavy residential fertilization, storm sewer drainage, and leaves (Cowen and Lee, 1973) can

TABLE 13-8 Concentrations of Nitrogen and Phosphorus in Runoff from Miami Silt Loam of Differing Gradients

| GRADIENT (%) | $g\ m^{-2}\ year^{-1}$ | |
	Nitrogen	Phosphorus
8	2.0	0.06
20	4.25	0.2

After data cited in Vollenweider, R. A.: Scientific Fundamentals of the Eutrophication of Lakes and Flowing Waters, with Particular Reference to Nitrogen and Phosphorus as Factors in Eutrophication. OECD Report no. DAS/CSI/68.27, Paris, OECD, 1968.

increase phosphorus inputs to surface drainage. For example, in the largest lake of southwestern Michigan, Gull Lake, approximately 24 per cent of the total phosphorus entering this rapidly eutrophicating lake was from fertilization of lakeside lawns. The soils of 75 per cent of the lawns, however, were saturated with phosphorus, making additions unnecessary (Moss, 1972b).

Surface runoff receives very large quantities of phosphorus from domestic sewage. Of course, loading of these waters varies greatly with population density, treatment of the sewage for nutrient removal, and points of discharge of effluents. Average values are given in Table 13-11. Industrial inputs, especially those associated with food processing, can be exceedingly high.

Ironically, cleaning detergents are major sources of phosphorus that contribute markedly to fertilization effects of many fresh waters. Where legally permitted, synthetic detergents include phosphate builders as a major constituent, mainly sodium pyrophosphate and polyphosphates, to complex and inactivate cations of water supplies and permit more effective cleaning action. Until very recently, when detergent phosphorus content was reduced or eliminated, from 7 to 12 per cent of the weight of deter-

TABLE 13-9 Relative Erosion from Soil in Relation of Vegetative Cover and Land Use, Pacific Northwest of United States

CROP OR PRACTICE	RELATIVE EROSION
Forest duff	0.001-1
Pastures, humid region, or irrigated	0.001-1
Range or poor pasture	5-10
Grass/legume hayland	5
Lucerne	10
Orchards, vineyards with cover crops	20
Wheat, fallow, stubble not burned	60
Wheat, fallow, stubble burned	75
Orchards, vineyards, without cover crops	90
Row crops and fallow	100

From data cited in Biggar and Corey, 1969, after Musgrave.

TABLE 13–10 Grouping of Soils of Natural and Agricultural Areas on the Basis of Their General Productivity and Export of Nitrogen and Phosphorus

	EXPORTS, $g\ m^{-2}\ year^{-1}$		
		PHOSPHORUS	
GENERAL SOIL PRODUCTIVITY	INORGANIC NITROGEN	PO_4-P	Total
Low	<0.5	<0.01	<0.02
Medium	0.5–2.5	0.01–0.025	0.02–0.05
High	>2.5	>0.025	>0.05

After Vollenweider, 1968.

gents was phosphorus. The amount of phosphorus used in the production of detergents is staggering; well over 2 million tons are produced annually in the United States alone. Although the percentage of phosphorus in detergents has been reduced somewhat, the phosphorus loading to many water treatment facilities is still extremely high, and originates from a source that could be eliminated with relative ease. In a majority of existing treatment facilities, phosphorus has been reduced significantly, but not sufficiently to prevent accelerated productivity in many recipient lakes. Practical technology for nearly complete removal of phosphorus is available (cf. Vollenweider, 1968; Rohlich, 1969).

The amounts of phosphorus in flowing waters with only slight loading are generally less than 100 mg m^{-3} total phosphorus (Vollenweider, 1968; Keup, 1968). Phosphorus concentrations depend on waste-water discharges, however, and levels of several hundred mg m^{-3} are not uncommon; in serious cases, concentrations of phosphorus can

TABLE 13–11 Surface Water Loadings of Nitrogen and Phosphorus per Unit Area from Human Excrement and Other Sources Based on an Average of 12 g N Caput^{-1} Day^{-1} and 2.25 g P Caput^{-1} Day^{-1}

POPULATION DENSITY (caput km^{-2})	NITROGEN (g m^{-2} year^{-1})	PHOSPHORUS (g m^{-2} year^{-1})
50	0.22	0.04
100	0.44	0.08
150	0.66	0.12
200	0.88	0.16
300	1.32	0.24
500	2.20	0.40
1000	4.40	0.80
2500	11.0	2.00
5000	22.0	4.05

After Vollenweider, R. A.: Scientific Fundamentals of the Eutrophication of Lakes and Flowing Waters, with Particular Reference to Nitrogen and Phosphorus as Factors in Eutrophication. OECD Report no. DAS/CSI/68.27, Paris, OECD, 1968.

exceed 1000 mg m^{-3}. Assimilation rates of phosphorus and uptake by stream biota are high, but a majority of the load is carried downstream in dissolved, particulate suspensoids, and scoured sediment deposits to receiving lakes or reservoirs in which lower flow velocities occur.

GENERAL EUTROPHICATION

The word *eutrophy* (from the German adjective *eutrophe*) in general refers to nutrient-rich and is a word greatly misused in contemporary dialogue.* Naumann (1919) introduced the general concepts of oligotrophy and eutrophy, and distinguished them on the basis of phytoplanktonic populations. Oligotrophic lakes contained few planktonic algae, and were common in regions dominated by primary rocks. Eutrophic lakes contained more phytoplankton, and were common among more naturally fertile lowland regions in which human activity provided an increased supply of nutrients. Although, at that time, chemical methodology was crude or nonexistent for certain substances, Naumann emphasized that chemical factors, particularly phosphorus, combined nitrogen, and calcium, were primary determining factors. Limnological evidence since that time has shown a strong relationship between the biological dynamics of a lake and its nutritional level (concentrations of nutrients).

Shortly after the turn of the present century, Thienemann (summarized in 1925) found that in alpine and subalpine lakes, midge larvae, largely of the genus *Tanytarsus*, were characteristic of unproductive, deep lakes in which the hypolimnetic water lost little of its oxygen content during summer stratification. Eutrophic lakes (he later adopted Naumann's terms) were shallower, richer in plankton, and the oxygen-reduced hypolimnetic water was dominated by fauna, such as the midge *Chironomus*, which can tolerate very low oxygen concentrations. The extensive studies and conceptualizations of these earlier workers, when coupled with those of numerous others, formed a foundation upon which much regional limnology of the 1930s was based.

A plethora of subsequent limnological studies was directed towards evaluation of the characteristics of different lake types. The terminology for different lake types that developed in the ensuing two decades was indeed phenomenal. Many lake types were differentiated on the basis of indicator species. The basis for many of these descriptive classifications was founded in controlling physical and chemical parameters, while the basis for others was not. The terminology and dichotomies of classification that developed, while instructive, were excessive. Lakes were categorized in nearly every limnological aspect—geomorphological, physical, chemical—and by indicator species or aggregations of nearly every group of organisms from bacteria to fish. Even the extremes of lake types based on the waterfowl that commonly are associated with them were analyzed. As more lakes were studied, exceptions were found which led to further splitting and name generation. Early integrative analyses, such as those of Thienemann (1925) and Naumann (1932), became less common as the terminology became more complex. If the details of classifying are carried to an

---

*Hutchinson (1973) has written an excellent general review of the subject and the development of the term in limnology. The superb semi-popular book *The Algal Bowl—Lakes and Man* by Vallentyne (1974) complements collections of studies on eutrophication: *Eutrophication: Causes, Consequences, Correctives* (National Academy of Sciences, 1969), and *Nutrients and Eutrophication*, edited by Likens (1972).

extreme, lakes are so individualistic that one would soon require taxonomic keys for lake types (which in fact was proposed by Zafar, 1959).

The grouping of lakes must be left as a spectrum of states of lake metabolism between oligotrophy and eutrophy. Historical studies show that in many small lakes of temperate, glaciated areas, a succession is common from more inorganic sediment, containing oligotrophic-indicator fossils, to a more organic sediment, containing eutrophic-indicator fossils (Hutchinson, 1973). Evidence indicates that once organic sedimentation has become established after the initial oligotrophic phase, a type of trophic equilibrium or reasonably stable steady state occurs. This implies that nutrient inputs from the drainage basin are relatively constant over long periods of time, and undergo only minor changes with oscillations in climate and inputs from vegetative cover and erosion. The systems can, and commonly do, regress in productivity as the surficial soils become reduced in nutrient content by leaching. Conversely, productivity can be greatly accelerated by increased nutrient inputs.

Effects of Phosphorous Concentrations on Lake Productivity

Although many definitions of lake eutrophication exist, based on an array of attendant conditions associated with increased productivity, the consensus among limnologists is that the term *eutrophication* is synonymous with increased growth of the biota of lakes, and that the rate of increasing productivity is accelerated over that rate that would have occurred in the absence of perturbations to the system. The most conspicuous, basic, and measurable criterion of accelerated productivity is an increased quantity of carbon assimilated by algae and larger plants per given area.

Under a large majority of lake conditions, the most important nutrient factors causing the shift from a lesser to a more productive state are phosphorus and nitrogen. Typical plant organic matter of aquatic algae and macrophytes contains phosphorus, nitrogen, and carbon in approximately the ratios (Vallentyne, 1974):

$$1 \, P : 7N : 40C \text{ per 100 dry weight or}$$
$$1 \, P : 7N : 40C \text{ per 500 wet weight.}$$

If one of the three elements is limiting and all other elements are present in excess of physical needs, phosphorus can theoretically generate 500 times its weight in living algae, nitrogen 71 (500:7) times, and carbon 12 (500:40) times.

Comparison of the relative amounts of different elements required for algal growth with supplies available in fresh waters illustrates the general importance of phosphorus and nitrogen (Table 13-12). Similar proportional ratios were demonstrated by comparison of the demand among terrestrial plants with the accessible supply of elements from the lithosphere (Hutchinson, 1973). Even though variations in conditions of solubility or availability may at times make very abundant elements (e.g., silicon, iron, and certain micronutrients) almost unobtainable, phosphorus, and secondarily nitrogen, are generally the first to impose limitation on the system. This relationship is emphasized further by consideration of average demand–supply ratios in late winter prior to the spring algal maximum common in lakes of temperate regions, and in midsummer during maximum sustained algal productivity (Table 13-13).

Oligotrophic lakes are often limited by phosphorus and contain an excess of nitrogen. As the lakes become more productive, the primary effecting agent is increased loading of phosphorus. As discussed earlier, the instantaneous concentrations of phosphorus usually decrease and are quite variable, but the turnover rate increases markedly. Rates of loss increase, and high algal productivity requires sustained phosphorus loading of the system from allochthonous and littoral sources.

Combined nitrogen, although found in large quantities in the lithosphere, is largely unavailable to plants. Nitrogen supplies to fresh waters are augmented more readily by inputs from external sources (Chapter 12) than are those of phosphorus. Under extremely eutrophic conditions, planktonic utilization of combined nitrogen can exceed inputs and literally deplete combined nitrogen supplies in the trophogenic zone. When this occurs, molecular nitrogen fixation by heterocystous blue-green algae can augment the income of nitrogen, but only at a high metabolic cost.

The potential for limitation of eutrophication by inorganic carbon availability has been a subject of much discussion (cf. Chapter 11). In a few exceedingly productive situations, such as sewage lagoons, in which phosphorus and nitrogen compounds are available in excess of any demands, carbon can become limiting to algal growth (Kerr, et al., 1972). While there is some evidence for inorganic carbon limitation in exceedingly soft lakes (see, for example, Allen, 1972), diffusion of atmospheric CO_2 is generally

TABLE 13-12 Proportions of Essential Elements for Growth in Living Tissues of Freshwater Plants (Requirements), in the Mean World River Water (Supply), and the Approximate Ratio of Concentrations Required to Those Available

ELEMENT	AVERAGE PLANT CONTENT OR REQUIREMENTS (%)	AVERAGE SUPPLY IN WATER (%)	RATIO OF PLANT CONTENT TO SUPPLY AVAILABLE
Oxygen	80.5	89	1
Hydrogen	9.7	11	1
Carbon	6.5	0.0012	5000
Silicon	1.3	0.00065	2000
Nitrogen	0.7	0.000023	30,000
Calcium	0.4	0.0015	<1000
Potassium	0.3	0.00023	1300
Phosphorus	0.08	0.000001	80,000
Magnesium	0.07	0.0004	<1000
Sulfur	0.06	0.0004	<1000
Chlorine	0.06	0.0008	<1000
Sodium	0.04	0.0006	<1000
Iron	0.02	0.00007	<1000
Boron	0.001	0.00001	<1000
Manganese	0.0007	0.0000015	<1000
Zinc	0.0003	0.000001	<1000
Copper	0.0001	0.000001	<1000
Molybdenum	0.00005	0.0000003	<1000
Cobalt	0.000002	0.000000005	<1000

After Vallentyne, J.R.: The Algal Bowl—Lakes and Man. Miscellaneous Special Publication 22, Ottawa, Dept. of the Environment, 1974.

TABLE 13-13 Comparisons of the Ratios of
Required Concentrations of Inorganic
Nutrients to Average Supplies Available in
Fresh Waters

	RATIO OF DEMAND TO AVAILABLE SUPPLIES	
ELEMENT	Late Winter	Midsummer
Phosphorus	80,000	up to 800,000
Nitrogen	30,000	up to 300,000
Carbon	5000	up to 6000
Iron, silicon	Generally low, but variable	
All other elements	<1000	<1000

After Vallentyne, J.R.: The Algal Bowl—Lakes and Man. Miscellaneous Special Publication 22, Ottawa, Dept. of the Environment, 1974.

adequate to sustain the carbon requirements of phytoplanktonic populations. It is extremely unlikely that carbon limitation is of significance in the limitation of algal populations in a majority of harder waters, in which dense algae develop under eutrophic conditions.

The importance of phosphorus in comparison to nitrogen and carbon has been particularly well illustrated by large-scale fertilization experiments (Schindler, 1974). Lake 226, located in northwestern Ontario in Precambrian Shield bedrock, was chemically and biologically similar to more than 50 percent of the waters draining to the Laurentian Great Lakes. The trophogenic zone of Lake 226 was partitioned into two lakes at a constriction in the basin (Fig. 13-8). One basin was fertilized with phosphorus, nitrogen, and carbon, and the other with equivalent concentrations of nitrogen and carbon.* The phosphate-enriched basin quickly became highly eutrophic, while the basin receiving only nitrogen and carbon remained at prefertilization conditions (Fig. 13-8). In this phosphate-enriched basin, and in several other lakes receiving analogous treatments over a period of several years, algal biomass increased to two orders of magnitude over that of lakes receiving only nitrogen and carbon enrichment. Recovery to near prefertilization levels was very rapid when phosphate additions were discontinued.

The amount of evidence for these relationships is so overwhelming, both from experimental and applied investigations, that it is difficult to appreciate how the bitter controversy on the importance of carbon rather than phosphorus as a limiting nutrient in fresh waters developed in Canada in the late 1960s. As a result of a combination of the magnification of a few findings from blue-green algal cultures and a few rare cases of discovered low carbon availability in natural habitats in which excessive amounts

---

*Additions were equivalent to 3.16 g of NO_3-N and 6.05 g of sucrose C per m^2 per year in both basins, made in 20 equal weekly increments. The northeast basin additionally received 0.6 g m^{-2} yr^{-1} of PO_4-P. The N/P and C/P ratios were greater than in treated sewage, but the quantity of P added was not exceptionally high for lakes that commonly receive pollution from domestic sources.

Figure 13–8. Lake 226 of northwestern Ontario, which was partitioned at the constriction of the basin, 4 September 1973. The far northeastern basin, fertilized with phosphorus, nitrogen, and carbon, was covered by a dense algal bloom within two months. No increases in algae or species changes were observed in the near basin, which received similar quantities of nitrogen and carbon but no phosphorus. (From Schindler, D. W.: Eutrophication and recovery in experimental lakes: Implications for lake management. Science, *184*:May 24, 897–899, 1974. Copyright 1974 by the American Association for the Advancement of Science. Photograph courtesy of D. W. Schindler.)

of nitrogen and phosphorus occurred, phosphorus was implied to be of less importance in the acceleration of eutrophication than carbon. So convincing was the misinterpretation of results by a few scientists, and so effective was the exploitation of the situation by the detergent industry and irresponsible journalists, that urgently needed legislation to reduce the effluent loading of surface waters with phosphorus was seriously impeded. The matter was finally put to rest by the rebuttal of Vallentyne (1970), numerous subsequent investigations, and a national symposium on the subject (Likens, 1972). The effects of the controversy, however, continue to be felt both in research on the subject and in the efforts to reduce the loading of fresh waters with millions of tons of phosphorus each year, so that ultimately lakes will not become, in the words of J. R. Vallentyne, algal bowls.

Phosphorus and Nitrogen Loading and Algal Productivity

When phosphorus is added as a pulse to unproductive lakes or ponds, either experimentally, for purposes of intentional fertilization, or in effluents resulting from man's activities, the usual response is a very rapid increase in algal productivity (e.g., Einsele,

1941; Maciolek, 1954; Mortimer and Hickling, 1954; Vinberg and Liakhnovich, 1965; Vollenweider, 1968). The increased productivity is not sustained, however, but decreases rather rapidly in a few weeks or months to levels near those prior to the addition. Losses in the colloidal fraction and from sedimentation of particulate phosphorus result in continuous losses of phosphorus from the trophogenic zone. Inputs to the system must be maintained in a continuous or pulsed manner in order to sustain the increased productivity. In other words, steady phosphorus loading of the system is critical to sustaining increased productivity in a majority of lakes of low or medium biological productivity.

Conversely, in order to reduce the productivity of a lake that is receiving a continuous loading of nutrients, algal growth is generally decreased most effectively by reduction of phosphorus inputs to below the level of losses within the lake. Reduction of total phosphorus of the water is the objective, since phosphorus is the major nutrient in greatest demand in relation to supply. Phosphorus is chemically reactive, technologically easier to remove from water than nitrogen, and does not have major reservoirs in the atmosphere.

NUTRIENT LOADING

The nutrient loading concept implies that a relationship exists between the quantity of nutrient entering a water body and its response to that nutrient input. The effects of this relationship can be expressed by some quantifiable index of productivity or water-quality parameter (e.g., chlorophyll concentrations, water transparency).

The loading concept was recognized some time ago as applicable to lake changes in response to phosphorus and nitrogen enrichment (e.g., Rawson, 1939, 1955; Ohle, 1956; Edmondson, 1961). Sawyer (1947) suggested that if critical levels of dissolved inorganic nitrogen (300 μg N l^{-1}) and phosphorus (10 μg P l^{-1}) measured at the time of spring turnover in Wisconsin lakes were exceeded, nuisance populations of phytoplankton would likely develop during the growing season. Vollenweider (1966, 1968) first formulated definitive quantitative loading criteria for phosphorus and nitrogen and expected trophic conditions in water bodies. He defined boundaries between oligotrophic and eutrophic lakes by relating nutrient loadings to mean depth (as a measure of lake volume), and later refined these relationships (Vollenweider, 1975). Because phosphorus is commonly the initial limiting nutrient to algae, as discussed earlier, and because the many processes of nitrogen cycling (nitrification, denitrification, nitrogen fixation; Chapter 12) complicate accurate measurements of nitrogen loading of lakes, the loading criteria and models emphasized phosphorus relationships.

The loading relationships are all based on the mass balance equation of a substance M between its sources and losses (sinks) (Vollenweider, et al., 1980):

$$\Delta M/\Delta t = I - O - (S - R) \tag{1}$$

where $\Delta M/\Delta t$ = storage gain or loss of nutrient M over time Δt
       I = external nutrient load
      O = nutrient loss by outflow
      S = nutrient loss to sediments
      R = nutrient regeneration from sediments (internal loading).

While I and O can often be measured directly with accuracy, $(S - R)$ is more difficult to measure and is often obtained by difference. Under steady-state conditions, $\Delta M/\Delta t = 0$. While in reality steady state does not exist in a lake, lakes that oscillate around relatively constant nutrient loads over periods of several years exhibit a set of trophic characteristics and may be viewed as in a repetitive steady state. Often this steady state is defined as the nutrient storage content as measured at spring turnover or as the annual storage content. The models based on mass balance further assume that the nutrient load is instantaneously and completely mixed throughout the lake water, an assumption that is only partially true for most lakes. The models, consequently, are particularly applicable to a large population of lakes, rather than for describing the specific behavior of an individual lake.

Assuming steady-state conditions, the simple mass balance equation accounts for volumnar loading of phosphorus as well as losses by flushing and sedimentation (Vollenweider and Kerekes, 1980):

$$\frac{d\overline{[P]}_\lambda}{dt} = \left(\frac{1}{\overline{\tau}_w}\right) \overline{[P]}_i - \left(\frac{1}{\overline{\tau}_p}\right) \overline{[P]}_\lambda \tag{2}$$

where $\overline{[P]}_\lambda$ = average (total) lake concentration (both dissolved and particulate phosphorus components)

$\overline{[P]}_i$ = average inflow concentration of total phosphorus

$\overline{\tau}_p$ = average residence time of phosphorus

$\overline{\tau}_w$ = average residence time of water

in which the right-hand terms represent the average rate of supply to and the average rate of loss of total phosphorus from the lake, respectively, and the left-hand term represents the corresponding temporal variations of the average lake concentration.

Then, simplifying Equation 2:

$$\overline{[P]}_\lambda = (\overline{\tau}_p/\overline{\tau}_w) \, \overline{[P]}_i \tag{3}$$

The premise of Equation 3 can also be expressed as:

$$\overline{[P]}_\lambda = (1 - R) \, \overline{[P]}_i \tag{4}$$

where R = the phosphorus-retention coefficient:

$$R = \frac{\overline{[P]}_i - \overline{[P]}_\lambda}{\overline{[P]}_i} = 1 - \overline{\tau}_p/\overline{\tau}_w \tag{5}$$

or statistically approximately $\overline{\tau}_p/\overline{\tau}_w = \dfrac{1}{1 + \sqrt{\overline{\tau}_w}}$ (Vollenweider, 1976; Larsen and Mercer, 1975). An approximate retention coefficient as a function of hydraulic load (q_s) was found to be (Dillon, 1974; Kirchner and Dillon, 1975):

$$R = 0.426 \, e^{-0.271q_s} + 0.574 \, e^{-0.00949q_s}$$

where q_s = areal water load (hydraulic load) to the lake (m year^{-1}), calculated as the lake outflow volume divided by the lake surface area.

These two retention values, while not identical, yield similar results for lakes of mean depths between 10 and 30 m and hydraulic loads between 3 and 90 m year^{-1} (Vollenweider and Kerekes, 1980).

When either of these models were used and data plotted to show the extent of prediction of spring-turnover phosphorus concentration from loading, the models describe the average statistical behavior of lakes in terms of the relationship between phosphorus load and phosphorus accumulation within the lake (Vollenweider, 1976, 1979; Rast and Lee, 1978). Deviant lakes could be shown to have mechanisms operating to overestimate or underestimate the loading (e.g., high sedimentation rates, high regeneration of phosphorus from the sediments).

BIOLOGICAL RESPONSES TO NUTRIENT LOADING

Based on these nutrient loading relationships, the loadings to a lake should be related to biological responses such as phytoplankton biomass (e.g., chlorophyll concentrations) and productivity. Early predictions of chlorophyll levels from phosphorus concentrations in lakes at spring turnover or during the growing season (Sakamoto, 1966; Dillon and Rigler, 1974; Jones and Bachmann, 1976; Schindler, 1978) were later extended further with larger worldwide data sets (Vollenweider, 1976; Rast and Lee, 1978; Schindler, et al., 1977; Oglesby and Schaffner, 1978; Canfield and Bachmann, 1981). Within certain boundaries, the regression between phosphorus loading and average annual chlorophyll concentrations is approximated by:

$$\overline{[\text{chl. a}]} = 0.55 \, \{[P]_i / (1 + \sqrt{\tau_w})\}^{0.76} \tag{6}$$

Examples of this relationship are given in Figure 13-9; similar relationships have been found for many other lakes with only minor deviations.

The annual rate of primary productivity of the algae has also been related to the predicted phosphorus concentration (Fig. 13-10). The nonlinearity of the log–log plot results from the light-reducing, self-shading effects of dense algal populations (high biomass) at high levels of productivity (see Chapter 15). The relationship can be represented by:

$$\Sigma C(\text{g m}^{-2} \text{y}^{-1}) = 7 \left[\frac{\{[P]_i / (1 + \sqrt{\tau_w})\}^{0.76}}{0.3 + 0.011 \, \{[P]_i / (1 + \sqrt{\tau_w})\}^{0.76}} \right] \tag{7}$$

which operates similarly to the integral of daily photosynthesis, and is based on average chlorophyll concentrations and light extinction resulting from turbidity and dissolved organic substances (Vollenweider and Kerekes, 1980).

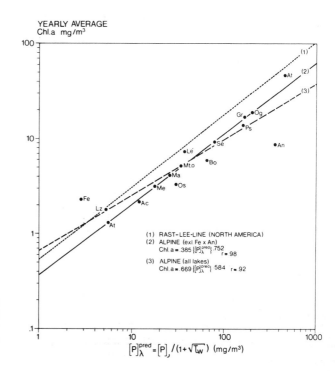

Figure 13-9. Statistical relationship between predicted total phosphorus concentration and annual mean chlorophyll a concentration for selected alpine lakes (modified slightly from Vollenweider, 1979).

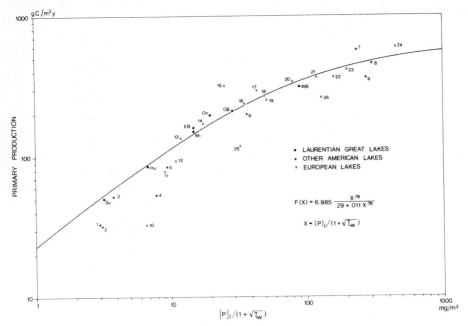

Figure 13-10. Annual primary productivity (g C m^{-2} y^{-1}) as a function of predicted total phosphorus concentration (From Vollenweider, 1979).

APPLICATION OF PREDICTIVE MODELS AND DATA

The evaluation of the trophic status of a lake has great practical importance. The eutrophicational status must be known before remedial corrective measures can be implemented in relation to the desired uses for any lake.

Based upon many data obtained during an international program on eutrophication of the Organization for Economic Cooperation and Development (OECD), nutrient load–eutrophication response relationships for lakes and reservoirs were quantified and evaluated (Vollenweider, 1968; Rast and Lee, 1978; Canfield and Bachmann, 1981; and others). The analyses provide a potential basis by which predictions can be made of the changes in water quality that will occur from changes in the phosphorus loadings to phosphorus-limited water bodies. Analyses of over 200 water bodies permitted a general classification of lakes based on water transparency and concentrations of phosphorus, nitrogen, and phytoplankton pigment (Table 13–14). Although any attempt at rigid classification of water bodies is subjective and fraught with exceptions, the general classification is useful.

The models permit predictions of the probability of oligotrophic, intermediate, or eutrophic conditions developing in phosphorus-limited lakes in response to various loading regimes (Fig. 13–11). The literature on this subject is very large, for the models have been continuously tested, refined, and improved (e.g., Reckhow, 1979; Chapra and Reckhow, 1981). As with all modeling efforts of natural and disturbed ecosystems, one

TABLE 13-14   General Trophic Classification of Lakes and Reservoirs in Relation to Phosphorus and Nitrogen

PARAMETER (ANNUAL MEAN VALUES)	OLIGOTROPHIC	MESOTROPHIC	EUTROPHIC	HYPEREUTROPHIC
Total phosphorus (mg m^{-3})				
Mean	8.0	26.7	84.4	–
Range	3.0–17.7	10.9–95.6	16–386	750–1200
N	21	19	71	2
Total nitrogen (mg m^{-3})				
Mean	661	753	1875	–
Range	307–1630	361–1387	393–6100	–
N	11	8	37	–
Chlorophyll a (mg m^{-3}) of phytoplankton				
Mean	1.7	4.7	14.3	–
Range	0.3–4.5	3–11	3–78	100–150
N	22	16	70	2
Chlorophyll a peaks (mg m^{-3}) ("worst case")				
Mean	4.2	16.1	42.6	–
Range	1.3–10.6	4.9–49.5	9.5–275	–
N	16	12	46	–
Secchi Transparency Depth (m)				
Mean	9.9	4.2	2.45	–
Range	5.4–28.3	1.5–8.1	0.8–7.0	0.4–0.5
N	13	20	70	2

Based on data of an international eutrophication program. Trophic status based on the opinions of the experienced investigators of each lake. (Modified from Vollenweider, 1979.)

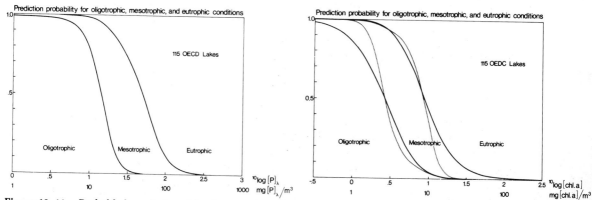

Figure 13-11.   Probable boundaries of the degrees of trophy of waterbodies with differing annual mean values of total phosphorus concentrations *(left)* and chlorophyll a concentrations *(right)* (modified from Vollenweider, 1979).

must not permit the mathematical manipulations to become an end in themselves (e.g., Pielou, 1981).

Models based on many data permit estimates of permissible loading rates of phosphorus and nitrogen while still allowing tolerable conditions of productivity. Considering a combination of influencing conditions, Vollenweider (1968) estimated very approximate provisional loading rates of nitrogen and phosphorus required to maintain lakes in a steady state (Table 13–15). Rapid eutrophication is likely to result as loading is increased. Although the predictive models have improved considerably since this analysis appeared, the tolerable loading level of 10 mg P m^{-3} remains unaltered, while the excessive loading level has been increased slightly to 25 mg P m^{-3} (Vollenweider, and Kerekes, 1980).

Reduction of the productivity of lakes by decreasing phosphorus loading can be very effective. Many examples are given in Vollenweider's (1968) review. Among the most frequently discussed examples is Lake Washington, Seattle, Washington (Edmondson, 1972). Lake Washington was enriched with increasing volumes of effluent from secondary sewage treatment facilities during the period from 1941 to 1953. Production increased markedly, and the algae became more abundant. Phosphate concentrations also increased proportionally much more than those of nitrate or carbon dioxide. Effluent was diverted away from the lake in 1963, and by 1969 phosphate had decreased to 28 percent of its 1963 value. Summer chlorophyll concentrations had decreased about as much, but nitrate and total CO_2 fluctuated from year to year at relatively high values. Reduction in phytoplankton and phosphorus is continuing (Edmondson and Lehman, 1981). Other examples are given by Edmondson (1969).

The rate at which the productivity of a lake will revert towards conditions existing prior to increased loading is variable, and depends upon basin morphometry, water chemistry, and the nature of the phosphorus sources (diffuse or concentrated at point sources). The loading of lakes with nitrogen and phosphorus is further influenced by the ratio of the surface area of the drainage basin to that of the lake (Ohle, 1965).

TABLE 13–15 Provisional Permissible Loading Levels for Total Nitrogen and Total Phosphorus (Biochemically Active) in g m^{-2} Year^{-1}

MEAN DEPTH (m)	PERMISSIBLE LOADING		DANGEROUS LOADING	
	N	P	N	P
5	1.0	0.07	2.0	0.13
10	1.5	0.10	3.0	0.20
50	4.0	0.25	8.0	0.50
100	6.0	0.40	12.0	0.80
150	7.5	0.50	15.0	1.00
200	9.0	0.60	18.0	1.20

After Vollenweider, R. A.: Scientific Fundamentals of the Eutrophication of Lakes and Flowing Waters, with Particular Reference to Nitrogen and Phosphorus as Factors in Eutrophication. OECD Report no. DAS/CSI/68.27, Paris, OECD, 1968.

Depending on the percentage losses from the land, some of which may be under cultivation, inputs increase roughly in proportion to the "surroundings" ratio.

Upon reduction of loading rates to or below those of the lake prior to accelerated inputs, recovery will require 2 to 10 years for lakes of average size and average hydrological replenishment time. The rate of recovery, however, is much slower in shallower lakes where the "internal loading" of phosphorus from sediment sources is high. In eutrophic Shagawa Lake, Minnesota, for example, where external phosphorus loading was reduced by more than 70 percent, predicted reduced equilibrium levels were not achieved (observed levels of phosphorus concentrations after diversion were over 100 percent in excess of those predicted) (Larsen, et al., 1975). Shagawa Lake is shallow, has an anaerobic hypolimnion, and has an extensively developed littoral zone of submersed macrophytes, all of which contribute to enhanced release of sediment phosphorus. Internal loading from sediment phosphorus sources has been observed in a number of other shallow lakes (e.g., Cooke, et al., 1977), which makes lake recovery by reduction of phosphorus loading from external allochthonous sources a much longer process. In a majority of lakes, however, the contribution of the "internal loading" is a relatively small proportion of the external phosphorus loading (e.g., Bannerman, et al., 1974; Edmondson and Lehman, 1981), so that regulation of loading rates is often effective in controlling lake trophic status.

SUMMARY

1. Phosphorus plays a major role in biological metabolism. In comparison to other macronutrients required by biota, phosphorus is least abundant and commonly is the first element to limit biological productivity.
2. Many quantitative data exist on the seasonal and spatial distribution of phosphorus in streams and lakes and the loading rates to recipient waters from drainage basins.
   a. Orthophosphate (PO_4^{-3}) is the only directly utilizable form of soluble inorganic phosphorus. Phosphate is extremely reactive and interacts with many cations (e.g., Fe, Ca) to form, especially under oxidizing conditions, relatively insoluble compounds that precipitate out of the water. Availability of phosphate is also reduced by adsorption to inorganic colloids and particulate compounds (e.g., clays, carbonates, hydroxides).
   b. A large proportion of phosphorus in fresh waters is bound in organic phosphates and cellular constituents of organisms, both living and dead, and within or adsorbed to organic colloids.
   c. The range of total phosphorus in fresh waters is large, from <5 μg l^{-1} in very unproductive waters to >100 μg l^{-1} in highly eutrophic waters. Most uncontaminated fresh waters contain between 10 and 50 μg total P l^{-1}.
   d. While concentrations of total and soluble phosphorus of oligotrophic lakes exhibit little variation with increasing depth, eutrophic lakes with strongly clinograde oxygen profiles commonly show a marked increase in phosphorus content in the lower hypolimnion (Fig. 13–1). Much of the

      hypolimnetic increase is from soluble phosphorus near the sediment–water interface.

3. Exchanges of phosphorus across the sediment–water interface are regulated by oxidation-reduction (redox) interactions dependent on oxygen supply, mineral solubility and sorptive mechanisms, the metabolic activities of bacteria and fungi, and turbulence from physical and biotic activities.

    a. In all but the upper few millimeters of sediment, exchange is slow and controlled by low diffusion rates.

    b. If water above the sediments is oxygenated (approximately >1 mg O_2 l^{-1}), an oxidized microzone is formed below the sediment–water interface (0 to -5 mm), below which the sediments usually become extremely reducing (Fig. 13–2). The oxidized microzone effectively prevents phosphorus (which is solublized under reducing conditions in the sediments) from migrating by diffusion upward into the water column (Fig. 13–3, *left*).

    c. As the hypolimnion becomes anoxic in productive lakes, the oxidized microzone is lost. The release of phosphate and ferrous iron into the water occurs readily when reducing conditions reach a redox potential (E_7) of about $+200$ mv (Fig. 13–3, *right*).

    d. Soluble phosphorus can accumulate in large quantities in anaerobic hypolimnia. With the advent of autumnal circulation, ferrous iron is rapidly oxidized and precipitates much of the phosphate as ferric phosphate.

    e. Bacterial metabolism of organic matter is the primary mechanism by which organic phosphorus is converted to phosphate in the sediments and for creating the reducing conditions required for release of phosphate to the water.

    f. Movement of phosphorus from sediment interstitial water can be accelerated by physical turbulence and by biota.

       i. Rooted aquatic macrophytes often obtain phosphorus from the sediments and can release large amounts into the water both during active growth and upon senescence and death.

      ii. High population densities of sediment-dwelling invertebrates, such as midge larvae, can increase the exchange of phosphorus across the sediment–water interface.

4. Recent studies of phosphorus cycling in the trophogenic zone have demonstrated that the exchange of phosphorus between its various forms is often rapid, and involves numerous complex pathways.

    a. A large portion, often >95 percent, of the phosphorus is bound in the particulate phase of living biota, especially algae.

    b. Organic phosphorus of the seston of the open water consists of at least two major fractions:

       i. A rapidly cycled fraction which is exchanged with soluble forms. In this fraction, phosphate is transferred rapidly through the particulate phase to low-molecular-weight compounds.

      ii. A fraction of dissolved organic and colloidal phosphorus material that is released and cycled more slowly.

c. Phosphorus uptake and turnover kinetics in numerous lakes show that turnover rates are extremely rapid (5 to 100 min) during summer periods of high demand and low loading inputs. During winter periods, turnover rates become as much as two orders of magnitude slower. Phosphorus turnover rates are generally faster under more oligotrophic conditions of greater phosphorus deficiency.

d. Zooplankton that feed on seston excrete soluble phosphorus and ammonia. These nutrients are rapidly utilized by algae and bacteria. When supplies of phosphorus are low, this source of recycled phosphorus can be critical to the growth and succession of phytoplankton.

e. Algal requirements for phosphorus are variable among species and can lead to selective advantages of certain algae as phosphorus supplies change seasonally or as phosphorus loadings to a fresh water change over a longer period of time. Algae must compete with bacteria for phosphorus sources; if organic substrates for bacterial growth are high, algal growth may be seriously impeded.

5. Sedimentation of particles results in constant losses of phosphorus from the trophogenic zone. As a result, new phosphorus supplies must enter the ecosystem in order to maintain or increase productivity.

a. Phosphorus enters fresh waters from atmospheric precipitation and from groundwater and surface runoff. The loading rates of phosphorus vary greatly with patterns of land use, geology and morphology of the drainage basin, soil productivity, human activities, pollution, and other factors.

b. When phosphorus is added to unproductive fresh waters, either experimentally or as a result of man's activities, the usual response is a rapid increase in algal productivity. Inputs must be more or less continuous, however, to maintain a higher level of productivity.

c. Numerous mass balance models have been developed to predict, on the basis of phosphorus loading and retention times, the anticipated responses of algal biomass and productivity. The models predict a reasonably accurate estimation of permissible phosphorus loading needed to achieve a certain level of productivity, if loading is lowered.

d. In certain shallow lakes with greater than average turbulence, large littoral areas, and small, anaerobic hypolimnia, reduced productivity does not always occur as rapidly in response to decreased loading as predicted from the models. In these lakes, phosphorus release from sediment sources ("internal loading") is much greater than values (10 to 30 percent of total loading) for deeper, more stratified lakes.

CHAPTER FOURTEEN
IRON, SULFUR, AND SILICA CYCLES

BIOGEOCHEMICAL CYCLING OF ESSENTIAL MICRONUTRIENTS

The biogeochemical cycling of essential micronutrients is complex, and is regulated to a large extent by variations in oxidation-reduction states, which in turn are mediated by photosynthetic and bacterial metabolism. Many of the reactions among different elemental nutrients are coupled, and the state of one can influence the availability of another. While the coupled reactions must be viewed simultaneously, the following discussion separates the components as much as possible.

Oxidation-Reduction Potentials in Freshwater Systems

In pure inorganic chemical systems, oxidation-reduction (redox) reactions proceed with a flow of electrons between the oxidized and reduced states until equilibrium is attained. There is a tendency for the reduced phase to lose electrons and be transformed to an oxidized state, e.g., $Fe^{+++} + e^- \rightleftarrows Fe^{++}$. Free electrons, however, usually inhibit this process, so that large quantities of free ions in reduced and oxidized states can exist together. If electrons are removed, as for example by an immersed platinum electrode, then a transformation of the reduced state to the oxidized state (Fe^{++} to Fe^{+++}) occurs, and a current of electrons passes through the electrodes. By applying a current (electrons) in the opposite direction, reduction can be induced; the reactions are reversible. Such potentiometric changes in electron flow can be accomplished with any solution containing ionic species, including that composite system of lake water, by drawing off electrons with a calomel*-platinum electrode. The resulting measurement, taken against standard conditions, is an expression of the oxidizing or reducing intensity or condition of the solution. The current, referenced against the hydrogen electrode, is termed the redox potential or the electrode potential.

---

*A calomel electrode consists of a chloride solution (usually KCl), solid Hg_2Cl_2 (calomel), and metallic mercury.

THE REDOX POTENTIAL

Redox potential is proportional to the equivalent free energy change per mole of electrons associated with a given reduction (Stumm, 1966; Morris and Stumm, 1967). Although aqueous solutions do not contain free protons and electrons, it is possible to define proton activity [pH = $-\log (H^+)$] and electron activity [pE = $-\log (e^-)$]. pE is large and positive in strongly oxidizing solutions (low electron activity), just as pH is high in strongly alkaline solutions (low proton activity). Thus both pH and pE are intensity factors of free energy levels and are not related to capacity or condition (e.g., alkalinity, acidity).

When oxygen is dissolved in water, a redox potential is generated according to the reaction (cf. discussion in Stumm, 1966, and Stumm and Morgan, 1981):

$$H_2O \rightleftharpoons \tfrac{1}{2} O_2 + 2H^+ + 2 e^-$$

The pE of water at equilibrium is relatively insensitive to change in oxygen concentration and extent of saturation, for it is influenced by the 4th root of the partial pressure of oxygen. If the oxygen content is decreased by 99 per cent, the redox potential would be reduced by only about 30 mv. The activity of the hydroxyl ions, however, influences the activity of the hydrogen ions. Therefore the redox potential is significantly changed by alterations of H^+ and is reflected in the pH. It is customary to express pE of redox reactions in natural waters at their activities in neutral water at pH 7 rather than at unity. Thus, the redox potential is expressed as E_h or as E_7, in which a correction is made for the change in redox at the pH of the sample to a pH of 7.† A change in pH of one unit is accompanied by a change in redox potential of 58 mv. Hence, a common practice is to correct potentials to pH 7 by subtracting 58 mv for every pH unit on the acid side of neutrality, and by adding 58 mv for every pH unit on the alkaline side of neutrality. E_h is influenced to only a small extent by temperature, e.g., E_h of water = 860 mv at 0°C, and 800 at 30°C at pH 7.

The preceding discussion applies to an ideal situation under conditions of equilibrium, in which oxidation-reduction systems are reversible (Stumm, 1966). True redox equilibrium is not found in any natural aquatic system because of the extreme slowness of most redox reactions in the absence of appropriate biochemical catalysis. There is, in addition, a continuous cyclic input of photosynthetic energy that disturbs the trend towards equilibrium conditions. In addition to nonequilibrium redox conditions, which are related to the highly dynamic state of biotic activities, electrochemical measurements of redox depend on the nature and composite rates of reactions at the electrode surface. In practice, the redox measurements reflect irreversible electrochemical redox potentials. Thus, in neutral, fully oxygenated water equilibrated with air, redox potentials slightly greater than 500 mv are obtained, which are considerably less than the theoretical E_h of 800 mv.

Few elements (C, O, N, S, Fe, Mn) are predominant reactants in redox processes

---

†E_h is compared to the hydrogen half-cell at any designated pH; usually it is pH O([H$^+$] = 1). Thus, E_h is usually, but not necessarily always, corrected to neutral pH.

in natural waters. By conversion of energy into chemical bonds, photosynthesis produces reduced states (negative E_h) of high free energy and results in nonequilibrium concentrations of C, N, and S compounds (Stumm, 1966). Respiratory, fermentative, and other nonphotosynthetic reactions of organisms tend to restore equilibrium by catalytically decomposing, through energy-yielding redox reactions, the thermodynamically unstable products of photosynthesis. It is through such reactions that nonphotosynthetic organisms obtain a source of free energy for their metabolic demands. The mean E_h is increased by these combined processes. It should be noted that organisms act as redox catalysts by mediating the reactions and transfer of electrons; the organisms themselves do not oxidize substrates or reduce compounds.

Reduced and oxidized iron (Fe^{++} and Fe^{+++}) are among the most electroactive redox reactants in natural water systems. Other than iron and to a certain extent manganese, the redox components of organic carbon, nitrogen, and sulfur are not electronegative and yield reversible potentials only following enzyme-mediated changes. As a result, redox measurements in natural waters do not lend themselves to quantitative interpretation and comparison. Qualitative, relative comparisons of E_h, representing mixed composite potentials, however, can be extremely instructive, as was demonstrated by the monographic comparison of pH and E_h limits in different environments and among organisms (Baas Becking, et al., 1960).

From this discussion, one would anticipate little change in redox potential with increasing depth as long as the water contained dissolved oxygen. Even though a distinctly clinograde oxygen curve may be found, as long as the water is not near anoxia, the E_h will remain positive and fairly high (300 to 500 mv). Such conditions have been found repeatedly in detailed studies of E_h in lake waters of widely diverse types (Kuznetsov, 1935; Hutchinson, et al., 1939; Allgeier, et al., 1941; Mortimer, 1942; Kjensmo, 1970). As oxygen concentrations approach zero and anoxic conditions appear, as in the lower hypolimnion and near the sediments, the E_h decreases precipitously. Similar relationships at the sediment-water interface were discussed in some detail in the previous chapter on phosphorus release from the sediments (cf. Figs. 13–2, 13–3, and 13–4). Within the sediments, reducing conditions prevail, and the E_h declines to 0 mv or below within a few millimeters of the interface (Figs. 13–2, 13–3). Of the many reducing compounds that contribute to the reductions in E_h, ferrous iron of the sediments is the most important (Fig. 13–3).

Lower redox potentials are generally observed in lake systems containing relatively high concentrations of dissolved humic compounds (Allgeier, et al., 1941; Kjensmo, 1970). Humic acids, especially those derived from the moss *Sphagnum* of bogs and bog lakes, have an E_h of about 350 mv (Visser, 1964b) and their reducing properties lead to a metal enrichment by complexing and adsorption to the acid molecules (Szilágyi, 1973; Zimmerman, 1981). The redox potentials of the deep monimolimnia of meromictic lakes are generally extremely low, among the lowest found in natural waters (Fig. 14–1). Extremely high concentrations of soluble total iron are frequently found in the monimolimnia of these permanently stratified lakes of high dissolved organic matter content. In shallow meromictic lakes where the water of the monimolimnion is slowly (over several years) exchanged with the overlying water strata, high monimolimnetic iron concentrations are maintained through the oxidation of ferrous iron in the chemolimnion (Campbell and Torgersen, 1980). Up to 90 per cent of the iron entering the chemolimnion is oxidized, and is returned to the monimolimnion by sedimentation and subsequent resolubilization.

Iron and Manganese Cycling in Lakes

The similarities in chemical reactivity between iron and manganese permit us to discuss them together. Although clear differences exist between the two metals, and although much more is known about the dynamics and cycling of iron, the two metals behave in similar fashion in freshwater systems. A strong interaction exists between the cycling of these metals, especially iron, and sulfur. The biogeochemical fluxes of iron and manganese reflect the combined spatial and temporal variations in physical chemistry, which are almost totally controlled by the dynamic conditions regulating bacterial metabolism.

CHEMICAL EQUILIBRIA AND FORMS OF IRON AND MANGANESE

Iron exists in solution in either the ferrous (Fe^{++}) or ferric (Fe^{+++}) state. (Detailed reviews are given in Hem and Cropper, 1959; Hem, 1960; Stumm and Lee, 1960; Doyle, 1968a; and Stumm and Morgan, 1981). Amounts of iron in solution in natural water and the rate of oxidation of Fe^{++} to Fe^{+++} in oxygenated water are dependent primarily on pH, E_h, and temperature. Ferrous constituents tend to be more soluble than ferric constituents.

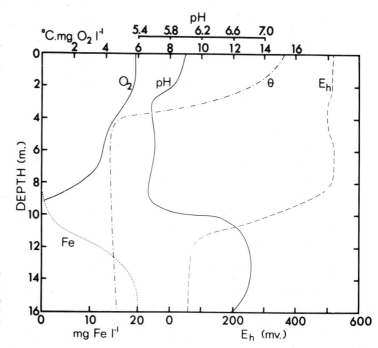

Figure 14–1. Vertical distribution of temperature, oxygen, pH, total iron, and redox potential E_h in permanently meromictic Lake Skjennungen, Norway, June 1967. (After Kjensmo, 1970.)

Soluble ferrous iron occurs mainly as hydrated Fe^{++} and hydrated hydroxo ions, the solubility of which is determined largely by the solubility of ferrous hydroxide [$Fe(OH)_2$], ferrous carbonate ($FeCO_3$), and ferrous sulfide (FeS). $Fe(OH)_2$ is exceedingly insoluble within the normal pH range of oxygenated natural waters. The solubility of ferrous iron is generally controlled by the solubility of $FeCO_3$. Even at low pH, carbonate concentrations are usually sufficient to limit solubilization. If water is saturated with $CaCO_3$, $FeCO_3$ is about 200 times less soluble than $CaCO_3$, unless very high pH values (>10) occur. Soft waters with very low concentrations of bicarbonate usually contain somewhat higher concentrations of Fe^{++}, although most is oxidized to Fe^{+++}. For example, a water free of bicarbonate at pH 7 would contain more than 1000 times as much Fe^{++} as a water containing 2 meq l^{-1} alkalinity, an average value for many waters. FeS is also exceedingly insoluble (Doyle, 1968b), and forms a stable, crystalline compound that darkens the color of anaerobic sediments. At low redox potentials in the anaerobic hypolimnia of productive or meromictic lakes, bacterial reduction of sulfate to sulfides is common. The resulting excess of sulfide can decrease concentrations of Fe^{++} through the formation of insoluble FeS.

IRON COMPLEXES

The most common species of ferric iron in natural waters is hydrated ferric hydroxide, $Fe(OH)_3$. At equilibrium in the pH range of 5 to 8, $Fe(OH)_3$ is in the solid state, because its solubility is very low (equilibrium constant approximately 10^{-36} at 25°C). Other insoluble ferric salts are of less significance. For example, phosphate does not influence the solubility of Fe^{+++} when inorganic phosphorus concentration is less than 10^{-4} mol l^{-1}, as is usually the case. It is apparent, therefore, that in the absence of organic matter and based only on solubility criteria as mediated by pH and E_h, a number of forms of inorganic iron can exist in natural waters. The simultaneous influence of hydrogen ions and electrons on the equilibria of aqueous iron is illustrated in Figure 14-2. All of the oxygenation reactions of Fe^{++} to Fe^{+++} are exergonic and capable of supplying energy; some of these reactions serve as an energy source for microorganisms.

Much of the iron of normal lake water is present as suspensions of flocculent ferric hydroxide, removable by filtration with membranes of a pore size of 0.5 μm (Hutchinson, 1957; Hem and Cropper, 1959; Hem and Skougstad, 1960). Under some conditions, a very finely divided precipitate of ferric hydroxide, which has colloidal properties (0.001 to 0.5 μm, charged), may form. Colloidal particles of $Fe(OH)_3$ are commonly positively charged, although a negatively charged sol can occur at high pH. Ions in solution, and negatively charged clay particles, organic colloids, and other suspended solids can neutralize the charges on the hydroxide colloidal particles. The uncharged aggregates join to form a rapidly settling precipitate. Other metals, such as copper ions, can be adsorbed by and coprecipitated with the ferric hydroxide precipitate.

Although ferrous iron in solution occurs in extremely low concentrations in oxygenated water, some apparently occurs in particulate and colloidal form. Such suspended ferrous iron is presumably associated with soil and sediment particles. A portion of total iron is contained in the living plankton, but this quantity is a very small part of the whole. The iron content of higher aquatic plants averages 5 mg g^{-1} dry weight, about an order of magnitude greater than the content of terrestrial plants (Oborn, 1960). Plant roots contain a higher proportion of iron than do stems or leaves.

Iron complexes with certain organic molecules greatly alter iron solubility and availability to organisms. Many organic bases form strong soluble iron complexes with ferrous (Gjessing, 1964) and ferric ions. An enrichment of iron is commonly found in surface waters with a high content of dissolved organic matter. These high concentrations of complexed soluble iron are associated with high levels of humic acids (Shapiro, 1957), tannic acids (Hem, 1960b), and other lignin derivatives. The intense yellow-brown color of bog waters is, in part, a result of these complexes. Natural and synthetic organic compounds, as, for example, citrate and ethylenediamine tetraacetic acid (EDTA), of high chelating capacity for metals, are commonly used in studies of plant and especially algal nutrition because they can maintain iron available for assimilation. Shapiro (1964, 1966, 1969) extended his earlier work on the complexing of iron with humic derivatives, especially with yellow organic acids of low molecular weight. The primary mechanism apparently is peptization, in which iron is dispersed in a solubilized form of the $Fe(OH)_3$ precipitate as a result of adsorption of organic acids onto the surfaces of the particles. Some iron is chelated with organic acids by weak chemical bonding, although chelation is not the primary complexing mechanism (cf. Steinberg, 1980). In acidic (pH 4 to 5) bog lakes, ferric iron may exist not as an inorganic complex but as reactive ferric iron by complexation with colloidal organic acids (Koenings, 1976).

Total iron found in oxygenated surface waters of pH 5 to 8 typically ranges from about 50 to 200 μg l^{-1}, almost none of which occurs in ionic form (Wetzel, 1972). Much higher levels occur in lakes heavily stained with dissolved humic compounds, in acidic volcanic lakes, acidic bog waters, and certain alkaline, closed lakes rich in organic matter.

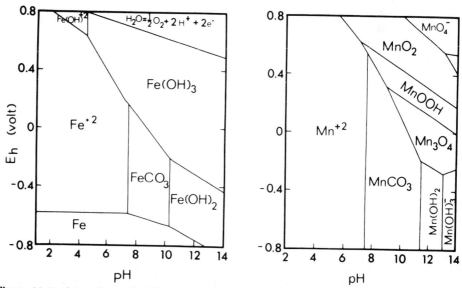

Figure 14–2 Approximate distribution of species of iron and manganese in relation to pH and redox potential E_h. Alkalinity is assumed to be equal to 2 meq l^{-1}. Lines denote points at which the activities of soluble Fe and Mn are 10^{-5} mol l^{-1}. (Modified from Stumm and Lee, 1960, and Stumm and Morgan, 1981.)

MANGANESE COMPLEXES

The theoretical thermodynamic redox equilibria of manganese have been studied in some detail (Hem, 1963, 1964; Stumm and Morgan, 1981), and permit an evaluation of manganese species and equilibria under various conditions of oxidation-reduction found in natural surface waters. In general, the behavior of manganese in lakes follows these predictions.

Although Mn occurs in several valence states, Mn^{+3} is thermodynamically unstable in aqueous solutions under normal conditions and Mn^{+4} compounds are insoluble at most environmental pH values. As with ferrous iron, ionic divalent Mn^{+2} occurs at low redox potentials and pH (Fig. 14-2). Some form of oxidized Mn will be in equilibrium with Mn^{+2} under oxidizing conditions of high pH and E_h, and some form of Mn^{+2} may be in equilibrium with manganese carbonate under reducing conditions of low pH and E_h. Supersaturation of both ferrous and manganese carbonate has been observed in anoxic hypolimnia of eutrophic lakes (Verdouw and Dekkers, 1980). The redox equilibrium E_h values are higher and the rates of oxidation slower than for iron; as a result, detectable quantities of Mn are commonly observed longer than comparable quantities of iron under lake conditions. Above a pH of 8.5, an intermediate complex forms in which Mn^{+2} is adsorbed onto manganese oxides. The Mn^{+2} of these oxide complexes can react relatively rapidly with other anions and precipitate as manganous carbonate ($MnCO_3$), manganous sulfide (MnS), and manganous hydroxide ($Mn(OH)_2$). Manganese is also adsorbed to iron oxides and coprecipitates with ferric hydroxide when the pH exceeds 7.

Manganese has a solubility of about 1 mg l^{-1} in distilled water of E_h of 550 mv and pH 7; solubility decreases markedly with increases in E_h and pH. Manganese forms soluble complexes with bicarbonate and sulfate. Increased bicarbonate activity decreases Mn solubility. At high concentrations, bicarbonate has been found to reduce the oxidation rate of manganese. In a manner analogous to iron, organic molecules can form stable complexes with Mn^{+2}, although their operation in aquatic systems is poorly understood. Manganese occurs in relative abundance in alkaline soils as hydrated oxides, and no doubt organic complexing plays a role in retention of Mn in a complexed dissolved form for transport once redox conditions are altered by microbial and plant metabolism. Igneous rock contains about 0.01 per cent manganese, which is involved in biogeochemical cycling following weathering and subsequent mobilization. Drainage from forest litter, especially from more acidic coniferous forests, is often high in manganese. This observation again points to the probability of the major role of organic complexes in the effective transport of easily oxidized metal ions.

DISTRIBUTION OF IRON AND MANGANESE IN LAKES

Under oxidized conditions, as in the epilimnia of lakes, large amounts of iron are found only in acidic water (pH < 3), such as in lakes of volcanic origin (Yoshimura, 1936) or in acid drainage from strip mining operations. In both cases, organic content is relatively low, and acidity usually results from sulfuric acid. The quantity of total iron found in most typical neutral or alkaline lakes, 50 to 200 μg l^{-1}, is dominated by $Fe(OH)_3$, organically complexed iron, and adsorbed sestonic iron in particulate form.

The world average distribution is variable among continents (Table 10–1), and higher than would be generally expected among most oxygenated surface waters of lakes. The range of manganese concentrations (approximately 10 to 850 μg l^{-1}) is also highly variable in relation to lithology and drainage of the lake basins (Hutchinson, 1957; Livingstone, 1963). The average quantity of manganese is about 35 μg l^{-1}, somewhat less than that of iron. The ratio of Fe to Mn in water is generally considerably lower than that of the lithosphere (50:1), and indicates the relative enrichment of Mn with respect to Fe; this is in agreement with the reaction equilibria already discussed.

The vertical distribution of iron and manganese is reflected in the distribution of redox potentials. Concentrations of ionic iron of oxygenated waters of oligotrophic lakes, epilimnia of more productive lakes, and of circulating waters are exceedingly low. Manganese is somewhat more soluble (Fig. 14–3). Ferrous ions diffuse readily from the sediments when redox potentials decline to about 200 mv; migration of Mn^{+2} from the sediments occurs at somewhat greater redox potentials (Robbins and Callender, 1975; Ostendorp and Frevert, 1979). These relationships were demonstrated very well in Mortimer's studies of the sediment interface, both experimentally (see Fig. 13–3) and in a eutrophic lake (see Fig 13–4). Here it can be seen that the release of Mn precedes that of iron. Released manganese will remain soluble if the oxygen saturation is less than about 50 per cent (Burns and Nriagu, 1976).

While the seasonal distribution of iron has been studied in some detail (reviewed in Hutchinson, 1957; McMahon, 1969), that of manganese has been investigated less extensively (Delfino and Lee, 1968, 1971; Brezonik, et al., 1969; Howard and Chisholm, 1975). The general pattern of seasonal distribution is apparent from the examples given of a mesotrophic lake that undergoes hypolimnetic oxygen reduction in the later phases of summer stratification (Fig. 14–4), and an interconnected eutrophic lake (Fig. 14–5). The eutrophic lake, Little Crooked Lake, is a small, protected eutrophic lake that frequently undergoes partial, temporary meromixis, as was the case in the year shown here when autumnal circulation was incomplete. Of the total manganese, nearly all

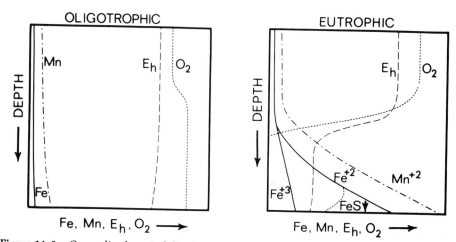

Figure 14–3. Generalized vertical distribution of iron, manganese, and redox potential (E_h) in stratified lakes of very low and high productivity.

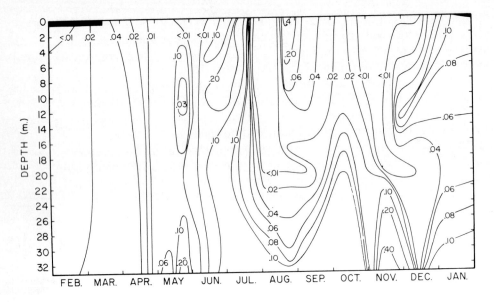

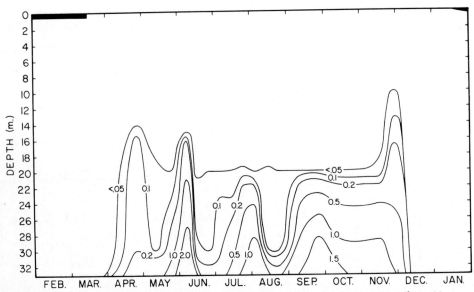

Figure 14-4. Depth-time diagrams of isopleths of total iron *(upper)* and manganese *(lower)* in mg l^{-1} of mesotrophic hardwater Crooked Lake, northeastern Indiana, 1963. (From Wetzel, unpublished data.)

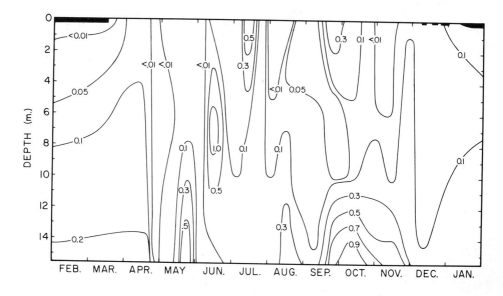

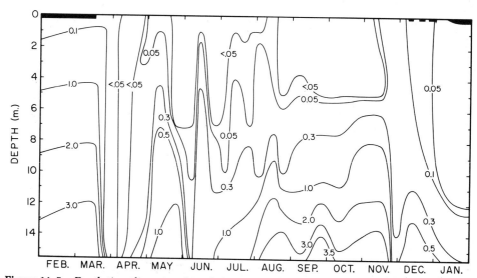

Figure 14–5. Depth-time diagrams of isopleths of total iron (upper) and manganese (lower) in mg l^{-1} of eutrophic hardwater Little Crooked Lake, northeastern Indiana, 1963. (From Wetzel, unpublished data.)

was in soluble form (passing through a 0.45-μm pore size membrane filter) in anaerobic water of the hypolimnion of a small eutrophic lake in New York (Howard and Chisholm, 1975). Less than 13 per cent of the total manganese was in soluble form in aerobic water strata of this lake.

As decomposition proceeds in the hypolimnion of very productive, thermally stratified lakes, the redox potential of hypolimnetic waters can decline to well below 100 mv. At an E_h below 100 mv, sulfate is reduced to hydrogen sulfide. Hydrogen sulfide is also produced by bacterial decomposition of sulfur-containing organic compounds. Since ferrous iron is released in significant quantities from the sediments at a higher E_h of about 250 mv, much Fe^{+2} is present in the hypolimnion at the time of sulfide formation. The formation of FeS and other metal sulfides (cuprous sulfide, CuS, cadmium sulfide, CdS, etc.), all of which are very insoluble under normal lake conditions, can result in a significant reduction of iron and other metals towards the end of summer stratification (Davison and Heaney, 1978). Manganous sulfide, on the other hand, is much more soluble and has little effect on the Mn^{++} concentrations under normal lake conditions.

Iron concentrations in the hypolimnia of soft-water lakes can reach very high levels under conditions that prevail in small, deep basins, especially in bog waters receiving high concentrations of humic organic matter. Levels of sulfate are low in such waters, and sulfide concentrations seldom become sufficient to precipitate iron as FeS. Kjensmo (1962, 1967, 1968) demonstrated that the hypolimnetic iron accumulations in protected lakes can reach such levels (> 250 mg l^{-1}) that the salinity gradient becomes adequate to render the lakes permanently meromictic. Precipitation of iron sulfides is inadequate to reduce such high iron concentrations appreciably, and contributes to meromictic conditions by allowing iron to accumulate.

Stages along the continuum of declining redox potentials can be divided into four phases of hypolimnetic conditions in stratified lakes of increasing productivity, or within a very productive lake during the period of summer stratification (Table 14–1).

UTILIZATION AND TRANSFORMATIONS OF IRON AND MANGANESE

The metabolic demands for iron and manganese are usually sufficiently low so that the biota do not materially deplete the concentrations of these metals in the environment. Both iron and manganese, however, are essential micronutrients of microflora, plants, and animals (Oborn, 1960b; Wangersky, 1963; Coughlan, 1971). Iron is required in the enzymatic pathways of chlorophyll and protein synthesis, and in respiratory enzymes of all living organisms. The function of iron in cytochromes and as the basic

TABLE 14–1 Changes in the Iron during the Continuum of Declining Redox Potentials in the Hypolimnia of Stratified Lakes of Increasing Productivity

LAKE STATUS	[O_2]	E_h	Fe^{+2}	H_2S	PO_4^{-3}
Oligotrophic ↓	High (orthograde) ↓	400–500 mv ↓	Absent ↓	Absent ↓	Very low ↓
↓	Much reduced (clinograde) ↓	400–500 mv ↓	Absent ↓	Absent ↓	Very low ↓
Eutrophic ↓	Much reduced (clinograde) ↓	approx. 250 mv ↓	High ↓	Absent ↓	High ↓
Hypereutrophic	Much reduced (or absent)	<100 mv	Decreasing	High	Very high

After discussion of Hutchinson, 1957.

component of hemoglobin in higher animals is well known. Manganese and iron are functional components of nitrate assimilation (e.g., Verstreate, et al., 1980), in the Hill reaction of photosynthesis, and are essential catalysts of numerous enzyme systems in animals and bacteria.

Although the requirements for these two micronutrients are low, their reactivity, very low concentrations, and restricted availability (especially of iron) in the trophogenic zones of lakes and in streams suggest that under certain conditions, availability of Fe and Mn may limit photosynthetic productivity. The mechanisms of assimilation of iron from the forms available in oxygenated natural waters are unclear. Some evidence, not completely satisfactory, indicates that ferric hydroxide is adsorbed onto algae and dead particles, from which iron is assimilated. The peptizing and chelating properties of organic acids for iron, which have been found in a large number of lakes (Shapiro, 1966, 1969), appear to be the primary source of iron, and demonstrate the importance of organic compounds in maintaining assimilable iron.

The available iron content of hard-water calcareous lakes is extremely low. For example, the reactive iron of Lawrence Lake, Michigan, seldom exceeded 5 μg l^{-1} over an annual period (Wetzel, 1972). In this and similar lakes, it is evident that iron availability is so low that high sustained primary productivity is limited by an effective iron deficiency (Schelske, 1962; Schelske, et al., 1962; Wetzel, 1965a, 1966b, 1972). The addition of iron in complexed form, either by synthetic chelating or natural complexing organic compounds, resulted in immediate increases in photosynthetic rates. Additions of complexing organic compounds were less effective but presumably increased the availability of iron already present in the water. The effectiveness of natural organic compounds from the hypolimnion and amino compounds in maintaining solubility of iron in hard-water lakes has also been demonstrated (Wetzel, 1972).

High concentrations of manganese (> 1 mg l^{-1}) are very inhibitory to blue-green and green algae and can induce marked changes in development and morphology (Gerloff and Skoog, 1957; Patrick, et al., 1969; Lorch, 1978). An antagonistic response was demonstrated in which increasing calcium concentrations progressively reduced the inhibitory effects of high manganese. This relationship indicates that high levels of manganese, for example at the time of fall circulation, could be inhibitory to natural populations of blue-green and green algae. Manganese concentrations of <50 μg l^{-1} were found to inhibit the development of green and blue-green algae in streams and to strongly favor diatom growth. This response by diatoms may be ubiquitous; if so, the moderate levels of Mn and high levels of Ca in hard-water lakes may contribute both to the general dominance of diatoms and to the stimulatory effects of primary productivity in response to small additions of chelated manganese observed by Wetzel (1966b) in these lakes.

In Lake Superior, an apparent synergistic effect between low concentrations of manganese and phosphorus can occur (Shapiro and Glass, 1978). Phytoplanktonic photosynthesis was enhanced much more by low-level enrichments of combined manganese and phosphate than by addition of either substance alone.

BACTERIAL TRANSFORMATIONS OF IRON AND MANGANESE

The cycling of iron and manganese is largely dictated by the oxidation-reduction conditions of lakes. While bacterial and photosynthetic metabolism greatly influence these controlling conditions, which indirectly regulate the states of these metals and their fluxes, certain bacteria utilize iron and manganese directly in energetic transformations (Lundgren and Dean, 1979). These transformations are minor, however, in comparison to heterotrophic metabolism of organic substrates in most natural systems.

Chemosynthetic utilization of energy from inorganic oxidations is relatively inefficient, especially in the case of the oxidation of iron and manganese. For example, the oxidation of Fe^{+2} to Fe^{+3} by iron bacteria releases only about 11 kcal per mole Fe.

The cycling of iron and manganese is influenced by two processes (Kuznetsov, 1970). First, as pointed out earlier, reduction of the oxidized combined metal occurs under appropriate redox conditions as ferrous bicarbonate or is precipitated and sedimented as a sulfide. Example reactions are:

$$Fe_2O_3 + 3H_2S \rightleftharpoons 2FeS + 3H_2O + S$$
$$FeS + 2H_2CO_3 \rightleftharpoons Fe(HCO_3)_2 + H_2S$$

Second, sheathed and stalked bacteria, algae, protozoan flagellates, and some true bacteria precipitate ferric and manganic oxides on their cells. The true iron bacteria occur in iron-rich waters of neutral or alkaline pH. Characteristic reactions of the few chemoautotrophic bacteria that deposit hydroxides and oxides are:

$$4Fe(HCO_3)_2 + O_2 + 6H_2O \rightarrow 4Fe(OH)_3 + 4H_2CO_3 + 4CO_2 + 58 \text{ kcal}$$
$$4MnCO_3 + O_2 \rightarrow 2Mn_2O_3 + 4CO_2 + 76 \text{ kcal}$$

Some species of *Leptothrix* are facultative iron bacteria that can oxidize both ferrous and manganous salts, whereas *Gallionella (Spirophyllum)* is restricted obligately to iron. Since at neutral pH and in the presence of oxygen Fe^{+2} is spontaneously oxidized, iron-oxidizing bacteria are restricted to zones of steep redox gradients, in which they can compete effectively with oxygen for reduced iron. Therefore, the iron bacteria are restricted to the interface regions of iron-bearing rock seeps, swamps, and bogs, and to upper hypolimnetic areas, where the redox potential is sufficiently low for reduced iron to occur. Over 220 g of ferrous iron are required to produce 0.5 g of cellular carbon. As a result, much oxidized iron will be precipitated on the sheaths of the bacteria and extruded materials. While this iron may not be important from the standpoint of synthesis of organic carbon, it is often important economically, for it causes corrosion and clogging of pipes. These examples refer to strict autotrophic bacteria that satisfy their CO_2 requirements without organic matter; they obtain all required energy from the oxidation of some specific, incompletely oxidized inorganic substance. Iron or manganese thereby enters the cells. Certain facultative autotrophic bacteria (mixotrophic) can develop in water containing only inorganic substances; other facultative autotrophic species are able to utilize organic substances as well.

Other groups of bacteria involved in the cycling of iron and manganese are heterotrophic. Certain filamentous forms (*Cladothrix*, some *Leptothrix*), in particular, deposit iron and manganese on the cell in sheaths during the process of metabolizing organic compounds. The colonial, coccoid cells or short rods of *Siderocapsa* of the Eubacteriales are a common form occurring at the oxic-anoxic interface zone of hypolimnion-metalimnion, especially in iron meromictic lakes (Dubinina, et al., 1973). *Siderocapsa* also is widely distributed in oxygenated zones of streams and lakes (Hardman and Henrici, 1939), and during periods of circulation (Sokalova, 1961). Species of this genus can mineralize humates and increase in numbers coincident with increases in iron humates during periods of high rainfall. *Spirothrix* is another dominant iron bacterium found in both aerobic and anoxic zones of Lake Glubok, USSR, whereas large populations of *Gallionella* developed only at the interface zone of the metalimnion. Over the period of an annual cycle in this lake, the number of iron bacteria reached a very high percentage (19 per cent) of the total bacteria during the later portion of summer stratification. At other times of the year, the percentage was about 5 per cent of the total bacteria. It is apparent that the iron bacteria significantly influence the iron cycle of more productive lakes.

Another widely distributed iron-oxidizing bacterium is the autotrophic *Thiobacillus thiooxidans*, which oxidizes iron disulfides to ferric sulfate (ZoBell, 1973):

$$FeS_2 + 3\frac{1}{2}O_2 + H_2O \rightarrow FeSO_4 + H_2SO_4$$
$$2FeSO_4 + \frac{1}{2}O_2 + H_2SO_4 \rightarrow Fe_2(SO_4)_3 + H_2O$$

Closely related to *Thiobacillus* is *Ferrobacillus ferrooxidans*, which is abundant in very acidic (pH < 3) mine waters, in which ferrous iron is soluble and stable. *Ferrobacillus* oxidizes ferrous carbonate to ferric hydroxide and CO_2.

$$4\ FeCO_3 + O_2 + 6\ H_2O \rightarrow 4\ Fe(OH)_3 + 4\ CO_2$$

A few heterotrophic species of iron-oxidizing bacteria of the genera *Sphaerotilis*, *Leptothrix*, *Clonothrix*, and *Siderobacter* deposit oxidized manganese along with iron on capsules in their sheaths. Some species of *Metallogenium* obtain part of their energy requirements from the oxidation of manganous oxide (MnO), manganous sulfate ($MnSO_4$), or manganous carbonate ($MnCO_3$) to manganese sesquioxide (Mn_2O_3) and manganese dioxide (MnO_2); others are heterotrophic, and along with several other manganese- and iron-oxidizing bacteria, contribute to the formation of manganese and iron oxides in lake sediments (Oborn, 1964; Perfil'ev and Gabe, 1969; Kuznetsov, 1970). *Metallogenium* is undoubtedly one of the dominant microorganisms involved in the deposition of manganese nodules in lakes (Sokalova, 1961; Sorokin, 1970).

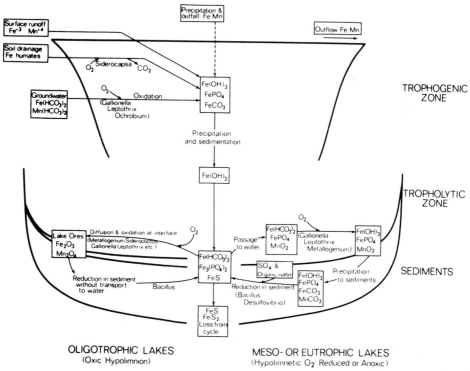

Figure 14-6. General iron and manganese cycles in lakes of low and high productivity emphasizing microbial interactions. (Modified from Kuznetsov, 1970.)

A general summary of these inorganic and bacterial relationships is given diagrammatically in Figure 14-6. The primary distinction between the fluxes of iron and manganese in hypolimnia of lakes of increasing productivity centers on the reduction of redox potential and pH. Our knowledge of the iron and manganese cycles at the bacterial level is qualitative owing to the paucity of quantitative information concerning both between-component flux rates and whole-lake metabolism of these two materials. Very little is known about the role of the littoral flora in this cycling. Iron, for example, is translocated from the sediments to the leaves of submersed vegetation (Oborn and Hem, 1962; DeMarte and Hartman, 1974). The magnitude of this transport from the sediments to the water in organic-bound phases of plant material as the plants decay is unknown.

Minor Metals of Nutritional Value

Minor metallic elements, collectively referred to as *micronutrients*, include Fe, Mn, zinc (Zn), copper (Cu), boron (B), cobalt (Co), molybdenum (Mo), and vanadium (V), all of which are required for the nutrition of plants and many animals. The universality of an absolute requirement for all of these elements is not clear. In some cases, as with iron and manganese, the essentiality of the element is established. In others, such as vanadium, it is known that one element can substitute for another, e.g., vanadium can replace molybdenum and that growth is enhanced by vanadium among certain algae (e.g., Meisch, et al., 1977; Patrick, 1978) but not among others (Gerloff, 1963; Holm-Hansen, 1968).

Selenium is another trace element of unknown biochemical significance among aquatic organisms. The dinoflagellate alga *Peridinium cinctum* requires a small amount of selenium (approximately 50 ng $Se^{+4}l^{-1}$) for optimal growth and reproduction (Lindström and Rodhe, 1978). In a Swedish lake, the concentrations of total and biologically available selenium showed considerable seasonal and vertical variation, which likely resulted from variations in loading from the atmosphere and the drainage basin, as well as algal uptake (Lindström, 1980; cf. Suzuki, et al., 1981). Although the potential for regulation exists, the significance of a trace element such as selenium in the regulation of algal growth in natural systems is unclear at the present time.

Most of the information on micronutrient requirements is obtained from studies of physiological deficiencies in cultures, which present an array of experimental difficulties, particularly involving contamination (Weissner, 1962). Requirements are further complicated by observed differences that depend on the composition and concentration of other minerals, and an antagonism among these elements. These effects, often found under the conditions of a controlled culture, become much more significant in the inorganic solution of natural waters, which are rendered heterogeneous because of constant seasonal and spatial variations. Hence, one finds a large number of analyses of concentrations of ionic and, less frequently, particulate micronutrients in water that may bear little relationship to actual metabolic requirements. The question of requirements concerns the *availability* of the micronutrient within the constraints of other ionic components of salinity, the extent of organic complexing of the micronutrient, and the metabolic demands of varying species. In general, concentrations and availability of micronutrients in most natural waters are adequate to sustain active populations of algae within constraints of light, temperature, and macronutrient availability. There are, however, clear cases in which micronutrients can limit photosynthesis to a degree. A true deficiency of micronutrients is found in some oligotrophic aquatic systems of granitic arctic, alpine, and volcanic areas, in which a paucity of these elements is well known (reviewed by Goldman, 1972). In other situations, such as in hard-water calcareous lakes, the micronutrients are not deficient in the system but are present in forms unavailable for assimilation (Wetzel, 1972). In both situations, the control of productivity can be quite transitory and effective on only certain algal species, thereby indirectly influencing algal succession and productivity.

The importance of dissolved organic matter in regulating micronutrient availability and growth, although long known to be effective in cultures, has been emphasized only recently as a major controlling factor of productivity in lakes (Wetzel, 1968, 1979; Wetzel and Allen, 1970; Wetzel, et al., 1972). Although involved in lake metabolism in many ways (cf. Chapter 22), complexation of micronutrients by organic compounds increases the physiological availability of many micronutrients when the ratio of organic matter to micronutrient concentration is reasonably high. The effects of chelated iron and manganese on algal productivity, discussed earlier, provide only one example of this process.

It is important to note that the relative concentrations of trace elements in fresh waters can have a significant effect on the competitive abilities of species in algal communities. For example, low concentrations of vanadium and hexavalent chromium favor the development of diatoms; at higher concentrations green and blue-green algae can dominate the communities (Patrick, 1978). Low concentrations of manganese favored the development of blue-green algae, whereas diatom communities dominated at concentrations of $> 40 \ \mu g$ Mn l^{-1} (Patrick, et al., 1969).

Most metallic micronutrients are very toxic when in excess or when complexed organically to the point when their availability exceeds tolerance limits. High concentrations of copper (usually as $CuSO_4$) have been used repeatedly as an herbicide to control algal blooms and growth of larger aquatic plants (Hodson, et al., 1979; McKnight, 1981). Fogg and Westlake (1955) have demonstrated that polypeptides secreted by blue-green algae can effectively complex copper ions and reduce toxic effects. This example is but one of many ways in which organic complexing can influence the availability of micronutrients. The effects of micronutrients and their availability to animals are less well understood. It is generally assumed that requirements are met by means of uptake from food sources. Positive and negative correlations between fluctuations of micronutrient concentrations of the water and animal populations (see Parker and Hazelwood, 1962) yield little insight into this question, but suggest that food quality can influence, in part, the succession and development of the fauna.

The general average concentrations of the soluble form of minor metallic micronutrients for a range of lakes and streams are given in Table 14–2. However, the ranges of concentration are extreme not only seasonally, but particularly in relation to contamination from industrial and other sources of pollution. Examples are discussed at length in the references given in Table 14–2, Chawla and Chau (1969), Robbins, et al. (1972), Mills and Oglesby (1971), Cowgill (1976, 1977a, 1977b), Hegi (1976), Baccini (1976), Boyle (1979), Leckie and Davis (1979), Martin, et al. (1980), and many others.

TABLE 14-2 Average Concentrations ($\mu g\ l^{-1}$) of Soluble Minor Metallic Elements of Natural Waters (Surface)

WATER	Fe	Mn	Cu	Zn	Co	Mo	V
World surface lakes and rivers (Livingstone, 1963)	approx. 40	35	10	10	0.9	0.8	approx. 0.1
Alpine lakes, California (Bradford, et al., 1968)	1.3	0.3	1.2	1.5	<0.3	0.4	–
Northern German lakes (Groth, 1971)	31.5	28.6	2.9	6.6	0.05	0.39	–
Northeastern Indiana lakes (Wetzel, 1966b)	15.0	21.3	<2	12.9	<2	30.0	<3
Linsley Pond, Connecticut (Hutchinson, 1957; Cowgill, 1976b, 1977b)	350	140	53	–	0.05	0.19	0.03
South American lakes (Groth, 1971)	533	15.6	1.7	8.7	0.11	0.6	–
Lake Constance, Southern Germany (Hegi, 1976)	10.9	11.7	1.7	6.2	–	–	–

DISTRIBUTION OF THE MINOR ELEMENTS IN LAKES

The distribution of the minor elements among ionic (soluble), organically complexed, and adsorbed fractions, as well as the amounts in living biota, is poorly understood. Quantitative rates of flux between the living and abiotic organic and inorganic phases are even less well delineated; generally much is inferred from analogous chemical species, such as iron and manganese. Fragmentary evidence on the cycling of the minor elements indicates that these similarities are real and that their cycling is comparable.

The amount of Cu, Zn, Co, and Mo in ionic solution is generally very small in aerated surface waters. Transport of these trace metals in flowing waters can be analytically partitioned into (1) ionic form, (2) that complexed in organic materials, (3) that adsorbed and precipitated on solids, and (4) that incorporated in crystalline structures (Gibbs, 1973). The solubilities of the various elements vary somewhat (cf. Groth, 1971), but in most cases, a large majority (> 70 per cent) of each is transported as crystalline solids or adsorbed solid phases. Much of the remainder is in organic complexes and dead seston, with very little in solution.

A summary of older literature on the cycling of Cu, Zn, Mo, and Co is given in Hutchinson (1957). The detailed analyses of Groth (1971) on Schöhsee in northern Germany highlight the basic phases of the dissolved and particulate fractions over an annual cycle. The total amounts of Co, Mo, and Zn, as well as of Fe and Mn, accumulated in hypolimnetic waters during summer stratification as shown in Table 14-3. The concentrations of Fe and Mn were found to be strongly related to redox conditions, while the primary source of the hypolimnetic accumulation of Co, Mo, and Zn was

TABLE 14-3 Average Concentrations (μg l^{-1}) of Minor Elements in the Epilimnion and Hypolimnion of Schöhsee, Germany

STRATUM	Mn	Fe	Cu	Zn	Co	Mo
Epilimnion (E)	4.5	15	1.0	1.8	0.03	0.21
Hypolimnion (H)	590	425	0.9	1.9	0.07	0.30
Enrichment ratio of H/E	130	28	0.9	1.1	2.3	1.4

Data after Groth, 1971.

from release and mineralization of sedimenting organic detritus (cf. also Baccini, 1976). No significant vertical distribution of Cu was found in Schöhsee during summer stratification, although moderate increases in hypolimnetic sestonic Cu have been observed elsewhere in eutrophic lakes (Riley, 1939).

Co, Cu, and Zn, like Fe and Mn, form stable complexes with organic compounds, so that losses of free ions by formation of insoluble hydroxides, sulfides, phosphates, and carbonates can be reduced appreciably (Groth, 1971). Molybdenum, however, exhibits greater mobility than the other ions (Mo > Cu > Zn > Co).

Phytoplankton tend to accumulate these minor metallic elements in the order: Fe > Zn > Cu > Co > Mn > Mo (Table 14-4). During decomposition and mineralization of plankton, release to a soluble phase occurs in the order of Mo > Co > Cu > Zn > Fe > Mn, approximately opposite the order of amounts of the elements found to be

TABLE 14-4 Average Accumulation of Minor Metallic Elements in the Plankton and Sediment in Comparison to the Average Concentrations in the Epilimnion of Schöhsee, Germany, 1968–1969

ELEMENT	DISSOLVED IN EPILIMNETIC WATER (μg l^{-1})	IN PLANKTON (μg g^{-1})	IN SEDIMENT (μg g^{-1})
Iron			
Concentration	15	950	58000
Enrichment factor	1	63×10^3	$3,900 \times 10^3$
Manganese			
Concentration	4.5	130	1600
Enrichment factor	1	29×10^3	355×10^3
Cobalt			
Concentration	0.03	1.1	8.3
Enrichment factor	1	37×10^3	280×10^3
Copper			
Concentration	1.0	60	95
Enrichment factor	1	60×10^3	95×10^3
Zinc			
Concentration	1.8	110	350
Enrichment factor	1	61×10^3	195×10^3
Molybdenum			
Concentration	0.21	4.2	1.4
Enrichment factor	1	20×10^3	7×10^3

Data after Groth, 1971.

transported to the sediments by sedimenting organic detritus: Fe > Mn > Co > Zn > Cu > Mo. Coprecipitation of these trace elements with Fe(OH)$_3$ was greatest with Cu, in the order of Cu > Mo > Co > Zn. Therefore, during summer stratification, algal uptake and sedimenting detritus play a major role in the cycling of Cu, Co, and Zn, whereas concentrations of Fe and Mn are regulated largely by redox conditions. Although variations among lakes are great, amounts of minor elements (especially copper) often increase during fall circulation and during winter. Although ionic concentrations increase somewhat in winter, most of the peak is in the organic fractions (Riley, 1939; Kimball, 1973). In general, the activity of zinc (Bachmann, 1963) and cobalt (Parker and Hasler, 1969; Benoit, 1957) follows the conclusions summarized above. Molybdenum exhibits much greater mobility than the other minor elements discussed, and does not show analogous patterns of distribution (cf. Dumont, 1972; Cowgill, 1977b).

In conclusion, it should be noted that the inputs of many trace elements to fresh waters are increasing. These increases often result from industrial and combustion emissions, which are subsequently deposited onto the drainage basins of rivers and lakes (e.g., Nriagu, 1979; Nriagu and Davidson, 1980). In other cases, the widespread acidification of rain and snow falling on poorly buffered soils and fresh waters can result in high rates of leaching of trace metals. Acidic precipitation is causing high concentrations of aluminum to be leached from poorly buffered soils in many continental regions (Almer, et al., 1978; Cronan and Schofield, 1979). Concentrations of aluminum often become so great that much of the fauna is eliminated. Moreover, as lakes become more acidic from anthropogenic pollution, the concentrations of many heavy metals increase as their solubilities increase and as sedimentary binding mechanisms decline (e.g., Jackson, et al., 1980).

The Sulfur Cycle

Sulfur is utilized by all living organisms in both inorganic and organic forms. Sulfate is reduced to sulfhydryl (−SH) groups in protein synthesis, with a concomitant production of oxygen that is utilized in oxidative metabolic reactions. Interest in the sulfur cycle of fresh waters, however, extends beyond nutritional demands of the biota, which are almost always met by the abundance and widespread distribution of sulfate, sulfide, and organic sulfur-containing compounds. Decomposition of organic matter containing proteinaceous sulfur and the anaerobic reduction of sulfate in stratified waters both contribute to altered conditions that markedly affect the cycling of other nutrients, ecosystem productivity, and distribution of the biota.

FORMS AND SOURCES OF SULFUR

Sources of sulfur compounds to natural waters include rocks, fertilizers, and atmospheric precipitation and dry deposition. At the present time, atmospheric sources, augmented greatly by combustion products of industry, dominate all other sources.

Sulfate is released during geochemical weathering of rocks and soils containing either sulfides or free sulfur, which are oxidized in the presence of water to form sulfuric acid (ZoBell, 1973):

$$FeS_2 + 3\tfrac{1}{2}O_2 + H_2O \rightarrow FeSO_4 + H_2SO_4$$
(pyrite)
$$2S + 3O_2 + 2H_2O \rightarrow 2H_2SO_4.$$

These two reactions tend to lower both the pH and E_h, which affects the oxidative weathering reactions of numerous other minerals. Calcium sulfate, which is moderately soluble in water, is a common constituent of sedimentary rocks, so drainage from calcareous regions generally contains higher than average concentrations of sulfate (Nriagu and Hem, 1978). As will be discussed further on, bacteria contribute to the oxidation of sulfides and elementary sulfur, both in soil and in water.

Large quantities of reduced sulfur, as hydrogen sulfide, are added in large quantities to the atmosphere from volcanic gases and biogenic and industrial sources (Kuznetsov, 1964; Kellogg, et al., 1972). H_2S undergoes a number of oxidative reactions to sulfur dioxide (SO_2), sulfur trioxide (SO_3), and sulfuric acid (H_2SO_4). Sulfur dioxide constitutes about 95 per cent of the sulfur compounds resulting from the burning of sulfur-containing fossil fuels. Although the oxidation of SO_2 to H_2SO_4 in air is slow (hours to days), SO_2 is rapidly oxidized to sulfuric acid as it dissolves in atmospheric water. The global cycling of sulfur compounds (Table 14–5) indicates clearly that in industrialized regions, sulfur inputs from the activities of man are rapidly exceeding inputs from natural sources. In eastern North America, for example, industrial emissions exceed natural ones by a factor of 10 (Table 14–6). The removal processes over land are sufficiently slow (several days) to result in markedly increased concentrations in areas hundreds to thousands of kilometers downwind. In nonindustrial areas, the primary source of sulfate (SO_4^-) in rain and snow is atmospherically oxidized H_2S that is pro-

TABLE 14–5 Sources, Sinks, and Residence Times of Atmospheric Sulfur Compounds (Units are 10^6 tons as sulfate per year.)

SOURCES	10^6 TON YR^{-1}	APPROXIMATE RESIDENCE TIME
Windblown sea salt SO_4^- in precipitation, 10% of global total deposited on all land	130	
Bacterial and plant production of H_2S (SO_2)	268	0.5–6 days
Man-made production of SO_2 and SO_4^- from fossil fuels (80% deposited on land; 93% of global production in the Northern Hemisphere)	150	0.5–6 days
Volcanic sources (H_2S, SO_2, SO_4^-)	2	<1 day
DEPOSITION		
Rain over oceans (SO_2, SO_4^-)	217	
Rain over land (SO_2, SO_4^-)	258	
Plant uptake (SO_2, SO_4^-)	45	
Dry fallout deposition (SO_4^-)	30	

From data of Kellogg, et al., 1972.

TABLE 14-6   Atmospheric Sulfur Budget for Eastern North America

COMPONENT	EASTERN CANADA	EASTERN U.S.A.	TOTAL, EASTERN NORTH AMERICA
	10^{12} g S YR^{-1}		
Inputs			
Man-made emissions	2.1	14	16.1
Natural emissions			
Sea spray	0.06	–	0.06
Terrestrial biogenic	0.06	0.04	0.10
Marine biogenic	0.2	0.4	0.6
Inflow from oceans	0.04	0.02	0.06
Inflow from west	0.1	0.4	0.5
Inflow to U.S. from Canada	–	0.7	–
Inflow to Canada from U.S.	2.0	–	–
TOTAL	4.6	15.6	17.4
Outputs			
Wet deposition	3.0	2.5	5.5
Dry deposition	1.2	3.3	4.5
Outflow to oceans	0.4	3.9	4.3
Outflow from Canada to U.S.	0.7	–	–
Outflow from U.S. to Canada	–	2.0	–
TOTAL	5.3	11.7	14.3

Modified from Galloway and Whelpdale (1980).

duced along coastal regions by anaerobic bacteria (Jensen and Nakai, 1961). The $SO_4^=$ derived from sea spray is largely returned to the ocean; over land, this source is minor and is generally restricted to coastal lakes (Table 14-6).

The relative contribution of sulfur compounds to natural waters of course varies with the regional lithology, agricultural application of sulfate-containing fertilizers, and atmospheric sources in relation to other sources. In calcareous areas of sedimentary rock, atmospheric contributions can be a small portion of the total. By contrast, in crystalline rock areas of northeastern America, precipitation supplies nearly all of the sulfate of natural waters (Fisher, et al., 1968; Kramer, 1978). Such acidic, sulfatic precipitation has significance in geochemical processes, especially weathering, equal to or greater than the carbonic acid system in these regions (Johnson, et al., 1972).

DISTRIBUTION OF SULFUR IN NATURAL WATERS

The sulfur content of biota ranges from 0.05 to nearly 5 per cent of dry weight in a few bacteria; the average content is 0.2 per cent. The amounts in the biota and detritus, although significant, are generally small in comparison to the inorganic sulfur components of aquatic systems. The distribution and cycling of sulfur therefore involve (a) the chemical species under various conditions, (b) biotic influences on the transformations of sulfur species, and (c) sulfur transport in the system.

The lowest concentrations of sulfate in oxic waters (often slightly less than 1 mg l^{-1}) appear to be a common feature of numerous African lakes situated in crystalline rock drainage basins (Talling and Talling, 1965). The other extreme is found in sulfate saline lakes (> 50 g l^{-1}). The usual range is about 5 to 30 mg l^{-1}, with an average of about 11 mg $SO_4^=$ l^{-1}. Low levels of $SO_4^=$ were implicated in the limitation of algal productivity in Lake Victoria, Africa (Fish, 1956), but more recently this role of $SO_4^=$ has been shown to be unlikely (Evans, 1961, 1962).

The predominant form of dissolved sulfur in water is sulfate (Fig. 14–7). Nearly all assimilation of sulfur is as sulfate, but during decomposition of organic matter, sulfur is released largely as hydrogen sulfide. Under oxic conditions, H_2S is oxidized rapidly. Therefore, little H_2S would be anticipated in aerated regions of aquatic systems. However, strict application of redox and pH to evaluate reactions involving sulfur is difficult, because some are chemically slow and mediated by bacterial metabolism. Although $SO_4^=$ and H_2S dominate, HS^- and very low concentrations of $S^=$ occur in strongly alkaline solutions, because H_2S, which is very soluble in water, dissociates weakly (k_1 10^{-7}; k_2 10^{-15}) (Hutchinson, 1957; Hem, 1960c). Under certain conditions of low redox and pH, partial oxidation of sulfides occurs and free $S°$ may be formed.

Metal sulfides are exceedingly insoluble at the neutral or alkaline pH values commonly encountered in a majority of natural waters. The equilibrium ion activity product of FeS of anaerobic lake sediments is $10^{-17.7}$ (Doyle, 1968b). Therefore, Fe^{+2} released from sediments reacts vigorously with H_2S to form FeS. In an anaerobic hypolimnion, the water must be somewhat acidic in order for appreciable H_2S to accumulate. If the water is alkaline, H_2S will accumulate only after most of the Fe^{+2} has been precipitated as FeS. The removal of sulfide by the release of ferrous ions permits an increase in the migration of other metals from the sediments, such as Cu, Zn, and lead (Pb), which form even more insoluble sulfides than does iron.

A generalized vertical distribution of sulfate and hydrogen sulfide for stratified lakes is depicted in Figure 14–8. Under oxic conditions, as is the case in many oligotrophic and mesotrophic lakes, and during periods of circulation, H_2S is absent and

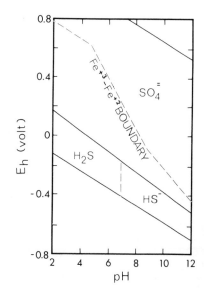

Figure 14–7. Approximate redox-pH fields of stability of dissolved sulfur species likely to occur in natural water. (Modified from Hem, 1960, with Chen and Morris, 1972.)

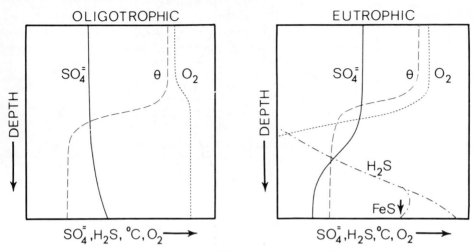

Figure 14-8. Generalized distribution of sulfate and hydrogen sulfide in lakes of very low and very high productivity.

SO_4^- concentrations change little with depth. Some release of SO_4^- occurs from the sediments, and this increase in sulfate in the hypolimnion can become more pronounced in hypolimnia of mesotrophic or eutrophic lakes in the earlier phases of summer stratification (Fig. 14-9). Reduction of sulfate to H_2S occurs as the redox potential declines to less than about 100 mv as a result of bacterial decomposition. Especially near the sediments, much of the H_2S reacts with Fe^{+2} ions to form insoluble FeS. In this way, considerable quantities of sulfur can be lost to the sediments. Lakes receiving rich sources of sulfate from inflowing water, such as meromictic lakes of crenogenic formation, often contain immense concentrations of H_2S in their anoxic monimolimnia. Analogous situations occur in certain anoxic stretches of rivers that are polluted with sulfate-rich organic wastes. Effluents from paper-producing industries are a common source of such pollution. Horizontal variations in concentrations of sulfate and sulfides can be large, especially in reservoirs, and are complicated by flow patterns (e.g., Hanušová, 1962).

The reduction of SO_4^- to sulfide, some of which is lost to the sediments as insoluble metallic sulfides, and oxidation of H_2S to sulfate, play a significant role in the modification of conditions for mobilization of phosphate and numerous other nutrients. In an excellent discussion of these reactions and their mediation by various bacterial groups, Ohle (1954) characterizes sulfate as a 'catalyzer' of limnetic nutrient cycling.

The sulfur cycle in Linsley Pond, Connecticut, was studied during summer stratification in appreciable detail using ^{35}S-labeled sulfuric acid (Stuiver, 1967). In this eutrophic lake, large quantities of sulfate were lost from the metalimnion and hypolimnion, especially at the sediment–water interface. The rates of reduction of sulfate in the metalimnion and hypolimnion differed little between regions that were anoxic or partly devoid of oxygen, but were about 10 times faster than in the fully aerobic epilimnion. Vertical diffusion of sulfate in the metalimnion and upper layers of the hypolimnion was very small during stratification. Horizontal diffusion at a given depth within

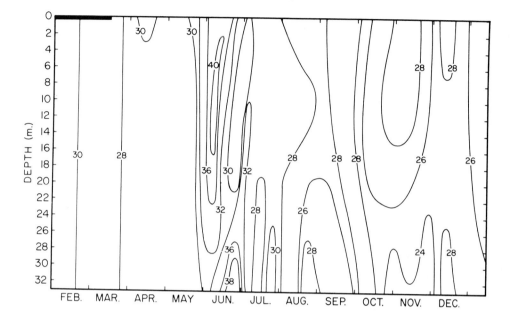

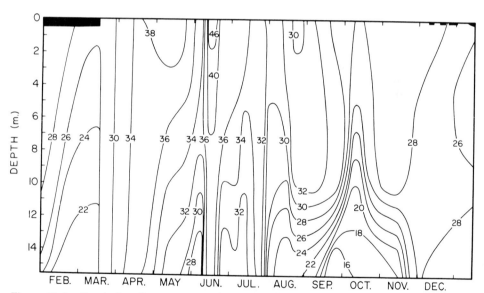

Figure 14-9. Depth-time diagrams of isopleths of sulfate concentrations (mg l^{-1}) of mesotrophic hardwater Crooked Lake *(upper)* and interconnected eutrophic Little Crooked Lake *(lower)*, Noble-Whitley counties, northeastern Indiana. Opaque areas = ice cover to scale. (From Wetzel, unpublished data.)

the lower strata of water was sufficient to transport sulfates to the surrounding sediments, where reduction occurs. Transport of sulfur by sedimenting biological material was negligible, and influenced the sulfur cycle of this lake only by providing organic substrates for bacterial metabolism in the hypolimnion and sediments.

This analysis permitted calculation of a total sulfur budget, estimated from the ^{35}S activities of the water and organic compounds (Fig. 14–10). Most of the sulfur was stored as sulfate and sulfide in the water, and as sulfide in the sediments. That fraction utilized by organisms was small, and did not materially influence the total cycle. The H_2S in the hypolimnion was oxidized in the epilimnion; escape of H_2S to the atmosphere by diffusion or gas bubbles was low.

Organic sulfur compounds were not studied extensively in this budgetary analysis. Most of the sulfur of the seston occurs as ester sulfates and protein sulfur. The bulk (up to 80 per cent) of sulfur in the sediments of productive lakes consists of organic sulfur compounds (ester sulfates, protein sulfur), and the remainder as pyritic sulfur, acid-volatile sulfides, sulfides dissolved in interstitial water, elemental sulfur, and dissolved sulfates (Table 14–7) (Doyle, 1968a, 1968b; Nriagu, 1968; King and Klug, 1980, 1982; Mitchell, et al., 1981; Smith and Klug, 1981). At least 40 per cent of the acid-soluble sulfide of Linsley Pond was identical to tetragonal FeS (mackinawite).

Sulfur-containing organic compounds are degraded more slowly than other organic compounds. In Lake Mendota, Wisconsin, about 45 per cent of the sulfur precipitated as sulfide was estimated to be derived from mineralization of organic matter, and the remainder (55 per cent) originated from bacterial reduction of sulfates (Nriagu, 1968). Estimates of net mineralization of sestonic sulfur inputs to the sediments in hypereutrophic Wintergreen Lake, Michigan, indicated that only about 45 to 50 per cent of the total and ester sulfate sulfur inputs, and 75 per cent of the protein sulfur inputs, were mineralized (King and Klug, 1982). About 3 per cent of the total water column sulfur was permanently lost to the sediments each year.

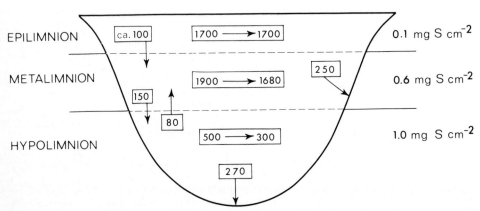

Figure 14–10. The total sulfate budget of Linsley Pond, Connecticut, estimated for an early stage of thermal stratification *(lefthand numbers in central squares)* and final stages *(righthand numbers in central squares)*. The other numbers give the transfer rates of sulfate, in kg S, between the different strata. At the right is the total amount of sulfate reduced and stored per cm² of sediment during the 4-month period. The total amount of dissolved H_2S at the end of stratification was about 15 kg S in the hypolimnion. (After Stuiver, 1967.)

TABLE 14-7 Distribution of Forms of Sulfur in Surface Sediments (0-10 cm)

SULFUR FORM	CONCENTRATION (moles)	PERCENTAGE OF TOTAL	SITE	PERCENTAGE ESTER SULFATE OF TOTAL
Wintergreen Lake, Michigan*			Wintergreen Lake, Michigan†	
Total sulfur	204	–	Littoral	22.1
Ester sulfate sulfur	65	35	Littoriprofundal	34.2
Pyritic sulfur	21	10	Profundal	32.3
Acid volatile sulfur	8	4	Oneida Lake, New York‡	80
Elemental sulfur	3	1	South Lake, New York‡	45
Dissolved sulfide	2	1	Deer Lake, New York‡	79
Dissolved sulfate	0.4	0.2		

*Smith and Klug, 1981; King and Klug, 1982.
†King and Klug, 1980.
‡Mitchell, et al., 1981.

BACTERIAL METABOLISM AND THE SULFUR CYCLE

Sulfate is reduced to the sulfhydryl ($-SH$) form during the synthesis of proteins by plants and animals. Further reduction of HS^- to H_2S occurs upon decomposition of this organic material by heterotrophic bacterial metabolism (Fig. 14–11). The most important of a large number of protein-decomposing bacteria belong to the genus *Proteus*, which are gram-negative, nonsporeforming rods that usually possess large numbers of flagella (Butlin, 1953). *Proteus* species are particularly active in soil systems. The dominant bacteria involved in proteinaceous decomposition to form H_2S in lakes of varying

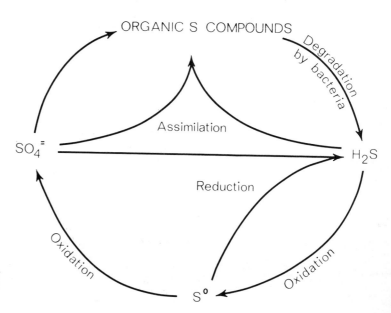

Figure 14-11. General sulfur cycle in nature. (Modified from Butlin, 1953.)

productivity (Table 14–8) are discussed in Kuznetsov (1970) and Zinder and Brock (1978). Bacterial densities in the water are 1 to 3 orders of magnitude lower than in the surface sediments.

A number of bacteria reduce sulfate, sulfite, thiosulfate, hyposulfite, and elemental sulfur to hydrogen sulfide. These *sulfate-reducing bacteria* are heterotrophic and anaerobic, and use the sulfur compound as a hydrogen acceptor during oxidative metabolism. Finally, several groups of bacteria oxidize sulfide to sulfur and sulfur to sulfate.

The *sulfur-reducing bacteria* of the genera *Desulfovibrio* and *Desulfotomaculum* are strictly anaerobic, and derive oxygen from sulfate for the oxidation of either organic matter or molecular hydrogen:

$$H_2SO_4 + 2(CH_2O) \rightarrow 2CO_2 + 2H_2O + H_2S$$
$$H_2SO_4 + 4H_2 \rightarrow 4H_2O + H_2S \ (\Delta G_0' = -60 \text{ kcal mole}^{-1})$$

While these reactions do not consume oxygen directly, the H_2S generated by sulfate-reducing bacteria is readily oxidized, and consumes oxygen upon intrusion into, or transport to, aerobic regions. The sulfate-reducing bacteria can reduce sulfite more rapidly, and thiosulfate less rapidly, than sulfate. Colloidal sulfur, but not pure noncolloidal sulfur, is reduced very slowly. Over a concentration range of 20 to 130 mg SO_4 l^{-1}, the rate of production of H_2S by bacteria is roughly proportional to SO_4 concentration (Ohle, 1954). The biological oxygen demand of oxidizable organic matter can theoretically be satisfied with about 1.6 g SO_4 g^{-1} by this reduction.

The *sulfur-oxidizing bacteria* are commonly differentiated into two groups. The *chemosynthetic (colorless) sulfur-oxidizing bacteria* are mostly aerobic forms that oxidize H_2S, and are of two types. The first deposits sulfur *inside* the cell:

$$H_2S + \frac{1}{2}O_2 \rightarrow S^\circ + H_2O (\Delta G_0' = -41 \text{ kcal mole}^{-1}),$$

which accumulates as long as H_2S is available. As sulfide sources are depleted, the internally stored sulfur is oxidized and sulfate is released:

$$S^0 + 1\frac{1}{2} O_2 + H_2O \rightarrow H_2SO_4 (\Delta G_0' = -118 \text{ kcal mole}^{-1})$$

TABLE 14–8 Predominant Bacteria that Form Hydrogen Sulfide from the Decomposition of Proteinaceous Organic Matter in Different Types of Lakes

TYPE OF LAKE	DOMINANT BACTERIA
Oligotrophic	*Mycobacterium phlei, Mycobacterium filiforme*
Mesotrophic	*Bacterium nitrificans, Pseudomonas liquefaciens, Chromobacter aurantiacum*
Eutrophic	*Pseudomonas liquefaciens, Bacterium delicatum*
Bog lakes of high organic matter	*Pseudomonas fluorescens, Bacillus pituitans*
Saline lakes and estuaries	*Mycobacterium luteum, Micrococcus nitrificans, Achromobacter halophilum, Flavobacterium halophilum, Bacterium albo-luteum, Vibrio hydrosulfureus*

After data of Kuznetsov, 1970.

Beggiatoa, a long, filamentous bacterium, and *Thiothrix* are common bacteria that oxidize H_2S with deposition of sulfur intracellularly. These two genera occur in areas where H_2S is being formed, e.g., canals, swamps, and sulfur springs.

By similar reactions, a second type of chemosynthetic sulfur-oxidizing bacteria deposits sulfur *outside* of the cell. This assemblage is represented best by the genus *Thiobacillus*, which oxidizes sulfide, $S°$, and other reduced sulfur compounds such as thiosulfate:

$$2 Na_2S_2O_3 + O_2 \rightarrow 2 S° + 2 Na_2SO_4$$

Of the many species of *Thiobacillus*, some (e.g., *T. thiooxidans*) are restricted to acidic waters (pH 1 to 5), while others such as *T. thioparus* grow optimally at neutral or alkaline pH values. The anaerobe *T. denitrificans* oxidizes thiosulfate in alkaline waters by reduction of nitrate to N_2:

$$5 S_2O_3^= + 8NO_3^- + 2HCO_3^- \rightarrow 10 SO_4^= + 2 CO_2 + H_2O + 4 N_2$$

or,

$$5 S° + 6 NO_3^- + 2 CO_3^= \rightarrow 5 SO_4^= + 2 CO_2 + 3 N_2 (\Delta G_0' = -179 \text{ kcal mole}^{-1}).$$

The elemental sulfur-oxidizing bacteria commonly adhere to sulfur granules, continuously utilizing a little at a time in the formation of sulfate.

The other major group of sulfur-oxidizing bacteria is the *photosynthetic (colored) sulfur bacteria*, anaerobes that can be divided conveniently into the *green sulfur bacteria* (Chlorobacteriaceae) and the *purple sulfur bacteria* (Thiorhodaceae). Excellent detailed reviews of the photosynthetic bacteria and their metabolism are given by Gest, et al. (1963), Vernon (1964), Kondrat'eva (1965), Pfennig (1967), and Fjerdingstad (1979). The green sulfur bacteria require light as an energy source, and use sulfur of H_2S as an electron donor in the photosynthetic reduction of CO_2:

$$CO_2 + 2 H_2S \xrightarrow{\text{light}} (CH_2O) + H_2O + 2S$$

$$2CO_2 + 2H_2O + H_2S \xrightarrow{\text{light}} 2(CH_2O) + H_2SO_4$$

A few species can utilize molecular hydrogen alone. The green sulfur bacteria, notably the genera *Chlorobium* and *Pelodictyon*, are generally unicellular, nonmotile, and produce sulfur granules outside of their cell membranes. At least four bacteriochlorophylls occur in the photosynthetic bacteria which differ from chlorophyll *a* in a primary absorption maximum at higher wavelengths (770 to 780 nm vs. 665 nm for chlorophyll *a*). The green sulfur bacteria can tolerate fairly high concentrations of H_2S, whereas the purple sulfur bacteria are less tolerant of H_2S and grow optimally at high pH values (9.5).

The *purple sulfur bacteria* require light energy for the oxidation of H_2S and other reduced sulfur compounds, especially thiosulfate, to sulfate in the photosynthetic reduction of CO_2. Members of this group are generally large (5 to 10 μm), actively motile, and deposit free $S°$ intracellularly. Sulfide is oxidized to S and $SO_4^=$ by the same reactions described for the green sulfur bacteria. In addition, some of the purple sulfur bacteria are able to grow photoautotrophically, with thiosulfate as the electron donor (*Thiopedia, Thiocapsa, Thiocystis, Rhodothece*). Other important genera (*Chromatium, Chlorobium, Thiospirillum*) are unable to utilize significant amounts of thiosulfate. Like the green sulfur bacteria, many purple bacteria can utilize hydrogen as the only electron acceptor, simultaneously with an assimilatory sulfate reduction system.

The Calvin cycle has been shown to be operational in all photosynthetic bacteria studied. Many strains can utilize low-molecular-weight organic substrates, especially fatty acids, as their carbon source, singly or in combination with CO_2. Nearly all green and purple sulfur bacteria require vitamin B_{12} from exogenous sources. This vitamin, as will be discussed further on, is an organic micronutrient that influences photosynthetic productivity under some limnological conditions.

A third group, the *purple nonsulfur bacteria* (Athiorhodaceae), is included here among the other pigmented photosynthetic bacteria because of its many metabolic and distributional similarities. The nonsulfur purple bacteria, including *Rhodopseudomonas, Rhodospirillum*, and *Rhodomicrobium*, are facultative photoautotrophs that grow photosynthetically or heterotrophically either aerobically or anaerobically in the dark on organic substrates. Some *Rhodopseudomonas* can utilize thiosulfate anaerobically as a hydrogen donor:

$$\text{light}$$
$$2\ CO_2 + Na_2S_2O_3 + 3\ H_2O \rightarrow 2(CH_2O) + Na_2SO_4 + H_2SO_4$$

In this group, sulfur is not stored intracellularly, and hydrogen sulfide inhibits growth.

The occurrence and distribution of the various sulfur-oxidizing or -reducing bacteria are restricted by the redox and pH conditions in relation to oxygen and the state of sulfur compounds (Fig. 14–12). The reducing conditions required by strictly anaerobic photosynthetic sulfur bacteria, for example, must coincide with adequate light of high wavelength before large populations can develop. Often conditions required for optimal growth of sulfur bacteria occur in stratified lakes as sharply defined layers with steep physical and chemical gradients, and result in thin layers or strata of bacterial populations.

Microbial processes involved in the sulfur cycle of lakes are diagrammatically represented in Figure 14–13. The processes on the lefthand side of the figure would be more characteristic of a lake with relatively high concentrations of sulfur in various forms. The gradient between the oxic upper strata and the lower H_2S-rich strata would be steep, but with a diffusion interface zone where both oxygen and H_2S occurred. Those processes on the righthand side of Figure 14-13 are more representative of lakes with lower sulfate content.

The rather specific requirements of the sulfate-reducing bacteria, photosynthetic and colorless sulfur bacteria, and sulfur-oxidizing bacteria would lead one to expect development of massive populations only in localized strata. Such is indeed the case in many situations, particularly in meromictic lakes where gradients within the chemolimnion are steep. Green sulfur bacteria are commonly found in profusion in a thin layer immediately below a dense population of purple sulfur bacteria at the interface of the oxic-anoxic layer of very productive lakes. Light levels at this region are almost always

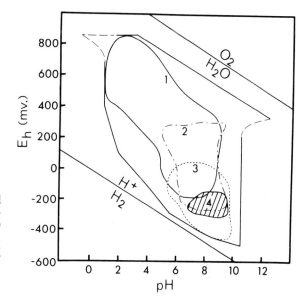

Figure 14-12. General E_h-pH environmental limits of *1*, chemosynthetic (colorless) sulfur-oxidizing bacteria, *2*, photosynthetic purple bacteria, *3*, sulfate-reducing bacteria, and *4*, green sulfur bacteria, all within the composite distributional field of E_h-pH measurements of habitats of organisms. (After data of Baas Becking, et al., 1960.)

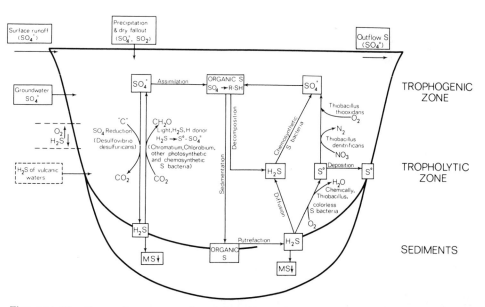

Figure 14-13. Composite representation of the sulfur cycle in a lake, with emphasis on the microbiological processes. *MS* = metallic, primarily iron, sulfides. (Greatly modified from Kuznetsov, 1970.)

TABLE 14-9 Comparison of Algal and Bacterial Photosynthetic Rates and Rates of Bacterial Chemosynthesis in Several Lakes (PS = photosynthesis)

LAKE	UNITS OF MEASUREMENT	ALGAL PS	BACTERIAL PS	BACTERIAL PS/ALGAL PS × 100 (%)	CHEMOSYNTHESIS	CHEMOSYNTHESIS/TOTAL PS × 100 (%)	REMARKS
Hiruga, Japan 21 July 66 (Takahashi and Ichimura, 1968)	mg C m^{-2} hr^{-1}	29.7	0.9	3.0	4.3	14.1	
Waku-ike, Japan 4 Aug. 65	mg C m^{-2} hr^{-1}	19.8	1.8	9.1	—	—	
Kisaratsu, Japan 16 Aug. 66	mg C m^{-2} hr^{-1}	22.1	128	579	3.5	2.3	
2 Sept. 65		12.0	10.6	88	2.5	11.1	
Smith Hole, Indiana Annual period 1963–1964 (Wetzel, 1973)	annual mean mg C m^{-2} day^{-1}	194	91	47	6	<10	Massive late summer population of *Chromatium*
14 Aug. 63 Wadolek, Poland	mg C m^{-2} day^{-1}	540	5960	1104	—	—	
June 66	mg C m^{-2} day^{-1}	32.8	55.3	169	—	—	*Chlorobium*
July 66		144.5	19.4	13	—	—	in metalimnion
Sept. 66 (Czeczuga, 1967, 1968a, 1968b)		24.6	19.1	78	—	—	
Muliczne, Poland 19 Aug. 67	mg C m^{-2} day^{-1}	478	157	32.8	—	—	*Thiopedia* in hypolimnion
16 Sept. 67		258	136	52.7	—	—	
20 Oct. 67 (Czeczuga, 1968c)		281	28	10.0	—	—	
Belovod, USSR July (Sorokin, 1970)	mg C m^{-2} day^{-1}	500	55	11	15	2.7	
13 m	mg C m^{-3} day^{-1}	0	210	—	74	35	
14 m		0	79	—	44	56	
18 m		0	2.7	—	16.4	607	
24 m		0	0	—	5.5	—	

low, usually less than 10 per cent of intensities at the surface. The seasonal development of optimal conditions for the specific groups of sulfur bacteria can be transitory, so that their contribution to the total annual productivity of the lake may be short-lived. Within meromictic lakes, bacterial plates can persist more or less continuously (cf. Guerrero, et al., 1980; Abella, et al., 1980).

While descriptions of the distribution of the sulfur bacteria in aquatic systems are fairly common, and much is known about the physiology of these interesting organisms, little is known of their contribution to the total productivity of lakes. It is clear that at certain periods in productive dimictic and in meromictic lakes, bacterial photosynthesis easily can exceed that of algae and macrophytes (Table 14-9). Most of the data given in Table 14-9, however, are taken from periods when the photosynthesis of bacteria was maximal. Occasionally, when the algal photosynthesis is low, as in the example of Smith Hole Lake, which receives a high proportion of allochthonous organic matter, the brief but very productive photosynthetic bacterial development represents a major part of the lake's annual photosynthetic productivity (Wetzel, 1973). In most situations, however, the contribution of bacterial photosynthesis to the entire system over an annual period is small, even though spectacular localized populations may develop periodically.

Rates of bacterial sulfur metabolism in situ in aquatic systems have been studied in only a few cases, (e.g., Sorokin, 1964, 1966, 1970). In several systems, as illustrated in Figure 14-14 for Lake Gek Gel, an oligotrophic meromictic lake of crenogenic origin, and in the deep, meromictic Black Sea, the most intensive rates of bacterial reduction of sulfates occurred near sediments and at the water layers on the littoral slopes at the upper boundary of the anaerobic H_2S zone. The higher rates of sulfate reduction in these cases are apparently related to higher inputs of organic matter from the littoral zone brought in allochthonously from the drainage basin. Higher H_2S concentrations along these sediment interfaces are then dispersed by weak water currents to the open water (Sorokin, 1970). Furthermore, high rates of sulfate reduction occur below the zone of most active bacterial chemosynthesis (Fig. 14-14), where again the concentrations of organic substrates are presumably higher.

A similar dichotomous distribution of sulfate-reducing bacteria below the zone of most active chemosynthesis is seen even more clearly from data of Lake Belovod (Fig. 14-15). The major bacteria in the zone of chemosynthesis were several species of *Thiobacillus*, over which lay a dense population of purple sulfur bacteria. Sorokin has provided evidence on the movements of zooplankton populations in response to changes in bacteria stratification, and studies of feeding indicate that zooplankton, especially the cladoceran microcrustacea, actively feed on the dense populations of bacteria.

Rates of sulfate reduction within the sediments are variable among different lakes (Table 14-10) but are generally rather low in comparison to those within the overlying water. Sediment sulfate concentrations varied little over time, which suggests a constant turnover time for the sulfate (Smith and Klug, 1981). Organic sulfate esters, entering the sediments from settling seston, form a major source of sulfur; they are mineralized by an active sulfhydrolase system in the surficial sediments (King and Klug, 1980, 1982).

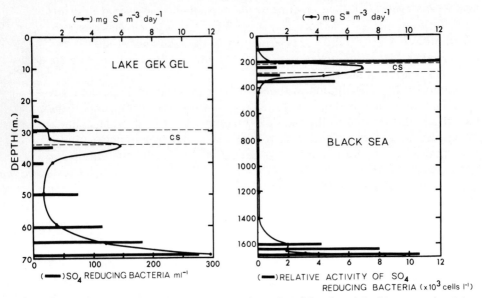

Figure 14-14. Rates of sulfate reduction (mg H_2S formed m^{-3} day^{-1}) and the biomass and activity of sulfate-reducing bacteria in meromictic Lake Gek Gel and Black Sea, USSR. CS = zone of active chemosynthetic bacterial metabolism. (After data of Sorokin, 1964, 1970.)

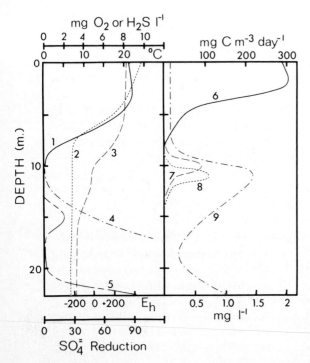

Figure 14-15. Distribution of midsummer characteristics and intensity of biological processes in the central depression of meromictic Lake Belovod, USSR. (After Sorokin, 1970.) 1, oxygen, mg l^{-1}; 2, °C; 3, E_h in mv; 4, H_2S, mg l^{-1}; 5, rate of sulfate reduction, mg H_2S formed m^{-3} day^{-1}; 6, photosynthesis by algae, mg C m^{-3} day^{-1}; 7, chemosynthesis, mg C m^{-3} day^{-1}; 8, photosynthesis by purple sulfur bacteria, mg C m^{-3} day^{-1}; 9, biomass of bacteria, mg l^{-1}.

TABLE 14-10 Rates of Sulfate and Elemental Sulfur Reduction in Various Sediments

LAKE	SO_4^{-2} CONCENTRATION (mg l^{-1})	REDUCTION RATE (mg H_2S $l^{-1}d^{-1}$)	SOURCE
Linsley Pond, Connecticut	7.5–17.5	0.8–1.5	Stuiver, 1967
Lake Beloe, USSR	48–205	$2-6 \times 10^{-4}$	Ivanov, 1968, in Zinder and Brock, 1978
Gor'kii Reservoir, USSR	32–132	0.04–2.94	
Lake Belovod, USSR	428–625	0.067–0.124	
Wintergreen Lake, Michigan	6–48		Molongoski and Klug, 1980a
Sulfate reduction		0.6–5.8	Smith and Klug, 1981
Elemental sulfur reduction		0.28	

The Silica Cycle

Silica (SiO_2) is usually moderately abundant in fresh waters and, although it is relatively unreactive, it is of major significance to diatomaceous algae. Diatoms assimilate large quantities of silicon in the synthesis of their frustules. Silicon is a major factor influencing algal production in many lakes, and diatom utilization of silica greatly modifies the flux rates of silica in lakes and streams. Availability of silica can have a strong influence on the overall pattern of algal succession and productivity in lakes.

FORMS AND SOURCES OF SILICON

Silicon occurs in fresh waters in two major forms of silicon dioxide or silica (SiO_2). (1) *Dissolved silicic acids* form stable solutions of H_4SiO_4 at much higher concentrations than are encountered in fresh waters (60 to 80 mg SiO_2 l^{-1} at 0°C to 100 to 140 mg SiO_2 l^{-1} at 25°C at commonly occurring pH values; Krauskopf, 1956). Unreactive silicon is not generally found; polymeric silicon is unstable and depolymerizes rather rapidly (within hours) (Burton, et al., 1970). Solution of silica from various rock sources is modified, however, by surface adsorption of silicic acid, which reduces solubility and leads to a general situation in which nearly all natural waters are greatly undersaturated with respect to silica (Stöber, 1967; Tessenow, 1966). (2) *Particulate silica* is found in two forms—that in biotic material, in particular in diatoms and a few other organisms that use large amounts of silica, and that adsorbed to inorganic particles or complexed organically. Silicate complexes with iron and aluminum hydroxides which decrease the solubility of silicates in sediments, especially in interstitial waters at pH values above 7 (Ohle, 1964). Solubility of silica is increased by humic compounds and through the formation of iron and aluminum-silicate-humic complexes.

The silica content of drainage to natural waters is less variable than many of the other major inorganic constituents. The world average is about 13 mg SiO_2 l^{-1}, with relatively little variation among the continents; the average of groundwater is somewhat higher than that of surface drainage (Davis, 1964). The major source of silica is from the degradation of aluminosilicate minerals. Greatest concentrations of silica are found in groundwater in contact with volcanic rocks; intermediate amounts occur in association with plutonic rocks and sediments containing feldspar and volcanic rock fragments; and small amounts originate from marine sandstones. The lowest amounts of silica are found in water draining from carbonate rocks. Silica forms aggregations and becomes relatively immobile at pH values below 3, but its mobility increases somewhat in the range of pH 4 to 9. Adsorption is the only significant mechanism of inorganic precipitation; flocculation of colloidal silica could not be demonstrated (Tessenow, 1966).

Carbonic acid originating from dissolved CO_2 reacts with silicates to form carbonates and silica. Concentrations of dissolved silica in soil water increase slightly with higher temperature and decrease with increasing soil pH values (McKeague and Cline, 1963a, 1963b) as adsorption increases within pH 4 to 9. Above a pH of 10, adsorption decreases sharply. Mineralization of silicates is presumed to be largely or entirely a nonenzymatic hydrolysis (Golterman, 1960). However, silicate bacteria are known to play a role in the weathering of rocks and minerals of arid tropical regions in the absence of organic substrates (Savostin, 1972), and benthic-living diatoms are known to attack silicious materials of sediments. Mechanisms of this mineralization are unclear, however.

The silica content of river waters tends to be remarkably uniform and shows little response to change in discharge rates (Edwards and Liss, 1973). This situation is in distinct contrast to other major constituents of river water, which commonly show an inverse relationship between concentration and discharge rates. Although rapid diurnal changes in the silica content of river waters are known and can be associated with division rates and growth of diatoms (Müller-Haeckel, 1965), these and other biological factors are insufficient to explain the relative stability of silica concentrations. An abiological buffering mechanism is apparently also operational, with adsorption reactions between dissolved silica and silica in the solid phase (i.e., in association with hydrated oxides). Adsorption and desorption equilibria can buffer changes in concentration over a period of several days. The effects of dissolved silica and degraded silicates assume much more significance in the oceans, where reactions occur in which silica and alkali metal cations are fixed. Hydrogen ions are released, and result in an effective buffering of both pH and silica concentrations (Garrels, 1965; Mackenzie et al., 1965, 1967).

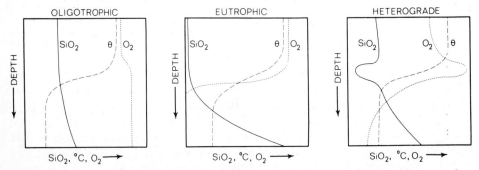

Figure 14-16. Generalized distribution of silica concentrations in unproductive and very productive lakes, and in a lake exhibiting a metalimnetic development of diatom algae and a negative heterograde silica curve.

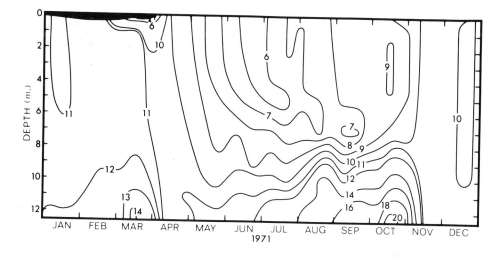

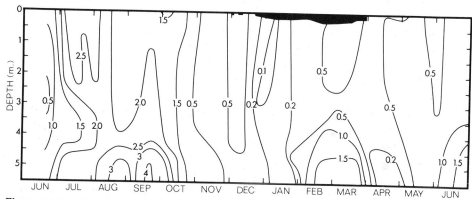

Figure 14-17. Depth-time diagram of isopleths of silica concentrations (mg SiO_2 l^{-1}) in oligotrophic hardwater Lawrence Lake, 1971 (upper), and hypereutrophic Wintergreen Lake, 1971-72 (lower), southwestern Michigan. Opaque areas = ice-cover to scale. (From Wetzel, unpublished data.)

DISTRIBUTION OF SILICA IN LAKES

Concentrations of silica within lakes frequently exhibit marked seasonal and spatial variations. Even in oligotrophic waters, a conspicuous decrease in silica is often found in the epilimnetic strata during early winter, as well as in the spring during circulation and during thermal stratification (Fig. 14-16). In eutrophic lakes, silica concentrations in the trophogenic zone are commonly near analytical undetectability. Reductions in silica concentrations within the epilimnetic and metalimnetic zones result in a negative heterograde silica curve against depth (Figs. 14-16 and 14-17). The heterograde silica distribution is clearly associated with intensive assimilation of silica by diatoms, which sediment from the trophogenic zone more rapidly than silica is replaced by inputs to the system from surface water and groundwater. Other sinks of silica are minor relative to quantities transported by diatoms. Abiogenic precipitation in the open water is relatively unimportant in the cycle of silica. Dilution by water low in silica can occur, as for example, by rain percolating through decaying ice (March–April, Lawrence Lake, Fig. 14–17).

The seasonal cycle of silica, demonstrated in Figure 14–17, has been observed frequently. By far the most detailed work on experimental investigations of mechanisms controlling silica dynamics in lakes is that of Tessenow (1966), from which many of the following statements are drawn. Utilization of silica by diatoms occurs during photosynthesis, and increases somewhat in darkness. Adsorption of SiO_2 to dead cells under some conditions can lead to reduction of silica content even below the trophogenic zone. Silica concentrations usually increase in the tropholytic zone of a lake during both summer and winter periods of stratification. The silica gradient becomes more steep in eutrophic lakes that exhibit an anaerobic clinograde oxygen curve. In most stratified lakes, silica concentrations increase in water immediately above the sediments (Fig. 14–17; cf. Conway, et al., 1977).

The amorphous silica of diatoms often settles to the sediments. This biochemical condensation of dissolved silica and sedimentation greatly exceeds inputs to the sediments from abiogenic sources, and reaches the sediments with variable periodicity. Diatom production is typically greatest in the spring and early winter, but often proceeds more or less continuously, particularly in oligotrophic lakes.

> Interstitial water is enriched in dissolved silica at concentrations far in excess of those of water entering the lake (Tessenow, 1966; Harriss, 1967). Interstitial concentrations increase as the pH declines below 7, decrease between pH 7 to 9, and greatly increase above pH 9. Concentrations also increase at higher temperatures within the range found in fresh waters. The dissolved silica of interstitial water is not in equilibrium with amorphous silica, but rather with chemically bound or adsorbed silica. Silica concentration of the interstitial water is controlled by dissolution of ferroaluminum silicate. This complex is formed in the sediments by the reaction of silica from diatoms with aluminum and ferric oxyhydroxides, or through the hydrolysis of clay minerals (Nriagu, 1978). Liberation of silica to the overlying water in relatively isolated lake sediments is governed by these equilibria, which are stable only in the presence of the solid phase of adsorbed silica. Exchange between sediment and water decreases concentrations of interstitial water and results in greater redissolution from the sediments. The rate of silica release from the sediments is influenced by temperature and the difference in silica concentrations between the sediments and the overlying water. Equilibrium is not attained because of the slowness of diffusion (weeks). The difference between silica of interstitial water and in the overlying water is influenced by currents, movements produced by benthic organisms (e.g., larvae of chironormid insects; Tessenow, 1966), and by gas bubbles escaping from the sediments.

In lakes dominated by diatom algae, large numbers of sedimenting diatom frustules can accumulate within the sediments and be lost permanently from the system. The extent of this permanent loss depends on rates of diatom productivity, on the morphometry of the lake basin and upon the percentage of the sediments located in the quiescent waters of the deep hypolimnion. In deep lakes, many of the diatom frustules undergo partial dissolution before reaching the sediments. The dissolution of suspended silica can be accelerated by consumption and fragmentation of diatom frustules by zooplankton, a process that is apparently of significance in both productive shallow ponds and in deep lakes (e.g., Ferrante and Parker, 1978). Silica sedimented from biogenic sources to shallower sediments is returned more rapidly to the overlying and circulating waters. Therefore, the metalimnetic and upper hypolimnetic waters that experience greater movement than the deeper waters are commonly enriched with silica in relation to strata above and below. A resulting positive heterograde silica curve has often been observed (numerous examples are given in Tessenow, 1966). The increased exchange induced by circulation over littoral sediments in shallow lakes can lead to silica concentrations that are sometimes greater than those of drainage inlet sources, and can, in addition, result in silica losses from the system by outflow. The silica cycle and economy of most lakes are regulated largely by autochthonous metabolism within the lake, but losses are balanced strongly by allochthonous inputs.

DIATOM UTILIZATION AND ROLE OF SILICA

Silicified structures occur in many aquatic organisms, but none approaches the importance of the diatoms (Bacillariophyceae). All diatoms are enclosed in a silica wall or frustule in which silicic acid has been dehydrated and polymerized to form silica particles (J. C. Lewin, 1962). The vegetative cells of some species of yellow-brown algae (Chrysophyceae) bear discrete siliceous scales and form cysts with silicified walls. These algae, as well as certain silicoflagellates, however, probably have only an insignificant impact on silica cycling in comparison to the active utilization and more extensive distribution of the diatoms. Similarly, utilization of silicon by certain aquatic macrophytes (*Equisetum*) and siliceous sponges, while of major importance to the organisms per se and their development and productivity, is rarely sufficient to alter the quantitative cycling of silica in lakes.

The succession and productivity of algal populations will be discussed in detail further on, but it is appropriate here to stress the importance of diatoms on the silica cycle and their effects on their own population dynamics. Of all of the aspects of chemical determination of succession and productivity, the relation between diatoms and silica concentrations is among the most apparent. The data, upheld by the detailed investigations of Lund (1949, 1950, 1954, 1955; Lund, et al., 1963; Heron, 1961), are irrefutable, and have been corroborated by experimental work (cf. Chapter 15).

Phytoplankton in temperate waters usually undergo a spring maximum. This population development may begin beneath the ice as light conditions improve, but is most conspicuous during and following spring circulation when the water is relatively rich in nutrients as the winter accumulations are mixed throughout the water column. One or several algal species usually dominate this exponential growth maximum for several weeks, and in a large number of lakes, diatoms constitute the predominant algae of the spring maximum. Increasing light, and to a lesser extent a rise in water temperatures, are major factors initiating the development of diatom populations from small residual planktonic populations. Circulation and turbulence are much higher in the spring than later in the season, and assist in maintaining the relatively dense diatom cells in optimal

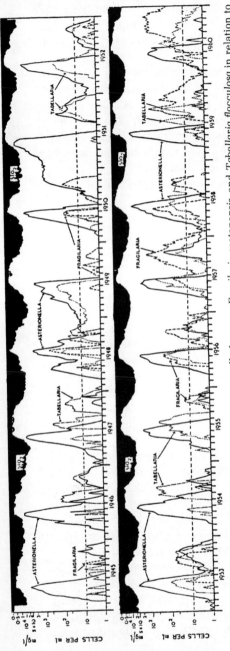

Figure 14-18. The periodicity of the diatom algae *Asterionella formosa, Fragilaria crotonensis,* and *Tabellaria flocculosa* in relation to fluctuations in the concentration of dissolved silica, 0–5 m in Lake Windermere, England, 1945–1960. (From Lund, J. W. G.: Verhandlungen Int. Ver. Limnol., 15:37, 1964.)

intensities of light. Other factors influencing the development of succession and development of the algal populations will be treated later (Chapter 15).

The diatom *Asterionella* commonly precedes other diatoms such as *Cyclotella*, *Fragilaria*, and *Tabellaria* in the spring maximum. Collectively, the spring maximum often declines abruptly as the silica concentrations fall below 0.5 mg l^{-1} (Fig. 14-18). The same annual pattern is illustrated for Lake Windermere, where it has been demonstrated continuously for over 20 years. Factors such as light intensity, temperature, grazing by zooplankton, fungal parasitism, and changes in other nutrients, especially nitrogen and phosphate, could not be shown to be associated with the decline of the maximum. Experimental results on silica requirements showed that the silica reduction was clearly the major factor contributing to the decline of the diatoms.

To generalize from this simple situation would be misleading, since even within the English Lake District, where this sequence occurs rather commonly, numerous other interacting factors are involved. Silica concentrations and their biogenic reduction from epilimnetic waters, however, are certainly major factors in diatom-community regulation. In lakes where silica levels remain high, even though severely reduced during the summer productive period (e.g., Lawrence Lake, Fig. 14-17, upper), the spring diatom maximum persists longer into the summer and is overtaken gradually by a predominance of green algae. In very productive lakes, a maximum diatom peak can be found in the fall and early winter, as for example in Wintergreen Lake (Fig. 14-17, lower), during which time a competitive advantage to diatoms is apparent until the silica levels decline to very low concentrations (< 100 μg l^{-1}). After spring overturn, diatom growth is very short-lived in this lake, followed by a brief dominance of green algae until inorganic combined nitrogen sources are depleted. Nitrogen-fixing algae then quickly dominate and persist until the combined effects of increased inorganic nitrogen, reduced light and temperature, and renewed silica reoccur during circulation in late summer and early autumn.

The sequence of succession of different species of diatoms can be seen not only within a composite spring maximum, but also in the patterns of diatom periodicity as lakes become more productive. In lakes in which silica concentrations are moderate to low (e.g., < 5 mg l^{-1}), progressive long-term enrichment with phosphorus and nitrogen can lead to rapid biogenic reduction in silica levels so that diatoms cannot effectively compete, and are replaced by nonsiliceous phytoplankton (Kilham, 1971). The current eutrophication of Lake Michigan is imposing such circumstances on the diatom populations, and as a result, they are being excluded gradually by green and blue-green algae during silica depletion in the summer period (Schelske and Stoermer, 1971).

A strong interaction can exist between the population level of littoral diatoms and the development of planktonic diatoms. For example, in Furesø, the spring maximum of the planktonic diatoms, primarily *Stephanodiscus*, reduced the silica concentrations to < 40 μg l^{-1}, levels experimentally demonstrated to inhibit the growth of these algae (Jørgensen, 1957) (Fig. 14-19). An immediate increase in epiphytic diatoms growing on submersed portions of the emergent macrophyte *Phragmites* accompanied this decrease. The *Phragmites* stems were shown to possess large quantities of easily dissolved silica, and their silica content decreased during the development of the epiphytes.

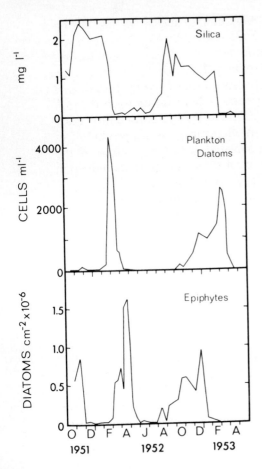

Figure 14–19. Variations in numbers of the dominant phytoplanktonic diatom *(Stephanodiscus hantzschii)*, silica concentrations at the surface, and epiphytic diatoms per cm^{-2} of *Phragmites* stems at a depth of 20 to 30 cm, Furesø, Denmark. (After data of Jørgensen, 1957).

SUMMARY

1. The biogeochemical cycling of essential micronutrients is regulated to a large extent by changes in oxidation-reduction (redox) states, which are governed by photosynthetic and bacterial metabolism.

   a. By conversion of light energy into chemical bonds, photosynthesis produces reduced states (negative E_h) of high free energy and nonequilibrium concentrations of carbon, nitrogen, and sulfur compounds. Respiratory and fermentative reactions of nonphotosynthetic organisms tend to restore equilibrium by catalytically decomposing the products of photosynthesis. The nonphotosynthetic organisms obtain free energy for their metabolism from this process.

   b. Complete redox equilibrium conditions do not occur in natural aquatic ecosystems because most redox reactions are slow and consist of a composite of many redox reactions of different reaction rates. In addition, continuously changing inputs of photosynthetic energy occur, which disrupt the tendency toward equilibrium.

c. The predominant reactants in the redox processes in natural waters are carbon, oxygen, nitrogen, sulfur, iron, and manganese.

d. Redox remains positive (300 to 500 mv) as long as some dissolved oxygen (> 1 mg l^{-1}) is present. Temperature and pH changes have only minor effects on redox. As oxygen concentrations approach zero, E_h decreases precipitously.

2. Concentrations of ionic iron are exceedingly low in aerated water. Most iron in oxygenated water occurs as ferric hydroxide in particulate and colloidal form and as complexes with organic, especially humic, compounds. The solubility of manganese is considerably higher than that of iron, but it is analogous to iron in the way it reacts.

a. Under conditions of both low pH and low redox potential (approximately 250 mv), ferrous and manganous ions diffuse from the sediments and accumulate in anaerobic hypolimnetic water of productive lakes (Fig. 14–3).

b. Under strongly reducing conditions of very low redox potentials (< 100 mv), sulfate is reduced to sulfide. Highly insoluble metallic sulfides, particularly ferrous sulfide (FeS), form under these conditions. Therefore, in hypereutrophic lakes, high concentrations of hydrogen sulfide resulting from bacterial decomposition of sulfur-containing organic matter and from reduction of sulfate can lead to significant decreases in dissolved iron (but not the more soluble manganese) as sulfides form during the latter portion of stratification.

3. Iron and manganese are micronutrients that are essential to freshwater flora and fauna.

a. Under certain conditions of restricted availability, photosynthetic productivity can be limited by these elements. Manganese is clearly involved in the seasonal succession of certain algal populations.

b. Certain chemosynthetic bacteria can utilize the energy of inorganic oxidations of ferrous and manganous salts in relatively inefficient reactions involving CO_2 fixation. Other autotrophic and heterotrophic iron-oxidizing bacteria deposit oxidized iron and manganese. These bacteria are restricted to zones of steep redox gradients between reduced metal ions and oxygenated water. (Figure 14–6).

4. Quantitative information on the dynamics of other essential metallic micronutrients—zinc, copper, cobalt, molybdenum, vanadium, and selenium—is limited. Micronutrient availability is governed by redox conditions and the extent of complexing with dissolved organic compounds and other inorganic ions.

a. In most natural waters, micronutrient concentrations and availability are usually adequate to meet metabolic requirements within the constraints of light, temperature, and macronutrient availability. A few clear cases of limitations to photosynthetic productivity by micronutrient deficiencies or physiological unavailability have been demonstrated.

b.   The dynamics of copper are strongly affected by redox conditions similar to those of iron. Cobalt, zinc, and molybdenum dynamics are more closely related to microbial metabolism and transport of the seston, and to the effectiveness of complexing with organic compounds. Molybdenum exhibits greater mobility than do the other micronutrient ions.

c.   Inputs of many trace elements to fresh waters are increasing as a result of pollution from industrial and combustion emissions to the atmosphere and subsequent deposition via precipitation.

5.   Sulfur is nearly always present in quantities adequate to meet high requirements for protein and sulfate ester synthesis. The dynamics of sulfate and the hydrogen sulfide produced by decomposition of organic matter, however, alter conditions in stratified, productive waters that affect the cycling of other nutrients, productivity, and biotic distribution.

a.   Atmospheric sulfur compounds originating primarily from combustion of coal and oil return to land by precipitation and dry fallout of particles. This sulfur constitutes a major global source of sulfur to fresh waters that in many natural waters exceeds inputs from rock and soil weathering and transport in surface runoff and groundwater.

b.   Sulfate is the primary dissolved form of sulfur in oxic waters; hydrogen sulfide accumulates in anoxic zones of intensive decomposition in productive lakes where the redox potential is reduced below about 100 mv.

c.   Most of the sulfur in lake water is stored as dissolved sulfate and hydrogen sulfide. Sestonic sulfur-containing proteins and sulfate esters and dissolved sulfides are primary constituents of the sediments.

d.   Although oxygen is obtained from sulfate by sulfate-reducing bacteria, the H_2S generated readily oxidizes and utilizes oxygen upon transport or intrusion into aerobic strata.

e.   Sulfur-oxidizing bacteria consist of two general types:

i.   Chemosynthetic aerobes that oxidize reduced sulfur compounds and elemental sulfur to sulfate, and

ii. Photosynthetic sulfur bacteria that utilize light as an energy source and reduced sulfur compounds as electron donors in the photosynthetic reduction of CO_2.

iii. The redox requirements of the sulfur-oxidizing bacteria, especially those requiring light, are rather specific, and distribution of species is often restricted to zones of steep gradients between anoxic and aerated water strata.

iv. When conditions are optimal, the photosynthetic bacteria often develop in extreme profusion and may contribute significantly to the annual productivity of lakes.

6.   Silica occurs in relative abundance in natural waters as dissolved silicic acid and particulate silica.

a.   Diatom algae assimilate large quantities of silica and markedly modify the flux rates of silica in lakes and streams.

b. Utilization of silica in the trophogenic zone of lakes by diatoms often reduces the epilimnetic concentrations (Fig. 14–16) and induces, along with other factors, a seasonal succession of diatom species.

c. When the concentration of silica is reduced below about 0.5 mg l^{-1}, many diatom species cannot compete effectively with nonsiliceous algae, and their growth rates decline until silica supplies are renewed, usually during autumnal circulation.

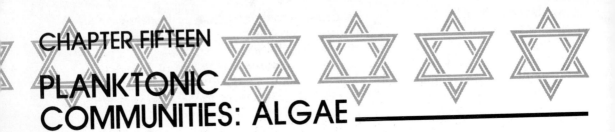

CHAPTER FIFTEEN
PLANKTONIC COMMUNITIES: ALGAE

The algae of the open water of lakes and large streams, the *phytoplankton*, consist of a diverse assemblage of nearly all major taxonomic groups. Many of these forms have different physiological requirements and vary in response to physical and chemical parameters such as light, temperature, and nutrient regimen. Despite these diversities, both taxonomic and physiological, many algal species coexist in the same water volume. However, dominant genera in algal groupings change not only spatially (vertically and horizontally within a lake), but seasonally, as physical, chemical, and biological conditions in the water body change. A general pattern of seasonal algal succession correlated with environmental factors has been described for many lakes, although the precise reasons for most of these changes are not well known.

Phytoplankton ecology has been one of the most popular areas of aquatic ecology in recent years and many inroads have been made in understanding algal light, temperature, and nutrient requirements, buoyancy regulation, competition, productivity, and the effects of predation and parasitism on phytoplankton. However, large voids still exist in our knowledge of the many complex mechanisms that result in the wide array of planktonic algal communities observed worldwide. Further investigations, particularly in the areas of algal-microbial, algal-algal, and algal-herbivore interactions, are much needed.

COMPOSITION OF THE ALGAE OF PHYTOPLANKTONIC ASSOCIATIONS

Morphological and Physiological Characteristics

Before summarizing the algal associations and population dynamics of phytoplanktonic communities, a few of the major morphological and physiological characteristics of the primary algal groups should be emphasized. The systematics of algae comprising the phytoplankton is a subject that has been treated well in a number of detailed works but is continually being revised and improved. The emphasis here is functional, from the standpoint of interactions of physiological characteristics with environmental variables, and the effects of these interactions on primary productivity and resultant population successions within communities.

In this physiologically functional and ecological treatment, it is defeating to become engrossed in discussion of whether certain groups are truly algae in a strict sense. In many respects, the blue-green algae (Myxophyceae) are physiologically similar to the bacteria. Many motile and colonial flagellates are classified as Protozoa on the basis of certain morphological and reproductive characteristics. The important characteristic, however, is that autotrophic photosynthesis is the primary mode of nutrition, and results in the synthesis of new organic matter. Therefore these photoautotrophic microbiota are treated collectively with the true algae of the phytoplankton.

PIGMENTS

A primary algal characteristic is the presence of photosynthetic pigments—the chlorophylls, carotenoids, and biliproteins. Chlorophyll a is the primary photosynthetic pigment of all oxygen-evolving photosynthetic organisms, and is present in all algae (Table 15-1) and photosynthetic organisms other than the photosynthetic bacteria (discussed separately in Chapter 14). Chlorophyll a has two in vitro absorption bands, as discussed on page 68, in the red-light region at 660 to 665 nm and at lower wavelengths near 430 nm.

Chlorophyll b, although commonly present in higher plants, is found only in the green algae and the euglenophytes (Table 15-1). Chlorophyll b is a light-gathering pigment that transfers absorbed light energy to chlorophyll a for primary photochemistry (reviewed in Govindjee and Brown, 1974). Maximum absorption bands are approximately 645 nm and 435 nm. Chlorophyll c, consisting of two spectrally distinct components, is probably an accessory pigment to photosystem II. Maximum extraction absorption bands are at about 630 to 635 nm, and in the dominant blue portion of the spectrum at about 583 to 586 nm and 444 to 452 nm (Meeks, 1974). Chlorophyll d, of no known function, is a minor pigment component found only in certain red algae. A fifth chlorophyll (e) has been isolated from a xanthophycean alga, but whether it exists in vivo is uncertain (Strain, 1951).

Among the many carotenoids of algae, the carotenes are linear unsaturated hydrocarbons, and the xanthophylls are oxygenated derivatives of carotenes (Goodwin, 1974). As is the case with chlorophyll b, light energy absorbed by carotenoids and biliproteins is transferred to chlorophyll a, leading to fluorescence and excitation of chlorophyll a molecules. β-carotene is the most widely distributed of the carotenes, and is replaced by α-carotene only in certain green algae and in the Cryptophyceae. The biliproteins are water-soluble, pigment-protein complexes occurring in the blue-green algae, and to a lesser extent, in certain cryptophytes and red algae.

BLUE-GREEN ALGAE

The blue-green algae (cyanobacteria) have been among the most studied of all algal groups. The Cyanophyta (cyano [Greek] = blue-green) or Myxophyceae (myx [Greek] = slime) is a primitive group which is procaryotic in cell structure like the bacteria. Procaryotic cells lack certain membranous structures including a nuclear membrane, mitochondria, and chloroplasts. The cells do contain cellular inclusions, which in some cases assume similar functions to eucaryotic organelles such as pigment-

TABLE 15-1 Distribution of Photosynthetic Pigments Among the Algae

PIGMENTS	CYANO-PHYCEAE (MYXO-PHYCEAE)	CHLORO-PHYTA	XANTHO-PHYCEAE	CHRYSO-PHYCEAE	BACILLARIO-PHYCEAE	CRYPTO-PHYCEAE	DINO-PHYCEAE (PYRRO-PHYTA)	EUGLENO-PHYCEAE	PHAEO-PHYCEAE	RHODO-PHYCEAE
Chlorophylls:										
chlorophyll a	+	+	+	+	+	+	+	+	+	+
chlorophyll b	−	+	−	−	−	−	−	+	−	−
chlorophyll c	−	−	+	+	+	+	+	−	+	−
chlorophyll d	−	−	−	−	−	−	−	−	−	+
Carotenoids:										
Carotenes:										
α-carotene	−	+	−	−	−	+	−	−	−	+
β-carotene	+	+	+	+	+	−	+	+	+	+
γ-carotene	−	+	−	−	−	−	−	−	−	−
ε-carotene	−	−	−	−	+	+	−	−	−	−
Xanthophylls										
lutein	+	+	−	+	−	−	−	−	−	+
violaxanthin	−	+	+	−	−	−	−	+	+	+
fucoxanthin	−	−	−	+	+	−	−	−	+	−
neoxanthin	−	+	−	−	−	−	−	+	−	−
astaxanthin	−	−	−	−	−	−	−	−	−	−
diatoxanthin	−	−	+	+	+	−	+	−	+	−
diadinoxanthin	−	−	+	+	+	−	+	+	+	−
peridinin	−	−	−	−	−	−	+	−	−	−
dinoxanthin	−	−	−	−	−	−	+	−	−	−
teraxanthin	−	−	−	−	−	−	−	−	−	−
antheraxanthin	−	−	−	−	−	−	−	+	−	+
myxoxanthin	+	−	−	−	−	−	−	−	−	−
myxoxanthophyll	+	−	−	−	−	−	−	−	−	−
oscilloxanthin	+	−	−	−	−	−	−	−	−	−
echinenone	+	−	−	−	−	−	−	+	−	−
Biliproteins:										
phycocyanin*	+	−	−	−	−	+	−	−	−	+
phycoerythrin*	+	−	−	−	−	+	−	−	−	+

From Morris (1967), Meeks (1974), and Goodwin (1974).
*These chromoproteins consist of distinctly different types in the algal groups in which they occur. Differentiation is based on their absorption spectra (cf. Goodwin, 1974).

bearing lamellae, a plasma membrane, and a nuclear region containing chromosomal material. Like bacteria, blue-green algae have murein in the cell wall, reproduce by binary fission, and do not divide by mitosis as other algae and higher organisms do. However, blue-green algae are distinguished from bacteria by the presence of chlorophyll *a*, which is common to eucaryotic algae and higher plants and different structurally from bacteriochlorophyll. Blue-green algae are also able to use water as an electron donor in photosynthesis, which is evolutionarily more advanced than bacterial photosynthesis. Therefore, blue-green algae possess the enzymatic and pigment capability to photosynthesize under oxygenated conditions as higher plants. In effect, blue-green algae are structurally and physiologically like bacteria, but functionally like plants in aquatic systems.

Blue-green algae occur in unicellular, filamentous, and colonial forms, and most are enclosed in mucilaginous sheaths either individually or in colonies. A majority of the planktonic blue-green algae consists of members of the coccoid family Chroococcaceae (e.g., *Anacystis* = *Microcystis*, *Gomphosphaeria* = *Coelosphaerium*, and *Coccochloris*) and filamentous families Oscillatoriaceae, Nostocaceae, and Rivulariaceae (e.g., *Oscillatoria, Lyngbya, Aphanizomenon, Anabaena*).

In filamentous blue-green algae, cells are arranged end to end to form long trichomes. Trichomes are usually contained within a mucilaginous sheath. Although vegetative reproduction by fragmentation of trichomes is common, some cells are differentiated and specifically involved in reproduction and perennation. Spores or akinetes form in a number of species as enlarged, thick-walled cells that accumulate proteinaceous reserves in the form of cyanophycin granules (cf. Fogg, et al., 1973; Stanier and Cohen-Bazire, 1977). Under favorable conditions, akinetes germinate directly into a trichome or into hormogonia. Hormogonia are short, slightly modified pieces of trichome that fragment from the parent trichome and move away by means of gliding motions to develop into a new filament. In a few species, hormogonia develop within the parent sheath as multiple trichomes. A few other spore structures develop in species that do not form hormogonia; in these species, development occurs without resting stages.

Heterocysts are differentiated cells unique to blue-green algae. Heterocysts can occur in all filamentous species except those in the Oscillatoriaceae and are believed to be the major sites of nitrogen fixation (Chapter 12). Heterocysts develop from vegetative cells, which form a thick envelope over the cell wall, except at the polar regions, in which the heterocyst is connected to adjacent vegetative cells by a pore channel. It is through this pore that the plasma membranes are adjoined and exchange of metabolic products occurs. In contrast to vegetative cells, heterocysts lack phycobilins, which are the primary light absorbers in blue-green algae, lack an oxygen-evolving photosystem, and show higher reducing activity. Therefore, the internal environment of the heterocyst is ideal for nitrogen fixation, since the nitrogenase enzyme is inactivated by oxygen. Organic carbon from adjacent cells is transferred into the heterocyst and used as an energy source in nitrogen fixation; reduced nitrogen is, in turn, transferred out of the heterocyst and into vegetative cells (Wolk, 1968, 1973). The interaction of vegetative cells and heterocysts is an example of structural separation of metabolic processes (photosynthesis and nitrogen fixation) in a relatively primitive organism. Nitrogen fixation can also occur in some nonheterocystous blue-green algae (e.g., *Gloeocapsa*). In this case photosynthesis and nitrogen fixation may be separated temporally rather than structurally (Gallon, et al., 1974).

GREEN ALGAE

The Chlorophyta are an extremely large and morphologically diverse group of algae that is almost totally freshwater in distribution. A majority of the planktonic green algae belong to the orders Volvocales (e.g., *Chlamydomonas, Sphaerocystis, Eudorina, Volvox*) and Chlorococcales (e.g., *Scenedesmus, Ankistrodesmus, Selenastrum, Pediastrum*). Many members are flagellated (2 or 4, rarely more) at least in the gamete stages; in the desmids (Conjugales or Desmidiales), the gametes are amoeboid.

Asexual reproduction by vegetative division is common to most of the green algae, but is lacking in most of the Chlorococcales and Siphonales. Cell division very often is synchronized, so that nuclear and cell division occurs at night. Cell division in colonial species results in enlargement of the colony; new colonies are formed only by fragmentation of the colony. Among filamentous species, such as *Spirogyra*, fragmentation of the filament is common. Vegetative formation of single or multiple zoospores within a cell is also common. When liberated from the parent cell, the flagellated zoospores are actively motile and positively phototactic for varying periods of time before losing the flagellae and entering a resting spore stage.

Sexual reproduction in the green algae is diverse. In the simplest case of isogamy, the flagellated male and female gametes are morphologically similar in size and structure. In anisogamy, the flagellated female gamete is larger than the male gamete. Among more specialized algae, oogamy is common, in which union occurs between a large, nonflagellated female gamete and a small, flagellated male antherozoid. Some green algae are sexually monoecious (gametes are derived from the same cell), and others are dioecious (male and female gametes are derived from different cells).

Whereas the planktonic Volvocales and Chlorococcales are ubiquitous in distribution among waters of differing salinity within the normal limnological range, the distribution of most species of desmids of the Conjugales is limited to low concentrations of the divalent cations, calcium and magnesium. Although not totally restricted to waters of low salinity, the desmids are most common and the species diversity is greatest in soft waters draining land forms developed in granitic or other igneous rocks, and especially in waters with a high content of dissolved organic matter. Their abundance and diversity are often greatest in bog waters that drain through deposits of the moss *Sphagnum*. Many desmids are distributed widely, but as a whole they are less cosmopolitan than most unicellular algae (cf. Hutchinson, 1967; Brook, 1981).

YELLOW-GREEN ALGAE

The Xanthophyceae are unicellular, colonial, or filamentous algae that are characterized by conspicuous amounts of carotenoids in comparison to chlorophylls that result in their predominantly yellow-green coloration. Nearly all of the motile cells possess two flagella, one of which is longer than the other (hence the older name, Heterokontae). A cell wall is often absent; when present, it contains large amounts of pectin and in many species may be silicified. Asexual reproduction is usually by fission, with the formation of zoospores. Sexual reproduction is poorly understood in this group, but is most often isogamous.

A majority of the xanthophycean algae are associated with substrata, and many are epiphytic on larger aquatic plants. A few members are planktonic and include common genera such as *Chlorobotrys*, *Gloeobotrys*, and *Gloeochloris*.

GOLDEN-BROWN ALGAE

The chromatophores of the Chrysophyceae often yield a distinctive golden-brown coloration because of the dominance of β-carotene and specific xanthophyll carotenoids, in addition to chlorophyll *a*. Most of the chrysophycean algae are unicellular. A few are colonial; they are rarely filamentous. Flagellation is variable; most often cells are uniflagellate, but some species possess two flagella, usually of equal length. Many species lack a cell wall, and are bounded only by a cytoplasmic membrane; others possess a cell surface covered by delicate siliceous or calcareous plates or scales. Vegetative reproduction by longitudinal cell division is most common, especially among the motile unicellular species. Although it has rarely been observed, sexual reproduction is isogamous; the resulting cystlike zygote undergoes reductive division under favorable conditions.

A number of chrysophycean algae form important components of the phytoplankton. The unicellular species with a single flagellum (e.g., *Chromulina*, *Chrysococcus*, *Mallomonas*) are usually very small algal constituents of the nannoplankton.* Larger colonial forms such as *Synura*, *Chrysosphaerella*, *Uroglena*, and particularly *Dinobryon* are widely distributed and may become major components of the phytoplankton under certain environmental conditions. The conspicuous development of populations of certain species of *Dinobryon* and *Uroglena* in lakes of very low phosphorus concentrations (mentioned earlier; p. 276) may be due to their ability to effectively take up phosphate at extremely low ambient concentrations (cf. Lehman, 1976). In contrast, other species of *Dinobryon* and *Synura* have high phosphorus requirements.

DIATOMS

A most important group of algae of the phytoplankton are the diatoms (Bacillariophyceae), even though most species are sessile and associated with littoral substrata. Their primary characteristic, silicified cell walls, has been discussed in the preceding chapter. Both unicellular and colonial forms are common among the diatoms. The group is commonly divided into the centric diatoms (Centrales), which have radial sym-

---

*The term *nannoplankton* was introduced long ago in reference to small plankton components (bioseston) of the total particulate material (seston) that pass through the older plankton nets of a mesh size of about 50 to 60μm. The generally accepted size ranges, as commonly used (see, for example, Strickland, 1960), are:

Macroplankton	>500 μm
Microplankton (net plankton)	approx. 50 to 500 μm
Nannoplankton (= nanoplankton)	10 to approx. 50 μm
Ultraplankton	0.5 to 10 μm

Hutchinson (1967) discusses further separations within these divisions and gives a more complex terminology. Since so many phytoplankton algae are smaller than the common porosity of plankton nets (50 to 60 μm), nets are relatively useless for even qualitative studies. Plankton nets have been relegated to limnological museums for algal work, and are employed now only for macrozooplankton studies.

metry, and the pennate diatoms (Pennales), which exhibit essentially bilateral symmetry. The cell wall or frustule of diatoms consists of two lidlike valves, one of which fits within the other; the overlapping area of the valves is connected by bands that constitute the girdle. The beautiful structures of the siliceous cell walls of diatoms are complex and used as taxonomic characteristics.

The valves of pennate diatoms exhibit various areas of cell thickenings and dilations. In some species a slit, termed a *raphe*, traverses all or part of the cell wall; in others, a depression in the axial areas of the cell wall, termed a *pseudoraphe*, is found. The four major groups of pennate diatoms are differentiated on the basis of these structures: (a) the Araphidineae, which possess a pseudoraphe (e.g., *Asterionella*, *Diatoma*, *Fragilaria*, *Synedra*); (b) the Raphidioidineae, in which a rudimentary raphe occurs at the cell ends (e.g., *Actinella*, *Eunotia*); (c) the Monoraphidineae, which have a raphe on one valve and a pseudoraphe on the other (e.g., *Achnanthes*, *Cocconeis*); and (d) the Biraphidineae, in which the raphe occurs on both valves (e.g., *Amphora*, *Cymbella*, *Gomphonema*, *Navicula*, *Nitzschia*, *Pinnularia*, *Surirella*). These divisions are of more than taxonomic interest since, as will be pointed out subsequently, distinct nutritional requirements favor the growth of one group over another.

Vegetative reproduction by cell division, usually at night, is the most common mode of multiplication (cf. Werner, 1977). Sexual reproduction occurs periodically when the cells reach a minimum critical size through repeated cell division in the course of asexual reproduction. Amoeboid gametes occur in isogamous reproduction in the Pennales, whereas in the Centrales reproduction is oogamous with flagellated spermatozoids. After fusion of the gametes, an auxospore develops and divides; some of the resulting cells give rise to vegetative cells that are of the largest dimensions of the species.

CRYPTOMONADS

Most of the cryptophycean algae are naked, unicellular, and motile. This class is very small and most of the planktonic members belong to the Cryptomonadineae (e.g., *Cryptomonas*, *Rhodomonas*, *Chroomonas*). All members are small and not a great deal is known about their ecology or physiology. Their significance in the metabolism of lakes is probably much greater than generally is believed, especially because dense populations of these algae often develop during cold periods of the year under relatively low light conditions.

The cryptomonads usually are flattened dorsoventrally, with an anterior invagination that bears two equal or subequal flagella. The one or usually two chromatophores contain a variety of pigments (Table 15–1) that can yield a spectrum of apparent colors including olive green, blue, red, or brown. Reproduction is by longitudinal division; sexual reproduction is unknown in the group.

DINOFLAGELLATES

The dinoflagellates (Dinophyceae of the Pyrrophyta) are unicellular flagellated algae, many of which are motile. Although a few species are naked or without a cell wall (e.g., *Gymnodinium* of the Gymnodiniales), most develop a conspicuous cell wall that often is sculptured and bears large spines and elaborate cell-wall processes (the

Peridiniales, e.g., *Ceratium, Glenodinium, Peridinium*). In both naked and armored types, the cell surface has transverse and longitudinal furrows that connect and contain the flagella, whose movements create water currents providing for weak locomotion and disrupting of chemical gradients at the cell surface. Although sexual reproduction occurs, asexual reproduction by the formation of aplanospores (in which a motile phase is omitted) predominates. These asexual resting stages or cysts can undergo considerable periods of diapause. In the large *Ceratium*, for example, the autumn decline of summer populations in temperate regions results in the production of overwintering cysts. The distribution of dinoflagellates in relation to major chemical characteristics shows that, while some species are widely tolerant and ubiquitous, especially among species of *Ceratium* and *Peridinium*, most dinoflagellate species have restricted ranges with respect to calcium, pH, dissolved organic matter, and temperature.

Although many zooplankton change size and form seasonally (Chapter 16), only a few phytoplankton undergo seasonal polymorphism or cyclomorphosis. One example of phytoplanktonic cyclomorphosis is the lengthening of cellular extensions or horns in the dinoflagellate *Ceratium* as temperatures increase from spring into midsummer (Hutchinson, 1967). In *Ceratium*, the conspicuous separation of the cell wall by the transverse furrow into the upper part (epitheca) and lower part (hypotheca) is accentuated further by the development of horns. The epitheca always has one horn; the hypotheca usually has one to three horns. As the water warms, the cells of populations exhibit an increase in the length of the horns, a decrease in the width and general size of the cell, and a lessening of the angle of divergence of the hypothecal horns (Fig. 15–1). In midsummer, with even warmer temperatures, the development of a short, fourth horn is common. Field and experimental evidence indicates that these changes are induced primarily by increasing water temperatures. Some experimental evidence points to the fact that the sinking rate is reduced appreciably in less viscous warmer water by the reduced cell dimensions, increased divergence of the horns, and the

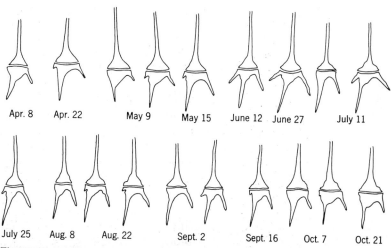

| Apr. 8 | Apr. 22 | May 9 | May 15 | June 12 | June 27 | July 11 |

| July 25 | Aug. 8 | Aug. 22 | Sept. 2 | Sept. 16 | Oct. 7 | Oct. 21 |

Figure 15–1. Seasonal polymorphism in *Ceratium hirundinella* in Kirchberg-teich, Darmstadt, Germany. (From Hutchinson, G. E.: A Treatise on Limnology. Vol. 2, New York, John Wiley & Sons, Inc., 1967, after work of List.)

development of the fourth horn. Presumably, these changes are of adaptive significance to *Ceratium* in that they reduce the rate of sinking out of the photic zone.

EUGLENOIDS

Although the euglenoid algae (Euglenophyceae) are a relatively large and diverse group, few species are truly planktonic. Nonetheless, when conditions are favorable, the euglenoids can develop in great profusion. Almost all euglenoids are unicellular, lack a distinct cell wall, and possess one, two, or three flagella that arise from an invagination in the cell membrane. Reproduction occurs by longitudinal division of the motile cell; sexual reproduction has not yet been substantiated.

Although some euglenoids are unpigmented and phagotrophic (able to ingest solid particles) and are best treated as Protozoa, most are photosynthetic and facultatively heterotrophic. Even when phagotrophy constitutes the dominant mode of carbon assimilation among certain nonpigmented members of the Dinophyceae, Chrysophyceae, and especially the Euglenophyceae, none of these algae depend solely on phagotrophy for their major carbon requirements (Droop, 1974). Nutrition is supplemented by the uptake of dissolved organic compounds. Ammonia and dissolved organic nitrogen compounds are the dominant sources of nitrogen among most euglenoid algae. Their development in the phytoplankton occurs most often in seasons, strata, or lake systems in which concentrations of ammonia and especially dissolved organic matter are high. For example, detailed seasonal analyses of the phytoplanktonic populations at weekly intervals at each meter of depth for 16 months in hard-water Lawrence Lake of southwestern Michigan showed euglenoid algae developing for only a brief two-week period immediately prior to autumnal turnover and in a small stratum, one meter above the deep sediments, that was anoxic during the last two months of summer stratification (Wetzel, unpublished). However, euglenoids are found most often in shallow water rich in organic matter such as farm ponds.

BROWN AND RED ALGAE

The brown algae (Phaeophyta) are mostly filamentous or thalloid algae, which, as a group, are almost exclusively marine. Only a few genera of this large, primitive group are represented in fresh water; those that are found in fresh water are attached to substrata, such as rocks. No species is planktonic.

The red algae (Rhodophyta) are also very sparsely represented in fresh water, and none is planktonic. A majority of the genera occurring in fresh water are found only in fresh water. The thalloid or filamentous species (e.g., *Batrachospermum*) are nearly all restricted to fast-flowing streams of well-oxygenated, cool waters.

PHYTOPLANKTONIC COMMUNITIES

An outstanding feature of phytoplanktonic communities in lacustrine habitats is the coexistence of a number of algal species. In some cases, one species is found in much greater abundance than others; more often, two or more algal species are codominant in the phytoplanktonic assemblage. A large number of rarer species can always be found among the dominant or subdominant algae. The algae co-occur even though each species has a specific niche based on its physiological requirements and constraints of the environment. Theoretically, niche overlap can lead to competitive exclusion, and should result in dominance by a single species (cf. Chapter 8). This problem has been formally referred to as the "paradox of the plankton" (Hutchinson, 1961): An apparent multispecific, rather than unispecific, equilibrium occurs in what seem to be physically uniform conditions of the turbulent open water of lakes.

A number of explanations have been advanced to account for species diversity in phytoplanktonic communities that is much higher than would be anticipated from theory and mathematical derivations. First, in order for the relatively slow process of competitive exclusion to occur, the physical conditions must be uniform for a sufficient period of time. If environmental conditions change rapidly, the advantages gained by one species that is a better competitor may not exist long enough to result in the competitive exclusion of other species. Stated another way, differences in the efficiencies of resource utilization among species may be too small for competitive exclusion to occur before conditions change.

Second, competition can be negated partially if commensalism and symbiotic relations exist among species. As discussed later, a number of organic compounds released extracellularly by one alga can influence the metabolism of another species, although most of these compounds are inhibitory and antibiotic in nature. Also, a majority of those algae requiring exogenous sources of vitamin micronutrients are motile and small (Chrysophyceae, Cryptophyceae, Euglenophyceae, and Dinophyceae). The differences in organic micronutrient requirements and the release of the excess by species capable of their synthesis would serve as a means of encouraging mixed equilibrium populations.

Greater consumption of one species of algae than another can also promote coexistence, if the species being preyed upon has a much greater competitive advantage. Selective grazing, largely on the basis of algal size, by zooplankton is an important phenomenon and is discussed in some detail in Chapter 16.

Some species of algae are exclusively planktonic. As far as is known, their populations oscillate temporally in abundance, dominating for a period and then becoming extremely rare but nonetheless remaining planktonic (Hutchinson, 1964). Alternatively, some species enter resting stages and thereby leave the competitive arena for a period of time. Some of these species are quiescent in the littoral sediments, and later develop sufficiently to form a significant component of the phytoplankton (e.g., Livingstone and Jaworski, 1980). Evidence for benthic algal populations in the littoral zone serving as a "seed" inoculum for some phytoplankton species is sparse (cf. Chapter 19), but when this occurs, phytoplankton species diversity is increased. This mode of opportunistic expansion into the pelagial region is certainly more common in shallow waters with extensive littoral development.

None of these mechanisms for explaining coexistence are mutually exclusive, and all factors probably operate simultaneously to a greater or lesser degree. Which one dominates at a particular time in a given body of water surely varies. There is some evidence that the distribution of phytoplankton is not necessarily as uniform as the concepts of multispecies equilibrium would imply. For example, the phytoplankton of an alpine lake showed a high degree of patchiness for many species, which indicated that the rate of mixing was sufficiently slow relative to the reproductive rate of the algae and allowed communities of many different species to exist simultaneously (Richerson, et al., 1970). Under these hypothetical conditions of "contemporaneous disequilibrium," at any given time, many patches of water exist in which one species has a competitive advantage relative to the others, but the patches are obliterated frequently enough to prevent exclusive occupation by a single species.

Types of Algal Associations

A great deal of descriptive work has been devoted to the associations of algal species among differing fresh waters. Algae are extremely diverse, and many exhibit a very wide tolerance to environmental conditions found under natural limnological situations. Nonetheless, certain characteristic phytoplanktonic associations occur repeatedly in lakes of increasing nutrient enrichment. Some of the commonly observed major associations are set out in Table 15-2 based on the detailed discussion of Hutchinson (1967). Such categorizations are not very satisfactory because of the wide spectrum of intergradations often observed. Additionally, shifts occur seasonally from one type to another, especially among more productive waters. Even though such characterizations yield little insight into regulating environmental factors, they are useful from the standpoint of general correlations between qualitative and quantitative abundance of the algae and available nutrients.

A number of phytoplankton indices have been developed, particularly by Thunmark (1945) and by Nygaard (1955), in admirable attempts to quantify more precisely the relationships of rarely occurring and dominant algal species in relation to lake productivity. Their use is revived periodically with limited and varying degrees of success. The indices developed by Nygaard (1949) are based on the ratios or quotients of numbers of species within general groups, such as those of blue-green algae to desmids, chlorococcalean algae to desmids, centric to pennate diatoms, and euglenophytes to blue-green and green algae. His compound index was the ratio of all species of blue-green, chlorococcalean green, centric diatoms, and euglenoid algae to the species of desmid algae. A general relationship was found between low compound index values of less than one and oligotrophic waters. Lowest values occurred in desmid-rich dystrophic lakes enriched by dissolved humic compounds. The values of the compound index increased in more productive waters, and approached 50 in highly enriched eutrophic lakes.

Comparisons of these indices with more modern measurements of algal productivity show weak positive correlations, but exceptions are many. Great variations are found among different regions of the world and among lake districts. It is apparent that these indices, while having some value in determining species relationships, are much too superficial in physiological foundation to be of significant use in evaluating productivity among lakes, or in elucidating causal mechanisms underlying the composite growth of algae.

TABLE 15-2 Characteristics of Common Major Algal Associations of the Phytoplankton in Relation to Increasing Lake Fertility

GENERAL LAKE TROPHY	WATER CHARACTERISTICS	DOMINANT ALGAE	OTHER COMMONLY OCCURRING ALGAE
Oligotrophic	Slightly acidic; very low salinity	Desmids *Staurodesmus*, *Staurastrum*	*Sphaerocystis*, *Gloeocystis*, *Rhizosolenia*, *Tabellaria*
Oligotrophic	Neutral to slightly alkaline; nutrient-poor lakes	Diatoms, especially *Cyclotella* and *Tabellaria*	Some *Asterionella* spp., some *Melosira* spp., *Dinobryon*
Oligotrophic	Neutral to slightly alkaline; nutrient-poor lakes or more productive lakes at seasons of nutrient reduction	Chrysophycean algae, especially *Dinobryon*, some *Mallomonas*	Other chrysophyceans, e.g., *Synura*, *Uroglena*; diatom *Tabellaria*
Oligotrophic	Neutral to slightly alkaline; nutrient-poor lakes	Chlorococcal *Oocystis* or chrysophycean *Botryococcus*	Oligotrophic diatoms
Oligotrophic	Neutral to slightly alkaline; generally nutrient poor; common in shallow Arctic lakes	Dinoflagellates, especially some *Peridinium* and *Ceratium* spp.	Small chrysophytes, cryptophytes, and diatoms
Mesotrophic or Eutrophic	Neutral to slightly alkaline; annual dominants or in eutrophic lakes at certain seasons	Dinoflagellates, some *Peridinium* and *Ceratium* spp.	*Glenodinium* and many other algae
Eutrophic	Usually alkaline lakes with nutrient enrichment	Diatoms much of year, especially *Asterionella* spp., *Fragilaria crotonensis*, *Synedra*, *Stephanodiscus*, and *Melosira granulata*.	Many other algae, especially greens and blue-greens during warmer periods of year; desmids if dissolved organic matter is fairly high
Eutrophic	Usually alkaline; nutrient enriched; common in warmer periods of temperate lakes or perennially in enriched tropical lakes	Blue-green algae, especially *Anacystis* (= *Microcystis*), *Aphanizomenon*, *Anabaena*	Other blue-green algae; euglenophytes if organically enriched or polluted

After Hutchinson, 1967.

Growth Characteristics of Phytoplankton

Analyses and evaluations of seasonal and spatial growth characteristics of phytoplankton are sometimes difficult because of the array of environmental factors involved, the individual physiological properties of each algal species and the magnitude of change that can occur in both. Clearly the organisms and environment are both highly dynamic. Some important factors regulating growth and succession are (a) light and temperature, (b) buoyancy regulation, i.e., means of remaining within the photic zone by alterations of sinking rates, (c) inorganic nutrient factors, (d) organic micronutrient factors and interactions of organic compounds with inorganic nutrient availability, and (e) biological factors of competition for available resources and predation by other organisms. Each species of algae possesses a range of tolerance to these factors, and population growth proceeds most rapidly at some optimal combination of the interacting factors. The optimal combination of factors required for greatest growth and productivity is probably seldom achieved under most natural conditions. Competitive advantage of one species over another is relative, and can change with changes in the physical and biotic conditions that affect growth.

LIGHT AND TEMPERATURE

The ecological effects of light and temperature on the photosynthesis and growth of algae are inseparable because of interrelationships between metabolism and light saturation. Qualitative factors of the selective attenuation of light energy with increasing depth have been discussed earlier (Chapter 5) in relation to photoreceptors and environmental parameters such as dissolved organic matter. Light intensity affects both rate of photosynthesis and algal growth. Response to light intensity, however, is species specific and in many cases, a considerable degree of adaptation to changing light intensities occurs.

Commonly, the intensity of light required to saturate algal photosynthesis increases as water temperature increases (Fig. 15-2A). In this type of response, exemplified by *Chlorella*, adaptation to higher or lower light intensities occurs mainly by changing the amount of pigment per cell (Steemann Nielsen and Jørgensen, 1968a; Jørgensen, 1969). Cells adapted to high light intensity have a lower chlorophyll *a* content per cell than those adapted to low light intensities. In *Chlorella pyrenoidosa*, for example, cells grown at 1 klux* have about 10 times more chlorophyll per cell than those grown at 21 klux. At high light intensities, but below light inhibition, the rate of photosynthesis per cell is not much greater than it is at low intensities (Fig. 15-2B).

In other algae, such as many diatoms, adaptation to changes in light intensity occurs by changing the light-saturated photosynthetic rate; the chlorophyll content of cells grown at low and high light intensities remains the same. The light-saturated rate of photosynthesis is generally much greater in algae grown under higher light intensities than it is for the same species grown under lower illuminance. Some algae are intermediate between the extreme types discussed here, or exhibit a combination of photoadaptive strategies (cf. Prézelin and Sweeney, 1978).

In light saturation curves, as exemplified in Figure 15-2, the upward sloping portion is the result of the physiological photochemical response of pigments to increasing light. At light saturation, in the horizontal portions of the family of curves, enzymatic reactions occur whose rates are dependent on the concentrations of active enzymes and temperature. Some algae, such as *Skeletonema*, adapt to lower temperatures by increasing enzyme concentrations; consequently the same rate of photosynthesis is achieved per given light intensity at high and low temperatures (see, for example, Jørgensen, 1968). In *Skeletonema* adapted to low temperatures, the amount of protein per cell was twice as great at 7°C as at 20°C, and the size of cells was greater at low temperatures. While rates of respiration are affected very little by changes in light intensity,

---

*See Appendix for conversions among units of illuminance.

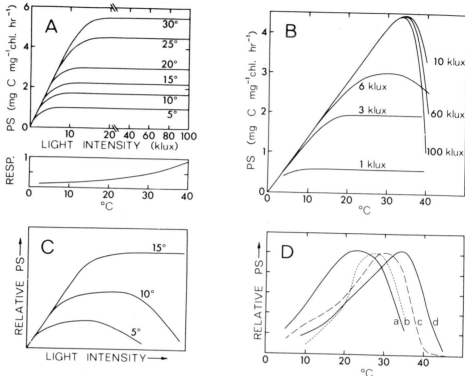

Figure 15–2. Interactions between photosynthesis (*PS*) and light intensity and temperature in plankton algae. *A*, Generalized photosynthetic rates (mg C fixed per mg chlorophyll per hour) in relation to increasing light intensities and rates of respiration at different temperatures; *B*, Commonly observed increasing inhibitory effects of high light intensities that are accentuated at higher temperatures; *C*, Effects of high light intensity on photosynthetic rates of algae adapted to cold temperatures, as in arctic lakes; and *D*, Optimum temperatures for photosynthesis of different algae: *a*, *Synedra*, *b*, *Anabaena*, *c*, *Chlorella*, and *d*, *Scenedesmus*. (After data of Aruga, 1965.)

they do increase with increasing temperatures (Fig. 15–2A). Photosynthetic rates of planktonic algae adapted to light and temperature conditions of arctic regions can be nearly as high as those observed in temperate regions during summer under relatively high light (see, for example, Steemann Nielsen and Jørgensen, 1968b).

Light of high intensity is detrimental to many algae (Fig. 15–2B, C). Ultraviolet light is particularly harmful to planktonic algae (Gessner and Diehl, 1952; Godward, 1962; van Baalen, 1968), but the effects are partly reversible if exposure is not too long, e.g., if plankton exposed to more intensive UV radiation at the immediate surface waters are circulated to deeper waters to which little UV penetrates (cf. Chapter 5). Effects of high light intensities depend upon the species in question, and some adaptation occurs with time. On bright days it is very common to observe a distinct depression in rates of in situ photosynthesis near the surface. The decrease in photosynthetic rates is associated with photo-oxidative destruction of enzymes, and not chlorophyll (Steemann Nielsen, 1962; Steemann Nielsen and Jørgensen, 1962). The depression of enzymatic processes at high light intensities takes some time to occur. If exposure is of short duration (several hours), reactivation can occur and permanent damage is avoided. A depression of pigment concentrations often occurs at midday on bright days. Some algae, such as the diatom *Cyclotella*, adapt rapidly (<24 hrs) to changing light intensities and are not inhibited at very high intensities (Jørgensen, 1964). Other algae, such as natural populations of phytoplankton in Antarctic lakes exposed to continuous light in summer or in Arctic lakes after loss of icecover in summer, are severely inhibited by high light intensities (Goldman, et al., 1963; Hobbie, 1964). In Arctic ponds, the phytoplankton exhibit increasing susceptibility to photoinhibition at cold (2 to 8°C) temperatures (Fig. 15-2C; Alexander, et al., 1980; Hobbie, personal communication). As water temperatures increase during the brief, 100-day period when the ponds are not frozen solid, adaptation to high illuminance occurs.

Interactions of light and temperature frequently result in one of several possible vertical profiles of photosynthesis in planktonic environments. In some cases, there is a zone of maximum photosynthetic rates at light saturation, which overlies a zone of near-exponential decline of rates with increasing depth (Fig. 15-3A). When light intensities exceed the photoinhibition threshold of the most productive species in the algal assemblage, a surface photoinhibition often occurs (Fig. 15-3B). The depth at which maximum rates of photosynthesis occur varies with the transparency of the water, which is governed by the concentration of dissolved and particulate organic matter and abiotic turbidity. When densities of phytoplankton, i.e., the biogenic turbidity, are high, self-shading effects can greatly reduce light penetration and the trophogenic zone is reduced accordingly (Fig. 15-3C). Although surface photoinhibitory effects occur, rapid light attenuation through self-shading of the algal populations reduces the severity of enzymatic damage. In some cases, the algal populations adapted to relatively low light intensities of deeper layers develop greatly and have higher photosynthetic rates than those in the epilimnion (Fig. 15-3D; cf. also Fig. 9-4A).

There is great diversity in tolerances to variations in temperature among the algae (Fig. 15-2D). The minimal temperature at which photosynthesis can occur depends upon the algal species. For many diatoms, the critical temperature is about 5°C; for others, about 15°C (e.g., Rodhe, 1948). In many green and blue-green algae, even higher temperatures are needed for incipient photosynthesis. The blue-green algae as a group are generally much more tolerant of higher temperatures than are other algae. Many species of blue-green algae have higher temperature optima than eukaryotic algae of the same waters. Further, thermophilic algae with optimal growth at temperatures above 45°C are almost exclusively blue-greens, and photosynthesis is known to occur at constant temperatures as high as at least 74°C (Castenholz, 1969). Respiration and especially photorespiration increase at high temperatures (Döhler and Przybylla, 1973). In the ensuing discussion of seasonal succession of algal populations, we will see again the frequently observed preponderance of diatoms at colder temperatures and the increasing diversity of algae as the waters warm.

FLOTATION MECHANISMS AND WATER TURBULENCE

The importance of water movements in the transport of particulate organic matter, in this case planktonic algae, has been emphasized at length earlier (Chapter 7). Not only are water movements important in the physical movement of algae into and out

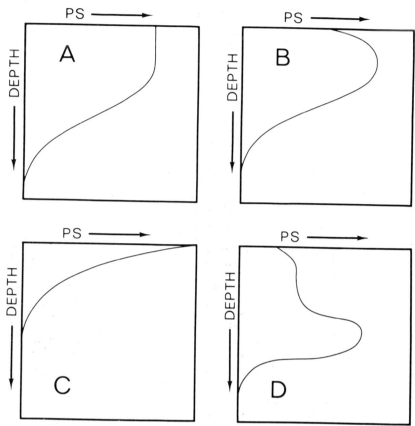

Figure 15-3. Generalized variations in vertical profiles of rates of photosynthesis (*PS*) among phytoplankton. *A,* Light saturation in surface waters, without inhibition, underlain with decreasing rates as light is reduced with depth; *B,* Profile when severe surface photoinhibition occurs; *C,* Photosynthetic rates when biogenic turbidity is very high; *D,* Photosynthetic maximum in the metalimnion.

of the photic zone, but they are critical in the vertical transport of mineralized matter from lower depths and littoral regions to the open water. The degree of turbulence and water movement are obviously critical in the regulation of algal periodicity and production.

The importance of basin morphometry to water movements must be emphasized again in relation to composition of planktonic algae aside from nutrient considerations. An excellent example is seen in the analyses of the spatial distribution of predominantly littoral desmid algae in several lakes of Minnesota (Bland and Brook, 1974). Many desmids occur throughout the year on and among submersed macrophytes of the littoral zone. During spring circulation, many desmid algae are dispersed, particularly to other areas of the littoral and can colonize newly growing annual macrophytes. In small lakes, circulation in the littoral is not as intense as it is in larger lakes. Larger desmids are carried out of the littoral less frequently in these smaller lakes. The more active circulation of larger lakes, however, transports a variety of desmids into the pelagic regions from the littoral zone, and this input increases the diversity of the planktonic communities.

The density of most freshwater planktonic organisms is 1.01 to 1.03 times that of the water. This slightly greater density causes them to sink when in undisturbed water. Spherical particles of a mean diameter of not more than 0.5 mm fall according to Stokes' law: the velocity of sinking varies inversely with the viscosity of the medium (cf. Chapter 2), directly with the excess density of the particle over that of the medium, and directly as the square of the diameter or some appropriate linear dimension (reviewed in Hutchinson, 1967). Thus, in planktonic habitats, velocity of sinking is greatest for relatively large, dense particles in open water (cf. Titman, 1975). It is apparent that sinking out of the photic zone is a distinct disadvantage for photosynthetic algae. However, movement of a cell through water, either by its own motility or by sinking, has the advantage of disrupting nutrient gradients around the cell and increasing the chances, because diffusion is so slow, of contact with nutrient molecules. Consequently, the disadvantage of sinking out of the photic zone, which is usually fatal, is offset to some extent by the advantage of increased nutrient acquisition.

Various characteristics are found among planktonic algae that improve flotation or reduce sinking rates (Lund, 1959, 1965; Fogg, 1965; Hutchinson, 1967). (a) Sinking rate is reduced by increasing *surface-to-volume ratios* (Table 15–3). The sinking rate of an algal cell is almost always less than a sphere whose volume and density are equal to those of the alga. The reduction in sinking rate due to cell morphology is called "form resistance." While the rough surface texture of the cell wall or membrane has relatively little effect on sinking, elongation into cylindrical or discoid shapes distinctly decreases settling rates. Projections or more elaborate protrusions such as spines decrease the settling rate by frictional resistance, and shift the orientation of the cell during descent. The physiological state of algal populations is of more importance to differential sinking rates of phytoplankton than is frictional resistance. Sinking rates can be directly related to nutrient assimilation and growth capabilities of the algae (e.g., Pasciak and Gavis, 1974; Titman and Kilham, 1976). Stationary phase populations sink several times more rapidly than exponentially growing populations (Table 15–3). Thus, in seasonal succession of algal populations, the demise of certain algae can be caused by changes in physiological states which *result* in accelerated sedimentation rates of a population. Hence, sedimentation per se may not be the cause of the decline (e.g. Knoechel and Kalff, 1975; cf. Lehman, 1979). (b) Production of *mucilage* reduces the sinking rate. Gelatinous sheaths occur in nearly all blue-green phytoplankton, some diatoms and green algae, and most desmids with rather elaborate projections and irregular cell morphology (Boney, 1981). The mucilage generally has a density less than that of the cell and increases both cell size and frictional resistance somewhat. The sheath, however, reduces the efficiency of nutrient uptake by creating an additional diffusional barrier. (c) *Gas vacuoles.* These vacuoles are found in the protoplasm of bacteria and blue-green algae. Each gas vacuole contains numerous hollow, cylindrical gas vesicles, the walls of which are composed of protein. The gas vacuole membranes and the walls of the vesicles are freely permeable to gases, and the gas contained in the vacuoles and vesicles is in equilibrium with gases dissolved in the surrounding water. Water is prevented from entering the gas vacuole by surface tension at the outer surface of the membrane (cf. reviews of Walsby, 1972, 1975; Fogg, et al., 1974; Reynolds and Walsby, 1975).

TABLE 15-3   Sinking Rates of Several Freshwater and Marine Algae in Relation to Area:Volume Ratios and Physiological State of Cells

CELL TYPE	AREA:VOLUME (μm^2:μm^3)	CELL SIZE (μm DIAMETER OR WIDTH $\times$ LENGTH)	SINKING RATE (m DAY^{-1})
Centric, Unicellular			
Thalassiosira nana	0.88–1.20	4.3–5.2	0.10–0.28
T. pseudonana	—	—	0.09
Cyclotella meneghiniana			
Exponential phase	—	2.0	0.08
Stationary phase	—	2.0	0.24
Coscinodiscus lineatus			
Exponential phase	—	50	1.9
Stationary phase	—	50	6.8
Centric, Elongate			
Rhizosolenia setigera			
Normal preauxospore	—	6 × 245	0.19–0.44
Spineless preauxospore	0.62–0.75	—	0.22–0.63
Postauxospore	0.10–0.16	33 × 363	0.79–1.94
Total population	—	—	0.63–0.95
R. robusta			
Exponential phase	—	84	1.1
Stationary phase	—	84	4.7
Centric, Chain-Forming			
Thalassiosira rotula	0.23–0.29	19–34	0.39–2.10
Skeletonema costatum	0.81–1.01	5 × 20	0.31–1.35
Bacteriastrum hyalinum	0.29–0.33	10 × 30	0.39–1.27
Chaetoceros lauderi	0.19–0.41	20 × 34	0.46–1.54
Pennate, Chain-Forming			
Nitzschia seriata	1.18–1.65	4 × 40	0.26–0.50
Asterionella formosa			
Exponential phase	—	25.0	0.20
Stationary phase	—	25.0	1.48
Melosira agassizii			
Exponential phase	—	54.8	0.67
Stationary phase	—	54.8	1.87
Fragilaria, natural populations			
Germany	—	—	0.5–1.0
England	—	—	0.2–0.65
Tabellaria fenestrata, natural populations	—	—	0–0.85
Scenedesmus quadricauda			
Exponential phase	—	8.4	0.27
Stationary phase	—	8.4	0.89

After data of Grim (1952), Smayda and Boleyn (1965, 1966a, 1966b), Pasciak and Gavis (1974), Knoechel and Kalff (1975), Reynolds (1976), Titman and Kilham (1976), and Beinfang (1979).

The gas vacuoles decrease the density of blue-green algae to below that of water, even though the volume of the vacuoles often occupies less than one per cent of the total cell volume. Blue-green algae possessing gas vacuoles float towards the surface and are not dependent only on turbulence, which is required by other algae to remain in the photic zone. Changes in the gas vacuole:cell-volume ratio determine whether the alga sinks or floats. The ratio changes as vacuole volume becomes proportionately less of the total volume in exponential growth phases. Gas vacuoles become more abundant when light is reduced and growth rate slows. Rises in turgor pressure of the cells as a result of the accumulation of photosynthate cause a collapse in existing gas vesicles and a reduction in buoyancy (Dinsdale and Walsby, 1972; Walsby and Booker, 1980). Additionally, the "dilution" of newly produced vacuoles during active cell division can reduce buoyancy. By these mechanisms, blue-green algae are able to regulate buoyancy and undergo limited vertical migration to poise themselves within physical and chemical gradients favorable to growth, usually towards the bottom of the euphotic zone (Fogg and Walsby, 1971). The population maximum, which results from population growth and movement downward, apparently often occurs as epilimnetic nutrient concentrations are depleted in summer (Klemer, 1978; Konopka, 1981). Vertical movement has the added advantage of disrupting nutrient gradients surrounding actively growing cells. The scattering of light by gas vacuoles is of slight benefit to blue-green algae in that they shield the cells from excessively bright light (cf. Walsby, 1975). Large colonial blue-green algae occur mainly in turbulent polymictic and epilimnetic waters, since these forms sink more slowly than helically shaped cells (Booker and Walsby, 1979). As turbulence subsides, these algae can adjust their position rapidly. As a result of these movements, large colonial blue-green algae commonly form algal blooms (cf. Reynolds, 1978). (d) The density of some algae is decreased by the *accumulation of fats.* Some algae, such as *Botryococcus*, can contain lipoids up to 30 to 40 per cent of dry weight and can float because of this fat accumulation (Fogg, 1965). Usually, however, fat accumulation occurs in senescent cells that are in various stages of degradation. (e) Finally, buoyancy alteration may also be achieved in some algae by means of ion regulation (cf. Anderson and Sweeney, 1977).

INORGANIC NUTRIENT FACTORS

The importance of inorganic macro- and micronutrients, particularly the two major elements, phosphorus and nitrogen, to algal nutrition has been emphasized in preceding chapters on biogeochemical cycling. Reiteration is not necessary here. Facets of algal nutrition and nutrient interactions will be discussed further in the sections on the periodicity and succession of phytoplanktonic populations.

The limiting nutrient concept, originally developed by Liebig in the middle of the last century as the "Law of the Minimum," has experienced considerable misuse, largely from attempts at over-simplification. Simply paraphrased, the law states that yield of any organism will be determined by the abundance of the substance that, in relation to the needs of the organism, is least abundant in the environment (cf. Hutchinson, 1973; Talling, 1979). Since yield is a result of growth, rate of growth has been substituted for yield in many subsequent analyses, the most important of which is the well-known Monod model for a single nutrient limitation in the growth of microorganisms (cf. Droop, 1973):

$$\mu/\mu_m = s/(K_s + s),$$

when

μ = specific growth rate (increase in biomass per unit biomass per unit time),

μ_m = maximum specific growth rate at infinite external substrate concentration, s = external substrate concentration (mass per unit volume),

K_s = saturation constant when the external substrate concentration results in half the maximal rate of uptake.

Irrespective of the internal nutrient concentration, the specific rate of uptake depends upon substrate concentration in a Michaelis function; the rate of uptake increases with increasing substrate concentration to a specific substrate level beyond which no further change in rate of uptake occurs.

Under conditions of steady state with nutrient concentrations that are not limiting, the specific rate of nutrient uptake (u, mass per unit biomass per unit time) equals the product of the specific growth rate (μ) and the cell nutrient quota (Droop, 1973):

$$u = \mu q$$

where q is a demand coefficient or cell nutrient quota (content) in the absence of nutrient excretion [(internal substrate concentration) (mass per unit biomass)]. The specific rate of uptake and the cell nutrient quota have a linear relationship:

$$\mu q = \mu_m(q - k_q) \text{ or } \mu/\mu_{m'} = 1 - k_q/q$$

where

$\mu_{m'}$ = maximum specific growth rate at nonlimiting internal substrate concentration,

k_q = the subsistence quota, i.e., the q intercept for zero μ.

The importance of these models, which are confirmed by many studies on planktonic algae (e.g., J. Goldman, 1977), lies in their applicability to obtaining specific growth values from more easily determined uptake rates (Eppley and Thomas, 1969).

It is important to view nutrient uptake as it really takes place, since uptake of several nutrients occurs simultaneously. Nutrient limitations of photosynthesis can be highly dynamic both spatially and temporally (Wetzel, 1972). Of importance among the many interacting, potentially limiting nutrients is the relative position of limitations within an *intensity* spectrum which undergoes constant change spatially and temporally.

The model of Droop given above can be expanded to the multinutrient condition, based on the fact that a limiting parameter (in this case μ/μ'_m) is proportional to the product of two or more functions if it is independently proportional to each of them. Therefore,

$$\mu/\mu_{m'} = (1 - k_{qA}/q_A)(1 - k_{qB}/q_B)(1 - k_{qC}/q_C)(\ldots\ldots,)$$

is a simple polynomial in which subscripts $A, B, C, \ldots$, refer to various nutrients. In the latter equation, all nutrients are formulated equally; the significance of a particular nutrient in establishing the composite limiting parameter (μ/μ'_m) is determined by relative magnitude of k_q compared with q. The parameter μ'_m is an abstract proportionality constant in this case, whose value is determined by all factors (physical, nutritional, etc.) excluded from the equation.

The effectiveness of individual nutrient limitations has been shown in a number of studies (e.g., Droop, 1974, 1977; Rhee, 1974, 1978). Frequently, a relatively sudden shift occurs in the transition from limitation by one nutrient to limitation by another nutrient. These results demonstrate the restricted usefulness of the simple multiplicative polynominal formulation given above as a descriptor of simultaneous limitation of growth or nutrient uptake by two or more nutrients. Although the shift from one nutrient limitation to another can be expressed mathematically (Droop, 1977), the predicted growth rates are likely serious underestimates for a given algal population (Talling, 1979).

Complications in simple multiplicative predictive analyses of algal population growth also result when synergistic effects occur between nutrients. An example mentioned earlier was an enhanced rate of photosynthesis resulting from enrichment with both phosphorus and manganese, but not with each nutrient alone (Shapiro and Glass, 1975). As will be discussed later (Chapter 17), among the bacteria, nutrient uptake can be influenced by the previous substrate (nutrient) concentration. If lack of a nutrient previously has limited growth severely, the uptake sites for that nutrient can be reduced. Replenishment of the limiting nutrient may not result in an immediate response of uptake and growth. This fact has important implications for short-term enrichment bioassays since effects of the added nutrient may not be realized in population growth for a considerable period, possibly longer than the duration of the experiment.

In fresh waters of low nutrient concentrations and slow turnover rates, it is metabolically advantageous to be both small and motile. Small size increases the ratio of absorptive surface area to cell volume. Since growth rates are slow and may not be adequate to offset mortality by sedimentation and predation, small size is further advantageous in reducing settling rate and possibly predation by grazers. Motility also is effective in disrupting nutrient gradients surrounding cell membranes during active growth. These relationships suggest that one would anticipate a greater percentage of the algal populations of oligotrophic lakes to be small and motile. In most oligotrophic lakes, this is indeed the case. Usually a much larger percentage of the total rates of primary productivity derives from nanno- and ultraplanktonic algae of oligotrophic lakes than from larger algae. In some extreme cases, all primary productivity is by algae smaller than 30 μm.

Organic Matter and Algal Growth

Dissolved organic matter is intimately coupled with the nutritional requirements of algae for organic growth factors. In addition, the availability of certain organic compounds is dependent upon the quantity and types of dissolved inorganic compounds that are present. Furthermore, a number of algae can supplement primary photosynthetic autotrophy by the uptake and utilization of organic substrates. Algae capable of synthesizing required organic growth factors, e.g., vitamins, have a distinct competitive advantage over algae that must rely on exogenous supplies synthesized by other microorganisms.

In addition to direct utilization as a micronutrient or utilization as supplementary energy and carbon sources, organic compounds can function as nonessential accessory substances that may stimulate growth of the algae. Growth substances or nonessential organic micronutrients include a large group of hormones known to be produced by algae and effective in growth regulation and cell development (Bentley, 1958a, 1958b; Conrad and Saltman, 1962; Provasoli and Carlucci, 1974). Little is known of the physiological mechanisms of these substances in freshwater algae, although they are probably of some significance in the success and succession of certain more specialized algae under natural conditions.

ORGANIC MICRONUTRIENT REQUIREMENTS

Knowledge of the specific requirements of phytoplanktonic algae for growth factors or essential organic micronutrients, especially the vitamins, is extensive (cf. reviews of Provasoli, 1958; Hutner and Provasoli, 1964; Provasoli and Carlucci, 1974). Although most higher plants do not require vitamins, many algae do; species that require vitamins for growth are termed *auxotrophic*. In the case of auxotrophic growth, the requirements for specific organic compounds are low and these compounds do not contribute significantly to the total cell carbon.

Only three water-soluble vitamins are known to be required by algae: vitamin B_{12} (cobalamine), thiamine, and biotin. Vitamin B_{12} and thiamine are required alone or in combination by a majority of the auxotrophic algae, and B_{12} is required more often than thiamine. Biotin is known to be required by a few chrysomonads, dinoflagellates, and euglenoid algae.

There is no strict taxonomic correspondence to the known distribution of the requirement for external vitamins (auxotrophy) among different algae. The differences, however, are sufficient to permit some generalizations. On the basis of relatively few analyses in pure cultures, a significant number of blue-green algae, diatoms, green algae, and dinoflagellates require exogenous sources of vitamin B_{12}, i.e., they are auxotrophic (Table 15-4). Only a few groups show any significant requirements for thia-

TABLE 15-4 General Requirements for Vitamins Among the Algae

ALGAL GROUPS	BIOTIN	THIAMINE	VITAMIN B_{12}	PREDOMINANT VITAMIN REQUIREMENTS
Cyanophyceae	0	0	+ +	B_{12}
Rhodophyceae	0	0	+ +	B_{12}
Bacillariophyceae	0	+	+ +	B_{12}
Xanthophyceae	0	0	0	None
Phaeophyceae	0	0	+	None
Chlorophyceae	0	+	+ +	B_{12}
Chrysophyceae and Haptophyceae	−	+ +	+	Thiamine
Cryptophyceae	−	−	+	None
Dinophyceae	+	0	+ +	B_{12}
Euglenophyceae	−	−	+	None

After discussion of Provasoli and Carlucci, 1974.
+ + = required in many species
+ = few species
− = requirement rare
0 = no known requirement

TABLE 15-5 Range of Concentrations of Organic Micronutrients Found in Fresh Waters (in $\mu g\ l^{-1}$ or mg m^{-3})

LAKE	VITAMIN B$_{12}$ (CYANOCOBALAMIN)	THIAMINE	BIOTIN	NIACIN (NICOTINIC ACIDAMIDE)	PANTOTHENIC ACID	FOLIC ACID
Linsley Pond, Conn. (Hutchinson, 1967; Benoit, 1957)	0.06–0.075	0.008–0.077	0.0001–0.004	0.15–0.89	—	—
Northern German lakes (Hagedorn, 1971)	—	0.05–12	—	—	—	—
Sagami, Japan (Ohwada and Taga, 1972)	0.005–0.85	0.001–0.38	0.010–0.068	—	—	—
Tsukui, Japan (surface) (Ohwada, et al., 1972)						
Dissolved	0.0005–0.0042	0.075–0.436	0.013–0.058	—	—	—
Particulate	0.0019–0.0203	0.031–0.159	0.0005–0.0042	—	—	—
Kasumigaura, Japan (Kashiwada, et al., 1963)	0.005–0.028		0.0021–0.050	0.30–3.3	0.01–0.26	0.040–0.244
Small Swiss Ponds (Clémencon, 1963)	—	<0.001–1.0	<0.001–0.004	<0.001–3.0	<0.01–0.034	<0.01–0.48
Kojima, Japan (Nishijima and Hata, 1977)	0.0021–0.036	0.020–2.30	0.0023–0.1016	—	—	—
Mergozzo, Italy (Kurata, et al., 1976)	0.0004–0.0025	0.047–0.334	0.005–0.019	—	—	—
Biwa, Japan (Kurata, et al., 1979a)						
Pelagial zone	0.001–0.006	0.008–0.095	0.003–0.015	—	—	—
Littoral zone	0.002–0.018	0.015–0.250	0.004–0.025	—	—	—
Mean of Precipitation, Missouri (Parker and Wachtel, 1971)						
April–November	0.001	—	0.004	2.0	—	—
November–March	0.0004	—	0.0008	0.42	—	—

mine. Only the xanthophycean algae exhibit no apparent need for these major water-soluble vitamin growth factors, although in several other groups the requirements are rare (Table 15-4). The Cyanophyceae, Chlorophyceae, Xanthophyceae, and Phaeophyceae exhibit the least number of species requiring vitamins. Most of the species in these groups are strictly autotrophic in metabolism. In contrast, a clear predominance of auxotrophic species occurs in the Chrysophyceae, Dinophyceae, Cryptophyceae, and Euglenophyceae.

When algae are unable to synthesize an organic micronutrient, the need for an exogenous source of the correct chemical moiety is obvious. Most of the vitamins are synthesized by bacteria and certain algae and are released to the water during active growth and upon death (Carlucci and Bowes, 1970; Aaronson, et al., 1977; Grieco and Desrochers, 1978; Nishijima, et al., 1979). Appreciable concentrations of vitamins derived from airborne soil particles, pollen, or active microorganisms (Parker and Wachel, 1971) can enter aquatic ecosystems from precipitation (Table 15-5). However, very low concentrations of organic micronutrients, especially vitamin B_{12}, which is in widespread demand, exist in fresh waters (Table 15-5). These concentrations, however, yield little insight into rates of turnover between supply sources and demand. All assays of the absolute vitamin requirements of algae are necessarily based on bacteria-free cultures, and in most cases indicate that the requirements are much lower than observed concentrations in fresh waters.

Studies of seasonal variations in concentrations of vitamin B_{12}, thiamine, and biotin have demonstrated large changes with time (Tal, 1962; Daisley, 1969; Kashiwada, et al., 1963; Kurata, et al., 1976, 1979a; Cavari and Grossowicz, 1977; Nishijima and Hata, 1977; Parker, 1977). In many cases, concentrations of vitamin B_{12} and thiamine, but seldom biotin, were negatively correlated with densities of phytoplankton, particularly small algae (<30 μm). Concentrations of the water-soluble vitamins commonly decrease with increasing water depth. In the surficial profundal sediments, vitamin concentrations are 100 to 1000 times greater than in the overlying water (e.g., Nishijima and Hata, 1978; Kurata, et al., 1979a). Concentrations of vitamin B_{12}, thiamine, and biotin also change diurnally, with higher concentrations occurring in the morning than in the evening (Kurata, et al., 1979b). Presumably these differences result from algal utilization in daylight exceeding inputs from synthesizing organisms.

Concentrations of the water-soluble vitamins have been found to be consistently higher in the littoral zone than in the open water (Kurata, et al., 1979a). The data suggest that epiphytic microorganisms associated with submersed macrophytes are major net exporters of vitamins, especially B_{12}, to the open water.

Concentrations in natural waters are so low that they must be determined by bioassay organisms of known growth responses to concentrations of the micronutrients. It must be emphasized that concentrations utilizable by the assay organism may not be equally utilized by algae. Moreover, evidence indicates that some or many of the natural organic micronutrients may not be readily available for assimilation. Photosynthetic rates and cell numbers of natural phytoplanktonic populations frequently increase markedly by additions of vitamin B_{12} or thiamine (see, for example, Wetzel, 1965a, 1966b, 1972; Hagedorn, 1971). Concentrations needed, especially in hard-water lakes, are often greatly in excess of those generally known to be required by algae in pure culture. White and Wetzel (unpublished) have shown experimentally that much vitamin B_{12} is adsorbed to particulate calcium carbonate, and is removed from the trophogenic zone by coprecipitation and sedimentation. Other losses are likely. It is doubtful whether the rates of synthesis generally are limiting to total growth and productivity of phytoplankton of lakes. However, large vertical seasonal fluctuations occur, and these organic micronutrients probably play a significant role in the succession and competitive success of phytoplankton populations.

HETEROTROPHY OF ORGANIC CARBON BY ALGAE

By far the dominant mode of metabolism among the algae is photosynthesis, in which cell carbon is obtained by reduction of carbon dioxide. Thus, a majority of algae are obligate *photoautotrophs*: they require light energy for these transformations, and have evolved light-receptor pigment systems.

In pure culture, an appreciable number of algae assimilate and utilize dissolved organic compounds as a source of carbon and energy both in the dark and the light. *Heterotrophy* or chemo-organotrophy in algae implies the capacity for sustained growth and cell division in the dark; energy and cell carbon are obtained from the metabolism of an organic substrate(s). Heterotrophy in algae occurs by means of aerobic dissimilation, and carbon dioxide may or may not be required. Excellent reviews on the physiology and biochemistry of heterotrophy by algae are given by Droop (1974) and Neilson and Lewin (1974).

A few algae are *mixotrophic*, and assimilate carbon dioxide in small amounts simultaneously with organic compounds both in the light and especially in darkness. In other words, photoautotrophy in the light can be supplemented by the assimilation of organic compounds in the dark. True *photoheterotrophy*, in which organic substrates serve as a significant source of cell carbon during growth, is more restrictive in that different metabolic pathways are involved.

In comparison to bacteria and fungi, heterotrophic algae can utilize only a few substrates, such as acetate and related compounds (pyruvate, ethanol, lactate, higher fatty acids), glycolate, hexose sugars, and amino acids. None of the algae known to grow heterotrophically can do so under anaerobic conditions. The ability to utilize organic substrates requires specific enzymes for transport across cell membranes, and many algae are deficient in these enzymes. In certain species that possess the enzymes, organic substrates still are not utilized because of an apparent impermeability of the cells or an inability to couple their dissimilation to the generation of adenosine 5'-triphosphate (ATP).

It must be emphasized that in a majority of cases in which algal heterotrophic utilization of organic substrates has been demonstrated, the concentrations of substrates were very high, often several orders of magnitude greater than those found in most natural waters. Furthermore, culture conditions have been bacteria-free, which eliminates the competitive interactions of bacteria that possess much more efficient active uptake mechanisms. Nearly all analyses in situ with natural populations have led to the following conclusions: (a) At naturally occurring substrate concentrations, the low affinity of algae for simple organic substrates makes heterotrophy a relatively unimportant process in comparison to photoautotrophy. (b) Algae cannot compete effectively with bacteria for available substrates (e.g., Wright and Hobbie, 1966). This subject is discussed at greater length later on (Chapter 17).

While it is obvious that photoautotrophic metabolism is the primary means of synthesis among algae in lake systems, it is incorrect to dismiss heterotrophic metabolism in natural populations of algae as unimportant. Indeed, there is evidence that heterotrophy and photoheterotrophy can be instrumental in subtle but important ways in the survival, competition, and succession of algal populations. Under conditions of light limitation, as occur deep in the hypolimnion, in turbid reservoirs, under heavy snow and ice cover, at extremely high latitudes at which total darkness exists for a period, or when very dense populations lead to shading, an ability to be facultatively heterotrophic (mixotrophy) is competitively advantageous. Observations of apparently viable and even reproducing phytoplanktonic algae under ice in northern Scandinavia during almost total darkness are incompletely explained (Rodhe, 1955). Some evidence exists that phytoplankton can compete with bacteria for acetate, but not glucose, under low light conditions (Maeda and Ichimura, 1973; Vincent, 1980; Vincent and Goldman, 1980). Certain deep-water blue-green algal populations, especially *Oscillatoria*, are facultatively heterotrophic under very low light conditions (Saunders, 1972c). Earlier remarks should be recalled on the possible utilization by these populations of extremely low light intensities to generate ATP by photophosphorylation without concomitant CO_2 reduction, which could be potentially important in maintaining viability without significant growth.

When organic loading is very high, such as in sewage oxidation ponds, algae such as *Chlamydomonas* can assimilate most of the acetate (generated by anaerobic bacteria) by photoheterotrophy, which involves photosynthetic production of ATP and reducing power (Eppley and MaciasR, 1963). While photoheterotrophy is generally not of quantitative importance in most natural waters, evidence indicates that it can amount to at least 20 per cent of total inorganic carbon fixation at low light intensities in certain oligotrophic lakes (McKinley and Wetzel, 1979). Two periods of relatively high photo-heterotrophic uptake activity were observed: one during spring turnover and a second during the late summer stratified period. Greatest photoheterotrophic contributions to overall carbon cycling occurred during morning-to-midday periods and at intermediate depths of the lower metalimnion. Microautoradiographic techniques have demonstrated selective utilization of organic substrates by algal and bacterial species (cf. Ellis and Stanford, 1982).

OTHER EFFECTS OF DISSOLVED ORGANIC MATTER

The extracellular production and release of antibiotics or growth inhibitors have been suggested in numerous cases from observations of the inhibitory effects of one alga on another in both mixed cultures and natural populations (Proctor, 1957; Krauss, 1962; Lefèvre, 1964; Hellebust, 1974). The chemical composition of these compounds is poorly understood. Inhibitory compounds include peroxides of fatty acids and possibly polyphenolic substances. Stimulatory compounds include a number of weak organic acids, especially glycolic acid. The effectiveness of antibiotics is governed in part by species specificity, relative concentrations, and rates of bacterial degradation. Such substances may play important roles in the succession of species under natural conditions, and are a fertile field for investigation.

Other effects of dissolved organic compounds include a number of ways in which the organic substances affect the availability of inorganic micronutrients (Saunders, 1957; Wetzel, 1968). Chelation of metal ions, an equilibrium reaction between a metal ion and an organic chelating agent, results in the formation of a stable ring structure incorporating the metal ion. Chelation can function in several complex ways in accordance with environmental conditions. The extent to which chelation of metallic ions occurs within a heterogeneous solution, such as lake water, is governed by the bonding characteristics of the organic compounds, the ratio of chelator to metal ions, and stability constants of chelates for different ions. Temperature, hydrogen ion activity, and the concentration and ionic strength of various anions further influence the extent of complexing. Other organic and inorganic compounds present in fresh waters function as sequestering agents for metals and cations (cf. Chapters 10 and 14). Complex formation occurs with pyrophosphate, and metals are bound by the complex formation with macromolecules such as proteins (Povoledo, 1961), and by the formation of peptidized metal hydroxides of yellow substances (Shapiro, 1964).

Functions of organic complexing of metal ions and major cations are only beginning to be appreciated as factors influencing the specific composition and succession of algal populations. Examples of some of the mechanisms involved include: (a) increases in the availability of inorganically reactive ions, such as iron and manganese (Chapter 14); (b) modifications of membrane permeability and osmoregulation by complexing of cations, which results in changes in monovalent:divalent cation ratios (Chapter 10); (c) if a trace metal is toxic to a particular organism, its availability may be reduced to a concentration below the threshold of toxicity by excessive complexing of the metal ions by organic compounds; (d) under certain circumstances, organic compounds may complex and effectively compete for a metal ion that is antagonistic to a toxicant, and thereby increase the concentration of the toxicant to harmful levels. (e) In addition, as discussed earlier (p. 273), evidence exists for the formation of organophosphorus-humic compounds that may reduce phosphorus availability to phytoplankton (Stewart and Wetzel, 1982).

SEASONAL SUCCESSION OF PHYTOPLANKTON

In spite of the number of pervasive generalizations about the common seasonal succession of phytoplanktonic algal populations in fresh waters, close inspection of existing data shows a great diversity of patterns. Some of the disparity is related to the choice of study methods employed. For example, many of the older analyses were based solely on number of organisms, which is an extremely biased indicator in comparison to biomass, because of great differences in size among algae. Furthermore, early investigators often used plankton nets of fairly large mesh size, through which significant and variable portions of the algae were lost.

Descriptions of seasonal succession of phytoplankton in north temperate, dimictic lakes frequently include correlations of changes in dominant algal species, biomass, numbers, and rates of photosynthesis with alterations in physical and biotic factors. From these descriptions, a general pattern of phytoplankton succession has emerged. However, many variations on the general-succession theme exist, particularly in lakes of other geographical regions (tropical, alpine, and polar). Although correlations are useful in pointing to possible areas of interaction for intensive study, they do not provide specific information on causal mechanisms. Briefly, the temperate pattern of succession involves a winter minimum of small flagellates adapted to low light and temperature, a spring burst of diatom activity and biomass, followed rapidly by a smaller development of green algae and a transitional lull between spring and summer. Summer populations vary in relation to the trophic status of the lakes, but can include either another diatom development in less productive lakes by late summer and early autumn or increases in nitrogen-fixing blue-green algae in eutrophic lakes.

The discussion that follows integrates a detailed description of phytoplankton succession, correlations with changes in environmental parameters, and recent information on mechanistic explanations of observed patterns. In this latter area, our knowledge is still incomplete.

Seasonal Patterns and Periodicity

A distinct periodicity in the biomass of phytoplankton is observed in polar and temperate fresh waters. Growth is greatly reduced or negligible during the winter period of low light and temperatures. Phytoplankton numbers and biomass normally increase in the spring under improved light conditions, and build up to a spring maximum. The spring maximum can begin under the ice in late winter, and often consists predominantly of diatoms adapted to low temperatures. In many dimictic lakes, the spring maximum does not develop fully until after spring circulation and the onset of summer stratification.

In many tropical lakes surrounded by extensive, nutrient-buffering wetlands and receiving perennial river inflow, total phytoplanktonic biomass and productivity are often both larger and more constant seasonally than those found in temperate lakes. Abrupt changes in abiotic factors (e.g., wind-induced vertical mixing; marked seasonality in rainfall and associated nutrient loading and turbidity) can be more frequent in tropical lakes than in stratified temperate lakes, which leads to a series of numerous episodic changes in the annual phytoplanktonic succession and productivity (e.g., Talling, 1969; Lewis, 1978a, 1978b; Melack, 1979). The episodic changes are often very regular in their seasonality from year to year. Occasionally, major changes (for example, increased salinity from unclear causes) can lead to abrupt alterations in algal assemblages and types of photosynthetic activity. Phytoplankton succession within any given episode, however, is similar in tropical and in typical temperate lakes.

The spring maximum of phytoplanktonic biomass is generally shortlived, usually less than one to two months in duration. This maximum often is followed by a period of low algal numbers and biomass that may extend throughout the summer. Among more eutrophic lakes of the temperate region, the summer minimum is often brief and phases into a late summer profusion of blue-green algae that persists into the autumn, until thermal stratification is disrupted. Populations of phytoplankton are often low throughout the summer in temperate oligotrophic lakes, but may develop a second maximum in the autumn period. This second maximum of the autumn, which often consists predominantly of diatoms, is generally not as strongly developed as that of the spring period. Decline of the autumnal populations into the winter minimum is often abrupt and more irregular than the decline after the spring development.

The limited growing season of lakes of high altitudes and in polar regions often is reflected in a conspicuous, single, summer maximum of phytoplanktonic biomass. By contrast, the maximum in tropical lakes often is observed during the winter.

Generalizations are difficult to make because of the great variability observed among phytoplanktonic numbers and biomass from lake to lake. Several points are reasonably consistent, however: (a) The seasonal periodicity of phytoplanktonic biomass is reasonably constant from year to year. If the freshwater system is not perturbed by outside influences, such as the activities of man in modifications of the watershed, nutrient loading, etc., the seasonal changes in the phytoplanktonic populations are similar from year to year. (b) The extent of change in phytoplanktonic numbers and biomass through the seasons is usually very great, on the order of a thousandfold in temperate and polar fresh waters. In keeping with the relatively constant environmental conditions, the seasonal variation in tropical waters is much lower, often as little as fivefold (Fogg, 1965). (c) The maxima and minima observed in numbers and biomass of phytoplankton often are quite out of phase with rates of primary production. Primary productivity usually follows the annual cycle of incident solar radiation in temperate, less productive lakes more closely (see following discussion). The spring maximum, if conspicuously developed, often is composed of larger algae such as diatoms, which have slower turnover rates than summer algae, which occur in warmer and more favorable light conditions.

Along with seasonal changes in total biomass, a periodicity of species composition occurs continuously. The species composition fluctuates in a regular way from year to year, as long as the system is not seriously perturbed. On an annual basis, algal communities consist of a composite of perennial species (holoplanktonic) that are present throughout the year, and of intermittent species (meroplanktonic) that periodically undergo some type of diapause in resting stages. In both cases, but much more markedly in the latter type, an interplay of environmental conditions results in fluctuations in growth and competition among other species. Among the perennial species, population numbers can decrease to extremely low levels, but cells are present in the water in sufficient numbers to reinoculate the community when growth conditions improve.

Winter Populations and the Spring Maximum

In temperate and arctic lakes, both the biomass and productivity of phytoplanktonic algae are usually low during winter. If nutrients are adequate, growth is limited to species that are adapted to low temperatures and low light irradiance. Population accrual by slow growth often is offset by respiratory losses, secretion of organic compounds, and sedimentation of cells from the limited photic zone under ice cover. In polar regions, inappreciable or zero growth exists under conditions of midwinter near-total darkness and very heavy ice and snow cover. At lower latitudes, within the temperate zone or at latitudes influenced by maritime amelioration of winter conditions, snow and ice are less thick and opaque, allowing more penetration of light.

The community of winter algae beneath ice is usually dominated by small and often motile algae. Cryptophyceans, such as *Rhodomonas* and *Cryptomonas*, are particularly common, as well as pyrrophyceans *(Gymnodinium)*, small green algae *(Chlamydomonas)*, chrysophyceans *(Dinobryon, Mallomonas, Chrysococcus, Synura)*, some diatoms *(Synedra, Tabellaria, Fragilaria)*, and euglenophyceans *(Trachelomonas)* (Wright, 1964; Rodhe, 1955; Verduin, 1959; Tilzer, 1972; and Maeda and Ichimura, 1973, among many others). The cold-water, low-light-adapted species are evidently quite species-specific in their tolerances to light quantity and quality, and are found in narrow strata in the photic zone beneath the ice cover. For example, over 95 per cent of the phytoplankton occurred in layers of the water column of an ice-covered lake of eastern Massachusetts receiving 0.5 to 20 percent of incident radiation (Wright, 1964). As discussed earlier, there is some evidence for mixotrophic growth in which photoautotrophy is supplemented somewhat by heterotrophic assimilation of simple organic substrates under darkness or very low irradiance. However, photoautotrophy clearly dominates the metabolism of these algae, even under exceedingly low light conditions. Changes in light, as occur with snowfall on ice, result in a rapid vertical shift in the distribution and rates of photosynthesis (Fig. 15–4).

Rates of primary production under ice cover can constitute a very significant portion of the total annual primary productivity of the phytoplankton. Nearly one-quarter of the annual phytoplanktonic productivity of hypereutrophic Sylvan Lake in northeastern Indiana occurred under the three-month period of ice cover (Wetzel, 1966a). Similar values have been found since in numerous lakes in the temperate zone, emphasizing that the widely held assumption that winter productivity is insignificant is not universally valid.

As ice cover is lost in the spring and the season progresses, circulation of the water column results in mixing of nutrient-laden water from lower depths with surface strata, which have more light. In shallow lakes in which the light penetrates throughout much

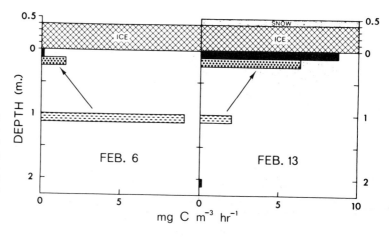

Figure 15–4. Rates of primary production beneath ice (6 February) and below ice and snow (13 February). Rectangles with dots are rates obtained when samples taken from 1 m were incubated alongside the surface samples. (Modified from Wright, 1964.)

of the water column, the spring algal maximum is often initiated by cold-water, poly-photic* species under ice in late winter. These species flourish and develop throughout the period of circulation and may persist for a time after summer stratification begins. Circulation of small lakes is commonly weaker and of shorter duration than it is in larger lakes. Among large lakes, net growth of phytoplankton can be significantly sup-pressed during circulation if mixing occurs at such a rate and to such a depth that the algae are carried out of the photic zone faster than they can multiply. The effects of circulation are accentuated in regions, such as much of the United Kingdom, in which lakes rarely freeze or do so for only brief periods. In these areas, circulation can con-tinue practically all winter, with a significant loss of phytoplankton occurring as a result of outflow and other mechanisms (e.g., Lund, et al., 1963).

There is little question that increasing light in the spring is the dominant factor contributing to the development of the spring "outburst," because water temperatures are still low. The spring maximum is frequently dominated by one species, a diatom, such as *Asterionella* discussed earlier (Fig. 14–18), *Cyclotella*, or *Stephanodiscus* (see, for example, Pechlaner, 1970). While a lag phase often is not seen before the dominants reach exponential growth phases, it obviously exists and is observed if one samples frequently (e.g., Lund, 1950). In Lake Erken, Sweden, some lag in the response of the algae constituting the spring maximum to the increase in radiation was caused by the necessity for light adaptation, even when the intensities of photosynthetically usable energy were still low (Pechlaner, 1970).

The relative rates of increase† of the dominant algae during the spring maxi-mum of exponential growth are usually much less than rates of increase observed in cultures grown under "optimal" conditions of light and temperatures (Fogg, 1965). This difference is variously attributable to restrictions of light, temperature, nutrients, and losses of cells by sedimentation and other causes.

The combined effects of nutrient limitations and suboptimal temperatures on growth of diatoms and green algae in cultures showed that the algae required more nutrient as temperatures declined from their optimal temperature range (Rhee and Gotham, 1980, 1981a). The optimum atomic ratio of nitrogen and phosphorus con-tained in the cells (i.e., the ratio at which one nutrient limitation shifts to another) increased at suboptimal temperatures. The combined effects of nutrient limitation and suboptimal temperature on growth were greater than the sum of the individual effects, and were not multiplicative. The larger species, such as major spring diatoms, have generation times† of about 4 to 8 days, which is somewhat longer than those of smaller species under natural conditions. The specific growth rates of the dominant species are considerably higher than those of the entire algal assemblage combined (e.g., Pechlaner, 1970). Continuous cultures simultaneously limited by light and nutrients also showed that the combined effects of these two factors on algal growth were greater than the sum of individual effects, and that the effects were not multi-plicative (Rhee and Gotham, 1981b). Under both nutrient-limited and nutrient-suffi-

---

*Adapted to a wide range of light intensities.

†The generation time (G) is the mean doubling time if the cells divide by two or $G = \dfrac{\log 2}{k'} = \dfrac{0.301}{k'}$, when

k' = the relative growth constant under exponential conditions, or $k' = \dfrac{\log_{10} N - \log_{10} N_0}{t}$ when N_0 is the

initial concentration of cells, N is the final cell concentration, and t the time interval.

cient conditions, the cell-nutrient quota needed for growth increased as irradiance decreased, indicating that nutrient requirements increase as irradiance decreases.

It should be noted that cold-, low-light-adapted species of algae that are well developed under winter conditions can be severely limited in the spring, or develop only in deeper strata in which temperatures and irradiance remain low. Population development in deep strata may result from actual migratory movement, as occurs among the dinoflagellates of alpine lakes, or the development of populations in lower strata, through sedimentation of cells from shallower strata, as is the case among some ubiquitous blue-green algae, e.g., *Oscillatoria rubescens*. Findenegg (1943), and others since, have demonstrated the characteristic weak development of *Oscillatoria* in upper layers under winter conditions. Before, during, and after spring circulation, the *Oscillatoria* population weakens in the upper strata and increases in the deep, cold layers of the lower metalimnion and upper hypolimnion.

In high-mountain lakes above the tree line, the nannoplankton dominate the phytoplankton, and are autotrophic throughout the long winter (Pechlaner, 1971). These algae are adapted to the low light intensities that penetrate heavy ice and snow cover. Both the concentration of algae and rates of in situ photosynthesis shift from the surface layers in winter to deep water strata during spring and summer. Immediately before ice breakup, the increasing light induces a downward migration of predominating dinoflagellates.

Spring Decline and Summer Populations

The decline of the spring maximum of phytoplankton and onset of summer populations in temperate lakes also is associated with a complex interaction of physical and biotic parameters. In many straightforward cases, reduction of nutrients in the photic zone of the epilimnion is responsible for slowing the growth of populations of the dominant as well as rarer algae. Since diatoms are often the dominant component of the spring maximum, especially in temperate fresh waters, silica concentrations are often reduced after extensive growth. The reduction of silica concentrations to limiting levels (<0.5 mg l^{-1}; cf. S. S. Kilham, 1975) has been shown repeatedly in many fresh waters, but is especially well documented in the northern British lakes, in which maximum concentrations are low (2 to 3 mg l^{-1}) (Fig. 15–5; see also Fig. 14–18). Reduction of silica to limiting concentrations occurs in less than two months when turbulence is low and sedimentation of diatom frustules occurs rapidly (Fig. 15–5).

In calcareous lakes in which silica concentrations are naturally high (>10 mg l^{-1}) silica levels can still be reduced appreciably during the spring and summer in the epilimnion (e.g., Fig. 14–17, upper). Utilization by diatoms continues throughout the period of the summer stratification. As hard-water lakes become more productive, the rapidity of silica reduction increases. For example, in extremely eutrophic Wintergreen Lake (Fig. 14–17, *lower*), concentrations are reduced to and below generally accepted inhibiting levels in both winter and spring periods.

The correlation between the observed decline of the diatom maximum with diminishing silica concentrations and the experimental results using bioassays are so distinct that the relationship appears to be predominantly causal. Factors of light intensity, temperature, nitrogen, phosphorus, zooplankton grazing, and fungal parasitism do not appear to be instrumental in the decline. However, much more complex interactions of both physicochemical and biotic factors undoubtedly exist in the seasonal succession of diatom populations when silica concentrations are not reduced to levels limiting to exponential growth. For example, the decline of the dominant spring diatom, *Asterionella formosa*, of a small mesotrophic lake was correlated more with combined nitrogen limitation than with depletion of silica (Lehman and Sandgren, 1978; Lehman, 1979).

Much insight into the competitive interactions among dominant spring blooms of algae has been gained from detailed experimentation in cultures (general methodology of competitive nutrient kinetic analyses is reviewed by S. S. Kilham, 1978). Nutrient kinetic parameters of two common spring diatoms, *Asterionella formosa* and *Cyclotella meneghiniana*, showed that maximal growth rates of the two species were not significantly different (Tilman and Kilham, 1976). *Asterionella* was able to grow significantly better, however, at low concentrations of extracellular phosphate and *Cyclotella* grew better at low concentrations of silica. Using a resource-based competition model (Titman, 1976; Tilman, 1977, 1978), the outcome of nutrient competition between these two species showed that (a) under conditions in which both species were phosphate-limited, *Asterionella* would be dominant; (b) under conditions in which both species were silica-limited, *Cyclotella* would be the competitive dom-

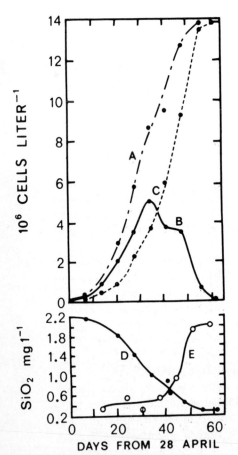

Figure 15-5. Production and loss of *Asterionella formosa* in the epilimnion of Lake Windermere, England, from 28 April to 30 June 1947. *A*, Cumulative total of cell production computed from silica uptake; *B*, Mean concentration of cells in epilimnion; *C*, Cumulative loss of cells from the epilimnion (*A* minus *B*); *D*, Epilimnetic silica concentration; and *E*, Relative rates of loss of cells from the population. (Modified from Lund, et al., 1963.)

inant; and (c) under intermediate conditions in which the growth rates of each species were limited by different nutrients (*Asterionella* by silica and *Cyclotella* by phosphate), both species would coexist stably. The model was verified for these species both experimentally and in natural populations along nutrient gradients (Kilham and Titman, 1977; Stoermer, et al., 1978). These studies indicate that a species is growth-limited by only one nutrient at a time, and that where two species are growth-limited by different nutrients simultaneously, they can coexist.

Gradients in nutrients and in the ratios of different nutrients exist both spatially (e.g., from near-shore to open-water areas) and seasonally when utilization or losses exceed inputs. Concentration of a nutrient is important to growth and phytoplanktonic community structure, but so is the supply rate of one nutrient relative to other potentially limiting nutrients. Competitive dominance of species populations will develop along nutrient gradients according to the species' maximal growth rates, their half-maximal growth rates (half-saturation constants of nutrient uptake), and their species-specific mortality rates (Kilham and Kilham, 1978; Kilham and Tilman, 1979; Tilman, 1978, 1980). For example, *Asterionella* is a successful competitor at high Si/P ratios, *Fragilaria* can dominate at intermediate ratios, and *Stephanodiscus* grows well when Si/P ratios are low.

As silica concentrations are reduced in productive lakes, diatom populations are often succeeded by a preponderance of first green algae and later blue-green algae. Growth in these eutrophic lakes can be so intense that combined nitrogen (NO_3, NH_4^+) sources are reduced to below detectable concentrations in the trophogenic zone (e.g., Wintergreen Lake, Chapter 12). When this happens, often by midsummer when the warmest epilimnetic temperatures occur, blue-green algae with efficient capabilities for fixing molecular nitrogen have a competitive advantage and can predominate (cf. Chapter 12). These lakes require, as a general rule, a reasonably sustained and heavy loading of phosphorus (cf. for example, Moss, 1973c).

Gradients in nutrient ratios are commonly observed during and after the spring diatom outburst in eutrophic lakes, because phosphorus and nitrate concentrations decline at different rates (e.g., Pechlaner, 1970). As a result, the Si/P ratio increases and the N/P ratio decreases markedly. At high Si/P ratios, diatoms can effectively outcompete blue-green algae (e.g., Holm and Armstrong, 1981). As concentrations of silica are reduced, and later combined nitrogen declines, green algae can compete less effectively with blue-green algae that possess N_2-fixing capabilities.

Similar relationships have been demonstrated in many lakes (e.g., Reynolds, 1976) and occur on a much larger scale in the Laurentian Great Lakes (Schelske and Stoermer, 1971, 1972; Schelske, et al., 1972). Phytoplanktonic algae in the upper Great Lakes are distinctly phosphate-limited. As the loading of phosphorus to the lakes has increased, diatoms reduce the available silica (ambient concentrations about 2 to 3 mg l^{-1}) to limiting levels more rapidly with each successive year. The diatoms then are replaced rapidly by green and blue-green algae. Increased loading with trace metals, vitamins, and dissolved organic matter further accelerates the algal succession. Experimental additions of nitrogen and phosphorus to ponds also demonstrate the influence of these nutrients on phytoplankton succession; with nitrogen and phosphorus enrichment, chrysophyte species were replaced by green and blue-green algae (DeNoyelles and O'Brien, 1978).

A series of papers by Moss (1972c, 1973a, 1973b, 1973c) is instructive in relation to both the seasonal succession and the distribution of phytoplankton. Changes in nutrient concentrations are just as important to seasonal periodicity within a lake as different nutrient concentrations among lakes are important in determining algal distribution. Working with 16 species of diverse but widely distributed phytoplanktonic algae, Moss found little effect of calcium ions above a concentration of 1 mg l^{-1} among both oligotrophic and eutrophic species. No evidence was found to support the contention that oligotrophic desmids are calciphobic. Certain oligotrophic algae were unable to grow at pH values greater than 8.6, whereas eutrophic species able to utilize both CO_2 and bicarbonate grew well at pH values above 9. Temperature optima for growth were similar to those discussed earlier: diatoms have an extended range into lower temperatures, whereas green and blue-green algae grow well at temperatures above 15°C. Temperatures above 15°C are common in many temperate waters throughout much of the summer and in tropical waters for the entire year. Temperature and nutrient interactions generally could be verified in experimental bioassays using a variety of natural lake waters ranging from oligotrophic to extremely eutrophic.

Parasitism and Grazing

Only recently has parasitism of phytoplanktonic algae been shown to be significant. Although the best-known parasites of algae are chytridiaceous or biflagellated phycomycetous fungi, viral and bacterial infection of algae surely occurs under natural conditions (Shilo, 1971). Viral lysis of specific blue-green algae has been demonstrated (Saffermann and Morris, 1963; Shilo, 1971), but attempts to exploit this specificity for the control of blue-green algal blooms have not been very successful (V. Sládeček, personal communication).

The best studies of chytrid fungal parasitism, centered in the English Lake District, indicate that infection of desmids, diatoms, and blue-green algae is fairly common. Even though the fungal percentage infection can be very high (>70 per cent) and can parasitize healthy, rapidly growing cells as well as declining senescent algae, parasitism usually does not greatly alter the overall seasonal pattern of periodicity of diatoms and desmids (Canter and Lund, 1948, 1969). Fungal parasitism, however, can have a significant effect on interspecific competition of algae because of the degree of specificity in the algae attacked. The reduction of dominant desmids apparently is influenced to a greater extent by parasitism, whereas the diatom decline, particularly of *Asterionella*, results from a combination of nutrient and parasite interactions. Certain diatom clones react hypersensitively to chytrid parasites, which results in both the death of the host alga and the fungi and causes the demise of the parasite population (Canter and Jaworski, 1979). Algal infection by some chytrid fungi occurs only at low temperatures (Blinn and Button, 1973). Parasitism probably increases in eutrophic waters, but little is known of its quantitative significance. Much further study in this area is needed.

Grazing of phytoplankton by animals, particularly by the microcrustacea, often has been invoked as a significant factor in the decline of the algal populations, and as contributory to seasonal succession. Although most of this discussion will be deferred until the subsequent chapter on zooplankton, a few points are appropriate here.

Evidence for grazing as a cause of quantitative changes in phytoplanktonic abundance and species composition is conflicting. As temperatures of the water increase in temperate waters, the reproduction and feeding of zooplankton increase markedly. In some cases, the population and grazing maxima of zooplankton coincide with the decline of algal maxima (e.g. Crumpton and Wetzel, 1982); in other cases, inverse correlations between the two are poor, or changes in zooplankton population correlate better with bacterial or detrital concentrations. Although certain recent studies indicate that, at certain times of the year, rotifers and microcrustaceans can totally graze the trophogenic zone in a day (cf. Chapter 16), much lower rates are more usual. A degree of size and species specificity also exists among the grazers. Such selectivity can lead to a competitive advantage by less effectively grazed algal species, and influences seasonal succession of algae within prevailing physical and nutrient constraints. In this way, invertebrates and fish possessing size-selective feeding habits with respect to zooplankton can influence zooplankton grazing effectiveness, and in turn, algal succession (e.g. Lampert, 1978).

An example of seasonal succession of phytoplankton and the interaction of growth rates and mortality is seen in detailed studies of a small, oligotrophic temperate lake (Crumpton and Wetzel, 1982). Weekly measurements were made of growth rates, sedimentation rates, and population densities for each of the dominant phytoplankton species, and of diurnal zooplankton grazing rates on the algae. The diatom *Cyclotella michiganiana* was the dominant alga through the end of June, at which time *C. comensis* began to increase and became dominant by August (Fig. 15-6). The combined effect of greater growth rates and lower loss rates due to sedimentation of *C. comensis* resulted in its dominance over *C. michiganiana*. In late August, high grazing pressure by herbivorous zooplankton caused rapid declines of both *Cyclotella* species, which were succeeded by the larger green alga *Sphaerocystis schroeteri*. The *Sphaerocystis* species was too large to be fed upon by the zooplankton dominating at that time. The *C. comensis–S. schroeteri* succession clearly resulted from differential mortality rather than from differences in growth rates.

Size-selective feeding is particularly conspicuous among the protozoans. Certain herbivorous protozoans feeding on small colonial algae have been shown to decimate over 99 per cent of certain chlorophycean planktonic algal populations over short periods of time (7 to 14 days) (Canter and Lund, 1968). Some of these protozoans, such as *Pseudospora*, have been shown to consume only certain species of algae. The effects on seasonal succession of algal populations are obvious. Numerous other examples are given in the following chapter.

Competitive Interactions and Successional Diversity

The phytoplanktonic community consists of populations of a diverse assemblage of algal species. Each of these species components and its growth dynamics are influenced by an array of environmental parameters (physical, chemical, and biotic) that undergo constant temporal variations. Some of the parameters, such as temperature, conservative nutrients, etc., undergo slow, periodic oscillations over an annual period. Other parameters, such as light intensity and quality, nutrients in rapid flux, etc., exhibit very rapid temporal, physiologically important changes in a matter of minutes or hours. The intensities of these forces greatly affect algal physiology and growth, some with regularity and others in a highly irregular fashion.

It is doubtful that an ideal species hypervolume is ever met under natural conditions. The fit of a species within an array of competing species is relative and changes constantly. The ability of an alga to store a critical nutrient imposes another time factor interaction that stresses the importance of past environmental conditions as well as

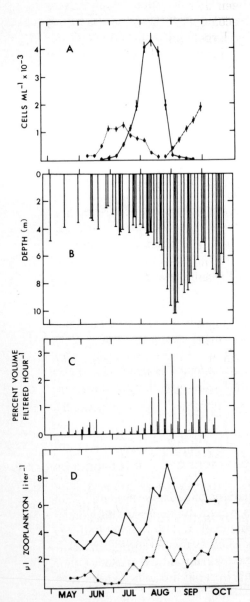

Figure 15-6. Dominant phytoplankton, water transparency, zooplankton grazing rates, and zooplankton biomass, Lawrence Lake, Michigan, 1979. (A) Population curves for *Cyclotella michiganiana* (-----), *Cyclotella comensis* (———), and *Sphaerocystis schroeteri* (- - - -). Error bars denote 95% confidence limits. (B) Transparency (Secchi depth) in meters. (C) Grazing rates as percentage of epilimnion filtered per hour. The first of each pair of bars represents daytime rates; the second represents nighttime rates. (D) Daytime (-----) and nighttime (———) epilimnetic zooplankton biomass. Dominant zooplankton were copepods (> 90%) from May through July and *Daphnia* spp. (> 90%) from August through October. (From Crumpton and Wetzel, 1982.)

contemporary ones. On the basis of the number of interacting parameters and ranges of variations tolerated by algae, it is not too surprising that (a) a species may be able to survive indefinitely provided that a suitable combination occurs with sufficient frequency and duration, and (b) the interactions are so complex and variable that, in spite of species overlap in their niche hypervolumes, coexistence can occur even though competitive interactions are in effect. The beginnings of mathematical formulations of these competitive interactions are set forth in the stimulating theoretical discussions of Grenney, Bella, and Curl (1973) and Tilman (1980).

Diversity of phytoplanktonic algal species has been determined in a number of ways because of observed species changes as lakes become more eutrophic. The best indices of species diversity are probably those that are largely independent of sample size (cf. Chapter 8).* Diversity indices are often determined primarily by the proportions of the more common species (equitability), and only secondarily by the number of species according to derivatives of the Shannon formula (Sager and Hasler, 1969; Moss, 1973d; Hallegraeff and Ringelberg, 1978). Predominance of one or two species results in low diversity values; high values occur when populations of several species each form moderate proportions of the whole.

Although data are far from satisfactory because of the variations found, there is a general tendency for species diversity to decrease with increasing fertility of the water. Presumably, the slower growth rates of algae in oligotrophic lakes permit a greater number of species with reasonably similar requirements (high degree of niche overlap) to coexist than would be found in more eutrophic waters.

In eutrophic temperate lakes, the tendency is for algal diversity to increase in summer and decline during winter (Moss, 1973d). However, when the relationship between species diversity and season is examined closely, one can find considerable variability (cf. Hallegraeff and Ringelberg, 1978). Nutrient availability and water temperature may account for much of the observed pattern; during and immediately after spring circulation, nutrient concentrations are high, and progressive depletion of nutrients usually occurs throughout the summer. Relatively few algal species are adapted to the low water temperatures found immediately after ice loss in spring, and these species rapidly come to dominate the algal community. During summer, warmer temperatures are more favorable for growth of a larger number of species but competition for limiting nutrients is likely to increase. Obviously, factors such as parasitism and predation are also involved, but the extent of influence of each of these factors on the overall pattern of succession is variable and poorly studied.

Congruous with these findings are the results of a measure of successional rate, or a seasonal rate of change of species composition. Jassby and Goldman (1974), employing a mathematically more involved but more realistic expression, showed that the rate of change in species composition per unit time decreases precipitously following the spring maximum and midsummer plateaus. Thus, when the community is disturbed by the rapid changes of spring circulation as in this case, or is perturbed by some cata-

---

*Increasing sample size tends to reveal a greater number of species, as more and more of the rarer species are encountered. Many of the rarer species are of littoral-zone origin, and not true members of the phytoplankton (cf. Moss, 1973d; Symons, 1972). Algal species diversity is much greater in the littoral zone than it is in pelagial areas.

strophic event (e.g., fertilization by man's activities), one would expect rapid changes in species composition and increased successional rates.

PERIODICITY AND DISTRIBUTION OF PRODUCTIVITY

The distribution of phytoplanktonic biomass and in situ rates of primary productivity has received a great deal of attention and represents perhaps the most thoroughly investigated aspect of lake biota. In contrast, the growth and productivity of phytoplankton of streams and rivers is poorly understood (cf. reviews of Hynes, 1970 and Wetzel, 1975b).

The distribution patterns of phytoplanktonic biomass within a lake usually change markedly throughout the year, and lake-to-lake variations are also extreme. Phytoplanktonic productivity is even more variable, for it is influenced by factors that can change within a matter of minutes (e.g., light intensity). However, in spite of these differences, there are some aspects of the distribution patterns that appear to be consistent. In addition, the productive capacity of a planktonic system is bounded by certain constraints, beyond which, productivity cannot be increased. These considerations are addressed in the following discussion.

VERTICAL DISTRIBUTION OF BIOMASS

The vertical distribution of phytoplanktonic biomass varies greatly from season to season with shifts in species composition. If the depth–time distribution of phytoplanktonic pigments of an oligotrophic hard-water lake of southern Michigan is taken as an example, conspicuous temporal changes can be seen. In the seasonal depth distribution of chlorophyll a concentrations (Fig. 15–7), corrected for pigment degradation products, values were low in winter, increased conspicuously in the spring maximum, and then increased in different strata during the period of summer stratification. Detailed analyses of the phytoplankton community (some 150 major species) revealed that cryptomonads and small green algae predominated in the winter. The composition then changed to predominantly diatoms in the spring, and green algae in June and late August during this particular year. The metalimnetic maximum in summer was largely composed of green and nonnitrogen-fixing blue-green algal populations with lesser percentages of diatoms. The development of a deep chlorophyll maximum (10 to 11 m) in the later portion of summer stratification was related to very dense populations of euglenoid algae just at the aerobic–anaerobic interface about one meter above the sediments. Concentrations of chlorophyll b were very much lower throughout the year and were irregularly distributed, except in the lower hypolimnion during late summer and autumn. Chlorophyll c concentrations were most abundant in algal populations dominated by diatoms. The total composite plant carotenoids, which are somewhat more stable than the chlorophyllous pigments, exhibited a vertical depth distribution seasonally similar to that of chlorophyll a (Fig. 15–8).

In using pigments as estimates of phytoplanktonic biomass, it is important to emphasize that a significant portion of the algal cells within the water column is nonviable, i.e., is particulate detritus in various stages of decomposition. The chlorophyl-

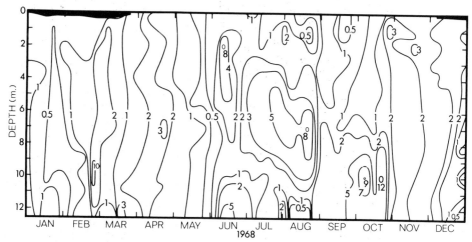

Figure 15-7. Depth-time distribution of phytoplanktonic chlorophyll *a* concentrations (mg m^{-3}), Lawrence Lake, Michigan, 1968. Opaque areas = ice cover to scale. Sampled at 7-day intervals at each meter of depth. (From Wetzel, unpublished data.)

lous pigments can be corrected for degradation products (e.g., Wetzel and Likens, 1979). Concentrations of these heterogeneous phaeopigments give some insight into the magnitude of nonfunctional pigments in the particulate fractions. Using the same data for Lawrence Lake, Figure 15-9 sets forth the concentrations of phaeopigments with depth over the year. At many times of the year, the phaeopigment concentrations equal or exceed those of chlorophyll *a*. Moreover, the phaeopigment concentrations are often displaced to slightly greater depths and to slightly later times in relation to the func-

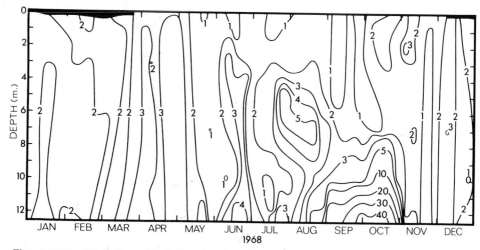

Figure 15-8. Depth-time distribution of phytoplanktonic total plant carotenoids (ca. mg m^{-3}), Lawrence Lake, Michigan, 1968. Sampled at meter depth intervals each 7 days; opaque areas = ice-cover to scale. (From Wetzel, unpublished data.)

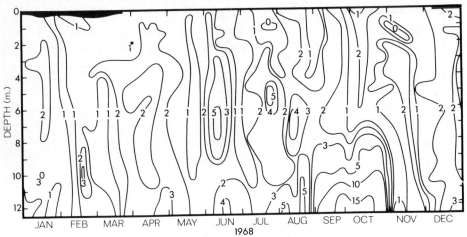

Figure 15-9. Depth-time distribution of phytoplanktonic phaeopigment concentrations (mg m^{-3}), Lawrence Lake, Michigan, 1968. Sampled at meter depths at 7-day intervals; opaque areas = ice-cover to scale. (From Wetzel, unpublished data.)

tional pigments. The importance of this detrital particulate matter of algal origin to lake metabolism is discussed at length later (Chapter 22). It is important to realize that cell integrity and pigment concentrations can persist for several days after cell death (e.g., Gusev and Nikitina, 1974). Chlorophyll a degrades most rapidly upon death, and when corrected for more stable phaeopigment degradation products, is an estimate of biomass (cf. Desortová, 1981).

VERTICAL DISTRIBUTION AND MAXIMUM GROWTH

The pigment-biomass values discussed above are low, which is typical of a rather oligotrophic lake. Considerably lower concentrations, about 1 mg chlorophyll a m^{-3}, are not unusual among pristine arctic and alpine oligotrophic lakes.

Among alpine oligotrophic lakes, and larger oligotrophic lakes in general, high concentrations of chlorophyll are commonly found deep into the upper hypolimnion and into aphotic zones (e.g., Brook and Torke, 1977; Tilzer, et al., 1977; Richerson, et al., 1978). Analyses have shown that much of the phytoplankton in these deep layers is adapted to low light intensities and that, during circulation periods, aphotic algae can augment phytoplankton growth of the euphotic zone upon re-entry into the upper waters.

At the other extreme of maximum growth and pigment concentrations, light limitations restrict productivity, and the biomass of the algae themselves imposes controls upon further growth by self-shading. Photosynthetic yield can continue to increase, given adequate nutrient availability, to a theoretical point at which total absorption of photosynthetically active incident radiation occurs. This absorption is distinct from that absorbed by the water itself, by dissolved organic matter, and by abiogenic particulate matter (cf. Talling, 1960, 1971). The theoretical maximum, as indicated by culture analyses, is approximately 1 g chlorophyll per m^3 (1 mg l^{-1}), but varies among algal types, being higher for blue-green algae and less for green algae and diatoms (e.g., Steemann Nielsen 1962b). As the algal populations increase biogenic turbidity in progressively more fertile lakes, the productivity per volume of water increases greatly (Fig. 15–10). However, the productivity per square meter of water column does not increase proportionately, because biogenic turbidity due to the increasing density of algal cells reduces light penetration. The depth of the trophogenic zone and, hence, the total volume occupied by photosynthesizing algae, are consequently reduced.

Therefore, as the composite extinction coefficient of light from biogenic sources increases, the productivity per square meter of water column increases to the maximum imposed by self-shading effects of the algal densities (Fig. 15–11). Increasing the extinc-

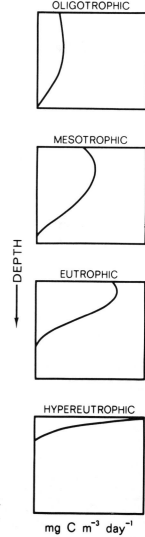

Figure 15–10. Generalized increases in productivity of phytoplankton per unit volume of water, with simultaneous reduction of the thickness of the trophogenic zone, in a series of lakes of increasingly greater fertility.

mg C m^{-3} day^{-1}

tion coefficient further, such as by high concentrations of dissolved humic compounds or inorganic particulate and colloidal turbidity, can similarly result in a progressive depression of maximum possible productivity.

The highest recorded concentrations of chlorophyll of phytoplanktonic origin emanate from tropical, eutrophic lakes. For example, shallow (2.5 m) but large (250 km^2) equatorial Lake George of western Uganda is a highly productive lake that supports very dense phytoplanktonic populations; these populations consist primarily of blue-green algae, with very little seasonal change in either species composition or densities. Based on in situ estimates of photosynthesis, the predicted concentrations of chlorophyll a in this lake stabilized at an algal density of about 500 mg chlorophyll a m^{-2} (Ganf, 1974a). Because of the shallow nature of this large lake, frequent daily mixing, and disturbance of the flocculent sediments, light penetration is reduced further by nonalgal components. Therefore, the maximum quantity of chlorophyll a actually measured within the euphotic zone is between 230 and 310 mg m^{-2}. However, many of the phytoplankton settling to the sediments are viable, and are resuspended each day (Ganf, 1974b), so that a more realistic mean algal biomass may be as high as 1000 mg chlorophyll a m^{-2}.

Among the maximum recorded values of algal biomass on the basis of chlorophyll content are those from two Ethiopian soda lakes (Talling, et al., 1973). In these extremely productive lakes, favored by high temperatures, abundant light, phosphorus, carbon, and other nutrients, the euphotic zone was less than 0.6 m in thickness and the algal populations were limited severely by self-shading. Although concentrations of chlorophyll a exceeded 2000 mg m^{-3} in one lake, which appears to be about the maximum possible under natural planktonic conditions (Fig. 15–12), the content of chlorophyll a per unit area of the euphotic trophogenic zone for these lakes was in the range of 180 to 325 mg m^{-2}. These values are similar to those maxima (180 to 450 mg chlorophyll a m^{-2}) that have been estimated indirectly or directly to occur in nature (cf. Westlake, et al., 1980).

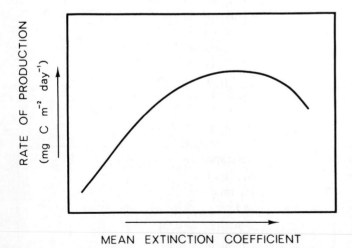

RATE OF PRODUCTION (mg C m^{-2} day^{-1})

MEAN EXTINCTION COEFFICIENT

Figure 15–11. Generalized maximum possible productivity of phytoplankton in relation to increasing extinction of light, resulting from biogenic turbidity initially, and then combined with increasing effects of extinction by dissolved organic matter and/ or particulate and colloidal inorganic matter.

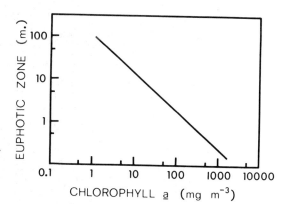

Figure 15–12. Generalized relationship of the depth of the euphotic zone in relation to the maximum concentrations of chlorophyll *a* per volume of water. (After data of several authors and Talling, et al., 1973.)

SEASONAL RATES OF PHOTOSYNTHESIS

Rates of in situ photosynthesis of the phytoplankton must be evaluated over an annual period to compensate for major differences in the length of active growing season with changing latitude and altitude. Although summer productivity of an arctic or high mountain lake may be just as high as productivity per unit volume of a lake at much lower latitude or elevation, the length of the growing season is considerably less. Since evaluations of phytoplanktonic productivity must be made on an annual basis, it is necessary to look first at the magnitude of these seasonal variations.

Further, it is important to keep in mind the ecological significance of gross versus net productivity. As discussed earlier, rates of the opposing processes of photosynthesis and respiration are very difficult to evaluate in heterogeneous planktonic populations of phytoplankton, bacteria, and zooplankton. The widely used ^{14}C light and dark bottle generally results in values close to net productivity under a majority of in situ conditions. This method permits a further evaluation of the extent of extracellular release of soluble organic compounds during growth of the algae. Respiration cannot, however, be evaluated by the ^{14}C method. The sensitive ^{14}C method permits an estimation of the in situ rates of photosynthesis approximating net productivity, which is useful since it is this particulate matter that is potentially available for consumption by higher organisms. More elaborate methods, however, are required to ascertain the fate of the total productivity, since a majority of the synthesized organic matter is not consumed by animals but instead enters the detrital food chain (cf. Chapter 22).

The less sensitive oxygen-change techniques measure community metabolism, although oxygen production over time is most often the result of phytoplanktonic algae. The importance of the effects of light-mediated photorespiration on the assumption that dark-bottle estimates of respiration are similar to those in the light have not been evaluated satisfactorily. Similarly, bacterial respiration is assumed to be small in relation to that of the planktonic algae, and to be the same in the light as it is in the dark.

It can only be concluded that at best, contemporary methods provide estimates of in situ rates of photosynthesis that are comparable only in a general way (cf. Peterson, 1980). Nonetheless, the advantages of being able to measure rates directly in situ are great, and are exceedingly valuable if viewed in correct perspective.

Several examples of seasonal primary productivity illustrate the types of variations that can be encountered. The changes in productivity of the phytoplankton at depth over the season set out in Figure 15-13 for a hard-water lake of low productivity in Michigan show a common pattern. Rates are low in the winter with periodic minor surges from near-surface populations under ice, particularly toward the end of winter. Rates often decrease during spring circulation, which in this lake for this year was very brief (until early April). Rates of production increased during the spring diatom maximum, entered a low period in May, and then succeeded through a series of pulses through the summer when values were higher than during the spring. Metalimnetic maxima are conspicuous in July and late August prior to a general slow decline in the autumn. These rates of production by phytoplankton are composite values for the community of many species populations, which show a great regularity in periodicity from year to year (monitored continuously for over a decade).

When these data at each depth interval are integrated over a square meter of water column, one can calculate the rates of in situ productivity per surface area of the pelagial zone (Fig. 15-14). It is usual to integrate the areal productivity under such curves for an annual period to obtain the estimate of the total annual productivity by phytoplankton for a year. Division by 365 results in an estimate of the mean daily productivity for the lake. Obviously, conically shaped lake basins have much greater volume for phytoplanktonic productivity in upper strata than at greater depth within the trophogenic zone. Therefore, it is advisable to determine the annual productivity for each stratum, say at meter intervals, over the year and integrate these values before summing productivity of each stratum for the whole. In this way, the changes in volume available for productivity are taken into account.

A feature of the total phytoplanktonic productivity and the mean daily productivity for the year is their stability from year to year as long as the system is not perturbed. In spite of variations in the seasonal rates of observed productivity and numerous sources

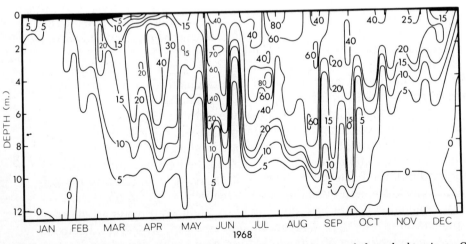

Figure 15-13. Depth-time distribution of the in situ rates of production of phytoplankton in mg C m^{-3} day $^{-1}$, Lawrence Lake, Michigan, 1968. Measured at meter depth intervals every 7 days; opaque areas = ice-cover to scale.

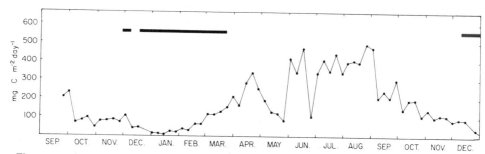

Figure 15-14. Integrated areal primary productivity per square meter of the pelagial zone of Lawrence Lake, Michigan, 1967-68. Bars indicate periods of ice-cover. (From Wetzel, unpublished data.)

of potential error in measurement and sampling, the values are reasonably constant. Examples of this consistency can be seen in Table 15-6. Some variations are to be expected. In this case, for example, 1972 was a particularly variant year with abnormally high cloud cover and rainfall. Within this type of variation, the annual primary productivity values serve as one of the best available criteria for following changes in the metabolism of lake systems in response to alterations by human activities.

Looking at the primary productivity of phytoplankton of more eutrophic lakes of the same latitude as the previous example, three points are conspicuous (Fig. 15-15). First, the thickness of the trophogenic zone is reduced markedly to the point where most of the productivity occurs in the first two meters of water. Second, the seasonal succession of productivity is much more irregular and exhibits marked fluctuations in comparison to less productive lakes. As discussed earlier, algal utilization and reduction of critical nutrients (Si, combined N) to levels limiting for certain algal groups progress very rapidly and contribute to the marked oscillations in composite photosynthetic

TABLE 15-6   Rates of In Situ Production of Phytoplankton of Lawrence Lake, Michigan, over Several Years

YEAR	g C m^{-2} YEAR^{-1}	MEAN mg C m^{-2} DAY^{-1}
1968	45.43	124.5
1969	41.76	114.4
1970	43.48	119.1
1971	36.13	99.0
1972	29.58	81.0
1973	39.17	107.3
1974	38.87	106.5
1975	43.01	117.8
1976	32.31	88.5
1977	30.93	84.7
1978	29.06	79.6
1979	29.03	79.5
1980	29.74	81.5
1981	39.13	107.2
Mean 14 years	36.26	99.3

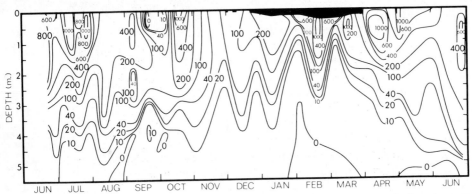

Figure 15–15. Depth-time distribution of the in situ rates of production of phytoplankton in mg C m^{-3} day^{-1}, Wintergreen Lake, Michigan, 1971–72. Opaque area = ice-cover to scale. (From Wetzel, et al., unpublished.)

rates. Finally, the data of Figure 15–15 again emphasize the importance of primary productivity under ice and winter conditions, which should not be ignored in contemporary studies.

Because the primary productivity of hypereutrophic Wintergreen Lake is attenuated so rapidly with increasing depth by the self-shading effects of the algae, the majority of the productivity is restricted to the uppermost portions of the photic zone. Therefore evaluation of the productivity of the lake on a per-square-meter-of-water-column basis throughout the trophogenic zone is similar to values for the entire lake, in which compensation for the decreasing volume of the lower strata is taken into account (Fig. 15–16).

In tropical regions when seasonal climatic changes are minimal, the length of the active growing season is extended to the entire year. The annual rates of primary pro-

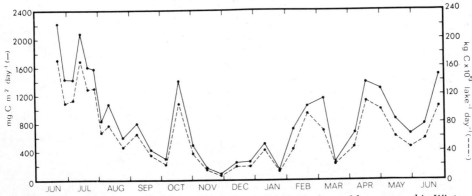

Figure 15–16 Integrated primary productivity of the phytoplankton of hypereutrophic Wintergreen Lake, Michigan, 1971–72, per square meter of the trophogenic zone at the deepest point and for the lake volume with morphometric corrections. (From Wetzel, et al., unpublished.)

duction of phytoplankton increase accordingly and when nutrient limitations are not severe, the equatorial lakes exhibit the highest values of phytoplanktonic productivity recorded in aquatic systems (for example, Talling, 1965; Lewis, 1974b; Westlake, et al., 1980). Despite nearly continuous optimal conditions for sustained high growth rates, definite seasonality persists even though it is not nearly as marked as that found at higher latitudes. In shallow, nonstratified tropical lakes that have high nutrient inputs, the productivity values are not only very high, but are relatively uniform throughout the year. An example is Lake George on the equator in Uganda (Talling, 1965; Ganf, 1974). Deeper tropical lakes that stratify thermally, even though weakly so, tend to exhibit some periodicity in productivity. For example, in equatorial Lake Victoria of Africa, somewhat higher rates occur in the early months of the year and in June and July, when the breakdown of stratification occurs. The net primary productivity of weakly monomictic Lake Lanao (8°N) in the southern Philippines was relatively high throughout the period of stratification (May–November) but decreased precipitously during the winter period when water temperatures were lowest and deepest seasonal mixing occurred (Fig. 15–17). Respiration rates as a percentage of gross photosynthesis generally are higher (towards 50 per cent) than the average for temperate waters (20 to 30 per cent of total fixed carbon).

Therefore, when viewing the annual cycles of primary productivity in lakes in relation to latitude, seasonal variation in input of solar radiation at higher latitudes is the general controlling mechanism. Near the equator this seasonal differentiation is muted, and the incidence of vertical mixing is more important in any observed periodicity (Talling, 1969). In some tropical lakes, as in Lake Victoria, seasonal cooling and mixing, regulated chiefly by atmospheric factors like wind and humidity, are decisive events. Nutrient supply has been implicated as a dominant controlling factor of primary productivity of Lake Lanao during stratification (Lewis, 1974b). Nutrient depletion is relieved at frequent intervals by changes in the depth of mixing associated with irregular storms.

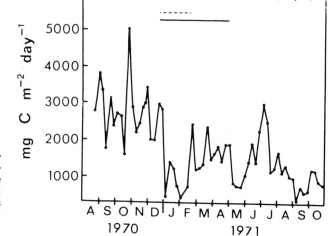

Figure 15–17. Net rates of primary production of the phytoplankton of Lake Lanao, Philippines, 1971–72. The dotted line marks the period of minimum water temperature and deepest seasonal mixing; the solid line marks the period when the lake lacked stable stratification. (Modified from Lewis, 1974b.)

EFFICIENCY OF LIGHT UTILIZATION

The efficiency of light utilization by algae can be estimated in several ways. All methods relate the amount of irradiance available at depth to the rate of transformation of light energy to chemical energy by photosynthesis. Existing values can be viewed only as estimates because of the difficulties in measuring the caloric equivalents of photosynthesis per unit volume at depth, or integrated for the water column and the variable light inputs of photosynthetically active radiation at depth. In spite of these limitations, it is of great interest to see the range of values encountered in fresh waters because of the desires of some to utilize aquatic systems as potential food sources.

Phytosynthetic efficiencies of light utilization can be defined as an ecological efficiency (ϵ'), in per cent, equal to the amount of photosynthetically stored energy (PSR)* in a volume of water $(z_1 + z_2)m^3$, divided by the photosynthetically available radiation (PAR; 350 to 700 nm) absorbed or dissipated in the water layer z_1 to z_2, regardless of the variable fraction of PAR absorbed by algal pigments (Dubinsky and Berman, 1976, 1981a; Morel, 1978):

$$\epsilon' = \frac{PSR}{PAR_{z_1} - PAR_{z_2}} \times 100 \; (\%).$$

The difference in down-welling irradiance flux between successive water planes (z_1 and z_2) approximates light absorbance in the water column, and is expressed in calories (or Joules) $m^{-3}h^{-1}$. Backscattered light from lower depths affects the estimate of PAR dissipation in the overlying stratum only slightly. Thus, the photosynthetic efficiency is expressed as a ratio of the caloric equivalent of photosynthesis to radiation inputs.

Photosynthetic efficiency of light utilization by phytoplankton can also be expressed on an areal basis, in which the efficiencies of all the water strata are integrated throughout the euphotic zone and related to the PAR at the surface (cf. Morel, 1978; Dubinsky and Berman, 1981a). The areal efficiency permits more direct comparisons with photosynthetic efficiencies in other aquatic, portions of aquatic (e.g., littoral, wetland), and terrestrial ecosystems.

It is well known that the areal efficiencies of light-energy utilization by phytoplankton in aquatic ecosystems are generally much lower than photosynthetic efficiencies of terrestrial systems. Much of the irradiance is selectively and rapidly absorbed by water itself, and by dissolved and particulate matter contained in it. The amount of light absorbed by the water and solutes is usually greater than the amount absorbed by plant pigments. Nearly all of the estimated efficiencies are less than 1 per cent, and the highest values (approximately 1 to 4 per cent) reported for phytoplankton are from eutrophic lakes and from tropical areas (Table 15–7).

Variations in photosynthetic efficiencies are great, both seasonally and vertically within the euphotic zone. As light intensities increase near the surface of the water, both light-utilization efficiencies and ratios of photosynthetically stored energy to the amount of PAR absorbed by algal pigments generally decrease, probably because photosynthetic reaction sites become light-saturated (Fig. 15–18; cf. Tilzer, et al., 1975;

---

*An approximate conversion of assimilated carbon to energy equivalents is 1 mg C assimilated = 9.33 cal, assuming all phytosynthate is glucose (Morel, 1978).

TABLE 15-7 Estimates of Photosynthetic Efficiencies of Utilization of Photosynthetically Active Radiation by Phytoplankton in Lakes of Approximately Increasing Productivity

LAKE	PERCENTAGE EFFICIENCY
Tahoe, Calif.-Nev.	0.035
Castle, Calif.*	0.040
Finstertaler, Austria*	0.068
Oliver, Ind.	0.19
Olin, Ind.	0.23
Walters, Ind. (mean of 4 basins)	0.26
Pretty, Ind.	0.26
Chad, Chad, Africa	0.26
Crooked, Ind.	0.32
Little Crooked, Ind.	0.33
Martin, Ind.	0.34
Kinneret, Israel	0.35
Smith Hole, Ind.	0.38
Sammamish, Wash.	0.42
Wingra, Wisc.*	0.45
Goose, Ind.	0.57
Sylvan, Ind. (mean of 3 interconnected basins)	0.98
Leven, Scotland	1.76

After data of Wetzel, 1966b, Tilzer, et al., 1975, and Dubinsky and Berman, 1976. Estimates are based on different criteria of conversion; values are only approximately comparable but should be within about 30 per cent (cf. Dubinsky and Berman, 1981b).

*Ice-free period only.

Dubinsky and Berman, 1981a). As irradiance decreases to values below photosynthetic light saturation, photosynthetic efficiencies usually increase. Although high efficiencies of light utilization are common at the bottom of the euphotic zone, photosynthetic rates are so low that these phytoplankton contribute little to the total primary productivity of the lake, since the absolute amount of light energy available at these depths is small. As a result, the overall areal efficiency is increased little.

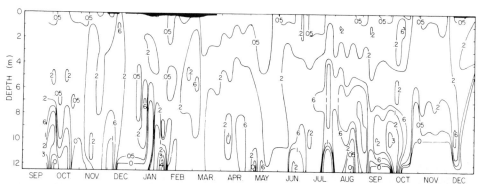

Figure 15-18. Depth-time distribution of estimates of the photosynthetic efficiency of utilization of photosynthetically active light income by phytoplankton of Lawrence Lake, Michigan, 1967-68. Values expressed as per cent, most of which are much less than one percent. (From Wetzel, unpublished data.)

As the quantity of phytoplanktonic biomass increases, the integral photosynthetic efficiencies generally increase until the maximum levels are restricted by light limitations imposed by self-shading. Therefore, the maximum areal values of photosynthetic efficiency are found in very productive waters containing dense algal populations. Exemplary ranges of photosynthetic efficiency encountered in lakes of increasing productivity are set out in Table 15–7.

It should be emphasized again that the productivity of the littoral zone and surrounding wetlands have usually been ignored as a component of aquatic ecosystems. Littoral and wetland productivity has been critically evaluated in only a few aquatic ecosystems; in every case examined, the combined productivity of these sessile plants and their photosynthetic efficiencies of light utilization are much higher than those of the phytoplankton and constitute a major productivity component of most freshwater ecosystems (cf. chapters 18 and 19).

EXTRACELLULAR RELEASE OF ORGANIC COMPOUNDS

Algae release a large number of organic compounds into the water extracellularly. These soluble compounds include glycolic acid, carbohydrates, polysaccharides, amino acids, peptides, organic phosphates, volatile substances, enzymes, vitamins, hormonal substances, inhibitors, and toxins. Literature on this subject is large and has been reviewed by Fogg (1971) and Hellebust (1974).

The release of organic compounds extracellularly represents first a significant loss of carbon fixed in photosynthesis. Second, the release of such organic compounds undoubtedly is of much greater importance than is generally recognized, and in a number of subtle but significant ways may modify growth, behavior of organisms, and the successional dynamics of algal populations. For example, a number of allelopathic interactions have been found among algae, in which organic compounds released by one alga are inhibitory or stimulatory to others (e.g., Proctor, 1957; Lefèvre, 1964; Keating, 1977; Wolfe and Rice, 1979; Jüttner, 1981). The release of biologically active organic compounds, especially by blue-green algae, is clearly implicated as one mechanism, among many others discussed earlier, that can influence fine adjustments in species succession and seasonal productivity. While causality is highly probable, quantitative evaluation of organic allelopathy has not yet been demonstrated in detail among natural populations of algae.

Two types of extracellular products are generally recognized (Fogg, 1971). (a) Metabolic intermediate compounds of low molecular weight. Glycolic acid, as an intermediate compound in photosynthesis, is known to be released under physiological stress conditions, especially under oligotrophic nutrient conditions and when photosynthesis is light-inhibited. Glycolic acid and polysaccharides released during active growth are utilized readily by bacteria (Nalewajko, et al., 1980; cf. Chapter 17). Intermediate compounds of respiration that are released include organic acids, organic phosphates, and, to a lesser extent, amino acids and peptides. (b) End products of metabolism, usually of higher molecular weight, the liberation of which does not depend on equilibrium and which are approximately proportional to the amount of growth. Included in this miscellaneous group are carbohydrates, peptides, volatile compounds such as aldehydes and ketones, enzymes, and a number of growth-promoting and -inhibiting substances.

The release of simple compounds, e.g., sugars, amino acids, and organic acids, by actively growing cells probably occurs mainly by diffusion through the plasmalemma (Hellebust, 1974). The rates of release depend on the concentration gradient of the substance across the membrane, and the permeability constant of the membrane for the compound. Although active excretion of small compounds is also possible, no convincing evidence for such processes in algae exists. Large molecules, such as polysaccharides, proteins, and polyphenolic substances, probably are excreted by means of more complex processes such as fusion of intracellular vesicles containing the compounds with the plasmalemma. The rates of extracellular release depend on the physiological and environmental factors affecting membrane permeability, and on intracellular concentrations (e.g., Mague, et al., 1980).

The rates of extracellular release under in situ conditions are quite variable. Although rates have been reported ranging from those equal to the rates of carbon fixation into cellular constituents down to less than 1 per cent of rates of carbon fixation, on the average most values are less than 20 per cent. For example, in the unproductive hardwater Lawrence Lake, in situ rates of release of organic compounds by phytoplankton monitored continuously over nearly two years ranged from 0.0 to 22.5 mg C m^{-2} day^{-1}, with a mean of 7.3 mg C m^{-2} day^{-1}. The annual mean percentage secretion was 5.7 per cent of net phytoplanktonic primary productivity (Wetzel, et al., 1972). Rates of release were greatest in April when production rates were low but increasing (Fig. 15–19). Secretion reached another maximum during the latter portion of summer stratification as primary productivity began to decrease while particulate organic carbon remained high.

The mean annual amount of released organic carbon in Lawrence Lake was highest at 1 m and decreased with increasing depth except below 10 m (Fig. 15–19). In these lower strata of very low light and low rates of photosynthesis, release rates, although extremely low in absolute value (cf. Fig. 11 of Wetzel, et al., 1972), were relatively high in relation to rates of primary production. Therefore, expressed as the mean percentage of extracellular release of all dates and samples, 23.5 per cent of the phytoplanktonic productivity was released to the dissolved phase. Similar results have been found in many other lakes (e.g., Berman, 1976).

As lakes become more productive, the absolute amounts of organic carbon released increase markedly. For example, the annual range in hypereutrophic Wintergreen Lake, Michigan, was between 2 and 100 mg C m^{-3} day^{-1}, with the highest values in the epilimnion (Fig. 15–20). However, because of the extremely high rates of photosynthetic carbon fixation and the rapid seasonal oscillations in photosynthesis within the compressed trophogenic zone, the percentage release values are generally lower than in oligotrophic waters (Fig. 15–21).

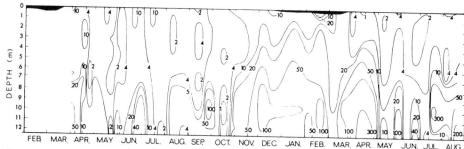

Figure 15–19. Depth-time isopleths of extracellular release of organic carbon by phytoplankton as a percentage of rates of carbon fixed photosynthetically, Lawrence Lake, Michigan, 1968-69. (From Wetzel, et al., 1972.)

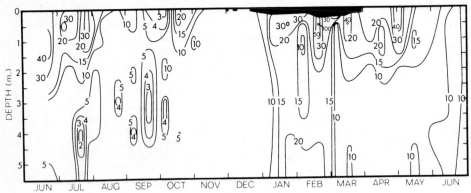

Figure 15–20. Depth-time isopleths of the rates of extracellular release of organic carbon by phytoplankton (mg C m^{-3} day^{-1}), hypereutrophic Wintergreen Lake, Michigan, 1971–72. (From Wetzel, et al., unpublished.)

High rates of release of extracellular products of algal photosynthesis have been shown to be a function of CO_2 limitation (e.g., by high pH), high population densities, inhibiting light intensities, low light intensities, and low cell densities and growth rates. The high percentage released at both extremes of the light and pH continuum indicates that the release of relatively large quantities of organic carbon compounds is favored by any environmental condition that inhibits cell multiplication but still permits photoassimilation to occur. Membrane permeability alteration or damage is probably the case in situations of high light intensities and other distinctly detrimental conditions.

During blue-green algal blooms, population growth of other algae can be strongly suppressed. Murphy, et al. (1976), have demonstrated that blue-green algae, which have high requirements for iron, can release hydroxamate siderchrome compounds, which complex iron and increase its availability, thereby favoring growth of the blue-green algae. Simultaneously, the hydroxamate chelators have growth-suppressing effects on non-blue-green algae.

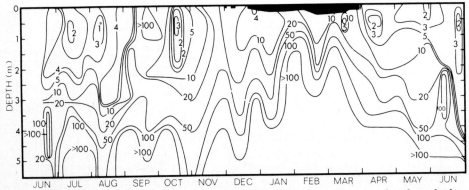

Figure 15–21. Depth-time isopleths of extracellular release of organic carbon by phytoplankton as a percentage of rates of carbon fixation, Wintergreen Lake, Michigan, 1971–72. (From Wetzel, et. al., unpublished.)

It should be noted that bacterial utilization of excreted organic compounds is extremely rapid (e.g., Nalewajko and Lean, 1972; Herbst and Overbeck, 1978; Nalewajko, et al., 1980; Coveney, 1981). As a result, the simpler low-molecular-weight compounds are utilized and degraded rapidly and large-molecular-weight substances dominate in time because of their much slower rates of degradation by bacteria (cf. Chapter 17).

DIURNAL CHANGES IN PHYTOPLANKTONIC PRODUCTION RATES

Evaluation of the methodology of oxygen change or carbon uptake for measuring in situ rates of production of phytoplankton has shown that the diurnal periodicity of photosynthesis is often not proportional to the daily insolation curve. Most studies show maximum photosynthetic rates in the morning hours, with a subsequent reduction at midday; sometimes recovery occurs in the afternoon. This diurnal periodicity is particularly acute at the surface, and undoubtedly is partly related to photoinhibition and increased rates of extracellular release of recently synthesized organic compounds. However, much of this surface midday depression dissipates with increasing depth, often in less than 1 meter, as light is attenuated and photosynthesis is undersaturated with light (Vollenweider, 1965; Jones and Ilmavirta, 1978).

Diurnal synchrony of cell division and growth has been well established in cultures of algae, and there is every reason to expect it to take place in the same way in natural populations under certain environmental conditions (Soeder, 1965). As a result, the composite phytoplanktonic populations consist of various combinations of algae in (a) active, strongly anabolic growth stages in which cell numbers are increasing and overriding losses by sinking, death, and grazing; (b) "neutrally" active stages that are slightly anabolic, not undergoing significant positive or negative changes in biomass; and (c) inactive, catabolically metabolizing stages in which cells in resting, decaying, or degenerating stages are decreasing. Thus, depending on the combination of algae in various physiological stages of population development, and the dominant stage within the composite phytoplanktonic community, diurnal, synchronized, predawn cell division often results in a pronounced photosynthetic periodicity from rapid early morning photosynthesis. For example, cell division in natural populations of the dinoflagellate *Ceratium* is phased to a brief period of the diurnal cycle (Heller, 1977; Heaney and Talling, 1980). The highest frequency of dividing cells occurred late in the dark period, with some carryover into the beginning of the following daylight hours.

It should also be mentioned that photorespiration increases at midday in submersed angiosperms (Chapter 18). Photorespiration certainly occurs in algae as well, but little is known as yet of its magnitude and diurnal periodicity under natural conditions.

An excellent example of diurnal biochemical coupling is seen in the bloom-forming blue-green algae, especially one species of *Anabaena*. During daylight, carbon dioxide fixation in *Anabaena oscillarioides* is greatest in the morning period, while N_2 fixation is maximum during the afternoon hours when oxygen concentrations of the water often become supersaturated (Paerl, 1979; Paerl and Kellar, 1979). High oxygen concentrations can inactivate nitrogenase enzyme activity of N_2 fixation. Nitrogenase-produced H_2, however, serves to remove excess intracellular oxygen by hydrogenase, leading to the formation of water and ATP (Paerl, 1980). The ATP is used in other metabolic reactions while protecting the N_2 fixation reactions from oxygen.

Vertical migration of phytoflagellates can contribute significantly to observed diurnal periodicity in stratification and photosynthesis of phytoplankton (e.g., Nauwerck, 1963; Tilzer, 1973; Ilmavirta, 1975; Happy-Wood, 1976). In high mountain lakes, for example, the dominant flagellates (Gymnodinium) usually ascend in the afternoon and evening and migrate downward with increasing light intensities (maximum rates approximately 1 m hr^{-1}). Photosynthesis is reduced near the surface during the midday hours as a result of light saturation, photoinhibition, and downward migration, but at greater depths, utilization of light energy usually increases at this time.

HORIZONTAL VARIATIONS IN PRODUCTIVITY

It is commonly assumed when measuring phytoplanktonic primary productivity that horizontal variations within the pelagial zone are relatively minor. This assumption is probably valid in many small- to moderate-sized lakes in which variations in different parts of the lakes are relatively small ($<$10 per cent). In such cases, horizontal variations in phytoplanktonic activity may be masked by larger variation associated with sampling and experimental errors of measurement. Horizontal variations become much more significant, however, in very small lakes, extremely large lakes such as the Great Lakes, in proximity to the littoral zone of most lakes, and as the morphometric complexity of the lake basins increases.

Even in very productive tropical lakes, phytoplanktonic productivity is often greater in the pelagial zone closer to the perimeter of the lake than in the central portion of the pelagial (e.g., Talling, 1965). In very large lakes, this horizontal disparity becomes more acute (Glooschenko, et al., 1973, 1974a, 1974b). Nutrient loading from specific regions of the drainage basins of such lakes can be sufficient to cause significantly higher, often several-fold higher, rates of primary productivity in near-shore areas or sectors of the pelagial zone. Phytoplanktonic productivity within the littoral zone or in the pelagial immediately adjacent to it is generally reduced significantly; the causes of this are discussed in some detail in Chapter 18.

Horizontal patchiness in phytoplankton distribution has also been observed in relation to wind-induced current patterns (Heaney, 1976; George and Heaney, 1978; Harris, et al., 1979; Heaney and Talling, 1980). Light wind stress (approx. 3 m s^{-1}) can cause horizontal flow of surface layers in small lakes. These currents induce up-welling of subsurface concentrations of phytoplankton at the windward end of a lake, or downwind transport of algae when concentrated in the surface waters. Moderate-to-strong wind conditions ($>$4 m s^{-1}) can cause sufficient turbulent mixing to eliminate heterogeneity within the epilimnion. In large, shallow lakes, wind-induced circulation can cause concentric distribution patterns, with maximum phytoplankton concentrations occurring toward the center (e.g., Ganf, 1974c).

Many lakes have complex morphometry, and may contain dendritic valleys with numerous large, long bays. Primary productivity within these relatively isolated bays can be quite different from that of the major basin (Table 15–8). Often the productivity of these isolated arms is higher than that of the primary basin, for they receive proportionally higher loading of nutrients per unit volume.

The primary productivity of phytoplankton of reservoirs is variable seasonally, and can vary considerably from year to year as well. During periods of high runoff from the drainage basin, abiogenic turbidity can be very high in the section of the basin that receives the primary influents. In this region, primary production rates often are greatly reduced. As much of the inorganic turbidity sediments out of the euphotic zone in the progression downstream (along the length of the reservoir) towards the dam and outlet, the depth and productivity of the trophogenic zone increase. However, if the retention time of the reservoir is short, i.e., the flow-through rate is high, there may be insufficient time for populations to develop fully before being flushed out of the reservoir.

Some reservoirs are so constructed that outflow is drawn from the hypolimnion and lower strata; the increased retention time of surface waters can then lead to a more typically lake-type of phytoplanktonic community development and primary productivity. Removal of hypolimnetic water, laden with nutrients from sedimenting organic matter and decomposition, can reduce the nutrient loading of the basin.

Annual Rates of Primary Production

The number of measurements of the primary productivity of phytoplankton in lake systems is prodigious. Only a few examples have been selected for general comparative purposes in Table 15-9, based on a rather subjective division of lakes into trophic categories. Further examples are listed in Westlake, et al. (1980).

Among the lowest annual productivity values for phytoplankton of natural freshwater systems are those from Char Lake at latitude 74° in the Canadian Arctic. For this lake, the annual primary productivity of phytoplankton is about 4 g C m^{-2} year^{-1} but productivity of mosses and epilithic algae increases this figure to about 20 g C m^{-2} year^{-1} (Kalff and Welch, 1974; Kalff and Wetzel, unpublished). The low productivity results from low light and temperatures as well as from low nutrient availability. For example, Meretta Lake, near Char Lake, receives domestic pollution from a small village in quantities sufficient to increase primary productivity of phytoplankton by five times (Table 15-9). Lower values exist (Table 15-9). To date all waters analyzed, even amictic, permanently ice-covered lakes, contain a limited algal flora and exhibit some net primary productivity.

TABLE 15-8 Variations in the Annual Primary Productivity of Phytoplankton in Different Sections of Sylvan Lake, Indiana, a Basin of Complex Morphometry

REGION OF LAKE	ANNUAL MEAN (mg C m^{-2} DAY^{-1})	ANNUAL PRODUCTIVITY (kg C ha^{-1})
Shallow arm near inlet	1475	5383
Large, isolated bay	1509	5509
Central main basin	1583	5777
Lower arm near outlet	1691	6171
Mean for lake	1564	5710

After data cited in Wetzel (1966a).

TABLE 15-9 Comparison of Rates of Primary Production of Phytoplankton in Selected Fresh Waters of Varying Fertility

LAKE	REMARKS	MEAN DAILY PRODUCTIVITY FOR ENTIRE YEAR (mg C m^{-2} DAY^{-1})	RANGE OBSERVED (mg C m^{-2} DAY^{-1})	ANNUAL PRODUCTIVITY (g C m^{-2} YEAR^{-1})
Oligotrophic				
Tundra ponds, Alaskan arctic (Alexander, et al., 1980)	Mean depth 20 cm; most production by benthic flora	1.6	(0–260)	0.6
Peters, Alaskan arctic (Hobbie, 1964)		2.5	(0–30)	0.9
Schrader, Alaskan arctic (Hobbie, 1964)		19.2	(0–40)	7.0
Char, N.W.T., Canadian arctic (lat. 74°) (Kalff and Welch, 1974)	80% of total production by benthic flora	11.5	0–35	4.2
Meretta, N.W.T., Canada (Kalff and Welch, 1974)	Polluted by sewage	8.5	0–170	11.3
Castle, Calif. (Goldman, pers. comm.)	Deep, alpine	98	6–317	36
Lunzer Untersee, Austria (Jónasson, 1972)	Small, alpine	(123)	—	45
Lawrence, Mich. (Wetzel; cf. Table 15-6)	Small, hardwater; 14-year average	99.3	5–497	36.3
Ransaren, Sweden (Rodhe, 1958)	Small Lapplandic	—	23–66	—
Lake Superior, USA–Canada (Putnam and Olson, 1961)	Most unproductive of Laurentian Great Lakes	—	50–260	—
Lake Huron, USA–Canada (Vollenweider, et al., 1974)	Offshore stations	—	150–700	approx. 100
Borax, Calif. (Wetzel, 1964)	Large, shallow saline lake (25% of total productivity)	249	10–524	91
Lake Michigan, USA (Vollenweider, et al., 1974)	Offshore stations	—	70–1030	approx. 130
Lake Ontario, USA–Canada (Vollenweider, et al., 1974)	Offshore stations	—	60–1400	approx. 180
Mesotrophic				
Erken, Sweden (Rodhe, 1958)	Large, deep, naturally productive	285	40–2205	104
Clear, Calif. (Goldman and Wetzel, 1963)	Very large, shallow	438	2–2440	160

Lake	Description			
Esrom, Denmark (Jónasson and Mathiesen, 1959)	Large, moderately deep	370	23–422	260
Furesø, Denmark (Jónasson and Mathiesen, 1959)	Large, deep, many macrophytes	462	0–1380	168
Walter, Ind. (Wetzel, 1973)	Series of 4 interconnected marl lakes			
Basin A		418	102–1395	153
Basin B		210	12–535	77
Basin C		276	30–1048	101
Basin D		437	98–1458	160
Oliver, Ind. (Wetzel, 1973)	Large, deep, marl lake	336	32–775	123
Olin, Ind. (Wetzel, 1973)	Large, deep, marl lake	374	89–996	137
Martin, Ind. (Wetzel, 1973)	Deep, stained marl lake	561	27–1708	205
Pretty, Ind. (Wetzel, 1966b)	Moderate-sized, deep, marl lake			
1963		440	68–1850	161
1964		305	6–895	111
Crooked, Ind. (Wetzel, 1966b)	Large, deep, hardwater lake			
1963		469	142–1364	171
1964		359	23–870	131
Little Crooked, Ind. (Wetzel, 1966b)	Small, deep, kettle lake			
1963		618	263–1903	226
1964		598	9–2451	218
Goose, Ind. (Wetzel, 1966a)	Small kettle lake	729	166–1753	266
Lake Erie, USA-Canada (Vollenweider, et al., 1974)				
Western stations		—	30–4760	(310)
Central stations		—	120–1690	(210)
Eastern stations		—	140–1440	(160)
Eutrophic				
Wintergreen, Mich. (Wetzel, et al., unpublished)	Shallow, extensive nutrient loading	1012	60–2240	369
Frederiksborg Slotssø Denmark (Nygaard, 1955)	Shallow, enriched	1030	12–4160	376
Minnetonka, Minn. (Megard, 1972)	Extremely complex basin, large, deep	(820)	—	(300)
Sollerød Sø, Denmark (Steemann Nielsen, 1955)		1430	0–3800	522

TABLE 15-9 Continued on Following Page.

TABLE 15-9 Comparison of Rates of Primary Production of Phytoplankton in Selected Fresh Waters of Varying Fertility (Continued)

LAKE	REMARKS	MEAN DAILY PRODUCTIVITY FOR ENTIRE YEAR (mg C m^{-2} DAY^{-1})	RANGE OBSERVED (mg C m^{-2} DAY^{-1})	ANNUAL PRODUCTIVITY (g C m^{-2} YEAR^{-1})
Sylvan, Ind. (Wetzel, 1966a; cf. Table 15-8)	Complex basin, large, shallow	1564	9–4959	570
Lanao, Philippines (Lewis, 1974b)	Large, deep tropical lake	1700	400–5000	620
Victoria, Africa (Talling, 1965)	Large, deep, equatorial lake	1750	1700–3800	640
Dystrophic				
Kattehale Mose, Denmark (Nygaard, 1955)	Very shallow, acidic, peat bog	80	0–400	29
Smith Hole, Ind. (Wetzel, 1973)	Shallow, humic stained	194	24–5960	71
Store Gribsø, Denmark (Nygaard, 1955)	Deep, acidic, humic stained	230	4–680	84
Grane Langsø, Denmark (Nygaard, 1955)	Deep, acidic	248	20–880	91

Data approximated in some cases; estimates given in parentheses.

A general categorization of the characteristics of lakes between the extremes of very oligotrophic and very eutrophic systems has been attempted many times. The criteria employed, largely resulting from man's innate desire for orderliness in the natural world, have led to classification of lakes on the basis of nearly every parameter, including geomorphology, chemical constituents, nearly every type of organism from bacteria to fish, and even the waterfowl associated with different types of lakes. The most realistic parameter for such categorization, at least in a general way, is one that can be quantitatively determined as a rate of growth, and one that integrates the host of environmental parameters controlling the synthesis of organic matter that enters the system. Rates of autochthonous primary production are basic to such an evaluation, and those of the phytoplankton have been proposed many times as the best existing criterion (see, for example, Rodhe, 1969).

The ranges of primary productivity of phytoplankton commonly associated with oligotrophy and eutrophy are given in Table 15–10, along with several related characteristics of the phytoplankton. Such groupings can only be used in a general way, since there are many exceptions. The relationships, however, predominate in a majority of inland waters. The dystrophic lake category, discussed at length in Chapter 24, is very deviant and variable. Productivity of phytoplankton in these lakes is frequently attenuated by complex interactions between light, high concentrations of dissolved humic compounds, and inorganic nutrient availability.

Primary productivity of phytoplankton is sometimes used as the criterion for determining a lake's trophic state. The validity of this criterion, however, depends upon the assumption that organic matter inputs from the littoral and allochthonous sources are small relative to those of the phytoplankton. However, in many cases, littoral productivity and allochthonous inputs are large relative to phytoplanktonic productivity; in such cases, phytoplanktonic productivity alone is a poor estimator of the trophic state. One of the major objectives of Chapters 18 and 19 is to place these components of aquatic systems in proper perspective with the real world. This does not decrease the importance of phytoplankton; rather, the primary productivity of phytoplankton must be viewed as a variably important contributor to overall lake and stream metabolism.

The importance of size relationships among algae and environmental parameters has been emphasized for some time (cf. particularly Margalef, 1955, 1959) and earlier discussion pointed out the relative preponderance of small algal species of the phytoplankton in oligotrophic lakes. In general, the size of dominant algae increases in more fertile lakes. A number of analyses of primary productivity in which the relative contribution of differently sized algae were determined show that a greater proportion of the total primary productivity is by smaller forms in less productive waters (Rodhe, et al., 1958; Goldman and Wetzel, 1963; Wetzel, 1964, 1965a, and others). The contributions of the smaller algae also shift seasonally to a point at which the nannoplankton and ultraplankton (<10 μm) are responsible for almost all of the planktonic primary productivity.

An inverse relationship generally exists between total algal biomass and productivity per unit biomass in the phytoplankton. Accordingly, communities dominated by populations of species with small cells usually have a greater primary productivity per unit algal biomass than communities dominated by populations of large algal species (Findenegg, 1965). As demonstrated by microautoradiography, the renewal time (gen-

TABLE 15–10 General Ranges of Primary Productivity of Phytoplankton and Related Characteristics of Lakes of Different Trophic Categories

TROPHIC TYPE	MEAN PRIMARY PRODUCTIVITY (mg C m^{-2} DAY^{-1})*	PHYTOPLANKTON DENSITY (cm^3 m^{-3})	PHYTOPLANKTON BIOMASS (mg C m^{-3})	CHLOROPHYLL a (mg m^{-3})	DOMINANT PHYTOPLANKTON	LIGHT EXTINCTION COEFFICIENTS (η m^{-1})	TOTAL ORGANIC CARBON (mg l^{-1})	TOTAL P (µg l^{-1})	TOTAL N (µg l^{-1})	TOTAL INORGANIC SOLIDS (mg l^{-1})
Ultraoligotrophic	<50	<1	<50	0.01–0.5		0.03–0.8	<1–3	<1–5	<1–250	2–15
Oligotrophic	50–300		20–100	0.3–3	Chrysophyceae, Cryptophyceae, Dinophyceae, Bacillariophyceae	0.05–1.0		5–10	250–600	10–200
Oligomesotrophic		1–3								
Mesotrophic	250–1000		100–300	2–15		0.1–2.0	<1–5	10–30	500–1100	100–500
Mesoeutrophic		3–5			Bacillariophyceae, Cyanophyceae, Chlorophyceae, Euglenophyceae	0.5–4.0	5–30			
Eutrophic	>1000		>300	10–500				30–>5000	500–>15000	400–60000
Hypereutrophic		>10								
Dystrophic	<50–500		<50–200	0.1–10		1.0–4.0	3–30	<1–10	<1–500	5–200

Modified from Likens (1975), after many authors and sources.
*Referring to approximately net primary productivity, such as measured by the ^{14}C method.

eration time) is related inversely to cell size; cells that are small with a low carbon content tend to have shorter renewal times and are more metabolically active than larger algae (Stull, et al., 1973). Therefore, some small species of relatively minor contribution to the algal community biomass turn over rapidly and contribute more to the total primary productivity than do larger species.

_____ SUMMARY

1. The freshwater phytoplankton comprise diverse algae of almost every major taxonomic group. Photosynthetic autotrophic metabolism is the sole metabolic pathway for the synthesis of organic matter among most algae. Although certain microbiota possess supplementary nonphotosynthetic means of nutrition, photosynthesis dominates and these organisms can be functionally included with the phytoplankton.
2. Light-energy-absorbing pigments are diverse among the algae. Chlorophyll *a* occurs in all algae, and is the primary photosynthetic pigment. Chlorophyll *b* is found only in green algae and euglenophytes, and chlorophyll *c* occurs in many algal groups. Other accessory chlorophylls, of uncertain functions, are restricted to specialized algae. Light energy absorbed by some carotenoids and the biliproteins (phycoerythrin and phycocyanin of the blue-green, cryptomonad, and red algae), as with chlorophylls *b* and *c*, is transferred to chlorophyll *a*.
3. Most phytoplanktonic algae lack or have very limited powers of locomotion. As a result, most phytoplankton are dispersed by turbulent water movements. Since the density of most phytoplankton is greater than that of water, they tend to sink from the lighted, trophogenic zone. Sinking has the advantage of disrupting nutrient gradients surrounding the cells. The disadvantage of sedimenting out of the photic zone is offset to varying degrees by upward movement and turbulent water transport.

    Improved flotation or reduced sinking rates are accomplished in various ways:
    a. Reduction in sinking rate is enhanced by increasing *surface-to-volume ratios*. Deviations in cell morphology from the spherical form by protrusions and projections decrease sinking rates by increasing frictional resistance with the water.

        Seasonal polymorphism (cyclomorphosis) is rare among phytoplankton, but is conspicuous among certain dinoflagellates. *Ceratium* alters its form and reduces its sinking rate as the viscosity of water decreases during warmer seasons (Fig. 15–1).
    b. *Production of mucilage* reduces sinking rates. Gelatinous sheaths are common among blue-green and green algae, diatoms, and certain desmids. The sheaths are less dense than the cell, and increase frictional resistance.
    c. *Gas vacuoles* are common in blue-green algae and decrease the density of the cells to below the density of water. Many blue-green algae

regulate buoyancy and position in the photic zone to poise the populations within vertical physical and chemical gradients favorable to growth.

    d.   A few algae reduce density by *accumulation of fats.*

    e.   Certain algae can alter density by *regulation of cellular ion content.*

4.   Coexistence of a number of phytoplankton species is a conspicuous feature of fresh waters. Although a few species commonly dominate a phytoplanktonic assemblage, a number of rarer algae coexist among the dominant species. Many differences in algal physiological characteristics, requirements, and tolerances, as well as seasonal and spatial variations in environmental parameters, permit the apparent multispecific equilibrium to exist for short periods.

    a.   Differences in the efficiency of utilization of resources may be too small for competitive exclusion to occur before environmental conditions change.

    b.   Partial commensalism or symbiotic dependency (e.g., through production and utilization of organic micronutrients) or both can reduce competition.

    c.   Selective herbivory of a more successful species can reduce competition with a less successful species.

    d.   Certain algae are meroplanktonic, enter resting stages, and do not compete for periods of varying duration with holoplanktonic (continually planktonic) species.

    e.   Many or all of these coexistence mechanisms can function simultaneously.

5.   Many environmental factors interact to regulate spatial and seasonal growth and succession of phytoplankton populations.

    a.   Light and temperature synergistically affect photosynthesis. Although photosynthetic rates and algal growth are directly related to irradiance intensity, the response to light intensity, especially at light saturation, is temperature-dependent (Fig. 15–2) and variable among species. A considerable degree of adaptation to changing light intensities occurs among algae, often by regulation of pigment concentrations per unit biomass.

    b.   High light intensities inhibit photosynthesis of many algae; effects are partly reversible if exposure is not too long (several hours) and phytoplankton are transported to regions of lower light intensities.

    c.   The vertical distribution of photosynthesis is strongly related to available light. A near-exponential decline with increasing depth underlies a surficial zone of maximum photosynthesis (Fig. 15–3). Photoinhibition by excessive light and ultraviolet radiation is common, with depression of photosynthesis at the surface and a subsurface photosynthetic maximum. As photosynthetic rates per unit volume of water increase in nutrient enriched waters, the biogenic turbidity resulting from dense algal populations constricts the thickness of the trophogenic zone towards the surface.

d. Algae have definite temperature optima and tolerance ranges, which interact with other parameters to cause seasonal succession. For example, many diatoms can photosynthesize well at cooler water temperatures, whereas the temperature optima of many green and blue-green algae are higher.

e. Major changes in the seasonal succession of phytoplankton in temperate lakes are related to changes in availability of phosphorus, nitrogen, and silica (cf. Chapters 12, 13 and 14). Growth of a population under conditions of adequate light and temperature is often limited by a single nutrient. Limitation can shift rapidly from nutrient to nutrient as their availabilities change on a diurnal, daily, and seasonal basis.

f. Many algae require, but are unable to synthesize, organic micronutrients, especially vitamin B_{12}. These substances have been found to be actively assimilated by phytoplankton and, at certain times, photosynthesis is enhanced by organic enrichment; they likely play a significant role in succession and competitive success of certain algal populations.

g. Most algae are obligate photoautotrophs in which inorganic carbon is reduced biochemically with light energy. Utilization of organic substrates in the dark (heterotrophy) or in the light (photoheterotrophy) by aerobic dissimilation occurs in certain algae, especially the blue-green algae. Heterotrophic augmentation occurs in some algae, but under most natural conditions of low substrate concentrations, bacteria are much better competitors for available substrates.

h. Dissolved organic compounds, largely of terrestrial and wetland plant origin, can influence phytoplankton metabolism by interacting with macro- and micronutrients and influencing their availability, either directly (e.g., complexation) or indirectly (e.g., altering other inorganic regulatory mechanisms).

6. Distinct seasonal patterns and successions of algal populations and biomass are observed in phytoplankton communities.

a. In temperate fresh waters, growth is greatly reduced during winter when both light and temperatures are low. A spring biomass maximum is commonly observed as light conditions improve, and often consists predominantly of diatoms and cryptophytes adapted to low light conditions. Lower biomass often occurs during the summer. Among more eutrophic waters with high phosphorus loading, silica concentrations of the trophogenic zone are commonly reduced in the spring to levels inadequate to support large diatom populations. Summer populations of nonsiliceous green algae then flourish until concentrations of combined nitrogen are reduced below replacement. Under these conditions, nitrogen-fixing blue-green algae have competitive advantages and often proliferate.

b. Summer phytoplankton populations of temperate oligotrophic lakes are often low. A second autumnal maximum, usually predominantly of diatoms, often develops during and after fall circulation. The limited growing season in high altitude and subpolar regions often results in a single summer maximum of phytoplankton biomass. In tropical lakes, total phytoplankton biomass is more constant (maximum often in winter) and larger than in temperate lakes. Frequent changes in abiotic factors occur in tropical lakes that lead to more episodic changes in phytoplankton succession than in stratified temperate lakes. Succession of algae within the episodes, however, is similar in both tropical and temperate lakes.

  c. Although variability is large, the general patterns of seasonal succession of phytoplanktonic biomass are reasonably constant from year to year, if the drainage basin and lake are not perturbed.

  d. Seasonal changes in phytoplanktonic numbers and biomass are usually very large (approximately a thousandfold) in temperate and subpolar fresh waters. In tropical lakes, seasonal variations are much lower (approximately fivefold).

  e. Changes in phytoplanktonic numbers and biomass are often out of phase with rates of photosynthesis. Photosynthetic productivity usually follows more closely the annual cycle of solar irradiance in low or moderately productive lakes, especially in temperate and subpolar areas. Smaller algae with faster turnover rates often predominate at warmer temperatures.

7. Parasitism of phytoplankton, particularly by chytrid fungi, is common and can increase as algae senesce in aerobic waters. Effects of selective parasites on species succession and algal productivity are unclear.

8. Grazing of phytoplankton by rotifers and microcrustacea can influence algal populations and their succession.

  a. On an annual basis, grazing losses of phytoplankton are usually minor in comparison to losses by sedimentation out of the photic zone.

  b. Grazing losses can be highly significant during certain periods of the year, during which time the algal populations are severely reduced. Much of the food ingested by zooplankton is not assimilated, but is egested and enters the detrital pool as dissolved and particulate organic matter.

  c. Selective grazing by microcrustacea, largely on the basis of algal cell size, can alter seasonal succession of phytoplankton.

  d. Nutrient regeneration, particularly of phosphorus, can be accelerated by zooplanktonic grazing; the increase in nutrient regeneration can enhance primary productivity of the algae.

9. As nutrient limitations of the phytoplankton of infertile waters are increasingly relieved by nutrient inputs, rates of algal production increase.

  a. The densities of phytoplanktonic community progressively reduce the available light and decrease the thickness of the trophogenic zone. A point is reached at which self-shading inhibits further increases in productivity, regardless of nutrient availability. Productivity per unit surface area of the freshwater system is usually lower under these hypereutrophic conditions than under less productive conditions where a thicker trophogenic zone exists.

b. Maximum photosynthetic efficiencies of light utilization by phytoplankton are low, usually less than one per cent of radiation incident on the water. These efficiencies are considerably less than utilization by wetland and terrestrial plants.

c. An inverse relationship generally exists between algal biomass and productivity per unit biomass in the phytoplankton. Often small species of relatively minor contribution to the algal community biomass have short generation times and contribute more to the total primary productivity than do larger species.

10. A significant amount of the carbon fixed photosynthetically by phytoplankton is released extracellularly as dissolved organic compounds.

   a. Certain low-molecular-weight compounds of intermediary metabolism are released in greater amounts under conditions of stress (e.g., light inhibition, low nutrient availability).

   b. Extracellular dissolved organic carbon released as a percentage of total carbon fixed photosynthetically is usually low (1–5%) in healthy, actively growing phytoplankton. Percentage release increases under very high and very low light conditions, and during senescence.

11. Horizontal variations in primary productivity of phytoplankton can be large. Spatial variations increase in significance in very small lakes, very large lakes (e.g., the Laurentian Great Lakes), in proximity to the littoral zone and inlet areas of most lakes, and as the morphometric complexity of lake basins increases.

CHAPTER SIXTEEN

PLANKTONIC COMMUNITIES: ZOOPLANKTON AND THEIR INTERACTIONS WITH FISH

The animals found in fresh waters are extremely diverse, and are represented by nearly all phyla. Evaluation of their functional roles within aquatic systems requires a balanced understanding between the general modes and timing of growth and reproduction in relation to the availability and utilization of food. The population dynamics and certain important adaptive behavioral characteristics that influence these dynamics regulate the productivity of individual species populations and entire communities. Underlying all evaluations of productivity of the animals are their food or trophic relations with plants and other animals, and competitive and predatory interactions that lead to a greater success of one species over another.

ZOOPLANKTON

Truly planktonic animals are dominated by three major groups: the rotifers, and two subclasses of the Crustacea, the Cladocera and Copepoda. These groups have been studied in some detail and form the bulk of the ensuing discussion. Under most circumstances, the rotifers and especially the limnetic crustaceans are overwhelmingly the dominants of zooplanktonic productivity.

A few coelenterates, larval trematode flatworms, gastrotrichs, mites, and the larval stages of insects and fish occasionally occur among the true zooplankton, if only for a portion of their life cycles. With a few exceptions discussed later, however, the quantitative significance of these organisms is largely obscure.

Protozoa

The population dynamics and productivity of the protozoa are poorly understood. Although generally a minor part of the zooplankton both numerically and in biomass, at times the protozoa can constitute a significant component of the zooplanktonic productivity (e.g., Pace and Orcutt, 1981).

> For example, in midsummer in large Lake Dalnee, Kamchatka, USSR, the flagellate and ciliate protozoans made up a substantial part of the pelagial zooplanktonic community (Sorokin and Paveljeva, 1972). Their maximal biomass was observed during the period of the decline of early summer algal populations and the simultaneous intensive development of bacterial populations. In this lake, protozoan maxima occurred in distinct layers of the water column, generally between 10 and 20 meters; their biomass (about 3 g m^{-3}) approached the total biomass of other zooplankton at their maximum development. Protozoan utilization of planktonic bacteria in this lake in July constituted a major pathway of energy flux among the fauna in the pelagial zone (discussed further in Chapter 17).
>
> In Lake Tanganyika in equatorial Africa, the biomass of protozoan zooplankton often exceeded that of the phytoplankton in the euphotic zone while the lake was stably stratified (Hecky and Kling, 1981). The dominant protozoan (*Strombidium*) was, however, probably symbiotic with zoochlorellae algae, and imposed lower demands on the available food.
>
> The lobosan rhizopod *Difflugia* is a rather common planktonic protozoan. An example of a common life cycle is seen in the dynamics of *D. limnetica* in several German lakes (Schönborn, 1962). Populations increased in early summer and attained a maximum in later summer (Fig. 16–1). Much of the population then sinks as fat globules are metabolized and density increases. Many die and decompose, some encyst, and others remain active throughout the winter in the littoral sediments. The benthic populations of *D. limnetica* slowly increase, and then in late spring become planktonic by reducing their density through formation of fat inclusions and gas bubbles. The tests of this species and some others are covered with minute sand grains; in the planktonic phase the tests may be covered instead with diatom frustules or quartz grains that are circulated into the upper strata (Fig. 16–1). Many other protozoans exhibit this type of meroplanktonic existence, in which only a portion of their life cycle is planktonic.

A number of ciliates are common to the zooplankton, although they usually do not dominate except in certain situations, e.g., in very shallow lakes or in the deeper strata of nearly or completely anaerobic hypolimnia. Ciliates can move much more rapidly (200 to 1000 μm sec^{-1})* than other Protozoa (0.5 to 3 μm sec^{-1} among those with pseudopodia; 15 to 300 μm sec^{-1} among those with flagella), which contributes significantly to their greater dispersal and higher feeding rates. Although a few ciliates are mixotrophic and supplement nutrition by photosynthesis, most are holozoic and feed on bacteria, algae, particulate detritus, and other protozoans. A few are carnivorous, and feed on small metazoans.

Smaller ciliated protozoans ($<$ 30 μm) graze upon bacteria-sized particles, whereas rotifers (see below) utilize both bacteria and small algae. Larger crustacean zooplankton are primarily phytophagous on small algae (Crisman, et al., 1981). Nutrients released when macrozooplankton graze on algae can enhance bacterial growth, and thereby provide food for ciliated protozoans and small rotifers. As such, the ciliated protozoans can serve as functional links in freshwater planktonic food chains; they uti-

---

*The swimming rates are temperature-dependent, e.g., two species of *Loxodes* averaged 270 μm sec^{-1} at 10°C and 430 μm sec^{-1} at 20°C (Jones and Goulder, 1973).

lize bacteria and very small particulate detritus and provide macrozooplankton with larger particles that can be more efficiently grazed (cf. Porter, et al., 1979). There is some evidence that the body size of macrozooplankton (especially cladocerans) is smaller in subtropical than in temperate lakes (e.g., Crisman, 1980). Ciliated protozoans and rotifers become more important in the zooplankton among eutrophic, subtropical lakes.

Several protozoans have been shown to feed actively on algae. The importance of planktonic amoeboid forms, known to consume diatoms (e.g., Canter, 1973), is not fully understood. Two species of benthic ciliates of the genus *Loxodes* were studied in a shallow eutrophic lake in relation to feeding and digestion of algae (Goulder, 1972; 1974a). Feeding rates were low (0.4 to 13 diatom cells per hour for the larger species), but there was an indication that the protozoan was able to distinguish among three species of the dominating alga *Scenedesmus* and feed selectively. Based on estimates of the feeding rates and the biomass of the algal and protozoan populations, maximum grazing by the ciliates resulted in consumption of less than 1 per cent of the algal pop-

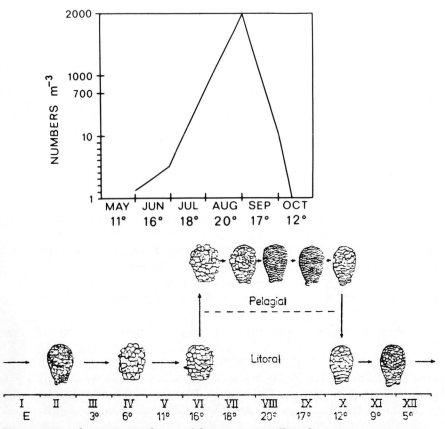

Figure 16-1. Changes in populations of the protozoan *Difflugia limnetica* in representative German lakes. *Upper:* Population densities per cubic meter during the planktonic phase. *Lower:* Generalized scheme of morphological changes during the population transitions between the littoral and planktonic habitats. E = Ice period. (Modified slightly from Schönborn, W.: *Limnologica*, 1:21–34, 1962.)

ulations per day, and had no appreciable effects on them. No evidence was found for interspecific competition between the two species; changes in environmental conditions affected both species in similar ways (Goulder, 1974b, 1980). These species are strongly negatively phototactic, and did not migrate from the anaerobic hypolimnion to the oxygenated epilimnion. Most individuals constituting the hypolimnetic population died as the oxygen was depleted.

The nature and quantity of available food have been implicated as major controlling factors in the population dynamics of ciliates. For example, in a shallow pond in Pennsylvania, three population surges of ciliates occurred and these were correlated with different nutritional habits (Bamforth, 1958). The small holophyid *Urotricha* developed rapidly under ice and fed actively on a dense population of chrysomonads. Oxytrichid ciliates appeared as the *Chlamydomonas* algal populations were declining rapidly in spring. Ciliate populations, mainly of the genus *Holophyra*, appeared following spring maxima of phytoplankton and persisted during the summer, coincident with euglenoid algal blooms.

Although nearly all Protozoa are aerobic, a majority can grow very well even when oxygen concentrations are very low (see, for example, Bragg, 1960). This microaerophilic ability is conspicuous among the planktonic and benthic ciliates, and is attested to by their major development in organic-rich and polluted waters (cf., for example, the monographic review of saprobic organisms by Sládeček, 1973). Populations of ciliates often develop in strata greatly reduced in or devoid of oxygen in which bacterial populations tend to be dense, such as the monimolimnion of meromictic lakes.

GENERAL CHARACTERISTICS OF ROTIFERS, CLADOCERA, AND COPEPODS

Rotifers

The Rotifera (Rotatoria) is a large class of the pseudocoelomate Phylum Aschelminthes, clearly arisen in fresh water; only two significant genera and a few species are marine. About three-quarters of the rotifers are sessile and associated with littoral substrates. Approximately 100 species are completely planktonic, and these rotifers form a significant component of the zooplankton. Rotifers are the most important soft-bodied invertebrates of the plankton. The general characteristics of the group have been treated in some detail by Pennak (1978), Hyman (1951), Hutchinson (1967), Ruttner-Kolisko (1972), and Dumont and Green (1980).

The rotifers exhibit a very wide range of morphological variations and adaptations. In a majority, the body shape tends to be elongated, and regions of the head, trunk, and foot usually are distinguishable (Fig. 16–2). The cuticle is generally thin and flexible, but in some rotifers it is thickened and more rigid and is termed a lorica; the lorica is of taxonomic importance in some groups. The anterior end or corona of rotifers is ciliated; in some species the periphery is ciliated as well. The movement of the cilia functions both in locomotion, especially among planktonic forms, and in movement of food particles towards the mouth. The mouth, although variously located, is generally anterior. The digestive system contains a complex mastax, a set of sclerotized jaws or trophi unique to the rotifers that functions to seize and disrupt food particles.

Most rotifers, both sessile and planktonic, are nonpredatory. Omnivorous feeding occurs by means of ciliary movement of living and detrital particulate organic matter into the mouth cavity. Predatory species, such as the common *Asplanchna*, are usually large and prey upon protozoa, other rotifers, and other micrometazoa of appropriate size.

Most rotifers are not planktonic, but are sessile and associated with littoral substrata. Population numbers are highest in association with submersed macrophytes, especially plants with finely divided leaves; densities commonly reach 25,000 per liter (Edmondson, 1944, 1945, 1946). Even greater densities are found in the interstitial water of beach sand at or slightly above the waterline (Pennak, 1940). With reduced sites for attachment and presumably less protection from predation, planktonic rotifer populations are much less dense. Densities of planktonic rotifers of 200 to 300 l^{-1} are common

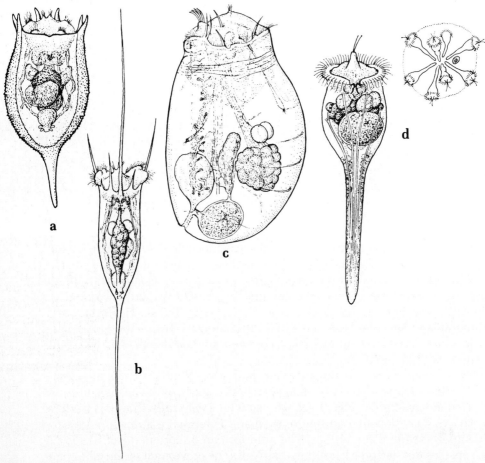

Figure 16-2. Exemplary planktonic rotifers: *a*, *Keratella cochlearis*; *b*, *Kellicottia longispina*; *c*, *Asplanchna girodi*; *d*, *Conochilus unicornis*, singly and in a colony. (From Ruttner-Kolisko, A.: III. Rotatoria. Das Zooplankton der Binnengewässer. I. Teil. Die Binnengewässer, 26:99–234, 1972, from various sources.)

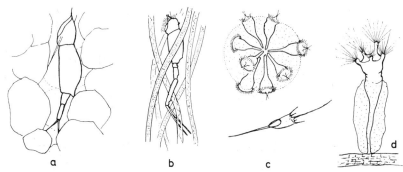

Figure 16-3. Exemplary types of rotifers of different habitats. a, Psammic rotifer (*Bryceella*) among sand grains; b, A littoral form (*Scaridium*) among algal filaments; c, Planktonic forms (*Conochilus* and *Kellicottia*); d, *Collotheca* epiphytic on the stem of a macrophyte. (From Ruttner-Kolisko, A.: III. Rotatoria. Das Zooplankton der Binnengewässer. I. Teil. Die Binnengewässer, 26:99–234, 1972.)

and occasionally reach 1000 l^{-1}; densities rarely exceed 5000 l^{-1} under natural conditions.

Several changes characterize the transition from the predominantly sessile to the planktonic life forms (Fig. 16-3). Weight reduction is common as a result of diminution of the lorica and enlargement of body volume with gelatinous materials. Planktonic species tend to have suspension processes and swimming organs in the form of immovable spines or movable setae. A reduction of attachment organs as a result of diminution or total loss of the foot structures also takes place. Adaptations that reduce the sinking rates of reproductive products also occur, e.g., attachment of eggs to the adult, production of lipid-rich eggs that may be extensively ornamented, and vivipary.

Crustacean Zooplankton

The crustacean arthropods are almost entirely aquatic; most are marine. Respiration is accomplished through the body surface or gills. The body generally is separated into three distinct regions, but the tendency is towards fusion of abdominal and thoracic segments until, in the Cladocera and Ostracoda, apparent body segmentation has been lost. In many crustaceans, the body bears paired, usually biramous, jointed appendages, and is covered wholly or in part by a carapace.

In fresh water, the truly planktonic Crustacea are dominated almost completely by the cladocerans and Copepoda, to which most of the ensuing discussion is devoted. Only a few insects are planktonic in immature stages; the larvae of *Chaoborus* (Diptera) is a notable example of major importance in the zooplankton that demands special consideration. Most of the Ostracoda are benthic (cf. Chapter 21). A few species of *Cypria* are apparently partly planktonic, but very little is known of their ecology.

The freshwater Branchiopoda (fairy and clam shrimps) are common inhabitants of shallow lakes, particularly of temporary, saline inland waters. All members of this primitive group are distinctly segmented and bear many pairs of swimming and respiratory appendages. The tadpole shrimps (Notostraca) are essentially benthic and most often restricted to shallow, temporary lakes of arid regions. The fairy shrimps (Anostraca), lacking a carapace, and the clam shrimps (Conchostraca), compressed laterally with a bivalved flexible carapace, are common in the plankton in shallow playa lakes and vernal ponds of semiarid regions. The members of the latter groups are bisexual, although parthenogenesis is known to occur in brine shrimp *Artemia* under some conditions. Resistant eggs can be subjected to long periods of desiccation and hatch when wetted again, probably as a result of a reduction in osmotic pressure at the egg surface (Hutchinson, 1967). In semipermanent lakes of more humid regions, hatching and reproductive rates are related strongly to temperature.

CLADOCERA

The suborder Cladocera (Branchiopoda; Order Phyllopoda) includes mainly microzooplankton. With the exception of two species, nearly all the cladocerans range in size from 0.2 to 3.0 mm. All have a distinct head, and the body is covered by a bivalve cuticular carapace. Light-sensitive organs usually consist of a large, compound eye and a smaller ocellus. The second antennae are large swimming appendages and constitute the primary organs of locomotion. The mouthparts consist of: (1) large, chitinized mandibles that grind food particles, (2) a pair of small maxillules, used to push food between the mandibles, and (3) a median labrum that covers the other mouthparts.

COPEPODA

The free-living copepods of this order of the class Crustacea are separable into three distinct groups: the suborders Calanoida, Cyclopoida, and Harpacticoida. Although accurate identification is based largely on morphological details of appendages, several general characteristics delineate the major groups. The body consists of

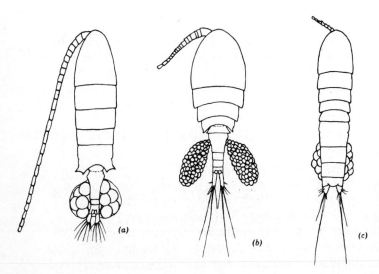

(a)

(b)

(c)

Figure 16-4. Diagrams of the three major types of free-living freshwater copepods (females). *a,* Calanoid. *b,* Cyclopoid. *c,* Harpacticoid. (From Wilson, M. S., and Yeatman, R. C.: Free-living copepoda. *In* Edmonson, W. T., ed., Freshwater Biology, 2nd ed., copyright © 1959 by John Wiley & Sons, Inc. New York, 1959. Reprinted by permission of John Wiley & Sons, Inc.)

TABLE 16-1 Some Characteristics of the Three Suborders of Free-Living Freshwater Copepoda

CALANOIDA	CYCLOPOIDA	HARPACTICOIDA
Anterior part of body much broader than posterior	Anterior part of body much broader than posterior	Anterior part of body a little broader than posterior
Marked constriction between somite of 5th leg and genital segment	Marked constriction between somites of 4th and 5th legs	Slight or no constriction between somites of 4th and 5th legs
One egg sac, carried medially	2 egg sacs, carried laterally	Usually 1 egg sac, carried medially
First antennae long, extend from end of metasome to near end of caudal setae, 23–25 segments in female	First antennae short, extend from proximal 3rd of head segment to near end of metasome, 6–17 segments in female	First antennae very short, extend from proximal 5th to end of head segment, 5–9 segments in female
5th leg similar to other legs	5th leg vestigial	5th leg vestigial
Planktonic, rarely littoral	Littoral, a few species planktonic	Exclusively littoral, on macrovegetation and sediments

Modified from the summary of Wilson and Yeatman, 1959.

the anterior metasome (cephalothorax), which is divided into the head region, bearing five pairs of appendages representing antennae and mouthparts, and the thorax, with six pairs of mainly swimming legs. The posterior urosome consists of abdominal segments, the first of which is modified in females as the genital segment, and terminal caudal rami bearing setae.

The three suborders of free-living copepods can be distinguished by the general structure of the first antennae, urosome, and fifth leg (Fig. 16-4 and Table 16-1). The harpacticoid copepods are almost exclusively littoral, habitating macrovegetation, mosses in particular, and the littoral sediments. Certain species have life histories with a diapause similar to that discussed further on for the cyclopoid copepods. Although the cyclopoid copepods are primarily littoral benthic species, those few members that are predominantly planktonic form major components of the copepod zooplankton, especially in small, shallow lakes. The calanoid copepods are almost exclusively planktonic.

FOOD, FEEDING, AND FOOD SELECTIVITY

ROTIFERS

The planktonic rotifers feed largely by sedimenting seston particles into the mouth orifice by means of the pulsating action of the coronal cilia (reviewed at length by Pourriot, 1965, and Hutchinson, 1967). Size of the food consumed is quite variable. Most food particles taken are small, less than about 12 μm in diameter, although larger cells, up to approximately 50 μm, are sometimes seized, ruptured, and their particulate parts ingested. Feeding behavior of suspension-feeding rotifers is related to the type of particles, the food size and shape, and food density. Food selectivity clearly occurs in some rotifers, and is governed by several rejection mechanisms, either by screening particles or by rejection of particles once they have been ingested (Starkweather, 1980a, 1980b).

Detailed analyses of the feeding behavior of *Brachionus calyciflorus* on varying quantities and mixtures of different types of foods (bacteria, yeast, and algae) demonstrated complex relationships between ingestion rates, food density, and selectivity. Several mechanisms regulate the ingestion of suspended particles (Gilbert and Starkweather, 1977): (a) Cirri of the pseudotrochus may be extended, allowing particles to enter the funnel-shaped buccal field, or they may be bent over to form a screen that prevents even some very small particles (e.g., yeast) from entering the buccal field. (b) Particles in the buccal field may be rejected by ciliary movements before entering the oral canal. (c) Particles in the oral canal can be rejected by the jaws and pushed back into the buccal field.

Different food types induce different ingestion rates in *Brachionus* (Starkweather and Gilbert, 1977a, 1977b, 1978; Gilbert and Starkweather, 1978). When feeding on yeast, ingestion increased continuously with increasing food-cell densities without the use of pseudotrochal screening. When feeding on certain bacteria and algae, however, ingestion rates were constant over a considerable range of particle densities. The use of pseudotrochal screening increased with larger-sized particles and with increasing particle densities. Screening was also influenced by past feeding history. Animals selected food of the type they had grown on if exposed to a mixed type of food, and previously starved rotifers used screening less than well-fed animals. The time required for ingested food to pass entirely through the alimentary canal is from 3 to 20 minutes; this time varies greatly among species and with external conditions.

Some rotifers are raptorial, seizing and ingesting whole prey organisms or drawing in the cell contents after the cell or body wall has been punctured. The largest rotifer *Asplanchna,* which is predatory on algae, rotifers, and small planktonic crustaceans, has the ability to shift size in response to changes in food size and prey densities, as discussed later (cf. Gilbert, 1980c). Although a large range in size of food particles is consumed by any given rotifer species, in many cases, there is a reasonable separation of rotifer species along a food-particle-size gradient. This separation is congruent with the observed co-occurrence of several species within the pelagial zone of lakes (e.g., Makarowicz and Likens, 1975). It is probable that fairly discrete niches have evolved within the zooplankton, corresponding to specialized utilization of part of the range of available particulate matter. These specializations are adequate to permit co-occurrence without severe competitive interactions.

CLADOCERA

Cladocerans usually have five pairs of legs attached to the ventral part of the thorax. The legs are flattened and bear numerous hairs and long setae. Complex movements of these setose legs create a current of water through the valves. This current oxygenates the body surface, and forces a stream of food particles anteriorly. The food particles, filtered by the setae, collect in a ventral food groove between the bases of the legs and are impelled forward toward the mouth to be mixed there with oral secretions.

The importance of size of food particles in relation to morphological limitations of the filtering apparatus and food selectivity will be discussed later.

The primarily littoral chydorid cladocerans have modified legs that are somewhat prehensile in scraping up larger pieces of detrital material. Feeding by filtration occurs as well. Two other common cladocerans, *Polyphemus* and *Leptodora,* contain members that are predaceous and feed mainly by seizing relatively large particles, such as protozoa, rotifers, and small crustaceans, with their prehensile legs.

COPEPODS

The mouthparts of harpacticoids are adapted for seizing and scraping particles from the sediments and macrovegetation. The food and feeding behavior of the cyclopoid copepods have been studied in detail in two masterful works by Fryer (1957a, 1957b). No filtration mechanisms occur in the free-living Cyclopoida. Feeding is raptorial; plant or animal food particles are seized by mouthparts and brought to the mouth. The maxillules hold and pierce the prey and force particles between the mandibles; intermittent oscillating movements macerate some of the food. Some particles are swallowed intact and are differentially digested. Diatoms tend to be digested, while some green algae, if they are not ruptured, pass through the gut undigested.

Many species of the major cyclopoid genera *Macrocyclops, Acanthocyclops, Cyclops,* and *Mesocyclops* are carnivorous. The food of these carnivores includes microcrustaceans, dipteran larvae, and oligochaetes, many of which are larger than the copepod that preys on them. Herbivorous cyclopoids include many species of *Eucyclops,* some *Acanthocyclops,* and *Microcyclops,* which feed on a variety of algae ranging from unicellular diatoms to long strands of filamentous species. Carnivorous cyclopoid species tend to be larger than herbivorous species. Random encounter appears to be the dominant mode of finding food in both carnivorous and herbivorous species, which search by discontinuous, irregular movements in the water or over the substratum. Herbivorous species apparently employ gustatory chemoreceptor organs, which may help in food seeking if only to facilitate discrimination between inorganic and organic particles encountered by chance. The filtering rate of *Diaptomus tyrrelli* is reduced in the presence of a predacious copepod *Epischura* (Folt and Goldman, 1981). The alteration in feeding rate is in response to a chemical released into the water by the predator. Presumably, the decreased movement functions as an evasive action taken by *Diaptomus* to reduce predatory success of *Epischura.*

While locomotion in most copepods is in the form of short, jerky swimming movements, the animals being propelled by rapid movement of most appendages simultaneously, swimming is more continuous in the Calanoida copepods. Their gliding movement results from rotary motion of the antennae and mouth appendages. The movements set up small vortical currents that carry particles to the maxillae, which are modified to remove the water. High-speed motion pictures have revealed that calanoid copepods do not strain particles out of the water, but instead propel water past the body by flapping four pairs of feeding appendages (Koehl and Strickler, 1981). The second maxillae actively capture "parcels" of that water containing food particles. Particles are then pushed into the mouth by the first maxillae. Selective feeding also exists among the calanoid copepods. For example, two closely related calanoids, *Diaptomus laticeps* and *D. gracilis*,* were found to coexist in the plankton of Lake Windermere, England (Fryer, 1954; cf. Maly and Maly, 1974). The species were separated by size differences, which were correlated with differences in food consumed. The larger *D. laticeps* fed chiefly on *Melosira*, and the smaller *D. gracilis* consumed mainly minute spherical green algae and particulate detritus. Neither species fed upon the then-abundant diatom, *Asterionella*. Competition for food by these two calanoid species was virtually nonexistent. Such small differences in feeding habits, based on morphological and behaviorial variations, are common in the calanoids and may be sufficient to separate species into different food niches, even though they occupy the same volume of water.

Similar results were found for *Diaptomus* feeding on natural phytoplanktonic populations (McQueen, 1970). Filtration rates were low for cells smaller than 100 μm^3, increased to a maximum of 12.9 ml per animal per day for cells ranging from 102 to 333 μm^3, and remained constant with increased cell volume and decreased cell concentrations.

FILTRATION AND FOOD SELECTIVITY

The entire subject of feeding in Cladocera and in copepods is an area currently under intensive investigation. Recent results indicate not only a number of physical and biotic factors that affect rates of ingestion, but also that some organisms possess at least minimal abilities to select food for consumption. The implications of these studies are major not only from the standpoint of effectiveness of food ingestion and assimilation by the animals, but also because they point out the effects of the consumption on the food populations. This latter subject is most complex. However, it is apparent that at times zooplankton, as a result of direct cropping, can have appreciable effects on phytoplanktonic populations, and, through selective grazing, can influence the seasonal succession of the phytoplankton. This area of research has great potential for demonstrating subtle but major community interactions of significant impact on the overall productivity of lake systems.

Filtration of water to remove particulate organic matter is the dominant mode by which most cladocerans collect and ingest food. The *filtration rate* of a zooplankter refers to volume per unit time, and is defined as the volume of water containing food particles that is filtered by the animal in a given time (cf. Rigler, 1971). This term does not imply that the volume of water passed over the filtering appendages is known, that all particles of any given type have been removed from the water, or that all particles retained by the filtration apparatus have been consumed. The terms *filtration rate, grazing rate,* and *filtration capacity* have been used synonymously with *filtering rate*. In contrast, *feeding rate* is a measure of the quantity of food ingested by an animal in a given time (Rigler, 1971), measured in terms of number of cells ingested, volume, dry weight, carbon, nitrogen, or some other relevant aspect of the food.

The size of particles that can be cleared from the water is a function of the morphology of the setae of the moving appendages or of entrapment as locomotion of the animal brings particle-laden water to the setae. Measurements of particle removal are

---

*Hutchinson (1967) differentiates these species into *Arctodiaptomus laticeps* and *Eudiaptomus gracilis*, respectively.

made by observing changes in the number of particles in different size classes caused by grazing over time (cf. Harbison and McAlister, 1980). More recently, the rate of removal of radioactively labeled food particles has been used to measure zooplankton feeding rates (see the excellent review of Rigler, 1971). The latter approach has been extended effectively to natural populations by use of an in situ grazing chamber (Haney, 1971) that permits estimates of feeding rates over short periods of time (< 10 minutes) so that ingested food is not excreted. These methods measure food actually taken into the gut after losses, either through active rejection of food or losses of food particles during the maceration process. Ingestion rates give little insight into assimilation rates of incorporated food, which can be highly variable with food type and concentrations.

A number of studies indicate that the rate of feeding stabilizes or decreases as the concentration of food particles increases. For example, in *Daphnia magna*, feeding rate is proportional to concentration of food particles below a critical level (McMahon and Rigler, 1963, 1965). Feeding rate is constant above the incipient limiting concentration of food particles. At concentrations of food above this limiting level, rates of movement of the thoracic appendages that collect the food decrease, movements of mandibles and swallowing of food remain about the same or decrease slightly, and the rate of rejection of food increases (Burns, 1968b, and personal communication). For adult *D. rosea* feeding on a yeast, the critical concentration for feeding was between 0.75×10^5 and 1.0×10^5 cells ml^{-1} (Burns and Rigler, 1967). Incipient limiting concentrations (cells ml^{-1}) for feeding rate appear to increase with decreasing particle size, although other factors may be involved as well.

Within a given range of food concentrations, filtering efficiency was found to be independent of size of food particles (0.9 μm^3 to 1.8×10^4 μm^3 for *D. magna*) (McMahon and Rigler, 1963, 1965; see also Kersting and Holterman, 1973; Geller, 1975). At concentrations of yeast cells above 0.25×10^5 cells ml^{-1}, filtering rates of *D. rosea* decreased.

The filtering rate of *Ceriodaphnia reticulata* decreased linearly with increasing concentrations of phytoplankton within a size range of 3 to 24 μm from low to high densities (10^4 to 10^6 ml^{-1}) (O'Brien and deNoyelles, 1974). This relationship was not evident in cultures of *Ceriodaphnia* exposed to low densities (10^4 to 10^5 ml^{-1}) of algae (Czeczuga and Bobiatyńska-Ksok, 1970), although the latter methods (very high zooplanktonic densities in a small volume with long incubation periods) do not permit comparison.

Filtering rates have been found to increase with increasing body length and with increasing temperatures (Fig. 16–5). Temperatures above a given point result in a decrease in the filtration rate. The temperature of maximum filtration rates apparently differs among species; for *D. rosea*, the maximum was found to be about 20°C (Figure 16–5), for *D. magna* 28°C, and for *D. galeata mendotae* and *D. pulex* 25°C or above (Burns, 1969b; Burns and Rigler, 1967; McMahon, 1965; Geller, 1975).

Brooks and Dodson (1965) have emphasized that the filtering rate should be proportional to the square of the body length in filter-feeding Cladocera. The increase in filtration rate relative to body length, however, varies among species, for example, about to the square in *D. magna* and to the cube of the length in smaller *D. rosea* (Burns, 1969b). Although the area of the filtering setae of these two *Daphnia* species

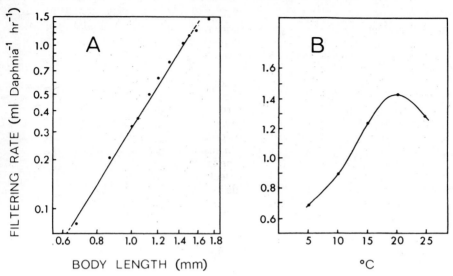

Figure 16-5. Relationship between filtering rate and *A*, body length at a constant food supply at 20°C, and *B*, water temperature (body length 1.65–1.85 mm) of *Daphnia rosea*. (Modified from Burns and Rigler, 1967.)

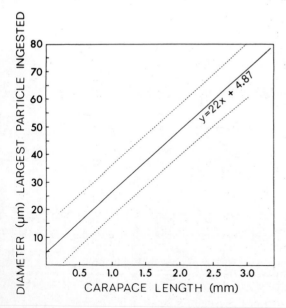

Figure 16-6. Relationship between body size and the diameter of the largest particle ingested by seven species of Cladocera. Broken lines equal 95 per cent confidence limits. (Modified from Burns, 1968a.)

increases approximately as the square of the body length, the filtering area and the filtering rates of *D. rosea* become proportionately greater than those of *D. magna* as the body length increases (Egloff and Palmer, 1971). The difference in the calculated rate of flow of water through the filtering setae, however, remains nearly constant at all body lengths.

The relationship between body size and the maximum size of particles that could be ingested was studied in six species of *Daphnia* and the smaller *Bosmina longirostris* using spherical beads within a size range from < 1 to 80 μm in diameter (Burns, 1968a, 1969a). A strong positive correlation between increasing body size and increasing size of the largest particle ingested was evident for the species assayed (Fig. 16-6). Experiments in which particles were kept in suspension or were allowed to sediment to the bottom of flasks showed that under the latter conditions, some species, when forced to forage for particles, were able to select particles of different sizes better than others. Leverage against the substratum aids in the ingestion of somewhat larger-sized particles.

Extension of these and other results from laboratory feeding studies to natural populations by observations of in situ grazing rates has demonstrated some striking relationships. In analyses from small, eutrophic Heart Lake of southern Ontario, seasonal grazing patterns (Fig. 16-7) were found to be similar over a two-year period (Haney, 1973). At certain depths during periods of the summer, grazing rates exceeded 100 per cent of the water volume per day, but during much of the year, values were less than complete filtration per day and became less than 10 per cent day^{-1} during the winter (Fig. 16-7; Table 16-2). The lower vertical boundary of zooplanktonic filter feeding was found to be closely defined by the 1 mg l^{-1} isopleth of dissolved oxygen concentration,

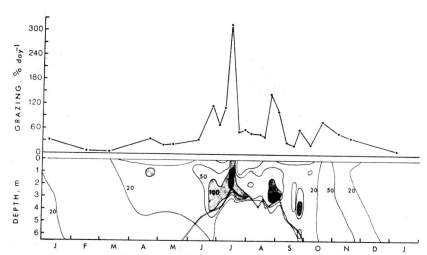

Figure 16-7. Seasonal and vertical changes in grazing by the zooplankton community in Heart Lake, Ontario, 1969. Mean grazing rates of upper figure calculated for the aerobic stratum. Isopleths of grazing rates (lower figure) are at 20, 50, 100, and 200 per cent per day. Broken line is the 1 mg l^{-1} dissolved oxygen isopleth. (After Haney, J. F.: An in situ examination of the grazing activities of natural zooplankton communities. Arch. Hydrobiol., 72:87–132, 1973.)

below which filtering rates declined precipitously. There were seasonal patterns in the rates of grazing on different food items. Smaller bacteria-sized particles apparently were ingested most rapidly in summer, whereas in the autumn the larger-sized algal particles were eaten most rapidly.

Since, as indicated earlier, the grazing rates of cladoceran zooplankton vary greatly with body size, food, and temperature, the in situ rates of filtration vary considerably seasonally. The average rates given in Table 16-3, however, show the general relationship to body size. Returning to the example from Heart Lake, the grazing contribution of each species can be calculated by multiplying filtering rates of each species at a given time by its population density at that time. In Heart Lake, *Daphnia rosea* and *D. galeata* were the most important grazers, and together accounted for about 80 per cent of the total annual grazing activity (Haney, 1973). Cladocerans (several *Daphnia* species) were also the dominant grazers in moderately productive Lawrence Lake, Michigan (Crumpton and Wetzel, 1982). Grazing of the phytoplankton in this lake became significant only during late summer (Table 16-2; cf. Figure 15-6). Zooplankton filtered less than 5 per cent of the volume of the epilimnion per day throughout most of the year, and had a minor impact on phytoplankton mortality. In August and September of 1979, however, grazing had a significant effect on both algal productivity and species succession (cf. Chapter 15). During the night, when many cladocerans and copepods migrate to the epilimnetic waters, grazing rates are about five times greater than during daylight hours (Table 16-2; cf. Hart, 1977). Often zooplankton encounter strata of reduced oxygen concentrations as they migrate to deep, poorly illuminated water during daylight hours.

TABLE 16-2   Average Grazing Rates of In Situ Phytoplankton

		MEAN GRAZING RATE ($\%$ DAY^{-1})			$\%$ TOTAL GRAZING OVER PERIOD
	TOTAL DAYS	DAY	NIGHT	TOTAL	
Lawrence, Michigan*					
May	31	0.2	3.0	3.2	4.2
Jun.	30	1.6	3.0	4.6	6.0
July	31	0.4	1.9	2.3	3.0
Aug.	31	4.1	20.2	24.3	31.7
Sept.	30	4.2	21.9	26.1	34.0
Oct.	15	4.2	12.0	16.2	21.1
Heart, Ontario†					
Jan.–May	151	—	—	19.2	17.4
Jun.–Sept.	122	—	—	80.1	61.4
Oct.–Jan.	92	—	—	35.2	20.9
Vechten, The Netherlands‡					
Winter (Dec.–Feb.)	90	—	—	2.1	4.5
Spring (Mar.–Apr.)	61	—	—	9.8	22.0
Summer (May–Sept.)	153	—	—	24	69.0
Autumn (Oct.–Nov.)	61	—	—	5.5	4.5
Annual Mean	365	—	—	13	100

*After data of Crumpton and Wetzel, 1982; see Figure 15-6.
†After data of Haney, 1973.
‡After data of Gulati, 1978.

TABLE 16-3   Filtering Rates of Various Cladoceran Zooplankters

SPECIES	ANIMAL SIZE RANGE (LENGTH IN mm)	AVERAGE FILTERING RATE (ml ANIMAL^{-1} DAY^{-1})*	SOURCE
Daphnia			
D. rosea	1.3–1.6	5.5	Haney, 1973
		3.6	Burns and Rigler, 1967
D. galeata	1.5–1.7	6.4	Haney, 1973
		3.7	Burns and Rigler, 1967
D. parvula	0.7–1.2	3.8	Haney, 1973
D. longispina		2.3	Nauwerck, 1963
Ceriodaphnia			
C. quadrangula	0.7–0.9	4.6	Haney, 1973
Diaphanosoma			
D. brachyurum	0.9–1.4	1.6	Haney, 1973
Bosmina			
B. longirostris	0.4–0.6	0.44	Haney, 1973
Chydorus			
C. sphaericus	0.1–0.2	0.18	Haney, 1973

*Type of food = in situ phytoplankton, in most cases.

Filtering and respiration rates decrease rapidly at oxygen concentrations below 3 mg l^{-1} (Kring and O'Brien, 1976; Heisey and Porter, 1977).

Filtering rates of the major zooplankton were also studied, although less intensively, in two much less productive lakes of southern Ontario: Halls Lake, a deep oligotrophic lake, and Drowned Bog Lake, a typical *Sphagnum* acidic bog lake (cf. Chapter 24) with high concentrations of dissolved humic matter. The filtration rates of the two dominant zooplankters, *Bosmina* and *Holopedium*, were very different, but the small *Bosmina* formed a dominant percentage of the total grazing in the autumn because of its very large population densities (Table 16-4). In contrast to the bog lake and to eutrophic Heart Lake, in which intense grazing occurred in the upper three meters, the

TABLE 16-4   Filtering Rates and Contribution to Total Grazing of Species-Dominant Zooplankton of Acidic Drowned Bog Lake, Ontario, in Early September

SPECIES	FILTERING RATES (ml ANIMAL^{-1} DAY^{-1})		SPECIES CONTRIBUTION TO TOTAL GRAZING (%)	
	Sept. 1968	Sept. 1969	Sept. 1968	Sept. 1969
Bosmina longirostris	0.46	0.45	85	44.8
Holopedium gibberum	—	9.4	12	46.2
Daphnia parvula	—	1.6	0.1	7.5
Diaptomus oregonensis	—	2.1	2.0	0.9
Diaphanosoma brachyurum	—	1.2	0.1	0.4

Extracted from data of Haney, 1973.

grazing rates in oligotrophic Halls Lake were extremely low and distributed uniformly throughout the water column. The dominant zooplankter of this oligotrophic lake was the copepod *Diaptomus*, and in general, nonpredatory copepods have much lower grazing rates than the Cladocera (Tables 16–3 and 16–5).

In studies in which species of widely different algae were fed to cladoceran and copepod zooplankton at concentrations above critical limitations to feeding, filtering rates decreased markedly with increases in the food concentration (Fig. 16–8). Ingestion rates were calculated as the product of the filtration rate (ml animal^{-1} day^{-1}) and the food concentration (cells ml^{-1}) (Infante, 1973). These results, when coupled with measurements of radioactively labeled carbon from the algae that were incorporated into the animals as an estimate of assimilation, showed marked differences with algal species. The diatom *Asterionella* was ingested and assimilated more readily than *Nitzschia*. Algae with heavy cell walls, such as *Scenedesmus* and *Stichococcus*, were ingested rapidly (Fig. 16–8) but were not utilized well. The Cladocera digested some of the protoplasm of such cells without visible changes in the cell walls, whereas the cope-

TABLE 16–5   Comparison of Filtering Rates of Various Copepods

SPECIES	TYPE OF FOOD	PARTICLE CONCENTRATION (1000 × CELLS ml^{-1})	FILTERING RATE (ml ANIMAL^{-1} DAY^{-1})	SOURCE
Diaptomus				
D. graciloides	Natural phytoplankton	—	0.3–2.8	Nauwerck, 1959
	Scenedesmus	13.6	4.1	Malovitskaia and Sorokin, 1961
D. siciloides	Pandorina and Chlamydomonas	—	2.0	Comita, 1964
D. oregonensis	Chlamydomonas	1.5–25.0	2.5	Richman, 1966
		25.0–52.0	2.5–1.4	
	Chlorella	52.0–198.0	1.4–0.3	
	In situ phytoplankton	—	1.4–0.00	Haney, 1973
D. gracilis	Melosira and Asterionella	24.2–52.0	1.92–1.96	
		198.0	0.68	Malovitskaia and Sorokin, 1961
	Chlorella	<30.0	0.61 (5°C)	Kibby, 1971
			1.51 (12°C)	
			2.40 (20°C)	
	Scenedesmus	<30.0	0.94 (12°C)	
			1.32 (20°C)	
	Diplosphaeria	<30.0	1.76 (12°C)	
			2.54 (20°C)	
	Ankistrodesmus	<30.0	1.61 (12°C)	
			2.45 (20°C)	
	Carteria	<30.0	0.87 (20°C)	
	Nitzschia	<30.0	1.96 (20°C)	
	Pediastrum	<30.0	0.02 (20°C)	
	Haematococcus	<30.0	2.16 (20°C)	
	Bacteria	<30.0	0.19 (20°C)	
Limnocalanus				
L. macrurus	Scenedesmus	—	2.45 (<5°C)	Kibby and Rigler, 1973
	Chlamydomonas	—	1.24 (<5°C)	
	Rhodoturula (yeast)	—	0.1 (<5°C)	

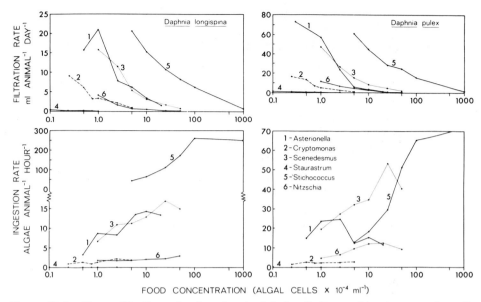

Figure 16-8. *Upper:* filtration rates (in ml animal^{-1} day^{-1}); *Lower:* ingestion rates (in cells animal^{-1} hour^{-1} as the filtration rate times food concentration in cells ml^{-1}), in *Daphnia longispina* and *D. pulex* exposed to different algal species. Ingestion rate of *Stichococcus* (5) for *D. pulex* is drawn × 0.1 that found. (Drawn from data of Infante, 1973.)

pods assayed (*Eudiaptomus* and *Cyclops*) were unable to digest them and the algae that left the gut were essentially intact. *Staurastrum* and *Cryptomonas* were poorly utilized. Differences also occurred among zooplanktonic species. *Daphnia longispina,* for example, incorporated nearly twice the number of *Nitzschia* that *D. pulex* did, but only one-half the amount of *Scenedesmus.*

Certain zooplankton feed on a wide variety of algae of different sizes and shapes, and with or without sheaths (Gliwicz, 1980; McNaught, et al., 1980; Porter and Orcutt, 1980). Other zooplankton are highly selective in the algal types ingested. Cellular forms are ingested more readily than filamentous or spinuosus forms, and zooplanktonic filtration rates, growth, and survivorship are greater when feeding on cellular forms. Circumstantial evidence also suggests that algae releasing toxic organic compounds are selected against by zooplankton, regardless of suitability of food size and shape, which implies chemoreception (taste).

Similarly, when the numbers of grazing Cladocera and copepods were increased experimentally among natural phytoplanktonic populations, small algae such as cryptomonads, certain diatoms, and other nannoplankton decreased, gelatinous green algae such as *Sphaerocystis* increased, and large or unpalatable species such as the blue-green *Anabaena* remained unchanged (Porter, 1973). While the size of particles grazed is limited by the morphological structure of the setae (Monakov and Sorokin, 1961), large or unpalatable algae result in reduced filtering rates. For example, the ingestion, assimilation, survivorship, and reproduction rates of *Daphnia* that were fed blue-green algae were lower than those fed green algae (Arnold, 1971). Therefore, even though food species of a usable size may be present, they may be too rare, or may be selectively

TABLE 16-6 Energy Budget of Adult *Daphnia pulex* After 18 Instars (40 days) of Growth at Four Food Concentrations of *Chlamydomonas*

FOOD CONCENTRATION		ENERGY CONSUMED (cal)	ENERGY OF GROWTH AND YOUNG (cal)	ENERGY OF RESPIRATION (cal)	ENERGY OF EGESTION (cal; BY DIFFERENCE)	% ASSIMILATION	% ENERGY CONSUMED AS GROWTH AND YOUNG	% ENERGY ASSIMILATED AS GROWTH AND YOUNG	% ENERGY OF GROWTH AND YOUNG AS RESPIRATION
cells x 10⁴ ml⁻¹ day⁻¹	mg ml⁻¹								
25	6.2	6.14	1.07	0.84	4.23	31.1	17.4	56.0	78.6
50	12.4	13.59	1.77	0.94	10.87	20.0	13.1	65.3	53.0
75	18.6	20.74	2.35	1.02	17.37	16.3	11.4	69.8	43.3
100	24.8	29.24	2.93	1.08	25.23	13.7	10.0	73.0	37.0

After data of Richman, 1958.

rejected by the grazer at high algal densities. This selection is apparently at least partly related to relative abundance of particles of different biomass. Selectivity by *Daphnia* increases with increasing particle concentrations until particle consumption is maximized when particles of one size dominate in mixed food (Berman and Richman, 1974). When concentrations of all particles were approximately equal, selection favored small, abundant particles.

Ingestion of algae does not necessarily result in their utilization and assimilation by the grazers. In some grazer species, algae with durable cell walls, gelatinous sheaths, or masses of colonial cells can pass through the gut intact and remain viable (Porter, 1975, 1976). Other species can utilize broken colonies or partially decayed cells of colonies by breaking them into smaller pieces, while intact colonies pass through relatively unaffected. Increases in the growth rates of certain algae after ingestion and passage through grazers may be the result of several interrelated mechanisms. Nutrients may be obtained from the degradation and digestion of other algae; phosphorus, in particular, is probably a major nutrient that is made more available during this process.

Selective grazing and utilization can remove species or reduce population size in the algal community, which can alter competition for the given resource base. Alternatively, grazer utilization of an algal species can result in enhancement of primary productivity of that species by increased selection for faster growing genotypes and by increased nutrient supply (Crumpton and Wetzel, 1982).

A number of studies have demonstrated the effects of food quality and concentrations on assimilation efficiencies. Among the better studies is that of *Daphnia pulex* (Richman, 1958), in which assimilation efficiency was shown to decrease markedly with increasing food concentrations (Table 16-6). Growth as average length, number of young per brood, and total young increased with increasing food concentrations at the levels used. The energy used for growth was nearly constant at all food levels in preadults, because the energy of egestion increased and assimilation efficiency declined as food consumption increased. In adults, however, the major portion of stored energy went into the production of young. Similar results have been found for a number of other cladocerans (cf. especially Lampert, 1977a, 1977b).

In addition to the increase in assimilation rates with increasing temperature, several studies show that the assimilation efficiency increases with higher energy content of the food (Schindler, 1968; Pechen'-Finenko, 1971). Foods of low caloric value, such as detritus of plant origin in the range of 2 to 4 cal mg^{-1} dry weight, are assimilated more slowly than algae and bacteria in a caloric range of 5 to 6 cal mg^{-1}. Although the assimilation efficiency of detritus is low, that of algae and bacteria by zooplankton is variable. The few studies on the subject show that in general, algae are assimilated more efficiently (15 to 90 per cent) than bacteria (3 to 50 per cent) (Monakov and Sorokin, 1961; Saunders, 1969; Winberg, et al., 1973). The percentage of assimilation varies seasonally with shifts in algal composition. In contrast, when given mixed diets of green algae and a photosynthetic bacterium, *Ceriodaphnia* assimilated less of the algae than when fed on algae alone (Gophen, et al., 1974). Utilization of bacteria was relatively unchanged when the algae were also included in the diet. There are several indications that algae alone do not meet the nutritional requirements of cladocerans (see, for example, Taub and Dollar, 1968). In spite of the relatively high assimilation rates of algae and bacteria and low assimilation rates of detritus, as will be pointed out later, it is unusual for bacteria to dominate the energy income of zooplankton. Moreover, in the few cases in which it has been analyzed in some detail (cf. Chapter 22), detritus can form a major energy resource over much of the year.

REPRODUCTION AND LIFE HISTORIES

A number of similarities exist among the common life histories and reproductive processes of rotifers and cladocerans. For many successive generations during the main growing season, males are absent and reproduction occurs via parthenogenesis of diploid females. Under certain conditions of environmental stress (e.g., crowding), males develop, meiosis occurs, and sexual (mictic) reproduction occurs, leading to the formation of mictic haploid eggs (Fig. 16-9). If fertilized, the now-diploid eggs ("resting eggs") develop a heavy cell wall and become resistant to environmental extremes. The resting eggs commonly enter diapause with hatching delayed for long periods, and, upon return of favorable conditions, will develop into amictic diploid females.

In contrast, freshwater copepods are bisexual and have a much longer life cycle than do the rotifers and cladocerans.

ROTIFERS

The reproductive life history of typical planktonic rotifers, all in the subclass Monogononta, is characterized by a large number of generations in which reproduction is parthenogenetic. These amictic females are all diploid and produce amictic eggs that develop into amictic females (Fig. 16-9). There may be as many as 20 to 40 of such amictic generations. With an egg-development time of about one day under warm, optimal conditions, and without the need to encounter males for fertilization, a population of amictic females can develop rapidly in 2 to 5 days.

This nonmeiotic egg production is intermittently broken, often only once or twice per year, by the development of morphologically indistinguishable mictic females. The eggs of the mictic females are produced by normal double meiotic division, and are thus haploid (Fig. 16-9). If eggs of these mictic females are fertilized, they develop into thick-walled encysted embryos called resting eggs. The resting eggs undergo prolonged diapause, and are highly resistant to adverse environmental conditions. If the mictic eggs are not fertilized, however, they develop rapidly into males. The males are greatly reduced in size and complexity. The males are short-lived and extremely active, and are capable of copulation within an hour of hatching. Hatching of resting eggs can be induced by biochemical factors, but induction under natural conditions is poorly understood. The diapause extends over a period of several weeks or months. Hatching is related to changes in temperature, osmotic pressure, water chemistry, and oxygen content. The resting eggs always result in parthenogenetic amictic females.

The mechanisms underlying the production of mictic females are unclear, although a number of environmental conditions have been proposed (reviewed by Hutchinson, 1967, and Ruttner-Kolisko, 1972). Although the stimuli are quite species-specific, significant factors include crowding of amictic females in relation to food and accumulation of substances such as pheromones, which are produced by the females. In clones

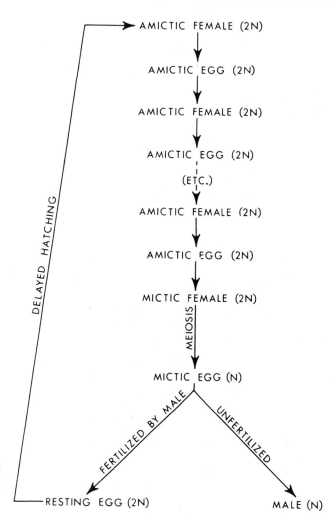

Figure 16-9. Diagrammatic sequence of the reproductive life history characteristic of most planktonic and many benthic rotifers. (Redrawn from Pennak, 1953.)

of *Brachionus* reproducing amictically at 20°C, brief exposure to a shock temperature of 6°C resulted in nearly 50 per cent of the clones becoming mictic in less than two days (Ruttner-Kolisko, 1964; Gilbert, 1977).

Asplanchna, the large predatory rotifer, is certainly among best-studied examples in this regard. Gilbert and co-workers, in a series of superb papers (1967a, 1968, 1972, 1973, 1975, 1976; Gilbert and Birky, 1971; Gilbert and Thompson, 1968; Riggs and Gilbert, 1972; Gilbert and Litton, 1975; reviewed by Gilbert, 1980a), has demonstrated the subtle reproductive controls in this species that have far-reaching implications among the zooplankton. When fed the protozoan *Paramecium*, *Asplanchna* produced only amictic females. Increasing the population densities of the rotifer to extraordinary levels of crowding produced no mictic females, whereas adding low populations of the algae *Chlamydomonas* or *Euglena* to the dense paramecia resulted in the production of a significant number of mictic females. It was also shown that the algal cells had to be eaten to induce the reproductive change in *Asplanchna*; extracellular algal products would not induce the change. It was shown subsequently that the important dietary component from the plants was d-α-tocopherol (vitamin E), which induced the transition from parthenogenetic to sexual reproduction and is essential for spermatogenesis or male fertility. The mictic female offspring were also larger and changed in morphology, exhibiting four partially retractile outgrowths of the body wall. These "humps" were elicited by d-α-tocopherol within a range of 5×10^{-13} mol (0.2 ng) in a linear fashion to a maximum of 5×10^{-11} mol (20 ng) per female. Tocopherol, as well as cannibalism, also induced a third polymorph, a very large campanulate type.

The adaptive significance of control of sexual reproduction and nongenetic female polymorphisms is complex (reviewed by Gilbert, 1980a). The initial response to tocopherol is a general increase in growth that permits the carnivorous *Asplanchna* to more easily ingest larger, possibly more nutritious, herbivorous prey. Tocopherol can act as an adaptive cue for this growth, for it is synthesized only by photosynthetic organisms and is readily assimilated by herbivores. Upon ingestion of rotifer and crustacean prey, tocopherol is readily assimilated by *Asplanchna*, and is transmitted to its embryos without degradation. Tocopherol can promote differential polymorphism in some species of *Asplanchna* through cell enlargement and increased cell division. Population densities of *Asplanchna* increase when herbivorous zooplankton are more abundant; the resulting increase in contacts between males and mictic females facilitates production of resting eggs.

Body-wall outgrowths, which reduce cannibalism by the large individuals, likely evolved with the growth response to tocopherol (Gilbert, 1980a). Two tocopherol-dependent, female morphotypes evolved in some species. The larger, campanulate morphotype is adapted for ingestion of very large prey and, in one species of *Asplanchna*, for parthenogenetic reproduction only. The well-protected cruciform morphotype, with prominent body-wall outgrowths, is adapted for ensuring sexual reproduction and avoiding capture by cannibalistic conspecific *Asplanchna*. Growth responses to tocopherol evolved only in those species which (1) do not depend on the direct utilization of relatively small tocopherol-rich algae, and (2) can ingest the copious quantities of zooplankton prey needed for sustaining populations of large individuals.

Other rotifers, such as the sessile bdelloids, none of which are planktonic, reproduce exclusively by parthenogenesis. Resting eggs are unknown, but these adults can survive drying.

The reproductive rate of rotifers is related strongly to the quality and abundance of food as well as to temperature (Edmondson, 1946, 1965; King, 1967; Halbach and Halbach-Keup, 1974; Baker, 1979; Sterzyński, 1979). It is clear that these factors are of major importance in determining seasonal fluctuations in the populations resulting from changes in the balance between increase by reproduction and mortality.

In addition to effects on the rate of development of eggs, temperature obviously influences the rates of biochemical reactions, feeding, movement, longevity, and fecundity. The composite effects can be seen in the relative rates of changes in concentrations of rotifers when plotted against temperature, both in natural and culture conditions (Fig. 16-10). A number of other factors affect natural populations, but temperature is clearly a major factor affecting birth rate. It is also apparent that species are variable in their responses; some are very restricted stenotherms, while others are fairly eurythermal.

A number of studies under both natural and laboratory conditions have shown that food type and quality exert a major influence on the population dynamics of rotifers. For example, in investigations of the dynamics and reproduction of two clones of *Euchlanis dilatata* fed different algal species at various concentrations, King (1967) found several relationships that also have been shown to varying degrees by others. Two genetically identical clones of *Euchlanis* were grown from a single parent by parthogenesis, so that in succeeding generations, the individuals of one clone, the young orthoclone, were always derived from young parents and those of the other orthoclone from old parents. The clones were fed different concentrations of three algal species, *Chlamydomonas reinhardi*, *Euglena gracilis*, and *E. geniculata*. Growth rates of individual animals were found to depend on food concentration and not on animal age or algal species. The rate of population increase was related directly to the density of food for concentrations up to 16.4 μg ml^{-1}, beyond which no further changes were observed. In *Brachionus*, filtration rate decreased at high algal densities; this was related to clogging of the filtration apparatus, reduced digestion, or inactivation of the corona by algal toxicants (Halbach and Halbach-Keup, 1974). Within the population, however, growth rates were related to both food species and food concentration (Fig. 16–11). *Euchlanis* fed on *Chlamydomonas* had higher rates of population increase than when feeding upon *Euglena gracilis*. *E. gracilis*, on the other hand, led to higher rates of population increase than *E. geniculata*. Individuals of the young orthoclone exhibited greater rates of population increase than those of the old orthoclone. The different rates of increase resulted more from corresponding differences in net reproduction (the number of eggs laid by the average female in her lifetime) than from the frequency of egg laying or survivorship. The results, confirmed by others (e.g., Snell, 1980), imply differences in assimilation, differences in the chemical quality of the food, or both.

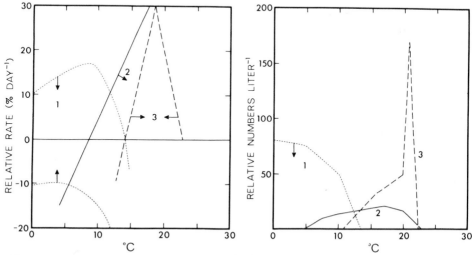

Figure 16–10. Range of response (between arrows) in relative rate of change of concentration of rotifers (*left*), and concentration per liter (*right*), in relation to temperature. *1, Synchaeta lakowitziana; 2, Euchlanis dilatata; 3, Pompholyx sulcata.* (Drawn from data given in Edmonson, 1946, after Carlin.)

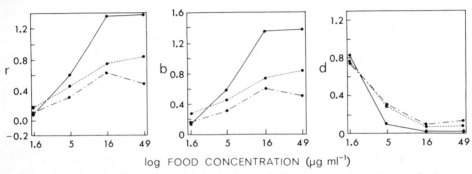

Figure 16-11. The instantaneous rate of increase (r), birth rate (b), and death rate (d) of young orthoclones of the rotifer *Euchlanis dilatata* as a function of food level and food species. ———— = *Chlamydomonas reinhardi*; - - - - - - = *Euglena gracilis*; —·—·— ·—·—· = *Euglena geniculata*. (Drawn from data of King, 1967.)

CLADOCERA

Somewhat analogous to that of the rotifers, reproduction in the cladocerans is parthenogenetic during a greater part of the year. Until interrupted by sexual reproduction, females produce eggs that develop into more parthenogenetic females. The number of eggs produced per clutch varies from two or more in the Chydoridae to as many as 40 in the larger Daphnidae. The eggs are deposited into a brood chamber, or pouch, a cavity dorsal to the body bounded by the valves of the carapace. The eggs develop in the brood pouch and hatch as a small form of the parent. As a result, in contrast to the copepods, there are, with the exception of *Leptodora*, no free-living larval forms among the Cladocera. One clutch of eggs is normally released into the brood pouch during each adult instar.

The number of molts is variable. Four preadult instars are common within a range of 2 to about 8 in some *Daphnia*. Longevity, from the time of egg release into the brood pouch until death of the adult, varies widely with species and environmental conditions. Longevity and time between molts are approximately inversely related to temperature. For example, longevity in *Daphnia magna* averaged 108, 88, 42, and 26 days at 8, 10, 18, and 28°C, respectively (MacArthur and Baille, 1929). Longevity also is affected by food availability, and commonly increases with a decrease in food consumption, short of starvation which results in rapid death. Lipid reserves in the Cladocera accumulate in the body coelom when food is abundant, and can be used as an energy reserve when food is scarce. The quantity of lipid that can be stored by individuals in a species may be related to competitive success when food availability declines (Goulden and Hornig, 1980). The importance of temperature, however, can be seen in studies of *Daphnia galeata mendotae* (Table 16-7) that were fed differing concentrations of mixed green algae at different temperatures (see Hall, 1964).

In adult cladocerans, the number of molts is more variable than in juveniles, and can range from a few to well over 20 instars. Each instar ends with the release of young from the brood pouch, molting, rapid increase in size, and deposition of a new clutch of eggs into the pouch, all of which occurs in a matter of minutes. Therefore, an individual can produce several hundred progeny in a life span under favorable growing conditions.

Increases in temperature result in an immediate effect by increasing the rate of molting and brood production, whereas increases in food supply, within limits, affect the rate of development of the population by increasing survivorship and fecundity, i.e., the number of eggs per brood. The latter process is not immediate and this time lag can contribute to population oscillations. When the growth of the population exceeds available food supply, reproduction decreases, with reduced replacement of adults. Reduction of the population below the environmental carrying capacity set by available food results in an increase in the population after a lag period. The extent of population oscillations depends greatly on the duration of the time lag. These relationships have been demonstrated in detailed studies under culture conditions (Frank, 1952, 1957; Slobodkin, 1954), and will be discussed in greater detail later on. When several zooplankton species are competing for limited food resources, the decline of some or all of the populations is responsive to the interactions of food quality and supply, feeding capabilities, and the effect of temperature on reproductive responses of each zooplankton species. Decline of a population does not necessarily eliminate that species, which may shift to an inactive state by the formation of resting stages.

Parthenogenesis continues until unfavorable conditions arise, whether they are physical, such as temperature reductions, drying, or possibly photoperiod (cf. Shan, 1974; Pourriot and Clement, 1975; Bunner and Halcrow, 1977), or biotic, i.e., induced through crowding by competition for food supply, or decrease in quality-size of food organisms. As the production of parthenogenetic eggs declines, some of the eggs develop into males. Some females then produce sexual (haploid) eggs, as in the rotifers. Though morphologically similar to parthenogenetic females, sexual females usually produce only one or a few resting eggs. Following copulation and fertilization, the carapace surrounding the brood pouch thickens and encloses the egg(s), forming a semi-elliptical, saddle-shaped *ephippium*. During the subsequent molt, the ephippium either is shed, as in the daphnids, or remains attached, as in the chydorids. If females carrying resting eggs are not fertilized, the resting eggs are resorbed. One species, the arctic *Daphnia middendorffiana*, is known to be able to produce ephippial eggs parthenogenetically.

The ephippia either sink or float, and can withstand severe conditions such as freezing or drying. It is not unusual to find large accumulations of ephippia along wind-

TABLE 16-7 Instantaneous Rates of Population Increase per Day (r) ± S.E. of *Daphnia galeata mendotae* at Different Temperatures and Levels of Food Supply

RELATIVE FOOD LEVELS	TEMPERATURE		
	11°C	20°C	25°C
Low (¼)	0.07 ± 0.005	0.23 ± 0.002	0.36 ± 0.015
Medium (1)	0.10 ± 0.003	0.30 ± 0.002	0.46 ± 0.013
High (16)	0.12 ± 0.006	0.33 ± 0.016	0.51 ± 0.006

After data of Hall, 1964.

ward shorelines. It is easy to envision entanglement and transport of ephippia by birds to other water bodies. A large percentage of the ephippia hatch under favorable but poorly understood conditions, always into parthenogenetic females. In a laboratory-cultured strain of *Daphnia pulex*, light was required for termination of diapause regardless of temperature or duration of ephippial storage (Stross, 1966). In naturally occurring ephippia of a lake population of the same species, diapause also was broken by light, but prolonged storage in darkness eliminated the need for light. The time required was 5 to 6 months at 3.5°C and only 3 to 6 weeks at 22°C.

Males are smaller and only slightly modified in morphology from females (larger antennules and clasping modifications of the first leg for copulation). Mechanisms underlying the periodic induction of males and ephippial females are unclear, although evidence indicates that the stimuli are different (reviewed by Hutchinson, 1967). Male production is correlated with crowding and a rapid reduction of food supply, whereas a constant low food supply simply inhibits reproduction (Slobodkin, 1954). Short-day photoperiods (12 hr. light, 12 hr. dark) increased the production of ephippia markedly in *Daphnia pulex*, in contrast to longer light photoperiods, which would be characteristic of midsummer in temperate latitudes (Fig. 16–12). If universal, this photoperiodic response must vary among species.

COPEPODS

Reproduction in free-living copepods is similar in spite of widely varying species differences in sexual behavior and periodicity of breeding. Some species reproduce throughout the year, others only briefly at specific times of the year. Copepod copulation occurs by a male briefly clasping the female with transfer of spermatophores to the ventral side of the female genital segment. Fertilization may take place immediately or several months after copulation. Fertilized eggs are carried by the female in 1 or 2 egg sacs (Fig. 16–4). The number of eggs in cyclopoid copepods, carried in two lateral egg sacs, is variable; differences observed between populations in different lakes and between seasons, have not been associated clearly with changing environmental conditions. Egg number, up to 72, is generally low in summer and increases in the autumn.

In the calanoid copepods, eggs are carried in a single egg sac and are variable in number (from about 1 to 30). Eggs are differentiated into subitaneous and resting eggs, both being fertilized in many species. Resting eggs usually are dropped to the sediments, where they undergo a period of diapause. Clutch size varies greatly in calanoids; often it is maximal in the spring and the fall, and these two peaks may be separated by a summer minimum. Larger clutch size is correlated with lakes of greater primary productivity and to spring conditions that favor more abundant food supplies. As in the Cladocera, food availability directly influences the clutch size, while temperature determines the rate of egg production (Elster, 1954; Comita and Anderson, 1959; Burgis, 1970; Geiling and Campbell, 1972). In this way, a small number of females carrying a few eggs for a short period at high temperatures can produce more young than many females with large egg clutches that are exposed to long periods at low temperatures.

Copepod eggs hatch into small, free-swimming larvae, termed *nauplii*, and then develop by molting through a number of subsequent larval stages. The initial nauplius has three pairs of reduced appendages (first and second antennae, mandibles). The successive naupliar stages, six in all, feed, grow, molt, and acquire further appendages.

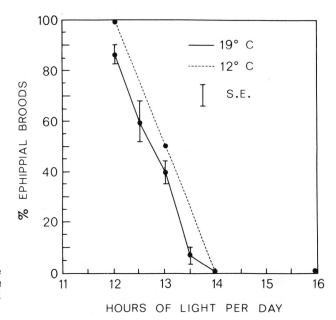

Figure 16-12. Influence of photoperiod on the production of ephippial egg broods in *Daphnia pulex* at high population densities. (Modified from Stross and Hill, 1965.)

After six naupliar molts, the next molt results in an enlarged and more elongated form, the first copepodite instar. There are five copepodite stages, during which additional appendages and body segments develop. The sixth and final copepodite stage is the adult. The time required to complete the juvenile stages is highly variable among species and depends upon seasonal conditions.

SEASONAL POPULATION DYNAMICS

GENERAL CONSIDERATIONS

More useful information on zooplankton population dynamics can be obtained if the stages in the life history of species are analyzed separately than if only the total population is measured.* Analyses of egg production and instar dynamics yield information on growth rates and mortality, as well as the duration of each stage and the interval between generations.

Instar analyses are easier with copepods than with cladocerans because the naupliar and immature and adult copepodite stages can be distinguished morphologically (cf. Wetzel and Likens, 1979:352–355). Among the cladocerans, and to some extent the rotifers, instars can sometimes be evaluated on the basis of size.

Analyses of the population dynamics of zooplankton require sound sampling methods with known statistics on sampling variance and on population heterogeneity within the aquatic system. This subject is outside the purview of the present work and, more-

---

*The generous counsel of W. T. Edmondson on the population analyses of zooplankton is gratefully acknowledged.

over, has been treated excellently in methods manuals edited by Edmondson and Winberg (1971) and Winberg (1971) and in the reviews by Bottrell, et al. (1976) and Waters (1977). Once representative samples of the animal population are obtained, a number of methods have been employed to evaluate population size and fluctuations. Enumeration can be adequate to address certain questions concerning population dynamics and interactions, but in order to evaluate secondary productivity, precise knowledge of changes in biomass throughout all stages of the organisms' life cycles is mandatory. Development times must be known and, although they are variable among organisms and environmental conditions, time between samplings must be less than the shortest developmental period.

Assuming stable egg distribution, the number of eggs present in a sample of a zooplankton population at some instant in time represents the increment that would have been added to the population over a period of time equal to the duration of development of the eggs (time from laying to hatching). The population would increase by the number hatched during the time period if no deaths occurred.

The population dynamics of rotifers are somewhat simplified because of the very short time between hatching and attainment of reproductive capacity. Analyses of birth and death rates may be estimated from reproductive rates and rates of population change. The reproductive rate can be determined indirectly from the ratio of eggs per female (Edmondson, 1960, 1965, 1968, 1974). The ratio of eggs per female (E) observable in a sample at any given moment can be used to calculate a finite population birth rate (B) as eggs per female per day, by:

$$B = \frac{E}{D} \tag{1}$$

when D is the duration of development of the eggs, or the mean time an individual spends in the egg stage. This relationship assumes that eggs present at a given time will be added to the population during the next period of time, D. If the age distribution of the eggs is stable, the fraction of eggs hatched during a day is $1/D$. If the population is changing size, the value of B is exact only if $D = 1$; the bias introduced by longer durations can be estimated (Edmondson, 1968).

From the reproductive rate B, the instantaneous growth rate of the population, r, can be calculated based on the conventional exponential growth model (cf. Chapter 8), in which positive or negative growth over a short time interval is:

$$N_t = N_0 e^{rt} \text{ and } r = b - d \tag{2}$$

in which
N_0 = the size of the initial population at time zero,
N_t = the population size at a later time t,
r = the growth rate coefficient or intrinsic rate of growth,
d = the instantaneous rate of mortality, and
b = the instantaneous birth rate (calculated from B).

The effective rate of population increase is the difference between the natural logarithms divided by the time increment $(t - 0)$:

$$r = \frac{\ln N_t - \ln N_0}{t} \tag{3}$$

which is in essence the difference between natality (b) and mortality (d) over the time interval (t).

Alternatively, birth rate, estimated by the egg-ratio method, gives the daily increment to the population. For a population growing by the amount E/D in one day with no deaths, its growth rate would be:

$$b = \ln (B + 1) \tag{4}$$

Here B is estimated at the beginning of the time interval with the assumption that for each female in the population there will be $(E/D) + 1 = B + 1$ females one day later. Assuming no mortality, the population would grow in one day at a rate of:

$$\ln (B + 1) - \ln (1) = \ln (B + 1) \tag{5}$$

Thus by Equation 4, a finite per capita birth rate, B, can be used to estimate the instantaneous rate, b. The calculations assume that B is small; under such circumstances both B and the E/D are approximately equal to the instantaneous birth rate b.

As pointed out by Edmondson (1968) and rederived by Paloheimo (1974), for moderately large values of r and bD, the finite per capita birth rate and the egg-ratio/D can diverge considerably (see Lynch, 1982). Further, the relationships are most applicable to planktonic animals that carry their eggs until hatching, at which time free swimming progeny are liberated, i.e., the eggs are subjected to the same mortality as the adults. Paloheimo (1974) has shown that:

$$E/D = (e^{bD} - 1)/D \tag{6}$$

when D is the development time of the eggs, which is algebraically equivalent to:

$$b = \ln [(C_t/N_t) + 1]/D \tag{7}$$

when C_t is the total number of eggs counted at time t, which estimates the instantaneous birth rate b from egg counts C_t where C_t/N_t is the egg ratio. This formula assumes that steady state conditions prevail, which is not usually the case. Development time is temperature-dependent, and in analyses of natural populations this primary factor must be estimated from good experimentally determined changes with temperature. Examples of estimates of instantaneous birth rates are given in Table 16–8, in which the C_t/N_t

TABLE 16–8 **Estimates of Instantaneous Birth Rates at Differing Development Times in Days by Equation 7 when the Egg Ratio $C_t/N_t = 1$**

DEVELOPMENT TIME, D (DAYS)	INSTANTANEOUS BIRTH RATE, b
0.5	1.386
1	0.693
2	0.346
3	0.231
4	0.173
5	0.139
10	0.069

After data of Paloheimo, 1974.

ratio has been held at 1. The egg-ratio method has been used in several detailed population analyses of planktonic rotifers (especially by Edmondson, 1960, 1968), from which the following examples are drawn, as well as for cladoceran and copepod zooplanktonic populations to be discussed later.

DEVELOPMENT TIME

The rate of development of eggs, which is required to determine the reproductive rate B, has been shown to be a function of temperature among a large number of rotifers and planktonic crustaceans. Examples of this relationship for a number of rotifer species are seen in Figure 16–13. The development time is quite independent of the type and quantitative nutrition of the adult female (see, for example, King, 1967) carrying the eggs.

The instantaneous growth rate, r, is the effective rate of population increase per unit time. A measure of death rate, d, is calculated by difference and varies with the abundance of predators.* The procedure assumes continuity of birth and death rates during the sampling interval (cf. Keen and Nassar, 1981; Taylor and Slatkin, 1981). The method demands that the sampling interval be short in relation to the life cycle and immaturity time of the animals. Although natural populations do not strictly conform to these assumptions, the discrepancies introduced are usually rather small (see Caswell, 1972; Edmondson, 1974; Paloheimo, 1974; Threlkeld, 1979; Polishchuk and Ghilarov, 1981; and Lynch, 1982 for a rigorous discussion of these assumptions). The method is very useful for analyzing factors that control birth and death rates in natural populations of zooplankton.

ROTIFER POPULATION DYNAMICS

Changes in the seasonal distribution of planktonic rotifer populations are complex and generalizations are difficult to make. A number of perennial species occur that commonly exhibit maximal densities in early summer in temperate regions. Other species are distinctly seasonal, of two general types: (a) cold stenotherms that develop greatest populations in winter and early spring, and (b) species that develop maxima in summer, often with two or more maxima, especially in late summer in conjunction with the development of certain blue-green algae populations. Some of the variations in seasonal succession can be seen in certain well-studied examples from several lakes. Additional examples from several older works are discussed critically in Hutchinson (1967).

The zooplanktonic community of Vorderer Finstertaler See, a deep, alpine drainage lake above the timber line in the Central Alps of Austria, was† extremely simple.

---

*In the calanoid copepod *Diaptomus clavipes*, egg size can differ, and larger eggs give rise to larger nauplii, each of which presumably has a greater likelihood of survival compared with smaller nauplii (Cooney and Gehrs, 1980). In some species of cladocerans, lipid reserves deposited in the eggs by the adult depend upon the nutritional status of the adult; neonates with lower lipid reserves may not survive as well as neonates with extensive lipid reserves (Goulden and Henry, 1982). The extent to which differential survival of newly hatched or neonate zooplankton is influenced by nutritional history of the parent animal is not fully understood, and may alter our view of zooplankton life strategies when more data become available.

†The lake has since been modified drastically by a large hydroelectric project.

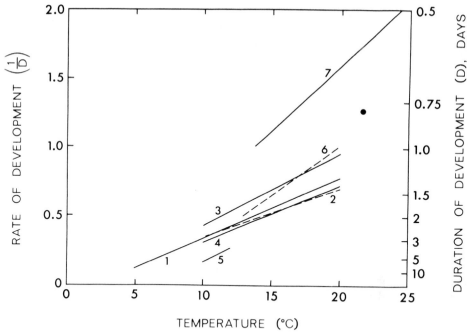

Figure 16-13. Average rates of development per day of rotifer eggs in relation to temperature. Line *1*, Least squares regression line for *Keratella cochlearis* and *Kellicottia longispina*; *2*, *Polyarthra vulgaris*; *3*, *Keratella quadrata*; *4*, *Kellicottia bostoniensis*; *5*, *Ploesoma truncatum*; *6*, *Ploesoma hudsoni*; *7*, *Hexarthra fennica*, •, *Euchlanis dilatata*. (From data of Edmondson, 1960, 1965, and King, 1967.)

Two rotifers (*Keratella hiemalis* and *Polyarthra dolichoptera*) and a single copepod (*Cyclops abyssorum*) constituted over 99 percent of the zooplankton. The species of *Keratella*, known as a *hypolimnetic cold stenotherm*, developed maximal population densities in midwinter, decreasing conspicuously during spring and early summer (Fig. 16-14). The species of *Polyarthra* is also a coldwater form that tolerates reduced oxygen concentrations, and is commonly found in hypolimnia. The *Polyarthra* maximum was seen in midsummer in this lake.

A number of other species of rotifers are similarly cold stenotherms with varying tolerances to reduced oxygen concentrations (Ruttner-Kolisko, 1975, 1980; Elliott, 1977; Herzig, 1980; May, 1980). These capabilities permit the rotifers to occupy niches both spatially (metalimnion, hypolimnion) and seasonally (winter throughout much of the water column; spring period) that are unsuitable for other rotifers. Often bacteria are more abundant in these regions than in the epilimnion, and apparently serve as significant food sources. Other rotifers, such as *Notholca*, can feed on the cell contents of large diatoms (e.g., *Asterionella*), and may increase in abundance during the spring diatom bloom.

Studies of the rotifer populations in two small lakes in southern Ontario showed that the seasonal distribution varies considerably among lakes, and is synchronous

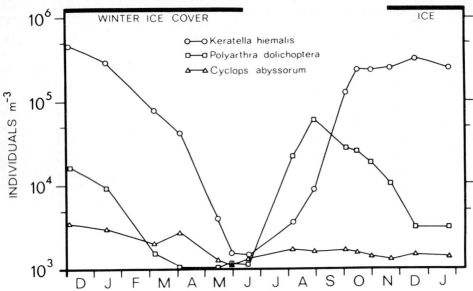

Figure 16–14. Seasonal variations in the population densities of three dominating zoo-plankton species (> 99% of the total) of Vorderer Finstertaler See, Central Alps of Austria, 1969. (Modified from Pechlaner, et al., 1970.)

among lakes only in a general way (George and Fernando, 1969). Paradise Lake, for example, is a small (7 ha), shallow (6 m), dimictic lake that exhibits reduced oxygen concentrations in the hypolimnion in late summer, but does not become anaerobic. Summer stratification extended from early June through September in the year of the investigation. Of the four species† of *Keratella* found in this lake, *K. canadensis* is a cold stenotherm found in the upper strata in early winter. *K. quadrata*, also a cold form, occurred from early winter in the upper layers and moved to the cooler hypo-limnion in midsummer before declining. *K. cochlearis* is also a cold stenotherm, and exhibited two maxima in this lake, one in winter and another in late spring, but the seasonal distribution of this species varies greatly among different lakes. *K. hiemalis*, known as a winter and early spring species, was common during colder periods but was reduced abruptly as temperatures reached approximately 15°C.

Polyarthra vulgaris, *P. euryptera*, *Conochilus unicornia*, *Gastropus stylifer*, and *Keratella crassa* are largely summer and autumn species, although a considerable displacement and spatial separation of the populations occurred seasonally with depth. The predatory species of *Asplanchna* generally was a minor component of the rotifer community, and was relatively uniformly distributed in winter and early sum-mer. The two species of *Kellicottia*, both perennial, were distributed uniformly with depth throughout much of the year; *K. bostoniensis*, however, was restricted to cooler areas during much of the summer. Both species exhibited summer population max-ima.

These examples serve to illustrate some of the variations found in seasonal succes-sion. Variations are very common, and in warm temperate regions, very little is known about seasonal succession among rotifers. Interactions among species of rotifers are

---

†*Keratella canadensis*, which may not be a valid species, is very similar to *K. quadrata*.

poorly understood; a notable exception is the predation of *Asplanchna* on *Brachionus*, discussed earlier. Clearly, temperature and food quality and quantity are dominant factors in the regulation of rotifer reproductive rates and population succession. Analyses of interactions in situ under natural conditions are rare. A detailed regression analysis of the seasonal dynamics of reproductive rates of planktonic rotifers of four lakes of the English Lake District in relation to food abundance verified the results of culture analyses (Edmondson, 1965). The abundance of food at a given time is an imperfect measure of food supply to the feeding organisms; the food consumed during a period of time depends upon the grazers' capabilities, as well as the predators' reproductive rate, which is under the influence of still other environmental variables. Among the three dominant rotifer populations analyzed by Edmondson, reproductive rates were strongly related to temperature, especially in the case of *Keratella*. A small flagellate, *Chrysochromulina*, was of major importance in the success of both *Keratella* and *Kellicottia*. Limited data on bacterial densities showed no significant correlations with reproductive rates of the rotifers. The reproductive rate of *Polyarthra* was directly correlated with the abundance of the small alga *Cryptomonas*.

The significance of food organisms to the rotifers was found to be partly but not wholly related to size. Some food organisms within an acceptable size range were of little significance. The green alga *Chlorella* was consistently not utilized as a food organism. Similarly, *Eudorina* was not eaten and may exert an inhibitory effect. Some rotifer species do not discriminate between feeding upon living and dead algal cells (Starkweather and Bogdan, 1980; Bogdan, et al., 1980). *Keratella cochlearis* was found to feed selectively on detritus when given both dead and live algae but, as with several other rotifers, would ingest bacteria, yeast, and algae in mixed diets, although at different rates. Other rotifers (e.g., *Polyarthra*) ingested only algal cells.

Food availability was found to be instrumental in the population dynamics of *Keratella cochlearis* in a eutrophic Danish Lake (Bosselmann, 1979a, 1979b). A small, nonreproducing population occurred in the open water during late autumn and in winter; mortality was low. The existing population started to reproduce and the population was augmented by the hatching of resting eggs. With low mortality in the spring, the population grew exponentially (Fig. 16–15); reproduction was limited only by low water temperatures. During May, the turnover rate of biomass of this rotifer species was 7.7 days. In early summer, the population was large but constant, had a moderate mortality, and a low birth rate because of low food availability. Food availability improved in late summer and, along with high water temperatures, resulted in more rapid biomass turnover rates (5.0 days). Mortality was higher in late summer and the population numbers fluctuated widely. In the autumn, reproduction decreased because of declining temperatures and food availability.

A detailed analysis of the zooplanktonic population dynamics and productivity of Mirror Lake, New Hampshire, also demonstrates the spatial and temporal separation of the individual populations (Makarewicz and Likens, 1975, 1979). As illustrated in the depth–time productivity data of Figure 16–34, the rotifers were major components of the zooplankton community in this lake. Although copepods constituted 55 per cent of the total zooplankton biomass, they represented only 19 per cent of the zooplankton productivity because of their slow growth over an entire year. In contrast, despite lower biomass, the rotifers accounted for 40 per cent of the annual zooplankton production

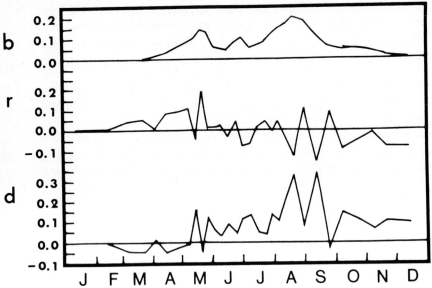

Figure 16-15. Estimates of the birth rate (b), rate of growth (r), and the rate of mortality (d) of *Keratella cochlearis* in Lake Esrom, Denmark, 1973. (Modified from Bosselmann, 1979a.)

primarily because of faster growth rates and generation times of 8 to 10 days (at 17°C). As discussed later (p. 476), the niche hyperspace of the limnetic zooplankton community was subdivided by dispersion of the populations of rotifers, cladocerans, and copepods along gradients of food size, depth, and time. Differences in the life histories in the timing of development of different populations, feeding behavior and food size and types utilized, tolerances of vertical and seasonal variations in temperature and oxygen, and vertical migratory behavior all contributed to minimizing niche overlap and permitting coexistence of species within the annual zooplankton community.

CLADOCERAN POPULATION DYNAMICS

The seasonal succession of the Cladocera is quite variable, both among species and within a species living in different lake conditions. Some species are perennial, and overwinter in low population densities as adults (parthenogenetic females) rather than as resting eggs. These species may exhibit one, two, or more irregular maxima. Some perennial species exhibit maxima in surface layers only during colder periods in the spring, and in the cooler hypolimnetic and metalimnetic strata during summer stratification. The aestival species that have a distinct diapause in a resting egg stage commonly develop population maxima in the spring and summer when the water is relatively warm. Although one population maximum is general, a second population peak often occurs in the autumn.

An example of a typical case of population dynamics can be taken from the well-studied zooplankton of Base Line Lake, Michigan (Hall, 1964), consisting mainly of *Daphnia*. Two maxima occurred in the population of *Daphnia galeata mendotae*, one

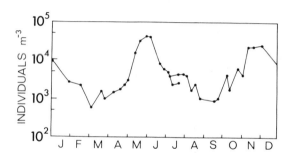

Figure 16-16. Population density of *Daphnia galeata mendotae* in Base Line Lake, Michigan, 1960–61. (Redrawn from Hall, 1964.)

in the spring and a second in late autumn (Fig. 16–16). Although a few ephippia were produced, this species, in contrast to *D. pulex* in the same lake, overwintered in the free-swimming stage with little or no reproduction. The initial spring maximum in May–June and that in the fall were preceded by a maximum in brood size (ratio of eggs to mature *Daphnia*) (Fig. 16–17). Degenerate eggs were observed frequently in the brood pouches of *Daphnia* and had no apparent relationship to normal brood size.

Employing the instantaneous rate of increase of the population discussed earlier, the population dynamics of *D. galeata mendotae* were evaluated in Base Line Lake. The instantaneous birth rate (b) was estimated from the finite birth rate of newborn occurring over a time interval in relation to the existing population size. The development time of eggs was determined experimentally in relation to temperature. The instantaneous rate of population increase, r (see Equation 2), and birth rate were both positive and nearly equal during the spring and fall months (Fig. 16–18). Maximum values of b occurred in the summer months, but r was negative during this period of population decrease because of increased death rate, d. During the winter, b was zero as reproduction ceased and r was negative as a slow death rate, d, gradually reduced the population until March.

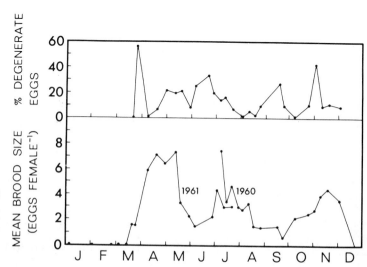

Figure 16-17. *Upper,* percentage of degenerate eggs in the population, and *lower,* mean brood size (mean number of eggs per adult) of *Daphnia galeata mendotae,* Base Line Lake, Michigan, 1960–61. (Redrawn from Hall, 1964.)

The summer minimum, with a loss rate of 28 per cent per day, apparently was associated with predation on the juvenile stages of *Daphnia galeata* by the large cladoceran *Leptodora* which reached its maximum in summer (Fig. 16–18). *D. retrocurva* reached its maximum density, greatly exceeding *D. galeata*, in late summer, well after water temperatures exceeded 20°C. Its successful competition for available food in fall was most likely contributory to the reduced birth rate of *D. galeata* at that time. Analyses of fish consumption showed that fish fed extensively on the large *D. pulex* in spring, but not significantly on the smaller *D. galeata*. *Daphnia catawba*, which like *D. retrocurva* exhibits one late summer maximum and overwinters with resting eggs, is also known to coexist with *D. galeata* in late summer (Tappa, 1965). The presence of *D.*

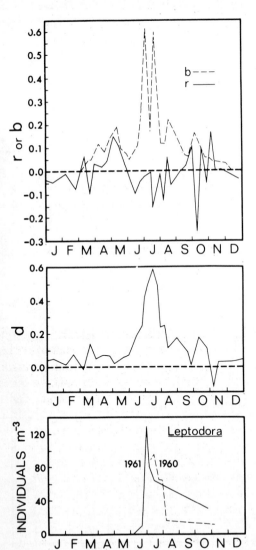

Figure 16–18. *Upper*, instantaneous birth rate (natality), *b*, and rate of population change (increase or decrease), *r*, *Daphnia galeata mendotae; middle*, instantaneous death rate, *d* (*b* minus *r*), of *Daphnia galeata mendotae; Lower*, population density of the predatory *Leptodora*, Base Line Lake, Michigan, 1960–61. (Redrawn from Hall, 1964.)

catawba and its competition for the same food base appeared to be instrumental in delaying the autumnal maximum of *D. galeata*.

In a detailed study of the zooplanktonic populations of shallow ($\bar{z}$ = approx. 1 m), extremely productive Sanctuary Lake in northwestern Pennsylvania, three species of *Daphnia* occurred and exhibited a large spring maximum and a small autumn peak (Cummins, et al., 1969), similar to that discussed in the Base Line Lake example. *D. galeata mendotae* dominated these populations and *D. retrocurva* became a codominant in late summer. With the decline of the spring maximum, a massive development of the predacious *Leptodora kindtii* occurred and reached a biomass maxima in June in one year and in June, in July, and in August the following year. Only the larger *Leptodora* of from 6 to 12 mm are predaceous, and ingest the fluids of their prey. The population dynamics of *Leptodora* were correlated with loss rates of prey; *Leptodora* apparently shifted seasonally to preying on *Daphnia*, *Ceriodaphnia*, *Bosmina*, and *Chydorus* and the copepods *Cyclops* and *Diaptomus*.

A few cladocerans are distinctly cold-water species, and inhabit shallow lakes of northern regions. These species develop large populations only in the early spring and in cooler meta- and hypolimnetic strata in more temperate lakes. These species generally are perennial, and tend to tolerate lower oxygen concentrations than the more common warm-water species.

Among cladoceran zooplanktonic populations in general, low populations in winter among perennial species and a near absence in winter among aestival species are common. With increasing food supply from algae, detritus, and bacteria with rising temperatures in spring, cladoceran populations increase from overwintering adults or resting eggs. Temperature increases the rate of molting and brood production, while rising food supply increases the number of eggs per brood. Population cycles in the summer are more variable, being influenced by reduced food supply, shifts in the quality of food to less palatable species, and predation by other zooplankton, fishes, and other organisms. Mortality of cladocerans, especially juveniles, is normally highest in summer, as is competition for available food resources among coinhabiting species.

Food availability also influences the size structure of zooplankton communities by age-specific differential mortality. Under conditions of starvation, respiration increases and body weight decreases before death (Threlkeld, 1976). Survival time (t) is greater in larger zooplankton, and is a function of body weight (w, in μg dry weight): $t = 2.95\ w^{0.25}$. Older individuals, however, survive for shorter periods of time than younger zooplankton under nonstarvation conditions. Juveniles tend to be more vulnerable to food limitations than adults (Neill, 1975). The impact of periods of size-related starvation on community size structure may be augmented or reversed by the timing of the onset of food limitation on its members (Threlkeld, 1976), and by the extent to which starvation influences prey utilization by the zooplankton (Williamson, 1980).

COPEPOD POPULATION DYNAMICS

A number of cyclopoid copepods are known to enter various periods of diapause, either at the egg stage or in the copepodite stages with or without encystment, upon the sediments (see, for example, Wierzbicka, 1966). Consequently, for reasons that are not very clear, the annual cycle of the copepod populations may be interrupted by a diapause that persists from one to several months. The resting stage may occur in midwinter, as in *Cyclops bicolor,* or in summer, as is seen in *Cyclops strenuus strenuus* (Fig. 16-19). Nauplii develop from egg-bearing females in this species in a bimodal population during winter and spring. Then, in summer, the larger population of copepodite stage IV aestivates in the sediments for about two months in midsummer before becoming pelagic again in the autumn and developing into adults. In contrast, *Cyclops strenuus abyssorum* exhibits only one effective generation in each year in several lakes of the English Lake District (Smyly, 1973). In the deepest of the four lakes studied in great detail, individuals of this generation hatch from eggs laid in the spring and reach the adult stage in early winter. The adults pass the winter in the planktonic zone, and start laying eggs early in the following year. In the three other lakes, most individuals of the spring generation reach the fifth copepodite stage by midsummer, and spend the next eight months aestivating in the profundal zone. They leave this zone in February or March to return to the planktonic zone, where they become adults and start breeding. Water temperature, oxygen concentrations, light intensity, and day length are believed to be associated with the initiation and termination of this diapause (Smyly, 1973). Co-occurring species of cyclopoid copepods are known to have their maxima and diapause periods at different times. This alternation presumably minimizes competitive interactions for the same food resources.

Although the general pattern of diapause is similar among the cyclopoid copepods, marked variations in the timing of instar stages and their distribution in the water and the sediments have been observed (e.g., Elgmork, 1959; Smyly, 1973; George, 1976; Nilssen and Elgmork, 1977; Vijverberg, 1977; Elgmork, et al., 1978; Boers and Carter, 1978; Elgmork and Langeland, 1980). The basic life-cycle pattern in limnetic cyclopoid cope-

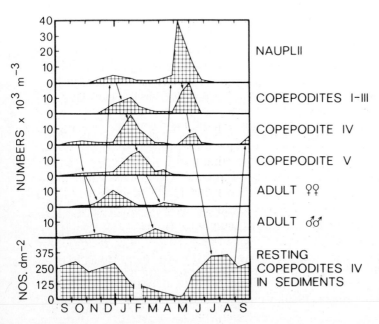

Figure 16-19. Development and resting stages of *Cyclops strenuus* in Bergstjern, Norway. Midsummer diapause is indicated by aestivation of copepodites IV in the sediments. (Drawn from data of Elgmork, 1959.)

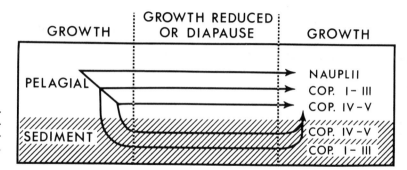

Figure 16-20. Generalized life-cycle patterns of limnetic cyclopoid copepods. COP. = copepodid instar stages as indicated. (Adapted from Nilssen, 1978.)

pods can be divided into two parts (Nilssen, 1978; Elgmork and Nilssen, 1978): (a) a period of growth, followed by (b) a period of retarded growth or diapause, the latter resulting from adverse environmental conditions (Fig. 16-20). Arrested development is induced by a combination of factors: decreasing water temperatures, photoperiod (on the copepodid I instar; Watson and Smallman, 1971), reduced food availability, anoxia, and increased competition and predation.

The intensity of the lines and pathways shown in Figure 16-20 differ in different lake ecosystems. (1) In arctic and alpine lakes, diapause as copepodites in the sediments is less common. Adult stages frequently predominate in pelagial waters just above the sediments, perhaps to avoid fish predation. The adults feed heavily on rotifers and their own nauplii during winter. (2) Large differences in life cycles of limnetic cyclopoid copepods occur in temperate environments. In less predictable environments (eutrophic lakes, temporary ponds), many cyclopoid copepods persist during periods of reduced growth in naupliar or advanced copepodid stages in the water; a relatively small proportion of the population enters copepodid IV-V diapause in the sediments. In more predictable and less productive lakes, during colder periods of reduced or arrested growth the tendency is for most of the population to diapause in naupliar stages or resting stages in the sediments. This response is believed to result largely as an adaptation to avoid predation by fish and predatory copepods during periods of slow growth when the population is most vulnerable (Nilssen, 1978; Strickler and Twomby, 1975). (3) In tropical lakes showing no great seasonal changes in environmental parameters, resting stages have not been observed.

The life cycles of the calanoid copepods are generally somewhat longer than those of the cyclopoids. In the temperate region, most species exhibit prolonged reproductive periods with several generations per year that are indistinguishable from one another. An example of these seasonal dynamics is seen in the populations of *Diaptomus* in a small, shallow beaver pond of southern Ontario (Fig. 16-21). The *Diaptomus* hatched from resting eggs in early May and produced four complete generations, the last maturing in late autumn (Carter, 1974). During spring and summer, generation times were about four weeks, while the time was about six weeks in autumn generations. Population characteristics were similar over a three-year period. In one year, water levels were low, which resulted in increased mean heat content per unit volume; during this year development time of the generation was reduced and population production increased significantly.

A number of studies have demonstrated that most adult cyclopoid copepods are carnivorous, and that their predatory activities can play a significant role in the population dynamics of other copepod species. For example, *Mesocyclops* was found to be selectively predacious on copepodites of *Diaptomus* rather than on cladocerans (Confer, 1971). Similarly, adult *Cyclops* preys heavily on nauplii of *Diaptomus* and its own species (McQueen, 1969). Some 30 per cent of the naupliar recruitment has been observed to be lost by copepod predation and by cannibalism. Juvenile mortality often is very high and is density-dependent, and can increase conspicuously at high population levels (Elster, 1954).

The coexistence of several congeneric species of copepods in the same volume of

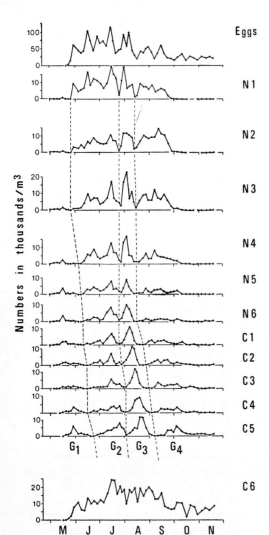

Figure 16-21. Seasonal cycles of *Diaptomus reighardi* in Black Pond, Ontario. Broken lines separate the estimated limits of successive generations, G. (From Carter, J. C. H.: Life cycles of three limnetic copepods in a beaver pond. J. Fish. Res. Bd. Canada, 31: 421–434, 1974. Reprinted by permission of the Journal of the Fisheries Research Board of Canada.)

water has been a subject of much interest. It is assumed that when coexistence does occur, the species have slightly differing tolerances and optima to existing environmental conditions. Mechanisms promoting coexistence include: (a) seasonal separation, discussed earlier; (b) vertical separation in relation to stratification and the distribution of food resources; and (c) size differences in relation to available food particles utilized. All of these mechanisms were implicated as factors permitting the coexistence of three species of *Diaptomus* in an Ontario lake (Sandercock, 1967; see also Hammer and Sawchyn, 1968). *D. minutus* was separated vertically from *D. sanguineus*, and the latter is distinctly smaller than the former species. *D. minutus* and *D. oregonensis* also were separated to some degree by size differences and exhibited different seasonal maxima. *D. sanguineus* and *D. oregonensis* were separated vertically and had different seasonal maxima. Therefore, in this lake, differences in two of the three factors provided the mutual separation among these three congeneric species that is believed necessary to coexistence.

VERTICAL MIGRATION AND SPATIAL DISTRIBUTION

CLADOCERA AND COPEPODS

One of the most conspicuous features of the cladoceran zooplankton, and to a lesser extent the copepods, is the marked vertical migration of these small animals over large distances on a daily basis.* There is no question of a light-mediated circadian rhythm in at least some cases of vertical migration (Siebeck, 1960; Ringelberg, 1964), but the adaptive significance of this movement is less clear. Movements vary considerably with lake conditions in relation to underwater light characteristics, season, and the age and sex of a species. However, certain generalizations do emerge.

Most species migrate upward from deeper waters to more surficial strata as darkness approaches, and return to deeper strata at dawn. Maximum numbers are found in the surface layers in *nocturnal* migration in a single maximum some time between sunset and sunrise. In other cases, *twilight* migration results in two maxima in surface layers, one at dawn and another at dusk. In a few species, *reverse* migration occurs with a single surface maximum during the daylight period.

The pattern of vertical migration commonly found may be seen in an example from relatively transparent Lake Michigan (Fig. 16–22). At dusk, the *Daphnia* population migrated from 24 m through the thermal discontinuity of the metalimnion at a rate of over 10 m hr^{-1}. Some sinking occurred after the initial ascent, a characteristic common to the twilight type of migration that could be associated with reduced nighttime activity and passive sinking. For example, narcotized *Daphnia galeata mendotae* sink passively

---

Diurnal refers to an event that occurs in a day (24-hour period) or recurs each day. *Diel* (= daily) is a more recent term (not yet in most dictionaries), that refers to events that recur at intervals of 24 hours or less, with no connotation of either daytime or nighttime (Odum, 1971). In studies of community periodicity, when whole groups of organisms exhibit synchronous activity patterns in the day-night cycle, *diurnal* is often used in a more restricted sense, referring to animals that are active only during the daytime (lighted period) as opposed to others that are active only during the period of darkness (= *nocturnal*) and still others only during twilight periods (= *crepuscular*).

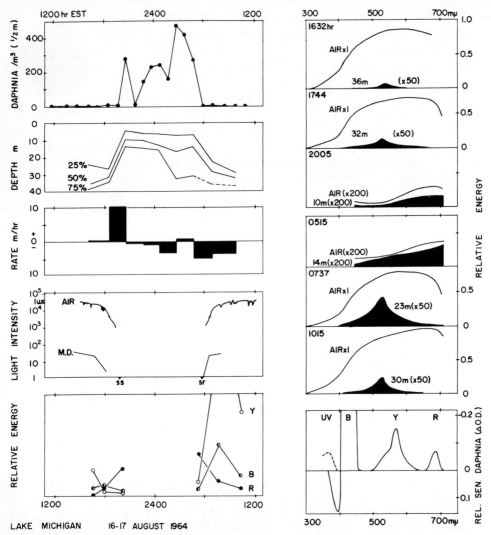

Figure 16–22. The vertical distribution and rate of movement of *Daphnia retrocurva* in relation to light quality at depth. *Left, top to bottom:* number of *Daphnia* m⁻³ at 0.5 m depth; quartile distribution; rate of movement (m hr⁻¹); light intensity (lux) in air and at mean population depth (*M.D.*); and relative energy received by blue (*B*), yellow (*Y*), and red (*R*) photosystems of *Daphnia*. *Right, top to bottom:* relative energy versus wavelength in air and at mean population depths at different times of the diurnal period; spectral sensitivity of *Daphnia*, expressed as change in optical density (*O.D.*) of visual pigments extracted with aqueous digitonin upon bleaching, *sr* = sunrise; *ss* = sunset. (From McNaught, D. C., and Hasler, A. D.: Photoenvironments of planktonic Crustacea in Lake Michigan. Verhandlungen Int. Ver. Limnol., 16:194–203, 1966.)

between 3.6 and 14.4 m hr^{-1}, at a rate inversely related to body length (Brooks and Hutchinson, 1950). The in situ populations generally do not sink that rapidly during the night period (usually less than 0.5 m hr^{-1}), which indicates that they are sufficiently active for sinking to only slightly exceed upward movements. After another surface maximum before sunrise, the daphnids left the upper waters abruptly about sunrise at velocities of about 5 m hr^{-1}. Both the evening ascent and morning descent occurred when light intensity at the mean depth of the population was changing rapidly (Fig. 16–22).

Daphnia has trichromatic vision with spectral peaks of sensitivity in the ultraviolet (370 nm), deep blue-violet (435 nm), yellow (570 nm), and possibly in the red (685 nm) (Fig. 16–22). At the midday depth of habitation, the photoenvironment is predominantly blue-green. The photosystem is most sensitive in the blue, and may be best adapted for detecting changes in light intensity at low energy levels (McNaught, 1966). All photosystems may function in the detection of changes in wavelength and intensity at higher levels during daytime vision. The responses to light changes are conspicuous, but there is no evidence that the zooplankters are attempting to remain in a constant photoenvironment.

The vertical migrations also occur under much less transparent conditions of more productive lakes. Under these situations, even though the extent of movement is much less and often only a meter or two, the pattern is conspicuously similar (Table 16–9; Fig. 16–23). The mean morning maximum at the surface occurred 60 minutes (SD ± 37 minutes) after sunrise and the mean evening maximum at the surface occurred 19 ± 61 minutes before sunset. Deviations from the mean pattern increased under less distinct light conditions of moderate to heavy cloud cover. But clearly, the cue is light intensity. Response can be shown experimentally and under abnormal natural light conditions, e.g., the response is identical under twilight conditions during a solar eclipse (see, for example, Bright, et al., 1972).

Variations in vertical migration are great. Generally, young stages have been found to migrate further than adults both in cladocerans and copepods. Differences also occur in the migratory pattern of females and males, especially among copepods. Males tend to be less migratory than females, although the pattern is not consistent. The amplitude and rates of migration vary considerably during the ice-free season. Some of this change may be related to a progressive decrease in oxygen content of lower layers as well as to changing light conditions resulting from seasonal changes in planktonic turbidity. Although temperature influences rates of locomotion, metabolism, and regeneration of photochemical visual pigments, it is minor in effecting vertical migrations in comparison to light. Migrations continue under ice cover and are strongly influenced by light reductions from snow cover. Reverse migration is more common under ice cover, especially among copepods (see, for example, Cunningham, 1972). Zooplanktonic concentrations are high in surface layers below heavy ice and snow where light levels are low. In lakes of tropical regions, little seasonal variation in diurnal vertical migration has been observed (e.g., Hart and Allanson, 1976).

Another conspicuous feature of the movements of pelagic cladocerans and copepods is a distinct "avoidance of shore." At the shore, the littoral and near-littoral waters of lakes are almost free of pelagic crustaceans. When released within the littoral zone,

TABLE 16-9 Examples of the Estimated Velocity and Distance of Vertical Migration of Several Zooplankters

GROUP/SPECIES	LAKE	MAXIMUM OBSERVED RATE (m hr⁻¹)		MAXIMUM OBSERVED MIGRATION (m)	SOURCE
		ASCENT	DESCENT		
Rotifera					
Polyarthra vulgaris	Sunfish, Ontario	0.25	0.18	5.6	George and Fernando, 1970
Filinia terminalis		0.65	0.32	4.0	
Keratella quadrata		0.36	0.50	3.0	
K. cochlearis	La Caldera, Spain	1.76	1.18	7.0	Cruz-Pizarro, 1978
Hexarthra bulgarica		0.90	2.05	8.9	
Euchlanis sp.		0.40	0.43	10.0	
Cladocera					
Daphnia galeata mendotae	Michigan	—	—	10	Wells, 1960
	Mendota, WI	1.4	1.4	1.5	McNaught and Hasler, 1964
D. longispina	Grand, CO	—	—	5.8	Pennak, 1944
	Lucerne	—	—	31	Worthington, 1931
	Victoria	—	—	50	
D. retrocurva	Michigan	10.6	5.0	24.3	McNaught and Hasler, 1966
D. pulex	Summit, CO	—	—	2.7	Pennak, 1944
	La Caldera, Spain	1.3	0.6	9.2	Cruz-Pizarro, 1978
D. schoedleri	Mendota, WI	1.4	1.4	1.5	McNaught and Hasler, 1964
Bosmina longirostris	Michigan	19	12	20	McNaught and Hasler, 1966
				1.8	Pennak, 1944
Polyphemus pediculus	Silver, CO	3.6	2.0	13	McNaught and Hasler, 1966
Holopedium gibberum	Michigan	6.9	—	13.7	
Leptodora kindtii		4.2	2.0	8.5	
Copepoda					
Limnocalanus macrurus	Michigan	18.0	9.8	24	McNaught and Hasler, 1966
Diaptomus shoshone	Summit, CO	—	—	8.8	Pennak, 1944
Cyclops bicuspidatus	Silver, CO	—	—	3.2	
Mixodiaptomus laciniatus	La Caldera, Spain				
Nauplii		0.4	0.3	6.8	Cruz-Pizarro, 1978
Copepodites		0.5	0.7	7.6	
Adults (♂)		0.6	0.9	8.5	
Adults (♀)		1.4	0.5	8.8	
Pseudodiaptomus hessei	Sibaya, S. Africa				
Nauplii		8.3	3.9	25	Hart and Allanson, 1976
Copepodites (CI-CIII)		5.5	9.4	40	
Copepodites (CIV-CV)		14.2	10.7	40	
Adults (♂)		16.2	14.9	40	
Adults (♀)		26.0	13.0	40	
Insecta					
Chaoborus trivittatus	Eunice, British Columbia	3	1	9	Swift, 1976
Amphipoda/Mysids					
Pontoporeia affinis	Michigan	11.7	13.9	40	McNaught and Hasler, 1966
		18	13	40	
Mysis relicta		30–48	30–48	76	Beeton, 1960

these animals migrate horizontally away from the shore area if illumination is adequate. This aspect of spatial orientation has been the subject of intensive experimentation (especially Ringelberg, 1964; Siebeck, 1968; Siebeck and Ringelberg, 1969; Daan and Ringelberg, 1969).

The behavioral analyses showed that these crustaceans orient themselves toward the elevation of the horizon and the position of the sun. The direction of migration coincides with the plane of symmetrical light stimuli in the underwater angular light distribution. Near the shore, this plane of symmetry of angular light depends primarily on the position of the elevation of the horizon. With increasing distance from the shore, the effect of the elevation of the horizon on the angular light distribution decreases and finally becomes insignificant, at which point the animals no longer select a directional

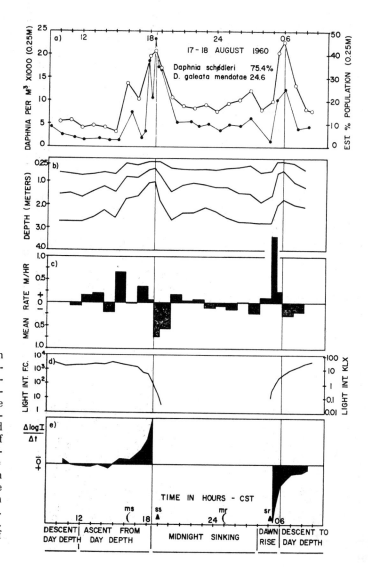

Figure 16-23. Diurnal vertical migration of two *Daphnia* populations in Lake Mendota in August in relation to light conditions. *a*, Number (●) and estimated population (○) at 0.25 m depth; *b*, quartile distribution; *c*, net rate of vertical movement (m hr⁻¹); *d*, light intensity (lux and foot-candles) at 0.25 m depth; and *e*, rate of change in the logarithm of the light intensity. *ss* = sunset; *sr* = sunrise; *ms* = moonset; *mr* = moonrise. (From McNaught, D. C., and Hasler, A. D.: Rate of movement of populations of *Daphnia* in relation to changes in light intensity. J. Fish. Res. Bd., Canada, 21:291–318, 1964. Reprinted by permission of the Journal of the Fisheries Research Board of Canada.)

movement away from shore. The higher the elevation of the horizon, as in mountain lakes, the wider the zone near shore containing a reduced concentration of pelagic crustacean zooplankton.

Light from the sun and sky incident on water is refracted in the direction of the vertical axis as it penetrates downward into the water. In effect, the angular light distribution operates in an inverted conical fashion, in which light impinging at angles greater than 49 degrees from the vertical is absorbed selectively so rapidly that a light–dark boundary or contrast forms at an angle of about 49 degrees to the vertical axis. The contrast or light gradient is used in orientation of the body axis in swimming. For example, *Daphnia magna* is unable to swim normally or maintain its normal body position when light from every direction is of equal intensity. Normal swimming reappears when a light gradient is introduced. The body and swimming positions are oriented by distinct pairs of ommatidia of the compound eye, which are capable of detecting contrasts. The eye is turned in a dorsal body direction so that the light contrast can be projected on the same dorsal part of the eye. The body is then turned in order to maintain a physiologically fixed angle between the eye axis and the body axis.

In the littoral zone, the angular light distribution consists of a dark area along the landward side within the cone of illumination. This distribution results in one of the two contrasts present at 49 degrees in the pelagial being shifted to the vertical, and swimming would be away from the shore (Fig. 16–24). In a plane parallel to the shoreline, this contrast difference is absent and the contrasts are the same as in the open water.

Evidence for changes in rate of feeding during the diurnal migration of crustacean zooplankton is meager. At certain times of the year in the seasonal succession of zooplanktonic species, the grazing rates are several times greater during the dark period (Haney, 1973; Haney and Hall, 1975; Starkweather, 1975; Crumpton and Wetzel, 1982).

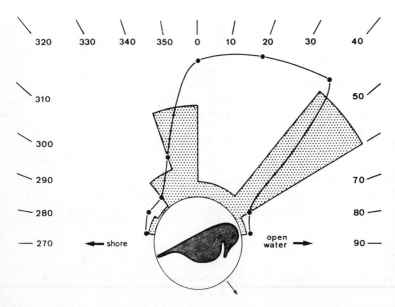

Figure 16–24. Orientation of a daphnid zooplankter along a plane perpendicular to the shore. Angular light distribution (solid line), magnitude of contrast expressed as the difference between two succeeding measurements (shaded bars), and the most probable orientation and movement of the daphnid. (From Siebeck, O., and Ringelberg, J.: Spatial orientation of planktonic crustaceans. Verhandlungen Int. Ver. Limnol., *17*:831–847, 1969.)

At other times, the zooplankton undergo diurnal migration without altering their grazing rates. However, simply because of the massive diurnal movements of major components of the zooplanktonic populations to surface strata at night, a marked increase in grazing pressure on epilimnetic algal, detrital, and bacterial populations occurs at night. Certain algae, such as motile unicellular flagellates, can move downward at night, and this may represent a survival adaptation to increased predation pressure.

Study of the adaptive significance of nocturnal vertical migration has led to a number of hypotheses, all of which probably interact to varying degrees. First, predation by fish and other predators is largely a visual process, and requires light. Movement upward into the trophogenic zone in darkness or periods of low light intensities would avoid much of this predation pressure (Zaret and Suffern, 1976; Wright, et al., 1980; Zaret, 1980). Second, the food quality of algae is variable diurnally, as is discussed further in the following chapter (see Fig. 17–10). During the day, algae synthesize predominantly carbohydrates; protein synthesis reaches a maximum during the night period (cf. Enright, 1977b). Third, growth efficiency is somewhat greater at low temperatures, such as are found in the lower strata when summer thermal stratification occurs (McLaren, 1963; 1974). Similarly, feeding rates and efficiency are greater at higher temperatures. According to this differential growth–feeding hypothesis, a significant advantage would be obtained with only one degree of temperature difference between the water strata. Since the photoperiod would have little influence on advantages conferred by these mechanisms, the adaptations most likely are coupled with the advantages of reduced predation and higher food quality that are related to the photoperiod. The energy expenditure in such migrations is uncertain but variable among the various swimming and drag characteristics of migrating zooplankton (cf. Vlymen, 1970; Haury and Weihs, 1976; Enright, 1977a; Strickler, 1977; Lehman 1977).

A number of zooplankton, particularly copepods, possess a deep red coloration. The red coloration is caused by several carotenoids synthesized from β-carotene present in the algal diet (Ringelberg, 1980). The carotenoid pigment content increases under conditions of high light intensity (e.g., in clear or shallow lakes, especially high altitude lakes) and, among many copepods, increases in summer. Carotenoid content also often increases several-fold in a diurnal rhythm, showing minima around midday and maxima around midnight (Ringelberg, 1980, 1981). Darkly pigmented copepods survived significantly longer than pale copepods at the same species when both were exposed to natural intensities of visible blue light. While carotenoid photoprotection is established among some species, the adaptive significance of irregular carotenoid distribution and diurnal patterns is not obvious at this time, although it is associated with diurnal vertical migration, energy utilization, and feeding periodicity (cf. Ringelberg, 1981; Hairston, 1976, 1980).

In arctic lakes, the predacious copepod *Heterocope* occurs in distinct dark red and pale green morphs (Luecke and O'Brien, 1981). The dark red form survived better in bright sunlight, but the pale green form was less susceptible to predation by fish that fed visually. In cases where fish predation was intense, the green forms predominated.

ROTIFERS

Although features of diurnal migration are known in some detail for the cladocerans and copepods, relatively little information exists on daily movements of planktonic

rotifers. Commensurate with their relatively small size in comparison with the crustacean zooplankton, the movements of rotifers are relatively slow (Table 16-9). The maximum observed velocity of ascent in *Filinia* of 0.018 cm sec^{-1} is nearly equivalent to
that of some of the slowest movements among the crustacean zooplankton (George and
Fernando, 1970). The maximum velocities of vertical movement, as well as the magnitude, vary among species with descent or ascent and with the season. The general tendency is toward upward movement during the daylight period, although limited evidence indicates an upward nocturnal migration at one season and the reverse at
another season (cf. Pennak, 1944). No clear-cut migration pattern is seen among the
rotifers, as occurs among the cladocerans and copepods. Presumably, the smaller size
of most rotifers removes much of the pressure of visually oriented predation by fishes
that probably is associated with nocturnal ascent patterns among Cladocera and some
copepods. Movement of rotifers into the surface waters at midday (e.g., Adeniji, 1978)
may, however, reflect an adaptive response to avoid predation by limnetic crustaceans.

HORIZONTAL VARIATIONS IN DISTRIBUTION

Insight into horizontal variations in the spatial distribution of rotifers is provided
by the detailed analyses of zooplankton in a single cross-section across the northern
arm of Skärshultsjön, southern Sweden (Berzins, 1958). The eastern shore of the basin
along this 145-meter transect is steepsided, and drops precipitously to the maximum
depth of about 13 m. The western, more gently sloping shore supported a well-developed littoral zone of submergents *Myriophyllum* and *Potamogeton*, grading through
waterlilies *(Nuphar)* to the emergent macrophytes *Phragmites, Equisetum,* and *Juncus.*
The numerous samples were collected within a 2-hour period in midafternoon on 7
August 1950. At this time the lake was stratified, with a freely circulating epilimnion
(19°C) to 3 m, below which the metalimnetic thermal discontinuity layer extended to 7
m, at which point the hypolimnion of 8 to 10°C extended to the bottom. The oxygen
content of the hypolimnion was reduced (<2 mg l^{-1}, <20 per cent saturation) but the
hypolimnion was not anoxic.

Within the 41 species of rotifers found, several conspicuous patterns of distribution
were noted and are summarized in Figure 16-25, although only a few of the dominant
species are considered. The epilimnetic rotifers tended to be more concentrated near
the littoral regions on both sides or either side of the basin (*Kellicottia longispina, Collotheca mutabilis, Polyarthra vulgaris,* and *P. euryptera*). The dominant epilimnetic *Kellicottia* was most concentrated in the open water. Two major species, *Polyarthra major*
and *Trichocerca similis,* were restricted to the metalimnion, while several others had
distinctly hypolimnetic populations (*Polyarthra longiremis, Keratella hiemalis, Filinia
longiseta*).

Within the population of *Keratella longispina,* Berzins found a distinct separation
according to size. Larger individuals of the population occurred at greater depth, which
suggests a greater sinking rate, although this species has some power of vertical movement, which helps it avoid sedimentation to deeper waters. A similar separation
according to size was found in the vertical distribution of several species of *Polyarthra.*

Certain pelagic rotifers (e.g., *Asplanchna*), when placed in littoral areas, moved
away from the shore towards the open water; others did not demonstrate such horizon-

tal movement (Preissler, 1977). How significant "avoidance of the shore" behavior is in planktonic rotifers is unclear.

In addition to the avoidance of the littoral regions by Cladocera and copepods discussed earlier, a number of detailed analyses have shown a nonrandom horizontal distribution of zooplankton in the pelagic zone of lakes. An example is seen in the intensive 1-day analysis of Berzins (1958) across Skärshultsjön in Sweden (Fig. 16–26). While *Polyphemus* is distinctly littoral, many of the cladocerans and copepods avoid the littoral and are not uniformly distributed across the pelagic zone.

Nonrandom dispersion of phytoplankton and zooplankton within the pelagial zone is in many cases caused by hydrographic water movements. Microcrustacea tend to accumulate in the epilimnion at the leeward end of the basin and to be rare at the windward end whenever a fairly strong wind persists for an appreciable period of time (see, for example, Langford and Jermolajev, 1966). Once at the lee side of the basin, weak locomotion is presumably adequate for most of the animals to maintain their depth distribution. Only under exceptionally strong winds and epilimnetic currents can they be swept back along the lower epilimnion to the windward side.

Part of the nonrandom dispersion of zooplankton is associated with Langmuir convectional helices whose parallel clockwise and counterclockwise rotations in the epilimnion create linear alternations of divergences and convergences (see pp. 106ff and Fig. 7–6). Zooplankton clearly concentrate in these areas of up-welling, which occur midway between zones of convergence (George and Edwards, 1973).

This concentration of zooplankton in linear patches affects the feeding behavior of certain fishes. An excellent example is that of the white bass (*Roccus chrysops*), a pelagic fish that, in its younger stages in summer, is largely planktivorous on *Daphnia*. *Daphnia* were found to be concentrated in the linear convergences (McNaught and

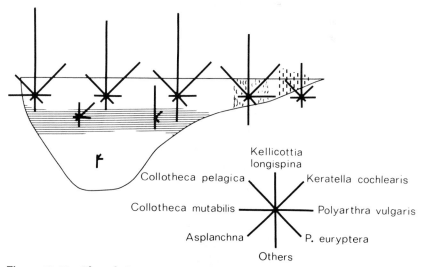

Figure 16–25. The relative quantitative distribution of dominant rotifers in per cent in different strata along a transect across Skärshultsjön, southern Sweden, 7 August 1950. Horizontal shading indicates the position of the thermal discontinuity. (Modified from Berzins, 1958.)

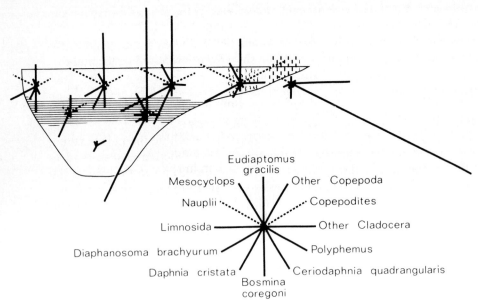

Figure 16-26. The relative quantitative distribution of crustacean zooplankton in per cent in different strata long a transect across Skärshultsjön, southern Sweden, midafternoon, 7 August 1950. (Modified from Berzins, 1958.)

Hasler, 1961). The schooling white bass were able to locate and feed actively on these concentrations of *Daphnia* and also were keyed to feed most actively in the early morning and evening periods, which coincided precisely with the maximal surface concentrations of the vertically migrating *Daphnia*. Similarly, several species of juvenile cyprinid fishes migrate from the littoral to the pelagial zone only during darkness, to feed on zooplankton concentrated in the surface waters by vertical migration (Bohl, 1980; Werner, et al., 1980).

CYCLOMORPHOSIS AMONG THE ZOOPLANKTON AND PREDATION

ROTIFERS

Cyclomorphosis, or seasonal polymorphism, is a phenomenon that is particularly conspicuous among planktonic Cladocera, but is also fairly common among the protozoa, the dinoflagellates (cf. p. 349), and the rotifers. The only comprehensive treatment of the subject of cyclomorphosis among plankton is the detailed résumé of Hutchinson (1967). Although it is more difficult to study the genetic continuity of seasonal forms of the same species of rotifers than of the planktonic crustaceans, cyclomorphosis in several rotifers has been analyzed in some depth.

Seasonal polymorphism is defined by the marked change in shape of some part of an organism in relation to some standard dimension of its size. The common changes in growth form among some rotifers include the following. (a) *Elongation in relation to body width.* In some species of *Asplanchna* midsummer populations can be about five times as long as wide, markedly changed from their nearly spherical morphology in late spring. These elongated forms are nearly always sterile and die back, and do not reappear until the next spring. (b) *Enlargement,* with the formation of body-wall outgrowths or humps. As discussed earlier, at least in *Asplanchna sieboldi,* this seasonal change in growth appears to be caused by tocopherol content of plant-derived food and most likely is an adaptive response to cope with larger sized food in summer. (c) *Reduction in size,* usually at higher temperatures in summer, with a disproportionate reduction in length of lorical spines. This form change is common among several species of *Keratella.* Its adaptive significance is unclear, since reduction in spine length increases sinking speed at higher summer temperatures, and is compensated for only partly by the reduction in body size. The causal mechanisms for the reduction in spine length are unclear. (d) *Production of lateral spines.* The best studied case of this type of cyclomorphosis is in *Brachionus calyciflorus,* in which the posterior spines elongate in the presence of its major predator, the large rotifer *Asplanchna. B. calyciflorus* always has two pairs of anterior spines and one pair of posteromedian spines. A pair of posterolateral spines may or may not be present (Fig. 16–27).

Gilbert (1967b; cf. also Pourriot, 1974) has demonstrated that extension of preexisting anterior and posteromedian spines and the de novo induction of large posterolateral spines are caused by a substance released into the medium by the predatory *Asplanchna.* These form changes cannot be explained by temperature induction or by an allometric growth response in which the substance works indirectly on the spines by influencing body size. Changes in body size were found to be independent of posterolateral spine production. The substance was shown to affect the spineform change only at the egg stage before cleavage. The substance released by *Asplanchna* is a relatively thermolabile proteinaceous compound of unknown composition.

The shape and movements of long posterolateral spines in *B. calyciflorus* significantly decrease predation by *Asplanchna.* Adult *A. girodi,* which captured about 25 per cent of newly hatched, spineless *Brachionus* with which they made direct contact, were completely unable to capture newly hatched, long-spined individuals. Adult *A. sieboldi* can capture nearly 100 per cent of adult, spineless *B. calyciflorus* contacted, but only about 78 per cent of the adult, long-spined forms. The capture rate drops to less than 15 per cent of the spined form by young *A. sieboldi.* The spines, however, provide no protection against predation by the copepod *Mesocyclops* (Gilbert, 1980b).

A number of analyses of natural populations substantiate the fact that the substance released by *Asplanchna* can induce posterolateral spine production in *B. calyciflorus* populations (Gilbert and Waage, 1967; Green and Lan, 1974). Posterolateral spine lengths were found to be unrelated to water temperature, but varied directly both with the density of *Asplanchna* and the presence of threshold quantities of *Asplanchna*-released substance.

The significance of predation upon rotifers varies depending on the species represented in the predator–prey interaction. For example, the predatory copepod *Mesocyclops* feeds effectively on the rotifers *Asplanchna* and *Polyarthra* but not on the roti-

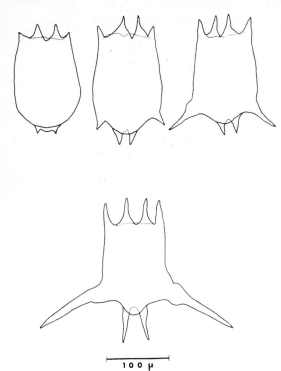

Figure 16-27. Spine variation in a clone of the rotifer *Brachionus calyciflorus* induced by increasing concentrations of a proteinaceous substance produced by its predator *Asplanchna* (From Gilbert, J. J: Asplanchna and posterolateral spine production in *Brachionus calyciflorus*. Arch. Hydrobiol., 64:1–62, 1967b.)

100 μ

fer *Keratella* (Gilbert and Williamson, 1978). *Keratella* is released by *Mesocyclops* after it has been captured, since the predator cannot remove soft parts of this rotifer. *Asplanchna* regularly eats *Keratella*, but cannot capture *Polyarthra* because of its effective escape behavior. The abundance of *Polyarthra* and *Keratella* in natural communities can be greatly affected by interactions with and between *Asplanchna* and predatory copepods.

CLADOCERA

Seasonal polymorphism occurs in the Cladocera perhaps more conspicuously than in any other group. The subject has been studied extensively, especially among the genus *Daphnia* (reviewed in detail by Hutchinson, 1967; cf. also Kerfoot, 1980a).

With increasing water temperatures in the spring, the most common seasonal pattern of variation among successive generations in temperate lakes is a gradual extension of the anterior part of the head to form a crest or helmet (Fig. 16–28). Carapace length, on the other hand, changes little or decreases only slightly between the spring and summer, and then increases again somewhat in autumn. The number of parthenogenetic eggs per individual generally decreases markedly in the strongly helmeted summer forms, although the number of instars increases with higher temperatures. The extent of head development and the change of head shape are extremely variable among species and within the same species under different environmental conditions.

In subtropical and tropical waters, cyclomorphosis is weak or does not occur, except when there is a large temperature change between winter and summer (e.g., Zago, 1976; Mitchell, 1978). An increase in the length of the tail spine also is observed in some species during summer. Cyclomorphosis in the genera *Bosmina*, *Ceriodaphnia*, and *Chydorus* is much less distinct than in *Daphnia*, and consists of slight reductions in length of the body and of the antennule in summer. In *Bosmina*, changes often occur by means of the formation of transparent dorsal humps, with no increase in length; reductions in antennule length and number of segments, and in mucro (caudal spine) length and number of sutures may also occur (Black, 1980a, 1980b; Kerfoot, 1980b).

The mechanisms causing cyclomorphosis in cladocerans have received much study. It is clear that temperature is a primary stimulus affecting the height of the head helmet in *Daphnia*, and that it is effective during the middle of embryogenesis. In a number of species, helmet extension does not occur at water temperatures below 10 to 15°C. Carapace length is maximal in winter, and the ratio of head length to carapace length increases with increasing temperatures. Transfers of animals reared at one temperature to a lower or higher temperature result in a reduction or increase in head length in relation to that of the carapace in later instars. Food supply also affects the specific growth rate and has a relatively greater effect on growth of the head.

Water turbulence also has been shown to be a significant factor in cladoceran

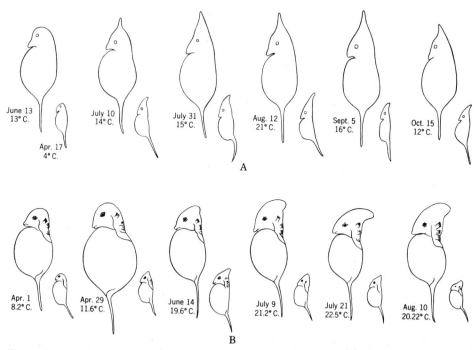

Figure 16-28. Cyclomorphosis in *A*, *Daphnia cucullata* of Esrom Sö, Denmark, and *B*, *Daphnia retrocurva* of Bantam Lake, Connecticut. The small individuals at the right are first instar juveniles drawn to the same scale as the adults. (*A*: From Hutchinson, G. E.: A Treatise on Limnology, Vol. 2, New York, John Wiley & Sons, Inc., 1967; *B*: From Brooks, John Langdon: Cyclomorphosis in *Daphnia*: Ecological monographs, 16:1946. Copyright 1946 by Duke University Press.)

cyclomorphosis (see, for example, Brooks, 1947; Jacobs, 1962). In turbulent cultures and in naturally turbulent environments such as the epilimnion of larger lakes, helmet length relative to that of the carapace is increased significantly in the light in comparison to quiescent situations. Antennal beating rate also increases significantly in turbulent conditions, probably in order to maintain a vertical orientation to light. Helmet development is distinctly more conspicuous in species typically inhabiting epilimnetic turbulent strata (Brooks, 1964). Populations genetically capable of considerable phenotypic variation do not exhibit their extreme phenotype when they live in lower water strata in which growth is slower.

Turbulence is virtually ineffective in cyclomorphic induction in the dark (Jacobs, 1961; Hazelwood, 1966). Temperature and turbulence influence relative growth rates from the beginning of the second half of embryogenesis; food becomes effective immediately after birth (Jacobs, 1980). In some cases, predators can induce cyclomorphic growth in *Daphnia* more effectively than can temperature, turbulence, and oxygen (Grant and Bayly, 1981). Daphnid cladocerans that are warm, well fed, illuminated, and exposed to turbulence, produce the largest helmets. All factors are influential as long as the animals are capable of growth.

A compensatory mechanism exists in daphnid growth: the larger (longer) the helmet resulting from earlier influences, the slower it will grow relative to the body in later instars (Jacobs, 1980). Certain body proportions are maintained regardless of previous environmental conditions. Since there is no single receptor mechanism for the effective environmental factors, one or more hormones likely interact to regulate specific differential growth.

Morphological variability in cladocerans has been ascribed only to phenotypic plasticity. It is important to note that some cyclomorphic species are, in reality, composed of many genetically differentiated species (e.g., Hebert, 1978a, 1978b; Black, 1980a; Kerfoot, 1980b). A number of these species often occur in the same community, and seasonal changes in their relative frequencies can cause polymorphic cycles that mimic true cyclomorphosis. Intraspecific genetic variation in head shape likely occurs as well in some species.

COPEPODS

Seasonal polymorphism among adult copepods is minor in comparison with that found among the parthenogenetic Cladocera and rotifers.The most conspicuous feature is a slight inverse relationship of body size to increasing temperature, so that, in the few cases studied, animals of summer populations tend to be somewhat smaller than animals living in colder seasons.

ADAPTIVE SIGNIFICANCE OF CYCLOMORPHOSIS

Seasonal polymorphism in cladocerans and rotifers has been assumed to be of major adaptive significance to these organisms. Early investigators assumed that cyclomorphic forms would have greater resistance to sinking because the viscosity of the water decreases at higher epilimnetic temperatures. Cyclomorphosis is confined almost completely to species that inhabit the epilimnion, or at least migrate to it daily. Epilimnetic turbulence, however, decreases the small advantage gained from form resistance

(Jacobs, 1967). Furthermore, such resistance would be somewhat disadvantageous to cladocerans swimming during vertical diurnal migration.

The adaptive significance of cyclomorphic growth appears to center on minimizing two aspects of predatory avoidance. (1) Predation of cladocerans and other large zooplankton by non-filter-feeding fish, which is a visual process (see p. 466). Cyclomorphic growth keeps the central portion of the body (thorax, abdomen, and basal portion of the head) that is visible to predators small. Growth continues by expansion of transparent peripheral structures (Brooks, 1965, 1966; Jacobs, 1966). Thus, when size-dependent planktivory is most intense in summer, the central visible portion of the daphnid body is smallest and the organism is less susceptible to being eaten by fish. Cyclomorphomic organisms are, through selection for high rates of growth in less visible peripheral structures, able to maintain assimilation rates during midsummer without the disadvantage of greater visible size when predation is intense. In the winter, when the rate of predation is negligible and turbulence essentially is eliminated under ice cover, evidence indicates that cessation of molting in juveniles and adults results from the lack of a turbulence cue. With the loss of ice cover, even at low temperatures, molting, growth, and egg production immediately begin from the overwintering population of large females. (2) Reduction of capture success by invertebrate predators. Many cladocerans of small size (<1.5 to 2.0 mm) exhibit more cyclomorphic growth than larger species (reviewed by Zaret, 1980). Invertebrate predation by copepods and phantom midge larvae (Chaoborus) is more effective on smaller zooplankton (<2.0 mm).

The lack of cyclomorphosis in copepods in comparison to marked polymorphism in cladocerans and rotifers is likely related to the fundamentally different ways in which these groups have developed different defenses from invertebrate predation (Kerfoot, 1980b). The thoracic legs of copepods are modified for locomotion, while those of cladocerans are modified for filter feeding. Nauplii, copepodite, and adult stages of copepods have evolved precontact defenses, in which they are capable of swift evasive swimming maneuvers (10 to 35 cm sec^{-1}; Strickler, 1975) away from predators. Locomotion in cladocerans by the pair of second antennae results in only feeble, slow escape reactions (e.g., Kerfoot, 1978). Although some rotifers (e.g., Polyarthra) have evasive propelling motions, most rotifer locomotion by ciliary movement is slow. Therefore, the cladocerans and rotifers have evolved postcontact defenses, including (a) larger size, (b) elaboration of spines and other body shapes to frustrate handling by predators, (c) hard, chitinized head shield and carapace, (d) protection of eggs and embryos internally in brood pouches, and (e) reduction of development time of immature stages (in contrast to relatively long naupliar and copepodite stages in copepods).

PREDATION BY FISHES AND SIZE SELECTIVITY

Thus far in this discussion, competitive interactions among zooplanktonic populations have centered on differences in spatial and temporal distributions and on food-size requirements that permit coexistence in the same body of water. The succession and competitive success of zooplanktonic populations can be materially influenced through predation by other zooplankters, a point which was demonstrated earlier among the rotifers and the large cladoceran Leptodora. A number of studies have demonstrated the importance of planktivorous fish in regulating zooplanktonic populations and in causing a distinct shift favoring survival of species that are smaller in size. As noted briefly earlier, Hrbáček and collaborators were the first to demonstrate unequivocally the importance of fish in the regulation of the size and species composition of zooplankters (Hrbáček, 1958, 1961, 1962, 1965; Novotná and Kořínek, 1966). When large, shallow ponds were stocked heavily with cyprinid fish, the zooplankton consisted of small cladocerans, *Bosmina*, *Ceriodaphnia*, and rotifers. Water transparency decreased with the predominant development of small nannoplanktonic algae. Upon selective removal of the fish, the zooplankton changed suddenly to larger cladocerans, particularly *Daphnia longispina*, and rotifer abundance decreased. Transparency increased as the phytoplankton shifted to smaller densities of larger species.

Another series of remarkably simple but definitive studies conducted by these Czechoslovakian workers involved experimental manipulation of portions of the same small lakes by fencing off portions of the littoral from fish and ducks (Straškraba, 1963, 1965, 1967; Poštolková, 1967). Within the protected littoral enclosures, submersed vegetation increased and afforded zooplankton greater protection from fish predation than in the open water. Zooplanktonic biomass increased in the littoral as the small *Chydorus* and other cladocerans were replaced by larger *Daphnia*, *Simocephalus*, and many copepods. Differences could not be related to changes in food supply but rather to changes in fish predation.

All planktivorous fish have closely spaced gill rakers. The example of the alewife of Figure 16–29 shows the closely spaced gill rakers of the first branchial arch of the planktivorous *Alosa pseudoharengus* in comparison to those of the closely related *Alosa mediocris*, which feeds principally on small fish. Studies of the pharyngeal sieves of the planktivorous yellow perch *(Perca flavescens)* and the rainbow trout *(Salmo gairdneri)* demonstrated that the trout and perch could remove few zooplankters of a size smaller than 1.3 mm (Galbraith, 1967; Wong and Wood, 1972; Nilsson and Pejler, 1973). It must be emphasized, however, that all freshwater planktivorous fish examined so far actively search for and visually select each plankter that they ingest (Brooks, 1968; Seghers, 1975). Some fish (e.g., ciscoes, *Coregonus artedii*; bluegills, *Lepomis macrochirus*) gulp water by constantly opening and closing the mouth while swimming (Janssen, 1980; Werner, et al., 1981). These actions ingest aggregations of prey when prey are at high densities. However, in the normal respiratory behavior of passing water through the mouth and across the gills, some zooplankters can be collected by gill rakers. Fish such as the alewife are more or less obligately planktivorous, while others such as the trout and perch eat large zooplankters, and if these are unavailable they readily shift to alternate food sources in the lake. Size selection of prey appears to be characteristic of both the obligate planktivores such as some of the alewives, and those species which feed facultatively on the plankton.

The relation between size selection of prey and foraging efficiency was shown in the bluegill sunfish *(Lepomis macrochirus)*, a common centrarchid of many temperate waters that is known to select prey on the basis of size (Werner, 1974; Werner and Hall, 1974; O'Brien, et al., 1976; Werner, et al., 1981; Mittelbach, 1981). Growth rates are known to increase significantly in direct relation to food size (see, for example, LeCren, 1958; Hall, et al., 1970), and much of this differential growth has been attributed to the efficiency of foraging, that is, the expenditure of energy or metabolic cost in obtaining food relative to the return (cf. Pyke, et al., 1977). In the bluegill, size selection of prey is related to the optimal allocation of time spent searching for and handling the prey. In other words, the size range of prey that maximizes the energy return per unit of energy expended depends on the costs of searching for and handling times of different prey in the environment. Both searching ability and prey-handling efficiency increase with increasing fish size (Mittelbach, 1981). The handling time per prey increases exponentially as the ratio of prey size to mouth size increases (Werner and Mittelbach, 1981). At times of low prey abundance, when search time is long, prey of different size are eaten as encountered. As prey abundance is increased and search time decreases, smaller-sized classes are eaten less frequently or ignored, such that overall return per time increases (see also Brooks, 1968).

The earlier Czechoslovakian observations on the effects of planktivorous fish on the zooplanktonic species composition have been demonstrated excellently by a study of the changes in the zooplanktonic populations of Crystal Lake, Connecticut (Brooks and Dodson, 1965). Prior to the introduction of the alewife into the lake, zooplankton included the large calanoid copepod *Epischura, Daphnia,* and the cyclopoid *Mesocy-*

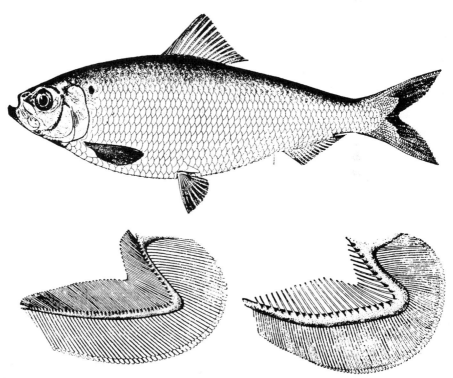

Figure 16–29. The alewife, *Alosa pseudoharengus. Upper:* A mature specimen, 300 mm in length. *Lower left:* First branchial arch with closely spaced gill rakers that act as a plankton sieve. *Lower right:* First branchial arch with widely spaced gill rakers of *A. mediocris,* a species that feeds primarily on small fish. (From Brooks, J. L., and Dodson, S. I.: Predation, body size, and composition of plankton. Science, 150(3692):28–35, 1965. Copyright 1965 by the American Association for the Advancement of Science.)

clops, as well as numerous smaller *Diaptomus* and *Cyclops* copepods (Fig. 16–30). Some 10 years after the alewife invaded the lake, the larger forms of zooplankters had been replaced and dominants included very much smaller forms, the cladoceran *Bosmina* and two cyclopoid copepods *Tropocyclops* and *Cyclops* (Fig. 16–30). The modal length of the numerically dominant forms had shifted from 0.8 to 0.3 mm in the zooplanktonic assemblages. Larger forms were found only in the littoral zone and near the sediments, areas that are avoided by the pelagic *Alosa.*

Similar predatory elimination of larger zooplankters, particularly *Daphnia,* from lakes by other planktivorous fish has been shown. The planktivorous smelt (*Osmerus mordax*) is particularly effective in this regard (see, for example, Reif and Tappa, 1966, and especially Galbraith, 1967). However, the relationship of size and planktivore predation is not always that simple. If prey are large but relatively transparent, such as *Leptodora* or planktonic larvae of the dipteran *Chaoborus,* they often are overlooked and predation is reduced (see, for example, Costa and Cummins, 1972). Furthermore, the slower, steady movements of cladocerans render them more vulnerable to predation than the jerky, irregular movements of copepods.

The importance of visibility in planktivorous predation by fish is demonstrated by the tropical *Ceriodaphnia cornuta,* which shows two distinct polymorphic forms of the same body length within the same lake (Zaret, 1972a, 1972b). One form has pointed, hornlike extensions of the exoskeleton on the head, body, and tail regions with a small area of black pigmentation in the compound eye. The other phenotype is unhorned but possesses a large, pigmented eye. Predation by the dominant planktivore fish, the silverside (Atherinidae; *Melaniris chagresi*) was more intense on the form with the large, pigmented eye. This form of *Ceriodaphnia cornuta* has a superior reproductive potential and a more rapid population growth with greater longevity than that of the horned, small-eyed form. Without fish predation, the large-eyed form can rapidly outcompete the less conspicuous form, but under predation pressure, the form with the small eye, although growing more slowly, can coexist because of reduced visibility to the predator. Similar results were found in the effects of fish predation on *Bosmina* (Zaret and Kerfoot, 1975). Prey selection was found to be related to the large, black pigmented eye, and body size was of negligible importance.

As lakes become more eutrophic, a greater proportion of the phytoplankton biomass and productivity often results from large algae (mostly colonial or filamentous; cf. Chapter 15). The larger algae interfere with food collection to a greater extent in larger cladocerans, causing reduced growth and fecundity, than in smaller cladoceran species that feed on small particles (Gliwicz, 1980; Gliwicz and Siedlar, 1980). In this way, larger cladocerans can experience a reduced efficiency of food collection under eutrophic conditions, and may consequently be selected against. Such interspecific competition, in addition to size-selective predation, could contribute to reduction of larger zooplankton.

When size selection by fish is not in effect and large zooplankters are present, a common observation is that the smaller-sized zooplankton are not generally found to co-occur with the larger forms. Brooks and Dodson (1965) developed the size-efficiency hypothesis in an attempt to explain the commonly observed inverse relationship between the abundances of small- and of large-bodied herbivorous zooplankton in lakes. According to this hypothesis (Hall, et al., 1976): (a) Planktonic herbivores all compete for fine particulate matter (1 to 15 μm) of the open waters. (b) Larger zooplankton do so more efficiently and can also take larger particles. This greater effectiveness of food collection leads to relatively reduced metabolic demands per unit mass, permitting more assimilation to go into egg production by the larger herbivores. (c) Therefore, when predation is of low intensity, the small planktonic herbivores will be competi-

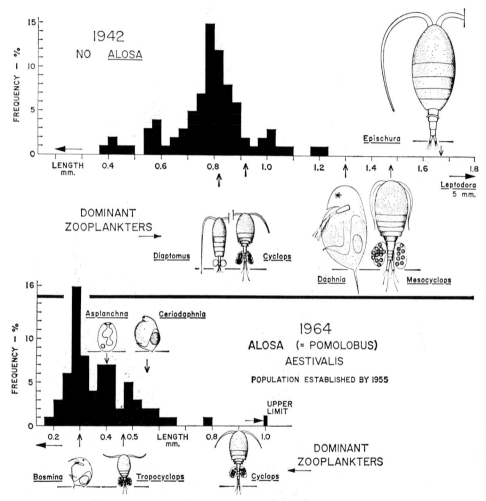

Figure 16-30. The composition of the crustacean zooplankton of Crystal Lake, Connecticut, before (1942) and after (1964) a population of *Alosa aestivalis* became well established. Each square of the histograms indicates that one per cent of the total sample counted was within that size range. Larger zooplankters are not represented because they were relatively rare. The specimens depicted represent the mean size (length from posterior base lines to the anterior end) of the smallest mature instar. The arrows indicate the position of the smallest mature instar of each dominant species in relation to the histograms. The predaceous rotifer *Asplanchna* is the only noncrustacean included in this study. (From Brooks, J. L., and Dodson, S. I.: Predation, body size, and composition of plankton. Science, 150(3692):28–35. Copyright 1965 by the American Association for the Advancement of Science.)

tively eliminated by large forms (dominance of large Cladocera and calanoid cope-pods). (d) When predation is intense, size-dependent predation will eliminate the large forms, allowing the small zooplankton (rotifers, small Cladocera) that escape predation to become the dominants. (e) When predation is moderate, it will, by removing more of the larger species, keep the populations of these more effective feeders sufficiently low so that smaller competitors are not eliminated by competition.

The basic assumptions of the size-efficiency hypothesis imply a complex succes-sional pattern in which optimal body size increases while the range of persisting sizes decreases with decreasing food concentration (critically analyzed by Hall, et al., 1976; Zaret, 1980). Vertebrate predation restricts the maximum adult body size of zooplank-ton, and invertebrate predation may restrict the minimum size. Both of these effects can augment declines of zooplankton productivity caused by food limitations. For many reasons, evidence in support of the size-efficiency hypothesis is meager or contradic-tory. In particular, food limitation and food competition have not been demonstrated to be strongly in effect in natural populations. Food competition is uncommon because of (1) the rarity of periods when zooplanktonic populations can even approach the potential of completely removing particulate material from the water of certain strata by filtration as discussed earlier, and (2) the presence of an excess of algae, bacteria, and particulate detrital organic matter beyond what can ever be consumed (most syn-thesized organic matter is decomposed in nonpredatory pathways; see Chapter 22). Experimental evidence on these food relationships is weak or negative (see, for exam-ple, Sprules, 1972; Dodson, 1974a; Neill, 1975; Gliwicz and Prejs, 1977), but does not exclude possible food-quality interactions.

A much more plausible explanation is again a size-selective predation, but a pre-dation by the larger zooplankters rather than by planktivorous fish. Such predation occurs among the size range of zooplankton that is smaller than can be utilized effec-tively by fish. A large number of studies have shown that invertebrate predation can be sufficient to eliminate certain smaller species, and that this predation, as with some fish, is size-selective (cf. review of Zaret, 1980). Larger copepods, especially, and other predators such as *Chaoborus* and *Leptodora* are particularly effective in causing sig-nigicant mortalities of small zooplanktonic species (see, for example, McQueen, 1969; Anderson, 1970; Confer, 1971; Dodson, 1970, 1972; Confer and Cooley, 1977; Confer and Applegate, 1979; Murtaugh, 1981). While fish planktivory is visual and size-selective, invertebrate predation is more tactile since eyes of rotifers, crustaceans, and most insects detect light intensity and movements, but do not form images. This relationship of size-selectivity and tactile responses in consumption suggests to Dodson (1974b) that shape of prey within the correct size range can influence whether the prey is taken or rejected. Thus cyclomorphic polymorphism would have an adaptive advantage for prey in both invertebrate predation by morphology and vertebrate predation by reduction of visibility while still permitting growth.

It should also be noted that in the absence of intense size-selective predation by fish, large cladoceran species could dominate because they possess greater reproduc-tive rates (intrinsic rates of increase, r) (Goulden, et al., 1978). Even though smaller species mature earlier than larger species, the larger forms have higher fecundities (number of young produced per female per day, b) which results in higher rates of increase.

ZOOPLANKTONIC PRODUCTIVITY

The production rates of specific populations of zooplankton refer to the net productivity or the sum of the growth increments of all specimens of the population. This net productivity excludes maintenance losses (respiration, excretion), and includes the growth increments of the animal itself as well as the biomass produced as gametes and as exuviae during molting. In some larger invertebrates and some vertebrates, productivity values of specific animal populations are influenced markedly by emigration and immigration. The productivity measurements are complicated when predation causes removal of a significant portion of the population; this loss to the population can be difficult to evaluate accurately.

The methods of estimating the rates of production of a specific population demand an accurate evaluation of the distribution of the organisms, the different stages of development and age, and the generation times. All of these parameters vary among species, seasonally, and under changing environmental conditions. These factors require alterations in methods of sampling, analyses of biomass within each age group, and frequency of sampling. Sampling intervals must be kept shorter than generation time of the animals.

The manner in which production rates of a specific population are estimated depends on the particular life cycle, reproductive characteristics, and generation times (discussed at length in Edmondson and Winberg, 1971; Winberg, 1971; Edmondson, 1974; Bottrell, et al., 1976). The productivity of a species with a long life cycle and a short period of reproduction is determined relatively easily if individuals are of the same age, or if separate cohorts can be recognized and their biomass determined. In such a situation, after the brief period of reproduction, no recruitment occurs to confound estimates of growth and mortality. When reproduction and recruitment to the population are continuous, estimation of production is more involved. In this latter case, which applies to most zooplanktonic populations, cohorts overlap, which makes it difficult to separate changes in their temporal abundance and biomass. Consequently, it is necessary to estimate finite birth and growth rates of individuals and evaluate changes in biomass over the life cycle from birth to death. Several simple examples were cited for growth rates of specific cohorts in earlier discussions.

An index of productivity that has become widespread in European and Russian literature in recent years is the P/B coefficient, the ratio of production to biomass (cf. Chapter 8). In populations with a constant age structure and biomass, which are a rarity in nature, the P/B ratio is relatively constant. The ratio varies in most situations when age structure, biomass, and growth are discontinuous and what is obtained is an average ratio over an interval of time. Multiplication of the P/B ratio by biomass yields an estimation of production during that interval of time. This ratio is analogous to the turnover rate, and also its reciprocal, the turnover time, which is the average duration of life of a species under a given set of conditions.

The P/B ratio is an expression of some value in comparing annual productivity among water bodies. However, it masks the large variations that occur within the population dynamics over a year. When used indiscriminately, the ratio relates few of the internal population characteristics or, more importantly, of the causal interrelationships leading to these dynamics and productivity. Over brief periods of time, under relatively

TABLE 16-10 Examples of the Production of Herbivorous and Predatory Zooplankton Communities

LAKE TYPE	PERIOD OF INVESTIGATION	BIOMASS (B)* g m⁻³	BIOMASS (B)* kcal m⁻³	PRODUCTION (P)* g m⁻³	PRODUCTION (P)* kcal m⁻³	BIOMASS TURNOVER TIME (DAYS) AVERAGE	BIOMASS TURNOVER TIME (DAYS) RANGE	REMARKS AND SOURCE
Oligotrophic								
Lake Baikal, USSR	June–July	0.136						Primarily *Epischura*; Moskalenko and Votinsev, 1970
	Sept.	0.43						
	Annual (0–50 m)	—		3.44	—			
Clear Lake, Ontario	Annual	0.20		3.02	16.435	25	12–333	Herbivorous zooplankton (rotifers, *Holopedium*, *Daphnia*, *Bosmina*, *Diaptomus*); Schindler, 1970
Lake 239, Ontario	May–Nov.	—		0.61	3.331	22	10–91	Arctic lake; Winberg, 1970
Lake Krugloe, USSR	Annual		0.405	0.94	5.116	29		
Mesotrophic								
Taltowisko Lake, Poland*	May–Oct.							29% of total lake area in littoral zone; Kajak, et al., 1970; Kajak, 1970
Herbivores		0.12		3.04	25.43	14.3		
Predators				4.68	2.50	10.2		
Lake Naroch, USSR	Annual	0.07	0.38	0.46	6.11	9.1		Winberg, 1970
Lake Krasnoe, USSR	Annual	0.14	0.76	1.12	16.82	22.4		
				3.09		16.5		
Eutrophic								
Lake Mikolajskie, Poland	May–Oct.							Kajak, et al., 1970; Hillbricht-Ilkowska, et al., 1970
Herbivores				6.45	35.09	9.2	4.0–12.5	
Predators				1.32	7.18	25.0		
Lake Sniardwy, Poland	May–Oct.							
Herbivores				3.08	16.78	14.9	8.3–33	
Predators				0.50	2.71	14.3		
Kiev Reservoir, USSR								Winberg, 1970
Herbivores		0.35	1.9	9.15	49.8	13.9		
Predators		0.022	0.12	1.16	6.3			
Severson Lake, Minnesota	May–Oct.							Comita, 1972
Herbivores				2.51	13.66			
Predators				0.11	0.60			
North Pine Dam, Australia	Annual (mean 3 years)							King, 1979
Rotifers		—		—	1.42	—		
Crustaceans		—		—	2.60	—		
Total Zooplankton		—		—	3.87			
Dystrophic (high dissolved organic matter)†								
Lake Flosek, Poland	May–Oct.							Sphagnum bog; high littoral and allochthonous organic inputs; Kajak, et al., 1970; Hillbricht-Ilkowska, et al., 1970
Herbivores				25.68	139.7	6.3		
Predators				1.16	6.3	25.0		

*Estimated using the mean caloric value for microconsumers (Cummins and Wuycheck, 1971).

†See discussion in Chapter 24.

TABLE 16–11   Correlations between Biomass and Production of Major Components in Fresh Waters

COMPONENT	SAMPLE NUMBER (N)	CORRELATION COEFFICIENT (r)
Phytoplankton	27	0.77
Herbivorous zooplankton	26	0.84
Carnivorous zooplankton	25	0.84
Herbivorous benthic organisms	15	0.61
Carnivorous benthic organisms	14	0.72

After data of Brylinsky (1980). All values significant at the 99% level.

steady environmental conditions, the turnover rates reflect relative growth of the populations.

Many studies have estimated zooplanktonic productivity. The number of estimates is increasing as the results of the International Biological Program are summarized (cf. Morgan, et al., 1980). A few examples will illustrate the ranges encountered in lakes of differing primary productivity (Table 16-10).

In general, a positive correlation exists between the rates of production of phytoplankton and of the zooplankton (Table 16-11; see, for example, Winberg, et al., 1970; Makarewicz and Likens, 1979; Brylinsky, 1980). However, this statement does not necessarily imply that the herbivorous zooplankters are consuming the algae directly in proportion to their biomass and growth. As was discussed earlier, size, quality, and other factors influence both the ingestion and assimilation of algae. Moreover, much of the algal production enters the detrital pathways of nonpredatory particulate and dissolved organic matter (cf. Chapter 22).

In a detailed and perceptive study of the relationships between phytoplankton and zooplankton of mesotrophic Lake Erken, southern Sweden, Nauwerck* (1963) demonstrated that algal productivity was inadequate to support the herbivorous zooplanktonic productivity. The filter-feeding zooplankton, which accounted for a total productivity of 0.22 g m^{-2} day^{-1} on an annual mean basis, consisted of primarily *Eudiaptomus* (60 per cent), with a turnover time of about 36 days (annual mean; 24 days during the ice-free period). Utilization of bacteria and particulate detrital organic matter was demonstrated in numerous subsequent studies, even though the caloric content of the detrital material is relatively low. The result, then, is a considerably modified trophic pyramid in which the transfer energy to higher food levels is short-circuited, to use the words of Nauwerck, via the bacteria and particulate detrital organic matter (Fig. 16–31). In similar fashion, the greater part of the herbivorous zooplanktonic productivity is not utilized by predatory zooplankton and fish, but instead enters the nonpredatory detrital pathways of decomposition and is partially reutilized by filter feeders as bacteria and detritus. Therefore, Lindeman's classic conceptualization of the generalized lacustrine food-cycle relationships must be modified to emphasize the importance of

---

*Certain computational errors have been found in the original paper, but these do not alter the general conclusions.

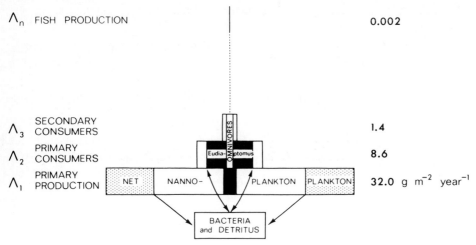

Figure 16-31. Diagrammatic representation of the trophic relationships in Lake Erken, Sweden. (Modified from Nauwerck, 1963.) Productivity estimates in g dry weight per m² per year of pelagic zone only.

nonpredatory detrital and bacterial pathways (Fig. 16–32); these pathways were underestimated because inadequate data were available when the cycle was formulated (cf. Rich and Wetzel, 1978). This change does not greatly alter the fundamental contribution of Lindeman, but instead properly emphasizes the major contributions of bacteria and detritus in the trophic relationships of primary consuming animals, and major losses of organic matter through decomposition that occur without animal consumption (cf. Chapter 22).

Other comparisons of the proportions of herbivorous zooplankton utilized by predators demonstrate similar results. In detailed studies of nine very different lakes in the Soviet Union, the biomass of predators varied from 6.5 to 52 per cent of the total biomass of zooplankton, averaging about 30 per cent (Winberg, 1970). The productivity of the filter-feeding zooplankters is distinctly higher than that of the predacious zooplankters (Table 16–12). Variation is great, however, particularly in relation to temperature, food quantity and quality, and other factors. Comparisons over two or more years with good sampling show the extent of variation that can occur with the same species in the same lake (Table 16–12; Figure 16–33). These differences are much greater than can be attributed to sampling and measurement errors. Perhaps the lowest production rates of zooplanktonic populations in a permanent lake are those of polar Char Lake at latitude 74° in the Canadian Arctic (Rigler, 1974). The dominant zooplankters *Limnocalanus macrurus* and *Keratella* produced 261 mg ash-free dry weight per m² per year and 3 mg m⁻² yr⁻¹, respectively. This is equivalent to an annual average of about 0.00006 g m⁻³ day⁻¹, an exceedingly low value.

In the particular case of *Daphnia hyalina* in a eutrophic reservoir (Fig. 16–33), maximum biomass was found nearly two months before highest numerical density in 1970, but numerical and biomass maxima were coincident the following year (George and

Edwards, 1974). Maximum spring biomass occurred during blooms of green algae in late April and early May, when many *Daphnia* survived and grew for some time after reaching maturity. Increases in production coincided with, or lagged somewhat behind, increases in biomass. In the second year, the production was about one-half that of the first year, and the average biomass turnover time also decreased (Table 16–12).

From an intensive analysis of the zooplankton productivity of oligotrophic Mirror Lake, New Hampshire (Makarewicz and Likens, 1979), two major relationships emerged. (a) Rotifers are a major component in energy transfer and intrasystem

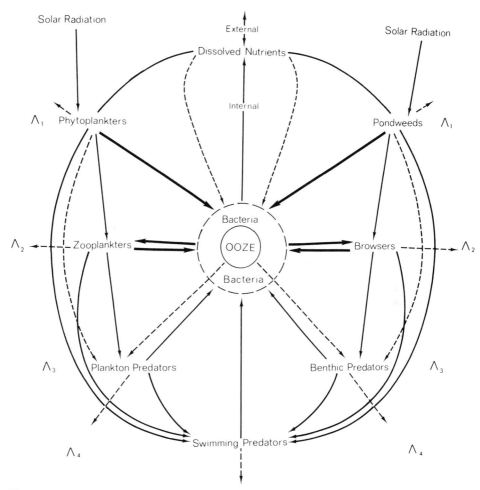

Figure 16–32. Generalized lacustrine food-cycle relationships after Lindeman (1942). Added darkened arrows indicate pathways ignored or underestimated in Lindeman's actual calculations, a consequence of accepting existing data on chemical assay of crude fiber in place of direct assimilation or egestion measurements. (From Wetzel, R. G., Rich, P. H., Miller, M. C., and Allen, H. L.: Metabolism of dissolved and particulate detrital carbon in a temperate hard-water lake. Mem. Ist. Ital. Idrobiol., 29: Suppl., 185–243, 1972.)

TABLE 16-12 Examples of Productivity of Herbivorous and Predatory Forms of Zooplankton

TYPE/SPECIES	LAKE/GENERAL PRODUCTIVITY	PERIOD OF INVESTIGATION	PRODUCTION ESTIMATES* g m^{-3} DAY^{-1}	kcal m^{-2} DAY^{-1}	BIOMASS TURN-OVER TIME (DAYS) AVERAGE	RANGE	SOURCE
Filter Feeders							
Cladocera							
Daphnia hyalina	Eglwys Nynydd Reservoir, Wales; eutrophic	Annual, 1970 Annual, 1971	0.57 0.32		21.3 15.9	3.8–333	George and Edwards, 1974
D. parvula	Severson Lake, Minn.; eutrophic	Annual	0.010	0.102	—		Comita, 1972
D. galeata mendota	Sanctuary Lake, Pa.; eutrophic reservoir	May–Nov., 1966 May–Nov., 1967	0.407 0.030	3.026 0.223			Cummins, et al., 1969
	Canyon Ferry Reservoir, Mont.; eutrophic	April–Sept.	0.114		10.0		Wright, 1965
D. schoedleri	Canyon Ferry Reservoir, Mont.; eutrophic	April–Sept.	0.227		6.7		Wright, 1965
D. longispina	Lake Sevan, southern USSR	Annual	0.006		58.9		Meshkova, 1952, in Winberg, 1971
Bosmina longirostris	Severson Lake, Minn.	Annual	0.007	0.071	—		Comita, 1972
B. longirostris and *B. coregoni*	Sanctuary Lake, Pa.	May–Nov., 1966 May–Nov., 1967	0.183 0.067	1.361 0.498			Cummins, et al., 1969
Ceriodaphnia reticulata		July–Nov.	0.031	0.154			
Chydorus sphaericus		July–Aug., 1966 July–Aug., 1967	0.004 0.047	0.020 0.233			
Cladocera	Naroch Lake, USSR	May–Oct.	0.0026	0.117	13.7		Winberg, et al. 1970
	Myastro Lake, USSR	May–Oct.	0.015	0.403	10.9		
	Batorin Lake, USSR	May–Oct.	0.033	0.484	10.5		
Copepods							
Cyclops strenuus	Buttermere, England; oligotrophic	Annual	0.0004				Smyly, 1973
	Rydal Water, England; eutrophic	Annual	0.0005				
	Grasmere, England; eutrophic	Annual	0.0006				
	Esthwaite Water, England; eutrophic	Annual	0.0017				
	Lake Sevan, southern USSR	Annual	0.0007		79.3		Meshkova, 1952, in Winberg, 1971
C. vicinus	Eglwys Nynydd Reservoir, Wales; eutrophic	Annual, 1970 Annual, 1971	0.13 0.12		22.5 19.4		George, 1976
Calamoecia lucasi	Lake Otota, New Zealand; oligotrophic	Annual	0.006		11.3		Green, 1976

Species	Location	Period				Reference
Boeckella dilatata	Lake Hayes, New Zealand; eutrophic	Annual	0.032		6.4	Burns, 1979
Eudiaptomus graciloides	Naroch Lake, USSR	May–Oct.	0.0010	0.044	24.7	Winberg, et al., 1970
	Myastro Lake, USSR	May–Oct.	0.0065	0.174	20.2	
	Batorin Lake, USSR	May–Oct.	0.0070	0.104	15.4	
	Esrom, Denmark	May–Oct.	—	0.10	6.8	Bosselmann, 1975
Mesocyclops edax	Severson Lake, Minn.	Annual	0.0046	0.045	—	Comita, 1972
Diaphanosoma leuchtenbergianum			0.0067	0.067	—	
Diaptomus siciloides	Waldsea Lake, Saskatchewan; mesotrophic	Annual	0.0342	0.341	—	Swift and Hammer, 1979
D. connexus			0.29	—	—	
Arthodiaptomus (2 species)	Lake Sevan, southern USSR	Annual	0.0014	—	162	Meshkova, 1952, in Winberg, 1971
Rotifers						
Keratella quadrata	Severson Lake, Minn.	Annual	0.0021	0.021	—	Comita, 1972
K. cochlearis			0.0007	0.0074	—	
Filinia longiseta			0.0011	0.0112	—	
Brachionus sp.			0.0075	0.0752	—	
Polyarthra sp.			0.0010	0.0103	—	
Rotifers	Naroch Lake, USSR	May–Oct.	0.0027	0.120	1.7	Winberg, et al., 1970
	Myastro Lake, USSR	May–Oct.	0.0024	0.065	3.9	
	Batorin Lake, USSR	May–Oct	0.0105	0.156	2.6	
Predatory Feeders:						
Cladocera						
Leptodora kindtii	Sanctuary Lake, Pa; eutrophic, shallow	May–Nov., 1966	0.003	0.022	—	Cummins, et al., 1969
	Lake George, NY; deep oligotrophic	May–Nov., 1967	0.013	0.097	—	LaRow, 1975
		June–August	0.006	—	—	
Cladocera	Naroch Lake, USSR	May–Oct.	0.0003	0.013	11.3	Winberg, et al., 1972
	Myastro Lake, USSR		0.0009	0.023	3.9	
	Batorin Lake, USSR		0.0002	0.003	10.8	
Copepods						
Cyclops sp.	Naroch Lake, USSR	May–Oct.	0.0008	0.034	9.7	
	Myastro Lake, USSR		0.0023	0.062	19.4	
	Batorin Lake, USSR		0.0094	0.140	14.2	
Rotifers						
Asplanchna priodonta	Naroch Lake, USSR	May–Oct.	0.0014	0.061	2.9	
	Myastro Lake, USSR		0.0061	0.163	2.5	
	Batorin Lake, USSR		0.0105	0.156	3.2	
Asplanchna sp.	Severson Lake, Minn.	Annual	0.0031	0.031	—	Comita, 1972
Synchaeta sp.			0.00009	0.0009	—	
Insect larvae						
Chaborus punctipennis			0.0001	0.001	—	

*Conversions estimated using the caloric mean of microconsumers (Cummins and Wuycheck, 1971), when mean depths available.

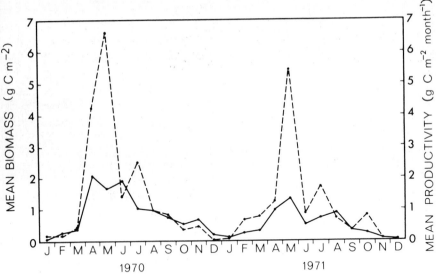

Figure 16–33. Seasonal monthly mean biomass (———) and productivity (- - - - -) of *Daphnia hyalina* in g C m^{-2} month^{-1}, calculated from daily measurements, of Eglwys Nyndd, Great Britain. (Drawn from data of George and Edwards, 1974.)

nutrient cycling. The major importance of rotifers in the transfer of energy is evident in a number of lakes of varying trophic status, from oligotrophic to eutrophic (Table 16–13). One third of the annual phosphorus flux through the zooplankton of Mirror Lake was by the rotifers. (b) The productivity of the different species of zooplankton, shown in Figure 16–34, indicates how the interacting populations are structured. The population positions or niches are dispersed in relation to complex gradients of time, depth, food, reproduction, and predation (cf., for example, Zaret, 1975, 1980; Kerfoot and Pastorak, 1978; McNaught, 1978; Seitz, 1980; Kerfoot, 1980a).

Separation of the niche hyperspace with relatively small regions of species overlap minimizes interspecific competition. Throughout the discussion in this chapter, emphasis has been on the general similarities and often subtle but important differences and

TABLE 16–13 Percentage of Production of Rotifer, Cladoceran, and Copepod Zooplankton in Various Lakes

LAKE	ROTIFERS	CLADOCERA	COPEPODS
Mirror, New Hampshire	39.8	40.9	19.3
Krivoe, USSR	15.2	36.1	48.7
Krugloe, USSR	19.2	71.8	8.9
Naroch, USSR	43.5	31.2	25.2
Myastro, USSR	23.9	47.6	28.6
Batorin, USSR	29.3	45.5	25.1
239, Ontario	67.2	5.4	27.4

After data of Makarewicz and Likens (1979), Alimov, et al. (1970), Winberg, et al. (1970), and Schindler (1970).

separations that occur in feeding, food utilization, reproduction, growth, and behavioral characteristics, the temporal (daily, seasonal) and spatial differences in distribution and growth, seasonal changes in morphology, and various characteristics of predator avoidance and predation effects. All of these factors contribute to the large diversity of population interactions that have evolved to permit coexistence in limnetic zooplankton communities.

The efficiency of energy transformations between components of animal populations is an area of intensive investigation. At present, few cases have been studied in detail in natural populations; much of the necessary information on ingestion, respiration, and egestion is determined under controlled laboratory conditions that have been applied variously to natural populations.

A certain percentage of the food ingested is not utilized. As a percentage, this egestion rate varies greatly with age, food quantity and quality, and other factors within an observed range of about 30 to 90 per cent loss (Table 16–14). How much of this egested detrital material is reingested is unclear, at least among the pelagic organisms. Consumption of feces after varying periods of colonization and decomposition by bacteria and fungi is a fairly common phenomenon among benthic fauna of lakes and streams.

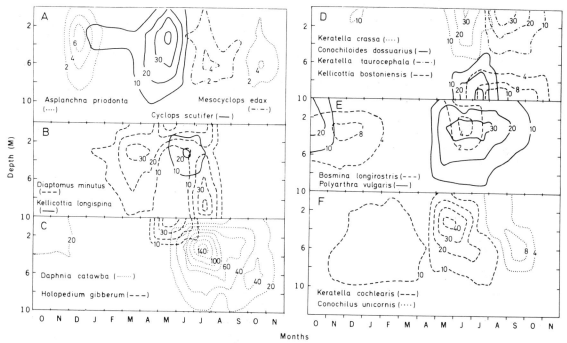

Figure 16–34. Productivity response surfaces (from mean production values, μg dry weight l⁻¹ month⁻¹) of zooplankton populations, Mirror Lake, New Hampshire.
A, Predators: Rotifer *Asplanchna* and two copepods; B, Macroconsumer herbivores: Rotifer *Kellicottia* and copepod *Diaptomus*; C, Microconsumer herbivores: Cladocerans; D, Microconsumer herbivores: Rotifers; E, Microconsumer herbivores: Rotifer *Polyarthra* and small cladoceran *Bosmina*; F, Microconsumer herbivores: Rotifers. (From Makarewitz, J. C. and G. E. Likens. Niche analysis of a zooplankton community. Science 190(4218):1000–1003, 1975. Copyright 1975 by the American Association for the Advancement of Science.)

TABLE 16-14 Exemplary Estimates of Efficiencies of Food Utilization by Various Animals

ORGANISMS	% OF INGESTED FOOD UTILIZED IN:				% OF ASSIMILATED ENERGY EXPENDED IN:		SOURCE
	EGESTION	ASSIMILATION	RESPIRATION	GROWTH AND REPRODUCTION	RESPIRATION	GROWTH AND REPRODUCTION	
Zooplankton							
Daphnia pulex	69–86	14–31	4–14	10–17	27–44	56–73	Richman, 1958
Daphnia ambigua	71.8	28.2	17.1	8.4	61	39	Lei and Armitage, 1980
Ceriodaphnia reticulata	—	10.6	1.8	—	—	—	Czeczuga and Bobiatyńska-Ksok, 1970
Simocephalus vetulus Juveniles (♀)	27.6	72.4	19.5	52.9	26.9	73.1	Klekowski, 1970
Reproducing (♀)	68.3	31.7	11.2	20.5	35.3	64.7	Cummins, et al., 1969; Moshiri, et al., 1969
Leptodora kindtii	60.0	40.0	—	—	92.7	7.3	Klekowski and Shushkina, 1966
Mesocyclops albidus	—	20–75	approx. 20	approx. 25	approx. 50	approx. 50	Comita, 1972
10 Herbivores	52.4	47.6	40.1	7.5	71–82	18–29	Klekowski, 1970
Benthic Animals							
Asellus aquaticus	69.7	30.3	24.7	5.6	81.8	18.2	Klekowski, et al., 1970
Lestes sponsa	63.4	36.6	13.2	23.5	36.0	64.0	
Fish							
Phytophagous carp, Ctenopharyngodon	86.0	14.0	12.2	1.9	86.0	14.0	Fischer, 1970
Predatory perch, Perca fluviatilis	64.2	35.8	16.2	19.5	45.5	54.5	Klekowski, et al., 1970

Methods of analysis and experimental conditions vary greatly and are comparable only approximately.

The rates of assimilation and respiratory loss for biochemical maintenance also vary widely with environmental conditions (Table 16–14). Assimilation efficiency is generally greater among young animals and decreases with age; average values, mostly from experimental conditions better than those in situ, are in the range of 10 to 50 per cent. Respiratory costs are high, and, as would be anticipated, increase greatly relative to assimilation rates under adverse conditions. For example, the population respiration of the dominant zooplankter of the Canadian arctic Char Lake, with a maximum of about 4°C for one month of the year, far exceeded net production (Table 16–15). The ratio of production to assimilation was about 13 per cent.

Of the production that is utilized for growth and reproduction, usually a small portion is utilized in the diapause stages such as in overwintering eggs. Most of the production enters the nonpredatory pool of detrital organic matter and is decomposed by microorganisms. A variable proportion is utilized by predators. The percentage of the primary consumers to appear as net production of secondary consumers also varies, but generally averages around 10 per cent or less (e.g., Comita, 1972) on an annual basis. As was discussed earlier, however, at certain periods of the year, the impact of herbivores and predators can be quite significant, and they can literally decimate populations of the primary producers.

In discussion of the roles of zooplankton in the general trophic-level concept (cf. Chapter 8), it should be recalled that trophic levels consist of population groups of different species of organisms that utilize the same mode of nutrition, and many of the species change their mode of feeding to another trophic level in the course of their life cycles. As was emphasized earlier in this chapter, animal nutrition is strongly dependent upon food density, quality, and many other factors. In turn, the efficiency of food utilization for growth and maintenance varies with the rate of food consumption and population densities. These efficiencies are only reasonably accurate when the populations studied are in a fairly fixed state of equilibrium (Slobodkin, 1960); however, this condition rarely exists in natural populations, since they frequently undergo large fluctuations.

While under controlled conditions, the metabolic efficiencies of food utilization by particular species have rigorous meanings from a physiological point of view, extrap-

TABLE 16–15 Mean Production and Respiration of the *Limnocalanus macrurus* Population in Polar Char Lake, Canadian Arctic, Over a 4-Year Period

PRODUCTION AND RESPIRATION	MEAN mg m^{-2} YEAR^{-1}	% OF TOTAL
Growth	220	11.1
Eggs	5.6	0.3
Exuviae	34.9	1.8
Total production	261	13.1
Respiration	1660	83.4
Assimilation	1990	

After data of Rigler, et al., 1974.

TABLE 16-16 Utilization and Transfer of Energy from Trophic Levels of Several Freshwater Systems, Expressed in g cal cm^{-2} Year^{-1}

PARAMETER*	LAKE MENDOTA, WISC.				TEMPERATE COLD SPRING			SILVER SPRINGS, FLA.				CEDAR BOG LAKE, MINN.			SEVERSON LAKE, MINN.		
	Λ_1	Λ_2	Λ_3	Λ_4	Λ_1	Λ_2	Λ_3	Λ_1	Λ_2	Λ_3	Λ_4	Λ_1	Λ_2	Λ_3	Λ_1	Λ_2	Λ_3
Ingestion (I)	118,872	55.4	3.4	0.3	1,095,000	—	—	1,700,000	—	—	—	118,872	19.6	3.4	89,586	57.7	—
Assimilation (A)	501.7	44.3	3.1	0.3	710	2318	242	20,810	3368	383	21	120	16.8	3.1	581.4	27.4	1.2
Egestion (nonassimilation)	118,370	11.1	0.3	0.0	1,094,290	—	—	1,679,190	—	—	—	118,752	2.8	0.3	89,005	30.2	—
Respiration (R)	125.3	16.9	1.8	0.2	55	1746	89	11,977	1890	316	13	30	6.4	1.8	309.1	23.1	0.8
Net productivity (A-R)	376.4	27.4	1.3	0.1	655	576	155	8833	1478	67	6	90	10.4	1.3	272.4	4.3	0.4
Production	55.4	3.4	0.3	0.0	(655)	208	—	—	—	—	—	19.6	3.4	0.0	57.7	—	—
Decomposition losses	321.0	24.0	1.0	0.1	—	—	—	—	—	—	—	70.4	7.0	1.3	—	—	—

From data of Juday (1940), Teal (1957); Odum (1957), and Lindeman (1942), as interpreted by Kozlovsky (1968), and from Comita (1972). The lake systems are planktonic only and do not consider littoral and allochthonous inputs.

*I = insolation for producers or food ingestion for organisms. Egestion includes nonutilized light and egestion. R includes all energy losses; also urine in higher organisms. Production refers here to that portion of productivity passed on to next trophic level, some of net productivity being dissipated in decomposition, tissue accumulation, and loss from the system.

TABLE 16-17 Phytoplankton Photosynthetic Efficiency and Secondary Production Energy Transfer Efficiencies

COMPONENT	SAMPLE SIZE (N)	MEAN PERCENTAGE EFFICIENCY	RANGE OF EFFICIENCIES (%)
Phytoplankton	93	0.34	0.002–1.0
Herbivorous zooplankton	27	7.1	0.10–27.4
Carnivorous zooplankton	24	1.2	0.17–5.0
Herbivorous benthic animals	17	2.3	0.16–11.1
Carnivorous benthic animals	16	0.3	0.35–1.8

After data of Brylinsky (1980).

olation to the complex dynamics of natural systems can only be done in a general way. The evaluation of ecological efficiencies among trophic levels is really only a theoretical way of viewing the interrelationships among organisms. Such an evaluation loses precision as soon as it is extended beyond measurements of solar radiation in studies of primary productivity. Such analyses can be instructive, however, if one can keep in mind the limitations imposed by their simplicity. By way of examples, five freshwater systems are compared in Table 16–16, based on mostly incomplete, rather limited data. In a detailed analysis of the numerous types of efficiency relationships developed by many workers, Kozlovsky (1968) used most of these limited data to illustrate changes taking place in the passage from one trophic level to the next. Some of his conclusions follow, the most important of which is that when the net productivity and assimilation of one level are compared to the net productivity and assimilation of the previous level, the efficiency of transfer is practically constant at about 10 per cent (Table 16–17). In more physiologically based comparisons of food utilization in animals in relation to food ingested, assimilation and respiration increase at higher trophic levels. It then follows that in relation to ingestion and assimilation in animals, production decreases at higher trophic levels.

_____ **SUMMARY**

1. Freshwater zooplankton are dominated by four major groups of animals: protozoa, rotifers, and two subclasses of crustaceans, the cladocerans and copepods.

2. Little is known about the population dynamics and productivity of planktonic protozoans. Under certain circumstances, flagellate, rhizopod, and ciliate protozoans make up a substantial component of pelagial zooplankton communities.

   a. Many pelagial protozoa are meroplanktonic, in that only a portion, usually in the summer, of their life cycle is planktonic. These forms spend the rest of their life cycle in the sediments, often encysted throughout the winter period.

b. Many protozoans feed on bacteria-sized particles, and thereby utilize a size class of bacteria and detritus generally not utilized by larger zooplankton.

3. Although most rotifers are sessile and are associated with the littoral zone, some are completely planktonic; these species can form major components of the zooplankton.

a. Most rotifers are nonpredatory, and omnivorously feed on bacteria, small algae, and detrital particulate organic matter. Most food particles eaten are small ($< 12\ \mu m$ in diameter).

b. A few rotifers are predatory on protozoa, rotifers, and small crustaceans, and can alter their size in response to changes in food size.

4. Most cladoceran zooplankton are small (0.2 to 3.0 mm) and have a distinct head; the body is covered by a bivalve carapace. Locomotion is accomplished mainly by means of the large second antennae.

a. Most cladocerans feed on particles filtered from the water by means of setae and hairs on five pairs of legs. Particles collect and move in a ventral food groove towards the mouth.

b. A few cladocerans are predaceous, and seize other zooplankton and detrital particles with prehensile legs.

5. Planktonic copepods consist of two major groups, the calanoids and the cyclopoids. These two groups are separated on the basis of body structure, length of antennae, and legs (Fig. 16–4; Table 16–1).

a. Cyclopoid copepods are raptorial; they seize food particles and draw them to the mouth. Many cyclopoids are carnivorous on other zooplankton; some are herbivorous on a variety of unicellular and filamentous algae. Locomotion is by movement of appendages, which results in short, rapid swimming movements.

b. Calanoid copepods swim more continuously in rotary motions, which set up currents that carry particles to modified structures of the maxillae. Particles that are retained are governed by the capture efficiency of the maxillae, and differ among calanoid species.

6. Filtration of particles is the dominant means of food collection by rotifers and cladocerans.

a. *Filtration* or *grazing rate* is the volume of water containing particles that is filtered by the animal in a given time, regardless of whether or not the particles are retained or ingested. *Feeding rate* is the quantity of food ingested in a given time.

b. Many zooplankton possess some capacity for selective feeding.

i. Among suspension-feeding rotifers, food selectivity occurs in some species by several food-rejection mechanisms (screening of particles by cilia, or rejection of particles once ingested).

ii. Among cladocerans, the size of particles cleared from the water is a function of the morphology of the setae of the moving appendages. Feeding rates commonly stabilize or decrease as concentrations of food particles increase.

    c. Filtering rates tend to increase with both increasing body length and increasing temperatures. Size of particles ingested is generally proportional to body size.

    d. The effectiveness of zooplankton grazing varies greatly seasonally and among lakes. Throughout much of the year, zooplankton grazing only filters a small proportion of the water volume (less than 15 per cent per day). At certain times of the year, grazing can remove large portions of the phytoplankton and can cause marked reduction in phytoplankton productivity.

    e. Algal species succession can also be altered by intensive, selective (usually size-specific) grazing and concommitant regeneration of nutrients. Certain algae can survive gut passage and their growth can be enhanced by contact with high nutrient levels within the gut of zooplankton.

7. Assimilation efficiency is variable among zooplankton species, but is usually less than 50 per cent.

    a. Efficiency of assimilation increases somewhat with higher temperatures and decreases markedly with increasing food concentrations.

    b. Food quality also influences assimilation efficiencies. Rates of assimilation are low when zooplankton are feeding on detritus particles, higher with bacteria, and generally highest when they are feeding on algae of acceptable size and type.

8. Reproductive rates and life histories of zooplankton are diverse.

    a. Rotifers and most cladocerans reproduce by diploid female parthenogenesis for many successive generations during the main growing season. Under certain conditions of environmental stress, meiosis occurs, which leads to the production of haploid males and females. The fertilized eggs (resting eggs) are diploid, and are encased in a heavy cell wall. These resting eggs often enter diapause, and hatching may be delayed for long periods. Under favorable conditions, resting eggs develop into amictic diploid females.

    b. A number of environmental conditions and mechanisms influence the production of sexual mictic females. Reductions in temperature, reduced food availability under crowded population conditions, and reduced availability of various dietary compounds (e.g., vitamin E) can induce sexual production in certain rotifers.

    c. Rates of growth and development of instars in Cladocera are generally proportional to temperature, and are often related to food availability. Food supply is, however, of secondary importance to temperature; increased food supply usually increases the rate of population development by increasing fecundity (number of eggs per brood).

    d. Copepods are bisexual. Fertilized eggs are carried externally in egg sacs.

        i. Temperature determines the rate of egg production, and food availability directly influences clutch size.

    ii. Copepod eggs hatch into small, free-swimming larvae (= *nauplii*) and molt successively through six naupliar stages. Nauplii then enlarge and form *copepodites,* and molt through five additional instars before forming an adult. Development time to complete all of the immature stages is far more extended than in the rotifers and Cladocera, and varies greatly among species and under different environmental conditions.

9. Analyses of the rates of egg production and the dynamics of instar populations yield information on growth rates, birth rates, mortality, and duration of generations.

    a. The seasonal distribution of planktonic rotifer populations is complex. Many perennial species exhibit maximal densities in early summer in temperate lakes. Other species are more seasonal, and include (i) cold stenotherms with greatest populations in winter and early spring, and (ii) species that develop two or more maxima during summer. Certain rotifer species also tolerate low oxygen concentrations, and populations may be separated spatially as well as seasonally.

    b. The seasonal succession and population dynamics of Cladocera are also variable among species and under different lake conditions.

        i. Some species are perennial (overwinter as adults rather than as resting eggs) and may exhibit one, two, or more irregular maxima.

        ii. Aestival species have a distinct diapause (resting egg stage), and usually develop one population maximum in the spring, and occasionally a second maximum in the late summer or autumn.

        iii. Temperature increases in the spring enhance rates of molting and brood production; increased food supply increases egg number per brood. Shifts in food quality and availability (often size-restrictive), and especially predation by invertebrates and fishes, can cause rapid changes in populations of cladocerans. Mortality by predation is normally highest in the summer.

    c. The life cycle of limnetic cyclopoid copepods is separated into two periods: (i) a period of growth followed by (ii) a period of retarded growth or diapause induced by a combination of decreasing water temperatures, photoperiod, reduced food availability, anoxia, and increased predation. The extent to which the population enters a resting stage, and the instar stage at which diapause occurs, is extremely variable. Similarly, the site where the resting stage occurs (water vs. sediment) and diapause duration, can vary greatly from lake to lake in arctic, alpine, and temperate areas. In tropical lakes, resting stages have not been observed. The seasonal life cycles of the calanoid copepods are usually longer than those of the cyclopoids.

        i. Most cyclopoid copepods are carnivorous and influence the population dynamics of the other copepods by predation. They influence their own dynamics by cannibalism, especially of juveniles.

        ii. Competition between copepods is reduced by (a) variations in timing and duration of diapause, (b) seasonal and vertical (spatial) separation, and (c) differences in utilization of available food particles.

10. Many zooplankton, particularly the Cladocera, exhibit marked diurnal vertical migrations.
    a. Most species migrate upward from deeper strata to more surficial regions as darkness approaches, and return to the deeper areas at dawn.
    b. Rates of movement over distances in excess of 50 meters in certain clear lakes are variable from less than 2 m per hour among rotifers to greater than 20 m per hour by certain Cladocera and copepods.
    c. Light intensity is the primary stimulus of vertical migration.
    d. Grazing rates of suspension feeders are usually several times greater during the dark period when they have migrated to surface strata.
    e. The adaptive significance of diurnal migrations is unclear but likely evolved as a mechanism to avoid predation by fish, much of which is a visual process requiring light.

11. The horizontal spatial distribution of zooplankton in lakes is often uneven and patchy.
    a. Pelagial cladocerans and copepods also migrate away from littoral areas (avoidance-of-shore movements) by behavioral swimming responses to angular light distributions.
    b. In many cases, nonrandom dispersion of zooplankton is caused by water movements, in particular Langmuir circulations and metalimnetic entrainment of epilimnetic water.

12. Seasonal polymorphism, or *cyclomorphosis,* is found among many zooplankton, but is most conspicuous among the Cladocera.
    a. Changes in rotifer growth form include elongation in relation to body width, enlargement, reduction in size, and production of lateral spines. A predator-produced organic substance can induce spine production in certain prey rotifers, which reduces predation success.
    b. Cyclomorphosis in Cladocera often results in extension of the head (helmet) with little change or only a slight decrease in carapace length (Fig. 16–28). Caudal spine length often increases.
        i. A combination of environmental parameters has been shown to induce internal growth factors (hormones) that influence differential growth: increased temperature, turbulence, photoperiod, and food enhance cyclomorphosis in daphnid cladocerans.
        ii. Adaptive significance of cyclomorphic growth likely centers on reducing predation by allowing continued growth of peripheral transparent structures without enlarging the central portion of the body visible to fish. Small cladocerans that increase size by cyclomorphic growth reduce capture success by invertebrate predators (e.g., copepods).
    c. Cyclomorphosis is lacking in copepods, which, by means of rapid, evasive swimming movements, can defend themselves better from invertebrate predators than can most rotifers and cladocerans.

13. Planktivorous fish can be important in regulating the abundance and size structure of zooplankton populations. Prey are visually selected, in most cases, on an individual basis, although the gill rakers of certain fish collect some zooplankton as water passes through the mouth and across the gills.
    a. Size selection of prey by fish is governed by energy return obtained per unit of energy expended in foraging, and by the abundance of prey. When prey are abundant, only larger prey are consumed; as prey abundance decreases, smaller prey are taken.
    b. Planktivorous fish select large zooplankters and can eliminate large cladocerans from lakes.
    c. When size selection by fish is not in effect, and when large zooplankters are present, smaller-sized zooplankton are generally not found to co-occur with the larger forms. The cause is likely a result of size-selective predation of smaller zooplankton by invertebrates (copepods, phantom midge larvae, and predaceous Cladocera).
14. The production rate (= net productivity) of zooplankton is the sum of all biomass produced in growth, including gametes and exuviae of molting, less maintenance losses from respiration and excretion. Emigration (e.g., outflow losses) and immigration from streams and other lakes of zooplankton are usually negligible.
    a. A general, positive correlation exists between the rates of production of phytoplankton and of zooplankton (Table 16–11).
    b. Much of autotrophic production is not utilized by herbivorous zooplankton, but instead enters detrital pathways as nonpredatory particulate and dissolved organic matter. Although particulate detritus has less energy content than living algae, detritus often augments the diet of suspension-feeding zooplankton.
    c. The productivity of suspension-feeding zooplankton is higher than that of predaceous zooplankton.
    d. Efficiency of assimilation of ingested food is somewhat higher in juvenile stages than in adults, but is nearly always less than 50 per cent in zooplankton. Most of the food is egested and enters detrital pathways. Assimilated energy expended in respiration is usually less than 50 per cent; the remainder is used for growth and reproduction. Assimilation and respiration rates generally increase at higher trophic levels, and production decreases.

CHAPTER SEVENTEEN

PLANKTONIC COMMUNITIES: PELAGIAL BACTERIA

Biochemical transformations of particulate and dissolved detrital organic matter by bacteria and fungi are fundamental to the structure and dynamics of nutrient cycling and energy flux within aquatic ecosystems. The entire subject of aquatic microbial ecology, including the consumption of particulate detritus by higher organisms, is a neglected frontier in limnology. The functioning of aquatic ecosystems is closely coupled to the metabolic transformations of organic matter by bacteria and fungi. Yet, few areas of understanding in aquatic ecology are in such comparative infancy.

Organic matter is synthesized either autochthonously within the lake, or allochthonously and is imported to the lake as water carries organic matter to the lake basin. As we will see later, organic matter entering lakes from the landscape is mostly in dissolved form, and is chemically modified by the flora and microflora of streams and wetlands before reaching the lake per se (Chapter 22). A major source of autochthonous organic matter is the phytoplankton. This organic matter is augmented with that synthesized by littoral flora (Chapters 18 and 19). These sources of organic matter form the base upon which all other lake-dwelling organisms depend for nutrients and energy.

Most photosynthetic organisms simply die and add their newly synthesized organic matter to the detrital components of the system, where they may be decomposed. The reduced carbon of organic matter is eventually transformed by microbes to the oxidized state, CO_2. A small portion of this organic matter, 1 to 10 per cent on the average, is consumed by animals, and is decomposed by their heterotrophic metabolism. In aquatic ecosystems, most organic detritus is metabolized by the bacteria, and to a lesser extent, the fungi. The bacteria are effective. The organic matter brought to a lake would quickly fill the basin if it were not decomposed, and the lake would cease to exist. Since decomposition is not complete, residual quantity remains; gradually the basin fills in, and the lake is transformed from an aquatic to a terrestrial ecosystem (Chapter 24). The rate of this development, or ontogeny, is governed by the capacity of bacteria to mineralize organic matter, which is in turn influenced by the geomorphology and morphometry of the system and particularly the type (quality) of organic matter brought to it (Chapter 22).

The ensuing discussion addresses only the planktonic microflora and their metabolism, of which we know little. As we will see later (Chapter 22), much of the organic matter synthesized within the lake or brought to it from the drainage basin is not decomposed in the water, but instead is decomposed on or in the sediments. The biochemical transformations of organic matter displaced to the sediments are crucial to the operation of freshwater ecosystems, and consequently deserve separate treatment (chapters 20 and 22).

It is perhaps most appropriate to begin discussion with the generalized composite organic carbon cycle of a lake (Fig. 17-1), modeled after the general discussions of Kuznetsov (1959, 1970), and to attempt systematically to provide a quantitative range for the reactions and processes involved among lakes of varying productivity. The sources and general composition of allochthonous and autochthonous organic matter in dissolved and particulate form are treated in some detail in subsequent chapters. Dead organic matter forming the detritus is separable into dissolved and particulate fractions, but it must be emphasized that this demarcation is only an operational division made by the investigators. From a functional standpoint of the system, the energetic transformations of organic carbon are similar whether they are operating on a large particle or a dissolved organic compound. Only the rates of transformation differ.

Bacteria and fungi assimilate dissolved organic compounds, some of which they obtain through enzymatic degradation of particulate organic matter. The decomposition rate of organic substances is greatly dependent on solubility (Vallentyne, 1962). Decom-

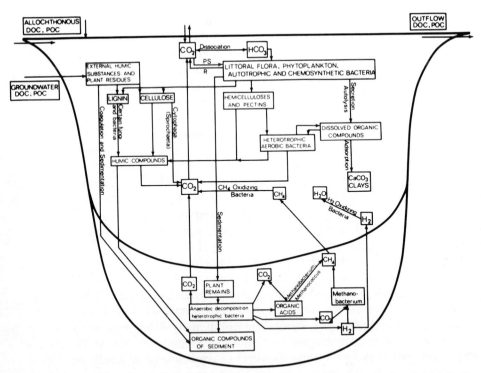

Figure 17-1. Simplified organic carbon cycle of a typical freshwater lake. *DOC* and *POC* = dissolved and particulate organic carbon; *PS* = photosynthesis; *R* = respiration. (Modified from Kuznetsov, 1959, 1970.)

position will be lowest for those organic compounds or complexes that occur in environmental concentrations exceeding the saturation level in the surrounding water. Degradation rates of soluble organic compounds vary, but for the compound to be preserved and removed from the organic carbon cycle, sedimentation must occur. The prerequisite for sedimentation by gravity is insolubility. Sedimented, relatively insoluble compounds become buried in anaerobic sediments, where rates of degradation are low. Although cellulose can be decomposed very rapidly by anaerobic bacteria under certain conditions, for example, in sewage digesters and in rumen, in anaerobic aquatic sediments, acidic fermentation products rapidly accumulate and lower the pH sufficiently to reduce or inhibit bacterial metabolism (Brock, 1966). Additionally, temperatures are generally low in many aquatic sediments. Lignin is even more resistant to degradation than cellulose, particularly under anaerobic conditions, and both lignin and cellulose form humic substances. The humic materials of the water coagulate to some extent, sink, and form a portion of aquatic sediments.

The primary intermediate products of anaerobic decomposition of sedimented organic matter (especially cellulose) are fatty acids, which are further degraded to CO_2 and hydrogen. Methane is biogenically generated from CO_2 and from hydrogen, and then is lost as bubbles or is reoxidized to CO_2 in less reduced strata of the sediments or water.

Much of the decomposition of organic compounds occurs in aerobic waters prior to sedimentation to the bottom of the basin. The extent of degradation en route is governed by an array of physical (morphometry, stratification patterns, temperature) and chemical conditions as well as the magnitude of allochthonous and autochthonous inputs of organic matter. Organic inputs to oligotrophic lakes are generally small, and organic matter is exposed to oxic conditions for long distances (greater time). Consequently, degradation of sedimenting organic matter is relatively complete, and organic sediment accumulation is slow. Massive inputs of organic matter in eutrophic lakes result in rapid sedimentation (shorter distances), less volume of aerobic water, and rapid accumulation of organic matter in anaerobic hypolimnia and sediments.

Distribution of Bacteria

Aquatic bacteria are typically substrate-limited. Consequently, one would expect a general increase in bacterial biomass as a lake system is increasingly loaded with organic substrates. The most direct approach to evaluation of microbial biomass is enumeration, either by isolation with cultures or by direct microscopic observation (cf. detailed discussions of techniques by Kuznetsov and Romanenko, 1963; Rodina, 1965, 1972; Sorokin and Kadota, 1972; Rosswall, 1973; and Romanenko and Kuznetsov, 1974).

The culturing of bacteria in media, while mandatory for species identification, is extremely selective for specific bacteria under most conditions. These methods underestimate natural heterogeneous bacterial populations by factors from 1000 to 100,000 times (Fig. 17–2). Direct enumeration by microscopy has the inherent difficulty of discriminating viable cells from detritus and cells that are dead, but recent techniques, such as epifluorescence microscopy (Francisco, et al., 1973; Hobbie, et al., 1977; Coleman, 1980; Porter and Feig, 1980), provide great improvements.

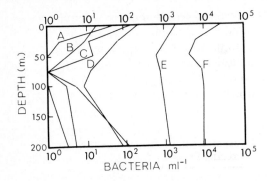

Figure 17-2. Bacterial populations of California coastal waters obtained by different methods: *A*, Agar pour plate method, *B*, Macrocolonies on membrane filters, *C*, Extinction dilution method, *D*, Microcolonies on membrane filters, *E*, Direct microscopic enumeration on membrane filters, and *F*, Cholodny method of direct enumeration of microorganisms on glass slides after filtration. (After Jannasch and Jones, 1959.)

Measurements of bacterial biomass separate from nonliving detrital organic matter can also be estimated by measurement of a uniformly distributed cellular constituent. Adenosine triphosphate (ATP) fulfills these criteria, since it decomposes readily on death, and can be measured by bioluminescent reactions at extremely low concentrations. Measurements of ATP concentrations permit an estimation of total biomass in terms of organic carbon or dry weight, because the ATP levels of microbial cells are fairly uniform (Holm-Hansen and Paerl, 1972).

In general, the numbers and biomass of bacteria increase with increasing productivity and concentrations of inorganic and organic compounds among lakes (Table 17-1). In spite of great seasonal variations, bacterial numbers and biomass increase from oligotrophic to eutrophic lakes. Highest concentrations have been observed in tropical alkaline, saline lakes (e.g., Kilham, 1981), and in eutrophic reservoirs, probably because the large reservoirs studied tend to occur on rivers that receive both industrial and municipal wastes. Numbers of bacteria are markedly lower in acidic dystrophic lakes which contain high concentrations of humic matter. In a similar manner, production of bacteria as calculated from changes in biomass over short time periods increases with increasing productivity of lakes and the average generation time decreases among more eutrophic lakes. A positive relationship has been demonstrated between the average rates of phytoplanktonic productivity and bacterial numbers (Table 17-2)* (cf. Overbeck, 1965; Godlewska-Lipowa, 1975, 1976; Jones, 1977; Straškrabová and Komárková, 1979).

SEASONAL AND VERTICAL DISTRIBUTION

The seasonal distribution of bacterial populations is highly variable from lake to lake and from year to year within a lake. A few generalizations, however, can be advanced even from population data determined with sampling frequencies which are far greater than the generation time of the bacteria. The most conspicuous generaliza-

---

*The term *saprophytic bacteria*, as used in these and similar studies, refers to those bacteria that grow on organic-enriched agar such as Difco nutrient agar or casein-glucose agar. These techniques are highly selective but reflect differences in concentration of organic substances easily assimilable by bacteria. Saprophytes generally require organic substances at high concentrations for development; in contrast, "oligotrophic bacteria" develop at minimal organic substrate concentrations (1 to 15 mg C l^{-1}), even though they can grow on richer media (Kuznetsov, et al., 1979).

tion is the close correlation between the numbers of phytoplanktonic algae and heterotrophic bacteria both seasonally (Fig. 17-3) and vertically (Fig. 17-4). Bacterial numbers and biomass are commonly highest in the epilimnion, decrease to a minimum in the metalimnion and upper hypolimnion, and increase in the lower hypolimnion, especially if anoxic (e.g., Niewolak, 1974; Jones, 1978; Kato and Sakamoto, 1979, 1981). But while there is no doubt that the heterotrophic metabolism of the bacteria depends to a

TABLE 17-1 Numbers, Volumes, and Biomass of Bacteria in Lakes of Differing Productivity

LAKE AND TYPE	NUMBER (10^6 ml^{-1})	VOLUME (μm^3)	BIOMASS $(g \text{ m}^{-3})$
Oligotrophic			
Zelenetskoye (1970, 1971)	0.175	0.265	0.06
Baikal	0.20	—	—
Krivoye (1968, 1969)	0.67	0.43	0.21
Onezhskoe	0.29	—	—
Ladozhskoe	0.35	—	—
Pert	0.13	—	—
Mesotrophic			
Krasnoye (1964–1970)	0.70	0.43	0.30
Naroch (1968–1970)	0.64	0.50	0.32
Dal'nee (1970, 1971)	1.50	0.67	1.00
Sevan (1952, 1962, 1966)	0.39	1.12	0.43
Glubokoe (1932)	1.2	—	0.97
Eutrophic			
Drivyaty (1964)	1.84	0.76	1.40
Myastro (1968–1970)	2.20	0.50	1.10
Batorin (1969, 1970)	6.40	0.50	3.20
Beloe (1932)	2.23	—	—
Chernoe (1932)	4.00	—	—
B. Krivoe (saline)	12.3	—	—
Four African saline lakes	3.7–360	—	—
Reservoirs			
Rybinsk (1964–1968)	1.70	0.60	1.00
Bratsk (1965–1972)	0.85	0.90	0.77
Kiev (1967–1968)	4.10	0.84	3.35
Kremenchug (1968)	3.50	0.60	2.10
Kakhov (1968)	4.00	0.47	1.90
Dneprodzerzhin (1968)	3.40	0.65	2.20
Kashkorenskoe	7.8–57.9	—	—
Rybovodnye (pond)	1.0–40.0	—	—
Kramet-Niyaz (pond)			
Without fertilization	2.0–6.0	—	2.0–6.0
With fertilization	5.0–20.0	—	5.0–25.0
Dystrophic			
Chernoe	1.07	—	—
Piyarochnoe	0.43	—	—
Serpovidnoe	0.1–0.5	—	—
Average Values			
Oligotrophic	0.50	0.2–0.4	0.15
Mesotrophic	1.00	0.4–1.2	0.70
Eutrophic	3.70	0.5–0.9	2.30

Often summer values in the trophogenic zone; after data of Kuznetsov (1970), Saunders, et al. (1980), and Kilham (1981).

TABLE 17-2 Relationship Between Average Rates of Phytoplanktonic Primary Production and Bacterial Numbers at 1 m Depth in Four Lakes of Northern Germany

LAKE	PRIMARY PRODUCTIVITY (mg C m^{-3} DAY^{-1})	SAPROPHYTES (NO. ml^{-1})	TOTAL BACTERIA ($\times$ 10^6 ml^{-1})
Schöhsee	16	300	0.5
Schluënsee	17	350	0.5
Grosser Plöner See	35	650	1.0
Pluss-See	57	1000	1.1

Modified from Overbeck, 1965.

significant extent on organic substrates produced by the phytoplankton, such correlations are often weak. Often increases in bacterial populations are delayed, and follow phytoplankton maxima by as much as 10 days (Straškrabová and Komárková, 1979). More frequent sampling has demonstrated that the vertical distribution of bacteria in dimictic lakes can change very rapidly (Rasumov, 1962; Saunders, 1971). For example, the rapid changes observed in Figure 17-5 became particularly acute in mid-June, following a phytoplanktonic maximum; bacterial numbers increased sharply, then decreased again in a few days. The rapid inversions in vertical distribution of the waxing and waning bacterial populations (Fig. 17-5) indicate that short-term variables and processes control bacterial abundance and distribution (Saunders, 1971; Saunders, et al., 1980). Correlations have been found between pulses of bacteria and changes in sources of organic matter, such as death and decomposition of phytoplankton and of littoral plants, as well as changes in precipitation and runoff, temperature, and, at times, grazing by zooplankton (e.g., Lane, 1977; Tanaka, et al., 1977; Goulder, 1980). However, correlations alone yield little insight into mechanisms controlling the apparent interaction. Causal relationships cannot be evaluated unless the sampling interval is considerably shorter than the life span of the organisms sampled. Studies of this kind on in situ populations are very few.

Planktonic bacterial biomass is generally lower during winter than during summer in temperate lakes; this relationship is correlated with low winter temperatures and reduced loading of particulate and dissolved organic substrates from autochthonous (phytoplankton, littoral plants) and allochthonous sources. It is possible that a significant portion of the planktonic bacterial community is physiologically dormant when conditions are cold and limited substrate is available (e.g., Stevenson, 1978; Wright, 1978). Bacteria are often found attached to detrital particulate matter (e.g., Hutchinson, 1967; Paerl, 1975), but most bacteria are freely planktonic (e.g., Riemann, 1978; Geesey and Costerton, 1979; Saunders, et al., 1980). Microbial attachment to particles can enhance availability of substrates relative to concentrations existing in the water.

Horizontal spatial variability in bacterial biomass is usually low in lakes that are not morphometrically complex or disturbed by high effluent loading (e.g., Jones, 1977; Zmyslowska and Sobierajska, 1977; Jones and Simon, 1980). Patchiness in bacterial distribution is found, however, upon critical sampling and analysis (e.g., Palmer, et al., 1976); this variance is often related to large-scale differences in organic loading from

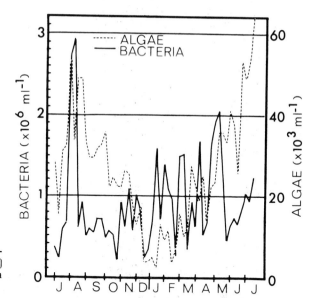

Figure 17-3. Annual cycle of heterotrophic bacteria and planktonic algae in the surface water (0.5 m) of a pond in Potsdam, Germany, 1958–59. (After data of Overbeck and Babenzien, 1964.)

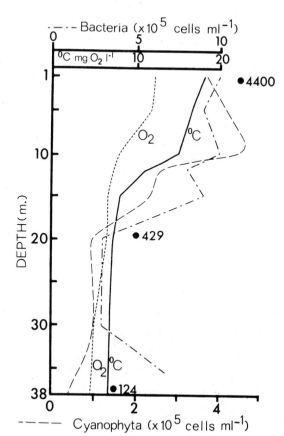

Figure 17-4. Vertical distribution of heterotrophic bacteria in relation to planktonic algae in Schluënsee, northern Germany, 18 August 1965. Circles = saprophytes ml⁻¹. (After data of Overbeck, 1968.)

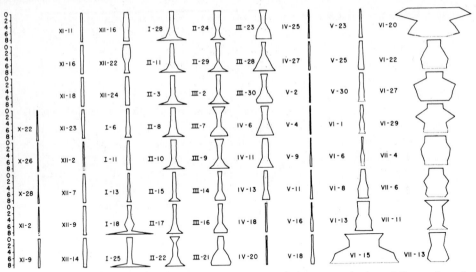

Figure 17–5. Relative vertical distribution of bacteria by direct enumeration on different dates in Frains Lake, southeastern Michigan, 1959–60. Ordinate: depth in meters; abscissa: relative numbers. (From Saunders, G. W.: Carbon flow in the aquatic system. *In* Cairns, J., Jr., ed., The Structure and Function of Fresh-Water Microbial Communities. Blacksburg, Va., Virginia Polytechnic Institute, 1971.)

human effluents (e.g., Rao and Burnison, 1976). Sinking rates of bacteria are so low (approximately 1 mm day^{-1}; Jassby, 1975) that horizontal differences are likely related to differences in water turbulence and variations in availability of organic compounds (e.g., phytoplankton).

Decomposition of Dissolved Organic Matter

AMINO ACIDS AND RELATED COMPOUNDS

Insight into the complex differential utilization of relatively labile organic compounds, such as amino acids, is gained from several detailed analyses of the amino acid and carbohydrate cycling in lakes. Concentrations of dissolved "free" amino acids of lake water and interstitial water of sediment generally are very low (<10 μg l^{-1}) at any given time (Brehm, 1967; Gardner and Lee, 1975), although they can increase to considerably higher levels in eutrophic pond waters (Zygmuntowa, 1972). All of the protein amino acids as well as several nonprotein amino acids have been found. Over an annual period, and in several different types of waters, concentrations of serine, glycine, and alanine were relatively high, those of aspartic acid and threonine occurred in less abundance, and those of all remaining amino acids and glutamic acid were very low. Maximum epilimnetic concentrations commonly occur in the winter, especially just below the ice, and during the summer period of maximum algal production. Concentrations in the metalimnion are generally very low but may increase somewhat in the hypolimnion. Just above the sediments, concentrations of amino acids often decline again. Much of the dissolved amino acids was found to be adsorbed to colloidal carbohydrates liberated by the microflora. These colloids are most abundant in the epilimnion during the summer and are reduced to extremely low concentrations in the hypolimnion, where intensive bacterial metabolism occurs.

High-molecular-weight peptides of the free water and interstitial water of the sediments were associated with humic substances and increased in surface waters during autumnal influxes of allochthonous plant material (Brehm, 1967). These compounds and, additionally, mucopeptides of blue-green algal and bacterial origin, undergo relatively slow bacterial degradation. Amino acids are adsorbed selectively by these humic substances of dead particulate matter, and by the cell walls of bacteria and blue-green algae. These polypeptides remain relatively undegraded after autolysis of algae. In contrast, certain amino acids, especially aspartic and glutamic acids, are released very rapidly during autolysis and are assimilated immediately by heterotrophic bacteria. Other amino acids are released through autolysis more slowly, resulting in a higher relative concentration in the dead plankton. Bacterial decomposition of plankton occurs first on the autolytically soluble products. Most of the cell decomposition occurs in the epilimnion; more resistant cell components are degraded more slowly after the soluble autolytic substances are utilized.

Concentrations of amino acids found are nearly always <50 μg l^{-1} in a variety of aquatic situations. Measurements with individual amino acids of uptake kinetics demonstrated that those in least abundance were assimilated most rapidly (<20 hours), and conversely that those compounds found in higher concentrations exhibited longer (30 to 90 hours) turnover times (the time required for the plankton to remove all the substrate) (Hobbie, et al., 1968). These results are reinforced by Gocke's (1970) demonstration of the rapid bacterial decomposition of free amino acids (reduced to 2 per cent within 6 days) and peptides (to 13 per cent in 6 days) by natural bacterial populations, while amino-colloidal complexes are resistant to degradation and accumulate. The differences in rates of degradation of amino components are reflected in the absolute concentrations found (Table 17–3).

TABLE 17–3 Dissolved Organic Nitrogen Compounds in the Epilimnion of Two Lakes of Northern Germany, 27 June 1967

DISSOLVED ORGANIC NITROGEN COMPOUNDS	SCHUËNSEE (MESOTROPHIC)	PLUSS-SEE (EUTROPHIC)
Extracellular Amino Compounds		
Free amino acids	39 μg l^{-1}	71 μg l^{-1}
Peptides	226	250
Colloids	118	414
Glucosamines	31	126
Total	414 μg l^{-1}	861 μg l^{-1}
Amino-N as a percentage of the total extracellular organic nitrogen	32.2%	23.7%
Amino acids and amino sugars as a percentage of the total dissolved organic matter	10.3%	5.5%
C:N of dissolved organic matter	7.5:1	10.7:1

From data of Gocke, 1970.

Simple organic nitrogen compounds are utilized more effectively in more productive lakes (faster turnover; lower instantaneous concentrations as a percentage of the total organic nitrogen). This observation suggests that the enzymatic systems of bacteria of eutrophic lakes are adapted or keyed to respond to a much greater variety of substrates than those of oligotrophic waters exposed to conditions of less frequent substrate encounters. The oligotrophic lake species, upon receipt of an influx of a specific substrate for which it does not possess an active enzymatic uptake system, must adapt to that substrate (cf. Hollibaugh, 1979). Hence, the observed assimilation rates are initially low but can increase markedly as adaptation occurs. Those species of eutrophic waters, which are exposed more frequently or continuously to a variety of substrates, respond to and assimilate the substrates immediately.

CARBOHYDRATES AND SIMPLE ORGANIC ACIDS

The distribution of carbohydrates and simple organic acids has been investigated in several lake systems (see, for example, Saunders, 1963; Walsh, 1965a, 1965b, 1966; and especially Weinmann, 1970). As with simple amino compounds, the instantaneous concentrations of individual organic acids and carbohydrate compounds are very low (usually <30 μg l^{-1}) and represent a small portion of the total dissolved organic matter (cf. Hama and Handa, 1980). Free monosaccharides and oligosaccharides are assimilated readily by bacteria and occur at very low concentrations (<10 μg l^{-1}). Dissolved polysaccharides, consisting of numerous hexoses, pentoses, and methylpentoses, and especially phenolic and cresolic compounds, occur in greater abundance. Various organic acids (lactic, citric, oxalic, malic, glycolic, and short-chain fatty acids) have also been found but only at low levels.

It is clear that major sources of these compounds are secretion products of phytoplankton (Chapter 15) and littoral flora (Chapter 18), autolysis of the plants and microflora, and intermediary products of microbial degradation. In bacteria-free algal cultures, carbohydrates, especially mono- and disaccharides, and total dissolved organic matter increase in proportion to algal growth. In the presence of a natural bacterial flora, the extracellular products are degraded rapidly (hours). This reduction is selective: The labile carbohydrates, organic acids, and amino acids decrease much more rapidly than the total dissolved organic matter (Weinmann, 1970; Gocke, 1970; Poltz, 1972).

The relative rates of microbial degradation, while understood in some detail from the wealth of physiological information on bacterial utilization of organic substrates, are exceedingly difficult to quantify in natural populations of aquatic bacteria. In natural systems, bacterial populations are heterogeneous, and each species may have different abilities to assimilate specific substrates. Consequently, measurements of substrate assimilation by bacteria are fraught with methodological problems (cf. Jannasch, 1969, 1974; but see Bell, 1980, and Bell and Sakshaug, 1980). Nevertheless, the kinetics of organic substrate utilization by heterogeneous planktonic populations, determined by the uptake rates of radioactivity labeled organic compounds *at substrate concentrations that occur naturally*, are instructive in a relative if not absolute way.

The uptake of inorganic ^{14}C-labeled carbon by natural populations over short intervals of time has been used widely as a measure of rates of carbon fixation by photosynthesis. Parsons and Strickland (1962) employed labeled organic substrates to measure uptake by natural heterotrophic populations in the dark by the relationship:

$$v = \frac{c \cdot f \cdot (S_n + A)}{C \mu t}$$

when:

v = rate of uptake (mg C m^{-3} hr^{-1}),

c = radioactivity of the filtered organisms (cpm, counts per minute),

f = a correction for isotopic discrimination (1.05 or 5 per cent slower uptake of ^{14}C which has a greater mass than ^{12}C),

S_n = in situ concentration (mg l^{-1}) of the organic substrate,

A = concentration (mg l^{-1}) of added substrate (labeled and unlabeled),

C = cpm from 1 μCi of ^{14}C-labeled substrate on the radioassay instrumentation used,

μ = quantity of ^{14}C added to the sample,

t = incubation time (hours).

This equation assumes that natural substrate concentrations (S_n) are much less than A. When the uptake of a solute is mediated by a transport system located on or in the cell membrane, the rate of uptake can be described by Michaelis-Menten kinetics:

$$v = \frac{V(S)}{K_m + S}$$

when:

v = velocity at a given substrate concentration **S**,

V = maximum velocity, attained when uptake sites are continually saturated with substrate,

K_m = Michaelis constant, by definition the substrate concentration at which the velocity is one-half the maximum velocity V. Also called K_t, the transport constant, K_m is a measure of the affinity of the uptake system for the substrate.

As developed by Wright and Hobbie (1966; cf. also Allen, 1969), this non-linear uptake relation over substrate concentration can be transformed into a linear relationship by the Lineweaver-Burk equation to yield:

$$\frac{C \mu t}{c} = \frac{(K_t + S_n)}{V} + \frac{A}{V}$$

With this equation, data from uptake measurements from algae and bacteria at low substrate concentrations can be plotted as $C\mu t/c$ versus A, giving values for $(K_t + S_n)$ and V (Fig. 17–6). The negative intercept on the abscissa is equal to $(K_t + S_n)$, and the reciprocal of the slope is V, the maximum rate of uptake. The ordinate intercept is equivalent to the turnover time (T_t), which is the time required for complete removal of the natural substrate by the microflora (see Hobbie, 1967). The $(K_t + S_n)$ approximates the maximum natural substrate concentration (S_n) if K_t is very small, as is often, but not always, the case. These relationships assume that a constant rate of regeneration of the organic solute is occurring in situ, or that steady state conditions exist. An appreciable portion of ^{14}C organic substrate is respired rapidly by the microflora, and corrections for this loss must be made (Hobbie and Crawford, 1969).

In contrast to the already discussed nonlinear active uptake velocities at low substrate concentrations, uptake of organic compounds by natural planktonic populations at high substrate concentrations (> approximately 0.5 mg l^{-1}) does not exhibit rate-limitation kinetics or saturation of uptake sites. Passive uptake velocity continually increases with rising substrate concentrations. The slope of the response line is constant (K_d), derived from diffusion kinetics, and has been used to estimate diffusion uptake by natural populations of algae and bacteria.

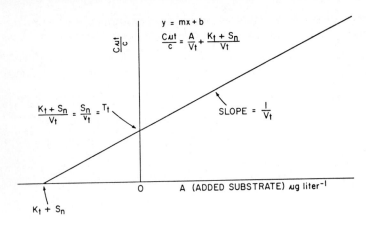

Figure 17-6. Graphical analysis of bacterial uptake at low organic substrate concentrations following Michaelis-Menten enzyme kinetics. A plot of $C\mu t/c$ against increasing substrate concentrations, S, illustrating derivation of (1) maximum natural substrate concentrations, $K_t + S_n$, as $\mu g\ l^{-1}$, (2) maximum velocity, V, as $\mu g\ l^{-1}\ hr^{-1}$, and (3) turnover time for substrate regeneration (T_t) in hours. (After Allen, H. L.: Chemo-organotrophic utilization of dissolved organic compounds by planktic algae and bacteria in a pond. Int. Rev. ges. Hydrobiol., 54:1–33, 1969.)

Analyses of utilization rates of organic substrates under in situ conditions with natural populations have been applied in only a few cases, and usually employ only a few simple substrates such as glucose, other sugars and isomers of glucose, acetate, glycolate, and amino acids. Only a few analyses have been extended over an annual period (Wetzel, 1967; Allen, 1969; Hobbie, 1967, 1971; Overbeck, 1975; Rai, 1978, 1979; Robarts, 1979; Gillespie and Spencer, 1980). However, despite inherent physiological limitations, the analyses do yield insight into utilization of dissolved organic substrates. Foremost, they demonstrate that dark diffusive algal uptake of simple organic substrates for use in heterotrophic growth at low natural substrate concentrations is almost always less than 10 per cent of active permease-mediated uptake by bacteria. Stated in another way, the algae are generally ineffective in competing for available organic substrates at substrate concentrations maintained by active bacterial heterotrophic uptake.

Maximum velocities of uptake are quite variable among lakes and different organic substrates, but generally occur after spring or following late summer algal maxima. Rates decrease by about an order of magnitude during winter in temperate lakes. Concentrations of substrates such as glucose, acetate, and amino acids remain low throughout the year, for the rate of inputs to the labile dissolved organic pool is balanced roughly by the rate of removal by bacteria. Vertical distribution of uptake velocities indicates trends toward maxima during and following algal maxima in the trophogenic zone, minima in the lower metalimnion–upper hypolimnion, and marked increases in rates near the sediments. Within the sediments, utilization velocities of simple organic solutes are several orders of magnitude greater than in the overlying water (Harrison, et al., 1971).

Utilization rates of simple substrates generally increase with greater productivity in an approximately direct relationship (Table 17–4). Great variations are observed in these rates both seasonally and vertically with changes in depth. Uptake rates of simple carbohydrates generally increase with greater bacterial densities and biomass, and with increasing concentrations of total inorganic nitrogen and total phosphorus (e.g., Bowie and Gillespie, 1976; Spencer, 1978). Glucose assimilation rates by planktonic bacteria were found to be nearly an order of magnitude greater in littoral areas where submersed macrophytes were abundant than in the pelagial areas (Gillespie and Spen-

TABLE 17-4 Comparison of Approximate Rates of Turnover of Organic Substrates by Natural Bacterioplankton in Lakes of Increasing Productivity Based on the Range of Maximum Phytoplanktonic Photosynthetic Rates of Carbon Fixation (P_{max})

LAKE	SUBSTRATES	P_{max} (mg C m^{-3} DAY^{-1})	T_t (HOURS)	SOURCE
Oligotrophic				
Lapplandic lakes, Sweden (summer)	Glucose, acetate	1–30	>10,000	Rodhe, et al., 1966
Lawrence, Michigan (annual)	Glucose	1–80	40–300	Wetzel, et al., 1972
	Acetate		10–120	
Tupé, Brazil (annual)	Glucose	—	105->30,000	Rai, 1978, 1979
Mesotrophic				
Crooked, Indiana (annual)	Glucose	63–110	80–470	Wetzel, 1967, 1968
	Acetate		20–350	
Klamath, Oregon (summer)	Water, glucose	—	220	Harrison, et al, 1971
	acetate	—	250	
	Sediments, glucose	—	2.25	
	acetate	—	0.75	
Gravelly Pond, Massachusetts (summer)	Glycolate	—	60–200	Wright, 1975
Kinneret, Israel	Glucose	—		Cavari, et al., 1978
0–10m		—	88	
20–40m		—	133	
Janauari, Brazil (annual)	Glucose	—	15–900	Rai, 1978
Four Várzea lakes, Brazil (annual)	Glucose	—		Rai, 1979
Low water		Higher	40–101	
High water		Lower	105->10,000	
Eutrophic				
Erken, Sweden				
Summer	Glucose	40–130	10–100	Hobbie, 1967
Winter	Glucose	2–20	100–1000	
Wintergreen, Michigan (summer)	Glucose	80–120	<1–20	Saunders, pers. comm., Wetzel, unpublished
Little Crooked, Indiana (annual)	Glucose	190–205	36–232	Wetzel, 1967, 1968
	Acetate		24–190	
Duck, Michigan (annual)	Glucose	10–320	8–50	Miller, 1972
	Acetate		4–40	
Pamlico River (estuary), North Carolina	Amino acids	—	1.5–26	Hobbie, 1971
Lötsjön, Sweden				
Summer	Glucose	<100	0.4–5	Allen, 1969
Winter	Glucose	<20	20–300	
Upper Klamath, Oregon (summer)	Glucose	—	2.4	Wright, 1975
	Acetate	—	2.3	
	Glycine	—	8	
	Glycolate	—	26	
Pluss, Northern Germany (annual)	Glucose	20–150	6–202	Overbeck, 1975

cer, 1980). Heterotrophic utilization of glucose by neustonic bacteria at the surface air-water interface was found to be much less than by planktonic bacteria (Dietz, et al., 1976). It is impossible to view these data in any way other than general *direct* measurements of the expected positive correlation between increased bacterial metabolism of ubiquitous simple substrates and waters with larger inputs of organic matter. Methodological problems, such as competitive inhibition of one substrate on the uptake of another (Burnison and Morita, 1973), substrate affinities (Button, 1978; Moaledj and Overbeck, 1980), variable substrate uptake in relation to phosphorus availability (Overbeck and Tóth, 1978), and the need for corrections for respiratory losses of CO_2, preclude a more detailed explanation of interactions.

FATTY ACIDS AND RELATED COMPOUNDS

Although relatively little is known about the distribution and metabolism of lipid compounds in fresh waters, the detailed investigations of Poltz (1972) on lakes of northern Germany demonstrate relationships that probably occur frequently. It is not known at this time how generally applicable these results are.

Total lipid, fatty acid, and triglyceride content declines rapidly when cells die and settle out of the trophogenic zone; concentrations of these materials in the sediments are much lower than in living net plankton (Table 17–5). Lipid content variations of net plankton, sedimenting particulate organic matter, and lake sediment were relatively large, because of rapid changes in species composition of the phytoplankton and zooplankton. Vertical migration and population oscillations both contributed to fluctuations in short-term lipid content in these particle types. Changes in the lipid content and fatty acid compositional patterns of the plankton over an annual period were not as great, however, because of opposing changes in lipid content of the phyto- and zooplankton. The relative amount of short-chained fatty acids, especially of the 16:1 isomer, increased from summer to winter so that the mean chain length of the fatty acids declined. The same fatty acids were found in the triglycerides of the plankton and in the total fatty acid fractions, but the relative concentrations of highly unsaturated fatty acids in the triglycerides were lower.

Decomposition of organic matter and fatty acids was most rapid under the warmer,

TABLE 17–5 Average Percentage of Lipid Fractions in the Net Plankton, Sedimenting Particulate Organic Matter, and Sediments of Several Lakes of Northern Germany

FRACTIONS	NET PLANKTON (>20 μm)	SEDIMENTING PARTICULATE MATTER	SEDIMENT*
Total organic matter per dry weight	83.1%	36.7%	16.0%
Total lipids per organic matter	22.3	9.0	6.8
Total fatty acids per total lipids	61.2	32.5	12.1
Triglycerides of total lipids	30.8	8.0	1.8

After Poltz, 1972.
*Grosser Plöner See only; mean values; values greatest near surficial interface and decreased progressively within first meter.

aerobic conditions of the epilimnion. The mineralization of the planktonic particulate organic matter of the pelagial zone was rapid and usually complete (>85 per cent). The intensity of decomposition of triglycerides was greatest (>98 per cent) in the epilimnion and followed the general sequence of degradation: triglycerides > total fatty acids > lipids > total organic substances (Poltz, 1972). Rates of decomposition decreased markedly in the hypolimnion. Only small amounts of the substances produced in the plankton were found settling to the sediments: 4 to 10 per cent of particulate organic matter, 1 to 2.5 per cent of lipids, 0.1 to 0.5 per cent of total fatty acids, and less than 0.1 per cent of triglycerides. After deposition, further degradation rates were most rapid at the surface interface zone of the sediments, although the rates were slower than those occurring in epilimnetic waters during the process of sedimentation.

These analyses were limited to the pelagic zone of lakes, and consequently do not reflect the massive amounts of particulate organic matter that can reach the sediments of many lakes when the littoral flora dominates production. The sediment interface is clearly the site of most intensive respiratory metabolism in many lakes (see Chapter 20), and much of the organic matter undergoing degradation there originates from littoral and allochthonous sources. Part of the lipid derivatives originate from the bacteria and fungi; these derivatives may be intermediate metabolic products formed during the degradation of this accumulated organic matter (cf. Schulz and Quinn, 1973).

HUMIC COMPOUNDS

Humic compounds derived from either allochthonous or autochthonous sources are structurally complex (phenolic linkages) and tend to have long residence times in lakes. In general, their degradation by aquatic microflora proceeds slowly. A portion of the humic materials present in a lake is truly dissolved, but under certain conditions, humic substances can aggregate into colloids or flocs. High-molecular-weight dissolved humic materials readily adsorb to particulate matter (Davis and Gloor, 1981), which may further alter their rates of degradation. Rates of glucose mineralization are also very low in tropical lakes that receive high loading of dissolved humic materials (Rai, 1978, 1979).

Haan (1972) demonstrated that a species of *Arthrobacter* was able to grow slowly on humic fractions and that it was able to hydrolyze the phenolic ether linkages. Studies of the degradation of dissolved humic compounds in heavily stained Finnish lake waters also provide evidence that these compounds, although relatively resistant, are mineralized slowly (Ryhänen, 1968; Sederholm, et al., 1973). Increased concentrations of inorganic nitrogen and phosphorus accelerated the rate of degradation. In some cases, sunlight may be sufficient to partially degrade dissolved humic materials, so that bacterial growth can proceed more rapidly (cf. Strome and Miller, 1978; Stewart and Wetzel, 1981a).

Studies of the effects of a low-molecular-weight humic fraction on the growth of a bacterial species of *Pseudomonas* showed that the presence of the humic fraction caused an increase in the cell yield and in the cell yield per unit of respiratory consumption of oxygen as compared to growth in organic media lacking the fraction (Haan, 1974). These results suggest that the stimulating effect of low-molecular-weight humic material on bacterial metabolism operates by coupling to the metabolism of less refractory organic substrates. Similar results were found by Stabel, et al. (1979). Cometabolism of many very refractory compounds, e.g., herbicides, is well known (Horvath, 1972; Hulbert and Kraviec, 1977), and is probably an important mechanism for the degradation of the more refractory compounds of fresh waters. The mechanisms of such cometabolism deserve greatly increased investigation.

Utilization of Biotically Released Dissolved Organic Matter

Dissolved organic compounds are released by organisms during active growth and by autolysis upon senescence and death.

AUTOLYSIS

As organisms senesce and die, *autolysis* of cytoplasmic cellular contents by deterioration of cellular membranes rapidly releases significant amounts of dissolved organic matter (DOM). This release is followed by rapid utilization of energy-, carbon-, and nitrogen-rich substrates by bacteria. The loss of total substance is very high immediately after "death" in all organisms (Table 17-6).

The autolytic release of DOM initially by algae, aquatic plants, and zooplankton, and its progressive decomposition thereafter has been studied in a few cases (Otsuki and Hanya, 1968, 1972a, 1972b; Golterman, 1971; Otsuki and Wetzel, 1974; Krause, 1964; Botan, et al., 1960; Godshalk and Wetzel, 1978a, 1978b, 1978c). Data from these studies illustrate the general relationships believed to prevail. Organic carbon of dead green algal cells decomposing under aerobic conditions declined about 55 per cent in 5 days

TABLE 17-6   Average Percentage Loss of Total Substance of Different Aquatic Organisms Immediately and 24 Hours After Death, Aerobic, 20° or 25°C

ORGANISMS	IMMEDIATE LOSS (%)	LOSS AFTER 24 HOURS (%)
Planktonic green algae	11	33
Planktonic diatoms	11	15
Leaves of aquatic plants:		
Callitriche hamulata	9	16
Scirpus subterminalis	35	—
Scirpus acutus	—	4
Najas flexilis	—	20
Myriophyllum heterophyllum	—	15
Nuphar variegatum	—	23
Mixed zooplankton:	15	52
Cladocera *(Daphnia)*	29	—
Copepods *(Cyclops, Diaptomus)*	8	—
Sediment-living worms *(Tubifex)*	5	68
Small fish *(Lebistes)*	7	28

After Krause, 1962; Otsuki and Wetzel, 1974b; Godshalk and Wetzel, 1978b.

TABLE 17-7   Changes in Composition of
Dissolved Organic Products During the
Aerobic Decomposition of *Scenedesmus* Cells
at 20°C

TIME (DAYS)	ORGANIC C (mg l^{-1})	ORGANIC N (mg l^{-1})	C:N RATIO
0	213	22	11
35	184	21	10
185	87	17	6

From Otsuki, A., and Hanya, T.: Production of
dissolved organic matter from dead green algal cells.
I. Aerobic microbial decomposition. Limnol. Oceanogr., 17:248–257, 1972.

(20°C), and decreased more slowly thereafter. Cell morphology changed little during
the first 50 days, indicating that although the cellular contents were decomposed
quickly after death, the cell wall was relatively resistant to microbial degradation.
Organic nitrogen of the cells was reduced by about 70 per cent during the first 30 days;
thereafter, decomposition of both organic C and N was slower. Dissolved organic C and
N released by autolysis, and subsequently from intermediate bacterial products of
decomposition, are degraded rapidly at first; rates of degradation then decrease with
time (Table 17-7). Similar results were found with the decomposition of submersed
vascular aquatic plants (Chapter 20).

RELEASE AND UTILIZATION OF ORGANIC COMPOUNDS.

Dissolved organic compounds also originate by *secretion* (extracellular release)
from actively growing algae (discussed in Chapter 15, p. 392ff) and larger aquatic plants
(Chapter 18). The quantities and qualitative composition of extracellularly released
organic matter are variable, and depend upon the species and their physiological status
under existing environmental conditions. Simple organic acids, sugars, and more complex carbohydrates, amino acids, peptides, pigments, and probably enzymes (e.g., Jüttner and Friz, 1974) are among the compounds known to be liberated during active
growth. Most of these secreted compounds are assimilated rapidly by the microbial
flora, so their concentrations in the water would always be expected to be low. Some
extracellular products, such as polypeptides, are more resistant to rapid degradation
and can form a more significant portion of the total dissolved organic matter at any
given time. Extracellular release of organic compounds during active growth represents
an elimination of metabolic products and is a relatively small loss of synthesized
organic matter for the plants.

Dissolved organic compounds are released by secretion from zooplankton, and by
leaching of their feces. For example, a significant amount (up to 17 per cent) of the algal
carbon ingested by cladoceran zooplankton can be immediately lost as dissolved
organic carbon as the algae are damaged during feeding (Lampert, 1978a). More than
10 percent of the algal particulate carbon removed from suspension by grazing can be

transformed to dissolved organic carbon by these processes. Compounds released through these processes can be of significance to the planktonic bacteria, since substrate availability in pelagial waters is usually limiting.

Extracellular organic products of phytoplankton are utilized in different ways (Nalewajko, 1977). Low-molecular-weight substances are assimilated rapidly by bacteria, essentially simultaneously with excretion (Nalewajko, et al., 1980), or by blue-green algae (Chang, 1981). Binding of low-molecular-weight compounds to colloids or to the surfaces of particulate matter may also occur, and can result in abiotic removal of these substances from solution through subsequent sedimentation. Estimates of bacterial utilization of secreted organic compounds from natural populations are few, but all evidence indicates rapid rates of assimilation. For example, in two eutrophic lakes of southern Sweden, from 28 to 80 per cent of the released dissolved organic carbon was recovered in bacterial particulate fractions after about four hours (Coveney, 1981). Variations in utilization were positively correlated with temperature. In the autumn, from 32 to 95 per cent of heterotrophic bacterial production was from the uptake of released algal carbon. Similar results have been found in both freshwater and marine pelagial systems (e.g., Saunders and Storch, 1971; Meffert and Overbeck, 1979; Larsson and Hagström, 1979; Bell and Satzshang, 1980).

Decomposition of Particulate Organic Detritus

Kinetic evaluations of aerobic decomposition of algae by bacteria approximate first-order reaction rates during the initial 30 days of degradation. Decomposition of cellular nitrogen and the production of dissolved organic nitrogen demonstrate that dead algal organic nitrogen is separable into labile and refractory constituents according to their relative resistance to bacterial degradation. Additional dissolved organic nitrogen is generated by bacteria through reassimilation of mineralized nitrogen. The general sequence of decomposition of dead algal material, the major organisms involved in this degradation, and the simultaneous regeneration of organic nitrogen are indicated in Figure 17-7. Following proteolysis by means of various enzyme systems, amino acids are deaminated by deaminases to ammonia; the ammonia then proceeds through a series of nitrification stages to hydroxylamine, nitrite, and nitrate. Organic nitrogen is synthesized during the processes of nitrate reduction, denitrification, and nitrogen fixation (see Chapter 12), each of which varies with numerous environmental parameters. In turn, lysis of the microbial flora generates further dissolved organic nitrogenous compounds.

Decomposition rates of organic matter are quite different under anaerobic conditions. For example, after 60 days of anaerobic decomposition of green algal cells at 20°C, 30 per cent of algal cell carbon was transformed into dissolved organic carbon (DOC) and 20 per cent was mineralized to CO_2; 50 per cent remained as particulate matter. Of the algal cell nitrogen, only 8 per cent was transformed to soluble form, 48 per cent was mineralized, and 44 per cent remained in particulate form after 60 days. The DOC in solution consisted largely of yellow organic acids that were resistant to further degradation. The decomposition rates and production of dissolved organic compounds from dead algal cells were only one-fourth as great under anaerobic as under aerobic conditions, and the rate constant of cell carbon and nitrogen decomposition decreased to less than one-half under anoxic conditions. Similar results have been found for the decomposition of aquatic angiosperms under anaerobic conditions (cf. Godshalk and Wetzel, 1978a, 1978b, 1978c).

The rates of decomposition under both aerobic and anaerobic conditions are strongly influenced by temperature (usually higher rates with increasing temperatures in the natural range), pH (commonly lower at acidic pH values), the major cation concentrations and cationic ratios, and other factors. The rates of cycling and utilization of the numerous dissolved organic intermediate products of bacterial and fungal degradation, such as amino acids, keto acids, various fatty and nonvolatile organic acids, phenolic derivatives, mercaptans, and others, are highly variable (see, for example Krause, et al., 1961).

Experimental studies on the transformation of artificially generated radioactive algal detritus under conditions approaching a natural community are instructive in understanding the relationships involved (Saunders, 1972). The living components of the water constitute a very small portion of the total organic matter (Fig. 17–8); most organic matter occurs as dissolved organic matter (DOM) and as detrital particulate organic mattter (POM). For example, in the surface water of a productive lake, the summer POM averaged 57 per cent of the total seston (inorganic and organic; Table 17–8.) The phytoplanktonic dry weight ranged from between 5 and 25 per cent of the organic particulate detritus most of the time, and only occasionally exceeded 40 to 50 per cent.

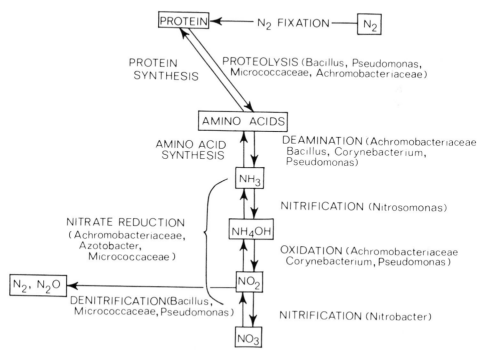

Figure 17–7. Decomposition and utilization stages of nitrogenous organic matter of aquatic organisms. (Modified from Botan, et al., 1960.)

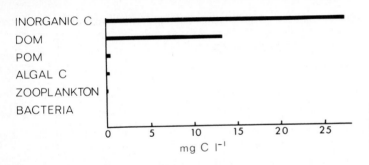

Figure 17-8. Distribution of carbon in inorganic and organic fractions at 1 m in Frains Lake, Michigan, 22 July 1968. (Modified from Saunders, 1972.)

Similar results have been shown for the seston composition of Lake Erie (Leach, 1975) and several Russian lakes (Winberg, et al., 1970). Considering that the POM constitutes only about 10 per cent of the total detrital organic matter (approx. 90 per cent as DOM), and that living organic matter makes up only a small fraction of the particulate component, it is clear that dissolved and particulate organic detritus is a major component of aquatic systems.

Most of the soluble components of DOM are utilized rapidly by bacteria in a period of usually less than 5 days. Decomposition of the particulate organic detritus, generated from planktonic communities of surface lake water, occurs more slowly at rates on the order of 10 per cent per day (Table 17-9). Radiolabeled dissolved organic carbon derived from the particulate organic detritus amounted to about 1 per cent of the initial radioactivity after one day; these values were about 2 to 6 times greater than the amount of dissolved organic matter released extracellularly by phytoplankton in the trophogenic zone.

Decomposition rates of detrital POM of course are influenced by an array of compositional and environmental factors. Certainly differences in the organic composition of algal cell walls lead to differences in degradation rates. For example, POM of dead blue-green algae decomposes much more rapidly than that of green algae and diatoms;

TABLE 17-8 Ratio of Dry Weight of Organic Particulate Detritus to Phytoplankton, and Organic Particulate Detritus as a Percentage of Seston, Frains Lake, Michigan

TIME PERIOD	RATIO OF DETRITAL POM TO PHYTOPLANKTON	ORGANIC PARTICULATE DETRITUS AS A % OF SESTON
21 Jan. 1967	5.7	—
27 Apr. 1967	4.6	45
3 May 1967	6.6	72
9 May 1967	16.9	65
16 May 1967	15.1	67
23 May 1967	2.3	21
30 May 1967	12.0	58
6 June 1967	4.5	57
14 June 1967	4.9	61
20 June 1967	8.4	68
27 June 1967	4.8	62
22 June 1968	1.3	44

After Saunders, 1972a.

TABLE 17-9   Decomposition of Fine Particulate Organic Matter in Lake Water

SUBSTRATE	PER CENT LOSS PER DAY
Phytoplankton	14
Mixed dead phytoplankton and detritus	5–20
Detritus (algal)	8
Large filamentous algae	7–25
Zooplankton	1–33

After data of Krause (1959), Saunders (1972a), and Pieczyńska (1972a).

desmid algal POM is especially resistant (Mills and Alexander, 1974; Gunnison and Alexander, 1975). The concentrations and species diversity of bacteria also are involved. Among environmental parameters, temperature and oxygen are paramount, and availability of inorganic nutrients, especially nitrogen, can also be an important variable.

Bacterial decomposition of organic detritus (soluble or particulate) can be considered simply as the rate of change between bacteria and detrital substrates, compounds, or particles, and can be expressed as a biomolecular second-order reaction (Saunders, 1972b):

$$(m_1, m_2. \ldots) \, (T, [N], n_1, n_2. \ldots)$$

$$\frac{-d \, [S]}{dt} = C_1 \, [S] \times C_2 \, [B]$$

when:

$$[S] = \text{concentration of detrital substrate,}$$
$$[B] = \text{concentration of bacteria,}$$
$$T = \text{temperature,}$$
$$[N] = \text{concentration of inorganic nutrients,}$$
$$m_1, m_2. \ldots; n_1, n_2. \ldots = \text{other variables,}$$
$$C_1, C_2 = \text{coefficients.}$$

The rate of decomposition of detrital substrates is a function of their concentration and of the enzymatic activity of the bacteria dispersed in the water or attached to surfaces of detrital particles (Saunders, 1972b). The coefficients C_1 and C_2 are obviously highly dynamic and variable operators on the detrital–bacterial interactions. Control variables that operate on the substrate coefficient C_1 include various biodegradability factors (Alexander, 1965b; 1975). Examples are many. A structural characteristic of the substrate molecule or of the particle surface parameters (that is, area, particle size, adsorption sites) can inhibit the enzymes. Certain enzymes can be inactivated by adsorption to surfaces of minerals or colloids, or inhibited by phenolic and polyaromatic substrates or their derivatives. Substrates may be rendered inaccessible not only by surface characteristics, but also by transport of the substrates to microenvironments in which conditions preclude or greatly reduce bacterial or enzymatic activities.

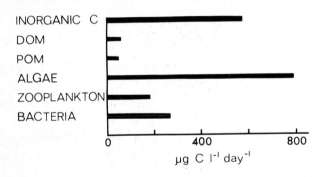

Figure 17–9. Assimilation rates of carbon in μg C l^{-1} day^{-1} by different microbial components at 1 m in Frains Lake, Michigan, on 22 July 1968. (Modified from Saunders, G. W.: Carbon flow in the aquatic system. *In* Cairns, J., Jr., ed., The Structure and Function of Fresh-Water Microbial Communities. Blacksburg, Va., Virginia Polytechnic Institute, 1971.)

Control variables that operate on the bacterial component C_2 include obvious parameters such as temperature T, inorganic nutrient concentrations [N], as well as other essential growth factors or biologically utilizable terminal electron acceptors. The presence of biologically generated organic inhibitors, microbially formed inorganic inhibitors, high salt concentrations, temperature extremes, acidity, or other environmental conditions outside the range suitable for microbial proliferation can all inhibit bacterial growth (Alexander, 1965b; 1975). A microbial community lacking the appropriate enzyme system may not be able to metabolize certain compounds, or the substrate may not be able to penetrate into cells that possess the appropriate enzymes. Thus, the decomposition rates of organic substrates in natural systems are a function of the physical and biological structure at any instant in time, as well as of the substrate characteristics and physiological state of the microflora and their enzymatic systems.

Within this general framework, it is frequently desirable to know the flux rates of organic carbon among the different components of the system. As we have seen earlier (Fig. 17–8), the system structure is dominated by two very large pools of nonliving carbon, the inorganic and dissolved organic carbon. Algae and particulate organic detritus contain much less carbon, zooplankton and other fauna contain even less, and bacteria contain about 10 per cent of that of the zooplankton (Saunders, 1971). This approximate general distribution of carbon mass of the trophogenic zone of lakes bears little relation to carbon flux rates, however.

A one-day intensive study undertaken to evaluate relationships between the carbon structure of a lake's pelagic system and various metabolic flux rates indicates that the microflora control the movement of carbon (Fig. 17–9). The biota utilize the carbon of the various pools within the limits of the environmental conditions. The environmental variables change constantly, resulting in continually changing rates of transfer among the carbon pools. The changes in carbon concentrations and metabolic rates in the surface waters over a 24-hour period during the summer can be generalized in a theoretical way in Figure 17–10. Photosynthetic rates of algae are coupled with light quantity. Usually photosynthetic rates are skewed toward the morning period of high light intensity, and decline precipitously with decreasing light in evening (Saunders and Storch, 1971). Coupled with the cycle of extracellular release of dissolved organic compounds by algae, bacterial assimilation rates increase and reach maximal levels about three hours following the secretion maximum. The division activity of bacterial populations is highest during the night and early morning, when the output of organic substances by the algae is low, whereas volume growth is highest during the second half of the day, when release of photosynthetic algal products is highest (Krambeck, 1978). The ratio of protein to carbohydrate of the algae also shifts diurnally; maximum carbohydrate levels are reached at the conclusion of the daily photosynthetic period. Zooplanktonic grazing rates on the epilimnetic system increase dramatically at dusk with a rapid diurnal migration of many zooplankters to the trophogenic zone. Assimilation rates of algae by *Daphnia* are greater in the night than during the day. Thus, we can envision a series of coupled oscillating reaction systems among the major components of the pelagic system in the trophogenic zone, which responds to an array of environmental variables.

Sedimentation and Decomposition of Pelagic Particulate Organic Matter

The vertical distribution of bacterial numbers and activity in stratified lakes commonly shows a maximum in the trophogenic zone, a minimum in the metalimnion, and an increase in the lower hypolimnion. The decomposition of sedimenting organic matter produced in the epilimnion follows a similar distribution. A number of devices have

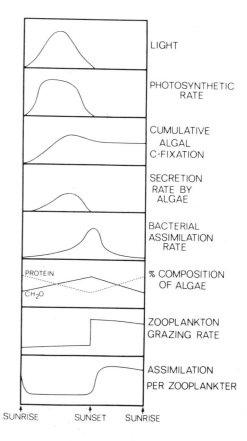

Figure 17–10. Generalized diagrams of the daily cycles of various processes and concentrations in the upper trophogenic zone of a lake system in summer. (Modified from Saunders, G. W.: Carbon flow in the aquatic system. *In* Cairns, J., Jr., ed., The Structure and Function of Fresh-Water Microbial Communities. Blacksburg, Va., Virginia Polytechnic Institute, 1971.)

TABLE 17-10 Percentage of Organic Carbon Produced by Phytoplanktonic Primary Production That Was Mineralized by Bacterial Decomposition per Area at Depth and at the Bottom of the Water Column in 3 Lakes of Northern Germany

SCHLUÈNSEE Dates	Depth (m)	% Mineralization Per Area	Total	SCHÖHSEE Dates	Depth (m)	% Mineralization Per Area	Total	PLUSS-SEE Dates	Depth (m)	% Mineralization Per Area	Total
9 June–6 July	20 30 40	58.8 33.5 5.0	97.3	11 April–18 May	10 20	52.8 30.7	83.5	11 May–30 May	5 15 25	93.1 5.0 0.9	98.9
6 July–16 Aug.	20 30 40	73.7 12.2 11.7	97.6	18 May–4 July	10 20	75.1 16.9	92.0	30 May–23 June	5 15 25	90.6 7.6 1.3	99.5
16 Aug.–7 Sept.	20 30 40	77.6 9.8 1.5	98.5	4 July–10 Aug.	10 20	26.8 54.0	80.8	23 June–8 Aug.	5 15 25	89.8 9.1 0.9	99.7
7 Sept.–5 Oct.	20 30 40	87.8 4.2 6.8	98.8	10 Aug.–5 Sept.	10 20	75.3 18.8	94.1	8 Aug.–16 Sept.	5 15 25	83.7 13.5 2.6	99.7
			Mean = 98.1%	5 Sept.–10 Oct.	10 20	17.2 77.2	94.4	16 Sept.–4 Nov.	5 15 25	70.0 26.5 2.5	99.1
				10 Oct.–17 Nov.	10 20	12.9 80.7	93.6				Mean = 99.4%
				17 Nov.–5 Dec.	10 20	8.8 74.3	83.1				
							Mean = 88.8%				

After Ohle, 1962.

been used to measure the amount of tripton, the term for all nonliving suspended matter (cf. methods discussed in Edmondson and Winberg, 1971 and Bloesch and Burns, 1980). Containers of various types are suspended at several depths to entrap sedimenting materials over a period of one to several weeks, depending on the sedimentation rates. After a period of collection, kept as short as possible to reduce complicating interferences, chemical analyses of the collected seston permit an approximate indirect evaluation of decomposition-induced changes in the organic matter that have occurred during sedimentation.

Most (>75 per cent) soluble organic matter released by secretion and autolysis of phytoplankton decomposes rapidly to a major extent at the site of its generation and release in the trophogenic zone. The particulate organic detritus undergoes slower degradation. During stratification periods in lakes of moderate depth, 75 to > 99 per cent of the particulate organic matter synthesized in the trophogenic zone is decomposed in the water column by the time it reaches the sediment interface (Table 17–10). In Lawrence Lake, Michigan, an annual mean of 88 per cent of the sedimenting particulate organic carbon was decomposed by the time it reached the lower epilimnion (5 m; Wetzel, et al., 1972). Little additional decomposition occurred on further hypolimnetic sedimentation; an annual mean of 10 per cent of planktonic particulate organic carbon of the trophogenic zone was found at 10.5 m. Similar results were obtained by Lawacz (1969) using different methods.

Even though the degradation of particulate organic matter is fairly complete with increasing depth, seasonal variations are great (Fig. 17–11). The increasing concentrations of sedimenting organic matter with depth must be considered in relation to the decreasing area of each stratum as the basin morphometry constricts with depth. This relationship is illustrated diagrammatically in Figure 17–12 for Schluënsee; in this lake, the funneling effects of the basin morphology increased the concentration of organic carbon per unit area, even though the total amount of synthesized particulate organic matter was greatly reduced (> 97 per cent) by decomposition at a cumulative depth of 40 m (Ohle, 1962). As the size of the basin decreases, the significance of the funneling effects of concentration increases.

As lakes become shallower and the importance of littoral production in relation to that of the phytoplankton increases, a greater proportion of the synthesized organic matter is shifted to the sediments. We would expect decomposition to be more complete

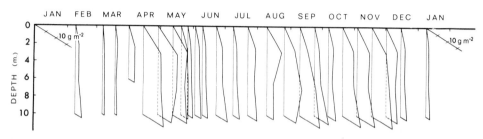

Figure 17–11. Seasonal changes in the organic content of sedimenting seston (g m^{-2}), Lawrence Lake, Michigan, 1972. (From White and Wetzel, unpublished.)

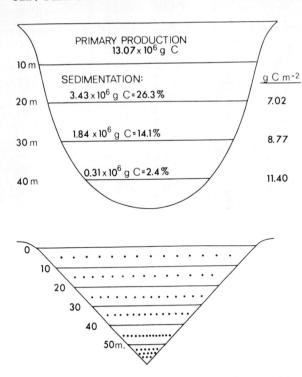

Figure 17-12. Schematic diagrams of (upper) the relationships between planktonic primary production and sedimentation of organic particulate carbon in Schluënsee, Germany, 16 July–16 August 1960, and (lower) the morphometric funnelling effect on sedimentation in lakes. (Modified from Ohle, 1962.) The numbers to the right of the upper figure (g C m⁻²) indicate the concentrating effect of basin morphometry on the quantity of particulate carbon per unit area in each of the strata at increasing depths. The percentage of the particulate carbon produced in the trophogenic zone and sedimenting to each stratum (inside the diagram) is markedly decreased by decomposition at each step.

in the water column of larger, oligotrophic lakes because of the amount of time required for decomposition of refractory materials of algal cells (cf. Jewell and McCarty, 1971) and the distance (and time) through which a particle travels during sedimentation. Evidence supports these expectations.

For example, in a shallow (z_m 6.5 m), hypereutrophic lake in southern Michigan, the high primary productivity of phytoplankton, relatively short distances that algal particulate organic matter must fall to reach the sediments, and the close proximity of the anaerobic hypolimnion (anaerobic below 4 m) to the photic zone resulted in high sedimentation of incompletely decomposed algal particulate matter to the sediments (Molongowski and Klug, 1980b; cf. also Gasith, 1975, 1976). The quantity of organic seston reaching the sediments was not only high in this shallow lake, but was relatively uniform throughout the thermally stratified summer period of high algal productivity. The residual sedimenting seston had a high protein content (50 to 60 per cent of that of phytoplankton) and, as a result, a low C/N ratio (4 to 6). In contrast, in deeper lakes, proteinaceous materials in sedimenting seston are rapidly degraded in the upper water strata (Brehm, 1967; Matsuyama, 1973; Lastein, 1976). The carbohydrate fraction of the sedimenting seston reaching the sediments of this shallow lake consisted primarily of glucose (19.2%), other soluble monomers (12.4%), and soluble polysaccharides (43.4%). Cellulosic polysaccharides represented only 7.2 per cent of the total carbohydrate fraction, which indicates that most of the carbohydrates were of algal origin, even though this lake has a massive littoral development that covers over a third of the lake area. As will be discussed later (chapters 20 and 22), much of the littoral particulate plant production decomposes on or within the sediments.

RESUSPENSION AND REDEPOSITION

A conspicuous feature of the annual cycle of sedimentation rates of seston among dimictic lakes is the marked increase during periods of spring and autumnal circulation (Fig. 17–13). The extent of resuspension of surficial sediments into the water column by the circulating water is of course variable among lakes of differing morphology. The extent of disturbance is also influenced by variations in meteorological conditions prior to initiation of stratification (Pennington, 1974). However, even the deepwater sediments of many lakes are disturbed to depths of from several millimeters to several cen-

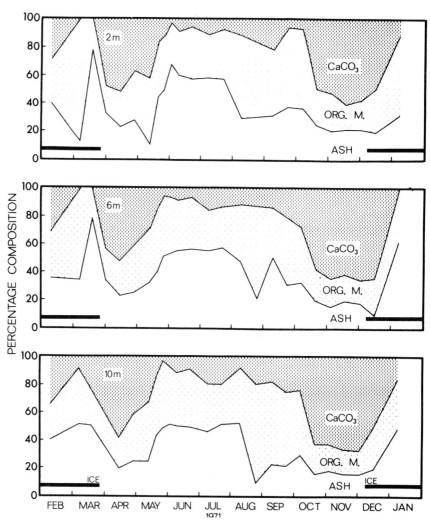

Figure 17–13. Seasonal changes in the percentage composition of calcium carbonate, particulate organic matter, and ash sedimenting to 2, 6, and 10 meter depths. Lawrence Lake, Michigan, 1971. (From White and Wetzel, unpublished.)

timeters (Davis, 1968, 1973; Wetzel, et al., 1972). This fine-grained particulate matter is swept into the water column where it is subject to additional degradation by bacteria under aerobic conditions. In calcareous lakes, as exemplified in Figure 17–13, calcium carbonate forms a major constituent of the seston during the periods of circulation. Resuspension of surface sediments occurs at irregular intervals throughout the year in nonstratified lakes.

Horizontal differential sedimentation of particulate matter has been demonstrated in studies of the movement, deposition, and redeposition of pollen grains of different sizes and densities by Davis and Brubaker (1973). Larger particles with rapid sinking rates are deposited fairly evenly onto the sediments throughout the lake basin. Smaller-sized particles with slower sinking rates in water are kept in suspension in the turbulent waters of the epilimnion, are carried across the lake in wind-induced water currents, and are deposited preferentially onto littoral sediments. Later, particularly during fall circulation, these fine particles are resuspended in the littoral areas, mixed in the lake water, and are redeposited over the entire basin. Larger particles of organic matter or seston are not extensively resuspended and transported from the littoral to the center of the lake, although they are frequently shifted about in the littoral region (Wetzel, et al., 1972). The result is a size-selective sorting and a transport of the finest-sized particles from the littoral, which sediment into the quiescent waters below the metalimnion (cf. also Sebestyén, 1949). This resuspension and redistribution of organic matter has important consequences for the rates of decomposition of particulate organic matter deposited in littoral areas from either littoral production or allochthonous matter brought to the lake basin (cf. chapters 18, 20, and 22).

Planktonic Bacterial Productivity

Comprehensive evaluations of planktonic bacterial productivity, in sufficient detail for functional analyses of carbon flux, are exceedingly few. At best, the situation is one of fragments of information on flux rates within bacterial communities, and many estimates of variable accuracy on metabolic rates. Most of the estimates of bacterial productivity will be discussed in Chapter 22 on the organic carbon cycling in aquatic ecosystems. A few statements are appropriate here, however, in relation to the discussion on the distribution of pelagial bacteria and their growth on specific organic compounds.

CHEMOSYNTHESIS

Microbial production by chemosynthesis occurs to a marked extent in layers having contact with anaerobic zones of an aquatic system, especially in boundary layers between anaerobic and aerobic zones. As indicated in earlier discussions of cycling of several elements, the anaerobic processes of decomposition of organic matter provide reduced inorganic compounds that serve as energy substrates for the chemoautotrophic bacteria. Chemosynthetic production, a type of secondary production, becomes significant primarily in steep gradient of redox potential (Sorokin, 1964a, 1965, 1970). Outside of these layers, chemosynthesis is very low in relation to total bacterial production (Romanenko, 1966; Jordan and Likens, 1980). Chemosynthesis by bacteria is normally very low in infertile waters and lakes of intermediate productivity, and becomes a significant contribution to the whole only in productive or meromictic lakes exhibiting steep redox gradients.

The ratio of dark CO_2 fixation (almost all by heterotrophic metabolism of bacteria; see Gerletti, 1968; Romanenko, 1973) to photosynthetic fixation of CO_2 and bacterial chemosynthesis generally is small in oligotrophic waters. For example, the 14-year average of dark CO_2 fixation was 13.3 per cent of phytoplanktonic fixation in the light in a small Michigan lake (Table 17–11). This percentage is much less for the lake, about halved, if the littoral photosynthesis of this lake is considered in addition to that of the phytoplankton. The ratio of photosynthetic to dark CO_2 fixation decreases in the transition to planktonic eutrophy and hypereutrophy. With further transition of the system or in lakes with a predominance of productivity by emergent macroflora and associated attached and eulittoral microflora, the relative contribution of dark CO_2 fixation to the DOC pool apparently decreases.

BACTERIAL PRODUCTIVITY

Chemosynthetic bacterial fixation rates of CO_2 in eutrophic Plusssee, northern Germany, fluctuated between 0.4 and 30.0 per cent of the total bacterial productivity over an annual period (Overbeck, 1979). Based on results using several methods, total pelagial bacterial production in oligotrophic Mirror Lake, New Hampshire, was 3 to 8 g C m^{-2} $year^{-1}$, which was equivalent to 8–22 per cent of phytoplankton productivity (Jordan and Likens, 1980). In Mirror Lake, 63 per cent of CO_2 uptake in the dark was bacterial in the epilimnion, and 15 per cent in the hypolimnion. In eutrophic Plusssee,

TABLE 17–11   Dark CO_2 Fixation in the Pelagial Zone of Lawrence Lake, Michigan, Integrated for the Water Column at the Central Depression, 1968-81

YEARS	g C m^{-2} $YEAR^{-1}$	MEAN mg C m^{-2} DAY^{-1}	% OF LIGHT CO_2 FIXATION
1968	4.7	12.9	10.4
1969	10.2	27.9	24.4
1970	8.2	22.4	18.8
1971	5.0	13.8	14.0
1972	3.6	9.9	12.2
1973	1.8	4.9	4.5
1974	0.7	1.9	1.8
1975	3.2	8.9	7.5
1976	3.7	10.2	11.5
1977	3.4	9.3	11.0
1978	3.5	9.6	12.1
1979	6.2	17.0	21.4
1980	5.9	16.2	19.8
1981	6.5	17.8	16.6
Average (14 years)	4.7	13.0	13.3

Data of Wetzel (unpublished).

bacterial dark uptake of CO_2 was 40 per cent in the epilimnion and 60 per cent in the hypolimnion. The total carbon flux through planktonic bacteria in Mirror Lake was between 16 and 43 per cent of phytoplankton net production, or 11 to 31 per cent of all autochthonous and allochthonous organic carbon inputs. In lakes that are deeper than Mirror Lake (z_m 10 m), planktonic bacterial production probably accounts for a greater proportion of the total decomposition simply because particulate detritus would have a longer residence time in the water column.

SUMMARY

1. Much decomposition of particulate and dissolved organic matter occurs in the aerobic pelagial waters prior to sedimentation of particulate detritus to the bottom of the basin. Most of this degradation is accomplished by planktonic bacteria, which enzymatically assimilate soluble organic substrates. Rates of degradation are governed by an array of conditions: quality (chemical composition) and quantity of organic matter, physical parameters (e.g., temperature, stratification patterns, basin morphometry), and chemical parameters (e.g., availability of terminal electron acceptors, particularly oxygen; inorganic nutrients).

2. Numbers, biomass, and productivity of planktonic bacteria generally increase with increasing photosynthetic productivity of fresh waters.
    a. A close correlation often exists between seasonal changes in biomass of phytoplankton and heterotrophic bacteria. Often bacterial increases lag behind pulses of phytoplankton by 5 to 10 days.
    b. In thermally stratified lakes, bacterial biomass is commonly highest in the epilimnion, decreases to a minimum in the metalimnion and upper hypolimnion, and increases in the lower hypolimnion.
    c. Seasonal and vertical distributions in bacterial populations can change very rapidly (e.g., a few days).

3. Simple organic substrates (amino acids, mono- and oligosaccharides, simple organic acids, and short-chained unsaturated fatty acids) are assimilated and mineralized by planktonic bacteria in aerobic waters more rapidly than larger, more complex soluble organic compounds.
    a. Dissolved organic substrates are utilized more effectively (faster turnover times) in warm, more productive, nutrient-rich fresh waters than they are in oligotrophic systems at lower temperatures.
    b. Humic compounds usually possess phenolic linkages that are more recalcitrant to degradation than simple substrates, and have longer residence times in fresh waters. These compounds are often cometabolized with simpler substrates in a coupled metabolic process.

4. Dissolved organic compounds released by phytoplankton and other biota during active growth, and upon senescence and death (autolysis), undergo rapid initial decomposition; declining mineralization at progressively slower rates of degradation then follows.

a. Upon death, large amounts (5 to 35 per cent of the total biomass) of organic matter are released as soluble organic compounds.

b. Extracellular release of dissolved organic substrates from phytoplankton can form a major source (to 95 per cent at certain times) of carbon and energy for planktonic bacteria.

c. Usually greater than 75 per cent of the soluble organic matter released by secretion and autolysis of phytoplankton is decomposed in the trophogenic zone.

d. In the pelagial trophogenic zone, decomposition of the algal dissolved organic matter can operate in a coupled oscillating manner (Fig. 17–10). Photosynthetic rates are coupled to light, often with high rates in the morning and increased extracellular release of dissolved organic compounds in the afternoon. Bacterial assimilation and growth attain maximum rates during and slightly after (about three hours) the maximum of algal product release.

5. Decomposition rates of particulate organic detritus approximate first-order reaction rates during the initial stages of degradation, and then decline as more refractory compounds accumulate.

a. Most pelagial organic matter occurs as dissolved organic matter. The particulate organic matter, most of which is nonliving, usually constitutes less than 10 per cent of the total organic matter.

b. Decomposition of fine particulate organic matter occurs at rates from 1 to about 15 per cent per day in warm, aerobic lake water, and is much slower than degradation of dissolved organic matter released by secretion and autolysis.

c. The rapidity and completeness of decomposition of particulate organic matter declines under anaerobic conditions, such as in anoxic hypolimnia of productive lakes.

d. Nitrogen-containing amino and proteinaceous compounds are generally utilized more rapidly than carbohydrate-based compounds. As a result, organic C/N ratios generally increase with time.

e. As particulate organic matter sinks from the trophogenic zone in stratified lakes, the extent of decomposition before reaching the sediments is controlled by many factors.

    i. In oligotrophic to moderately productive lakes of moderate depth (>10 to 20 m), 75 to 99 per cent of the particulate organic matter synthesized in the trophogenic zone is decomposed in the water column before reaching the sediments.

    ii. In shallower and more productive lakes, the amount of organic matter sedimenting out of the trophogenic zone increases, the residence time in the water column decreases (shorter settling distance), and close proximity of anaerobic hypolimnetic conditions results in large amounts of partially decomposed organic seston reaching the sediments.

    iii. A portion of the organic matter of surficial sediments is resuspended into the water column during periods of water circulation and undergoes further degradation in the pelagial water.

6. Chemosynthetic bacterial metabolism, in which CO_2 is assimilated in the presence of utilizable dissolved organic substrates, is significant only in strata of steep redox gradients, such as between an anaerobic hypolimnion and the aerobic trophogenic zone. Bacterial chemosynthesis is usually insignificant in aerobic waters, and is normally very low in oligotrophic and moderately productive lakes.

7. Few direct measurements of bacterial productivity exist from natural communities. Carbon flux through planktonic bacteria is usually less than 50 per cent of net phytoplankton production. Planktonic bacterial productivity as a percentage of all decomposition in lakes increases in deeper lakes that have a longer residence time for degradation of sedimenting particulate organic matter.

LITTORAL COMMUNITIES: LARGER PLANTS

The littoral region is an interface zone between the land of the drainage basin and the open water of lakes. The size of the littoral zone in relation to the size of the pelagial region varies greatly among lakes, and depends on the geomorphology of the basin and the rates of sedimentation that have occurred since the inception of the lake. Most lakes of the world are relatively small in area and shallow. In such lakes, the littoral flora contributes significantly to the productivity, and may regulate metabolism of the entire lake ecosystem.

The general zonation of the littoral region has been introduced earlier (Chapter 8; Fig. 8-2). The physiological and ecological adaptations of freshwater aquatic angiosperms influence their distribution and often permit very high productivity. These plants also provide excellent habitats for photosynthetic and heterotrophic microflora, as well as many zooplankton (Chapter 19) and larger invertebrates (Chapter 21). In addition, the wetland and littoral flora synthesize large quantities of organic matter, most of which is displaced to the sediments before degradation occurs (Chapter 22).

AQUATIC MACROPHYTES OF THE LITTORAL ZONE

CLASSIFICATION OF AQUATIC MACROPHYTES

In any inquiry into the botanical aspects of lakes, separate from the phytoplankton, one is confronted with an array of rather arbitrary definitions of the sessile flora of aquatic systems (Arber, 1920; Gessner, 1955, 1959; Sculthorpe, 1967; cf. Hutchinson, 1975). Many definitions are based on reproductive characteristics in relation to the aquatic stages in the life history of groups of angiosperms. From a strictly botanical viewpoint, such definitions are quite satisfactory. From an ecological standpoint, however, such species-oriented categorizations are unrealistic and ignore major system interrelationships. Alternatively, words such as *hydrophytes* are ambiguous; even though the term *hydrophyte* generally refers to vascular aquatic plants, it can include any form of aquatic plant. The term *aquatic macrophyte*, as it is commonly used, including in this work, refers to the macroscopic forms of aquatic vegetation, and encompasses macroalgae (e.g., the alga *Cladophora*, the stoneworts such as *Chara*), the few species of mosses and ferns adapted to the aquatic habitat, as well as true angiosperms.

Division on the basis of size is admittedly also arbitrary but, as discussed further on, when combined with the definition of the attached microflora, it permits a meaningful separation of primary producers in the littoral zone.

Numerous lines of evidence indicate that aquatic angiosperms originated on the land. Adaptation and specialization to the aquatic habitat have been achieved by only a few angiosperms (<1 per cent) and pteridophytes (<2 per cent). Consequently, the richness of plant species in aquatic habitats is relatively low compared with that of most terrestrial communities. Many aquatic angiosperms possess relics of their terrestrial heritage, such as a cuticle, stomata, and a lignified xylem tracheary structure. Most are rooted, but a few species float freely in the water. Aquatic macrophytes have evolved from many diverse groups, and often demonstrate extreme plasticity in structure and morphology in relation to changing environmental conditions. These factors, in combination with the very heterogeneous conditions of their littoral habitat, make difficult the precise ecological classification of this group into growth forms.

Numerous classification systems have been proposed and used. The primary groups of aquatic angiosperms, rooted and nonrooted, have been subdivided according to types of foliage and inflorescence, and whether these organs are emergent, floating on the water surface, or submersed (Arber, 1920). Differences in the extent of emergence or submergence and the manner of attachment or rooting to the substratum have led to the creation of numerous complex subclassifications and a corresponding terminology (cf. Hejný, 1960; Luther, 1949; den Hartog and Segal, 1964). Although useful for certain phytosociological analyses, this terminology is cumbersome, and lines of demarcation are rarely distinct.

The following simple classification of aquatic macrophytes, after Arber (1920) and Sculthorpe (1967), is based on attachment, but has also proven useful in morphological, physiological, and ecological studies.

(A) *Aquatic macrophytes attached to the substratum*

1. Emergent Macrophytes: These occur on water-saturated or submersed soils, from the point at which the water table is about 0.5 m below the soil surface to where the sediment is covered with approximately 1.5 m of water; they are primarily rhizomatous or cormous perennials (e.g., *Glyceria, Eleocharis, Phragmites, Scirpus, Typha, Zizania*); in heterophyllous species, submersed and/or floating leaves precede mature aerial leaves; many species may exist as (usually sterile) submersed forms; all produce aerial reproductive organs.

2. Floating-Leaved Macrophytes: These are primarily angiosperms that occur attached to submersed sediments at water depths from about 0.5 to 3 m; in heterophyllous species, submersed leaves precede or accompany the floating leaves; reproductive organs are floating or aerial; floating leaves are on long, flexible petioles (e.g., the water lilies *Nuphar* and *Nymphaea*), or on short petioles from long ascending stems (e.g., *Brasenia, Potamogeton natans*).

3. Submersed Macrophytes: These comprise a few pteridophytes (e.g., the quillwort *Isoetes*), numerous mosses and charophytes (stonewort algae *Chara, Nitella*), and many angiosperms. They occur at all depths within the photic zone, but vascular angiosperms occur only to about 10 m (1 atm hydrostatic pressure); leaf morphology is highly variable, from finely divided to broad; reproductive organs are aerial, floating, or submersed.

(B) *Freely floating macrophytes*

An extremely diverse group, freely floating macrophytes are typically not rooted to the substratum, but live unattached within or upon the water. They are diverse in form and habit, ranging from large plants with rosettes of aerial and/or floating leaves and well-developed submersed roots (e.g., *Eichhornia, Trapa, Hydrocharis*), to minute surface-floating or submersed plants with few or no roots (e.g., Lemnaceae, *Azolla, Salvinia*); reproductive organs are floating or aerial (e.g., aquatic *Utricularia*) but rarely submersed (e.g., *Ceratophyllum*).

Because of the major impact of the littoral vegetation on the metabolism of a majority of lake systems, a necessarily brief résumé of the main aspects of their morphological and physiological adaptations is required in order to appreciate their role in the productivity of fresh waters. Excellent reference works include Arber (1920), Gessner (1955, 1959), Sculthorpe (1967), and Hutchinson (1975).

Emergent Flora

Aerial stems and leaves of emergent macrophytes possess many similarities in both morphology and physiology to related terrestrial plants. The emergent monocotyledons such as *Phragmites* and the cattail *Typha* produce erect, approximately linear leaves from an extensive anchoring system of rhizomes. Epidermal cells are elongated parallel to the long axis of the leaf, which allows flexibility for bending. The cell walls are heavily thickened with cellulose, which provides the necessary rigidity. The mesophyll is generally undifferentiated, and contains large air spaces (lacunae) traversed at intervals by diaphragms that are porous to gases but not to water. These lacunae are separated from each other by thin walls of parenchyma cells. The anatomy of vascular bundles is similar to that of typical terrestrial plants: The xylem consists of scattered tracheids and parenchyma cells; the phloem, which consists of sieve tubes, companion, and parenchyma cells, is ensheathed with supporting sclerenchyma fibers. Emergent dicots produce erect, leafy stems, which show greater anatomical differentiation. The mesophyll tissue of leaves is divided into typically dicotyledonous upper palisade and lower spongy layers.

The root and rhizome systems of these plants exist in permanently anaerobic sediments, and must obtain oxygen from the aerial organs for sustained development (e.g., Sale and Wetzel, 1982). Similarly, the young foliage under water must be capable of respiring anaerobically for a brief period until the aerial habitat is reached, since the oxygen content of the water is extremely low in comparison to that of the air. Once the foliage has emerged into the aerial habitat, the intercellular lacunae increase in size, thus facilitating gaseous exchange between the photosynthetic cells and the atmosphere.

Oxygen concentration in the gases of the intercellular lacunae declines as one moves from the aerial foliage to the underground organs. The rate at which oxygen is transported through the lacunal system is influenced by the rate of net photosynthetic oxygen production in the foliage, which undergoes diurnal fluctuation, by the respiratory demands for oxygen of the root system, and by the resistance to diffusion afforded by diaphragms of the lacunae. Experimental analyses have demonstrated that roots and rhizomes can tolerate appreciable periods, as long as a month, of very low oxygen supply or anaerobiosis without any apparent harmful effects.

TRANSPIRATION

The epidermal cuticle of emergent and floating leaves is generally well developed. Cuticular transpiration is much lower in these plants than in land plants (Table 18–1). Cuticular transpiration is generally greater from floating-leaved macrophytes than from aerial foliage of emergent macrophytes. As a result, transpiratory water loss to the atmosphere by emergent macrophytes occurs largely via the stomata.

Rates of transpiration by emergent macrophytes are extremely high, and result in an efflux of water vapor from the leaves that is much greater than evaporation from an equivalent area of water (cf. Chapter 4, Table 4–3). Evaporation from the water surface is lower when the water surface is densely covered by macrophytes, such as water lilies or duckweeds (Lemna). These plants restrict contact of moving air with the water surface. However, this effect is counterbalanced by plant transpiration rates that increase markedly when the vegetation is exposed to moderate wind velocities (e.g., van der Weert and Kamerling, 1974; Benton, et al., 1978). As a result, the relative humidity of the atmosphere in and above stands of emergent or floating-leaved vegetation is greatly increased in comparison to that of air overlying open water. The differential increase in relative humidity of air over a macrophyte stand in relation to that over open water changes diurnally as well, and is maximal at the midday peak of net photosynthesis. Stomata of aquatic macrophytes generally do not close to nearly the extent observed in terrestrial flora during the light period, and higher transpiration rates occur in aquatic macrophytes than are found among land plants. In both aquatic plants and terrestrial flora, stomata prevent desiccation when water availability declines.

> Many aquatic plants with emergent and floating leaves have developed means of increasing surface area and evapotranspiration. Such structures include papillae, either single-celled or multicellular extensions of the epidermis, or perforations into the leaf blade, termed *stomatodes*. Many emergents exude water (guttation) from enlarged ends of veins (hydathodes) around the margins of their leaves.

The very high transpiration rates among emergent vegetation also vary greatly diurnally, and change seasonally with the morphological development of the plant species (Bernatowicz, et al., 1976; Królikowska, 1978). For example, among the reeds *Phragmites* and *Glyceria*, daily transpiration patterns are positively correlated with increasing temperature, decreasing relative humidity, and increasing evaporation rates. Transpiration is usually lower in the basal leaves than in apical foliage, resulting in a general tendency toward xerophytic properties with increasing age of the leaves.

Among aquatic macrophytes, population density is important in the regulation of transpiration rates. Evapotranspirational water losses by emergent macrophytes in the littoral zone of lakes and marshes can be so effective that the plants can significantly reduce the water levels of the surrounding terrestrial area, and can result in diminished growth of nearby terrestrial plants.

TABLE 18-1 Comparison of the Ratio of Maximal to Minimal Cuticular Transpiration among Aquatic Macrophytes and Terrestrial Flora

AQUATIC FLORA		TERRESTRIAL FLORA	
Potamogeton natans	1.4	Quercus ilex	4.6
Eichhornia crassipes	1.5	Syringa vulgaris	4.9
Alisma plantago	1.7	Viola tricolor	7.6
Nymphaea marliacea	3.2	Laurus nobilis	8.4

Modified from Gessner, 1959.

Increases in water level, to the point when large portions of or the entire emergent plant are submerged, result in a varied number of morphological and physiological changes. Many species can survive submergence for long periods of time, but then growth is greatly reduced because of conditions of diminished light and oxygen. Leaves are generally reduced in size, and tend to elongate and lose rigidity. Mesophyll tissue often is reduced, and there is a corresponding increase in spongy tissue and intercellular air spaces. The xylem and lignified portions of the phloem fibers also generally decrease, concomitant with an increase in stem diameter and lacunal air spaces.

Floating-Leaved Macrophytes

The macrophytes that are attached to the substratum and possess leaves that float on the water surface are nearly all angiosperms, most conspicuously represented by the ubiquitous water lilies. The surface of the water is a habitat subject to severe mechanical stresses from wind and water movements. Adaptations to these stresses by floating-leaved macrophytes include the tendency towards peltate leaves that are strong, leathery, and circular in shape with an entire margin. The leaves usually have hydrophobic surfaces, and long, pliable petioles. A similar suite of adaptations is exhibited by floating-leaved macrophytes from many taxonomically unrelated groups. In spite of these adaptations, severe winds and water movements restrict these macrophytes to relatively sheltered habitats in which there is little water movement.

Floating leaves exhibit well-developed dorsoventral organization, in which the mesophyll usually is differentiated into an upper photosynthetic palisade tissue and an extensive lacunate tissue. Localized masses of spongy tissue aid buoyancy and, in combination with vascular tissues, offer resistance to tearing. The venous network of the leaves of aquatic macrophytes is much less extensive than among terrestrial plants, and is least developed among submersed angiosperms (Table 18-2).

Positioning of leaf surfaces parallel to the water surface creates a vigorous competition for space to expose maximum leaf area to incident light. Leaf growth some distance from the root or rhizome system is accommodated by long, very pliable petioles. Between 10 cm and 4 m water depth, a complete proportionality is found between water depth and the length of the leaf petioles of water lilies. Petioles are about 20 cm longer than the water depth, which permits leaves to remain on the surface

TABLE 18-2   Vein Length Densities (per unit leaf area, cm cm^{-2}) in Different Plant Types

AQUATIC PLANTS		TERRESTRIAL PLANTS	
Floating Leaves		*Rosa canina*	108
Nelumbo		*Acer campestris*	102
Aerial leaf		*Fraxinus excelsior*	88
Edge	104	*Quercus cerris*	86
Middle	91	*Salix rubrum*	52
Floating leaf			
Edge	68		
Middle	71		
Nymphaea mexicana	44		
Salvinia auriculata	27		
Submersed Leaves			
Potamogeton praelongus	14		
Nuphar luteum	7		

After Gessner, 1959.

among undulating waves. Stems of flowers, separate from the leaves in many species, grow to the surface in the same way. If gaseous exchange to the leaf surface is restricted experimentally, elongation of the petioles continues, and the leaves become aerial instead of floating. Although the control mechanisms of elongation are poorly known, it appears that neither oxygen nor carbon dioxide availability is involved, but rather that petiole or stem growth is arrested at the surface when the leaves begin to lose ethylene to the atmosphere (see page 526). Under crowded conditions, elongation frequently continues until leaves extend somewhat above the water surface.

The presence and abundance of stomata vary among different species of floating-leaved angiosperms. The most common pattern is a restriction of stomata to the upper-leaf epidermis. A few stomata are found on the undersides of leaves; these are considered relict features and have no known function.

The epidermis of the leaf underside of many floating-leaved and submersed angiosperms contains groups of many smaller cells nested around a pore about 0.05 μm in diameter; an extremely thin cutin lamella lies over the pore. These "organs," termed *hydropoten*, assume a three-celled lenticular form in some water lilies or shieldlike hairs in certain submersed angiosperms. The hydropoten function in ion absorption in both submersed and floating leaves in contact with water by an active mechanism analogous to that of root absorption (Lüttge, 1964). Absorbed ions are then translocated to veins of the mesophyll.

Submersed Macrophytes

The submersed macrophytes are a heterogeneous group of plants that include (a) filamentous algae (e.g., *Cladophora*) that under certain conditions develop into profuse mats and may become loosely attached to the substrata of the littoral zone, (b) certain macroalgae (e.g., the calcareous stoneworts, Charales) that can dominate littoral macrophyte communities in hard-water lakes, (c) mosses that are occasionally the major macroflora of soft-water lakes and streams, (d) a few totally submersed nonvascular plants (e.g., the lycopsid quillwort, *Isoetes*) that are most abundant in clear, soft-water lakes, and (e) vascular submersed macrophytes of slightly less than 20 diverse families, mostly monocotyledons.

Among the vascular submersed macrophytes, numerous morphological and physiological modifications are found that allow existence in a totally aqueous environment. Stems, petioles, and leaves usually contain little or no lignin, even in vascular tissues. Sclerenchyma and collenchyma are usually absent. No secondary growth occurs, and no cambium can be recognized. Conditions of reduced illumination under the water are reflected in numerous characteristics: an extremely thin cuticle, leaves only a few cells in thickness, and an increase in the number of chloroplasts in epidermal tissue. Leaves tend to be much more divided and reticulated than those of terrestrial or other aquatic plants. The vascular system is greatly reduced, and all major conducting vessels have been lost from the stems. As a rule, conducting bundles have coalesced to axial vascular bundles in both mono- and dicotyledon submersed species. Phloem and xylem cannot be differentiated, since phloem and woody parenchyma are nearly absent.

Leaves of submersed macrophytes occur in three main types: entire, fenestrated, and dissected. Entire leaf form is the most common throughout all groups and habitats. Entire leaves are often elongated to ribbonlike and filiform (threadlike) morphology in which length greatly exceeds width, even among more lanceolate-type leaves. Elongated, pliable leaves resist tearing in moving water, maximally utilize the reduced available light, increase the ratio of surface area to volume, and therefore presumably increase the efficiency of gaseous exchange and nutrient absorption.

Fenestrated leaves are rare among submergents and occur only among a few tropical monocots. The adaptive advantages of such perforated, lacelike foliage are unclear. Dissected leaves are common among submersed dicotyledons. The most common form is extreme dissection with segments in whorls radiating from the petiole. Both the fenestrated and dissected leaf forms greatly increase the surface-area-to-volume ratios.

HETEROPHYLLY AND DORMANCY

Although vegetative polymorphism or *heterophylly* is not unique to submersed macrophytes, extremes in foliar plasticity are particularly conspicuous among aquatic macrophytes that normally grow in shallow, submersed habitats that undergo fluctuations in water level. Such species can readily produce floating or aerial foliage. A marked polymorphism of leaves may often be found on the same stem or petiole. There is a tendency to shift from finely divided submersed leaves in the submersed habitat to more coalesced entire leaves in the transition to surface-floating or aerial leaves. Leaf form and anatomy can vary widely with age, water depth, current velocities, nutrient supplies, light intensity and day length, temperature, and other factors.

Heterophylly is particularly conspicuous among such common genera as *Ranunculus*, *Callitriche*, *Hippuris*, and *Potamogeton*, and creates chaos when the taxonomy is based on leaf morphology. For example, leaf morphogenesis of the amphibious buttercup *Ranunculus* responds conspicuously to submergence and water temperature (Fig. 18–1). The lobate tripartite aerial leaf of the semi-aquatic phase is approximated in the submersed phase at high temperatures. Both lobe numbers and blade lengths increase conspicuously with growth at decreasing temperatures. Analogous morphological differentiation has been shown to be directly related to carbon dioxide concentrations in at least three taxonomically distinct amphibious species (Bristow, 1969). Under conditions of high aqueous CO_2 concentrations, the plants developed dissected submersed leaves, whereas when the CO_2 content was reduced, lobate aerial foliage developed rapidly. These plants were unable to utilize bicarbonate as a source of inorganic carbon for photosynthesis. Among water lilies, continued elongation of petiole length occurs under conditions of low CO_2 of the water until the leaves reach the relatively high levels of atmospheric CO_2 (Gessner, 1959).

Transitions between submersed and aerial-type leaves of *Hippuris* are reversible by altering the ratio of red (660 nm) and far red (730 nm) light, implying direct phytochromal control (Bodkin, et al., 1980; cf. Stross, 1981). A low ratio of red to far red light, such as occurs in very shallow water (far red light is selectively attenuated with increasing depth), induces aerial leaf formation on submersed shoots.

Recent evidence has demonstrated that when leaves of several aquatic angiosperms are submersed, ethylene concentrations reach high levels in the internal

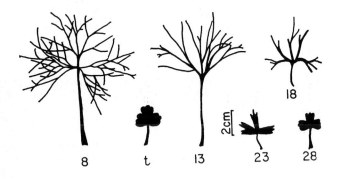

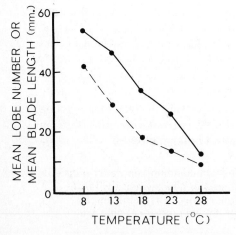

Figure 18–1. *Upper:* Silhouettes of leaves of a single clone of *Ranunculus flabellaris* grown in the semi-aquatic phase (t) and submersed at temperatures of 8, 13, 18, 23, and 28°C. *Lower:* Mean blade length (———) and mean lobe number (- - - -) in relation to temperature. (Modified from Johnson, M. P.: Temperature dependent leaf morphogenesis in *Ranunculus flabellaris.* Nature (London), 214:1354–1355, 1967.)

intercellular lacunae (Musgrave, et al., 1972; Musgrave and Walters, 1973). These high levels of ethylene enhance the sensitivity of the tissue to the hormone gibberellic acid and consequently promote petiole elongation. When the tissues reach the water surface, gaseous contact is made between the intercellular air and the atmosphere, and the accumulated ethylene rapidly dissipates into the atmosphere. Gibberellic acid activity and cell elongation then decline. The plant growth regulator abscisic acid also induces the formation of floating leaves in totally submersed germinating tubers of *Potamogeton nodosus* (Anderson, 1978).

Numerous other factors influence leaf form among aquatic macrophytes (cf. Sculthorpe, 1967; Gaudet, 1968; Hutchinson, 1975). Generalizations are difficult to make, however, because among some plants polymorphism occurs irrespective of external factors; in such cases, polymorphism is distinguishable at the earliest primordial stages, and is apparently determined either genetically or by the nutritional status of the shoot meristem.

Further morphological variations are found among temperate aquatic macrophytes that overwinter by the formation of *winter buds (turions)*. In the fall, many common macrophytes (e.g., *Utricularia, Myriophyllum, Potamogeton*) form masses of aborted leaves with very short internodes in the axils of lower leaves (Weber and Noodén, 1974, 1976a, 1976b; Winston and Gorham, 1979; Aiken and Walz, 1979; Sastroutomo, 1980a, 1980b). These *turions* separate from the mother plant and sink or float some distance away, and serve as a means of vegetative propagation (Weber, 1972). Many submergents persist for years by this means, without a sexual cycle. Turion formation is absent among the same species in the tropics, but can be easily induced when plants near maturity are exposed to low water temperatures (less than 10 to 15°C) and short daily photoperiods (e.g., Van, et al., 1978).

Freely Floating Macrophytes

The freely floating macrophytes, which occur submersed or on the surface, also exhibit great diversity in morphology and habit. Several of these plants, notably *Lemna, Pistia, Salvinia, Eichhornia,* and *Trapa,* develop so profusely in some waterways and lakes that they inhibit the commercial use of these systems (Hillman, 1961; Moore, 1969; Mitchell and Thomas, 1972).

The most elaborate of the free-floating life forms consists of a rosette of aerial and surface-floating leaves, a greatly condensed stem, and pendulous submerged roots. Most rosette species are perennials and are free-floating throughout growth, except for initial stages of seedling development. Among some free-floating groups, there is a strong tendency for reduction from the rosette habit. In the duckweeds (Lemnaceae), extreme reduction is seen in the trend toward elimination of roots and lack of separation between stem and leaf. Similarly, in the submersed free-floating "carnivorous" *Utricularia,* stems and leaves are not differentiated, and vegetative organs are greatly modified for entrapment of microfauna. Some species of *Lemna* and *Utricularia* are, like the ubiquitous *Ceratophyllum,* submersed during vegetative growth and rootless. Free-floating macrophytes generally are restricted to sheltered habitats and slow-flowing rivers. Their nutrient absorption is completely from the water; most of these macrophytes are found in waters rich in dissolved salts.

Most floating plants possess little lignified tissue. Rigidity and buoyancy of the leaves are maintained by turgor of living cells and extensively developed lacunate mesophyll tissue (often >70 per cent gas by volume). Vascular tissues of the leaves are very poorly differentiated; in most, the protoxylem is represented by a lacuna. Vegetative propagation by the production of lateral stolons that develop into new rosettes is a common mode of rapid proliferation in this group. All free-floating rosette plants form well-developed adventitious roots, lateral roots, and epidermal hairs. The root system of the water hyacinth (Eichhornia), for example, represents 20 to 50 per cent of the plant biomass.

Gaseous Metabolism by Aquatic Macrophytes

CARBON ASSIMILATION

Atmospheric carbon dioxide is clearly the dominant source of inorganic carbon among emergent, floating-leaved (rooted), and freely floating macrophytes. A few species of freely floating angiosperms, such as some duckweeds (Lemna), utilize both atmospheric and aqueous carbon sources (Wohler, 1966; Wetzel and Manny, 1972; Ultsch and Anthony, 1973). Heterophyllous aquatic macrophytes presumably are similarly adapted to use both atmospheric CO_2 and aqueous $CO_2 - HCO_3$.

Rates of diffusion of gases in water are several orders of magnitude slower in water than in air (cf. Chapter 11) and reduce the availability of CO_2 from aqueous solution. Morphological adaptations among submersed leaves, stems, and some petioles include thin leaves (1 to 3 cell layers), reduced cuticle development, and extreme reduction or elimination of mesophyll, and chloroplasts densely distributed in epidermal cells. These adaptations all increase utilization and exchange of gases. Massive intercellular gas spaces in leaves, stems, and petioles facilitate rapid internal diffusion.

The question of limitation of photosynthesis by the slow diffusion of CO_2 in water has received much attention, especially in relation to the alternate utilization of bicarbonate ions. In large submerged angiosperms and algae, the diffusion path through the cells of the thallus or leaf is long (up to 50 μm), to which must be added an effective gradient zone of about 10 μm of unstirred water at the cellular surfaces; further, diffusion rates through the larger cells are much slower than among microalgae (Steemann Nielsen, 1947; Raven, 1970; Smith and Walker, 1980). Low-velocity currents increase photosynthetic rates among submersed angiosperms by disrupting the stagnant boundary layer and increasing diffusion rates (e.g., Barth, 1957; Westlake, 1967).

The ability to assimilate bicarbonate ions as a carbon source supplementary to carbon dioxide has been demonstrated in a number of investigations. Space does not permit detailed discussion of the conflicting results of some of these studies (cf. Raven, 1970; Helder and Zanstra, 1977; Lucas, et al., 1978; Browse, et al., 1979; Helder, et al., 1980; Prins, et al., 1980). Physiological mechanisms of bicarbonate uptake are not fully understood. However, among certain submersed angiosperms, the process is as follows: (a) Carbon dioxide and bicarbonate ions are taken up through both leaf surfaces. (b) Carbon dioxide (both as free CO_2 and as CO_2 derived from the dehydration of bicarbonate) is fixed, and hydroxyl ions equivalent to the amount of bicarbonate used pass out through the adaxial upper leaf surface. (c) A quantity of cations, primarily calcium, equivalent to the quantity of bicarbonate taken in through the abaxial (lower) leaf surface, is transported from the abaxial to the adaxial leaf surface, thereby achieving stoichiometry and charge balance in all compartments. (d) The passage of bicarbonate, calcium, and hydroxyl ions to the adaxial leaf surface usually results in precipitation of $CaCO_3$ on that surface. The carbonate deposits encrusting the submersed parts of macrophytes in calcareous hard waters often exceed the weight of the plant material (Wetzel, 1960). Among emergent and floating macrophytes that utilize atmospheric carbon dioxide, encrustation on submersed plant portions was found to be highly variable and proportional to the extent of development of epiphytic algae. Deposits on submersed species were much heavier and were correlated with morphological variations and the ability to utilize bicarbonate ions.

The ability to assimilate bicarbonate ions as well as carbon dioxide can be regarded as an adaptation by those plants possessing a long diffusion path from the medium (i.e., from an unstirred layer) to sites of biochemical utilization, when they are living in waters containing more bicarbonate ions than free carbon dioxide. In algae, especially larger macroalgae, the ability to utilize bicarbonate ions in photosynthesis is widespread, but not universal. The extensive work on this subject among mosses and angiosperms by Steemann Nielsen (1944, 1947), Ruttner (1947, 1948, 1960), and others (e.g., Bain and Proctor, 1980) has shown that the freshwater red alga *Batrachospermum* and all assayed genera of freshwater submersed mosses utilize only free CO_2 and cannot assimilate bicarbonate. These latter plants are restricted almost universally to soft waters of relatively low pH, or to streams in which CO_2 concentrations are relatively high. Utilization of HCO_3^- under natural conditions has been reported for a number of freshwater angiosperms (e.g., Raven, 1970; Helder and Zanstra, 1977; Lucas, et al., 1978; Browse, et al., 1979; Prins, et al., 1980; Beer and Wetzel, 1981), while other species were found to utilize primarily CO_2 (Wetzel, 1969; Brown, et al., 1974; Van, et al., 1976; Moeller, 1978; Winter, 1978; Kadono, 1980).

AERATION AND METABOLISM IN SUBMERSED ORGANS

It has been assumed that as photosynthesis proceeds in submersed macrophytes, oxygen is released from the plants into the surrounding water in quantities proportional to rates of photosynthesis. Such a relationship is approximately accurate for micro- and macroalgae and apparently also for bryophytes. The presence of large lacunae in aquatic angiosperms complicates the relationship between oxygen released to the water and rates of photosynthesis by introducing variable time lags. During the photoperiod, oxygen accumulates rapidly in the lacunal intercellular atmosphere on a diurnal basis and only slowly diffuses out into the surrounding water (Hartman and Brown, 1966). Consequently, diffusion rates of oxygen into the water are not directly correlated with the intensity of photosynthesis, and accumulations of oxygen in the intercellular lacunae can be utilized for respiration during both the photoperiod and in darkness, without affecting the oxygen concentrations of the medium. Although release of bubbles, largely oxygen, from submerged plants can be significant during periods of intensive photosynthesis in warm water, this release should not be used as a measure of photosynthesis (as has been done, e.g., Odum, 1957, among others) because it can represent only a small proportion of the oxygen actually produced. Most of the oxygen diffuses into the water, or is retained within the plant and used for respiration, especially at low light intensities (and would measure something less than net productivity). At high rates of water flow over the plants, exchange of oxygen with the water is fairly rapid (e.g., Moeslund, et al., 1981).

The magnitude of the internal lacunal system is highly variable among species, but is almost universally extensive and constitutes a major portion (often exceeding 70 per cent) of the total plant volume. The fragility of the lacunal tissue and the possibility of its becoming filled with water if the plant is damaged at any point are protected against by a number of types of lateral plates and watertight diaphragms, permeable to gases, that interrupt the lacunae at intervals (Arber, 1920; Sculthorpe, 1967). Although experimental evidence is very meager, it is known that much of the oxygen produced during photosynthesis and retained in the lacunal system diffuses from the leaves through the petioles and stems to underground root and rhizome systems, where respiratory demands are high.

The gases within the intercellular lacunae of submersed angiosperms diffuse along gas–partial-pressure gradients. In floating-leaved plants (e.g., the yellow water lily *Nuphar*), however, evidence indicates that the internal gas spaces function as a pressurized flow-through system (Dacey, 1981). Ambient air enters the youngest emergent leaves against a small gas-pressure gradient as a result of physical processes driven by gradients in temperature (thermal transpiration) and water vapor (hygrometric pressure) between the atmosphere and the lacunae. The lacunal gas spaces are continuous through young emergent leaves, petioles, rhizomes, and petioles of the older emergent leaves. The older leaves vent the elevated pressure generated by the younger leaves. The resulting flow-through ventilation system accelerates both the rate of oxygen supply from the atmosphere to the root tissue, and the rate of CO_2 and methane (Dacey and Klug, 1979) transport from the roots to the atmosphere.

The movement of oxygen (from the atmosphere in emergent and floating-leaved plants or from sites of oxygen production in submersed plants) to roots is essential to prevent the accumulation of toxic endproducts (e.g., ethanol produced during glycolysis). Wetland and submersed plants often increase the volume of the lacunal gas system when sediments become more reducing (e.g., Katayama, 1961; Armstrong, 1978; Penhale and Wetzel, 1982). In addition, a number of aquatic plants have developed metabolic adaptations under anaerobic conditions: diverse nontoxic endproducts (malate, shikimic acid) can be produced during glycolysis (Crawford, 1978; Penhale and Wetzel, 1982; Sale and Wetzel, 1982), or ethanol can be released to the environment (Bertani, et al., 1980).

Bacterial endproducts of fermentation (H_2S, volatile fatty acids) in anoxic, reducing sediments are toxic to many plants. Some oxygen diffuses from the roots and forms an oxidized microzone in the surrounding sediments (reviewed by Armstrong, 1978). This oxidized microzone reduces the toxicity of fermentation products, but simultaneously reduces the availability of certain nutrients, such as iron and manganese (Tessenow and Baynes, 1978). Certain sulfur bacteria (*Beggiatoa*) of the plant rhizosphere can oxidize hydrogen sulfide to sulfur granules, and consequently lower H_2S concentrations adjacent to root tissues (Joshi and Hollis, 1977).

Photosynthetic carbon fixation of aquatic macrophytes is offset by growth, losses of respiratory CO_2 (mitochondrial and photorespiration), and secretion of soluble organic compounds. Photosynthetic efficiency is influenced directly by rates of these processes.

In calculations of primary productivity, the respiration rate has usually been assumed to be the same in light as in the dark. Physiological evidence, however, indicates that such an assumption is erroneous. Mitochondrial (dark) respiration may be inhibited in the light in some plants, perhaps by suppression of glycolysis (Jackson and Volk, 1970). Efficient refixation of respired CO_2 in the light restricts the loss of respiratory CO_2 from submersed aquatic plants (Søndergaard, 1979; Søndergaard and Wetzel, 1980).

Photorespiration, well known in terrestrial plants (Goldsworthy, 1970; Jackson and Volk, 1970; Hatch, et al., 1971), enhances loss of CO_2 and reduces photosynthetic efficiency of aquatic macrophytes. In this process, CO_2 is generated in the light from glycolic acid, a direct product of C_3–Calvin-cycle photosynthesis. The rate of this reaction is influenced by, and is proportional to, oxygen concentration, light intensity, and temperature. Glycolate metabolism is also enhanced when low CO_2 limits photosynthesis. The rate at which photorespired CO_2 is lost from the plant depends on the efficiency of CO_2 refixation.

Terrestrial plants in which all cells photosynthesize by the C_3–Calvin-cycle can lose up to 50 per cent of fixed carbon immediately by photorespiration, depending on environmental conditions. In contrast, little or no CO_2 is lost in the light from plants in which photosynthesis proceeds through the C_4–β-carboxylation pathway in mesophyll cells. These plants efficiently refix CO_2 both from dark respiration and from photorespiration in the C_3 bundle sheath cells.

C_3 and C_4 plants can be distinguished by the following combination of characteristics, any one of which is often reasonable evidence for making the distinction (Black, 1971): (a) C_4 plants have highly developed bundle sheath cells in leaf cross sections, with unusually high concentrations of organelles and starch accumulation (elucidated by use of iodine–potassium iodide stain); C_3 leaves lack this differentiation. (b) Photosynthesis is difficult to light-saturate in C_4 plants, while in C_3 plants, saturation illuminance is in the range of 10,000 to 45,000 lux. (c) Photosynthetic temperature optima are 30 to 40°C in C_4 plants, 10 to 25°C in C_3 plants. (d) Photosynthetic CO_2 compensation points are low for C_4 plants (0 to 10 ppm CO_2), and high for C_3 plants (30 to 70 ppm CO_2). (e) Response of apparent photosynthesis to O_2 concentration is not detectable in C_4 plants, and shows increasing inhibition above 1 per cent O_2 in C_3 plants. Response of photorespiration (CO_2 release in light) to O_2 concentration is not detectable in C_4 plants, but shows increasing enhancement with increasing O_2 in C_3 plants. (f) Glycolate synthesis and glycolate oxidase activity are low in C_4 plants as compared to C_3 plants.

Emergent hydrophytes are partially exposed to environmental conditions similar to those of terrestrial plants, and photorespiration can undoubtedly be extensive. The C_4 photosynthetic system thus is very likely of adaptive value in many emergent hydrophytes, particularly in regions of high temperature and high light intensity, or in situations in which the salt content of the environment adversely affects internal CO_2 and water balance (Hatch et al., 1971). In the latter context, *Spartina*, the dominant plant of the salt marshes of the east coast of the United States, and papyrus (*Cyperus papyrus*) of the tropics appear to be C_4 plants (Black, 1971; Jones and Milburn, 1978). There is some evidence that *Typha latifolia* is a C_3 plant with relatively low rates of photorespiration (McNaughton, 1966a, 1969; McNaughton and Fullem, 1969); limiting mechanisms are not known in this case.

Submersed macrophytes are exposed to lower concentrations of ambient oxygen, lower levels of light, and lower summer temperatures than are terrestrial and emergent aquatic plants. Furthermore, submersed and emergent aquatic plants are never subjected to the water stress to which terrestrial C_4 plants of arid environments have become adapted. The slower diffusion of CO_2 in water than in air, and the presence of massive internal gas lacunae can retard loss of CO_2 from submersed angiosperms and facilitate refixation of respired CO_2 (Carr, 1969; Hough and Wetzel, 1972; Hough, 1974; Søndergaard, 1979; Søndergaard and Wetzel, 1980). In soft-water lakes, concentrations of total inorganic carbon in the water are very low; certain soft-water submersed angiosperms utilize CO_2 of the sediment-interstitial water to supplement CO_2 assimilated from the water (Wium-Andersen, 1971; Søndergaard and Sand-Jensen, 1979a). Uptake of CO_2 by root tissue for photosynthetic fixation by submersed plants of hard-water lakes, in which concentrations of inorganic carbon were high, could not be demonstrated (Wetzel and Beer, in preparation).

The photosynthetic CO_2 compensation points of submersed angiosperms are generally very high (Helder, et al., 1974; Brown, et al., 1974; Van, et al., 1976; Lloyd, et al., 1977; Hough and Wetzel, 1978; Bowes, et al., 1979). In addition, photorespiration is enhanced under conditions of increasing concentrations of dissolved oxygen. Although evidence from early studies of the biochemical pathways of carbon fixation in submersed angiosperms was conflicting, the results of recent investigations indicate that these angiosperms possess the C_3–Calvin type of photosynthetic pathway (Hough and Wetzel, 1977; Winter, 1978; Browse, et al., 1979; Valanne, et al., 1982; critically discussed in Beer and Wetzel, 1980, 1982). Carbon incorporation into malate (a characteristic of C_4 plants) in submersed angiosperms is often a stable product rather than an intermediary carbon source for net photosynthetic fixation. In some aquatic species, malate accumulates in the light, and may function as an ion to balance excess cation uptake (Browse, et al., 1980). However, malate may also accumulate in the dark for subsequent decarboxylation and refixation of CO_2 in the light, complementary to fixation of exogenous carbon (Keeley, 1981, 1982; Beer and Wetzel, 1981).

Rates of photosynthetic carbon fixation in submersed macrophytes are generally correlated with the intensity of solar radiation on a daily basis. However, a midday and afternoon depression of carbon fixation is often observed so that highest rates of photosynthesis are skewed toward morning (Wetzel, 1965b; Sculthorpe, 1967; Goulder, 1970; Hough, 1974, 1979). In situ experiments with the submergent Najas showed that the afternoon decrease in net photosynthesis was correlated with an increase in photorespiration. The results of numerous studies suggest that photorespiration may increase through the day with increasing light intensity, with increasing oxygen tension of photosynthetic origin, increasing temperature, and possibly decreasing CO_2 availability (Fig. 18–2). Conditions conducive to accelerated photorespiration can develop in the littoral zone, particularly among dense macrophytic populations during calm weather with little turbulence. An increase in photorespiration would reduce net photosynthesis.

With cessation of photosynthesis during darkness, internal oxygen content decreases rapidly and carbon dioxide increases; then respiration is limited by the rate at which oxygen diffuses from the water to the cells. Dark respiration rates increase with rising temperatures and are generally, but not in all cases, enhanced by high concentrations of oxygen in the surrounding water, especially if the water is agitated slightly by low-velocity currents (Pannier, 1957, 1958; Owens and Maris, 1964; McIntire,

1966; Westlake, 1967; McDonnell, 1971; Hough and Wetzel, 1972; Stanley, 1972; Prins and Wolff, 1974). During the daytime, the effect of oxygen on photorespiration and dark respiration would further increase the release of CO_2.

Photorespiration and dark respiration were found to be 10-fold greater during the fall than during summer as the annual submergent *Najas* was entering senescence (Hough, 1974). Photorespiration in the submersed perennial *Scirpus subterminalis* was relatively low during summer and increased under the ice in late winter.

Physiological evidence indicates that freshwater submersed angiosperms possess C_3–Calvin-type photosynthetic metabolism and high rates of photorespiration, which collectively reduce rates of net photosynthesis. Efficient internal recycling of respired CO_2, utilization of bicarbonate as well as dissolved CO_2, and augmentation of exogenous inorganic CO_2 by decarboxylation of malate formed during darkness are mechanisms by which certain submersed species adapt to the generally restricted availability of CO_2. There is no question that the evolution of an extensive internal lacunal gas system in submersed angiosperms constitutes an array of interrelated morphological adaptations related to (a) enhanced efficiency of carbon fixation, (b) flexibility to withstand water movements, and (c) positive buoyancy and positioning of leaves toward greater available light.

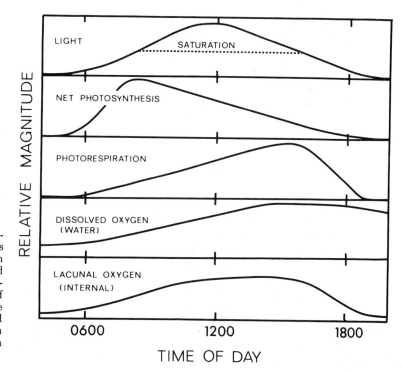

Figure 18–2. Commonly observed relative diurnal variations in light intensity, dissolved oxygen concentrations of the water and the internal lacunal spaces of submergents, net photosynthesis of submersed macrophyte (and some phytoplankton) populations, and proposed concurrent variations in photorespiration. (Modified from Hough, 1974.)

Secretion of Dissolved Organic Matter

The loss of recently synthesized organic carbon and nitrogen as secreted dissolved organic compounds has been demonstrated for a few submersed and freely floating angiosperms (Wetzel, 1969; Wetzel and Manny, 1972; Hough and Wetzel, 1972, 1975; Hough, 1979; Søndergaard, 1981). Well known among the planktonic algae (see Chapter 15), this loss of organic carbon during active photosynthesis in submersed macrophytes can represent a significant reduction in photosynthetic efficiency. Secretion rates of dissolved organic carbon vary greatly (0.05 to over 100 per cent of carbon photosynthetically fixed) under an array of experimental conditions of varying light and ionic composition of the medium. Most values, however, including those from in situ analyses, were between 1 to 10 per cent of photosynthetically fixed carbon, increasing somewhat during the course of the daylight period.

The secretion may represent an incomplete adaptation of submersed macrophytes to a totally aqueous medium. Functionally, however, the secretion of labile organic compounds by submersed macrophytes enhances the development of a highly productive epiphytic community of microflora (Wetzel and Allen, 1970). Nutrient interactions, both inorganic and organic, take place between the submersed macrophyte and epiphytic algal and bacterial populations (cf. Allen, 1971; Allanson, 1973; Coveney and Søndergaard, 1982; Chapter 19). These interactions are lacking or greatly reduced in the diluted planktonic regime. This apparent decrease in photosynthetic efficiency by loss of dissolved organic carbon could be viewed as a symbiotic interaction between the macrophyte and its epiphytic bacteria and algae.

Absorption of Ions and Nutrition

The functional significance of the rooting systems in aquatic macrophytes, primarily angiosperms, has been a long-standing question in limnology (reviewed in Gessner, 1959; Wetzel, 1964; Sculthorpe, 1967; Bristow, 1975; Hutchinson, 1975). There are two views held, each with supporting evidence: first, that the rooting systems function in absorption of nutrients from the substratum, and second, that the roots function merely as organs of attachment. Among emergent and floating-leaved angiosperms with active transpiration-mediated root-pressure systems, nutrient absorption and translocation from the roots to the foliage are clearly operational. The anaerobic sediments, while necessitating a well-developed aeration system from aerial organs, are highly reducing and consequently contain relatively large quantities of nutrients in available form. However, macroalgae, liverworts, mosses, ferns, and other macrophytes that float on the surface or in the water or are variously attached to the substratum presumably obtain nutrients by foliar absorption rather than by rhizoidal structures. An exception is the rhizoid-bearing macroalga *Chara*, which absorbs phosphorus equally well in all parts; phosphorus absorbed by the rhizoids is translocated to other parts of the plant (Littlefield and Forsberg, 1965).

In rooted submergents, a positive absorptive capacity is present and guttation always occurs in small quantities, increasing from root tip to the basal portions of the plant and increasing from the apical parts to the base (reviewed by Stocking, 1956).

Water exits via hydathodes on the upper portions of many submergents, and nutrients can enter apically through the numerous porous hydropoten. The extensive development of roots and root hairs among aquatic plants ($>$95 per cent of 200 species of 105 genera and 54 families) emphasizes the probable dependency of these plants on roots for absorption of solutes as well as for anchorage (Shannon, 1953).

The absorption of phosphorus by roots of submergents both from the water and from the sediments is well known (cf. Chapter 13, and reviews of Gessner, 1959; Schwoerbel and Tillmanns, 1964a; Bristow, 1975). Among most submergents, phosphate absorption rates by foliage are proportional to and dependent upon the concentrations in the water. Very large quantities of several mg l^{-1} are rapidly assimilated in excess of requirements until concentrations in the water are reduced to about 10 μg l^{-1}. These relationships of luxury consumption, known especially for phosphorus but also for nitrogen, have led to the application of elemental analysis of tissues for evaluating nutrient supplies for algae and aquatic angiosperms, a technique long used in agriculture (Gerloff, 1969; Fitzgerald, 1972; Schmitt and Adams, 1981). Tissue analysis is based on the assumption that concentration of an element in an organism varies over a wide range in response to the concentration in the environment, and that over part of this range yield (biomass) of the organisms is related to tissue content of the element. Tissue analysis depends on establishing, for each species, the critical concentration for each element of interest, that concentration which is just adequate for maximum growth (Fig. 18–3). Below this point in the "deficient zone," plant yield is dependent on the supply of the element, but the elemental concentration in the organism changes little. Above this point in the "zone of luxury consumption" the content of the element of the organism increases but plant growth does not. The method assumes that below a critical concentration, plant growth is being limited by the supply of that element. Such an assumption is of course true in a general way, and readily demonstrable in crop plants; some data additionally suggest a similar relationship among algae and aquatic plants grown hydroponically in completely aqueous media. The correlation between nutrient concentration of the water and tissue content among submergents rooted in sediments, however, is less clear.

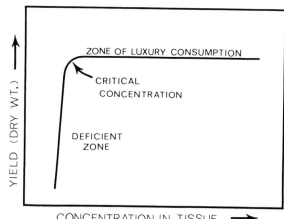

Figure 18–3. General relationship between aquatic plant biomass and the concentration of an essential element in plant tissue. (Modified from Gerloff, 1969.)

Experimental analyses have demonstrated that most rooted submersed angiosperms obtain most of their phosphorus from the interstitial water of the sediments (Bristow and Whitcombe, 1971; Schults and Malueg, 1971; DeMarte and Hartman, 1974; Bole and Allan, 1978; Welsh and Denny, 1979; Barko and Smart, 1980; Carignan and Kalff, 1980). Although some evidence is contradictory (e.g., Seadler and Alldridge, 1977; Swanepoel and Vermaak, 1977), most studies show an active absorption by the roots and translocation to the leaves. When concentrations of phosphate and sulfate of the water are low, absorption of these nutrients by the leaves of *Elodea* is limited by the rate of diffusion through the concentration gradient of the microlayer surrounding the leaves (Jeschke and Simonis, 1965). Absorption of phosphate is limited by the process of active uptake at intermediate exogenous concentrations, and by the rate of diffusive influx into the leaves at high external concentrations (Grunsfelder and Simonis, 1973; Grunsfelder, 1974). Much of the phosphorus uptake from the water by the macrophyte-epiphyte community is utilized by the epiphytic algae (e.g., Howard-Williams, 1981). High phosphorus concentrations in water can encourage the development of epiphytic algae, to the detriment of submersed macrophytes (cf. Chapter 19).

In a floating-leaved water lily (*Nuphar luteum*), phosphorus absorption rates differed with the absorbing tissue (roots > submersed leaves > floating leaves) (Twilley, et al., 1977). About four times more phosphorus moved from the roots to leaves than from the leaves to the roots during summer.

Among emergent aquatic macrophytes, phosphorus requirements are met by uptake of inorganic and organic phosphorus compounds from the sediments (e.g., Islam, et al., 1979; Atwell, et al., 1980). Detailed studies of the growth and reproduction of the cattail (*Typha latifolia*) in marshes of differing successional maturity demonstrated that phosphorus limitations affect not only overall growth, but also the allocation of photosynthate to sexual and vegetative reproduction, early spring mortality, and competitive success (Grace and Wetzel, 1981a).

Rates of nitrate assimilation by the foliage of several submersed macrophytes are considerably less than are rates of ammonia assimilation, especially at high pH values (Schwoerbel and Tillmanns, 1964b, 1964c, 1977; Toetz, 1973b, 1974; Cole and Toetz, 1975; Nichols and Keeney, 1976; Holst and Yopp, 1979; Best, 1980). Ammonium ions are readily absorbed by roots and translocated to apical tissues. Furthermore, nitrogen from N_2-fixing bacteria of the sediments and plant rhizosphere can serve as a major nitrogen source for aquatic plants (e.g., Bristow, 1974a; Tjepkema and Evans, 1976; Purchase, 1977; Kana and Tjepkema, 1978; Ogan, 1979; Watanabe, et al., 1979; Blotnick, et al., 1980).

Considerable information exists on the inorganic chemical composition of aquatic plants (cf. review of Hutchinson, 1975). There are large variations in chemical composition, even among closely related species and from site to site in the same species. Metals and alkalies are absorbed from the water by leaves and interstitial water of sediment by roots (e.g., DeMarte and Hartman, 1974; Marquenie-van der Werff and Ernst, 1979), and many elements are concentrated in the macrophytes in great excess of metabolic requirements. As a result, submersed and emergent plants are often suggested as scavengers of contaminants in surface waters (e.g., Wolverton and McDonald, 1978, 1979).

It is therefore apparent that ion absorption in submergents occurs both from the water by foliage and from the sediments by root and rhizoid systems. Translocation occurs in both directions. Certainly in a majority of cases, roots serve as a primary site of nutrient absorption from the sediments (cf. Denny, 1972). This fact significantly negates the value of presently used tissue analyses of element concentrations in aquatic macrophytes as an index of the fertility of the lake or river water, since nutrient content of the water can be quite unrelated to plant growth of those species having ready access to the abundant nutrient supply in the sediments. Also, as was indicated earlier (Chapter 13), the rooted plants can function as a "pump" of nutrients from the sediments; those nutrients are then lost to the water during both active growth and decomposition.

Rates of Photosynthesis and Depth Distribution of Macrophytes

LIGHT AND TEMPERATURE

The inhibitory effects of high light intensities in the surface waters, commonly observed among planktonic algae, are not generally observed among submersed macrophytes. Results from the few studies on in situ rates of photosynthesis in relation to light indicate a tendency toward increased photosynthesis at 1 to 2 meters below the surface on bright, clear days, where surface light intensities are reduced from 25 to 50 per cent. On the other hand, exposure of some submergents to extremely high light intensities (greater than 80,000 lux) did not significantly alter rates of photosynthesis; other species are clearly adapted to growth at low light levels (Gessner, 1955). Many submergents exhibit distinct morphological variations in relation to light intensity. Shade-adapted leaves are finely divided, whereas on the same plant growing at higher light intensities (shallower depths) leaves can be larger and much more lobate. Photosynthetic rates can be significantly higher near the surface in cases where overlying foliage of dense populations shades the lower leaves. Shade-adapted plants have also been shown to be sensitive to ultraviolet light, which is rapidly removed below the first meter of water, whereas photosynthetic rates of plants adapted to high light intensities were unaffected by ultraviolet light that is found in surface layers of water.

Gessner (1955) attempted to clarify the many physiological reactions of submersed macrophytes to light by grouping the plants into adaptation types either genetically fixed or phenotypically plastic. (a) A strictly shade-adapted form, such as the fenestrated tropical *Aponogeton,* is physiologically active only at low light intensities. (b) Numerous submergents are strictly light-adapted and require high light intensities for optimal metabolism. (c) Some submersed macrophytes exhibit a broad adaptation to light intensities, including some (d) that adapt to light or shade conditions depending on the area of development, others (e) that are generally tolerant of high light but develop best in weak light, and still others (f) that are shade-adapted but photosynthesize optimally at intermediate littoral light intensities. While such a scheme does not explain the physiological and morphological bases underlying such adaptations, it does demonstrate the extreme plasticity and adaptability of submersed macrophytes to the highly variable underwater light conditions.

Some insight into the mechanisms of photosynthetic adaptation of submersed macrophytes has been gained by the investigations of Spence and Chrystal (1970a, 1970b) on two species of *Potamogeton.* The rates of net oxygen production at various irradiances, dark respiratory oxygen uptake, chlorophyll content, specific leaf area (cm² leaf area per mg leaf dry weight), and leaf thickness were measured for a shallow-water, "sun-adapted" species (*P. polygonifolius*) and a deepwater, "shade-adapted" species (*P. obtusifolius*). From estimates of relative rates of photosynthesis and respiration per leaf area and per pigment content of the species, and sun and shade leaves of each species, it was concluded that (a) surface leaf area increases and respiration and leaf thickness decrease with depth and light reduction, and (b) higher net photosynthetic capacity per unit area of shade-adapted leaves and shade species of *Potamogeton* at low irradiances (approximately 1 per cent of summer daylight) is achieved by lowered respiration per area. The latter may result from a reduction in leaf weight per unit area. Potential complications introduced by photorespiration, discussed earlier, were not considered in these studies.

Other studies on the photosynthetic capacities and compensation points of submersed plants growing at low light intensities indicate marked variability. Compensation points commonly occur at 1 to 3 per cent of full sunlight (Wilkinson, 1963; Spence, 1976, 1982; Bowes, et al., 1977; Agami, et al., 1980; Moeller, 1980). However, considerable metabolic adaptation is found. For example, in an oligotrophic Danish lake, *Littorella* occurred between 0 and 2 m, where *Isoetes* grew between 2 and 4.5-m depths (Sand-Jensen, 1978). Light-saturated photosynthesis of *Littorella* was higher than that of *Isoetes*, and the difference increased with rising temperatures within the range of 5 to 20°C. At low irradiance, photosynthetic rates of *Isoetes* were similar to those of *Littorella*, despite greater chlorophyll content in *Isoetes*. Respiration rates were lower in *Isoetes* than in *Littorella*.

Two other common competing submersed angiosperms, *Myriophyllum spicatum* and *Vallisneria americana*, exhibit different growth forms; most of the biomass of *Myriophyllum* is near the water surface, whereas most of the leaf biomass of *Vallisneria* is near the sediment (Titus and Adams, 1979). *Vallisneria* was shade-adapted, and photosynthetic fixation rates were much more efficient than those of *Myriophyllum* at low light intensities. Optimum temperatures for photosynthesis (33°C) were nearly identical for the two species, but at 10°C, photosynthetic rates of *Myriophyllum* were nearly twice those of *Vallisneria*. Similar light and temperature interactions have been found among several other submersed angiosperms (Barko and Smart, 1981).

PRESSURE

Normal growth can be abruptly inhibited when certain submersed angiosperms are exposed to even very moderate increases in hydrostatic pressure (Fig. 18-4). Leaves become shorter, stems are much thinner and possess greatly reduced lacunal intercellular spaces, and growth of adventitious roots is inhibited (Ferling, 1957; Gessner, 1961; Hutchinson, 1975). Increased pressure of as little as 0.5 atm, equivalent to about 5 m depth, also induces increased growth of internodes, similar to the effects of low light intensities. Increased pressure also can inhibit flower formation; most submersed angiosperms produce aerial flowers that are wind-pollinated.

Growth inhibition is not usually proportional to pressure, and many species exhibit an irreversible "all or none" effect at an increase of about 1 atm pressure above surface atmospheric pressure (approximately 10 m depth equivalency). Little can be said about the physiological effects of increased pressure. One effect is the inhibition of movement of intercellular gases to the root system. Short-term effects include increased secretion of soluble organic matter by excessive pressure. Such increased cellular permeability may lead to losses of nutrients and hormonal growth substances. Golubić (1963) demonstrated in a very clear, transparent lake that a hydrostatic pressure of 0.8 atm limited photosynthesis of submersed angiosperms. The macroalgae *Chara* and *Nitella*, lacking intercellular gas systems, are unaffected by pressure differences.

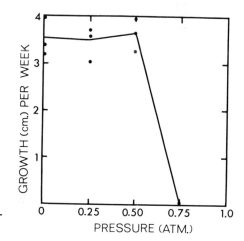

GROWTH (cm) PER WEEK

PRESSURE (ATM.)

Figure 18-4. Average growth (cm) per week of *Hippurus* in relation to hydrostatic pressure. (Modified from Gessner, 1955.)

Growth inhibition by hydrostatic pressure is apparently not universal among all submersed angiosperms. For example, marine angiosperms commonly grow at depths of several atmospheres of hydrostatic pressure in clear waters, and rates of photosynthesis are unaffected in these plants by pressure changes (Payne, Beer, Wetzel, and Iverson; in preparation). Pressure effects (to 2.3 atm., corresponding to a depth of 13.3 m) on growth of the freshwater angiosperm *Hippuris* could be overcome by warm temperatures (>15°C) and high light intensities (>100 μE m^{-2}s^{-1}; Bodkin, et al., 1980). The physiological ramifications of various pressures remain obscure, but it is likely that hydrostatic pressure interacts with other parameters, especially light, to restrict the distribution of freshwater angiosperms to depths* of less than 10 m. Shoots of *Myriophyllum spicatum* are not affected by increased hydrostatic pressure (Dale, 1981; Payne, 1982). Payne (1982) demonstrated that a lacunar arch system in stems of *Myriophyllum* expands the lacunae against external pressure by increasing cellular turgor pressure. Older stems and roots were weaker and were limited by low hydrostatic pressures.

Primary Productivity of Macrophytes

CHANGES IN BIOMASS

Macrophyte productivity is most commonly evaluated by measuring changes in biomass (Westlake, 1965b). The initial biomass of seeds of annual macrophytes is negligible. The biomass of macrophytes typically increases in a sigmoid fashion during the growing season (Fig. 18-5). In this idealized example, gross productivity reaches a plateau and later declines in older tissues; net productivity decreases and becomes nega-

---

Potamogeton strictus has been reported to a depth of 11 m in Lake Titicaca, Peru-Bolivia (Hutchinson, 1975). However, at an altitude of 3815 m, pressure at 11 m would be equivalent to a depth of 7.5 m in a lake at sea level. *Elodea canadensis* has been reported at a depth of 12 m in Lake George, New York (elevation approximately 100 m) (Sheldon and Boylen, 1977), although the range of this species in this lake is predominantly between 1 and 7 m.

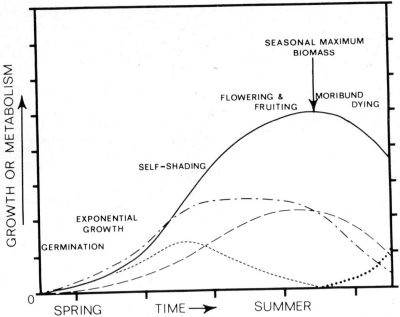

Figure 18-5. Generalized growth and metabolic patterns for a typical annual aquatic macrophyte. ———— = total biomass; — · — · = current gross productivity; · · · · = current net productivity; – – – = current respiration rate; · · · · · = death losses. (After Westlake, 1965b.)

tive because respiration continues to increase. Maximum biomass, the maximum cumulative net production, is reached when the current daily net productivity becomes zero. Subsequently, both the biomass and cumulative net production decrease, and terminal net production is equal to the quantity of material that has not been respired by the time the whole plant is dead. Although numerous variations are found among natural populations, this general pattern of growth and metabolism prevails among annual macrophyte species in temperate and subtropical communities.

If the initial biomass (e.g., seeds of an annual plant) is negligible, or if the plant dies before the seasonal maximum is reached, and losses other than respiration are negligible, as is commonly the case, the seasonal maximum biomass is equal to the maximum annual net production (Fig. 18-6, curves A and B) (Westlake, 1965b). This annual regrowth is characteristic of many plants that die down to a small quantity of perennating organs, but the extent of regrowth becomes less obvious when much of the plant survives until the next spring growing season. If the initial biomass persists without appreciable losses until after the seasonal maximum, the annual net production may be obtained as the difference between the final and initial biomass values for curve C (Fig. 18-6). Most commonly, a variable proportion of the initial biomass is lost (curve D). The maximum seasonal biomass in the latter case, as seen for example in populations of the macroalga *Chara*, is commonly only 50 to 80 per cent of the annual net production. When losses of the current season's production are appreciable, the estimation of production from changes in biomass is complicated (Fig. 18-6, curve E). Plant biomass remains more or less constant in extreme cases, such as in tropical communities, despite continuous growth. In populations with a high mortality, e.g., in some reed-swamp emergent macrophytes (Westlake, 1966), or in populations with a high turnover of leaves (Dickerman and Wetzel, 1982a, 1982b, 1982c), much of the production does not survive to be measured as terminal maximum biomass. Similar situations have been found among submersed macrophytes (Borutskii, 1950; Rich, et al., 1971). Demographic methods developed for analyses of fish populations subject to high mortality have been applied successfully to dense macrophyte populations, in which, for example among *Glyceria* populations, net production was some 3.4 times greater than seasonal maximum biomass (Mathews and Westlake, 1969; cf. Dickerman and Wetzel, 1982c).

UNDERGROUND BIOMASS

The roots or rhizoids of submersed freshwater macrophytes generally constitute a small but significant proportion of the total plant biomass (Table 18-3). Among floating-leaved and emergent macrophytes, however, the extensive system of roots, rootstocks, and rhizomes constitutes a major portion of the total biomass. Clearly, accurate analyses of productivity must include an evaluation of growth of the root system; measurements of only the above-ground foliage (= standing crop, see definitions in Chapter 8) are inadequate and can be quite misleading.

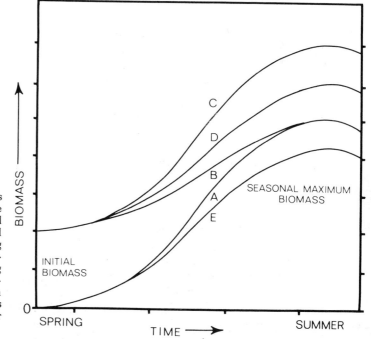

Figure 18-6. Types of growth curves among aquatic macrophytes. *A,* True annual or a plant with manifest annual regrowth; *B,* Plant with obscured annual regrowth; *C,* Plant with spring biomass persisting until seasonal maximum; *D,* Plant with only part of spring biomass persisting until seasonal maximum; *E,* Plant with annual regrowth and losses from current year's biomass before seasonal maximum. (After Westlake, 1965b.)

TABLE 18-3   Percentage of Total Biomass Found in Underground Tissues (Roots and Rhizomes) of Mature Aquatic Macrophytes

TYPE AND SPECIES	% OF TOTAL BIOMASS
Submersed	
Chara	<10
Ceratophyllum	< 5
Elodea canadensis	2.6
Myriophyllum sp.	11
Potamogeton richardsonii	12
P. perfoliatus	39
P. lucens	49
Heterantheria dubia	19
Vallisneria americana	48
Isoetes lacustris	25–40
Lobelia dortmanna	21–26
Littorella uniflora	46–55
Thalassia testudinum (marine)	75–85
Floating	
Eichhornia crassipes (submersed roots)	10–48
Floating-Leaved	
Nuphar lutea and N. pumilum	50–80
N. variegatum	68
Nymphaea candida	48–80
Emergent	
Zizania aquatica	7–8
Cyperus fuscus	7–8
Alisma plantago-aquatica	40
Equisetum fluviatile	40–83
Pontederia cordata	67
Glyceria maxima	>30–67
Phragmites communis	>36–96
Schoenoplectus lacustris	>46–90
Sparganium spp.	>25–66
Typha angustifolia	>32–67
Typha hybrid	64
Typha latifolia	43–67

After Westlake, 1965b, 1968; Nicholson and Best, 1974; Kansanen and Niemi, 1974; Ozimek, et al., 1976; Fiala, 1976, 1978; Szczepańska, 1976; Grace and Wetzel, 1981a.

The biomass of underground organs is difficult to measure, but can be estimated by randomized coring and quadrat excavation (Westlake, 1968; Fiala, et al., 1968; Fiala, 1971, 1973, 1978). Growth patterns and discrimination of annual growth among perennial root systems can be determined using a combination of tagging and growth experiments in which biomass increments of different age classes are analyzed. The ratio of biomass of underground parts to above-sediment parts is relatively constant among certain species at maturity in a particular habitat (e.g., Szczepański, 1969), but usually increases markedly as the season progresses (e.g., Szczepańska, 1976; Fiala, 1976, 1978). Photosynthetic allocation to above- and below-ground tissues can also be evaluated by in situ pulse labeling of photosynthetic tissue with radioactive CO_2 and determination of the distribution to plant parts over time (Grace and Wetzel, 1981a; cf. also Tietema, 1980).

LOSSES DURING GROWTH

Losses of plant biomass from damage (e.g., from water movement) and grazing before the seasonal maximum biomass are generally considered small among temperate annual macrophytes (Westlake, 1965b; Sculthorpe, 1967). Senescence and death during an annual period of active growth estimated for several submersed macrophytes ranged between 2 to 10 per cent of the maximum biomass. However, it is apparent from detailed analyses of perennial emergent macrophytes (e.g., the cattail *Typha*) that the composite annual population can consist of several cohorts, each of which undergoes biomass maxima with phased cohort senescence and decline (Dickerman and Wetzel, 1982a, 1982b). As a result, the composite population experiences continual senescence after the first cohort reaches maturity and starts to decline. Further, early shoot mortality can also account for a significant loss of population biomass.

In a cattail *(Typha latifolia)* population in southern Michigan, 19 per cent of the shoots had completely senesced prior to maximum above-ground biomass, and 95 per cent of these "prematurely" senesced shoots had entirely disappeared from the population before the maximum biomass occurred (Dickerman and Wetzel, 1982a, 1982b). Other significant biomass losses resulted from high leaf turnover. On the average, 50 per cent of the total number of leaves produced by a cattail shoot had undergone senescence and abscission by the time maximum above-ground biomass had been attained.

Pathogenic decimation of entire macrophyte populations is rare; natural diseases usually only cause gaps in existing populations, which result in a more heterogeneous distribution of individuals (see, for example, Klötzli, 1971; cf. Zettler and Freeman, 1972; Freeman, 1977). While many animals, especially immature insects, eat live aquatic macrophytes, few cause extensive damage (see especially the extensive review of Gaevskaya, 1966). Fish, birds, and a few mammals can, under exceptional circumstances, cause extensive damage or consume significant portions of the annual production. Some herbivorous fish, such as the grass carp *(Ctenopharyngodon idella)*, can decimate submersed and floating macrophytes (e.g., Sutton, 1977). These fish have been used as a management tool to control aquatic macrophytes in semitropical and tropical fresh waters. Under natural conditions, values of grazing loss, however, usually range from about 0.5 to 8 per cent of the total. Under most circumstances, most of macrophyte production is consumed by bacteria and fungi when the plants undergo senescence.

ORGANIC CHEMICAL COMPOSITION OF AQUATIC MACROPHYTES

An extensive collation of existing data on the general organic chemical composition of aquatic macrophytes (Straškraba, 1968; Hutchinson, 1975; Muztar, et al., 1978a, 1978b) provides useful information on the relative potential contribution of these plants to production and decomposition processes in aquatic systems. Morphologic and physiologic differences among the different plant types are reflected in a higher content of water, ash, and protein and a lower content of fiber in submergents as compared to emergent macrophytes (Table 18–4). Content of carbohydrate and fats is similar within the two groups. Floating-leaved macrophytes exhibit approximately intermediate composition values but contain significantly more lipids, which is reflected in somewhat

TABLE 18-4   Average and Range of Chemical Composition and Caloric Values of the Three Primary Types of Aquatic Plants

| ECOLOGICAL TYPE | WATER (%) | ASH (% OF DRY WEIGHT) | % OF ORGANIC MATTER | | | | CALORIC VALUE (cal g⁻¹) |
			Protein	Ether Extract	Crude Fiber	Carbo-hydrate	
Emergent	79	12	13	2.1	32	50	4480
	(70–85)	(5–25)	(3.5–26.7)	(0.4–6.1)	(17.6–45.4)	(23–69)	(4207–4987)
Plants with floating leaves	82	16	26.5	4.0	27	42	4770
	(80–85)	(10–25)	(18–44)	(2.8–5.7)	(14–38)	(31–64)	(4560–5140)
Submersed	88	21	22	2.2	27	51	4580
	(85–92)	(9–25)	(7.5–34.7)	(4.4–5.1)	(14.0–61.6)	(20–70)	(4165–5201)
Average all groups	83	18	19	2.4	29	50	4570

From Straškraba, 1968, from numerous sources.

higher average caloric values. The caloric values of aquatic macrophytes (Table 18-4) are approximately 20 per cent lower than the mean for phytoplankton (approximately 6000 cal g⁻¹ organic matter).

COMPARATIVE MACROPHYTE PRODUCTIVITY AMONG HABITATS

Comparison of productivity among freshwater ecosystems is difficult because of the variations in analytical and sampling techniques employed. Variability is especially great for evaluations of macrophyte productivity. Sampling of total biomass, including the rooting systems, has been done seriously only in recent years. Species distribution and production depend on an array of physical and chemical characteristics of both the water and the sediment. The result is extreme heterogeneity in distribution and productivity, both spatially and temporally. Further, results are commonly biased toward the growing season, which is usually much shorter than a year. A year is a well-defined unit in which the variability of annual productivity is much less than for shorter periods, since most communities have definite natural periodicities related to the year.

Despite these difficulties, approximate comparisons of macrophyte productivity are possible (see especially Westlake, 1963; Sculthorpe, 1967; Květ, 1971). The few exemplary figures given in Table 18-5 indicate the general range of above-ground maximum annual biomass and productivity among macrophyte communities. Submersed macrophytes generally exhibit conspicuously lower productivity than plants of floating or emergent communities. Annual net primary productivity of submersed macrophytes ranges from 1 to 7 metric tons per hectare (100 to 700 g m⁻² year⁻¹), the lower values being more common in oligotrophic and moderately fertile waters, and ranging from 4 to 7 mt ha⁻¹ year⁻¹ in fertile lakes. Values of greater productivity are found among submersed communities in subtropical regions, from 10 to as high as 30 mt ha⁻¹ year⁻¹ in the marine *Thalassia* community of Puerto Rico. Few accurate data are available for productivities of floating or floating-leaved macrophytes, although it is apparent that this community reaches levels of 11 to 33 mt ha⁻¹ year⁻¹ in warmer climates, and can easily achieve greater productivity under tropical conditions (Westlake, 1963).

Numerous investigations of the productivity of emergent macrophytes of fertile

TABLE 18-5 Representative Estimations of Annual Maximum Biomass and Productivity of Aquatic Macrophytes

TYPE AND LAKE	SEASONAL MAXIMUM BIOMASS OR ABOVE-GROUND BIOMASS (g dry m^{-2})	PRODUCTIVITY* (g m^{-2} year^{-1})	SOURCE
Submergents Dominating			
Trout L., Wisc. (soft-water)	0.07	—	Wilson, 1941
Sweeney L., Wisc. (soft-water)	1.73	—	Wilson, 1937
Weber L., Wisc. (soft-water)	16.8	—	Potzger and Engel, 1942
Mirror L., New Hampshire *Lobelia dortmanna*	80	55	Moeller, 1978a
L. Opinicon, Ontario	496	—	Crowder, et al., 1977
West Blue L., Manitoba	—	6.4	Love and Robinson, 1977
Lowes L., Scotland (dystrophic)	32	—	Spence, et al., 1971
Spiggie L., Scotland (dystrophic)	100	—	
L. Mendota, Wisc. (hard-water)	202	—	Rickett, 1921
Lawrence L., Mich. (hard-water)			
Submersed *Scirpus subterminalis*	338	565	Rich, Wetzel, and Thuy, 1971
Chara	110	155	
Annuals	130	199	
Croispol, Scotland	400	—	Spence, et al., 1971
Borax L., Calif. (saline lake, *Ruppia*)	60	64	Wetzel, 1964
River Ivel, England (*Berula*; very fertile)	500	—	Edwards and Owens, 1960
River Test, England (*Ranunculus*)	100–400	—	Owens and Edwards, 1961
River Yare, England (*Potamogeton*)	380	—	Owens and Edwards, 1962
Saline channels, Puerto Rico (*Thalassia*)	700–7300	—	Burkholder, et al., 1959
Floating			
New Orleans, La. (*Eichhornia*)	630–1472	1500–4400	Penfound and Earle, 1948
Emergents Dominating			
Ladoga L., USSR	0.4–10.7	2.4	Raspopov, 1971
Onega L., USSR	0.1–33	0.6	Raspopov, et al., 1977
Kubenskoje L., USSR	—	29	
Voze L., USSR	—	70	
Laca L., USSR	—	120	
Czechoslovakian ponds			
Phragmites australis	2980	—	Dykyjová, 1971
Typha angustifolia	4040	—	
Scirpus lacustris	3000	—	
Southern Swedish lake, *Phragmites australis*	2380	—	Björk, 1967
L. Arreso, Denmark (eutrophic)	2900–9900	—	Andersen, 1976
Temperate wetlands, N. America *Zizania aquatica*	—	630–1450	Whigham and Simpson, 1977
Alberta oxbow lakes	465	—	van der Valk and Bliss, 1971
Michigan Hollow, New York, *Carex lacustris*			
Above ground	1145	965	Bernard and Solsky, 1977
Below ground	575	208	

(Continued)

TABLE 18-5 Representative Estimations of Annual Maximum Biomass and Productivity of Aquatic Macrophytes (continued)

TYPE AND LAKE	SEASONAL MAXIMUM BIOMASS OR ABOVE-GROUND BIOMASS (g dry m^{-2})	PRODUCTIVITY* (g m^{-2} year^{-1})	SOURCE
Emergents Dominating (continued)			
Blanket bog, England (Sphagnum)			
on hummocks	—	180	Clymo and Reddaway,
in pools	—	290	1971, 1974
on "lawns"	—	340	
Polish lakes (emergent species)	440–830	—	Szczepańska, 1973
Minnesota wetlands (Carex)	850	738	Bernard, 1973
Surlingham Broad, England (Glyceria, Typha, and Phragmites)	800–1100	—	Buttery and Lambert, 1965
Opatovicky Pond, Czechoslovakia (Phragmites)			
Aerial	1100–2200	—	Dykyjová and
Below ground	6000–8560	—	Hradecká, 1973
Cedar Creek, Minn. (Typha)	4640	2500	Bray, et al., 1959
Lawrence L., Michigan (Typha latifolia, above ground)	770	6000	Dickerman and Wetzel, 1982a, 1982b

*Where maximum seasonal biomass values are higher than productivity estimates, the biomass values include a carryover of biomass from the previous season.

reedswamps and marshes indicate very high values, even though the range is quite large. Temperate-zone emergents can attain productivity levels of 20 to 45 mt ha^{-1} year^{-1} and, in subtropical and tropical regions, can reach 40 to at least 75 mt ha^{-1} year^{-1}.

Inspection of the annual net productivity of the plant communities (Table 18-6) shows that the emergent reed communities are as productive as terrestrial communities. Under fertile conditions with a continuous growing season, emergent aquatic macrophyte communities can be more productive than terrestrial communities. In the temperate zones, the organic productivity of emergent communities is comparable to that of perennial agricultural crops and coniferous forests, and appreciably greater than that of deciduous forests. In tropical zones, emergent communities are among the most productive of all aquatic macrophytes. By comparison, the submersed macrophytes, with the possible exception of tropical marine angiosperms, are relatively unproductive, and are similar to the composite areal productivity of phytoplankton. In spite of the number of adaptations to aqueous conditions found among submersed macrophytes, the reductions in rates of diffusion and spectrally selective attenuation of light by the water preclude intensive productivity. In certain respects, the submersed habitat is less severe than the terrestrial, e.g., in terms of water availability and thermal stability. These ameliorated parameters permit some growth of perennial populations of submersed macrophytes to continue all year (Rich, et al., 1971; Hough and Wetzel, 1975; Boylen and Sheldon, 1976; Stickey, et al., 1978). However, these advantages are not sufficient to offset the strictures of being totally submersed.

It should be emphasized in comparative analyses of productivity of macrophyte communities, as done briefly above and by others (e.g., Keefe, 1972), that geographically diverse ecosystems are not fully distinguished. The importance of geographic and genetic differences within a species may be reflected in their productivity and other growth characteristics, which vary in a given environment from population to population as well as among populations from one environment to another. These multidimensional variations among ecotypes of the same species are well characterized in detailed studies of *Typha* populations in which phenotypic adaptations to numerous environmental parameters, especially temperature, photoperiod, light availability, and nutrients, are seen in marked differences in productivity (McNaughton, 1966b, 1970; Grace and Wetzel, 1981a, 1981b, 1981c, 1982a, 1982b).

PRODUCTIVITY OF AQUATIC MACROPHYTES WITHIN HABITATS

The literature on qualitative phytosociological composition of aquatic macrophyte communities is prodigious (e.g., Hejný, 1960; Hutchinson, 1975). While these analyses of changes in dominance yield much information on the floristic structure of the communities, the criteria used, e.g., percentage areal cover, do not allow much insight into the causal mechanisms underlying competitive interactions.

Seasonal changes in species dominance (e.g., percentage areal cover) are commonly observed among submersed, floating-leaved, and emergent communities, but a clear tendency to form extensive monospecific populations is also commonly observed.

TABLE 18-6   Annual Net Primary Productivity of Aquatic Communities on Fertile Sites in Comparison to That of Other Communities

TYPE OF ECOSYSTEM	APPROXIMATE ORGANIC (DRY) PRODUCTIVITY (mt ha^{-1} year^{-1})*	RANGE (mt ha^{-1} year^{-1})
Marine phytoplankton	2	1–4.5
Lake phytoplankton	2	1–9
Freshwater submersed macrophytes		
Temperate	6	1–7
Tropical	17	12–20
Marine submersed macrophytes		
Temperate	29	25–35
Tropical	35	30–40
Marine emergent macrophytes		
(salt marsh)	30	25–85
Freshwater emergent macrophytes		
Temperate	38	30–45
Tropical	75	65–85^{+}
Desert, arid	1	0–2
Temperate forest		
Deciduous	12	9–15
Coniferous	28	21–35
Temperate herbs	20	15–25
Temperate annuals	22	19–25
Tropical annuals	30	24–36
Rain forest	50	40–60

After Westlake, 1963, 1965b.

*mt = metric tons. Values × 100 = g m^{-2} year^{-1} (approximately) and × 50 = g C m^{-2} year^{-1} (approximately).

Vegetative reproduction is prevalent among aquatic macrophytes, and rapid, relatively complete expansion into favorable habitats usually occurs.

An example of competitive interactions has been observed between two emergent reeds, Glyceria and Phragmites (Buttery, et al., 1965a, 1965b). Glyceria grows rapidly in the spring, and under optimal growing conditions it can completely suppress Phragmites by producing dense foliage that effectively shades its slower-growing competitor. When sediments are water-saturated and anaerobic, however, Glyceria growth is reduced and Phragmites, with taller and more erect foliage that absorbs more incident light, can increase more effectively and eventually exclude Glyceria. Experimental manipulation of nutrient supply had no effect on this competitive interaction.

Two species of cattail, Typha latifolia and T. angustifolia, commonly segregate along a gradient of increasing water depth, with T. latifolia restricted to depths of less than 80 cm and T. angustifolia to depths greater than 15 cm (Grace and Wetzel, 1981b, 1981c). In these two species, habitat partitioning resulted from differences in morphology; T. latifolia was prevented from growing in deep water because of the higher metabolic cost of producing broader leaves, but was able to outcompete T. angustifolia for light in shallow water owing to its greater leaf surface area.

Among submersed macrophytes of calcareous hard-water lakes, the ratio of organic matter to calcium carbonate of the sediments has been implicated in the competition between the macroalga Chara and rooted angiosperms (Wohlschlag, 1950). Higher organic matter content of manipulated lake plots favored the production of rooted angiosperms.

Other experimental and correlational in situ studies indicate that, in general, growth and productivity of both submersed and emergent macrophytes increase on organic-rich sediments (Szczepańska and Szczepański, 1976; Barko and Smart, 1979; Sand-Jensen and Søndergaard, 1979b; Aiken and Picard, 1980; Grace and Wetzel, 1981a, 1982a).

The seasonal growth patterns of aquatic macrophytes can be exemplified by the common emergents Phragmites and Typha, a submersed annual Myriophyllum, and submersed perennial macrophytes in a hard-water lake. Many other examples exist and variations from these typical temperate zone patterns can be found.

Growth and production analyses of emergent macrophyte communities are commonly based on measurements of total plant biomass, biomass of component species (or plant parts), or leaf area index. The leaf area index (LAI) is simply the ratio of leaf area per unit ground area (Kvĕt, 1971). The reedswamp plants exemplified in Figure 18–7 demonstrate the marked differences in growth rates found among perennial emergent plant species. In Phragmites, maximum biomass (W) occurred in midsummer, whereas the maximum biomass of Typha increased over the entire growing season with little difference between the maximum values (1810 and 1620 g m^{-2}, respectively) for these two species. The average rate of increase in shoot biomass was 9.4 g m^{-2} day^{-1} in the cattail population, and nearly 20 g m^{-2} day^{-1} in the Phragmites population. In both species, stem biomass (W$_s$) formed an increasingly greater proportion of the total (W), with a concommitant decrease in the proportion of leaf biomass (W$_1$). In Phragmites, however, both stem and leaf biomass decreased in the latter half of the growing season, whereas they continued to increase in Typha. The leaf area index followed the changes in leaf biomass, and, coupled with changes in shoot density, it is clear that shoots of Typha that emerged later in the growing season had a better chance to survive in sparse stands than in the dense Phragmites stand. In the dense Phragmites stand, shorter shoots of Typha were handicapped by competition for light.

Monodominant stands of emergent macrophytes (e.g., Phragmites, Typha) are common; invasion by other species can be resisted for long periods of time. Detailed population studies of nearly monospecific stands of macrophytes revealed pulses of shoot emergence throughout the growing season (Bernard and Solsky, 1977; Dickerman and Wetzel, 1982a). The presence of such cohorts results in a mixture of senescing and young shoots, which allows effective exploitation of the available habitat. Emergence of cohorts throughout the growing season (Fig. 18–8) frequently more than offsets shoot mortality; a constant or increasing shoot density may help exclude potential competitor species. Density decreases with increasing accumulations of litter, increases when the area is flooded by rising water levels and increased inorganic nutrients, and decreases with grazing of buds and moderate increases in salinity (Rudescu, et al., 1965; Björk, 1967; Haslam, 1971a; Rodewald-Rudescu, 1974). High light, warmer temperatures, and higher water levels promote growth in dense, monospecific stands. Among mixed

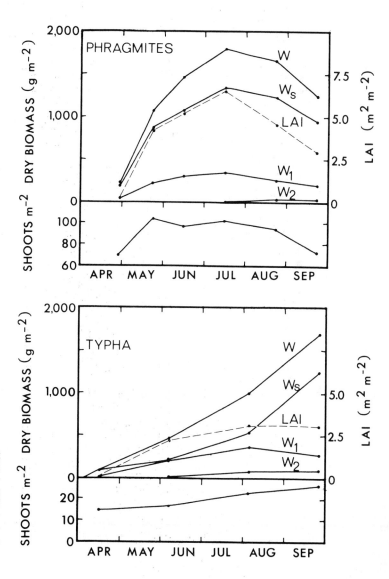

Figure 18-7. Changes in total plant biomass (dry, W); stem, sheaths, and stubble (W_s); leaf laminae (W_1); inflorescence (W_2); leaf area index (LAI); and the stand densities (shoot m^{-2}) of *Phragmites communis* and *Typha latifolia*, southern Czechoslovakia. (Modified from Květ, Svoboda, and Fiala, 1969.)

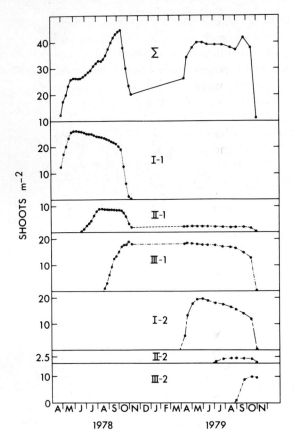

Figure 18-8. Shoot density of the composite population (Σ) and individual cohorts (roman numerals) during the first (1) and second (2) years of analysis of the common cattail (*Typha latifolia*), Lawrence Lake, Michigan. (After Dickerman and Wetzel, 1982a.)

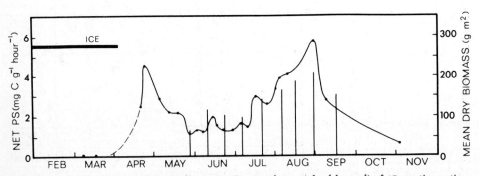

Figure 18-9. Rates of net photosynthesis (mg C gram dry weight^{-1} hour^{-1}) of 15-cm tip sections of *Myriophyllum spicatum* (curved line) and mean ash-free organic weight (g m^{-2}) (vertical bars) in eutrophic Lake Wingra, Wisconsin, 1971. (From data of Adams and McCracken, 1974.)

stands of emergents, the final balance of dominance is determined by a combination of (a) physical factors and inherent properties of the plants, such as soil and water regime, growth form, size, and abundance of sexual and vegetative organs, (b) physiological characteristics of the species, such as shade tolerance, requirements for germination and establishment, survival in relation to temperature, water availability, and litter cover, and (c) physical events such as opening of the litter mats, grazing, and diseases (Haslam, 1971b, 1973).

The littoral community of eutrophic Lake Wingra, Wisconsin, is dominated by the submersed annual macrophyte *Myriophyllum* (Adams and McCracken, 1974). In situ

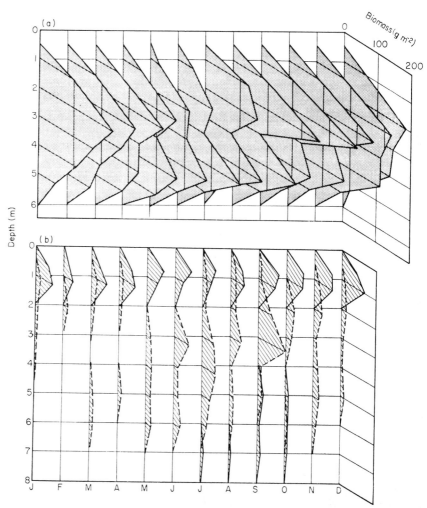

Figure 18-10. Annual biomass (g m^{-2} ash-free dry organic weight) depth distribution of (a) *Scirpus subterminalis*, and (b) Characeae (———) and annual macrophytes (- - - - -) Lawrence Lake, Michigan. (After Rich, P. H., Wetzel, R. G., and Thuy, N. V.: Distribution, production and role of aquatic macrophytes in a southern Michigan marl lake. Freshwater Biol., 1:3–21, 1971.)

measurements of photosynthetic rates demonstrated a rapidly formed spring maximum, followed by a marked summer reduction and subsequent maximum in late summer (Fig. 18–9). The spring maximum of photosynthetic rates did not coincide with the rates of biomass accumulation, a phenomenon similarly observed for the single submergent (*Ruppia maritima*) of a saline lake (Wetzel, 1964). The biomass maxima occurred at the time of flowering in late summer (see review of Grace and Wetzel, 1978). A cumulative amount of organic matter approximately equal to 1.6 times the maximum biomass of *Myriophyllum* was lost by night-time respiration and sloughing of leaves.

Within the *Myriophyllum* community, most of the total biomass, leaf surface area, and photosynthetic activity occurred in the upper portion of the water column. Light reduction is definitely a major factor in the marked decrease in photosynthetic rates with depth observed by many investigators. Surface photoinhibition, so frequently found in phytoplankton photosynthesis, is seen only rarely among submersed macrophytes. Other factors, such as temperature reduction, greater age of tissue, and encrustations of epiphytic algae and carbonates on older leaf surfaces, contribute to lower rates of photosynthesis in deeper strata.

The importance of light to submersed macrophytes is further indicated by the depth distribution of biomass of the dominating perennial macrophyte *Scirpus subterminalis* (76 per cent of total biomass) in oligotrophic Lawrence Lake, Michigan (Fig. 18–10). Two annual maxima are evident from two peaks at different depths. The larger fall maximum consisted of a major peak at 2 m in September, and an additional peak occurred at 4 m in midsummer when insolation was greatest. A fall maximum of the macroalgae of the Characeae (16 per cent of total biomass) was evident even though its seasonal biomass was relatively constant; the Characeae of this lake is similarly best described as a perennial population (Figs. 18–10 and 18–11).

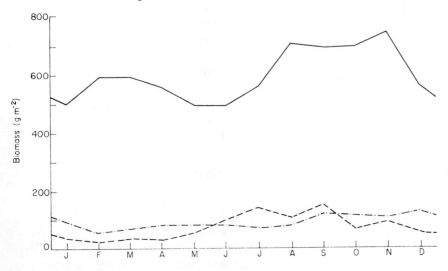

Figure 18–11. Annual biomass (g m⁻² ash-free dry organic weight) of the important macrophytic groups summed with respect to depth, Lawrence Lake, Michigan. ——— = *Scirpus subterminalis*, · · · · · · · = Characeae, - - - - - = annuals. (After Rich, P. H., Wetzel, R. G., and Thuy, N.V.: Distribution, production and role of aquatic macrophytes in a southern Michigan marl lake. Freshwater Biol., 1:3–21, 1971.)

TABLE 18-7   Comparison of Annual Primary Productivity of Submersed Macrophytes in Several Lake Systems

LAKE	g C m^{-2} (OF LAKE) YEAR^{-1}	REMARKS
Wingra, Wisconsin	117	*Myriophyllum*; eutrophic lake (Adams and McCracken, 1974)
Lawrence, Michigan	87.9	Perennials; oligotrophic lake (Rich, Wetzel, and Thuy, 1971)
Narock, USSR	59.4	Largely perennial charophytes; mesotrophic lake (Winberg, et al., 1972)
Marion, British Columbia	50.8	Annuals; mesotrophic lake (Davies, 1970)
Kalgaard, Denmark	21.0	Perennial isoetids; oligotrophic, soft-water lake (Søndergaard and Sand-Jensen, 1978)
Mirror, New Hampshire	1.4	Perennials; oligotrophic, soft-water lake (Moeller, 1975)
Borax, California	1.2	Single annual *Ruppia*; saline lake (Wetzel, 1964)

The submersed annual macrophytes (8 per cent of total macrophytic biomass) of Lawrence Lake were dominated by two species of *Potamogeton*. These macrophytes exhibited two maxima, and died back in the autumn (Figs. 18-10 and 18-11).

The relatively constant biomass of perennial macrophytes throughout the year presented in this example is probably a much more common characteristic of lakes than is generally recognized. Turnover rates (ratio of production to biomass) of emergent and submergent annual plants are relatively low (1.02 to 1.2 times maximum biomass on an annual basis). Among perennial populations, turnover rates are higher (range 1.5 to 5.0 times the maximum biomass). Even among oligotrophic lakes, productivity of perennial populations of submersed macrophytes can approach or exceed that of the commonly more productive annual macrophytes (Table 18-7). Comparisons of the relative contribution of the macrophytes to the primary productivity of attached algae and phytoplankton of lakes will be deferred until the conclusion of the discussion of littoral microalgae (Chapter 19).

The fate of organic matter produced by aquatic macrophytes as they undergo senescence, death, and decomposition is treated separately with a discussion of the metabolic processes in sediments (cf. Chapters 20 and 22).

INHIBITION OF PHYTOPLANKTON AND ZOOPLANKTON BY SUBMERSED MACROPHYTES

A number of investigations have demonstrated a distinct inhibition of phytoplanktonic development in the littoral zones containing heavy stands of emergent and submersed macrophytes (e.g., Schreiter, 1928; Hasler and Jones, 1949; Hogetsu, et al., 1960; Stangenberg, 1968; Goulder, 1969; Dokulil, 1973). Circumstantially, these investigations implied that organic compounds excreted by the macrophytes inhibited the growth of phytoplankton. However, competitive reduction of nutrients, light, and other substances by the macrophytes as factors inhibiting phytoplanktonic growth was not satisfactorily ruled out in these studies.

A detailed study of the influence of four species of submersed macrophytes on phytoplankton within the stands showed that decreases in rates of phytoplankton photosynthesis were caused by shading (Brandl, et al., 1970). Average reductions over three years in the rates of phytoplankton production among the plant stands as compared to rates in the open water were: *Potamogeton pectinatus*, 60 per cent; *P. lucens*, 46 per cent; *Batrachium aquatile*, 40 per cent; and *Elodea canadensis*, 12 per cent. Phytoplankton production was markedly reduced when the phytoplankton were taken from the pelagial zone and incubated among the macrophytes, and increased greatly when littoral phytoplanktonic algae were incubated at the same depth in the open water (Table 18–8).

The presence of membrane filtratable (1.2 μm pore size) organic substances capable of inhibiting the photosynthesis of phytoplankton could not be confirmed. However, increasing the pH of the water, as occurs with a reduction in available free CO_2 in dense stands of actively photosynthesizing submersed macrophytes (Figure 18–12), reduced the rates of phytoplankton productivity. This decrease could be reversed completely by small additions of CO_2. Brammer (1979) similarly found that the submersed macrophyte *Stratiotes* inhibited the growth of contiguous populations of phytoplankton. Competition for nutrients and reductions in the concentrations of ions in the littoral water by the macrophytes were likely responsible for the decline of the phytoplankton.

Dense stands of macrophytes can markedly alter other immediate environmental conditions in comparison to those of the open water. Under floating macrophytes (e.g., duckweeds; hyacinths), water temperature increases, pH values decline with increased CO_2 levels, and oxygen concentrations decrease (Ultsch, 1973; Dale and Gillespie, 1976; Rai and Munshi, 1979; Schreiner, 1980). In reed *(Phragmites)* stands, conductance and calcium and bicarbonate concentrations can be higher than in the open water (Planter, 1970). In all cases, light is severely and selectively attenuated by the macrovegetation (Gessner, 1955; Dykyjová and Květ, 1978).

While much of the apparent inhibition of phytoplankton growth among dense stands of macrophytes is likely related to competition for light and nutrients, the production of biotically inhibitory or stimulatory organic substances cannot be totally excluded. For example, the presence of macrophytes favors the dominance of some blue-green algae (Guseva and Goncharova, 1965) but is inhibitory to other groups

TABLE 18–8 Changes in Rates of Primary Production of Phytoplankton from the Pelagial Zone Incubated among Submerged Macrophyte Stands of the Littoral and Littoral Phytoplankton Incubated in the Pelagial Zone, Czechoslovakian Shallow Lakes

SUBMERSED MACROPHYTE SPECIES DOMINATING LITTORAL AREA	PERCENTAGE REDUCTION IN LITTORAL ZONE	PERCENTAGE INCREASE IN PELAGIAL ZONE
Potamogeton pectinatus	56	250
Potamogeton lucens	46	270
Batrachium aquatile	39	230
Elodea canadensis	27	500

From data of Brandl, Brandlová, and Poštolková, 1970.

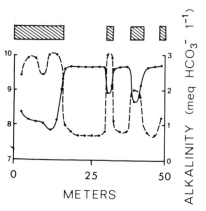

Figure 18-12 Changes in the pH ($\cdot$ - - - $\cdot$) and alkalinity ($\cdot$ ——— $\cdot$) in the surface water along a transect across Sangwin Pond, in the afternoon (June) through intermittent areas of dense cover of the submersed macrophyte *Ceratophyllum demersum* (indicated by hatching). (Modified from Goulder, 1969.)

(Kogan and Chinnova, 1972). Macrophytes may influence the competition of blue-green algae with other algal groups by blue-green algal heterotrophic utilization of certain excreted organic compounds. Zooplankton of many groups are inhibited strongly or repelled by substances secreted into the water by submersed angiosperms and macroalgae (much of the scattered literature is reviewed by Pennak, 1973 and Dorgelo and Koning, 1980). For example, *Daphnia* is strongly repelled by water containing *Elodea*, *Myriophyllum*, or *Nitella*. The magnitude of the negative responses varies with the plant species and is more pronounced at higher temperatures. Sexually mature *Daphnia* respond more rapidly than immature instars. The inhibitory or insecticidal effects of organic compounds secreted by macrophytes, especially the macroalgae Characeae, on mosquito larvae have been reported widely (and critically reviewed by Hutchinson, 1975). The active compounds are probably of the general allyl sulfide group, analogous to those derived enzymatically from garlic oil.

Retention and Release of Nutrients

The surface area of wetland and littoral plants increases greatly in the transition from emergent to submersed macrophytes. As a result, colonization of surfaces by epiphytic algae and bacteria increases along the littoral gradient from emergent to submersed macrophytes (Chapter 19). Both the macrophytes and epiphytic microflora utilize and release inorganic nutrients and dissolved organic compounds (Wetzel and Allen, 1970; Allanson, 1973; Mickle and Wetzel, 1978a, 1978b, 1979). The photosynthetic productivity of epiphytic algae can be very high, often exceeding that of the submersed macrophytes (Chapter 19).

As water of the drainage basin containing inorganic nutrients and organic compounds passes through the complex vegetation and their epiphytes, marked alterations can occur in the chemical composition of the influent water. Submersed vegetation-epiphytic complexes can be extremely effective in removing such inorganic nutrients as Ca, K, combined N, and P of inflowing water, either by direct assimilation or via adsorption onto inorganic precipitates (such as $CaCO_3$) induced by photosynthetic activity (Mickle and Wetzel, 1978a). Labile organic compounds of low molecular weight (<1000 Daltons) can be effectively removed by the epiphytic microflora. The fate of high molecular weight compounds (>1000 to 30,000 Daltons), which dominate from terrestrial and wetland sources, is various: they may be selectively removed or, alternately, pass through relatively unaffected, and their concentration may increase owing to release from the macrophytes themselves (Mickle and Wetzel, 1978b, 1979).

During senescence and after death of the macrophytes, a massive release of inorganic nutrients and dissolved organic matter occurs (Chapter 20). The release rates are governed by the population growth dynamics of the macrophytes, their mortality during and following the growth period, and the conditions prevailing at the time of death and decomposition. Release rates are also influenced by the type of vegetation; decomposition and release of nutrients and organic compounds are much faster from submersed and floating-leaved macrophytes than from emergent macrophytes (Chapter 20). Depending upon when these nutrients and organic compounds enter the influent water, and how they are selectively modified by the metabolic sieve of the submersed macrophyte–epiphyte complex, the inputs can have marked influence on phytoplanktonic growth in the recipient stream or lake (Chapter 15). It should be recalled from earlier discussions that phytoplanktonic responses can be stimulated or inhibited by the inorganic–organic loadings, since the stimulatory effects of some inorganic nutrients can be counteracted by inhibitory effects of certain organic compounds of macrophyte origin.

SUMMARY

1. The size of the littoral zone of lakes and streams varies greatly in relation to the size of the open-water pelagial region. Since most lakes of the world are small in area and shallow, the wetland and littoral flora and its epiphytes are commonly a dominant source of synthesized organic matter to the recipient water.

2. Four groups of aquatic macrophytes can be distinguished on the basis of morphology and physiology; each of the groups dominates a major region of the littoral zone (cf. Fig. 8–2):

    a. *Emergent* macrophytes grow on water-saturated or submersed soils from where the water table is about 0.5 m below the soil surface (supralittoral) to where the sediment is covered with approximately 1.5 m of water (upper littoral).

    b. *Floating-leaved* macrophytes are rooted in submersed sediments in the middle littoral zone (water depths of approximately 0.5 to 3 m), and possess either floating or slightly aerial leaves.

    c. *Submersed* macrophytes occur at all depths within the photic zone. Vascular angiosperms occur only to about 10 m (1 atm hydrostatic pressure) within the lower littoral (infralittoral), and nonvascular macrophytes (e.g., macroalgae) occur to the lower limit of the photic zone (littoriprofundal).

    d. *Freely floating* macrophytes are not rooted to the substratum; they float freely on or in the water and are usually restricted to nonturbulent, protected areas.

3. Emergent macrophytes produce erect, linear leaves from an extensive rhizome/root base, and are physiologically similar to terrestrial plants.

    a. Rooting tissues grow in an anaerobic substratum and must obtain oxygen for respiration from aerial organs. The plants develop extensive intercellular gas lacunae that permit internal gas movement and exchange.

    b. Rates of transpiration by emergent and floating macrophytes are often extremely high, and result in water losses to the atmosphere that are greater than evaporation from an equivalent area of water. Vegetative evapotranspiratory losses of water can be sufficiently large to appreciably reduce water levels of the fresh waters and surrounding terrestrial areas.

4. Submersed angiosperms possess numerous morphological and physiological adaptations to their aqueous environments.

    a. Leaves are only a few cells in thickness, and photosynthetic pigments are concentrated in epidermal tissue. Leaves tend to be much more divided, with greater surface-to-volume ratios, than are leaves of other aquatic or terrestrial plants. Leaf morphology maximizes exposure of cellular surfaces to reduced light, gas, and nutrient availability under water.

    b. Many submersed angiosperms exhibit extreme vegetative polymorphism *(heterophylly)* in leaf structure, often on the same stem or petiole. Morphology shifts from finely divided submersed leaves to more coalesced, entire shapes in surface-floating or aerial leaves. Heterophyllous growth is regulated by interactions of temperature, CO_2, and hormonal action mediated by ethylene concentrations and light.

5. Carbon assimilation is more difficult in submersed macrophytes than in emergent and floating-leaved plants that have access to atmospheric CO_2, primarily because rates of diffusion of gases are several orders of magnitude slower in water than in air.

    a. The cuticle of submersed angiosperms is reduced or absent, photosynthetic tissue is thin (1 to 3 cells thick), and chloroplasts are densely distributed in outer epidermal cells—all enhance exchange of gases and nutrients that are dissolved in the water.

    b. Extensive intercellular gas lacunae (often >70 percent of plant volume) in leaves, stems, petioles, and roots facilitate rapid diffusion of internal gases. Microzones at the cellular surface of leaves rapidly develop in quiescent littoral waters; these microzones set up gradients of reduced gas availability.

    c. Certain species of submersed macrophytes assimilate and subsequently decarboxylate bicarbonate ions. This assimilation represents a physiological adaptation to limited availability of free CO_2.

d. Submersed angiosperms possess C_3–Calvin-type photosynthetic metabolism with high rates of photorespiration. CO_2 from mitochondrial respiration of foliage and rooting tissue, photorespiration, and, in a few soft-water plants, CO_2 from sediments, augment CO_2 and/or HCO_3^- assimilated from the water. Some CO_2 from respiration is efficiently recycled in the internal gas lacunae of leaves.

e. Some oxygen produced in photosynthesis, or entering emergent and floating plants from the atmosphere, moves downward in internal gas lacunae to rooting tissue. Some oxygen diffuses to the surrounding rhizosphere and offers some protection to the plant from toxic metabolic products (H_2S, volatile fatty acids) produced by fermentation in the anaerobic sediments.

6. Nutrients are assimilated from the sediments by emergent and rooted floating-leaved macrophytes, and from the water in free-floating macrophytes.

a. Most submersed rooted macrophytes obtain a majority of their nutrients from the interstitial water of sediments, where concentrations are much greater than in the water. When nutrient concentrations of the water are high, a greater proportion is assimilated from the water.

b. Some nutrients are released to the water by submersed macrophytes during active growth; large amounts are released upon senescence and death.

7. Light availability is a major factor regulating the growth and competitive interactions of aquatic macrophytes.

a. Among submersed macrophytes, light attenuation rapidly limits the vertical distribution of plant growth. Compensation points of photosynthesis are commonly at 1 to 3 per cent of full sunlight. Considerable photosynthetic and respiratory adaptation to low light and differing temperatures is found among submersed species.

b. Light availability and interspecific shading can be major factors influencing the competitive success of one species over another.

c. Hydrostatic pressure likely interacts with low light intensities to limit the distribution of freshwater angiosperms to depths of less than 10 m.

8. The productivity of aquatic macrophytes has been evaluated from changes in biomass over time and, in a few cases, by in situ determinations of rates of photosynthesis.

a. Many macrophytes are perennial, with varying degrees of biomass turnover from year to year.

b. Roots and rhizomes of submersed macrophytes make up a lesser portion (1–40 per cent) of total plant biomass than is the case for floating-leaved (30–70 per cent) or emergent (30–95 per cent) macrophytes.

c. Under most circumstances, losses of aquatic plant biomass from grazing and pathogens are generally small (0.5–8 per cent). Senescence and death of plant parts during active growth have been accurately evaluated in only a few cases. The losses via senescence depend upon the plant species and upon the environmental conditions; among perennial species, turnover of biomass can range from 1.5 to 3 or more times per year where populations pass through a number of distinct cohorts.

d. Emergent macrophytes contain a much higher content of structural tissue than do submersed or floating-leaved macrophytes. This structural tissue is of a chemical composition relatively refractory to rapid microbial decomposition.

e. Growth of aquatic macrophytes is generally higher on organic-rich sediments than it is on sandy sediments.

f. On a unit-area basis, the net primary productivity of aquatic macrophytes is among the highest of any community in the biosphere. Emergent macrophyte productivity is the highest ($1500–4500$ g C m^{-2} year^{-1}); submersed macrophytic productivity is considerably less ($50–1000$ g C m^{-2} year^{-1}) but often equals or exceeds that of phytoplankton ($50–450$ g C m^{-2} year^{-1}).

g. Seasonal variations in aquatic macrophyte productivity are great. In the temperate zone, productivity of aquatic macrophytes commonly follows increasing insolation and temperature; some submergents persist under ice cover, although net productivity is low or zero. In the tropics, growth can be nearly continuous.

9. Phytoplankton productivity is generally lower in littoral zones containing stands of aquatic macrophytes. Experimental evidence indicates that this phenomenon results largely from a competition for nutrients (including carbon) by submersed macrophytes, and by a reduction of light by macrophyte foliage.

10. Wetland and littoral regions of freshwater ecosystems are commonly intensely metabolically active owing to the presence of aquatic macrophytes.

a. The macrophytes are frequently the primary source of organic matter to fresh waters.

b. The synthesis of organic matter by emergent vegetation and its decomposition in and on wetland sediments result in high loading of dissolved organic matter and inorganic nutrients to recipient lakes. The timing of loading is strongly influenced by the stages of wetland plant growth, senescence, and decomposition.

c. As water laden with nutrients and dissolved organic compounds passes to the lake, the chemical loading is greatly modified by the metabolic activity of littoral submersed macrophytes and their epiphytic microflora. The littoral macrophyte–epiphyte complex can function as a selective metabolic sieve.

CHAPTER NINETEEN

LITTORAL COMMUNITIES: ALGAE AND ZOOPLANKTON

ALGAE OF THE LITTORAL ZONE

In contrast to the immense amount of study on the systematics, physiology, and ecology of the phytoplanktonic algae of aquatic systems, there is a dearth of information on the algae attached to substrata or loosely aggregated in the littoral regions of lakes or shallow zones of streams. Although the taxa of these largely sessile algae are somewhat better understood, little information exists on their geographical distribution, seasonal population dynamics, utilization of microhabitats, responses to parameters of water movement or water and substratum chemistry, or on their interactions with other organisms.

Much of the problem centers on the extreme heterogeneity in distribution of the algae across an exceptionally variegated spectrum of microhabitats, which in turn are subjected to much more variable environmental physicochemical and biotic parameters than usually occur in the open water. The problems of spatial and temporal heterogeneity of phytoplankton in pelagial habitats are minor in comparison to those of the attached algae. Nonetheless, despite the numerous problems associated with obtaining quantitative information, indications are that the algal populations specialized to these habitats sustain high rates of productivity. Our lack of knowledge of the complex interactions between the sessile flora and their substrata, and the contributions of the attached microorganisms to the total system productivity, represents a major void in contemporary limnology that warrants intensified study.

The limnological terminology applied to attached algae is complex. These terms and the general zonation of littoral algae have been introduced earlier (Chapter 8; cf. Fig. 8–3).

General Distribution of Littoral Algae

A floristic survey of the distribution of algae among subcommunities of the littoral zone results in long lists of species. The different associations reflect differences in parameters of light, temperature, nutrient availability, water movement, substratum, grazing pressure, and other factors. Detailed floristic analyses indicate specific characteristics of dominant taxa and designate the range of probable interactions between species, thereby providing a foundation upon which experimental approaches can be designed (Round, 1964a). The initial evaluations should lead to recognition of the various habitats and their algal associations, thereby enabling quantitative sampling and the separation of dominant and casual species. These initial evaluations are difficult to carry out, especially among algae of the sediments, but have been done in several detailed analyses of littoral communities.

The attached algae often dominate algal biomass in small streams and shallow lakes. Probably over 90 per cent of all algal species grow attached to a substratum; these species include practically all of the pennate diatoms, and a majority of the Conjugales, Cyanophyta, Euglenophyta, Xanthophyceae, and Chrysophaceae (Round, 1964a).

Epilithic and epiphytic communities have many algal species in common, but there is relatively little interchange of species between those of the mud-living epipelic habitat and the other substrata types. Epipelic algae are largely motile; their motility is essential to enable them to move to the surface after any disturbance of the sediments. The algal species adnate to sediments of streams and rivers with moderate to fast flow are almost exclusively motile.

Epipsammic algae, consisting largely of small diatoms and blue-green algae attached more or less firmly to crevices in the surface of sand grains, exhibit an extremely variegated, heterogeneous distribution (Fig 19–1). These algae tend to be less motile than epipelic algae, which usually grow on finer, organic sediments (Round, 1965a; Meadows and Anderson, 1966, 1968). In the shallows of a temperate English lake, Moss and Round (1967) found that both epipelic and epipsammic algae showed marked population peaks during the spring. The biomass of the epipsammic algae of the surface 5 cm of sediments was consistently greater than that of the epipelic algae.

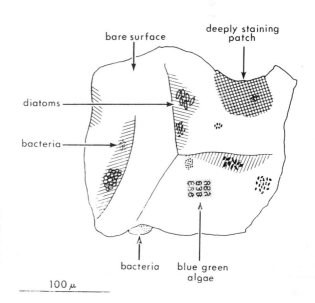

Figure 19–1. Diagrammatic representation of typical localized distribution of microorganisms and staining (carbol fuchsin) patches of organic materials on the surface of a sand grain. (From Meadows, P. S., and Anderson, J. G.: Micro-organisms attached to marine and freshwater sand grains. Nature (London), 212:1059–1060, 1966.)

Algal populations in interstitial water of a sandy beach of a southern Michigan lake were also large, and exhibited a spring maximum (Davies, 1971). Percentage water content was an important determinant of algal composition at the zone of the water's edge.

Within the sand, light is attenuated rapidly and is essentially extinguished within half a centimeter. Yet viable epipsammic algae have been found attached to sand grains as deep as 20 cm in sandy beach zones exposed to heavy wave action. It is believed that wave action is adequate to mix the sand grains and return attached algae to a lighted zone sufficiently often to permit active, although low, photosynthesis (Steele and Baird, 1968).

VERTICAL MIGRATION AND PHOTOSYNTHETIC RHYTHMS

Algae growing adnate to unconsolidated, shifting sediments are constantly in danger of being removed from light, which is attenuated to zero in a few millimeters.* It is not known whether these epipelic algae supplement photosynthetic growth by heterotrophic utilization of organic substrates. An important adaptation to the rigors of this habitat is an ability to move vertically within the sediments in response to light availability.

Persistent diurnal vertical migration rhythms have been shown in epipelic diatoms, flagellates, and blue-green algae from flowing waters and shallow lakes (Round and Happey, 1965; Round and Eaton, 1966). Cell numbers on the sediment surface start to increase before dawn, and reach a maximum about midmorning. Thereafter, surface cell numbers decrease and reach a minimum before the onset of darkness. This rhythm persists for a limited time in continuous darkness or continuous light, although the synchrony is lost more rapidly under continuous light conditions.

Studies of the diurnal rhythms of freshwater epipelic algae have shown that the maxima of cell emergence and photosynthetic capacity coincide, although the increase in photosynthesis preceded cell emergence (Brown, et al., 1972). The photosynthetic rhythm persisted at light intensities well below saturation; minimum photosynthetic values were found in midafternoon, after which the rate increased again, a rise that preceded the reemergence of cells. The maximum photosynthetic rate occurred when maximum cell numbers were present on the sediment surface.

SEASONAL POPULATION DYNAMICS

Quantitative evaluations of attached benthic algal populations are much more difficult than qualitative floristic analyses. Recognition of the various habitats and their associations is a problem common to both analyses, and this problem is accentuated by ineffective sampling techniques that are difficult at best. It is difficult to separate epipelic algae from sediments, but separation is necessary for accurate enumeration. Phototactic responses of the algae can assist in this separation; alternatively, one can use other biomass methods such as pigment content (correcting for pigment degradation

---

*The attenuation of light by sediments is highly dependent upon the type of sediment. Light is attenuated much more rapidly by highly organic sediments than it is through sand. For example, Palmer and Round (1965) found that radiation was reduced to about 1 per cent of the incident value under about 3 mm of organic-rich sediment.

products common to sediments). Quantitative sampling of epilithic and epiphytic algal populations is even more difficult, but a number of devices and procedures have been invented (see the review of Sládečková, 1962) that permit some degree of quantification. The problems of quantitative sampling of attached algae are further compounded by the settling of casual species originating from the plankton and metaphyton into the benthic populations, especially during winter. Only detailed sampling of the different communities combined with metabolic assays of growth can resolve this problem.

Use of Artificial Substrata. Because of extreme habitat and community heterogeneity, artificial substrata of uniform composition and colonizable area are commonly used to estimate colonization and growth characteristics of attached algae. Substrate types that have been used include glass slides, concrete blocks, rocks and rock plates, sand, various plastic and metal plates, and wood.

There are several restrictions inherent in the use of artificial substrata. Marked differences in colonization rates and biomass accrual have been found in relation to the position of the supported substrata and their location within the aquatic system. For example, vertically positioned plates on which sedimenting organisms from the plankton or metaphyton could not accumulate contained less biomass, but more realistically simulated population characteristics found on natural substrata, than did those positioned horizontally (e.g., Castenholz, 1961; Pieczyńska and Spodniewska, 1963). The influence of water movement or simply the process of retrieving the slides following incubation can lead to significant losses of attached microflora. Some organisms will not adhere to artificial substrata such as glass until early successional forms have colonized the substratum, which may or may not occur within the period of incubation (e.g., Blinn, et al., 1980). Apparently glass is not seriously selective for the attachment of diatoms (Patrick, et al., 1954). Grazing can lead to alterations of attached communities that may influence the results, depending on whether the questions being addressed concern specific attached flora or the entire attached community.

Some investigators have positioned artificial substrata in the pelagic zone of lakes or main surficial flow of streams quite removed from sites of natural substrates. While these analyses may have some value in a relative sense, such as in evaluating a response to toxicants in different water types, they are of little value in evaluating natural attached communities.

The most serious criticism that can be directed against studies using artificial substrata centers on the implicit assumption that any metabolite, inorganic or organic, of the living or nonliving substrata has no appreciable effects on the attached community. Similarly, it is assumed that the metabolism of the attached microfloral community has no reciprocal effects upon the substratum. Although recent evidence for such synergistic effects is meager, it is sufficient to indicate that more than a passive relationship exists (see discussion on pp. 567–572). With these thoughts in mind, difficulties of quantitatively evaluating natural population growth on natural substrata become apparent. Even though the mutual metabolic relationships are real, especially among epiphytic algae and the supporting macrophytes, measuring the effects of the heterogeneous attached populations on the highly variable substrata often exceeds the capacities of existing methods. The problem is a technical one that must be approached without the use of artificial substrata.

On the other hand, when used critically, artificial substrata can be a meaningful tool for the approximate estimation of biomass accrual of many attached microorganisms. Much of our existing information on these communities is based on analyses using artificial substrata. Only a few investigations have addressed the in situ rates of algal productivity by means of metabolic techniques.

Factors Affecting Growth. The detailed investigations of the population dynamics of epipelic algae of several lakes of the English Lake District (Round, 1957a, 1957b, 1957c, 1960, 1961a, 1961b) serve to illustrate several characteristics of these populations, as well as the potential spatial, temporal, chemical, and biotic factors affecting growth. Diatoms completely dominated the epipelic community, and patterns of distribution could be discerned on the basis of general chemical features of the sediments and overlying water. Other algal groups were less well represented. The Volvocales were very sparse, and the Chlorococcales were represented only by the genera *Pediastrum* and *Scenedesmus.* Although not abundant, the desmid flora was richer in species than other groups. Euglenoids were much more common on more organic-rich sediments. Little correlation was found between the growth cycles of the epipelic and planktonic blue-green algae, and in some lakes, a conspicuous development occurred on the sediments but not in the plankton. Epipelic blue-green algae developed equally well in alkaline, neutral, and slightly acidic waters.

In general, a larger diatom biomass corresponded to a higher organic matter content of the sediments, and growth was better on sediments with more leachable silica. Moderate organic matter content of the sediments favored blue-green algal populations, while low populations were associated with sediments of very high and very low organic-matter content. Populations of the poorly represented Chlorophyceae and the flagellated forms were distinctly greater on organic-rich sediments. Although little correlation was found between population numbers and sediment content of sodium, potassium, iron, or manganese, the diatom, blue-green, and flagellated algal populations were directly correlated with calcium content. Algal growth was also greater on sediments with high phosphate content.

Although population numbers of epipelic algae exhibited large variations from lake to lake, and between sites within lakes, and with depth, certain consistent patterns emerged. The main growth period of epipelic diatoms and blue-green algae of lakes of this moderate temperate region* was in the spring, commencing about February (Fig. 19–2). Except for the wave-active shallow zone (0 to 1 m depth), population numbers

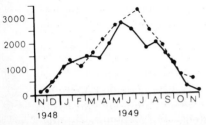
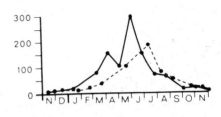

Figure 19–2. Seasonal cell counts of epipelic diatoms (*left*) and blue-green algae (*right*), in which all the depth stations are added together for each sampling date in the photosynthetic zone (1–6 m) of Lake Windermere (———) and Blelham Tarn (- - - - -), England. (Modified from Round, 1961b.)

*Ice-cover is rare and usually short-lived, typically only a few days in duration.

were relatively constant on sediments between 1 to 6 m depths; below this depth, numbers decreased precipitously to practically nil at depths greater than 8 m. Growth of the dominant benthic diatom *Tabellaria flocculosa* was shown experimentally to be largely related to light intensity, which decreased rapidly with increasing depth (Cannon, et al., 1961; Hillbricht-Ilkowska, et al., 1972). Strictly phototrophic diatoms are more sensitive to light than blue-green algae, whose depth distribution usually extends considerably deeper than that of the diatoms. The general seasonal biomass of epipelic algal populations in the photosynthetically illuminated zone generally followed the curves of incident light and water temperature.

In streams, light availability frequently limits benthic algal growth (e.g., Whitton, 1975; Wetzel, 1975b; Marker, 1976; Aizaki, 1978). Light availability can be altered markedly by seasonal changes in leaf canopy of deciduous trees adjacent to the stream. Succession of species within the overall annual cycle of epipelic algae is a much more elusive and complex matter. Among planktonic populations of algae, discussed in some detail in Chapter 15, the changes in physical and chemical parameters, particularly those associated with periods of circulation in stratified lakes, are very large and can influence dominance and competitive abilities of the species. The sudden collapse of populations at certain times of the year is related, in part, to these "shock" events.[†] Epipelic populations tend to occur largely above the metalimnion, so that the effects of disruption of stratification may be less on these populations than among the plankton. The seasonal succession of attached algal species is, however, just as pronounced as succession among the planktonic species (Round, 1972).

No temperate epipelic algal species is known to persist as a numerically dominant species over an entire year. Epipelic algal populations tend to increase quite rapidly, and then, decline equally rapidly. Often several successive species associations tend to dominate during an annual cycle, each phasing in to the next from live cells that have remained in the habitat for some time after cell division has ceased. The substrata provide innumerable microhabitats, certainly many more than exist under planktonic conditions. Therefore, one would expect a large number of coexisting species in the attached algal community at almost any time of the year. This distribution can be obscured by sampling techniques that tend to aggregate algae from many different microhabitats.

Annual and seasonal cycles of epipelic algae are marked and complex (e.g., Round, 1964a; Ilmavirta, 1977; Sheath and Hellebust, 1978; Romagoux, 1979). Diatom growth often commences in early spring and reaches maximum cell numbers in April to May in the temperate zone, after which a decline occurs in midsummer prior to a smaller autumnal peak that is over by November. While this pattern is similar to that of planktonic species, the epipelic algal pattern occurs more slowly than that found among planktonic algae. On shallow sediments, the epipelic algal populations tend to exhibit early winter and spring peaks, followed by a later midsummer maximum. Spring maxima are common among attached algal populations of streams, although these populations are subject to irregular decimation and disturbance by moving substrata during spates (see, for example, Douglas, 1958). In woodland streams, the seasonally changing light passing through the terrestrial canopy can shift the productivity pattern from one

---

†A concept discussed by Round (1971).

that is autotrophic in the spring to one of essentially heterotrophic dominance later in the season* (see Kobayasi, 1961, and also discussion in Chapter 22).

The seasonal population dynamics of epiphytic algae growing on submersed portions of macrophytes are much more involved because the surface area available for colonization and growth is continually changing as the macrophytes grow, senesce, and enter detrital phases. Computations of the population changes of the total epiphytic algal community are rare, but changes obviously are quite different among different supporting macrophyte communities (Kowalczewski, 1975a, 1975b; Cattaneo and Kalff, 1980). For example, changes in the biomass of attached algae associated with a zone of the emergent bulrush *Scirpus acutus* were more than an order of magnitude lower and out of phase with those of algae attached to substrates in a shallow zone dominated by submersed plants (*Najas flexilis* and *Chara*) (Fig. 19–3).

Only correlative data exist regarding potential factors affecting population structure or controlling seasonal population variations of benthic algae. Unlike the planktonic habitat, which is relatively isotropic, selective physical factors are involved in the benthic habitat. Within the overriding control factors of light and temperature, the complex and dynamic interactions of inorganic and organic chemical parameters of the substrata and surrounding water surely play major roles. Both epipelic and epipsammic algae are subject to burial by water movements and disturbance of the sediments by animals. If buried in the sediments, the algae may also be exposed not only to lower intensities of light, but also to anaerobic conditions within the sediments. Many benthic algae can survive anaerobiosis for several days, and may retain considerable photosynthetic potential in the absence of oxygen (Moss, 1977). So far, correlations between benthic algae and chemical parameters have centered on rel-

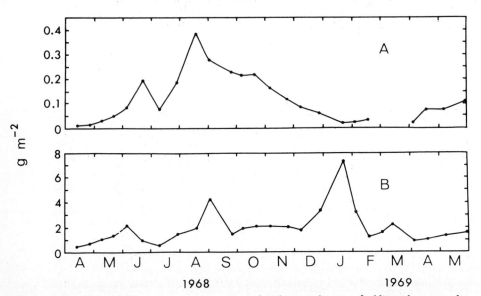

Figure 19–3. Chlorophyll *a* concentrations, extrapolated upward to g m^{-2} of littoral zone on the basis of estimated surface areas of the macrophytes, of algae attached to artificial substrates in dense stands of the littoral zone dominated by *A*, the bulrush *Scirpus acutus*, and *B*, the submergents *Najas flexilis* and *Chara*, Lawrence Lake, Michigan. (After Allen, 1971.)

*The author acknowledges the fruitful discussions on this subject of much unpublished work of Donna King of Central Michigan University.

atively extreme habitats that permit only crude generalizations. Some correlations drawn between changing water chemistry and benthic algae may be helpful, as in the case of epilithic algae that can obtain relatively meager nutrients from their substratum. An additional example is the previously discussed (see Chapter 14) results of Jørgenson (1957), who found that diatoms epiphytic on the reed *Phragmites* competed with plankton for silica of the water early in the major growing season. Later in summer and fall, apparent utilization by the diatoms of silica from the *Phragmites* stems gave them a competitive advantage over planktonic forms (see Fig. 14–19). Certain other littoral algae also have high requirements for silicon (e.g., Moore and Traquair, 1976; Hooper-Reid and Robinson, 1978). Spring and autumnal population maxima are common in epiphytic diatom communities on submersed macrophytes, and have been correlated with nutrient availability (e.g., Siver, 1978).

The general assumption that the chemistry of the overlying water has a direct controlling effect on epipelic algae has not been verified experimentally. For those species adapted to epipelic existence, the proximity to the totally different chemical milieu of the interstitial water within and immediately adjacent to the sediment–water interface can give them a distinct advantage that is quite unrelated to chemical events occurring in the overlying water. For example, the biomass of epipelic algae commonly increases as the organic carbon, nitrogen, and phosphorus content of the sediments increases (Skorik, et al., 1972). A large portion of the ammonia nitrogen of interstitial water of sediments was found to be assimilated by benthic algae (Jansson, 1980). Dissolved organic nitrogen compounds released to the overlying water originated largely from the algae. Nitrogen fixation by epipelic, as well as epiphytic, blue-green algae and bacteria is common (Moeller and Roskowski, 1978; Finke and Seeley, 1978) and, as indicated earlier (Chapter 12), can constitute a significant input of combined nitrogen to some lakes. Yet the epipelic forms are not completely exempt from characteristics and events of the overlying water. For example, a high content of dissolved organic matter or a seasonally variable development of planktonic algae or of macrophytic vegetation populations would markedly affect the quality and quantity of light reaching the sediments, the thermal regime of the sediments, water movements near and along the sediments, etc. While it is intuitively apparent that changing events in the overlying water can influence the species succession and productivity of epipelic algae, studies of these interactions are essentially nonexistent. The whole area of investigation is uncharted and difficult, but is one that is approachable by systematic application of modern research techniques.

Insight into the community structure of epiphytic microflora of submersed macrophytes can be gained from electron photomicrographs* of a typical association (Fig. 19–4). The surface structure of the epiphytic community (Fig. 19–4A) indicates the manner in which the loosely woven diatomaceous component is held in place by a stranded matrix formed largely from gelatinous stalks of mucoid substances (Fig. 19–4B), which in places extends as a fragmented membrane above the deposits of calcium carbonate covering the cell wall (Allanson, 1973; Roos, et al., 1981). While some algae (diatoms, blue-greens) and bacteria are loosely associated within and on the calcite and mucoid complex, others penetrate through the complex to the macrophyte itself, and attach themselves directly along its entire surface or by means of simple or branched mucilagenous stalks. Much of the matrix is associated with the mucopeptide cell walls of

---

*The scanning electron photomicrographs were kindly provided by Dr. B. R. Allanson, Rhodes University, South Africa; some are from his investigations reported in Allanson, 1973.

Figure 19–4. The structure of a microfloral community epiphytic upon the macroalga *Chara* from Wytham Pond, Oxford, England. *A*, Surface view of microflora on a stem (× 75); *B*, The epiphytic community after exposure to pH 4.5 to remove calcite deposits showing the dominant diatom *Achnanthes minutissima* and mucoid membranes attached to the wall of the host plant (× 500); *C*, Intimate

(cont. on next page)

the epiphytic bacteria and blue-green algae, an association that is lost in the techniques used to prepare these scanning electron photographs; this matrix can be seen, however, in transmission electron microscopy.

Most of this relatively rich community of microflora epiphytic upon macrophytes is destined to become detritus. Massive accumulation rates on the supporting plants, *Chara* and *Potamogeton* (Allanson, 1973), and differences seen in the fine structure of the epiphytic-carbonate-mucoid complex between parts of the plant as it grows and provides new surfaces for colonization indicate that little of the community is utilized by grazing animals. In places, however, evidence shows that some grazing occurs (Fig. 19-4D, E). Little is known of the in situ feeding processes, nutrition obtained from the epiphytic association, or effects of the grazing organisms on productivity of the epiphytes or macrophytes, although the feeding process may enhance growth of the epiphytes or macrophytes by solubilizing certain nutrients.

The growth and productivity of the sessile attached microbiota, particularly epiphytes on submersed portions of macrophytes, can be affected by grazing activities of animals, e.g., insect larvae, snails, crayfish, and certain phytophagous fishes (Flint and Goldman, 1975; Mason and Bryant, 1975; Bowen, 1978; Eichenberger and Schlatter, 1978; Kitchell, et al., 1978; Hunter, 1980; Hagashi, et al., 1981). In certain cases, when the epiphytic algal biomass is low, growth can be stimulated (turnover rates increased) by moderate levels of grazing. Heavy grazing pressures result in reductions in overall epiphytic productivity under the commonly used experimental conditions of high grazer density without natural predation pressures upon the herbivores. Very little evidence exists for selectivity of attached microflora by grazers; the animals are generally omnivorous, and randomly ingest algae, bacteria, particulate detritus, and inorganic deposits (e.g., $CaCO_3$). Certain fatty acid components of attached algae are distinctly inhibitory to grazing invertebrates (e.g., LaLonde, et al., 1979). Most of the microflora attached to macrophytes and other substrates is not consumed by animals, but instead enters detrital pathways within the ecosystem (cf. Chapter 22).

Metabolic Interactions

RELATIONSHIPS BETWEEN ALGAE AND BACTERIA

The metabolic relationships between the attached algae and bacteria and the substrata are complex. The detailed investigation of Allanson (1973) on the fine structure of the epiphytic microflora of macrophytes illustrated the structural proximity required for the metabolic interactions hypothesized earlier for these associations (Fig. 19-5) by Wetzel and Allen (1970) and Allen (1971). The macrophytes release large quantities of both inorganic compounds (e.g., oxygen, carbon dioxide, phosphates, silica, etc.) and organic compounds (secreted during active photosynthesis and released by autolysis during senescence; e.g., Howard-Williams, et al., 1978, Siver, 1980) into the immediate

---

association between the diatom component and supporting mucoid matrix, calcite deposits removed ($\times$ 1,000); D, Transverse feeding tracks left by nymphs of the baetid mayfly *Cloeon dipterum* ($\times$ 500); E, A chironomid larva, found enclosed in fronds of *Chara*, which may have been feeding in this position ($\times$ 90). (Photographs courtesy of B. R. Allanson, 1973. From Allanson, B. R.: Fine structure of the periphyton of *Chara* sp. and *Potamogeton natans* from Wytham Pond, Oxford, and its significance to the macrophyte periphyton metabolic model of R. G. Wetzel and H. L. Allen. Freshwater Biol., 3:535–541, 1973.)

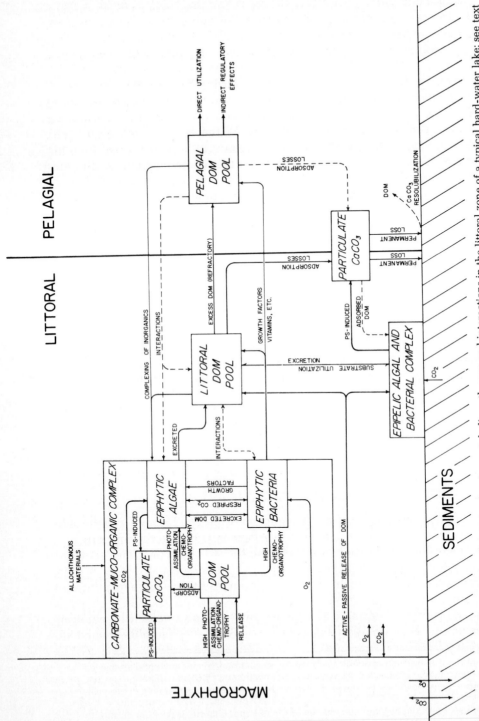

Figure 19-5. Diagrammatic representation of major metabolic pathways and interactions in the littoral zone of a typical hard-water lake; see text for discussion. DOM = dissolved organic matter; PS = photosynthesis. (Modified from Wetzel and Allen, 1970.)

proximity of adnate epiphytic microflora and littoral waters. This proximity alone is sufficient for the transfer and exchange of nutrients (PO_4, CO_2) between the epiphyte and the supporting plant (see, for example, Harlin, 1973). In hard waters, precipitation of calcium carbonate is induced by photosynthetic activity (Chapter 11) to form a matrix intermixed with the microflora and mucoid substances (predominantly polysaccharides and peptides). Organic substrates released by the macrophytes and algae are in part actively utilized by the epiphytic bacteria. Dissolved organic compounds not utilized by this association and not adsorbed within or to monocarbonate surfaces, enter the pool of littoral dissolved organic matter for further degradation by bacteria. In turn, epiphytic bacteria produce CO_2 and certain organic micronutrients, e.g., vitamin B_{12} (cf. Wetzel, 1969, 1972) that can serve as growth factors for the macrophytes or algae, or both. Photoassimilative active uptake of labile organic substrates has been demonstrated for a submersed angiosperm in axenic culture and for a species of the macroalga *Nitella* (Wetzel, unpublished; Smith, 1967).

There is evidence to support the conclusion that most phytoplanktonic algae cannot actively augment photosynthesis by heterotrophic utilization of organic substrates when in competition with bacteria that actively assimilate the same substrate (e.g., Mayfield and Inniss, 1978; cf. chapters 15 and 17). The immediate juxtaposition of attached algae to the metabolically active area of a living plant and bacterial concentrations can give them a distinct competitive advantage over their planktonic counterparts which live in a habitat where nutrients are much more dispersed (see, for example, the studies of Allen, 1971).

> Controversy has evolved in recent literature on whether or not the epiphytes acquire nutrients from supporting macrophytes (e.g., Cattaneo and Kalff, 1979; Gough and Gough, 1981). Despite marked differences in colonization patterns and densities of epiphytic algae on natural and artificial (plastic) plants (Cattaneo, 1978; Cattaneo and Kalff, 1978), metabolic differences between algal communities on the different substrata have been small. Acquisition of phosphorus by epiphytes from the macrophytes has been evident, but the effects on overall community productivity have been slight. Other evidence indicates a specificity of algal communities to particular macrophyte species (reviewed in Eminson and Moss, 1980). It is concluded that it is highly probable that significant interactions exist between epiphytes and the supporting macrophytes, and that these metabolic exchanges influence algal species composition, successional population dynamics, and resultant productivity. As pointed out by Eminson and Moss (1980), the influence of supporting macrophytes in determining the epiphytic algal community is greatest in infertile waters. As the surrounding water becomes more fertile, external environmental factors influence epiphytic growth more than do inputs from the host macrophytes. It is apparent that complex, coupled interactions exist between the epiphytes, supporting macrophytes, and adjacent environmental conditions; they remain to be resolved.

Analogous interactions certainly exist within the epipelic associations of algae and bacteria. Unfortunately, even less is known about the qualitative and quantitative metabolic interrelationships of these attached forms and their substrate than is known about the epiphytic microflora. Potential sources of inorganic nutrients and labile dissolved organic substrates are likely to be greater under reducing conditions of the sediment–water interface than at the surfaces of macrophytes.

Dissolved organic substrates emanating from littoral sources or from the drainage basin are degraded, flocculated, and adsorbed as they travel to the pelagic zone (cf.

chapters 18, 20, and 22). The rates and kinds of "processing" that occur en route are highly variable, and depend upon the dynamics of seasonal and individual differences in physical and chemical parameters of lakes. The littoral region can be viewed as functioning as an effective but selective metabolic sieve or sink for certain dissolved organic compounds and inorganic nutrients from littoral and external sources towards the pelagic zone (e.g., Lee, et al., 1975; Howard-Williams and Lenton, 1975; Klopatek, 1975; Gaudet, 1977, 1978, 1979; Mickle and Wetzel, 1978a, 1978b, 1979; Prentki, et al., 1978; Howard-Williams and Howard-Williams, 1978; Stewart and Wetzel, 1981a). During this transport, the more easily degradable organic substrates tend to be decomposed in the littoral zone, and compounds of greater resistance (i.e., slower turnover rates) tend to move towards the open water. Therefore, even though the amounts, rates of utilization, and fate of inorganic and organic compounds released from the littoral zone vary seasonally and differ from lake to lake, their direct and indirect effects on the metabolism of the pelagial microflora can be great in many lakes.

EFFECTS OF LIGHT AVAILABILITY

Within dense stands of emergent macrophytes, the rapid attenuation of light plays a major role in the reduction of photosynthesis of littoral algae. For example, in reed stands of *Phragmites*, Straškraba and Pieczyńska (1970) found a rectilinear relationship between the percentage transmission of light through the emergent parts of the reed and shoot density. As much as 96 per cent of the incoming radiation was removed above the water, whereas the density of the stand had little further influence on the light availability underwater. The rather low light intensities occurring in reed stands, especially those of an average or high density, substantially lowered photosynthesis and biomass of phytoplankton entering the stands from the pelagial region. Photosynthetic activity of attached algae is similarly reduced in these dense stands. Cutting experiments showed a large increase in photosynthetic activity of each algal group, and the increase was directly proportional to the percentage improvement of light conditions (Table 19-1). In less dense stands of emergent plants, as for example in low-growing

TABLE 19-1 Effect of Increased Light Availability on Photosynthetic Activity of Phytoplankton and Attached Algae in a Dense Stand of *Phragmites communis* Before and After Removal of Plant Shoots, Smyslov Pond, Czechoslovakia, in September

	WITH PHRAGMITES		PHRAGMITES CUT	
	0 cm	30 cm	0 cm	30 cm
Phytoplankton (mg O_2 l^{-1} day $^{-1}$)				
Gross production	0.14	0.06	7.9	3.0
Respiration	0.17	0.78	2.6	1.5
Attached algae				
Gross production (mg O_2 dm^{-2} day^{-1})		3.1		8.7

From data of Straškraba and Pieczyńska, 1970.

sedges, light reduction is not as serious, but a strong limitation of photosynthesis of phytoplankton still has been observed (Straškraba, 1963). Some decompositional products or compounds secreted from the plants were also believed to contribute to this reduction.

ATTACHED ALGAE AS A SOURCE OF PHYTOPLANKTON

The concept that the presence of littoral benthic algae gives rise to plankton algae emerged many years ago. It was assumed that plankton algae formed resting spores that overwinter in or upon the sediments, and that under conditions favorable to germination and population development, algae on the sediments served as the inoculum for the phytoplankton. In moderate-size to large lakes, there is no evidence for any conspicuous buildup of algae on the sediments before a plankton bloom, but only afterward when the dead plankton settle out onto the sediment (Round, 1964).

An alternative concept is that planktonic populations are always present in the water, although in extremely low concentrations. These few remnants of past populations can then serve as the inoculum for a rapid population increase under favorable growing conditions. This situation seems to be the case for most species.* In one of the better studies of algae, resting spores of the diatom *Asterionella formosa* were not observed, and no evidence exists that appreciable numbers originate from shallow areas (Lund, 1949). Massive population developments more commonly originate from exponential growth of live cells present in the pelagial waters.

Some phytoplankton do settle out during periods of reduced turbulence, and remain in a resting state on the sediments until water movements are again sufficient to reintroduce the resting cells back into the plankton.† *Melosira italica* is a large diatom that clearly does enter a resting stage on the sediments, particularly under ice cover and during summer stratification (Lund, 1954, 1955). There is no evidence, however, for active growth on the sediments, which are often below the photic zone. Resuspension of filaments during autumnal and spring circulation serves as a large inoculum for the rapidly developing pelagic population. Similar results have been found for blue-green algae (Preston, et al., 1980; Fallon and Brock, 1981).

In shallow lakes, the separation between epipelic algae and littoral metaphyton is less distinct. It is common to find strictly benthic algae or predominantly planktonic forms intermixed, especially following irregular storm turbulence (see, for example, Lehn, 1968; Moss, 1969a; Moss and Abdel Karim, 1969; Brown and Austin, 1973a). In shallow lakes in which thermal stratification is irregular and intermittent, shifts in the percentage composition of largely planktonic or of epipelic algae can be rapid. For example, in a pond of southern England, nonmotile species dominated planktonic populations only when the water column was relatively turbulent. Motile species persisted in the water column, and the largest biomass of any species occurred when the water column was stratified. In the shallow Obersee of the Lake of Constance, southern Germany, littoral diatoms and algae of the metaphyton are transported to the open water, especially during spring storms (Müller, 1967).

---

*Termed *holoplankton* in older literature, in contrast to *meroplankton* that have resting stages in sediments and are not observed in the open water for long periods of time, i.e., they are planktonic only at certain times in their life history.

†See page 370.

Productivity of Littoral Algae

QUANTITATIVE EVALUATION

Quantitative evaluation of attached algae is basic to analyses of population changes and productivity, but is confounded seriously by the simultaneous collection of much debris and dead cells in various stages of decomposition. Counting and identification, with all proper precautions and statistical evaluations, have a number of advantages for detailed analyses (Lund and Talling, 1957; Lund, et al., 1958; Vollenweider, 1969b). Although developed for lake plankton, these methods of analysis usually are equally applicable to studies of littoral algae. Species enumeration can be extended for productivity evaluations when coupled to estimations of more meaningful biomass parameters such as cell volumes or carbon at the species level (cf. Chapter 8). Cellular constituents such as pigment concentrations can be used effectively as an estimate of attached algal biomass (e.g., Szczepański and Szczepańska, 1966). It is particularly imperative, however, to correct for pigment degradation products (Wetzel and Westlake, 1969), and not to overextend these estimates, which vary significantly in response to an array of physiological conditions.

BIOMASS MEASUREMENTS

To a limited extent, a temporal series of biomass measurements permits estimates of rates of primary production. These estimates often underestimate actual productivity because of losses via excretion of organic compounds, respiration, mortality, decomposition, grazing, etc. The observed rates also are influenced by the rate of colonization of new propagula, which is generally faster when conditions for growth are better. Hence, the observed accumulation of biomass is a composite of colonization and production. The saturated accumulations of biomass on substrata yield little insight into the production dynamics that have occurred before the cumulative point.

The rate of community turnover is an average of several variable species rates, and is difficult to determine from biomass analyses. Subjective errors also occur in determinations of attached algae of lakes and streams; factors important include the frequency of sampling and estimation of the degree of colonization that represents a climax stage of the community (Sládeček and Sládečková, 1964). Higher turnover rates result as sampling interval is decreased, especially when the intervals are shorter than a month.

Productivity values of littoral algae are difficult to compare from one lake to another, or even seasonally within a lake, because of the diversity of techniques employed. Very few in situ metabolic measurements have been made on these communities. As mentioned earlier, many analyses have employed artificial substrata to reduce the natural heterogeneity of the algae. The resemblance of benthic algae on artificial substrates to those on adjacent natural substances is highly variable and deviates significantly, both qualitatively and quantitatively, with seemingly minor differences in substrata, environmental differences in microhabitats, and particularly in rates and duration of colonization. The discrepancies found between populations on natural and artificial substrata are sufficient to necessitate a critical evaluation of each study in which artificial substrata are contemplated (see, for example, Tippett, 1970; Warren and Davis, 1971). The extent of compromise must be determined in any ecological method, and this problem becomes especially acute among littoral communities.

The admixture of estimates of productivity values in the ensuing discussion should be viewed as representing the range encountered under natural conditions. Although no attempt has been made to summarize all of the literature, which is quite diverse and of variable precision, the examples are presented to point out primary characteristics and emphasize the major contributions that the attached flora can make to freshwater systems.

SPATIAL AND TEMPORAL VARIATIONS

The rates of primary production of attached algae are obviously dependent upon the substrate area available for colonization within the zone of adequate light. In some lakes, the littoral substratum may be relatively uniform for much of the perimeter, but in most cases, the variation is great. Moreover, the living macrophytic substrata are constantly changing, especially in the more seasonal temperate regions. An example of this spatial heterogeneity in productivity can be seen in a two-year analysis of littoral components—macrophytes, littoral phytoplankton, and attached, largely epiphytic, algae—in Mikolajskie Lake, northern Poland (Pieczyńska, 1965, 1968; Pieczyńska and Szczepańska, 1966; Kowalczewski, 1965). Simultaneous measurements in midsummer along numerous points of the littoral zone showed much variation, which was accentuated during sporadic mass appearances of sessile blue-green *Gloeotrichia* and the filamentous green alga *Cladophora* (Fig. 19-6). Similar spatial variations in numbers and species assemblages have been demonstrated in the littoral of a eutrophic lake of British Columbia (Brown and Austin, 1973b).

Seasonal fluctuations in primary productivity of attached algae are also quite variable, and are analogous to changes in biomass discussed earlier. In general, estimates of biomass accrual, such as chlorophyll, and in situ measurements of production rates by carbon uptake or oxygen production are fairly similar, particularly during periods

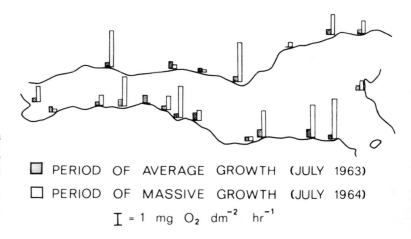

Figure 19-6. Spatial variations in the simultaneously measured rates of primary production of attached algae in different areas of the littoral zone of Mikolajskie Lake, central Poland. (After data of Pieczyńska and Szczepańska, 1966.)

☐ PERIOD OF AVERAGE GROWTH (JULY 1963)

☐ PERIOD OF MASSIVE GROWTH (JULY 1964)

$\text{I} = 1 \text{ mg } O_2 \text{ dm}^{-2} \text{ hr}^{-1}$

of active growth (see Pieczyńska and Szczepańska, 1966; Allen, 1971; Hunding and Hargrave, 1973; Hickman, 1971a). Reductions in temperature and light under winter conditions clearly are the dominant causal mechanisms limiting photosynthesis both seasonally and vertically with depth (see Figs. 19-7 and 19-8). The productivity of attached algae collected under ice and exposed to increased light was found to increase

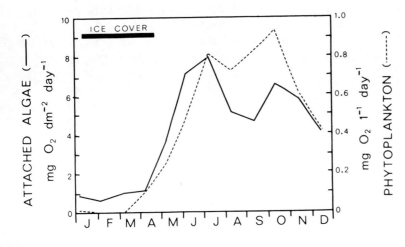

Figure 19-7. Rates of primary production of attached algae and phytoplankton in the littoral zone dominated by the emergent macrophyte *Phragmites*. (Modified from Pieczyńska and Szczepańska, 1966.)

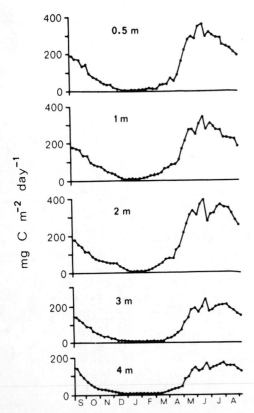

Figure 19-8. Weekly mean values of gross primary production rates of epipelic algae in shallow Marion Lake, a composite from 5 littoral stations, southern British Columbia. (Redrawn from Gruendling, 1971.)

greatly and approached values obtained under summer conditions. In the example from Mikolajskie Lake, productivity in the spring increased rapidly in both the attached algae and littoral phytoplankton, decreased in midsummer during maximum development of the littoral macrophytes, and then increased in the fall with a bloom of diatoms.

Photosynthetic enhancement of epipelic algae was shown experimentally when sediment samples were moved to shallower depths with improved light conditions (Table 19-2). Hunding (1971) has shown that benthic diatoms exhibit light-saturated photosynthesis over a wide range of light intensities. During summer, light-saturated photosynthesis was found even at very high irradiances, and no photoinhibition could be demonstrated up to 38 klux ($= 30.4$ μW cm^{-2}). The light-saturated photosynthetic value decreased in autumn, indicating that maximum photosynthetic rates are probably maintained under reduced prevailing light conditions at that time of year. Similar depth relationships can be seen in comparisons of pigment biomass and primary productivity of epipelic algae of a small English lake (Table 19-3), and in a clear, deep alpine lake (Capblancq, 1973).

Within sediments the rates of carbon fixation and the chlorophyll concentrations of epipelic algae are attenuated very rapidly with depths of a few millimeters (Table 19-4), whereas rates and pigment concentrations of epipsammic algae penetrate somewhat deeper. As discussed earlier, light-mediated metabolism of these algae is probably

TABLE 19-2 Enhancement of Photosynthesis of the Epipelic Algal Community of Marion Lake, British Columbia, in June by Moving Intact Sediments from Different Depths to 0.5 m

SAMPLE DEPTH (m)	SOLAR RADIATION (g cal cm^{-2} hr^{-1})	TEMPERATURE (°C)	INCUBATED IN SITU AT DEPTH (ml O$_2$ m^{-2} hr^{-1})	INCUBATED AT 0.5 m DEPTH (ml O$_2$ m^{-2} hr^{-1})
0.5	32.6	23.0	19.9	19.9
1.0	23.3	22.0	22.7	26.4
2.0	14.4	17.0	19.4	28.8
3.0	10.6	14.0	14.5	19.4
4.0	5.5	13.0	2.4	5.7

From data of Gruendling, 1971.

TABLE 19-3 Annual Mean Biomass and Rates of Primary Production of Epipelic Algae of Abbot's Pond, Southern England

YEAR	LITTORAL DEPTH CATEGORY (m)	MEAN BIOMASS (mg CHLOROPHYLL a m^{-2})	MEAN RATE OF PRIMARY PRODUCTION (mg C m^{-2} hr^{-1})
1966	0–1.0	11.7	5.50
	1.0–2.5	7.3	2.89
	2.5–4.0	4.0	0.56
1967	0–1.0	9.09	4.13
	1.0–2.5	1.38	0.75
	2.5–4.0	1.47	0.17
1968	0–1.0	13.6	6.23
	1.0–2.5	2.14	0.70
	2.5–4.0	1.47	0.23

From data of Moss, 1969b, and Hickman, 1971a.

TABLE 19-4   Relative Rates of Carbon Uptake by Epipelic and Epipsammic Algae at Different Depths within Sediment, Expressed as a Percentage of the Maximum Value, Shear Water, England

| DEPTH (cm) | % OF MAXIMUM VALUES | | | |
| | Epipelic Algae | Epipsammic Algae | | |
	17 Mar. 68	23 Nov. 67	17 Mar. 68	9 Sept. 68
0–1.0	100	100	100	100
1.0–2.0	11.6	57.4	42.1	6.6
2.0–3.0	0	33.4	5.7	0.2
3.0–4.0	0	30.5	3.1	1.2
4.0–5.0	0	14.3	2.5	0.3

All samples incubated at the same light conditions. From data of Hickman and Round, 1970.

intermittent and occurs predominantly when wave action circulates sand grains up to the sediment surface. Biomass and rates of production of epipsammic algae were consistently greater throughout the year than those of epipelic algae in the same lake system.

Seasonal cycles of primary productivity of epiphytic algae differ considerably in relation to the type of supporting macrophyte. In the Lawrence Lake example shown in Figure 19-9, the annual maxima occurred in midsummer when macrophyte biomass was greatest. At this time conservative estimates exceeded 8 g C m^{-2} littoral zone day^{-1} of carbon fixed by epiphytic algae in this oligotrophic lake. Even though the rates of production per substrate area were similar among the emergent bulrush *Scirpus acutus* site and that of the much more dissected submersed macrophytes *Najas* and *Chara*, the

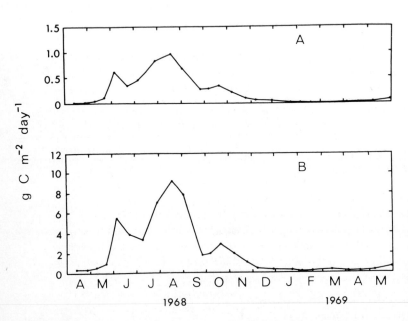

Figure 19-9. In situ primary production of attached algae in g C m^{-2} of the littoral zone day^{-1} among *A*, emergent *Scirpus acutus*, and *B*, submersed macrophytes *Najas flexilis* and *Chara*, Lawrence Lake, Michigan. (Modified from Allen, 1971.)

much greater area available for colonization in the submersed macrophyte area increased the rates of production by epiphytic algae in the littoral zone area dominated by the two more dissected macrophytes. In both cases, annual maxima were found in August near the sediments at the basal portions of the macrophytes. Although the winter rates of carbon fixation by attached algae among the *Najas-Chara* were 100 to 200 times greater than those among the *Scirpus*, the values per substrate area among the more open *Scirpus* littoral were consistently much higher, ranging from double to an order of magnitude higher per area of plant. Causes for this difference are unknown, but the more open *Scirpus* bed has greater light and water movement than occurs in the dense *Chara* beds.

> In an oligotrophic lake of New York, photosynthetic rates of epiphytic algae were approximately five per cent of the photosynthetic rates of the supporting macrophytes (Sheldon and Boylen, 1975). Epiphytic algal densities increased with the seasonal growth patterns of the macrophytes; maximal algal colonization of leaves of the plants, however, was similar per unit time. Epiphytic algal productivity from older leaves, lower on the macrophyte, was as much as 10-fold greater than that of epiphytes from younger top leaves.
>
> Rates of epiphytic bacterial utilization of simple organic substrates (glucose and acetate) followed first-order active transport kinetics (Allen, 1971). These rates were nearly an order of magnitude greater per unit area in the submersed macrophyte littoral region than among the emergent plants. This difference is potentially related to the greater quantities of organic substrates released by the submersed plants than by emergent macrophytes.

Comparison of production rates of attached algae among lakes and climatic regimes is difficult because of the variety of methods that have been used. However, there is a clear tendency for the rates of primary productivity of the attached algae to increase with decreasing latitude and longer growing season (Tables 19-5 and 19-6). The turnover times of the attached algal populations also tend, not unexpectedly, to decrease under the better growing conditions. The highest values recorded are from subtropical springs in which constant temperatures, high flow, and adequate nutrient supplies combine to enhance high rates of production. A similar relationship is seen among streams (cf. Wetzel, 1975b; Aizaki, 1978), although available data are few and very difficult to compare in a satisfactory manner.

PRODUCTIVITY RATES OF ATTACHED ALGAE VS. PHYTOPLANKTON AND MACROPHYTES

Only a handful of detailed studies exists in which in situ measurements of productivity of attached algae, phytoplankton, and macrophytes have been made simultaneously (Table 19-7). However, these studies are important in that they demonstrate the magnitude of effect that littoral productivity can have on lake systems. The range is great. In the large, shallow Borax Lake of northern California, epilithic algae are a major source of organic matter. Because the littoral area is small, when productivity per unit area is expanded for the whole lake surface, the attached algal contribution is reduced but still constitutes nearly one-half of total primary productivity. In Marion

TABLE 19-5 Rates of Production of Attached Algae and Estimated Turnover Rates of the Algal Populations of Several Exemplary Aquatic Systems

LAKE	TYPE OF ALGAE	AVERAGE BIOMASS (g dry m^{-2})	AVERAGE ANNUAL ESTIMATED NET PRODUCTIVITY (mg dry m^{-2} day^{-1})	ESTIMATED TURNOVER TIME (days)	SOURCE
Sodon Lake, Mich.	Glass slides, horizontal, open water	1.23	37.5	32.8	Newcombe, 1950
Falls Lake, Wash.	Glass slides, horizontal, at sediment interface (0.4 m)	3.24	148	21.8	Castenholz, 1960
Alkali Lake, Wash.	Glass slides, horizontal, at sediment interface (0.4 m)	2.94	131	22.4	Castenholz, 1960
Sedlice Reservoir, Czechoslovakia	Glass slides, vertical, open water, several depths	12.7	213	59.6	Sládeček and Sládečková, 1963
Lake Glubokoye, USSR	Epiphytic on *Equisetum*, biomass, and O_2 production	18.7	128.8	21.7	Assman, 1953
Silver Springs, Fla.	Epiphytic on *Sagittaria*, biomass, and O_2 production	177.	12,300.	14.4	Odum, 1957
Laboratory analyses, diatoms of Lake Balaton, Hungary	Epilithic diatoms, biomass, and O_2 production	—	—	approx. 1 (optimal conditions)	Felföldy, 1961
Jeziorak and Tynwald, northern Poland	Glass slides; in reed zones	—	—	4.5–16	Bohr and Luścińska, 1975
Crescent Pond, Manitoba	Epiphytic algae on macrophytes; ¹⁴C methods	40.	147	—	Hooper and Robinson, 1976, 1978
Suomunjärvi, Finland	Epipelic algae of an oligotrophic lake; ¹⁴C methods	—	1.1	—	Sarsa, 1979
Pääjärvi, Finland	Epipelic algae, ¹⁴C methods	—	60	—	Kairesalo, 1976, 1977
	Epilithic algae, ¹⁴C methods	—	6	—	
Changing, Antarctica	Epipelic algae on rocky debris; O_2 methods	—	18	—	Priddle, 1980
Sombre, Antarctica	Epipelic algae on rocky debris; O_2 methods	—	49	—	Priddle, 1980
Tundra, Ponds, Alaska	Epipelic algae of shallow ponds; ¹⁴C methods	—	22–55	—	Stanley, 1976
Tamagawa River, Japan	Glass slides; Bacteria	—	—	0.13–0.42	Aizaki, 1979

TABLE 19-6 Comparisons of Net Rates of Production of Attached Algae, Expressed as Organic Matter Accrual Over a 10-Day Period, of Several Different Fresh Waters

LAKE	MEAN NET PRODUCTION RATE (mg org. mat. dm^{-2} day^{-1})	DEPTH OF MAXIMAL NET PRODUCTION ACCRUAL	REMARKS
Sodon Lake, Mich.	0.5	0.2 m	Horizontal glass slides, 3 summer months; Newcombe, 1950
Falls Lake, Wash.	1.28	0.4 m	Horizontal glass slides, 22 months; Castenholz, 1960
Lenore Lake, Wash.	1.00	0.4 m	17 months
Alkali Lake, Wash.	1.76	0.4 m	12 months
Soap Lake, Wash.	1.67	0.4 m	17 months
Walnut Lake, Mich.	1.69	0.6 m	Horizontal glass slides, 2 summer months; Newcombe, 1950
Sedlice Reservoir, Czechoslovakia	2.13	3.0 m	Vertical slides, 10 months; Sládeček and Sládečková, 1964
Lake Tiberias, Israel	2.20	1.2 m	Vertical glass slides, 14 months; Dor, 1970
Shallow ponds, central Mich.	3.63	—	Glass slides, 2 summer months; Knight, et al., 1962
Borax Lake, Calif.	14.63	0.2 m	^{14}C methods, 12 months, in situ on epilithic algae; Wetzel, 1964
Red Cedar River, central Mich.	21.2	—	Plexiglas plates, summer months; King and Ball, 1966
Tiberias Hot Springs, Israel	73.0	0.1 m	Horizontal glass slides, 2 months; Dor, 1970
Silver Springs, Fla.	96.7	—	O$_2$ methods, epiphytic on *Sagittaria*, 12 months; Odum, 1957

Techniques employed vary widely and only an approximate comparison is possible.

Lake the primarily epipelic algae dominate the lake system. In both Lawrence Lake, Michigan, and Lake Wingra, Wisconsin, the submersed macrophytes are major components of the total primary productivity. Submersed macrophytes and largely epiphytic algae of Lawrence Lake constitute nearly three-fourths of the primary lake productivity, in spite of the limited littoral area of this small lake. In contrast, the littoral contribution to the total productivity of the two Ontario lakes constitutes only a few per cent because the shapes of these lakes do not lend themselves to extensive littoral development.

A few other less complete studies that do not permit full comparisons should be mentioned to relate the relative rates of photosynthesis of attached algae to those of phytoplankton. Slow-growing submerged mosses, almost exclusively *Marsupella aquatica*, cover some 40 per cent of the bottom from between 2 to 35 m of clear, low mountain Lake Latnajaure in Swedish Lappland (latitude 68°). This large, deep ($\bar{z} = 16.5$ m, $z_m = 43.5$ m) lake is covered with over a meter of ice for about 10 months of the year, and has a mean annual water temperature of 2°C. During the period of investigation in August and September, the average primary productivity for the whole lake was between 3.5 and 4.0 kg C per day (Bodin and Nauwerck, 1969). The average turnover time of the biomass of the perennial moss was about 30 years, between extremes of about 15 years at a depth of 5 m, and more than 125 years at depths greater than 30 m. Of the total primary productivity of the lake, the phyto-

TABLE 19-7 Examples of Annual Net Productivity of Phytoplankton, Littoral Algae, and Macrophytes of Several Lakes in Which Productivity Estimates of Attached Algae Were Made on Natural Substrata

LAKE	AREA (ha)	MEAN DEPTH (m)	ANNUAL MEAN (mg C m^{-2} day^{-1})	ANNUAL MEAN (kg C lake^{-1} day^{-1})	kg C ha^{-1} OF LAKE SURFACE year^{-1}	(%)	REMARKS
Borax, California	39.8	<0.5					Saline lake; benthic algae, primarily epilithic, some epiphytic and metaphyton; single macrophyte species *Ruppia maritima*; ^{14}C methods for all components (Wetzel, 1964)
Phytoplankton			249.3	101.0	926	[56.8]	
Littoral algae			731.5	75.5	692	[42.5]	
Macrophytes			76.5	1.36	12	[0.7]	
					1630		
Marion, British Columbia	13.3	2.2					Soft-water, oligotrophic lake; benthic algae, primarily epipelic; O$_2$ techniques, from which net production was estimated (Efford, 1967; Hargrave, 1969; Gruendling, 1971)
Phytoplankton			21.9	0.29	8	[1.6]	
Littoral algae			109.6	11.3	310	[62.2]	
Macrophytes			49.3	6.5	180	[36.1]	
					498		
Lake 239, Ontario	56.1	10.5					Soft-water, oligotrophic lake; probably underestimates since winter production is not included; benthic algae, primarily epilithic, macrophytes probably insignificant; CO$_2$ utilization methods (Schindler, et al., 1973)
Phytoplankton					823	[99.0]	
Littoral algae					8.1	[1.0]	
Macrophytes					N.D.		
					approx. 831		
Lake 240, Ontario	44.1	6.1					(Same as for Lake 239)
Phytoplankton					501	[98.2]	
Littoral algae					9.0	[1.8]	
Macrophytes					N.D.		
					approx. 510		
Lawrence, Michigan	5.0	5.9					Hard-water, oligotrophic marl lake; benthic algae, primarily epiphytic on sparse submersed macrophytes; ^{14}C methods (Wetzel, et al., 1972)
Phytoplankton			118.9	2153	434	[25.4]	
Littoral algae			2003.	1977	399	[23.3]	
Macrophytes			240.8	4360	879	[51.3]	
					1712		

						Remarks
Wingra, Wisconsin						Large, shallow hard-water eutrophic lake; large littoral zone with dominant submersed macrophyte *Myriophyllum* and metaphytic mats of macroalga *Oedogonium*; 14C methods for all components; mostly only summer values (McCracken, et al., 1974; Adams and McCracken, 1974; J. F. Koonce, personal communication)
Phytoplankton	139.6	1200.	1675.	4380.	(78.6)	
Metaphyton	approx. 2	3.0	4.2	11.1	(0.4)	
(*Oedogonium*)	(Summer, 1971) (Summer, 1972)	5.5	7.6	19.9		
Macrophytes		320.5	447	1170.	(21.0)	
				5581.		
Algal, Antarctica						Shallow oligotrophic lake; dense epipelic algal mats; 14C techniques (Goldman, et al., 1972)
Epipelic algae	—	468				
Phytoplankton	<0.5	8.2				
Kalgaard, Denmark						Small oligotrophic, soft-water lake; 14C techniques for phytoplankton and epiphytes; biomass of macrophytes (Søndergaard and Sand-Jensen, 1978)
Phytoplankton	10.5			241	(52.3)	
Littoral algae	4.7			5	(1.1)	
Macrophytes				215	(46.6)	
Skua, Antarctica						Shallow eutrophic lake; dense epipelic algal mats; 14C methods (Goldman, et al., 1972)
Epipelic algae	—	340				
Phytoplankton	<0.5	216				
Lača, USSR						Large, shallow, eutrophic lake; macrophytes colonizing 48 per cent of lake area; biomass methods (Raspopov, 1979)
Phytoplankton	16,600			996	(36.9)	
Littoral algae	—			725	(26.9)	
Macrophytes				979	(36.3)	
Eagle, California						Large, subalpine, hard-water, eutrophic lake; oxygen and biomass techniques (Huntsinger and Maslin, 1976)
Phytoplankton	12,150	356	50,860	1168	(85.8)	
Epilithic algae	7.0	274	3,800	95	(7.0)	
Epiphytic algae		1153	1,910	47	(3.5)	
Macrophytes (emergent)		1249	2,690	51	(3.8)	

plankton were responsible for 60 per cent, the moss for 20 per cent, the epipelic diatoms for 15 per cent, and the epiphytic algae on the moss for 5 per cent. Similar results have been found in Char Lake of the Canadian Arctic at latitude 75° N (Kalff and Wetzel, unpublished).

Evaluation of the productivity of phytoplankton and attached algae of a large deep alpine lake of the central Pyrenees Range also indicated the major contribution of the macroalga *Nitella* and of epilithic algae to the total primary productivity (Capblancq, 1973). The attached algae of the littoral zone (0 to 6 m) and *Nitella* (6 to 19 m) formed a biomass some 140 times greater than the mean biomass of the phytoplankton during summer, and *Nitella* constituted about 80 per cent of the total benthic algae. Based on in situ productivity measurements, the contribution of benthic algae to the total primary productivity of this lake was estimated to be 30 per cent.

The work of Straškraba (1963) on two large, shallow fishponds of southern Czechoslovakia deserves particular mention. Based on extensive annual cycles of the nitrogen content and biomass dynamics of all plant and animal components of these shallow lakes, it was determined that about 73 per cent of the primary productivity resulted from largely emergent macrophytes, 20 per cent from attached, mainly epiphytic algae, and 7 per cent from phytoplankton. It was shown that less than 10 per cent of the primary productivity was utilized by higher trophic levels.

Studies evaluating the relative photosynthetic rates of algae of the littoral in comparison to the rates of phytoplankton are few. But the suggestion that the most productive site is epiphytic on macrophytes repeatedly appears, as shown for example in Table 19-8 for a shallow English lake, Lawrence Lake, discussed earlier, and in the littoral of Mikolajskie Lake (Table 19-9). These results support earlier discussion on the importance of factors of light, wave action-sediment movement, and inorganic-organic nutrient exchanges between macrophytes and attached algae. It is apparent that growth conditions are often better in the littoral attached habitat than in the planktonic habitat. The significance of littoral productivity to a lake system depends to a great extent on the physical conditions of the lake morphometry and characteristics of the substrata available for algal attachment and growth.

TABLE 19-8 Mean Rates of Primary Production of Phytoplankton in Relation to Those of Epipelic Algae and Algae Epiphytic on the Horsetail (*Equisetum fluviatile* L.)

LAKE	PHYTOPLANKTON (mg C m^{-2} hr^{-1})	EPIPELIC ALGAE (mg C m^{-2} hr^{-1})	EPIPHYTIC ALGAE (mg C m^{-2} OF SUBSTRATUM hr^{-1})
Eutrophic, shallow; Priddy Pool, England*	1.55	1.71	63.9
Oligotrophic, deep, with limited littoral area; Lake Pääjärvi, Finland†			
0–1 m	5.35	3.68	3.6
2–4 m	3.85	1.68	—

*Mean annual values; based on data of Hickman, 1971b.
†Mean summer values; Kairesalo, 1980b.

TABLE 19-9 Percentage Contribution of Various Producers to the Annual Net Primary Productivity per m^2 in the Littoral Zone, Mikolajskie Lake, Poland

ZONE/COMPONENT	PERCENTAGE CONTRIBUTION
Eulittoral	
Macrophytes	28
Planktonic	10
Metaphyton	21
Attached algae	41
Littoral Overgrown with Emergent Vegetation	
Macrophytes	57.2
Planktonic	19.6
Metaphyton	0.1
Attached algae	23.1

After Pieczyńska, 1970.

Changing Littoral Productivity and Eutrophication

As the water of lakes receives increasingly large loads of nutrients, there is a strong tendency for phytoplankton to increase to the maximum capacity within existing limitations of temperature and available light (cf. Chapter 15). However, it is imperative that eutrophication of aquatic systems is not viewed in the restricted sense of phytoplanktonic productivity. Within obvious geomorphological restrictions on littoral development, the common situation is for littoral productivity to play a major role in the early and final stages of increasing fertility *of the lake system as a whole*. Exceptions certainly exist, but these conditions of important littoral production are clearly widespread in a large percentage of lakes of the world.

Submersed macrophytes often assume an increasingly greater importance to the total primary productivity of lakes, until, under high nutrient loading and lake fertility, the whole system is subject to severe light attenuation (Fig. 19-10). This light limitation is usually associated with intense phytoplanktonic and epiphytic algal productivity. Light attenuation is critical. Dense phytoplankton populations alone are sufficient to attenuate light to a point where it is inadequate to support growth of submersed macrophytes (e.g., Mulligan, et al., 1976; Jupp and Spence, 1977). Reductions in phytoplankton densities and improved light conditions (e.g., by reductions in nutrient or turbidity loading, or by selective fish predation on zooplankton which in turn alters phytoplankton composition; Chapters 15 and 16; Leah, et al., 1980) can result in recolonization by submersed macrophytes. In addition, increased nutrient loading can increase epiphytic growth on macrophytes to the point of severely reducing the amount of light reaching the supporting macrophytes (e.g., Sand-Jensen, 1977; Phillips, et al., 1978; Tsuchiya and Iwaka, 1979; Sand-Jensen and Søndergaard, 1981). As water fertility increases, enhanced growth of epiphytes, along with simultaneously increasing phytoplankton densities, can contribute to the demise of submersed macrophytes. As the macrophytes decline and are eliminated, phytoplankton and emergent macrophytes dominate lake productivity. Within a given latitude and climate, maximum growth of phytoplankton in fertile waters is determined by light reduction induced by self-shading; growth can

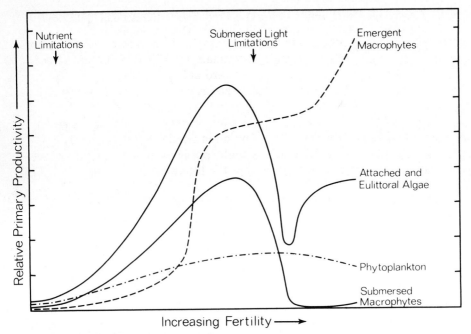

Figure 19-10. Generalized relationship of primary productivity of submersed and emergent macrophytic flora, attached algae and metaphyton, and phytoplankton of lakes of increasing fertility of the whole lake ecosystem. (After Wetzel and Hough, 1973.)

be increased beyond these limits only by increasing turbulence and light availability to levels greater than those normally found under natural conditions (Wetzel, 1966b).

As the emergent vegetation assumes greater dominance in a lake ecosystem, and eventually covers a majority of the lake basin (see Chapter 24), an exceedingly productive combination of littoral macrophytes and attendant microflora develops (Fig. 19-10). Attached, largely epiphytic algae, eulittoral algae, and metaphyton develop in strong association with the emergent flora. Natural changes in this general sequence are usually slow, and may extend over centuries and millennia, depending on the basin morphometry. The process can be accelerated greatly by increased nutrient loading, either artificially (see Hasler, 1947; Smith, 1969) or more gradually (see Mattern, 1970).

A large majority of the lakes of the world are small, and their morphometry is such that the ratio of colonizable littoral zone to pelagic zone of production is large. In these cases, there is little question that macrophytic vegetation, and importantly, its attendant microfloral community, have a major impact on the lake ecosystem. The distinction between what constitutes allochthonous production (e.g., eulittoral, terrestrial, etc.) and what is part of true autochthonous lake production is an extremely artificial one and has, indeed, led to an artificial treatment of mechanisms of ecosystem metabolism. All components of the drainage basin are influential in regulating lake metabolism. Integrated data on the *functional* impact of these components on the entire system are essentially nonexistent. The concept of a lake as a microcosm must finally be laid to rest.

Littoral Zooplankton Communities

Littoral zooplankton communities are made up of a diverse assemblage of protozoans, rotifers, and microcrustaceans. Many of these animals are sessile upon the sediments or macrophytes and are not truly planktonic, or are only intermittently planktonic. The number of species within littoral macrophyte zones is generally larger among all groups of organisms than in the open-water zone (e.g., Pennak, 1966).

Littoral zooplankton have been studied much less intensively than have pelagial forms. Many of the life history, reproductive, and population characteristics of pelagial zooplankton, discussed at length in Chapter 16, are equally applicable to littoral-inhabiting zooplankton. A few conspicuous characteristics of the littoral fauna are noteworthy.

Microcrustacean communities have been found to fall into three general groups in the littoral zone: (a) highly "plant-associated" and infrequently found away from aquatic macrophytes, (b) "free-swimming" among the large plants, and (c) sessile and living mainly in littoral sediments (e.g., Pennak, 1966; Whiteside, et al., 1978; Fairchild, 1981). Truly planktonic species are common in littoral zones devoid of vegetation, but are rare within macrophyte zones. Among the large plants, free zooplankton are predominantly plant browsers or sediment-inhabiting species that are temporarily swimming, or open-water species that enter vegetated areas only in small numbers.

Plant-associated microcrustacean taxa are often closely associated with macrophyte species or morphologically similar plant types (Quade, 1969, 1971), and are distributed according to available surface area (Fairchild, 1981). Numbers of Cladocera feeding on detrital periphyton were significantly correlated with epiphytic diatom density, whereas filter-feeding Cladocera attached to the plants were not. Similar animal-plant associations have been found among littoral rotifers (Edmondson, 1944, 1945; Wallace, 1978).

Among the common littoral microcrustaceans are the chydorids; most chydorid species increase in numbers during spring and autumn, and decrease during midsummer (Goulden, 1971; Keen, 1973; Whiteside, 1974). In a detailed study of the population dynamics of four species of a hard-water lake in southern Michigan (Keen, 1973), for example, *Chydorus sphaericus* reached its maximum population density in the spring, declined to a low level in the summer, and then increased to a smaller fall peak preceding a winter plateau (Fig. 19–11). It is common for *Chydorus* to become abundant in the open water in summer, coincident with the minimal littoral populations. As indicated in the upper portion of the figure, production of males and ephippial females was very small in this littorally perennial species. In contrast, *Graptoleberis, Acroperus,* and *Camptocerus* are aestival and after a winter absence appeared in spring, from ephippial eggs, and attained maximal densities in late summer and autumn (Fig. 19–11). The latter peaks terminated with the massive development of ephippial females and males prior to winter. The factors of natural death and emigration were negligible in this and several similar studies. High summer mortality was associated with active predation by small fishes ($<$ 4 cm in length) and nymphs of dragonflies (Odonata). In another analysis (Goulden, 1971), the midsummer population minima of chydorids were largely caused by predaceous tanypodine midge larvae. The midges were rare or absent in spring and autumn when the chydorids were most abundant.

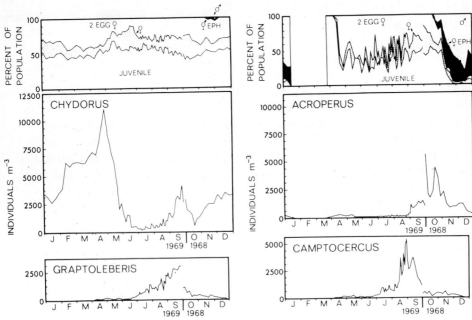

Figure 19–11. Annual population densities of the four major species of littoral chydorid cladocerans among submersed beds of *Scirpus subterminalis*, west littoral, Lawrence Lake, Michigan. Upper panels show percentage of the population of *Chydorus* and *Acroperus* seasonally in juveniles, females, females with eggs, males, and ephippial females. (Redrawn from Keen, 1973.)

Although the general relationships between temperature and egg and other developmental characteristics of littoral and benthic cladocerans and copepods are similar to those discussed earlier for planktonic zooplankton (Chapter 16), several differences have been found (e.g., Bottrell, 1975a, 1975b; Sarvala, 1979). For example, egg development time is increased markedly with decreasing temperature, but among many species of cladocerans and copepods, eggs of sessile species take longer to develop than those of planktonic species, regardless of temperature. Instar duration and frequency of molting also tend to be greater among littoral species than among planktonic species.

Diurnal migration has also been observed among zooplankton living among macrophytes. During the day, many populations, particularly copepods, aggregate among macrophytes near the sediments (Szlauer, 1963; Kairesalo, 1980a). Portions of the populations may then migrate to surface water strata during darkness.

SUMMARY

1. Submersed macrophytes and aqueous portions of emergent and floating macrophytes provide a large surface area that is colonized by microflora. In addition, all nonliving substrata, particularly within the photic zone, are colonized by sessile microflora.

a. Extremely diverse microhabitats occur in littoral areas among substrates of sand, rock, organic sediments, and macrophytes.

b. Complex interactions exist between the attached microflora and their substrates; these relationships are poorly understood. It is evident, however, that the proximity of microflora attached to or living in close juxtaposition with living and nonliving substrata can enhance acquisition of certain required nutrients and provide metabolic advantages not as prevalent among planktonic microflora living in the more dilute open water medium. Sessile populations tend to be more stable than their planktonic counterparts.

2. Attached algae of littoral communities often form the dominant algal biomass in streams and shallow lakes. Over 90 per cent of all algal species likely grow attached in submersed habitats.

a. Diatoms dominate on all substrata among moderately productive aquatic ecosystems.

b. Sessile blue-green algae and flagellates increase in abundance in organically enriched lakes.

3. Algae living on loose sediments that are subject to disturbance and displacement by water movements and animals often exhibit an ability to migrate vertically in response to changes in light. This vertical migration is phased with diurnal rhythms in cell division and photosynthetic capacity.

4. In temperate regions, the seasonal community biomass of attached algae commonly follows seasonal changes in incident light and temperature.

a. Epipelic diatoms frequently exhibit population maxima in the spring and in the autumn.

b. Seasonal population dynamics of epiphytic algae are more variable, and change as the surface area available for colonization of the supporting macrophytes changes seasonally.

5. Within the littoral zone, photosynthesis of attached algae is influenced to a great extent by light availability, although compensatory mechanisms exist that permit appreciable active growth at low levels of irradiance.

a. Biomass of epipelic algae commonly increases with increasing nutrient content of the sediments.

b. Epipelic algae can utilize significant quantities of combined nitrogen in the interstitial water of sediments and alter the flux of nutrients from the sediments to overlying water in the littoral zone.

6. Where comparisons of growth among algal populations associated with different substrates are possible, they indicate that photosynthesis is commonly higher for algae attached to sand grains than it is for those associated with organic sediments. Algae epiphytic on macrophytes are often more productive than algae associated with sediments. The meager available evidence indicates that a complex and highly dynamic metabolic relationship exists among the epiphytic algae, bacteria, and the supporting macrophyte.

7. Productivity of littoral algae is not much greater than that of phytoplankton, and turnover rates tend to be somewhat slower.

a. The large surface area available for colonization, however, particularly on submersed macrophytes, can result in very high contributions by attached littoral algae to the total primary productivity of many fresh-water ecosystems.

b. Spatial heterogeneity in rates of production by attached algae is usually very large because of the great variability in littoral habitats within fresh-water ecosystems.

c. Community rates of primary production of attached algae commonly follow the annual insolation and temperature curves, with maxima occurring in midsummer.

d. Most lakes of the world are shallow and possess large littoral zones. The productivity of the littoral algae and macrophytes tends to increase in relation to that of the phytoplankton as lakes become enriched. Under eutrophic conditions, excessive growth of phytoplankton and epiphytic algae can reduce light sufficiently to cause the decline of submersed macrophytes and attached algae; a consequent overall reduction in total productivity within the lake will then result.

8. Many of the protozoan, rotifer, and microcrustacean zooplankton of the littoral zone are sessile on the sediments or macrophytes, and are not truly (or are only intermittently) planktonic.

a. Plant-associated zooplankton are commonly sessile on specific macrophyte types. Population densities of animals feeding on epiphytic microflora and detritus are positively correlated with epiphytic algal density.

b. Many of the life history, reproductive, and population characteristics of littoral-inhabiting zooplankton are similar to those of their pelagial counterparts.

c. Population numbers of common littoral microcrustaceans increase during spring and autumn, and decrease during summer. Midsummer population minima often result from intense predation by insect larvae and small fishes.

CHAPTER TWENTY

SEDIMENTS AND MICROFLORA

Sediments of freshwater systems have been analyzed from a number of stand-points. Sediments of the littoral zone of lakes and of running waters are sorted by particle size by water movements. Gradients in substrate particle sizes form with changes in water velocity. Particle size of both inorganic and organic sediments is of major importance to the distribution and growth of many benthic invertebrates (cf. Cummins, 1962). Because of the overriding importance of microbial metabolism in mineralization of organic matter, in biogeochemical cycling of nutrients, and in the amount of matter deposited without appreciable further degradation in permanent sediments, the organic composition of sediments in lake systems has received much study.

GENERAL COMPOSITION

Sediments consist of three primary components: (a) organic matter in various stages of decomposition; (b) particulate mineral matter, including clays, carbonates, and non-clay silicates; and (c) an inorganic component of biogenic origin, e.g., diatom frustules and certain forms of calcium carbonate (cf. Jones and Bowser, 1978; Kelts and Hsü, 1978). In many respects, the organic sediments of lakes are similar to the uppermost A_o horizon in terrestrial soils (Hansen, 1959a). Heterogeneous humus in particulate form can be divided into two main types, acid humus and neutral humus (sapropel*). From a colloid-chemistry viewpoint, acid humus is largely an unsaturated sol whose particles have negatively charged surfaces, while neutral humus is largely a gel in which anions are adsorbed to the surfaces of the humus particles.

Acid humus prevails in peat bog systems. Decomposing peat plant materials, such as the moss *Sphagnum*, form unsaturated humus colloids (dopplerite). Aggregations of black, gelatinous dopplerite are very soluble in water unless dried; upon drying, they become insoluble. Nitrogen content of this humic material is very low (0 to 2 per cent).

---

*True sapropel is bluish-black, contains much hydrogen sulfide (H_2S) and methane (CH_4), and is deposited under anaerobic reducing conditions.

DY AND GYTTJA

Two words of Scandinavian origin are widely used in limnology to describe the general character of organic sediments. The terms dy (pronounced dē, nasally) and gyttja (pronounced yǐt'-ja) were introduced in the middle of the 19th century by von Post (Hansen, 1959a, 1959b). Gyttja is a coprogenous sediment containing the remains of all particulate organic matter, inorganic precipitations, and minerogenic matter. In a fresh state, gyttja is very soft and hydrous, with a dark greenish-grey to black color; it is never brown. In a dry state, some gyttjas are hard and black, while others are more friable and lighter in color, depending upon the main constitutent. The organic carbon content of gyttja is less than 50 per cent.

Dy* is a gyttja mixed with unsaturated humus colloids. Fresh dy is soft, hydrous, and brown in color. In a dry condition dy is very hard and dark brown. The organic carbon content of dy and peat is greater than 50 per cent.

Numerous comparative studies of the organic components in sediments have demonstrated that the protein and nitrogen content of acidic humus is low. The C:N ratio of *Sphagnum* peat is about 35; that of pure dopplerite varies from 46 to 52. Hansen concluded that if the C:N is less than 10, the humus is neutral humus and the sediment is gyttja. If the C:N exceeds 10, the gyttja is mixed with acid humus and the sediment is a dy.

The sources of particulate organic matter and their respective rates of decomposition during sedimentation in the pelagic zone were discussed elsewhere (Chapters 17 and 22). In large oligotrophic as well as large eutrophic lakes in which phytoplanktonic productivity dominates, humic inputs to the sediments will be relatively low. Gyttja sediments would be anticipated in these lakes, and this is generally borne out by empirical observations (cf., for example, Hansen, 1961). Dy sediments are found in some small lakes dominated by littoral productivity, especially by acid-producing *Sphagnum* mosses, or allochthonous inputs of humic organic matter. All sediments contain some humic matter, and there is a broad transition between gyttja and dy.

Comparisons of the changes in chemical composition of pelagial seston while it is sedimenting in the water column were discussed in Chapter 17. Diagenesis of this organic matter is faster within the water column than once it is buried in the sediments. Within the sediments, decomposition rates of organic matter follow the general sequence: carbohydrates–amino acids–amino sugars > humic compounds > lipids (e.g., Kayama, et al., 1973; Kemp and Johnston, 1979; Klug, personal communication).

HUMIC COMPOUNDS ORIGINATING FROM WETLAND AND LITTORAL FLORA

Products formed by microbial degradation of lignin can polymerize as humic compounds (Flaig, 1964; Larson and Hufnal, 1980; Wang, et al., 1980). As a result, it would be expected that the littoral macrophytes, and particularly the more highly lignified emergent flora, would be a greater source of humic compounds than submersed hydrophytes or planktonic microflora. Analyses of the successive degradation of macroflora cast some light on the decomposition products and the rates of their formation.

---

*Also termed *tyrfopel* in older literature, meaning *a fine-grained peat.*

The carbon and nitrogen contents of fractions of (1) fresh, living plant material of *Stratiotes aloides* were compared with those of (2) semi-decomposed, dead plant material that was morphologically distinct but undergoing intermediate phases of decomposition, (3) young, thin sapropel formed by partial mineralization, and (4), old, thick, stratified sapropel resulting from nearly complete mineralization and humification (Úlehlová, 1970, 1971, 1976). Results (Table 20-1) showed that the carbon and nitrogen content decreased markedly with progressive decomposition to sapropels, while little change occurred in the C:N ratio. Free humic substances exhibited a loss of carbon in the sapropel phases of decomposition; this loss occurred primarily in the fulvic acid fraction. In the chemically bound humic fractions, carbon content increased with progressive degradation; again, humic acid content increased at the expense of the fulvic acid component. Nitrogen content increased in the semidecomposed stage, and the quantity of old sapropel decreased in the bound humic fractions. Most of the nitrogen was found in the fulvic acid fractions of the humic substances. These results suggest that free humic substances present in the early stages of decomposition were microbiologically transformed to chemically bound humic substances in later stages of decomposition. As the nitrogen of humic compounds is selectively removed, C:N ratios increase markedly. Slower rates of degradation occur when nitrogen levels decline sufficiently, which then permits some net accumulation of organic matter in the sediments.

MICROFLORA OF SEDIMENTS AND RATES OF DECOMPOSITION

A conspicuous feature of microbial populations in lakes is the great increase in numbers in the transition from the overlying water to the diffuse, uncompacted zone of the surficial sediments (Fig. 20-1). Bacteria increase about 3 to 5 orders of magnitude from the water to the surface sediments and decrease rapidly with increasing depth in the sediments. Expressed in numbers per gram dry weight of sediment, bacterial populations in the surficial sediments can reach 6 to 7 orders of magnitude greater than in an equivalent weight of overlying water (Table 20-2). Saprophytic bacteria decrease much more rapidly than do total bacteria with increasing depth below the sediment-water interface, which suggests a depletion of readily assimilable organic substrates below the interface.

Bacteria within the surface sediments are unevenly distributed over the lake basin as well (Henrici and McCoy, 1938; Steinberg, 1980a). When the littoral zone is covered with a well-developed macrophyte community, bacterial numbers of the sediments are much higher by several orders of magnitude than those found in profundal sediments of deepwater areas. Numbers of bacteria in sandy, wave-swept littoral areas are lower than those in profundal sediments (e.g., Jones, 1980). A more detailed analysis of bacterial concentrations in sediments within the macrovegetated littoral zone as compared with those of sediments free of vegetation shows similar relationships (Table 20-3), but with marked shifts seasonally in relation to growth of the macrovegetation. Nitrogen-fixing bacteria *Azotobacter* and the obligate anaerobe *Clostridium* were more abundant in sediments with macrovegetation than in sediments without vegetation or in the overlying water. Similar relationships were found among populations of hydrocarbon- and hydrogen-oxidizing bacteria, cellulose-decomposing bacteria and actinomycetes, and

TABLE 20-1 Carbon, Nitrogen, and C:N Ratios Among Fractions of Decomposing *Stratiotes* Organic Matter in Lake Venematen, Netherlands (see text)

CARBON, NITROGEN, AND C:N RATIO	TOTAL ORGANIC MATTER	HUMIC ORGANIC MATTER	RESIDUAL ORGANIC MATTER	FREE HUMIC SUBSTANCES	FREE HUMIC ACIDS	FREE FULVIC ACIDS	SORBED HUMIC ACIDS	BOUND HUMIC SUBSTANCES	BOUND HUMIC ACIDS	BOUND FULVIC ACIDS
Carbon										
Fresh *Stratiotes*	57.0	40.4	16.7	20.6	6.5	14.1	8.9	10.8	4.3	6.5
Semi-decomposed plant material	60.6	38.6	22.0	20.6	10.2	10.4	3.1	14.9	9.3	5.6
Young, thin sapropel	34.0	34.1	—	14.4	10.7	3.8	3.7	16.0	16.0	0.00
Old, thick sapropel	34.6	34.6	—	14.7	10.3	4.4	3.9	16.0	15.2	0.7
Nitrogen										
Fresh *Stratiotes*	2.6	2.3	0.2	1.1	0.1	0.9	0.8	0.5	0.3	0.2
Semi-decomposed	3.0	1.4	1.6	0.8	0.2	0.6	0.2	0.4	0.3	0.1
Thin sapropel	1.3	0.6	0.7	0.2	0.1	0.06	0.3	0.2	0.2	0.00
Thick sapropel	1.7	0.6	1.1	0.1	0.1	0.04	0.2	0.2	0.2	0.03
C:N										
Fresh *Stratiotes*	22.3	17.5	69.4	19.7	46.7	15.5	11.7	21.7	14.3	32.7
Semi-decomposed	20.5	28.0	14.0	27.2	56.4	18.0	17.2	33.9	29.0	46.9
Thin sapropel	25.5	55.8	—	72.3	76.1	63.2	13.9	106.6	106.6	—
Thick sapropel	20.5	58.6	—	105.1	103.2	109.8	16.1	76.1	84.5	24.3

After data of Úlehlová, B.: Decomposition and humification of plant material in the vegetation of *Stratiotes aloides* in NW Overijssel-Holland. Hidrobiologia (Romania), 12:279–285, 1971.

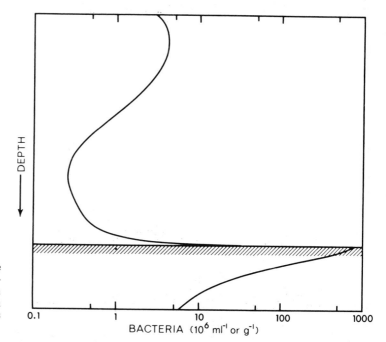

Figure 20-1. Generalized relative vertical distribution of bacterial numbers in water and sediments of a moderately productive lake. Depth scale in the water relative in meters and in the sediments in centimeters.

BACTERIA (10^6 ml^{-1} or g^{-1})

TABLE 20-2 Bacterial Numbers (Direct Enumeration) in the Surficial Layer of Sediments of Several Lakes of the Moscow Region, USSR

LAKE	ASH CONTENT OF SEDIMENTS (%)	BACTERIAL NUMBERS (10^6 g^{-1}) Fresh Sediment	Dry Sediment	ORGANIC SUBSTANCES OF BACTERIA IN % OF ORGANIC MATTER OF SEDIMENTS
Beloe (Kosino)	54.1	2326	54,219	7.8
B. Medvezh'e	58.3	1905	38,253	6.0
Chernoe (Kosino)	49.3	1285	35,109	4.5
M. Medvezh'e	47.3	1624	32,032	4.0
Sviatoe (Kosino)	18.6	922	29,790	2.3
Krugloe	79.2	1110	7991	2.5
Gab (Karelian region)	81.9	1883	15,680	5.8

Modified from Kuznetsov, 1970, after Khartulari.

fungi. Large numbers of methane-forming and sulfate-reducing bacteria developed under anaerobic conditions in sediments containing large quantities of decomposition products of higher aquatic plants.

Bacterial numbers and metabolic activities of the deepwater profundal sediments show a fair correspondence with the productivity of the lakes (Table 20-4); bacterial numbers in the open water tend to undergo large, rapid fluctuations, and consequently do not correlate to lake productivity as well (cf. Chapter 17; Jones, 1979; Jones, et al., 1979). In extremely oligotrophic lakes, total numbers of bacteria of the sediments can

TABLE 20-3 Comparison of Microbial Populations (Thousands per Gram of Sediment) of the Sediments Among Emergent Bulrush Vegetation (A) and of Open Water Free of Bulrushes (B) in Lake Mekhteb, USSR

Period	TOTAL SAPROPHYTES		ACTINOMYCETES		FUNGI		CLOSTRIDIUM PASTEURIANUM		HYDROGEN OXIDIZING		METHANE FORMING		SULFATE REDUCING	
	A	B	A	B	A	B	A	B	A	B	A	B	A	B
Dec.–Jan.: no bulrushes	590	790	0	80	0	30	10	10	1	10	0	1	0.1	2
Feb.: young bulrushes	1800	2700	200	0	400	400	1	10	1	10	40	40	8	1
Mar.: young bulrushes	4400	4800	400	300	0	0	1	10	10	10	5	240	40	10
Apr.: developing stands	5300	3100	800	0	0	300	100	100	10	1	1500	140	50	1300
May:	1800	1800	1000	100	200	300	1000	10	1000	10	2000	15,000	100	100
June: mature bulrushes	7900	4800	2900	300	100	0	10	10	1000	10	1200	15,000	400	1200
July:	60,200	3800	0	0	1000	200	100	100	10	100	1800	2000	240,000	5000
Aug.: declining bulrushes	16,400	5400	1500	0	2800	0	10	10	100	100	170	3000	1400	1500
Sept.:	3800	2700	300	0	0	0	10	10	10	10	70	13	1	1
Nov.:	230	80	0	0	0	9	1	1	10	1		10	0	0

From data of Aliverdieva-Gamidova, 1969.

TABLE 20-4 Comparison of Relative Bacterial Numbers (Plate Culture Techniques) of the Profundal Sediments Among Several Wisconsin Lakes of Differing Productivity

LAKE	TYPE	TYPE OF SEDIMENT	AVERAGE BACTERIA PER CM³ OF SURFACE SEDIMENTS	TOTAL BACTERIA PER CM³ OF FIRST 18 CM OF SEDIMENT
Crystal	Oligotrophic	Gyttja	2160	38,880
Weber	Oligotrophic	Gyttja	2350	42,300
Big Muskellunge	Mesotrophic	Gyttja	10,930	196,740
Trout	Oligotrophic	Gyttja	29,790	536,220
Little John	Mesotrophic	Gyttja	39,050	642,900
Mary	Dystrophic	Dy	39,450	710,100
Helmet	Dystrophic	Dy	120,300	2,165,400
Alexander	Eutrophic	Marl gyttja	144,240	2,599,320
Mendota	Eutrophic	Marl gyttja	609,300	>10,000,000

After data of Henrici and McCoy, 1938.

be less than the cumulative total of the entire overlying water, which indicates that in these lakes a majority of the readily decomposable organic matter has been processed before the particulate residues reach the sediments (cf. Chapter 17; Bengtsson, et al., 1977; Mothes, 1981).

BACTERIAL ACTIVITY OF SEDIMENTS

The relationship of increased bacterial populations and metabolic activity to greater organic matter in sediments may appear obvious, but in situ measurements are very few. A significant correlation ($r = 0.93$; $p = 0.01$) between dehydrogenase activity and organic content of surface sediments has been found in a eutrophic reservoir (Lenhard, et al., 1962). These correlations, however, would not be expected to hold in organic-rich sediments in which other conditions, such as acidity (as occurs in bogs), depress microbial activity.

Data on oxygen consumption by the benthic community of bacteria, algae, micro- and macrofauna offer some insight into sediment respiration rates, provided that oxygen utilization by these community components can be appropriately partitioned and corrected for abiotic oxygen uptake (Bowman and Delfino, 1980). Under aerobic conditions of shallow ($\bar{z} = 2.4$m) Marion Lake, Canada, respiratory oxygen consumption by sediment-dwelling bacteria over an annual period was found to be related primarily to temperature (Fig. 20-2; also Granéli, 1978). The proportion of community respiration resulting from bacterial respiration varied with season (Fig. 20-2), and was lowest during maximal community respiration during the summer period of higher temperatures. This percentage certainly increases under other conditions among lakes; in deepwater sediments, for example, nearly all metabolism is microbial.

Particulate organic detritus consumed up to 3 orders of magnitude more oxygen per unit dry weight than sand (Hargrave, 1972). Uptake rates were inversely related to particle size, and directly related to particle organic content of carbon and nitrogen.

In shallow lakes, water turbulence can frequently disturb and mix sediments to

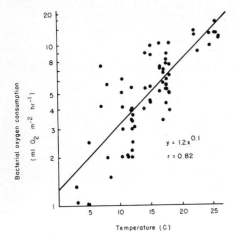

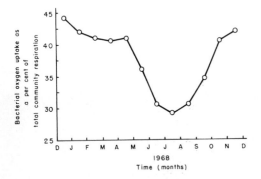

Figure 20-2. *Upper*, relationship of temperature and sediment bacterial respiration, and *lower*, bacterial respiration as a proportion of total community oxygen consumption in shallow water Marion Lake sediments during 1968. (After Hargrave, B. T.: Epibenthic algae production and community respiration in the sediments of Marion Lake. J. Fish. Res. Bd. Canada, 26:2003–2026, 1969.)

appreciable depths (e.g., to 10 cm; Viner, 1975). If reoxygenated briefly by turbulence or the activities of benthic animals, community respiration and oxygen consumption rates of highly organic sediments quickly (hours to a few days) return to values similar to those of undisturbed sediments of similar organic content and composition (Viner, 1975; Hargrave, 1975).

Assays of heterotrophic activity of benthic microorganisms of shallow water sediments by the uptake of labeled glucose, acetate, and glycine demonstrated that the greatest activity occurred in the summer months when the water temperatures exceeded 10°C (K. J. Hall, et al., 1972; Toerien and Cavari, 1982). Availability of substrates and slow diffusion rates exerted some control over the uptake and utilization of dissolved organic compounds of the interstitial water. The fraction of the substrate respired as CO_2 varied with the compound and was highest (average 63 per cent) with glycine in comparison to glucose (22 per cent) and acetate (13 per cent). Most of the heterotrophic activity occurred in the upper sediment layers, the sites of highest bacterial numbers in the sediments.

Glucose uptake rates by bacteria of surficial sediments of a small, eutrophic alpine lake were greater in littoral areas than in profundal sediments (Steinberg, 1978b). Similar results were found in a hypereutrophic lake of southern Michigan (King and Klug, 1982). The turnover times (T_t) of glucose were much more rapid in the anaerobic sediments (0.3 to 4 hours) than in the overlying water column (20 to 40 hours). About 40 per cent of the methane carbon generated in the sediments emanated from glucose. Turnover rates were faster in the littoral sediments than in the profundal sediments.

A number of studies have indicated enhanced growth of bacteria in sediments at temperatures greater than usually occur in profundal sediments of temperate lakes (e.g., Inniss and Mayfield, 1978a, 1978b, 1979; Tison, et al., 1980; Tison and Pope, 1980). Appreciable adaptation to cold temperatures apparently occurs by selection of bacterial species and by a physiological tolerance of temperature changes by individual species (e.g., Boylen and Brock, 1973; Leduc and Ferroni, 1979).

Anaerobic Decomposition in Sediments

Oxygen serves as the universal hydrogen acceptor for biochemical reactions of microbes under aerobic conditions. Under anaerobic conditions, however, the relationships are far more complicated, for various other substances and intermediate metabolic organic compounds become hydrogen acceptors. Often the same compound can serve as a hydrogen acceptor or donor, depending on the environmental conditions.

METHANOGENESIS

Degradation of large quantities of organic matter occurs under anaerobic conditions through methanogenesis in the sediments of lakes, ponds, and streams, as well as in waste-treatment facilities (oxidation ponds, anaerobic lagoons, and septic tanks). In the degradation process, organic matter is converted quantitatively to methane and carbon dioxide in two stages (Fig. 20–3). In the first stage, a heterogeneous group of facultative and obligate anaerobic bacteria, termed *acid formers*, converts proteins, carbohydrates, and fats primarily into fatty acids by hydrolysis and fermentation (Deyl, 1961; McCarty, 1964; Zeikus, 1977). The methane-producing bacteria then utilize the organic acids, converting them to carbon dioxide and methane. Certain alcohols from carbohydrate fermentation can also be converted to methane and carbon dioxide by methane-producing bacteria.

Utilization of organic compounds in suspension or solution by acid-forming bacteria in the first stage results only in the synthesis of bacterial cells and the generation of endproduct organic compounds, such as organic acids. Removal of oxidizable organic compounds occurs in the second stage, and is directly proportional to the quantity of methane produced. During most anaerobic metabolism, the formation of hydrogen occurs (see below); alternatively, the reduction of inorganic hydrogen acceptors, such as sulfates, nitrates, and nitrites, takes place (Chapters 12 and 14).

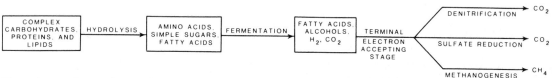

Figure 20–3. Stages of anaerobic metabolism of complex organic compounds. (Modified from Wolfe, 1971; Molongoski, 1978; and Zehnder, 1978.)

The first stage of acid formation results from hydrolysis and fermentation (McCarty, 1964). Proteins are first enzymatically hydrolyzed to polypeptides, and then to simple amino acids. Complex carbohydrates such as starch and cellulose are hydrolyzed to simple sugars, while fats and oils are hydrolyzed to glycerol and fatty acids. The amino acids, simple sugars, and glycerol formed by hydrolysis are soluble, and are fermented by acid-forming bacteria. In the absence of oxygen, one portion of the organic molecule is oxidized, while another portion of the same molecule (or sometimes another compound) is reduced. These fermentative energy-yielding oxidation-reduction reactions produce reduced (saturated) fatty-acid compounds and oxidized carbon as CO_2. Ammonia is produced as an end product during amino-acid fermentation.

The methane-producing bacteria are strictly anaerobic, and consist of four major genera: rod-shaped nonsporulating *Methanobacterium*, rod-shaped sporulating *Methanobacillus*, and the spherical *Methanococcus* and *Methanosarcina*. Species are differentiated on the basis of the substrates they are capable of using.

Methane is formed by two major processes. In the first, carbon dioxide serves as a hydrogen acceptor and is reduced to methane by enzymatic addition of hydrogen derived from the organic acids:

$$CO_2 + 8H \rightarrow CH_4 + 2H_2O.$$

In the second process, acetic acid, which is a major intermediate produced by the fermentation of complex organic compounds, is converted to CO_2 and methane. In this reaction, the carbon of the methane originates from the methyl carbon of acetic acid:

$$*CH_3COOH \rightarrow *CH_4 + CO_2.$$

Methane produced during the fermentation of mixed organic compounds originates largely (approximately 70 per cent) via the intermediate, acetic acid. Much of the remainder of the methane results from reduction of CO_2. Propionic acid, an intermediate compound produced in significant quantities from protein, carbohydrate, glycerol, and other compounds, also can serve as a substrate for methane production. Longer-chain acids are of less importance.

The concentrations of acetate, H_2, and CO_2, the major in situ substrates for methanogenesis, are variable and dependent upon the inputs and rates of fermentation of settling organic matter. Acetate is the preferred substrate for methanogenesis at low partial pressures of hydrogen; as the partial pressure of hydrogen increases, bicarbonate is the preferred substrate (e.g., Winfrey, et al., 1977; Strayer and Tiedje, 1978a; Lovley and Klug, 1982; Jones, et al., 1982; Lovley, et al., 1982). Nitrate, nitrite, and sulfate in sediment in interstitial water inhibit methanogenesis either directly by metabolic inhibition of methanogens, or indirectly by channeling electron flow from methane carbon precursors to these alternate electron acceptors (e.g., Chen, et al., 1972; MacGregor and Keeney, 1973; Cappenberg, 1974, 1975; Winfrey and Zeikus, 1977, 1979; Zeikus, 1977; Zehnder, 1978). The degree of inhibition depends upon the relative concentrations of alternate electron acceptors. Evidence suggests that certain sulfate-reducing bacteria can function commensally with methanogens: methanogens ferment acetate released by the sulfate-reducing bacteria (Cappenberg, 1975). When sulfate is not limiting, the lower half-saturation constant of sulfate-reducing bacteria enables them to inhibit methane production by lowering the hydrogen partial pressure below levels that methanogenic bacteria can effectively utilize (Lovley, et al., 1982). However, methanogenic bacteria can coexist with sulfate-reducing bacteria in the presence of sulfate; the outcome of competition at any time is a function of the rate of hydrogen production, the relative population sizes, and sulfate availability.

METHANE OXIDATION

Although the production of methane in sediments can be very intense and reach as much as 85 per cent of the total gas volume formed in the deposits, little of the methane escapes to the atmosphere in low to moderately productive lakes, owing to the presence and activity of methane-oxidizing bacteria in the overlying water (ZoBell, 1964). Methane oxidizers are widely distributed in most natural waters (cf. Higgins, et al., 1981). Among the best-known of many species of methane-oxidizing bacteria is *Methanomonas methanica*, which, in addition to methane, can oxidize numerous other hydrocarbon compounds. All members of this group appear to be strict aerobes, although they tolerate microaerobic conditions. The overall empirical equation for methane oxidation is:

$$5CH_4 + 8O_2 \rightarrow 2(CH_2O) + 3CO_2 + 8H_2O,$$

with a free-energy yield of 195 kcal per mole.

The distribution of methane-oxidizing bacteria in dimictic lakes indicates that these forms are present throughout the year in significant numbers. The methane oxidizers increase during summer stratification, the highest numbers ($>5 \times 10^2$ ml^{-1}) occurring in metalimnetic and hypolimnetic waters with reduced oxygen concentrations and maximum concentrations of methane (Cappenberg, 1972). During summer stratification, the highest rates of methane oxidation have been consistently found in the lower metalimnion where oxygen concentrations are reduced to approximately 1 mg l^{-1} or less (Rudd, et al., 1974, 1976; Rudd and Hamilton, 1975; Jannasch, 1975; Panganiban, et al., 1979). During the summer, when combined inorganic nitrogen sources in the epilimnion were low in the eutrophic lakes studied, the methane-oxidizing bacteria fixed molecular nitrogen. Methane oxidation in strata above the metalimnion declined by as much as 40 per cent, because N$_2$ fixation is inhibited by the higher concentrations of oxygen occurring there. High rates of methane oxidation, amounting to as much as 95 per cent of the annual quantity of methane oxidized (Rudd and Hamilton, 1975), have been found in spring and fall periods of lake circulation. During the periods of lake turnover, combined inorganic nitrogen is available and the dependency of methane oxidation on N$_2$ fixation found in summer months is alleviated. Under ice cover, high rates of methane oxidation can occur throughout the water column of eutrophic lakes, possibly lowering oxygen concentrations or even causing total anoxia (cf. Chapter 9).

Oxidation of methane to carbon dioxide in the absence of oxygen has been observed, but only at the sediment surface (Panganiban, et al., 1979). The methane-oxidizing organisms and the metabolic pathways involved are unclear, but are likely associated with the activity of sulfate-reducing bacteria (cf. Reeburgh and Heggie, 1977; Rudd and Taylor, 1980).

METHANE CYCLING IN LAKES

Several recent investigations have attempted to quantify methane cycling in eutrophic lakes in terms of overall carbon cycling, and to evaluate its effects on the entire lake ecosystem (reviewed in Rudd and Taylor, 1980). For example, the annual methane cycle was evaluated in Lake 227, a lake of western Ontario rendered eutrophic by experimental fertilization (Rudd and Hamilton, 1978). Lake 227 is small (5 ha.), moderately deep (z_m = 10 m), and during the investigation circulated only for about six weeks in the fall before ice formation. During summer stratification, most methane produced in hypolimnetic sediments accumulated in the hypolimnion. Vertical diffusion of methane from the hypolimnion to the metalimnion was very slow, and amounted to 11 per cent of the methane produced in the hypolimnion. Nearly all of the methane that diffused into the metalimnion was converted to bacterial-cell carbon and CO_2 by methane-oxidizing bacteria. Small amounts of methane observed in the epilimnion during summer emanated largely from methanogenesis in epilimnetic sediments. Epilimnetic methane slowly evaded to the atmosphere, since little was oxidized because of low combined inorganic nitrogen concentrations and the inhibitory effects of high oxygen on N_2-fixation by methane-oxidizing bacteria.

Oxidation rates of methane in Lake 227 increased abruptly during fall circulation as hypolimnetic water rich in accumulated methane and inorganic nitrogen was mixed and oxygenated. About 40 per cent of the methane that was transported into the upper strata escaped to the atmosphere. Under winter ice cover, methane production in hypolimnetic sediments was similar to that during the summer. Only about 10 per cent of the methane was consumed by methane-oxidizing bacteria at the aerobic–anaerobic interface; the remainder accumulated and extended into the summer hypolimnion, since this lake did not circulate completely in the spring following ice loss.

In Lake 227, the annual methane production, oxidation, and evasion were compared to the annual phytoplankton primary production and total carbon inputs from runoff and the atmosphere. This comparison showed that the amounts of CO_2 and bacterial-cell material produced by methane oxidizers were insignificant in terms of carbon flow in comparison to primary production (Table 20-5). Methane cycling was much more significant, however, when compared with the total carbon input into the lake. During 1974, the amount of carbon regenerated as methane was equivalent to about 55

TABLE 20-5 Annual Rates of Methane Production, Methane Oxidation, Phytoplanktonic Primary Production, and Total Carbon Inputs from the Atmosphere and Runoff in Lake 227, Ontario

	g C m^{-2} OF LAKE SURFACE*
CH_4 production	18
Total CH_4 oxidation	12
CH_4-oxidizing bacterial production	6
CO_2 production by CH_4-oxidizing bacteria	6
Phytoplanktonic primary production	138
Total carbon inputs	33

After Rudd and Hamilton (1978).

*Primary production and total carbon inputs occurred mostly during the summer. The methane production and oxidation year included the succeeding winter since regeneration of methane carbon fixed during the previous summer would be occurring. The large difference between the amount of carbon entering the lake and the amount fixed by phytoplankton presumably resulted from rapid aerobic recycling of carbon within the epilimnion during the summer months.

per cent of the total carbon input into the lake (Table 20–5). Methane-oxidizing bacteria recycled about two-thirds of the methane carbon. Similar values have been found in a number of other eutrophic and hypereutrophic lakes during summer periods of thermal stratification (Table 20–6). A general direct relationship has been found between the rates of particulate organic carbon sedimentation and rates of methane release from pelagial sediments in eutrophic lakes (Robertson, 1979).

The terminal processes involved in the decomposition of organic carbon compounds and electron flow in sediments can be divided into two groups (Fig. 20–4): (a) those processes requiring electron acceptors (e.g., O_2, NO_3^-, $SO_4^=$,) generated or regenerated externally (i.e., from aerobic overlying water) to the zone of anaerobic metabolism in the sediments, and (b) those processes utilizing electron acceptors (e.g., HCO_3^-), generated or regenerated internally within the zone of anaerobic metabolism (Klug, et al., 1982). Denitrification and sulfate reduction are examples of microbial processes dependent upon external electron acceptors. Fermentation and methanogenesis, in contrast, depend upon internally generated electron acceptors.

HYDROGEN METABOLISM

Molecular hydrogen is formed in the sediments, under anaerobic conditions, by fermentative degradation of carbohydrates, cellulose, and hemicelluloses to fatty acids, especially acetic acid. Hydrogen and CO_2 are generated by *Clostridium* from simple carbohydrates, and from cellulose by *Achromobacter* and *Bacillus*. Formate serves as the carbon source for hydrogen formation in some species of *Bacterium*.

Hydrogen is oxidized by several pathways, so that, as is the case with methane, little reaches the surface and escapes to the atmosphere as bubbles. Certain strictly

TABLE 20–6 Relative Amounts of Particulate Organic Carbon Converted to Methane Carbon in Profundal Sediments, and Losses of Methane Carbon by Evasion to the Atmosphere in Several Eutrophic Lakes

LAKE	PERIOD	% OF INPUT C CONVERTED TO CH$_4$	% OF INPUT C LOST FROM LAKE AS CH$_4$–C
Lake 227, Ontario (Rudd and Hamilton, 1978)	Annual	55	<10
Wintergreen Lake, Michigan (Strayer and Tiedje, 1978; Molongoski and Klug, 1980a)	Summer, 1976	34	14
	Summer, 1977	44	20
Third Sister Lake, Michigan (Robertson, 1979)	Summer	36	6
Frains Lake, Michigan (Robertson, 1979)	Summer	59	28
Lake Mendota, Wisconsin (Fallon, et al., 1980)	Summer	54	ca. 5

Values of particulate loading are based on phytoplanktonic sestonic sedimentation only; inputs of organic carbon from the wetlands and littoral productivity, which are extensive in many of these lakes and could be transported along the sediments to hypolimnetic areas and missed in pelagial sedimentation traps, were not evaluated. Therefore, the methanogenetic values could be overestimates.

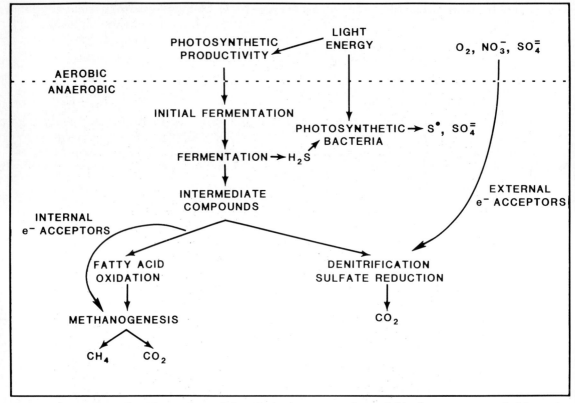

Figure 20-4. Relationship of main terminal degradation processes to the available electron acceptors in lakes sediments. (Modified from Klug, et al., 1982.)

anaerobic sulfate-reducing bacteria of the genus *Desulfovibrio* oxidize hydrogen while utilizing CO_2 to synthesize at least part of their cell substance (ZoBell, 1973):

$$H_2SO_4 + 4 H_2 \rightarrow H_2S + 4 H_2O \; (\Delta G_0' = -73,130 \text{ cal mole}^{-1})$$

Such sulfate reducers were found to fix an average of 0.24 g CO_2–C per gram of H_2 oxidized, and nearly all of the CO_2–C was accounted for as bacterial biomass. Oxygen for the oxidation is derived from the sulfate.

Certain species of all of the major genera of the methane-forming bacteria can oxidize molecular hydrogen.

CARBON MONOXIDE METABOLISM

Carbon monoxide results from certain microbial fermentations but even under exceptionally favorable conditions it generally does not exceed 3 per cent of the gases formed by bacterial fermentation of organic wastes (ZoBell, 1964, 1973). The chemical reactivity of CO and the relative ease with which various bacteria reduce it to methane or oxidize it to CO_2 probably account for its absence or very low concentrations in most environments.

Aerobic CO-oxidizing bacteria are distributed widely, especially in organic-rich sediments. Species of *Carboxydomonas, Hydrogenomonas, Bacillus,* and certain methane-oxidizing bacteria oxidize CO (Hubley, et al., 1974):

$$CO + \frac{1}{2} O_2 \rightarrow CO_2 \ (\Delta G'_0 = -66 \ \text{kcal mole}^{-1})$$

Additionally, several aerobic bacteria in mixed populations quantitatively convert carbon monoxide into methane in the presence of hydrogen:

$$CO + 3 H_2 \rightarrow CH_4 + H_2O \ (\Delta G'_0 = -46 \ \text{kcal mole}^{-1}).$$

In the absence of H_2, some anaerobic bacteria *(Methanosarcina)* produce CO_2 and methane:

$$4 CO + 2 H_2O \rightarrow 3 CO_2 + CH_4.$$

EFFECTS OF HYDROSTATIC PRESSURE ON GASEOUS METABOLISM IN SEDIMENTS

Analyses of the distribution of the gases contained in lake sediments and water show that the total quantity of nitrogen, methane, and hydrogen in lake sediments increases with water depth (Table 20-7; cf. also Fig. 12-1). The total pressure of the gases was usually slightly higher than the corresponding hydrostatic pressure (Koyama, 1964).

Experimental investigation of the metabolism of lake sediments under normal pressure demonstrated the following sequence (Koyama, 1955): (a) CO_2 production began immediately at high rates; (b) as the sediments became anaerobic, hydrogen was evolved, followed by increasing rates of methane production. The primary reaction of methane formation was the reduction of CO_2 by hydrogen and hydrogen donor compounds. Hydrostatic pressures that would be encountered in the sediments of most lakes do not greatly alter bacterial production of gases. Methane production decreased slightly as pressures were increased from 0 to 50 atmospheres (atm) and then increased up to 300 atm, after which gradual decreases in methane production were observed.

TABLE 20-7 Average Content of Gases in Sediments of Several Japanese Lakes in Relation to Gas and Hydrostatic Pressures

LAKE	DEPTH (m)	ml GAS l^{-1} OF INTERSTITIAL WATER OF SEDIMENTS					AVERAGE TOTAL GAS PRESSURE (atm)	HYDRO-STATIC PRESSURE OF THE BOTTOM (atm)
		CO_2	O_2	N_2	CH_4	H_2		
Nakatsuna-ko	12	234	0	12	72	7	2.68	2.08
Kizaki-ko	29	137	0	17	110	17	4.13	3.71
Aoki-ko	56	89	0	18	161	24	5.14	6.32

After Koyama, 1964.

PRODUCTION OF GASES IN SEDIMENTS, EBULLITION, AND RELATIONSHIPS TO LAKE PRODUCTIVITY

Concentrations of fermentation gases within the water column of most productive lakes are very low in the surface layers and markedly increase in the hypolimnion and near the sediments. If production rates and pressure conditions are such (cf. p. 159ff) as to permit bubble formation, the gases can rise rapidly to overlying strata. Usually, these reactive gases are redissolved and metabolized by microflora en route. Bubble release to the water surface occurs in shallow aquatic situations, such as marshes and littoral areas, where intense anaerobic fermentation in the surficial sediments occurs (e.g., Koyama, 1963; King and Wiebe, 1978; Barber and Ensign, 1979).

As lakes become very eutrophic, the anaerobic fermentation of organic matter in the sediments can reach levels sufficient to result in the steady ebullition of gases. For example, Strayer and Tiedje (1978) found that an average of 21 mmoles m^{-2} day^{-1} of methane left the anaerobic sediments by ebullition in hypereutrophic Wintergreen Lake, Michigan, during the period of late May through August. Ohle (1958) demonstrated by both chemical analyses and sonar echo-sounding that, under stratified conditions in very productive lakes, the ebullition is sufficient to induce "methane convection" currents by which sediment particles and dissolved substances of the hypolimnion can be transported to the trophogenic zone.

The return of nutrients from the sediment zone to the zone of primary production represents an *internal fertilization* of the system. The ebullition of gases from the sediments is closely related to lake productivity (Ohle, 1978). When the oxygen content of the deeper water is exhausted, ebullition of fermentation gases increases. These gases further reduce the oxygen content of water strata nearer the surface through bacterial oxidation and cyclically generate conditions for further intensification of gas generation and ebullition. The result can be observed in a jump-like increase in eutrophication (*rasante Seen-Eutrophierung*) (Fig. 20-5). The accelerated biogenic response intensifies

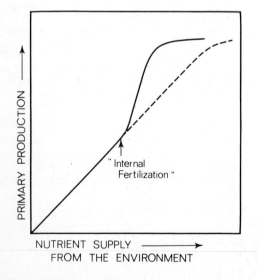

Figure 20-5. Accelerated primary production of lakes resulting from "internal fertilization" during the progression of increasing nutrient inputs from the surrounding environment. (After Ohle, 1958.)

production and inputs of organic matter to the sedimentary fermentation system and enhances rates of generation of gases.

Ohle (1958) further proposed a symbiotic relationship among the bacterial populations that utilize and generate fermentation gases. Sulfate-reducing bacteria of the sediments can utilize methane as an excellent energy and carbon source, but they are strict anaerobes. Their generation of hydrogen sulfide, even though inhibitory to many bacteria, serves as an efficient means to maintain anoxia against intrustions of aerated water.

Littoral Decomposition and Microbial Metabolism

Evaluations of the horizontal distribution patterns of bacteria in lakes demonstrate that bacterial densities are greater in the littoral zone than in the open water. Although comprehensive studies of these spatial distributions are few, examples from both small and large lakes are consistent. The quantity of total bacterioplankton in the littoral zone of Lake Balaton, Hungary, determined by direct enumeration, was found to be 2 to 3 times greater than that of the open water (Oláh, 1969a, 1969b). Saprophytic bacteria* of the littoral zone were 40 to 120 times more abundant than in the open water. The observed differentiation was greatest during the spring period of active growth of the emergent macrophyte reed zone (*Scirpus* and *Phragmites*) and in the autumn as the littoral foliage senesced. High numbers of bacteria from the littoral zone extended out into the open water for considerable distances (1500 meters) in this large lake during the spring, but they were restricted largely to the macrophyte belt during the summer.

Similarly, the microbial activity of large, shallow Lake Mekhtev, USSR, was strongly influenced by the aquatic macrophytes and associated sessile algal flora (Aliverdieva-Gamidova, 1969). While temperature plays an overriding role in bacterial growth on a seasonal basis, and maximal populations and shortest generation times occur in midsummer (Table 20-8), bacteria are clearly more abundant and on the aver-

TABLE 20-8   Distribution and Activity of Bacterioplankton in Water Among Emergent Bulrush Vegetation and in Open Water Without Macrovegetation in Lake Mekhteb, Dagestan, USSR

	WATER TEMPERATURE (°C)	WATER AMONG MACROPHYTES			OPEN WATER		
PERIOD		Total Bacteria $(\times 10^6 \text{ ml}^{-1})$	Generation Time (hours)	Total Saprophytes $(\times 10^3 \text{ ml}^{-1})$	Total Bacteria $(\times 10^6 \text{ ml}^{-1})$	Generation Time (hours)	Total Saprophytes $(\times 10^3 \text{ ml}^{-1})$
ec.–Jan.: no bulrushes	0–2	0.55	19–17	5–12	0.55	19–27	5–12
eb.–Mar.: young bulrushes	3–7	1.4	11–16	28–39	1.0	17	22–103
pr.–May: developing stands	17–23	1.3–1.8	10–13	11–31	1.4	10–12	14–21
ne: mature bulrushes	26	3.3	5.2	40	3.0	7.1	31
ly	28	5.3	4.4	232	2.5	6.5	28
ug: declining bulrushes	21	4.3	7.6	178	3.1	6.3	81
ct.	11	1.5	11	21	1.1	23	17
ov.	3	0.85	21	0.2	0.27	18	0.4

After data of Aliverdieva-Gamidova, 1969.

*See footnote p. 490.

age grow faster in water among macrophyte stands than in open water lacking littoral flora. While these studies indicate that massive amounts of organic substrates are being released by the macrophytes and their epiphytic microflora during growth and senescence, measurements of actual transfer rates are not yet available.

LOSSES OF MACROPHYTIC PARTICULATE ORGANIC MATTER

Earlier discussion indicated that the microbial utilization and degradation of dissolved organic matter released extracellularly during active plant growth and by autolysis was rapid and relatively complete. In contrast, the degradation of the particulate organic matter of the littoral zone is much slower. Studies are few and generally employ approximate techniques (e.g., following the rates of weight loss over periods of time in experimental or in situ situations).

Emergent macrophytic vegetation generally exhibits very slow rates of degradation in relation to those of floating-leaved and submersed aquatic plants (Table 20-9). Part of this greater resistance to decomposition is related to the greater quantity of lignified tissue found in emergent flora (Sculthorpe, 1967). With adaptation to a largely or totally submersed existence, a conspicuous reduction in lignification of supportive and conducting tissues is found in floating and submersed vegetation. Lignin, consisting of aromatic rings with side chains and $-OH$ and $-OCH_3$ groups, exhibits a high degree of chemical stability. Catabolism of lignin by oxidative depolymerization is restricted largely to fungi, especially the Hymenomycetales (Basidiomycetes), which possess the inductive exoenzymes for this cleavage (Trojanowski, 1969). The role of bacteria in the decomposition of lignin macromolecules is limited primarily to side chains or to monomers detached from the lignin by fungi (e.g., Hackett, et al., 1977; Healy, et al., 1980). Large amounts of paracoumaric esters and lignin derivatives are found to persist during decomposition of emergent macrophytes under different environmental conditions (Boon, et al., 1982) and can further retard decay of emergent plants in the littoral zone of lakes. Under conditions of aerobic decomposition, appreciable amounts (to 30 per cent) of the nitrogen can occur in nonprotein nitrogen compounds (Odum, et al., 1979). Much of this nitrogen is apparently chitinaceous and of fungal origin; these compounds are also relatively resistant to decay.

Planktonic fungi found in the open water likely originate mainly from wetland and littoral areas (Willoughby, 1965; Novotny and Tews, 1975), although some (e.g., many chytrid species) are highly specialized parasites of planktonic algae (cf. Chapter 15). Fungal propagules are transported in the water directly or in association with inflowing particulate plant debris. Little is known about their metabolic activities.

The sources of accumulated organic matter in the littoral zone are variable among lakes in relation to morphometry and production by the littoral algal and macrophyte components. In most situations, a majority of the particulate organic matter originates from the macrophytes (approximately 90 per cent), most conspicuously from the emergent forms. Lesser quantities are derived from attached algae, loosely attached littoral algae, and windrowed planktonic algae. Rates of decomposition of the littoral particulate organic matter are quite variable in relation to conditions of accumulation (Table 20-10). In the example from Mikolajskie Lake, decomposition rates were lowest among

TABLE 20-9 Percentage Loss of Dry Organic Matter of Aquatic Macrophytes Over Long Periods of Time

PLANT TYPE	% LOSS OF DRY WEIGHT								SOURCE
	7–14 Days	4 Weeks	6 Weeks	9 Weeks	12 Weeks	15–16 Weeks	24 Weeks	43 Weeks	
Emergent									
Juncus effusus	5	—	—	—	30	60	65	—	Boyd, 1971
Typha latifolia	5	—	—	—	25	30	45	—	Boyd, 1970
Spartina alterniflora	2	—	—	—	10	30	50	85	Burkholder and Bornside, 1957
Carex gracilis	—	42	42	55	—	—	—	—	Koreliakova, 1959
Sparganium simplex (overwintered)	22	—	48	—	58	100	—	—	Koreliakova, 1958
Salix leaves	17–56	—	—	—	—	—	—	—	Pieczyńska, 1972
Phragmites communis									
Overwintered plants	40	—	42	—	50	53	—	—	Koreliakova, 1958
Fresh plants	—	43	48	63	—	—	—	—	Koreliakova, 1959
Scirpus acutus									
Spring–summer	10–12	23	26	—	35	—	38	—	Godshalk, 1977
Fall–winter	5–10	15	18	—	—	18	—	—	
Floating-Leaved									
Lemnaceae	20	—	—	—	90	—	—	—	Laube and Wohler, 1973
Brasenia schreiberi									
Leaves in surface water	(2)	—	—	—	45	—	50	95	Kormondy, 1968
Leaves at sediment	20	—	—	—	20	—	45	90	
Nymphaea odorata									
Leaves in surface water	(15)	—	—	—	45	—	70	98	
Leaves at sediment	(20)	—	—	—	58	—	75	95	
Polygonum amphibium									
Overwintered plants	25	—	35	—	35	37	—	—	Koreliakova, 1958, 1959
Fresh plants	—	42	46	55	—	—	—	—	
Nuphar variegatum									
Spring–summer	23–53	70	80	—	88	—	90	—	Godshalk, 1977
Fall–winter	10–25	30	38	—	47	53	—	—	
Salvinia auriculata	32–34	—	—	45	—	47	50	90	Howard-Williams and Junk, 1976
Eichhornia crassipes	47–53	—	—	63	—	80	80	90	Junk, 1976
Submersed									
Potamogeton lucens	6–92	—	—	—	—	—	—	—	Pieczyńska, 1972; cf. also
P. perfoliatus	6–95	—	—	—	—	—	—	—	Bastardo, 1979
Scirpus subterminalis									
Spring–summer	4–10	33	55	—	62	—	70	—	Godshalk, 1977
Fall–winter	5–10	—	15	—	—	18	21	—	
Najas flexilis									
Spring–summer	15–33	55	85	—	95	—	99	—	
Fall–winter	10–15	20	25	—	—	25	—	—	
Myriophyllum heterophyllum									
Spring–summer	28–40	55	75	—	88	—	95	—	
Fall–winter	12–20	30	35	—	—	38	—	—	
Paspalum repens	57	—	—	—	—	80	90	95	Howard-Williams and
Leersia hexandra	47	—	—	—	—	62	72	95	Junk, 1976

plant material in the emergent part of the eulittoral zone on the surface of the sediments. Senescence and death of submersed species in the submersed littoral site was slow and delayed decomposition; when dead and decaying submersed plants occurred in pools and among reed heaps, decomposition was very rapid. Filamentous algae and accumulations of planktonic blue-green algae in the littoral decayed much more rapidly (> 95 per cent in 3 to 10 days) under comparable conditions (Pieczyńska, 1970).

TABLE 20–10 Decomposition Rates of Macrophyte Particulate Organic Matter in Various Habitats of the Littoral Zone of Mikolajskie Lake, Poland, 1967

PLANT MATERIAL	PERIOD	EMERGENT, 1 m FROM SHORE LINE	REED HEAPS ON THE SHORE LINE	SMALL POOLS ON SHORE PARTLY ISOLATED FROM LAKE	IN WATER, 0.5 m DEPTH, 2 m FROM SHORE LINE
Emergent					
Phragmites	Mid-June	4	25	43	21
communis leaves	Late July	4	34	57	14
	Early Sept.	6	27	38	24
Salix leaves	Mid-June	5	35	37	18
	Late July	7	28	56	25
	Early Sept.	4	30	39	17
Submersed					
Potamogeton	Mid-June	3	35	89	—
lucens	Late July	5	41	92	5
	Early Sept.	6	—	77	4
Potamogeton	Mid-June	1	38	83	7
perfoliatus	Late July	6	40	95	6
	Early Sept.	7	50	70	8

After Pieczyńska, 1970. Percentage losses of dry weight after 10 days of in situ exposure.

DEGRADATION PATTERNS OF LITTORAL FLORA

Lignified tissue of emergent hydrophytes is degraded slowly by a succession of fungal flora. For example, it has been observed that the mycoflora of the cattail *Typha latifolia* passes through early stages that are similar to those of terrestrial plants, with primary colonization by leaf-surface fungi, followed by a secondary phase in which many species of *Leptosphaeria* dominate moribund leaves (Pugh and Mulder, 1971). The succession is associated with aging of plant material and substrate changes. Final stages of decomposition are accompanied by a dominance of fungi predaceous on nematodes. Similar successional patterns have been found on analogous substrates (Sparrow, 1968; Mason, 1976). The roots and rhizomes are very poorly colonized by fungi.

The movement of fine organic detritus particles of decomposing macrophytes from the littoral zone to the open water can occur by water movements and currents returning from the littoral to the pelagic zones, both along the surface and near the sediments (cf. Schröder, 1973, 1975; Gaudet, 1976; Kistritz, 1978; Rho and Gunner, 1978). In addition to these fine particles, nutrients, particularly phosphorus, are transported in major quantities from the littoral flora to the open water. The maximum phosphorus inputs from the littoral occur in the summer period when the phosphate concentrations in the open lake water are minimal.

Very fine-sized (< 100 μm) particles of the emergent reed *Phragmites* were studied as they underwent decomposition in lake water (Oláh, 1972). The successional patterns of bacterial populations in the water containing leached dissolved organic matter from the particles and on the particles correlate with progressive utilization of organic substances (Fig. 20–6). Within a few hours, large rods dominated the liquid phase, presumably utilizing the leached dissolved organic substrates of the detritus. A large popula-

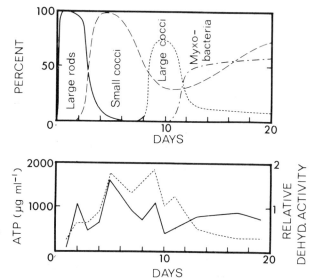

Figure 20-6. *Upper,* bacterial populations, and *lower,* activity associated with decomposition of fine particles of *Phragmites* detritus. ——— = ATP biomass carbon; - - - - = relative dehydrogenase activity. (Drawn from data of Oláh, 1972.)

tion of small cocci succeeded these bacteria; a few days later, the cocci were displaced by large cocci. Both of the coccal types were associated with the liquid phase and populated the surfaces of the small *Phragmites* fragments. Myxobacteria, well-known decomposers of living and dead particulate organic matter (Ruschke, 1968), colonized both the large cocci and the residual, stabilized detrital particles. Maximal bacterial biomass, measured as ATP content, coincided with the maximal development of the bacterial population succession, and corresponded to the time of the mass development of small cocci (Fig. 20-6). Bacterial metabolism, determined by dehydrogenase activity, behaved similarly and decreased after about 10 days. The decrease was related to a stabilizing condition of decreasing quantities of easily assimilable organic substances in the residual particulate detritus.

Analogous sequences of microbial succession were found during the decomposition of emergent reeds *(Typha, Phragmites, Sparganium)*, floating-leaved plants, and submersed macroflora *(Ceratophyllum, Myriophyllum, Najas)* (Gorbunov, 1953; Krasheninnikova, 1958). Degradation rates were greater for the submersed flora under similar conditions than for emergent flora.

Intensive investigation of the decomposition of the particulate organic matter from several species of emergent, floating-leaved, and submersed macrophytes, and of dissolved organic matter leached from them, was undertaken under different conditions of temperature and oxygen availability (Godshalk and Wetzel, 1978a, 1978b, 1978c). Resistance of particulate tissue to microbial decomposition was strongly influenced by the chemical composition of the plant species. The floating-leaved species decomposed faster than the submersed plants, which decomposed more rapidly than the emergent species. Decay rates were related to initial nitrogen and fiber contents, with high-nitrogen, low-fiber plants decomposing most rapidly. Unlike the leached dissolved organic matter of the plants (whose rate of decomposition was found to be mostly influenced by oxygen availability), the rate of conversion of particulate organic matter to carbon dioxide and/or dissolved organic compounds was regulated primarily by temperature, tissue nitrogen, and fiber content.

The relative rates of decomposition, k, of aquatic macrophytes have been variously defined mathematically (see Godshalk and Wetzel, 1978b; Carpenter, 1980) and are controlled by factors of the plant itself and the environment within which it is decaying (Godshalk and Wetzel, 1978c):

$$k \propto \frac{T \times O \times N}{R \times S}$$

where T = temperature, O = dissolved oxygen, N = mineral nutrients necessary for microbial proliferation (e.g., Waite and Kurucz, 1977; Almazan and Boyd, 1978; Andersen, 1978; Carpenter and Adams, 1979), R = initial tissue refractility, and S = particle size (i.e., particle volume/surface area).

The relative rate of decomposition, k, can be summarized graphically (Fig. 20-7) versus undefined units of time; in this case, k is defined as the amount of detrital dissolved and particulate organic carbon metabolized (e.g., POM → DOM, POM → microbial cellular material, DOM → CO_2, etc.) per unit time. While decomposition is a continuous process, the *rate* of decomposition varies with time, and can conceptually be viewed as three distinct phases.

Phase A (Fig. 20-7) is a period of increasing biomass loss from leaching and/or autolytic production of dissolved organic matter (DOM). Much of this DOM is rapidly (hours) metabolized, and the quantity and quality of DOM that persists are influenced by temperature and oxygen availability. Following the maximum rate of weight loss, decay rates decrease to varying degrees during the next phase (B, Fig. 20-7) of decomposition, which can last from several days to many months. Microbial activity and nitrogen content of the particulate detritus increases in the initial phases; factors controlling decay rates have their greatest influence in this phase. The relative resistance of the remaining particulate organic matter continually increases as the more labile compounds are metabolized. In Phase C, the rate of decomposition of the residual resistant particulate matter approaches an asymptotic limit of zero and is limited mostly by the high refractility of the particulate detritus. Some of this matter can enter the anaerobic sediments and become buried; if so, decay rates become so slow that, for all practical purposes, decomposition ceases.

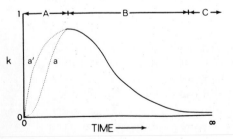

Figure 20-7. Generalized sequence of decay rates of dissolved and particulate organic matter with three phases of decay (see text). The logistic S-curve (line a) is found among plants beginning to decompose under natural conditions where decay is initially slow and is followed by increasing cellular senescence. Line a' is found in laboratory studies using pre-dried plant material at the start of experimental assays (cf. Rogers and Breen, 1982). (After Godshalk and Wetzel, 1978c.)

_____ SUMMARY

1. Sediments of fresh waters are sites of major biological activity.
   a. Sediments of running waters are size-sorted in relation to velocity of water movement. Inorganic- and organic-particle size markedly influences the distribution and growth of many benthic invertebrates.
   b. Lake sediments are the major sites of microbial degradation of detrital organic matter and biogeochemical recycling of nutrients.
2. Sediments are composed of organic matter in various stages of decomposition, particulate mineral matter, and an inorganic component of biogenic origin (e.g., diatom frustules, $CaCO_3$).
3. Particulate organic matter formed in the pelagial zone is decomposed more readily and completely during sedimentation than is the more refractory structural tissue of littoral flora. As senescing littoral plants are decomposed and fragmented by successive communities of fungi and bacteria, redistribution often occurs. The partly degraded material is displaced widely among both littoral and pelagial sediments.
4. Bacterial populations and metabolic activity in surficial sediments are several orders of magnitude larger than those in the overlying water column.
   a. Bacterial populations and metabolic activity decrease precipitously from the sediment–water interface to lower depths within the sediments.
   b. Bacterial numbers and metabolic activities of deepwater profundal sediments are often positively correlated with lake productivity and the quantity of particulate organic matter reaching the sediments.
   c. Respiratory rates of sediment-dwelling bacteria are commonly inversely related to surface area of particles (particle size), and directly related to organic nitrogen content.
5. Most sediments contain significant amounts of organic matter and experience limited oxygen intrusions. Bacterial metabolism rapidly produces anoxic, reducing conditions. Under anaerobic conditions, rates of mineralization are slower than under aerobic conditions. Under anoxic conditions, various substances and intermediate metabolic compounds function as electron acceptors and are reduced instead of molecular oxygen. Often the same compound can serve as a hydrogen acceptor or donor, depending on the environmental conditions.
   a. Large amounts of organic matter are anaerobically degraded by methane fermentation. Organic matter is converted to methane and CO_2 in two stages (Fig. 20–3).
   b. Facultative and obligate anaerobic bacteria ('acid formers') convert proteins, carbohydrates, and fats primarily to fatty acids (especially acetate) by hydrolysis and fermentation.
   c. Obligately anaerobic methanogenic bacteria then convert the organic acids to CH_4 and CO_2; alternately, they can reduce CO_2 to CH_4 by enzymatic addition of hydrogen derived from the organic acids.
6. Methane generation in sediments can be very intense in anaerobic sedi-

ments. Methane diffuses into anaerobic overlying hypolimnetic waters of productive lakes. At the aerobic–anaerobic water interface (often in the metalimnion), much of the methane is converted to CO_2 by aerobic methane-oxidizing bacteria.

a. During summer stratification, highest rates of methane oxidation often occur in the lower metalimnion where oxygen concentrations are reduced (< 1 mg l^{-1}). For the lake as a whole, the total quantity of methane oxidized is often low during summer, however, because the bacteria are limited by available combined inorganic nitrogen, and because oxygen present in the epilimnion is inhibitory to N_2-fixation. Nonetheless, little dissolved methane is found in the epilimnion and only small amounts evade to the atmosphere.

b. High rates of methane oxidation (amounting to as much as 95 per cent of the annual oxidation) occur in periods of lake circulation, especially during fall turnover. Considerable amounts of methane can diffuse and escape to the atmosphere during these periods.

c. Under ice cover, high rates of methane oxidation can occur throughout the water column of eutrophic lakes. Methane oxidation under ice cover can contribute to reduced levels of dissolved oxygen or even total lake anoxia.

d. The amounts of CO_2 and bacterial-cell material production are usually insignificant in comparison to primary production. In relation to the total carbon inputs from the atmosphere and runoff, however, methane cycling can be significant in eutrophic and hypereutrophic lakes. An amount of organic carbon regenerated as methane equivalent to as much as 60 per cent of the total carbon inputs has been found in several very productive lakes; methane-oxidizing bacteria can recycle about two-thirds of this carbon.

7. As lakes become very productive, anaerobic fermentation processes can evolve large quantities of gases, mainly CO_2, CH_4, and H_2. The quantities produced can exceed microbial oxidizing capacities and hydrostatic pressure constraints, which allows the formation of gas bubbles that rise to the lake surface and escape to the atmosphere.

a. In relatively shallow hypereutrophic lakes, a steady ebullition of gases can result.

b. Rising gas bubbles can generate small currents that are effective in redistributing hypolimnetic nutrients and degradation products to overlying water strata. These ebullition currents can enhance nutrient availability and primary productivity in the epilimnion, and thus represent a positively reinforcing, internal fertilization process.

CHAPTER TWENTY-ONE
BENTHIC ANIMALS AND
FISH COMMUNITIES

BENTHIC ANIMAL COMMUNITIES

Analyses of the complex interrelationships among the benthic animals of fresh waters have focused to a large extent on descriptions of species and their distributions within lakes and streams in relation to environmental variables. Although such analyses are essential to initial evaluations of the communities, physiologically oriented experimental analyses of regulating environmental parameters have not been utilized among benthic communities to nearly the extent that they have been in studies of planktonic communities. The population, productivity, and trophic interrelationships of the benthic fauna are poorly understood in lakes; they are somewhat better known in running waters (e.g., Hynes, 1970).

The distribution of the diverse fauna within lakes and streams is extremely heterogeneous, in part a product of variable requirements for feeding, growth, and reproduction. These requirements are strongly influenced by changes in the substratum and overlying water on a seasonal basis, e.g., changes in oxygen content, and in the inputs of living and dead organic matter for food. The benthic organisms either possess adaptive mechanisms to cope with these changes, enter relatively dormant stages until more physiologically amenable conditions return, move, or die. The adaptive capabilities of the benthic animals to the dynamics of environmental parameters are basic to their distribution, growth and productivity, and reproductive potential.

Several major problems must be overcome in order to analyze benthic animal communities effectively. First is the difficulty of obtaining quantitative samples. Substrate heterogeneity leads to a patchy, nonrandom distribution that requires extensive replicated sampling according to procedures that depend on the organisms and bottom substrata involved (cf. detailed discussions of sampling methods in Edmondson and Winberg, 1971; Holme and McIntyre, 1971; Brinkhurst, 1974; Elliott, 1977; and Wetzel and Likens, 1979). Organisms must be separated from samples obtained with benthic grabs, dredges, or cores, which retrieve organisms together with the substrate in which they live. The taxonomy of many animal groups is confusing to the nonspecialist; some groups are still very incompletely described. Emigration and immigration of members of the populations of certain groups, especially among the insects, necessitates more elaborate sampling methods. In spite of these problems, careful and detailed analyses

of some populations provide insight into the controlling environmental and biotic interactions within benthic animal communities.

Benthic animals are extremely diverse, and are represented by nearly all phyla
from protozoans through large macroinvertebrates and vertebrates. This fact, together
with their heterogeneous characteristics of habitats, feeding, growth, reproduction,
mortality, and behavior, makes it exceedingly difficult to treat these animals in an integrated, functional manner. In the brief space available, we can only point out some of
the major groups, giving examples of their population dynamics in relation to freshwater ecosystems. Important characteristics of the sediments have been discussed elsewhere in relation to oxygen (Chapter 9), redox (Chapter 14), microbial metabolism of
organic matter (Chapter 20), and organic composition of different sediments (Chapters
20, 22).

PROTOZOA

Perhaps the least understood groups of benthic animals that occur in massive numbers on and in surficial sediments are the Protozoa and the ostracod crustaceans. General characteristics of the protozoans were mentioned earlier in relation to planktonic
populations. In spite of the abundant information that we have on protozoan morphology, physiology, genetics, and behavior, very little is known concerning their population dynamics and contributions to productivity in the sediments. A majority of the Protozoa are attached to substrata, and they are particularly abundant in habitats of active
oxidative decomposition. Ciliate protozoans are particularly important in the metabolism of dissolved and particulate organic matter of sewage-treatment facilities and
organically polluted streams (Cairns, 1974). The diversity of species of Protozoa, their
possession of wide ranges of tolerance to environmental extremes and varied feeding
capabilities (including algae, bacteria, particulate detritus, and other protozoa), and
their large population densities on aerobic, organic-rich sediments, all point to a significant metabolic role in freshwater systems. Because of their low population biomass,
it generally has been assumed that their contributions are small. However, their short
generation times and rapid turnover under the optimal trophic conditions of the sediments suggest that the Protozoa contribute appreciably to the degradation processes in
lakes.

Many factors are involved in the distribution and growth of free-living Protozoa
(cf. review of Noland and Gojdics, 1967), but oxygen is of paramount importance. Few
species can tolerate the anaerobic conditions of the sediments for any appreciable
period. Distribution and abundance of chlorophyll-bearing forms are regulated to a
large extent in ways analogous to the regulation of algae, such as by inorganic and
organic nutrient availability and light. Among heterotrophic protozoa, the distribution
and quality of dissolved and particulate organic matter for food are important. The
nutritional requirements of only a few species are known. The significance of vitamins
and other growth substances to protozoan population dynamics is just beginning to be
appreciated.

Ciliates and rhizopods dominate protozoans in benthic communities. Most of the
protozoans occur in the upper 1 cm of sediment. The number of species within sed-

iments of a lake can be very large (50 to 150; e.g., Webb, 1961; Laminger, 1973; Finlay, et al., 1979). While coexistence in the same habitat is common, microspatial differences in distribution exist, and species separation by differences in food utilization, reproduction, and timing of population development is also common (Goulder, 1971, 1974a, 1974b; Taylor and Berger, 1980).

Most species of protozoans are intolerant of low oxygen concentrations and reducing conditions in sediments, but some species tolerate reducing conditions much better than others (e.g., Goulder, 1971, 1980a; Laminger, 1973; Finlay, 1980). As a result, protozoan species in the sediments are often segregated vertically in the lake according to the depth of the sediments. These depth distributions may change seasonally, owing to changes in oxygen concentrations of overlying water. Many species migrate to shallower areas as anoxia develops in the hypolimnion, or the populations gradually assume a different depth distribution as reproductive rates change. Some species are negatively phototactic and avoid high light intensities within epilimnetic sediments.

Many of the benthic protozoan populations exhibit pronounced summer maxima (Finlay, 1980; Bark, 1981; Rogerson, 1981). These changes have been positively correlated with summer increases in temperature, day length, and abundance of organic matter and microflora serving as food in the surficial sediments. As oxygen is depleted in the hypolimnion, many populations migrate upward to metalimnetic and epilimnetic regions; this migration also contributes to increases in the protozoan populations in these areas.

PORIFERA

Few sponges occur in fresh waters; most species are marine. Although some species endemic to Lake Baikal grow to magnificent lobed structures of about half a meter in size, most are small, inconspicuous, and morphologically variable. The flagellated chambers create weak currents that pass through the internal chambers, where bacteria and particulate detrital material are trapped for intracellular digestion. Zoochlorellae, ingested algae that persist inter- and intracellularly in the tissue, can grow and metabolize appreciably within certain sponges (Gilbert and Allen, 1973). Although some symbiotic relationships between the organisms exist (Frost and Williamson, 1980), the zoochlorellae are not essential to the survival of the sponge, at least not for *Spongilla*.

Sponge growth occurs by proliferation of the tissue over the substratum, with the addition of new chambers. Under unfavorable conditons, such as in autumn in the temperate zone, tissue is reduced by partial or total deterioration. Highly resistant resting structures termed gemmules, usually less than a millimeter in diameter, are formed in many species (Gilbert, 1975; Gilbert and Simpson, 1976a). Some species, particularly those in sublittoral environments of muted seasonality, do not form gemmules; instead, sexual reproduction continues into the autumn (Gilbert and Allen, 1973; Simpson and Gilbert, 1974; Gilbert and Hadžišče, 1975, 1977).

Gemmules contain undifferentiated mesenchymal cells, and have an outer layer of compacted spicules within a tough sclerotized membrane. They can withstand freezing and desiccation. Germination of the gemmules is usually associated with warmer temperatures above 15°C in the early spring (Harrison, 1974; Gilbert, et al., 1975). Syngamic sexual reproduction also occurs in warmer periods of the year. Sponges are usually dioecious, and the sex ratio is about 1:1 in natural populations. A common freshwater sponge, *Spongilla lacustris*, exhibits a type of alternate hermaphroditism, in which a sponge may be exclusively male or female during the period of sexual reproduction one year and the opposite sex the next year (Gilbert and Simpson, 1976b). This form of sexuality may facilitate larval production and thus dispersal following colonization of a new habitat.

Sponges are among the few animals requiring large amounts of silica for spicule development. Sponges are restricted largely to waters of moderate silica content (>0.5 mg l^{-1}) (Jewell, 1935). Species distribution is also correlated with the calcium content of the water (Jewell, 1939; Strekal and McDiffett, 1974); although causal mechanisms for this relationship remain obscure, they are probably related to food abundance rather than to a direct demand for calcium. Freshwater sponges, which usually occur only in relatively clear, unproductive waters, rarely, if ever, become a major component of benthic communities. Their significance in benthic productivity is minor in most situations (but cf. Frost, 1978).

COELENTERATA

The hydroids and "jellyfish" of the Class Hydrozoa belong to a predominately marine group that is poorly represented in fresh waters. The common *Hydra* occurs only as the tentacled polyp stage, while the rare *Craspedacusta* or freshwater jellyfish undergoes metagenesis with alternation between medusoid and polyp stages that is common to the marine Hydrozoa. Epidermal cells contain nematocysts or stinging cells that function in capture of prey; food particles are then moved to the mouth orifice by the tentacles. The hydroids are sessile and are almost exclusively carnivorous on zooplankters. One species of green hydra *(Chlorohydra)* contains unicellular green algae in a symbiotic relationship, and the algae live intracellularly in gastrodermal cells. About 10 percent of the carbon photosynthetically fixed by the algae is released and assimilated by the *Hydra* (Muscatine and Lenhoff, 1963). This photosynthate is clearly of nutritional significance to the hydra (Pardy and White, 1977; Phipps and Pardy, 1982). Development in quiet waters is rapid in spring, and greatest densities usually are found in early summer (Cuker and Mozley, 1981). Growth rates decrease markedly when populations are crowded (Thorp and Barthalmus, 1975). Only rarely do hydroid densities become appreciable, however, and their contribution to total benthic productivity is considered negligible. Under some conditions, dense populations of *Hydra* can significantly reduce population sizes of small zooplankton by predation (cf. Cuker and Mozley, 1981).

FLATWORMS

The only important free-living members of the Platyhelminthes are the turbellarian flatworms. Most members of the group, the tapeworms and the flukes, are entirely parasitic. The free-living flatworms generally possess abundant cilia that assist them in movement over the substrata of shallow lakes and streams. When the animal encounters small invertebrates or detrital organic matter, the ventral pharynx is extruded through the mouth to incorporate the material (Young, 1973). Some of the rhabdocoels contain symbiotic zoochlorellae in their parenchyma and gastrodermis cells.

Reproduction is either asexual by budding or fission (Pattee and Persat, 1978), or sexual. In the latter case, sex organs usually develop over winter and egg capsules are released in the later winter or spring in permanent bodies of water. Nearly all flatworms are hermaphroditic, with both sex organs occurring in the same individual. The eggs are encapsulated or sometimes enclosed in a stalked shell.

Winter eggs are heavily encapsulated and are able to withstand low temperatures, but not desiccation. Development is strongly temperature-dependent in both the egg and immature stages; no larval stages are found in most species (e.g., Pattee, 1975; Folsom and Clifford, 1978).

Most flatworms, especially the triclads, are negatively phototactic and occur in shaded areas among debris, rocky substrata, and macrophytes. Among four species of lake-dwelling triclads studied extensively by Reynoldson (1966), all occupied the same general substratum in the shallow littoral zone and fed on the prey organisms there. Feeding mechanisms of all the species were similar, and wide overlap existed in the type and size of prey eaten by young and adults of each species. The species populations exhibited restricted fluctuations in numbers and lived under conditions of food shortage for long periods of the year. Flatworms are able to avoid starvation because they can resorb and then regenerate tissues by reductions in size and metabolic rate (Calow, 1977), and because of low predation pressure by other organisms.

The distribution and abundance of individual triclad species are determined primarily by interspecific competition for food (Reynoldson and Bellamy, 1971, 1975; Reynoldson and Sefton, 1976; Reynoldson and Piearce, 1979; Reynoldson, et al., 1981; Adams, 1980; Young, 1981). Despite the great overlap of food prey species, each triclad has been found to feed on a particular kind of prey to a greater extent than others. Thus the competition depends largely on the pattern of distribution and abundance of important types of prey, such as gastropod mollusks, stonefly and mayfly nymphs, amphipods, and oligochaetes, each of which represents the dominant prey of particular triclad species.

The abundance of these triclad species has been directly correlated with the general productivity of the waters and amount of food available in the littoral zone, which in turn has been directly related to concentrations of calcium and total dissolved matter (Reynoldson, 1966, 1981). As a result, unproductive lakes receiving low nutrient loading support smaller populations and fewer species, owing to competition for food, than lakes of higher productivity, in which both the diversity and abundance of food species increase. As food supply is reduced, short-term reproductive effort generally increases (Callow and Woollhead, 1977).

Similar relationships have been found among stream-dwelling flatworms (Chandler, 1966). Shallow streams exhibit wide fluctuations in temperature, which influences triclad feeding activity, reproductive capacity, and acclimation capacity. The distribution of individual species is widely separated as a result. Additionally, triclad abundance is inversely related to velocity of water flow. The shifting substrates of rapidly flowing streams and wave-swept littoral zones create an unfavorable habitat for flatworms and many other benthic animals. Although adults may be able to tolerate the mechanical abrasion associated with shifting substrata, eggs or immature stages often cannot.

NEMATODES

The Nematoda or roundworms, which constitute a class of the pseudocoelomate Phylum Aschelminthes, are a significant component of the benthic fauna. Although most of the large number of nematode species are parasitic, the many free-living species are distributed widely in all types of freshwater habitats. The taxonomy of this group is difficult, incomplete, and has contributed to the relative paucity of knowledge on the ecology and productivity of nematodes.

The nematode's external proteinaceous cuticle is noncellular and layered, and is molted four times during its life. This ridged cuticle is interfaced with longitudinal muscle cells that permit active, snakelike movements along and through sediments.

Great diversity in feeding habits occurs among the nematodes. Some members are strictly detritivores, feeding solely on dead plant or animal particulate matter, or both. Others are herbivorous, and have specialized mouthparts for chewing living or dead plant material or for piercing and sucking cytoplasm from plants (Prejs, 1977a). Mouthparts are more specialized among carnivorous nematodes, enabling them to seize, rasp, and macerate small prey animals such as other nematodes, protozoans, gastrotrichs, tardigrades, and oligochaetes.

Reproduction is variable, with some species being parthenogenetic or hermaphroditic. Most species, however, are syngamic with separate females and males. Fertilized eggs, usually laid outside of the body, develop rapidly (1 to 10 days) and hatch into young, which are nearly fully developed except for the reproductive organs. Reproduction and egg laying by adults are more or less continuous. Little is known of the fecundity of free-living nematodes under natural conditions, but it is apparent that a large proportion of their productivity is diverted into reproduction in a normal life cycle. Several hundred eggs with a biomass several times greater than that of the producing female are apparently common. The duration of the life cycle is also variable, from about 2 to 40 days; most species, however, require 20 to 30 days for development and reproduction.

In a detailed study of the nematode populations of the sediments of a typical dimictic, relatively oligotrophic alpine lake, Bretschko (1973) found nematodes to be concentrated at depths of 3 to 4 cm in the sediments. Few penetrated below 6 cm into the sediments. Although the four molting instars could not be separated, analyses of the seasonal population dynamics of the many species demonstrated three generations per year: (a) The first hatched in September–October and matured in November–December just after ice cover formed. (b) The second generation was the most productive, and reached maximum abundance and biomass in January (Table 21–1). The peak of relative abundance of adults of this generation occurred in April. (c) Offspring of the winter generation formed the third generation, with a maximum in biomass and abundance in May (about 50 per cent of that peak in January). As the ice cover was lost in May and the lake circulated, the entire community of nematode populations collapsed for unknown reasons. Over a two-year period, the average population density was 235,000 m^{-2} under ice cover but only 60,000 m^{-2} during the ice-free period.

Nematode species distribution and biomass production were strongly correlated with the type of substratum (Table 21–1). Over 80 per cent of production occurred in the deeper zone of deep, fine sediments where gravel was scarce and boulders were absent. Within this zone, 90 percent of the biomass production was by one species of

TABLE 21-1 The Seasonal Production (kg dry wt) of Nematodes in Vorderer Finstertalersee, Austria, in Different Zones of Substratum, October 1968 to December 1969

SUBSTRATUM ZONE AND ITS AREA	SEASONAL PERIOD				PER CENT OF TOTAL PRODUCTION
	Oct.–Jan.	Feb.–May	June–Sept.	Oct.–Dec.	
Littoral-slope zone, fine silt; 58,868 m^2	1.24	0.16	0.34	1.74	13
Current-steep slope, reduced silt covering gravel and boulders; 47,213m^2	0.24	0.26	0.32	0.82	6
Flat-depth zone, below 20 m, deep fine mud, gravel rare, boulders absent; 51,520 m^2	8.64	1.56	0.44	10.64	81
Entire lake, 157,600 m^2	10.16	1.96	1.08	13.20	100

From data of Bretschko (1973); dry weight estimated from fresh-weight values given by a factor of 20%; totals do not sum parts exactly because of rounding.

Tobrilus. The reasons for the observed high productivity in early winter are unclear, but presumably were related to good food conditions that followed the ice-free period of higher primary productivity, and to reduced predation by fishes and possibly by certain predacious cyclopoid copepods.

Greatest densities of nematodes are found in oligotrophic lakes (Prejs, 1977b, 1977c). Nematode densities decrease in profundal sediments of eutrophic and dystrophic lakes; presumably low oxygen concentrations are a primary limitation in these lakes.

Nematodes can reach very high densities in the littoral zone, among the attached algae of dense emergent and submersed macrophytes. In a detailed study of the seasonal dynamics of nematode populations in the littoral zones of several lakes, Pieczyńska (1959, 1964) found that maxima occurred in May and June, when the populations numbered over one million individuals per square meter of substrate surface, a level brought about by mass reproduction of the dominant species. The same species reproduced throughout the year, although not as intensely. During the other seasons of the year, the nematodes occurred in lower, more or less constant densities. Variations in the development of the spring maximum from year to year were correlated with warming temperatures of the water and the duration of ice cover in the littoral.

Species of *Prochromadorella* and *Punctodora* dominated in littoral substrates, especially shoots of higher aquatic plants, to which they attached readily by the secretion of an adhesive substance from their caudal glands and by their ability to become entangled in colonies of littoral algae. Littoral ice cover was found to be especially destructive to the substrata and nematode populations. Wave action, in contrast, caused little mortality and was found to facilitate dispersal and subsequent rapid colonization of newly growing algal and macrophytic surfaces. The densities of nematodes were correlated directly with the development and density of attached algae.

BRYOZOANS

The freshwater colonial bryozoan members of the primarily marine Ectoprocta are rarely of quantitative importance. These sessile forms, however, occasionally form massive colonies that can become conspicuous members of shallow eutrophic lakes and open areas of swamps for brief periods. The microscopic zooids possess ciliated tentacular crowns that project into the water and create water currents. Particles as large as microcrustaceans are directed into the lophophore and mouth. Although hermaphroditic sexual reproduction occurs in summer, asexual reproduction permits rapid proliferation on the outer edge of the colonies, which occasionally form amorphous ball-like masses one-quarter of a meter in diameter. These colonies are common in small farm ponds in water less than a meter in depth, and their growth is influenced by fish predation (Dendy, 1963). When protected from fish predation, colonies were branched and more productive; when exposed to fish predation, unbranched, cropped colonies persisted. A large number of invertebrate animals, from protozoans to large insect larvae, particularly chironomids, live within the colonies of bryozoans. Predators of the bryozoans consume these coinhabitants as well as the bryozoans.

Extensive studies of 12 species of bryozoans in 122 lakes, streams, and shallow waters of Michigan showed correlations of species distribution with several general limnological parameters such as pH, current, and substratum type (Bushnell, 1966; cf. the detailed review of Bushnell, 1974). Some bryozoan species were associated with particular macrophyte substrata, and others were distributed more ubiquitously. Live colonies were found under ice at water temperatures of $<2°C$. Growth was often erratic, but luxuriant colonies of *Plumatella* were found to grow rapidly, doubling in 3.4 to 4.9 days in summer and from 4.0 to 7.4 days in the spring. Growth ceased when temperatures decreased below 9°C. A majority of the mature colonies broke up and died after a period of growth, for reasons not completely understood. Predation on actively growing colonies was often very high, especially by caddisfly larvae and snails. *Plumatella* longevity ranged from 4 to 53 days; death rate of individuals of the colony (the polypides) was highest (80 percent) between 21 and 36 days after statoblast germination, and few lived longer than six weeks.

AQUATIC WORMS

Two major groups of the annelids, or segmented worms, are represented in fresh waters. The first of these, the Oligochaeta or aquatic "earthworms," often form a major component of the benthic fauna, particularly of lakes. The other group, Hirudinea or leeches, is a diverse class of annelids of much biological interest. Although the significance of leeches as a component of total benthic animal productivity is unclear at this time, their predatory habits can materially influence the population dynamics of other benthic organisms, especially oligochaetes. By far the most comprehensive study of the aquatic oligochaetes is the work of Brinkhurst and Jamieson (1971). While largely a global taxonomic work, much of the presently rather limited knowledge of oligochaete anatomy, embryology, distribution, and ecology is reviewed in this book (cf. Learner, et al., 1978; Brinkhurst and Cook, 1980).

Oligochaetes. Oligochaetes are typically segmented, bilaterally symmetrical, hermaphroditic annelids with an anterior ventral mouth and a posterior anus. Size ranges from <1 mm to about 40 cm (maximum >200 cm), but most freshwater forms are less than 5 cm in length. Each segment, except for the first and a few terminal segments, bears four bundles of setae. Sexually mature Oligochaeta possess an anteriorly located thickened region in the body wall which is glandular and produces a cocoon at oviposition. Sexual reproduction predominates in many species under poor environmental conditions (Loden, 1981). Asexual reproduction is more common during favorable environmental conditions (e.g., McElhone, 1978).

Much of the information on oligochaete populations focuses on geographical distribution, habitat selection, and effects of organic pollution. Several species usually are found in each habitat and these species differ little between streams and lakes of the same region. The number of species is often greater in larger lakes, perhaps because of the greater number of different microhabitats. Some species are restricted to relatively oligotrophic waters while others are distributed widely in lakes of greatly differing productivity, from oligotrophic to extremely eutrophic (see, for example, Milbrink, 1973a, 1973b, 1978; Lang and Lang-Dobler, 1980; Särkkä and Aho, 1980). Obviously a number of interacting parameters influence species distribution within the physiological tolerances of individual species. Some of these factors are discussed below.

The distribution of oligochaetes in sediments at different depths within a lake also indicates that a large number of species coexist in the same substrata. Individual species, however, do reach maximal abundance at different depths (Fig. 21–1). Although differences in the composition of the sediments were not analyzed critically in this

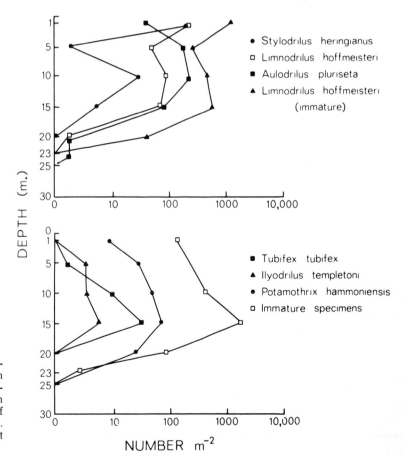

Figure 21–1. Distribution of various species of oligochaetes in relation to depth in hypereutrophic Rostherne Mere, southern England. The lake was devoid of oxygen below a depth of 25m. (Drawn from data of Brinkhurst and Walsh, 1967.)

• Stylodrilus heringianus
□ Limnodrilus hoffmeisteri
■ Aulodrilus pluriseta
▲ Limnodrilus hoffmeisteri
  (immature)

■ Tubifex tubifex
▲ Ilyodrilus templetoni
● Potamothrix hammoniensis
□ Immature specimens

DEPTH (m.)

NUMBER m^{-2}

study of a eutrophic lake, a number of studies suggest that the particulate composition and organic content of the sediments are correlated with the distribution and abundance of oligochaetes (Timm, 1962a, 1962b). In a large, oligotrophic Italian lake, the number of oligochaetes was generally less in sediments of relatively large particle size (0.11 to 0.12 mm) than in more organic-rich sediments of finer particle size (0.07 to 0.08 mm) (Della Croce, 1955). A similar relationship has been found in relation to the total organic content of the sediments (Brinkhurst, 1967; Wachs, 1967). Experimental studies on the ubiquitously distributed *Tubifex tubifex* verified these general relationships. Nutrition and the availability of food are primary factors influencing distribution and abundance of oligochaetes.

> Oligochaetes ingest surficial sediments containing organic matter of autochthonous and allochthonous origins colonized with bacteria and other microorganisms. Some species actively graze attached microorganisms growing epiphytically on macrophytic vegetation. Other oligochaetes actively ingest epipelic algae (e.g., Moore, 1978, 1981). Detailed analyses of the gut contents and feeding habits of oligochaetes living in the same habitat indicate that resource partitioning occurs by differences in worm morphology and by selective feeding on food items, particularly algae (Streit, 1977; McElhone, 1979). In studies on sediment feeding by this group (reviewed in Brinkhurst and Jamieson, 1971), it has been found that the organic carbon, caloric, and total nitrogen contents of the ingested sediments were only reduced slightly. However, critical experiments with mixed species cultures failed to demonstrate any reduction in the percentage of organic matter, nitrogen, or caloric content in feces as opposed to that in sediment provided as food (Brinkhurst, et al., 1972; Brinkhurst and Austin, 1979). These and other experimental studies on the utilization of bacterial species in sediments indicated that one species of oligochaete may be able to selectively utilize different microbial components of the sediment (Brinkhurst and Chua, 1969; Wavre and Brinkhurst, 1971; Harper, et al., 1981).

As lakes and streams become organically polluted, it is common to find an abundance of tubificid oligochaetes (Brinkhurst and Cook, 1974; Milbrink, 1980). Associated with such organic enrichment is acute reduction or elimination of dissolved oxygen. Acute reduction of oxygen is lethal to a majority of benthic animals. The number of species of tubificid worms also decreases with higher productivity and increased organic pollution. However, if some oxygen is available periodically and if toxic products of anaerobic metabolism do not accumulate, the combined conditions of a rich food supply and freedom from competing benthic animals permit rapid growth.

Many oligochaetes, especially the tubificids, burrow headfirst into the sediments but leave their caudal ends, which contain most of their gills or respiratory appendages, projecting and undulating in the water above the sediment–water interface. The caudal respiratory movements increase in frequency and vigor as the oxygen content of the water decreases. In general, tubificids are able to adjust their respiration to decreasing oxygen concentration down to a critical level in the range of 10 to 15 per cent saturation, below which hemoglobin can no longer be loaded with oxygen. Below this critical level, feeding and defecation are then greatly reduced or cease (cf. Brandt, 1978). Many tubificids can tolerate anaerobic conditions for at least a month or longer, but survive only if they are exposed intermittently to some oxygen, since they cannot respire anaerobically for any appreciable length of time. Water movement is apparently an important factor in removal of toxic metabolic wastes, since anaerobic conditions can be tolerated longer in moving situations than when the water is stagnant.

Reproduction in oligochaetes varies seasonally, and even when more or less continuous, some periodicity in breeding intensity is common. Oligochaetes lack discrete age classes, and production is often approximately continuous. The limited data from natural populations suggest that maturation takes as long as a year in some species, and up to 2 to 4 years in others. Variation is great, however, within the same species, and even from site to site in the same body of water. Breeding intensifies in late winter and spring as temperatures increase above about 10°C. Most adult oligochaetes die after sexual reproduction. Selective predation by other animals as a factor in species coexistence has been studied very little in natural populations. However, it is clear that selective breeding and metabolic adaptations by oligochaetes permit cohabitation. For example, when three species were combined, population respiration of mixed cultures decreased about 35 per cent, while increasing the population density of a single species proportionately had no such effect (Brinkhurst, et al., 1972). More energy was used for growth when the three species were mixed than when they were maintained separately in single-species culture, and the rate of respiration decreased in the mixed populations. Isolated single species spent more time burrowing and searching for their preferred food, which is feces of other oligochaete species, than they did in feeding (Chua and Brinkhurst, 1973; Brinkhurst, 1974). Such interspecific interactions are probably causal factors in the frequently observed clumped distribution of several species of oligochaetes.

In the deep profundal sediments of the Danish eutrophic Lake Esrom, some tubificid *Potamothrix hammoniensis* can first reproduce at three years of age, but most do not until they are four years old (Jónasson and Thorhauge, 1972). The life span of this species can reach five or more years. Mature *Potamothrix* in this lake laid eggs in May–July, and this process was related to moderate increases in sediment temperatures. From 2 to 13 embryos hatched from each cocoon about two months later, in August. Growth was most rapid in the autumn and spring, and no weight increase occurred during the summer period of stratification. Numbers of this species varied between 10,000 and 25,000 m^{-2} over a 6-year period. In spite of these large numbers, biomass varied between 0.75 and 1.80 g dry weight m^{-2}, but usually was greater than 1.0 g m^{-2}. These values are similar to those observed in other lakes (Table 21–2).

Mortality in the Lake Esrom populations was clearly related to death following reproduction. However, predation by the dipteran larvae of *Chironomus anthracinus*, studied in great detail in this lake, also caused significant reduction of the oligochaete populations. The reproduction of the oligochaete populations was depressed severely every second year as a result of the two-year life cycle of the *Chironomus* larvae (see also Loden, 1974).

Very few estimates of production rates of oligochaete populations exist (Brinkhurst, 1980). Detailed evaluations of the egg development times, mortality, cohort structure, and population dynamics of two dominant species of a reservoir in northern Italy demonstrated marked differences between the seasonal population characteristics of these oligochaetes (Bonomi, 1979; Adreani, et al., 1981). *Tubifex* had higher fecundity and shorter generation times, and was nearly twice as productive as *Limnodrilus* (Table 21–3). However, the annual production-to-biomass ratios (turnover times) of the two species were very similar.

Oligochaete species frequently segregate vertically within the sediments. Naidid oligochaetes are concentrated at the sediment–water interface, rarely more than 2 to 4

TABLE 21-2   Average Population Densities and Biomass of 29 Species of Oligochaetes of Chud-Pskov Lake, Estonia, USSR

	AVERAGE POPULATION DENSITY (No. m^{-2})	PER CENT OF TOTAL DENSITY OF BENTHIC ANIMALS	BIOMASS* (g m^{-2})	PER CENT OF TOTAL BIOMASS OF BENTHIC ANIMALS
Total lake	370–1400	—	—	—
June	859	47	1.307	23
Fine-grained silt of profundal	287	58	0.433	44
Silt with fibrous particulate detritus	1452	62	2.776	16
Sandy with little silt; no vegetation	1015	22	1.409	36
Upper littoral	1185	20	0.133	6

From data of Timms, 1962a.

*Assumed to be wet weight; methods not stated (see Jónasson and Thorhauge, 1972, for discussion of errors involved with preserved vs. fresh wet weights).

TABLE 21-3   Average Biomass, Annual Production, and Annual Production/Biomass Ratios of Oligochaete Populations in Pietra del Pertusillo Reservoir, Italy

		IMMATURE (to 1 mm)	YOUNG	MATURE ADULTS	EGG-BEARING ADULTS	TOTAL
Production (P) (g m^{-2} yr^{-1})	Tubifex	18.89	21.31	6.46	2.26	48.92
	Limnodrilus	4.47	14.24	10.31	−1.50	27.53
Mean Biomass (B̄) (g m^{-2})	Tubifex	0.69	3.59	1.13	3.91	9.32
	Limnodrilus	1.07	1.78	0.87	2.44	6.15
Annual P/B	Tubifex	27.4	5.9	5.7	0.6	5.25
	Limnodrilus	4.2	8.0	11.8	—	4.48

After Adreani, et al., 1981. Biomass estimates based on wet (formalin) weights.

cm below the surface (Milbrink, 1973c). Tubificid oligochaetes are most dense between 2 to 4 cm of sediment depth, and occasionally penetrate as deep as 15 cm. The movements of benthic organisms within the sediments are important because their activity can disrupt the oxidized microzone of the sediment–water interface and thereby alter rates of chemical exchange between the sediments and overlying water. Experiments with tubificids have shown that their activity can alter the stratigraphy of surface sediments (Davis, 1974; McCall and Fisher, 1980), which is of interest in micropaleontological work (Chapter 23). As was discussed in Chapter 20, it appears that the intensive microbial activity at the sediment–water interface is sufficient to reestablish redox conditions almost immediately after such disturbance.

Hirudinea.   Excellent general summaries on the structure, physiology, and ecology of leeches are given by Mann (1962) and especially Sawyer (1974). In general, leeches are ectoparasites that intermittently consume blood and body fluids of vertebrates greatly in excess of their own body weight. They then enter a period of fasting, during which the consumed food is utilized over a period that can be as long as 200

days, with progressive loss of body weight. Other leeches are predatory on invertebrates such as oligochaetes, amphipods, and chironomid larvae, and entirely consume
their prey (Elliott, 1973a; Green, 1974; Davies and Everett, 1975).

Reproduction is usually initiated in the spring or early summer, and apparently is
governed by temperature, the density of the populations, and age. Many leeches breed
at one year of age and die after breeding, although some pass through two generations
in one year (Tillman and Barnes, 1973; Davies and Reynoldson, 1975; Davies, 1978).
Other species live two years, overwinter and breed in the second, and die soon thereafter (Elliott, 1973b). In the first three months after hatching there is heavy mortality of
the young (approximately 95 per cent; see Table 21–4) from predators (carnivorous
invertebrates, amphibia, many fish, and birds). Thereafter mortality lessens, but is still
severe. When population densities become high, population size can be regulated by
high mortality of the eggs, which are consumed by adult leeches (Elliott, 1973a).

Among blood-consuming leeches, the time required for individuals to attain
maturity is dependent upon the number and frequency of blood meals taken (Wilkialis and Davies, 1980; Davies and Wilkialis, 1980). A seasonal migration has been
found in some species, in which populations move into shallow water areas, where
encounters with vertebrates would be more probable, during the ice-free period and
then back into deep water zones during the period of ice cover.

The abundance of leeches is highly variable among different habitats of lakes and
streams. A general direct correlation exists between leech abundance and lake productivity. This relationship probably is associated with the increasing diversity of substrate types among the macrophytes and sediments, with correspondingly greater
amounts of invertebrate food sources for the predacious leeches, and birds and other
vertebrates for the blood-consuming leeches (Sander and Wilkialis, 1972).

Estimates of the production rates of leeches are extremely rare. The figures cited
in Table 21–5 are for a moderately productive stream, and point out some of the population characteristics. Production of this leech species varied considerably between
years and year-classes. Production was always greatest in the first year of life cycle,

TABLE 21-4   Life History and Mortality of Two Common Freshwater Leeches in England

	GLOSSIPHONIA COMPLANATA	ERPOBDELLA OCTOCULATA
Breeding period	Mar.–May	June–Aug.
Average number of young per parent	26	23.5
Mortality in first 6–9 months (%)	97	91
Proportion of year-group breeding at 1 year old (%)	70 (or less)	87
Mortality from 6 to 18 months (%)	33	49
Proportion of year-group breeding at 2 years old (%)	100	100
Mortality from 18 to 30 months (%)	84–88	69
Proportion of total population surviving to 3 years (%)	5–6	4

After Mann, K. H.: Leeches (Hirudinea). Their Structure, Physiology, Ecology and Embryology. Oxford,
Pergamon Press, 1962, from several sources.

TABLE 21-5 Production, Mean Biomass, and the Ratio of Production: Biomass (P/B) for Populations of the Leech *Erpobdella octoculata* in an English Stream

POPULATION	PRODUCTION (g WET WEIGHT m^{-2})	BIOMASS (g WET WEIGHT m^{-2})	P/B RATIO
1964 year-class			
Second year	8.25	10.35	0.80
1965 year-class			
First year	23.42	10.02	2.34
Second year	8.52	11.99	0.71
Both years	31.94	11.01	2.90
1966 year-class			
First year	12.34	4.61	2.68
Second year	4.60	5.29	0.87
Both years	16.94	4.95	3.42
1967 year-class			
First year	6.54	2.47	2.65
Annual estimates			
1966	33.46	20.96	1.60
1967	19.77	14.36	1.38
1968	11.63	7.82	1.49

After data of Elliott, 1973b.

even though the mean biomass was very similar in both years of the life cycle. The rate of production was most rapid during March to July, after which production declined sharply.

Overwintering leeches of the species *Helobdella stagnalis* were found to reproduce in May in a shallow, eutrophic reservoir of South Wales (Learner and Potter, 1974; cf. Davies and Reynoldson, 1975). These postreproductive adults then died. The spring generation grew rapidly, and about one-third of this generation reproduced in July. The nonbreeding portion of the spring generation overwintered with the summer generation; both reproduced at the same rate in the following spring. The mean biomass was 96 mg (dry) m^{-2} in 1970, and 393 mg (dry) m^{-2} in 1971. Production was estimated to be 310 mg (dry) m^{-2} yr^{-1} in 1970 and 1190 mg (dry) m^{-2} yr^{-1} in 1971, with production to biomass ratios of 3.2 and 3.0 in the respective years.

OSTRACODS

The ostracods are small, bivalved crustaceans usually less than 1 mm in size, which are widespread in nearly all aquatic habitats. Their size, distribution in the surface sediments with few planktonic species, and difficult taxonomy have all contributed to a very poor understanding of their ecology, population dynamics, and productivity in fresh waters. Their valves, which superficially resemble clam shells, are held apart when undisturbed. Cladoceranlike appendages protrude between the open valves and are used for locomotion. Most ostracods move about on the sediments by beating movements of the antennae and the caudal ramus. Ostracods are omnivorous and like many cladoceran microcrustacea, they feed on bacteria, algae, detritus, and other microorganisms by means of filtration. Very little is known of the effects of ostracods on the turnover and metabolism of benthic microflora. The large population numbers of ostracods suggest that their role in the metabolism of surficial sediments could be considerable.

Ostracod reproduction is usually parthenogenetic for much of their life cycle. In some species, males have not been found; in others sexual reproduction is common. The time required for egg development is variable from days to months, and is strongly temperature-dependent. The larva hatches as a nauplius with a reduced number of appendages, and then undergoes a series of growth and molting stages; usually eight molts are needed to reach the ninth, adult stage, during which morphology becomes more complex and appendages develop.

Little is known of the population dynamics of ostracod populations, although many exhibit distinct seasonal periodicity. Some species exhibit a single generation per year, others two or three per year. The detailed study of the reproductive potential and life history of a species of *Darwinula* in a deep, dimictic, temperate lake illustrates some of the characteristics of this group (McGregor, 1969). *D. stevensoni* was abundant throughout the littoral sediments of this mesotrophic lake, and reached a maximum density at a depth of 6 m throughout the year. Most individuals occurred between depths of 3 and 6 m and densities decreased markedly between 6 and 9 m. Only a few individuals were found below a depth of 9 m, and none below 12 m. More than 95 per cent of the adults and juveniles present in the sediments occurred in the upper 5 cm of sediment.

The reproductive period of this *Darwinula* species in this Michigan lake began in May and was effectively completed by October (Fig. 21-2). The number of young per individual increased from a maximum of 3 in May to 15 in August. Although the reproductive potential of a given individual was correlated strongly with temperature and varied somewhat with depth, most adults produced only one brood per year of about 11 young per individual at depths of 3 to 6 m (13 at 9 m).

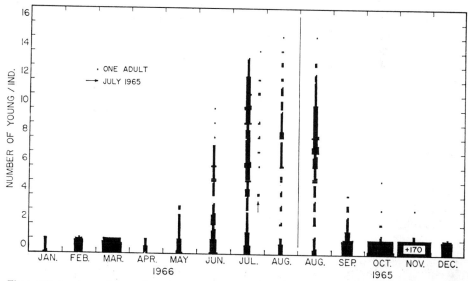

Figure 21-2. Annual reproductive cycle and reproductive potential of the benthic ostracod *Darwinula stevensoni* in Gull Lake, southwestern Michigan. (From McGregor, D. L.: The reproductive potential, life history and parasitism of the freshwater ostracod *Darwinula stevensoni* (Brady and Robertson). *In* Neale, J. W., ed., The Taxonomy, Morphology and Ecology of Recent Ostracoda. Edinburgh, Oliver & Boyd, 1969.)

Eggs of *Darwinula* are released into and develop within the carapace of the parent (McGregor, 1969). Juveniles produced by first and second reproductive season adults are released during the summer and early autumn, and overwinter as juveniles. These young mature and reproduce the following summer. Surviving members of this age class overwinter as adults during the second winter and reproduce again the following summer. Most adults of the second reproductive year die following release of young, and are replaced by overwintering juveniles entering their first reproductive period. Thus, a nearly complete turnover of the adult portion of the population occurs each year.

Although much of the *Darwinula* population was parasitized by ectoparasitic rotifers, this did not appear to affect the reproductive potential. Population consequences of predation by other benthic organisms and bottomfeeding fish are unclear in this and other species. Evaluations of densities of ostracod populations are few, and range from an average of 57 m^{-2} in oligotrophic Great Slave Lake in the Canadian subarctic (Tressler, 1957) to about 10,000 m^{-2} of the *Darwinula* of mesotrophic Gull Lake. Much higher population densities of $>$50,000 m^{-2} are common in more productive lakes but densities decrease, and species diversity is reduced, in hypereutrophic lakes (e.g., Delorme, 1978).

OTHER CRUSTACEA

Of the many species of the malacostracean crustaceans, four groups have distinct freshwater representatives of some interest and importance. All malacostraceans have a definite and fixed number of body segments.

Mysids. The mysids, or opossum shrimps, are morphologically similar to crayfish; they can attain a maximum length of about 3 cm. Their appendages, however, are elongated, contain abundant setae, and generally are greatly modified for active swimming. Their mode of feeding is by filtration of small zooplankton, phytoplankton, and particulate detritus via setose appendages, particularly two pairs of maxillipeds. Aggregations of food particles are carried anteriorly to the mouth by water currents set up by movements of the appendages. Gut analyses of *Mysis relicta* showed that at night adults were voracious predators, feeding on *Daphnia*, other cladocerans, and the rotifer *Kellicottia* during their vertical migration (Lasenby and Langford, 1973). *Mysis* had an assimilation efficiency of about 85 per cent when feeding on *Daphnia*, but this efficiency dropped to zero when the animal fed on debris from mosses growing in the lake.

Several studies have shown that mysids are size-selective in their predation of zooplankton (Bowers and Grossnickle, 1978; Cooper and Goldman, 1980; Murtaugh, 1981a, 1981b; Langeland, 1981). Large mysids prefer large-sized prey (*Daphnia*, *Epischura*) consistently, while small mysids select the smallest zooplankton available. Although freshwater mysids consume prey over a broader size range than most other invertebrate predators, the size-selective predation pressure can lead to marked alterations in the zooplankton community when mysids are introduced into lake ecosystems as a food source for fish (e.g., Richards, et al., 1975; Langeland, 1981). Mysids will consume algae and detritus, but are preferentially carnivores.

The freshwater mysids are distinctly stenothermal, cold-water forms, and in stratified lakes are restricted to hypolimnetic strata of less than 15°C. Reproduction occurs only during the colder periods of autumn, winter, and early spring. There are as many as 40 eggs per clutch, and the eggs are kept within the female brood pouch (marsupium).

After hatching, juveniles are retained within the female for considerable periods of time (up to 3 months); hence the name *opossum shrimps*.

A comparison of the life history, growth, and respiration of *Mysis relicta* in an arctic lake (Char Lake, Northwest Territories, Canada) and a temperate lake (Stony Lake) of southern Ontario is most instructive (Lasenby and Langford, 1972). In the arctic lake, *Mysis* takes two years to reach maturity, while those of the temperate lake mature in one year. *Mysis* from both lakes exhibited a trimodal distribution of size classes. In the arctic lake, the population consisted of 8-to-9-month old immature mysids, 20-month old adult males and females, and 32-month old females. In contrast, in the temperate lake the cohorts consisted of two-month old immatures, one-year old females, and a few two-year old females. Respiration measurements showed no differences between *Mysis* from the two lakes over its environmental temperature range. Over a two-year period from hatching to maturity, the energy used by the arctic population in growth and reproduction was approximately 68 per cent of that used by the temperate mysids in growth and reproduction over a one-year period from hatching to maturity (Table 21-6). Of much interest is the finding that the amount of energy (200 cal) used by each mysid for growth and respiration in order to become a reproducing adult was about the same for the arctic population in two years as for the temperate population in one year. Similarly, size-frequency distributions of *Mysis relicta* of the Superior, Huron, and Ontario basins of the Great Lakes showed

TABLE 21-6 Energy Budgets in Calories per Individual Female *Mysis relicta* from Two Canadian Lakes, Assuming an Assimilation Efficiency of 85 Per Cent

ENERGY BUDGET	ARCTIC CHAR LAKE	TEMPERATE STONY LAKE
Ingestion*		
First Year	62	245
Second Year	180	510
Total	242	755
Growth		
First Year	10	39
Second Year	18	39
Total	28	78
Reproduction		
First Year	0	14
Second Year	8	14
Total	8	28
Respiration		
First Year	43	156
Second Year	127	376
Total	170	532
Egestion		
First Year	9	36
Second Year	27	81
Total	36	117

Based on data of Lasenby and Langford (1972).
*Based on ingestion = growth + reproduction + respiration + egestion, when $0.85 \, I = G + R$.

differences in life cycles in the three lakes (Carpenter, et al., 1974). In the colder Lake Superior the generation time appeared to be two years, while in lakes Huron and Ontario the generations matured in 18 months. When lake productivity is relatively high, the life cycle of mysids takes from one to two years, but when productivity is low, it may last up to four years (Morgan, 1981).

Lacking gills, mysids respire by exchange through a thin carapace. The high respiratory demands for oxygen and the requirement for cold temperatures restrict their distribution to oligotrophic lakes in which the oxygen content of the hypolimnion is not reduced appreciably.

Members of the two major genera, *Mysis* and *Neomysis*, all exhibit distinct, rapid diurnal migrations. During the day, the mysids are on the sediments or in the strata immediately overlying the sediments (Beeton, 1960; Herman, 1963). *Mysis relicta* ascends when surface light intensities decrease to about 160 to 10 lux, and migrates into or below the metalimnion, in which temperatures are well below 15°C. This species rarely migrates into warmer surface waters. Descent occurred in populations of the Great Lakes when surface light intensities were increasing from 0.01 to 0.1 lux. Rates of the upward and downward migrations were similar at 0.5 to 0.8 meters per minute over distances of at least 75 meters, and represent the fastest rates known among small invertebrates (see Table 16-9). *Mysis relicta* is very well adapted to the photic environment of these relatively clear, oligotrophic lakes in which light in the range of 490 to 540 nm penetrates the deepest. The peak photosensitivity of *Mysis* is to light at 515 nm, with a threshold response of at least 0.00009 lux (Beeton, 1959). This low irradiance is well within the light intensities at 100 m in lakes of low extinction coefficients even late in the daylight period. Moonlight can inhibit ascent of *Mysis* at night (Bowers and Grossnickle, 1978). Langsby and Langford (1972) indicate that the respiratory expenditure of energy for the daily migration is very small, about 3 per cent of the total over a life span of two years.

Isopods. The isopods or sowbugs are largely marine or terrestrial; occasionally isopods become a significant part of the benthic community of lakes and streams. These small organisms (<2 cm) are flattened dorsoventrally and are well adapted to living on substrata exposed to water movements, as in streams or the littoral of lakes. Their seven pairs of walking legs are well developed. The first is modified for grasping particles. Isopods are omnivorous feeders on both plant and animal matter (Marcus, et al., 1978; Willoughby and Marcus, 1979). Isopods are generally not active predators on other benthic fauna.

Reproduction is more common in periods of warmer temperatures, but otherwise is similar to that mentioned for the mysids. The number of eggs per female is fairly high (to several hundred), and both eggs and young are retained in the brood pouch for about a month. Relatively little is known about their life cycle and population dynamics; generation time is about 8 to 12 months but quite variable. In an English lake, gravid females of *Asellus* were present in late winter–early spring and again during late summer. These females gave rise to spring and autumn cohorts (Adcock, 1979). Large *Asellus* of the spring cohort reproduced in the autumn, and the remainder in the following spring. The bimodal population growth characteristics showed that growth was faster, and productivity greater, in the overwintering spring cohort than was the case among the smaller individuals of the autumn cohort. Annual production (P) of the population was 3005 mg m^{-2} from a mean biomass (B) of 757 mg m^{-2}. The P/B ratio of 4.0 was double that found for the same species in colder Swedish lakes (Andersson, 1969). Reynoldson (1961) has shown a strong direct correlation between the distribution of *Asellus* and the concentrations of calcium and total dissolved matter. This isopod was usually absent in soft, unproductive waters (< 5 mg Ca l^{-1}; < 70 mg l^{-1} of total dissolved matter). Isopods are common among the benthic invertebrates of saline inland waters and can tolerate high salinity (see, for example, Eriksen, 1968; Ellis and Williams, 1970).

Decapods. The decapod Crustacea, of which the lobster is a familiar example, are largely marine. The few freshwater crayfish and shrimps are characterized by their approximately cylindrical body, heavily sclerotized translucent shell, and laterally compressed rostrum.

Their 19 pairs of appendages include well-developed antennae and five pairs of large walking legs, the first three of which are clawed and the first greatly enlarged with a strong pincer claw used for crushing food. While the few shrimps are more or less continuous swimmers, the crayfish move slowly over the sediments and can move backward quickly by flexing the abdomen. Crayfish are omnivorous but primarily herbivorous on algae and larger aquatic plants; occasionally, they are scavengers (Momot, 1967a; Jones and Momot, 1981).

The population dynamics and productivity of a common crayfish of temperate lakes have been studied in some detail (Momot, 1967a) and are exemplary for this group. In the crayfish population of a hard-water lake, the age structure exhibited a year-class fluctuation, and males had a higher growth rate than females. After the second year, mortality rates for females were greater than for males. Both sexes matured after the midsummer molting in the second year, then mated until early autumn. Eggs were laid the following spring. The average number of eggs carried by females in the spring, a measure of the reproductive capacity, was 58 per cent of the eggs carried in the preceding fall. Within the maximum life span of three years of this species, *Orconectes virilis*, two-year-old females produced most (92.5 per cent) of the eggs. Young remained in shallow water after hatching in spring or early summer, but adult females molted and then migrated to deeper waters (3 to 7 m), where most remained all summer. Adult males also migrated, somewhat later than females, to deeper water. Population size appeared to be regulated by natural mortality at molting and cannibalism of both males and females on each other at high population densities. Predation by fish was not an important population control mechanism in this and other lakes (Momot, 1967b; Momot and Gowing, 1977a; Gowing and Momot, 1979). A number of pollutants of lakes and streams particularly those that reduce oxygen content or alter the substratum, cause changes in the population distribution or seasonal migrations (Hobbs and Hall, 1974). The successive year-class estimates of the crayfish biomass of this hard-water lake are given in Table 21-7. Seasonal changes in biomass are small because of compensatory changes among age groups. The estimated net production of the crayfish population was 127.1 kg ha^{-1} between the summer of 1962 and that of 1963*. These values are about intermediate within the range of biomass values found among crayfish populations studied in other waters. (Momot and Gowing, 1977b; Flint and Goldman, 1977; Momot, 1978; Momot, et al., 1978).

In many lakes, crayfish can dominate the annual production of the benthic animal biomass, at times reaching 100 to 140 g m^{-2} and densities of 15 m^{-2} (Momot, et al., 1978). The rate of population growth, reproductive capacity, age at maturity, and life span are

---

*A corrected value; see footnote to Table 21-7.

TABLE 21-7   Successive Estimates of the Annual Biomass (in kg) of the Crayfish Populations of West Lost Lake, Michigan

AGE GROUP	SEX	SUMMER 1962	TOTAL ♂♂ AND ♀♀	SPRING 1963	TOTAL ♂♂ AND ♀♀	SUMMER 1963	TOTAL ♂♂ AND ♀♀
0	♂♂	8.7	—	6.2	—	4.2	—
0	♀♀	9.7	18.4	6.2	12.4	5.2	9.4
I (2nd year)	♂♂	36.2	—	23.7	—	41.9	—
I	♀♀	24.5	60.7	18.9	42.6	27.8	69.7
II (3rd year)	♂♂	49.7	—	25.9	—	39.0	—
II	♀♀	8.0	57.7	16.5	42.4	10.0	49.0
III (4th year)	♂♂	9.8	—	25.7	—	7.7	—
III	♀♀	3.7	13.5	1.6	27.3	1.8	9.5
Total, kg		150.3		124.7		137.6	
kg ha^{-1}		100.3		88.1		91.8	

Data of Momot (1967a) with corrections for computational errors that appeared in the original publication (Momot, personal communication). West Lost Lake has an area of 1.5 ha and a maximum depth of 13.4 m (mean depth, 8.5 m).

primarily density-dependent. Populations often regulate their development by control of brood stock size.

The annual production of crayfish populations ranges from 4 to 67 g m^{-2} year^{-1} (Table 21-8). Turnover rates, estimated from the ratios of production to mean biomass, are mostly about 1, but vary from 0.5 to 2.0 (Table 21-8).

Amphipods.   The amphipods or scuds, also chiefly marine, are represented in fresh waters by a few important species. With the exception of *Pontoporeia*, discussed further on, the amphipods live on the sediments. Most species are small (5 to 20 mm), with a laterally compressed, many-segmented body. At the base of their seven pairs of thoracic legs many have gills that, like the lateral gills on some species, are exposed to currents of water created by beating of pleopod appendages on the abdomen. Amphipods have high oxygen requirements and are usually restricted to waters of high dissolved oxygen concentrations (Franke, 1977). Amphipods are generally omnivorous substrate feeders that consume bacteria, algae, fungi, and animal and plant remains; only rarely are they predacious on living animals (Minckley and Cole, 1963; Marzolf, 1965a; Hargrave, 1970a; Bärlocher and Kendrick, 1975; Kostalos and Seymour, 1976; Moore, 1977).

In a detailed study of the population dynamics of the amphipod *Hyalella azteca* in a eutrophic lake of southeastern Michigan, Cooper (1965) was able to estimate production rates and biomass turnover of the natural populations (Table 21-9). The amphipod populations showed positive intrinsic rates of increase at temperatures above 10°C with maxima between 20 to 25°C. Reproduction resumed again when water temperatures reached 20°C in the spring. Estimates of mortality rates based on size and age structures of the population indicated that a size-specific mortality factor was operating on the large, reproductively mature amphipods during the summer months. It was established that the highly size-specific feeding behavior of the yellow perch accounted for most of the mortality in the adult amphipod size class. A similar

TABLE 21-8 Comparison of the Annual Production (P), Mean Biomass ($\overline{B}$), and Production/Biomass Ratios for Crayfish Populations

	SPECIES	PRODUCTION (g m^{-2} year^{-1})	MEAN BIOMASS (g m^{-2})	P/$\overline{B}$ RATIO
Dock Lake, Ontario	Orconectes virilis	5.25	2.87	1.8
Shallow Lake, Ontario	O. virilis	4.38	2.96	1.5
West Lost Lake, Michigan	O. virilis	3.87–13.4	5.17–14.3	0.7–1.4
North Twin Lake, Michigan	O. virilis	1.70–8.46	3.17–8.24	0.5–1.4
South Twin Lake, Michigan	O. virilis	11.0–11.2	6.44–12.7	1.2–1.5
Red Cedar River, Michigan	O. propinquus	41.5	44.0	0.9
Lake Berziukas, Lithuanian SSR	Astacus astacus	4.2	6.0	0.7
Berry Creek, Oregon	Pacifasticus leniusculus	13.7	14.3	0.9
Farm Pond, Australia	Cherax albidus	67.0	34.0	2.0*

Modified from Momot (1978), after numerous sources.
*Ratio given as P/B rather than P/$\overline{B}$.

TABLE 21-9 Estimated Mean Production Rates and Numerical Turnover of the Amphipod *Hyalella azteca* Population from Four Stations of Sugarloaf Lake, Southeastern Michigan, May to October, 1962

AGE GROUP	NUMBER OF ANIMALS m^{-2}	MEAN DRY WEIGHT (μg)	NUMERICAL YIELD m^{-2} DAY^{-1}	MEAN POPULATION DENSITY (mg m^{-2})	PER CENT NUMERICAL TURNOVER DAY^{-1}
11–13 antennal segments	1259.4	15.3	39.8	19.3	3.2
14–16 antennal segments	1383.0	35.6	24.8	49.2	1.8
New adults	2044.6	133.3	71.8	272.5	3.5
Old adults	183.8	200.0	4.8	36.8	2.6
Total	4870.8	—	141.2	—	2.9

Compiled from data of Cooper (1965).

situation was found in the benthic invertebrate populations of shallow experimental ponds (Hall, et al., 1970), in which the fish (*Lepomis*) preyed upon larger particles, selectively removing the larger adult stages of *Hyalella* and the terminal aquatic stages of many benthic insects. There is some evidence that heavy predation pressure by fish can lead to genetic selection for smaller-sized adult amphipods (Strong, 1972).

In spite of the relatively low biomass levels of the *Hyalella* populations, rapid growth rates and short generation times (33 days to maturation at 25°C; 98 days at 15°C) of the animals result in a mean production rate of about 0.13 kg ha^{-1} day^{-1}, which is significant in comparison to other components of the benthic fauna. The observed turnover rate of about 3 per cent day^{-1} is close to the maximum of 4 per cent day^{-1}, estimated from experimental studies, that the amphipod population could maintain over the course of the summer at temperatures above 20°C with any degree of population stability. Similar production rates were found for amphipods of temperate streams (Welton, 1979; Waters and Hokenstrom, 1980).

The extensive studies of *Hyalella azteca* in shallow (z_m = 5 m) Marion Lake, British Columbia, by Hargrave (1970a, 1970b, 1970c) demonstrated that growth, density, and body size of this deposit-feeder depended on the quantity of epipelic algae and sediment microflora. The highest concentrations of sedimentary chlorophyll and microflora, as well as the lowest concentrations of nondigestible ligninlike organic material, occurred in the upper 2 cm of sediment, which was also the limit of the vertical distribution of the *Hyalella* in the sediments. Increased growth of *Hyalella* during June was independent of temperature in Marion Lake, but was correlated closely with increased rates of epipelic primary productivity. Egg production began in May as growth rates increased, and the maximum density of *Hyalella* was reached in August.

Bacteria and algae, except for blue-green algae, were assimilated with an efficiency of about 50 per cent; cellulose and ligninlike compounds were not assimilated at all by *Hyalella* (Hargrave, 1970a). Overall, sediment and associated microflora were assimilated with a 6 to 15 per cent efficiency. In the estimated energy budget for adult *Hyalella* at 15°C in Marion Lake, 49 per cent of calories assimilated were respired, 36 per cent were lost as soluble excretory products, and 15 per cent were accumulated as growth, egg production, and molts (Table 21-10). The fecal pellets produced were colonized rapidly by heterotrophic microorganisms, and the dissolved organic compounds excreted by *Hyalella* significantly increased the rate of recolonization of the fecal material (Hargrave, 1970c). Epipelic algal production was stimulated somewhat by the feeding activities of the amphipod at natural densities, even though less than 10 per cent of the daily microfloral production was required to supply the energy necessary for observed rates of amphipod growth, respiration, and egg production. The mechanism of stimulation of algal productivity on the sediments is unclear. It may emanate directly from inorganic and organic nutrients from the amphipod excretions, indirectly from increased metabolism of the heterotrophic microorganisms of the sediments, or both. Alternatively, the mechanical disruption of the oxidized microzone of the sediment–water interface may be instrumental in encouraging release of certain reduced, and more soluble, nutrients into the proximity of the epipelic algae. While these compounds could not migrate far into overlying aerobic water before being oxidized and decreasing in solubility, the algae directly at the interface could benefit from such disturbance.

Fungi colonizing autumn-shed tree leaves are the preferred food of many woodland-stream-dwelling amphipods (Bärlocher and Kendrick, 1975; Kostalos and Seymour, 1976). Less than 20 percent of the ingested leaf material was assimilated, whereas over 75 percent of the fungi of decomposing leaves was assimilated in some cases.

Pontoporeia affinis is exceptional among the amphipods in that it migrates exten-

TABLE 21-10   The Estimated Energy Budget of Adult Amphipod *Hyalella azteca* of 700 μg Dry Weight at 15°C Feeding on the Surface Sediment of Marion Lake, British Columbia

ENERGY	CALORIES AMPHIPOD^{-1} HOUR^{-1}
Ingestion	0.0525
Production (growth, molts, and egg production)	0.0012
Respiration	0.0039
Egestion	0.0430
Excretion (soluble organic matter)	0.0029

From data of Hargrave (1971). The mean caloric value of adult *Hyalella* was 3850 cal g^{-1} ash-free dry weight; therefore a 700-μg amphipod contains 2.7 calories.

sively into the pelagic zone at night. The behavior and habitat requirements of *Pontoporeia* are similar to those of *Mysis* discussed earlier. *Pontoporeia* is restricted to cold, relatively oligotrophic waters and migrates only into the upper hypolimnion and metalimnion, in which temperatures are usually less than 15°C (Marzolf, 1965b; Wells, 1968). Only a portion of the predominantly benthic population migrates into the water. During daytime, *Pontoporeia* is largely on and in the sediments, or in the immediately overlying water.

Pontoporeia is a burrowing amphipod and a deposit feeder on the surficial sediments. This amphipod is the predominant macrobenthic invertebrate of the upper Great Lakes. In Lake Michigan, *Pontoporeia* showed no significant correlation with depth, particle size, or organic matter content of the sediments (Marzolf, 1965a). Distribution of *Pontoporeia* was, however, positively correlated with the number of bacteria in the sediments. In experiments with different-sized sediment particles, *Pontoporeia* selected sediments with a particle size smaller than 0.05 mm. Such sediments contained more bacteria and organic matter than coarser sediments.

Because bacteria of the sediments are the predominant food source of *Pontoporeia*, and there is no evidence for planktonic feeding, the adaptive significance of the nocturnal vertical migration of a small portion of the population is unclear. Marzolf (1965b) suggest that the migrations have adaptive value in maintaining genetic continuity across otherwise isolated benthic populations. Migration encourages dispersal and, possibly, copulation with individuals from populations from other areas within the community.

MOLLUSKS

The freshwater Mollusca are separable into two distinct groups, the univalve snails (Gastropoda) and the bivalve clams and mussels (Pelecypoda). Among the univalve mollusks, the shells of snails generally are spirally coiled, while those of the few freshwater limpets are conically shaped. The mollusk head is distinct, with a pair of contractile tentacles and a ventral mouth with a chitinous, sclerotized jaw and a chitinous internal radula containing numerous transverse teeth. The radula is extended from the mouth and moves back and forth rapidly, scraping and macerating food particles. Respiration in the snails occurs by gills (Prosobranchia) in many aquatic forms, and by pulmonary cavities or "lungs" in the pulmonate snails. The latter group can stay submersed for long periods of time. Some of the most common pulmonate snails, however, are able to stay submersed for a very long time or for their entire life cycle without filling their pulmonary cavities with air. Cutaneous respiration through the body membranes is common to all freshwater snails (Ghiretti, 1966). Even though the lung is retained in pulmonate freshwater snails, it is of secondary function and is used only when the oxygen concentration drops to low values. Then, the snails come to the surface to breathe air while moving or floating. Other pulmonates have water-filled lungs through which gaseous exhange occurs, much the same as it does with a gill. Locomotion in snails is by muscular movements of the ventral surface of the body, the "foot."

The body of clams or mussels is enclosed in two symmetrical, opposing shell valves and lacks a head, tentacles, eyes, jaws, and a radula. The body is enclosed by membranous tissue, the mantle, which secretes the shell valves. At the posterior end of the clam, the mantle has two openings, the siphons. The lower ciliated siphon draws water into the body cavity. Thus, flow aerates the gills and carries in food particles. Food particles are removed from the water by filtration through the gills and cilia. Water then exits from the upper siphon. The muscular foot can be extended from the valves in front of the clam, implanted in the sediment, and then contracted to draw the animal forward. The food of the Pelecypoda clams consists primarily of particulate detritus and microzooplankters of the sediments (see review of Fuller, 1974). There is no evidence to support the view that clams utilize living phytoplankton; living and active diatoms apparently pass through the alimentary canal unaffected.

The life cycle of freshwater snails in temperate regions, particularly the smaller species, tends to be annual (Harman, 1974). One reproductive period may occur in the spring or fall, or two or more reproductive periods may occur throughout the summer, during which the original cohort is replaced or supplemented. Some species overwinter as juveniles or adults and reach maturity the following spring or summer. Larger species tend to have life cycles that extend over two, three, or even four years. A few examples of the growth and population characteristics of some freshwater snails and clams illustrate these variations.

In the common pond snail *Physa gyrina* of the midtemperate region, copulation occurred in April (DeWitt, 1954). Egg masses adhere to substrata and the number of eggs laid, about 100, varies with the size of the snail. After about a week of embryological development at 20 to 30°C, the young snails hatch and grow very rapidly, attaining most of their adult size of approximately 12 mm in about 8 weeks. Beyond this period, growth is very slow for the remainder of the 12- to 13-month life span. Growth in *Physa* is not determinate, but continues throughout the life of the individual.

Analyses of the population-size classes of the pulmonate snail *Lymnaea* that lives along the shore line in littoral regions of fresh waters showed a similar pattern, in which growth of overwintering individuals was linear but slow, and then very rapid in the spring and early summer (McCraw, 1961, 1970). Little growth occurred before mid-April or after early July, when most of the production is shunted into reproductive development. Small individuals overwinter and almost all of the spring and summer population is derived from snails ovipositing in later summer or autumn of the previous year. Although changes in population densities of *Lymnaea stagnalis* did not affect rates of food consumption, egg production under experimental conditions was significantly lower at high population densities (Mooij-Vogelaar, et al., 1973). Eisenberg (1966) found in both natural and experimental populations of *Lymnaea elodes* that both the quantity and quality of food limited the population densities. Adult fecundity increased some 25-fold and numbers of young increased 4- to 9-fold when abundant food of high quality was available. Mortality among natural populations of young snails, caused especially by dipteran larvae, was very high (93 to 98 per cent), but food limitations on fecundity were considered to be the primary mechanism affecting the final population structure.

The efficiency of assimilation (expressed as the percentage of ingested organic carbon that is assimilated) of the stream snail *Potamopyrgus* varied from 3.7 to 9.0 per cent, depending on the type of sediment used as food (Heywood and Edwards, 1962). The average was 4 per cent for adult snails, in which about 7 µg organic carbon was assimilated per snail per day. Diatoms, a primary food of this snail, are crushed between sand grains, which are ingested with the food, and the strongly cuticularized stomach walls (Frenzel, 1979). In contrast, the tropical snail *Pila*, feeding on the macrophyte *Ceratophyllum*, had much higher assimilation rates, possibly* greater

---

*The methods employed in measurements of changes in plant matter are questionable, and the reported values of loss most likely are overestimates.

than 50 per cent (Vivekanandan, et al., 1974). Utilization efficiency of the plant food decreased as the supply of food increased. Assimilation efficiencies of diatom and green algae ingested by the gastropod *Ancylus* were between 43 and 65 per cent, and increased slightly with increased temperatures (Streit, 1975a, 1975b). Ingested blue-green algae *(Anabaena)* were not assimilated by *Ancylus*; similarly, the green alga *Chlorella* was not assimilated by the tropical snail *Bulinus* (van Aardt and Wolmarans, 1981).

The filtration rates and ingestion rates of a filter-feeding mussel *(Dreissena)* remained constant when the mussel was fed diatoms *(Nitzschia)* until high food concentrations were reached, above which rates declined (Walz, 1978). Assimilation efficiency of food carbon was about 40 per cent when the mussel was fed this alga.

The sperm of male clams is shed into the water and drawn into the female mantle cavity by the normal water current of the siphon system. Fertilized ova are incubated in portions of the gills, but mortality is very high from bacteria and protozoans. In mussels of the family Unionidae, the mature larvae hatch after periods of a few months to a year, to temporarily free-living forms, called glochidia, which have two chitinous valves bearing ventral hooks in some species. The glochidia are discharged in large numbers, fall to the sediments, and die within a few days if they do not come into contact with a suitable fish host. If they come into contact with a fish, the glochidia attach to surface membranes or to the gills; some species of clams are specific to a particular species of fish. During this parasitic stage, which lasts from about 10 to 30 days, internal development occurs although there is little change in size while the glochidia are encysted on the host fish. Juveniles then break encystment and fall to the sediments. Development ensues if substrate and food conditions are within tolerable limits. Large clams live 15 years or longer, while the smaller "fingernail clams" (Sphaeriidae) have a longevity of about a year or slightly longer. Mortality is exceedingly high in the egg, larval-glochidial, and juvenile stages. The adult populations and their distribution can be modified greatly by fish predation, certain birds, and a few mammals such as the muskrat (cf. Fuller, 1974).

Study of the distribution of five species of sphaeriid clams of the common genus *Pisidium* showed that four coexisted among the more heterogeneous substrata of Lunzer Untersee, Austria (Hadl, 1972). The profundal sediments were dominated by a single species. The littoral species were exposed to large seasonal fluctuations in temperature and were gravid in spring, released young in the summer during which rapid growth occurred. The new generation overwintered to reproduce the following spring. In contrast the deep-water species, which were exposed to relatively constant temperatures throughout the year, exhibited no definite reproductive cycle. Embryos were found throughout the year in mature females. The larger clams, for example the *Anodonta*, migrate seasonally on the sediments to different water depths (Burla, 1971). Species of this genus were found to move up to the upper littoral in spring and early summer where they remained until late in the year. In autumn and early winter most returned to deeper water. These movements were not correlated with temperature but rather with photoperiod, which may be related more to food availability.

In a very detailed study of the benthic fauna of a eutrophic, lowland lake of southern Norway, Ökland (1964) studied the mollusk fauna in particular (13 species of gastropod snails and 1 dominant bivalve clam, *Anodonta piscinalis*). The average quantitative data of Table 21–11 are included to point out a typical distribution with depth of the sediments. The greatest quantity of gastropods occurred in the zone of submersed plants at a depth of 1.5 m. The bivalve *Anodonta* was restricted to the littoral, and all mollusks were intolerant of the reduced oxygen concentrations of the deep strata during both summer and winter periods of stratification. The *Anodonta* constituted a trivial proportion of the numerical population density of the total benthic fauna (Table 21–11). Because of their large size and heavy shell, however, their mass can completely bias the quantitative biomass comparisons. Such data, while instructive in a comparative way, are meaningless for productivity estimates

TABLE 21-11 Average Numerical Density and Wet Weight Biomass of Mollusks of Eutrophic Lake Borrevann, Norway, in Relation to Other Benthic Macroinvertebrates at Different Depths

ZONE AND DEPTH	ANODONTA PISCINALIS		OTHER MOLLUSKS		OTHER BENTHIC FAUNA	
	Ind. m^{-2}(%)	g m^{-2}(%)	Ind. m^{-2}(%)	g m^{-2}(%)	Ind. m^{-2}(%)	g m^{-2}(%)
Littoral						
0.2 m	− (−)	− (−)	121 (5)	3.3 (18)	2422 (95)	14.6 (82)
1.5 m	11 (−)	186 (80)	2041 (57)	28.7 (12)	2705 (43)	17.0 (7)
2 m	20 (−)	361 (94)	1182 (26)	6.6 (2)	3398 (74)	14.6 (4)
3 m	31 (2)	1185 (99)	93 (5)	0.3 (−)	1648 (93)	6.1 (1)
Sublittoral						
5 m	24 (2)	500 (99)	12 (1)	0.2 (−)	1242 (97)	5.4 (1)
6 m	11 (1)	73 (96)	5 (1)	− (−)	785 (98)	2.8 (4)
7 m	− (−)	− (−)	− (−)	− (−)	810 (100)	3.3 (100)
Profundal						
10 m	− (−)	− (−)	4 (−)	− (−)	1143 (100)	7.0 (100)
15 m	− (−)	− (−)	− (−)	− (−)	1598 (100)	5.9 (100)

Extracted from data of Ökland (1964).

unless the rates of turnover of each of the components constituting the categories are known. The much smaller and more numerous nonmolluscan animals have a small instantaneous biomass, but in general a higher turnover rate (shorter turnover time). The gastropod snails reached maximum numbers and biomass in the fall (October). The sphaeriid clams attained maximum biomass per area in early summer (June) prior to maximum densities in later summer (mid-August).

Production rates and turnover values of freshwater mollusks have not been determined frequently. For example, detailed population and growth analyses of the snail *Lymnaea catascopium* of a lake in Quebec showed considerable variability in the littoral areas (Pinel-Alloul, 1978). Mean biomass of this species varied between 3.4 and 7.4 g m^{-2}; production rates ranged between 6.6 and 17.8 g m^{-2} year^{-1}. Turnover ratios (P/$\overline{\text{B}}$), however, were relatively constant at approximately 2 among the different populations, and were found to be similar to those of other snail populations (e.g., Browne, 1978; Frenzel, 1980). Of two annual cohorts of a snail (*Potamopyrgus*) in a southcentral European lake, mean annual biomass was 8.8 g fresh weight m^{-2}, which resulted in production to mean biomass ratios of 2.5 and 3.7 for the first and second cohorts, respectively.

Aquatic Insects

The insects are very abundant and diverse; most are terrestrial. Of those that are aquatic, nearly all are freshwater. Some orders are entirely aquatic; others inhabit fresh waters only during certain life stages (cf. Hutchinson, 1981). The characteristics of the insects are well known, and no attempt will be made here to summarize the group differences except to point out salient features of the life cycles that are important to feeding and reproduction as related to benthic productivity and distribution.

The chitinous exoskeleton of arthropods necessitates molting for continued growth. The true bugs (Hemiptera), dragonflies (Odonata), stoneflies (Plecoptera), and mayflies (Ephemeroptera) are orders of winged insects that undergo gradual metamorphosis. In these orders, the young are referred to as *nymphs*: their wings develop as external pads, and the organisms increase in size with each molt. The other orders, including the flies (Diptera), caddisflies (Trichoptera), the alderflies and dobsonflies (Megaloptera), beetles (Coleoptera), a few species of moths (Lepidoptera), and spongeflies (Neuroptera) undergo complete metamorphosis. In this development, the wing pads develop internally in early *larval* instars and then evert to the outside in the preadult instar *pupal* stage. In nearly all of the important aquatic insects only the immature stages live in the water; the adults and, in some groups, the pupae are terrestrial (cf. Hutchinson, 1981). Only some of the beetles and hemipterans have adapted to the point at which both the adults and immature stages can live in the water.

A brief summary of the major life history characteristics follows. In generalizing interrelationships among so diverse a group as the insects, less attention is given to taxonomic characteristics, and more attention is given to functional relations of feeding, reproduction, and productivity within aquatic systems. Because much of Hynes' (1970) compendium on running waters is devoted to the invertebrates of stream and river systems, remarks here are directed more to lakes and reservoirs.

Dragonflies and Damselflies.   After fertilization, the eggs of dragonflies and damselflies (Odonata) are deposited into the water, onto substrata in or near the water, or into submersed parts of macrovegetation. Egg development, as in most invertebrates, is temperature-dependent; nymphs hatch after about two to five weeks of egg development. Growth in the nymphal stages is quite variable, especially in relation to temperature and food supply. Within a range of about five weeks to five years, 10 to about 20 instar moltings occur. The mature nymphs often leave the water on some emergent substratum as aerial adults. The odonate nymphs are almost entirely littoral in habitat, living among macrovegetation and littoral sediments, and burrowing into surficial sediments. The nymphs have fairly high respiratory demands and oxygen requirements.

Mayflies.   The mayflies (Ephemeroptera) are almost totally aquatic (Edmunds, et al., 1976). The adult longevity is very brief (3 to 4 days), during which no feeding occurs. After mating in flight, fertilized eggs are laid in the water or on submersed objects; the time to hatching varies from a few days to many weeks. Growth is relatively rapid, from less than 1 mm in length at hatching to about 2 cm in length in nymphal stages of many species. From 10 to over 40 instars of molting occur over an average life cycle of about one year. A few species live as nymphs for two years or longer. All mayflies are restricted to waters of relatively high oxygen content but are widely distributed even in waters with moderate organic loading (Roback, 1974). Some species can regulate respiratory movements of gills in response to changing oxygen concentrations (Eriksen, 1963a). Mayfly nymphs are characteristic of shallow streams and littoral areas of lakes, and are distributed widely. However, many species are restricted to specific substrata of macrophytes, sediments of waveswept or moving stream areas, or sediments of specific sized particles (Eriksen, 1963b).

Stoneflies.   The stoneflies (Plecoptera) are terrestrial as adults, but in the nymphal stages they are strictly aquatic, and most are restricted to flowing waters of relatively high oxygen concentrations. Fertile eggs, laid over or in the water, require two to three weeks for hatching in many species, and several months among some larger forms. The nymphal instars, from 10 to over 30 moltings, occur in one to three years.

Hemiptera.   The Hemiptera or true bugs are essentially terrestrial; a few are semiaquatic and very few species have adapted to submersed conditions. Most hemipterans overwinter as adults in moist sediments or vegetation. Eggs are laid on semiaquatic substrata or in aquatic macrophytes and develop rapidly in one to four weeks. Nymphs also develop rapidly in one to two months, commonly with five instar stages, and generally have a one-year life cycle. Few hemipterans are truly aquatic, with cutaneous respiration; most are dependent on atmospheric oxygen for respiration. They inhabit substrata either near or above the water, possess adaptations for obtaining air while underwater, or carry entrapped pockets of air with them on underwater excursions. Several large families of these hemipterans are adapted to stride over the water surface without disrupting the surface tension (cf. page 13).

Diptera. The most important group of the aquatic insects with complete morphogenesis includes the flies, midges, and mosquitoes (Diptera). The dipterans are commonly dominant components of benthic invertebrate communities in many standing and running waters, and the chironomid larvae are ubiquitous, as will be seen in some of the examples that follow. Much variability is found in the morphology, reproductive biology, and respiration of dipteran larvae. Adults are essentially never aquatic, but most of their lives are spent as immature forms in fresh waters. The larval stage, with about three or four growth molts, extends from several weeks to at least two years in different species, and many overwinter in the larval stage. Most species have one generation per year, some have two per year, and a few of the species studied have a two-year life cycle (to seven years in arctic ponds; Butler, 1982). Most larvae respire cutaneously or by means of "blood gills," which also function in ionic regulation. Fewer larvae rely on air, and possess various adaptive structures to obtain air above the water or from the internal lacunae of aquatic angiosperms. Some chironomid larvae possess a type of hemoglobin in their blood that functions efficiently at low oxygen concentrations.

Caddisflies and Moths. The caddisflies or Trichoptera generally have a one-year cycle (Wiggins, 1977). Adults emerge in the warmer periods of the year, often from overlapping cohorts, from May to October. Eggs are dropped or placed on vegetation, or laid under water on submersed substrata and develop in about one to three weeks. Many caddisfly larvae build beautifully intricate cases from substrate particles of sand, small stones, leaf fragments, and the like, and are highly specific to types of substratum (cf. Cummins, 1964; Cummins and Lauff, 1969; Mackay and Wiggins, 1979; Wallace and Merritt, 1980). Some construct a net that traps microorganisms and detrital particles in flowing water. After five to seven larval instars, pupation occurs underwater within a cocoon. The pupal stage generally lasts only a few weeks, after which the pupa leaves the cocoon, moves to an aerial substratum, and emerges as an adult.

Few species of the moths (Lepidoptera) have aquatic larval stages; most aquatic moth species belong to the family *Pyralididae*. Many characteristics of the life history of the "aquatic caterpillars" are similar to those of the closely related caddisflies.

Spongeflies, Alderflies, Dobsonflies. The spongeflies (Neuroptera) and alderflies and dobsonflies (Megaloptera) are closely related and often are grouped together under the Neuroptera. Both groups are primarily terrestrial for all of their life cycles. Of the few freshwater species, however, some megalopteran larvae are so large that they may represent a significant portion of the biomass of some benthic communities. Eggs generally are laid on aerial substrates overhanging the water. After a rapid development period (1 to 2 weeks), the larvae drop into the water and feed actively. The aquatic neuropterans (Sisyridae) are restricted to and parasitic on freshwater sponges and can have multiple generations annually. The megalopterans have numerous molting instars and most of their life cycle of one to three years is spent in the larval stage. The pupal stage occurs in soil out of water and lasts up to one month. The megalopteran larvae are vicious predators on other insects, and some can reach a size of several centimeters.

Aquatic Beetles. Such great diversity exists among the aquatic beetles (Coleoptera) that even broad generalizations are difficult to make. Generally, the beetles adapted for larval and adult existence in water occur among the more phylogenetically primitive coleopteran groups. The life cycles of many (Gyrinidae, Haliplidae, Dytiscidae, and Hydrophilidae) are annual, with three to eight larval instars. Eggs laid on or in macrophytes or sediments hatch in one to three weeks. Larval development is variable (1 to 8 months), and pupation takes place on some nearby terrestrial or aerial substratum. Overwintering generally occurs in the adult stage. Nearly all adult beetles are dependent on atmospheric oxygen and carry an air supply with them on the ventral side of their bodies, or are variously adapted to obtain air from the surface or from aquatic angiosperms. Some larvae have gills to obtain oxygen directly from the water. Although beetles are generally omnivorous, their feeding habits often change markedly between the larval and adult stages.

TROPHIC MECHANISMS AND FOOD TYPES

The great diversity of food ingested by the aquatic insects and their various feeding mechanisms can be organized within a functional framework combining feeding mechanisms with the particle size of the dominant food utilized (Table 21–12). Distinctions can be made between herbivory, defined as the ingestion of living vascular plant tissue or algae, carnivory, the ingestion of living animal tissue, and detritivory, the intake of nonliving particulate organic matter and the microorganisms associated with it (Cummins, 1973). The organic matter in any of these categories can be ingested either in particulate form (by swallowing, biting, or chewing) or in dissolved form (by piercing or sucking). While association of aquatic insects with a particular substratum usually is related directly to feeding on that substratum or on the associated microflora, this is not always the case. For example, the insect may use aquatic angiosperms as a source of air or as a site for reproduction or protection from predation. However, the relationship between insects and certain substrates, e.g., angiosperms, is often highly specialized and even species-specific (cf. the compilation of Gaevskaia, 1966). Other insect groups, especially large ones such as the dipteran Chironomidae, show great diversity in feeding mechanisms and substrata ingested (Table 21–12). Most aquatic insects tend to be nonselective in their food habits.

Classification of the functional feeding groups among invertebrates, as has been done in Table 21–12 for aquatic insects, describes the morphological and behavioral capacities of these organisms to consume available food resources (see particularly the reviews of Cummins, 1973; Anderson, et al., 1978; Anderson and Sedell, 1979; Wallace and Merritt, 1980). Facultative-feeding invertebrates ingest a wider array of substrates (have a wider niche breadth) and inhabit a greater diversity of stream and lake habitats than more specialized feeders (Cummins and Klug, 1979).

TABLE 21-12   A General Categorization of the Trophic Mechanisms and Food Types of Aquatic Insects

GENERAL CATEGORY BASED ON FEEDING MECHANISM	GENERAL PARTICLE SIZE RANGE OF FOOD (μ m)	SUBDIVISION BASED ON FEEDING MECHANISMS	SUBDIVISION BASED ON DOMINANT FOOD	AQUATIC INSECT TAXA CONTAINING PREDOMINANT EXAMPLES
Shredders	$>10^3$	Chewers and miners	Herbivores: living vascular plant tissue	Trichoptera (Phryganeidae, Leptoceridae) Lepidoptera Coleoptera (Chrysomelidae) Diptera (Chironomidae, Ephydridae)
		Chewers and miners	Detritivores (large particle detritivores): decomposing vascular plant tissue	Plecoptera (Filipalpia) Trichoptera (Limnephilidae, Lepidostomatidae) Diptera (Tipulidae, Chironomidae)
Collectors	$<10^3$	Filter or suspension feeders (see also Wallace and Merritt, 1980)	Herbivore-detritivores: living algal cells, decomposing particulate organic matter	Ephemeroptera (Siphlonuridae) Trichoptera (Philopotamidae, Psychomyiidae, Hydropsychidae, Brachycentridae) Lepidoptera Diptera (Simuliidae, Chironomidae, Culicidae)
		Sediment or deposit (surface) feeders	Detritivores (fine particle detritivores): decomposing organic particulate matter	Ephemeroptera (Caenidae, Ephemeridae, Leptophlebiidae, Baetidae, Ephemerellidae, Heptageniidae) Hemiptera (Gerridae) Coleoptera (Hydrophilidae) Diptera (Chironomidae, Ceratopogonidae)
Scrapers	$<10^3$	Mineral scrapers	Herbivores: algae and associated microflora attached to living and nonliving substrates	Ephemeroptera (Heptageniidae, Baetidae, Ephemerellidae) Trichoptera (Glossosomatidae, Helicopsychidae, Molannidae, Odontoceridae, Goreridae) Lepidoptera Coleoptera (Elmidae, Psephenidae) Diptera (Chironomidae, Tabanidae)
		Organic scrapers	Herbivores: algae and associated attached microflora	Ephemeroptera (Caenidae, Leptophlebiidae, Heptageniidae, Baetidae) Hemiptera (Corixidae) Trichoptera (Leptoceridae) Diptera (Chironomidae)

TABLE 21-12   A General Categorization of the Trophic Mechanisms and Food Types of Aquatic Insects (Continued)

GENERAL CATEGORY BASED ON FEEDING MECHANISM	GENERAL PARTICLE SIZE RANGE OF FOOD (μ m)	SUBDIVISION BASED ON FEEDING MECHANISMS	SUBDIVISION BASED ON DOMINANT FOOD	AQUATIC INSECT TAXA CONTAINING PREDOMINANT EXAMPLES
		Swallowers	Carnivores: whole animals (or parts)	Odonata Plecoptera (Setipalpia) Megaloptera Trichoptera (Rhyacophilidae, Polycentropidae, Hydropsychidae) Coleoptera (Dytiscidae, Gyrinnidae) Diptera (Chironomidae)
Predators	$>10^3$	Piercers	Carnivores: cell and tissue fluids	Hemiptera (Belastomatidae, Nepidae, Notonectidae, Naucoridae) Diptera (Rhagionidae)

Modified slightly from Cummins, K. W.: Trophic relations of aquatic insects. Ann. Rev. Entomol., 18: 183–206, 1973.

The ingestion of food bears little relation to rates of assimilation. Assimilation rates may not differ with variations in the caloric content of food material. For example, fortuitous ingestion of inorganic sediment by fine-particle detritivores obviously has little nutritive value, but coatings of organic compounds and attached microflora may be of considerable value or small inorganic particles may aid in digestion by means of mechanical disruption of cells within the gut. Little is understood about the digestive capabilities or efficiencies of food utilization by aquatic insects (see reviews of Cummins, 1973; Anderson, et al., 1978; Cummins and Klug, 1979). Much of the material ingested by litter and deposit-feeding invertebrates probably is indigestible, and must be degraded by the enzymatic activity of symbiotic microflora and fauna of the gut. A number of carbohydrases have been found in the digestive tracts of several species of Trichoptera, the carnivorous amphipod *Gammarus*, and two species of sediment-feeding dipteran *Chironomus* (Bjarnov, 1972). All species degraded mono- and disaccharides and species differences were found only in polysaccharide degradation. Cellulose was found to be degraded only by the amphipod, and chitinase activity was found only in carnivorous species and in one detritivorous caddisfly. Species possessing digestive cellulase can live in a wide range of habitats because they can utilize a greater array of detritus and algal types than can organisms lacking this physiological adaptation.

Macroinvertebrates utilize various feeding modes to compensate for differing dietary sufficiency of ingested substrates (cf. reviews of Anderson and Cummins, 1979; Anderson and Sedell, 1979; Cummins and Klug, 1979). Shredders and some collectors feed preferentially on particulate organic detritus colonized by microorganisms, and utilize the associated microorganisms and partially hydrolyzed substrates. Collectors, scrapers, and facultative shredders may increase consumption of low-quality food (e.g., resistant detrital plant parts) to compensate for the lower nutritional benefit.

A number of immature freshwater insects that consume relatively resistant plant materials and particulate organic detritus may harbor microbial symbionts in their alimentary tracts (Cummins and Klug, 1979). These animals are analogous to ruminant mammals, which ingest food of high C:N ratios (>17) and absorb through their gutwall materials with lower C:N ratios (<17) produced by intestinal microbiota that degrade polymers and supply essential amino acids.

Most particulate organic matter that enters lakes and streams from allochthonous and autochthonous sources is not eaten by animals, and is eventually degraded to CO_2 by microflora attached to the particles. The feeding and shredding activities of benthic invertebrates, however, can accelerate reduction of particle size and increase surface areas exposed to the microbiota, and thereby stimulate microbial degradation rates.

BIOMASS AND FEEDING RATE

The biomass of aquatic insects is relatively constant when they are supplied with consistent and abundant food supplies of similar caloric and protein content throughout the year (Cummins, 1973; Anderson and Cummins, 1979). The rate of biomass turnover in insects is controlled primarily by temperature, and is mediated mainly by the positive relation between temperature, feeding rate, and respiration. The ratios of feeding and respiration to growth are relatively constant. In general, streams exhibit less seasonal fluctuation in temperature than do the littoral regions of lakes. Much of the feeding and growth of aquatic insects living in streams occurs during the fall and winter, even in the temperate zones. Assimilation efficiencies seem to be fairly constant over the broad range of foods consumed by insects. There is a suggestion (Welch, 1968), based on limited and variable data, that this efficiency* is somewhat higher in carnivores than in herbivores and detritivores. However, the net or tissue growth efficiencies† of carnivores tend to be lower than those of herbivores and detritivores.

Littoral and Profundal Benthic Communities

Some of the important relationships among the benthic macroinvertebrates can be seen by a comparison of the littoral fauna with that of the profundal. As lakes become more productive and the hypolimnetic water strata undergo periods of oxygen reduction and increases in the metabolic products of microbial decomposition, the number of animals adapted to these conditions decreases precipitously. Those adapted to these conditions are exposed to relatively homogeneous conditions of temperature and substratum throughout the year. Additionally, both competitive pressures for available resources and predation are reduced considerably. Therefore, a commonly observed community structure consists of a rich fauna with high oxygen demands in the littoral zone above the metalimnion. Substratum heterogeneity is much greater in the littoral, and species diversity and competitive interactions are more complex. By contrast, the

---

*Assimilation efficiency is defined here as assimilation/ingestion, or growth plus respiration/ingestion (see Chapter 8).

†Defined as growth/assimilation.

profundal zone is more homogeneous and becomes more so as lakes become more productive and species diversity decreases correspondingly (cf. Jónasson, 1969; 1978).

The production of phytoplankton and macrophytes is a major determinant of the subsequent conditions of food supply, oxygen, ionic composition, pH, and numerous other factors that delimit the range and competitive abilities of benthic fauna (cf. general reviews of Macan, 1961; Brinkhurst, 1974; Jónasson, 1978). The effects are seen not only in the qualitative distribution, but also in the quantitative aspects. Two maxima of abundance are typically observed with depth, one in the littoral zone and another in the profundal. Examples of this distribution are very numerous. The data presented in Table 21-13 show these two maxima both in numbers and biomass, as well as the reduction in species diversity in the progression from the littoral to the profundal zones. The same type of distribution is illustrated by the example from Lake Washington (Fig. 21-3). If lakes become extremely enriched, to the point that the population densities of the phytoplankton and epiphytes become so great that they shade out the submersed macrovegetation, then the habitat diversity of the littoral decreases. Correspondingly, the diversity and quantity of littoral macroinvertebrates decrease, and often a peak in animal biomass can be observed only in the lower profundal zone. With further increases in eutrophication, and lengthening of the period of hypolimnetic oxygen reduction and associated chemical changes, the rates of respiratory activity of the adapted benthic animals are reduced. Rates of growth and survival also decline, and some insect larvae increase their life cycles from one to two years. As hypolimnetic strata of hypereutrophic waters undergo extreme eutrophicational or pollutional loading of organic matter, essentially all of the aquatic insects may be eliminated. Practically the only group of benthic fauna adapted to conditions of extremely high organic loading is the oligochaete annelids. Variations in the vertical distribution of biomass of benthic invertebrates are discussed in some detail by Deevey (1941), Brinkhurst (1974), and Jónasson and Lindegaard (1979).

The changes in species composition, especially among the dipteran larvae, and the quantity of benthic fauna in the profundal zone of lakes as they become more eutrophic have been subjects of intensive study since shortly after the turn of the century. Prompted by the perceptive work of Thienemann and other German workers, especially Lenz and Lundbeck, an array of typological schemes of lake classifications was

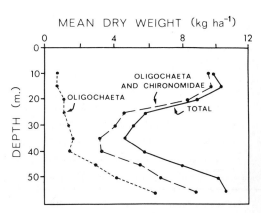

Figure 21-3. Vertical distribution of the mean biomass of oligochaetes, of oligochaetes and chironomids, and of the total benthic macroinvertebrates of Lake Washington, Washington. (Redrawn after Thut, 1969.)

TABLE 21-13 Average Numerical Density* and Wet Weight Biomass† of the Principal Groups of Benthic Invertebrates at Different Depths of Eutrophic Lake Borrevann, Southern Norway

BENTHIC INVERTEBRATES	LITTORAL								SUBLITTORAL						PROFUNDAL			
	0.2 m		1.5 m		2 m		3 m		5 m		6 m		7 m		10 m		15 m	
	No. m^{-2}	g m^{-2}	No. m^{-2}	g m^{-2}	g m^{-2}	No. m^{-2}	No. m^{-2}	g m^{-2}	No. m^{-2}	g m^{-2}	No. m^{-2}	g m^{-2}	No. m^{-2}	g m^{-2}	No. m^{-2}	g m^{-2}	No. m^{-2}	g m^{-2}
Oligochaeta	114	0.6	1370	8.9	8.4	1652	897	4.5	252	1.6	280	1.4	245	0.8	378	1.4	406	0.7
Hirudinea	126	2.9	163	2.1	2.2	368	63	0.5	2	0.02	—	—	—	—	—	—	—	—
Ephemeroptera	1416	5.0	150	1.1	0.04	18	9	0.01	—	—	—	—	—	—	—	—	—	—
Trichoptera	391	5.2	203	2.0	0.6	112	193	0.3	80	0.2	10	0.02	60	0.1	172	0.5	958	3.0
Chaoborus	—	—	—	—	0.01	2	3	0.01	—	—	5	0.02	60	0.1	172	0.5	958	3.0
Chironomidae	279	0.5	643	1.2	2.5	1138	463	0.7	900	3.5	490	1.4	505	2.5	575	5.0	299	2.2
Gastropoda	105	3.2	1964	27.9	5.2	902	60	0.2	6	0.2	—	—	—	—	2		—	—
Other Groups	112	0.5	253	2.5	2.3	388	53	0.2	14	0.1	5	0.02	—	—	20	0.1	5	0.01
Total	2543	17.9	4746	45.8	21.3	4580	1741	6.4	1254	5.6	790	2.9	810	3.4	1147	7.0	1668	5.9

From data of Ökland (1964); bivalve mollusk Anodonta excluded.

developed on the basis of indicator species of dominant benthic fauna. Particular emphasis was given to the distribution of species within the ubiquitous dipteran family of midges (Chironomidae) and to the oxygen content of the hypolimnion. Much of the early work on the classification of lakes in relation to the Chironomidae is reviewed in the book by Thienemann (1954; see also the critical discussion by Deevey, 1941 and Brinkhurst, 1974). While these analyses did much to stimulate study of the benthic fauna, such classification systems are of limited value since they do not consider fully the complex interactions of morphometry, the chemical differences among sediments and water, the biotic interactions such as predation, and an array of other factors.

As lakes become more eutrophic, shifts occur in percentage composition of the two dominant groups of benthic animals in the profundal zone of lakes, the Chironomidae and the oligochaetes. The few examples given in Table 21-14, which could be expanded greatly, simply show the general reduction in the number of chironomids and other benthic animals and a concomitant increase in oligochaete worms among more productive lakes. The transition can occur during natural development of a particular lake, and can be accelerated greatly by the eutrophicational activites of man, as was shown dramatically in the case of Lake Erie (Table 21-14) in a period of less than 30 years (cf. the general review of this subject by Wiederholm, 1980).

BENTHIC FAUNA OF LAKE ESROM

Certainly one of the foremost analyses of the profundal benthic fauna can be found in the work of Jónasson (1972, 1978, and numerous earlier works reviewed therein). Environmental factors, especially the phytoplanktonic productivity of the moderately large (17.3 km²) and deep (22 m), eutrophic Lake Esrom, a dimictic lake

TABLE 21-14 Comparison of the Relative Composition of the Dominant Benthic Macroinvertebrates of Several Lakes of Differing Productivity Based on Other Criteria

| | PERCENTAGES | | | | |
LAKE	CHIRONOMIDAE	OLIGOCHAETA	SPHAERIDAE	OTHERS	SOURCE
Oligotrophic					
Convict, Calif.	65.3	30.8	0.4	3.5	Reimers, et al. (1955)
Bright Dot, Calif.	77.5	3.1	19.1	0.3	
Dorothy, Calif.	69.5	23.3	3.5	3.7	
Constance, Calif.	56.9	20.5	20.5	2.1	
Cultus, British Columbia	65.0	24.0	—	—	Ricker (1952)
Lake Ontario	1.8	6.4	3.4	88.4*	Johnson and Brinkhurst (1971)
Lake Erie (1929–1930)	10	1	2	87†	Wright (1955)
Eutrophic					
Washington, Wash.	43	51	3	3	Thut (1969)
Lake Erie (1958)	27	60	5	8	Beeton (1961)
Glenora Bay, Lake Ontario	42.3	29.4	6.2	22.1	Johnson and Brinkhurst (1971)

Data are only approximately comparable because of different methods employed.
*Mostly amphipods.
†Mostly the mayfly *Hexagenia*.

in Denmark, were studied in relation to the food, growth, life cycles, and population dynamics of three profundal detritivores and two carnivores. Because of the completeness of these studies, it is instructive to discuss them in some detail.

The dipteran detritivore *Chironomus anthracinus* was found to feed at the sediment surface. Its growth was limited to two very short periods, one in spring during the phytoplanktonic maximum when the hypolimnion was oxygen-rich, and the other after autumnal circulation when oxygen was available again but food production was declining. Growth continued during winter as long as the lake was ice-free, but essentially ceased under ice cover. During the summer, growth stopped when the oxygen concentrations of the hypolimnion were reduced slightly below 1 mg O_2 l^{-1}. Biomass, protein content, and fat content of larvae of this species were positively correlated seasonally with the influx of oxygen and phytoplanktonic detritus to the hypolimnion.

Continuous quantitative data extending over 17 years showed that deep (20-m) populations of *C. anthracinus* exhibited a two-year life cycle. The two-year cycle is maintained by alternate years of high initial population densities in which a large percentage of the population does not emerge as adults in the first year. These larvae prevent a new generation from becoming established, because their feeding inflicts severe mortality on new eggs laid on the sediments. As a result, at any given time the population derives from a single generation of eggs. In alternate years, when the density of larvae declines below 2000 m^{-2} after the first year's emergence of adults, a new generation is able to establish itself, because now the larvae are not able to feed over the whole of the sediment surface and most eggs survive. During these years of recruitment of a new generation, the larval population is a mixture of two generations (Fig. 21–4).

The larvae of this species in the metalimnetic sediments (11 and 14 m) grew longer because of greater oxygen concentrations, and each individual attained greater biomass. There was more exposure to fish predation in the shallower metalimnetic sediments, however, and population numbers were lower than in the profundal zone. Those at 11-m depth had a one-year life cycle; the transitional zone between one and two year life cycles was between 14- and 17-m sediments. In sediments of the latter depth, recruitment of larvae every year was not successful (Fig. 21–4). At 14 m, the larvae succeeded in having a yearly recruitment in some years. Population size increased with depth, probably as a result of reduced fish predation.

Larvae of the carnivorous phantom midge *Chaoborus flavicans*, and of the midge *Procladius pectinatus*, had similar growth patterns and consistently exhibited a one-year life cycle. The detritivorous tubificid oligochaete *Ilyodrilus hammoniensis* lives at least five years and does not begin to reproduce until after three years; the detritivorous small clam *Pisidium* also exhibits a long life cycle.

Detailed analyses of the population dynamics, survivorship, emergence, and respiration permitted an evaluation of the net production of the profundal macroinvertebrates of eutrophic Lake Esrom (Table 21–15). About 83 per cent of the production was by the three benthic detritivores, mostly by the *Chironomus* species. Jónasson was able to extend these data, making a number of reasonable assumptions, to a partial energy flow budget. In this moderately eutrophic lake, an appreciable portion (perhaps 20 per cent) of the phytoplanktonic primary production reaches the profundal sediments and, while undergoing decomposition, is utilized by profundal invertebrate fauna along with the microflora (Table 21–16). In preparing this table from his data, the unknown category of littoral and allochthonous organic matter has been included intentionally. As discussed at length in earlier chapters, these inputs can be very significant and often exceed that of the phytoplankton. Jónasson himself emphasizes that these figures represent only part of the whole. In spite of the excellence of his studies, further study is needed to evaluate how significant these littoral and allochthonous inputs are to profundal invertebrate productivity in a lake of moderately large size such as Esrom.

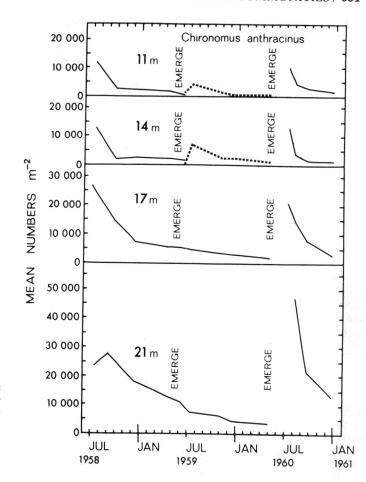

Figure 21-4. Seasonal fluctuations in populations of the midge larvae *Chironomus anthracinus* in sediments of different depths of Lake Esrom, Denmark. One-year life cycles were found at 11 and 14 m, two-year cycles at 17 and 21 m depths. (Modified from Jónasson, 1972.)

TABLE 21-15 **Estimates of the Average Annual Production, Mortality, Respiration, and Emergence of the Dominant Profundal Benthic Fauna of Lake Esrom, Denmark**

SPECIES	RESPIRATION (MINIMAL)	MORTALITY	EMERGENCE	NET PRODUCTION
Dipterans				
Chironomus anthracinus	361.8	35.9	39.4	75.3
Chaoborus flavicans	51.2	2.6	11.6	14.4
Procladius pectinatus	7.3	1.3	1.7	2.9
Oligochaetes				
Ilyodrilus hammoniensis	7.3	—	—	6.0
Mollusks				
Pisidium casertanum	4.8	—	—	1.5

Composite from data of Jónasson (1972). Data expressed as kcal m^{-2} $year^{-1}$.

TABLE 21-16   Estimates of Energy of Biomass and Flow from Components of Profundal of Lake Esrom, Denmark

COMPONENTS	KCAL m^{-2} YEAR^{-1}	PER CENT
Littoral and Allochthonous Organic Matter	?	—
Phytoplankton		
Net production	1630	73
Respiration	610	27
Zooplankton		
Net production	120	40
Respiration	180	60
Detritivorous Benthic Fauna		
Respiration		
Micro benthic fauna	210	28
Macro benthic fauna	374	50
Mortality-predation	43	6
Emergence	40	5
Net production (macro)	83	11
Carnivorous Benthic Fauna		
Respiration	58	63
Mortality-predation	4	4
Emergence	13	14
Net production	17	18

Generated from data of Jónasson (1972). Respiratory values of phytoplankton and zooplankton estimated.

Food supply is of major importance in the regulation of the population dynamics of detritivore species (cf. also Iyengar, et al., 1963), especially in the profundal zone, in which predation shifts from other animals to cannibalism of eggs. Among carnivorous chironomid larvae, the young (first instar) are largely detritivores, but beyond the second instar, they shift to feeding on small chironomid larvae, small crustacea, and some plant food. Although *Procladius* are carnivorous, their diet is highly variable and opportunistic (Kajak and Dusoge, 1970). Mature larvae consume about 10 per cent of their body weight per day and have a relatively high degree of utilization of animal food. Abundant food availability increases growth and development greatly.

The Chironomidae form the dominant macroinvertebrates of all arctic lakes studied (Hobbie, 1973; Butler, 1982). It is noteworthy that those few studied have two- to seven-year cycles and are detritivores. The carnivorous *Procladius* is also usually present.

PHANTOM MIDGES

In addition to the chironomid larvae, oligochaetes, and the small clam *Pisidium*, another major component of the profundal zone of lakes is the phantom midge *Chaoborus*. Extensive studies of *C. flavicans* demonstrate some of the major characteristics of this group (Parma, 1971). This midge dominates the benthic fauna of the profundal zone of Lake Vechten, a small (5 ha), moderately deep (12 m) lake in The Netherlands. Lake Vechten is thermally stratified from April to November and mixes all winter (very brief period of ice cover). The lake is moderately eutrophic and the hypolimnion becomes anaerobic during stratification.

The life span of an adult phantom midge is very brief ($<$ 6 days). Eggs (about 500 in number) are laid on the water in rafts and most (97 per cent) hatch in two to four days. The larvae can develop to the fourth instar in six to eight weeks. The first and second instars are always limnetic and positively phototactic, and they develop rapidly in a few weeks. The third instar, mostly limnetic but also occurring in the sediments, is of much longer duration; the larvae can overwinter in this stage (Stahl, 1966, Roth, 1968). After a variable period of up to several months, ecdysis to the fourth instar occurs; this instar is limnetic much of the time.

The fourth instar of many species of *Chaoborus* undergoes strong diurnal vertical migrations (see, for example, Northcote, 1964; Malueg and Hasler, 1966; LaRow, 1968; Goldspink and Scott, 1971). A typical rhythm is movement from the sediments or their daytime depth to the surface strata at sunset; they stay there until about sunrise, then descend. Rates of ascent and descent are about the same at 4 to 6 m hr^{-1}. Migration does not occur as actively at temperatures below 5°C. When the larvae are burrowed into the sediments, they push up to the surface of the mud as if to sample the light conditions. If light levels are sufficiently low, they emerge from the sediments and begin the migration (LaRow, 1969). In some interesting experiments, LaRow (1970) obtained evidence that the migration was influenced in part by oxygen tension of the water near the sediments. When the oxygen concentrations were high, most of the populations stayed in the sediments; when there was less than 1 mg O_2 l^{-1}, about one-third of the population migrated into the water above the sediments and about one-third migrated to the surface during the night. Migration varies with season, being highest in midsummer and lowest in winter.

Since the early instars are small and very transparent, predation pressure probably is low. With increasing size, predation pressure in the open water becomes greater and results in a more benthic habit. Oxygen stresses in the hypolimnion impose a demand for oxygen which can be alleviated by moving to the upper strata, but this activity is best accomplished at night when predation pressure from fish is reduced. Fish predation is greater upon the larger third and fourth instar larvae of *Chaoborus* than on the smaller instars. Smaller species, or species in which the larval eye size and visibility to fish is reduced, have less pronounced vertical migration than larger species. Some highly visible species are eliminated in lakes by fish predation, and do not occur when fish are present or introduced (Macan, 1977; Northcote, et al., 1978; Stenson, 1978a, 1978b; von Ende, 1979).

Pupation takes from 1 to 2 weeks, and occurs from May to October in *C. flavicans*. The pupae migrate daily. Pupation and emergence as adults result in a reduction in the total benthic population during summer. Highest densities of larvae, 1400 to 1800 m^{-2} in Lake Vechten, occurred in November. Mortality in the winter was high (65 per cent) due to reductions in food supply, and a total mortality of about 97 per cent occurred before the fourth larval stage was reached. Annual net production of the benthic part of the population was estimated to be 8 to 12.5 kg (dry) ha^{-1} [70 to 90 kg wet weight ha^{-1}] for *C. flavicans* in this lake.

Chaoborus larvae are predatory on pelagic zooplankton and certain small benthic animals. Their daily ration of food, amounting to about one crustacean zooplankter per hour, is variable but amounts to approximately 10 per cent of their body weight (Kajak and Ranke-Rybicka, 1970). It was estimated that *Chaoborus* could consume about 12 per cent of the macrozooplanktonic biomass on the average, and a much greater percentage at certain times of the year, such as late in the period of summer stratification.

It has been demonstrated further that *C. flavicans* is conspicuously selective in the food it consumes (Swuste, et al., 1973). Among zooplankton, copepods were actively selected over the cladoceran *Daphnia*; the pelagic ostracod *Cypria* was attacked but then immediately rejected. When given the choice between benthic oligochaetes and zooplankton, the oligochaetes were taken and predation on *Daphnia* ceased. Similar results have been found in a number of other studies (Swift, 1976; Lewis, 1978b; Smyly, 1980; Vinyard and Menger, 1980; Winner and Greber, 1980; Chimney, et al., 1981; Pastorok, 1981). Differences in prey size and shape, as well as in swimming, escape, and vertical migration behavior, influence the selection of prey by different instars of *Chaoborus*.

The coexistence of two or more species of *Chaoborus* without any apparent great differences in their characteristics or requirements that would reduce competition, has been noted often (Stahl, 1966a, 1966b). Previously, it was assumed that the abundance of food precluded serious competition. Roth (1968), however, observed severe competitive interactions among three species of *Chaoborus* in the same lake, and presented evidence to show that slight differences in morphology of the larvae and their distribution, especially during nocturnal migration, were adequate to permit the two quantitatively minor species to coexist with the dominant species. Similarly, two coexisting species of *Chaoborus*, the dominant profundal fauna of Lake George of tropical Africa, were found to have interspecific differences in the mean number of eggs per adult female, the nature of egg batches, the morphology of mouthparts, as well as differences in the vertical distribution of the larvae and seasonal occurrence of the adults (McGowan 1974).

Productivity of Benthic Fauna

The extreme heterogeneity in distribution and the seasonal population dynamics of the benthic invertebrate fauna of lakes and streams have impeded detailed analyses of their productivity. Moreover, mortality from natural causes and by predation is complex and related to changes in the types and intensity of predation, since the macroinvertebrates change their size and distribution during their life cycles. Several aspects of these predation-related changes have already been discussed, especially among the oligochaetes, chironomids, and the phantom midge *Chaoborus*.

Many of the existing quantitative data on the benthic fauna are expressed on the basis of biomass per unit area. A comparative set of examples of the average biomass among many lakes is set out in Table 21-17; the number of examples of this nature could be expanded many times over. Such data, however, are of quite limited value owing to great differences in sampling times and in sampling frequency and duration. As has been pointed out several times in the foregoing discussion, the summer period of low population densities and high predation pressure is the poorest time for analyses. Many studies are conducted only during the ice-free period. In some streams and especially in sediments from deeper parts of lakes, the maximum period of growth often occurs during the winter months in the temperate zone. More importantly, most of these studies yield insufficient data on size–frequency distribution and other population characteristics to permit an evaluation of turnover rate. Animals with small biomass but high turnover rates (shorter generation times) can obviously be more productive

TABLE 21-17   Comparisons of the Biomass of Benthic Macroinvertebrates in Lakes

LAKE	WET WEIGHT (kg ha⁻¹)	DRY WEIGHT (kg ha⁻¹)	NOTES AND SOURCE
Great Slave, N.W.T.	20.0	2.5	Rawson (1955)
Athabaska, Alberta-Saskatchewan	(32.8)	4.1	
Minnewanka, Alberta	(36.0)	4.5	
Simcoe, Ontario	(99.2)	12.4	
Waskesiu, Saskatchewan	(198.0)	24.7	
Wyland, Ind.			
Mean 1955	—	8.05	Gerking (1962)
Mean 1956	—	8.60	
July	56.0	10.3	
August	31.6	5.8	
75 Finnish lakes	23.6	(3.0)	Deevey (1941) and Hayes (1957) after numerous sources
5 Swedish lakes	31.1	(3.9)	
3 New Brunswick lakes	25.4	(3.2)	
10 Lakes of USSR	41.3	(5.2)	
38 Lakes of USA, mostly of	87.2	(10.9)	
Connecticut and New York	(10–348)	(1–44)	
13 Lakes of northern Canada	88.9	(11.1)	
43 European alpine lakes	76.1	(9.5)	
64 Lakes of northern Germany	115.0	(14.4)	(includes some mollusks)
19 Eutrophic Polish lakes	28	3.5	Pieczyńska, et al. (1963)
	(2–56)	(0.3–7.0)	
15 Pond-type Polish lakes	101	12.6	
	(20–370)	(2.5–46.3)	
Dystrophic Store Gribsø, Denmark	94	(11.8)	Berg and Peterson (1956)
Weber, Wisc. 1940	553	(69)	Juday (1942)
1941	147	(18)	
Nebish, Wisc. 1940	122	(15)	
1941	590	(74)	
Shallow fish culture ponds,	222–272	(28–34)	Lellák (1961)
Czechoslovakia	272		
Parvin Reservoir, Colorado	582	99.8	Buscemi (1961)
Green, Wisconsin			
0–1 m	36.6	8.7	Juday (1924)
0–10 m	82.2	15.9	
10–20 m	166.0	29.9	
20–40 m	171.1	33.5	
40–66 m	149.6	30.9	
Mendota, Wisconsin			
0–7 m	433.9	53.8	Juday (1942)
>20 m	696.8	76.6	
Average for lake	414		Juday (1921)
Memphremagog, Quebec–Vermont			Dermott, et al. (1977)
North basin (mesotrophic)	—	21.9	
South basin (eutrophic)	—	43.2	

Values are only approximately comparable because of the differences in methods and frequency and duration of sampling; all exclude mollusks unless stated to the contrary. Values in parentheses were estimated using the relationship: dry weight equals 12.5 per cent of wet weight (see discussion of Thut, 1969). Winberg, et al. (1971) recommended 10 per cent and Gerking (1962) and others found about 20 per cent more realistic.

and contribute as much or more to the system than can animals with high biomass but low turnover rates.

Thus, average data such as are summarized in Table 21-17 must be viewed with caution. One not too surprising trend is observed, namely, that in general, the biomass of benthic fauna increases as the productivity of the system increases. The biomass of benthic animals of oligotrophic lakes of northern latitudes and alpine regions is quantitatively less than that of lowland lakes of temperate and tropical regions. Within some rather obvious limits, as food supply by autochthonous productivity and the diversity of littoral habitats increases, the productivity of benthic animals increases with greater eutrophication. If the organic loading becomes so great that the conditions of hypolimnion are intolerable to even the most adapted fauna, as occurs in extremely hypereutrophic lakes, productivity of the animals decreases. Generalizations based only on biomass data weaken rapidly beyond this level.

> Changes in the total amount of biomass of benthic fauna with enrichment have been demonstrated dramatically many times in fertilization experiments. For example, high phosphate and nitrogen fertilization of fish culture ponds resulted in a 42 per cent greater yield of benthic invertebrates, and 3.3 times greater yield of zooplankton, than yields of unfertilized ponds (Ball, 1949). In a similar study on ponds in Michigan, fertilization increased biomass of benthic invertebrates by about 75 per cent and increased food organisms utilized by the common bluegill sunfish (Lepomis) by about 70 per cent. (Patriarche and Ball, 1949). However, in a more detailed analysis of fertilization relationships in these shallow waters, Hall, et al. (1970) demonstrated the complex interactions between shifting development of macrophytic vegetation and increasing densities of phytoplankton, and the effects of these substrata on species composition of benthic animals and on their predation by fish. Although increases in productivity of benthic invertebrates were found in more nutrient-rich waters, in large part the production was compensated for by shifts in species composition owing to changing substrata and increases in mortality as a result of greater fish production and predation.

In the foregoing discussion, examples of estimates of the production rates of benthic macrofauna have been given for individual groups or for specific regions of lakes, e.g., the profundal zone. Studies of the rates of production of the total invertebrate benthic fauna of whole-lake systems, for which sufficient detail is available to calculate turnover rates over the entire year, are rare. The data in Table 21-18 can be treated only as approximate estimates, but they do show the general relationships commonly observed between nonpredatory, herbivorous or detritivorous benthic fauna and the less productive predatory fauna. In plots of the frequency of occurrence of the annual production values among different benthic invertebrate groups (annelids, mollusks, crustaceans, and insects) in many lakes and rivers, the most common production values were between 1 and 10 kg ha^{-1} year^{-1} for all groups (Waters, 1977). The productivity estimates of the nonpredatory benthic fauna were found to be 5 to 35 times lower than those of nonpredatory zooplankton (Winberg, 1970). As would be anticipated, the productivity of predatory benthic fauna was consistently found to be much lower than that of the nonpredatory benthic fauna. In general, the productivity of the nonpredatory benthic fauna accounts for less than 10 per cent of the primary productivity, and decreases appreciably as the productivity of macrophytes increases and the loading of

TABLE 21-18   Estimates of Annual Production (Dry Weight) of Total Benthic Invertebrates

BENTHIC ORGANISMS	PRODUCTION (kg ha^{-1} year^{-1})	LAKE/STREAM	SOURCE
All benthic animals	29.5	Bodensee, Fed. Rep. of Germany	Streit and Schröder, 1978
	84	Wyland Lake, Indiana	Gerking, 1962
	152	Marion Lake, British Columbia	Hall and Hyatt, 1974
	283	Mikolajskie Lake, Poland	Kajak and Rybak, 1966
	367	Taltowisko Lake, Poland	
	507	Bay of Quinte, Lake Ontario	Johnson and Brinkhurst, 1971
	2000	Speed River, Ontario	Hynes and Coleman, 1968
Herbivores-Detritivores	3.7	Lake Krivoe, USSR	Alimov, et al., 1970
Carnivores	0.8		
Herbivores-Detritivores	9.0	Lake Krugloe, USSR	
Carnivores	1.4		
Herbivores	41.0	Volchja Reservoir, USSR	Pidgaiko, et al., 1970
Predators	6.1		
Herbivores-Detritivores	57.1	Lake Manitoba, Manitoba	Tudorancea, et al., 1979
Carnivores	10.1		
Herbivores-Detritivores	214.4	River Thames, England	Mann, et al., 1970
Predators	34.8		
Nonpredators	18.1	Average of three lakes, USSR	Winberg, et al., 1970
Predators	5.6		
Nonpredators	4.7	Lake Zelenetzkoye, USSR (subarctic)	Winberg, et al., 1973
Predators	0.2		
Nonpredators	341	Kiev Reservoir, USSR	Winberg, et al., 1970
Herbivores-Detritivores	556	Oconee River, Georgia	Nelson and Scott, 1962
Nonpredators	204	Five Polish lakes	Kajak, et al., 1970

Based on the *approximate* conversion coefficients for invertebrates of Waters (1977): 1 g dry weight = 6 g wet weight, = 5 kcal, = 0.9 g ash-free dry weight, = 0.5 g carbon.

allochthonous organic matter to the lake increases in proportion to phytoplanktonic productivity.

Annual turnover rates for benthic invertebrates, calculated as the ratio of annual production to mean biomass, vary considerably (see review of Waters, 1977). Most values, however, follow the general rule of larger P/$\overline{B}$ ratios with increasing number of generations per year (Table 21-19). Cohort P/$\overline{B}$ ratios show considerable constancy between 3 and 5, irrespective of the length of cohort life, similar to the annual P/B ratios for univoltine species. The relationship of the annual P/$\overline{B}$ ratio to voltinism is directly attributable to water temperature among benthic invertebrates, since water

TABLE 21-19   Ranges and Average Production/Biomass Ratios of Benthic Invertebrates

GENERATIONS OF LIFE CYCLE	AVERAGE P/$\overline{B}$ RATIO	P/$\overline{B}$ RATIO RANGE
Multivoltine (2–3 generations per year)	8.3	3–16
Bivoltine	5.4	2–9
Univoltine	4.5	<1–9
Hemivoltine (2–3-year life span)	2.2	<1–4
Long-lived Mollusks	1.7	<1–5

After discussion of Waters (1977).

temperature affects the number of generations per year. $P/\overline{B}$ ratios have been empirically approximated by $T^2/10$, where T = mean temperature (°C) at the bottom of lakes (Johnson and Brinkhurst, 1971).

Fish and Predation on Benthic Fauna

The subject of fish biology and the fish fauna of lakes and streams is an enormous field to which much attention has been devoted. The diversity in the morphology, physiology, behavior, reproductive capacity, growth, and feeding of freshwater fishes is a field of study that has been treated in detail in many recent works many times the size of this book. This separation of the treatment of fish biology, ecology, and population dynamics from limnology is perhaps justifiable from the standpoint of giving due attention to these organisms of major economic importance, and because the subject matter is so extensive. The schism is unfortunate in many other ways, however.

Fish are an integral component of freshwater systems. Although the amount of energy and carbon flux to the fish trophic level is exceedingly small in most aquatic systems, this does not justify their neglect as an operational component within the system. Their impact on the operation of the system in terms of carbon flux and nutrient regeneration at times can be quite significant, as has been discussed repeatedly in nearly every foregoing chapter. For example, the shift of fish species feeding on larger-sized food organisms to planktivorous species can have marked effects on zooplanktonic composition and productivity. This change can in turn influence the species composition of phytoplankton and consequently productivity at the primary level.

Another conspicuous example occurs during eutrophication of lake systems, in which the fish species change dramatically from salmonid and coregonid species of quite stringent low thermal and high oxygen requirements to warm-water species that are increasingly tolerant of eutrophic conditions. The warm-water fish are restricted to the epilimnion during summer stratification. Certain fish, such as the carp, have omnivorous feeding habits and can be very effective in modifying the littoral substrata to the point where many submersed macrophytes are eliminated. Such feeding activities often can so disturb the sediments that turbidity is increased, and transparency and phytoplanktonic productivity decline.

A large number of studies have indicated the major effects of fish predation on the population dynamics and productivity of benthic macroinvertebrates, as has been discussed already in this and previous chapters. Many other examples exist. In Third Sister Lake, Michigan, the littoral benthic fauna increased to a maximum concentration in early winter (Ball and Hayne, 1952). The invertebrate populations declined steadily after the formation of ice cover until spring loss of ice, when an abrupt reduction occurred as adult insects emerged. During summer stratification, the populations and their biomass progressively increased from this annual spring low to the autumn and winter maxima. Experimental removal of all of the fish from the lake not only resulted in a large increase in the biomass and numbers of benthic fauna, but the low populations characteristic of early summer were much less pronounced than under conditions of normal fish predation. Experimental manipulation of fish populations in half-hectare ponds showed that the presence of fish predation resulted in a very significant increase in the rate of production of benthic invertebrate fauna and a reduction in their biomass (Hayne and Ball, 1956). In the absence of fish, the apparent production rate of benthic fauna decreased and stabilized at a higher level of benthic faunal biomass relative to ponds in which fish predation occurred.

While there are great differences in the qualitative composition of invertebrate fauna and seasonal succession of fauna among littoral stands of emergent vegetation, little difference is found between the quantity of macrofauna and the production of higher emergent vegetation in productive lakes or ponds (e.g. Dvořák, 1970). The presence or absence of submersed vegetation, however, results in large differences. These plants can in turn be influenced markedly by fish and waterfowl. A number of studies have experimentally excluded the feeding and disturbance activities of fish and waterfowl from the littoral vegetation (see, for example, Kořínková, 1967; Kajak, 1970). The biomass and production rates of benthic macrofauna were severely reduced (usually by more than 50 per cent) when the populations were exposed to fish (carp) and birds (especially ducks). The reductions resulted from increased predation pressure and from losses of submersed vegetation (the latter provided both refugia and food). Although mortality of benthic invertebrates can be very high from fish predation (at least half), food supply is usually the dominant factor controlling the population dynamics and productivity of benthic invertebrates (see Lellák, 1965; Hall, et al., 1970).

Studies of the profundal macroinvertebrate fauna, for example in the early detailed analyses by Borutskii (1939), showed that predation mortality, chiefly by fish, was relatively low (Table 21-20). Only a small percentage of the profundal macroinvertebrates lived to the adult stage. His studies indicated that the turnover rate was about two times the mean biomass, which is in general agreement with other studies (production rates equal about 2 to 4 times the mean annual biomass). There is evidence that the turnover values are considerably higher in the littoral zone (4 to 8 times the mean annual biomass) (see, for example, Anderson and Hooper, 1956).

In a detailed study by Gerking (1962), the production rates of zooplankton and macroinvertebrate fauna were determined both directly and by measuring of utilization by the dominant predator, the bluegill sunfish (*Lepomis macrochirus*). Population dynamics and mortality analyses showed that the dense fish population grew slowly in this eutrophic Indiana lake, because individuals became overcrowded and undernourished. The diet of the fish was varied, but the dominant prey were midges (45 per cent) and *Daphnia* (26 per cent) among all size classes of fish (Table 21-21). Each year the diet shifted from a high proportion of *Daphnia* in July to a high proportion of midge larvae in August as the cladoceran populations decreased drastically during the summer. The estimated fraction of midge production that was consumed by fishes during the summer was 0.50 (Table 21-21), which indicates a high cropping efficiency of the benthic fauna, much higher than "losses" through emergence as adults.

The production rate of the bluegill fish was estimated by multiplying the instantaneous rate of growth by the average summer biomass, and adding the recruitment

TABLE 21-20 Estimated Rates of Production of Profundal Macroinvertebrates of Lake Beloie, USSR, Over an Annual Period, 1935-36, and the Fate of the Fauna Within the Lake

	kg (DRY WEIGHT) LAKE^{-1}	PER CENT
Net production	4010	100
Emerging insects	256	6
Predation mortality (mainly by fish)	542	14
Natural mortality and decomposition	2213	55
Surviving population at end of year	999	25

Generated from data of Borutskii (1939).

to this value (Table 21-21). The minimum summer production of benthic fauna was computed as the amount necessary to replace the losses by predation. Midges and other dipteran larvae accounted for more than one-half of the production of food that was derived from the bottom fauna. Therefore it was estimated that minimum food production was four to five times the production of bluegills during the summer, and probably near 10 times bluegill production on an annual basis.

As is common among fish of the temperate zone, both reproduction and growth of the bluegill sunfish are limited to the spring, summer, and early autumn. Spawning occurs for about a month in the spring, depending on the temperature. Growth occurs for about five summer months, during which time nearly all of the annual production

TABLE 21-21 Relationships Between the Bluegill Sunfish (*Lepomis macrochirus*) and Its Food Supply of Benthic Fauna and Zooplankton in Wyland Lake, Indiana

BLUEGILL SUNFISH	Weight (kg ha^{-1})		FOOD SUPPLY	Weight (kg ha^{-1})	
	Protein	Live Weight		Protein	Live Weight
Biomass:			July biomass:		
June	17.0	98	Benthic fauna	6.1	56
Average	12.2	72	Zooplankton	9.2	191
			Total	15.3	247
Percent diet constituents:			August biomass:		
Benthic fauna	—	74	Benthic fauna	3.4	33
Zooplankton	—	26	Zooplankton	1.1	23
Midges	—	45	Total	4.5	56
Monthly summer food turnover:			July-August decline:		
Benthic fauna	6.2	64	Benthic fauna	2.7	24
Zooplankton	2.2	23	Zooplankton	8.1	168
Total	8.4	87	Total	10.8	192
Annual production:			Minimum summer production rate:		
Growth	6.5	37	Benthic fauna	17.8	196
Recruitment	9.1	54	Midges	9.2	71
Total	15.6	91			

	PRODUCTION TURNOVER RATIO	
	Protein	Live Weight
Monthly summer food turnover compared with:		
July biomass:		
Benthic fauna	0.99	1.14
Zooplankton	0.24	0.12
July-August decline:		
Benthic fauna	2.30	2.67
Zooplankton	0.27	0.14
Summer efficiencies based on total production:		
Production biomass/intake	0.15	0.09
Production fish/production food	0.23	0.19
Intake by fish/production food:		
Benthic animals	0.48	0.47
Midges	0.50	0.55

Modified from Gerking (1962). Mahon (1976) reevaluated the fish production data of Gerking's study and found that production values of the younger bluegills were underestimated (corrected to approximately 625 kg ha^{-1} year^{-1} total annual production). Most of these younger-aged fish would prey more heavily on zooplankton than upon benthic organisms.

occurs, about 625 kg ha^{-1} in this example (Table 21-21). Growth and feeding decreased abruptly in the autumn, and remained low for the remainder of the year. In contrast, the production of benthic fauna continued for much of the year and increased greatly in the fall, under reduced predation pressure, until about the time ice cover formed. On an annual basis, the benthic faunal production is much larger (approximately 5 to 10 times) than that utilized by fish.

Fish Production Rates

Although an enormous literature exists on the yield of fish per given area of freshwater habitat, relatively few analyses exist of their growth and mortality over annual periods. Consequently, only a few reasonable estimates of production rates can be made. Growth is highly variable among different species of fish. In general, the instantaneous growth is correlated inversely with fish size. At greater sizes, most of the food energy is utilized for maintenance; in young fish, a larger proportion of the energy intake is diverted into growth. Hence, the ratio of production to biomass usually decreases with age and greater size. Reproductive effort by sexually mature members of a fish population can represent a high proportion of the annual fish production.

The summary estimates of net production rates of various fish groups set out in Table 21-22 demonstrate the general ranges encountered. Production rates are considerably higher in tropical waters than they are in temperate fresh waters, where growth is restricted to about half of the year. In fertile standing waters, the production rates of herbivorous fish in the tropics can reach several hundred g m^{-2} year^{-1} (Chapman, 1967, 1978). In standing waters of temperate regions, in which a single species often predominates, the range of rates is from 1 to about 20 g m^{-2} year^{-1}. Production rates in streams are usually higher than in standing waters (Table 21-22), and average about 50 g m^{-2} year^{-1} in temperate areas.

TABLE 21-22 Estimates of Rates of Production of Fish from Various Fresh Waters

FISH GROUP/COMMUNITIES	ANNUAL PRODUCTION (kg ha^{-1} year^{-1})	P/$\overline{\text{B}}$ RATIOS
Salmonidae (Trout, Salmon)		
From standing waters	0.21–66	0.62–2.0
From streams	11–300	1.0–5.0
Cottidae (Sculpins)	8.0–431	1.2–5.5
Percidae (Perch Family)	0.91–52	0.35–2.4
Esocidae (Pike Family)	0.75–14.2	0.32–0.7
Cyprinidae (Minnow and Carp Family)	0.1–915	0.18–1.94
Others	40–625	—
Total fish fauna, multispecies		
Temperate zone	90–1980	—
Tropical zone	1306–3468	—
Planktivores (USSR lakes)	9–24	0.7–0.8
Benthivores (USSR lakes)	13–60	0.4–0.5
Piscivores (USSR lakes)	9–14	0.3–0.4

After data from numerous sources, particularly Chapman (1967, 1978), Winberg, et al. (1970), and Waters (1977). Values are expressed as wet weight; *approximate* conversion values for fishes: 1 g wet weight = 0.2 g dry weight or approximately 1 kcal (Waters, 1977).

Annual P/$\overline{\text{B}}$ ratios for fishes are generally lower than for most invertebrates as a result of longer life spans (several years). Relatively high P/$\overline{\text{B}}$ ratios are found in the sculpins, for example, which have shorter life spans (Table 21–22), because the younger, faster-growing stages contribute disproportionately to the population productivity. Annual P/$\overline{\text{B}}$ ratios of a population in an expanding or colonizing stage, such as in a new reservoir or in a stream recovering from flood damage, will be high, because growth is high relative to mortality (Waters, 1977). A population that is overcrowded and stunted, as is commonly the case with sunfish populations in temperate lakes, have lower annual P/$\overline{\text{B}}$ ratios, slower growth, and longer life spans. A very approximate estimate of annual production can be obtained if the annual mean biomass of zooplankton is multiplied by 15 to 20, of benthic animals by 6 to 8, and of fish by 0.5 to 1.0 (1.2 for stream salmonids).

SUMMARY

1. The benthic animals of fresh waters are extremely diverse; representatives of nearly every animal phylum occur in or are associated with the sediments of lakes and streams. This diversity and extreme heterogeneity in distribution, modes of feeding, reproduction, and morphological and behavioral characteristics make it difficult to generalize for the entire benthic animal community. Nonetheless, complex general patterns of coexistence, interrelationships, and community productivity emerge. Invertebrate and vertebrate (primarily fish) predation upon benthic animals are important regulators of spatial and temporal population structure and dynamics.

2. Most freshwater Protozoa are attached to benthic substrata. Few protozoans tolerate low dissolved oxygen concentrations; most inhabit surficial sediments and migrate to shallower water when dissolved oxygen of deeper strata declines in stratified productive lakes. Protozoan species are segregated by differences in food utilization, reproduction, and timing of population development. Many protozoan populations exhibit summer maxima; their population densities are positively correlated with summer increases in temperature and abundance of food items on the surficial sediments. Little is known of natural protozoan productivity in fresh waters. Although protozoan community biomass is low, rates of turnover (number of generations) are high. Protozoan productivity is likely higher than is generally recognized.

3. Freshwater sponges are rarely abundant, and their contribution to total benthic productivity is usually minor.

4. The hydroid coelenterates are not common in fresh waters and rarely is their productivity a significant portion of the whole. Occasionally, dense populations of *Hydra* can impart significant predation pressures upon populations of small zooplankton.

5. The turbellarian flatworms are the only important free-living members of the Platyhelminthes in fresh waters. Most are restricted to quiescent shallow areas of lakes and low-velocity streams where water movement is reduced.

Flatworms prey upon other small invertebrates, and species segregate by the dominant type of prey consumed and by differences in reproductive capacities. Flatworm productivity is directly correlated with the general fertility and productivity of fresh waters.

6. Free-living nematodes, or roundworms, are widely distributed in fresh waters and can constitute a significant component of the benthic fauna. Nematode feeding habits are diverse; some species are strict herbivores, others are strict carnivores on other small animals, and still others are detritivores on dead particulate organic matter. A large proportion of production is expended in reproduction, inasmuch as mortality is high in the egg and juvenile stages.

   a. In the temperate zone, population dynamics of many nematode species exhibit three generations per year, with maximum production during winter and spring periods with reduced fish predation and increased food abundance.

   b. Highest densities and productivity of nematodes are commonly found in littoral substrata of productive lakes.

7. Colonial bryozoans are rarely quantitatively important in fresh waters. A number of invertebrates live commensally within bryozoan colonies. Bryozoans and these coinhabitants are heavily preyed upon by fish.

8. Two major groups of aquatic annelids or segmented worms form significant components of the benthic fauna.

   a. Oligochaete worms are diverse, and occur in a spectrum of fresh waters, from unproductive to extremely eutrophic lakes and rivers.

      i. A large number of species of oligochaetes coexist in sediments of lakes, but species abundance patterns differ at different water depths, since the particle size and organic content of the sediment change with depth. Resource partitioning occurs by differences in worm morphology and selective feeding on microbiota.

      ii. As lakes become organically polluted and dissolved oxygen concentrations become reduced or are eliminated, an abundance of tubificid oligochaetes is commonly found concomitant with a precipitous reduction and exclusion of most other benthic animals. As long as some oxygen is periodically available, and toxic products of anaerobic sedimentary metabolism do not accumulate, the rich food supply and freedom from competing benthic animals and predators permit rapid growth.

      iii. Oligochaete densities can be very large (many thousands per m²). Productivity can vary greatly from year to year because of changes in mortality associated with population dynamics of major long-lived predators (e.g., chironomid midge larvae).

   b. Leeches are primarily ectoparasites that intermittently consume vertebrate blood. A few species are predators on other invertebrates.

      i. Leech abundance is highly variable, but generally increases in more productive fresh waters.

      ii. Production rates of leeches are greatest in the first year of the typical two-year life cycle. Rates of growth and production are most rapid during spring and early summer.

9. The bivalved microcrustacean ostracods are widespread in fresh waters. Because of their small size (<1 mm), occurrence in surficial sediments, and difficult taxonomy, little is known of ostracod ecology, population dynamics, or productivity.
   a. Ostracods occur in the surficial sediments (0–5 cm) where they feed by filtration on bacteria, algae, detritus, and other microorganisms.
   b. Reproduction is parthenogenetic for much of the ostracod life cycle. Egg development is variable and is strongly temperature-dependent. One to three generations occur per year.
   c. Little is known about effects of invertebrate or fish predation on ostracod population dynamics.
   d. Ostracod densities increase in more productive lakes (to >50,000 m^{-2}). Little is known of their productivity rates or of their roles in benthic metabolism; both are probably more significant than is presently realized.

10. Representatives of four groups of malocostracean crustaceans can form major components of the benthic fauna of some fresh waters.
   a. The mysids, or opossum shrimps, feed on small zooplankton, phytoplankton, and particulate detritus. All mysids exhibit rapid diurnal migrations from the sediments into the metalimnion or lower hypolimnion (waters of <15°C) at night. Mysids are size-selective predators; their introduction into a lake can lead to marked alterations in the composition and productivity of the zooplankton community. When lake productivity is high, the mysid life cycle is one to two years in duration; when productivity or temperatures are low, mysids may require up to four years to complete their life cycle.
   b. The isopods, or sowbugs, occasionally become a significant part of the benthic animal community of lakes and streams. Populations commonly have two distinct cohorts (spring and autumn). Productivity is greatest in the spring from the overwintering cohort.
   c. The decapod crustacea, or freshwater crayfish and shrimps, are primarily herbivorous on algae and larger aquatic plants. Despite the long life cycle of crayfish (to 3 years), populations often reach high densities and can dominate the annual production of benthic animals in some lakes and streams.
   d. The amphipods, or scuds, are generally omnivorous substrate feeders on bacteria, algae, fungi, and particulate organic detritus; they are restricted to well-oxygenated waters. Size-selective fish predation commonly severely reduces the larger-sized adult amphipods.
      The amphipod *Pontoporeia* burrows into the sediments and is a surficial deposit feeder during the daylight period. At night, *Pontoporeia* migrates into the upper hypolimnion and metalimnion (<15°C), similarly to the mysids. In contrast to the mysids, however, *Pontoporeia* does not feed on plankton. The adaptive value of its nocturnal migration is unclear.

11. The freshwater mollusks consist of univalve snails (Gastropoda) and bivalve clams and mussels (Pelecypoda). The snails are grazers on attached micro-organisms, whereas clams and mussels are filter feeders on particulate detritus and microzooplankton of the sediments. In spite of the large biomass of mollusks, the life cycle is long (1 to 4 years in snails; 1 to 15 years in clams) and productivity is relatively low. Egg and juvenile mortality is commonly very high in both groups.

12. Aquatic insects are extremely diverse. Some orders of insects are entirely aquatic; others inhabit fresh waters only during certain life stages.

    a. Feeding mechanisms and types of food ingested by insects are extremely varied.

        i. Organic matter is ingested either in particulate form (by swallowing, biting, or chewing) or in dissolved form (by piercing or sucking) (see Table 21–12).

        ii. Association of aquatic insects with a particular substrate is usually directly related to feeding on that substratum or on the associated microflora.

        iii. Most aquatic insects tend to be nonselective in their food habits; a few species feed specifically on a given species of food substrate. Facultative-feeding invertebrates ingest a wider array of substrates and inhabit a greater diversity of stream and lake habitats than do more specialized feeders.

        iv. Some immature detritivorous insects harbor microbial symbionts in their alimentary tracts that degrade relatively resistant plant detritus. Species possessing these symbionts or digestive cellulase can utilize a wider range of habitats because they can use a greater array of detritus and algal types than can organisms that lack these adaptations.

    b. The biomass of aquatic insects is relatively constant if food supplies of similar nutritional content are supplied. Insect-biomass turnover is controlled primarily by water temperature, and is governed by the positive relation between temperature, food availability, feeding rate, and respiration. Growth efficiency of carnivores tends to be less than that of herbivores or detritivores.

13. Benthic community structure in lakes usually consists of a rich fauna with high oxygen demands in the littoral zone above the metalimnion. Heterogeneity of substrata is great in the littoral; benthic-animal species diversity is greater in the littoral than in the more homogeneous profundal zone. As lakes become more productive, the number of benthic animals adapted to hypolimnetic conditions of reduced oxygen and increased decompositional end-products declines.

    a. Two maxima in abundance and biomass of benthic animals are often observed: one in the littoral zone, the other in the lower profundal zone.

    b. As lakes become more fertile, submersed macrovegetation can be eliminated as a result of light attenuation. Maximum abundance and biomass of benthic animals may then shift to the profundal zone.

    c. With further eutrophication and intensive organic-matter decomposition in the profundal zone, much of the benthic fauna of the profundal zone can be eliminated.

   d. As lakes become more eutrophic, a shift occurs in the percentage composition of two dominant benthic groups, with a decrease in the dipteran chironomid larvae and an increase in the more tolerant oligochaete worms (e.g., tubificids).

   e. The dipteran phantom midge *Chaoborus* is another major component of the profundal benthic fauna of lakes. *Chaoborus* larvae migrate into the open water at night and prey heavily on zooplankton. Feeding on limnetic zooplankton by *Chaoborus* is highly selective.

14. Composite densities of benthic invertebrates are often lowest during summer, especially among insect-dominated communities, both in streams and in the profundal zone of lakes. In general, biomass and productivity of benthic fauna increases as the overall fertility and productivity of lakes and streams increase.

   a. Mortality of benthic invertebrates from fish predation commonly increases with rising ecosystem productivity, in part offsetting enhanced benthic productivity.

   b. The most common production values among different benthic invertebrate communities (annelids, mollusks, crustaceans, and insects) are between 1 and 10 kg ha^{-1} year^{-1}.

   c. Productivity of nonpredaceous benthic animals (herbivores and detritivores) is at least 5 to 10 times greater than that of predaceous (carnivorous) benthic animals.

   d. Ratios of annual production (P) to mean biomass ($\overline{B}$) for benthic invertebrates vary considerably, but in general P/$\overline{B}$ ratios increase with increasing number of generations per year (voltinism) (see Table 21–19). The relation of this annual P/$\overline{B}$ ratio to voltinism is, in part, related to water temperature, since temperature affects the number of generations in many of the dominant benthic organisms.

15. Although growth among different fish species is highly variable, growth rate and size are generally inversely correlated. In young fish, more energy is diverted into growth, and in larger fish, more food energy is utilized for maintenance.

   a. Production rates of fish are considerably higher in tropical fresh waters than in temperate waters, where growth is restricted to about half of the year. In temperate regions, production rates are usually higher in streams than they are in standing waters.

   b. Annual production-to-mean-biomass ratios are higher among fish with shorter life cycles than they are among fish species that have slower growth and longer life spans. Similarly, in cases where young stages contribute disproportionately to population productivity, annual P/$\overline{B}$ ratios are greater. In general, annual P/$\overline{B}$ ratios are highest among planktivorous fishes, lower for benthic feeders, and least for carnivores.

CHAPTER TWENTY-TWO

ORGANIC CARBON
CYCLING AND DETRITUS

Nearly all of the organic carbon of natural waters consists of dissolved organic carbon (DOC) and dead particulate organic carbon (POC). The ratio of DOC to POC often approximates 6:1 to 10:1 in both lacustrine and stream systems (Wetzel and Rich, 1973). DOC has been separated from POC by many techniques, such as sedimentation, centrifugation, and filtration. However, DOC is defined arbitrarily in most studies by the practical necessity of fractionation of POC from DOC by filtration at the 0.5-μm size level; hence, DOC concentrations frequently include a significant colloidal fraction, in addition to truly dissolved organic carbon (e.g., Lock, et al., 1977). Living POC of the biota constitutes a very small fraction of the total POC. The metabolism of the biota, however, creates a series of reversible fluxes between the dissolved and particulate phases of detrital carbon.

SOURCES OF ALLOCHTHONOUS ORGANIC MATTER

The sources and composition of organic matter are diverse and poorly understood. Production of dissolved and particulate organic carbon is a result of autotrophic and heterotrophic metabolism. Instantaneous measurements of the chemical mass of DOC and POC, however, are highly biased toward refractory compounds (that is, compounds that are relatively chemically stable, of low solubility, and resistant to rapid microbial degradation). These refractory components of detrital organic carbon persist in fresh waters for longer periods of time than do more labile organic compounds. The readily utilizable labile components cycle rapidly at low equilibrium concentrations, but represent major carbon pathways and energy fluxes. Moreover, in streams and most lakes, much of the detrital organic carbon, mostly dissolved, is of terrestrial origin. Most lakes are small, with a high proportion of their surface area as littoral zone. Allochthonous and littoral sources of dissolved and particulate detrital carbon form major inputs to the lacustrine system, and can markedly influence metabolism of organisms in the open-water pelagic zone. In streams, dissolved organic carbon is decomposed by both the benthic microflora and the planktonic bacteria (Cummins, et al., 1972; Wetzel and Manny, 1972b; Dahm, 1981). In the relatively static waters of lakes, gravity acts as an important selective agent by which sedimentation displaces a major portion of carbon and its metabolism to the sediments (Vallentyne, 1962; Wetzel, et al., 1972). To adequately assess the importance of the complex carbon cycle that is central to both the structure and function of lakes and streams, one must know of the productivity of all components of the ecosystem, and have an understanding of the origins and metabolism of detrital organic carbon.

Detrital DOC and POC have long been known to exceed by many times the amount of organic carbon present as living material in the form of bacteria, plankton, flora, and fauna (Birge and Juday, 1926, 1934; Saunders, 1969; Wetzel, et al., 1972). An example from a lake in southeastern Michigan demonstrates the general relationships found among many lakes (see Fig. 17-8). The amount of carbon in higher organisms is even smaller than the smallest amount indicated here, and is trivial in comparison to the whole amount of carbon in the lake. These relationships are instantaneous measures of chemical mass and, as indicated earlier, bear little relationship to the metabolic fluxes of carbon.

In spite of the relatively crude techniques for separation of particulate from dissolved organic matter and the approximate chemical analyses available to earlier workers on the subject, the massive data of Birge and Juday (1926, 1934) on a large number of Wisconsin lakes provide an initial orientation to the general relationships. Seston consisting of living and mostly dead particulate organic matter was separated by centrifugation from colloidal organic matter and true DOC. The organic matter was analyzed for proteinaceous nitrogen and other nitrogen-containing compounds, lipids present in ether extracts, and the residual carbohydrate. The range of total organic carbon content of natural waters is 1 to 30 mg l^{-1}; higher values usually are encountered only under polluted conditions. The average values from over 500 Wisconsin lakes were: dissolved organic matter, 15.2 mg l^{-1}, and particulate organic matter, 1.4 mg l^{-1}, a ratio of 11:1 (cf. Meybeck, 1982; world DOC average of rivers is 5.8 mg l^{-1}). The average composition of the dissolved organic matter (detailed in Table 12-8 of the earlier discussion on dissolved organic nitrogen) was:

Crude protein	15.6 percent
Lipid material	0.7 percent
Carbohydrate	83.7 percent

The importance of these and related studies to the present discussion lies in their relationship to the magnitude of allochthonous organic carbon entering most lakes from terrestrial and littoral marsh areas, and the qualitative changes in composition of this organic matter en route to the pelagic zone. As the total organic matter of lake water increases, the percentage in the dissolved fraction increases disproportionately to that of the particulate fraction. The nitrogen content of the organic matter progressively decreases with increasing DOC concentration, and similarly declines as the proportion of allochthonous to autochthonous organic matter increases. As a result, the organic C:N ratios increase—a reflection of an increased input of organic compounds low in nitrogen as well as decreased decomposition rates of the more refractory organic compounds. Dissolved organic matter of allochthonous origin contains a very low percentage of nitrogen (C:N about 50:1), while that produced autochthonously within the lake by algae and macrophytes has a much higher initial nitrogen content (C:N about 12:1). It is clear that much of the DOC originating allochthonously from terrestrial and marsh plants has humic characteristics (discussed further on) that impart a stained brown color to the water. Numerous workers have shown a nearly linear relationship between color units (Chapter 5) and dissolved organic matter or carbon, but some studies indicate that the color is affiliated primarily with the colloidal or high-molecular-weight fractions of truly dissolved organic carbon (cf. Black and Christman, 1963; Lock, et al., 1977; Stewart and Wetzel, 1981a).

The colloidal fraction of measured "dissolved" organic matter is probably small in most cases (Krough and Lange, 1932), but may constitute a greater fraction in heavily stained waters (cf. Lock, et al., 1977). Unfortunately, data pertaining to this point are very few. Nitrogen content decreases and carbon content increases in the progression from particulate to colloidal to truly dissolved organic matter fractions (Table 22-1). The colloidal fraction is higher in lakes rich in DOC such as bog waters, and in hard waters in which organic compounds are adsorbed onto carbonate particles (cf. Ohle, 1934b; White, 1974; White and Wetzel, 1975).

Both allochthonous and autochthonous sources of detrital particulate organic matter (POM) and dissolved organic matter (DOM), therefore, are instrumental as variable inputs to aquatic systems. Within the lake, autochthonous sources include: (1) littoral sources of POM and DOM by active secretion and autolysis of the macrophytes and attached microflora, and (2) primary producers of the pelagic zone, primarily the algal phytoplankton. Under some circumstances, photosynthetic and chemosynthetic bacteria are also significant. Rapid transformations between POC and DOC by heterotrophic microflora progressively degrade organic matter to CO_2 and heat. The amount of organic carbon utilized and transformed by animals is a quantitatively small portion of that of the whole system. Much of the heterotrophic metabolism, which is almost completely microbial, occurs in the open water; some heterotrophic decomposition is displaced to the sediments, and this site becomes the major place of transformation in many lakes, especially as depth and volume decrease.

TERRESTRIAL SOURCES

Allochthonous sources of organic matter to aquatic systems are primarily of terrestrial plant origin. The organic carbon of plant residues is transformed variously by animal utilization and microbial degradation while being transported by runoff water. Much of the input of terrestrial organic matter to streams is in the form of DOM. This input results from direct leaching from living vegetation or from soluble compounds

TABLE 22-1 Annual Average Particulate, Colloidal, and Dissolved Organic Fractions of Lake Fureso, Denmark

ORGANIC FRACTION*	TOTAL DRY WEIGHT† (mg l⁻¹)	ORGANIC WEIGHT† (mg l⁻¹)	CRUDE PROTEIN (%)	LIPID MATERIAL (%)	CARBOHYDRATE (%)
Particulate	2.1	1.56	53.2	9.9	36.9
Colloidal	3.6	0.67	41.5	11.7	46.8
Dissolved	approx. 70.	8.8	37.5	—	62.5

After Krogh and Lange, 1932.
*Colloidal and dissolved fractions separated by ultrafiltration.
†Based on a limited number of largely surface and near-sediment samples.

carried in runoff from dead plant material in various stages of decomposition. POM, again mostly of plant origin, can fall directly into stream water from overhanging canopies, be transported by runoff water, or be windblown into the stream. Foliage from trees and ground vegetation can provide very significant inputs of organic matter to streams, both as POM and as leached DOM from the dead POM. The large as well as fine particulate matter of terrestrial vegetation entering streams is leached of a significant portion of its organic content as dissolved compounds, the amount varying with the plant species (Fig. 22-1). This DOM leachate can be metabolized very rapidly by bacterial populations that increase markedly in response to the DOM loading. For example, in large experimental streams, leaf leachate was demonstrated to have a bacteriologically labile dissolved organic carbon (DOC) fraction that was rapidly decomposed ($T_{1/2}$ 2 days), and a refractory DOC fraction ($T_{1/2}$ 80 days) (Wetzel and Manny, 1972b). Most of the refractory dissolved organic nitrogen compounds persisted in a relatively unmodified state for at least 24 days. A quantity of the dissolved leachate precipitates (Lush and Hynes, 1973); the rate of precipitation and the size of the resulting particles depend upon leaf species and water chemistry.

Aggregations of large POM, such as leaf packs trapped among stream sediments and rocks, undergo colonization by fungi in complex successional patterns within the detrital microhabitats (Triska, et al., 1975; Triska and Sedell, 1976; Suberkropp and Klug, 1976, 1980; Suberkropp, et al., 1975, 1976). As the resistant plant material is degraded, solubilized products of decomposition are utilized by bacteria living in highly stratified populations in the steep redox gradients of the compacted plant mate-

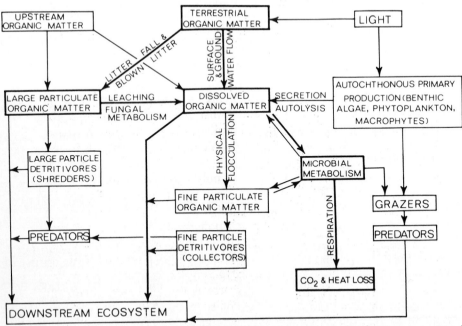

Figure 22-1. Simplified compartment model of the structure of an idealized stream ecosystem. Heavier lines indicate dominant transport and metabolic pathways of organic matter. (Composite of modified figures after Fisher and Likens, 1973, and Cummins, et al., 1973.)

rial (Fig. 22–1). The detrital material and its associated microflora serve as a major nutritive source for numerous aquatic invertebrates, especially the immature insect fauna (Kaushik and Hynes, 1971; Cummins, 1973; Cummins, et al., 1973; Iversen, 1973; Anderson and Sedell, 1979). There is no question that the shredding, collecting, and grazing activities of aquatic insects play a role in quickening the reduction of the size of POM and subsequent microbial degradation. It is also clear that a major portion of the faunal nutrition is obtained from the microflora attached to the resistant detrital POM. Only a few animals are adapted with symbiotic gut microflora to utilize the particulate plant material itself (cf. Cummins and Klug, 1979).

There are few detailed analyses of the metabolism of organic carbon in streams (e.g., Fisher and Likens, 1973; Wetzel and Otsuki, 1974; Fisher, 1977; Dahm, 1981; Mulholland, 1981). However, there is sufficient evidence to indicate that (1) Allochthonous inputs of terrestrial organic matter, in the form of detrital DOM and POM, commonly form a major source of material and energy for stream ecosystems. (2) Approximately 10 times more organic matter occurs as DOM than occurs in particulate form. (3) The rates of decomposition of DOM are rapid (days) in comparison to those for POM (leaves in weeks, woody material in years). (4) The labile components of DOM are decomposed rapidly, but more refractory components are exported to downstream ecosystems. Large POM is decomposed slowly, and has a longer retention time within a particular reach of the stream system.

Estimations of the rates and importance of autochthonous primary production in streams by the attached benthic algae, lotic phytoplankton, and larger aquatic plants are very difficult (reviewed by Wetzel, 1975b; Minshall, 1978, and Bott, 1982). It generally is assumed that terrestrial photosynthesis and importation of this organic matter to stream ecosystems are the primary carbon and energy sources of these systems; that is, that streams are largely heterotrophic. Such is indeed the case in heavily canopied woodland and forested streams in which autochthonous primary production is very low or negligible. In noncanopied streams and in rivers as they increase in size and decrease in velocity of flow, the significance of primary productivity of lotic phytoplankton and macrophytes increases (Fig. 22–2). Running waters can clearly vary in their proportion of heterotrophic and autotrophic metabolism. The relative significance of these types of metabolism is dynamic and varies considerably at local levels within a river system, and seasonally with shifts in many physicochemical parameters and shifts in loading with allochthonous (natural or artificially by pollution) organic matter.

The primary productivity within reservoirs is extremely variable because of individual characteristics of morphometry, seasonal changes in water-retention times, etc. In general, nutrient loading and trophic state are higher at the river end of reservoirs than at the deeper, dammed end. Although phytoplanktonic productivity per unit volume may be higher at the river end than at the dammed end, increased inorganic turbidity and light reduction at the river end can cause a reduction of the depth of the trophogenic zone and a decrease in total productivity (e.g., Fig. 22–2). The primary productivity per unit area in some reservoirs can be approximately the same over the length of the reservoir. In these cases, the higher trophic state and volumetric productivity at the inflow end of the reservoirs shift to lower epilimnetic trophic states, lower volumetric productivity, but increased depth of the trophogenic zone at the dammed end.

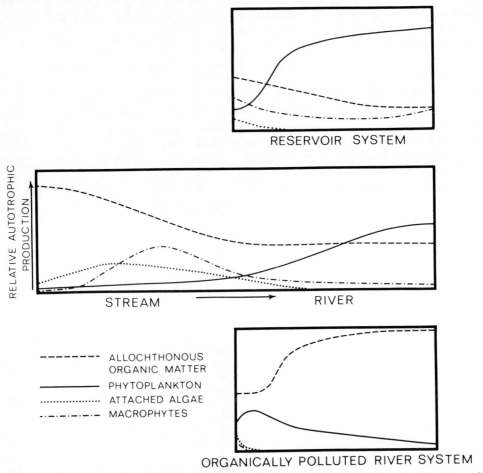

Figure 22-2. Generalized scheme of the relative contributions of allochthonous organic matter and auto-trophic production by attached algae, phytoplankton, and aquatic macrophytes in the transition of a stream to a river system, if it is impounded where velocity of flow is reduced or practically eliminated, or if it is organically polluted. (From Wetzel, R. G.: Primary production. *In* Whitton, B., ed., River Ecology. Oxford, Blackwell Sci. Publs., 1975.)

As will be emphasized further on, the trophic dynamic structure of aquatic eco-systems depends operationally on a dynamic detrital structure (Wetzel, et al., 1972; Wetzel and Rich, 1973; Rich and Wetzel, 1978). From the standpoint of carbon fluxes, most energy and organic carbon of the systems is dead, of autotrophic origin, and undergoing microbial degradation, the rates of which are variable and serially decrease with increasing refractility and resistance of the organic matter. In lakes in general, detrital organic matter is the main supportive metabolic base of carbon and energy. Since most lakes of the world are small to very small, much of the autochthonous pro-duction of detrital matter originates from benthic littoral and wetland vegetation. Sim-ilarly, detrital heterotrophic metabolism dominates in streams, where usually the major sources are allochthonously derived terrestrial plant material. Common to both lake and stream ecosystems is the dominance of detrital metabolism, which gives the ecosystems a fundamental metabolic stability. The trophic structure above the producer-decomposer level, with all of its complexities of population fluctuations, metabolism, and behavior, has a relatively minor impact on the total carbon flux of the system. The detrital system provides stability to streams; the slower, relatively consistent degradation of dissolved and particulate detritus by microorganisms underlies the more sporadic autochthonous metabolism that responds rapidly to, and depends to a greater extent on, environmental fluctuations. Autochthonous primary production is often minor and variable, but in combination with allochthonous, autotrophically produced detritus, drives the lotic system. The functional operation of lentic and lotic systems converges at this point of similarity in detrital metabolism.

Allochthonous Organic Matter Received by Lakes

The DOM of surface runoff is composed of relatively refractory organic compounds resistant to rapid microbial degradation. The amounts of DOC and POC reaching a lake and the chemical composition of these organic compounds change seasonally with the volume of flow in relation to time in the stream, the growth and decay cycles of the terrestrial and marsh vegetation through which runoff flows, and other factors, especially climatic changes.

The influxes of allochthonous detrital soluble and particulate organic carbon to a small temperate lake were analyzed over an annual period in relation to their fate within and losses from the lake (Wetzel and Otsuki, 1974). Detrital organic carbon influxes were determined in water from an inlet stream, both before and after its flow had briefly traversed a marsh adjacent to the lake, and from a second inlet stream, both at its headwaters and after it crossed the marsh. Similarly, measurements were made of organic content of groundwater where it entered the lake, and from the outlet. Concentrations of organic carbon in inflows and outflows were converted to values of total carbon loading and outflow using a detailed annual water budget (Fig. 22–3). The DOC increased significantly ($\times$ 2) as the inlet water of the streams passed through the marshes prior to entering the lake. A similar result was found in a larger stream as it passed through an extensive marsh system (Manny and Wetzel, 1973). Lowest inputs of DOC occurred during the summer, when there was little precipitation, and vegetation was growing rapidly. During the autumn and early winter, DOC increased markedly. The DOC content of the groundwater was very low, but because of the large volume of this inflow, it constituted one-third of the annual influx of allochthonous DOC (Table 22–2). A net loss of 15 g DOC m^{-2} year^{-1} from the outflow of this lake was observed, which represented a major pathway of organic carbon removal from the lake. Spectrophotometric analyses (absorption of ultraviolet light and fluorescence) of water flowing into and out of the lake indicated that a high percentage of the DOC was composed of humic compounds or yellow organic acids or both.

Concentrations of POC in the influxes, of water within the lake and of the outflow water, was consistently about 10 percent that of the DOC. Inputs of POC were minimal during the summer months, when marsh vegetation was growing rapidly, but increased markedly during periods of high autumnal rainfall and spring runoff (Fig. 22–4). Groundwater POC was very low, but on an annual basis constituted over one-fourth of the allochthonous POC input (Table 22–3). Allochthonous POC was small in relation to the quantities of littoral and pelagial organic carbon synthesized within the lake, and represented less than 5 per cent of the gross production of POC by the lake.

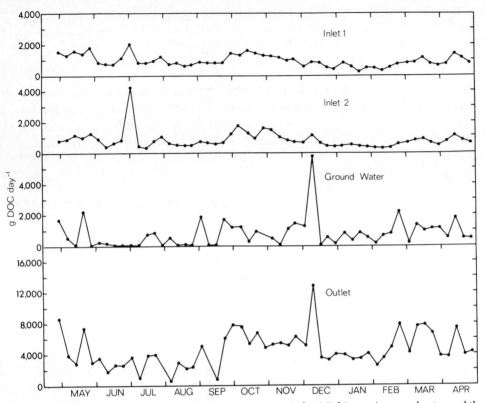

Figure 22-3. Dissolved organic carbon entering Lawrence Lake, Michigan, via groundwater and the two stream inlets, and leaving via the outlet, 1971-72. (From Wetzel, R. G., and Otsuki, A.: Allochthonous organic carbon of a marl lake. Arch. Hydrobiol., 73:31-56, 1974.)

TABLE 22-2 Annual Dissolved Organic Carbon Entering Allochthonously and Leaving Lawrence Lake, 1971-72

DOC	g C m^{-2} YEAR^{-1}
Influxes of DOC	
Inlet 1	7.00
Inlet 2	7.94
Groundwater and seepage	6.01
Total	20.95
Outflow DOC	35.82

From Wetzel, R. G., and Otsuki, A.: Allochthonous organic carbon of a marl lake. Arch. Hydrobiol., 73:31-56, 1974.

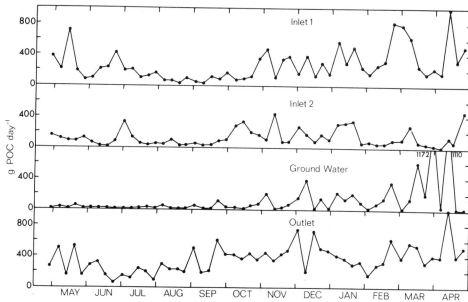

Figure 22-4. Particulate organic carbon of allochthonous inputs and the outflow, Lawrence Lake, Michigan, 1971-72. (From Wetzel, R. G., and Otsuki, A.: Allochthonous organic carbon of a marl lake. Arch. Hydrobiol., 73:31-56, 1974.)

TABLE 22-3   Annual Particulate Organic Carbon Entering Allochthonously and Leaving Lawrence Lake, Michigan, 1971-72

POC	g C m^{-2} YEAR^{-1}
Influxes of POC	
Inlet 1	1.99
Inlet 2	0.962
Groundwater and seepage	1.15
Windblown POC	0.00
Total	4.10
Outflow POC	2.75

From Wetzel, R. G., and Otsuki, A.: Allochthonous organic carbon of a marl lake. Arch. Hydrobiol., 73:31-56, 1974.

The quantity of windblown POM to Lawrence Lake was negligible. Leaf litter fall and material blown into a lake can become significant in small lakes or ponds in heavily forested areas, but generally this input is small (Szczepański, 1965) in comparison to the major import of organic matter from many streams. Domestic and industrial pollution of lakes by organic wastes is an exceedingly individualistic parameter that lends itself poorly to generalization. Pollution of this nature is normally transported to lakes and reservoirs via stream systems. The types of organic pollution and its processing and effects are treated excellently by Hynes (1960, 1970). As pointed out by Pieczyńska (1972; cf. Rai and Hill, 1981), the degradation of organic matter by microflora in lakes that receive a high proportion of allochthonous organic matter is generally more variable and requires a greater diversity of metabolic pathways than microbial degradation in predominantly autotrophic lakes. It should be stressed, however, that many lakes derive much of their organic carbon and many nutrients from terrestrial sources of the surrounding drainage basin. The magnitude of allochthonous inputs is difficult to measure and, more important, the metabolic role of allochthonous materials within the lake is difficult to separate from that of autochthonous inputs. This area is in great need of further investigation.

COMPOSITION OF ORGANIC MATTER

Before discussing our limited knowledge of the utilization and degradation of detrital DOM and POM in lake systems, a brief consideration of the heterogeneous composition of such materials must be presented. The detailed reviews of Vallentyne (1957) on the distribution and composition of organic matter in lakes, of Breger, et al. (1963) on organic geochemistry of all major components, and of Konanova (1966) on soil organic matter are especially recommended. Numerous compendia have appeared on organic matter in natural waters that are most relevant to this involved subject (e.g., the lengthy summations of marine and littoral organic compounds of Datsko, 1959; Duursma, 1961; Hood, 1970; and Khailov, 1971; and the freshwater symposia, e.g., Maistrenko, 1965; Shtegman, 1969; Trifonova, et al., 1969; and Faust and Hunter, 1971). Humic substances are reviewed comprehensively by Swain (1963), Haworth (1971), Schnitzer and Khan (1972), and Gjessing (1976).

The organic matter of soils and waters can be viewed as a mixture of plant and animal products in various stages of decomposition; it consists of compounds synthesized biologically and chemically from degradation products, and of microorganisms and their decomposing remains. This complex system may be simplified by separation into two categories: nonhumic and humic substances. *Nonhumic substances* are a class of compounds that includes carbohydrates, proteins, peptides, amino acids, fats, waxes, resins, pigments, and other low-molecular-weight organic substances (see discussion in Chapter 17). These substances generally are labile, that is, relatively easily utilized and degraded by microorganisms, and exhibit rapid flux rates. Because of rapid rates of utilization and turnover, the instantaneous concentrations of nonhumic substances in water are usually very low. *Humic substances* form most of the organic matter of soils and waters. Humic substances consist of dark-colored and acidic compounds of molecular weights ranging from hundreds to many thousands of daltons. Humic substances are formed largely as a result of microbial activity on plant and animal material, but further polymerization can occur abiotically (cf. Larson and Hufnal, 1980). The resulting compounds are relatively resistant to further microbial degradation and tend to persist in aquatic systems.

Humic Substances. Humic substances are traditionally separated into three categories, each of which is defined in terms of acid–base-solubility characteristics of the materials when extracted into an alkaline solution: *Humic acids* precipitate upon acidification, *fulvic acids* are soluble in both acid and base, and *humin* is insoluble in dilute solutions of acid or base. These three fractions have certain structural similarities, but differ from one another in molecular weight and functional group content. Fulvic acids are of lower molecular weight than humic acid or humin fractions, and have a higher proportion of oxygen-containing functional groups. This fractionation scheme is arbitrary; the individual fractions are all heterogeneous mixtures of various substances (sugars, carbohydrates, phenols, proteins, etc.), whose specific compositions vary with the source material and its state of degradation.

High-molecular-weight humic materials exhibit a colloidal structure that is important in the physical behavior of humic solutions. The colloidal particles (approximately 0.02 μm in size) can be fractionated by dialysis, gel permeation chromatography, ultracentrifugation, or electrophoresis, and exhibit Brownian movement. A basic characteristic of the humic colloids, as well as of truly dissolved humic and fulvic acid fractions, is their association with organic and inorganic materials via adsorption or peptization. The importance of the association of humic materials with Fe, P, Ca, and other ions has already been indicated in previous chapters. The pH and ionic strength of the solution governs both the extent of condensation and colloidal size, as well as the availability of adsorption sites.

Colloidal humic materials provide a very large surface area that is suitable for the adsorption of both inorganic and organic materials (e.g., Poirrier, et al., 1972; Stevenson, 1972; Seitz, 1982). Consequently, humic materials can fundamentally alter the availability of required or toxic metals and organic substances to aquatic biota. The binding of herbicides or pesticides (e.g., DDT) to humic materials, for example, can significantly change transport, fate, and bioaccumulation characteristics of these toxicants. Humic constituents can also link normally thermodynamically discrete redox couples by functioning as broad redox potential "flavoproteins" (Zimmerman, 1981). This feature can facilitate the oxidation or reduction of a number of biologically important substances (e.g., Miles and Brezonik, 1981; Francko and Heath, 1982).

> Humic material of the series humic acid-hymatomelanic acid (an alcohol-soluble component of humic acid)-fulvic acid-humin consists of polymeric micelles whose basic structure is aromatic rings which are variously bridged into condensed form by —O—, —NH—, —CH$_2$—, or —S— linkages (Swain, 1963; Schnitzer and Khan, 1972). Attached hydroxy and carboxyl groups provide acidity, hydrophilic properties, base-exchange capacity, and tanninlike character (for example, they react with proteins). Fulvic acid is the least polymerized of the fractions, while humin is the most condensed humic component. Fulvic acid is highly oxidized, stable, and water-soluble (Schnitzer, 1971). About 60 per cent of the weight of the fulvic acid fraction is composed of functional groups such as carboxyls, hydroxyls, and carbonyls attached to a predominantly aromatic nucleus ring structure. Fulvic acid is a naturally occurring metal complexing agent that can bring di- and trivalent metal ions into stable solution from practically insoluble hydroxides and oxides.

Humic substances contain a variable amount of nitrogen (1 to 6 per cent), which is largely hydrolyzable to amino compounds. Most of the combined amino acid nitrogen occurs in amino acid–peptide linkages. The remainder is in the form of amino sugars, ammonia, and amine linkages.

Formation of Humic Compounds.   The formation of humic substances occurs during the degradation of plant material; both microbiological and abiotic processes contribute to its production. Carbohydrates serve as the main microbial source of energy and carbon in the intercellular synthesis of protein and hemicellulose. Polyphenolic lignin of higher plant tissues is modified during this degradation, and humiclike substances of high molecular weight result (Zeikus, 1981). Further degradation of these compounds results in the generation of humic and fulvic acids and various products of decomposition (fats, amino compounds, and CO_2, H_2, CH_4, N_2, NH_3, and H_2S). Additional polymerization of the decomposition products can occur on the surfaces of clay particles (Larson and Hufnal, 1980; Wang, et al., 1978). Autolysis of the microorganisms themselves represents a major source of humiclike substances.

It is clear that the fungi play a major role in the degradation of plant material and the synthesis of humic substances. Fungi, as well as cellulolytic actinomycetes and myxobacteria, enzymatically oxidize lignin-derived polyphenols, which condense with proteinaceous degradation products to form humic acids. Humic substances of terrestrial plant and soil systems form largely from cellulose and lignin. The needles and other products of coniferous origin are decomposed mainly by fungi which degrade celluloses, hemi-celluloses, and lignin. This coniferous litter is easily leached, especially of organic acids (Nykvist, 1963). Litter of deciduous vegetation, however, is decomposed more readily by fungi and bacteria to lignoprotein complexes. When allochthonous inputs of particulate organic matter from higher plants are large, as in the case of many river systems and many lakes with extensive littoral vegetation, fungi probably play a major role in the degradation of lignified and cellulolytic tissue for subsequent bacterial decomposition. In the open water of lakes, in which autotrophic production by algae results in plant constituents composed mainly of hemicelluloses, proteins, and carbohydrates, cellulolytic activity is reduced. As would be anticipated, fungal populations are well developed in sediment accumulations, in littoral regions of higher plant densities, and in streams receiving large amounts of terrestrial plant debris. Sparrow (1968) provides an excellent review of the distribution and general ecology of the freshwater fungi.

The phenols, quinones, phenol carboxylic acids, and related compounds are common constituents of humic materials. These compounds can be inhibitory to bacterial fermentation and to fungi. These inhibitory compounds, and associated acidity and resultant low redox potentials, frequently lead to accumulations of humic substances undergoing very slow rates of degradation. The antiseptic qualities of humic substances from plant remains result in accumulations in topographical depressions, for example in the peat of bogs and lake sediments, over long periods of time without significant degradation. The superb preservation of human bodies placed into peat bogs in Europe over 2000 years ago (Glob, 1969) attests to the effectiveness of these combined properties and conditions.

DISTRIBUTION AND SOURCES OF ORGANIC MATTER IN LAKES

The implications of the preceding discussion are that a large portion of the particulate and dissolved detrital organic matter that enters lakes from allochthonous sources has undergone microbial stripping of more labile compounds. The bulk of organic detrital carbon is in the form of humic and other substances that are relatively resistant to further microbial degradation. It is then necessary to superimpose upon this variable allochthonous input of detrital organic matter the organic matter that is produced in the body of water itself. The autochthonous organic matter falls into two broad categories: that produced by plankton of the open water, and that produced in the littoral zone.

In previous discussions, we have systematically scrutinized the autochthonous synthesis of organic matter in both the open water (Chapter 15) and the littoral regions (chapters 18 and 19). The utilization of portions of this organic matter by animals was considered in chapters 16 and 21. The complexities of microbial degradation of the dissolved and particulate organic matter in the open water and the sediments were treated in chapters 17 and 20, respectively. These analyses provide the parts that can be fitted into an integrated whole that quantitatively combines the pathways of flow among producer, consumer, and decomposer components.

Historically, integration of the ecosystem components has been attempted in many different ways. Early attempts coupled the plant and animal components qualitatively as complex food webs, in which the feeding or trophic relationships among organisms were interconnected. While these analyses provide descriptors of the biotic components, the food-production and food-consumption processes are very dynamic and constantly change with differing environmental conditions and during the life histories of the organisms. So many organisms are involved that it is exceedingly difficult to quantify, in the words of Hutchinson, the roles of even the major actors of the ecological play.

Alternatively, the energy content of organisms and communities has been examined. Present technology is incapable of measuring directly the rates of energy transfer between individual organisms and communities. If the energy content of all organisms and their energy expenditures for all behavioral, maintenance, and reproductive costs were known, however, it would be possible to estimate the energy transfer between biotic components. The dearth of bioenergetic knowledge for many of the organisms occurring in fresh waters means that only very approximate estimates are possible at this time.

Quantitative carbon fluxes among communities of freshwater ecosystems have been used with some success. The underlying premise of this approach is that the synthesis of organic matter begins with the fixation (primarily photosynthetic) of CO_2. Much of this organic carbon is variously cycled within the ecosystem before being respired (primarily via decomposition) and returned as CO_2. Despite limitations of this approach, discussed below, much functional information on the operation of freshwater ecosystems has emerged from quantitative analyses of carbon cycling.

We initiated discussion throughout previous chapters with quantitative aspects of productivity and the dispersion of gross productivity to obtain estimates of net productivity. These values were expressed in terms of carbon wherever possible for comparative purposes under differing physical, chemical, and biotic (e.g., predation) environmental conditions. The biotic components were coupled qualitatively earlier, as, for example, in Figure 17-1. It is now instructive to examine the mass and flux of organic carbon between the components in as quantitative a manner as possible in representative lakes. None of these ecosystem analyses is complete; the analyses do, however, provide insight into the relative importance of the different communities and processes. Additionally, they emphasize areas in need of intensive further investigation.

DISTRIBUTION OF ORGANIC CARBON IN LAKES

DISSOLVED ORGANIC CARBON (DOC)

A conspicuous feature of the depth–time distribution of total dissolved organic carbon in a lake (e.g., Fig. 22–5) is the absence of strong vertical stratification or seasonal fluctuations, despite the parallel stratification and seasonal pulses of metabolic activity. It is apparent that the DOC pool consists primarily of carbon compounds relatively resistant to bacterial decomposition. Inputs of these organic substrates are approximately equal to their slow rates of microbial degradation. Highest concentrations occur during summer stratification in the epilimnion, and consistently fluctuate to a greater extent in the upper strata, especially near the surface, than do concentrations in the hypolimnion. It is evident from earlier discussion (see Chapters 15, 18, and 19), that secretion of DOC by phytoplanktonic algae and littoral flora is associated with a portion of the higher epilimnetic concentrations of DOC. Decomposition of largely labile secreted organic compounds often is very rapid (e.g., < 48 hours), and their dynamics would not be delineated by the sampling frequency employed for this generalized picture of the DOC pool. The constancy of the DOC pool is reflected further in annual average values per square meter and for the whole lake calculated as the depthwise integration of the products of DOC concentrations and volume of water of each water stratum (Table 22–4). Horizontal variations of DOC within this lake were very small (< 10 percent deviation from the mean of the central depression). In some lakes, however, and especially in reservoirs, point-source inputs of DOC, such as river influxes high in DOC, would cause greater horizontal variations.

It should be emphasized that the origins of DOC in aquatic ecosystems are largely photosynthetic; this DOC is either autotrophically synthesized within the water or allochthonously generated in terrestrial systems of the drainage basin and transformed

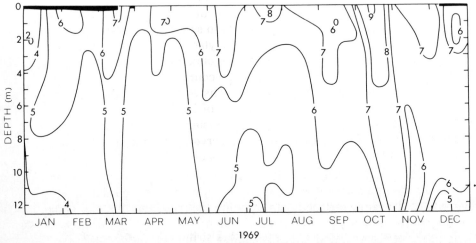

Figure 22–5. Depth–time diagram of isopleths of dissolved organic carbon (mg C l^{-1}), Lawrence Lake, Michigan, 1969. (From Wetzel, R. G., Rich, P. H., Miller, M. C., and Allen, H. L.: Metabolism of dissolved and particulate detrital carbon in a temperate hard-water lake. Mem. Ist. Ital. Idrobiol., 29:Suppl., 185–243, 1972.)

TABLE 22-4 Dissolved Organic Carbon of Lawrence Lake,
Michigan, 1968-79

YEARS	MEAN g C m^{-2}	MEAN kg C LAKE^{-1}
1968	30.93	1535
1969	34.73	1724
1970	27.80	1380
1971	26.71	1326
1972	20.19	1002
1973	25.15	1248
1974	23.25	1153
1975	25.87	1284
1976	25.47	1264
1977	21.54	1069
1978	22.12	1098
1979	28.70	1424
Average, 12 years	26.04	1292

Data of Wetzel, unpublished.

during transport to recipient basins. As indicated, the major sources to the DOC pool are (1) photosynthetic inputs of the littoral and pelagic flora added to the pool through secretions and autolysis of cellular contents, and (2) allochthonous DOC, composed largely of terrestrial humic substances refractory to rapid bacterial degradation. Very much smaller amounts, usually quantitatively negligible but qualitatively of potential importance, include (3) DOC from excretions of zooplankton and higher animals, and (4) bacterial chemosynthesis of organic matter with subsequent release of DOC.

Phytoplanktonic productivity and allochthonous sources from the drainage basin are the primary sources of the DOC pool of oligotrophic waters. In moderately large or very large bodies of water, phytoplanktonic photosynthesis alone can dominate inputs to the DOC pool. However, most lakes are small, and the ratio of littoral to pelagic photosynthetic productivity greatly increases as the mean depth of the basin decreases.

With increasing fertility, the relative contributions to system productivity by phytoplankton, submerged and emergent littoral macrophytes, and eulittoral algae are altered (Chapters 18, 19; see Fig. 19-10). The shift in dominance with respect to the productivity of these major groups results in differential contributions to the DOC pool (Fig. 22-6). If the drainage basin is not manipulated extensively by man, allochthonous inputs of DOC to a lake are defined by characteristics of its drainage basin and are relatively constant; their relative contribution to the ecosystem consequently decreases as autochthonous sources increase (Fig. 22-6). In reservoir systems, allochthonous inputs of DOC can exceed by several times the amounts of POC and DOC produced autochthonously (e.g. Romanenko, 1967).

PARTICULATE ORGANIC CARBON (POC)

Examination of the general spatial and temporal distribution of pelagic POC of lakes of varying degrees of productivity permits a few generalizations. In oligotrophic to moderately productive lakes, the observed depth-time distribution of pelagic POC follows the productivity and biomass distribution of phytoplankton rather closely dur-

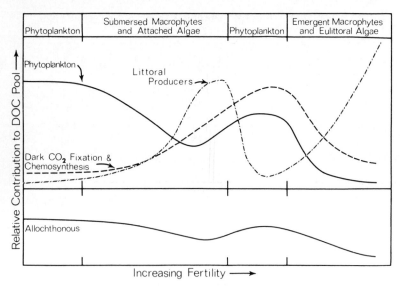

Figure 22–6. Generalized relative contributions of phytoplankton, littoral producers, dark and chemosynthetic CO_2 fixation, and allochthonous sources of dissolved organic carbon (DOC) to lakes of increasing fertility in the progression of dominating pelagic and littoral flora. Upper panel of generalized changes in productivity by autotrophic flora refers to those detailed in Figure 19–10.

ing stratified periods. When Lawrence Lake, for example, is stratified, the POC maxima lag behind peaks in productivity from several days to about two weeks, especially when sedimentation of plankton is slowed in thermal density gradients of the metalimnion (Fig. 22–7). During ice cover, one can frequently observe a period of reduced primary production, lower inputs of allochthonous POC, and a gradual reduction in concentrations of POC. An increase in POC generally is observed during periods of circulation when surficial sediments are disturbed and resuspended into the water column. In less productive lakes, hypolimnetic POC generally does not increase greatly unless the lower hypolimnion becomes anaerobic, for example, in the latter portions of summer

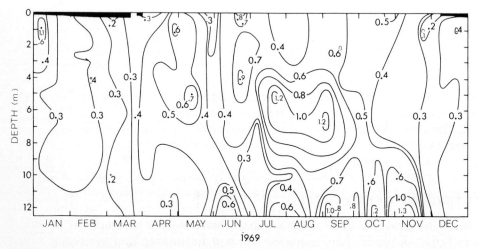

Figure 22–7. Depth–time distribution of pelagic particulate organic carbon (mg C l^{-1}), Lawrence Lake, Michigan, 1969. (From Wetzel, R. G., Rich, P. H., Miller, M. C., and Allen, H. L.: Metabolism of dissolved and particulate detrital carbon in a temperate hard-water lake. Mem. Ist. Ital. Idrobiol., 29: Suppl., 185–243, 1972.)

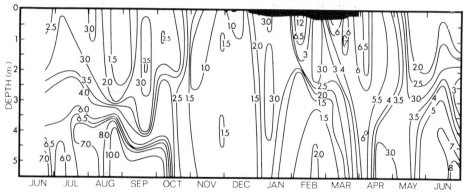

Figure 22–8. Depth–time distribution of pelagic particulate organic carbon (mg C l⁻¹) of hypereutrophic Wintergreen Lake, Michigan, 1971–72. (From Wetzel, et al., unpublished.)

TABLE 22–5   Particulate Organic Carbon of the Pelagic Zone of Lawrence Lake, Michigan, 1968–81

YEARS	MEAN g C m⁻²	MEAN kg C LAKE⁻¹
1968	2.03	101.0
1969	2.65	131.8
1970	2.18	108.4
1971	2.10	104.2
1972	1.73	86.0
1973	1.91	94.6
1974	2.06	102.2
1975	2.14	106.0
1976	2.25	111.8
1977	2.07	102.7
1978	2.72	135.2
1979	2.38	118.2
1980	2.42	120.1
1981	1.84	91.5
Average, 14 years	2.18	108.1

Data of Wetzel, unpublished.

stratification when specialized bacterial populations or algae (e.g., euglenophytes) may develop in profusion. In hypereutrophic lakes that receive large inputs of planktonic and littoral POC, the hypolimnia are rapidly rendered anoxic (as exemplified by Wintergreen Lake, Fig. 22–8) and bacterial productivity contributes to marked increases in POC.

The amounts of pelagic POC of a lake are relatively constant from year to year (Table 22–5), as long as the lake system is not disturbed by external influences such as enrichment activities of man. The ratio of DOC to POC is rather constant at about 10:1 in most unproductive to moderately productive lakes, but may be more variable in streams (e.g., Moeller, et al., 1979). Deviations from this 10:1 ratio with depth and season are small in less productive waters (compare, for example, Figs. 22–5 and 22–7). As lakes become more eutrophic, the DOC:POC ratio fluctuates greatly with season and depth. In the Wintergreen Lake example (Fig. 22–8), the annual average is about 5:1;

but during periods of intensive algal (epilimnetic) and bacterial (metalimnetic and hypolimnetic) growth, the ratio decreases to 1:1 or less, and then increases to about 10:1 during the fall period of circulation (see also Weinmann, 1970). In riverine lakes and in reservoirs with high allochthonous loading of POC, DOC:POC ratios fluctuate and have lower average values (e.g., 4:1; Rai and Hill, 1981). In another productive lake in Michigan, Saunders (1972) found that the particulate organic detritus constituted from 1.3 to 16.9 times the phytoplanktonic biomass and made up more than 50 per cent of the seston. The remainder of the seston was dominated by inorganic matter, such as particulate $CaCO_3$ and silica.

Estimates of the replacement of algal-cell carbon in the pelagic POC have been made from measurements of the biomass of algal-cell carbon and the rates of net primary production (Miller, 1972). Compensating for respiratory loss of carbon, the daily net accumulation of POC can be estimated from net primary production. Total suspended epilimnetic POC of Lawrence Lake had an average replacement time (turnover) of 40.7 days (range 8.1 to 544 days). In the POC pool, a mean of 83 mg C m^{-2} (range 30 to 241 mg C m^{-2}) was algal-cell carbon that was replaced by primary productivity in 1.1 days (range 0.30 to 2.55 days) during the ice-free seasons. The algal-cell carbon had an annual mean replacement time of 3.6 days.

Generalizations on the cycling of organic carbon among phytoplanktonic populations are difficult to make because of the paucity of data under natural conditions. POC

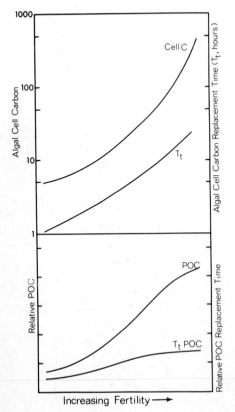

Figure 22-9. Generalized relationship of pelagic algal-cell carbon and particulate organic carbon (*POC*), and their relative replacement times (*T$_t$*) in lakes of increasing fertility. (From Wetzel, R. G., and Rich, P. H.: Carbon in freshwater systems. *In* Woodwell, G. M., and Pecan, E. V., eds., Carbon and the Biosphere. Springfield, Va., National Information Service, 1973.)

concentrations of the pelagic zones usually are considerably larger (double, or greater) in eutrophic lakes than in infertile waters (Fig. 22–9). Differences in concentrations of DOC are less marked in the transition from oligotrophic to highly eutrophic waters. Algal-cell carbon commonly increases nonlinearly with increasing fertility, and a slight tendency towards algal cells of greater size is found in eutrophic lakes. Replacement times of algal-cell carbon by net primary production are usually larger (slower turnover times) in eutrophic than in oligotrophic waters, but the increases in replacement times are not proportional to increases in cell carbon. Hence, in less fertile waters, the algal cells are photosynthesizing more per unit cell carbon, i.e., they have a greater carbon flux per cell.

All indications point to the importance of autochthonous primary production by the phytoplanktonic and littoral flora as the major contributor to the POC of natural lake systems. Allochthonous POC is relatively small in contrast to major inputs of DOC, except for special cases (for example, in small lakes or ponds in heavily forested areas, or reservoirs that are small in volume in relation to inputs and flow-through). The sources of POC shift in their relative contributions to the total POC pool as the fresh-water systems frequently progress through stages of increasing fertility when exposed to enhanced nutrient loading or as seen among a series of lakes at different stages of development (Fig. 22–10). Generalizations are again problematic because of the high degree of lake individuality. The importance of the littoral components to the production of POC in a given series of lakes increases greatly and changes markedly in the transition from nutrient-limited conditions of oligotrophic lakes, to light limitations imposed by biogenic turbidity, to dominance by emergent macrophytes and associated littoral microflora.

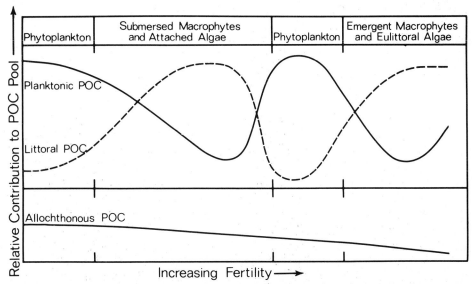

Figure 22–10. Generalized relative contributions of planktonic, littoral, and allochthonous sources of particulate organic carbon (*POC*) to lakes of increasing fertility as the dominance of pelagic and littoral primary productivity shifts (as detailed in Fig. 19-10). (From Wetzel, R. G., and Rich, P. H.: Carbon in freshwater systems. *In* Woodwell, G. M., and Pecan, E. V., eds., Carbon and the Biosphere. Springfield, Va., National Information Service, 1973.)

CONSEQUENCES OF METABOLISM DISPLACED TO THE SEDIMENTS

Sealing a lake off from its drainage basin would alter it drastically. Although lakes are frequently referred to as ecosystems, the important influences of the drainage basin (imported and exported detrital organic matter, inorganic inputs and losses, morphometry, and so forth) intimately link a lake with its surrounding terrestrial and wetland environments. A lake itself is only one component of a larger landscape unit, the lake ecosystem, which must include the entire drainage basin of the lake per se.

The Central Role of Detritus

Stokes's law states that spherical particles of a given density sediment through water at a rate directly proportional to the square of their radii. Both POC and DOC move with water, but POC ultimately will be deposited at the bottom of static water. If associated with inorganic particulate matter, DOC also will sediment such as, for example, when it is adsorbed to clay or $CaCO_3$ particles. These factors result in a displacement of some terrestrial production to the littoral zone of lakes with runoff. Littoral and pelagic production can also be largely displaced to the benthic sediments of lakes. In terms of energetics of the ecosystem, the medium of exchange is detritus; functionally, POC and DOC are equivalent. These two detrital fractions are merely arbitrary divisions of a smooth continuum of large particles to small molecules (Rich and Wetzel, 1978). Although the specific nature of the detritus modifies its behavior and its effects upon the environment (e.g., Bowen, 1979), detritus inevitably carries energy from its point of origin to its place of transformation.

Chemosynthesis and Nutrient Regeneration

In most cases, only low intensities of light reach the pelagial sediments; this fact, combined with the continual sedimentation of predominantly dead POC, ensure that benthic metabolism is primarily heterotrophic and detrital (Rich and Wetzel, 1978). The low rate of diffusion of oxygen into saturated sediments and its rapid utilization there mean that much metabolism is anaerobic, even when hypolimnetic oxygen depletion does not occur. Hutchinson (1941) terms this metabolism "pelometabolism," and describes its importance relative to "hydrometabolism" that occurs in the free water of the lake. Carbon dioxide produced by anaerobic detrital pelometabolism represents the escape of the oxidized product of an oxidation-reduction reaction (Fig. 22–11). The continued production of CO_2 during anaerobiosis indicates that the respiratory quotient (CO_2 release/O_2 uptake) of benthic metabolism is greater than 1, and that alternate electron acceptors are being reduced instead of molecular oxygen. Thus, alternate electron acceptors are receiving energy originally captured during photosynthesis and transferred to carbon as water is oxidized to oxygen and CO_2 is reduced to carbohydrate (cf. Chapter 20).

This accumulation of electrons in the sediments must be interpreted as functional with respect to the ecosystem (Wetzel, et al., 1972; Rich and Wetzel, 1978). The *detrital*

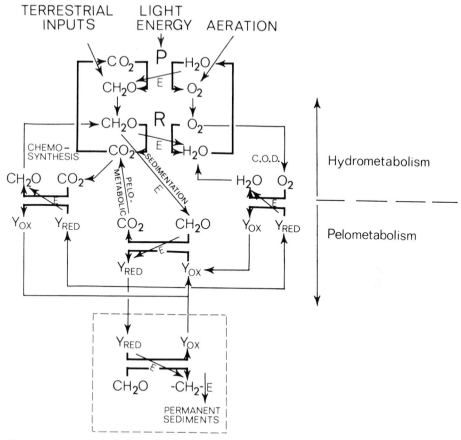

Figure 22–11. Oxidation-reduction reactions associated with benthic detrital electron flux from sediments. P = photosynthesis; R = respiration; E = electrons; red = reduced; ox = oxidized; $C.O.D.$ = chemical oxygen demand. (Modified from Rich and Wetzel, 1978.)

electron flux into and ultimately out of the sediments originates from the same photosynthetic energy source as does a predator's food and is a much larger proportion of the total source. Presumably, evolution, or at least integration, of the ecosystem has incorporated this energy flow into its overall function to the same extent as the more recognized flow through predator–prey systems. Reduced ions, radicals, and molecules are generally more soluble than their oxidized form; as a result, the reducing conditions favor an increase in the net rate of diffusion out of the sediments. Upon entry of these products into the oxidized layers of the lake, further oxidation may proceed either chemically or biologically. If the original detrital reduction involved a nutrient (sulfate, nitrate, or ferric phosphate) or a fermentation product, the detrital energy has effected both nutrient regeneration and translocation. If the subsequent "hydrometabolic" oxidation is biological (chemosynthesis), the detrital energy has been reintroduced into the biota. The energy thus transferred must be considered heterotrophic rather than synthetic in that it is originally derived from primary production (cf. Sorokin, 1964; 1965).

In order to further understand the consequences of the benthic detrital electron flux, some basic considerations of the possible alternatives of benthic metabolism are worth evaluating. Material enters the sediments by sedimentation because it is particulate—that is, it is relatively insoluble. If it remains insoluble it will enter the permanent sediments of the lake and decrease the volume of the lake—that is, reduce the life span of the lake. In order to escape the sediments it must become more soluble, gaseous, or both.

Presumably, cellulose and similar materials constitute much of the organic material that enters the sediments. The intermediary metabolic products of such compounds are increasingly soluble, and culminate in carbon dioxide which is both gaseous and highly soluble. Very reduced carbon compounds, for example, methane, are also gaseous and volatile, soluble, or both. The loss of dissolved organic carbon from the lake sediments is not well understood, but organic acids, acetic acid in particular, are significant products (Chapter 20). A carbohydrate (cellulose, for example) requires one molecule of oxygen for each carbon atom in order to be oxidized to CO_2 and water. The removal of a molecule of oxygen will release two carbon atoms when the product is methane. However, oxygen represents the heaviest atom in cellulose, and the simple removal of oxygen via the production of CO_2 reduces the mass of the molecule appreciably and results in less potential sediment.

As we have seen, CO_2 and methane are the major gases escaping from benthic metabolism. Highly reduced carbon compounds are also well-known products of long-term sedimentation, that is, coal and petroleum, and increasing concentrations of carbon, in excess of that found in carbohydrates, are measurable in lake sediments over much shorter time periods (Shiegl, 1972). Given this format, at least two additional ecosystem-level consequences of the benthic detrital electron flux may be hypothesized.

The redox potential in lake sediments brought about by anaerobic metabolism represents a measure of electron activity that has been dissociated from biochemical (enzymatic) "constraints" or reaction specificity (Rich and Wetzel, 1978). Consequently, these electrons may diffuse out of the sediments in association with a number of inorganic and organic compounds, as has already been described. The exhaustion of a particular class of electron acceptors causes the benthic redox potential to become more negative relative to molecular oxygen; this continues until new acceptors precipitate back into the sediments following hydrometabolic oxidation as highly insoluble compounds (for example, FeS, CuS, and so forth). On the other hand, the carbon and oxygen in the form of CO_2 are the relatively massive products of the oxidized component of the anaerobic oxidation-reduction reactions. Thus, the net diffusion of reduced compounds out of the sediments is controlled by redox potential, and the net diffusion controls the export of mass—that is, carbon and oxygen—from the lake sediments. In this manner, the life span of the lake is affected directly by reduction of mass through losses of large quantities of gases of carbon and oxygen. The rate of the detrital electron flux is fundamental to eutrophication rates of lake systems.

The importance of detrital electron flux in benthic metabolism is reflected further by the changes in ratios of CO_2 evolution to consumption of molecular oxygen in hypolimnia over an annual period. The oxygen uptake by the benthic community of sediments in lakes generally is considered to exceed the concomitant evolution of CO_2. Oxidations of proteins and fats result in respiratory quotients (RQ = CO_2 released/O_2 consumed) of less than unity, as has been demonstrated often by caloric combustions of benthic organisms. An RQ value of 0.85 is generally accepted as an average value (Ohle, 1952; Hutchinson, 1957) that has been used in estimates of lake productivity by hypolimnetic CO_2 accumulation (cf. Chapter 11). Anaerobic bacterial fermentative metabolism produces excess CO_2, a positive CO_2 anomaly, and volatile organic compounds, such as methane, that diffuse out of the sediments.

As Rich (1975) points out, lake sediments have an "oxygen debt" that is indicated by low and negative redox potentials and nonbiological chemical uptake of molecular oxygen by reduced substrates that were formed under anaerobic conditions. This suggests that microbial metabolism involving glycolysis and reactions in the tricarboxylic acid cycle are continuing, while oxidative reactions using molecular oxygen as the terminal electron acceptor (i.e., the respiratory cytochrome system) are impeded by limited O_2 diffusion (Wetzel, et al., 1972). Viewing the RQ concept at the benthic community level, which would include the anaerobic bacteria that utilize electron acceptors other than molecular O_2, the benthic respiratory quotients would be greater than those of organisms undergoing aerobic metabolism. Additionally, the in situ RQ values would be expected to vary seasonally with circulation periods and changes in redox gradients in response to varying rates of oxygen diffusion. Such was the case in Dunham Pond, Connecticut, in which Rich (1975) found that during stratification the respiratory quotients of the hypolimnion varied inversely with the availability of oxygen (Fig. 22–12), from less than 1 after spring circulation and oxygen renewal, but increased to nearly 3 under anoxic conditions during the latter portion of summer stratification. Similar results have been found in a number of other lakes (Rich, 1980; Rich and Devol, 1978). Under anaerobic conditions, high respiratory quotients are the result of the oxidation of organic carbon to CO_2 during the reduction of alternate electron acceptors, which then appear as oxidizable substrates.

Reductions in sediment mass also result from this process, because soluble and

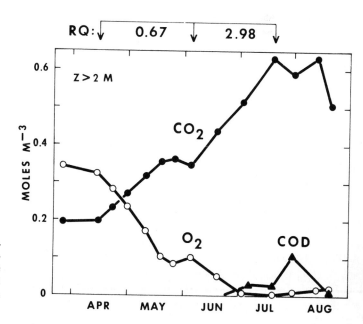

Figure 22–12. Changes in the hypolimnetic concentrations of CO_2, O_2, and chemical oxygen demand (COD) and the benthic respiratory quotients, Dunham Pond, Connecticut. (From data of Rich, 1975.)

gaseous products are formed which increase diffusion of reduced matter out of the sediments. Further, the removal of oxygen (reduction) from permanently sedimented carbon compounds represents the removal of more than half their mass.

To conclude, the alternate role of detrital electron flux suggests that existing lakes can be viewed as temporarily sedimenting energy rather than mass, which permits an extended existence. Simultaneously, benthic detrital electron flux tends to close the carbon and particularly the oxygen cycles of lakes.

Detritus: Organic Matter as a Component of the Ecosystem

Studies that address the problems of detrital origins and metabolism directly, including the nonplanktonic and terrestrial components of the lacustrine ecosystem, are exceedingly rare. The heterogeneity and diversity of lacustrine detritus are reflected in the compound nature of most lake ecosystems. The role of allochthonous inputs in the metabolism and trophy of streams has been emphasized by numerous investigations (Ross, 1963; Darnell, 1964; Hynes, 1963; Cummins, 1973). Similarly, pelagic metabolism can be strongly influenced by edaphic factors and terrestrial metabolism in small lakes with high drainage-area-to-volume ratios. The relatively recent invasion of fresh waters by angiosperms disputes Forbes's (1887) statement that a lake " . . . is an islet of older, lower life in the midst of the higher, more recent life of the surrounding region." Further, it is simply untrue, as Shelford (1918) proposed, that "one could probably remove all the larger plants from a lake and substitute glass structure of the same form and surface texture without greatly affecting the immediate food relations." A great majority of lakes are small with high shoreline-to-surface-area ratios, and pelagic metabolism is modified by littoral metabolic activities and inputs. Moreover, sedimentation in the relatively static waters of lakes results in a displacement of much of the lake's metabolism to the sediments. Prerequisite to a study of lacustrine ecosystem structure is a complete representation of the productivity inputs of *all* components of the lake. The metabolism of detrital organic matter results in a complex carbon cycle that dominates both the structure and function of lake systems.

DEFINITIONS OF DETRITUS AND ITS FUNCTIONS

Organic detritus or "biodetritus" was described by Odum and de la Cruz (1963) as dead particulate organic matter inhabited by decomposer microorganisms. The existence and importance of detritus in many habitats have since become the subject of a large and widespread literature. However, the position of detritus in relation to the trophic dynamic concept of Lindeman (cf. Chapter 8), was not clarified until recently.

Balogh (1958) demonstrated that detritus, as egested material, can constitute an important fraction of total metabolism in several terrestrial soil and litter communities. On the other hand, Odum (1962, 1963) and others have emphasized that detritus originating as ungrazed primary production supports a "detritus food chain" which is essentially parallel to the conventional "grazer food chain" at succeeding trophic levels.

The importance of the divergence between the concepts of the grazer food chain and the detritus food chain has been emphasized and clarified by Wetzel and coworkers (1972; Rich and Wetzel, 1978). Part of the difficulty stems from the concept of ecological efficiencies as it grew out of the original statement of the trophic dynamic model by Lindeman (1942; cf. Kozlovsky, 1968). Lindeman defined trophic efficiency as the ratio between assimilation by one trophic level and the assimilation of the preceding trophic level, that is, $\wedge_n / \wedge_{n-1}$. This formulation recognizes neither the existence of material egested or otherwise lost by trophic level n nor postassimilatory, nonpredatory losses at $\wedge_{n-1}$ which may be lost to those trophic levels and to the grazer food chain, but which *are not lost to the ecosystem as a whole*. Although later studies on ecological efficiencies have used different formulae, for example, Ingestion$_n$/Ingestion$_{n-1}$ (Slobodkin, 1960, 1962), a failure to distinguish between egestion and other nonpredatory losses from respiration tends to persist. Therefore, most empirically derived aquatic ecological efficiencies that do not specify respiration are in reality agricultural or grazer food chain efficiencies, and not applicable to complete ecosystems in which detrital organic matter is the significant trophic pathway.

A more realistic system operation is integrated diagrammatically in Figure 22–13. Here autochthonous organic matter represents all carbon fixed by autotrophs living within the lake ecosystem, minus their respiratory losses. The actual immediate form of productivity may be living cellular material (primary POM), which may become detrital secondarily, or dead or dissolved organic carbon, or both.

A three-phase system of organic carbon is operational: dead organic matter, both particulate and dissolved (Fig. 22–13, *left*); living particulate organic material *(right)*, which has the potential of entering either the POC or DOC pool upon death, and whose

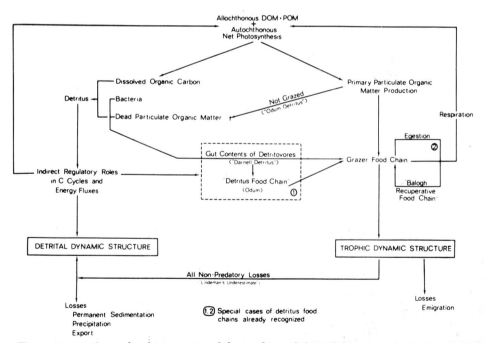

Figure 22–13. Generalized integration of the trophic and detrital dynamics of aquatic ecosystems. (From Wetzel, et al., 1972.)

metabolism may either create or destroy all three phases; and CO_2. Nonliving equilibrium forces are also present, for example, photooxidation and hydrolysis of organic compounds, but under most conditions, these processes probably represent turnover rates several orders of magnitude smaller than biological metabolism. Therefore, metabolism represents the organizing force in the structure of the organic system, and it is metabolism that determines the kinds and magnitudes of transformations that occur between the oxidized and reduced forms of carbon. The top of the diagram represents the zone of reduced carbon inputs, the middle portion the zone of oxidation to CO_2, and the lower portion represents losses of organic matter from boundaries of the system. The two parallel dynamic pools of organic carbon (dead, *left*; living, *right*) originate photosynthetically and are depleted by oxidation to CO_2 and exports. Although bacteria are shown in association with detritus, they function as biota. Odum's detritus food chain is really a link between the two pools in the direction of the biota.

A major point to be emphasized is that frequently more than 90 per cent of the organic matter entering the system from photosynthetic sources is not utilized in the grazer and detrital food chains. Among studies of the trophic dynamic food-cycle relationships, lack of appropriate recognition of nonpredatory losses from autotrophs ("Odum detritus") and nonpredatory losses from heterotrophs ("Lindeman's underestimate," see Fig. 22–13) represents a major void in contemporary system analyses.

Detritus and associated terms have been redefined (Wetzel, et al., 1972; Rich and Wetzel, 1978) in an attempt to make current terminology consistent with the ecosystem concept and to avoid perpetuating the oversight of nonpredatory pathways. The new definitions do not contradict the old, but are defined more broadly so as to avoid some of the ambiguity inherent in the concepts of the ecosystem and trophic level efficiencies espoused by earlier investigators and perpetuated by many recent ones.

Detritus consists of organic carbon lost by nonpredatory means from any trophic level (includes egestion, excretion, secretion, and so forth) or inputs from sources external to the ecosystem that enter and cycle in the system (allochthonous organic carbon). This definition removes the highly arbitrary "particulate" restriction from existing definitions of detritus. In terms of the ecosystem, there is no energetic difference between DOC lost from a phytoplankter or feces or other exudates lost from an animal. Detritus is all dead organic carbon, distinguishable from living organic and inorganic carbon. Secondly, the bacterial component is not combined with detritus ("biodetritus" = POM plus bacteria of Odum and de la Cruz, 1963) because this interferes with the general applicability of the term to situations in which detritus is not simply ingested. Such cases include the use of detrital energy in the regeneration of nutrients such as CO_2, N, and P, algal heterotrophy, losses by adsorption, flocculation, precipitation with $CaCO_3$, chelation, and so forth.

The detritus food chain is any route by which chemical energy contained within detrital organic carbon becomes available to the biota. Detrital food chains must include the cycling of detrital organic carbon, both dissolved and particulate, to the biota by direct heterotrophy of DOC, chemoorganotrophy, or absorption and ingestion. The definition of "detritus food chain" emphasizes the actual trophic linkage between the nonliving detritus and living organisms, and recognizes the metabolic activities of bacteria attached to detrital substrates as a trophic transfer.

The special case of a detrital food chain, in which detrital energy is subsequently transferred by noncarbon substrates in an anaerobic environment, has been termed *detrital electron flux*. As discussed in the previous section, such energy may reenter the biota (chemosynthesis), or mediate chemical or physical phenomena, or both, such as increasing the availability of inorganic nutrients in an anaerobic hypolimnion or in the sediments. The term "flux" specifies the flow of electrons, not carbon, to alternate electron acceptors in the absence of molecular oxygen.

Detritus, as a component of the environment, can also affect facets of the chemical and physical milieu without clearly defined energetic transformations of the detritus itself. In this situation, reference is made to effects of detritus that do not involve oxidation-reduction reactions. Examples include several indirect effects by which detrital organic carbon can influence and regulate the total energy and carbon flux of an aquatic system. Adsorption onto the surfaces of and coprecipitation of dissolved organic compounds with inorganic particulate matter (clays, $CaCO_3$, and so forth), and complexing of inorganic nutrients by dissolved organic substances, such as humic compounds, are examples of this interaction that have been discussed at some length earlier.

Detrital Dynamic Structure of Carbon in Lakes

There have been very few attempts to examine the dynamics of detrital organic carbon on a functional basis; such studies require measurements of all nonpredatory losses of organic carbon from any trophic level within the ecosystem as well as inputs from sources external to the ecosystem that enter and cycle in the system. Obviously, a number of simultaneously interacting components are operational. It is clear, nonetheless, that most of the organic metabolism of lake systems involves the detrital dynamic structure, and that an understanding of this functional system is a prerequisite for meaningful evaluation of control mechanisms of the whole ecosystem. Stated another way, we must understand carbon flux rates among all components at the same time that we seek an understanding of parameters important in regulating metabolism.

Quantitative analyses of the major organic carbon transformations in lake systems have shown the following general characteristics: (1) The central pool of organic carbon is in the dissolved form. (2) Three major sources of particulate organic carbon occur: allochthonous, and two distinct major zones of autochthonous carbon flux, the littoral and the pelagic. (3) Allochthonous inputs from the drainage basin and exports from the lake occur largely as dissolved organic carbon, and represent a major flow of carbon through the metabolism of the lake. (4) Detrital metabolism occurs principally in the benthic region, where a majority of POC is decomposed in many lakes, and the pelagial area during sedimentation. Secondary metabolism thus is displaced from sites of production and input, by sedimentation in the case of POC, and by aggregation and coprecipitation with inorganic matter in the case of DOC.

Supportive information for these generalizations can be obtained from analyses of an arctic tundra pond and of the pelagic zone of a eutrophic temperate lake for a single day, of a large reservoir over a 30-day period of the summer, and of a softwater and a hardwater temperate lake of low productivity over an annual period. Because of the complexity of these studies and the insights they offer into the central questions on metabolism, these analyses warrant closer scrutiny than has been given to other examples in this discussion.

The instantaneous carbon mass of the major components and transfer rates of carbon between them were evaluated for a typical midsummer day in a small arctic tundra pond by Hobbie, et al. (1972, as modified and corrected in Hobbie, 1980). These small ponds cover the northern tundra by the thousands, lying in depressions between low ridges formed above ice wedges; beneath them lies permafrost, which prevents any outflow. Most of the ponds are very shallow (less than 40 cm) and are frozen solid for 9 months of the year. The tundra vegetation surrounding the ponds is dominated by two sedges (*Carex aquatilis, Eriophorum angustifolium*) and a grass (*Dupontia fischeri*). The dominant herbivore of the area is the brown lemming, which clips this vegetation close to the ground. Litter from these clippings and lemming feces form a significant source of particulate and dissolved matter reaching the ponds.

The DOC pool represented most of the total carbon present in the aqueous portion of the system, because the inorganic carbon content was quite low (Fig. 22–14). Although the primary productivity on an annual basis was very low, 50 per cent of the total was contributed by the macrophytes, 47 per cent by the benthic algae, and 3 per cent by the phytoplankton. The most active transfer path was the movement of detrital organic carbon from the aquatic plants to the sediments, where decomposition results in major respiratory loss as CO_2. The flux rates are much smaller in the water than in the sediments, and the transfer of organic carbon between POC and the zooplanktonic crustacea is significant. The photooxidation of dissolved organic compounds to CO_2, although small, is significant in this shallow habitat, as has been indicated for certain streams and lakes (cf. Gjessing, 1970; Gjessing and Gjerdahl, 1970; Stewart and Wetzel, 1981a). In most systems, however, such abiotic transformations are likely very small in relation to other fluxes.

The carbon flux in the pelagic zone at midsummer in the surface water of eutrophic Frains Lake, southeastern Michigan (Saunders, 1972a), demonstrates an analogous relationship (Fig. 22–15). Within a 24-hour period, fluctuations in DOC ranged from two to six times more than that released extracellularly by the phytoplanktonic algae. Particulate organic detritus, ordinarily five to ten times greater than that of the phytoplankton, constituted a much larger potential source of DOC than living phytoplankton, and was dispersed both in the photic and aphotic zones. Similarly, detrital particulate carbon was, at this time and depth, much more important energetically than phytoplankton as a food source to the crustacean zooplankton. Clearly, the organic detrital matter is quantitatively a major constituent of this system, in terms of both mass and metabolism. Its importance increases markedly from the photic to the tropholytic zone.

Another example of the relationships among pools of organic matter, living components, and fluxes among components emerges from investigations of the pelagial zone of Dalnee Lake, eastern USSR (Sorokin and Paveljeva, 1972). This lake receives a major influx of organic matter from allochthonous sources, equal to about one-third of the total organic input (Fig. 22–16). During the month of July, the average flux rates of organic matter, expressed as calories m^{-2} per 30 days, indicated that the planktonic bacterial populations were consumed mainly by protozoa. Most of the protozoan organic matter entered the nonpredatory organic matter pool and underwent mineralization. Only a very small portion of the total organic base of detritus and microflora was utilized by zooplankton and higher organisms. About one-third of the zooplanktonic utilization of the microflora was by the rotifer *Asplanchna*, which was predatory on protozoa at this time of year. It was estimated that well over 70 per cent of the organic matter entering the system from autochthonous photosynthetic and allochthonous sources became part of the nonpredatory organic matter pool, and was sub-

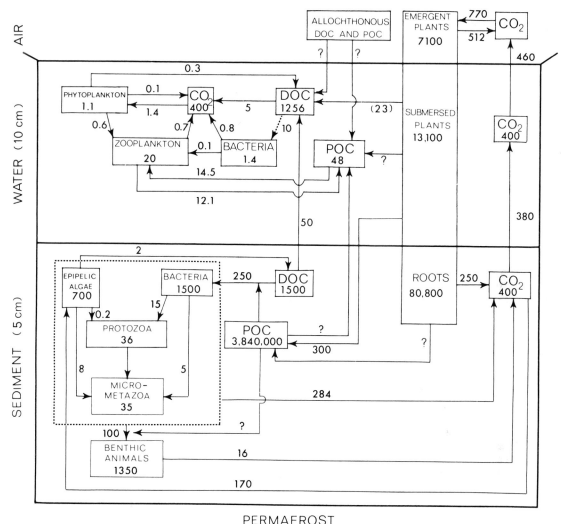

Figure 22–14. Carbon flow among components of a tundra pond ecosystem representative of approximately average midsummer conditions at coastal Barrow, Alaska (lat. 71°N). Data from Pond B on 12 July 1971 are used in the water portion of the diagram and those for Pond J on the same date for the sediment portion. Numbers in boxes refer to standing crops (mg C m^{-2}), while arrows indicate the transfer rates (mg C m^{-2} day^{-1}). DOC = dissolved organic carbon; POC = particulate organic carbon. (Modified from Hobbie, et al., 1972, as corrected by Hobbie, 1980.)

sequently decomposed or lost from the system by sedimentation and outflow without entering the animal components of the system.

A detailed evaluation of the carbon budgets of a whole lake system at meter depth intervals over several years was undertaken on a hard-water lake in southwestern Michigan (Wetzel, et al., 1972). In this system, allochthonous inputs, exports, and dynamics in the littoral, pelagic, and benthic zones were evaluated over an annual period (Fig. 22–17). Even though the littoral flora of this lake is not well developed in comparison to most lakes of similar morphometry, the single largest flow of

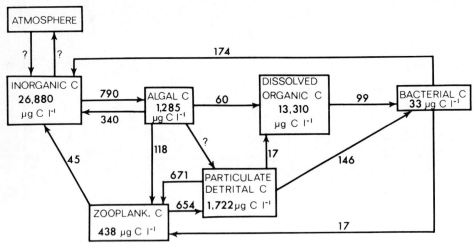

Figure 22–15. Pelagic carbon flow at 1-meter depth in Frains Lake, Michigan, on 22 July 1968. Numbers on arrows refer to transfer rates in μg C l^{-1} 24 hours^{-1}; numbers in blocks refer to carbon concentration in μg C l^{-1} at sunrise. (Modified from Saunders, 1972a.)

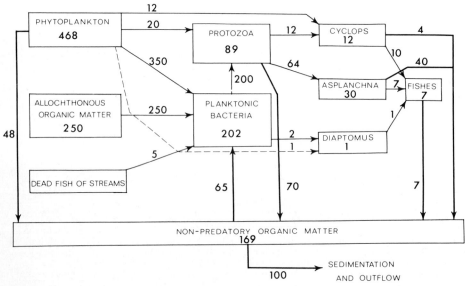

Figure 22–16. Generalized relationships of the average energy flow in the pelagic zone of Dalnee Lake, Kamchatka, USSR, during the month of July. Data expressed in calories m^{-2} per 30 days. (From data of Sorokin and Paveljeva, 1972.)

organic carbon originated as primary particulate production of the littoral zone (Table 22-6). This material was transported to the deeper sediments, where it was supplemented by additional particulate organic carbon sedimenting from the pelagic zone (Table 22-7). Allochthonous POC input to the sediments was a fraction of the POC sedimenting in the pelagic zone, and is included there. Benthic microbial metabolism of the sediments released a major amount of the sedimented organic car-

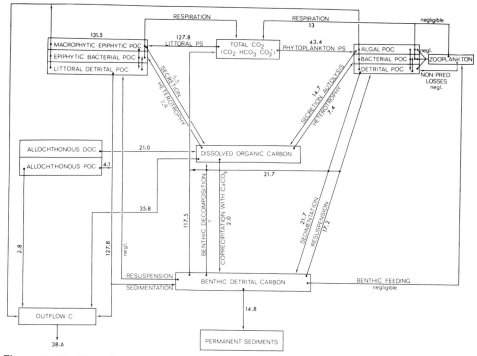

Figure 22-17. Detrital structure and flux of organic carbon (g C m⁻² year⁻¹) of Lawrence Lake, southwestern Michigan. *DOC* = dissolved organic carbon; *POC* = particulate organic carbon; *PS* = photosynthesis. (After Wetzel, et al., 1972.)

TABLE 22-6 Littoral Carbon Budget, Lawrence Lake, Michigan

CARBON	g C m⁻² YEAR⁻¹
Inputs	
Macrophytes	87.9
Epiphytic algae	37.9
Epipelic algae	2.0
Heterotrophy	2.8
Resuspension	0.0
Total	130.6
Outputs	
Secretion	5.5
Gross productivity	125.1
Benthic respiration	117.5
Net productivity	+7.6

From Wetzel, R. G., et al.: Metabolism of dissolved and particulate detrital carbon in a temperate hard-water lake. Mem. Ist. Ital. Idrobiol., 29, Suppl.: 185–243, 1972. Estimates are based on a lakewide areal basis rather than on an arbitrarily defined littoral zone.

TABLE 22-7   Pelagic Organic Carbon Budget, Lawrence Lake, Michigan

DISSOLVED ORGANIC CARBON	g C m^{-2} YEAR^{-1}	PARTICULATE ORGANIC CARBON	g C m^{-2} YEAR^{-1}
Inputs		**Inputs**	
		Phytoplankton	43.4
Algal secretion ⎱	14.7	Resuspension	17.2
Algal autolysis ⎰		Allochthonous	4.1
Littoral secretion:		Bacteria:	
Macrophytes	3.5	Chemosynthesis	7.1
Epiphytic algae	1.9	Heterotrophy of	
Epipelic algae	0.1	DOC	7.4
Allochthonous DOC	21.0	Total	79.2
Total	41.2	**Losses**	
Losses		Outflow POC	2.8
Outflow	35.8	Sedimentation	21.7
Coprecipitation with CaCO$_3$	2.0	Total	24.5
Total	37.8	**Gross production**	+54.7
Gross production	+3.4	**Respiration**	
Respiration		Algae	13.0
Bacteria	−20.6	Sedimenting POC	21.7
(Estimated as 50% of DOC		Total	−34.7
production)	−17.2	**Net production**	+20.0
Net production			

Total net production (DOC + POC) = +2.8 g C m^{-2} year^{-1}

From Wetzel, R. G., et al.: Metabolism of dissolved and particulate detrital carbon in a temperate hard-water lake. Mem. Ist. Ital. Idrobiol., 29, Suppl.: 185–243, 1972.

bon to the dissolved CO_2—HCO_3^-—$CO_3^=$ pool in respiration (Table 22-8); CO_2 released by microbial respiration was the primary flow of carbon to the photic zone. A much smaller amount of benthic organic carbon was directly utilized by the pelagic community, and approximately an equivalent amount was lost by permanent sedimentation. Utilization of benthic detrital carbon (resuspension of Fig. 22-17) was almost equivalent to sedimentation losses from the pelagic zone, and represented an important contribution to pelagic productivity of material originating from the littoral zone.

TABLE 22-8   Annual Benthic Particulate Carbon Budget of Lawrence Lake, Michigan

PARTICULATE CARBON	g C m^{-2} YEAR^{-1}
Inputs	
Submersed macrophytes	+ 87.9
Epiphytic algae	+ 37.9
Epipelic algae	+ 0.0
Loess	+ 0.0
Precipitation of DOC with CaCO$_3$	+ 2.0
Sedimentation	+ 21.7
Outputs	
Benthic respiration	−117.5
Permanent sedimentation	− 14.8
Balance (utilization)	+ 17.2

From Wetzel, R. G., et al.: Metabolism of dissolved and particulate detrital carbon in a temperate hard-water lake. Mem. Ist. Ital. Idrobiol., 29, Suppl.: 185–243, 1972.

An integration of these budgets of inputs and outputs permits an evaluation of the annual carbon fluxes in a whole lake system in which the allochthonous, littoral, benthic, and pelagic organic carbon flows are considered (Table 22–9). The whole lake budget is within 5 per cent of being balanced, which is very good considering the magnitude of variances involved in the analyses under in situ conditions. Analyses of zooplanktonic grazing by serially increasing animal concentrations to 200-fold in situ (Wetzel, et al., 1972) and in situ measurements of grazing of phytoplankton (Haney, Wetzel, Manny, unpublished; Crumpton and Wetzel, 1982) yielded maximal values of 25 per cent (nighttime) and < 5 per cent (daytime) of the epilimnetic volume grazed per day during the period of maximum development of zooplanktonic populations. At other times of the year, the values were lower and therefore were considered negligible in this lake. Without question, zooplanktonic grazing can cause marked alterations in algal species composition and succession within the phytoplanktonic community at certain times of the year (see critical analyses of Lawrence Lake algal populations in Crumpton and Wetzel, 1982). These analyses, as well as others (e.g., Jewson, et al., 1981), however, demonstrated the importance of algal losses by sedimentation out of the photic zone; on an annual basis, sedimentation losses were much greater than losses from grazing. Sedimenting algae, as well as detritus egested by zooplankton, enter the detrital portion of the carbon cycle. Analyses of the dynamics of the annual carbon fluxes in Lawrence Lake, of which only summary values are given in Tables 22–6 through 22–9, demonstrate the major inputs from the littoral flora and the dominance

TABLE 22-9 Total Annual Budget of Carbon Fluxes in g C m^{-2} Year^{-1} for Lawrence Lake, Michigan

COMPONENTS	g C m^{-2} YEAR^{-1}	PER CENT
Inputs		
Macrophytes	87.9	38.8
Epiphytic algae	37.9	16.8
Epipelic algae	2.0	0.9
Heterotrophy	2.8	1.2
Algal secretion and autolysis	14.7	6.5
Phytoplankton	43.4	19.1
Allochthonous particulate	4.1	1.8
Allochthonous dissolved	21.0	9.3
Littoral plant secretion	5.5	2.5
Bacterial chemosynthesis	7.1	3.1
	226.4	100.0
Outputs		
Benthic respiration	117.5	54.6
Outflow dissolved	35.8	16.5
Outflow particulate	2.8	1.3
Permanent particulate sedimentation	14.8	6.9
Coprecipitation with CaCO$_3$	2.0	1.0
Bacterial respiration dissolved organic matter	20.6	9.6
Bacterial respiration particulate organic matter	8.6	4.0
Algal respiration	13.0	6.1
	215.1	100.0

Derived from Wetzel, et al., 1972.

of respiration in the sediments, where this material is decomposed. When macrophytes are minor contributors to the total autochthonous organic inputs, as is the case in Mirror Lake, New Hampshire (Table 22-10), the respiratory metabolism of the sediments is still a major site of decomposition. In Mirror Lake (Table 22-10), phytoplankton dominate the inputs of organic carbon; the output percentages of Mirror Lake are nearly identical to those of Lawrence Lake (Table 22-9), in which the littoral inputs dominate. The importance of dissolved organic detritus in the inputs from allochthonous sources and in the outflow is in the same relationship to the whole in both lakes. The decomposition of photosynthetically fixed carbon by microflora of the littoral and the sediments, which probably dominates in most lakes of the world since a vast majority of lakes are small to very small, relegates the role of animals as "decomposers" to a lesser category, and the burden of metabolism is shifted to the microflora. In large lakes in which the inputs from littoral sources are proportionately low, or in hypereutrophic lakes in which phytoplanktonic densities reduce light to the point of excluding much

TABLE 22-10 Annual Organic Carbon Fluxes in Mirror Lake, New Hampshire

COMPONENTS	g C m^{-2} YEAR^{-1}	PER CENT OF SUBTOTAL
Inputs		
Autochthonous		
Phytoplankton	56.5	69.5
Epilithic algae	2.2	2.7
Epipelic algae	0.6	0.7
Epiphytic algae	0.06	0.07
Macrophytes	2.5	3.1
Dark CO_2 fixation	2.1	2.6
Allochthonous		
With precipitation	1.4	1.7
Shoreline litter	4.3	5.3
Stream DOC	10.5	12.9
Stream POC	1.15	1.4
	81.31	100.0
Outputs		
Respiration		
Phytoplankton	19.1	23.5
Zooplankton	12.0	14.8
Macrophytes	1.0	1.2
Attached algae	1.16	1.4
Benthic invertebrates	2.8	3.4
Fish	0.2	0.2
Sediment bacteria	17.3	21.3
Planktonic bacteria	4.9	6.0
Permanent sedimentation	10.7*	13.2
Outflow		
Dissolved	10.87	13.4
Particulate	0.78	1.0
Insect emergence	0.5	0.6
	81.31	100.0

From unpublished data of M. Jordan, G. E. Likens, and B. Petersen (1982).
*Estimated by difference.

of the littoral flora, animals may assume a somewhat greater importance in the degradation of POC. The bulk of organic carbon is dissolved, however, and decomposition of that organic matter is almost completely microbial.

The structure of biomass or carbon fluxes in flowing waters is much less well documented. From a detailed budget of the average daily biomass flow in a temperate woodland stream, it was estimated that 25 to 30 per cent of the particulate biomass input flowed through, and was decomposed by the benthic animals, assuming an average assimilation rate of 40 per cent for nonpredators (Cummins, 1972). Allochthonous and autochthonous inputs of dissolved organic matter, which constitute at least 5 times and probably 10 times that of the POC, were not considered in this budget. Experimental analyses in artificial streams have demonstrated that a large portion ($>$ 50 per cent) of this DOM can be rapidly degraded (24 to 48 hours) by planktonic and attached bacteria under optimal conditions. The amount of DOC that is converted to POM via flocculation, especially in hard-water streams as discussed earlier, or enters bacterial biomass that is then partially degraded by animals is unclear. This is an area of great interest in need of further research (cf. Willoughby, 1974). Less than 1 per cent of detrital organic inputs was estimated to be processed by macroinvertebrate fauna of a mountain stream of New England (Fisher and Likens, 1973). The role of animals in comparison to microflora in decomposition of POM varies widely, and may be even more variable in streams than in lakes.

It is of the utmost importance to emphasize that these examples of functional structure of lake and stream ecosystems are only a prelude to the important underlying questions of control mechanisms that operate to determine the observed structural integrity of the system. Analyses of structure and dynamics of ecosystems are a prerequisite to analyses of factors regulating the observed and changing structure. But such analyses must not become an end in themselves. Sufficient underlying similarities between the structure and dynamics of ecosystems probably exist that will permit generalizations to emerge with less rigorous analyses. As a result, more effort could be devoted to experimental evaluation of control mechanisms. It is critical that the regulatory mechanisms of natural ecosystems be understood before manmade perturbations of the systems can be effectively evaluated and minimized. Meaningful applied research and wise management of aquatic resources cannot be undertaken effectively without a basic understanding of functional control mechanisms.

Net Ecosystem Production

The dissolved organic carbon pool of lakes receives inputs from both pelagic and littoral photosynthesis. In the Lawrence Lake example, an equivalent amount was received allochthonously from the drainage basin. The major loss of dissolved organic carbon through outflow represented most of the net ecosystem production of Lawrence Lake relative to the surrounding environment (Table 22–11). This estimate of *net ecosystem production* yields essentially a measure of export from the system, for example by outflow, sedimentation, and emigration.

TABLE 22-11 Net Ecosystem Production of Organic Carbon, Lawrence Lake, Michigan

	DISSOLVED ORGANIC CARBON	PARTICULATE ORGANIC CARBON	TOTAL (g C m^{-2} YEAR^{-1})
Inputs			
Inlet 1	7.0	2.0	9.0
Inlet 2	7.9	1.0	8.9
Groundwater	6.0	1.2	7.2
Total	20.9	4.2	25.1
Outputs			
Outflow	35.8	2.8	38.6
Sedimentation	–	14.8	14.8
Total	35.8	17.6	53.4

Net ecosystem production (NEP): + 28.3 g C m^{-2} year^{-1}

From Wetzel, R. G., et al.: Metabolism of dissolved and particulate detrital carbon in a temperate hardwater lake. Mem. Ist. Ital. Idrobiol., 29, Suppl.: 185–243, 1972.

Woodwell and Whittaker (1968) originally defined net ecosystem production (NEP) as:

$$NEP = GP - [Rs_{(A)} + Rs_{(H)}],$$

when GP   = gross production,

$Rs_{(A)}$ = respiration of autotrophs,

$Rs_{(H)}$ = respiration of heterotrophs,

or simply NEP = (gross production) − (respiration of the ecosystem). The essential feature of NEP is that it accounts for importation and losses via respiration.

The original expression was modified by Woodwell, et al. (1972), to account for the import and export of organic material:

$$NEP = (GP + NP_{in}) - [Rs_{(A)} + Rs_{(H)} + NP_{out}],$$

when NP_{in} = net production imported from another ecosystem,

NP_{out} = net production exported.

Both formulations interpret NEP as the positive or negative increment of organic matter, either as living biota or dead organic storage, after total respiration within the ecosystem. The latter definition of NEP, which specifically subtracts organic exports, limits application of the term to only that material which remains inside ecosystem boundaries.

Biotic Stability and Succession of Productivity

The mechanisms and couplings of detrital metabolism, as elaborated in the three-phase system (allochthonous, littoral, and pelagic) and exemplified in the dynamics of a whole lake system (Lawrence Lake), provide a fundamental stability to the entire biotic dynamics of lakes and streams. Living components of the system generally undergo rapid oscillations of productivity in an opportunistic series of competitive responses to changes in availability of constraining nutritional and physical factors governing their growth. However, the overall metabolism of both lakes and streams is dependent to a very major extent on the detritus components of dead DOC and POC that form the primary source of utilizable energy.

Detrital organic carbon is the dominant pool of organic carbon, and forms an excess detrital reserve that is refractory to rapid metabolism. The relatively slow utilization rate of this detrital reservoir gives stability to aquatic systems, tiding the system over during periods of low detrital carbon inputs and recharging the reserve during excessive inputs.

The stability is afforded in part by the refractory chemical structure of the organic substrates and in part by the displacement of much of the organic matter to anoxic environments. When oxygen is absent or its diffusion rate into a habitat is insufficient to fulfill the requirements of aerobic microflora, the rate of carbon mineralization is slower and the number of microbial cells formed per unit of substrate is less than when oxygen is abundant (Alexander, 1971; J. B. Hall, 1971). Complex and often simple organic compounds such as fatty acids and amines can accumulate when the quantity of readily fermentable carbon is large, as in the sediments. The low redox conditions lead to the generation of anaerobic fermentative products, including gases (CO_2, CH_4, H_2, H_2S), that simultaneously inhibit many forms of microbial growth and under aerobic conditions serve as a major sink for molecular oxygen. Degradation under these conditions is slowed and limited largely to utilization of intermediate compounds of anaerobic microflora. When oxic conditions are in close juxtaposition to anaerobic degradation, intermediate products are more completely and efficiently degraded. Inhibitory conditions caused by low E_h and toxic gaseous products are greatly diminished.

The overriding importance of detrital metabolism, within which the trophic dynamic structure operates and utilizes a small portion of the available organic matter, has been alluded to in earlier works (see, for example, Lindeman, 1942; Odum and de la Cruz, 1963; Wetzel, 1968, 1972; Wetzel and Allen, 1970; Rich and Wetzel, 1972; Wetzel, et al., 1972; Saunders, 1972a). But its demonstration for whole lake systems had not been available until recently. This detrital metabolism and the stability it affords to the trophic dynamic structure must be operational in all biotic systems, both terrestrial and aquatic. The trophic dynamic structure is almost totally dependent energetically upon the detrital dynamic structure. The manner in which it is manifested differs in fine detail among systems, but functionally, it must be the same in all ecosystems.

Within the operation of the detrital dynamic structure, the sources, quantities, and composition of detrital inputs are important, and are related to lake eutrophication. As has been discussed in detail earlier (Chapters 18 and 19), the littoral plant complex of macrophytes and their associated microflora dominate autochthonous productivity of many lake systems. The contributions by phytoplankton, submersed and emergent macrophytes, and attached and eulittoral algae change with increasing fertility. Dominance by the sessile flora is usually the case, except for certain extremely oligotrophic lakes, lakes of very large area and volume, or in hypereutrophic lakes, where biogenic turbidity severely attenuates submersed light conditions (see Fig. 19–10).

The detrital dynamic structure functionally controls metabolism in a majority of lakes, in part because nutrient regeneration is controlled by detrital dynamics. Wetland and littoral plants are implicated as a major metabolic force driving detrital dynamics in most lakes, where they are dominant sources of allochthonous and autochthonous organic matter. Moreover, the detrital inputs of particulate and dissolved organic substrates vary with the flora adapted to individual lake conditions. For example, in hardwater marl lakes, DOC interacts with inorganic components such as $CaCO_3$, which effectively suppresses subsequent pelo- and hydrometabolism. These suppressing mechanisms can be loaded to saturation, resulting in rapid transition to highly eutrophic conditions, geologically speaking. In other cases, the succession of certain macrophytes can lead to a community dominated by *Sphagnum* and other mosses that function as effective ion exchangers with simultaneous release of refractory organic compounds (cf. Chapter 24), creating conditions in which detrital metabolism is slowed and permits accumulation of organic matter in particulate form. Detrital metabolism and the rates and cycling of dissolved organic matter underlie the entire metabolic and trophic structures of all ecosystems.

SUMMARY

1. Nearly all of the organic carbon of natural waters consists of dissolved organic carbon (DOC) and dead particulate organic carbon (POC). The ratio of DOC to POC is commonly between 6:1 and 10:1, both in lacustrine and running-water ecosystems.
2. Living POC of the biota constitutes a very small portion of the total POC. The metabolism of the biota, however, is of primary importance in mediating reversible fluxes of carbon between the dissolved and particulate phases of detrital organic carbon.
3. The distribution of DOC within lakes and streams is relatively constant with changes in season and in depth. Much of the DOC pool consists of refractory compounds that are chemically complex, and resistant to rapid (days to weeks) microbial degradation. The readily utilizable, labile dissolved organic compounds are cycled rapidly (hours) at low concentrations, and represent major carbon pathways of energy transformations.
4. DOC of allochthonous origin (terrestrial and wetland marsh sources) constitutes the primary source of external organic carbon loading to fresh waters. This DOC can be extensively modified by microbial metabolism in streams and in wetlands as it flows towards recipient lakes. Dissolved organic matter (DOM) that enters lakes via surface runoff or groundwater seepage contains a low percentage of nitrogen (C:N approximately 50:1). The high C:N ratio results from a high proportion of relatively resistant structural compounds in terrestrial and emergent macrophytes, and from the selective decomposition of more labile compounds by microflora as the DOM is transported by water flow.
5. There are three major sources of particulate organic carbon (POC):
   a. Allochthonous sources of POC, which consist primarily of plant materials. This POC is transported by water drainage and by air movements to recipient lakes or streams. Rates of decomposition of POC in recipient

waters vary greatly with the organic composition of the material, but are usually much slower (e.g., leaves in weeks; woody material in years) than those of DOC (days to weeks).

    b.   Most ($> 90\%$) of the POC in lakes originates from autochthonous production in (i) the pelagic zone, and (ii) the littoral zone. In reservoirs, the proportion of autochthonous POC production to allochthonous POC loading is more equal, although it can be quite variable (particularly seasonally).

6.   The spatial and temporal distribution of pelagic particulate organic carbon is commonly closely correlated with the productivity and biomass distribution of phytoplankton and bacteria during periods of thermal stratification. The inputs from littoral productivity to the POC pool of lakes increase in relative importance in the transition (e.g., in a series of lakes) from the nutrient-limited conditions of oligotrophic lakes to more fertile conditions of eutrophic lakes (see Fig. 22–10).

7.   Most of the organic carbon of aquatic ecosystems occurs as dead organic matter, both in dissolved and particulate forms, and this organic matter or detritus functions in several pivotal roles.

    a.   Dissolved and particulate organic carbon move with the water, and POC is ultimately deposited at the bottom of static water. Dissolved organic carbon will also sediment if it adsorbs to inorganic or organic particulate matter or if it undergoes polymerization to particulate form.

    b.   The major areas of detrital metabolism are the benthic region, where most particulate organic carbon is decomposed in many lakes, and the pelagic area during POC sedimentation. Secondary metabolism is thus displaced from sites of production to sites of decomposition by sedimentation in the case of POC, and by aggregation and coprecipitation with particulate matter in the case of DOC.

    c.   Although the specific nature of detritus molecules modifies their behavior and effects upon the environment, detritus inevitably carries most of the energy of the ecosystem from its points of origin to its places of transformation.

    d.   Most benthic metabolism is heterotrophic, microbial, and anaerobic, even when hypolimnetic oxygen depletion does not occur. The continued production of CO_2 by anaerobic benthic metabolism, and release of this oxidized product (synthesized using alternate electron acceptors) to the overlying water, results in benthic and hypolimnetic respiratory quotients (CO_2 released/O_2 uptake) greater than one.

    e.   The reduction of alternate electron acceptors (nitrate, sulfate, or various organic substrates) instead of molecular oxygen results in an accumulation of electrons in the sediments and the receipt of energy originally transferred during photosynthesis (where CO_2 was reduced to carbohydrate and water was oxidized to oxygen). This flow, termed *detrital electron flux*, into and ultimately out of the sediments (i) derives its energy from photosynthesis, just as a predatory animal does from its food, and (ii) is a much larger proportion of the total energy inputs and transfer processes because most organic matter becomes detrital in the ecosystem without ever passing through an animal. Thus, the ecosystems incorporate detrital energy flow into their overall functioning to an extent that is similar to, or usually greater than, that occurring through the well-recognized energy flow of predator-prey processes.

f. From a functional standpoint of ecosystem operation, detritus consists of all organic carbon (DOC and POC) lost by nonpredatory means from any trophic level (by egestion, excretion, secretion. autolysis, etc.) or inputs from sources external to the ecosystem that enter and cycle within the system (allochthonous organic carbon). The detritus food chain is any pathway by which the chemical energy contained within detrital organic carbon compounds becomes available to the biota.

g. Lack of appropriate quantitative recognition of nonpredatory losses of organic carbon from autotrophs and nonpredatory losses from heterotrophs represents a major void in most contemporary ecosystem analyses.

8. Much functional insight into the operation of freshwater ecosystems has been obtained from quantitative analyses of the dynamics of carbon cycling among the biotic communities and the inorganic and organic carbon reservoirs. From several examples discussed in detail, those general properties discussed above were verified. In addition:

a. Increases in the magnitude of autochthonous organic carbon inputs in relation to allochthonous inputs result in progressively greater decomposition in the sediments than in the water column.

b. Detrital dissolved and particulate organic carbon from allochthonous, littoral, and pelagic components is often metabolized largely in the sediments, and provides a fundamental stability to ecosystems. Many components of the biotic (living), trophic dynamic structure of freshwater ecosystems generally undergo rapid oscillations in relation to an array of dynamic factors governing growth and reproduction, while detrital metabolism is slower and more evenly sustained by a much larger organic reserve.

c. The trophic dynamic structure of freshwater ecosystems is almost totally dependent for its energy upon the detrital dynamic structure. In many lakes and reservoirs, the littoral and allochthonous components are the dominant sources of synthesized organic carbon; these two components also dominate inputs of detrital organic carbon, and are implicated as the major driving force of the ecosystems.

CHAPTER TWENTY-THREE

HISTORICAL RECORDS OF PRODUCTIVITY: PALEOLIMNOLOGY

In the past, alterations of aquatic ecosystems have been brought about by changes in climatic conditions; more recently, changes have been initiated by man's activities. These changes have affected conditions of the drainage basins, water budgets, nutrient loadings and budgets, and the resultant productivity of fresh waters. Records of these changes have been left in the sediments, which are relatively static derivatives of the dynamic ecosystems. Paleolimnology focuses on the interpretation of sedimentary sequences, and on diagenetic processes that can alter that record. The ultimate goal is to gain insights into past conditions, productivity, and changes in regulatory parameters that have caused the lake ecosystems to enter a different stage of productivity.

Sedimentary Record

The fossil record of sediments includes both the biochemical substances produced by organisms, or resulting from their degradation, and morphological remnants of specific organisms. The subject is large and the literature is voluminous. Particularly instructive reviews on paleolimnology, from which much of the following résumé was drawn, include Frey, 1964, 1969a, 1969b, 1974; Juse, 1966; Korde, 1966; Swain, 1965; Vallentyne, 1960.

Sediments that accumulate in lake basins consist of numerous source materials. The composition is influenced to a great extent by the geomorphology of the lake basin and the drainage basin. Although many lakes were formed between the intervals of cyclic glaciation and deglaciation, most lakes of earlier glacial periods have been obliterated by succeeding glaciations, by filling with sedimentary materials, or by draining. It is only because of the very recent retreat of the last major glaciation phase that so many lakes exist at the present time. With the notable exceptions of some very ancient lakes, particularly in rift areas in Africa and Asia (Livingstone, 1975), most lakes are very young ($< 20,000$ years).

The primary sedimented materials are controlled by regional geology and climate, and are modified by biological processes of both the drainage basin of the lake and the lake itself. Water movements such as wave action and currents sort particulate matter

707

according to its size and density, and the energy available for displacement. The result is a general gradient from coarser particles near shore regions to finer particles in sediments under deeper waters. This gradient can be disrupted by various mechanisms that often are related to the morphology of the lake basin. For example, deposits of sediments in littoral regions or river deltas can accumulate until they become gravitationally unstable and slump in landslide fashion, or move as turbidity flows to deeper portions of the basin where they come to overlie finer sediments.

DATING OF SEDIMENTS

Several methods have been used to determine the age of sedimented materials in lake deposits. In postglacial lakes, the most important method has been to evaluate the age of organic material by measuring its content of radioactive carbon (^{14}C).

Radiocarbon is formed in the upper atmosphere by the reaction of nitrogen (^{14}N) with neutrons (n): $^{14}N + n \rightarrow {}^{14}C + {}^{1}H$. The production of ^{14}C has been nearly constant over time, but has varied because of changes in the supply of neutrons produced by cosmic radiation. Neutron quantity is thus dependent upon the intensity of cosmic radiation. The cosmic-ray flux in the upper atmosphere is influenced by (a) the intensity of the earth's magnetic field, and (b) short-term changes in solar wind magnetic properties (Ralph and Michael, 1974; Stuiver and Quay, 1979). When the earth's geomagnetic field is strong, fewer cosmic rays reach the upper atmosphere; when it is weak, the inverse occurs. These geomagnetic changes cause about an 8 per cent change in atmospheric ^{14}C over a period of approximately 6000 years. The short-term solar magnetic variations of ^{14}C production occur in cycles of a few hundred years or less, and superimpose small deviations upon the long-term geomagnetic trends. Radioactive carbon decays back to ^{14}N with emission of a β^- particle, and has a half-life of 5568 years.

The ^{14}C of the atmosphere is a very small part of the earth's CO_2 reservoir (approximately 1 in 10^{12} parts). For purposes of ^{14}C dating, it is assumed that ^{14}C produced in the atmosphere is in rapid equilibrium with these CO_2 reservoirs and that the balance among the carbon reservoirs has been constant over time.

During photosynthesis, $^{14}CO_2$ is photosynthetically incorporated into organic matter in proportion to its availability in the atmosphere. When the organisms die and are interred in sediments, the ^{14}C contained in the organic matter continues to decay back to ^{14}N with the emission of a β^- particle. The half-life of ^{14}C disintegration is constant, and when the residual specific activity of ^{14}C in the carbon of old organic matter is accurately radioassayed, it provides an estimate of the age of the organic matter. The decay rate of ^{14}C should permit age determinations back to a limit of about 75,000 years. Technical detection limitations of assaying for negative beta radiation of ^{14}C above background radiation usually limit ^{14}C-age determinations to 40,000 years, although techniques are improving (e.g., Stuiver, 1978b; Grootes, 1978).

Much effort has been devoted to evaluating the accuracy of ^{14}C dating of organic deposits, including potential errors caused by variations of ^{14}C production within the atmosphere and ^{14}C incorporation into plant organic matter (e.g., reviews of Ralph and Michael, 1974; Stuiver, 1978a; Krishnaswami and Lal, 1978; Stuiver and Quay, 1979). Recent changes in past atmospheric ^{14}C levels can be determined by measuring the present-day ^{14}C activity of wood laid down in trees (e.g., Douglas fir, Sequoia redwoods, bristlecone pine) and accurately dated by counting tree rings. Calibrations of ^{14}C dating measurements of older samples have been made by careful comparisons with other dating techniques ($^{230}Th/^{234}U$ ratios, thermoluminescence, magnetic dating). In hard-water lakes containing appreciable quantities of old, ^{14}C-deficient carbonates, inorganic carbon fixed in photosynthesis is not in equilibrium with atmospheric ^{14}C activity, but includes carbon dissolved from carbonates of the drainage basin. This old ^{14}C-deficient carbon dilutes contemporary carbon originating from atmospheric sources. Organic matter from sediments of calcareous lakes thus gives dates that are spuriously old. Corrections are only approximate.

Dating of very recent sediments can be problematic with ^{14}C methodology. More recent dates of sediment deposits have been obtained by analyses of lead-210 (^{210}Pb). Radium-226 in soils decays to radon-222, which escapes to the atmosphere, where it decays to ^{210}Pb. This ^{210}Pb enters a lake via precipitation and is eventually incorporated into the sediments; it is termed "unsupported" ^{210}Pb, since it was produced from radium located outside the sediments. Dating by ^{210}Pb relies on the difference in the total ^{210}Pb in the sediment and the "supported" ^{210}Pb resulting from the presence of ^{226}Ra and produced within the sediments. The difference is the concentration of "unsupported" ^{210}Pb resulting from atmospheric deposition, which will decrease with depth in the sediments because of radioactive decay, provided the input of ^{210}Pb to a lake is constant, its residence time is constant, and there is no significant migration within the sediment (Pennington, et al., 1976; Robbins and Edgington, 1975). The age of the sediment at any particular depth can be calculated from its activity of unsupported ^{210}Pb relative to that at the surface, since the radioactive half-life (22.26 years) is constant. Because of the short half-life of ^{210}Pb, this dating technique is limited to about a 150-year time span.

The distribution in sediments of cesium-137, which has been present in the atmosphere since 1954 because of atomic bomb testing, has been a useful method of dating sediments deposited within the last 25 years. This method assumes that ^{137}Cs fallout with rain becomes attached to particles and that these particles are quickly (< 1 year) transported from the drainage basin to lake deposits (Ritchie, et al., 1973). ^{137}Cs falling directly upon the lake surface is adsorbed onto suspended particulate matter and sedimented.

In both the ^{210}Pb and ^{137}Cs dating techniques for recently deposited sediments, it is assumed that disturbance of the sediment stratigraphy is small. In some cases, disturbance by water movements and biological redistribution of sediments by benthic macroinvertebrates can be sufficient to obscure dating chronology (e.g., Robbins, et al., 1977; McCall and Fischer, 1980). In other cases, careful interpretation of recent dating analyses, combined with other paleolimnological data, provides much insight into reconstruction of recent lake events (e.g., Pennington, et al., 1976; Von Damm, et al., 1979).

After sediments are deposited in a lake, properties of magnetic intensity and geomagnetic declination (direction) are retained in the sediments as a *remnant magnetism*. These magnetic characteristics reflect past, relatively short (approximately 1000 years) variations in the direction of the earth's magnetic field. Much of the magnetic remanence is carried by hematite (Fe_2O_3) (Mackereth, 1971, Creer, et al., 1972). Initial studies carried out in Lake Windermere, England, showed that the direction of the horizontal magnetization oscillates about a mean direction with an amplitude of approximately ± 20° and a frequency of 2700 years. Remnant magnetism was similarly demonstrated in sediments of a number of other lakes, although with somewhat different periods of oscillation (e.g., Yaskawa, et al., 1973; Tolonen, et al., 1975; Creer, et al., 1976; Oldfield, et al., 1978a, 1978b; Thompson, et al., 1980). This method assists in rapidly dating sediments by nondestructive methods once the detailed chronology has been established within a given lake region.

A further magnetic property of sediments, their *magnetic susceptibility*, is useful in evaluating probable sources of sedimented materials in lakes (Thompson, et al., 1975; Thompson, 1978; Oldfield, et al., 1978). Magnetic susceptibility is simply the magnetizability of a sediment, i.e., its degree of attraction to a magnet. The most magnetically susceptible sediments contain the highest amount of allochthonous inorganic material washed into the lake from the drainage basin. Magnetic susceptibility also assists in evaluating spatial differences in stratigraphy within a lake basin. These differences result from variations in the deposition in different places, including the slumping of deposits from shallower areas over deeper deposits.

Some lake deposits exhibit distinct *laminations* in which temporal differences in the composition and quantities of suspended matter laid down create fine alternations of light- and dark-colored layers. When a pair of laminae (a dark and a light) represent a time period of one year, the lamination is called a *varve* (DeGeer, 1912; Nipkow, 1920; Sturm, 1979). Varves are generally restricted to sediments of lakes that have not been physically disturbed during deposition. In such lakes, they provide a means of dating sediments at annual intervals and can be used to interpret former conditions of climate, mineralogy of the drainage basin, water level, the succession of certain organisms, and trophic conditions as evaluated by other sediment parameters and from biotic remains (e.g., Merkt, 1971; Renberg, 1978; Satake and Saijo, 1978; Sturm and Matter, 1978).

Whether the deposition of sediments occurs in distinct laminations or varves depends upon the timing of lake stratification in relation to the timing of inputs of suspended matter from sources to the basin (from river runoff, deposition from the atmosphere, autochthonous production of organic matter or inorganic precipitates, e.g., $CaCO_3$, resuspension of deposited matter, or importation of suspended matter from groundwater) (Sturm, 1979). Often inputs are not sufficiently coordinated with quiescent stratified conditions; as a result, deposition is chaotic and distinct annual laminations are not formed. Bimodal sedimentary varves occur if the seasonal (discontinuous) influx of suspended matter coincides with thermal stratification of the water column in annually stratified lakes, and no bioturbation subsequently erases the pattern.

In glacial environments, relatively coarse sediments are brought to recipient lakes with the inflow of melt water in the spring and summer, and are deposited during the period of thermal stratification. Finer sediments are deposited over these during the winter when the lake is ice-covered. Similar bimodal deposition is often found in temperate lakes, particularly in meromictic lakes in which the sediments are not disturbed by water circulation (e.g., Tippett, 1964; Brunskill, 1969; Digerfeldt, et al., 1975; Ludlam, 1976; Simola, 1977). In these cases, records of seasonal differences can be found in the deposition of diatoms, tree pollen, clay, iron oxides, and/or calcium carbonate.

INORGANIC CHEMISTRY

External source materials in dissolved and particulate forms that leave the drainage basin and enter the recipient lake basin are influenced markedly by the vegetative cover. Numerous examples were discussed quantitatively in the preceding chapters, and need not be reiterated here. The long-term effects on the productivity of lakes are influenced by climatic variations as well as by man-induced changes in the vegetation and weathering processes (e.g., Mackereth, 1965, 1966; Bradbury, et al., 1975; McColl and Grigal, 1975; Dean and Gorham, 1976; Likens, et al., 1977; Bormann and Likens, 1979). An example of the effect of climate on weathering was demonstrated in some endorheic lakes of East Africa (Hecky and Kilham, 1972). Lakes in humid areas that had high weathering rates of the drainage basin received greater inputs of silica and were largely bicarbonate-dominated. In more arid regions, sodium chloride originating from precipitation dominated the inorganic composition, and inputs of silica are low. The chemistry and climate changes of these endorheic lakes were reflected in the sedimentary record as well as in the extant diatom populations.

Chemical constituents of the sediments have been used variously to interpret rates of both limnological activities of the lake as well as changes in climate and alterations of the drainage basin. The analyses assume that the chemical chronology represents, at least on a long-term basis, the composite changes in inputs to the lake and metabolic transformations that have occurred within the lake. For example, comparison of the chemical composition of numerous elements, mostly biologically nonessential, in lake sediments as well as in rock and soils of the drainage basins permits an effective evaluation of the chronology of erosion and rates of leaching (Cowgill, Hutchinson, and collaborators, 1963, 1966a, 1966b, 1970, 1973). Inputs of extremely inert, conservative elements, such as titanium, permit an evaluation of erosion rates of the lake basins. Depending on the geology of the area, other elements may indicate disturbances of drainage basins. For example, in Laguna de Petenxil, a small lake in Guatemala, sediment concentrations of calcium, strontium, potassium, and, to some extent, sodium, increased during periods of active Mayan agricultural activity in the drainage basin. Manganese, iron, and phosphorus followed the fluctuations of agriculturally determined alkalies and alkaline earths in an analogous way (cf. Deevey, et al., 1979).

Detailed investigations of the paleolimnology of a small, closed lake of volcanic origin in Italy, Lago di Monterosi, demonstrated the dramatic effects of alterations to the catchment area by the construction of a road alongside the lake in early historical times (Hutchinson, et al., 1970). The lake was quite productive shortly after its formation for a period of about 2000 years, probably related to initial easy leaching of materials rich in potassium and of fairly high phosphorus content. By 20,000 years B.P.,* sedimentation rates were very low. Low rates of deposition of inorganic constituents and organic matter, indicative of very low productivity, continued until at least about 5000 years B.P. and the Roman period. This interval was marked by a period of cold, dry conditions following the last glacial retreat from the area. A slight increase in productivity was discernible during the latter portion of this interval and a more recent period, most likely related to slight climatic amelioration and some human activity. A very large increase in nutrients and productivity occurred slightly

---

*B.P. = Before Present

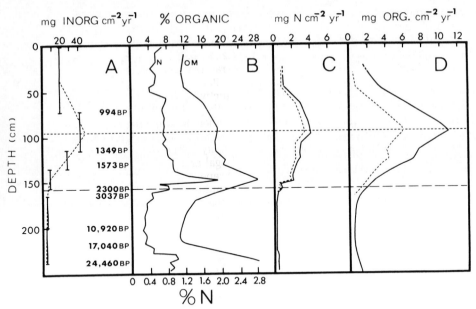

Figure 23-1. Stratigraphy of organic matter and nitrogen in the sediments of Lago di Monterosi, Italy. *A*, Estimation of sedimentation rates as a function of sediment depth and radiocarbon dates; *B*, Ignitable organic matter and nitrogen as percentages of dry sediment; *C*, Estimated rate of deposition of nitrogen, the broken line being corrected for organic nitrogen brought in by erosion; *D*, Estimated rate of deposition of organic matter, the broken line being corrected for organic matter brought in by erosion. (Redrawn from Hutchinson, et al., 1970.)

over 2000 years B.P. (Fig. 23–1). The changes in the sedimentary records coincided rather precisely with the major disturbance of the drainage basin by the construction of a Roman road, the Via Cassia, in about 171 B.C. A great increase in inputs of phosphorus and alkaline earths, particularly calcium, resulted and the lake rapidly became eutrophic, causing the deposition of much organic matter. Productivity of the lake since the period of major disturbance has declined as land uses in the area have changed.

Sedimentary records of such disturbances in the drainage basin are common. For example, the productivity of Grosser Plöner See in northern Germany was increased markedly in the early 13th century by the construction of a mill dam that raised the water depth a few meters and flooded portions of the surrounding catchment basin (Ohle, 1972, 1979; Averdieck, 1978). In more recent times in the New World, the total clearance of forests over large portions of North America has lead to dramatic changes in productivity of lakes that are recorded in the chemical stratigraphy of sediments deposited during the last two centuries. Often the increased leaching and nutrient loading leads to a marked acceleration of productivity. In contrast, when this increased leaching occurs in calcareous regions, extreme loading of calcium and carbonates can result in a negative effect on productivity owing to carbonate interactions and sedimentation of essential inorganic and organic nutrients (Wetzel, 1970; Manny, et al., 1978; Wetzel and Manny, 1978; cf. Livingstone, 1957; Livingstone and Boykin, 1962).

The paleolimnological conditions and productivity of a eutrophic hard-water lake of southern Michigan were reconstructed by combined analyses of the sedimentation rates of organic carbon, nutrients, fossil pigments, pollen, and diatoms (Manny, et al., 1978). During the late glacial period (14,075 to 10,200 years B.P.), algal productivity was low under boreal conditions of low temperatures and competition for nutrients, particularly phosphorus, which was coprecipitated with $CaCO_3$. A long period of relatively constant but slowly increasing productivity followed from about 10,200 to 5900 years B.P. During this period, the surrounding vegetation shifted from pine to oak dominance as the climate ameliorated. From about 5900 to 140 years B.P., several lines of evidence indicated a gradual increase in areas suitable for the growth of submersed littoral macrophytes, until littoral productivity likely exceeded that of the phytoplankton. Alternating shifts between littoral algal and macrophytic dominance and then to phytoplanktonic dominance occurred repeatedly, which suggested fluctuating water levels in response to alternating moist and dry periods. During this period, evidence also indicated that the hypolimnion became anoxic each year. As the land surrounding the lake was deforested by man settling into the region approximately 140 years B.P., sedimentation rates increased sevenfold. The increase in sedimentation rates occurred concomitant with markedly increased sediment concentrations of carbonates and phosphorus, and decreased organic matter, nitrogen, and plant pigments. Littoral diatoms increased over planktonic diatoms during this period. In 1926 (some 55 years B.P.), a bird sanctuary with extensive waterfowl populations was established on the lake in conjunction with a reforestation program designed to stabilize eroding soils around the lake. These actions rapidly resulted in reduced carbonate loading and markedly increased concentrations of organic nitrogen, pigments, organic matter, and particularly of phosphorus. During this period, diatoms of the sediments were dominated by eutrophic, planktonic species. The hypereutrophic conditions continue to the present time, with exceedingly high sedimentation rates (approximately 6 mm year^{-1}).

The old concept that lakes progress from states of oligotrophy to eutrophy is not necessarily universal (Deevey, 1955; Hutchinson, 1973). As has been demonstrated repeatedly from chemical analyses of sediment stratigraphy, coupled with other indices of past productivity, lake productivity responds to changing nutrient incomes, climate, and morphometry. Accumulation rates of nitrogen, phosphorus, and organic matter in sediments are commonly high in early postglacial time, a period of high availability of nutrients. Accumulation rates typically then decline as the drainage basin gradually is impoverished by leaching. In recent times, increases or decreases in lake productivity are usually directly related to disturbances caused by the activities of man.

ORGANIC CONSTITUENTS

Chemical analyses of sedimentary organic constituents must address the problem of origin: determining whether the organic compound originated inside or outside of the lake basin. Further, many organic compounds undergo degradation both during and after sedimentation; the extent of preservation is not always uniform with geologically changing conditions at and within the sediments. However, certain organic constituents of sediments, especially pigment degradation products, are very helpful in the interpretation of past episodes of productivity, particularly when coupled with other indices.

Total amino acid, carbohydrate, and chlorophyllous and flavinoid pigment residues in lake and bog sediments typically increase toward the surface (Swain, 1965). Carotenoid pigments also commonly increase upward in more recent lake sediments, but decrease markedly in the developmental transition from a lake to a bog (see discussions further on). Maxima of many organic compounds occur just beneath the surface of the sediment, probably associated both with increased microbial metabolism at the interface and with redox-mediated adsorption phenomena.

Much information is available on organic compounds and their distribution in sediments (Vallentyne, 1957b, 1960, 1969; Cranwell, 1976; Philp, et al., 1976). At the present time, a shift is occurring from simple measurements, such as the amount and distribution of total organic matter, carbon, or nitrogen, to molecular characterization of individual compounds. Many organic compounds are not very stable, and tend to be hydrolyzed enzymatically and chemically; they may then be microbially metabolized. It is often very difficult to distinguish the time, origin, and mechanism of synthesis of many organic compounds. Significant progress has been made in the association of specific compounds with specific organisms, particularly with certain pigment products (see the following discussion) and with lipid derivatives (e.g. Mermoud, et al., 1981).

> For example, analyses of carbon-chain lengths of n-alkanes derived from plant lipids showed that sediment could be characterized by the type of humic material from which it was derived (Cranwell, 1973). In higher plants, n-alkanes consist of largely odd-numbered carbon chain lengths of C_{23} to C_{35}, whereas lower plants contain n-alkanes of both odd- and even-numbered carbon chains. Humic derivatives in sediments from acidic peat areas around a lake contained predominantly C_{31} n-alkanes, whereas those from deciduous forested areas had mostly C_{27} and C_{29} n-alkanes. Investigation of monocarboxylic unsaturated fatty acid components in lake sediments, soil, and peat indicated that autochthonous material was the main source of C_{12}–C_{18} n-alkanoic acids, while terrestrial sources provided the C_{22}–C_{28} n-alkanoic acids (Cranwell, 1974, 1979). The relative contribution of these two sources could be related to the trophic conditions (greater C_{12}–C_{18} acid content in the sediment of productive lakes) and to the rate of erosion of the drainage basin. High abundance of branched/cyclic alkanoic acids was correlated with a large autochthonous contribution from algae and bacteria; a low abundance was correlated with a high terrestrial contribution to the n-alkanoic acids. High amounts of alkanoic acids were also found in productive, eutrophic lake sediments that were strongly reducing.

These and similar organic chemical approaches hold much potential and deserve intensified investigation. Problems associated with diagenetic changes in organic compounds after deposition, about which little is known, are serious because these transformations can be confused with those resulting from succession or other changes in the organisms that synthesize the compounds. Controlled experimental analyses, however, can provide insight into the mechanisms and rates of diagenesis involved.

PIGMENTS

By far the most promising organic constituents of sedimentary stratigraphy that thus far have been investigated in detail are fossil pigments, a subject reviewed at length by Vallentyne (1957b, 1960) and Brown (1969). Upon plant senescence and death, the photosynthetic pigments undergo molecular transformations in which ions (such as the magnesium ion of chlorophylls) and side groups are lost progressively during physical or biological degradation. There is a tendency for the degradation products to increase in stability—that is, decrease in solubility and increase in relative resistance to further microbial and physical decomposition. Chlorophyll a, for example, degrades to pheophytin a with the loss of magnesium; pheophytin a then degrades to pheophorbide a with the loss of the phytyl group. The alternate sequence in which the phytyl group is lost first leads to the formation of the intermediate product, chlorophyllide a.

The number of pigments found in sediments is large. Nondegraded chlorophylls are rare, but pheophytins, chlorophyllides, pheophorbides, and bacteriochlorophyll degradation products have been identified and quantified. Additionally, a number of carotenoids have been found, some of which are highly specific to groups of organisms such as the blue-green algae or certain families of organisms. The stratigraphic distribution of pigments, therefore, offers the possibility of identification and estimation of past plant populations as well as of environmental conditions of the water and sediment at the time of their deposition.

In the 1950s, Vallentyne and colleagues pioneered studies of fossil pigments in sediments of many different lake systems. These data, when correlated with the stratigraphy of microfossils of organisms, allow interpretation of the ecological records of the various lakes. The potential information that pigments can provide, with critical interpretation, is exemplified in analyses of Bethany Bog, Connecticut (Vallentyne, 1956). Myxoxanthin, a carotenoid specific to the blue-green algae, was present only in the sediments of the eutrophic phase of the lake and was not found in the more recent sediments of the lake as it entered its bog phase.

A number of analyses on several different types of lakes have demonstrated general correlations between sedimentary pigments and productivity. Most studies show maximum concentrations of pigments in early postglacial time after a period of tundra conditions associated with retreating glaciation (Fig. 23–2). The peaks generally are attributed to early leaching of nutrients from the drainage basin during the ameliorated thermal period, which is characterized by transitions from tundra to coniferous and then deciduous vegetation. This initial surge in productivity usually is followed by an intermediate period of relative stability or slowly declining productivity, with minor fluctuations associated with variations in climate, rainfall, and nutrient conditions. Subsequent pigment concentrations, and by inference phytoplanktonic productivity, are variable but commonly have been lower until very recent times. This reduction can be associated with alterations of the catchment area that increase the ratio of allochthonously derived organic matter (including pigment degradation products) to autochthonous organic matter sedimenting from production within the lake (Fogg and Belcher, 1961).

The rapid eutrophication of many lakes in recent times, often within the last century, generally is reflected in large increases in the pigment concentrations of recent sediments. This recent increase often coincides with renewed leaching of nutrients as the land was deforested for agriculture. In highly calcareous drainage basins, deforestation can accelerate inputs of calcium and bicarbonate, which interact with numerous inorganic and organic factors to suppress productivity. The inverse relationship between $CaCO_3$ and pigment stratigraphy has been demonstrated in several hard-water lakes (Wetzel, 1970; Wetzel and Manny, 1978; Manny, et al., 1978). The abrupt decline in pigment concentrations within 200 years B.P. (Fig. 23–2, *upper*) coincided precisely with deforestation, massive increases in $CaCO_3$, and the initiation of *Ambrosia* (ragweed) pollen associated with agriculture.

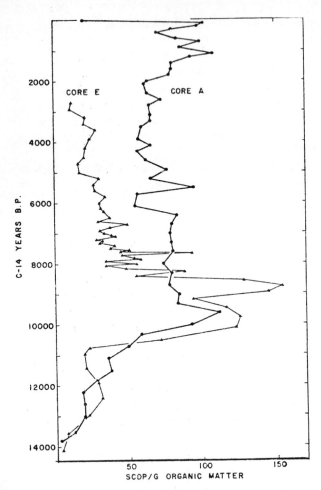

Figure 23-2. Sedimentary chlorophyll degradation products (*SCDP*) per gram organic matter in two cores from Pretty Lake, northeastern Indiana. *Core A* was from the deepwater central depression of the lake at a water depth of 25 m; *Core E* was from a shallow marl lakemount near shore that ceased accretion of sediment because of water movements at a water depth of about 1 m below the surface 2740 years B.P. (Wetzel, 1970, as modified by Frey, D. G.: Paleolimnology. Mitteilungen Int. Ver. Limnol., 20:95–123, 1974.)

Morphologically recognizable parts of a number of algae (diatoms, certain desmids and green algae, and cysts of some dinoflagellates and chrysophytes) are well preserved in freshwater sediments. Parts of other algae are preserved less well or not at all (see following discussion). Therefore, if pigments that are preserved are restricted to specific taxonomic groups that are characteristic of relatively distinct ecological conditions, their stratigraphy can be most helpful in interpretation of past conditions in the lake's ontogeny. The cosmopolitan distribution of chlorophylls among plants limits their specificity to protistan groups. Certain bacterial carotenoids and bacteriochlorophyllous degradation products of purple photosynthetic and green sulfur bacteria have been found in sediment stratigraphy and associated with events that led to eutrophication and meromixis (Brown, 1968; Czeczuga and Czerpak, 1968). Myxoxanthin, myxoxanthophyll, oscillaxanthin and other carotenoids have been used in this manner to infer the time of invasion and development of blue-green and other algae (Züllig, 1961, 1981; Brown and Coleman, 1963; Griffiths, et al., 1969; Griffiths, 1978). For example, quantitative measurements of myxoxanthophyll in sediment cores of five Swiss lakes provided a clear record of the first appearance of blue-green algae, in the absence of morphological fossils, during the marked recent eutrophical changes in three of these lakes (Züllig, 1961). The appearance of oscillaxanthin, which is specific to blue-green algae of the family Oscillatoriaceae, in recent sediments of Lake Washington in Seattle, was related directly to the development of dense populations of this algal group as the lake became excessively eutrophic owing to enrichment with sewage (Griffiths, et al., 1969; Griffiths and Edmondson, 1975).

Interpretation of fossil pigments as a measure of qualitative and quantitative changes of former protistan populations must be done critically. Several assumptions are often made that are never completely valid, and that vary with specific conditions of individual lake systems (cf. reviews of Brown, 1969; Moss, 1968; Wetzel, 1970). The extent to which pigment products of terrestrial origin dilute those formed within the lake is not always clear. Chlorophyllous pigments decompose readily to pheopigments in soils and in senescent leaves (Hoyt, 1966). The relative allochthonous input of inorganic and organic matter is largely a function of morphometry of the lake basin in relation to its drainage basin and the rate of erosion (cf. Mackereth, 1966). Although allochthonous inputs of organic matter can be quite significant in small, shallow lakes, the distribution and decomposition rates of chlorophyll and carotenoid derivatives in woodland soils, swamps, ponds, and lakes indicate that allochthonous contributions usually are small in relation to autochthonous sources (Gorham and Sanger, 1964, 1972, 1975, 1976; Sanger and Gorham, 1970, 1973).

The latter workers introduced a pigment diversity index, based on the enumeration of pigment spots on two-dimensional thin-layer chromatograms of acetone extracts, as a means of assessing the origins of sedimentary organic matter. Particulate detritus of terrestrial origin contains three dominant pigments: pheophytin a, β-carotene, and lutein, even though quantities of these pigments in leaves of various deciduous trees are very low at the time of abscission (Sanger, 1971). The carotenoids in aerobic soils were degraded more rapidly than pheophytins. Therefore, the terrestrial vegetation has a low pigment diversity which decreases further on degradation in soils. Examinations of pigment diversities of upland vegetation, aquatic macrophytes, algae, and sediments showed a marked increase in diversity with the transition to lakes and autochthonous sources (Table 23–1). The pigment diversity in

TABLE 23–1   Pigment Maxima and Diversity Indices in Woodland Soil Detritus, Swamp Peats, and Sediments of a Series of Dimictic Lakes and a Meromictic Lake

| SYSTEM | RELATIVE UNITS PER GRAM ORGANIC MATTER | | | |
	Chlorophyll Derivatives	Epiphasic (xanthophyll) Carotenoids	Hypophasic (carotenes) Carotenoids	Pigment Diversity
Woodland maximum	1.3	0.2	0.4	10
Swamp maximum	12.6	2.4	5.0	23
Dimictic lakes, minimum	1.1	0.4	0.8	24
Dimictic lakes, maximum	16.3	26.6	31.4	47
Meromictic lake, maximum	59.2	40.5	64.7	50

Modified from Sanger and Gorham, 1973, and Gorham and Sanger, 1972.

lake sediments was greater than that of other materials as a result of decomposition of the phytoplankton. Thus, the pigment diversity index is primarily a measure of pigment diagenesis. The initially high pigment diversity of algae is further increased during decomposition (cf. Daley, 1973). The ranges commonly observed were 7 to 8 for upland vegetation, 12 to 15 for aquatic macrophytes, 10 to 21 for algae, and 24 to 47 for lake sediments. The highest diversity was found under ideal conditions for preservation of pigments in sediments of a meromictic lake.

The results indicate that during oligotrophic conditions, much of the organic matter preserved in sediments is allochthonous in origin, while under eutrophic conditions pigments of autochthonous origins greatly exceed those sedimented from external sources.

Another assumption generally made is that the pigment concentrations of living algal populations represent a reasonable estimate of algal productivity, so that the sedimentary record reflects past levels of algal productivity. Although chlorophylls have been used widely as a measure of contemporary productivity (Chapters 15, 19), the pigment content of living populations is influenced by numerous environmental factors. Over time, the succession of species and changes in nutrient status may result in changes in the ratio of pigment to organic matter in sedimenting materials. In a general and relative way, however, pigment concentrations do reflect contemporary plant productivity.

Differential degradation of pigments during and after sedimentation is a more serious problem that is difficult to resolve. If pigment diagenesis is greatly variable, it is difficult to distinguish between high productivity and rapid rates of degradation or low productivity and low rates of diagenesis. Detailed experimental investigations of the destruction of chlorophyll and the formation of derivatives from lacustrine pigments demonstrated that several factors are involved (Daley, 1973; Daley and Brown, 1973). Photooxidative destruction of chlorophyll occurred in senescent phytoplanktonic cells and at an accelerated rate in cells lysed artificially, by bacteria, or by a virus. Chlorophyll a was degraded faster than chlorophyll b (cf. Brown, et al., 1977). In prolonged darkness, no destruction was observed. Enzyme-mediated chlorophyll degradation could not be detected, and from a comparison of available information, it was concluded that photooxidation is one of the principal causes of chlorophyll destruction in situ.

Destruction of pigments and derivative formation were found to be unrelated processes. Pheophytins and pheophorbides accumulated only in lysed cells, and concentrations of these two derivatives increased in the presence of dilute acids and oxygen. Derivative formation is strictly chemical and governed by oxygen concentration and pH. The rate of derivative formation decreases precipitously at pH values above neutrality. Ingestion of algae by herbivores, such as *Daphnia* and the chrysophycean flagellate *Ochromonas*, resulted in the destruction of chlorophyll, and the formation of pheophytins and pheophorbides.

Preservation of pigments is favored by the low temperatures found in hypolimnetic waters. Thus one would anticipate that the best conditions for preservation occur in the anoxic, cold, and relatively dark monimolimnia of meromictic lakes (cf. Table 23–1). In more typical lakes, however, the effects of circulation should be kept in mind. Pigments deposited in surficial sediments may be resuspended during circulation of the water column, and thereby reexposed to oxygenated, lighted conditions. Temperatures are relatively low in hypolimnia, during circulation periods in the temperate regions, and during winter. Hence, the conditions for chemically mediated formation of pigment derivatives may be fairly similar over much of the history of the lake development. As the basin fills in and becomes too shallow to permit stratification or becomes meromictic, however, conditions for preservation and diagenesis can change markedly (Daley, et al., 1977). Therefore it must be clearly recognized that use of pigment concentrations and constituents as paleolimnological indices requires critical evaluation (Brown, et al, 1977; Daley, et al., 1977). The methods have merit in interpreting past lacustrine events, but should complement morphological and other reconstructions rather than serve as a sole approach.

Morphological Remains

Paleolimnology has a long history that began with analyses of morphological remains of aquatic and terrestrial organisms. Preservation of dead organisms in the sediments is typically incomplete, and the extent of preservation is dependent both upon the type of organism and the environmental conditions that prevail at the time of sedimentation. Pollen and spores of terrestrial plants are most abundant, as are frustules of diatoms and chrysomonad cysts (Frey, 1974). Some other groups of algae can occur, but many are not represented. The Cladocera and chironomid midges have the most abundant and diversified of animal remains, although all groups of animals are represented to some extent. Many remains can be matched to species, particularly among the diatoms, desmids, Cladocera, ostracods, and beetles. When only a stage in the life cycle is represented, e.g. cysts of chrysomonads or resting eggs of rotifers, features allowing specific identification are lacking and differentiation often is possible only to the point of the genus or family level.

POLLEN

Plant pollen or spores are produced in great abundance, but only a few of those produced fulfill their reproductive function. As pollen types are dispersed, they are well mixed by atmospheric turbulence in proportion to the quantity and composition of the parent vegetation in and surrounding lakes. Most pollen grains are resistant to decay if deposited in nonoxidizing sites, such as occur in the sediments of most standing waters. The taxonomy of pollen is relatively well known in many parts of the world. Since pollen is relatively abundant in sediments, small aliquots of sediment are sufficient for analysis. As a result, analyses can be performed at very close intervals within the stratigraphy of sedimentary deposits. The pollen assemblage in sediments of known age, which can be determined by various dating techniques, provides an index of vegetation in and surrounding the lake in the past, and also reflects changes that have occurred in the vegetation through time.

A number of detailed works are available on the principles of pollen analysis, including pollen identification and interpretation of pollen data. A good summary of these analyses is given in Birks and Birks (1980). Although many earlier analyses depicted the stratigraphy of pollen as relative percentages of the different pollen types, much greater interpretative power is gained from determinations of the absolute pollen frequency (number of pollen grains cm^{-3}). Then, with determinations of the sediment accumulation rates obtained from dating measurements, the absolute pollen accumulation rate (grains cm^{-3} $year^{-1}$) can be determined for the different types or species. Absolute pollen accumulation rates (= pollen influx), unlike percentages, provide independent rates for each species in proportion to its densities and its past population dynamics. Careful interpretation is required, however, because pollen influx into lakes and their sediments can vary with factors not associated with the population and community dynamics of the surrounding vegetation. For example, some lakes are more efficient than others in collecting pollen because of differences in their morphology, their position within the landscape, and the quantities of pollen influx received from inflowing streams vs. that received from the atmosphere (Davis, et al., 1973; Bonny, 1976; Pennington, 1973, 1979). Within a lake, sediments and pollen can be resuspended and moved from shallow to deep parts of the basin (Lundqvist, 1927; Davis, 1968, 1973). This process of sediment focusing can result in a greater net accumulation of sediment in the deeper parts of the lake basin (e.g., Likens and Davis, 1975; Lehman, 1975; Davis and Ford, 1982).

The intensity of sediment focusing can change over time. For example, it can decrease as the basin fills. This reduction may be erroneously interpreted as a decrease in apparent rates of accumulation per unit area (when measured at a single point), since more recent deposits are being spread more thinly over a greater area than previously.

The stratigraphy of pollen and spores provides a chronology of the vegetation from which presumptive evidence for climatic changes can be obtained. These vegetational changes of the landscape and catchment area provide insight into temperature and rainfall changes, soil development, and changes in the drainage basin caused by the activities of man. Knowledge of the detailed vegetational history of the drainage basin is indispensable for interpretation of changes in the developmental stratigraphy of the lake through its sediments.

Nearly all detailed paleolimnological analyses augment their interpretative power by means of pollen diagrams. Hence, the number of studies on pollen stratigraphy is very large. A few examples will demonstrate their usefulness.

> In a classical study of the now meromictic Längsee, Austria, Frey (1955) demonstrated that permanent meromixis began about 2000 years ago when man moved into the area and began clearing forest vegetation from the drainage basin for agriculture. Pollen stratigraphy showed a marked decrease in the dominant forest species and a sharp increase in pollen of smaller species, two cultivated plants, and a number of agricultural weeds. Evidence also was found for markedly increased erosion of fine clay particles that probably were sufficient to increase the density of lower water strata and initiate meromixis, which was subsequently maintained biogenically.

Numerous other examples exist in which the pollen stratigraphy reflects how sensitively lake metabolism responds to changes in the drainage basin. The response of the Italian Lago di Monterosi to construction of a Roman road, discussed earlier, also can be seen in the abrupt changes in pollen deposition from surrounding terrestrial vegetation. More recent changes in lake drainage basins resulting from agricultural clearing of land are seen in many parts of North America. As the land was denuded, erosion and leaching was increased markedly, usually resulting in accelerated nutrient loading and productivity. This sedimentary horizon is demonstrated by many chemical and biotic changes, the most conspicuous of which is the sudden rise in pollen of the agricultural weed, *Ambrosia,* or common ragweed. The importance of the erosion rate in the drainage basin also was shown to be of major importance in the stratigraphy of lake sediments in the English Lake District (Tutin, 1969; Pennington, 1981). Pollen spectra throughout late and postglacial times correspond closely with the chemical history of the drainage basin. During the late glacial and early postglacial period, the erosion rates were controlled by low temperatures. In the mid-postglacial period of deciduous forest, erosion was at a minimum. Within the last 5000 years, variation in or destruction of the deciduous forests surrounding many lakes in the English Lake District is evident from pollen stratigraphy and rates of erosional deposition.

ALGAL REMAINS

Several types of algae are well preserved in lake sediments. By far the most common algal microfossils are the siliceous frustules of diatoms (Korde, 1966; Juse, 1966; Round, 1964b; Bradbury, 1975). Diatoms are useful paleoecological indicators because their remains commonly occur in abundance, are often well preserved (see following discussion), and most can be identified to species from frustule morphology. Additionally, the ecological characteristics of many diatoms are distinct and well known, which permits reconstruction of probable past lake conditions from past diatom communities. Other algae or parts of algae are preserved less completely than diatoms. Certain green algae (e.g., *Pediastrum*), scales and cysts of chrysophytes, cysts of dinoflagellates, and heterocysts of blue-green algae are preserved in lake sediments (e.g., Leventhal, 1970; Brugam, 1978; Crisman, 1978; Elner and Happey-Wood, 1978; Munch, 1980; Smol, 1980). The stratigraphy of these algae, and of diatoms as well, can be useful in interpreting past conditions when coupled with changes in other parameters recorded in the sedimentary record. However, less is known of the physiology and ecology of these other algae than is known of many diatoms.

Diatom profiles in sediments are complicated by differential rates of dissolution of the frustules during and after sedimentation, which raises questions about how rigorously the remains found in sediments reflect the quantitative composition of producing populations. The few studies directed toward this question are conflicting, but it is apparent that dissolution rates vary among individual lake systems in relation to their chemical and morphometrically related history (cf. Chapter 14). For example, it has been found that frustules sedimented in the littoral and shallower profundal zones dissolved at higher rates than those sedimenting in the deep profundal zone (Tessenow, 1966). Under mildly acidic conditions, as in the hypolimnia of some meromictic lakes, dissolution of diatom frustules can be higher than in less acidic sediments (e.g., Meriläinen, 1969, 1971).

In spite of these interpretative problems, the stratigraphy of diatoms has been most useful in casting insight into past conditions and causes of change. Because of known ranges of ecological requirements for many species and knowledge of community composition from contemporary diatoms, inferences about past conditions are possible.

In very general terms, centric diatoms are associated with more oligotrophic waters, while planktonic pennate diatoms tend to be more characteristic of eutrophic waters. The ratio of these two indicator groups of diatoms in the stratigraphic record can offer some help in interpreting the chronology of positive or negative changes in productivity. Around Lake Washington in Seattle, early development of the human population extended over the period from 1850 to the early 1900s. From about 1910 to the mid-1920s, raw sewage entered the lake in increasing amounts. Between 1926 and 1941, much of the primary effluent was diverted away from the lake. In 1941, with progressively increasing population expansion, large amounts of secondarily treated sewage entered the lake. These effluents of treated sewage continued to enter the lake until the mid-1960s, when all sewage was diverted to Puget Sound. Examination of the diatoms preserved in the recent sediments indicated detectable responses to the known sequence of enrichment, reduction, and renewed enrichment (Stockner and Benson, 1967). Major changes in the diatom community occurred (Figs. 23-3 and 23-4), related largely to reciprocal fluctuations in the abundance of the centric *Melosira italica* and the pennate *Fragilaria crotonensis*. Relative composition of the diatoms deposited, as well as species diversity redundancy values, was constant for the period prior to enrichment. Algal dominance could be correlated with the pattern of sewage discharge into the lake. Many of the species changed in proportion in accordance with their generally known nutritional behavior as indicator species. It was concluded that use of indicator groups (in this case, the ratio of centric to araphid pennate diatoms) was more reliable than use of single indicator species.

The increased ratio of Araphidineae (e.g., *Fragilaria*, *Asterionella*) to centric diatoms (A/C) has been successfully applied to a number of other recent sediments as an

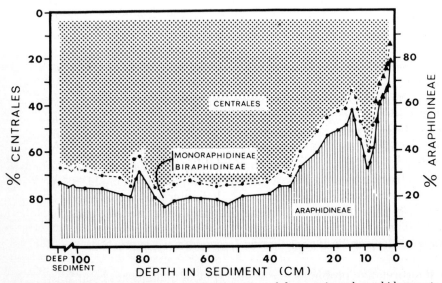

Figure 23-3. Changes in the percentage composition of the centric and araphid pennate diatoms in the surficial sediments of Lake Washington. The changes around 80 cm are associated with a clay turbidite layer peculiar to this core but not generally present throughout the lake. (Redrawn and modified from Stockner and Benson, 1967.)

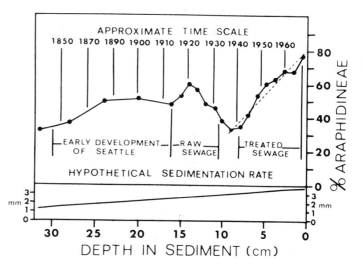

Figure 23-4. Changes in the relative abundance of araphid pennate diatoms in the recent sediments of Lake Washington in relation to the developmental history of the region surrounding the lake. The time scale is derived from an estimated average sedimentation rate of 2.5 mm per year over the upper 30 cm. (Redrawn and modified from Stockner and Benson, 1967.)

indicator of eutrophication of lakes (e.g., Stockner, 1971, 1972, 1975; Bailey and Davis, 1978; Brugam, 1978; Culver, et al., 1981). It is clear, however, that high A/C indices of eutrophic conditions are not universal, and are characteristic only in certain lakes of moderate total phosphorus concentrations and low alkalinities. In other lakes, even though araphidinate diatoms such as *Fragilaria* commonly increase in response to human settlement and increased nutrient loading, certain centric diatoms (particularly *Stephanodiscus*) increased even more vigorously (Haworth, 1972; Bradbury and Megard, 1972; Bradbury, 1975; Bradbury and Winter, 1976; Battarbee, 1978; Koivo and Ritchie, 1978; Brugam, 1979). Eutrophic lakes dominated by the centric diatom *Stephanodiscus* are nearly always of high alkalinity (> 1.5 meq l^{-1}). Caution is clearly warranted in any use of simple A/C ratios or of indicatory species as an index of trophic conditions.

Experimentally determined tolerances of diatoms to ranges of alkalinity and other chemical parameters have permitted reconstructions of past chemical conditions in lakes based on past diatom communities. For example, in certain shallow African lakes wide fluctuations in total ion content, lake morphometry, and paleoclimate have been inferred from stratigraphic changes in diatom communities (Richardson, et al., 1978).

The predominant habitats of many diatoms are well understood. Therefore, stratigraphic separation of diatom communities into littoral (epiphytic or benthic) vs. planktonic-dwelling species can yield further insight into past conditions of lake morphometry, water-level fluctuations, and shifts between littoral vs. planktonic dominance of primary productivity (e.g., Round, 1961c; Evans, 1970; Manny, et al., 1978).

PLANT MACROFOSSILS

Remnants of large plants are frequently preserved in lake sediments. Plant macrofossils usually consist of fruits, seeds, megaspores, and fragments of leaves, rhizomes, flowers, and woody tissue (see reviews of Dilcher, 1974; Birks, 1980; Birks and Birks, 1980). Seeds and other reproductive parts are often well preserved and identifiable to species. Such macrofossils supplement the pollen record. Many pollen grains cannot be identified to the species level, and some pollen types are poorly preserved. In addition, some common plants produce so little pollen that they are not routinely represented in pollen counts. Because of their larger size, plant macrofossils are usually deposited close to their sites of origin; in contrast, pollen is easily and widely dispersed. The limited transport of plant macrofossils can result in heterogeneous distributions within sediments in different areas of a lake. Moreover, because the quantity of plant macrofossils per volume of sediment is usually very low, one must examine large quantities of sediment.

Quantitative analyses of the stratigraphy of plant macrofossils assist in reconstruction of past lake development and changes in primary productivity (see the Wintergreen Lake example discussed earlier). Changes in lake levels, water chemistry, and regional climate have been evaluated, in part, from plant remains and inferences from known physiological and ecological characteristics of the plants (e.g., Watts and Winter, 1966; Birks, 1980). Macrofossils can also provide insight into the rates and types of plant succession that has occurred in and surrounding lakes.

ANIMAL REMAINS

Nearly all groups of animals leave at least some identifiable morphological remains in lake sediments. At the microscopic level, the most abundant animal remains in sediments are those of ostracods, Cladocera, and midges, and under favorable circumstances shells or cases of testaceous rhizopods, spicules of sponges, egg cocoons of neorhabdocoele Turbellaria, resting eggs of rotifers, bryozoan statoblasts, copepod spermatophores, and oribatid mites can also be found (Frey, 1964, 1969, 1976). At the macroscopic level, mollusk and beetle remains dominate. The remains of larger organisms (e.g., fish scales, bones) are generally less abundant and require larger volumes of sediments for the recovery of significant numbers. Therefore, their stratigraphy must be based on coarser intervals than that of microscopic fossils.

Numbers of species or groups are often expressed in relative percentages of the total assemblage or, preferably, are related to the weight of organic matter or some other parameter. Varying rates of sedimentation can be determined from radiocarbon dating and certain other stratigraphic criteria that permit expression as an absolute rate of accumulation, which can result in quite different interpretations than those based on percentage composition (cf. Davis and Deevey, 1964).

Certain shifts in populations of organisms would be anticipated with changes in the characteristics of a lake during its ontogeny from a primitive to an advanced condition. Therefore, as a lake becomes more eutrophic with decreasing oxygen content of the hypolimnion, one might expect, for example, a succession of profundal midge populations from those that require high oxygen to those that tolerate decreasing oxygen concentrations. In detailed studies of the stratigraphy of chironomid and *Chaoborus* midge faunas of several lakes of northern Indiana, Stahl (1959) demonstrated such a succession in midge populations and offered strong evidence for the transition from aerobic to anoxic hypolimnetic conditions among the more productive lakes. Additional evidence has been found for historical changes in eutrophication of lakes from marked changes in benthic invertebrate remains in sediments (e.g., Stahl, 1969; Carter, 1977; Hofman, 1978b; Wiederholm and Eriksson, 1979).

Exoskeletons of cladoceran zooplankton preserve reasonably well in lake sediments. This group has been studied in the fossil record in great detail. Analyses of past populations also have been made in which information theory at the community level has been applied, refining interpretations of responses to changing ecological conditions.

Species of chydorid Cladocera are largely littoral and occur on a variety of substrata. Considerable evidence exists, however, that the littoral faunal remains are redistributed by currents within lake basins, where they become integrated with remains of planktonic Cladocera (Mueller, 1964; DeCosta, 1968; Goulden, 1969a). The populations recorded in profundal sediments, therefore, offer a reasonable integration of community dynamics over habitats and seasons. Changes with time or among lakes can then reflect changes in the communities in response to such varying biotic conditions as shifts in food quantity and quality, competition, or predators (Frey, 1976; Hofmann, 1978a, 1978b; Crisman and Whitehead, 1978).

There are many examples of the usefulness of cladoceran remains in interpreting past conditions (Frey, 1969, 1974; Birks and Birks, 1980). Changes in species diversity and equitability of chydorid communities, as pioneered by Goulden (1969b), have been used to assess the stability and responses of the communities to disturbances. Equilibria of the population associations have been shown to be disrupted by the establishment of competing species, or by agricultural disturbances, climatic alterations, and shifts in the ratio of lake productivity contributed by littoral and phytoplanktonic sources (Whiteside, 1969). The ratio between planktonic and littoral cladoceran remains also is useful in interpreting past changes in water level and the extent of littoral development.

Freshwater mollusks are relatively poor indicators of past paleoclimatic fluctuations and lake conditions because of their generally broad environmental tolerances (Ouellet, 1975). In some cases, however, temporal changes in the molluscan fauna provide useful interpretations of postglacial conditions of lake-water chemistry and water-level fluctuations (e.g., Ohlhorst, et al., 1977).

_____ SUMMARY

1. The mineralogy and structure of sediments, their organic and inorganic chemical constituents, and the morphological remains of organisms preserved in the sediments permit interpretations about past states and conditions of the ecosystem that have led to positive or negative alterations of productivity. As in all paleontological records, gaps occur and the record is incomplete. Variable inputs from the drainage basin, redistribution of sediments within the lake basin, and differential preservation of interred remains demand critical interpretation of the record. The amount of information in the sediments, however, is large and has been appreciated fully only in

recent years. Accurate interpretation of the historical records depends upon a good understanding of the biology and regulatory physicochemical processes of contemporary ecosystems.

2. Evaluations of past events in the historical development of a lake require accurate determinations of the rates of sedimentation over time. Several methods have been used to determine the age of sedimented materials.

   a. Radioactive carbon-14 is being continuously generated from atmospheric nitrogen by cosmic radiation. This ^{14}C is also in equilibrium with CO_2 in the water. During photosynthesis, a small amount of the radioactive ^{14}C is incorporated proportionally into new organic matter. When organic matter is interred in the sediments, the ^{14}C decays back to ^{14}N at a constant, known rate. Accurate radioassay of the amount of ^{14}C remaining in old organic matter thereby provides an estimate of its age. Numerous variables in the synthesis of ^{14}C and its incorporation into organic matter require careful time corrections in order to provide accurate dates. The ^{14}C dating technique permits age estimates over the past 40,000 years or somewhat longer.

   b. Age of recent sediments (to 150 years) can be estimated by the content of more short-lived isotopes (e.g., lead-210).

   c. As sediments are deposited in a lake, they retain remnant properties of magnetic declination and intensity in minerals such as hematite (F_2O_3). The remnant magnetism is caused by the earth's magnetic field, which oscillates about a mean direction with a constant frequency at a given location (approximately 2700 years).

   d. Some lake deposits are laid down in distinct laminae. When annual laminations are distinguishable, they are called *varves*. Under ideal depositional circumstances, as is found in certain meromictic lakes and in lakes where annually pulsed loading (e.g., glacially fed streams) occurs during thermal stratification, varves permit direct age determinations.

3. Paleolimnological analyses of many chemical and biological remains in sediments indicate that the old concept that most temperate lakes progress from oligotrophy to eutrophy is not universal. Nutrient loading and productivity were commonly very low for long periods following the inception of lakes under the cold boreal conditions following glaciation. But the transition period from tundra to coniferous to deciduous-forested conditions, with its ameliorated temperatures, is characterized by high lake productivity. Productivity often declined and stabilized as land surrounding lakes became impoverished of nutrients by leaching of the soils and as nutrients were retained by forested vegetation surrounding lakes. As the vegetation of the drainage basin was disturbed by man's activities for agriculture and other purposes, nutrient loading commonly increased, resulting in enhanced productivity. In calcareous regions, drainage-basin disturbances increased cation and carbonate loading to recipient lakes, which counteracted the effects of simultaneous loading of important nutrients (e.g., phosphorus) and thereby reduced productivity.

4. Sediments of lakes contain numerous organic compounds of biological

origins produced either autochthonously or within the drainage basin and transported to the lake.

a. Some compounds, e.g., lipid derivatives, are characteristic not only of terrestrial plants but also of algae and bacteria. Analyses of the proportions of these compounds in sediments provide insight into the differences in the proportion of allochthonous vs. autochthonous sources of the organic matter found in sediments.

b. Degradation products of plant pigments are relatively stable after death of the organisms that synthesized these compounds. Many pigment degradation products preserve well in sediments, and their composition and remnant concentrations in sediments have been used to interpret past changes in algal communities and productivity.

    i. Certain carotenoids are synthesized only by specific algal or bacterial groups. The stratigraphy and quantities of these compounds have permitted inference about the time of invasion and development of certain blue-green and other algae and bacteria that do not leave morphologically recognizable fossils in the sediments.

    ii. In conjunction with other chemical and biological characteristics, the quantitative stratigraphy of chlorophyllous and carotenoid degradation products in sediments has been used in a number of paleolimnological studies to determine past general productivity. Critical chromatographic discrimination of the many pigment degradation products is necessary for the accurate evaluation of those relatively stable compounds that originated from autochthonous primary productivity. In many lakes, evidence indicates that most of the remnant pigment products originated from autochthonous primary producers.

5. Morphological remains of many plants and animals are fossilized in lake sediments. Since most lakes are young ($< 20,000$ years since inception), if the contemporary physiology and ecology of the organisms are known, inferences can be made about past ecological conditions.

a. Plant pollen and spores are commonly deposited into lakes in proportion to changes in the composition and quantity of parent plants in or surrounding lakes. Most pollen grains preserve well once they have settled into anoxic reducing sediments, and they can be identified, sometimes to species. Absolute pollen accumulation rates provide independent measures of each species in proportion to its density and past population dynamics.

    i. The stratigraphy of pollen and spores provides presumptive evidence of the chronology for vegetational changes in the drainage basin related to changes in temperature, rainfall, soil development, and disturbances caused by the activities of man.

    ii. Pollen stratigraphy is used in all comprehensive studies to augment other paleolimnological analyses; changes in the plant communities of the drainage basin markedly influence the nutrient loading characteristics and are reflected in subsequent chemical and biological indicators of lake productivity.

b. Although the siliceous frustules of diatoms are the most common algal microfossils preserved in lake sediments, some other algae or algal parts (e.g., cysts or scales of chrysophytes or dinoflagellates) are also found. Because diatoms generally preserve well, can be identified to species from the frustules, and because much is known of their physiological ecology, their stratigraphy has been studied extensively in conjunction with other paleolimnological parameters.

    i. Changes in diatom communities through the history of lake development have been related particularly to probable past changes in nutrient chemistry and salinity, and by inference to alterations in lake level and climate.

    ii. Changes in ratios of littoral-dwelling to planktonic diatoms have also been used to evaluate past changes in water levels and littoral development.

c. Plant macrofossils (fruits, seeds, megaspores, tissue fragments) are frequently found in sediments but are less abundant than are microfossils. Plant macrofossils assist in reconstruction of past changes in plant succession, water levels, water chemistry, and regional climate from known physiological and ecological characteristics of the plants.

d. Nearly all freshwater animals leave some identifiable remains fossilized in lake sediments. The most widely studied and abundant fossils are remains of Cladocera, ostracods, and midges, many of which can be identified to species from the exoskeletons.

    i. Based upon known population dynamics and factors controlling community composition and species diversity, past cladoceran communities provide presumptive evidence for past environmental conditions, including food quantity and quality, competition, and selective predation. Ratios of past littoral vs. planktonic cladocera can be useful in interpreting previous changes in lake-water levels and littoral development.

    ii. Study of past midge community structure has proven particularly useful in reconstruction of the historical transitions of increased productivity and resulting depletion of oxygen from the hypolimnia of lakes in more productive stages.

CHAPTER TWENTY-FOUR

ONTOGENY OF LAKE ECOSYSTEMS

The ontogeny (successional development) of lake ecosystems is governed by many interacting causal mechanisms that affect the rates of autotrophic productivity. Nearly all of the preceding discussion of this book has been directed to these mechanisms and their integration. As we have seen, a number of complex factors influence productivity of fresh waters. A fundamental property encompasses the multitude of dynamic factors that regulate the productivity of individual ecosystems at any given time: in general, lakes are formed, usually by catastrophic geological events, and proceed through a series of stages to eventual filling, until they are obliterated as lakes per se and are incorporated into the terrestrial landscape. As we have seen, particularly in Chapter 23, the *rate* at which lakes fill varies greatly. Filling by inorganic materials is governed to a great extent by the basin morphometry, characteristics of the drainage basin, and climatic factors. Organic loading is influenced to a greater extent by factors regulating autotrophic productivity and the decomposition of the organic matter produced.

Although the general ontogeny of lakes is from low to higher productivity, we have seen many cases (Chapter 23) in which past productivity was higher in the earlier ontogeny of lakes than in subsequent periods. In very recent times, however, many lakes are experiencing accelerated productivity or *eutrophication*, often as a direct result of increased loading of nutrients (see particularly Chapter 13). The enhanced productivity is desired and encouraged by certain human cultures to increase production of protein sources for consumption. In western cultures, eutrophication is held less desirable, at least at the present time, because lakes are used primarily for water supply and recreational purposes, and these uses are more compatible with oligotrophic or mesotrophic characteristics.

Throughout the previous discussions, I have attempted to integrate littoral and wetland productivity into overall lake functions. In the late stages of lake ontogeny, autochthonous productivity shifts from phytoplanktonic dominance to total dominance by littoral components. In shallow lake basins, which predominate the earth, and in lake systems in which sedimentation has reduced water depth sufficiently for littoral dominance, rates of lake ontogeny greatly accelerate. The ensuing brief discussion attempts to summarize and integrate the earlier treatments.

ONTOGENY OF AQUATIC ECOSYSTEMS

Throughout most of this text, the dynamics of aquatic ecosystems have been considered in a comparative, biogeochemical context. Rationale for this input–output approach has focused on its power to provide insight into the mechanisms governing contemporary productivity. Clearly, the physiological and population characteristics of the heterogeneous biota of aquatic ecosystems must be integrated with the dynamics of the physical and chemical parameters that influence growth and behavior. However, aquatic ecosystems commonly are not in steady state, except in a restricted short-term sense. The differences among aquatic ecosystems, as is evident from comparative studies, are obviously related to an array of dynamic factors that change over time. Short-term changes, keyed to diel and seasonal fluctuations, are reasonably repetitive and can be accommodated within a longer-term steady state. The preceding summary of paleolimnology (Chapter 23), however, should make it clear that the long-term progression cannot be adequately viewed as a steady-state process.

Aquatic systems change. Erosion is a dominant geomorphological feature in running water systems and lotic metabolism changes with progressive physical and chemical changes that occur in the headwater regions. Geomorphology also is critical to the ontogeny of lake ecosystems. Geochemical inputs to lake basins, coupled with morphometric characteristics of the depression and changes in both the morphology and drainage patterns, influence lake productivity.

A primary characteristic of this ontogeny is the process of sedimentation, which gradually fills the lake basin. Inorganic inputs to the sediments vary greatly with the parent materials of the surrounding catchment area. Sedimentation is generally quite slow; a majority of organic inputs are mineralized prior to and soon after initial sedimentation. The ratio of organic to inorganic deposition can shift markedly in response to changes in the nutrient incomes, drainage-basin characteristics, and basin morphometry. A time is reached at which the rate of organic deposition exceeds the capacity for its decomposition. Sedimentation rates can increase very rapidly once such a combination of characteristics is reached, and can result in an accelerated conversion of the lake basin toward a terrestrial landscape. Although the precise mechanisms involved in this ontogeny are highly individualistic, certain similarities among lakes emerge.

Inputs of organic matter are crucial in determining lake ontogeny, and in order to view development of natural lake systems within an ontogenetical framework of succession, one must have a solid understanding of basic factors that control both production and decomposition. The productivity of lakes is fundamentally autotrophic, and is controlled to a great extent by aspects of inorganic and organic biogeochemical cycling. Since biogeochemically cycled materials of necessity link the activities of decomposers and primary producers, aspects of biogeochemical cycling have been emphasized repeatedly in earlier chapters. Central to this concept, however, must be an awareness that any consideration of lake productivity must include material produced in littoral and wetland regions, as well as production by pelagic communities.

Although the boundary of the lake per se can often be delineated relatively easily by the shoreline and the supralittoral area, it is in many cases much more diffuse. From a functional standpoint, both in terms of the pelagic region and of the entire lake, saturated marshy areas that surround many lakes constitute a major source of inorganic and organic inputs that can radically alter lake metabolic processes and ontogeny. Separation of these massive sources of organic matter from consideration of the metabolic processes that occur in the open-water "lake" is functionally incorrect. Similarly, one cannot ignore the sensitivity of the lake per se to differences in the types, quantities, and timing of allochthonous inputs of nutrients and organic materials.

To reiterate earlier statements, lakes cannot be treated as separate microcosms. While the importance of allochthonous inputs has long been known, especially from British work, the importance of littoral inputs has been advanced more recently. There remains an erroneous, persistent tendency to treat the pelagic community as an isolated, nonintegrated component of the lake system.

The importance of dissolved organic compounds in recycling of nutrients in natural systems is only beginning to be fully appreciated. Dissolved organic matter is important in changing the trophic states of lakes by accelerating the cyclic regeneration of nutrients through bacterial metabolism, which increases the availability of inorganic nutrients required for photosynthesis.

Major causal pathways governing the eutrophicational ontogeny of lakes are based on interacting mechanisms that regulate autotrophic metabolism. Our discussion is directed primarily at the majority of temperate lakes of glacial, tectonic, and volcanic origin that are of moderate size and depth. Exceptions certainly exist. In many cases, the following discussion is applicable to certain lakes since their inception; in other cases, the pathways described are applicable only to recent times, e.g., since the recent disturbances by man or those changes that could be anticipated with changes induced by man's activities that affect the loading rates of nutrients and organic matter.

INCREASING EUTROPHICATION

Low rates of productivity of oligotrophic lakes are frequently determined by low inputs of inorganic nutrients from external sources. Morphometric characteristics of relatively large size and depth that yield high ratios of hypolimnion to epilimnion volume are commonly characteristic to oligotrophic lakes and affect nutrient cycling. The low production of organic matter, resultant low rates of decomposition, and oxidizing hypolimnetic conditions result in relatively low rates of nutrient release from the sediments in a cyclical causal system (Fig. 24-1). Low concentrations of relatively readily decomposable dissolved organic substrates from both planktonic and littoral sources result in small bacterial populations and slow rates of microbial metabolism. Synthesis of organic micronutrients, which are essential to most planktonic algae, would be correspondingly limited. Complexing of essential inorganic micronutrients by dissolved organic compounds, by which solubility and physiological availability could be partially maintained under oxidizing conditions, would be less effective under oligotrophic conditions of low synthesis of organic matter and high rates of degradation of available substrates. The importance of phosphorus and nitrogen limitations to maintenance of low productivity in a large percentage of lakes of the world was discussed in detail earlier (cf. particularly Chapter 13, pp. 280–295).

The circumstances of the Shield lakes of Canada are ideal for demonstrating the importance of phosphorus limitation and rapid recovery from eutrophic conditions once the inputs were controlled. The Lake Washington and Swiss lake examples, dis-

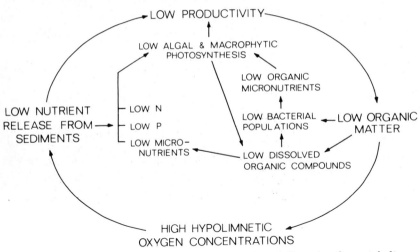

Figure 24-1. Major inorganic and organic interactions influencing the metabolism of phytoplankton of oligotrophic lakes. (Modified from Wetzel and Allen, 1970.)

cussed earlier (Chapter 23), are analogous examples of the success in the abatement of eutrophication that is possible once phosphorus-control measures are instituted. In spite of the obvious caution that must be exercised in making generalizations among such complex and varied systems as lakes, it is important to emphasize once more that phosphorus abatement will not return all lakes to preenrichment conditions. As discussed in Chapter 13, recycling of phosphorus from internal stores in the sediments can sustain high rates of production for many years after influent loading has been reduced. The importance of phosphorus demand and supply for plant growth is such that its reduction in inputs to fresh waters is the first place to begin, and likely to succeed in a majority of cases. Further, if phosphorus inputs to surface water are to be reduced most effectively, point sources should be eliminated as rapidly as possible. The millions of tons of phosphate currently introduced by synthetic detergents can be eliminated relatively easily; technologically available methods exist for phosphate removal during wastewater treatment. The scientific basis for the importance of phosphate and nitrogen in eutrophication is so overwhelming that an international resolution was ratified at the 19th International Congress of Theoretical and Applied Limnology in 1974 (Wetzel, 1975a). This resolution emphasizes the critical role of phosphorus in the rapid eutrophication of inland waters and the need to control the addition of this element to any inland water by any means available. In addition to secondary treatment of sewage, methods of control include: (a) restrictions on the use of cleaning products that contain phosphates or other ecologically harmful substances, (b) removal of phosphate at sewage treatment facilities discharging effluents into such water, and (c) control of drainage from feedlots, agricultural areas, septic tanks, and other diffuse sources of phosphorus. Control measures for nitrogen should also be considered in basins in which there is evidence that such controls are appropriate. Implementation of such control measures is socially complex, politically controversial, but technologically attainable.

Under eutrophic conditions, the loading rates of phosphorus and nitrogen, as well

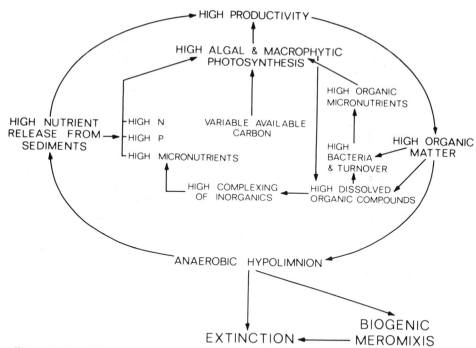

Figure 24-2. Major inorganic and organic interactions influencing the metabolism of phytoplankton of eutrophic lakes. (Modified from Wetzel and Allen, 1970.)

as of other nutrients under less acute demand, are relatively high (Fig. 24-2). As the rates of photosynthetic productivity increase during eutrophication, the cyclic interactions involving regeneration of inorganic nutrients and organic compounds increase in intensity. Planktonic productivity increases markedly and results in a light-limited compression of the trophogenic zone. Reduction in depth of the trophogenic zone continues with intensification of eutrophication. Eventually, planktonic populations impose self shading to such an extent that further increases are not possible under natural conditions of incident solar irradiance (Fig. 24-3). The rate of planktonic production then reaches a plateau, with a gradual progression toward eventual extinction through sedimentation of organic matter that accrues faster than decomposition can remove it. Under certain conditions of morphometry, meteorology, and productivity, it is not rare for mesotrophic to eutrophic lakes of small surface area and moderate depth to undergo biogenically induced meromixis (Chapter 6). A reduction, usually temporary, in productivity can occur when redistribution of nutrient-rich hypolimnetic waters is prevented.

HARD-WATER CALCAREOUS LAKES

The ontogeny of an oligotrophic lake toward eutrophic conditions can be altered markedly by natural inputs of carbonates and associated cations from surrounding sedimentary bedrock and glacial till. Calcareous hard-water lakes are common to large regions of the world, and in such lakes, oligotrophic states are maintained by high cal-

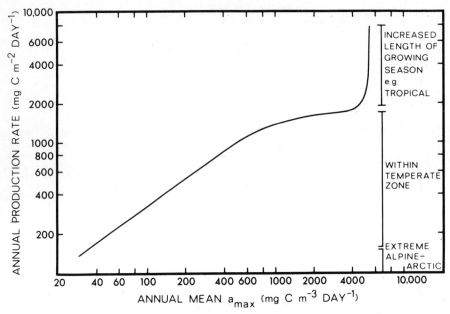

Figure 24-3. General relationship between the annual mean volumetric production rate at the depth of optimal growth and the annual mean areal production rates of phytoplankton. (After Wetzel, 1966a, and others.)

careous inputs sustained over long periods of time. In these waters, reduced productivity is maintained by decreased nutrient availability rather than the type of oligotrophy persistent in soft-water lakes of low productivity, in which nutrient inputs to the systems are deficient.

In calcareous and in noncalcareous lakes, early postglacial eutrophication generally proceeded rapidly as the climate ameliorated. These initial high rates of eutrophication were favored by inputs of readily leached nutrients that entered lakes from adjacent terrestrial areas. In calcareous regions, however, high inputs of carbonates were maintained over long periods of time. Under excessively buffered bicarbonate conditions, productivity is suppressed by an array of inorganic–organic interactions affecting the metabolism of the micro- and macroflora (Fig. 24-4). Major interactions have been discussed in earlier chapters and need be mentioned only briefly here.

Phosphate and essential inorganic micronutrients, particularly iron and manganese, form highly insoluble compounds and precipitate from the trophogenic zone of marl lakes (Wetzel, 1972). Combined nitrogen compounds tend to occur in high concentrations as a result of the relatively high inputs common in calcareous regions and the low rates of biotic utilization within the lake. In extremely calcareous lakes, it is not rare to find inorganic nitrogen concentrations that approach levels considered toxic for human consumption. Calcium and magnesium concentrations can be exceedingly high; calcium is often in apparent supersaturation and approaches or exceeds 100 mg l^{-1}. Intensive biogenically induced decalcification is characteristic of these hard waters, during which massive amounts of particulate carbonate precipitate from the trophogenic zone. The sedimenting particulate and colloidal $CaCO_3$ adsorbs inorganic

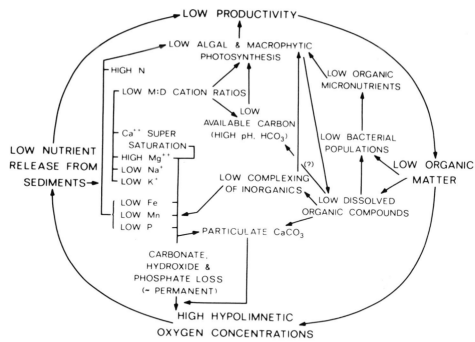

Figure 24-4. Major inorganic and organic interactions influencing the metabolism of producers in hardwater calcareous lakes. (From Wetzel, 1972.)

nutrients and certain labile and recalcitrant dissolved organic materials, and effectively removes them from the trophogenic zone. The major monovalent cations, sodium and potassium, commonly occur at low concentrations. Low levels of sodium, an essential element of certain blue-green algae, have been implicated as a contributory factor in reducing the growth of heterocystous blue-green algae in marl lakes (Ward and Wetzel, 1975). Although inorganic carbon concentrations occur at very high levels, the low availability of free CO_2 can reduce photosynthesis of certain plants under stagnated conditions when the pH is excessively high (cf. Chapter 11). The extent to which photosynthetic utilization of organic carbon sources—for example, carbamino carboxylic acids—occurs under natural conditions is unclear. Recent evidence indicates that certain species of natural phytoplanktonic algae of a marl lake are capable of significantly augmenting photosynthesis by photoheterotrophy of simple organic substrates and compounds produced extracellularly by aquatic macrophytes (Chapter 15).

Low photosynthetic productivity of calcareous lakes contributes to reduced inputs of dissolved organic compounds, such as extracellular losses from phytoplankton and macrovegetation. This situation is compounded by effective adsorptive losses of dissolved organic matter to monocarbonates. Bacterial rates of metabolism are low because of the low concentrations and turnover of easily decomposable organic compounds, which can lead further to concomitant reduced synthesis of organic micronutrients. Reduced concentrations of certain organic micronutrients such as vitamin B_{12} have been shown to be contributory to reduced phytoplanktonic productivity in marl lakes; a portion can be metabolically inactivated by adsorption to $CaCO_3$. The more reactive organic substrates are either utilized rapidly by bacteria or are inactivated by adsorption to sedimenting $CaCO_3$. The loss of the more labile organic compounds reduces the complexation capacity of the water, so that certain inorganic nutrients cannot be maintained in solution. In addition, the selective removal of the more labile organic constituents leads to a relative accumulation of more recalcitrant dissolved organic materials, the consequences of which are not fully known. It is common to find very hard lakes with a stained coloration (brown) and high concentrations of dissolved humic materials; these compounds degrade slowly, and may inhibit additional epilimnetic decalcification by interfering with the formation of $CaCO_3$ particles, thereby sustaining an oligotrophic condition.

Factors that reduce the buffering capacity and carbonate reservoirs of marl lakes promote eutrophication in these ecosystems. Two obvious means of altering the major controlling aspects of calcareous inputs are reductions in or depletion of bicarbonate and cation sources in the drainage basin, or increases in the loading of dissolved organic matter. At some point in this transition, the inhibitory effects of the calcareous state on nutrient availability are reduced, and the stimulatory effects (both direct and indirect) of dissolved organic matter increase. At that point, eutrophication can proceed relatively rapidly. Morphometric changes that accompany eutrophication-induced increases in sedimentation rates and infilling also have significant effects on stratification patterns, hypolimnetic capacity of maintaining oxidizing conditions, and nutrient recycling. These changes accelerate as productivity increases, and operate to further enhance eutrophication.

Where large areas of sedimentary rock have been glaciated, it is common to find former marl lakes underlying bog lakes* or bogs (discussed in some detail further on). It is apparent that numerous marl lakes have shifted relatively rapidly, in the range of a millennium, from highly calcareous, alkaline conditions to a very acidic, organic-rich state, markedly depleted in divalent cations and bicarbonate. In nearly closed basins of moderate to small size, the ontogeny of initially calcareous lakes can be altered at a rapid rate by the development of specialized littoral flora, particularly the mosses such as *Sphagnum*. These plants and associated encroaching vegetation function as particularly effective cationic sieves (Clymo, 1963, 1964). It is not uncommon to find small pioneering hummocks of *Sphagnum* growing in the back-littoral reaches of marl lakes directly on a thin organic deposit overlying sediments containing more than 50 per cent of $CaCO_3$ (e.g., Glime, et al., 1982). The eulittoral of highly stained alkaline hard-water lakes is often interspersed with stretches of *Sphagnum* in dense mats. These alkaline bog lakes represent a transitional stage between certain marl lakes and true *Sphagnum* bogs. The ionic exchange mechanisms of these mosses, accompanied by the simultaneous release of organic acids (e.g., polyuronic acids), are effective in reducing the cation influxes from surface sources. As the littoral sieving vegetation circumscribes the basin, the buffering capacity of the marl lake system is progressively reduced. The subsequent vegetative development then can proceed relatively rapidly, in a classical pattern of bog succession from a lake to a terrestrial ecosystem, via accelerated accumulation of organic matter under acidic, reducing conditions that do not favor rapid decomposition.

---

*Bog lakes have been variously described; they commonly refer to strongly basic alkaline lakes surrounded by an extensive acid-forming bog mat (e.g., Welch, 1936).

DYSTROPHY AND BOG SYSTEMS

Early workers formulating relationships in studies of comparative regional limnology were acutely aware of the important differences among lakes in regard to the proportion of allochthonous and autochthonous organic matter.* *Trophy* of a lake refers to the rate at which organic matter is supplied by or to the lake per unit time.† Trophy, then, is an expression of the combined effects of organic matter supplied to the lake.

As has already been indicated (Chapter 22), under natural conditions, the relatively resistant humic substances of largely terrestrial plant origin represent the most common component of allochthonous organic matter. Lakes that receive large amounts of their organic matter supply from allochthonous sources commonly are heavily stained and have been referred to as "brown-water lakes." These lakes were termed *dystrophic*, in reference to their high content of humic organic matter. Productivity of most dystrophic lakes classically has been described as low. It must be emphasized, however, that this low to moderate productivity criterion *refers to the planktonic productivity and, as it was developed, ignores the littoral plant components of the lake system*. As we shall see, in dystrophic lakes that develop bog flora, the littoral plants completely dominate the metabolism of these lake systems as sources of dissolved and particulate organic matter.

As a consequence of uncritical use of the terms associated with dystrophy, bogs, bog lakes, and dystrophic lakes have been loosely, and incorrectly, equated. Much of the unfortunate confusion emanates from inappropriate generalizations of ontogeny of lake systems from specific lake investigations. For example, the famous successional scheme put forth by Lindeman for Cedar Bog Lake, Minnesota, has been more than once uncritically proposed in general ecology texts as the universal situation. In this way, the erroneous concept that all lakes become bogs and then land may become widely accepted. While some lakes do progress through this sequence, it is far from the rule. A large, heavily stained Finnish lake on a primary bedrock basin with no characteristic bog vegetation can be correctly termed *dystrophic* just as effectively as the *open pelagial water* of a quaking *Sphagnum* bog. Trophy refers to the rate of supply of organic matter, and dystrophy denotes a high loading of allochthonous organic matter (Fig. 24–5). Dystrophy is a subset of trophy (oligotrophy to eutrophy), rather than a parallel concept.

These examples emphasize a most important point. As developed originally and as largely used today, the trophic concept refers to the pelagial-zone–planktonic *portion* of the lake ecosystem. The differentiation was between rate of organic matter input to the system from autotrophic phytoplanktonic sources and from allochthonous sources of the catchment basin (allotrophy) (Fig. 24–5). The littoral flora and its often dominating supply of autochthonous organic matter to the system, was and usually still is, ignored. Although conceptually it is easy to place the littoral productivity within such

---

*Birge and Juday (1927), for example, in their extensive studies on the origin and supplies of dissolved organic matter, differentiated between autotrophic lakes dominated by autotrophic inputs, and allotrophic lakes that receive a majority of their soluble organic matter from the drainage basin.

†A concise review of the development of the trophic concept is given by Rodhe (1969). A number of other terms have been used to indicate the direction of the state of trophy, e.g., *eutrophication* in the broad sense of an increase in productivity (also *auxotrophication*; Thunmark, 1948; Lillieroth, 1950), which refers to an increase in trophic standard, and the contradictory term *oligotrophication* (also *meiotrophication*; Quennerstedt, 1955), which refers to a decrease in trophic standard.

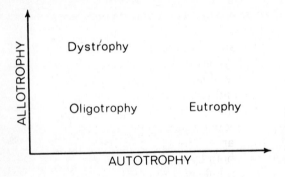

Figure 24-5. Classic trophic types of lakes based on the rate of supply of organic matter from autotrophic and allochthonous sources. In the original development of these concepts, only phytoplanktonic productivity was considered. (Modified from Rodhe, 1969.)

trophic schemes, there are few quantitative evaluations of its contribution. Sufficient information is available, however, to indicate that it cannot be ignored in most cases.

Photosynthetic rates of phytoplankton are generally low in highly stained open basins that receive copious quantities of allochthonous dissolved organic matter; they are similarly low in the open-water zones of closed bog lakes that receive inputs of dissolved organic matter largely from littoral vegetation (Fig. 24-6). The abundance of dissolved humic materials and the relatively low pH values in these systems are not conducive to bacterial metabolism. Additionally, monovalent-to-divalent cation ratios tend to be high because of low inputs from the drainage basin or because of effective cation exchange mechanisms of the littoral flora. The combination of acidity and low

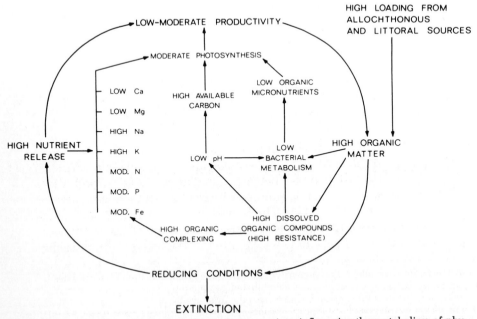

Figure 24-6. Dominant inorganic and organic interactions influencing the metabolism of phytoplankton in the pelagial waters of dystrophic and bog lakes. (Modified from Wetzel and Allen, 1970.)

nutrient availability usually restricts phytoplanktonic productivity even though inorganic carbon in the form of free CO_2 is readily available.*

The relationships among the major causal mechanisms regulating the phytoplanktonic trophic states in the four main types of lakes may be further connected in an overall sequence of development (Fig. 24-7). This scheme of lake ontogeny emphasizes the importance of changes in functional relationships and environmental parameters that regulate algal, and to some extent, macrophytic, productivity. The designated pathways of ontogeny (Fig. 24-7) can obviously vary greatly, depending upon geomorphology of the specific system and the macroclimate of the area. The rate at which a lake proceeds to extinction via sedimentation is determined both by interactions that occur within the pelagic zone and by certain aspects of geomorphology of the drainage basin. Finally, as has been stressed repeatedly throughout this text, one must have an appreciation for the importance of littoral macrovegetation in the culmination of this sequence.

LITTORAL DEVELOPMENT

Shallow water bodies, correctly termed *lakes* but often referred to in older literature as *ponds*, are usually characterized by an abundance of aquatic macrovegetation and associated microflora attached to all surfaces. It is not unusual for the vegetation

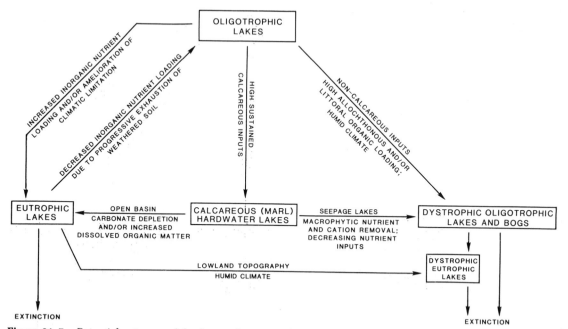

Figure 24-7. Potential ontogeny of the four main types of lakes. See text for discussion.

---

*High productivity of phytoplankton in heavily stained dystrophic lakes is known, especially in lowland areas of Scandinavia. Järnefelt (1925) called these lakes mixotrophic, but introduction of another term in this conceptual discussion is unnecessary and confusing.

to extend over the entire basin, provided the depth does not exceed plant tolerances, primarily for adequate light (cf. Chapter 18). Shallow lakes include basins that were never preceded by a larger, deeper lake, as well as those basins which represent a terminal stage in the extinction of deeper lakes.

From a hydrological viewpoint, the shallow water bodies can be separated into those that are *permanent*, containing some water at all times of the year, and those that are *temporary*, in which the basin periodically becomes dry. Vernal lakes are common in shallow lowland areas following high water inputs from spring runoff and precipitation, but dry out during summer in the temperate region. Aestival lakes are shallow lakes that contain water permanently, but freeze completely during winter periods. It is apparent that a nearly infinite variety of conditions lead to the development and persistence of small, shallow lakes in relation to seasonal and long-term climatic changes in the water budget. Inflow and precipitation are counterbalanced by outflow (surface and subsurface), evaporation, and retention, including water retained by accumulations of organic matter of plant origin.

The terminal stages of the transition from lake to terrestrial ecosystems is characterized by an accumulation of organic matter in excess of degradation. Partially decayed organic matter, mainly of plant origin, termed *peat**, accumulates in aquatic systems under a wide variety of conditions in all except the driest macroclimatic regions of the world. Mire systems form: (1) in basins or depressions (primary mire systems); (2) beyond the physical confines of the basin or depression (secondary), the peat itself acting as a reservoir and increasing the surface retention of water; and (3) above the physical limits of the groundwater (tertiary), the peat functioning as a reservoir holding a volume of water by capillarity above the level of the main groundwater mass of the region (Moore and Bellamy, 1974). This last mire system obtains much of its water directly from precipitation and forms a slightly raised (perched) water table.

Although dense stands of small mosses and herbs are the dominant components of contemporary active peat formations over immense areas of subarctic and temperate regions, shallow depressions of lake systems commonly gain excessive organic matter deposits through submersed, floating-leaved, and emergent macrophytes. The successional sequence may or may not terminate in moss and associated bog vegetation. Many highly fertile eutrophic lakes develop increasing amounts of emergent littoral vegetation (Fig. 19–10), to the point at which organic matter accumulation increases the level of flora and sediments above the water table, which allows terrestrial vegetation to encroach.

---

*Excellent reviews and interpretation of the widely dispersed literature on peat-producing ecosystems or *mires* are given by Gorham (1957), and especially in the book *Peatlands*, by Moore and Bellamy (1974). The terminology and synonymity of this area of study are complex; only a brief résumé of major features of the systems is given here. It should be noted that mire is a collective term which includes both *bog* (= "moss" in British literature) and *fen*, differentiated according to water and nutrient sources, as well as by rather subtle floristic variations. Mire is equivalent to the Swedish *myr* and the German *Moor*. All mires are in areas where the groundwater table is permanently at or near the surface of the ground. Mires are commonly separated into *minerotrophic mires* (= fens) in which water and nutrients are supplied from groundwater, surface sources, and atmospheric precipitation, and *ombrotrophic mires* (= bogs) in which water-table recharge and nutrient inputs originate mainly from atmospheric precipitation. A *carr* refers to a mire or portion of a mire that is dominated by woody shrubs or trees among the herbaceous vegetation on a relatively stable peat mass (e.g., a fen carr could be dominated by trees, such as species of *Alnus*, *Fraxinus*, and *Betula*, with a dense understory of fen herbaceous plants).

A common successional pattern associated with the ontogeny of shallow basins follows along the pathway of excessive macrophytic and sessile algal production:

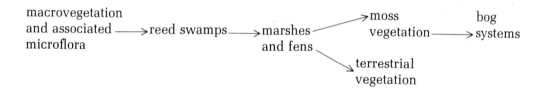

Reed swamps are characterized by closing vegetation of the littoral, often of polycorm, tall graminoid emergent plants—for example, *Phragmites, Cyperus papyrus, Scirpus,* and tall *Carex* species—with only occasional mixtures of submersed and floating-leaved macrophytes. Usually species diversity is low, and two herbaceous layers commonly dominate within the main vegetative structure.

As organic accumulations increase to the displacement of the standing water, the water-logged peat habitat often is characterized by moderate-sized graminoid vegetation and small herbs in two or three layers. This vegetation is referred to as marsh or fen in areas of slightly flowing surficial water, or bogs in areas of subsurface seepage drainage. The distinction between the terms *marsh* and *fen* is largely made on the basis of phytosociological differences in floristic associations. As marshes and fens dry, and depositions of organic matter exceed the mean water table, very tall terrestrial dicotyledonous vegetation increasingly colonizes and develops over the basin and displaces the emergent aquatic macrophytes. In nutrient-poor areas that are humid and receive abundant rainfall, marshes and fens may possess well-developed bryophyte floras. In many instances, mosses (especially species of *Sphagnum*) dominate the system, and can lead to the formation of bogs (e.g., Glime, et al., 1982).

Early distinctions of various bog systems and types were made mostly on the basis of the primary, secondary, and tertiary stages of mire development discussed above in relation to water sources and floristic differences. The ontogeny of mire systems is now generally viewed with respect to a combination of hydrological, phytosociological, and chemical processes. A sequence under ideal geomorphological and hydrological conditions illustrating common stages in the succession of mire systems in shallow surface depressions is portrayed in Fig. 24-8 (Moore and Bellamy, 1974 after Kulczynski). Water supply to the model lake system of Figure 24-8 consists of direct rainfall, runoff, and seepage from the immediate catchment of the lake basin, and a continuous flow of groundwater that enters by an inflow stream, of sufficient volume to affect the whole lake.

When these mire systems initially develop in a shallow drainage lake (Fig. 24-8), the system is under the influence of continuously or intermittently flowing groundwater. The ionic composition of the water of these "rheophilous mires"* is dominated by calcium and bicarbonate. Allochthonous organic-matter inputs are high and result in accrual of peat within the central portion of the basin (Stage 2). In some cases, this peat

---

*Synonymous with terms "low moor" and *Niedermoore.*

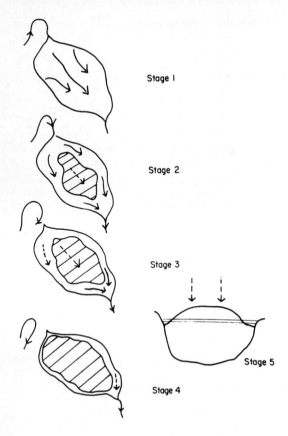

Stage 1

Stage 2

Stage 3

Stage 5

Stage 4

Figure 24-8. Idealized successional stages in mire systems from a small, shallow drainage lake (1), through a seepage system (2, 3), to a raised bog deriving most of its water supply from precipitation directly on its surface (4, 5). Hatched area = peat accumulation. (From Moore, P. D., and Bellamy, D. J.: Peatlands. London, Paul Elek Scientific Books Ltd., 1974.)

is sufficiently buoyant to float as a mat. The main flow of water tends to be channelized to peripheral areas. In the third transitional stage of mire development,† continued accrual of peat deposits diverts inflow from the basin. Water supply is derived largely from the immediate catchment of the mire and from direct precipitation on the surface. Ionic composition usually is dominated by calcium and sulphate. Further accrual of peat results in large areas of the mire surface that are not affected directly by moving water (Stage 4), but are inundated periodically when the water level rises during heavy rainfall.

Ombrogenous or ombrophilous (ombro = rain) mires* (Stage 5) are no longer subject to the influence of flowing groundwater. Nearly all of the water, which is dominated by sulphate and hydrogen ions, is received from direct rainfall. The surface of the mire system is too high to be materially influenced by fluctuations in the level of groundwater.

The change in ionic composition of water from dominance by calcium and bicarbonate to that of sulphate and hydrogen ions is a conspicuous feature characterizing these mire systems (Fig. 24-9). The loss of ions from surface water supplies results in

---

†*Ubergangsmoore* of older German literature.
*Termed "high moors" or *Hochmoore* in early works.

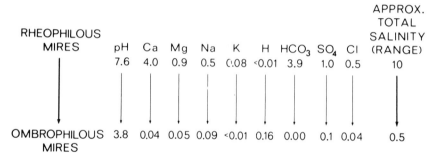

Figure 24-9. General shifts in ionic composition (milliequivalents l^{-1}) of the water in the transition from rheophilous to ombrophilous mire systems of Western European mires. (From average values of Bellamy and of Sjörs given in Moore and Bellamy, 1974.)

very low salinities and greatly reduced buffering capacity. Under these conditions, relatively small additions of acids can result in a considerable reduction in pH.

The origin of acidity and control of cation exchanges of mire systems result from a combination of factors. The increase in acidity and reduction of cations is particularly marked in communities dominated by the mosses of the genus *Sphagnum*. Water surrounding *Sphagnum*, a dominant genus of mosses over immense areas of the temperate, boreal, and subarctic regions of the world (see, for example, Sjörs, 1961), has a pH usually below 4.5 and sometimes below 3.0. Much of the acidity of the aquatic environment surrounding *Sphagnum* can be attributed to cation exchange (Anschütz and Gessner, 1954; Clymo, 1963, 1967; Brehm, 1970; Glime, et al., 1982; and others). *Sphagnum* and other mosses behave as cation exchangers even when dead, because of their high concentrations (up to 30 per cent of the plant's dry weight) of polymerized unesterified uronic acids.* Although small amounts of free organic acids are released from *Sphagnum*, evidence indicates that most of the acidity of *Sphagnum*-dominated mire systems originates from the release of carboxyl-associated H^+ ions by freshly produced plant growth; the hydrogen ions are exchanged for nutritionally important cations that occur in rain or groundwater. Some of the acidity of bog pools may be derived from H^+ ions released during the metabolic activities of anaerobic sulfur-metabolizing bacteria (Gorham, 1966; Clymo, 1965), but this source of acidity is clearly minor.

BOGS AND QUAKING BOGS

The littoral vegetation is clearly critical in the senescence of a lake as the dominant source of organic matter produced in excess of decomposition. As mentioned earlier, bogs are acidic, dystrophic aquatic ecosystems that possess an exceptionally characteristic littoral flora. Bogs do not represent an ontogenetic stage through which all lakes must pass in their route to extinction: The development of a bog system requires climatic conditions of abundant precipitation and relatively high humidity over much of the annual cycle. The littoral vegetation of bogs completely dominates key character-

---

*Similar to sugars, but the sixth carbon is part of a carboxyl group. These long-chain molecules may be mixed polymers containing both sugars and uronic acids.

istics of the system: acidic pH, heavy staining resulting from the presence of large amounts of dissolved and colloidal humic materials, and low concentrations of nutritionally important cations.

Commonly, the transition of a small lake to a bog or a terrestrial ecosystem occurs sequentially as depicted in Figures 24–8 and 24–10. Depth of the open water decreases continually because of the sedimentation of organic matter. As the basin progressively fills, the littoral vegetation simultaneously advances toward the center of the lake basin. Eventually, the entire area is covered with what is loosely referred to as *swamp conditions*, with standing water occurring among the vegetation. Production of littoral vegetation increases because of the high productivity of emergent macrophytes, and sediments eventually reach the surface of the original lake. This stage can be referred to as a *marsh*, for the sediments are continually saturated but there is little, if any, standing water among the vegetation. Under these conditions, decomposition proceeds rapidly. However, for a number of reasons (discussed in detail in Chapter 20, and summarized in Fig. 24–11), inputs of relatively recalcitrant organic matter from the emergent macro-

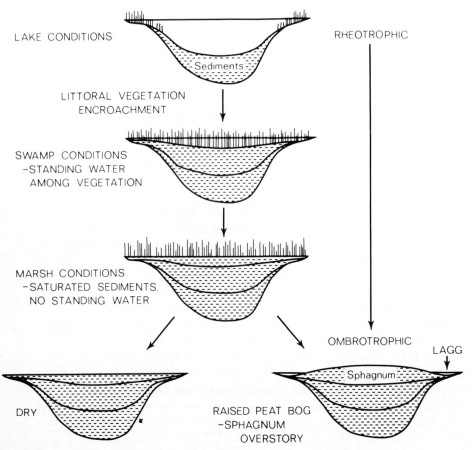

Figure 24–10. Frequently observed ontogeny of shallow lake systems through swamp and marsh stages to terrestrial conditions or to raised peat bogs. See text for discussion.

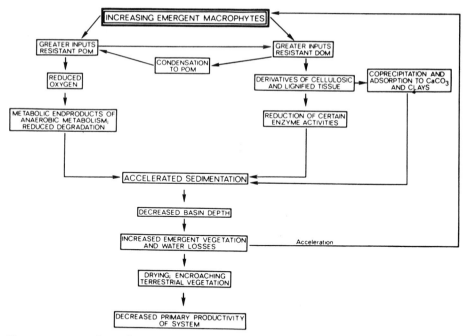

Figure 24-11. Relationships of increased loading of relatively resistant organic matter from dominating emergent littoral and wetland flora to controls of decomposition rates, acceleration of emergent macrophyte development, and the ontogeny of lake systems. POM and DOM = particulate and dissolved organic matter, respectively. (From Wetzel, 1979.)

phytes occur even more rapidly. The mean pH decreases substantially, and concentrations of dissolved humic materials in the water increase. Changes in the dominant vegetation also occur, and acidophilic flora (especially sedges and grasses) becomes more abundant.

The formation of a bog depends greatly on the prevailing climate. In moderately dry regions, the transition is from marsh conditions to terrestrial vegetation, without the characteristic development of mosses. Over much of the temperate zone, however, conditions are sufficiently moist to encourage the colonization and eventual dominance of *Sphagnum* and other mosses among the often predominantly sedge flora. Through a succession of mosses, especially *Sphagnum* species, cations decrease and acidity increases, which effectively decreases rates of decomposition (Chapter 20) and accelerates net accumulation of organic matter. The accumulation of particulate organic matter in the under-story, continued growth of plants at the surface, and excellent capillarity of the dead and living mass of mosses raise the water level within the mat above the original level of the lake. In this way, the vegetative mat can develop slightly (usually <50 cm) above the water table; hence, the term *raised bog*.*

---

*Raised bogs occur in depressions, often former lake basins, where losses of water from runoff are not as great as in more upland areas. In areas of high rainfall (> 130 cm year^{-1}) and low evaporation, bog vegetation can extend onto low-lying landscapes that have slopes of < 15°, forming *blanket bogs*. Blanket bogs cover large regions, notably in northern Great Britain, western Norway, and boreal continental areas in general (Sjörs, 1961).

Moatlike areas of shallow water characteristically surround the central mat of peat; these areas, which are called *lagg* zones, are the remnants of flowing groundwater whose path has been diverted around the central peat mat (cf. Fig. 24-8). Organic matter accumulates more slowly in the lagg zones than it does in the raised portions of the bog because of greater movements of water and better oxidizing conditions. Vegetation of the lagg zone is commonly dominated by small graminoid species, bryophytes, and foliaceous liverworts.

A dominant feature during the succession of lakes to bogs is a shift in origin of nutrients. Initially, most of the nutrients are derived from soil in the drainage basin; in later successional stages, most of the nutrients enter the bog ecosystem via atmospheric precipitation and particulate fallout. In mire systems in general, the terms *rheotrophic* and *ombrotrophic* have been introduced† to designate systems whose nutrients are derived predominantly from surface water/groundwater or from atmospheric sources, respectively. These two terms emphasize the interplay of geomorphological, chemical, climatic, and biotic factors that are involved in the succession of bogs. In regions where the ionic composition of rainfall is strongly influenced by oceanic contributions (cf. Chapter 10), such as in western Britain and much of Ireland, ombrotrophic nutrition is only slightly less effective than weak rheotrophic nutrition, and little difference in vegetation results.

Quaking bogs represent a particular development of bogs within a relatively deep lake basin of small surface area. The development of *Sphagnum* mosses occurs early in the succession of littoral flora, and in geological context leads to a very rapid accumulation of organic matter in the littoral areas (Fig. 24-12). In more advanced conditions, the littoral vegetation encroaches more rapidly on the open water than deposition of peat deposits occurs beneath it. This growth results in a thick mat (several meters) that floats above the open water and sediments consisting of loosely aggregated, flocculant organic matter. Growth proceeds all along the periphery toward the lake center, and eventually covers the entire surface.

A marked feature of quaking bogs is the concentric zonation of plant communities. *Sphagnum* is most often the pioneering organism, and dominates the floating mat formation closest to the center of the lake. Within the open water of the bog, certain submersed angiosperms (e.g., *Potamogeton* and *Utricularia* spp.) and floating-leaved macrophytes (such as *Nuphar* and *Nymphaea*) may be abundant. Lakeward portions of

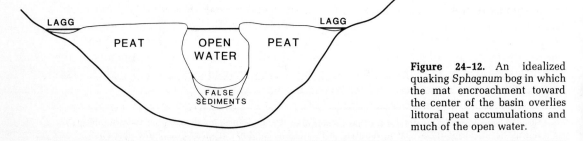

Figure 24-12. An idealized quaking *Sphagnum* bog in which the mat encroachment toward the center of the basin overlies littoral peat accumulations and much of the open water.

---

†Originally proposed as "minerotrophic" and "ombrotrophic" by DuRietz (Sjörs, 1961). A detailed discussion of the interrelationships and terminology is given by Moore and Bellamy (1974).

the mat may contain sedges (especially *Carex* sp.) and low shrubs such as leather leaf (*Chamaedaphne*) and blueberry (*Vaccinium* sp.). Farther landward, tall shrubs and, finally, bog trees, such as black spruce and tamarack, are common (Gates, 1942).

The older portions of the mat are grounded firmly in the basin by underlying deposits of peat. The floating mat, growing considerably above the level of water-saturated flocculant organic matter often is attached only by the vegetative mass, and can be displaced by the added weight of a person and "quake." The sensation is that of walking on an immense floating, saturated sponge. The sedimentation of much particulate, littorally derived organic matter in the open water areas results in loosely aggregated, flocculant sediment that is not compacted sufficiently to support small weights (for example, an anchor), often for many meters; hence, the term "false bottom" or "false sediments" (Fig. 24–12).

Quantitative information on the productivity and population dynamics of biota of the pelagic and profundal regions of bog lakes and quaking bogs is extremely sparse. Much more knowledge exists on qualitative aspects simply because of the extremely diverse and interesting biota that are adapted to the extreme conditions of bogs. The microflora of the plankton and of small pools among hummocks of *Sphagnum* and larger plants are characterized by a great diversity of algal species, although few are ubiquitous or abundant. Most abundant are desmids, often represented by several hundred species. Species of blue-green algae, chrysomonads, dinophyceans, and diatoms are adapted to bog conditions but seldom are found in abundance. The microfauna is dominated by testaceous rhizopod protozoans and rotifers; very few cladoceran and copepod zooplankters are tolerant of the bog milieu.

Larger animals have not adapted well to the extreme acidity and low salinity of bog waters. Species diversity is very low and entire groups are lacking or poorly represented. For example, in acidic, quaking *Sphagnum* bogs, sponges, coelenterates, ostracods, hydracarnid mites, oligochaetes, Ephemeroptera, Malacostraca, mollusks, nematodes, flatworms, and fish are completely absent or very poorly represented.

The genesis and ontogeny of bog systems can be viewed as a transitional succession of lake basins of greatly differing geomorphology to wetlands in which organic matter of macrophytic vegetation accumulates in excess of complete degradation. The causal processes regulating productivity can be effectively viewed from a short-term biogeochemical standpoint as operating in a reasonably steady-state system. However, the very nature of excessive accumulation of organic matter, the successional pathways of which are profoundly influenced by climatic and hydrological factors, indicates that the systems are not balanced even over very brief geological time periods.

SUMMARY

1. The interacting mechanisms regulating autotrophic productivity follow several ontogenetic pathways as lake basins undergo succession and are ultimately obliterated as aquatic ecosystems and transformed into terrestrial ecosystems. Many rate-regulating mechanisms are influenced by the general transition from a dominance of allochthonous loading of nutrients and

organic matter to accelerated loading of organic matter from littoral and wetland productivity.

   a. In the early stages of lake ontogeny, particularly among lakes formed by glaciation, autochthonous and allochthonous productivity is severely limited by climatic conditions, especially cold temperatures.

   b. In the intermediate stages of ontogeny, autochthonous autotrophy is first dominated by phytoplanktonic productivity, with some contributions by littoral production.

   c. In the later and terminal stages, however, a rapidly accelerating shift occurs to a total dominance of productivity by littoral components.

   d. The planktonic-dominated stages of lake development are relatively lengthy; during these stages, organic sedimentation and deposition are nearly balanced by decomposition.

   e. In shallow basins and in lakes in which gradual sedimentation has reduced the depth sufficiently, massive production by littoral plants and wetland *emergent* macrophytes dominates inputs of photosynthetically accrued organic matter.

      i. The increasing contribution of derivatives of lignified and cellulosic tissues results in reduced rates of and less complete decomposition, and

      ii. Leads to accelerated rates of sedimentation and favors conditions conducive to enhanced emergent macrophyte development in a cyclically reinforcing process (cf. Fig. 24–11).

2. Increased inorganic nutrient loading, particularly of phosphorus and combined nitrogen, is fundamental to initial eutrophication and to maintenance of high sustained productivity by the phytoplankton community.

   a. Low rates of productivity in oligotrophic lakes are maintained to a large extent by low inputs of inorganic nutrients from external sources. Low production of organic matter, concomitant low rates of decomposition, and oxidizing hypolimnetic conditions result in low rates of nutrient release from the sediments in a cyclical causal system (Fig. 24–1). Dissolved organic compounds from both internal and external sources are usually low; the results are limited availability of organic micronutrients and limited complexing capability for essential inorganic micronutrients.

   b. Under eutrophic conditions, the loading rates of inorganic nutrients, especially of phosphorus and combined nitrogen, are relatively high (Fig. 24–2). As rates of photosynthetic productivity and organic loading to lower strata increase, nutrients are released from sediments into anoxic hypolimnia, increasing recycling rates.

   c. Phytoplanktonic productivity of eutrophic lakes increases markedly. In increasingly eutrophic lakes, dense algal communities reduce light penetration, and thereby compress the depth of the trophogenic zone. Light limitations caused by self-shading set an upper boundary to the total photosynthetic productivity, beyond which further increases are not possible, regardless of increased nutrient availability (Fig. 24–3). Further increases in photosynthetic productivity are possible only by extending the length of the growing season (e.g., in the tropics) or by increasing turbulence and frequency of exposure to available light (as in artificially mixed sewage lagoons).

d. In calcareous hard-water lakes, availability of certain organic micronutrients (such as vitamins) and inorganic nutrients (particularly phosphorus and metallic micronutrients) can be suppressed by inactivation, i.e., chemical competition, or sedimentation with inorganic particulate materials (e.g., coprecipitation with $CaCO_3$) (Fig. 24-4). Maintenance of reduced productivity in calcareous lakes by lowered nutrient availability depends upon high carbonate loading from the drainage basin. Calcareous loading may be reduced through long-term leaching or overcome by increased loading of dissolved organic matter. The result is a relatively rapid transition to eutrophic conditions. Under certain conditions, high cationic (e.g., Ca, Mg) loading also can be counterbalanced and reduced by cationic exchange mechanisms of littoral–wetland plants (e.g., bryophytes), which may result in the transition of calcareous lakes to acidic bogs.

3. Many lakes receive large amounts of dissolved and particulate organic matter from allochthonous and surrounding wetland–littoral sources. These lakes are commonly heavily stained as a result of the abundance of dissolved humic compounds of plant origin; such lakes are termed *dystrophic.*

a. Although phytoplanktonic productivity is usually low in dystrophic lakes (Fig. 24-6), productivity of the surrounding littoral vegetation is moderate to high.

b. The surrounding macrovegetation can function as effective nutrient scavengers, and frequently reduces the nutrient loading that reaches the open water. In some cases, particularly in bogs, the vegetation can effectively shift the lake per se from an open to a closed basin, so that nutrient inputs to the lake are mainly from atmospheric precipitation.

4. Terminal stages in the successional sequence of lake ontogeny are dependent upon the types of macrovegetation in the littoral zone and wetland areas surrounding the basin. The development of the vegetation is, in turn, strongly influenced by prevailing climatic conditions (rainfall, humidity) of the region.

a. Many fertile eutrophic lakes develop ever increasing amounts of emergent littoral vegetation, until littoral vegetation encroaches over the entire lake basin (Fig. 24-10). The resultant *swamp conditions,* with standing water among the emergent vegetation, gradually succeed to *marsh* conditions with little or no standing water over the water-saturated sediments. Evapotranspiration from the vegetation can gradually exceed water income, permitting invasion of a terrestrial flora. Ecosystem productivity decreases in the transition to terrestrial conditions.

b. Under climatic conditions of high rainfall and humidity, various types of *mire* ecosystems develop; in these systems, partially decayed plant organic matter *(peat)* accumulates in abundance.

i. *Rheophilous mires* can develop from shallow drainage lake basins in which large amounts of organic matter accumulate in the depression. With time, the main water flow through the mire tends to be channelized to peripheral areas around the central portion of the organically filled basin.

ii. Eventually the mire is no longer under the influence of flowing groundwater. In these *ombrophilous mires,* water and nutrients are received primarily from rainfall. Salinity, buffering capacity, and pH of the water are low.

iii. The acidity of the water often increases during mire development. Mosses, particularly *Sphagnum,* often develop in profusion in *bog* mires. Mosses have effective cation exchange mechanisms, in which divalent cations are retained with a commensurate release of H^+ ions and organic acids. *Sphagnum* and other acidophilic plants can develop above the mean water table (raised bog), yet still acquire sufficient water by capillarity.

iv. Bog mats and underlying peat accumulations can gradually encroach upon, and eventually eliminate, the open water of the lake remnant within the bog.

v.   Individual bog succession proceeds at rates and along developmental pathways that are controlled by a complex interplay of geomorphological, chemical, climatic, and biotic factors.

CHAPTER TWENTY-FIVE
EPILOGUE

The foregoing pages represent only a brief summary of contemporary knowledge of freshwater systems. In spite of concerted efforts for brevity, this book demonstrates that diversity in the biota is large and that interactions among biotic populations and environmental parameters are numerous. Furthermore, it has been emphasized repeatedly how the metabolism of the biota often affects the total cycling and characteristics of the system. The entire thrust of this treatment, however, has focused on generalized operational similarities among fresh waters, whether they are standing in basins or running in channels. Despite the heterogeneity and individuality from site to site, one must appreciate that there is underlying unity of function.

An attempt has been made to state these generalized relationships in simple terms, without resorting to mathematical expressions and their attendant disadvantages of oversimplification. General principles in limnology are, and will be, modified little by mathematical manipulations. Nonetheless, mathematical modeling and systems simulation possess intrinsic manipulative and predictive power if based soundly on adequate quantitative data. In no way, however, will such manipulation improve poor data or conceptualizations.

In the comparative approach taken, a concerted attempt was made to integrate the massive number of dynamic variables that govern and effect the resultant biotic productivity. Throughout this treatment, the dominant factors relating to the productivity of oligotrophic and eutrophic systems were contrasted. The underlying rationale was to emphasize the basic control mechanisms that operate on the progressive changes that lead to enrichment and increases in productivity per given quantity of water. A further intent was to emphasize what happens as aquatic systems are loaded excessively from man-induced changes.

As has been succinctly described by Vallentyne (1974), a common result of misuse of the drainage basin and excessive nutrient loading of fresh waters is an accelerated eutrophication; our lakes are literally turning into "algal bowls." In the present treatment, I have repeatedly emphasized that the metabolism of all aquatic systems, and indeed of a major portion of the biosphere, is dominated by detrital metabolism. Accelerated eutrophication leads to accelerated pelagial and littoral primary productivity with progressive intensification of detrital metabolism, effectively relegating lakes to "detrital bowls" in an operational sense. Metabolically mediated changes in the environment leading to strata of prolonged anoxia and attendant reductions in catabolism of detrital organic matter result in decreased efficiencies of utilization and degradation of organic matter.

A conscientious individual must view these changes in his natural environment with concern. As the exploitative pressures of demophoric growth increase, man's concern must involve more than simply his aesthetic values and those of future generations of humans. The very survival of man centers on the wise utilization of finite freshwater resources; to think otherwise is naive and myopic.

Use of fresh water for purposes of aquaculture, to provide protein for burgeoning human demands, is not a very promising endeavor because of the relatively low efficiencies of conversion of solar to potential chemical energy. Terrestrial agriculture, in spite of its high freshwater requirements, is in general more efficient. The greatest demands for fresh water, however, are in the area of technology. Much of this water is returned to the environment in a seriously degraded form or removed from its source. Groundwater is particularly susceptible to misuse because of its commonly slow renewal times.

The demands on water resources now are increasing exponentially, and necessitate a corresponding increase in management, as well as a degree of manipulation. Unless we comprehend the rudiments of their functional operation, any hope of using them effectively is lost. Contemporary management of freshwater resources is often performed in erratic and irrational ways. The rationale behind many practices is commonly purely hydrological, without the slightest interest in or appreciation of the biotic interactions of the water bodies themselves, or the long-term related effects on the climate of large regional areas, and, in the end, on the biosphere. The metabolism of aquatic systems *does* matter and must be considered and understood in all manipulative measures taken.

Analyses of dynamic aquatic ecosystems require a deep understanding of the physiology and biochemistry of organisms, evaluated within the context of physical and chemical variables that determine the external environment of these organisms. From these evaluations, the functional control of contemporary metabolic states of organisms, populations, and aquatic systems can emerge; this represents a major objective of limnologists today. The ability to predict changes in the operation of aquatic ecosystems in response to perturbations is essential to limnological efforts.

The effectiveness of our predictive capability can be increased in many ways. Certainly among the most informative approaches is to determine the functional state of an aquatic ecosystem, and then observe the responses of the system and its biota when some specific parameter is manipulated. This method is far from new, and has been practiced in various forms for centuries, for example in the fertilization of ponds to increase the growth rates and biomass of a particular species of fish. The early methods were largely trial and error, without appreciable cognizance of controlling factors. The same approach permeates contemporary manipulation of water resources. Because of existing voids in the operational biology of aquatic ecosystems, engineering methods of water utilization are often implemented blindly. Some of these methods are marginally operational, most are inefficient, and many are blatantly destructive of water resources on a long-term basis. Insights into parameters that regulate metabolic processes and population dynamics also have been obtained under more controlled conditions; in this category can be included studies on laboratory cultures, small to very large isolated portions of aquatic systems containing natural biota, and the manipulation of whole lake systems. Much more of this integrated physiological and large-scale manipulative research is needed, and certainly these efforts represent a major direction in contemporary limnology.

Our knowledge of the operation of freshwater ecosystems has come a great distance in the last 60 years. However, if we really scrutinize the existing information in an unbiased way, the only conclusion is that we remain quite ignorant of this beautifully complex integration. Accrual of the needed information must be greatly accelerated. There is no alternative to devoting a greater percentage of our intellectual and financial resources to fresh waters. Everything allocated to this critical resource thus far has been utter tokenism. We have passed the point at which we can continue to take unabatedly without casting back some comprehensive efforts in return. Nature is remarkably resilient to human insults. Man must learn what are nature's dynamic capacities, however, because excessive violation without harmony will only unleash her intolerable vengeance.

APPENDICES

APPENDIX I Units and Conversion Factors for Solar Radiation Data (See Chapter 5)

1 watt = 1 joule s^{-1}
1 watt cm^{-2} = 1 joule s^{-1} cm^{-2}
1 watt = 14.3 cal min^{-1}
1 watt cm^{-2} = 14.3 langley min^{-1}
1 ly min^{-1} = 0.0698 watt cm^{-2}
1 ft-candle = 10.764 lux
1 lux = 0.0929 ft-candle
1 ly = 2.11 × 10^{15} × Å quanta cm^{-2} (where Å = wavelength of the quanta in Å units)
1 ly min^{-1} = 2.11 × 10^{15} × Å quanta cm^{-2} min^{-1}
1 ly min^{-1} = × 3.50 × 10^{-9} × Å einsteins cm^{-2} min^{-1}

A: ENERGY [DIMENSIONS: MASS LENGTH^2 TIME^{-2}]

	Erg	Cal	Kcal	J (= Ws)	Wh	kWh	Btu*
erg	1	2.39 × 10^{-8}	2.39 × 10^{-11}	10^{-7}	2.78 × 10^{-11}	2.78 × 10^{-14}	9.48 × 10^{-11}
cal	4.19 × 10^{7}	1	10^{-3}	4.19	1.16 × 10^{-3}	1.16 × 10^{-6}	3.97 × 10^{-3}
kcal	4.19 × 10^{10}	10^{3}	1	4.19 × 10^{3}	1.16	1.16 × 10^{-3}	3.97
J (= Ws)	10^{7}	0.239	2.39 × 10^{-4}	1	2.78 × 10^{-4}	2.78 × 10^{-7}	9.48 × 10^{-4}
Wh	3.6 × 10^{10}	8.60 × 10^{2}	8.60 × 10^{-1}	3.6 × 10^{3}	1	10^{-3}	3.41
kWh	3.6 × 10^{13}	8.60 × 10^{5}	8.60 × 10^{2}	3.6 × 10^{6}	10^{3}	1	3.41 × 10^{3}
Btu*	1.06 × 10^{10}	2.52 × 10^{2}	2.52 × 10^{-1}	1.06 × 10^{3}	2.93 × 10^{-1}	2.93 × 10^{-4}	1

B: POWER (ENERGY PER UNIT TIME) [DIMENSIONS: MASS LENGTH^2 TIME^{-3}]

	Erg s^{-1}	Cal s^{-1}	Cal min^{-1}	W (= J s^{-1})	Btu* min^{-1}	Btu* h^{-1}
erg s^{-1}	1	2.39 × 10^{-8}	1.43 × 10^{-6}	1.00 × 10^{-7}	5.69 × 10^{-9}	9.48 × 10^{-11}
cal s^{-1}	4.19 × 10^{7}	1	60	4.19	2.38 × 10^{-1}	3.97 × 10^{-3}
cal min^{-1}	6.97 × 10^{5}	1.67 × 10^{-2}	1	6.97 × 10^{-2}	3.97 × 10^{-3}	6.62 × 10^{-5}
W(= J s^{-1})	1.00 × 10^{7}	2.39 × 10^{-1}	14.3	1	5.69 × 10^{-2}	9.48 × 10^{-4}
Btu* min^{-1}	1.76 × 10^{8}	4.20	2.52 × 10^{2}	17.6	1	1.67 × 10^{-2}
Btu* h^{-1}	1.06 × 10^{10}	2.52 × 10^{2}	1.52 × 10^{4}	1.06 × 10^{3}	60	1

C: RADIANT FLUX DENSITY AND IRRADIANCE UNITS ([DIMENSIONS: MASS TIME^{-3}][λ in cm])

	Quantum cm^{-2} s^{-1}	Einstein cm^{-2} s^{-1}	Erg cm^{-2} s^{-1}	Cal cm^{-2} s^{-1}	Cal cm^{-2} min^{-1}	Cal cm^{-2} h^{-1}
quantum cm^{-2} s^{-1}	1	1.66×10^{-24}	$1.99 \times 10^{-16} \lambda^{-1}$	$4.75 \times 10^{-24} \lambda^{-1}$	$2.85 \times 10^{-22} \lambda^{-1}$	$1.71 \times 10^{-20} \lambda^{-1}$
einstein cm^{-2} s^{-1}	6.02×10^{23}	1	$1.20 \times 10^{8} \lambda^{-1}$	$2.86 \lambda^{-1}$	$1.72 \times 10^{2} \lambda^{-1}$	$1.03 \times 10^{4} \lambda^{-1}$
erg cm^{-2} s^{-1}	$5.03 \times 10^{15} \lambda$	$8.35 \times 10^{-9} \lambda$	1	2.39×10^{-8}	1.43×10^{-6}	8.6×10^{-5}
cal cm^{-2} s^{-1}	$2.11 \times 10^{23} \lambda$	$3.50 \times 10^{-1} \lambda$	4.19×10^{7}	1	60	3.6×10^{3}
cal cm^{-2} min^{-1}	$3.51 \times 10^{21} \lambda$	$5.83 \times 10^{-3} \lambda$	6.98×10^{5}	1.67×10^{-2}	1	60
cal cm^{-2} h^{-1}	$5.85 \times 10^{19} \lambda$	$9.71 \times 10^{-5} \lambda$	1.16×10^{4}	2.78×10^{-4}	1.67×10^{-2}	1
cal dm^{-2} h^{-1}	$5.85 \times 10^{17} \lambda$	$9.71 \times 10^{-7} \lambda$	1.16×10^{2}	2.78×10^{-6}	1.67×10^{-4}	10^{-2}
kcal m^{-2} h^{-1}	$5.85 \times 10^{18} \lambda$	$9.71 \times 10^{-6} \lambda$	1.16×10^{3}	2.78×10^{-5}	1.67×10^{-3}	10^{-1}
μW cm^{-2}	$5.03 \times 10^{16} \lambda$	$8.35 \times 10^{-8} \lambda$	10	2.39×10^{-7}	1.43×10^{-5}	8.6×10^{-4}
W cm^{-2}	$5.03 \times 10^{22} \lambda$	$8.35 \times 10^{-2} \lambda$	10^{7}	2.39×10^{-1}	14.33	8.6×10^{2}
W m^{-2}	$5.03 \times 10^{18} \lambda$	$8.35 \times 10^{-6} \lambda$	10^{3}	2.39×10^{-5}	1.43×10^{-3}	8.6×10^{-2}
Btu* ft^{-2} min^{-1}	$0.95 \times 10^{21} \lambda$	$1.58 \times 10^{-3} \lambda$	1.89×10^{5}	4.53×10^{-3}	0.27	16.2

	cal dm^{-2} h^{-1}	kcal m^{-2} h^{-1}	μW cm^{-2}	W cm^{-2}	W m^{-2}	Btu* ft^{-2} min^{-1}
quantum cm^{-2} s^{-1}	$1.71 \times 10^{-18} \lambda^{-1}$	$1.71 \times 10^{-19} \lambda^{-1}$	$1.99 \times 10^{-17} \lambda^{-1}$	$1.99 \times 10^{-23} \lambda^{-1}$	$1.99 \times 10^{-19} \lambda^{-1}$	$1.05 \times 10^{-21} \lambda^{-1}$
einstein cm^{-2} s^{-1}	$1.03 \times 10^{6} \lambda^{-1}$	$1.03 \times 10^{5} \lambda^{-1}$	$1.20 \times 10^{7} \lambda^{-1}$	$12.0 \lambda^{-1}$	$1.20 \times 10^{5} \lambda^{-1}$	$6.34 \times 10^{2} \lambda^{-1}$
erg cm^{-2} s^{-1}	8.6×10^{-3}	8.6×10^{-4}	10^{-1}	10^{-7}	10^{-3}	5.28×10^{-6}
cal cm^{-2} s^{-1}	3.6×10^{5}	3.6×10^{4}	4.19×10^{6}	4.19	4.19×10^{4}	2.21×10^{2}
cal cm^{-2} min^{-1}	6×10^{3}	6×10^{2}	6.98×10^{4}	6.98×10^{-2}	6.98×10^{2}	3.69
cal cm^{-2} h^{-1}	10^{2}	10	1.16×10^{3}	1.16×10^{-3}	11.63	6.17×10^{-2}
cal dm^{-2} h^{-1}	1	10^{-1}	11.63	1.16×10^{-5}	1.16×10^{-1}	6.17×10^{-4}
kcal m^{-2} h^{-1}	10	1	1.16×10^{2}	1.16×10^{-4}	1.163	6.17×10^{-3}
μW cm^{-2}	8.6×10^{-2}	8.6×10^{-3}	1	10^{-6}	10^{-2}	5.28×10^{-5}
W cm^{-2}	8.6×10^{4}	8.6×10^{3}	10^{6}	1	10^{4}	5.28×10
W m^{-2}	8.6	8.6×10^{-1}	10^{2}	10^{-4}	1	5.28×10^{-3}
Btu* ft^{-2} min^{-1}	1.62×10^{3}	1.62×10^{2}	1.89×10^{4}	1.89×10^{-2}	1.89×10^{2}	1

Data modified from Strickland, 1958, and Šesták, Čatský, and Jarvis, 1971.

*Btu = British thermal unit

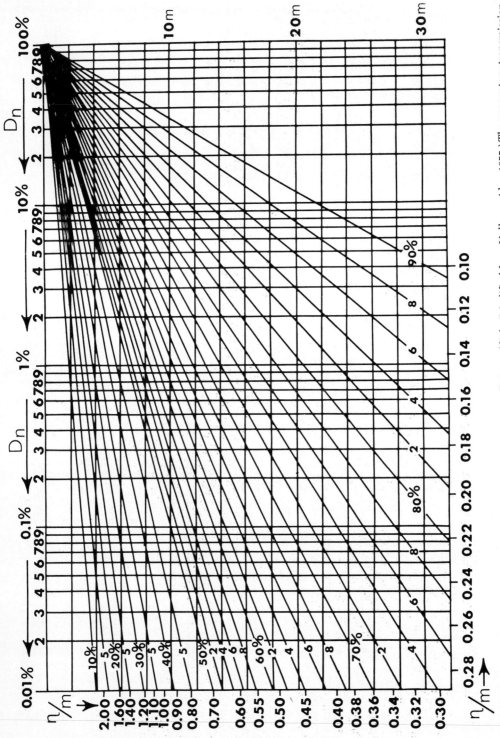

Appendix II. Nomogram for the estimation of vertical extinction coefficient of light (Modified from Vollenweider, 1955.) The percentage transmission of surface light along the upper abscissa is coordinated with depth (right ordinate) to diagonal line and followed to η along the left ordinate and lower abscissa.

APPENDIX III   Units and Conversion Factors of Length and Weight

Length, meter, m
1 inch, in = 25.4 mm = 0.0254 m
1 foot, ft = 12 in = 304.8 mm = 0.3048 m
1 yard, yd = 36 in = 0.9144 m
1 mile = 1760 yd = 1.60934 km
1 international nautical mile (n mile) = 1.852 km

Area, m^2
$1\ in^2 = 6.4516\ cm^2 = 6.4516 \times 10^{-4}\ m^2$
$1\ ft^2 = 929.0304\ cm^2 = 9.290304 \times 10^{-2}\ m^2$
$1\ yd^2 = 8361.2736\ cm^2 = 0.83612736\ m^2$
$1\ acre = 4046.86\ m^2 = 0.40486\ ha$

Volume, m^3 (1 liter, l = 1 dm^3 = $10^{-3}\ m^3$ and 1.0 ml = 1 cm^3)
$1\ in^3 = 16.3871\ cm^3 = 1.63871 \times 10^{-5}\ m^3$
$1\ ft^3 = 28.3168\ dm^3$ (l) = 0.0283168 m^3
1 pint (UK), pt = 0.568261 dm^3 (l) = $5.68261 \times 10^{-4}\ m^3$ [$\times 1.201$ = pints (US)]
1 quart (UK), qt = 1.13652 dm^3 (l) = $1.13652 \times 10^{-3}\ m^3$ [$\times 1.201$ = quarts (US)]
1 gallon (US), US gal = 3.78541 dm^3 (l) = $3.78541 \times 10^{-3}\ m^3$
1 gallon (UK), gal = 4.54609 dm^3 (l) = $4.54609 \times 10^{-3}\ m^3$

Mass, kg
1 ounce (av), oz = 28.3495 g = 0.0283495 kg
1 ounce (UK), fluid; fl oz = 28.413 cm^3 = $2.84131 \times 10^{-5}\ m^3$
1 ounce (US), fluid; fl oz = 29.57 cm^3
1 pound, lb = 453.59239 g = 0.45359239 kg
1 quarter (UK, long) = 12.7006 kg
1 hundred weight (long), cwt = 50.8023 kg
1 ton (long) = 1016.05 kg

1 dz (Doppelzentner, German)
1 ctr (centner, Russian and Polish)   } all = 100 kg
1 m.q. (metric quintal, French and Polish)

The quintal in the U.S.A. and Britain is a variable unit, which can equal 100 or 112 lbs.

	Multiplying factors	
	to kg/m^2	from kg/m^2
mt/ha (metric tons or tonnes)	0.1000	10.00
mt/acre (metric tons)	0.2471	4.047
lt/acre (long or British tons)	0.2511	3.983
st/acre (short or United States tons)	0.2242	4.460
lb/yd^2	0.5429	1.842

Pounds per acre are best converted to short tons by dividing by 2000.

(Appendix III continued on p. 762.)

APPENDIX IV   Oxygen Saturation Nomogram (See Chapter 9)

The nomogram provides estimates of: (a) solubility of oxygen in fresh water in equilibrium with the atmosphere (at 100% humidity), and (b) per cent saturation of oxygen in samples at observed temperatures and oxygen content.

Scale 1:   Bottom scale = water temperature (°C);
Upper scale = corresponding saturated concentration (C^*) of oxygen in equilibrium with the atmosphere (at 100% humidity and standard pressure of 760 mm Hg or 101.325 kPa).

Scale 2:   Oxygen concentration (s) in mg l^{-1} (lower) and μmol l^{-1} (upper).

Scale 3:   Per cent saturation of dissolved oxygen.

Scale 4:   Scales to estimate equilibrium concentrations at pressures (P) other than standard (P_{st}). Place one end (point) of a scale (or calipers) at zero h and extend the other end (point) into the fan of scales corresponding to the altitude (in km) or pressure (lower scale) or pressure (in mm Hg, upper scale) to intersect with the water temperature among the temperature lines extending from the fan ends.

The distance from zero h to the pressure (altitude)-temperature point within the fan Scale 4 is then brought down onto Scale 1. This distance is extended from the temperature of the water from which oxygen concentration was determined to the right.† Where a straight line between this point and the oxygen concentration (Scale 1) intersects Scale 3 gives the per cent saturation of dissolved oxygen.

If the oxygen concentration (s) is below the range of Scale 2, multiply s by a convenient factor, e.g., 10. Determine the percentage saturation as described above, but divide that estimate by the factor to give the correct result.

From Mortimer, C. H. 1981. The oxygen content of air-saturated fresh waters over ranges of temperature and atmospheric pressure of limnological interest. Mitt. Int. Ver. Limnol. 22. 23 pp.

†Extended to the left of the water temperature on Scale 1 if the lake surface is below sea level.

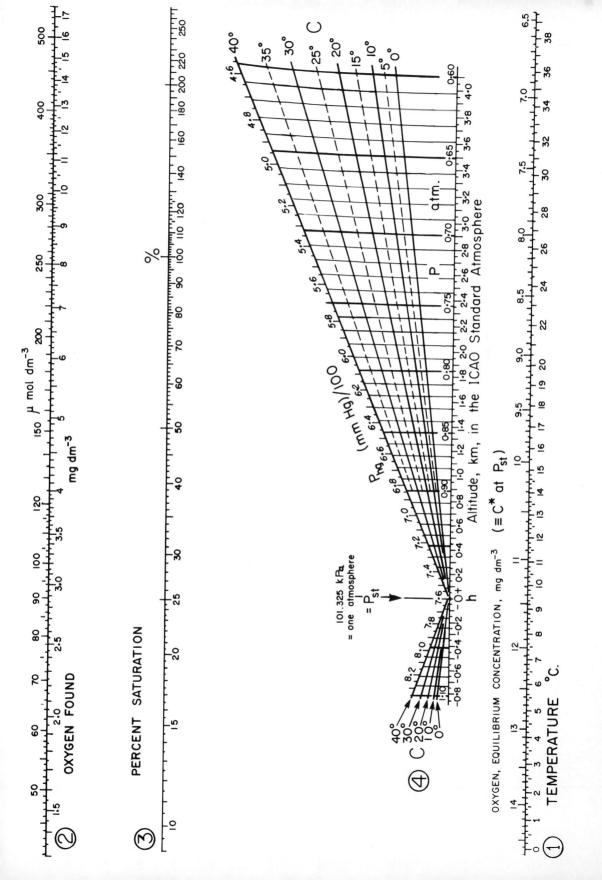

APPENDIX III (Continued
To Convert the Values in Units of the Top Line, Multiply by the Relevant Factors in Order to Obtain Corresponding Values of the Lefthand Column.

	ANGSTROMS	NANOMETERS (MILLIMICRONS)	MICROMETERS (MICRONS)	MILLIMETERS	CENTIMETERS	KILOMETERS	METERS	INCHES	FEET	MILES
Angstroms	1	10	10^4	10^7	10^8	10^{13}	10^{10}	2.540×10^8	3.048×10^9	1.609×10^{13}
Nanometers (millimicrons)	10^{-1}	1	10^3	10^6	10^7	10^{12}	10^9	2.540×10^7	3.048×10^8	1.609×10^{12}
Micrometers (microns)	10^{-4}	10^{-3}	1	10^3	10^4	10^9	10^6	2.540×10^4	3.048×10^5	1.609×10^9
Millimeters	10^{-7}	10^{-6}	10^{-3}	1	10	10^6	10^3	2.540×10	3.048×10^2	1.609×10^6
Centimeters	10^{-8}	10^{-7}	10^{-4}	0.1	1	10^5	10^2	2.540	3.048×10	1.609×10^5
Kilometers	10^{-13}	10^{-12}	10^{-9}	10^{-6}	10^{-5}	1	10^3	2.540×10^{-5}	3.048×10^{-4}	1.609
Meters	10^{-10}	10^{-9}	10^{-6}	10^{-3}	10^{-2}	10^3	1	2.540×10^{-2}	3.048×10^{-1}	1.609×10^3
Inches	3.937×10^{-9}	3.937×10^{-8}	3.937×10^{-5}	3.937×10^{-2}	3.937×10^{-1}	3.937×10^4	3.937×10	1	12	6.336×10^4
Feet	3.281×10^{-10}	3.281×10^{-9}	3.281×10^{-6}	3.281×10^{-3}	3.281×10^{-2}	3.281×10^3	3.281	8.333×10^{-2}	1	5.280×10^3
Miles	6.214×10^{-14}	6.214×10^{-13}	6.214×10^{-10}	6.214×10^{-7}	6.214×10^{-6}	6.214×10^{-1}	6.214×10^{-4}	1.578×10^{-5}	1.894×10^{-4}	1

Modified from Bickford and Dunn, 1972, Westlake, 1963, and Šesták, et al., 1971.

SYMBOLS IN CHAPTER TITLES

Many readers of the first edition of LIMNOLOGY have written to ask about the meaning of the symbols used on the first page of each chapter. The symbols were originally used by alchemists and have been modernized slightly by the publisher's artist. While the symbols have some relationship to the content of the chapters, some were chosen with tongue in cheek. Listed here are the symbols and their meanings.

CHAPTER 1   **Prologue**

Water

CHAPTER 2   **Water as a Substance**

Water

CHAPTER 3   **Lakes—Their Distribution, Origins, Forms**

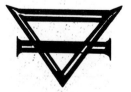

Earth

CHAPTER 4   **Water Economy**

Air

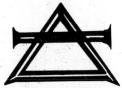

CHAPTER 5   **Light in Lakes**

Sol

CHAPTER 6   **Fate of Heat**

Fire

CHAPTER 7   **Water Movements**

Projection

CHAPTER 8   **Aquatic Ecosystems: Structure and Productivity**

Amalgam;
a mixture
of
different
things

CHAPTER 9   **Oxygen**

Sublimation

Digestion

CHAPTER 10   **Salinity of Inland Waters**

salt   lime

CHAPTER 11   **Inorganic Carbon**

alkali   precipitation   precipitate

CHAPTER 12   **Nitrogen Cycle**

nitric acid    sal ammoniac

CHAPTER 13   **Phosphorus Cycle**

apiapite

CHAPTER 14   **Iron, Sulfur, and Silica Cycles**

iron   sulfur   glass

CHAPTER 15   **Planktonic Communities: Algae**

Materia prima

CHAPTER 16   **Planktonic Communities: Zooplankton and Their Interactions with Fish**

bolus

CHAPTER 17   **Planktonic Communities: Bacteria**

Process (Prozess)

CHAPTER 18   **Littoral Communities: Larger Plants**

Essential oil

CHAPTER 19   **Littoral Communities: Algae and Zooplankton**

Mashing Rod (Magischer Stab)

CHAPTER 20   **Littoral Communities: Sediments and Microflora**

Distillation

CHAPTER 21   **Benthic Animals and Fish Communities**

sublimate

CHAPTER 22   **Detritus: The Organic Carbon Cycle**

fermentation    acetic acid   wax

CHAPTER 23   **Historical Records of Productivity: Paleolimnology**

lasting or
perennial
plants
(ausdauernde
Pflanze)

CHAPTER 24   **Ontogeny of Lake Ecosystems**

annealing

CHAPTER 25   **Epilogue**

minium

REFERENCES CITED

van Aardt, W. J. and C. T. Wolmarans. 1981. Evidence for non-assimilation of *Chlorella* by the African freshwater snail *Bulinus (Physopsis) globosus*. S. African J. Sci. 77:319–320.

Aaronson, S., S. W. Dhawale, N. J. Patni, B. DeAngelis, O. Frank, and H. Baker. 1977. The cell content and secretion of water-soluble vitamins by several freshwater algae. Arch. Microbiol. 112:57–59.

Abella, C., E. Montesinos, and R. Guerrero. 1980. Field studies on the competition between purple and green sulfur bacteria for available light (Lake Siso, Spain). Developments Hydrobiol. 3:173–181.

Åberg, B., and W. Rodhe. 1942. Über die Milieufaktoren in einigen südschwedischen Seen. Symbol. Bot. Upsalien. 5(3), 256 pp.

Adam, N. K. 1937. A rapid method for determining the lowering of tension of exposed water surfaces, with some observations on the surface tension of the sea and of inland waters. Proc. Roy. Soc. London (Ser. B) 122:134–139.

Adams, J. 1980. The role of competition in the population dynamics of a freshwater flatworm, *Bdellocephala punctata* (Turbellaria, Tricladida). J. Anim. Ecol. 49:565–579.

Adams, M. S., and M. D. McCracken. 1974. Seasonal production of the *Myriophyllum* component of the littoral of Lake Wingra, Wisconsin. J. Ecol. 62:457–465.

Adams, W. A. 1978. Effects of ice cover on the solar radiation regime in Canadian lakes. Verh. Int. Ver. Limnol. 20:141–149.

Adams, W. P. and D. C. Lasenby. 1978. The role of ice and snow in lake heat budgets. Limnol. Oceanogr. 23:1025–1028.

Adams, W. P. and T. D. Prowse. 1978. Observations on special characteristics of lake snowcover. Proc. Eastern Snow Conf. 1978: 117–128.

Adams, W. P., M. C. English, and D. C. Lasenby. 1979. Snow and ice in the phosphorus budget of a lake in south central Ontario. Water Res. 13:213–215.

Adcock, J. A. 1979. Energetics of a population of the isopod *Asellus aquaticus*: Life history and production. Freshwat. Biol. 9:343–355.

Adeniji, H. A. 1978. Diurnal vertical distribution of zooplankton during stratification in Kainji Lake, Nigeria. Verh. Int. Ver. Limnol. 20:1677–1683.

Adreani, L., C. Bonacina, and G. Bonomi. 1981. Production and population dynamics in profundal lacustrine Oligochaeta. Verh. Int. Ver. Limnol. 21:967–974.

Agami, M., S. Beer, and Y. Waisel. 1980. Growth and photosynthesis of *Najas marina* L. as affected by light intensity. Aquatic Bot. 9:285–289.

Ahl, T. 1980. Variability in ionic composition of Swedish lakes and rivers. Arch. Hydrobiol. 89:5–16.

Aiken, S. G. and K. F. Walz. 1979. Turions of *Myriophyllum exalbescens*. Aquatic Bot. 6:357–363.

Aiken, S. G. and R. R. Picard. 1980. The influence of substrate on the growth and morphology of *Myriophyllum exalbescens* and *Myriophyllum spicatum*. Can. J. Bot. 58:1111–1118.

Aizaki, M. 1978. Seasonal changes in standing crop and production of periphyton in the Tamagawa River. Jap. J. Ecol. 28:123–134.

NOTE: A few discrepancies occur in the transliteration of Slavic languages. When the work of an author is translated into another language and I have cited only one of his or her works, that citation is recorded as printed even if the transliteration is incorrect. When more than one work of the same author is cited, but some are from the mother language, the name is cited consistently as it should be transliterated. An exception is Vinberg in the Russian; because of his European ancestry, the author uses Winberg in several important writings in the English. Conforming to the transliteration in this case could lead to confusion if the references were sought.

Aizaki, M. 1979. Growth rates of microorganisms in a periphyton community. Jap. J. Limnol. 40:1–10.

Albers, J. 1963. Interaction of Color. New Haven, Yale Univ. Press, 75 pp.

Albrecht. M. -L. 1964. Die Lichtdurchlässigkeit von Eis und Schnee und ihre Bedeutung für die Sauerstoffproduktion im Wasser. Deutsche Fischerei-Zeitung 11:371–376.

Alexander, M. 1961. Introduction to Soil Microbiology. New York, John Wiley & Sons, Inc., 472 pp.

Alexander, M. 1965a. Nitrification. Agronomy 10:307–343.

Alexander, M. 1965b. Biodegradation: Problems of molecular recalcitrance and microbial fallibility. Adv. Appl. Microbiol. 7:35–80.

Alexander, M. 1971. Microbial Ecology. New York, John Wiley & Sons, Inc., 511 pp.

Alexander, M. 1975. Environmental and microbiological problems arising from recalcitrant molecules. Microbial Ecol. 2:17–27.

Alexander, V., D. W. Stanley, R. J. Daley, and C. P. McRoy. 1980. Primary producers. In J. E. Hobbie, ed. Limnology of Tundra Ponds. Stroudsburg, Dowden, Hutchinson, & Ross, Inc., pp. 179–250.

Alimov, A. F., et al. 1970. Biological productivity of lakes Krivoe and Krugloe. In Z. Kajak and A. Hillbricht-Ilkowska, eds. Productivity Problems of Freshwaters. Warsaw, PWN Polish Scientific Publishers, pp. 39–56.

Aliverdieva-Gamidova, L. A. 1969. Microbiological processes in Lake Mekhteb. Mikrobiologiya, 38:1096–1100. (Translated into English, Consultants Bureau, New York, 1970.)

Allanson, B. R. 1973. The fine structure of the periphyton of Chara sp. and Potamogeton natans from Wytham Pond, Oxford, and its significance to the macrophyte-periphyton metabolic model of R. G. Wetzel and H. L. Allen. Freshwat. Biol. 3:535–541.

Allen, E. D. and D. H. N. Spence. 1981. The differential ability of aquatic plants to utilize the inorganic carbon supply in fresh waters. New Phytol. 87:269–283.

Allen, H. L. 1969. Chemo-organotrophic utilization of dissolved organic compounds by planktic algae and bacteria in a pond. Int. Rev. ges. Hydrobiol. 54:1–33.

Allen, H. L. 1971. Primary productivity, chemoorganotrophy, and nutritional interactions of epiphytic algae and bacteria on macrophytes in the littoral of a lake. Ecol. Monogr. 41:97–127.

Allen. H. L. 1972. Phytoplankton photosynthesis, micronutrient interactions, and inorganic carbon availability in a soft-water Vermont lake. In G. E. Likens, ed. Nutrients and Eutrophication: The Limiting-Nutrient Controversy. Special Symposium, Amer. Soc. Limnol. Oceanogr. 1:63–83.

Allen, M. B. 1952. The cultivation of Myxophyceae. Arch. Mikrobiol. 17:34–53.

Allen, M. B., and D. I. Arnon. 1955. Studies on nitrogen-fixing blue-green algae. II. The sodium requirement of Anabaena cylindrica. Physiol. Plant. 8:653–660.

Allgeier, R. J., B. C. Hafford, and C. Juday. 1941. Oxidation-reduction potentials and pH of lake waters and of lake sediments. Trans. Wis. Acad. Sci. Arts Lett., 33:115–133.

Almazan, G. and C. E. Boyd. 1978. Effects of nitrogen levels on rates of oxygen consumption during decay of aquatic plants. Aquatic Bot. 5:119–126.

Almer, B., W. Dickson, C. Ekström, and E. Hörnström. 1978. Sulfur pollution and the aquaticst ecosystem. In J. O. Nriagu, ed. Sulfur in the Environment. II. Ecological Impacts. New York, John Wiley & Sons, Inc., pp. 271–311.

Alsterberg, G. 1927. Die Sauerstoffschichtung der Seen. Bot. Not. 1927:255–274.

American Public Health Association. 1976. Standard Methods for the Examination of Water and Wastewater. 14th ed. Washington, D.C., American Public Health Association, 1193 pp.

Andersen, F. O. 1976. Primary production in a shallow water lake with special reference to a reed swamp. Oikos 27:243–250.

Andersen, F. O. 1978. Effects of nutrient level on the decomposition of Phragmites communis Trin. Arch. Hydrobiol. 84:42–54.

Andersen, J. M. 1975. Influence of pH on release of phosphorus from lake sediments. Arch. Hydrobiol. 76:411–419.

Andersen, J. M. 1977. Rates of denitrification of undisturbed sediment from six lakes as a function of nitrate concentration, oxygen and temperature. Arch. Hydrobiol. 80:147–159.

Anderson, D. V. 1961. A note on the morphology of the basins of the Great Lakes. J. Fish. Res. Bd. Canada 18:273–277.

Anderson, G. C. 1958. Some limnological features of a shallow saline meromictic lake. Limnol. Oceanogr. 3:259–270.

Anderson, L. W. J. 1978. Abscisic acid induces formation of floating leaves in the heterophyllous aquatic angiosperm Potamogeton nodusus. Science 201:1135–1138.

Anderson, L. W. J. and B. M. Sweeney. 1977. Diel changes in sedimentation characteristics of Ditylum brightwelli: Changes in cellular lipid and effects of respiratory inhibitors and iontransport modifiers. Limnol. Oceanogr. 22:539–552.

Anderson, N. H. and K. W. Cummins. 1979. Influences of diet on the life histories of aquatic insects. J. Fish. Res. Board Can. 36:335–342.

Anderson, N. H. and J. R. Sedell. 1979. Detritus processing by macroinvertebrates in stream ecosystems. Ann. Rev. Entomol. 24:351–377.

Anderson, N. H., J. R. Sedell, L. M. Roberts, and F. J. Triska. 1978. The role of aquatic invertebrates in processing of wood debris in coniferous forest streams. Amer. Midland Nat. 100:64–82.

Anderson, R. O., and F. F. Hooper. 1956. Seasonal

abundance and production of littoral bottom fauna in a southern Michigan lake. Trans. Amer. Microsc. Soc. 75:259-270.

Anderson, R. S. 1970. Predator-prey relationships and predation rates for crustacean zooplankters from some lakes in western Canada. Can. J. Zool. 48:1229-1240.

Andersson, E. 1969. Life cycle and growth of *Asellus aquaticus* (L.). Rept. Inst. Freshwat. Res. Drottningholm 49:5-26.

Andronikova, I. N., V. G. Drabkova, K. N. Kuzmenko, N. F. Michailova, and E. A. Stravinskaya. 1970. Biological productivity of the main communities of the Red Lake. *In* Z. Kajak and A. Hillbricht-Ilkowska, eds. Productivity Problems of Freshwaters. Warsaw, PWN Polish Scientific Publishers, pp. 57-71.

Anschütz, I., and F. Gessner. 1954. Der Ionenaustausch bei Torfmoosen *(Sphagnum)*. Flora 141:178-236.

Anthony, E. H., and F. R. Hayes. 1964. Lake water and sediment. VII. Chemical and optical properties of water in relation to the bacterial counts in the sediments of twenty-five North American lakes. Limnol. Oceanogr. 9:35-41.

Arber, A. 1920. Water Plants. A Study of Aquatic Angiosperms. Cambridge, Cambridge University Press, 436 pp.

Armstrong, W. 1978. Root aeration in the wetland condition. *In* D. D. Hook and R. M. M. Crawford, eds. Plant Life in Anaerobic Environments. Ann Arbor, Ann Arbor Science Publishers, Inc., pp. 269-297.

Arnold, D. E. 1971. Ingestion, assimilation, survival, and reproduction by *Daphnia pulex* fed seven species of blue-green algae. Limnol. Oceanogr. 16:906-920.

Aruga, Y. 1965. Ecological studies of photosynthesis and matter production of phytoplankton. II. Photosynthesis of algae in relation to light intensity and temperature. Bot. Mag. Tokyo 78:360-365.

Assman, A. V. 1953. Rol vodoroslevykh obrastanii v obrazovanii organicheskogo veshchestva v Glubokom Ozere. Trudy Vsesoiuznogo Gidrobiol. Obshchestva 5:138-157.

Atwell, B. J., M. T. Veerkamp, B. Stuiver, and P. J. C. Kuiper. 1980. The uptake of phosphate by *Carex* species from oligotrophic to eutrophic swamp habitats. Physiol. Plant. 49:487-494.

Averdieck, F. -R. 1978. Palynologischer Beitrag zur Entwicklungsgeschichte des Grossen Plöner Sees und der Vegetation seiner Umgebung. Arch. Hydrobiol. 83:1-46.

Ayers, J. C., D. C. Chandler, G. H. Lauff, C. F. Powers, and E. B. Henson. 1958. Currents and Water Masses of Lake Michigan. Publ. Great Lakes Res. Div., Univ. Mich., 3, 169 pp.

Baas Becking, L. G. M., I. R. Kaplan, and D. Moore. 1960. Limits of the natural environment in terms of pH and oxidation-reduction potentials. J. Geol. 68:243-284.

Baccini, P. 1976. Untersuchungen über den Schwermetallhaushalt in Seen. Schweiz. Z. Hydrol. 38:121-158.

Bachmann. R. W. 1963. Zinc-65 in studies of the freshwater zinc cycle. *In* V. Schultz and A. W. Klement, Jr., eds. Radioecology. First National Symposium on Radioecology. New York, Reinhold Publishing Corp., pp. 485-496.

Bachmann, R. W., and C. R. Goldman. 1965. Hypolimnetic heating in Castle Lake, California. Limnol. Oceanogr. 10:233-239.

Bailey, J. H. and R. B. Davis. 1978. Quantitative comparisons of lake water quality and surface sediment diatom assemblages in Maine, U.S.A. lakes. Verh. Int. Ver. Limnol. 20:531.

Bain, J. T. and M. C. F. Proctor. 1980. The requirement of aquatic bryophytes for free CO_2 as an inorganic carbon source: Some experimental evidence. New Phytol. 86:393-400.

Baker, R. L. 1979. Birth rate of planktonic rotifers in relation to food concentration in a shallow, eutrophic lake in western Canada. Can. J. Zool. 57:1206-1214.

Ball, R. C. 1949. Experimental use of fertilizer in the production of fish-food organisms and fish. Tech. Bull. Mich. State Univ. Agricult. Exp. Stat., Sec. Zool. 210, 28 pp.

Ball, R. C., and D. W. Hayne. 1952. Effects of the removal of the fish population on the fish-food organisms of a lake. Ecology 33:41-48.

Ball, R. C., T. A. Wojtalik, and F. F. Hooper. 1963a. Upstream dispersion of radiophosphorus in a Michigan trout stream. Pap. Mich. Acad. Sci. Arts Lett. 48:57-64.

Ball, R. C., and F. F. Hooper. 1963b. Translocation of phosphorus in a trout stream ecosystem. *In* V. Schultz and A. W. Klement, Jr., eds. Radioecology. First National Symposium on Radioecology. New York, Reinhold Publishing Corp., pp. 217-228.

Balogh, J. 1958. On some problems of production biology. Acta Zool. Acad. Sci. Hungaricae 4:89-114.

Bamforth, S. S. 1958. Ecological studies on the planktonic protozoa of a small artifical pond. Limnol. Oceanogr. 3:398-412.

Bandurski. R. S. 1965. Biological reduction of sulfate and nitrate. *In* J. Bonner and J. E. Varner, eds. Plant Biochemistry. New York, Academic Press, pp. 467-490.

Bannerman, R. T., D. E. Armstrong, G. C. Holdren, and R. F. Harris. 1974. Phosphorus mobility in Lake Ontario sediments. Proc. Conf. Great Lakes Res. 17:158-178.

Barber, L. E., and J. C. Ensign. 1979. Methane formation and release in a small Wisconsin lake. Geomicrobiol. J. 1:341-353.

Barica, J. 1970. Untersuchungen über den Stickstoff-Kreislauf des Titisees und seiner Quellen. Arch. Hydrobiol. (suppl.) 38:212-235.

Barica, J., and F. A. J. Armstrong. 1971. Contribution by snow to the nutrient budget of some small northwest Ontario lakes. Limnol. Oceanogr. 16:891–899.

Barica, J., and J. A. Mathias. 1979. Oxygen depletion and winterkill risk in small prairie lakes under extended ice cover. J. Fish. Res. Board Can. 36:980–986.

Barlow, J. P., and J. W. Bishop. 1965. Phosphate regeneration by zooplankton in Cayuga Lake. Limnol. Oceanogr. 10(Suppl.):R15–R25.

Bark, A. W. 1981. The temporal and spatial distribution of planktonic and benthic protozoan communities in a small productive lake. Hydrobiologia 85:239–255.

Barko. J. W., and R. M. Smart. 1979. The nutritional ecology of Cyperus esculentus, an emergent aquatic plant, grown on different sediments. Aquatic Bot. 6:13–28.

Barko, J. W., and R. M. Smart. 1980. Mobilization of sediment phosphorus by submersed freshwater macrophytes. Freshw. Biol. 10:229–238.

Barko, J. W., and R. M. Smart. 1981. Comparative influence of light and temperature on the growth and metabolism of selected submersed freshwater macrophytes. Ecol. Monogr. 51:219–235.

Barko, J. W., P. G. Murphy, and R. G. Wetzel. 1977. An investigation of primary production and ecosystem metabolism in a Lake Michigan dune pond. Arch. Hydrobiol. 81:155–187.

Bärlocher, F. and B. Kendrick. 1975. Assimilation efficiency of Gammarus pseudolimnaeus (Amphipoda) feeding on fungal mycelium or autumn-shed leaves. Oikos 26:55–59.

Barrett, P. H. 1957. Potassium concentrations in fertilized trout lakes. Limnol. Oceanogr. 2:287–294.

Bartell, S. M., and J. F. Kitchell. 1978. Seasonal impact of planktivory on phosphorus release by Lake Wingra zooplankton. Verh. Int. Ver. Limnol. 20:466–474.

Barth, H. 1957. Aufnahme und Abgabe von CO_2 and O_2 bei submersen Wasserpflanzen. Gewässer Abwässer 4(17/18):18–81.

Bastardo, H. 1979. Laboratory studies on decomposition of littoral plants. Polskie Arch. Hydrobiol. 26:267–299.

Battarbee, R. W. 1978. Relative composition, concentration, and calculated influx of diatoms from a sediment core from Lough Erne, Northern Ireland. Polskie Arch. Hydrobiol. 25:9–16.

Baumeister, W. 1943. Der Einfluss des Bors auf die Photosynthese und Atmung submerser Pflanzen. Jb. wiss. Bot. 91:242–277.

Baxter, R. M., R. B. Wood, and M. V. Prosser. 1973. The probable occurrence of hydroxylamine in the water of an Ethiopian lake. Limnol. Oceanogr. 18:470–472.

Bayly, I. A. E., and W. D. Williams. 1973. Inland Waters and Their Ecology. Victoria, Longman Australia Pty. Ltd., 316 pp.

Bazin, M., and G. W. Saunders. 1971. The hypolimnetic oxygen deficit as an index of eutrophication in Douglas Lake, Michigan. Mich. Academician 3(Pt. 4):91–106.

Beadle, L. C. 1943. Osmotic regulation and the faunas of inland waters. Biol. Rev. 18:172–183.

Beadle, L. C. 1957. Comparative physiology: osmotic and ionic regulation in aquatic animals. Ann. Rev. Physiol. 19:329–358.

Beadle, L. C. 1959. Osmotic and ionic regulation in relation to the classification of brackish and inland saline waters. Arch. Oceanogr. Limnol. (suppl.) 11:143–151.

Beadle, L. C. 1969. Osmotic regulation and the adaptation of freshwater animals to inland saline waters. Verh. Int. Ver. Limnol. 17:421–429.

Beadle, L. C. 1974. The Inland Waters of Tropical Africa: An Introduction to Tropical Limnology. London, Longman Group Ltd. 365 pp.

Beer, S., and R. G. Wetzel. 1981. Photosynthetic carbon metabolism in the submerged aquatic angiosperm Scirpus subterminalis. Plant Sci. Lett. 21:199–207.

Beer, S., and R. G. Wetzel. 1982. Photosynthesis in submersed macrophytes of a temperate lake. Plant Physiol. 70: 488–492.

Beeton, A. M. 1958. Relationship between Secchi disc readings and light penetration in Lake Huron. Trans. Amer. Fish. Soc. 87(1957):73–79.

Beeton, A. M. 1959. Photoreception in the opossum shrimp, Mysis relicta Lovén. Biol. Bull. 116:204–216.

Beeton, A. M. 1960. The vertical migration of Mysis relicta in lakes Huron and Michigan. J. Fish. Res. Bd. Canada 17:517–539.

Beeton, A. M. 1961. Environmental changes in Lake Erie. Trans. Amer. Fish. Soc. 90:153–159.

Behre, K. 1956. Die Algenbesiedlung einiger Seen um Bremen und Bremerhaven. Ver. Inst. Meeresforsch. Bremerhaven 4:221–283.

Bell, W. H. 1980. Bacterial utilization of algal extracellular products. I. The kinetic approach. Limnol. Oceanogr. 25:1007–1020.

Bell, W. H., and E. Sakshaug. 1980. Bacterial utilization of algal extracellular products. 2. A kinetic study of natural populations. Limnol. Oceanogr. 25:1021–1033.

Bengtsson, L., T. Brorson, S. Fleischer, and C. Aström. 1977. Beschaffenheitsänderungen des Sestons in Seen nach dem Sedimentieren. Acta Hydrochim. Hydrobiol. 5:153–165.

Benoit, R. J. 1957. Preliminary observations on cobalt and vitamin B_{12} in fresh water. Limnol. Oceanogr. 2:233–240.

Benoit, R. J. 1969. Geochemistry of eutrophication. In Eutrophication: Causes, Consequences, Correctives. Washington, D.C., National Academy of Sciences, pp. 614–630.

Benson, B. B., and D. Krause, Jr. 1980. The concentration and isotopic fractionation of gases dissolved in freshwater in equilibrium with the atmosphere. 1. Oxygen. Limnol. Oceanogr. 25:662–671.

Bentley, J. A. 1958a. Role of plant hormones in algal metabolism and ecology. Nature 181:1499–1502.

Bentley, J. A. 1958b. The naturally-occurring auxins and inhibitors. Ann. Rev. Plant Physiol. 9:47–80.

Benton, A. R., Jr., W. P. James, and J. W. Rouse, Jr. 1978. Evapotranspiration from water hyacinth (*Eichhornia crassipes* [Mart]. Solms) in Texas reservoirs. Water Resour. Bull. 14:919–930.

Berg, K. 1938. Studies on the bottom animals of Esrom Lake. K. Danske Vidensk. Selsk. Skr. Nat. Mat. Afd. 9, 8, 255 pp.

Berg, K., and I. C. Petersen. 1956. Studies on the humic, acid Lake Gribsø. Folia Limnol. Scandinavica 8, 273 pp.

Berger, F. 1955. Die Dichte natürlicher Wässer und die Konzentrations-Stabilität in Seen. Arch. Hydrobiol. (suppl.) 22:286–294.

Bergstein, T., Y. Henis, and B. Z. Cavari. 1981. Nitrogen fixation by the photosynthetic sulfur bacterium *Chlorobium phaeobacteroides* from Lake Kinneret. Appl. Environ. Microbiol. 41:542–544.

Berman, M. S., and S. Richman. 1974. The feeding behavior of *Daphnia pulex* from Lake Winnebago, Wisconsin. Limnol. Oceanogr. 19:105–109.

Berman, T. 1970. Alkaline phosphatases and phosphorus availability in Lake Kinneret. Limnol. Oceanogr. 15:663–674.

Berman, T. 1976. Release of dissolved organic matter by photosynthesizing algae in Lake Kinneret, Israel. Freshw. Biol. 6:13–18.

Bernard, J. M. 1973. Production ecology of wetland sedges: the genus *Carex*. Pol. Arch. Hydrobiol. 20:207–214.

Bernard, J. M., and B. A. Solsky. 1977. Nutrient cycling in a *Carex lacustris* wetland. Can. J. Bot. 55:630–638.

Bernatowicz, S., S. Leszczynski, and S. Tyczynska. 1976. The influence of transpiration by emergent plants on the water balance in lakes. Aquatic Bot. 2:275–288.

Bertani, A., I. Brambilla, and F. Menegus. 1980. Effect of anaerobiosis on rice seedlings: Growth, metabolic rate, and fate of fermentation products. J. Exp. Bot. 31:325–331.

Berzins, B. 1958. Ein planktologisches Querprofil. Rept. Inst. Freshwat. Res. Drottningholm 39:5–22.

Best, E. P. H. 1980. Effect of nitrogen on the growth and nitrogenous compounds of *Ceratophyllum demersum*. Aquatic Bot. 8:197–206.

Bickford, E. D., and S. Dunn. 1972. Lighting for Plant Growth. Kent, Ohio, Kent State University Press, 221 pp.

Bienfang, P. K. 1979. A new phytoplankton sinking rate method suitable for field use. Deep-Sea Res. 26:719–729.

Biggar, J. W., and R. B. Corey. 1969. Agricultural drainage and eutrophication. *In* Eutrophication: Causes, Consequences, Correctives. Washington, D.C., National Academy of Sciences, pp. 404–445.

Billaud, V. A. 1968. Nitrogen fixation and the utilization of other inorganic nitrogen sources in a subarctic lake. J. Fish. Res. Bd. Canada 25:2101–2110.

Birge, E. A. 1915. The heat budgets of American and European lakes. Trans. Wis. Acad. Sci. Arts Lett. 18(Pt. 1):166–213.

Birge, E. A. 1916. The work of the wind in warming a lake. Trans. Wis. Acad. Sci. Arts Lett. 18(Pt. 2):341–391.

Birge, E. A., and C. Juday. 1911. The inland lakes of Wisconsin. The dissolved gases of the water and their biological significance. Bull. Wis. Geol. Nat. Hist. Survey 22, Sci. Ser. 7, 259 pp.

Birge, E. A., and C. Juday. 1914. A limnological study of the Finger Lakes of New York. Bull. U.S. Bur. Fish. 32:525–609.

Birge, E. A., and C. Juday. 1926. Organic content of lake water. Bull. U.S. Bur. Fish. 42:185–205.

Birge, E. A., and C. Juday. 1927. The organic content of the water of small lakes. Proc. Amer. Phil. Soc. 66:357–372.

Birge, E. A., and C. Juday. 1934. Particulate and dissolved organic matter in inland lakes. Ecol. Monogr. 4:440–474.

Birge, E. A., C. Juday, and H. W. March. 1927. The temperature of the bottom deposits of Lake Mendota. A chapter in the heat exchanges in the lake. Trans. Wis. Acad. Sci. Arts Lett. 23:187–231.

Birks, H. J. B., and H. H. Birks. 1980. Quaternary Palaeoecology. Edward Arnold Publ., London. 289 pp.

Birks, H. H. 1980. Plant macrofossils in quaternary lake sediments. Arch. Hydrobiol. Ergeb. Limnol. 15. 60 pp.

Bjarnov, N. 1972. Carbohydrases in *Chironomus, Gammarus* and some Trichoptera larvae. Oikos 23:261–263.

Bjerke, G., A. H. Erlandsen, and K. Vennerød. 1979. The temperature of maximum density. Temperature profiles and circulation in deep, temperate lakes. Arch. Hydrobiol. 86:87–94.

Björk, S. 1967. Ecologic investigations of *Phragmites communis*. Studies in theoretic and applied limnology. Folia Limnol. Scand. 14, 248 pp.

Black, A. P., and R. F. Christman. 1968. Chemical characteristics of fulvic acids. J. Amer. Water Works Assoc. 55:897–912.

Black, C. C. 1971. Ecological implications of dividing plants into groups with distinct photosynthetic production capacities. Adv. Ecol. Res. 7:87–114.

Black, R. W. 1980a. The nature and causes of cyclomorphosis in a species of the *Bosmina longirostris* complex. Ecology 61:1122–1132.

Black, R. W. 1980b. The genetic component of cyclomorphosis in *Bosmina*. *In* W. C. Kerfoot, ed. Evolution and Ecology of Zooplankton Communities. Hanover, NH, Univ. Press New England, pp. 456–469.

Bland, R. D., and A. J. Brook. 1974. The spatial distribution of desmids in lakes in northern Minnesota, U.S.A. Freshwat. Biol. 4:543–556.

Blanton, J. O. 1973. Vertical entrainment into the

epilimnia of stratified lakes. Limnol. Oceanogr. 18:697–704.

Blinn, D. W., A. Fredericksen, and V. Korte. 1980. Colonization rates and community structure of diatoms on three different rock substrata in a lotic system. Br. Phycol. J. 15:303–310.

Blinn, D. W., and K. S. Button. 1973. The effect of temperature on parasitism of *Pandorina* sp. by *Dangeardia mammillata* B. Schröder in an Arizona mountain lake. J. Phycol. 9:323–326.

Bloesch, J., and N. M. Burns. 1980. A critical review of sedimentation trap technique. Schweiz. Z. Hydrol. 42:15–55.

Blotnick, J. R., J. Rho, and H. B. Gunner. 1980. Ecological characteristics of the rhizosphere microflora of *Myriophyllum heterophyllum*. J. Environ. Qual. 9:207–210.

Bodin, K., and A. Nauwerck. 1969. Produktionsbiologische Studien über die Moosvegetation eines klaren Gebirgssees. Schweiz. Z. Hydrol. 30:318–352.

Bodkin, P. C., D. H. N. Spence, and D. C. Weeks. 1980. Photoreversible control of heterophylly in *Hippuris vulgaris* L. New Phytol. 84:533–542.

Bodkin, P. C., U. Posluszny, and H. M. Dale. 1980. Light and pressure in two freshwater lakes and their influence on the growth, morphology and depth limits of *Hippuris vulgaris*. Freshw. Biol. 10:545–552.

Boers, J. J., and J. C. H. Carter. 1978. The life history of *Cyclops scutifer* Sars (Copepoda: Cyclopoida) in a small lake of the Matamek River System, Quebec. Can. J. Zool. 56:2603–2607.

Bogdan, K. G., J. J. Gilbert, and P. L. Starkweather. 1980. In situ clearance rates of planktonic rotifers. Hydrobiologia 73:73–77.

Bohl, E. 1980. Diel pattern of pelagic distribution and feeding in planktivorous fish. Oecologia 44:368–375.

Bohr, R., and M. Luścińska. 1975. The primary production of the periphyton association *Oedogonio-Epithemietum litoralae*. Verh. Int. Ver. Limnol. 19:1309–1312.

Bole, J. B., and J. R. Allan. 1978. Uptake of phosphorus from sediment by aquatic plants, *Myriophyllum spicatum* and *Hydrilla verticillata*. Water Res. 12:353–358.

Boney, A. D. 1981. Mucilage: The ubiquitous algal attribute. Br. Phycol. J. 16:115–132.

Bonomi, G. 1979. Ponderal production of *Tubifex tubifex* Müller and *Limnodrilus hoffmeisteri* Claparède (Oligochaeta, Tubificidae), benthic cohabitants of an artificial lake. Boll. Zool. 46:153–161.

Bonny, A. P. 1976. Recruitment of pollen to the seston and sediment of some Lake District lakes. J. Ecol. 64:859–887.

Booker, M. J. and A. E. Walsby. 1979. The relative form resistance of straight and helical blue-green algal filaments. Br. Phycol. J. 14:141–150.

Boon, J. J., R. G. Wetzel, and G. L. Godshalk. 1982. Pyrolysis mass spectrometry of some *Scirpus* species and their decomposition products. Limnol. Oceanogr. 27: 839–848.

Bormann, F. H., and G. E. Likens. 1979. Pattern and process in a forested ecosystem. New York, Springer-Verlag, 253 pp.

Borutskii, E. V. 1939. Dynamics of the total benthic biomass in the profundal of Lake Beloie. Proc. Kossino Limnol. Stat. Hydrometeorol. Serv. USSR 22:196–218. (Translated by M. Ovchynnyk, Michigan State University.)

Borutskii, E. V. 1950. Dinamika organicheskogo veshchestva v vodoeme. Trudy Vses. Gidrobiol. Obshchestva 2:43–68.

Bosselmann, S. 1975. Production of *Eudiaptomus graciloides* in Lake Esrom, 1970. Arch. Hydrobiol. 76:43–64.

Bosselmann, S. 1979a. Population dynamics of *Keratella cochlearis* in Lake Esrom. Arch. Hydrobiol. 87:152–165.

Bosselmann, S. 1979b. Production of *Keratella cochlearis* in Lake Esrom. Arch. Hydrobiol. 87:304–313.

Botan, E. A., J. J. Miller, and H. Kleerekoper. 1960. A study of the microbiological decomposition of nitrogenous matter in fresh water. Arch. Hydrobiol. 56:334–354.

Bott, T. L. 1982. Primary productivity in streams. In G. W. Minshall and J. R. Barnes, eds. Stream Ecology: The Testing of General Ecological Theory in Stream Ecosystems (In press)

Bottrell, H. H. 1975a. The relationship between temperature and duration of egg development in some epiphytic Cladocera and Copepoda from the River Thames, Reading, with a discussion of temperature functions. Oecologia 18:63–84.

Bottrell, H. H. 1975b. Generation time, length of life, instar duration and frequency of moulting, and their relationship to temperature in eight species of Cladocera from the River Thames, Reading. Oecologia 19:129–140.

Bottrell, H. H., A. Duncan, Z. M. Gliwicz, E. Grygierek, A. Herzig, A. Hillbricht-Ilkowska, H. Kurasawa, P. Larsson, and T. Weglenska. 1976. A review of some problems in zooplankton production studies. Norw. J. Zool. 24:419–456.

Bowen, S. H. 1978. Benthic diatom distribution and grazing by *Sarotherodon mossambicus* in Lake Sibaya, South Africa. Freshwat. Biol. 8:449–453.

Bowen, S. H. 1979. Determinants of the chemical composition of periphytic detrital aggregate in a tropical lake (Lake Valencia, Venezuela). Arch. Hydrobiol. 87:166–177.

Bowers, J. A., and N. E. Grossnickle. 1978. The herbivorous habits of *Mysis relicta* in Lake Michigan. Limnol. Oceanogr. 23:767–776.

Bowes, G., A. S. Holaday, and W. T. Haller. 1979. Seasonal variation in the biomass, tuber density, and photosynthetic metabolism of *Hydrilla* in three Florida lakes. J. Aquat. Plant Manage. 17:61–65.

Bowes, G., T. K. Van, L. A. Garrard, and W. T. Haller. 1977. Adaptation to low light levels by *Hydrilla.* J. Aquat. Plant Manage. *15*:32–35.

Bowie, I. S., and P. A. Gillespie. 1976. Microbial parameters and trophic status of ten New Zealand lakes. N.Z. J. Mar. Freshw. Res. *10*:343–354.

Bowman, G. T., and J. J. Delfino. 1980. Sediment oxygen demand techniques: A review and comparison of laboratory and in situ systems. Water Res. *14*:491–499.

Boyd, C. E. 1970. Losses of mineral nutrients during decomposition of *Typha latifolia.* Arch. Hydrobiol. *66*:511–517.

Boyd, C. E. 1971. The dynamics of dry matter and chemical substances in a *Juncus effusus* population. Amer. Midland Nat. *86*:28–45.

Boyle, E. A. 1979. Copper in natural waters. *In* J. O. Nriagu, ed. Copper in the Environment. I. Ecological Cycling. New York, John Wiley & Sons, Inc., pp. 77–88.

Boylen, C. W., and T. D. Brock. 1973. Bacterial decomposition processes in Lake Wingra sediments during winter. Limnol. Oceanogr. *18*:628–634.

Boylen, C. W. and R. B. Sheldon. 1976. Submergent macrophytes: Growth under winter ice cover. Science *194*:841–842.

Bradbury, J. P. 1975. Diatom stratigraphy and human settlement in Minnesota. Special Pap. Geol. Soc. Amer. *171.* 74 pp.

Bradbury, J. P., and R. O. Megard. 1972. Stratigraphic record of pollution in Shagawa Lake, northeastern Minnesota, Bull. Geol. Soc. Amer. *85*:2639–2648.

Bradbury, J. P., S. J. Tarapchak, J. C. B. Waddington, and R. F. Wright. 1975. The impact of a forest fire on a wilderness lake in northeastern Minnesota. Verh. Int. Ver. Limnol. *19*:875–883.

Bradbury, J. P., and T. C. Winter. 1976. Areal distribution and stratigraphy of diatoms in the sediments of Lake Sallie, Minnesota. Ecology *57*:1005–1014.

Bradford, G. R., F. I. Bair, and V. Hunsaker. 1968. Trace and major element content of 170 high Sierra lakes in California. Limnol. Oceanogr. *13*:526–530.

Bragg, A. N. 1960. An ecological study of the protozoa of Crystal Lake, Norman, Oklahoma. Wasmann J. Biol. *18*:37–85.

Brammer, E. S. 1979. Exclusion of phytoplankton in the proximity of dominant water-soldier *(Stratiotes aloides).* Freshw. Biol. *9*:233–249.

Brandl, Z., J. Brandlová, and M. Poštolková. 1970. The influence of submerged vegetation on the photosynthesis of phytoplankton in ponds. Rozpravy Českosl. Akad. Věd, Řada Matem. Přír. Věd. *80*(6):33–62.

Brandt, E. 1978. Anpassungen von *Tubifex tubifex* Müller (Annelida, Oligochaeta) an die Temperatur, den Sauerstoffgehalt und den Ernährungszustand. Arch. Hydrobiol. *84*:302–338.

Bray, J. R., D. B. Lawrence, and L. C. Pearson. 1959. Primary production in some Minnesota terrestrial communities for 1957. Oikos *10*:38–49.

Breger, I. A., ed. 1963. Organic Geochemistry. New York, Macmillan Company, 658 pp.

Brehm, J. 1967. Untersuchungen über den Aminosäure-Haushalt holsteinischer Gewässer, insbesondere des Pluss-Sees. Arch Hydrobiol. (suppl.) *32*:313–435.

Brehm, K. 1970. Kationenaustausch bei Hochmoorsphagnen: Die Wirkung von an den Austauscher gebundenen Kationen in Kulturversuchen. Beitr. Biol. Pflanzen *47*:91–116.

Bretschko, G. 1973. Benthos production of a high-mountain lake: Nematoda. Verh. Int. Ver. Limnol. *18*:1421–1428.

Brezny, O., I. Mehta, and R. K. Sharma. 1973. Studies on evapotranspiration of some aquatic weeds. Weed Sci. *21*:197–204.

Brezonik, P. L., J. J. Delfino, and G. F. Lee. 1969. Chemistry of N and Mn in Cox Hollow Lake, Wisc., following destratification. J. Sanit. Eng. Div. Proc. Amer. Soc. Civil Eng. *SA-5*:929–940.

Brezonik, P. L., and C. L. Harper. 1969. Nitrogen fixation in some aquatic lacustrine environments. Science *164*:1277–1279.

Brezonik, P. L., and G. F. Lee. 1968. Denitrification as a nitrogen sink in Lake Mendota, Wisconsin. Environ. Sci. Technol. *2*:120–125.

Bright, T., F. Ferrari, D. Martin, and G. A. Franceschini. 1972. Effects of a total solar eclipse on the vertical distribution of certain oceanic zooplankters. Limnol. Oceanogr. *17*:296–301.

Brinkhurst, R. O. 1967. The distribution of aquatic oligochaetes in Saginaw Bay, Lake Huron. Limnol. Oceanogr. *12*:137–143.

Brinkhurst, R. O. 1974a. Factors mediating interspecific aggregation of tubificid oligochaetes. J. Fish. Res. Bd. Canada *31*:460–462.

Brinkhurst, R. O. 1974b. The Benthos of Lakes. New York, St. Martin's Press, 190 pp.

Brinkhurst, R. O. 1980. The production biology of the Tubificidae (Oligochaeta). *In* R. O. Brinkhurst and D. G. Cook, eds. Aquatic Oligochaete Biology. New York, Plenum Press, pp. 205–209.

Brinkhurst, R. O., and M. J. Austin. 1979. Assimilation by aquatic Oligochaeta. Int. Rev. ges. Hydrobiol. *64*:245–250.

Brinkhurst, R. O., and K. E. Chua. 1969. Preliminary investigation of the exploitation of some potential nutritional resources by three sympatric tubificid oligochaetes. J. Fish. Res. Bd. Canada *26*:2659–2668.

Brinkhurst, R. O., K. E. Chua, and N. K. Kaushik. 1972. Interspecific interactions and selective feeding by tubificid oligochaetes. Limnol. Oceanogr. *17*:122–133.

Brinkhurst, R. O., and D. G. Cook. 1974. Aquatic earthworms (Annelida: Oligochaeta). *In* C. W. Hart, Jr., and S. L. H. Fuller, eds. Pollution Ecology of Freshwater Invertebrates. New York, Academic Press, pp. 143–156.

Brinkhurst, R. O., and D. G. Cook, eds. 1980. Aquatic Oligochaete Biology. Plenum Press, New York.

Brinkhurst, R. O., and B. G. M. Jamieson. 1971. Aquatic Oligochaeta of the World. Toronto, University of Toronto Press, 860 pp.

Brinkhurst, R. O., and B. Walsh. 1967. Rostherne Mere, England, a further instance of guanotrophy. J. Fish. Res. Bd. Canada 24:1299–1309.

Bristow, J. M. 1969. The effects of carbon dioxide on the growth and development of amphibious plants. Can. J. Bot. 47:1803–1807.

Bristow, J. M. 1974. Nitrogen fixation in the rhizosphere of freshwater angiosperms. Can. J. Bot. 52:217–221.

Bristow, J. M. 1975. The structure and function of roots in aquatic vascular plants. In J. G. Torrey and D. Clarkson, eds. The Development and Function of Roots. London, Academic Press, pp. 221–236.

Bristow, J. M., and M. Whitcombe. 1971. The role of roots in the nutrition of aquatic vascular plants. Am. J. Bot. 58:8–13.

Broch, E. S., and W. Yake. 1969. A modification of Maucha's ionic diagram to include ionic concentrations. Limnol. Oceanogr. 14:933–935.

Brock, T. D. 1966. Principles of Microbial Ecology. Englewood Cliffs, N.J., Prentice-Hall, Inc., 306 pp.

Broecker, W. S. 1965. An application of natural radon to problems in ocean circulation. In Diffusion in Oceans and Freshwaters. Symposium Proceedings, New York, Lamont Geological Observatory, Columbia University, pp. 116–145.

Broecker, W. S. 1973. Factors controlling CO_2 content in the oceans and atmosphere. In G. M. Woodwell and E. V. Pecan, eds. Carbon and the Biosphere. Brookhaven, N.Y., Proc. Brookhaven Symp. in Biol. 24. Tech. Information Center, U.S. Atomic Energy Commission CONF-720510, pp. 32–50.

Broecker, W. S., J. Cromwell, and Y. H. Li. 1968. Rates of vertical eddy diffusion near the ocean floor based on measurement of the distribution of excess ^{222}Rn. Earth Planet. Sci. Lett. 5:101–105.

Broecker, W. S., and T.-H. Peng. 1971. The vertical distribution of radon in the Bonex area. Earth Planet. Sci. Lett. 11:99–108.

Broecker, W. S., T. Takahashi, H. J. Simpson, and T.-H. Peng. 1979. Fate of fossil fuel carbon dioxide and the global carbon budget. Science 206:409–418.

Bronowski, J. 1956. Science and human values. New York, Harper & Brothers Publs., 94 pp.

Brook, A. J. 1981. The Biology of Desmids. Berkeley, Univ. California Press, 276 pp.

Brooks, A. S. and B. G. Torke. 1977. Vertical and seasonal distribution of chlorophyll a in Lake Michigan. J. Fish. Res. Bd. Can. 34:2280–2287.

Brooks, J. L. 1947. Turbulence as an environmental determinant of relative growth in Daphnia. Proc. Nat. Acad. Sci. 33:141–148.

Brooks, J. L. 1964. The relationship between the vertical distribution and seasonal variation of limnetic species of Daphnia. Verh. Int. Ver. Limnol. 15:684–690.

Brooks, J. L. 1965. Predation and relative helmet size in cyclomorphic Daphnia. Proc. Nat. Acad. Sci. 53:119–126.

Brooks, J. L. 1966. Cyclomorphosis, turbulence and overwintering in Daphnia. Verh. Int. Ver. Limnol. 16:1653–1659.

Brooks, J. L. 1968. The effects of prey size selection by lake planktivores. Syst. Zool. 17:273–291.

Brooks, J. L., and S. I. Dodson. 1965. Predation, body size, and composition of plankton. Science 150:28–35.

Brooks, J. L., and G. E. Hutchinson. 1950. On the rate of passive sinking in Daphnia. Proc. Nat. Acad. Sci. 36:272–277.

Brown, D. H., C. E. Gibby, and M. Hickman. 1972. Photosynthetic rhythms in epipelic algal populations. Brit. Phycol. Bull. 7:37–44.

Brown, J. M. A., F. I. Dromgoole, M. W. Towsey, and J. Browse. 1974. Photosynthesis and photorespiration in aquatic macrophytes. In R. L. Bieleski, A. R. Ferguson, and M. M. Cresswell, eds. Mechanisms of Regulation of Plant Growth. Bull. 12, Royal Soc. New Zealand, Wellington pp. 243–249.

Brown, S. -D., and A. P. Austin. 1973a. Diatom succession and interaction in littoral periphyton and plankton. Hydrobiologia 43:333–356.

Brown, S. -D., and A. P. Austin. 1973b. Spatial and temporal variation in periphyton and physicochemical conditions in the littoral of a lake. Arch. Hydrobiol. 71:183–232.

Brown, S. R. 1968. Bacterial carotenoids from freshwater sediments. Limnol. Oceanogr. 13:233–241.

Brown, S. R. 1969. Paleolimnological evidence from fossil pigments. Mitt. Int. Ver. Limnol. 17:95–103.

Brown, S. R., R. J. Daley, and R. N. McNeely. 1977. Composition and stratigraphy of the fossil phorbin derivatives of Little Round Lake, Ontario. Limnol. Oceanogr. 22:336–348.

Brown, S., and B. Coleman. 1963. Oscillaxanthin in lake sediments. Limnol. Oceanogr. 8:352–353.

Browne, R. A. 1978. Growth, mortality, fecundity, biomass and productivity of four lake populations of the prosobranch snail, Viviparus georgianus. Ecology 59:742–750.

Browse, J. A., J. M. A. Brown, and F. I. Dromgoole. 1979. Photosynthesis in the aquatic macrophyte Egeria densa. II. Effects of inorganic carbon conditions on ^{14}C fixation. Aust. J. Plant Physiol. 6:1–9.

Browse, J. A., J. M. A. Brown, and F. I. Dromgoole. 1980. Malate synthesis and metabolism during photosynthesis in Egeria densa Planch. Aquatic Bot. 8:295–305.

Browse, J. A., F. I. Dromgoole, and J. M. A. Brown. 1979. Photosynthesis in the aquatic macrophyte *Egeria densa*. III. Gas exchange studies. Aust. J. Plant Physiol. 6:499–512.

Brugam, R. B. 1978. Human disturbance and the historical development of Linsley Pond. Ecology 59:19–36.

Brugam, R. B. 1979. A re-evaluation of the Araphidineae/Centrales index as an indicator of lake trophic status. Freshwat. Biol. 9:451–460.

Brunskill, G. J. 1969. Fayetteville Green Lake, New York. II. Precipitation and sedimentation of calcite in a meromictic lake with laminated sediments. Limnol. Oceanogr. 14:830–847.

Brunskill, G. J., and S. D. Ludlam. 1969. Fayetteville Green Lake, New York. I. Physical and chemical limnology. Limnol. Oceanogr. 14:817–829.

Brylinsky, M. 1980. Estimating the productivity of lakes and reservoirs. *In* E. D. Le Cren and R. H. Lowe-McConnell, Eds. The Functioning of Freshwater Ecosystems. Cambridge, Cambridge Univ. Press, pp. 411–453.

Bryson, R. A. and T. J. Murray. 1977. Climates of hunger. Mankind and the world's changing weather. Madison, Univ. Wisconsin Press, 171 pp.

Bryson, R. A., and R. A. Ragotzkie. 1960. On internal waves in lakes. Limnol. Oceanogr. 5:397–408.

Bunner, H. C. and K. Halcrow. 1977. Experimental induction of the production of ephippia by *Daphnia magna* Straus (Cladocera). Crustaceana 32:77–86.

Burgis, M. J. 1970. The effect of temperature on the development time of eggs of *Thermocyclops* sp., a tropical cyclopoid copepod from Lake George, Uganda. Limnol. Oceanogr. 15:742–747.

Burkholder, P. R., and G. H. Bornside. 1957. Decomposition of marsh grass by aerobic marine bacteria. Bull. Torr. Bot. Club 84:366–383.

Burkholder, P. R., L. M. Burkholder, and J. A. Rivero. 1959. Some chemical constituents of turtlegrass, *Thalassia testudinum*. Bull. Torr. Bot. Club 86:88–93.

Burla, H. 1971. Gerichtete Ortsveränderung bei Muscheln der Gattung *Anodonta* im Zürichsee. Vierteljahrsschr. Naturf. Gesellschaft Zürich 116:181–194.

Burnison, B. K., and R. Y. Morita. 1973. Competitive inhibition for amino acid uptake by the indigenous microflora of Upper Klamath Lake. Appl. Microbiol. 25:103–106.

Burns, C. W. 1968a. The relationship between body size of filter-feeding Cladocera and the maximum size of particle ingested. Limnol. Oceanogr. 13:675–678.

Burns, C. W. 1968b. Direct observations of mechanisms regulating feeding behavior of *Daphnia* in lakewater. Int. Rev. ges. Hydrobiol. 53:83–100.

Burns, C. W. 1969a. Particle size and sedimentation in the feeding behavior of two species of *Daphnia*. Limnol. Oceanogr. 14:392–402.

Burns, C. W. 1969b. Relation between filtering rate, temperature, and body size in four species of *Daphnia*. Limnol. Oceanogr. 14:696–700.

Burns, C. W. 1979. Population dynamics and production of *Boeckella dilatata* (Copepoda: Calanoida) in Lake Hayes, New Zealand. Arch. Hydrobiol. (suppl.) 54:409–465.

Burns, C. W., and F. H. Rigler. 1967. Comparison of filtering rates of *Daphnia rosea* in lake water and in suspensions of yeast. Limnol. Oceanogr., 12:492–502.

Burns, N. M. and J. O. Nriagu. 1976. Forms of iron and manganese in Lake Erie waters. J. Fish. Res. Bd. Can. 33:463–470.

Burns, N. M., and C. Ross. 1971. Nutrient relationships in a stratified eutrophic lake. Proc. Conf. Great Lakes Res., Int. Assoc. Great Lakes Res. 14:749–760.

Burris, R. H., F. J. Eppling, H. B. Wahlin, and P. W. Wilson. 1943. Detection of nitrogen-fixation with isotopic nitrogen. J. Biol. Chem. 148:349–357.

Burton, J. D., T. M. Leatherland, and P. S. Liss. 1970. The reactivity of dissolved silicon in some natural waters. Limnol. Oceanogr. 15:473–476.

Buscemi, P. A. 1958. Littoral oxygen depletion produced by a cover of *Elodea canadensis*. Oikos 9:239–245.

Buscemi, P. A. 1961. Ecology of the bottom fauna of Parvin Lake, Colorado. Trans. Amer. Microsc. Soc. 80:266–307.

Bushnell, J. H., Jr. 1966. Environmental relations of Michigan Ectoprocta, and dynamics of natural populations of *Plumatella repens*. Ecol. Monogr. 36:95–123.

Bushnell, J. H. 1974. Bryozoans (Ectoprocta). *In* C. W. Hart, Jr., and S. L. H. Fuller, eds. Pollution Ecology of Freshwater Invertebrates. New York, Academic Press pp. 157–194.

Butler, M. G. 1982. A 7-year life cycle for two *Chironomus* species in arctic Alaskan tundra ponds (Diptera: Chironomidae). Can. J. Zool. 60:58–70.

Butlin, K. R. 1953. The bacterial sulphur cycle. Research 6:184–191.

Buttery, B. R., and J. M. Lambert. 1965. Competition between *Glyceria maxima* and *Phragmites communis* in the region of Surlingham Broad. I. The competition mechanism. J. Ecol. 53:163–181.

Buttery, B. R., W. T. Williams, and J. M. Lambert. 1965. Competition between *Glyceria maxima* and *Phragmites communis* in the region of Surlingham Broad. II. The fen gradient. J. Ecol. 53:183–195.

Button, D. K. 1978. On the theory of control of microbial growth kinetics by limiting nutrient concentrations. Deep-Sea Res. 25:1163–1177.

Cairns, J., Jr. 1974. Protozoans (Protozoa). *In* C. W. Hart, Jr., and S. L. H. Fuller, eds. Pollution Ecology of Freshwater Invertebrates. New York, Academic Press, pp. 1–28.

Caldwell, D. E., and J. M. Tiedje 1975. The structure of anaerobic bacterial communities in the hypolimnia of several Michigan lakes. Can. J. Microbiol., 21:377–385.

Caldwell, D. R., J. M. Brubaker, and V. T. Neal. 1978. Thermal microstructure on a lake slope. Limnol. Oceanogr. 23:372–374.

Calow, P. 1977. The joint effect of temperature and starvation on the metabolism of triclads. Oikos 29:87–92.

Calow, P. and A. S. Woollhead. 1977. The relationship between ration, reproductive effort and age-specific mortality in the evolution of life-history strategies—some observations on freshwater triclads. J. Anim. Ecol. 46:765–781.

Campbell, P. and T. Torgersen. 1980. Maintenance of iron meromixis by iron redeposition in a rapidly flushed monimolimnion. Can. J. Fish. Aquat. Sci. 37:1303–1313.

Canfield, D. E., Jr. and R. W. Bachmann. 1981. Prediction of total phosphorus concentrations, chlorophyll a, and Secchi depths in natural and artificial lakes. Can. J. Fish. Aquat. Sci. 38:414–423.

Cannon, D., J. W. G. Lund, and J. Sieminska. 1961. The growth of Tabellaria flocculosa (Roth) Kütz. var. flocculosa (Roth) Knuds. under natural conditions of light and temperature. J. Ecol. 49:277–287.

Canter, H. M. 1973. A new primitive protozoan devouring centric diatoms in the plankton. Zool. J. Linn. Soc. 52:63–83.

Canter, H. M. and G. H. M. Jaworski. 1979. The occurrence of a hypersensitive reaction in the planktonic diatom Asterionella formosa Hassall parasitized by the chytrid Rhizophydium planktonicum Canter Emend., in culture. New Phytol. 82:187–206.

Canter, H. M., and J. W. G. Lund. 1948. Studies on plankton parasites. I. Fluctuations in the numbers of Asterionella formosa Hass. in relation to fungal epidemics. New Phytol. 47:238–261.

Canter, H. M., and J. W. G. Lund. 1968. The importance of Protozoa in controlling the abundance of planktonic algae in lakes. Proc. Linn. Soc. Lond. 179:203–219.

Canter, H. M., and J. W. G. Lund. 1969. The parasitism of planktonic desmids by fungi. Österr. Bot. Z. 116:351–377.

Capblancq, J. 1973. Phytobenthos et productivité primaire d'un lac de haute montagne dans les Pyrénées centrales. Ann. Limnol. 9:193–230.

Cappenberg, T. E. 1972. Ecological observations on heterotrophic, methane oxidizing and sulfate reducing bacteria in a pond. Hydrobiologia 40:471–485.

Cappenberg, T. E. 1974. Interrelations between sulfate-reducing and methane-producing bacteria in bottom deposits of a fresh-water lake. II. Inhibition experiments. Antonie van Leeuwenhoek 40:297–306.

Cappenberg, T. E. 1975. A study of mixed continuous cultures of sulfate-reducing and methane-producing bacteria. Microbial Ecol. 2:60–72.

Carignan, R. and J. Kalff. 1980. Phosphorus sources for aquatic weeds: Water or sediments? Science 207:987–989.

Carignan, R., and J. Kalff. 1982. Phosphorus release by submerged macrophytes: Significance to epiphyton and phytoplankton. Limnol. Oceanogr. 27:419–427.

Carlucci, A. F. and P. M. Bowes. 1970. Production of vitamin B_{12}, thiamine, and biotin by phytoplankton. J. Phycol. 6:351–357.

Carpenter, G. F., E. L. Mansey, and N. H. F. Watson. 1974. Abundance and life history of Mysis relicta in the St. Lawrence Great Lakes. J. Fish. Res. Bd. Canada 31:319–325.

Carpenter, S. R. 1980. Enrichment of Lake Wingra, Wisconsin, by submersed macrophyte decay. Ecology 61:1145–1155.

Carpenter, S. R. and M. S. Adams. 1979. Effects of nutrients and temperature on decomposition of Myriophyllum spicatum L. in a hard-water eutrophic lake. Limnol. Oceanogr. 24:520–528.

Carr, J. F. 1962. Dissolved oxygen in Lake Erie, past and present. Publ. Great Lakes Res. Div., Univ. Mich. 9:1–14.

Carr, J. L. 1969. The primary productivity and physiology of Ceratophyllum demersum. II. Micro primary productivity, pH, and the P/R ratio. Austral. J. Mar. Freshwat. Res. 20:127–142.

Carroll, D. 1962. Rainwater as a chemical agent of geologic processes—a review. U.S. Geol. Surv. Water-Supply Pap. 1535-G, 18 pp.

Carter, C. E. 1977. The recent history of the chironomid fauna of Lough Neagh, from the analysis of remains in sediment cores. Freshwat. Biol. 7:415–423.

Carter, J. C. H. 1974. Life cycles of three limnetic copepods in a beaver pond. J. Fish. Res. Bd. Canada, 31:421–434.

Casselman, J. M. and H. H. Harvey. 1975. Selective fish mortality resulting from low winter oxygen. Verh. Int. Verein. Limnol. 19:2418–2429.

Castenholz, R. W. 1960. Seasonal changes in the attached algae of freshwater and saline lakes in the Lower Grand Coulee, Washington. Limnol. Oceanogr. 5:1–28.

Castenholz, R. W. 1961. An evaluation of a submerged glass method of estimating production of attached algae. Verh. Int. Ver. Limnol. 14:155–159.

Castenholz, R. W. 1969. Thermophilic blue-green algae and the thermal environment. Bacteriol. Rev. 33:476–504.

Caswell, H. 1972. On instantaneous and finite birth rates. Limnol. Oceanogr. 17:787–791.

Cattaneo, A. 1978. The microdistribution of epiphytes on the leaves of natural and artificial macrophytes. Br. Phycol. J. 13:183–188.

Cattaneo, A. and J. Kalff. 1978. Seasonal changes in the epiphyte community of natural and artificial macrophytes in Lake Memphremagog (Que. & Vt.). Hydrobiologia 60:135–144.

Cattaneo, A. and J. Kalff. 1979. Primary production of algae growing on natural and artificial aquatic plants: A study of interactions between epiphytes and their substrate. Limnol. Oceanogr. 24:1031–1037.

Cattaneo, A. and J. Kalff. 1980. The relative contribution of aquatic macrophytes and their epiphytes to the production of macrophyte beds. Limnol. Oceanogr. 25:280–289.

Cavari, B. Z. 1977. Nitrification potential and factors governing the rate of nitrification in Lake Kinneret. Oikos 28:285–290.

Cavari, B. and N. Grossowicz. 1977. Seasonal distribution of vitamin B_{12} in Lake Kinneret. Appl. Environ. Microbiol. 34:120–124.

Cavari, B. Z., G. Phelps, and O. Hadas. 1978. Glucose concentrations and heterotrophic activity in Lake Kinneret. Verh. Int. Ver. Limnol. 20:2249–2254.

Chamberlain, W. M. 1968. A preliminary investigation of the nature and importance of soluble organic phosphorus in the phosphorus cycle of lakes. Ph.D. Diss., University of Toronto, Ontario, 232 pp.

Chan, Y. K. and N. E. R. Campbell. 1978. Phytoplankton uptake and excretion of assimilated nitrate in a small Canadian Shield lake. Appl. Environ. Microbiol. 35:1052–1060.

Chan, Y. K. and N. E. R. Campbell. 1980. Denitrification in Lake 227 during summer stratification. Can. J. Fish. Aquat. Sci. 37:506–512.

Chan, Y. K. and R. Knowles. 1979. Measurement of denitrification in two freshwater sediments by an in situ acetylene inhibition method. Appl. Environ. Microbiol. 37:1067–1072.

Chandler, C. M. 1966. Environmental factors affecting the local distribution and abundance of four speces of stream-dwelling triclads. Invest. Indiana Lakes Streams 7:1–56.

Chandler, D. C. 1937. Fate of typical lake plankton in streams. Ecol. Monogr. 7:445–479.

Chang, T.-P. 1980. Mucilage sheath as a barrier to carbon uptake in a cyanophyte, Oscillatoria rubescens D.C. Arch. Hydrobiol. 88:128–133.

Chang, T.-P. 1981. Excretion and DOC utilization by Oscillatoria rubescens D. C. and its accompanying micro-organisms. Arch. Hydrobiol. 91:509–520.

Chapin, J. D., and P. D. Uttormark. 1973. Atmospheric Contributions of Nitrogen and Phosphorus. Tech. Rep. Wat. Resources Ctr. Univ. Wis. 73-2, 35 pp.

Chapman, D. W. 1967. Production in fish populations. In Gerking, S. D., ed. The Biological Basis of Freshwater Fish Production. New York, John Wiley & Sons, Inc., pp. 3–29.

Chapman, D. W. 1978. Production in fish populations. In S. D. Gerking, ed. Ecology of Freshwater Fish Production. New York, John Wiley & Sons, pp. 5–25.

Chapra, S. C. and K. H. Reckhow. 1979. Expressing the phosphorus loading concept in probabilistic terms. J. Fish. Res. Bd. Can. 36:225–229.

Charlton, M. N. 1980. Hypolimnion oxygen consumption in lakes: Discussion of productivity and morphometry effects. Can. J. Fish. Aquat. Sci. 37:1531–1539.

Chatarpaul, L., J. B. Robinson, and N. K. Kaushik. 1980. Effects of tubificid worms on denitrification and nitrification in stream sediment. Can. J. Fish. Aquat. Sci. 37:656–663.

Chave, K. E. 1965. Carbonates: Association with organic matter in surface seawater. Science 148:1723–1724.

Chave, K. E., and E. Suess. 1970. Calcium carbonate saturation in seawater: Effects of dissolved organic matter. Limnol. Oceanogr. 15:633–637.

Chawla, V. K., and Y. K. Chau. 1969. Trace elements in Lake Erie. Proc. Conf. Great Lakes Res., Int. Assoc. Great Lakes Res. 12:760–765.

Chen, K. Y., and J. C. Morris. 1972. Kinetics of oxidation of aqueous sulfide by O_2. Environ. Sci. Technol. 6:529–537.

Chen, R. L., D. R. Keeney, and J. G. Konrad. 1972. Nitrification in lake sediments. J. Environ. Quality 1:151–154.

Chen, R. L., D. R. Keeney, D. A. Graetz, and A. J. Holding. 1972. Denitrification and nitrate reduction in lake sediments. J. Environ. Quality 1:158–162.

Chen, R. L., D. R. Keeney, J. G. Konrad, A. J. Holding, and D. A. Graetz. 1972. Gas production in sediments of Lake Mendota, Wisconsin. J. Environ. Quality 1:155–158.

Chimney, M. J., R. W. Winner, and S. K. Seilkop. 1981. Prey utilization by Chaoborus punctipennis Say in a small, eutrophic reservoir. Hydrobiologia 85:193–199.

Chow, V. T., ed. 1964. Handbook of Applied Hydrology. New York, McGraw-Hill Book Co.

Chu, S. P. 1942. The influence of the mineral composition of the medium on the growth of planktonic algae. I. Methods and culture media. J. Ecol. 30:284–325.

Chu, S. P. 1943. The influence of the mineral composition of the medium on the growth of planktonic algae. II. The influence of the concentration of inorganic nitrogen and phosphate phosphorus. J. Ecol. 31:109–148.

Chua, K. E., and R. O. Brinkhurst. 1973. Evidence of interspecific interactions in the respiration of tubificid oligochaetes. J. Fish. Res. Bd. Canada 30:617–622.

Clark, C. 1967. Population Growth and Land Use. London, Macmillan Press Ltd., 406 pp.

Clarke, F. W. 1924. The Data of Geochemistry. 5th ed. Bull. U.S. Geol. Surv. 770, 841 pp.

Clarke, G. L. 1939. The utilization of solar energy by aquatic organisms. In Problems in lake biology. Publ. Amer. Assoc. Adv. Sci. 10:27–38.

Clémencon, H. 1963. Verbreitung einiger B-Vitamine in Kleingewässern aus der Umgebung von Bern und Untersuchungen über die Abgabe von B-Vitaminen durch Algen. Schweiz. Z. Hydrol. 25:157–165.

Clymo, R. S. 1963. Ion exchange in Sphagnum and

its relation to bog ecology. Ann. Bot., N.S. 27:309-324.

Clymo, R. S. 1964. The origin of acidity in *Sphagnum* bogs. Bryologist 67:427-431.

Clymo, R. S. 1965. Experiments on breakdown of *Sphagnum* in two bogs. J. Ecol. 53:747-758.

Clymo, R. S. 1967. Control of cation concentrations, and in particular of pH, in *Sphagnum* dominated communities. *In* H. L. Golterman and R. S. Clymo, eds. Chemical Environment in the Aquatic Habitat. Amsterdam, N.V. Noord-Hollandsche Uitgevers Maatschappij, pp. 273-284.

Clymo, R. S., and E. J. F. Reddaway. 1971. Productivity of *Sphagnum* (bog moss) and peat accumulation. Hidrobiologia 12:181-192.

Clymo, R. S., and E. J. F. Reddaway. 1974. Growth rate of *Sphagnum rubellum* Wils, on pennine blanket bog. J. Ecol. 62:191-196.

Coats, R. N., R. L. Leonard, and C. R. Goldman. 1976. Nitrogen uptake and release in a forested watershed, Lake Tahoe basin, California. Ecology 57:995-1004.

Coffin, C. C., F. R. Hayes, L. H. Jodrey, and S. G. Whiteway. 1949. Exchange of materials in a lake as studied by the addition of radioactive phosphorus. Can. J. Res. (ser. D) 27:207-222.

Cole, B. S. and D. W. Toetz. 1975. Utilization of sedimentary ammonia by *Potamogeton nodosus* and *Scirpus*. Verh. Int. Ver. Limnol. 19:2765-2772.

Coleman, A. W. 1980. Enhanced detection of bacteria in natural environments by fluorochrome staining of DNA. Limnol. Oceanogr. 25:948-951.

Comita, G. W. 1964. The energy budget of *Diaptomus siciloides*, Lilljeborg. Verh. Int. Ver. Limnol. 15:646-653.

Comita, G. W. 1972. The seasonal zooplankton cycles, production and transformations of energy in Severson Lake, Minnesota. Arch. Hydrobiol. 70:14-66.

Comita, G. W., and G. C. Anderson. 1959. The seasonal development of a population of *Diaptomus ashlandi* Marsh, and related phytoplankton cycles in Lake Washington. Limnol. Oceanogr. 4:37-52.

Confer, J. L. 1971. Intrazooplankton predation by *Mesocyclops edax* at natural prey densities. Limnol. Oceanogr. 16:663-666.

Confer, J. L. 1972. Interrelations among plankton, attached algae, and the phosphorus cycle in artificial open systems. Ecol. Monogr. 42:1-23.

Confer, J. L. and G. Applegate. 1979. Size-selective predation by zooplankton. Amer. Midl. Nat. 102:378-383.

Confer, J. L. and J. M. Cooley. 1977. Copepod instar survival and predation by zooplankton. J. Fish. Res. Bd. Canada 34:703-706.

Conrad, H. M., and P. Saltman. 1962. Growth substances. *In* R. A. Lewin, ed. Physiology and Biochemistry of Algae. New York, Academic Press, pp. 663-671.

Conway, E. J. 1942. Mean geochemical data in relation to oceanic evolution. Proc. Royal Irish Acad. (ser. B) 48:119-159.

Conway, H. L., J. I. Parker, E. M. Yaguchi, and D. L. Mellinger. 1977. Biological utilization and regeneration of silicon in Lake Michigan. J. Fish. Res. Bd. Can. 34:537-544.

Cooke, G. D., M. R. McComas, D. W. Waller, and R. H. Kennedy. 1977. The occurrence of internal phosphorous loading in two small, eutrophic glacial lakes in northeastern Ohio. Hydrobiol. 56:129-135.

Cooke, W. B. 1956. Colonization of artificial bare areas by microorganisms. Bot. Rev. 22:613-638.

Cooney, J. D. and C. W. Gehrs. 1980. The relationship between egg size and naupliar size in the calanoid copepod *Diaptomus clavipes* Schacht. Limnol. Oceanogr. 25:549-552.

Cooper, C. F. 1969. Nutrient output from managed forests. *In* Eutrophication: Causes, Consequences, Correctives. Washington, D.C., National Academy of Sciences, pp. 446-463.

Cooper, S. D. and C. R. Goldman. 1980. Opossum shrimp *(Mysis relicta)* predation on zooplankton. Can. J. Fish. Aquat. Sci. 37:909-919.

Cooper, W. E. 1965. Dynamics and production of a natural population of a fresh-water amphipod, *Hyalella azteca*. Ecol. Monogr. 35:377-394.

Cornett, R. J. and F. H. Rigler. 1979. Hypolimnetic oxygen deficits: Their prediction and interpretation. Science 205:580-581.

Cornett, R. J. and F. H. Rigler. 1980. The areal hypolimnetic oxygen deficit: An empirical test of the model. Limnol. Oceanogr. 25:672-679.

Costa, R. R., and K. W. Cummins. 1972. The contribution of *Leptodora* and other zooplankton to the diet of various fish. Amer. Midland Nat. 87:559-564.

Coughlan, M. P. 1971. The role of iron in microbial metabolism. Sci. Progress, Oxford 59:1-23.

Coulson, K. L. 1975. Solar and terrestrial radiation. Methods and measurements. Academic Press, New York. 322 pp.

Coveney, M. F. 1982. Bacterial uptake of photosynthetic carbon from freshwater phytoplankton. Oikos 38:8-20.

Coveney, M. F., G. Cronberg, M. Enell, K. Larsson, and L. Olofsson. 1977. Phytoplankton, zooplankton and bacteria—standing crop and production relationships in a eutrophic lake. Oikos 29:5-21.

Coveney, M. F. and M. Søndergaard. 1982. Formation of particulate carbon from dissolved organic carbon released by aquatic plants. Aquatic Bot. (In press)

Cowen, W. F., and G. F. Lee. 1973. Leaves as source of phosphorus. Environ. Sci. Technol. 7:853-854.

Cowgill, U. M. 1976. The hydrogeochemistry of Linsley Pond, North Branford, Connecticut. IV. Minor constituents by optical emission spectroscopy. Arch. Hydrobiol. 78:279-309.

Cowgill, U. M. 1977a. The hydrogeochemistry of Linsley Pond, North Branford, Connecticut. VI.

Summary, Part 1: Geochemical relationships within zones of the lake. Arch. Hydrobiol. Suppl. 50:401–438.

Cowgill, U. M. 1977b. The hydrogeochemistry of Linsley Pond, North Branford, Connecticut. VI. Summary, Part 2: Geochemical relationships between zones of the lake. Arch. Hydrobiol. Suppl. 53:1–47.

Cowgill, U. M. 1977c. The molybdenum cycle in Linsley Pond, North Branford, Connecticut. In W. R. Chappell and K. K. Petersen, Editors. Molybdenum in the Environment. II. M. Dekker, New York. pp. 705–723.

Cowgill, U. M. and G. E. Hutchinson. 1966. El Bajo de Santa Fé. Trans. Amer. Phil. Soc., N.S. 53(7),51 pp.

Cowgill, U. M., and G. E. Hutchinson. 1963. The history of a pond in Guatemala. Arch. Hydrobiol. 62:355–372.

Cowgill, U. M., C. E. Goulden, G. E. Hutchinson, R. Patrick, A. A. Raček, and M. Tsukada. 1966. The History of Laguna de Petenxil. A Small Lake in Northern Guatemala. Mem. Conn. Acad. Arts Sci. 17, 126 pp.

Craik, A. D. D. 1977. The generation of Langmuir circulations by an instability mechanism. J. Fluid Mech. 81:209–223.

Craik, A. D. D. and S. Leibovich. 1976. A rational model for Langmuir circulations. J. Fluid Mech. 73:401–426.

Cranwell, P. A. 1973. Chain-length distribution of n-alkanes from lake sediments in relation to postglacial environmental change. Freshwat. Biol. 3:259–265.

Cranwell, P. A. 1974. Monocarboxylic acids in lake sediments: Indicators, derived from terrestrial and aquatic biota, of paleoenvironmental trophic levels. Chem. Geol. 14:1–14.

Cranwell, P. A. 1976. Organic geochemistry of lake sediments. In J. O. Nriagu, Editor. Environmental Biogeochemistry. I. Carbon, Nitrogen, Phosphorus, Sulfur and Selenium Cycles. Ann Arbor Sci. Publs., Inc., Ann Arbor, Michigan. pp. 75–88.

Cranwell, P. A. 1979. Decomposition of aquatic biota and sediment formation: Bound lipids in algal detritus and lake sediments. Freshwat. Biol. 9:305–313.

Crawford, R. M. M. 1978. Metabolic adaptations to anoxia. In D. D. Hook and R. M. M. Crawford, Eds. Plant Life in Anaerobic Environments. Ann Arbor Science Publ. Inc., Ann Arbor, Michigan. pp. 119–136.

Creer, K. M., D. L. Gross, and J. A. Lineback. 1976. Origin of regional geomagnetic variations recorded by Wisconsinan and Holocene sediments from Lake Michigan, U.S.A., and Lake Windermere, England. Geol. Soc. Amer. Bull. 87:531–540.

Creer, K. M., R. Thompson, and L. Molyneux. 1972. Geomagnetic secular variation recorded in the stable magnetic remanence of recent sediments. Earth Planetary Sci. Lett. 14:115–127.

Crisman, T. L. 1978. Algal remains in Minnesota lake types: A comparison of modern and late-glacial distributions. Verh. Int. Ver. Limnol. 20:445–451.

Crisman, T. L. 1980. Algal control through trophic-level interactions: A subtropical perspective. U.S.E.P.A. Workshop on Algal Management and Control. 20 pp.

Crisman, T. L., J. R. Beaver, and J. S. Bays. 1981. Examination of the relative impact of microzooplankton and macrozooplankton on bacteria in Florida lakes. Verh. Int. Ver. Limnol. 21:359–362.

Crisman, T. L. and D. R. Whitehead. 1978. Paleolimnological studies on small New England (U.S.A.) ponds. Part II. Cladoceran community responses to trophic oscillations. Polskie Arch. Hydrobiol. 25:75–86.

Cronan, C. S., and C. L. Schofield. 1979. Aluminum leaching response to acid precipitation: Effects on high-elevation watersheds in the Northeast. Science 204:304–306.

Crowder, A. A., J. M. Bristow, M. R. King, and S. Vanderkloet. 1977. Distribution, seasonality, and biomass of aquatic macrophytes in Lake Opinicon (Eastern Ontario). Naturaliste Can. 104:441–456.

Crumpton, W. G. and R. G. Wetzel. 1982. Effects of differential growth and mortality in the seasonal succession of phytoplankton populations. Ecology (In press)

Cruz-Pizarro, L. 1978. Comparative vertical zonation and diurnal migration among Crustacea and Rotifera in the small mountain lake LaCaldera (Granada, Spain). Verh. Int. Ver. Limnol. 20:1026–1032.

Cuker, B. E. and S. C. Mozley. 1981. Summer population fluctuations, feeding, and growth of Hydra in an arctic lake. Limnol. Oceanogr. 26:697–708.

Culver, D. A., R. M. Vaga, C. S. Munch, and S. M. Harris. 1981. Paleoecology of Hall Lake, Washington: A history of meromixis and disturbance. Ecology 62:848–863.

Cummins, K. W. 1962. An evaluation of some techniques for the collection and analysis of benthic samples with special emphasis on lotic waters. Amer. Midland Nat. 67:477–504.

Cummins, K. W. 1964. Factors limiting the microdistribution of larvae of the caddisflies Pycnopsyche lepida (Hagen) and Pycnopsyche guttifer (Walker) in a Michigan stream (Trichoptera: Limnephilidae). Ecol. Monogr. 34:271–295.

Cummins, K. W. 1972. Predicting variations in energy flow through a semi-controlled lotic ecosystem. Tech. Rept. Inst. Water Res. Mich. State Univ. 19, 21 pp.

Cummins, K. W. 1973. Trophic relations of aquatic insects. Ann. Rev. Entomol. 18:183–206.

Cummins, K. W. and M. J. Klug. 1979. Feeding ecology of stream invertebrates. Ann. Rev. Ecol. Syst. 10:147–172.

Cummins, K. W., and G. H. Lauff. 1969. The influence of substrate particle size on the microdistribution of stream macrobenthos. Hydrobiologia 34:145–181.

Cummins, K. W., R. R. Costa, R. E. Rowe, G. A. Moshiri, R. M. Scanlon, and R. K. Zajdel. 1969. Ecological energetics of a natural population of the predaceous zooplankter *Leptodora kindtii* Focke (Cladocera). Oikos 20:189–223.

Cummins, K. W., and J. C. Wuycheck. 1971. Caloric equivalents for investigations in ecological energetics. Mitt. Int. Ver. Limnol. 18, 158 pp.

Cummins, K. W., M. J. Klug, R. G. Wetzel, R. C. Petersen, K. F. Suberkropp, B. A. Manny, J. C. Wuycheck, and F. O. Howard. 1972. Organic enrichment with leaf leachate in experimental lotic ecosystems. BioScience 22:719–722.

Cummins, K. W., R. C. Petersen, F. O. Howard, J. C. Wuycheck, and V. I. Holt. 1973. The utilization of leaf litter by stream detritivores. Ecology 54:336–345.

Cunningham, L. 1972. Vertical migrations of *Daphnia* and copepods under the ice. Limnol. Oceanogr. 17:301–303.

Czeczuga, B. 1959. On oxygen minimum and maximum in the metalimnion of Rajgród lakes. Acta Hydrobiol. 1:109–122.

Czeczuga, B. 1968a. An attempt to determine the primary production of the green sulphur bacteria, Chlorobium limicola Nads, (Chlorobacteriaceae). Hydrobiologia 31:317–333.

Czeczuga, B. 1968b. Primary production of the green hydrosulphuric bacteria, *Chlorobium limicola* Nads. (Chlorobacteriaceae). Photosynthetica 2:11–15.

Czeczuga, B. 1968c. Primary production of the purple sulphuric bacteria, *Thiopedia rosea* Winogr. (Thiorhodaceae). Photosynthetica 2:161–166.

Czeczuga, B., and E. Bobiatyńska-ksok. 1970. The extent of consumption of the energy contained in the food suspension by *Ceriodaphnia reticulata* (Jurine). In Z. Kajak and A. Hillbricht-Ilkowska, eds. Productivity Problems of Freshwaters. Warsaw, PWN Polish Scientific Publishers, pp. 739–748.

Czeczuga, B., and R. Czerpak. 1967. Obserwacje nad bakterioplanktonem jezior legińskich. Zesz. Nauk WSR Olszt. 23:35–44.

Czeczuga, B., and R. Czerpak. 1968. Investigations on vegetable pigments in post-glacial bed sediments of lakes. Schweiz. A. Hydrol. 30:217–231.

Daan, N., and J. Ringelberg. 1969. Further studies on the positive and negative phototactic reaction of *Daphnia magna* Straus. Netherlands J. Zool. 19:525–540.

Dacey, J. W. H. 1981. Pressurized ventilation in the yellow waterlily. Ecology 62:1137–1147.

Dacey, J. W. H. and M. J. Klug. 1979. Methane efflux from lake sediments through water lilies. Science 203:1253–1255.

Dahm, C. N. 1981. Pathways and mechanisms for removal of dissolved organic carbon from leaf leachate in streams. Can. J. Fish. Aquat. Sci. 38:68–76.

Daisley, K. W. 1969. Monthly survey of vitamin B_{12} concentrations in some waters of the English Lake District. Limnol. Oceanogr. 14:224–228.

Dale, H. M. 1981. Hydrostatic pressure as the controlling factor in the depth distribution of Eurasian watermilfoil, *Myriophyllum spicatum* L. Hydrobiologia 79:239–241.

Dale, H. M. and T. Gillespie. 1976. The influence of floating vascular plants on the diurnal fluctuations of temperature near the water surface in early spring. Hydrobiologia 49:245–256.

Dale, H. M. and T. Gillespie. 1977. Diurnal fluctuations of temperature near the bottom of shallow water bodies as affected by solar radiation, bottom color and water circulation. Hydrobiologia 55:87–92.

Daley, R. J. 1973. Experimental characterization of lacustrine chlorophyll diagenesis. II. Bacterial, viral and herbivore grazing effects. Arch. Hydrobiol. 72:409–439.

Daley, R. J., and S. R. Brown. 1973. Experimental characterization of lacustrine chlorophyll diagenesis. I. Physiological and environmental effects. Arch. Hydrobiol. 72:277–304.

Daley, R. J., S. R. Brown, and R. N. McNeely. 1977. Chromatographic and SCDP measurements of fossil phorbins and the postglacial history of Little Round Lake, Ontario. Limnol. Oceaongr. 22:349–360.

Danforth, W. F. 1962. Substrate assimilation and heterotrophy. In R. A. Lewin, ed. Physiology and Biochemistry of Algae. New York, Academic Press, pp. 99–123.

Darnell, R. M. 1964. Organic detritus in relation to secondary production in aquatic communities. Verh. Int. Ver. Limnol. 15:462–470.

Datsko, V. G. 1959. Organischeskoe Veshchestvo v Vodakh Iuzhnykh Morei SSSR. Moscow, Izdatel'stvo Akademii Nauk SSSR, 271 pp.

Davies, G. S. 1970. Productivity of macrophytes in Marion Lake, British Columbia. J. Fish. Res. Bd. Canada 27:71–81.

Davies, R. W. 1978. Reproductive strategies shown by freshwater Hirudinoidea. Verh. Int. Ver. Limnol. 20:2378–2381.

Davies, R. W. and R. P. Everett. 1975. The feeding of four species of freshwater Hirudinoidea in southern Alberta. Verh. Int. Ver. Limnol. 19:2816–2827.

Davies, R. W. and T. B. Reynoldson. 1975. Life history of *Helobdella stagnalis* (L.) in Alberta. Verh. Int. Ver. Limnol. 19:2828–2839.

Davies, R. W. and J. Wilkialis. 1980. The population ecology of the leech (Hirudinoidea: Glossiphoniidae) *Theromyzon rude*. Can. J. Zool. 58:913–916.

Davies, W. 1971. The phytopsammon of a sandy beach transect. Amer. Midland Nat. 86:292–308.

Davis, C. C. 1963. On questions of production and productivity in ecology. Arch. Hydrobiol. 59:145–161.

Davis, J. A. and R. Gloor. 1981. Adsorption of dissolved organics in lake water by aluminum oxide. Effect of molecular weight. Environ. Sci. Technol. 15:1223–1229.

Davis, M. B. 1968. Pollen grains in lake sediments: Redeposition caused by seasonal water circulation. Science 162:796–799.

Davis, M. B. 1973. Redeposition of pollen grains in lake sediment. Limnol. Oceanogr. 18:44–52.

Davis, M. B., and L. B. Brubaker. 1973. Differential sedimentation of pollen grains in lakes. Limnol. Oceanogr. 18:635–646.

Davis, M. B., L. B. Brubaker, and T. Webb. 1973. Calibration of absolute pollen influx. In H. J. B. Birks and R. G. West, Editors. Quaternary Plant Ecology. Blackwell Sci. Publs., Oxford. pp. 9–25.

Davis, M. B., and E. S. Deevey, Jr. 1964. Pollen accumulation rates: Estimates from late-glacial sediment of Rogers Lake. Science 145:1293–1295.

Davis, M. B. and M. S. J. Ford. 1982. Sediment focusing in Mirror Lake, New Hampshire. Limnol. Oceanogr. 27:137–150.

Davis, R. B. 1974. Stratigraphic effects of tubificids in profundal lake sediments. Limnol. Oceanogr. 19:466–488.

Davis, R. B., D. L. Thurlow, and F. E. Brewster. 1975. Effects of burrowing tubificid worms on the exchange of phosphorus between lake sediment and overlying water. Verh. Int. Ver. Limnol. 19:382–394.

Davis, S. N. 1964. Silica in streams and ground water. Amer. J. Sci. 262:870–891.

Davis, S. N., and R. J. M. DeWiest. 1966. Hydrogeology. New York, John Wiley & Sons, Inc., 463 pp.

Davis, W. M. 1933. The lakes of California. Calif. J. Mines Geology 29:175–236.

Davison, W. and S. I. Heaney. 1978. Ferrous iron-sulfide interactions in anoxic hypolimnetic waters. Limnol. Oceanogr. 23:1194–1200.

Dean, W. E. and E. Gorham. 1976. Major chemical and mineral components of profundal surface sediments in Minnesota lakes. Limnol. Oceanogr. 21:259–284.

DeCosta, J. 1968. Species diversity of chydorid fossil communities in the Mississippi Valley. Hydrobiologia 32:497–512.

Deevey, E. S., Jr. 1941. Limnological studies in Connecticut. VI. The quantity and composition of the bottom fauna of thirty-six Connecticut and New York lakes. Ecol. Monogr. 11:414–455.

Deevey, E. S., Jr. 1955. The obliteration of the hypolimnion. Mem. Ist. Ital. Idrobiol. Suppl. 8:9–38.

Deevey, E. S., D. S. Rice, P. M. Rice, H. H. Vaughan, M. Brenner, and M. S. Flannery. 1979. Mayan urbanism: Impact on a tropical karst environment. Science 206:298–306.

DeGeer, G. 1912. A geochronology of the last 12,000 years. Proc. 11th Int. Geol. Congr., Stockholm. Part 1, pp. 241–253.

Delfino, J. J., and G. F. Lee. 1968. Chemistry of manganese in Lake Mendota, Wisconsin. Environ. Sci. Technol. 2:1094–1100.

Delfino, J. J., and G. F. Lee. 1971. Variation of manganese, dissolved oxygen and related chemical parameters in the bottom waters of Lake Mendota, Wisconsin. Water Res. 5:1207–1217.

Della Croce, N. 1955. The conditions of sedimentation and their relations with Oligochaeta populations of Lake Maggiore. Mem. Ist. Ital. Idrobiol. 8(Suppl.):39–62.

Delorme, L. D. 1978. Distribution of freshwater ostracodes in Lake Erie. J. Great Lakes Res. 4:216–220.

DeMarte, J. A., and R. T. Hartman. 1974. Studies on absorption of ^{32}P, ^{59}Fe, and ^{45}Ca by water-milfoil (*Myriophyllum exalbescens* Fernald). Ecology 55:188–194.

Dendy, J. S. 1963. Observations on bryozoan ecology in farm ponds. Limnol. Oceanogr. 8:478–482.

DeNoyelles, F., Jr., and W. J. O'Brien. 1978. Phytoplankton succession in nutrient enriched experimental ponds as related to changing carbon, nitrogen and phosphorus conditions. Arch. Hydrobiol. 84:137–165.

Denny, P. 1972. Sites of nutrient absorption in aquatic macrophytes. J. Ecol. 60:819–829.

Dermott, R. M., J. Kalff, W. C. Leggett, and J. Spence. 1977. Production of *Chironomus*, *Procladius*, and *Chaoborus* at different levels of phytoplankton biomass in Lake Memphremagog, Quebec-Vermont. J. Fish. Res. Board Can. 34:2001–2007.

Desortová, B. 1981. Relationship between chlorophyll-*a* concentration and phytoplankton biomass in several reservoirs in Czechoslovakia. Int. Rev. ges. Hydrobiol. 66:153–169.

DeWitt, R. M. 1954. Reproduction, embryonic development, and growth in the pond snail, *Physa gyrina* Say. Trans. Amer. Microsc. Soc. 73:124–137.

Deyl, Z. 1961. Anaerobic fermentations. Sci. Pap. Inst. Chem. Technol. Praque, Technol. Water 5(2):131–234.

Dickerman, J. A. and R. G. Wetzel. 1982a. Life history and population dynamics of *Typha latifolia*. Ecol. Monogr. (Submitted)

Dickerman, J. A. and R. G. Wetzel. 1982b. Biomass turnover in a *Typha latifolia* population. Ecology (submitted)

Dickerman, J. A. and R. G. Wetzel. 1982c. Shoot production in a *Typha latifolia* population: A comparison of sampling interval sensitivity for seven techniques. Ecology (submitted)

Dickman, M. D. and J. S. Hartman. 1979. A rationale for the subclassification of biogenic meromictic lakes. Int. Revue ges. Hydrobiol. 64:189–192.

Di Cola, G., M. Dilingenti, and I. Guerri. 1977. Identification of the thermal eddy diffusivity in a stratified lake. Mem. Ist. Ital. Idrobiol. 34:175–185

Dietz, A. S., L. J. Albright, and T. Tuominen. 1976.

Heterotrophic activities of bacterioneuston and bacterioplankton. Can. J. Microbiol. 22:1699–1709.

Digerfeldt, G., R. W. Battarbee, and L. Bengtsson. 1975. Report on annually laminated sediment in Lake Järlasjön. Geol. Fören. Förhandl. Stockholm 97:29–40.

Dilcher, D. L. 1974. Approaches to the identification of angiosperm leaf remains. Bot. Rev. 40:1–157.

Dillon, P. J. 1974. The prediction of phosphorus and chlorophyll concentrations in lakes. Ph.D. Dissertation, Univ. Toronto. 330 pp.

Dillon, P. J. and W. B. Kirchner. 1975. The effects of geology and land use on the export of phosphorus from watersheds. Water Res. 9:135–148.

Dillon, P. J. and F. H. Rigler. 1974. The phosphorus-chlorophyll relationship in lakes. Limnol. Oceanogr. 19:767–773.

Dillon, P. J. and F. H. Rigler. 1974. A test of a simple nutrient budget model predicting the phosphorus concentration in lake water. J. Fish. Res. Bd. Can. 31:1771–1778.

Dinsdale, M. T., and A. E. Walsby. 1972. The interrelations of cell turgor pressure, gas-vacuolation, and buoyancy in a blue-green alga. J. Exp. Bot. 23:561–570.

Dodson, S. I. 1970. Complementary feeding niches sustained by size-selective predation. Limnol. Oceanogr. 15:131–137.

Dodson, S. I. 1972. Mortality in a population of Daphnia rosea. Ecology 53:1011–1023.

Dodson, S. I. 1974a. Zooplankton competition and predation: An experimental test of the size-efficiency hypothesis. Ecology 55:605–613.

Dodson, S. I. 1974b. Adaptive change in plankton morphology in response to size-selective predation: A new hypothesis of cyclomorphosis. Limnol. Oceanogr. 19:721–729.

Döhler, G., and K. -R. Przybylla. 1973. Einfluss der Temperatur auf die Lichtatmung der Blaualge Anacystis nidulans. Planta 110:153–158.

Dokulil, M. 1973. Planktonic primary production within the Phragmites community of Lake Neusiedlersee (Austria). Pol. Arch. Hydrobiol. 20:175–180.

Domogalla, B. P., and E. B. Fred. 1926. Ammonia and nitrate studies of lakes near Madison, Wisconsin. J. Amer. Soc. Agronomy 18:897–911.

Domogalla, B. P., E. B. Fred, and W. H. Peterson. 1926. Seasonal variations in the ammonia and nitrate content of lake waters. J. Amer. Water Works Assoc. 15:369–385.

Dor, I. 1970. Production rate of the periphyton in Lake Tiberias as measured by the glass-slide method. Israel J. Bot. 19:1–15.

Dorgelo, J. and K. Koning. 1980. Avoidance of macrophytes and additional notes on avoidance of the shore by 'Acanthodiaptomus denticornis' (Wierzejski, 1887) from Lake Pavin (Auvergne, France). Hydrobiol. Bull. 14:196–208.

Douglas, B. 1958. The ecology of the attached diatoms and other algae in a small stony stream. J. Ecol. 46:295–322.

Downes, M. T. and H. W. Paerl. 1978. Separation of two dissolved reactive phosphorus fractions in lakewater. J. Fish. Res. Bd. Can. 35:1636–1639.

Doyle, R. W. 1968a. The origin of the ferrous ion-ferric oxide Nernst potential in environments containing dissolved ferrous iron. Amer. J. Sci. 266:840–859.

Doyle, R. W. 1968b. Identification and solubility of iron sulfide in anaerobic lake sediment. Amer. J. Sci. 266:980–994.

Droop, M. R. 1973. Some thoughts on nutrient limitation in algae. J. Phycol. 9:264–272.

Droop, M. R. 1974a. Heterotrophy of carbon. In W. D. P. Stewart, ed. Algal Physiology and Biochemistry. Berkeley, Univ. of California Press, pp. 530–559.

Droop, M. R. 1974b. The nutrient status of algal cells in continuous culture. J. Mar. Biol. Assoc. U. K. 54:825–855.

Droop, M. R. 1977. An approach to quantitative nutrition of phytoplankton. J. Protozool. 24:528–532.

Drummond, A. J. 1971. Recent measurements of the solar radiation incident on the atmosphere. In Space Research XI. Berlin, Akademie-Verlag, pp. 681–693.

Dubay, C. I. and G. M. Simmons, Jr. 1979. The contribution of macrophytes to the metalimnetic oxygen maximum in a montane, oligotrophic lake. Amer. Midland Nat. 101:108–117.

Dubinina, G. A., V. M. Gorlenko, and J. I. Suleimanov. 1973. A study of microorganisms involved in the circulation of manganese, iron, and sulfur in meromictic Lake Gek-Gel'. Mikrobiologiya 42:918–924.

Dubinsky, Z. and T. Berman. 1976. Light utilization efficiencies of phytoplankton in Lake Kinneret (Sea of Galilee). Limnol. Oceanogr. 21:226–230.

Dubinsky, Z. and T. Berman. 1981a. Light utilization by phytoplankton in Lake Kinneret (Israel). Limnol. Oceanogr. 26:660–670.

Dubinsky, Z. and T. Berman. 1981b. Photosynthetic efficiencies in aquatic ecosystems. Verh. Int. Ver. Limnol. 21:237–243.

Dugdale, R. C., and V. A. Dugdale. 1961. Sources of phosphorus and nitrogen for lakes on Afognak Island. Limnol. Oceanogr. 6:13–23.

Dugdale, V. A., and R. C. Dugdale. 1962. Nitrogen metabolism in lakes. II. Role of nitrogen fixation in Sanctuary Lake, Pennsylvania. Limnol. Oceanogr. 7:170–177.

Dumont, H. J. 1972. The biological cycle of molybdenum in relation to primary production and waterbloom formation in a eutrophic pond. Verh. Int. Ver. Limnol. 18:84–92.

Dumont, H. J. and J. Green, Editors. 1980. Rotatoria. Hydrobiologia 73:1–262. (Also as Developments in Hydrobiology, vol. 1)

Dunne, T. 1978. Field studies of hillslope flow processes. In M. J. Kirkby, Ed. Hillslope Hydrology. John Wiley & Sons, New York. pp. 227–293.

Dunne, T., and R. D. Black. 1970a. An experimental investigation of runoff production in permeable soils. Water Resources Res. 6:478–490.

Dunne, T., and R. D. Black. 1970b. Partial area contributions to storm runoff in a small New England watershed. Water Resources Res. 6:1296–1311.

Dunne, T., and R. D. Black. 1971. Runoff processes during snowmelt. Water Resources Res. 7:1160–1172.

Duong, T. P. 1972. Nitrogen fixation and productivity in a eutrophic hard-water lake: In situ and laboratory studies. Ph. D. Disser., Michigan State University, 241 pp.

Durum, W. H., and J. Haffty. 1961. Occurrence of minor elements in water. Circ. U.S. Geol. Surv. 445, 11 pp.

Dussart, B. 1966. Limnologie. L'Etude des Eaux Continentales. Paris, Gauthier-Villars, 677 pp.

Dussart, B. H., K. F. Lagler, P. A. Larkin, T. Scudder, K. Szesztay, and G. F. White. 1972. Man-made lakes as modified ecosystems. SCOPE Rept. 2. Int. Council Sci. Unions, Paris. 76 pp.

Dutton, J. A., and R. A. Bryson. 1962. Heat flux in Lake Mendota. Limnol. Oceanogr. 7:80–97.

Duursma, E. K. 1961. Dissolved organic carbon, nitrogen and phosphorus in the sea. Netherlands J. Mar. Res. 1:1–148.

Duursma, E. K. 1967. The mobility of compounds in sediments in relation to exchange between bottom and supernatant water. In H. L. Golterman and R. S. Clymo, eds. Chemical Environment in the Aquatic Habitat. Amsterdam, N. V. Noord-Hollandsche Uitgevers Maatsahappij, pp. 288–296.

Dvořák, J. 1970. A quantitative study on the macrofauna of stands of emergent vegetation in a carp pond of south-west Bohemia. Rozpravy Československ. Akad. Věd, Řada Matem. Přírod. Věd 80(6):63–110.

Dykyjová, D. 1971. Production, vertical structure and light profiles in littoral stands of reed-bed species. Hidrobiologia (Romania) 12:361–376.

Dykyjová, D., and D. Hradecká. 1973. Productivity of reed-bed stands in relation to the ecotype, microclimate and trophic conditions of the habitat. Pol. Arch. Hydrobiol. 20:111–119.

Dykyjová, D. and J. Květ, Editors. 1978. Pond Littoral Ecosystems. Springer-Verlag, New York. 464 pp.

Eberly, W. R. 1959. The metalimnetic oxygen maximum in Myers Lake. Invest. Indiana Lakes Streams 5:1–46.

Eberly, W. R. 1963. Oxygen production in some northern Indiana lakes. Proc. Indust. Wastes Conf., Purdue Univ. 17:733–747.

Eberly, W. R. 1964. Further studies on the metalimnetic oxygen maximum, with special reference to its occurrence throughout the world. Invest. Indiana Lakes Streams 6:103–139.

Edmondson, W. T. 1944. Ecological studies of sessile Rotatoria. Part I. Factors affecting distribution. Ecol. Monogr. 14:31–66.

Edmondson, W. T. 1945. Ecological studies of sessile Rotatoria. Part II. Dynamics of populations and social structures. Ecol. Monogr. 15:141–172.

Edmondson, W. T. 1946. Factors in the dynamics of rotifer populations. Ecol. Monogr. 16:357–372.

Edmondson, W. T. 1948. Ecological applications of Lansing's physiological work on longevity in Rotatoria. Science 108:123–126.

Edmondson, W. T. 1960. Reproductive rates of rotifers in natural populations. Mem. Ist. Ital. Idrobiol. 12:21–77.

Edmondson, W. T. 1961. Changes in Lake Washington following an increase in the nutrient income. Verh. Int. Ver. Limnol. 14:167–175.

Edmondson, W. T. 1965. Reproductive rate of planktonic rotifers as related to food and temperature in nature. Ecol. Monogr. 35:61–111.

Edmondson, W. T. 1966. Changes in the oxygen deficit of Lake Washington. Verh. Int. Verein. Limnol. 16:153–158.

Edmondson, W. T. 1968. A graphical model for evaluating the use of the egg ratio for measuring birth and death rates. Oecologia 1:1–37.

Edmondson, W. T. 1969. Eutrophication in North America. In Eutrophication: Causes, Consequences, Correctives. Washington, D.C., National Academy of Sciences pp. 124–149.

Edmondson, W. T. 1972. Nutrients and phytoplankton in Lake Washington. In G. E. Likens, ed. Nutrients and Eutrophication: The Limiting-Nutrient Controversy. Special Symposium, Amer. Soc. Limnol. Oceanogr. 1:172–193.

Edmondson, W. T. 1974. Secondary productivity. Mitt. Int. Ver. Limnol. 20:229–272.

Edmondson, W. T. and J. T. Lehman. 1981. The effect of changes in the nutrient income on the condition of Lake Washington. Limnol. Oceanogr. 26:1–29.

Edmondson, W. T., and G. G. Winberg, eds. 1971. A Manual on Methods for the Assessment of Secondary Productivity in Fresh Waters. Int. Biol. Program Handbook, 17. Oxford, Blackwell Scientific Publications, 358 pp.

Edmunds, G. F., Jr., S. L. Jensen, and L. Berner. 1976. The mayflies of North and Central America. Univ. Minnesota Press, Minneapolis. 330 pp.

Edwards, A. M. C., and P. S. Liss. 1973. Evidence for buffering of dissolved silicon in fresh waters. Nature 243:341–342.

Edwards, R. W., and M. Owens. 1960. The effects of plants on river conditions. I. Summer Crops and estimates of net productivity of macrophytes in a chalk stream. J. Ecol. 48:151–160.

Efford, I. E. 1967. Temporal and spatial differences in phytoplankton productivity in Marion Lake, British Columbia. J. Fish. Res. Bd. Canada 24:2283–2307.

Eggleton, F. E. 1956. Limnology of a meromictic, interglacial, plunge-basin lake. Trans. Amer. Microsc. Soc. 75:334–378.

Egloff, D. A., and D. S. Palmer. 1971. Size relations of the filtering area of two *Daphnia* species. Limnol. Oceanogr. 16:900–905.

Eichenberger, E. and A. Schlatter. 1978. Effect of herbivorous insects on the production of benthic algal vegetation in outdoor channels. Verh. Int. Ver. Limnol. 20:1806–1810.

Einsele, W. 1936. Über die Beziehungen des Eisenkreislaufs zum Phosphatkreislauf im eutrophen See. Arch. Hydrobiol. 29:664–686.

Einsele, W. 1941. Die Umsetzung von zugeführtem, anorganischen Phosphat im eutrophen See und ihre Rüchwirkungen auf seinen Gesamthaushalt. Zeitsch. f. Fischerei 39:407–488.

Eisenberg, D., and W. Kauzmann. 1969. The Structure and Properties of Water. New York, Oxford University Press, 296 pp.

Eisenberg, R. M. 1966. The regulation of density in a natural population of the pond snail, *Lymnaea elodes*. Ecology 47:889–906.

Eklund, H. 1963. Fresh water: Temperature of maximum density calculated from compressibility. Science 142:1457–1458.

Eklund, H. 1965. Stability of lakes near the temperature of maximum density. Science 149:632–633.

Elgmork, K. 1959. Seasonal occurrence of *Cyclops strenuus strenuus* in relation to environment in small water bodies in southern Norway. Folia Limnol. Scandinavica 11, 196 pp.

Elgmork, K. and A. Langeland. 1980. *Cyclops scutifer* Sars: One and two-year life cycles with diapause in the meromictic Lake Blankvatn. Arch. Hydrobiol. 88:178–201.

Elgmork, K. and J. P. Nilssen. 1978. Equivalence of copepod and insect diapause. Verh. Int. Ver. Limnol. 20:2511–2517.

Elliott, J. I. 1977. Seasonal changes in the abundance and distribution of planktonic rotifers in Grasmere (English Lake District). Freshw. Biol. 7:147–166.

Elliott, J. M. 1973a. The diel activity pattern, drifting and food of the leech *Erpobdella octoculata* (L.) (Hirudinea: Erpobdellidae) in a Lake District stream. J. Anim. Ecol. 42:449–459.

Elliott, J. M. 1973b. The life cycle and production of the leech *Erpobdella octoculata* (L.) (Hirudinea: Erpobdellidae) in a Lake District stream. J. Anim. Ecol. 42:435–448.

Elliott, J. M. 1977. Some methods for the statistical analysis of samples of benthic invertebrates. 2nd Ed. Sci. Publ. Freshwat. Biol. Assoc. 25. 156 pp.

Ellis, B. K. and J. A. Stanford. 1982. Comparative photoheterotrophy, chemoheterotrophy, and photolithotrophy in a eutrophic reservoir and an oligotrophic lake.Limnol. Oceanogr. 27:440–454.

Ellis, J., and S. Kanamori. 1973. An evaluation of the Miller method for dissolved oxygen analysis. Limnol. Oceanogr. 18:1002–1005.

Ellis, P., and W. D. Williams. 1970. The biology of *Haloniscus searlei* Chilton, an oniscoid isopod living in Australian salt lakes. Austral. J. Mar. Freshw. Res. 21:51–69.

Elner, J. K. and C. M. Happey-Wood. 1978. Diatom and chrysophycean cyst profiles in sediment cores from two linked but contrasting Welch lakes. Brit. Phycol. J. 13:341–360.

Elster, H.-J. 1954a. Einige Gedanken zur Systematik, Terminologie und Zielsetzung der dynamischen Limnologie. Arch. Hydrobiol. (suppl.) 20:487–523.

Elster, H.-J. 1954b. Über die Populationsdynamik von *Eudiaptomus gracilis* Sars und *Heterocope borealis* Fischer in Bodensee-Obersee. Arch. Hydrobiol. (suppl.) 20:546–614.

Elster, H.-J., and M. Stěpánek. 1967. Eine neue Modifikation der Secchischeibe. Arch. Hydrobiol. Suppl. 33:101–106.

Emerson, K., R. C. Russo, R. E. Lund, and R. V. Thurston. 1975. Aqueous ammonia equilibrium calculations: Effect of pH and temperature. J. Fish. Res. Bd. Canada 32:2379–2383.

Emerson, S. 1975. Chemically enhanced CO_2 gas exchange in a eutrophic lake: A general model. Limnol. Oceanogr. 20:743–753.

Emerson, S., W. Broecker, and D. W. Schindler. 1973. Gas-exchange rates in a small lake as determined by the radon method. J. Fish. Res. Bd. Canada 30:1475–1484.

Eminson, D. and B. Moss. 1980. The composition and ecology of periphyton communities in freshwaters. I. The influence of host type and external environment on community composition. Br. Phycol. J. 15:429–446.

von Ende, C. N. 1979. Fish predation, interspecific predation, and the distribution of two *Chaoborus* species. Ecology 60:119–128.

von Engeln, C. D. 1961. The Finger Lakes Region: Its Origin and Nature. Ithaca, N.Y., Cornell University Press, 156 pp.

English, M. C. 1978. The magnitude and significance of the terrestrial snowpack and white ice contribution to the phosphorus budget of a lake in the Canadian Shield Region of the Kawartha lakes. Proc. Eastern Snow Conf. 35:173–189.

Enright, J. T. 1977a. Problems in estimating copepod velocity. Limnol. Oceanogr. 22:160–162.

Enright, J. T. 1977b. Diurnal vertical migration: Adaptive significance and timing. Part 1. Selective advantage: A metabolic model. Limnol. Oceanogr. 22:856–872.

Eppley, R. W. 1962. Major cations. In R. A. Lewin, ed. Physiology and Biochemistry of Algae. New York, Academic Press, pp. 255–266.

Eppley, R. W., and F. M. MaciasR. 1963. Role of the alga *Chlamydomonas mundana* in anaerobic waste stabilization lagoons. Limnol. Oceanogr. 8:411–416.

Eppley, R. W., and W. H. Thomas. 1969. Comparison of half-saturation constants for growth and nitrate uptake of marine phytoplankton. J. Phycol. 5:375–379.

Epstein, E. 1965. Mineral metabolism. In J. Bonner and J. E. Varner, eds. Plant Biochemistry. New York, Academic Press, pp. 438–466.

Eriksen, C. H. 1963a. Respiratory regulation in Ephemera simulans Walker and Hexagenia limbata (Serville) (Ephemeroptera). J. Exp. Biol. 40:455–467.

Eriksen, C. H. 1963b. The relation of oxygen consumption to substrate particle size in two burrowing mayflies. J. Exp. Biol. 40:447–453.

Eriksen, C. H. 1968. Aspects of the limno-ecology of Corophium spinicorne Stimpson (Amphipoda) and Gnorimosphaeroma oregonensis (Dana) (Isopoda). Crustaceana 14:1–12.

Eugster, H. P. and L. A. Hardie. 1978. Saline lakes. In A. Lerman, Ed. Lakes: Chemistry, Geology, Physics. Springer-Verlag, New York. pp. 237–293.

Evans, G. H. 1970. Pollen and diatom analyses of late-Quaternary deposits in the Blelham Basin, North Lancashire. New Phytol. 69:821–874.

Evans, J. H. 1961. Growth of Lake Victoria phytoplankton in enriched cultures. Nature 189:417.

Evans, J. H. 1962. The distribution of phytoplankton in some central East African waters. Hydrobiologia 19:299–315.

Fairchild, G. W. 1981. Movement and microdistribution of Sida crystallina and other littoral microcrustacea. Ecology 62:1341–1352.

Faller, A. J. 1969. The generation of Langmuir circulations by the eddy pressure of surface waves. Limnol. Oceanogr. 14:504–513.

Faller, A. J. 1971. Oceanic turbulence and the Langmuir circulations. Ann. Rev. Ecol. Syst. 2:201–236.

Faller, A. J. 1978. Experiments with controlled Langmuir circulations. Science 201:618–620.

Faller, A. J. and E. A. Caponi. 1978. Laboratory studies of wind-driven Langmuir circulations. J. Geophys. Res. 83:3617–3633.

Fallon, R. D. and T. D. Brock. 1981. Overwintering of Microcystis in Lake Mendota. Freshwat. Biol. 11:217–226.

Fallon, R. D., S. Harrits, R. S. Hanson, and T. D. Brock. 1980. The role of methane in internal carbon cycling in Lake Mendota during summer stratification. Limnol. Oceanogr. 25:357–360.

Faust, S. J., and J. V. Hunter, eds. 1971. Organic Compounds in Aquatic Environments. New York, Marcel Dekker, Inc., 638 pp.

Felföldy, L. J. M. 1960. Experiments on the carbonate assimilation of some unicellular algae by Ruttner's conductiometric method. Acta Biol. Hung. 11:67–75.

Felföldy, L. J. M. 1961. On the chlorophyll content and biological productivity of periphytic diatom communities on the stony shores of Lake Balaton. Ann. Biol. Tihany 28:99–104.

Fenchel, T. and T. H. Blackburn. 1979. Bacteria and mineral cycling. New York, Academic Press, 225 pp.

Ferling, E. 1957. Die Wirkungen des erhöhten hydroctatischen Druckes auf Wachstum und Differenzierung submerser Blütenpflanzen. Planta 49:235–270.

Ferrante, J. G. 1976. The role of zooplankton in the intrabiocoenotic phosphorus cycle and factors affecting phosphorus excretion in a lake. Hydrobiologia 49:203–214.

Ferrante, J. G. and J. I. Parker. 1978. The influence of planktonic and benthic crustaceans on silicon cycling in Lake Michigan, U.S.A. Verh. Int. Ver. Limnol. 20:324–328.

Feth, J. H. 1971. Mechanisms controlling world water chemistry: Evaporation-crystallization process. Science 172:870.

Fiala, K. 1971. Seasonal changes in the growth of clones of Typha latifolia L. in natural conditions. Folia Geobot. Phytotax. Praha 6:255–270.

Fiala, K. 1973. Growth and production of underground organs of Typha angustifolia L., Typha latifolia L. and Phragmites communis Trin. Pol. Arch. Hydrobiol. 20:59–66.

Fiala, K. 1976. Underground organs of Phragmites communis, their growth, biomass and net production. Folia Geobot. Phytotax. Praha 11:225–259.

Fiala, K. 1978. Underground organs of Typha angustifolia and Typha latifolia, their growth, propagation and production. Acta Sci. Nat. Brno 12(8):1–43.

Fiala, K., D. Dykyjová, J. Květ and J. Svoboda. 1968. Methods of assessing rhizome and root production in reed-bed stands. In Methods of Productivity Studies in Root Systems and Rhizosphere Organisms. Leningrad, Publishing House Nauka, pp. 36–47.

Ficke, E. R., and J. F. Ficke. 1977. Ice on Rivers and Lakes: A Bibliographic Essay. U.S. Geol. Surv. Water-Resources Invest. Report 75–95. 173 pp.

Findenegg, I. 1935. Limnologische Untersuchungen im Kärntner Seengebiete. Ein Beitrag zur Kenntnis des Stoffhaushaltes in Alpenseen. Int. Rev. ges. Hydrobiol. 32:369–423.

Findenegg, I. 1937. Holomiktische und meromiktische Seen. Int. Rev. ges. Hydrobiol. 35:586–610.

Findenegg, I. 1943. Untersuchungen über die Ökologie und die Produktionsverhältnisse des Planktons im Kärntner Seengebiete. Int. Rev. ges. Hydrobiol. 43:368–429.

Findenegg, I. 1965. Relationship between standing crop and primary productivity. Mem. Ist. Ital. Idrobiol. 18(Suppl.):271–289.

Finke, L. R. and H. W. Seeley, Jr. 1978. Nitrogen fixation (acetylene reduction) by epiphytes of freshwater macrophytes. Appl. Environ. Microbiol. 36:129–138.

Finlay, B. J. 1980. Temporal and vertical distribution of ciliophoran communities in the benthos of a small eutrophic loch with particular reference to the redox profile. Freshwat. Biol. 10:15–34.

Finlay, B., P. Bannister, and J. Stewart. 1979. Temporal variation in benthic ciliates and the application of association analysis. Freshwat. Biol. 9:45–53.

Fischer, Z. 1970. Elements of energy balance in grass carp Ctenopharyngodon idella Val. Pol. Arch. Hydrobiol. 17:421–434.

Fish, G. R. 1956. Chemical factors limiting growth of phytoplankton in Lake Victoria. East African Agricult. J. 21:152–158.

Fisher, D. W., A. W. Gambell, G. E. Likens, and F. H. Bormann. 1968. Atmospheric contributions to water quality of streams in the Hubbard Brook Experimental Forest, New Hampshire. Water Resources Res. 4:1115–1126.

Fisher, S. G. 1977. Organic matter processing by a stream-segment ecosystem: Fort River, Massachusetts, U.S.A. Int. Rev. ges. Hydrobiol. 62:701–727.

Fisher, S. G., and G. E. Likens. 1973. Energy flow in Bear Brook, New Hampshire: An integrative approach to stream ecosystem metabolism. Ecol. Monogr. 43:421–439.

Fitzgerald, G. P. 1972. Bioassay analysis of nutrient availability. In Allen, H. E., and J. R. Kramer, eds. Nutrients in Natural Waters. New York, John Wiley & Sons, Inc., pp. 147–169.

Fjerdingstad, E. 1979. Sulfur Bacteria. Tech. Publ. 650, Philadelphia, Am. Soc. Testing Materials, 121 pp.

Flaig, W. 1964. Effects of micro-organisms in the transformation of lignin to humic substances. Geochim. Cosmochim. Acta 28:1523–1535.

Fleischer, S. 1978. Evidence for the anaerobic release of phosphorus from lake sediments as a biological process. Naturwissenschaften 65:109.

Flett, R. J., D. W. Schindler, R. D. Hamilton, and N. E. R. Campbell. 1980. Nitrogen fixation in Canadian Precambrian Shield lakes. Can. J. Fish. Aquat. Sci. 37:494–505.

Flint, R. W. and C. R. Goldman. 1975. The effects of a benthic grazer on the primary productivity of the littoral zone of Lake Tahoe. Limnol. Oceanogr. 20:935–944.

Flint, R. W. and C. R. Goldman. 1977. Crayfish growth in Lake Tahoe: Effects of habitat variation. J. Fish. Res. Board Can. 34:155–159.

Flohn, H. 1973. Der Wasserhaushalt der Erde Schwankungen und Eingriffe. Naturwissenschaften 60:340–348.

Florin, M. -B. and H. E. Wright, Jr. 1969. Diatom evidence for the persistence of stagnant glacial ice in Minnesota. Bull. Geol. Soc. Amer. 80:695–704.

Fogg, G. E. 1963. The role of algae in organic production in aquatic environments. Brit. Phycol. Bull. 2:195–205.

Fogg, G. E. 1965. Algal Cultures and Phytoplankton Ecology. Madison, University of Wisconsin Press, 126 pp.

Fogg, G. E. 1971a. Nitrogen fixation in lakes. Plant Soil (Spec. Vol.) 1971:393–401.

Fogg, G. E. 1971b. Extracellular products of algae in freshwater. Arch. Hydrobiol. Beih. Ergebn. Limnol. 5, 25 pp.

Fogg, G. E. 1974. Nitrogen fixation. In: W. D. P. Stewart, Editor. Algal Physiology and Biochemistry. Univ. California Press, Berkeley. pp. 560–582.

Fogg, G. E., and J. H. Belcher. 1961. Pigments from the bottom deposits of an English lake. New Phytol. 60:129–142.

Fogg, G. E., and A. E. Walsby. 1971. Buoyancy regulation and the growth of planktonic blue-green algae. Mitt. Int. Ver. Limnol. 19:182–188.

Fogg, G. E., and D. F. Westlake. 1955. The importance of extracellular products of algae in freshwater. Verh. Int. Ver. Limnol. 12:219–232.

Fogg, G. E., W. D. P. Stewart, P. Fay, and A. E. Walsby. 1973. The Blue-Green Algae. New York, Academic Press, 459 pp.

Folsom, T. C. and H. F. Clifford. 1978. The population biology of Dugesia tigrina (Platyhelminthes: Turbellaria) in a thermally enriched Alberta, Canada lake. Ecology 59:966–975.

Folt, C. and C. R. Goldman. 1981. Allelopathy between zooplankton: A mechanism for interference competition. Science 213:1133–1135.

Forbes, S. A. 1887. The lake as a microcosm. Bull. Peoria (Ill.) Sci. Assoc. 1887. Reprinted in Bull. Ill. Nat. Hist. Surv. 15:537–550 (1925).

Forel, F. -A. 1892. Le Léman: Monographie Limnologique. Tome I. Géographie, Hydrographie, Géologie, Climatologie, Hydrologie. Lausanne, F. Rouge, 543 pp. (Reprinted Genève, Slatkine Reprints, 1969).

Forel, F. -A. 1895. Le Léman: Monographie Limnologique. Tome II. Mécanique, Hydraulique, Thermique, Optique, Acoustique, Chemie. Lausanne, F. Rouge, 651 pp. (Reprinted Genève, Slatkine Reprints, 1969).

Forel, F. -A. 1904. Le Léman: Monographie Limnologique. Tome III. Biologie, Histoire, Navigation, Pêche. Lausanne, F. Rouge, 715 pp. (Reprinted Genève, Slatkine Reprints, 1969.)

Francisco, D. E., R. A. Mah, and A. C. Rabin. 1973. Acridine orange-epifluorescence technique for counting bacteria in natural waters. Trans. Amer. Microsc. Soc. 92:416–421.

Francko, D. A. and R. T. Heath. 1979. Functionally distinct classes of complex phosphorus compounds in lake water. Limnol. Oceanogr. 24:463–473.

Francko, D. A., and R. T. Heath. 1982. UV-sensitive complex phosphorus: Association with dissolved humic material and iron in a bog lake. Limnol. Oceanogr. 27:564–569.

Francko, D. A. and R. G. Wetzel. 1980. Cyclic adenosine-3':5'-monophosphate: Production and extracellular release from green and blue-green algae. Physiol. Plant. 49:65–67.

Francko, D. A. and R. G. Wetzel. 1981. Dynamics of cellular and extracellular cAMP in *Anabaena flos-aquae* (Cyanophyta): Intrinsic culture variability and correlation with metabolic variables. J. Phycol. 17:129–134.

Francko, D. A. and R. G. Wetzel. 1982. The isolation of cyclic adenosine 3':5'-monophosphate (cAMP) from lakes of differing trophic status: Correlation with planktonic metabolic variables. Limnol. Oceanogr. 27:27–38.

Frank, P. W. 1952. A laboratory study of intraspecific and interspecific competition in *Daphnia pulicaria* (Forbes) and *Simocephalus vetulus* O. F. Müller. Physiol. Zool. 25:178–204.

Frank, P. W. 1957. Coactions in laboratory populations of two species of *Daphnia*. Ecology 38:510–519.

Franke, U. 1977. Experimentelle Untersuchungen zur Respiration von *Gammarus fossarum* Koch 1835 (Crustacea-Amphipoda) in Abhängigkeit von Temperatur, Sauerstoffkonzentration und Wasserbewegung. Arch. Hydrobiol. (suppl.) 48:369–411.

Freeman, T. E. 1977. Biological control of aquatic weeds with plant pathogens. Aquatic Bot. 3:175–184.

Frenzel, P. 1979. Untersuchungen zur Biologie und Populationsdynamik von *Potamopyrgus jenkinsi* (Smith) (Gastropoda: Prosobranchia) im Litoral des Bodensees. Arch. Hydrobiol. 85:448–464.

Frenzel, P. 1980. Die Produktion von *Potamopyrgus jenkinsi* Smith (Gastropoda, Prosobranchia) im Bodensee. Hydrobiologia 74:141–144.

Frevert, T. 1979a. Phosphorus and iron concentrations in the interstitial water and dry substance of sediments of Lake Constance (Obersee). I. General discussion. Arch. Hydrobiol. (suppl.) 55:298–323.

Frevert, T. 1979b. The pe redox concept in natural sediment-water systems; its role in controlling phosphorus release from lake sediments. Arch. Hydrobiol. (suppl.) 55:278–297.

Frevert, T. 1980. Dissolved oxygen dependent phosphorus release from profundal sediments of Lake Constance (Obersee). Hydrobiologia 74:17–28.

Frey, D. G. 1955a. Längsee: A history of meromixis. Mem. Ist. Ital. Idrobiol. 8(Suppl.):141–164.

Frey, D. G. 1955b. Distributional ecology of the cisco (*Coregonus artedii*) in Indiana. Invest. Indiana Lakes Streams 4:177–228.

Frey, D. G. 1964. Remains of animals in Quaternary lake and bog sediments and their interpretation. Arch. Hydrobiol. Beih. Ergebn. Limnol. 2, 114 pp.

Frey, D. G. 1969a. The rationale of paleolimnology. Mitt. Int. Ver. Limnol. 17:7–18.

Frey, D. G. 1969b. Evidence for eutrophication from remains of organisms in sediments. *In* Eutrophication: Causes, Consequences, Correctives. Washington, D.C., National Academy of Sciences, pp. 594–613.

Frey, D. G. 1974. Paleolimnology. Mitt. Int. Ver. Limnol. 20:95–123.

Frey, D. G. 1976. Interpretation of Quaternary paleoecology from Caldocera and midges, and prognosis regarding usability of other organisms. Can. J. Zool. 54:2208–2226.

Frink, C. R. 1967. Nutrient budget: Rational analysis of eutrophication in a Connecticut lake. Environ. Sci. Technol. 1:425–428.

Frost, T. M. 1978. Impact of the freshwater sponge *Spongilla lacustris* on a *Sphagnum* bog-pond. Verh. Int. Ver. Limnol. 20:2368–2371.

Frost, T. M. and C. E. Williamson. 1980. In situ determination of the effect of symbiotic algae on the growth of the freshwater sponge *Spongilla lacustris*. Ecology 61:1361–1370.

Fryer, G. 1954. Contributions to our knowledge of the biology and systematics of the freshwater Copepoda. Schweiz. Z. Hydrol. 16:64–77.

Fryer, G. 1957a. The food of some freshwater cyclopoid copepods and its ecological significance. J. Anim. Ecol. 26:263–286.

Fryer, G. 1957b. The feeding mechanism of some freshwater cyclopoid copepods. Proc. Zool. Soc. London 129:1–25.

Fuhs, G. W., S. D. Demmerle, E. Canelli, and M. Chen. 1972. Characterization of phosphorus-limited plankton algae (with reflections on the limiting-nutrient concept). *In* G. E. Likens, ed. Nutrients and Eutrophication: The Limiting-Nutrient Controversy. Special Symposium, Amer. Soc. Limnol. Oceanogr. 1:113–133.

Fuller, S. L. H. 1974. Clams and mussels (Mollusca: Bivalvia). *In* C. W. Hart, Jr., and S. L. H. Fuller, eds. Pollution Ecology of Freshwater Invertebrates. New York, Academic Press, pp. 215–273.

Gaevskaia, N. S. 1966. Rol' vysshikh vodnykh rastenii v pitanii zhivotnykh presnykh vodoemov. Moskva, Izdatel'stvo Nauka, 327 pp. (Translated into English, Nat. Lending Libr. Sci. Technol., Yorkshire, England, 1969, as: The Role of Higher Aquatic Plants in the Nutrition of the Animals of Fresh-Water Basins.)

Gak, D. Z. 1959. Fiziologicheskaia aktivnost' i sistematicheskoe polozhenie mobilizuiushchikh fosfor mikroorganizmov, vydelennykh iz vodoemov pribaltiki. Mikrobiologiya, 28:551–556.

Gak, D. Z. 1963. Vertikal'noe raspredelenie mobilizuiushchikh fosfor bakterii v gruntakh Latviiskikh vodoemov. Mikrobiologiya, 32:838–842.

Galbraith, M. G., Jr. 1967. Size-selective predation on *Daphnia* by rainbow trout and yellow perch. Trans. Amer. Fish. Soc. 96:1–10.

Gallepp, G. W. 1979. Chironomid influence on phosphorus release in sediment-water microcosms. Ecology 60:547–556.

Gallepp, G. W., J. F. Kitchell, and S. M. Bartell. 1978. Phosphorus release from lake sediments as affected by chironomids. Verh. Int. Ver. Limnol. 20:458–465.

Gallon, J. R., T. A. LaRue, and W. G. W. Kurz. 1974. Photosynthesis and nitrogenase activity in the blue-green alga *Gloeocapsa*. Can. J. Microbiol. 20:1633–1637.

Galloway, J. N. and D. M. Whelpdale. 1980. An atmospheric sulfur budget for eastern North America. Atmospheric Environ. 14:409–417.

Ganf, G. G. 1974a. Incident solar irradiance and underwater light penetration as factors controlling the chlorophyll *a* content of a shallow equatorial lake (Lake George, Uganda). J. Ecol. 62:593–609.

Ganf, G. G. 1974b. Diurnal mixing and the vertical distribution of phytoplankton in a shallow equatorial lake (Lake George, Uganda). J. Ecol. 62:611–629.

Ganf, G. G. 1974c. Phytoplankton biomass and distribution in a shallow eutrophic lake (Lake George, Uganda). Oecologia 16:9–29.

Gardner, W. S. and G. F. Lee. 1975. The role of amino acids in the nitrogen cycle of Lake Mendota. Limnol. Oceanogr. 20:379–388.

Garrels, R. M. 1965. Silica: Role in the buffering of natural waters. Science 148:69.

Gasith, A. 1975. Tripton sedimentation in eutrophic lakes - simple correction for the resuspended matter. Verh. Int. Ver. Limnol. 19:116–122.

Gasith, A. 1976. Seston dynamics and tripton sedimentation in the pelagic zone of a shallow eutrophic lake. Hydrobiologia 51:225–231.

Gates, D. M. 1962. Energy Exchange in the Biosphere. New York, Harper & Row, 151 pp.

Gates, F. C. 1942. The bogs of northern Lower Michigan. Ecol. Monogr. 12:213–254.

Gaudet, J. J. 1968. The correlation of physiological differences and various leaf forms of an aquatic plant. Physiol. Plant 21:594–601.

Gaudet, J. J. 1976. Nutrient relationships in the detritus of a tropical swamp. Arch. Hydrobiol. 78:213–239.

Gaudet, J. J. 1977. Uptake, accumulation, and loss of nutrients by papyrus in tropical swamps. Ecology 58:415–422.

Gaudet, J. J. 1978. Effect of a tropical swamp on water quality. Verh. Int. Ver. Limnol. 20:2202–2206.

Gaudet, J. J. 1979. Seasonal changes in nutrients in a tropical swamp: North Swamp, Lake Naivasha, Kenya. J. Ecol. 67:953–981.

Geesey, G. G. and J. W. Costerton. 1979. Microbiology of a northern river: Bacterial distribution and relationship to suspended sediment and organic carbon. Can. J. Microbiol. 25:1058–1062.

Geiling, W. T., and R. S. Campbell. 1972. The effect of temperature on the development rate of the major life stages of *Diaptomus pallidus* Herrick. Limnol. Oceanogr. 17:304–307.

Geller, W. 1975. Die Nahrungsaufnahme von *Daphnia pulex* in Abhängigkeit von der Futterkonzentration, der Temperatur, der Körpergrösse und dem Hungerzustand der Tiere. Arch. Hydrobiol. Suppl. 48:47–107.

George, D. G. 1976. Life cycle and production of *Cyclops vicinus* in a shallow eutrophic reservoir. Oikos 27:101–110.

George, D. G., and R. W. Edwards. 1973. *Daphnia* distribution within Langmuir circulations. Limnol. Oceanogr. 18:798–800.

George, D. G., and R. W. Edwards. 1974. Population dynamics and production of *Daphnia hyalina* in a eutrophic reservoir. Freshwat. Biol. 4:445–465.

George, D. G. and S. I. Heaney. 1978. Factors influencing the spatial distribution of phytoplankton in a small productive lake. J. Ecol. 66:133–155.

George, M. G., and C. H. Fernando. 1969. Seasonal distribution and vertical migration of planktonic rotifers in two lakes of eastern Canada. Verh. Int. Ver. Limnol. 17:817–829.

George, M. G., and C. H. Fernando. 1970. Diurnal migration in three species of rotifers in Sunfish Lake, Ontario. Limnol. Oceanogr. 15:218–223.

Gerking, S. D. 1962. Production and food utilization in a population of bluegill sunfish. Ecol. Monogr. 32:31–78.

Gerletti, M. 1968. Dark bottle measurements in primary productivity studies. Mem. Ist. Ital. Idrobiol. 23:197–208.

Gerloff, G. C. 1963. Comparative mineral nutrition of plants. Ann. Rev. Plant Physiol. 14:107–123.

Gerloff, G. C. 1969. Evaluating nutrient supplies for the growth of aquatic plants in natural waters. *In* Eutrophication: Causes, Consequences, Correctives. Washington, D. C., National Academy of Sciences, pp. 537–555.

Gerloff, G. C., G. P. Fitzgerald, and F. Skoog. 1952. The mineral nutrition of *Microcystis aeruginosa*. Amer. J. Bot. 39:26–32.

Gerloff, G. C., and F. Skoog. 1957. Availability of iron and manganese in southern Wisconsin lakes for the growth of *Microcystis aeruginosa*. Ecology 38:551–556.

Gessner, F. 1955. Hydrobotanik. Die Physiologischen Grundlagen der Pflanzenverbreitung im Wasser. I. Energiehaushalt. Berlin, VEB Deutscher Verlag der Wissenschaften, 517 pp.

Gessner, F. 1959. Hydrobotanik. Die Physiologischen Grundlagen der Pflanzenverbreitung im Wasser. II. Stoffhaushalt. Berlin, VEB Deutscher Verlag der Wissenschaften, 701 pp.

Gessner, F. 1961. Hydrostatischer Druck und Pflanzenwachstum. *In* W. Ruhland, Ed. Handbuch der Pflanzenphysiologie. Vol. 16. Heidelberg, Springer-Verlag, pp. 668–690.

Gessner, F., and A. Diehl. 1951. Die Wirkung natürlicher Ultraviolettstrahlung auf die Chlorophyllzerstörung von Planktonalgen. Arch. Mikrobiol. 15:439–453.

Gest, H., A. San Pietro, and L. P. Vernon, eds. 1963. Bacterial Photosynthesis. Yellow Springs, Ohio, Antioch Press, 523 pp.

Ghiretti, F. 1966. Respiration. *In* Wilbur, K. M., and C. M. Yonge, eds. Physiology of Mollusca, Vol. 2. New York, Academic Press, pp. 175–208.

Gibbs, R. J. 1970. Mechanisms controlling world water chemistry. Science 170:1088–1090.

Gibbs, R. J. 1973. Mechanisms of trace metal transport in rivers. Science. 180:71–73.

Gilbert, J. J. 1967a. Control of sexuality in the rotifer *Asplanchna brightwelli* by dietary lipids of plant origin. Proc. Nat. Acad. Sci. 57:1218–1225.

Gilbert, J. J. 1967b. *Asplanchna* and postero-lateral spine production in *Brachionus calyciflorus*. Arch. Hydrobiol. 64:1–62.

Gilbert, J. J. 1968. Dietary control of sexuality in the rotifer *Asplanchna brightwelli* Gosse. Physiol. Zool. 41:14–43.

Gilbert, J. J. 1972. α-tocopherol in males of the rotifer *Asplanchna sieboldi*: Its metabolism and its distribution in the testis and rudimentary gut. J. Exp. Zool. 181:117–128.

Gilbert, J. J. 1973. Induction and ecological significance of gigantism in the rotifer *Asplanchna sieboldi*. Science 181:63–66.

Gilbert, J. J. 1975a. Polymorphism and sexuality in the rotifer *Asplanchna*, with special reference to the effects of prey-type and clonal variation. Arch. Hydrobiol. 75:442–483.

Gilbert, J. J. 1975b. Field experiments on gemmulation in the fresh-water sponge *Spongilla lacustris*. Trans. Amer. Microsc. Soc. 94:347–356.

Gilbert, J. J. 1976. Polymorphism in the rotifer *Asplanchna sieboldi*: Biomass, growth, and reproductive rate of the saccate and campanulate morphotypes. Ecology 57:542–551.

Gilbert, J. J. 1977. A note on the effect of cold shock on mictic-female production in *Brachionus calyciflorus*. Arch. Hydrobiol. Beih. Ergebn. Limnol. 8:158–160.

Gilbert, J. J. 1980a. Female polymorphism and sexual reproduction in the rotifer *Asplanchna*: Evolution of their relationship and control by dietary tocopherol. Amer. Nat. 116:409–431.

Gilbert, J. J. 1980b. Further observations on developmental polymorphism and its evolution in the rotifer *Brachionus calyciflorus*. Freshwat. Biol. 10:281–294.

Gilbert, J. J. 1980c. Feeding in the rotifer *Asplanchna*: Behavior, cannibalism, selectivity, prey defenses, and impact on rotifer communities. *In* W. C. Kerfoot, ed. Evolution and Ecology of Zooplanktonic Communities. Hanover, NH, Univ. Press New England, pp. 158–172.

Gilbert, J. J., and H. L. Allen. 1973. Studies on the physiology of the green freshwater sponge *Spongilla lacustris*: Primary productivity, organic matter, and chlorophyll content. Verh. Int. Ver. Limnol. 18:1413–1420.

Gilbert, J. J., and C. W. Birky, Jr. 1971. Sensitivity and specificity of the *Asplanchna* response to dietary α-tocopherol. J. Nutrition 101:113–126.

Gilbert, J. J. and S. Hadžišče. 1975. Sexual reproduction in the freshwater sponge *Ochridaspongia rotunda*. Verh. Int. Ver. Limnol. 19:2785–2792.

Gilbert, J. J. and S. Hadžišče. 1977. Life cycle of the freshwater sponge *Ochridaspongia rotunda* Arndt. Arch. Hydrobiol. 79:285–318.

Gilbert, J. J. and J. R. Litton, Jr. 1975. Dietary tocopherol and sexual reproduction in the rotifers *Brachionus calyciflorus* and *Asplanchna sieboldi*. J. Exp. Zool. 194:485–494.

Gilbert, J. J. and T. L. Simpson. 1976a. Gemmule polymorphism in the freshwater sponge *Spongilla lacustris*. Arch. Hydrobiol. 78:268–277.

Gilbert, J. J. and T. L. Simpson. 1976b. Sex reversal in a freshwater sponge. J. Exp. Zool. 195:145–151.

Gilbert, J. J., T. L. Simpson, and G. S. de Nagy. 1975. Field experiments on egg production in the fresh-water sponge *Spongilla lacustris*. Hydrobiologia 46:17–27.

Gilbert, J. J. and P. L. Starkweather. 1977. Feeding in the rotifer *Brachionus calyciflorus*. I. Regulatory mechanisms. Oecologia 28:125–131.

Gilbert, J. J. and P. L. Starkweather. 1978. Feeding in the rotifer *Brachionus calyciflorus*. III. Direct observations on the effects of food type, food density, change in food type, and starvation on the incidence of pseudotrochal screening. Verh. Int. Ver. Limnol. 20:2382–2388.

Gilbert, J. J., and G. A. Thompson, Jr. 1968. Alpha tocopherol control of sexuality and polymorphism in the rotifer *Asplanchna*. Science 159:734–736.

Gilbert, J. J., and J. K. Waage. 1967. *Asplanchna*, *Asplanchna*-substance, and posterolateral spine length variation of the rotifer *Brachionus calyciflorus* in a natural environment. Ecology 48:1027–1031.

Gilbert, J. J. and C. E. Williamson. 1978. Predator-prey behavior and its effect on rotifer survival in associations of *Mesocyclops edax*, *Asplanchna girodi*, *Polyarthra vulgaris*, and *Keratella cochlearis*. Oecologia 37:13–22.

Gillespie, P. A. and M. J. Spencer. 1980. Seasonal variation of heterotrophic potential in Lake Rotoroa. N. Z. J. Mar. Freshw. Res. 14:15–21.

Gjessing, E. T. 1964. Ferrous iron in water. Limnol. Oceanogr. 9:272–274.

Gjessing, E. T. 1970. Reduction of aquatic humus in streams. Vatten 26:14–23.

Gjessing, E. T. 1976. Physical and Chemical Characteristics of Aquatic Humus. Ann Arbor, Ann Arbor Science Publs., 120 pp.

Gjessing, E. T., and T. Gjerdahl. 1970. Influence of ultra-violet radiation on aquatic humus. Vatten 26:144–145.

Glime, J. M., R. G. Wetzel, and B. J. Kennedy. 1982. The effects of bryophytes on succession from alkaline marsh to *Sphagnum* bog. Amer. Midland Nat. *108:* 209–223.

Gliwicz, Z. M. 1979. Metalimnetic gradients and trophic state of lake epilimnia. Mem. Ist. Ital. Idrobiol. *37:*121–143.

Gliwicz, Z. M. 1980. Filtering rates, food size selection, and feeding rates in cladocerans—Another aspect of interspecific competition in filter-feeding zooplankton. *In* W. C. Kerfoot, ed. Evolution and Ecology of Zooplankton Communities. Hanover, NH, Univ. Press New England, pp. 282–291.

Gliwicz, Z. M. and A. Prejs. 1977. Can planktivorous fish keep in check planktonic crustacean populations? A test of size-efficiency hypothesis in typical Polish lakes. Ekol. Polska 25:567–591.

Gliwicz, Z. M. and E. Siedlar. 1980. Food size limitation and algae interfering with food collection in *Daphnia.* Arch. Hydrobiol. 88:155–177.

Glob, P. V. 1969. The Bog People. Iron-Age Man Preserved. Ithaca, N.Y., Cornell University Press, 200 pp.

Glooschenko. W. A., J. E. Moore, and R. A. Vollenweider. 1973. Chlorophyll *a* distribution in Lake Huron and its relationship to primary production. Proc. Conf. Great Lakes Res., Int. Assoc. Great Lakes Res. 16:40–49.

Glooschenko, W. A., J. E. Moore, M. Munawata, and R. A. Vollenweider. 1974a. Primary production in lakes Ontario and Erie: A comparative study. J. Fish. Res. Bd. Canada 31:253–263.

Glooschenko, W. A., J. E. Moore, and R. A. Vollenweider. 1974b. Spatial and temporal distribution of chlorophyll *a* and pheopigments in surface waters of Lake Erie. J. Fish. Res. Bd. Canada 31:265–274.

Gocke, K. 1970. Untersuchungen über Abgabe und Aufnahme von Aminosäuren und Polypeptiden durch Planktonorganismen. Arch. Hydrobiol. 67:285–367.

Gode, P., and J. Overbeck. 1972. Untersuchungen zur heterotrophen Nitrifikation ím See. Z. Allg. Mikrobiol. 12:567–574.

Godlewska-Lipowa, W. A. 1975. Ecosystem of the Mikolajskie Lake. The role of heterotrophic bacteria in the pelagial. Polskie Arch. Hydrobiol. 22:79–87.

Godlewska-Lipowa, W. A. 1976. Bacteria as indicators of the degree of eutrophication and degradation of lakes. Polskie Arch. Hydrobiol. 23:341–356.

Godshalk, G. L. 1977. Decomposition of aquatic plants in lakes. Ph.D. Dissertation, Michigan State University. 139 pp. + 340 figs.

Godshalk, G. L. and R. G. Wetzel. 1978a. Decomposition of aquatic angiosperms. I. Dissolved components. Aquatic Bot. 5:281–300.

Godshalk, G. L. and R. G. Wetzel. 1978b. Decomposition of aquatic angiosperms. II. Particulate components. Aquatic Bot. 5:301–327.

Godshalk, G. L. and R. G. Wetzel. 1978c. Decomposition of aquatic angiosperms. III. *Zostera marina* L. and a conceptual model of decomposition. Aquatic Bot. 5:329–354.

Godward, M. B. E. 1962. Invisible radiations. *In* R. A. Lewin, ed. Physiology and Biochemistry of Algae. New York, Academic Press, pp. 551–566.

Goering, J. J., and V. A. Dugdale. 1966. Estimates of the rates of denitrification in a subarctic lake. Limnol. Oceanogr. 11:113–117.

Goering, J. J., and J. C. Neess. 1964. Nitrogen fixation in two Wisconsin lakes. Limnol. Oceanogr. 9:530–539.

Golachowska, J. B. 1971. The pathways of phosphorus in lake water. Pol. Arch. Hydrobiol. 18:325–345.

Goldman, C. R. 1960. Primary productivity and limiting factors in three lakes of the Alaska Peninsula. Ecol. Monogr. 30:207–230.

Goldman, C. R. 1960. Molybdenum as a factor limiting primary productivity in Castle Lake, California. Science 132:1016–1017.

Goldman, C. R. 1961. The contribution of alder trees (*Alnus tenuifolia*) to the primary productivity of Castle Lake, California. Ecology 42:282–288.

Goldman, C. R. 1970. Antarctic freshwater ecosystems. *In* M. W. Holdgate, ed. Antarctic Ecology. New York, Academic Press, pp. 609–627.

Goldman, C. R. 1972. The role of minor nutrients in limiting the productivity of aquatic ecosystems. *In* G. E. Likens, ed. Nutrients and Eutrophication: The Limiting-Nutrient Controversy. Special Symposium, Amer. Soc. Limnol. Oceanogr. 1:21–38.

Goldman, C. R., D. T. Mason, and B. J. B. Wood. 1963. Light injury and inhibition in Antarctic freshwater phytoplankton. Limnol. Oceanogr. 8:313–322.

Goldman, C. R., D. T. Mason, and B. J. B. Wood. 1972. Comparative study of the limnology of two small lakes on Ross Island, Antarctica. *In* G. A. Llano, ed. Antarctic Terrestrial Biology. Washington, American Geophysical Union, pp. 1–50.

Goldman, C. R., and R. G. Wetzel. 1963. A study of the primary productivity of Clear Lake, Lake County, California. Ecology 44:283–294.

Goldman, J. C. 1977. Steady state growth of phytoplankton in continuous culture: Comparison of internal and external nutrient equations. J. Phycol. 13:251–258.

Goldman, J. C. and S. J. Graham. 1980. Inorganic carbon limitation and chemical composition of two freshwater green microalgae. Appl. Environ. Microbiol. 41:60–70.

Goldman, J. C., D. B. Porcella, E. J. Middlebrooks, and D. F. Toerien. 1972. The effect of carbon on algal growth—its relationship to eutrophication. Water Res. 6:637–679.

Goldspink, C. R., and D. B. C. Scott. 1971. Vertical

migration of *Chaoborus flavicans* in a Scottish loch. Freshwat. Biol. *1*:411–421.

Goldsworthy, A. 1970. Photorespiration. Bot. Rev. *36*:321–340.

Golterman, H. L. 1960. Studies on the cycle of elements in fresh water. Acta Bot. Neerlandica 9:1–58.

Golterman, H. L., ed. 1969. Methods for Chemical Analysis of Fresh Waters. Int. Biol. Program Handbook 8. Oxford, Blackwell Scientific Publications, 172 pp.

Golterman, H. L. 1971. The determination of mineralization losses in correlation with the estimation of net primary production with the oxygen method and chemical inhibitors. Freshwat. Biol. *1*:249–256.

Golterman, H. L., C. C. Bakels, and J. Jakobs-Mögelin. 1969. Availability of mud phosphates for the growth of algae. Verh. Int. Ver. Limnol. *17*:467–479.

Golubić, S. 1963. Hydrostatischer Druck, Licht und submerse Vegetation im Vrana-See. Int. Rev. ges. Hydrobiol. *48*:1–7.

Golubić, S. 1967. Algenvegetation der Felsen. Eine ökologische Algenstudie im dinarischen Karstgebiet. Die Binnengewässer *23*. 183 pp.

Golubić, S. 1969. Cyclic and noncyclic mechanisms in the formation of travertine. Verh. Int. Ver. Limnol. *17*:956–961.

Golubić, S. 1973. The relationship between blue-green algae and carbonate deposits. *In* N. G. Carr and B. A. Whitton, eds. The Biology of Blue-Green Algae. Berkeley, University of California Press, pp. 434–472.

Goodwin, T. W. 1974. Carotenoids and biliproteins. *In* W. D. P. Stewart, ed. Algal Physiology and Biochemistry. Berkeley, University of California Press, pp. 176–205.

Gophen, M., B. Z. Cavari, and T. Berman. 1974. Zooplankton feeding on differentially labelled algae and bacteria. Nature *247*:393–394.

Gorbunov, K. V. 1953. Raspad ostatkov vysshikh vodnykh rastenii i ego ekologicheskaia rol' v vodoemakh nizhnei zony Del 'ty Volgi. Trudy Vses. Gidrobiol. Obshschestva 5:158–202.

Gorham, E. 1955. On some factors affecting the chemical composition of Swedish fresh waters. Geochim. Cosmochim. Acta *7*:129–150.

Gorham, E. 1956. On the chemical composition of some waters from the Moor House Nature Reserve. J. Ecol. *44*:377–382.

Gorham, E. 1957. The development of peat lands. Quart. Rev. Biol. *32*:145–166.

Gorham, E. 1958. Observations on the formation and breakdown of the oxidized microzone at the mud surface in lakes. Limnol. Oceanogr. *3*:291–298.

Gorham, E. 1961. Factors influencing supply of major ions to inland waters, with special reference to the atmosphere. Bull. Geol. Soc. Amer. *72*:795–840.

Gorham, E. 1964. Morphometric control of annual heat budgets in temperate lakes. Limnol. Oceanogr. *9*:525–529.

Gorham, E., and J. Sanger. 1964. Chlorophyll derivatives in woodland, swamp, and pond soils of Cedar Creek Natural History Area, Minnesota, U.S.A. *In* Recent Researches in the Fields of Hydrosphere, Atmosphere and Nuclear Geochemistry. Tokyo, Mauzen Co., Ltd., pp. 1–12.

Gorham, E., and J. E. Sanger. 1972. Fossil pigments in the surface sediments of a meromictic lake. Limnol. Oceanogr. *17*:618–622.

Gorham, E. and J. E. Sanger. 1975. Fossil pigments in Minnesota lake sediments and their bearing upon the balance between terrestrial and aquatic inputs to sedimentary organic matter. Verh. Int. Ver. Limnol. *19*:2267–2273.

Gorham, E. and J. E. Sanger. 1976. Fossilized pigments as stratigraphic indicators of cultural eutrophication in Shagawa Lake, northeastern Minnesota. Geol. Soc. Amer. Bull. *87*:1638–1642.

Gough, S. B. and L. P. Gough. 1981. Comment on "Primary production of algae growing on natural and artificial aquatic plants: A study of interactions between epiphytes and their substrate" (Cattaneo and Kalff). Limnol. Oceanogr. *26*:987–989.

Goulden, C. E. 1969a. Interpretative studies of cladoceran microfossils in lake sediments. Mitt. Int. Ver. Limnol. *17*:43–55.

Goulden, C. E. 1969b. Temporal changes in diversity. *In* G. M. Woodwell and H. H. Smith, eds. Diversity and Stability in Ecological Systems. Brookhaven Symposia in Biology *22*:96–102.

Goulden, C. E. 1971. Environmental control of the abundance and distribution of the chydorid Cladocera. Limnol. Oceanogr. *16*:320–331.

Goulden, C. E. and L. L. Henry. 1982. Lipid energy reserves and their role in Cladocera. Spec. Symposium, Amer. Assoc. Advancement Science (In press)

Goulden, C. E. and L. Hornig. 1980. Population oscillations and energy reserves in plankton Cladocera and their consequences to competition. Proc. Nat. Acad. Sci. *77*:1716–1720.

Goulden, C. E., L. Hornig, and C. Wilson. 1978. Why do large zooplankton species dominate? Verh. Int. Ver. Limnol. *20*:2457–2460.

Goulder, R. 1969. Interactions between the rates of production of a freshwater macrophyte and phytoplankton in a pond. Oikos *20*:300–309.

Goulder, R. 1970. Day-time variations in the rates of production by two natural communities of submerged freshwater macrophytes. J. Ecol. *58*:521–528.

Goulder, R. 1972. Grazing by the ciliated protozoon *Loxodes magnus* on the alga *Scenedesmus* in a eutrophic pond. Oikos *23*:109–115.

Goulder, R. 1974a. The seasonal and spatial distribution of some benthic ciliated Protozoa in Esthwaite Water. Freshw. Biol. *4*:127–147.

Goulder, R. 1974b. Relationships between natural populations of the ciliated protozoa *Loxodes magnus* Stokes and *L. striatus* Penard. J. Ecol. 43:429–438.

Goulder, R. 1980a. The ecology of two species of primitive ciliated protozoa commonly found in standing fresh waters (*Loxodes magnus* Stokes and *L. striatus* Penard). Hydrobiologia 72:131–158.

Goulder, R. 1980b. Seasonal variation in heterotrophic activity and population density of planktonic bacteria in a clean river. J. Ecol. 68:349–363.

Govindjee, and B. Z. Braun. 1974. Light absorption, emission and photosynthesis. *In* W. D. P. Stewart, ed. Algal Physiology and Biochemistry. Berkeley, University of California Press, pp. 346–390.

Gowing, H. and W. T. Momot. 1979. Impact of brook trout *(Salvelinus fontinalis)* predation on the crayfish *Orconectes virilis* in three Michigan lakes. J. Fish. Res. Board. Can. 36:1191–1196.

Grace, J. B. and R. G. Wetzel. 1978. The production biology of Eurasian waterfilfoil (*Myriophyllum spicatum* L.): A review. J. Aquatic Plant Manage. 16:1–11.

Grace, J. B. and R. G. Wetzel. 1981a. Phenotypic and genotypic components of growth and reproduction in *Typha latifolia*: Experimental studies in marshes of differing successional maturity. Ecology 62:789–801.

Grace, J. B. and R. G. Wetzel. 1981b. Effects of size and growth rate on vegetative reproduction in *Typha*. Oecologia 50:158–161.

Grace, J. B. and R. G. Wetzel. 1981c. Habitat partitioning and competitive displacement in cattails *(Typha)*: Experimental field studies. Amer. Nat. 118:463–474.

Grace, J. B. and R. G. Wetzel. 1982a. Niche differentiation between two rhizomatous plant species: *Typha latifolia* and *Typha angustifolia*. Can. J. Bot. 60:46–57.

Grace, J. B. and R. G. Wetzel. 1982b. Variations in growth and reproduction within populations of two rhizomatous plant species: *Typha latifolia* and *Typha angustifolia*. Oecologia 53:258–263.

Graham, W. F., S. R. Piotrowicz, and R. A. Duce. 1979. The sea as a source of atmospheric phosphorus. Mar. Chem. 7:325–342.

Granéli, W. 1978. Sediment oxygen uptake in south Swedish lakes. Oikos 30:7–16.

Grant, J. W. G. and I. A. E. Bayly. 1981. Predator induction of crests in morphs of the *Daphnia carinata* King complex. Limnol. Oceanogr. 26:201–208.

Green, E. J., and D. E. Carritt. 1967. New tables for oxygen saturation of seawater. J. Mar. Res. 25:140–147.

Green, J., and O. B. Lan. 1974. *Asplanchna* and the spines of *Brachionus calyciflorus* in two Javanese sewage ponds. Freshwat. Biol. 4:223–226.

Green, J. D. 1976. Population dynamics and production of the calanoid copepod *Calamoecia lacasi* in a northern New Zealand lake. Arch. Hydrobiol. (suppl.) 50:313–400.

Greenbank, J. 1945. Limnological conditions in ice-covered lakes, especially as related to winter-kill of fish. Ecol. Monogr. 15:343–392.

Greene, K. L. 1974. Experiments and observations on the feeding behavior of the freshwater leech *Erpobdella octoculata* (L.) (Hirudinea: Erpobdellidae). Arch. Hydrobiol. 74:87–99.

Greeney, W. J., D. A. Bella, and H. C. Curl, Jr. 1973. A theoretical approach to interspecific competition in phytoplankton communities. Amer. Nat. 107:405–425.

Grey, D. M., ed. 1970. Handbook on the Principles of Hydrology. Ottawa, National Research Council of Canada, 676 pp.

Grieco, E. and R. Desrochers. 1978. Production de vitamin B_{12} par une algue bleue. Can. J. Microbiol. 24:1562–1566.

Griffith, E. J., A. Beeton, J. M. Spencer, and D. T. Mitchell. 1973. Environmental Phosphorus Handbook. New York, John Wiley & Sons, 718 pp.

Griffiths, M. 1978. Specific blue-green algal carotenoids in sediments of Esthwaite Water. Limnol. Oceanogr. 23:777–784.

Griffiths, M. and W. T. Edmondson. 1975. Burial of oscillaxanthin in the sediment of Lake Washington. Limnol. Oceanogr. 20:945–952.

Griffiths, M., P. S. Perrott, and W. T. Edmondson. 1969. Oscillaxanthin in the sediment of Lake Washington. Limnol. Oceanogr. 14:317–326.

Grim, J. 1952. Vermehrungsleistungen planktischer Algenpopulationen in Gleichgewichtsperioden. Arch. Hydrobiol. Suppl. 20:238–260.

Grootes, P. M. 1978. Carbon-14 time scale extended: Comparison of chronologies. Science 200:11–21.

Groth, P. 1971. Untersuchungen über einige Spurenelemente in Seen. Arch. Hydrobiol. 68:305–375.

Gruendling, G. K. 1971. Ecology of the epipelic algal communities in Marion Lake, British Columbia. J. Phycol. 7:239–249.

Grünsfelder, M. 1974. Die Wirkung von Licht auf die Phosphataufnahme in Blattzellen von *Elodea densa*. Ber. Deutsch. Bot. Ges. 87:501–508.

Grünsfelder, M. and W. Simonis. 1973. Aktive und inaktive Phosphataufnahme in Blattzellen von *Elodea densa* bei hohen Phosphat-Aussenkonzentrationen. Planta 115:173–186.

Guerrero, R., E. Montesinos, I. Esteve, and C. Abella. 1980. Physiological adaptation and growth of purple and green sulfur bacteria in a meromictic lake (Vila) as compared to a holomictic lake (Siso). Developments Hydrobiol. 3:161–171.

Gulati, R. D. 1978. Vertical changes in the filtering, feeding and assimilation rates of dominant zooplankters in a stratified lake. Verh. Int. Ver. Limnol. 20:950–956.

Gunnison, D., and M. Alexander. 1975. Resistance and susceptibility of algae to decomposition by natural microbial communities. Limnol. Oceanogr. 20:64-70.

Gusev, M. V., and K. A. Nikitina. 1974. A study of the death of blue-green algae under conditions of darkness. (Trans. Consultants Bureau, Plenum Publ. Corp.) Mikrobiologiya 43:333-337.

Guseva, K. A., and S. P. Goncharova. 1965. O vliianii vysshei vodnoi rastitel 'nosti na razvitie planktonnykh sinezelenykh vodoroslei. In Ekologiia i Fiziologiia Sinezelenykh Vodoroslei. Leningrad, pp. 230-234.

Haan, H. de 1972. Some structural and ecological studies on soluble humic compounds from Tjeukemeer. Verh. Int. Ver. Limnol. 18:685-695.

Haan, H. de 1974. Effect of a fulvic acid fraction on the growth of a Pseudomonas from Tjeukemeer (The Netherlands). Freshwat. Biol. 4:301-309.

Hackett, W. F., W. J. Connors, T. K. Kirk, and J. G. Zeikus. 1977. Microbial decomposition of synthetic ^{14}C-labeled lignins in nature: Lignin biodegradation in a variety of natural materials. Appl. Environ. Microbiol. 33:43-51.

Hadl, G. 1972. Zur Ökologie und Biologie der Pisidien (Bivalvia: Sphaeriidae) im Lunzer Untersee. Sitzungsber. Österr. Akad. Wissensch. Math.- nat. Kl., Abt. I 180:317-338.

Hagedorn, H. 1971. Experimentelle Untersuchungen über den Einfluss des Thiamins auf die natürliche Algenpopulation des Pelagials. Arch. Hydrobiol. 68:382-399.

Haines, D. A., and R. A. Bryson. 1961. An empirical study of wind factor in Lake Mendota. Limnol. Oceanogr. 6:356-364.

Hairston, H. G., Jr. 1976. Photoprotection by carotenoid pigments in the copepod Diaptomus nevadensis. Proc. Natl. Acad. Sci. 73:971-974.

Hairston, N. G., Jr. 1980. The vertical distribution of diaptomid copepods in relation to body pigmentation. In W. C. Kerfoot, eds. Evolution and Ecology of Zooplankton Communities. Hanover, NH, Univ. Press New England, pp. 98-110.

Håknson, L. 1981. A Manual of Lake Morphometry. Springer-Verlag, New York. 78 pp.

Halbach, U., and G. Halbach-Keup. 1974. Quantitative Beziehungen zwischen Phytoplankton und der Populationsdynamik des Rotators Brachionus calyciflorus Pallas. Befunde aus Laboratoriumsexperimenten und Freilanduntersuchungen. Arch. Hydrobiol. 73:273-309.

Hall, D. J. 1964. An experimental approach to the dynamics of a natural population of Daphnia galeata mendotae. Ecology 45:94-112.

Hall, D. J., W. E. Cooper, and E. E. Werner. 1970. An experimental approach to the production dynamics and structure of freshwater animal communities. Limnol. Oceanogr. 15:839-928.

Hall, D. J., S. T. Threlkeld, C. W. Burns, and P. H. Crowley. 1976. The size-efficiency hypothesis and the size structure of zooplankton communities. Ann. Rev. Ecol. Syst. 7:177-208.

Hall, D. J., E. E. Werner, J. F. Gilliam, G. G. Mittelbach, D. Howard, C. G. Doner, J. A. Dickerman, and A. J. Stewart. 1979. Diel foraging behavior and prey selection in the golden shiner (Notemigonus crysoleucas). J. Fish. Res. Bd. Can. 36:1029-1039.

Hall, J. B. 1971. Evolution of the prokaryotes. J. Theor. Biol. 30:429-454.

Hall, K. J., and K. D. Hyatt. 1974. Marion Lake (IBP)—From bacteria to fish. J. Fish. Res. Board Can. 31:893-911.

Hall, K. J., P. M. Kleiber, and I. Yesaki. 1972. Heterotrophic uptake of organic solutes by microorganisms in the sediment. Mem. Ist. Ital. Idrobiol. 29(suppl.):441-471.

Halldal, P. 1967. Ultraviolet action spectra in algology. A review. Photochem. Photobiol. 6:445-460.

Hallegraeff, G. M. and J. Ringelberg. 1978. Characterization of species diversity of phytoplankton assemblages by dominance-diversity curves. Verh. Int. Ver. Limnol. 20:939-949.

Halmann, M. and A. Elgavish. 1975. The role of phosphate in eutrophication. Stimulation of plankton growth and residence times of inorganic phosphate in Lake Kinneret water. Verh. Int. Ver. Limnol. 19:1351-1356.

Halmann, M. and M. Stiller. 1974. Turnover and uptake of dissolved phosphate in freshwater. A study in Lake Kinneret. Limnol. Oceanogr. 19:774-783.

Hama, T. and N. Handa. 1980. Molecular weight distribution and characterization of dissolved organic matter from lake waters. Arch. Hydrobiol. 90:106-120.

Hammer, U. T. 1978. The saline lakes of Saskatchewan. III. Chemical characterization. Int. Revue ges. Hydrobiol. 63:311-335.

Hammer, U. T., and W. W. Sawchyn. 1968. Seasonal succession and congeneric associations of Diaptomus spp. (Copepoda) in some Saskatchewan ponds. Limnol. Oceanogr. 13:476-484.

Haney, J. F. 1971. An in situ method for the measurement of zooplankton grazing rates. Limnol. Oceanogr. 16:970-977.

Haney, J. F. 1973. An in situ examination of the grazing activities of natural zooplankton communities. Arch. Hydrobiol. 72:87-132.

Haney, J. F. and D. J. Hall. 1975. Diel vertical migration and filter-feeding activities of Daphnia. Arch. Hydrobiol. 75:413-441.

Hansen, K. 1959a. Sediments from Danish lakes. J. Sediment. Petrol. 29:38-46.

Hansen, K. 1959b. The terms gyttja and dy. Hydrobiologia 13:309-315.

Hansen, K. 1961. Lake types and lake sediments. Verh. Int. Ver. Limnol. 14:285-290.

Hanušová, J. 1962. Ein Beitrag zum Studium des Schwefelkreislaufes während der Sommerstagnation in der Talsperre Sedlice. Sci. Pap. Inst.

Chem. Technol. Prague Technol. Water 6:177–191.

Happey-Wood, C. 1976. Vertical migration patterns in phytoplankton of mixed species composition. Br. Phycol. J. 11:355–369.

Harbison, G. R. and V. L. McAlister. 1980. Fact and artifact in copepod feeding experiments. Limnol. Oceanogr. 25:971–981.

Hardman, Y. 1941. The surface tension of Wisconsin lake waters. Trans. Wis. Acad. Sci. Arts Lett. 33:395–404.

Hardman, Y., and A. T. Henrici. 1939. Studies of freshwater bacteria. V. The distribution of Siderocapsa treubii in some lakes and streams. J. Bact. 37:97–104.

Hardy, J. T. 1973. Phytoneuston ecology of a temperate marine lagoon. Limnol. Oceanogr. 18:525–533.

Hardy, R. W. F., R. C. Burns, and R. D. Holsten. 1973. Applications of the acetylene-ethylene assay for measurement of nitrogen fixation. Soil Biol. Biochem. 5:47–81.

Hargrave, B. T. 1969. Epibenthic algal production and community respiration in the sediments of Marion Lake. J. Fish Res. Bd. Canada 26:2003–2026.

Hargrave, B. T. 1970a. The utilization of benthic microflora by Hyalella azteca (Amphipoda). J. Anim. Ecol. 39:427–437.

Hargrave, B. T. 1970b. Distribution, growth, and seasonal abundance of Hyalella azteca (Amphipoda) in relation to sediment microflora. J. Fish. Res. Bd. Canada 27:685–699.

Hargrave, B. T. 1970c. The effect of a deposit-feeding amphipod on the metabolism of benthic microflora. Limnol. Oceanogr. 15:21–30.

Hargrave, B. T. 1971. An energy budget for a deposit-feeding amphipod. Limnol. Oceanogr. 16:99–103.

Hargrave, B. T. 1972a. A comparison of sediment oxygen uptake, hypolimnetic oxygen deficit and primary production in Lake Esrom, Denmark. Verh. Int. Ver. Limnol. 18:134–139.

Hargrave, B. T. 1972b. Oxidation-reduction potentials, oxygen concentration and oxygen uptake of profundal sediments in a eutrophic lake. Oikos 23:167–177.

Hargrave, B. T. 1972c. Aerobic decomposition of sediment and detritus as a function of particle surface area and organic content. Limnol. Oceanogr. 17:583–596.

Hargrave, B. T. 1975. Stability in structure and function of the mud-water interface. Verh. Int. Ver. Limnol. 19:1073–1079.

Hargrave, B. T., and G. H. Geen. 1968. Phosphorus excretion by zooplankton. Limnol. Oceanogr. 13:332–342.

Harlin, M. M. 1973. Transfer of products between epiphytic marine algae and host plants. J. Phycol. 9:243–248.

Harman, W. N. 1974. Snails (Mollusca: Gastropoda). In C. W. Hart, Jr., and S. L. H. Fuller, eds. Pollution Ecology of Freshwater Invertebrates. New York, Academic Press, pp. 275–312.

Harper, R. M., J. C. Fry, and M. A. Learner. 1981. A bacteriological investigation to elucidate the feeding biology of Nais variabilis (Oligochaeta: Naididae). Freshwat. Biol. 11:227–236.

Harris, G. P., S. I. Heaney, and J. F. Talling. 1979. Physiological and environmental constraints in the ecology of the planktonic dinoflagellate Ceratium hirundinella. Freshwat. Biol. 9:413–428.

Harris, G. P. and J. N. A. Lott. 1973. Observations of Langmuir circulations in Lake Ontario. Limnol. Oceanogr. 18:584–589.

Harrison, F. W. 1974. Sponges (Porifera: Spongillidae). In C. W. Hart, Jr., and S. L. H. Fuller, eds. Pollution Ecology of Freshwater Invertebrates. New York, Academic Press, pp. 29–66.

Harrison, M. J., R. T. Wright, and R. Y. Morita. 1971. Method for measuring mineralization in lake sediments. Appl. Microbiol. 21:698–702.

Harriss, R. C. 1967. Silica and chloride in interstitial waters of river and lake sediments. Limnol. Oceanogr. 12:8–12.

Hart, R. C. 1977. Feeding rhythmicity in a migratory copepod (Pseudodiaptomus hessei (Mrazek)). Freshwat. Biol. 7:1–8.

Hart, R. C. and B. R. Allanson. 1976. The distribution and diel vertical migration of Pseudodiaptomus hessei (Mrázek) (Calanoida: Copepoda) in a subtropical lake in southern Africa. Freshwat. Biol. 6:183–198.

Harter, R. D. 1968. Adsorption of phosphorus by lake sediment. Proc. Soil Sci. Soc. Amer. 32:514–518.

Hartman, R. T., and D. L. Brown. 1967. Changes in internal atmosphere of submersed vascular hydrophytes in relation to photosynthesis. Ecology 48:252–258.

den Hartog, C., and S. Segal. 1964. A new classification of the water-plant communities. Acta Bot. Nerl. 13:367–393.

Haslam, S. M. 1971a. Community regulation in Phragmites communis Trin. I. Monodominant stands. J. Ecol. 59:65–73.

Haslam, S. M. 1971b. Community regulation in Phragmites communis Trin. II. Mixed stands. J. Ecol. 59:75–88.

Haslam, S. M. 1973. Some aspects of the life history and autecology of Phragmites communis Trin. A review. Pol. Arch. Hydrobiol. 20:79–100.

Hasler, A. D. 1947. Eutrophication of lakes by domestic drainage. Ecology 28:383–395.

Hasler, A. D., and W. G. Einsele. 1948. Fertilization for increasing productivity of natural inland waters. Trans. N. Amer. Wildl. Conf. 13:527–555.

Hasler, A. D., and E. Jones. 1949. Demonstration of the antagonistic action of large aquatic plants on algae and rotifers. Ecology 30:359–364.

Hatch, M. D., C. B. Osmond, and R. O. Slatyer, eds. 1971. Photosynthesis and Photorespiration. New York, John Wiley & Sons, Inc., 565 pp.

Haury, L. and D. Weihs. 1976. Energetically efficient swimming behavior of negatively buoyant zooplankton. Limnol. Oceanogr. 21:797–803.

Haworth, E. Y. 1972. The recent diatom history of Loch Leven, Kinross. Freshwat. Biol. 2:131–141.

Haworth, R. D. 1971. The chemical nature of humic acid. Soil Sci. 111:71–79.

Hayes, F. R. 1955. The effect of bacteria on the exchange of radiophosphorus at the mud-water interface. Verh. Int. Ver. Limnol. 12:111–116.

Hayes, F. R. 1957. On the variation in bottom fauna and fish yield in relation to trophic level and lake dimensions. J. Fish. Res. Bd. Canada 14:1–32.

Hayes, F. R. 1964. The mud-water interface. Oceanogr. Mar. Biol. Ann. Rev. 2:121–145.

Hayes, F. R., and E. H. Anthony. 1959. Lake water and sediment. VI. The standing crop of bacteria in lake sediments and its place in the classification of lakes. Limnol. Oceanogr. 4:299–315.

Hayes, F. R., and E. H. Anthony. 1964. Productive capacity of North American lakes as related to the quantity and the trophic level of fish, the lake dimensions, and the water chemistry. Trans. Amer. Fish. Soc. 93:53–57.

Hayes, F. R., and C. C. Coffin. 1951. Radioactive phosphorus and the exchange of lake nutrients. Endeavor 10:78–81.

Hayes, F. R., and M. A. MacAulay. 1959. Lake water and sediment. V. Oxygen consumed in water over sediment cores. Limnol. Oceanogr. 4:291–298.

Hayes, F. R., J. A. McCarter, M. L. Cameron, and D. A. Livingstone. 1952. On the kinetics of phosphorus exchange in lakes. J. Ecol. 40:202–216.

Hayes, F. R., and J. E. Phillips. 1958. Lake water and sediment. IV. Radiophosphorus equilibrium with mud, plants, and bacteria under oxidized and reduced conditions. Limnol. Oceanogr. 3:459–475.

Hayes, F. R., B. L. Reid, and M. L. Cameron. 1958. Lake water and sediment. II. Oxidation-reduction relations at the mud-water interface. Limnol. Oceanogr. 3:308–317.

Hayne, D. W., and R. C. Ball. 1956. Benthic productivity as influenced by fish predation. Limnol. Oceanogr. 1:162–175.

Hazelwood, D. H. 1966. Illumination and turbulence effects on relative growth in Daphnia. Limnol. Oceanogr. 11:212–216.

Healey, F. P. and L. L. Hendzel. 1975. Effect of phosphorus deficiency on two algae growing in chemostats. J. Phycol. 11:303–309.

Healy, J. B., Jr., L. Y. Young, and M. Reinhard. 1980. Methanogenic decomposition of ferulic acid, a model lignin derivative. Appl. Environ. Microbiol. 39:436–444.

Heaney, S. I. 1976. Temporal and spatial distribution of the dinoflagellate Ceratium hirundinella O. F. Müller within a small productive lake. Freshwat. Biol. 6:531–542.

Heaney, S. I. and J. F. Talling. 1980a. Dynamic aspects of dinoflagellate distribution patterns in a small productive lake. J. Ecol. 68:75–94.

Heaney, S. I. and J. F. Talling. 1980b. Ceratium hirundinella—Ecology of a complex, mobile, and successful plant. Ann. Rept. Freshwat. Biol. Assoc. U.K. 48:26–40.

Heath, R. T. and G. D. Cooke. 1975. The significance of alkaline phosphatase in a eutrophic lake. Verh. Int. Ver. Limnol. 19:959–965.

Hebert, P. D. N. 1978a. Cyclomorphosis in natural populations of Daphnia cephalata King. Freshwat. Biol. 8:79–90.

Herbert, P. D. N. 1978b. The adaptive significance of cyclomorphosis in Daphnia: More possibilities. Freshwat. Biol. 8:313–320.

Hecky, R. E., and P. Kilham. 1973. Diatoms in alkaline, saline lakes: Ecology and geochemical implications. Limnol. Oceanogr. 18:53–71.

Hecky, R. E. and H. J. Kling. 1981. The phytoplankton and protozooplankton of the euphotic zone of Lake Tanganyika: Species composition, biomass, chlorophyll content, and spatio-temporal distribution. Limnol. Oceanogr. 26:548–564.

Hegel, G. W. F. 1807. Phänomenologie des Geistes. System der Wissenschaft. Bamberg u. Würzburg, J. A. Goebhardt, p. 21.

Hegi, H. -R. 1976. Schwermetalle (Fe, Mn, Cd, Cr, Cu, Pb, Zn) im Pelagial des Bodensees (Obersee und Untersee) und des Greifensees. Schweiz. Z. Hydrol. 38:35–47.

Heisey, D. and K. G. Porter. 1977. The effect of ambient oxygen concentration on filtering and respiration rates of Daphnia galeata mendotae and Daphnia magna. Limnol. Oceanogr. 22:839–845.

Hejný, S. 1960. Ökologische Charakteristik der Wasser- und Sumpfpflanzen in den slowakischen Tiefebenen (Donau- und Theissgebiet). Bratislava, Verlag Slowakischen Akad. Wissenschaften, 487 pp.

Helder, R. J., H. B. A. Prins, and J. Schuurmans. 1974. Photorespiration in leaves of Valisneria spiralis. Proc. Koninklijke Nederl. Akad. Wetenschappen, Ser. C, Biol. Med. Sci. 77:338–344.

Helder, R. J. and P. E. Zanstra. 1977. Changes of the pH at the upper and lower surface of bicarbonate assimilating leaves of Potamogeton lucens L. Proc. Koninklijke Nederl. Akad. Wetenschappen, Ser. C, Biol. Med. Sci. 80:421–436.

Helder, R. J., J. Boerma, and P. E. Zanstra. 1980. Uptake of carbon dioxide and bicarbonate by leaves of Potamogeton lucens L. Proc. Koninklijke Nederl. Akad. Wetenschappen, Ser. C, Biol. Med. Sci. 83:151–166.

Hellebust, J. A. 1974. Extracellular products. In W.

D. P. Stewart, ed. Algal Physiology and Biochemistry. Berkeley, University of California Press, pp. 838–863.

Heller, M. D. 1977. The phased division of the freshwater dinoflagellate *Ceratium hirundinella* and its use as a method of assessing growth in natural populations. Freshwat. Biol. 7:527–533.

Hem, J. D. 1960a. Restraints on dissolved ferrous iron imposed by bicarbonate, redox potential, and pH. *In* Chemistry of Iron in Natural Water. U.S. Geol. Surv. Water-Supply Pap. *1459-B*:33–55.

Hem, J. D. 1960b. Complexes of ferrous iron with tannic acid. *In* Chemistry of Iron in Natural Water. U.S. Geol. Surv. Water-Supply Pap. *1459-D*:75–94.

Hem, J. D. 1960c. Some chemical relationships among sulfur species and dissolved ferrous iron. *In* Chemistry of Iron in Natural Water. U.S. Geol. Surv. Water-Supply Pap. *1459-C*:57–73.

Hem, J. D. 1963. Chemical equilibria and rates of manganese oxidation. *In* Chemistry of Manganese in Natural Water. U.S. Geol. Surv. Water-Supply Pap. *1667-A*, 64 pp.

Hem, J. D. 1964. Deposition and solution of manganese oxides. *In* Chemistry of Manganese in Natural Water. U.S. Geol. Surv. Water-Supply Pap *1667-B*, 42 pp.

Hem, J. D., and W. H. Cropper. 1959. Survey of ferrous-ferric chemical equilibria and redox potentials. *In* Chemistry of Iron in Natural Water. U.S. Geol. Surv. Water-Supply Pap. *1459-A*:1–30.

Hem, J. D., and M. W. Skougstad. 1960. Coprecipitation effects in solutions containing ferrous, ferric, and cupric ions. *In* Chemistry of Iron in Natural Water. U.S. Geol. Surv. Water-Supply Pap. *1459-E*:95–110.

Henrici, A. T., and E. McCoy. 1938. The distribution of heterotrophic bacteria in the bottom deposits of some lakes. Trans. Wis. Acad. Sci. Arts Lett. 31:323–361.

Hepher, B. 1966. Some aspects of the phosphorus cycle in fish ponds. Verh. Int. Ver. Limnol. 16:1293–1297.

Herbst, V. and J. Overbeck. 1978. Metabolic coupling between the alga *Oscillatoria redekei* and accompanying bacteria. Naturwissenschaften 65:598.

Herman, S. S. 1963. Vertical migration of the opossum shrimp, *Neomysis americana* Smith. Limnol. Oceanogr. 8:228–238.

Heron, J. 1961. The seasonal variation of phosphate, silicate, and nitrate in waters of the English Lake District. Limnol. Oceanogr. 6:338–346.

Herzig, A. 1980. Ten years quantitative data on a population of *Rhinoglena fertöensis* (Brachionidae, Monogononta). Hydrobiologia 73:161–167.

Hesslein, R., and P. Quay. 1973. Vertical eddy diffusion studies in the thermocline of a small stratified lake. J. Fish. Res. Bd. Canada *30*:1491–1500.

Heywood, J., and R. W. Edwards. 1962. Some aspects of the ecology of *Potamopyrgus jenkinsi* Smith. J. Anim. Ecol. 31:239–250.

Hickey, J. R., and others. 1980. Initial solar irradiance determinations from Nimbus 7 cavity radiometer measurements. Science *208*:281–283.

Hickman, M. 1971a. Standing crops and primary productivity of the epipelon of two small ponds in North Somerset, U.K. Oecologia 6:238–253.

Hickman, M. 1971b. The standing crop and primary productivity of the epiphyton attached to *Equisetum fluviatile* L. in Priddy Pool, North Somerset. Brit. Phycol. J. 6:51–59.

Hickman, M., and F. E. Round. 1970. Primary production and standing crops of epipsammic and epipelic algae. Brit. Phycol. J. 5:247–255.

Higashi, M., T. Miura, K. Tanimizu, and Y. Iwasa. 1981. Effect of the feeding activity of snails on the biomass and productivity of an algal community attached to a reed stem. Verh. Int. Ver. Limnol. 21:590–595.

Higgins, I. J., D. J. Best, R. C. Hammond, and D. Scott. 1981. Methane-oxidizing microorganisms. Microbiol. Rev. 45:556–590.

Hillbricht-Ilkowska, A., I. Spodniewska, T. Weglenska, and A. Karabin. 1970. The seasonal variation of some ecological efficiencies and production rates in the plankton community of several Polish lakes of different trophy. *In* Z. Kajak and A. Hillbricht-Ilkowska, eds. Productivity Problems of Freshwaters. Warsaw, PWN Polish Scientific Publishers, pp. 111–127.

Hillbricht-Ilkowska, A., A. Kowalczewski, and I. Spodniewska. 1972. Field experiment on the factors controlling primary production of the lake plankton and periphyton. Ekol. Polska 20:315–326.

Hillman, W. S. 1961. The Lemnaceae, or duckweeds. Bot. Rev. 27:221–287.

Hobbie, J. E. 1961. Summer temperatures in Lake Schrader, Alaska. Limnol. Oceanogr. 6:326–329.

Hobbie, J. E. 1964. Carbon-14 measurements of primary production in two arctic Alaskan lakes. Verh. Int. Ver. Limnol. 15:360–364.

Hobbie, J. E. 1967. Glucose and acetate in freshwater: Concentrations and turnover rates. *In* H. L. Golterman and R. S. Clymo, eds. Chemical Environment in the Aquatic Habitat. Amsterdam, N. V. Noord-Hollandsche Uitgevers Mattschappij, pp. 245–251.

Hobbie, J. E. 1971. Heterotrophic bacteria in aquatic ecosystems; Some results of studies with organic radioisotopes. *In* J. Cairns, Jr., ed. The Structure and Function of Fresh-Water Microbial Communities. Res. Div. Monogr. 3, Virginia Polytechnic Inst., pp. 181–194.

Hobbie, J. E. 1973. Arctic limnology: A review. *In* M. E. Britton, ed. Alaskan Arctic Tundra. Tech.

Paper No. 25, Arctic Inst. of North America, pp. 127-168.

Hobbie, J. E., Editor. 1980. Limnology of Tundra Ponds. Stroudsburg, PA, Dowden, Hutchinson, & Ross, Inc., 514 pp.

Hobbie, J. E., and C. C. Crawford. 1969. Bacterial uptake of organic substrate: New methods of study and application to eutrophication. Verh. Int. Ver. Limnol. 17:725-730.

Hobbie, J. E., C. C. Crawford, and K. L. Webb. 1968. Amino acid flux in an estuary. Science 159:1463-1464.

Hobbie, J. E., and R. T. Wright. 1965. Competition between planktonic bacteria and algae for organic solutes. Mem. Ist. Ital. Idrobiol. 18(Suppl.):175-185.

Hobbie, J. E., R. J. Barsdate, V. Alexander, D. W. Stanley, C. P. McRoy, R. G. Stross, D. A. Bierle, R. D. Dillon, and M. C. Miller. 1972. Carbon Flux Through a Tundra Pond Ecosystem at Barrow, Alaska. U.S. Tundra Biome Report 72-1, 28 pp.

Hobbie, J. E., R. J. Daley, and S. Jasper. 1977. Use of Nuclepore filters for counting bacteria by fluorescence microscopy. Appl. Environ. Microbiol. 33:1225-1228.

Hobbs, H. H., Jr., and E. T. Hall, Jr. 1974. Crayfishes (Decapoda: Astacidae). In C. W. Hart, Jr., and S. L. H. Fuller, eds. Pollution Ecology of Freshwater Invertebrates. New York, Academic Press, pp. 195-214.

Hochmüller, K. and H. Simoneth. 1980a. Zur Situation bei der Angabe von Ergebnissen im Bereich der Wasserchemie und -technologie. Teil I. Das Val wurde nicht ersatzlos gestrichen. Z. Wasser Abwasser Forsch. 13:153-158.

Hochmüller, K. and H. Simoneth. 1980b. Zur Situation bei der Angabe von Ergebnissen im Bereich der Wasserchemie und -technologie. Teil II. Gesetzliche Einheiten, Grössen, Terminologie. Z. Wasser Abwasser Forsch. 13:227-232.

Hodson, P. V., U. Borgmann, and H. Shear. 1979. Toxicity of copper to aquatic biota. In J. O. Nriagu, Editor. Copper in the Environment. II. Health Effects. New York, John Wiley & Sons, pp. 307-372.

Hofmann, W. 1978a. Bosmina (Eubosmina) populations of Grosser Plöner See and Schöhsee lakes during late-glacial and postglacial times. Polskie Arch. Hydrobiol. 25:167-176.

Hofmann, W. 1978b. Analysis of animal microfossils from the Grosser Segeberger See (F.R.G.). Arch. Hydrobiol. 82:316-346.

Hogetsu, K., Y. Okanishi, and H. Sugawara. 1960. Studies on the antagonistic relationship between phytoplankton and rooted aquatic plants. Japan. J. Limnol. 21:124-130.

Holden, A. V. 1961. The removal of dissolved phosphate from lake waters by bottom deposits. Verh. Int. Ver. Limnol. 14:247-251.

Holdren, G. C., Jr., D.E. Armstrong, and R. F. Harris. 1977. Interstitial inorganic phosphorus concentrations in lakes Mendota and Wingra. Water Res. 11:1041-1047.

Höll, K. 1972. Water: Examination, Assessment, Conditioning, Chemistry, Bacteriology, Biology. Berlin, Walder de Gruyter, 389 pp.

Hollibaugh, J. T. 1979. Metabolic adaptation in natural bacterial populations supplemented with selected amino acids. Estuarine Coastal Mar. Sci. 9:215-230.

Holm, N. P. and D. E. Armstrong. 1981. Role of nutrient limitation and competition in controlling the populations of Asterionella formosa and Microcystis aeruginosa in semicontinuous culture. Limnol. Oceanogr. 26:622-634.

Holme, N. A., and A. D. McIntyre, eds. 1971. Methods for the Study of Marine Benthos. Int. Biol. Program Handbook 16. Oxford, Blackwell Scientific Publications, 334 pp.

Holm-Hansen, O. 1968. Ecology, physiology and biochemistry of bluegreen algae. Ann. Rev. Microbiol. 22:47-70.

Holm-Hansen, O., and H. W. Paerl. 1972. The applicability of ATP determination for estimation of microbial biomass and metabolic activity. Mem. Ist. Ital. Idrobiol 29(suppl.):149-168.

Holst, R. W. and J. H. Yopp. 1979. Comparative utilization of inorganic and organic compounds as sole nitrogen sources by the submergent duckweek, Lemna trisulca L. Biol. Plantarum (Praha) 21:245-252.

Hood, D. W., ed. 1970. Organic matter in natural waters. Occas. Publ. Inst. Mar. Sci. Univ. Alaska 1, 625 pp.

Hooper, F. F., and A. M. Elliott. 1953. Release of inorganic phosphorus from extracts of lake mud by protozoa. Trans. Amer. Microsc. Soc. 72:276-281.

Hooper, N. M. and G. G. C. Robinson. 1976. Primary production of epiphytic algae in a marsh pond. Can. J. Bot. 54:2810-2815.

Hooper-Reid, N. M. and G. G. C. Robinson. 1978. Seasonal dynamics of epiphytic algal growth in a marsh pond: productivity, standing crop, and community composition. Can. J. Bot. 56:2434-2440.

Horie, S. 1962. Morphometric features and the classification of all the lakes in Japan. Mem. College Sci. Univ. Kyoto (ser. B) 29:191-262.

Horne, A. J. 1979. Nitrogen fixation in Clear Lake, California. 4. Diel studies on Aphanizomenon and Anabaena blooms. Limnol. Oceanogr. 24:329-341.

Horne, A. J. and J. W. Carmiggelt. 1975. Algal nitrogen fixation in Californian streams: Seasonal cycles. Freshwat. Biol. 5:461-470.

Horne, A. J., and G. E. Fogg. 1970. Nitrogen fixation in some English lakes. Proc. Roy. Soc. London (ser. B) 175:351-366.

Horne, A. J., and C. R. Goldman. 1972. Nitrogen fixation in Clear Lake, California. I. Seasonal variation and the role of heterocysts. Limnol. Oceanogr. 17:678–692.

Horne, A. J., J. E. Dillard, D. K. Fujita, and C. R. Goldman. 1972. Nitrogen fixation in Clear Lake, California. II. Synoptic studies on the autumn *Anabaena* bloom. Limnol. Oceanogr. 17:693–703.

Horne, A. J., J. C. Sandusky, and C. J. W. Carmiggelt. 1979. Nitrogen fixation in Clear Lake, California. 3. Repetitive synoptic sampling of the spring *Aphanizomenon* blooms. Limnol. Oceanogr. 24:316–328.

Horne, R. A., ed. 1972. Water and Aqueous Solutions. Structure, Thermodynamics, and Transport Processes. New York, Wiley-Interscience, 837 pp.

Horvath, R. S. 1972. Microbial co-metabolism and the degradation of organic compounds in nature. Bacteriol. Rev. 36:146–155.

Hough, J. L. 1958. Geology of the Great Lakes. Urbana, University of Illinois Press, 313 pp.

Hough, R. A. 1974. Photorespiration and productivity in submersed aquatic vascular plants. Limnol. Oceanogr. 19:912–927.

Hough, R. A. 1979. Photosynthesis, respiration, and organic carbon release in *Elodea canadensis* Michx. Aquatic Bot. 7:1–11.

Hough, R. A., and R. G. Wetzel. 1972. A ^{14}C-assay for photorespriation in aquatic plants. Plant Physiol. 49:987–990.

Hough, R. A., and R. G. Wetzel. 1975. The release of dissolved organic carbon from submersed aquatic macrophytes: Diel, seasonal, and community relationships. Verh. Int. Ver. Limnol. 19:939–948.

Hough, R. A. and R. G. Wetzel. 1977. Photosynthetic pathways of some aquatic plants. Aquatic Bot. 3:297–303.

Hough, R. A. and R. G. Wetzel. 1978. Photorespiration and CO_2 compensation point in *Najas flexilis*. Limnol. Oceanogr. 23:719–724.

Howard, H. H. and S. W. Chisholm. 1975. Seasonal variation of manganese in a eutrophic lake. Amer. Midland Nat. 93:188–197.

Howard-Williams, C. 1981. Studies on the ability of a *Potamogeton pectinatus* community to remove dissolved nitrogen and phosphorus compounds from lake water. J. Appl. Ecol. 18:619–637.

Howard-Williams, C. and B. R. Davies, 1978. The influence of periphyton on the surface structure of a *Potamogeton pectinatus* L. leaf (an hypothesis). Aquatic Bot. 5:87–91.

Howard-Williams, C. and W. Howard-Williams. 1978. Nutrient leaching from the swamp vegetation of Lake Chilwa, a shallow African lake. Aquatic Bot. 4:257–267.

Howard-Williams, C. and W. J. Junk. 1976. The decomposition of aquatic macrophytes in the floating meadows of a central Amazonian Várzea Lake. Biogeographica 7:115–123.

Howard-Williams, C. and G. M. Lenton. 1975. The role of the littoral zone in the functioning of a shallow tropical lake ecosystem. Freshwat. Biol. 5:445–459.

Hoyt, P. B. 1966. Chlorophyll-type compounds in soil. II. Their decomposition. Plant Soil 25:313–328.

Hrbáček, J. 1958. Typologie und Produktivität der teichartigen Gewässer. Verh. Int. Ver. Limnol. 13:394–399.

Hrbáček, J. 1962. Species composition and the amount of the zooplankton in relation to the fish stock. Rozpravy Českosl. Akad. Věd, Řada Matem. Přír. Věd 72:(10):1–114.

Hrbáček, J., M. Dvořakova, V. Kořínek, and L. Procházkóva. 1961. Demonstration of the effect of the fish stock on the species composition of zooplankton and the intensity of metabolism of the whole plankton association. Verh. Int. Ver. Limnol. 14:192–195.

Hrbáček, J., and M. Dvořáková-Novotná. 1965. Plankton of four backwaters related to their size and fish stock. Rozpravy Českosl. Akad. Věd, Řada Matem. Přír. Věd 75(13):1–65.

Hrbáček, J., and M. Straškraba. 1966. Horizontal and vertical distribution of temperature, oxygen, pH and water movements in Slapy Reservoir (1958–1960). Hydrobiol. Stud. 1:7–40.

Hubley. J. H., J. R. Mitton, and J. F. Wilkinson. 1974. The oxidation of carbon monoxide by methane-oxidizing bacteria. Arch. Microbiol. 95:365–368.

Hudec, P. P., and P. Sonnenfeld. 1974. Hot brines on Los Roques, Venezuela. Science 185:440–442.

Hulbert, M. H. and S. Krawiec. 1977. Cometabolism: A critique. J. Theor Biol. 69:287–291.

Hunding, C. 1971. Production of benthic microalgae in the littoral zone of a eutrophic lake. Oikos 22:389–397.

Hunding, C., and B. T. Hargrave, 1973. A comparison of benthic microalgal production measured by C^{14} and oxygen methods. J. Fish. Res. Bd. Canada 30:309–312.

Hunter, R. D. 1980. Effects of grazing on the quantity and quality of freshwater Aufwuchs. Hydrobiologia 69:251–259.

Huntsinger, K. R. G. and P. E. Maslin. 1976. Contribution of phytoplankton, periphyton, and macrophytes to primary production in Eagle Lake, California. Calif. Fish. Game 62:187–194.

Hutchinson, G. E. 1937. A contribution to the limnology of arid regions primarily founded on observations made in the Lahontan Basin. Trans. Conn. Acad. Arts Sci 33:47–132.

Hutchinson, G. E. 1938. Chemical stratification and lake morphology. Proc. Nat. Acad. Sci. 24:63–69.

Hutchinson, G. E. 1941. Limnological studies in Connecticut. IV. The mechanisms of intermediary metabolism in stratified lakes. Ecol. Monogr. 11:21–60.

Hutchinson, G. E. 1944a. Limnological studies in Connecticut: VII. A critical examination of the supposed relationship between phytoplankton periodicity and chemical changes in lake water. Ecology 25:3–26.

Hutchinson, G. E. 1944b. Nitrogen in the biogeochemistry of the atmosphere. Amer. Scientist. 32:178–195.

Hutchinson, G. E. 1950. The biogeochemistry of vertebrate excretion. Bull. Amer. Mus. Nat. Hist. 96, 554 pp.

Hutchinson, G. E. 1957a. A Treatise on Limnology. I. Geography, Physics, and Chemistry. New York, John Wiley & Sons, Inc., 1015 pp.

Hutchinson, G. E. 1957b. Concluding remarks. Cold Spring Harbor Symp. Quant. Biol. 22:415–427.

Hutchinson, G. E. 1961. The paradox of the plankton. Amer. Nat. 95:137–146.

Hutchinson, G. E. 1964. The lacustrine microcosm reconsidered. Amer. Sci. 52:334–341.

Hutchinson, G. E. 1967. A Treatise on Limnology. II. Introduction to Lake Biology and the Limnoplankton. New York, John Wiley & Sons, Inc., 1115 pp.

Hutchinson, G. E. 1973. Eutrophication. The scientific background of a contemporary practical problem. Amer. Sci. 61:269–279.

Hutchinson, G. E. 1974. De rebus planktonicis. Limnol. Oceanogr. 19:360–361.

Hutchinson, G. E. 1975. A treatise on limnology. Vol. III. Limnological botany. New York, John Wiley & Sons, 660 pp.

Hutchinson, G. E. 1978. An Introduction to Population Ecology. New Haven, Yale Univ. Press, 260 pp.

Hutchinson, G. E. 1981. Thoughts on aquatic insects. BioScience 31:495–500.

Hutchinson, G. E., and V. T. Bowen. 1947. A direct demonstration of the phosphorus cycle in a small lake. Proc. Nat. Acad. Sci. 33:148–153.

Hutchinson, G. E., and V. T. Bowen. 1950. Limnological studies in Connecticut. IX. A quantitative radiochemical study of the phosphorus cycle in Linsley Pond. Ecology 31:194–293.

Hutchinson, G. E., and U. M. Cowgill. 1973. The waters of Merom: A study of Lake Heleh. III. The major chemical constituents of a 54 m. core. Arch. Hydrobiol. 72:145–185.

Hutchinson, G. E., E. S. Deevey, Jr., and A. Wollack. 1939. The oxidation-reduction potentials of lake waters and their ecological significance. Proc. Nat. Acad. Sci. 25:87–90.

Hutchinson, G. E., and H. Löffler. 1956. The thermal classification of lakes. Proc. Nat. Acad. Sci. 42:84–86.

Hutchinson, G. E., et al. 1970. Ianula: An account of the history and development of the Lago di Monterosi, Latium, Italy. Trans. Amer. Phil. Soc. N. S., 60(4), 178 pp.

Hunter, S. H., and L. Provasoli. 1964. Nutrition of algae. Ann. Rev. Plant Physiol. 15:37–56.

Hyman, L. H. 1951. The Invertebrates: Acanthocephala, Aschelminthes, and Entoprocta. The Pseudocoelomate Bilateria, Vol. III. New York, McGraw-Hill Book Co., 572 pp.

Hynes, H. B. N. 1960. The Biology of Polluted Waters. Liverpool, Liverpool University Press, 202 pp.

Hynes, H. B. N. 1963. Imported organic matter and secondary productivity in streams. Proc. 16th Int. Congr. Zool. 4:324–329.

Hynes, H. B. N. 1970. The Ecology of Running Waters. Toronto, University of Toronto Press, 555 pp.

Hynes, H. B. N. and M. J. Coleman. 1968. A simple method of assessing the annual production of stream benthos. Limnol. Oceanogr. 13:569–573.

Hynes, H. B. N., and B. J. Greib. 1970. Movement of phosphate and other ions from and through lake muds. J. Fish. Res. Bd. Canada 27:653–668.

Idso, S. B. 1973. On the concept of lake stability. Limnol. Oceanogr. 18:681–683.

Idso, S. B. and R. G. Gilbert. 1974. On the universality of the Poole and Atkins Secchi disk-light extinction equation. J. Appl. Ecol. 11:399–401.

Ilmavirta, V. 1975. Diel periodicity in the phytoplankton community of the oligotrophic lake Pääjärvi, southern Finland. II. Late summer phytoplankton biomass. Ann. Bot. Fennici 12:37–44.

Ilmavirta, V., R. I. Jones, and T. Kairesalo. 1977. The structure and photosynthetic activity of pelagial and littoral plankton communities in Lake Pääjärvi, southern Finland. Ann. Bot. Fennici 14:7–16.

Imboden, D. M. and S. Emerson. 1978. Natural radon and phosphorus as limnologic tracers: Horizontal and vertical eddy diffusion in Greifensee. Limnol. Oceanogr. 23:77–90.

Incoll, L. D., S. P. Long, and M. R. Ashmore. 1977. SI units in publications in plant science. Current Adv. Plant Sci. 9:331–343.

Infante, A. 1973. Untersuchungen über die Ausnutzbarkeit verschiedener Algen durch das Zooplankton. Arch. Hydrobiol. (suppl.) 42:340–405.

Ingram, L. O., J. A. Calder, C. Van Baalen, F. E. Plucker, and P. L. Parker. 1973a. Role of reduced exogenous organic compounds in the physiology of the blue-green bacteria (algae): Photoheterotrophic growth of a "heterotrophic" blue-green bacterium. J. Bacteriol. 114:695–700.

Ingram, L. O., C. Van Baalen, and J. A. Calder. 1973b. Role of reduced exogenous organic compounds in the physiology of the blue-green bacteria (algae): Photoheterotrophic growth of an "autotrophic" blue-green bacterium. J. Bacteriol. 114:701–705.

Inniss, W. E. and C. I. Mayfield. 1978a. Growth rates of psychrotrophic sediment microorganisms. Water Res. 12:231–236.

Inniss, W. E. and C. I. Mayfield. 1978b. Psychrotrophic bacteria in sediments from the Great Lakes. Water Res. 12:237–241.

Inniss, W. E. and C. I. Mayfield. 1979. Seasonal variation of psychrotrophic bacteria in sediment from Lake Ontario. Water Res. *13*:481-484.

International Association of Limnology. 1959. Symposium on the classification of brackish waters. Arch. Oceanogr. Limnol. *11*(suppl.):1-248.

Isirimah, N. O., D. R. Keeney, and E. H. Dettmann. 1976. Nitrogen cycling in Lake Wingra. J. Environ. Qual. *5*:182-188.

Islam, A., R. Mandal, and K. T. Osman. 1979. Direct absorption of organic phosphate by rice and jute plants. Plant Soil *53*:49-54.

Iversen, H. W. 1952. Laboratory study of breakers. *In* Gravity Waves. Circ. U.S. Bur. Standards *521*:9-32.

Iversen, T. M. 1973. Decomposition of autumn-shed beech leaves in a springbrook and its significance for the fauna. Arch. Hydrobiol. *72*:305-312.

Iyengar, V. K. S., D. M. Davies, and H. Kleerekoper. 1963. Some relationships between Chironomidae and their substrate in nine freshwater lakes of southern Ontario, Canada. Arch. Hydrobiol. *59*:289-310.

Jackson, T. A., G. Kipphut, R. H. Hesslein, and D. W. Schindler. 1980. Experimental study of trace metal chemistry in soft-water lakes at different pH levels. Can. J. Fish. Aquat. Sci. *37*:387-402.

Jackson, W. A., and R. J. Volk. 1970. Photorespiration. Ann. Rev. Plant Physiol. *21*:385-432.

Jacobs, J. 1961. Cyclomorphosis in *Daphnia galeata mendotae*, a case of environmentally controlled allometry. Arch. Hydrobiol. *58*:7-71.

Jacobs, J. 1962. Light and turbulence as co-determinants of relative growth rates in cyclomorphic *Daphnia*. Int. Rev. ges. Hydrobiol. *47*:146-156.

Jacobs, J. 1966. Predation and rate of evolution in cyclomorphic *Daphnia*. Verh. Int. Ver. Limnol. *16*:1645-1652.

Jacobs, J. 1967. Untersuchungen zur Funktion und Evolution der Zyklomorphose bei *Daphnia*, mit besonderer Berücksichtigung der Selektion durch Fische. Arch. Hydrobiol. *62*:467-541.

Jacobs, J. 1980. Environmental control of cladoceran cyclomorphosis via target-specific growth factors in the animal. *In* W. C. Kerfoot, ed. Evolution and Ecology of Zooplankton Communities. Hanover, NH, Univ. Press New England, pp. 429-437.

Jacobsen, T. R. and G. W. Comita. 1976. Ammonia-nitrogen excretion in *Daphnia pulex*. Hydrobiologia *51*:195-200.

James, H. R., with E. A. Birge. 1938. A laboratory study of the absorption of light by lake waters. Trans. Wis. Acad. Sci. Arts Lett. *31*:1-154.

Jannasch, H. W. 1969. Current concepts in aquatic microbiology. Verh. Int. Ver. Limnol. *17*:25-39.

Jannasch, H. W. 1974. Steady state and the chemostat in ecology. Limnol. Oceanogr. *19*:716-720.

Jannasch, H. W. 1975. Methane oxidation in Lake Kivu (central Africa). Limnol. Oceanogr. *20*:860-864.

Jannasch, H. W., and G. E. Jones. 1959. Bacterial populations in sea water as determined by different methods of enumeration. Limnol. Oceanogr. *4*:128-139.

Janssen, J. 1980. Alewives (*Alosa pseudoharengus*) and ciscoes (*Coregonus artedii*) as selective and non-selective planktivores. *In* W. C. Kerfott, ed. Evolution and Ecology of Zooplankton Communities. Hanover, NH, Univ. Press New England, pp. 580-586.

Jansson, M. 1976. Phosphatases in lakewater: Characterization of enzymes from phytoplankton and zooplankton by gel filtration. Science *194*:320-321.

Jansson, M. 1980. Role of benthic algae in transport of nitrogen from sediment to lake water in a shallow clearwater lake. Arch. Hydrobiol. *89*:101-109.

Jansson, M., H. Olsson, and O. Broberg. 1981. Characterization of acid phosphatases in the acidified Lake Gårdsjön, Sweden. Arch. Hydrobiol. *92*:377-395.

Järnefelt, H. 1925. Zur Limnologie einiger Gewässer Finlands. Ann. Soc. Zool.-Bot. Fennicae Vanamo *2*:185-352.

Jarvis, N. L. 1967. Adsorption of surface-active material at the sea-air interface. Limnol. Oceanogr. *12*:213-221.

Jarvis, N. L., W. D. Garrett, M. A. Scheiman, and C. O. Timmons. 1967. Surface chemical characterization of surface-active material in seawater. Limnol. Oceanogr. *12*:88-96.

Jassby, A. D. 1975. The ecological significance of sinking to planktonic bacteria. Can. J. Microbiol. *21*:270-274.

Jassby, A. D., and C. R. Goldman. 1974. A quantitative measure of succession rate and its application to the phytoplankton of lakes. Amer. Nat. *108*:688-693.

Jassby, A. and T. Powell. 1975. Vertical patterns of eddy diffusion during stratification in Castle Lake, California. Limnol. Oceanogr. *20*:530-543.

Jensen, M. L., and N. Nakai. 1961. Sources and isotopic composition of atmospheric sulfur. Science *134*:2102-2104.

Jeschke, W. D., and W. Simonis. 1965. Über die Aufnahme von Phosphat- und Sulfationen durch Blätter von *Elodea densa* und ihre Beeinflussung durch Licht, Temperatur und Aussenkonzentration. Planta *67*:6-32.

Jewell, M. E. 1935. An ecological study of the freshwater sponges of northern Wisconsin. Ecol. Monogr. *5*:461-504.

Jewell, M. E. 1939. An ecological study of the freshwater sponges of Wisconsin. II. The influence of calcium. Ecology *20*:11-28.

Jewell, W. J., and P. L. McCarty. 1971. Aerobic decomposition of algae. Environ. Sci. Technol. *5*:1023-1031.

Jewson, D. H., B. H. Rippey, and W. K. Gilmore. 1981. Loss rates from sedimentation, parasitism, and grazing during the growth, nutrient limitation, and dormancy of a diatom crop. Limnol. Oceanogr. 26:1045–1056.

Johannes, R. E. 1964a. Uptake and release of phosphorus by a benthic marine amphipod. Limnol. Oceanogr. 9:235–242.

Johannes, R. E. 1964b. Uptake and release of dissolved organic phosphorus by representatives of a coastal marine ecosystem. Limnol. Oceanogr. 9:224–234.

Johannes, R. E. 1964c. Phosphorus excretion and body size in marine animals: Microzooplankton and nutrient regeneration. Science 146:923–924.

Johnson, L. 1964. Temperature regime of deep lakes. Science 144:1336–1337.

Johnson, L. 1966. Temperature of maximum density of fresh water and its effect on circulation in Great Bear Lake. J. Fish Res. Bd. Canada 23:963–973.

Johnson, M. G., and R. O. Brinkhurst. 1971. Production of benthic macroinvertebrates of Bay of Quinte and Lake Ontario. J. Fish. Res. Bd. Canada 28:1699–1714.

Johnson, M. P. 1967. Temperature dependent leaf morphogenesis in Ranunculus flabellaris. Nature 214:1354–1355.

Johnson, N. M., R. C. Reynolds, and G. E. Likens. 1972. Atmospheric sulfur: Its effect on the chemical weathering of New England. Science 177:514–516.

Johnson, N. M., J. S. Eaton, and J. E. Richey. 1978. Analysis of five North American lake ecosystems. II. Thermal energy and mechanical stability. Verh. Int. Ver. Limnol. 20:562–567.

Johnson, N. M. and D. H. Merritt. 1979. Convective and advective circulation of Lake Powell, Utah-Arizona during 1972–1975. Water Resour. Res. 15:873–884.

Jónasson, P. M. 1969. Bottom fauna and eutrophication. In Eutrophication: Causes, Consequences, Correctives. Washington, D.C., National Academy of Sciences, pp. 274–305.

Jónasson, P. M. 1972. Ecology and production of the profundal benthos in relation to phytoplankton in Lake Esrom. Oikos (suppl.) 14, 148 pp.

Jónasson, P. M. 1978. Zoobenthos of lakes. Verh. Int. Ver. Limnol. 20:13–37.

Jónasson, P. M. and C. Lindegaard, 1979. Zoobenthos and its contribution to the metabolism of shallow lakes. Arch. Hydrobiol. Beih. Ergebn. Limnol. 13:162–180.

Jónasson, P. M., and H. Mathiesen. 1959. Measurements of primary production in two Danish eutrophic lakes. Esrom Sø and Furesø. Oikos 10:137–167.

Jónasson, P. M., and F. Thorhauge. 1972. Life cycle of Potamothrix hammoniensis (Tubificidae) in the profundal of a eutrophic lake. Oikos 23:151–158.

Jones, B. F. and C. J. Bowser, 1978. The mineralogy and related chemistry of lake sediments. In A.

Lerman, ed. Lakes: Chemistry, Geology, Physics. New York, Springer-Verlag, pp. 179–235.

Jones, J. G. 1972. Studies on freshwater bacteria: Association with algae and alkaline phosphatase activity. J. Ecol. 60:59–75.

Jones, J. G. 1977. The effect of environmental factors on estimated viable and total populations of planktonic bacteria in lakes and experimental enclosures. Freshwat. Biol. 7:67–91.

Jones, J. G. 1979a. Microbial nitrate reduction in freshwater sediments. J. Gen. Microbiol. 115:27–35.

Jones, J. G. 1978. The distribution of some freshwater planktonic bacteria in two stratified eutrophic lakes. Freshwat. Biol. 8:127–140.

Jones, J. G. 1979b. Microbial activity in lake sediments with particular reference to electrode potential gradients. J. Gen. Microbiol. 115:19–26.

Jones, J. G. 1980. Some differences in the microbiology of profundal and littoral lake sediments. J. Gen. Microbiol. 117:285–292.

Jones, J. G., M. J. L. G. Orlandi, and B. M. Simon. 1979. A microbiological study of sediments from the Cumbrian lakes. J. Gen. Microbiol. 115:37–48.

Jones, J. G. and B. M. Simon. 1980. Variability in microbiological data from a stratified eutrophic lake. J. Appl. Bacteriol. 49:127–135.

Jones, J. G., B. M. Simon, and S. Gardener. 1982. Factors affecting methanogenesis and associated anaerobic processes in the sediments of a stratified eutrophic lake. J. Gen. Microbiol. 128:1–11.

Jones, J. R. and R. W. Bachmann. 1976. Prediction of phosphorus and chlorophyll levels in lakes. J. Water Poll. Control Fed. 48:2176–2182.

Jones, M. B. and T. R. Milburn. 1978. Photosynthesis in papyrus (Cyperus papyrus L.). Photosynthetica 12:197–199.

Jones, P. D. and W. T. Momot. 1981. Crayfish productivity, allochthony, and basin morphometry. Can. J. Fish. Aquat. Sci. 38:175–183.

Jones, R. I. and V. Ilmavirta. 1978. A diurnal study of the phytoplankton in the eutrophic Lake Lovojärvi, southern Finland. Arch. Hydrobiol. 83:494–514.

Jones, S. W., and R. Goulder. 1973. Swimming speeds of some ciliated Protozoa from a eutrophic pond. Naturalist (Hull, England) (No. 924): 33–35.

Jordan, M., and G. E. Likens. 1975. An organic carbon budget for an oligotrophic lake in New Hampshire, U.S.A. Verh. Int. Ver. Limnol. 19:994–1003.

Jordan, M. J. and G. E. Likens. 1980. Measurement of planktonic bacterial production in an oligotrophic lake. Limnol. Oceanogr. 25:719–732.

Jørgensen, E. G. 1957. Diatom periodicity and silicon assimilation. Dansk Bot. Arkiv. 18(1), 54 pp.

Jørgensen, E. G. 1964. Adaptation to different light intensities in the diatom Cyclotella meneghiniana Kütz. Physiol. Plant. 17:136–145.

Jørgensen, E. G. 1968. The adaptation of plankton algae. II. Aspects of the temperature adaptation of *Skeletonema costatum*. Physiol. Plant. 21:423–427.

Jørgensen, E. G. 1969. The adaptation of plankton algae. IV. Light adaptation in different algal species. Physiol. Plant. 22:1307–1315.

Joshi, M. M. and J. P. Hollis. 1977. Interaction of *Beggiatoa* and rice plant: Detoxification of hydrogen sulfide in the rice rhizosphere. Science 195:179–180.

Juday, C. 1921. Quantitative studies of the bottom fauna in the deeper waters of Lake Mendota. Trans. Wis. Acad. Sci. Arts Lett. 20:461–493.

Juday, C. 1924. The productivity of Green Lake, Wisconsin. Verh. Int. Ver. Limnol. 2:357–360.

Juday, C. 1940. The annual energy budget of an inland lake. Ecology 21:438–450.

Juday, C. 1942. The summer standing crop of plants and animals in four Wisconsin lakes. Trans. Wis. Acad. Sci. Arts Lett. 34:103–135.

Juday, C., and E. A. Birge. 1931. A second report on the phosphorus content of Wisconsin lake waters. Trans. Wis. Acad. Sci. Arts Lett. 26:353–382.

Juday, C., and E. A. Birge. 1933. The transparency, the color and the specific conductance of the lake waters of northeastern Wisconsin. Trans. Wis. Acad. Sci. Arts Lett. 28:205–259.

Juday, C., E. A. Birge, G. I. Kemmerer, and R. J. Robinson. 1927. Phosphorus content of lake waters of northeastern Wisconsin. Trans. Wis. Acad. Sci. Arts Lett. 23:233–248.

Juday, C., E. A. Birge, and V. W. Meloche. 1938. Mineral content of the lake waters of northeastern Wisconsin. Trans. Wis. Acad. Sci. Arts Lett. 31:223–276.

Judd, J. H. 1970. Lake stratification caused by runoff from street deicing. Water Res. 4:521–532.

Jupp, B. P. and D. H. N. Spence. 1977. Limitations on macrophytes in a eutrophic lake, Loch Leven. I. Effects of phytoplankton. J. Ecol. 65:175–186.

Juse, A. 1966. Diatomeen in Seesedimenten. Arch. Hydrobiol. Beih. Ergebn. Limnol. 4, 32 pp.

Jüttner, F. 1981. Biologically active compounds released during algal blooms. Verh. Int. Ver. Limnol. 21:227–230.

Jüttner, F. and R. Friz. 1974. Excretion products of *Ochromonas* with special reference to pyrrolidone carboxylic acid. Arch. Microbiol. 96:223–232.

Kadono, Y. 1980. Photosynthetic carbon sources in some *Potamogeton* species. Bot. Mag. Tokyo 93:185–193.

Kairesalo, T. 1976. Measurements of production of epilithiphyton and littoral plankton in Lake Pääjärvi, southern Finland. Ann. Bot. Fennici 13:114–118.

Kairesalo, T. 1977. On the production ecology of epipelic algae and littoral plankton communities in Lake Pääjärvi, southern Finland. Ann. Bot. Fennici 14:82–88.

Kairesalo, T. 1980a. Diurnal fluctuations within a littoral plankton community in oligotrophic Lake Pääjärvi, southern Finland. Freshwat. Biol. 10:533–537.

Kairesalo, T. 1980b. Comparison of in situ photosynthetic activity of epiphytic, epipelic and planktonic algal communities in an oligotrophic lake, southern Finland. J. Phycol. 16:57–62.

Kajak, Z. 1970a. Some remarks on the necessities and prospects of the studies on biological production of freshwater ecosystems. Pol. Arch. Hydrobiol. 17:43–54.

Kajak, Z. 1970b. Analysis of the influence of fish on benthos by the method of enclosures. In Z. Kajak and A. Hillbricht-Ilkowska, eds. Productivity Problems of Freshwaters. Warsaw, PWN Polish Scientific Publishers, pp. 781–793.

Kajak, Z. and K. Dusoge, 1970. Production efficiency of *Procladius choreus* MG (Chironomidae, Diptera) and its dependence on the trophic conditions. Pol. Arch. Hydrobiol. 17:217–224.

Kajak, Z., A. Hillbricht-Ilkowska, and E. Pieczyńska. 1970. The production processes in several Polish lakes. In Z. Kajak and A. Hillbricht-Ilkowska, eds. Productivity Problems of Freshwaters. Warsaw, PWN Polish Scientific Publishers, pp. 129–147.

Kajak, Z., and B. Ranke-Rybicka. 1970. Feeding and production efficiency of *Chaoborus flavicans* Meigen (Diptera, Culicidae) larvae in eutrophic and dystrophic lake. Pol. Arch. Hydrobiol. 17:225–232.

Kajak, Z. and J. I. Rybak. 1966. Production and some trophic dependences in benthos against primary production and zooplankton production of several Masurian lakes. Verh. Int. Ver. Limnol. 16:441–451.

Kajosaari, E. 1966. Estimation of the detention period of a lake. Verh. Int. Ver. Limnol. 16:139–143.

Kalff, J., and H. E. Welch. 1974. Phytoplankton production in Char Lake, a natural polar lake, and in Meretta Lake, a polluted polar lake, Cornwallis Island, Northwest Territories. J. Fish. Res. Bd. Canada 31:621–636.

Kalinin, G. P., and V. D. Bykov. 1969. The world's water resources, present and future. Impact of Science on Technology 19:135–150.

Kalk, M., A. J. McLachlan, and C.. Howard-Williams. 1979. Lake Chilwa: Studies of change in a tropical ecosystem. The Hague, Junk BV Publishers, 462 pp.

Kamiyama, K., S. Okuda, and A. Kawai. 1977. Studies on the release of ammonium nitrogen from the bottom sediments in freshwater regions. II. Ammonium nitrogen in dissolved and absorbed form in the sediments. Japan. J. Limnol. 38:100–106.

Kana, T. M. and J. D. Tjepkema. 1978. Nitrogen fixation associated with *Scripus atrovirens* and other nonnodulated plants in Massachusetts. Can. J. Bot. 56:2636–2640.

Kansanen, A. and R. Niemi. 1974. On the production ecology of isoetids, especially *Isoëtes lacustris* and *Lobelia dortmanna*, in Lake Pääjärvi, southern Finland, Ann. Bot. Fennici 11:178–187.

Karcher, F. H. 1939. Untersuchungen über den Stickstoff-haushalt in ostpreussischen Waldseen. Arch. Hydrobiol. 35:177–266.

Kashiwada, K., A. Kanazawa, and S. Tachibanazono. 1963. Studies on organic compounds in natural water. II. On the seasonal variations in the content of nicotinic acid, pantothenic acid, biotin, folic acid and vitamin B_{12} in the water of the Lake Kasumigaura. (In Japanese.) Mem. Fac. Fish., Kagoshima Univ. 12:153–157.

Katayama, T. 1961. Studies on the intercellular spaces in rice. Crop Sci. Soc. Japan Proc. 29:229–233.

Kato, K. and M. Sakamoto. 1979. Vertical distribution of carbohydrate utilizing bacteria in Lake Kizaki. Japan. J. Limnol. 40:211–214.

Kato, K. and M. Sakamoto. 1981. Vertical distribution of free-living and attached heterotrophic bacteria in Lake Kizaki. Japan. J. Limnol. 42:154–159.

Kaushik, N. K., and H. B. N. Hynes. 1971. The fate of the dead leaves that fall into streams. Arch. Hydrobiol. 68:465–515.

Keating, K. I. 1977. Allelopathic influence on blue-green bloom sequence in a eutrophic lake. Science 196:885–887.

Keefe, C. W. 1972. Marsh production: A summary of the literature. Contr. Mar. Sci. Univ. Texas 16:163–181.

Keeley, J. E. 1981. *Isoetes howellii*: A submerged aquatic CAM plant? Amer. J. Bot. 68:420–424.

Keeley, J. E. 1982. Distribution of diurnal acid metabolism in the genus *Isoetes*. Amer. J. Bot. (In press).

Keen, R. 1973. A probabilistic approach to the dynamics of natural populations of the Chydoridae (Cladocera, Crustacea). Ecology 54:524–534.

Keen, R. And R. Nassar. 1981. Confidence intervals for birth and death rates estimated with the egg-ratio technique for natural populations of zooplankton. Limnol. Oceanogr. 26:131–142.

Keeney, D. R. 1972. The fate of nitrogen in aquatic ecosystems. Literature Rev., 3, Water Resources Center, University of Wisconsin, 59 pp.

Keeney, D. R. 1973. The nitrogen cycle in sediment-water systems. J. Environ. Quality 2:15–29.

Keeney, D. R., R. L. Chen, and D. A. Graetz. 1971. Importance of denitrification and nitrate reduction in sediments to the nitrogen budgets of lakes. Nature 233:66–67.

Keeney, D. R., J. G. Konrad, and G. Chesters. 1970. Nitrogen distribution in some Wisconsin lake sediments. J. Water Poll. Control Fed. 42:411–417.

Kellogg, W. W., R. D. Cadle, E. R. Allen, A. L. Lazrus, and E. A. Martel. 1972. The sulfur cycle. Science 175:587–596.

Kelts, K. and K. J. Hsü. 1978. Freshwater carbonate sedimentation. *In:* A. Lerman, Ed. Lakes: Chemistry, Geology, Physics. New York, Springer Verlag, pp. 295–323.

Kemp, A. L. W. and L. M. Johnston. 1979. Diagenesis of organic matter in the sediments of lakes Ontario, Erie, and Huron. J. Great Lakes Res. 5:1–10.

Kerfoot, W. C. 1978. Combat between predatory copepods and their prey: *Cyclops*, *Epischura*, and *Bosmina*. Limnol. Oceanogr. 23:1089–1102.

Kerfoot, W. C., ed. 1980a. Evolution and Ecology of Zooplankton Communities. Hanover, NH, Univ. Press New England, 793 pp.

Kerfoot, W. C. 1980b. Perspectives on cyclomorphosis: Separation of phenotypes and genotypes. *In* W. C. Kerfoot, ed. Evolution and Ecology of Zooplankton Communities. Hanover, NH, Univ. Press New England, pp. 470–496.

Kerfoot, W. C. and R. A. Pastorok. 1978. Survival versus competition: Evolutionary compromises and diversity in the zooplankton. Verh. Int. Ver. Limnol. 20:362–374.

Kern, D. M. 1960. The hydration of carbon dioxide. J. Chem. Education 37:14–23.

Kerr, P. C., D. L. Brockway, D. F. Paris, and J. T. Barnett, Jr. 1972. The interrelation of carbon and phosphorus in regulating heterotrophic and autotrophic populations in an aquatic ecosystem, Shriner's Pond. *In* G. E. Likens, ed. Nutrients and Eutrophication: The Limiting-Nutrient Controversy. Special Symposium, Amer. Soc. Limnol. Oceanogr. 1:41–62.

Kersting, K., and W. Holterman. 1973. The feeding behaviour of *Daphnia magna*, studied with the Coulter Counter. Verh. Int. Verl. Limnol. 18:1434–1440.

van Kessel, J. F. 1978. The realtion between redox potential and denitrification in a water-sediment system. Water Res. 12:285–290.

Keup, L. E. 1968. Phosphorus in flowing waters. Water Res. 2:373–386.

Khailov, K. M. 1971. Ekologicheskii metabolizm v more. Kiev, Izdatel 'stvo Naukova Dumka 252 pp.

Kibby, H. V. 1971. Energetics and population dynamics of *Diaptomus gracilis*. Ecol. Monogr. 41:311–327.

Kibby, H. V., and F. H. Rigler. 1973. Filtering rates of *Limnocalanus*. Verh. Int. Ver. Limnol. 18:1457–1461.

Kilham, P. 1971. A hypothesis concerning silica and the freshwater planktonic diatoms. Limnol. Oceanogr. 16:10–18.

Kilham, P. 1975. Mechanisms controlling world water chemistry based on data for African lakes and rivers. Proc. Int. Symp. Geochemistry of Natural Waters. Burlington, Ontario. 5 pp.

Kilham, P. 1981. Pelagic bacteria: Extreme abundances in African saline lakes. Naturwissenschaften 67:380–381.

Kilham, P. and D. Tilman. 1979. The importance of

resource competition and nutrient gradients for phytoplankton ecology. Arch. Hydrobiol. Beih. Ergebn. Limnol. *13*:100–119.

Kilham, P. and D. Titman. 1976. Some biological effects of atmospheric inputs to lakes: Nutrient ratios and competitive interactions between phytoplankton. J. Great Lakes Res. Suppl. 2:187–191.

Kilham, S. S. 1975. Kinetics of silicon-limited growth in the freshwater diatom *Asterionella formosa*. J. Phycol. *11*:396–399.

Kilham, S. S. 1978. Nutrient kinetics of freshwater planktonic algae using batch and semicontinuous methods. Mitt. Int. Ver. Limnol. *21*:147–157.

Kilham, S. S. and P. Kilham. 1978. Natural community bioassays: Predictions of results based on nutrient physiology and competition. Verh. Int. Ver. Limnol. *20*:68–74.

Kimball, K. D. 1973. Seasonal fluctuations of ionic copper in Knights Pond, Massachusetts. Limnol. Oceanogr. *18*:169–172.

King, C. E. 1967. Food, age, and the dynamics of a laboratory population of rotifers. Ecology *48*:111–128.

King, C. R. 1979. Secondary productivity of the North Pine Dam. Tech. Pap. Australian Water Resour. Council No. 39. 75 pp.

King, D. L., and R. C. Ball. 1966. A qualitative and quantitative measure of *Aufwuchs* production. Trans. Amer. Microsc. Soc. *85*:232–240.

King, G. M. and M. J. Klug. 1980. Sulfhydrolase activity in sediments of Wintergreen Lake, Kalamazoo County, Michigan. Appl. Environ. Microbiol. *39*:950–956.

King. G. M. and M. J. Klug. 1982a. Comparative aspects of sulfur mineralization in sediments of a eutrophic lake basin. Appl. Environ. Microbiol. *43*:1406–1412.

King, G. M. and M. H. Klug. 1982b. Glucose metabolism in sediments of a eutrophic lakes: A tracer analysis of uptake and product formation. Appl. Environ. Microbiol. (In press)

King, G. M. and W. J. Wiebe. 1978. Methane release from soils of a Georgia salt marsh. Geochim. Cosmochim. Acta *42*:343–348.

Kingsland, S. 1982. The refractory model: The logistic curve and the history of population ecology. Quart. Rev. Biol. *57*:29–52.

Kirchner, W. B. and P. J. Dillon. 1975. An empirical method of estimating the retention of phosphorus in lakes. Water Resour. Res. *11*:182–183.

Kirby, M. J., Editor. 1978. Hillslope Hydrology. New York, John Wiley & Sons, 389 pp.

Kistritz, R. U. 1978. Recycling of nutrients in an enclosed aquatic community of decomposing macrophytes (*Myriophyllum spicatum*). Oikos *30*:561–569.

Kitchell, J. F., R. A. Stein, and B. Knežević. 1978. Utilization of filamentous algae by fishes in Skadar Lake, Yugoslavia. Verh. Int. Ver. Limnol. *20*:2159–2165.

Kjensmo, J. 1962. Some extreme features of the iron metabolism in lakes. Schweiz. Z. Hydrol. *24*:244–252.

Kjensmo, J. 1967. The development and some main features of "iron-meromictic" soft water lakes. Arch. Hydrobiol. (suppl.) *32*:137–312.

Kjensmo, J. 1968. Iron as the primary factor rendering lakes meromictic, and related problems. Mitt. Int. Ver. Limnol. *14*:83–93.

Kjensmo, J. 1970. The redox potentials in small oligo and meromictic lakes. Nordic Hydrol. *1*:56–65.

Klein, R. M. 1978. Plants and near-ultraviolet radiation. Bot. Rev. *44*:1–127.

Klekowski, R. Z. 1970. Bioenergetic budgets and their application for estimation of production efficiency. Pol. Arch. Hydrobiol. *17*:55–80.

Klekowski, R. Z., E. Fischer, Z. Fischer, M. B. Ivanova, T. Prus, E. A. Shushkina, T. Stachurska, Z. Stepien, and H. Zyromska-Rudzka. 1970. Energy budgets and energy transformation efficiencies of several animal species of different feeding types. *In* Z. Kajak and A. Hillbricht-Ilkowska, eds. Productivity Problems of Freshwaters. Warsaw, PWN Polish Scientific Publishers, pp. 749–763.

Klekowski, R. Z., and E. A. Shushkina. 1966. Ernährung, Atmung, Wachstum und Energie-Umformung in *Macrocyclops albidus* (Jurine). Verh. Int. Ver. Limnol. *16*:399–418.

Klemer, A. R. 1978. Nitrogen limitation of growth and gas vacuolation in *Oscillatoria rubescens*. Verh. Int. Ver. Limnol. *20*:2293–2297.

Klopatek, J. M. 1975. The role of emergent macrophytes in mineral cycling in a freshwater marsh. *In* F. G. Howell, J. B. Gentry, and M. H. Smith, eds. Mineral Cycling in Southeastern Ecosystems. Washington, DC, U.S. Energy Res. Development Admin., pp. 367–393.

Klötzli, F. 1971. Biogenous influence on aquatic macrophytes, especially *Phragmites communis*. Hidrobiologia *12*:107–111.

Klug, M. J., G. M. King, R. L. Smith, D. R. Lovley, and J. W. H. Dacey. 1982. Comparative aspects of anaerobic carbon and electron flow in freshwater sediments. (Manuscript)

Knight A., R. C. Ball, and F. F. Hooper. 1962. Some estimates of primary production rates in Michigan ponds. Pap. Mich. Acad. Sci. Arts Lett. *47*:219–233.

Knoechel, R. and J. Kalff. 1975. Algal sedimentation: The cause of a diatom-blue-green succession. Verh. Int. Ver. Limnol. *19*:745–754.

Kobayasi, H. 1961. Productivity in sessile algal community of Japanese mountain river. Bot. Mag. Tokyo *74*:331–341.

Koehl, M. A. R. and J. R. Strickler. 1981. Copepod feeding currents: Food capture at low Reynolds number. Limnol. Oceanogr. *26*:1062–1073.

Koenings, J. P. 1976. In situ experiments on the dissolved and colloidal state of iron in an acid bog lake. Limnol. Oceanogr. *21*:674–683.

Koenings, J. P. and F. F. Hooper. 1976. The influence

of colloidal organic matter on iron and iron-phosphorus cycling in an acid bog lake. Limnol. Oceanogr. 21:684–696.

Kogan, Sh. I., and G. A. Chinnova. 1972. Relations between *Ceratophyllum demersum* (L.) and some blue-green algae. Hydrobiol. J. (USSR; Translation Ser.), 8:14–25.

Koivo, L. K. and J. C. Ritchie. 1978. Modern diatom assemblages from lake sediments in the boreal-arctic transition region near the Mackenzie Delta, N.W.T., Canada. Can. J. Bot. 56:1010–1020.

Kondrat 'eva, E. N. 1965. Photosynthetic Bacteria. Moscow, Izdatel 'stvo Akademii Nauk SSSR 1963. (Translated into English, Israel Program for Scientific Translations, Jerusalem). 243 pp.

Kononova, M. M. 1966. Soil Organic Matter. Its Nature, its Role in Soil Formation and in Soil Fertility. 2nd ed. Oxford, Pergamon Press, 544 pp.

Konopka, A. 1981. Influence of temperature, oxygen, and pH on a metalimnetic population of *Oscillatoria rubescens*. Appl. Environ. Microbiol. 42:102–108.

Konrad, J. G., D. R. Keeney, G. Chesters, and K.-L. Chen. 1970. Nitrogen and carbon distribution in sediment cores of selected Wisconsin lakes. J. Water Poll. Control Fed. 42:2094–2101.

Korde, N. W. 1966. Algenreste in Seesedimenten. Zur Entwicklungsgeschichte der Seen und umliegenden Landschaften. Arch. Hydrobiol. Beih. Ergebn. Limnol. 3, 38 pp.

Koreliakova, I. L. 1958. Nekotorye nabluideniia nad raspodom perezimovavshei pribrezhno-vodnoi rastitel'nosti Rybinskogo Vodokhranilishcha. Bull. Inst. Biol. Vodokhranilishch 1:22–25.

Koreliakova, I. L. 1959. O raslade skoshennoi pribrezhno-vodnoi rastitel'nosti. Bull. Inst. Biol. Vodochranilishch 3:13–16.

Kořínková, J. 1967. Relations between predation pressure of carp, submerged plant development and littoral bottom-fauna of Pond Smyslov. Rozpravy Československ. Akad. Věd, Rada Matem. Přírod. Věd 77(11):35–62.

Kormondy, E. J. 1968. Weight loss of cellulose and aquatic macrophytes in a Carolina bay. Limnol. Oceanogr. 13:522–526.

Koshinsky, G. D. 1970. The morphometry of shield lakes in Saskatchewan. Limnol. Oceanogr. 15:695–701.

Kostalos, M. and R. L. Seymour. 1976. Role of microbial enriched detritus in the nutrition of *Gammarus minus* (Amphipoda). Oikos 27:512–516.

Kowalczewski, A. 1965. Changes in periphyton biomass of Mikolajskie Lake. Bull. Acad. Polon. Sci. (Cl. II) 13:395–398.

Kowalczewski, A. 1975a. Algal primary production in the zone of submerged vegetation of a eutrophic lake. Verh. Int. Ver. Limnol. 19:1305–1308.

Kowalczewski, A. 1975b. Periphyton primary pro-duction in the zone of submerged vegetation of Mikolajskie Lake. Ekol. Polska 23:509–543.

Koyama, T. 1955. Gaseous metabolism in lake muds and paddy soils. J. Earth Sci. Nagoya Univ. 3:65–76.

Koyama, T. 1963. Gaseous metabolism in lake sediments and paddy soils and the production of atmospheric methane and hydrogen. J. Geophys. Res. 68:3971–3973.

Koyama, T. 1964. Gaseous metabolism in lake sediments and paddy soils. *In* U. Colombo and G. D. Hobson, eds. Advances in Organic Geochemistry. New York, Macmillan Co., pp. 363–375.

Koyama, T., M. Nakaido, T. Tomino, and H. Hayakawa. 1973. Decomposition of organic matter in lake sediments. *In* E. Ingerson, ed. Proceedings of Symposium on Hydrogeochemistry and Biogeochemistry. Washington, DC, Clarke Company, pp. 512–535.

Kozhov, M. 1963. Lake Baikal and Its Life. The Hague, W. Junk, Publishers, 344 pp.

Kozlovsky, D. G. 1968. A critical evaluation of the trophic level concept. I. Ecological efficiencies. Ecology 49:48–60.

Kózmiński, Z., and J. Wisniewski. 1935. Über die Vorfrühlingthermik der Wigry-Seen. Arch. Hydrobiol. 28:198–235.

Krambeck, C. 1978. Changes in planktonic microbial populations—an analysis by scanning electron microscopy. Verh. Int. Ver. Limnol. 20:2255–2259.

Kramer, J. R. 1978. Acid precipitation. *In* J. O. Nriagu, ed. Sulfur in the Environment. I. The Atmospheric Cycle. New York, John Wiley & Sons, pp. 325–370.

Krasheninnikova, S. A. 1958. Mikrobiologicheskie protsessy raslada vodnoi rastitel'nosti v litorali Rybinskogo Vodokhanilishcha. Bull. Inst. Biol. Vodokhranilishch 2:3–6.

Kratz, W. A., and J. Myers. 1955. Nutrition and growth of several blue-green algae. Amer. J. Bot 42:282–287.

Krause, H. R. 1962. Investigation of the Decomposition of Organic Matter in Natural Waters. FAO Fish. Biol. Report 34(FB/R34), 19 pp.

Krause, H. R. 1964. Zur Chemie und Biochemie der Zersetzung von Süsswasserorganismen, unter besonderer Berücksichtigung des Abbaues der organischen Phosphorkomponenten. Verh. Int. Ver. Limnol. 15:549–561.

Krause, H. R., L. Möchel, and M. Stegmann. 1961. Organische Säuren als gelöste Intermediärprodukte des postmortalen Abbaues von Süsswasser-Zooplankton. Naturwissenschaften 48:434–435.

Krauskopf, K. B. 1956. Dissolution and precipitation of silca at low temperatures. Geochim. Cosmochim. Acta 10:1–26.

Krauss, R. W. 1958. Physiology of the fresh-water algae. Ann. Rev. Plant Physiol. 9:207–244.

Krauss, R. W. 1962. Inhibitors. *In* R. A. Lewin, ed.

Physiology and Biochemistry of Algae. New York, Academic Press, pp. 673–685.

Kretsinger, R. H. 1979. The informational role of calcium in cytosol. In P. Greengard and G. A. Robison, eds. Advances in Cyclic Nucleotide Research. Vol. 11. New York, Raven Press, pp. 2–26.

Kring, R. L. and W. J. O'Brien, 1976. Effect of varying oxygen concentrations on the filtering rate of Daphnia pulex. Ecology 57:808–814.

Krishnaswami, S. and D. Lal. 1978. Radionuclide limnochronolgy. In A. Lerman, ed. Lakes: Chemistry, Geology, Physics. New York, Springer Verlag, pp. 153–177.

Kriss, A. E., and R. Tomson. 1973. Origin of the warm water near the bottom of Lake Vanda in the Antarctic (25.5–27°):Microbiological data. Mikrobiologiya 42:942–943.

Krogh, A. 1939. Osmotic Regulation in Aquatic Animals. Cambridge, Cambridge University Press, 242 pp.

Krogh, A., and E. Lange. 1932. Quantitative Untersuchungen über Plankton, Kolloide und gelöste organische und anorganische Substanzen in dem Füresee. Int. Rev. ges. Hydrobiol. 26:20–53.

Królikowska, J. 1978. The transpriation of helophytes. Ekol. Polska 26:193–212.

Kuenzler, E. J. 1970. Dissolved organic phosphorus excretion by marine phytoplankton. J. Phycol. 6:7–13.

Kuhl, A. 1962. Inorganic phosphorus uptake and metabolism. In R. A. Lewin, ed. Physiology and Biochemistry of Algae. New York, Academic Press, pp. 211–229.

Kuhl, A. 1974. Phosphorus. In W. D. P. Stewart, ed. Algal Physiology and Biochemistry. Berkeley, Univ. California Press, pp. 636–654.

Kurata, A., C. Saraceni, and H. Kadota. 1979a. The status of B group vitamins in macrophyte and pelagic zones of Lake Biwa. Mem. Ist. Ital. Idrobiol. 37:63–85.

Kurata, A., C. Saraceni, and H. Kadota. 1979b. Diurnal changes of concentration in water of B group vitamins in macrophyte and pelagic zones of Lake Biwa. Mem. Ist. Ital. Idrobiol. 37:87–103.

Kurata, A., C. Saraceni, D. Ruggiu, M. Nakanishi, U. Melchiorri-Santolini, and H. Kadota. 1976. Relationship between B group vitamins and primary production and phytoplankton population in Lake Mergozzo (Northern Italy). Mem. Ist. Ital. Idrobiol. 33:257–284.

Kuznetsov, S. I. 1935. Microbiological researches in the study of the oxygenous regimen of lakes. Verh. Int. Ver. Limnol. 7:562–582.

Kuznetsov, S. I. 1959. Die Rolle der Mikroorganismen im Stoffkreislauf der Seen. Berlin, VEB Deutsch. Verlag Wissenschaften, 301 pp.

Kuznetsov, S. I. 1964. Biogeochemistry of sulphur. In Lo zolfo in agricoltura. Simposio Int. Agrochimica (Palermo, Italy) 5:312–330.

Kuznetsov, S. I. 1968. Recent studies on the role of

microorganisms in the cycling of substances in lakes. Limnol. Oceanogr. 13:211–224.

Kuznetsov, S. I. 1970. Mikroflora ozer i ee geokhimicheskaya deyatel'nost'. (Microflora of Lakes and Their Geochemical Activities.) (In Russian.) Leningrad, Izdatel'stvo Nauka, 440 pp.

Kuznetsov, S. I. and G. S. Karzinkin. 1931. Direct method for the quantitative study of bacteria in water and some considerations on causes which produce a zone of oxygen-minimum in Lake Glubokoje. Zbl. Bakt., Ser. II 83:169–174.

Kuznetsov, S. I., and E. M. Khartulari. 1941. Mikrobiologicheskaia kharakteristika protsessov anaerobnogo raspada organicheskogo veshchestva ila Belogo Ozera v Kosine. (Microbiological characteristics of the process of anaerobic decomposition of organic substances of sediments of Beloye Lake in Kosine.) Mikrobiologiya 10:834–849.

Kuznetsov, S. I., and V. I. Romanenko. 1963. Mikrobiologicheskoe izuchenie viutrennikh yodoemov. Laboratornoe rukobodstvo. Moscow, Izdatel'stvo Akademii Nauk SSSR, 129 pp.

Kuznetsov, S. I., G. A. Dubinina, and N. A. Lapteva. 1979. Biology of oligotrophic bacteria. Ann. Rev. Microbiol. 33:377–387.

Květ, J. 1971. Growth analysis approach to the production ecology of reedswamp plant communities. Hidrobiologia. 12:15–40.

Květ, J., J. Svoboda, and K. Fiala. 1969. Canopy development in stands of Typha latifolia L. and Phragmites communis Trin. in South Moravia. Hidrobiologia 10:63–75.

LaLonde, R. T., C. D. Morris, C. F. Wong, L. C. Gardner, D. J. Eckert, D. R. King, and R. H. Zimmerman. 1979. Response of Aedes triseriatus larvae to fatty acids of Cladophora. J. Chem. Ecol. 5:371–381.

Lam, C. W. Y., W. F. Vincent, and W. B. Silvester. 1979. Nitrogenase activity and estimates of nitrogen fixation by freshwater benthic blue-green algae. N. Z. J. Mar. Freshwat. Res. 13:187–192.

Laminger, H. 1973. Untersuchungen über Abundanz und Biomasse der sedimentbewohnenden Testaceen (Protozoa, Rhizopoda) in einem Hochgebirgssee (Vorderer Finstertaler See, Kühtai, Tirol). Int. Rev. ges. Hydrobiol. 58:543–568.

Lampert, W. 1977a. Studies on the carbon balance of Daphnia pulex de Geer as related to environmental conditions. II. The dependence of carbon assimilation on animal size, temperature, food concentration and diet species. Arch. Hydrobiol. (suppl.) 48:310–335.

Lampert, W. 1977b. Studies on the carbon balance of Daphnia pulex de Geer as related to environmental conditions. III. Production and production efficiency. Arch. Hydrobiol. (suppl.)48:336–360.

Lampert, W. 1978a. Climatic conditions and planktonic interactions as factors controlling the regular succession of spring algal bloom and extremely clear water in Lake Constance. Verh. Int. Ver. Limnol. 20:969–974.

Lampert, W. 1978b. Release of dissolved organic carbon by grazing zooplankton. Limnol. Oceanogr. 23:831–834.

Landers, D. H. 1982. Effects of naturally senescing aquatic macrophytes on nutrient chemistry and chlorophyll *a* of surrounding waters. Limnol. Oceanogr. 27:428–439.

Landner, L., and T. Larsson. 1973. Indications of disturbances in the nitrification process in a heavily nitrogen-polluted water body. Ambio 2:154–157.

Lane, L. S. 1977. Microbial community fluctuations in a meromictic, Antarctic lake. Hydrobiologia 55:187–190.

Lang, C. and B. Lang-Dobler. 1980. Structure of tubificid and lumbriculid worm communities, and three indices of trophy based upon these communities, as descriptors of eutrophication level of Lake Geneva (Switzerland). *In* R. O. Brinkhurst and D. G. Cook, eds. Aquatic Oligochaete Biology. New York, Plenum Press, pp. 457–470.

Langeland, A. 1981. Decreased zooplankton density in two Norwegian lakes caused by predation of recently introduced *Mysis relicta*. Verh. Int. Ver. Limnol. 21:926–937.

Langford, R. R., and E. G. Jermolajev. 1966. Direct effect of wind on plankton distribution. Verh. Int. Ver. Limnol. 16:188–193.

Langmuir, I. 1938. Surface motion of water induced by wind. Science 87:119–123.

LaRow, E. J. 1968. A persistent diurnal rhythm in *Chaoborus* larvae. I. The nature of the rhythmicity. Limnol. Oceanogr. 13:250–256.

LaRow, E. J. 1969. A persistent diurnal rhythm in *Chaoborus* larvae. II. Ecological significance. Limnol. Oceanogr. 14:213–218.

LaRow, E. J. 1970. The effect of oxygen tension on the vertical migration of *Chaoborus* larvae. Limnol. Oceanogr. 15:357–362.

LaRow, E. J. 1975. Secondary productivity of *Leptodora kindtii* in Lake George, N. Y. Am. Midl. Nat. 94:120–126.

Larsen, D. P., K. W. Malueg, D. W. Schults, and R. M. Brice. 1975. Response of eutrophic Shagawa Lake, Minnesota, U.S.A., to point-source phosphorus reduction. Verh. Int. Ver. Limnol. 19:884–892.

Larsen, D. P. and H. T. Mercier. 1976. Phosphorus retention capacity of lakes. J. Fish. Res. Bd. Can. 33:1742–1750.

Larson, R. A. and J. M. Hufnal, Jr. 1980. Oxidative polymerization of dissolved phenols by soluble and insoluble inorganic species. Limnol. Oceanogr. 25:505–512.

Larsson, U. and A. Hagström. 1979. Phytoplankton exudate release as an energy source for the growth of pelagic bacteria. Mar. Biol. 52:199–206.

Lasenby, D. C. 1975. Development of oxygen deficits in 14 southern Ontario lakes. Limnol. Oceanogr. 20:993–999.

Lasenby, D. C., and R. R. Langford. 1972. Growth, life history, and respiration of *Mysis relicta* in an arctic and temperate lake. J. Fish. Res. Bd. Canada 29:1701–1708.

Lasenby, D. C., and R. R. Langford. 1973. Feeding and assimilation of *Mysis relicta*. Limnol. Oceanogr. 18:280–285.

Lastein, E. 1976. Recent sedimentation and resuspension of organic matter in eutrophic Lake Esrom, Denmark. Oikos 27:44–49.

Latimer, J. R. 1972. Radiation measurement. Tech. Manual Ser. Int. Field Year Great Lakes 2, 53 pp.

Laube, H. R., and J. R. Wohler. 1973. Studies on the decomposition of a duckweed (Lemnaceae) community. Bull. Torr. Bot. Club. 100:238–240.

Laurent, M., and J. Badia. 1973. Étude comparative du cycle biologique de l'azote dans deux étangs. Ann. Hydrobiol. 4:77–102.

Lauwers, A. M., and W. Heinen. 1974. Bio-degradation and utilization of silica and quartz. Arch. Microbiol. 95:67–78.

Lawacz, W. 1969. The characteristics of sinking materials and the formation of bottom deposits in a eutrophic lake. Mitt. Int. Ver. Limnol. 17:319–331.

Leach, J. H. 1975. Seston composition in the Point Pelee Area of Lake Erie. (Unpublished manuscript)

Leah, R. T., B. Moss, and D. E. Forrest. 1980. The role of predation in causing major changes in the limnology of a hyper-eutrophic lake. Int. Rev. ges. Hydrobiol. 65:223–247.

Lean, D. R. S. 1973a. Phosphorus dynamics in lake water. Science 179:678–680.

Lean, D. R. S. 1973b. Movements of phosphorus between its biologically important forms in lake water. J. Fish. Res. Bd. Canada 30:1525–1536.

Lean, D. R. S. and C. Nalewajko. 1976. Phosphate exchange and organic phosphorus excretion by freshwater algae. J. Fish. Res. Bd. Can. 33:1312–1323.

Lean, D. R. S. and C. Nalewajko, 1979. Phosphorus turnover time and phosphorus demand in large and small lakes. Arch. Hydrobiol. Beih. Ergebn. Limnol. 13:120–132.

Lean, D. R. S., C. F.-H. Liao, T. P. Murphy, and D. S. Painter. 1978. The importance of nitrogen fixation in lakes. *In* Environmental Role of Nitrogen-Fixing Blue-Green Algae and Asymbiotic Bacteria. Ecol. Bull. (Stockholm) 26:41–51.

Lean, D. R. S., and F. H. Rigler. 1974. A test of the hypothesis that abiotic phosphate complexing influences phosphorus kinetics in epilimnetic lake water. Limnol. Oceanogr. 19:784–788.

Learner, M. A., and D. W. B. Potter. 1974. Life-history and production of the leech *Helobdella stagnalis* (L.) (Hirudinea) in a shallow eutrophic reservoir in South Wales. J. Anim. Ecol., 43:199–208.

Learner, M. A., G. Lochhead, and B. D. Hughes. 1978. A review of the biology of British Naididae (Oligochaeta) with emphasis on the lotic environment. Freshwat. Biol. 8:357–375.

Leckie, J. O. and J. A. Davis III. 1979. Aqueous environmental chemistry of copper. In J. O. Nriagu, ed. Copper in the Environment. I. Ecological Cycling. New York, John Wiley & Sons, pp. 89–121.

LeCren, E. D. 1958. Observations on the growth of perch (*Perca fluviatalis* L.) over twenty-two years with special reference to the effects of temperature and changes in population density. J. Anim. Ecol. 27:287–334.

Leduc, L. G. and G. D. Ferroni. 1979. Quantitative ecology of psychrophilic bacteria in an aquatic environment and characterization of heterotrophic bacteria from permanently cold sediments. Can. J. Microbiol. 25:1433–1442.

Lee, G. F., E. Bentley, and R. Amundson. 1975. Effects of marshes on water quality. In A. D. Hasler, ed. Coupling of Land and Water Systems. New York, Springer-Verlag, pp. 105–127.

Lefèvre. M. 1964. Extracellular products of algae. In D. F. Jackson, ed. Algae and Man. New York, Plenum Press, pp. 337–367.

Lehman, J. T. 1975. Reconstructing the rate of accumulation of lake sediment: The effect of sediment focusing. Quat. Res. 5:541–550.

Lehman, J. T. 1976. Ecological and nutritional studies on *Dinobryon* Ehrenb.: Seasonal periodicity and the phosphate toxicity problem. Limnol. Oceanogr. 21:646–658.

Lehman, J. T. 1977. On calculating drag characteristics for decelerating zooplankton. Limnol. Oceanogr. 22:170–172.

Lehman, J. T. 1979. Physical and chemical factors affecting the seasonal abundance of *Asterionella formosa* Hass. in a small temperate lake. Arch. Hydrobiol. 87:274–303.

Lehman, J. T. 1980a. Release and cycling of nutrients between planktonic algae and herbivores. Limnol. Oceanogr. 25:620–632.

Lehman, J. T. 1980b. Nutrient recycling as an interface between algae and grazers in freshwater communities. In W. C. Kerfoot, ed. Evolution and Ecology of Zooplankton Communities. Hanover, NH, Univ. Press New England, pp. 251–263.

Lehman, J. T., and C. D. Sandgren. 1978. Documenting a seasonal change from phosphorus to nitrogen limitation in a small temperate lake, and its impact on the population dynamics of *Asterionella*. Verh. Int. Ver. Limnol. 20:375–380.

Lehman, J. T., and D. Scavia. 1982. Microscale patchiness of nutrients in plankton communities. Science 216:729–730.

Lehn, H. 1965. Zur Durchsichtigkeitsmessung im Bodensee. Schrift. Ver. Geschichte Bodensees Umgebung 83:32–44.

Lehn, H. 1968. Litorale Aufwuchsalgen im Pelagial des Bodensee. Beitr. Naturk. Forsch. Südw.-Dlt. 27:97–100.

Lei, C.-H., and K. B. Armitage. 1980. Ecological energetics of a *Daphnia ambigua* population. Hydrobiologia 70:133–143.

Leibovich, S. 1977. Convective instability of stably stratified water in the ocean. J. Fluid Mech. 82:561–581.

Leighly, J. 1942. Effects of the Great Lakes on the annual march of air temperature in their vicinity. Pap. Mich. Acad. Sci. Arts Lett. 21:377–414.

Lellák, J. 1961. Zur Benthosproduktion und ihrer Dynamik in drei böhmischen Teichen. Verh. Int. Ver. Limnol. 14:213–219.

Lellák, J. 1965. The food supply as a factor regulating the population dynamics of bottom animals Mitt. Int. Ver. Limnol., 13:128–138.

Lenhard, G., W. R. Ross, and A. du Plooy. 1962. A study of methods for the classification of bottom deposits of natural waters. Hydrobiologia 20:223–240.

Leopold, L. B., M. G. Wolman, and J. P. Miller. 1964. Fluvial Processes in Geomorphology. San Francisco, W. H. Freeman and Co., 522 pp.

Lerman, A., and M. Stiller, 1969. Vertical eddy diffusion in Lake Tiberias. Verh. Int. Ver. Limnol. 17:323–333.

Leventhal, E. A. 1970. The Chrysomonadina. In G. E. Hutchinson, ed. Ianula: An Account of the History and Development of the Lago di Monterosi, Latium, Italy. Trans. Amer. Philos. Soc. 60(pt. 4):123–142.

Levine, S. 1975. Orthophosphate concentration and flux within the epilimnia of two Canadian Shield lakes. Verh. Int. Ver. Limnol. 19:624–629.

Lewin, J. C. 1962. Silicification. In R. A. Lewin, ed. Physiology and Biochemistry of Algae. New York, Academic Press, pp. 445–455.

Lewis, W. M., Jr. 1973. The thermal regime of Lake Lanao (Philippines) and its theoretical implications for tropical lakes. Limnol. Oceanogr. 18:200–217.

Lewis, W. M., Jr. 1974. Primary production in the plankton community of a tropical lake. Ecol. Monogr. 44:377–409.

Lewis, W. M., Jr. 1975. A theoretical comparison of the attenuation of light energy and quanta in waters of divergent optical properties. Arch. Hydrobiol. 75:285–296.

Lewis, W. M., Jr. 1977a. Ecological significance of the shapes of abundance-frequency distributions for coexisting phytoplankton species. Ecology 58:850–859.

Lewis, W. M., Jr. 1977b. Feeding selectivity of a tropical *Chaoborus* population. Freshwat. Biol. 7:311–325.

Lewis, W. M., Jr. 1978a. A compositional, phytogeographical, and elementary community structural analysis of the phytoplankton in a tropical lake. J. Ecol. *66*:213–226.

Lewis, W. M., Jr. 1978b. Dynamics and succession of the phytoplankton in a tropical lake: Lake Lanao, Philippines. J. Ecol. *66*:849–880.

Li, W. C., D. E. Armstrong, J. D. H. Williams, R. F. Harris, and J. K. Syers. 1972. Rate and extent of inorganic phosphate exchange in lake sediments. Proc. Soil Sci. Soc. Amer. *36*:279–285.

Liao, C. F.-H. and D. R. S. Lean. 1978. Nitrogen transformations within the trophogenic zone of lakes. J. Fish. Res. Bd. Can. *35*:1102–1108.

Likens, G. E., ed. 1972. Nutrients and Eutrophication: The Limiting-Nutrient Controversy. Special Symposium, Amer. Soc. Limnol. Oceanogr. *1*, 328 pp.

Likens, G. E. 1975. Primary production of inland aquatic ecosystems. *In* H. Lieth and R. W. Whittaker, eds. The Primary Productivity of the Biosphere. New York, Springer-Verlag, pp. 185–202.

Likens, G. E., and F. H. Bormann. 1972. Nutrient cycling in ecosystems. *In* J. A. Weins, ed. Ecosystem Structure and Function. Corvallis, Oregon State University Press, pp. 25–67.

Likens, G. E., and M. B. Davis. 1975. Post-glacial history of Mirror Lake and its watershed in New Hampshire, U.S.A.: An initial report. Verh. Int. Ver. Limnol. *19*:982–993.

Likens, G. E., F. H. Bormann, and N. M. Johnson. 1972. Acid rain. Environment *14*:33–40.

Likens, G. E., F. H. Bormann, N. M. Johnson, and R. S. Pierce. 1967. The calcium, magnesium, potassium, and sodium budgets for a small forested ecosystem. Ecology *48*:772–785.

Likens, G. E., F. H. Bormann, N. M. Johnson, D. W. Fisher, and R. S. Pierce. 1970. Effects of forest cutting and herbicide treatment on nutrient budgets in the Hubbard Brook watershed-ecosystem. Ecol. Monogr. *40*:23–47.

Likens, G. E., F. H. Bormann, R. S. Pierce, J. S. Eaton, and N. M. Johnson. 1977. Biogeochemistry of a forested ecosystem. New York, Springer-Verlag, 146 pp.

Likens, G. E., and A. D. Hasler. 1960. Movement of radiosodium in a chemically stratified lake. Science *131*:1676–1677.

Likens, G. E., and A. D. Hasler. 1962. Movements of radiosodium (Na²⁴) within an ice-covered lake. Limnol. Oceanogr. *7*:48–56.

Likens, G. E., and N. M. Johnson. 1969. Measurement and analysis of the annual heat budget for the sediments in two Wisconsin lakes. Limnol. Oceanogr. *14*:115–135.

Likens, G. E., and P. L. Johnson, 1966. A chemically stratified lake in Alaska. Science *153*:875–877.

Likens, G. E., and R. A. Ragotzkie. 1965. Vertical water motions in a small ice-covered lake. J. Geophys. Res. *70*:2333–2344.

Likens, G. E., and R. A. Ragotzkie. 1966. Rotary circulation of water in an ice-covered lake. Verh. Int. Ver. Limnol. *16*:126–133.

Likens, G. E., R. F. Wright, J. N. Galloway, and T. J. Butler. 1979. Acid rain. Sci. Amer. *241*:(4):43–51.

Lillieroth, S. 1950. Über Folgen kulturbedingter Wasserstandsenkungen für Makrophyten- und Planktongemeinschaften in seichten Seen des südschwedischen Oligotrophiegebietes. Acta Limnol. *3*. 288 pp.

Lind, O. T. 1978. Interdepression differences in the hypolimnetic areal relative oxygen deficits of Douglas Lake, Michigan. Verh. Int. Verein. Limnol. *20*:2689–2696.

Lindegaard, C. and P. M. Jónasson. 1979. Abundance, population dynamics and production of zoobenthos in Lake Mývatn, Iceland. Oikos *32*:202–227.

Lindeman, R. L. 1942. The trophic-dynamic aspect of ecology. Ecology *23*:399–418.

Lindström, K. 1980. *Peridinium cinctum* bioassays of Se in Lake Erken. Arch Hydrobiol. *89*:110–117.

Lindström, K. and W. Rodhe. 1978. Selenium as a micronutrient for the dinoflagellate *Peridinium cinctum* fa. *westii*. Mitt. Int. Ver. Limnol. *21*:168–173.

Lingeman, R., B. J. G. Flik, and J. Ringelberg. 1975. Stability of the oxygen stratification in a eutrophic lake. Verh. Int. Verein. Limnol. *19*:1193–1201.

Littlefield, L., and C. Forsberg. 1965. Absorption and transolcation of phosphorus-32 by *Chara globularis* Thuill. Physiol. Plant. *18*:291–296.

Livingstone, D. A. 1954. On the orientation of lake basins. Amer. J. Sci. *252*:547–554.

Livingstone, D. A. 1957. On the sigmoid growth phase in the history of Linsley Pond. Am. J. Sci. *255*:364–373.

Livingstone, D. A. 1963a. Chemical composition of rivers and lakes. Chap. G. Data of Geochemistry. 6th ed. Prof. Pap. U.S. Geol. Surv. *440–G*, 64 pp.

Livingstone, D. A. 1963b. Alaska, Yukon, Northwest Territories, and Greenland. *In* D. G. Frey, ed. Limnology in North America. Madison, University of Wisconsin Press, pp. 559–574.

Livingstone, D. A. 1975. Late Quaternary climatic change in Africa. Ann. Rev. Ecol. Syst. *6*:249–280.

Livingstone, D. A. and J. C. Boykin. 1962. Vertical distribution of phosphorus in Linsley Pond. mud. Limnol. Oceanogr. *7*:57–62.

Livingstone, D. A., K. Bryan, Jr., and R. G. Leahy. 1958. Effects of an arctic environment on the origin and development of freshwater lakes. Limnol. Oceanogr. *3*:192–214.

Livingstone, D., and G. H. M. Jaworski. 1980. The viability of akinetes of blue-green algae recovered from the sediments of Rostherne Mere. Br. Phycol. J. *15*:357–364.

Lloyd, N. D. H., D. T. Canvin, and J. M. Bristow.

1977. Photosynthesis and photerespiration in submerged aquatic vascular plants. Can. J. Bot. 55:3001–3005.

Lock, M. A., P. M. Wallis, and H. B. N. Hynes. 1977. Colloidal organic carbon in running waters. Oikos 29:1–4.

Loden, M. S. 1974. Predation by chironomid (Diptera) larvae on oligochaetes. Limnol. Oceanogr. 19:156–159.

Loden, M. S. 1981. Reproductive ecology of Naididae (Oligochaeta). Hydrobiologia 83:115–123.

Loehr, R. C., C. S. Martin, and W. Rast, eds. 1980. Phosphorus Management Strategies for Lakes. Ann Arbor, Ann Arbor Science Publ., Inc., 490 pp.

Lohuis, D., V. W. Meloche, and C. Juday. 1938. Sodium and potassium content of Wisconsin lake waters and their residues. Trans. Wis. Acad. Sci. Arts Lett. 31:285–304.

Lorch, D. W. 1978. Desmids and heavy metals. II. Manganese: Uptake and influence on growth and morphogenesis of selected species. Arch. Hydrobiol. 84:166–179.

Loucks, O. L., and W. E. Odum. 1978. Analysis of five North American lake ecosystems. I. A strategy for comparison. Verh. Int. Ver. Limnol. 20:556–561.

Love, R. J. R., and G. G. C. Robinson. 1977. The primary productivity of submerged macrophytes in West Blue Lake, Manitoba. Can. J. Bot. 55:118–127.

Lovley, D. R., and M. J. Klug, 1982. Intermediary metabolism of organic matter in the sediments of a eutrophic lake. Appl. Environ. Microbiol. 43:552–560.

Lovley, D. R., D. F. Dwyer, and M. J. Klug. 1982. Kinetics analysis of competition between sulfate reducers and methanogens for hydrogen in sediments. Appl. Environ. Microbiol. 43:1373–1379.

Lucas, W. J., M. T. Tyree, and A. Petrov. 1978. Characterization of photosynthetic ^{14}carbon assimilation by Potamogeton lucens L. J. Exp. Bot. 29:1409–1421.

Ludlam, S. D. 1976. Laminated sediments in holomictic Berkshire lakes. Limnol. Oceanogr. 21:743–746.

Luecke, C. and W. J. O'Brien. 1981. Phototoxicity and fish predation: Selective factors in color morphs in Heterocope. Limnol. Oceanogr. 26:454–460.

Lueschow, L. A., J. M. Helm, D. R. Winter, and G. W. Karl. 1970. Trophic nature of selected Wisconsin lakes. Trans. Wis. Acad. Sci. Arts Lett. 58:237–264.

Lund, J. W. G. 1949. Studies on Asterionella. I. The origin and nature of the cells producing seasonal maxima. J. Ecol. 37:389–419.

Lund, J. W. G. 1950. Studies on Asterionella formosa Hass. II. Nutrient depletion and the spring maximum. (parts I and II). J. Ecol. 38:1–35.

Lund, J. W. G. 1954. The seasonal cycle of the plankton diatom, Melosira italica (Ehr.) Kütz. subsp. subarctica O. Müll. J. Ecol. 42:151–179.

Lund, J. W. G. 1955. Further observations on the seasonal cycle of Melosira italica (Ehr.) Kütz. subsp. subarctica O. Müll. J. Ecol. 43:90–102.

Lund, J. W. G. 1959. Buoyancy in relation to the ecology of the freshwater phytoplankton. Brit. Phycol. Bull. 1:1–17.

Lund, J. W. G. 1964. Primary production and periodicity of phytoplankton. Verh. Int. Ver. Limnol. 15:37–56.

Lund, J. W. G. 1965. The ecology of the freshwater phytoplankton. Biol. Rev. 40:231–293.

Lund, J. W. G., C. Kipling, and E. D. LeCren. 1958. The inverted microscope method of estimating algal numbers and the statistical basis of estimations by counting. Hydrobiologia 11:143–170.

Lund, J. W. G., F. J. H. Mackereth, and C. H. Mortimer. 1963. Changes in depth and time of certain chemical and physical conditions and of the standing crop of Asterionella formosa Hass. in the North Basin of Windermere in 1947. Phil. Trans. Roy. Soc. London (ser. B) 246:255–290.

Lund, J. W. G., and J. F. Talling. 1957. Botanical limnological methods with special reference to the algae. Bot. Rev. 23:489–583.

Lundgren, A. 1978. Nitrogen fixation induced by phosphorus fertilization of a subarctic lake. In Environmental Role of Nitrogen-Fixing Blue-green Algae and Asynbiotic Bacteria. Ecol. Bull. (Stockholm) 26:52–59.

Lundgren, D. G., and W. Dean. 1979. Biogeochemistry of iron. In P. A. Trudinger and D. J. Swaine, ed. Biogeochemical Cycling of Mineral-forming Elements. Amsterdam, Elsevier Sci. Publ. Co., pp. 211–251.

Lundqvist, G. 1927. Bodenablagerungen und Entwicklungstypen der Seen. Die Binnengewässer 2:124 pp.

Lush, D. L., and H. B. N. Hynes. 1973. The formation of particles in freshwater leachates of dead leaves. Limnol. Oceanogr. 18:968–977.

Lüttge, U. 1964. Mikroautoradiographische Untersuchungen über die Funktion der Hydropoten von Nymphaea. Protoplasma 59:157–162.

Lvovitch, M. I. 1973. The global water balance. Trans. Amer. Geophys. Union 54:28–42.

Lynch, M. 1982. How well does the Edmondson-Paloheimo model approximate instantaneous birth rates? Ecology 63:12–18.

Macan, T. T. 1961. Factors that limit the range of freshwater animals. Biol. Rev. 36:151–198.

Macan, T. T. 1970. Biological Studies of the English Lakes. New York, American Elsevier Publishing Co., Inc., 260 pp.

Macan, T. T. 1977. The influence of predation on the composition of fresh-water animal communities. Biol. Rev. 52:45–70.

MacArthur, J. W., and W. H. T. Baillie. 1929. Metabolic activity and duration of life. I. Influence of temperature on longevity in *Daphnia magna*. J. Exp. Zool., 53:221–242.

MacFayden, A. 1948. The meaning of productivity in biological systems. J. Anim. Ecol. 17:75–80.

MacFayden, A. 1950. Biologische Produktivität. Arch. Hydrobiol. 43:166–170.

MacGregor, A. N., and D. R. Keeney. 1973a. Denitrification in lake sediments. Environ. Lett. 5:175–181.

MacGregor, A. N., and D. R. Keeney. 1973b. Methane formation by lake sediments during in vitro incubation. Water Resour. Bull. 9:1153–1158.

Machta, L. 1973. Prediction of CO_2 in the atmosphere. *In* G. M. Woodwell and E. V. Pecan, eds. Carbon and the Biosphere. Brookhaven, N.Y., Proc. Brookhaven Symp. in Biol. 24. Tech. Information Center, U.S. Atomic Energy Commission CONF-720510, pp. 21–31.

Maciolek, J. A. 1954. Artificial fertilization of lakes and ponds. A review of the literature. Spec. Sci. Rep. Fish. *113*, 41 pp.

Mackay, R. J., and G. B. Wiggins. 1979. Ecological diversity in Trichoptera. Ann. Rev. Entomol. 24:185–208.

Mackenzie, F. T., and R. M. Garrels. 1965. Silicates: Reactivity with sea water. Science 150:57–58.

Mackenzie, F. T., R. M. Garrels, O. P. Bricker, and F. Bickley. 1967. Silica in sea water: Control by silica minerals. Science 155:1404–1405.

Mackereth, F. J. H. 1953. Phosphorus utilization by *Asterionella formosa* Hass. J. Exp. Bot. 4:296–313.

Mackereth, F. J. H. 1966. Some chemical observations on postglacial lake sediments. Phil. Trans. Roy. Soc. London (ser. B) 250:165–213.

Mackereth, F. J. H. 1965. Chemical investigation of lake sediments and their interpretation. Proc. Roy. Soc. (ser. B) 161:295–309.

Mackereth, F. J. H. 1971. On the variation in direction of the horizontal component of remanent magnetisation in lake sediments. Earth Planetary Sci. Lett. 12:332–338.

Macpherson, L. B., N. R. Sinclair, and F. R. Hayes. 1958. Lake water and sediment. III. The effect of pH on the partition of inorganic phosphate between water and oxidized mud or its ash. Limnol. Oceanogr. 3:318–326.

Maeda, O., and S. Ichimura. 1973. On the high density of a phytoplankton population found in a lake under ice. Int. Rev. ges. Hydrobiol. 58:673–685.

Maeda, O., and H. Tomioka. 1977. Vertical distributions of particulate protein and nucleic acids in lakes in central Japan in late summer. Japan. J. Limnol. 38:109–115. (In Japanese)

Mague, T. H., E. Friberg, D. H. Hughes, and I. Morris. 1980. Extracellular release of carbon by marine phytoplankton: A physiological approach. Limnol. Oceanogr. 25:262–279.

Mahon, R. 1976. A second look at bluegill production in Wyland Lake, Indiana. Environ. Biol. Fish. 1:85–86.

Maistrenko, Iu. G. 1965. Organicheskoe Veshchestvo Vody i Donnykh Otlozhenii Rek i Vodoemov Ukrainy (Basseiny Dnepra i Dunaia). Kiev, Inst. Gidrobiologii, 239 pp.

Makarewicz, J. C., and G. E. Likens. 1975. Niche analysis of a zooplankton community. Science 190:1000–1003.

Makarewicz, J. C., and G. E. Likens. 1979. Structure and function of the zooplankton community of Mirror Lake, New Hampshire. Ecol. Monogr. 49:109–127.

Malovitskaia, L. M., and Ju. Sorokin. 1961. Eksperimental'noe issledovanie pitaniia *Diaptomus* (Crustacea, Copepoda) s pomoshch'iu C^{14}. Trudy Inst. Biol. Vodokhranilishch 4:262–272.

Malueg, K. W., and A. D. Hasler. 1966. Echo sounder studies on diel vertical movements of *Chaoborus* larvae in Wisconsin (U.S.A.) lakes. Verh. Int. Ver. Limnol. 16:1697–1708.

Maly, E. J., and M. P. Maly. 1974. Dietary differences between two co-occurring calanoid copepod species. Oecologia 17:325–333.

Mann, K. H. 1962. Leeches (Hirudinea). Their Structure, Physiology, Ecology and Embryology. Oxford, Pergamon Press, 201 pp.

Mann, K. H., R. H. Britton, A. Kowalczewski, T. J. Lack, C. P. Mathews, and I. McDonald. 1970. Productivity and energy flow at all trophic levels in the River Thames, England. *In* Z. Kajak and A. Hillbricht-Ilkowska, eds. Productivity Problems of Freshwaters. Warsaw, PWN Polish Sci. Publs. pp. 579–596.

Manny, B. A. 1972a. Seasonal changes in dissolved organic nitrogen in six Michigan lakes. Verh. Int. Ver. Limnol. 18:147–156.

Manny, B. A. 1972b. Seasonal changes in organic nitrogen content of net- and nannophytoplankton in two hardwater lakes. Arch. Hydrobiol. 71:103–123.

Manny, B. A., and R. G. Wetzel. 1973. Diurnal changes in dissolved organic and inorganic carbon and nitrogen in a hardwater stream. Freshwat. Biol. 3:31–43.

Manny, B. A., and R. G. Wetzel. 1982. Allochthonous dissolved organic and inorganic nitrogen budget of a marl lake. (In preparation).

Manny, B. A., R. G. Wetzel, and W. C. Johnson. 1975. Annual contribution of carbon, nitrogen, and phosphorus to a hard-water lake by migrant Canada geese. Verh. Int. Ver. Limnol. 19:949–951.

Manny, B. A., R. G. Wetzel, and R. E. Bailey. 1978. Paleolimnological sedimentation of organic carbon, nitrogen, phosphorus, fossil pigments, pollen, and diatoms in a hypereutrophic, hardwater lake: A case history of eutrophication. Polskie Arch. Hydrobiol. 25:243–267.

Marcus, J. H., D. W. Sutcliffe, and L. G. Willoughby.

1978. Feeding and growth of *Asellus aquaticus* (Isopoda) on food items from the littoral of Windermere, including green leaves of *Elodea canadensis*. Freshwat. Biol. 8:505–519.

Margalef, R. 1955. Temperature and morphology in freshwater organisms. Verh. Int. Ver. Limnol. 12:507–514.

Margalef, R. 1969. Size of centric diatoms as an ecological indicator. Mitt. Int. Ver. Limnol. 17:202–210.

Marker, A. F. H. 1976. The benthic algae of some streams in southern England. II. The primary production of the epilithon in a small chalkstream. J. Ecol. 64:359–373.

Marquenie-van der Werff, M., and W. H. O. Ernst. 1979. Kinetics of copper and zinc uptake by leaves and roots of aquatic plant, *Elodea nuttallii*. Z. Pflanzenphysiol. 92:1–10.

Martin, J. H., G. A. Knauer, and A. R. Flegal. 1980. Distribution of zinc in natural waters. *In* J. O. Nriagu, Ed. Zinc in the Environment. I. Ecological Cycling. New York, John Wiley & Sons, pp. 193–197.

Marx, J. L. 1980. Calmodulin: A protein for all seasons. Science 208:274–276.

Marzolf, G. R. 1965a. Substrate relations of the burrowing amphipod *Pontoporeia affinis* in Lake Michigan. Ecology 46:579–592.

Marzolf, G. R. 1965b. Vertical migration of *Pontoporeia affinis* (Amphipoda) in Lake Michigan. Publ. Great Lakes Res. Div., Univ. Mich. 13:133–140.

Mason, C. F. 1976. Relative importance of fungi and bacteria in the decomposition of *Phragmites* leaves. Hydrobiologia 51:65–69.

Mason, C. F., and R. J. Bryant. 1975. Periphyton production and grazing by chironomids in Alderfen Broad, Norfolk. Freshwat. Biol. 5:271–277.

Mason, M. A. 1952. Some observations of breaking waves. *In* Gravity Waves. Circ. U.S. Bur. Standards 521:215–220.

Mathews, C. P., and D. F. Westlake. 1969. Estimation of production by populations of higher plants subject to high mortality. Oikos 20:156–160.

Matsuyama, M. 1973. Organic substances in sediment and settling matter during spring in meromictic Lake Suigetsu. J. Oceanogr. Soc. Japan 29:53–60.

Mattern, H. 1970. Beobachtungen über die Algenflora im Uferbereich des Bodensees (Überlinger See und Gnadensee). Arch. Hydrobiol. (suppl.) 37:1–163.

Matveev, V. P. 1964. O vertikal'nom raspredelenii temperatury v donnykh otlozheniyakh Ozer Dolgogo (Pitkayarvi) i Volochaevskogo (Vuotyarvi). (On the vertical distribution of temperature in the bottom deposits of Lake Dolgom (Pitkayarvi) and Volochaevskom (Vuotyarvi). *In* Ozera Karel' skogo Ieresheika. Moscow, Izdatel' stvo Nauka, pp. 45–50.

Maucha, R. 1932. Hydrochemische Methoden in der Limnologie. Die Binnengewässer 12, 173 pp.

May, L. 1980. On the ecology of *Notholca squamula* Müller in Loch Leven, Kinross, Scotland. Hydrobiologia 73:177–180.

Mayfield, C. I., and W. E. Inniss. 1978. Interactions between freshwater bacteria and *Ankistrodesmus braunii* in batch and continuous culture. Microbial Ecol. 4:331–344.

McCall, P. L. and J. B. Fisher, 1980. Effects of tubificid oligochaetes on physical and chemical properties of Lake Erie sediments. *In* R. O. Brinkhurst, and D. G. Cook, eds. Aquatic Oligochaete Biology. New York, Plenum Press, pp. 253–317.

McCarter, J. A., F. R. Hayes, L. H. Jodrey, and M. L. Cameron. 1952. Movements of materials in the hypolimnion of a lake as studied by the addition of radioactive phosphorus. Can. J. Zool. 30:128–133.

McCarty, P. L. 1964. The methane fermentation. *In* H. Heukelekian and N. C. Dondero, eds. Principles and Applications in Aquatic Microbiology. New York, John Wiley & Sons, Inc., pp. 314–343.

McColl, J. G., and D. F. Grigal. 1975. Forest fires: Effects on phosphorus movement to lakes. Science 188:1109–1111.

McCracken, M. D., T. D. Gustafson, and M. S. Adams. 1974. Productivity of *Oedogonium* in Lake Wingra, Wisconsin. Amer. Midland Nat. 92:247–254.

McCraw, B. M. 1961. Life history and growth of the snail *Lymnaea humilis* Say. Trans. Amer. Microsc. Soc. 80:16–27.

McCraw, B. M. 1970. Aspects of the growth of the snail *Lymnaea palustris* (Müller). Malacologia 10:399–413.

McDonnell, A. J. 1971. Variations in oxygen consumption by aquatic macrophytes in a changing environment. Proc. Conf. Great Lakes Res., Int. Assoc. Great Lakes Res. 14:52–58.

McElhone, M. J. 1978. A population study of littoral dwelling Naididae (Oligochaeta) in a shallow mesotrophic lake in North Wales. J. Anim. Ecol. 47:615–626.

McElhone, M. J. 1979. A comparison of the gut contents of two co-existing lake-dwelling Naididae (Oligochaeta), *Nais pseudobtusa* and *Chaetogaster diastrophus*. Freshwat. Biol. 9:199–204.

McGowan, L. M. 1974. Ecological studies on *Chaoborus* (Diptera, Chaoboridae) in Lake George, Uganda. Freshwat. Biol. 4:483–505.

McGregor, D. L. 1969. The reproductive potential, life history and parasitism of the freshwater ostracod *Darwinula stevensoni* (Brady and Robertson). *In* J. W. Neale, ed. The Taxonomy, Morphology and Ecology of Recent Ostracoda. Edinburgh, Oliver & Boyd, pp. 194–221.

McIntire, C. D. 1966. Some factors affecting respiration of periphyton communities in lotic environments. Ecology 47:918–930.

McKeague, J. A., and M. G. Cline. 1963a. Silica in soil solutions. I. The form and concentration of dissolved silica in aqueous extracts of some soils. Can. J. Soil Sci. 43:70–82.

McKeague, J. A., and M. G. Cline. 1963b. Silica in soil solutions. II. The adsorption of monosilicic acid by soil and by other substances. Can. J. Soil Sci. 43:83–96.

McKinley, K. R. 1977. Light-mediated uptake of ³H-glucose in a small hard-water lake. Ecology 58:1356–1365.

McKinley, K. R., and R. G. Wetzel. 1979. Photolithotrophy, photoheterotrophy, and chemoheterotrophy: Patterns of resource utilization on an annual and a diurnal basis within a pelagic microbial community. Microbial Ecol. 5:1–15.

McKnight, D. M. 1979. Release of weak and strong copper-complexing agents by algae. Limnol. Oceanogr. 24:823–837.

McKnight, D. 1981. Chemical and biological processes controlling the response of a freshwater ecosystem to copper stress: A field study of the CuSO₄ treatment of Mill Pond Reservoir, Burlington, Massachusetts. Limnol. Oceanogr. 26:518–531.

McLaren, I. A. 1963. Effects of temperature on growth of zooplankton, and the adaptive value of vertical migration. J. Fish. Res. Bd. Can. 20:685–727.

McLaren, I. A. 1974. Demographic strategy of vertical migration by a marine copepod. Am. Nat. 108:91–102.

McMahon, J. W. 1965. Some physical factors influencing the feeding behavior of Daphnia magna Straus. Can. J. Zool. 43:603–612.

McMahon, J. W. 1969. The annual and diurnal variation in the vertical distribution of acid-soluble ferrous and total iron in a small dimictic lake. Limnol. Oceanogr. 14:357–367.

McMahon, J. W., and F. H. Rigler. 1963. Mechanisms regulating the feeding rate of Daphnia magna Straus. Can. J. Zool. 14:321–332.

McMahon, J. W., and F. H. Rigler. 1965. Feeding rate of Daphnia magna Straus in different foods labeled with radioactive phosphorus. Limnol. Oceanogr. 10:105–113.

McNaught, D. C. 1966. Depth control by planktonic cladocerans in Lake Michigan. Publ. Great Lakes Res. Div., Univ. Mich., 15:98–108.

McNaught, D. C. 1978. Spatial heterogeneity and niche differentiation in zooplankton of Lake Huron. Verh. Int. Ver. Limnol. 20:341–346.

McNaught, D. C., and A. D. Hasler. 1961. Surface schooling and feeding behavior in the white bass, Roccus chrysops (Rafinesque), in Lake Mendota. Limnol. Oceanogr. 6:53–60.

McNaught, D. C., and A. D. Hasler. 1964. Rate of movement of populations of Daphnia in relation to changes in light intensity. J. Fish. Res. Bd. Can. 21:291–318.

McNaught, D. C., and A. D. Hasler. 1966. Photoenvironments of planktonic Crustacea in Lake Michigan. Verh. Int. Ver. Limnol. 16:194–203.

McNaught, D. C., D. Griesmer, and M. Kennedy. 1980. Resource characteristics modifying selective grazing by copepods. In W. C. Kerfoot, ed. Evolution and Ecology of Zooplankton Communities. Hanover, NH, Univ. Press N. Engl. pp. 292–298.

McNaughton, S. J. 1966a. Light stimulated oxygen uptake and glycolic acid oxidase in Typha latifolia L. leaf discs. Science 211:1197–1198.

McNaughton S. J. 1966b. Ecotype function in the Typha community-type. Ecol. Monogr. 36:297–325.

McNaughton, S. J. 1969. Genetic and environmental control of glycolic acid oxidase activity in ecotypic populations of Typha latifolia. Amer. J. Bot. 56:37–41.

McNaughton, S. J. 1970. Fitness sets for Typha. Amer. Nat. 104:337–341.

McNaughton, S. J., and L. W. Fullem. 1969. Photosynthesis and photorespiration in Typha latifolia. Plant Physiol. 45:703–707.

McQueen, D. J. 1969. Reduction of zooplankton standing stocks by predaceous Cyclops bicuspidatus thomasi in Marion Lake, British Columbia. J. Fish. Res. Bd. Can. 26:1605–1618.

McQueen, D. J. 1970. Grazing rates and food selection in Diaptomus oregonensis (Copepoda) from Marion Lake, British Columbia. J. Fish. Res. Bd. Can. 27:13–20.

McRoy, C. P., R. J. Barsdate, and M. Nebert. 1972. Phosphorus cycling in an eelgrass (Zostera marina L.) ecosystem. Limnol. Oceanogr. 17:58–67.

Meadows, P. S., and J. G. Anderson. 1966. Microorganisms attached to marine and freshwater sand grains. Nature 212:1059–1060.

Meadows, P. S., and J. G. Anderson. 1968. Microorganisms attached to marine sand grains. J. Mar. Biol. Assoc. U. K. 48:161–175.

Meeks, J. C. 1974. Chlorophylls. In W. D. P. Stewart, ed. Algal Physiology and Biochemistry. Berkeley, University of California Press, pp. 161–175.

Meffert, M.-E., and J. Overbeck. 1979. Regulation of bacterial growth by algal release products. Arch. Hydrobiol. 87:118–121.

Meffert, M.-E., and H. Zimmermann-Telschow. 1979. Net release of nitrogenous compounds by axenic and bacteria-containing cultures of Oscillatoria redekei (Cyanophyta). Arch. Hydrobiol. 87:125–138.

Megard, R. O. 1972. Phytoplankton, photosynthesis, and phosphorus in Lake Minnetonka, Minnesota. Limnol. Oceanogr. 17:68–87.

Meinzer, O. E., ed. 1942. Hydrology. New York, McGraw-Hill Book Co., 712 pp.

Meisch, H.-U., H. Benzschawel, and H.-J. Bielig.

1977. The role of vanadium in green plants. II. Vanadium in green algae—two sites of action. Arch. Microbiol. 114:67–70.

Melack, J. M. 1979. Temporal variability of phytoplankton in tropical lakes. Oecologia 44:1–7.

Merkt, J. 1971. Zuverlässige Auszählungen von Jahresschichten in Seesedimenten mit Hilfe von Gross-Dünnschliffen. Arch. Hydrobiol. 69:145–154.

Meriläinen, J. 1969. Distribution of diatom frustules in recent sediments of some meromictic lakes. Mitt. Int. Ver. Limnol. 17:186–192.

Meriläinen, J. 1971. The recent sedimentation of diatom frustules in four meromictic lakes. Ann. Bot. Fenn. 8:160–176.

Mermoud, F., O. Clerc, F. O. Gülacar, and A. Buchs. 1981. Analyse des acides gras et des stérols dans le plancton du Lac Léman. Arch. Sc. Genève 34:367–382.

Meybeck, M. 1982. Carbon, nitrogen, and phosphorus transport by world rivers. Amer. J. Sci. 282:401–450.

Meyer, J. L., and G. E. Likens. 1979. Transport and transformation of phosphorus in a forest stream ecosystem. Ecology 60:1255–1269.

Meyers, P. A., and J. G. Quinn. 1971. Interaction between fatty acids and calcite in seawater. Limnol. Oceanogr. 16:922–997.

Mickle, A. M., and R. G. Wetzel. 1978a. Effectiveness of submersed angiosperm-epiphyte complexes on exchange of nutrients and organic carbon in littoral systems. I. Inorganic nutrients. Aquatic Bot. 4:303–316.

Mickle, A. M., and R. G. Wetzel. 1978b. Effectiveness of submersed angiosperm-epiphyte complexes on exchange of nutrients and organic carbon in littoral systems. II. Dissolved organic carbon. Aquatic Bot. 4:317–329.

Mickle, A. M., and R. G. Wetzel. 1979. Effectiveness of submersed angiosperm-epiphyte complexes on exchange of nutrients and organic carbon in littoral systems. III. Refractory organic carbon. Aquatic Bot. 6:339–355.

Milbrink, G. 1973a. Communities of Oligochaeta as indicators of the water quality in Lake Hjälmaren. Zoon 1:77–88.

Milbrink, G. 1973b. On the use of indicator communities of Tubificidae and some Lumbriculidae in the assessment of water pollution in Swedish lakes. Zoon 1:125–139.

Milbrink, G. 1973c. On the vertical distribution of oligochaetes in lake sediments. Rep. Inst. Freshw. Res. Drottningholm 53:34–50.

Milbrink, G. 1978. Indicator communities of oligochaetes in Scandinavian lakes. Verh. Int. Ver. Limnol. 20:2406–2411.

Milbrink, G. 1980. Oligochaete communities in pollution biology: The European situation with special reference to Scandinavia. In R. O. Brinkhurst and D. G. Cook, eds. Aquatic Oligochaete Biology. New York, Plenum Press, pp. 433–455.

Miles, C. J., and P. L. Brezonik. 1981. Oxygen consumption in humic-colored waters by a photochemical ferrous-ferric catalytic cycle. Environ. Sci. Technol. 15:1089–1095.

Miller, M. C. 1972. The carbon cycle in the epilimnion of two Michigan lakes. Ph.D. Diss., Michigan State University, 214 pp.

Mills, A. L., and M. Alexander. 1974. Microbial decomposition of species of freshwater planktonic algae. J. Environ. Quality 3:423–428.

Mills, E. L., and R. T. Oglesby. 1971. Five trace elements and vitamin B_{12} in Cayuga Lake, New York. Proc. Conf. Great Lakes Res., Int. Assoc. Great Lakes Res. 14:256–267.

Minckley, W. L., and G. A. Cole. 1963. Ecological and morphological studies on gammarid amphipods (Gammarus spp.) in spring-fed streams of northern Kentucky. Occas. Pap. Adams Center Ecol. Stud. 10, 35 pp.

Minder, L. 1922. Über biogene Entkalkung im Zürichsee. Verh. Int. Ver. Limnol. 1:20–32.

Minder, L. 1923. Studien über den Sauerstoffgehalt des Zürichsees. Arch. Hydrobiol. (suppl.) 3:107–155.

Minshall, G. W. 1978. Autotrophy in stream ecosystems. BioScience 28:767–771.

Mitchell, B. D. 1978. Cyclomorphosis in Daphnia carinata King (Crustacea: Cladocera) from two adjacent sewage lagoons in South Australia. Aust. J. Mar. Freshwat. Res. 29:565–576.

Mitchell, D. S., and P. A. Thomas. 1972. Ecology of water weeds in the neotropics: An ecological survey of the aquatic weeds Eichhornia crassipedes and Salvinia species, and their natural enemies in the neotropics. Tech. Pap. in Hydrology, UNESCO 12, 50 pp.

Mitchell, M. J., D. H. Landers, and D. F. Brodowski. 1981. Sulfur constituents of sediments and their relationship to lake acidification. Water, Air Soil Poll. (In press)

Mitchell, S. F., and C. W. Burns. 1979. Oxygen consumption in the epilimnia and hypolimnia of two eutrophic, warm-monomictic lakes. N. Z. J. Mar. Freshw. Res. 13:427–441.

Mittelbach, G. G. 1981. Foraging efficiency and body size: A study of optimal diet and habitat use by bluegills. Ecology 62:1370–1386.

Moaledj, K., and J. Overbeck. 1980. Studies on uptake kinetics of oligocarbophilic bacteria. Arch. Hydrobiol. 89:303–312.

Moeller, J. R., G. W. Minshall, K. W. Cummins, R. C. Petersen, C. E. Cushing, J. R. Sedell, R. A. Larson, and R. L. Vannote. 1979. Transport of dissolved organic carbon in streams of differing physiographic characteristics. Org. Geochem. 1:139–150.

Moeller, R. E. 1975. Hydrophyte biomass and community structure in a small, oligotrophic New Hampshire lake. Verh. Int. Ver. Limnol. 19:1004–1012.

Moeller, R. E. 1978a. Seasonal changes in biomass,

tissue chemistry, and net production of the ever-green hydrophyte, *Lobelia dortmanna*. Can. J. Bot. 56:1425–1433.

Moeller, R. E. 1978b. Carbon-uptake by the submerged hydrophyte *Utricularia purpurea*. Aquatic Bot. 5:209–216.

Moeller, R. E. 1980. The temperature-determined growing season of a submerged hydrophyte: Tissue chemistry and biomass turnover of *Utricularia purpurea*. Freshwat. Biol. 10:391–400.

Moeller, R. E. and J. P. Roskoski. 1978. Nitrogen-fixation in the littoral benthos of an oligotrophic lake. Hydrobiologia 60:13–16.

Moeslund, B., M. G. Kelly, and N. Thyssen. 1981. Storage of carbon and transport of oxygen in river macrophytes: Mass-balance, and the measurement of primary productivity in rivers. Arch. Hydrobiol. 93:45–51.

Molongoski, J. J. 1978. Sedimentation and anaerobic metabolism of particulate organic matter in the sediments of a hypereutrophic lake. Ph.D. Dissertation, Michigan State University. 143 pp.

Molongoski, J. J., and M. J. Klug. 1980a. Anaerobic metabolism of particulate organic matter in the sediments of a hypereutrophic lake. Freshw. Biol. 10:507–518.

Molongoski, J. J., and M. J. Klug. 1980b. Quantification and characterization of sedimenting particulate organic matter in a shallow hypereutrophic lake. Freshw. Biol. 10:497–506.

Momot, W. T. 1967a. Population dynamics and productivity of the crayfish, *Orconectes virilis*, in a marl lake. Amer. Midland Nat. 78:55–81.

Momot, W. T. 1967b. Effects of brook trout predation on a crayfish population. Trans. Amer. Fish. Soc. 96:202–209.

Momot, W. T. 1978. Annual production and production/biomass ratios of the crayfish, *Orconectes virilis*, in two northern Ontario lakes. Trans. Amer. Fish. Soc. 107:776–784.

Momot, W. T. and H. Gowing. 1977a. Response of the crayfish *Orconectes virilis* to exploitation. J. Fish. Res. Board Can. 34:1212–1219.

Momot, W. T., and H. Gowing. 1977b. Production and population dynamics of the crayfish *Orconectes virilis* in three Michigan lakes. J. Fish. Res. Board Can. 34:2041–2055.

Momot, W. T., H. Gowing, and P. D. Jones. 1978. The dynamics of crayfish and their role in ecosystems. Am. Midl. Nat. 99:10–35.

Monakov, A. B., and Ju. I. Sorokin. 1961. Kolichestvenn'ie dann'ie o pitanii Dafnii. Trudy Inst. Biol. Vodokhranilishch 4:251–261.

Mooij-Vogelaar, J. W., J. C. Jager, and W. J. van der Steen. 1973. Effects of density levels, and changes in density levels on reproduction, feeding and growth in the pond snail *Lymnaea stagnalis* (L.). Proc. Nederl. Akad. Wetensc. Amsterdam (ser. C) 76:245–256.

Moore, A. W. 1969. Azolla: Biology and agronomic significance. Bot. Rev. 35:17–34.

Moore, J. W. 1977. Importance of algae in the diet of subarctic populations of *Gammarus lacustris* and *Pontoporeia affnis*. Can. J. Zool. 55:637–641.

Moore, J. W. 1978. Importance of algae in the diet of the oligochaetes *Lumbriculus variegatus* (Müller) and *Rhyacodrilus sodalis* (Eisen). Oecologia 35:357–363.

Moore, J. W. 1981. Inter-species variability in the consumption of algea by oligochaetes. Hydrobiologia 83:241–244.

Moore, L. F. and J. A. Traquair. 1976. Silicon, a required nutrient for *Cladophora glomerata* (L) Kütz. (Chlorophyta). Planta 128:179–182.

Moore, P. D., and D. J. Bellamy. 1974. Peatlands. London, Elek Science, 221 pp.

Morel, A. 1978. Available, usable, and stored radiant energy in relation to marine photosynthesis. Deep-Sea Res. 25:673–688.

Morel, A., and R. C. Smith. 1974. Relation between total quanta and total energy for aquatic photosynthesis. Limnol. Oceanogr. 19:591–600.

Morgan, M. D. 1981. Abundance, life history, and growth of introduced populations of the opossum shrimp (*Mysis relicta*) in subalpine California lakes. Can. J. Fish. Aquat. Sci. 38:989–993.

Morgan, N. C., T. Backiel, G. Bretschko, A. Duncan, A. Hillbricht-Ilkowska, Z. Kajak, J. F. Kitchell, P. Larsson, C. Lévêque, A. Nauwerck, F. Schiemer, and J. E. Thorpe. 1980. Secondary production. *In* E. D. LeCren and R. H. Lowe-McConnell, eds. The Functioning of Freshwater Ecosystems. Cambridge, Cambridge Univ. Press, pp. 247–340.

Morikawa, M., Y. Fukuo, and F. Hirao. 1959. Limnological researches in Lake Biwa near the mouth of the River Ado. I. Density distribution of the lake water off the mouth of the river. (In Japanese.) Japan. J. Limnol. 20:10–20.

Morowitz, H. J. 1980. The dimensionality of niche space. J. Theor. Biol. 86:259–263.

Morris, I. 1967. An Introduction to the Algae. London, Hutchinson University Library, 189 pp.

Morris, J. C., and W. Stumm. 1967. Redox equilibria and measurements of potentials in the aquatic environment. Advances in Chemistry Series 67:270–285.

Mortimer, C. H. 1941. The exchange of dissolved substances between mud and water in lakes (Parts I and II). J. Ecol. 29:280–329.

Mortimer, C. H. 1942. The exchange of dissolved substances between mud and water in lakes (Parts III, IV, summary, and references). J. Ecol. 30:147–201.

Mortimer, C. H. 1951. The use of models in the study of water movement in stratified lakes. Verh. Int. Ver. Limnol. 11:254–260.

Mortimer, C. H. 1952. Water movements in lakes during summer stratification; Evidence from the distribution of temperature in Windermere. Proc. Roy. Soc. London (ser. B) 236:355–404.

Mortimer, C. H. 1953. The resonant response of

stratified lakes to wind. Schweiz. Z. Hydrol. 15:94–151.

Mortimer, C. H. 1954. Models of the flow-pattern in lakes. Weather 9:177–184.

Mortimer, C. H. 1955. Some effects of the earth's rotation on water movements in stratified lakes. Verh. Int. Ver. Limnol. 12:66–77.

Mortimer, C. H. 1961. Motion in thermoclines. Verh. Int. Ver. Limnol. 14:79–83.

Mortimer, C. H. 1963. Frontiers in physical limnology with particular reference to long waves in rotating basins. Publ. Great Lakes Res. Div., Univ. Mich. 10:9–42.

Mortimer, C. H. 1965. Spectra of long surface waves and tides in Lake Michigan and at Green Bay, Wisconsin. Publ. Great Lakes Res. Div., Univ. Mich. 13:304–325.

Mortimer, C. H. 1971. Chemical exchanges between sediments and water in the Great Lakes—speculations on probable regulatory mechanisms. Limnol. Oceanogr. 16:387–404.

Mortimer, C. H. 1971. Large-Scale Oscillatory Motions and Seasonal Temperature Changes in Lake Michigan and Lake Ontario. Spec. Rept. No. 12, Center for Great Lakes Studies, University of Wisconsin-Milwaukee. Part I, Text, 111 pp. and Part II, Illustrations, 106 pp.

Mortimer, C. H. 1974. Lake hydrodynamics. Mitt. Int. Ver. Limnol. 20:124–197.

Mortimer, C. H. 1975. Substantive corrections to SIL Communications (IVL Mitteilungen) numbers 6 and 20. Verh. Int. Ver. Limnol. 19:60–72.

Mortimer, C. H. 1981. The oxygen content of air-saturated fresh waters over ranges of temperature and atmospheric pressure of limnological interest. Mitt. Int. Ver. Limnol. 22: 23 pp.

Mortimer, C. H., and C. F. Hickling. 1954. Fertilizers in fishponds. Fish. Publ. U.K. Colonial Office, London 5, 155 pp.

Mortimer, C. H., and F. J. H. Mackereth. 1958. Convection and its consequences in ice-covered lakes. Verh. Int. Ver. Limnol. 13:923–932.

Mortimer, C. H., D. C. McNaught, and K. M. Stewart. 1968. Short internal waves near their high-frequency limit in central Lake Michigan. Proc. Conf. Great Lakes Res., Int. Assoc. Great Lakes Res. 11:454–469.

Morton, S. D., and G. F. Lee. 1968. Calcium carbonate equilibria in lakes. J. Chem. Educ. 45:511–513.

Mortonson, J. A., and A. S. Brooks. 1980. Occurrence of a deep nitrite maximum in Lake Michigan. Can. J. Fish. Aquat. Sci. 37:1025–1027.

Moshiri, G. A., K. W. Cummins, and R. R. Costa. 1969. Respiratory energy expenditure by the predaceous zooplankter *Leptodora kindtii* (Focke) (Crustacea: Cladocera). Limnol. Oceanogr. 14:475–484.

Moskalenko, B. K., and K. K. Votinsev. 1970. Biological productivity and balance of organic substance and energy in Lake Baikal. *In* Z. Kajak and A. Hillbricht-Ilkowska, eds. Productivity Problems of Freshwaters. Warsaw, PWN Polish Scientific Publishers, pp. 207–226.

Moss, B. 1968. Studies on the degradation of chlorophyll *a* and carotenoids in freshwaters. New Phytol. 67:49–59.

Moss, B. 1969a. Vertical heterogeneity in the water column of Abbot's Pond. II. The influence of physical and chemical conditions on the spatial and temporal distribution of the phytoplankton and of a community of epipelic algae. J. Ecol. 57:397–414.

Moss, B. 1969b. Algae of two Somersetshire pools: Standing crops of phytoplankton and epipelic algae as measured by cell numbers and chlorophyll *a*. J. Phycol. 5:158–168.

Moss, B. 1972a. Studies on Gull Lake, Michigan. I. Seasonal and depth distribution of phytoplankton. Freshwat. Biol. 2:289–307.

Moss, B. 1972b. Studies on Gull Lake, Michigan. II. Eutrophication—evidence and prognosis. Freshwat. Biol. 2:309–320.

Moss, B. 1972c. The influence of environmental factors on the distribution of freshwater algae: An experimental study. I. Introduction and the influence of calcium concentration. J. Ecol. 60:917–932.

Moss, B. 1973a. The influence of environmental factors on the distribution of freshwater algae: An experimental study, II. The role of pH and the carbon dioxide-bicarbonate system. J. Ecol. 61:157–177.

Moss, B. 1973b. The influence of environmental factors on the distribution of freshwater algae: An experimental study. III. Effects of temperature, vitamin requirements and inorganic nitrogen compounds on growth. J. Ecol. 61:179–192.

Moss, B. 1973c. The influence of environmental factors on the distribution of freshwater algae: An experimental study. IV. Growth of test species in natural lake waters, and conclusion. J. Ecol. 61:193–211.

Moss, B. 1973d. Diversity in fresh-water phytoplankton. Amer. Midland Nat. 90:341–355.

Moss, B. 1977. Adaptations of epipelic and epipsammic freshwater algae. Oecologia 28:103–108.

Moss, B., and A. G. Abdel Karim. 1969. Phytoplankton associations in two pools and their relationships with associated benthic flora. Hydrobiologia 33:587–600.

Moss, B., and J. Moss. 1969. Aspects of the limnology of an endorheic African lake (L. Chilwa, Malawi). Ecology 50:109–118.

Moss, B., and F. E. Round. 1967. Observations on standing crops of epipelic and epipsammic algal communities in Shear Water, Wilts. Brit. Phycol. Bull. 3:241–248.

Mothes, G. 1981. Sedimentation und Stoffbilanzen in Seen des Stechlinseegebiets. Limnologica 13:147–194.

Mueller, W. P. 1964. The distribution of cladoceran remains in surficial sediments from three northern Indiana lakes. Invest. Indiana Lakes Streams 6:1–63.

Mulholland, P. J. 1981. Organic carbon flow in a swamp-stream ecosystem. Ecol. Monogr. 51:307–322.

Müller, H. 1967. Eine neue qualitative Bestandsaufnahme des Phytoplanktons des Bodensee-Obersees mit besonderer Berücksichtigung der tychoplanktischen Diatomeen. Arch. Hydrobiol. (suppl.) 33:206–236.

Müller-Haeckel, A. 1965. Tagesperiodik des Siliziumgehaltes in einem Fliessgewässer. Oikos 16:232–233.

Mulligan, H. F., A. Baranowski, and R. Johnson. 1976. Nitrogen and phosphorus fertilization of aquatic vascular plants and algae in replicated ponds. I. Initial response to fertilization. Hydrobiologia 48:109–116.

Mullin, M. M., P. R. Sloan, and R. W. Eppley. 1966. Relationship between carbon content, cell volume, and area in phytoplankton. Limnol. Oceanogr. 11:307–311.

Munch, C. S. 1980. Fossil diatoms and scales of Chrysophyceae in the recent history of Hall Lake, Washington. Freshwat. Biol. 10:61–66.

Murphy, T. P., D. R. S. Lean, and C. Nalewajko. 1976. Blue-green algae: Their excretion of iron-selective chelators enables them to dominate other algae. Science 192:900–902.

Murtaugh, P. A. 1981a. Selective predation by Neomysis mercedis in Lake Washington. Limnol. Oceanogr. 26:445–453.

Murtaugh, P. A. 1981b. Size-selective predation on Daphnia by Neomysis mercedis. Ecology 62:894–900.

Murtaugh, P. A. 1981c. Inferring properties of mysid predation from injuries to Daphnia. Limnol. Oceanogr. 26:811–821.

Muscatine, L., and H. M. Lenhoff. 1963. Symbiosis: On the role of algae symbiotic with hydra. Science 142:956–958.

Musgrave, A., M. B. Jackson, and E. Ling. 1972. Callitriche stem elongation is controlled by ethylene and gibberellin. Nature (New Biol.) 238:93–96.

Musgrave, A., and J. Walters. 1973. Ethylene-stimulated growth and auxin transport in Ranunculus sceleratus petioles. New Phytol. 72:783–789.

Muztar, A. J., S. J. Slinger, and J. H. Burton. 1978a. Chemical composition of aquatic macrophytes. I. Investigation of organic constituents and nutritional potential. Can. J. Plant Sci. 58:829–841.

Muztar, A. J., S. J. Slinger, and J. H. Burton. 1978b. The chemical composition of aquatic macrophytes. II. Amino acid composition of the protein and non-protein fractions. Can. J. Plant Sci. 58:843–849.

Myer, G. E. 1969. A field study of Langmuir circulations. Proc. Conf. Great Lakes Res., Int. Assoc. Great Lakes Res. 12:652–663.

Nagasawa, M. 1959. On the dichotomous microstratification of pH in a lake. Japan. J. Limnol. 20:75–79.

Nakazawa, S. 1973. Artificial induction of lake balls. Naturwissenschaften 60:481.

Nalewajko, C. 1977. Extracellular release in freshwater algae and bacteria: Extracellular products of algae as a source of carbon for heterotrophs. In J. Cairns, Jr., ed. Aquatic Microbial Communities. New York, Garland Publ., Inc., pp. 589–624.

Nalewajko, C., and D. R. S. Lean. 1972. Growth and excretion in planktonic algae and bacteria. J. Phycol. 8:361–366.

Nalewajko, C., K. Lee, and P. Fay. 1980. Significance of algal extracellular products to bacteria in lakes and in cultures. Microbial Ecol. 6:199–207.

National Academy of Sciences. 1969. Eutrophication: Causes, Consequences, Correctives. Washington, D.C., National Academy of Sciences, 661 pp.

Naumann, E. 1919. Några synpunkter angående limnoplanktons ökologi med särskild hänsyn till fytoplankton. Svensk Bot. Tidskr., 13:129–163. (English translat. by the Freshwater Biological Association, No. 49.)

Naumann, E. 1931. Limnologische Terminologie. Handbuch der biologischen Arbeitsmethoden, Abt. IX, Teil 8. Berlin, Urban & Schwarzenberg, 776 pp.

Naumann, E. 1932. Grundzüge der regionalen Limnologie. Die Binnengewässer 11, 176 pp.

Nauwerck, A. 1959. Zur Bestimmung der Filterierrate limnischer Planktontiere. Arch. Hydrobiol. Suppl. 25:83–101.

Nauwerck, A. 1963. Die Beziehungen zwischen Zooplankton und Phytoplankton im See Erken. Symbol. Bot. Upsalien. 17(5), 163 pp.

Neal, J. T., Editor. 1975. Playas and dried lakes. Occurrence and development. Stroudsburg, PA, Dowden, Hutchinson & Ross, 411 pp.

Neill, W. E. 1975. Experimental studies of microcrustacean competition, community composition and efficiency of resource utilization. Ecology 56:809–826.

Neilson, A. H., and R. A. Lewin. 1974. The uptake and utilization of organic carbon by algae: An essay in comparative biochemistry. Phycologia 13:227–264.

Nelson, D. J., and D. C. Scott. 1962. Role of detritus in the productivity of a rock-outcrop community in a Piedmont stream. Limnol. Oceanogr. 7:396–413.

Neumann, J. 1959. Maximum depth and average depth of lakes. J. Fish. Res. Bd. Canada 16:923–927.

Newcombe, C. L. 1950. A quantitative study of attachment materials in Sodon Lake, Michigan. Ecology 31:204–215.

Nichols, D. S., and D. R. Keeney. 1973. Nitrogen and phosphorus release from decaying water milfoil. Hydrobiologia 42:509–525.

Nichols, D. S., and D. R. Keeney. 1976. Nitrogen nutrition of Myriophyllum spicatum: Uptake and translocation of ^{15}N by shoots and roots. Freshwat. Biol. 6:145–154.

Nicholson, S. A., and D. G. Best. 1974. Root:shoot and leaf area relationships of macrophyte communities in Chautauqua Lake, New York. Bull. Torrey Bot. Club 101:96–100.

Niewolak, S. 1970. Seasonal changes of nitrogen-fixing and nitrifying and denitrifying bacteria in the bottom deposits of Ilawa lakes. Pol. Arch. Hydrobiol. 17:89–103.

Niewolak, S. 1972. Fixation of atmospheric nitrogen by Azotobacter sp., and other heterotrophic oligonitrophilous bacteria in the Ilawa lakes. Acta Hydrobiol. 14:287–305.

Niewolak, S. 1974. Distribution of microorganisms in the waters of the Kortowskie Lake. Polskie Arch. Hydrobiol. 21:315–333.

Nilssen, J. P. 1978. On the evolution of life histories of limnetic cyclopoid copepods. Mem. Ist. Ital. Idrobiol. 36:193–214.

Nilsson, N.-A., and B. Pejler. 1973. On the relation between fish fauna and zooplankton composition in North Swedish lakes. Rep. Inst. Freshw. Res. Drottningholm 53:51–77.

Nilssen, J. P., and K. Elgmork. 1977. Cyclops abyssorum: Life cycle dynamics and habitat selection. Mem. Ist. Ital. Idrobiol. 34:197–238.

Nipkow, F. 1920. Vorläufige Mitteilungen über Untersuchungen des Schlammabsatzes im Zürichsee. Z. Hydrol. 1:1–27.

Nishijima, T., and Y. Hata. 1977. Distribution of thiamine, biotin, and vitamin B$_{12}$ in Lake Kojima. I. Distribution in the lake water. Bull. Japan. Soc. Sci. Fish. 43:1403–1410.

Nishijima, T., and Y. Hata. 1978. Distribution of thiamine, biotin, and vitamin B$_{12}$ in Lake Kojima. II. Distribution in the bottom sediments. Bull. Japan. Soc. Sci. Fish. 44:815–818.

Nishijima, T., R. Shiozaki, and Y. Hata. 1979. Production of vitamin B$_{12}$, thiamin, and biotin by freshwater phytoplankton. Bull. Japan. Soc. Sci. Fish. 45:199–204.

Nishimura, M., S. Nakaya, and K. Tanaka. 1973. Boron in the atmosphere and precipitation: Is the sea the source of atmospheric boron? In Proc. Symposium on Hydrogeochemistry and Biogeochemistry. I. Hydrogeochemistry. Washington, D.C., Clarke Company, pp. 547–557.

Noble, V. E. 1961. Measurement of horizontal diffusion in the Great Lakes. Publ. Great Lakes Res. Div., Univ. Mich. 7:85–95.

Noble, V. E. 1967. Evidences of geostrophically defined circulation in Lake Michigan. Proc. Conf. Great Lakes Res., Int. Assoc. Great Lakes Res. 10:289–298.

Noland, L. E., and M. Gojdics. 1967. Ecology of Free-Living Protozoa, In T. Chen, ed. Research in Protozoology. Vol. 2. Oxford, Pergamon Press, pp. 215–266.

Nordlie, F. G. 1972. Thermal stratification and annual heat budget of a Florida sinkhole lake. Hydrobiologia 40:183–200.

Northcote, T. G. 1964. Use of a high-frequency echo sounder to record distribution and migration of Chaoborus larvae. Limnol. Oceanogr. 9:87–91.

Northcote, T. G., C. J. Walters, and J. M. B. Hume. 1978. Initial impacts of experimental fish introductions on the macrozooplankton of small oligotrophic lakes. Verh. Int. Ver. Limnol. 20:2003–2012.

Novotná, M., and V. Kořínek. 1966. Effect of the fishstock on the quantity and species composition of the plankton of two backwaters. Hydrobiol. Stud. 1:297–322.

Novotny, I. H., and L. L. Tews. 1975. Lentic moulds of southern Lake Winnebago. Trans. Br. Mycol. Soc. 65:433–441.

Nriagu, J. O. 1968. Sulfur metabolism and sedimentary environment: Lake Mendota, Wisconsin. Limnol. Oceanogr. 13:430–439.

Nriagu, J. O. 1978. Dissolved silica in pore waters of lakes Ontario, Erie, and Superior sediments. Limnol. Oceanogr. 23:53–67.

Nriagu, J. O. 1979. Copper in the atmosphere and precipitation. In: J. O. Nriagu, ed. Copper in the Environment. I. Ecological Cycling. New York, John Wiley & Sons, Inc., pp. 43–75.

Nriagu, J. O., and J. D. Hem. 1978. Chemistry of pollutant sulfur in natural waters. In J. O. Nriagu, ed. Sulfur in the Environment. II. Ecological Impacts. New York, John Wiley & Sons, pp. 211–270.

Nriagu, J. O., and C. I. Davidson. 1980. Zinc in the atmosphere. In J. O. Nriagu, ed. Zinc in the Environment. I. Ecological Cycling, John Wiley & Sons, Inc., New York. pp. 113–159.

Nygaard, G. 1938. Hydrobiologische Studien über dänische Teiche und Seen. 1. Teil: Chemisch-Physikalische Untersuchungen und Planktonwägungen. Arch. Hydrobiol. 32:523–692.

Nygaard, G. 1949. Hydrobiological studies on some Danish ponds and lakes. Part II. The quotient hypothesis and some new or little known phytoplankton organisms. Kongel. Danske Vidensk. Selskab Biol. Skrift. 7(1), 293 pp.

Nygaard, G. 1955. On the productivity of five Danish waters. Verh. Int. Ver. Limnol. 12:123–133.

Nykvist, N. 1963. Leaching and decomposition of water-soluble organic substances from different types of leaf and needle litter. Stud. Forest. Suecica 3, 31 pp.

Oborn, E. T. 1960a. Iron content of selected water and land plants. In Chemistry of Iron in Natural Water. U. S. Geol. Surv. Water-Supply Pap. 1459-G:191–211.

Oborn, E. T. 1960b. A survey of pertinent biochemical literature. In Chemistry of Iron in Natural

Water. U. S. Geol. Surv. Water-Supply Pap. *1459-F*:111–190.

Oborn, E. T. 1964. Intracellular and extracellular concentration of manganese and other elements by aquatic organisms. *In* Chemistry of Manganese in Natural Water. U.S. Geol. Surv. Water-Supply Pap. *1667-C*, 18 pp.

Oborn, E. T., and J. D. Hem. 1962. Some effects of the larger types of aquatic vegetation on iron content of water. *In* Chemistry of Iron in Natural Water. U.S. Geol. Surv. Water-Supply Pap. *1459-I*:237–268.

O'Brien, W. J., and F. deNoyelles. 1974. Filtering rate of *Ceriodaphnia reticulata* in pond waters of varying phytoplankton concentrations. Amer. Midland Nat. *91*:509–512.

O'Brien, W. J., N. A. Slade, and G. L. Vinyard. 1976. Apparent size as the determinant of prey selection by bluegill sunfish *(Lepomis macrochirus).* Ecology *57*:1304–1310.

Odum, E. P. 1962. Relationships between structure and function in the ecosystem. Japan. J. Ecol. *12*:108–118.

Odum, E. P. 1963. Primary and secondary energy flow in relation to ecosystem structure. Proc. 16th Int. Congr. Zool. *4*:336–338.

Odum, E. P. 1971. Fundamentals of Ecology, ed. 3 Philadelphia, W. B. Saunders Co., 574 pp.

Odum, E. P., and A. A. de la Cruz. 1963. Detritus as a major component of ecosystems. Bull. Amer. Inst. Biol. Sci. *13*:39–40.

Odum, H. T. 1957. Trophic structure and productivity of Silver Springs, Florida. Ecol. Monogr. *27*:55–112.

Odum, W. E., P. W. Kirk, and J. C. Zieman. 1979. Non-protein nitrogen compounds associated with particles of vascular plant detritus. Oikos *32*:363–367.

Ogan, M. T. 1979. Potential for nitrogen fixation in the rhizosphere and habitat of natural stands of the wild rice *Zizania aquatica.* Can. J. Bot. *57*:1285–1291.

Ogawa, R. E., and J. F. Carr. 1969. The influence of nitrogen on heterocyst production in blue-green algae. Limnol. Oceanogr. *14*:342–351.

Oglesby, R. T., and W. R. Schaffner. 1978. Phosphorus loadings to lakes and some of their responses. Part 2. Regression models of summer phytoplankton standing crops, winter total P, and transparency of New York lakes with phosphorus loadings. Limnol. Oceanogr. *23*:135–145.

Ohle, W. 1934a. Chemische und physikalische Untersuchungen norddeutscher Seen. Arch. Hydrobiol. *26*:386–464 and 584–658.

Ohle, W. 1934b. Über organische Stoffe in Binnenseen. Verh. Int. Ver. Limnol. *6*(Pt. 2):249–262.

Ohle, W. 1938. Zur Vervollkommnung der hydrochemischen Analyse. III. Die Phosphorbestimmung. Angew. Chem. *51*:906–911.

Ohle, W. 1952. Die hypolimnische Kohlendioxyd-Akkumulation als produktionsbiologischer Indikator. Arch. Hydrobiol. *46*:153–285.

Ohle, W. 1953. Die chemisch und elektrochemische Bestimmung des molekular gelösten Sauerstoffs der Binnengewässer. Mitt. Int. Ver. Limnol. *3*:44 pp.

Ohle, W. 1954. Sulfat als "Katalysator" des limnischen Stoffkreislaufes. Vom Wasser *21*:13–32.

Ohle, W. 1955a. Ionenaustausch der Gewässersedimente. Mem. Ist. Ital. Idrobiol. *8*(Suppl.):221–245.

Ohle, W. 1955b. Beiträge zur Produktionsbiologie der Gewässer. Arch. Hydrobiol. (suppl.) *22*:456–479.

Ohle, W. 1956. Bioactivity, production, and energy utilization of lakes. Limnol. Oceanogr. *1*:139–149.

Ohle, W. 1958a. Die Stoffwechseldynamik der Seen in Abhängigkeit von der Gasausscheidung ihres Schlammes. Vom Wasser *25*:127–149.

Ohle, W. 1958b. Typologische Kennzeichnung der Gewässer auf Grund ihrer Bioaktivität. Verh. Int. Ver. Limnol. *13*:196–211.

Ohle, W. 1962. Der Stoffhaushalt der Seen als Grundlage einer allgemeinen Stoffwechseldynamik der Gewässer. Kieler Meeresforschungen *18*:107–120.

Ohle, W. 1964. Kolloidkomplexe als Kationen- und Anionenaustauscher in Binnengewässern. Vom Wasser *30* (1963):50–64.

Ohle, W. 1965. Nährstoffanreicherung der Gewässer durch Düngemittel und Meliorationen. Münchner Beiträge *12*:54–83.

Ohle, W. 1972. Die Sedimente des Grossen Plöner Sees als Dokumente der Zivilisation. Jahrb. Heimatkunde Plön *2*:7–27.

Ohle, W. 1978. Ebullition of gases from sediment, conditions, and relationship to primary production of lakes. Verh. Int. Ver. Limnol. *20*:957–962.

Ohle, W. 1979. Ontogeny of the lake Grosser Plöner See. *In* S. Horie, ed. Paleolimnology of Lake Biwa and the Japanese Pleistocene *7*:3–33.

Ohlhorst, S., G. E. Hutchinson, and J. G. J. Kuiper, 1977. The waters of Meron: A study of Lake Huleh. V. Temporal changes in the molluscan fauna. Arch. Hydrobiol. *80*:1–19.

Ohmori, M., and A. Hattori. 1972. Effect of nitrate on nitrogen-fixation by the blue-green alga *Anabaena cylindrica.* Plant Cell Physiol. *13*:589–599.

Ohwada, K., M. Otsuhata, and N. Taga. 1972. Seasonal cycles of vitamin B_{12}, thiamine and biotin in the surface water of Lake Tsukui. Bull. Japan. Soc. Sci. Fish. *38*:817–823.

Ohwada, K., and N. Taga. 1972. Vitamin B_{12}, thiamine, and biotin in Lake Sagami. Limnol. Oceanogr. *17*:315–320.

Ökland, J. 1964. The eutrophic Lake Borrevann (Norway)—an ecological study on shore and bottom fauna with special reference to gastropods, including a hydrographic survey. Folia Limnol. Scandinavica *13*, 337 pp.

Oláh, J. 1969a. The quantity, vertical and horizontal distribution of the total bacterioplankton of Lake Balaton in 1966/67. Annal. Biol. Tihany 36:185–195.

Oláh, J. 1969b. A quantitative study of the saprophytic and total bacterioplankton in the open water and the littoral zone of Lake Balaton in 1968. Annal. Biol. Tihany 36:197–212.

Oláh, J. 1972. Leaching, colonization and stabilization during detritus formation. Mem. Ist. Ital. Idrobiol. 29(suppl.):105–127.

Oldfield, F., J. A. Dearing, R. Thompson, and S. Garret-Jones. 1978a. Some magnetic properties of lake sediments and their possible links with erosion rates. Polskie Arch. Hydrobiol. 25:321–331.

Oldfield, F., R. Thompson, and K. E. Barber. 1978b. Changing atmospheric fallout of magnetic particles recorded in recent ombrotrophic peat sections. Science 199:679–680.

Olsen, S. 1958a. Phosphate adsorption and isotopic exchange in lake muds. Experiments with P 32; Preliminary report. Verh. Int. Ver. Limnol. 13:915–922.

Olsen, S. 1958b. Fosfatbalancen mellem bund og vand i Furesø. Forsøg med radioaktivt fosfor. Folia Limnol. Scandinavica 10:39–96.

Olsen, S. 1964. Phosphate equilibrium between reduced sediments and water. Laboratory experiments with radioactive phosphorus. Verh. Int. Ver. Limnol. 15:333–341.

Olson, F. C. W. 1960. A system of morphometry. Int. Hydrogr. Rev. 37:147–155.

Olson, F. C. W., and T. Ichiye. 1959. Horizontal diffusion. Science 130:1255.

Ondok, J. P. 1973a. Photosynthetically active radiation in a stand of Phragmites communis Trin. I. Distribution of irradiance and foliage structure. Photosynthetica 7:8–11.

Ondok, J. P. 1973b. Photosynthetically active radiation in a stand of Phragmites communis Trin. III. Distribution of irradiance on sunlit foliage area. Photosynthetica 7:311–319.

Ondok, J. P. 1978. Radiation climate in fishpond littoral plant communities. In D. Dykyjová and J. Květ, eds. Pond Littoral Ecosystems. Berlin, Springer-Verlag, pp. 113–125.

O'Neill, J. G., and J. F. Wilkinson. 1977. Oxidation of ammonia by methane-oxidizing bacteria and the effects of ammonia on methane oxidation. J. Gen. Microbiol. 100:407–41?

Orlov, D. S., I. A. Pivovarova, and N. I. Gorbunov. 1973. Interaction of humic substances with minerals and the nature of their bond—a review. Agrokhimiya, 1973(9):140–153. (Translat. in Soviet Soil Science 5:568–581.)

Ostendorp, W., and T. Frevert. 1979. Untersuchungen zur Manganfreisetzung und zum Mangangehalt der Sedimentoberschicht im Bodensee. Arch. Hydrobiol. (suppl.) 55:255–277.

Oswald, G. K. A., and G. de Q. Robin. 1973. Lakes beneath the Antarctic ice sheet. Nature 245:251–254.

Otsuki, A., and T. Hanya. 1968. On the production of dissolved nitrogen-rich organic matter. Limnol. Oceanogr. 13:183–185.

Otsuki, A., and T. Hanya. 1972a. Production of dissolved organic matter from dead green algal cells. I. Aerobic microbial decomposition. Limnol. Oceanogr. 17:248–257.

Otsuki, A., and T. Hanya. 1972b. Production of dissolved organic matter from dead green algal cells. II. Anaerobic microbial decomposition. Limnol. Oceanogr. 17:258–264.

Otsuki, A., and R. G. Wetzel. 1972. Coprecipitation of phosphate with carbonates in a marl lake. Limnol. Oceanogr. 17:763–767.

Otsuki, A., and R. G. Wetzel. 1973. Interaction of yellow organic acids with calcium carbonate in freshwater. Limnol. Oceanogr. 18:490–493.

Otsuki, A., and R. G. Wetzel. 1974a. Calcium and total alkalinity budgets and calcium carbonate precipitation of a small hard-water lake. Arch. Hydrobiol. 73:14–30.

Otsuki, A., and R. G. Wetzel. 1974b. Release of dissolved organic matter by autolysis of a submersed macrophyte, Scirpus subterminalis. Limnol. Oceanogr. 19:842–845.

Ottesen Hansen, N.-E. 1978. Mixing processes in lakes. Nordic Hydrol. 9:57–74.

Ouellet, M. H. 1975. Paleoclimatological implications of a Late-Quaternary molluscan fauna from Atkins Lake, Ontario. Verh. Int. Ver. Limnol. 19:2251–2258.

Overbeck, H.-J. 1968. Bakterien im Gewässer—Ein Beispiel für die Gegenwärtig Entwicklung der Limnologie. Mitt. Max-Planck-Ges. 3:165–182.

Overbeck, J. 1962. Untersuchungen zum Phosphathaushalt von Grünalgen. III. Das Verhalten der Zellfraktionen von Scenedesmus quadricauda (Turp.) Bréb. im Tagescyclus unter verschiedenen Belichtungsbedingungen und bei verschiedenen Phosphatverbindungen. Arch. Mikrobiol. 41:11–26.

Overbeck, J. 1963. Untersuchungen zum Phosphathaushalt von Grünalgen. VI. Ein Beitrag zum Polyphosphatstoffwechsel des Phytoplanktons. Ber. Deut. Bot. Gesellsch. 76:276–286.

Overbeck, J. 1965. Primärproduktion und Gewässerbakterien. Naturwissenschaften 51:145.

Overbeck, J. 1968. Prinzipielles zum Vorkommen der Bakterien im See. Mitt. Int. Ver. Limnol. 14:134–144.

Overbeck, J. 1975. Distribution pattern of uptake kinetic responses in a stratified eutrophic lake. Verh. Int. Ver. Limnol. 19:2600–2615.

Overbeck, J. 1979. Dark CO_2 uptake—biochemical background and its relevance to in situ bacterial production. Arch. Hydrobiol. Beih. Ergebn. Limnol. 12:38–47.

Overbeck, J., and H.-D. Babenzien. 1964. Bakterien

und Phytoplankton eines Kleingewässers im Jahreszyklus. Z. Allg. Mikrobiol. 4:59–76.

Overbeck, J. and D. Tóth. 1978. Einfluss des Phosphatgehalts auf die Glucoseaufnahme bei Bakterien. Arch. Hydrobiol. 82:114–122.

Owens, M., and R. W. Edwards. 1961. The effects of plants on river conditions. II. Further crop studies and estimates of net productivity of macrophytes in a chalk stream. J. Ecol. 49:119–126.

Owens, M., and R. W. Edwards. 1962. The effects of plants on river conditions. III. Crop studies and estimates of net productivity of macrophytes in four streams in southern England. J. Ecol. 50:157–162.

Owens, M., and P. J. Maris. 1964. Some factors affecting the respiration of some aquatic plants. Hydrobiologia, 23:533–543.

Ownbey, C. R., and D. A. Kee. 1967. Chlorides in Lake Erie. Proc. Conf. Great Lakes Res., Int. Assoc. Great Lakes Res. 10:382–389.

Ozimek, T., A. Prejs, and K. Prejs. 1976. Biomass and distribution of underground parts of *Potamogeton perfoliatus* L. and *P. lucens* L. in Mikolajskie Lake, Poland. Aquatic Bot. 2:309–316.

Pace, M. L. and J. D. Orcutt, Jr. 1981. The relative importance of protozoans, rotifers, and crustaceans in a freshwater zooplankton community. Limnol. Oceanogr. 26:822–830.

Paerl, H. W. 1975. Microbial attachment to particles in marine and freshwater populations. Microbial Ecol. 2:73–83.

Paerl, H. W. 1978. Role of heterotrophic bacteria in promoting N₂ fixation by *Anabaena* in aquatic habitats. Microbial Ecol. 4:215–231.

Paerl, H. W. 1979. Optimization of carbon dioxide and nitrogen fixation by the blue-green alga *Anabaena* in freshwater blooms. Oecologia 38:275–290.

Paerl, H. W. 1980. Ecological rationale for H₂ metabolism during aquatic blooms of the cyanobacterium *Anabaena*. Oecologia 47:43–45.

Paerl, H. W., and P. E. Kellar. 1979. Nitrogen-fixing *Anabaena*: Physiological adaptations instrumental in maintaining surface blooms. Science. 204:620–622.

Paerl, H. W., and D. R. S. Lean. 1976. Visual observations of phosphorus movement between algae, bacteria, and abiotic particles in lake water. J. Fish. Res. Bd. Can. 33:2805–2813.

Paerl, H. W., and M. T. Downes. 1978. Biological availability of low versus high molecular weight reactive phosphorus. J. Fish. Res. Bd. Can. 35:1639–1643.

Palmer, F. E., R. D. Methot, Jr., and J. T. Staley. 1976. Patchiness in the distribution of planktonic heterotrophic bacteria in lakes. Appl. Environ. Microbiol. 31:1003–1005.

Palmer, J. D., and F. E. Round. 1965. Persistent, vertical-migration rhythms in benthic microflora. I.

The effect of light and temperature on the rhythmic behaviour of *Euglena obtusa*. J. Mar. Biol. Assoc. U.K. 45:567–582.

Paloheimo, J. E. 1974. Calculation of instantaneous birth rate. Limnol. Oceanogr. 19:692–694.

Panganiban, A. T., Jr., T. E. Patt, W. Hart, and R. S. Hanson. 1979. Oxidation of methane in the absence of oxygen in lake water samples. Appl. Environ. Microbiol. 37:303–309.

Pannier, F. 1957. El consumo de oxigeno de plantas acuaticas en relacion a distintas concentraciones de oxigeno. Parte I. Acta Cient. Venez. 8:148–161.

Pannier, F. 1958. El consumo de oxigeno de plantas acuaticas en relacion a distintas concentraciones de oxigeno. Parte II. Acta Cient. Venez. 9:2–13.

Pardy, R. L. and B. N. White. 1977. Metabolic relationships between green *Hydra* and its symbiotic algae. Biol. Bull. 153:228–236.

Parker, B. C., and M. A. Wachtel. 1971. Seasonal distribution of cobalamins, biotin and niacin in rainwater. *In* Cairns, J., Jr., ed. The structure and function of freshwater microbial communities. Res. Div. Monogr., Virginia Polytech. Inst. 3:195–207.

Parker, M. 1977. Vitamin B₁₂ in Lake Washington, USA: Concentration and rate of uptake. Limnol. Oceanogr. 22:527–538.

Parker, M., and A. D. Hasler. 1969. Studies on the distribution of cobalt in lakes. Limnol. Oceanogr. 14:229–241.

Parker, R. A., and D. H. Hazelwood. 1962. Some possible effects of trace elements on fresh-water microcrustacean populations. Limnol. Oceanogr. 7:344–347.

Parma, S. 1971. *Chaoborus flavicans* (Meigen) (Diptera, Chaoboridae): an autecological study. Ph.D. Diss., University of Groningen. Rotterdam, Bronder-Offset n.v., 128 pp.

Parsons, T. R., and J. D. H. Strickland. 1962. On the production of particulate organic carbon by heterotrophic processes in sea water. Deep-Sea Res. 8:211–222.

Pasciak, W. J. and J. Gavis. 1974. Transport limitation of nutrient uptake in phytoplankton. Limnol. Oceanogr. 19:881–888.

Pastorok, R. A. 1981. Prey vulnerability and size selection by *Chaoborus* larvae. Ecology 62:1311–1324.

Patalas, K. 1961. Wind- und morphologiebedingte Wasserbewegungstypen als bestimmender Faktor für die Intensität des Stoffkreislaufes in nordpolnischen Seen. Verh. Int. Ver. Limnol. 14:59–64.

Patriarche, M. H., and R. C. Ball. 1949. An analysis of the bottom fauna production in fertilized and unfertilized ponds and its utilization by young-of-the-year fish. Tech. Bull. Mich. State Univ. Agricult. Exp. Stat., Sec. Zool. 207, 35 pp.

Patrick, R. 1978. Effects of trace metals in the aquatic ecosystem. Amer. Sci. 66:185–191.

Patrick, R., B. Crum, and J. Coles. 1969. Temperature and manganese as determining factors in the presence of diatom or blue-green algal floras in streams. Proc. Nat. Acad. Sci. 64:472–478.

Patrick, R., M. H. Hohn, and J. H. Wallace. 1954. A new method for determining the pattern of the diatom flora. Notulae Naturae Acad. Nat. Sci. Philadelphia, 259, 12 pp.

Pattee, E. 1975. Température stable et température fluctuante. I. Etude comparative de leurs effets sur le développement de certaines Planaires. Verh. Int. Ver. Limnol. 19:2795–2802.

Pattee, E. and H. Persat. 1978. Contribution of asexual reproduction to the distribution of triclad flatworms. Verh. Int. Ver. Limnol. 20:2372–2377.

Payne, F. C. 1982. Influence of hydrostatic pressure on gas balance and lacunar structure in *Myriophyllum spicatum* L. Ph.D. Dissertation, Michigan State Univ. 109 pp.

Pearsall, W. H. 1922. A suggestion as to factors influencing the distribution of free-floating vegetation. J. Ecol. 9:241–253.

Pearsall, W. H. 1932. Phytoplankton in the English lakes. II. The composition of the phytoplankton in relation to dissolved substances. J. Ecol. 20:241–262.

Pechan'-Finenko, G. A. 1971. Effectiveness of the assimilation of food by plankton crustaceans. (Translat. Consultants Bureau.) Ekologiia 2:64–72.

Pechlaner, R. 1970. The phytoplankton spring outburst and its conditions in Lake Erken (Sweden). Limnol. Oceanogr. 15:113–130.

Pechlaner, R. 1971. Factors that control the production rate and biomass of phytoplankton in high-mountain lakes. Mitt. Int. Ver. Limnol. 19:125–145.

Pechlaner, R., G. Bretschko, P. Gollmann, H. Pfeifer, M. Tilzer, and H. P. Weissenbach. 1970. The production processes in two high-mountain lakes (Vorderer and Hinterer Finstertaler See, Kühtai, Austria). In Z. Kajak and A. Hillbricht-Ilkowska, eds. Productivity Problems of Freshwaters. Warsaw, PWN Polish Scientific Publishers, pp. 239–269.

Penfound, W. T., and T. T. Earle. 1948. The biology of the water hyacinth. Ecol. Monogr. 18:447–472.

Penhale, P. A. and R. G. Wetzel. 1982. Structural and functional adaptations of eelgrass (*Zostera marina* L.) to the anaerobic sediment environment. Can. J. Bot. (In press)

Pennak, R. W. 1940. Ecology of the microscopic Metazoa inhabiting the sandy beaches of some Wisconsin lakes. Ecol. Monogr. 10:537–615.

Pennak, R. W. 1944. Diurnal movements of zooplankton organisms in some Colorado mountain lakes. Ecology 25:387–403.

Pennak, R. W. 1966. Structure of zooplankton populations in the littoral macrophyte zone of some Colorado lakes. Trans. Amer. Microsc. Soc. 85:329–349.

Pennak, R. W. 1973. Some evidence for aquatic macrophytes as repellents for a limnetic species of *Daphnia*. Int. Rev. ges. Hydrobiol. 58:569–576.

Pennak, R. W. 1978. Freshwater Invertebrates of the United States, 2nd ed. New York, Wiley-Interscience, 803 pp.

Pennington, W. 1973. Absolute pollen frequencies in the sediments of lakes of different morphometry. In H. J. B. Birks and R. G. West, eds. Quaternary Plant Ecology. Oxford. Blackwell Sci. Publs., pp. 79–104.

Pennington, W. 1974. Seston and sediment formation in five Lake District lakes. J. Ecol. 62:215–251.

Pennington, W. 1979. The origin of pollen in lake sediments: An enclosed lake compared with one receiving inflow streams. New Phytol. 83:189–213.

Pennington, W. 1981. Records of a lake's life in time: The sediments. Hydrobiologia 79:197–219.

Pennington, W., R. S. Cambray, J. D. Eakins, and D. D. Harkness. 1976. Radionuclide dating of the recent sediments of Blelham Tarn. Freshwat. Biol. 6:317–331.

Perfil'ev, B. V., and D. R. Gabe. 1969. Capillary Methods of Investigating Micro-Organisms. (Translat. of 1961 edition by J. M. Shewan.) Toronto, Univeristy of Toronto Press, 627 pp.

Peters, R. H. 1975. Orthophosphate turnover in central European lakes. Mem. Ist. Ital. Idrobiol. 32:297–311.

Peters, R. H. 1979. Concentrations and kinetics of phosphorus fractions along the trophic gradient of Lake Memphremagog. J. Fish. Res. Bd. Can. 36:970–979.

Peters, R. and D. Lean. 1973. The characterization of soluble phosphorus released by limnetic zooplankton. Limnol. Oceanogr. 18:270–279.

Peters, R. H. and S. MacIntyre. 1976. Orthophosphate turnover in East African lakes. Oecologia 25:313–319.

Peterson, B. J. 1980. Aquatic primary productivity and the ^{14}C-CO_2 method: A history of the productivity problem. Ann. Rev. Ecol. Syst. 11:359–385.

Peterson, W. H., E. B. Fred, and B. P. Domogalla. 1925. The occurrence of amino acids and other organic nitrogen compounds in lake water. J. Biol. Chem. 23:287–295.

Pettersson, K. 1980. Alkaline phosphatase activity and algal surplus phosphorus as phosphorus-deficiency indicators in Lake Erken. Arch. Hydrobiol. 89:54–87.

Pfennig. N. 1967. Photosynthetic bacteria. Ann. Rev. Microbiol. 21:285–324.

Phillips, G. L., D. Eminson, and B. Moss. 1978. A mechanism to account for macrophyte decline in progressively eutrophicated freshwaters. Aquatic Bot. 4:103–126.

Philp,, R. P., J. R. Maxwell, and G. Eglinton. 1976.

Envrionmental organic geochemistry of aquatic sediments. Sci. Prog. Oxford 63:521–545.

Phipps, D. W., Jr. and R. L. Pardy. 1982. Host enhancement of symbiont photosynthesis in the hydra-algae symbiosis. Biol. Bull. 162:83–94.

Pia, J. 1933. Kohlensäure und Kalk. Die Binnengewässer 13, 183 pp.

Pidgaiko, M. L., V. G. Grin, L. A. Kititsina, L. G. Lenchina, M. F. Polivannaya, O. A. Sergeeva, and T. A. Vinogradskaya. 1970. Biological productivity of Kurakhov's Power Station cooling reservoir. In Z. Kajak and A. Hillbricht-Ilkowska, Eds. Productivity Problems of Freshwaters. Warsaw, PWN Polish Sci. Publs., pp. 477–491.

Pieczyńska, E. 1959. Character of the occurrence of free-living Nematoda in various types of periphyton in Lake Tajty. Ekol. Polska (ser. A) 7:317–337. (Polish; English summary.)

Pieczyńska, E. 1964. Investigations on colonization of new substrates by nematodes (Nematoda) and some other periphyton organisms. Ekol. Polska (ser. A) 12:185–234.

Pieczyńska, E. 1965. Variations in the primary production of plankton and periphyton in the littoral zone of lakes. Bull. Acad. Polon. Sci. Cl. II, 13:219–225.

Pieczyńska, E. 1968. Dependence of the primary production of periphyton upon the substrate area suitable for colonization. Bull. Acad. Polon. Sci. Cl. II, 16:165–169.

Pieczyńska, E. 1970. Production and decomposition in the eulittoral zone of lakes. In Z. Kajak and A. Hillbricht-Ilkowska, eds. Productivity Problems of Freshwaters. Warsaw, PWN Polish Scientific Publishers, pp. 271–285.

Pieczyńska, E. 1972a. Ecology of the eulittoral zone of lakes. Ekol. Polska 20:637–732.

Pieczyńska, E. 1972b. Rola materii allochtonicznej w jeziorach. Wiadomości Ekologiczne 18:131–140.

Pierczyńska, E., E. Pieczyński, T. Prus, and K. Tarwid. 1963. The biomass of the bottom fauna of 42 lakes in the Wegorzewo District. Ekol. Polska (ser. A) 11:495–502.

Pieczyńska, E., and I. Spodniewska. 1963. Occurrence and colonization of periphyton organisms in accordance with the type of substrate. Ekol. Polska (ser. A) 11:533–545.

Pieczyńska, E., and W. Szczepańska. 1966. Primary production in the littoral of several Masurian lakes. Verh. Int. Ver. Limnol. 16:372–379.

Pielou, E. C. 1977. Mathematical Ecology. New York, John Wiley & Sons, 385 pp.

Pielou, E. C. 1981. The usefulness of ecological models: A stock-taking. Quart. Rev. Biol. 56:17–31.

Pimentel, G. C., and A. L. McClellan. 1960. The Hydrogen Bond. San Francisco, W. H. Freeman and Co., 475 pp.

Pinel-Alloul, B. 1978. Ecologie des populations de *Lymnaea catascopium catascopium* (Mollusques, Gastéropodes, Pulmonées) du lac St-Louis, près de Montréal, Québec. Verh. Int. Ver. Limnol. 20:2412–2426.

Pivovarov, A. A. 1973. Thermal Conditions in Freezing Lakes and Rivers. Israel Program for Scientific Translations of 1972 Russian edition. New York, John Wiley & Sons, 136 pp.

Planter, M. Physico-chemical properties of the water of reed-belts in Mikolajskie, Taltowisko, and Sniardwy lakes. Polskie Arch. Hydrobiol. 17:337–356.

Poirrier, M. A., B. R. Bordelon, and J. L. Laseter. 1972. Adsorption and concentration of dissolved carbon-14 DDT by coloring colloids in surface waters. Environ. Sci. Technol. 6:1033–1035.

Polishchuk, L. V., and A. M. Ghilarov. 1981. Comparison of two approaches used to calculate zooplankton mortality. Limnol. Oceanogr. 26:1162–1168.

Poltz, J. 1972. Untersuchungen über das Vorkommen und den Abbau von Fetten und Fettsäuren in Seen. Arch. Hydrobiol. (suppl.) 40:315–399.

Pomeroy, L. R., E. E. Smith, and C. M. Grant. 1965. The exchange of phosphate between estuarine water and sediments. Limnol. Oceanogr. 10:167–172.

Poole, H. H. and W. R. G. Atkins. 1929. Photo-electric measurements of submarine illumination throughout the year. J. Mar. Biol. Assoc. U.K. 16:297–324.

Porter, K. G. 1973. Selective grazing and differential digestion of algae by zooplankton. Nature 244:179–180.

Porter, K. G. 1975. Viable gut passage of gelatinous green algae ingested by *Daphnia*. Verh. Int. Ver. Limnol. 19:2840–2850.

Porter, K. G. 1976. Enhancement of algal growth and productivity by grazing zooplankton. Science 192:1332–1334.

Porter, K. G. and Y. S. Feig. 1980. The use of DAPI for identifying and counting aquatic microflora. Limnol. Oceanogr. 25:943–948.

Porter, K. G. and J. D. Orcutt, Jr. 1980. Nutritional adequacy, manageability, and toxicity as factors that determine the food quality of green and blue-green algae for *Daphnia*. In W. C. Kerfoot, ed. Evolution and Ecology of Zooplankton Communities. Hanover, NH, Univ. Press New England, pp. 268–281.

Porter, K. G., M. L. Pace, and J. F. Battey. 1979. Ciliate protozoans as links in freshwater planktonic food chains. Nature 277:563–565.

Poštolková, M. 1967. Comparison of the zooplankton amount and primary production of the fenced and unfenced littoral regions of Smyslov Pond. Rozpravy Československ. Akad. Věd, Řada Matem. Přír. Věd 77:(11):63–79.

Potts, W. T. W., and G. Parry. 1964. Osmotic and Ionic Regulation in Animals. Oxford, Pergamon Press, 423 pp.

Potzger, J. E., and W. A. van Engel. 1942. Study of the rooted aquatic vegetation of Weber Lake,

Vilas County, Wisconsin. Trans. Wis. Acad. Sci. Arts Lett. 34:149–166.

Pourriot, R. 1965. Recherches sur l'écologie des rotifères. Vie et Milieu (suppl.) 21, 224 pp.

Pourriot, R. 1974. Relations prédateur-proie chez les rotifères: influence du prédateur (Asplanchna brightwelli) sur la morphologie de la proie (Brachionus bidentata). Ann. Hydrobiol. 5:43–55.

Pourriot, R. and P. Clément. 1975. Influence de la durée de l'éclairement quotidien sur le taux de femelles mictiques chez Notommata copeus Ehr. (Rotifère). Oecologia 22:67–77.

Povoledo, D. 1961. Ulteriori studi sulle sostanze organiche disciolte nell'acqua del Lago Maggiore: Frazionamento e separazione delle proteine dai peptidi e dagli aminoacidi liberi. Mem. Ist. Ital. Idrobiol. 13:203–222.

Powell, T. and A. Jassby. 1974. The estimation of vertical eddy diffusivities below the thermocline in lakes. Water Resour. Res. 10:191–198.

Preissler, K. 1977. Do rotifers show "avoidance of the shore"? Oecologia 27:253–260.

Prejs, K. 1977a. The nematodes of the root region of aquatic macrophytes, with special consideration of nematode groupings penetrating the tissues of roots and rhizomes. Ekol. Polska 25:5–20.

Prejs, K. 1977b. The species diversity, numbers and biomass of benthic nematodes in central part of lakes with different trophy. Ekol. Polska 25:31–44.

Prejs, K. 1977c. The littoral and profundal benthic nematodes of lakes with different trophy. Ekol. Polska 25:21–30.

Prentki, R. T., T. D. Gustafson, and M. S. Adams. 1978. Nutrient movements in lakeshore marshes. In R. E. Good, D. F. Whigham, and R. L. Simpson. Freshwater Wetlands: Ecological Processes and Management Potential. New York, Academic Press, pp. 169–194.

Prepas, E. E., and F. H. Rigler. 1982. Improvements in quantifying the phosphorus concentration in lake water. Can. J. Fish. Aquat. Sci. 39:882–829.

Preston, T., W. D. P. Stewart, and C. S. Reynolds. 1980. Bloom-forming cyanobacterium Microcystis aeruginosa overwinters on sediment surface. Nature 288:365–367.

Prézelin, B. B. and B. M. Sweeney. 1978. Photoadaptation of photosynthesis in Gonyaulax polyedra. Mar. Biol. 48:27–35.

Priddle, J. 1980. The production ecology of benthic plants in some Antarctic lakes. I. In situ production studies. J. Ecol. 68:141–153.

Priddle, J. and R. B. Heywood. 1980. Evolution of Antarctic lake ecosystems. Biol. J. Linn. Soc. 14:51–66.

Prins, H. B. A., and R. W. Wolff. 1974. Photorespiration in leaves of Vallisneria spiralis. The effect of oxygen on the carbon dioxide compensation point. Proc. Akad. van Wetensc. Amsterdam (ser. C) 77:239–245.

Prins, H. B. A., J. F. H. Snel, R. J. Helder, and P. E. Zanstra. 1980. Photosynthetic HCO_3^- utilization and OH^- excretion in aquatic angiosperms. Light-induced pH changes at the leaf surface. Plant Physiol. 66:818–822.

Proctor, V. W. 1957. Studies of algal antibiosis using Haematococcus and Chlamydomonas. Limnol. Oceanogr. 2:125–139.

Provasoli, L. 1958. Nutrition and ecology of Protozoa and algae. Ann. Rev. Microbiol. 12:279–308.

Provasoli, L. 1963. Organic regulation of phytoplankton fertility. In M. N. Hill, ed. The Sea. Vol. 2. New York, Interscience, pp. 165–219.

Provasoli, L., and A. F. Carlucci. 1974. Vitamins and growth regulators. In W. D. P. Stewart, ed. Algal Physiology and Biochemistry. Berkeley, University of California Press, pp. 741–787.

Porvasoli, L., J. J. A. McLaughlin, and I. J. Pintner. 1954. Relative and limiting concentrations of major mineral constituents for the growth of algal flagellates. Trans. N.Y. Acad. Sci. (ser. 2) 16:412–417.

Pugh, G. J. F., and J. L. Mulder. 1971. Mycoflora associated with Typha latifolia. Trans. Br. Mycol. Soc. 57:273–282.

Purchase, B. S. 1977. Nitrogen fixation associated with Eichhornia crassipes. Plant Soil 46:283–286.

Putnam, H. D., and T. A. Olson. 1961. Studies on the Productivity and Plankton of Lake Superior. Rep. School Public Health, University of Minnesota, 24 pp.

Pyke, G. H., H. R. Pulliam, and E. L. Charnov. 1977. Optimal foraging: A selective review of theory and tests. Quart. Rev. Biol. 52:137–154.

Quade, H. W. 1969. Cladoceran faunas associated with aquatic macrophytes in some lakes in northwestern Minnesota. Ecology 50:170–179.

Quade, H. W. 1971. Niche specificity of littoral Cladocera: Habitat. Trans. Amer. Microsc. Soc. 90:104–105.

Quennerstedt, N. 1955. Diatoméerna i Långans Sjövetetation. Acta Phytogeographica Suecica 36. 208 pp.

Ragotzkie, R. A. 1978. Heat budgets of lakes. In A. Lerman, ed. Lakes: Chemistry, Geology, Physics. New York, Springer-Verlag, pp. 1–19.

Ragotzkie, R. A., and G. E. Likens. 1964. The heat budget of two Antarctic lakes. Limnol. Oceanogr. 9:412–425.

Rai, D. N. and J. D. Munshi. 1979. The influence of thick floating vegetation (water hyacinth: Eichhornia crassipes) on the physico-chemical environment of a freshwater wetland. Hydrobiologia 62:65–69.

Rai, H. 1978. Utilizacào de glicose por bactérias heterotróficas no ecossistema lacustre de Amazônia Central. Acta Amazonica 8:225–232.

Rai, H. 1979. Microbiology of central Amazon lakes. Amazoniana 6:583–599.

Rai, H. and G. Hill. 1981. Bacterial biodynamics in Lago Tupé, a central Amazonian black water

"Ria Lake". Arch. Hydrobiol. (suppl.) 58:420–468.

Rainwater, F. H., and L. L. Thatcher. 1960. Methods for collection and analysis of water samples. U.S. Geol. Surv. Water-Supply Pap. 1454, 301 pp.

Ralph, E. K. and H. N. Michael. 1974. Twenty-five years of radiocarbon dating. Am. Sci. 62:553–560.

Ramsey, W. L. 1962a. Bubble growth from dissolved oxygen near the sea surface. Limnol. Oceanogr. 7:1–7.

Ramsey, W. L. 1962b. Dissolved oxygen in shallow nearshore water and its relation to possible bubble formation. Limnol. Oceanogr. 7:453–461.

Rao, S. S. and B. K. Burnison. 1976. Bacterial distributions in Lake Erie (1967, 1970). J. Fish. Res. Bd. Canada 33:574–580.

Raspopov, I. M. 1971. Litoralvegatation der Onega- und Ladogaseen. Hidrobiologia 12:241–247.

Raspopov, I. M. 1979. Vegetation der grossen seichten Seen im Nordwesten der UdSSR und ihre Produktion. Arch. Hydrobiol. 86:242–253.

Raspopov, I. M., V. A. Ekzercev, and I. L. Koreljakova. 1977. Production by freshwater vascular plant (macrophyte) communities of lakes and reservoirs in the European part of the U.S.S.R. Folia Geobot. Phytotax. Praha 12:113–120.

Rast, W. and G. F. Lee. 1978. Summary analysis of the North American (U.S. portion) OECD Eutrophication Project: Nutrient loading-lake response relationships and trophic state indices. U.S. Environm. Protection Agency Rept. EPA-600/3-78-008. 455 pp.

Rasumov, A. S. 1962. Mikrobial'nyi plankton vody. Trudy Vsesoyusnogo Gidrobiol. Obshch. 12:60–190.

Raven, J. A. 1970. Exogenous inorganic carbon sources in plant photosynthesis. Biol. Rev. 45:167–221.

Rawson, D. S. 1939. Some physical and chemical factors in the metabolism of lakes. In Problems of Lake Biology. Publ. Amer. Assoc. Adv. Sci. 10:9–26.

Rawson, D. S. 1952. Mean depth and the fish production of large lakes. Ecology 33:515–521.

Rawson, D. S. 1955. Morphometry as a dominant factor in the productivity of large lakes. Verh. Int. Ver. Limnol. 12:164–175.

Rawson, D. S. 1956. Algal indicators of trophic lake types. Limnol. Oceanogr. 1:18–25.

Rawson, D. S., and J. E. Moore. 1944. The saline lakes of Saskatchewan. Can. J. Res. (sec. D), 22:141–201.

Reckhow, K. H. 1979. Empirical lake models for phosphorus: Development, applications, limitations and uncertainty. In D. Scavia and A. Robertson, eds. Perspectives on Lake Ecosystem Modeling. Ann Arbor, MI, Ann Arbor Sci. Publs., pp. 193–221.

Reeburgh, W. A. and D. T. Heggie. 1977. Microbial methane consumption reactions and their effect on methane distributions in freshwater and marine environments. Limnol. Oceanogr. 22:1–9.

Reichardt, W., J. Overbeck, and L. Steubing. 1967. Free dissolved enzymes in lake waters. Nature 216:1345–1347.

Reif, C. B. 1969. Temperature profiles and heat flow in sediments of Nuangola. Proc. Pennsylvania Acad. Sci. 43:98–100.

Reif, C. B., and D. W. Tappa. 1966. Selective predation: Smelt and cladocerans in Harveys Lake. Limnol. Oceanogr. 11:437–438.

Reimers, N., J. A. Maciolek, and E. P. Pister. 1955. Limnological study of the lakes in Convict Creek Basin, Mono County, California. Fish. Bull. U.S. Fish. Wildl. Serv. 56:437–503.

Remane, A., and C. Schlieper. 1971. Biology of brackish water. 2nd ed. Die Binnengewässer 25, 372 pp.

Renberg, I. 1978. Palaeolimnology and varve counts of the annually laminated sediment of Lake Rudetjärn, Northern Sweden. Early Norrland 11:63–92.

Reynolds, C. S. 1976a. Succession and vertical distribution of phytoplankton in response to thermal stratification in a lowland mere, with special reference to nutrient availability. J. Ecol. 64:529–551.

Reynolds, C. S. 1978. Stratification in natural populations of bloom-forming blue-green algae. Verh. Int. Ver. Limnol. 20:2285–2292.

Reynolds, C. S. and A. E. Walsby. 1975. Waterblooms. Biol. Rev. 50:437–481.

Reynolds, C. S. 1976b. Sinking movements of phytoplankton indicated by a simple trapping method. I. A Fragilaria population. Br. Phycol. J. 11:279–291.

Reynoldson, T. B. 1961. Observations on the occurrence of Asellus (Isopoda, Crustacea) in some lakes of northern Britain. Verh. Int. Ver. Limnol. 14:988–994.

Reynoldson, T. B. 1966. The distribution and abundance of lake-dwelling triclads—towards a hypothesis. Adv. Ecol. Res. 3:1–71.

Reynoldson, T. B. 1981. The ecology of the Turbellaria with special reference to the freshwater triclads. Hydrobiologia 84:87–90.

Reynoldson, T. B., and L. S. Bellamy. 1971. Intraspecific competition in lake-dwelling triclads. A laboratory study. Oikos 22:315–328.

Reynoldson, T. B. and P. Bellamy. 1975. Triclads (Turbellaria: Tricladida) as predators of lake-dwelling stonefly and mayfly nymphs. Freshwat. Biol. 5:305–312.

Reynoldson, T. B. and B. Piearce. 1979. Feeding on gastropods by lake-dwelling Polycelis in the absence and presence of Dugesia polychroa (Turbellaria, Tricladida). Freshwat. Biol. 9:357–367.

Reynoldson, T. B. and A. D. Sefton. 1976. The food

of *Planaria torva* (Müller) (Turbellaria-Tricladida), a laboratory and field study. Freshwat. Biol. 6:383–393.

Reynoldson, T. B., J. F. Gilliam, and R. M. Jaques. 1981. Competitive exclusion and co-existence in natural populations of *Polycelis nigra* and *P. tenuis* (Tricladida, Turbellaria). Arch. Hydrobiol. 92:71–113.

Rhee, G. -Y. 1972. Competition between an alga and an aquatic bacterium for phosphate. Limnol. Oceanogr. 17:505–514.

Rhee, G.-Y. 1973. A continuous culture study of phosphate uptake, growth rate and polyphosphate in *Scenedesmus* sp. J. Phycol. 9:495–506.

Rhee, G.-Y. 1974. Phosphate uptake under nitrate limitation by *Scenedesmus* sp. and its ecological implications. J. Phycol. 10:470–475.

Rhee, G.-Y. 1978. Effects of N:P atomic ratios and nitrate limitation on algal growth, cell composition, and nitrate uptake. Limnol. Oceanogr. 23:10–25.

Rhee, G.-Y. and I. J. Gotham. 1980. Optimum N:P ratios and coexistence of planktonic algae. J. Phycol. 16:486–489.

Rhee, G.-Y. and I. J. Gotham. 1981a. The effect of environmental factors on phytoplankton growth: Temperature and interactions of temperature with nutrient limitation. Limnol. Oceanogr. 26:635–648.

Rhee, G.-Y. and I. J. Gotham. 1981b. The effect of environmental factors on phytoplankton growth: Light and the interactions of light with nitrate limitation. Limnol. Oceanogr. 26:649–659.

Rho, J. and H. B. Gunner. 1978. Microfloral response to aquatic weed decomposition. Water Res. 12:165–170.

Rice, E. L., and S. K. Pancholy. 1972. Inhibition of nitrification by climax ecosystems. Amer. J. Bot. 59:1033–1040.

Rice, E. L., and S. K. Pancholy. 1973. Inhibition of nitrification by climax ecosystems. II. Additional evidence and possible role of tannins. Amer. J. Bot. 60:691–702.

Rich, P. H. 1975. Benthic metabolism of a soft-water lake. Verh. Int. Ver. Limnol. 19:1023–1028.

Rich, P. H. 1980. Hypolimnetic metabolism in three Cape Cod lakes. Amer. Midland Nat. 104:102–109.

Rich, P. H. and A. H. Devol. 1978. Analysis of five North American lake ecosystems. VII. Sediment processing. Verh. Int. Ver. Limnol. 20:598–604.

Rich, P. H., and R. G. Wetzel. 1969. A simple, sensitive underwater photometer. Limnol. Oceanogr. 14:611–613.

Rich, P. H., R. G. Wetzel, and N. V. Thuy. 1971. Distribution, production and role of aquatic macrophytes in a southern Michigan marl lake. Freshwat. Biol. 1:3–21.

Rich, P. H. and R. G. Wetzel. 1978. Detritus in lake ecosystems. Amer. Nat. 112:57–71.

Richardson, J. L., T. J. Harvey, and S. A. Holdship. 1978. Diatom in the history of shallow East African lakes. Polskie Arch. Hydrobiol. 25:341–353.

Richardson, L. F. 1925. Turbulence and vertical temperature difference near trees. Phil. Mag. 49:81–90.

Richerson, P., R. Armstrong, and C. R. Goldman. 1970. Contemporaneous disequilibrium, a new hypothesis to explain the "Paradox of the Plankton." Proc. Nat. Acad. Sci. 67:1710–1714.

Richerson, P. J., M. Lopez, and T. Coon. 1978. The deep chlorophyll maximum layer of Lake Tahoe. Verh. Int. Ver. Limnol. 20:426–433.

Richman, S. 1958. The transformation of energy by *Daphnia pulex*. Ecol. Monogr. 28:273–291.

Richman, S. 1966. The effect of phytoplankton concentration on the feeding rate of *Diaptomus oregonensis*. Verh. Int. Ver. Limnol. 16:392–398.

Ricker, W. E. 1934. A critical discussion of various measures of oxygen saturation in lakes. Ecology 15:348–363.

Ricker, W. E. 1937. Physical and chemical characteristics of Cultus Lake, British Columbia. J. Biol. Bd. Canada 3(4):363–402.

Ricker, W. E. 1952. The benthos of Cultus Lake. J. Fish. Res. Bd. Canada 9:204–212.

Rickett, H. W. 1921. A quantitative study of the larger aquatic plants of Lake Mendota, Wisconsin. Trans. Wis. Acad. Sci. Arts Lett. 20:501–527.

Ricklefs, R. E. 1979. Ecology, 2nd ed. New York, Chiron Press, Inc., 966 pp.

Riemann, B. 1978. Differentiation between heterotrophic and photosynthetic plankton by size fractionation, glucose uptake, ATP and chlorophyll content. Oikos 31:358–367.

Rigby, C. H., S. R. Craig, and K. Budd. 1980. Phosphate uptake by *Synechococcus leopoliensis* (Cyanophyceae): Enhancement by calcium ions. J. Phycol. 16:389–393.

Riggs, L. A., and J. J. Gilbert. 1972. The labile period for α-tocopherol-induced mictic female and body wall outgrowth responses in embryos of the rotifer *Asplanchna sieboldi*. Int. Rev. ges. Hydrobiol. 57:675–683.

Rigler, F. H. 1956. A tracer study of the phosphorus cycle in lake water. Ecology 37:550–562.

Rigler, F. H. 1961. The uptake and release of inorganic phosphorus by *Daphnia magna* Straus. Limnol. Oceanogr. 6:165–174.

Rigler, F. H. 1964a. The phosphorus fractions and the turnover time of inorganic phosphorus in different types of lakes. Limnol. Oceanogr. 9:511–518.

Rigler, F. H. 1964b. The contribution of zooplankton to the turnover of phosphorus in the epilimnion of lakes. Can. Fish Culturist 32:3–9.

Rigler, F. H. 1971. Feeding rates: zooplankton. *In* W. T. Edmondson and G. G. Winberg, eds. A Manual on Methods for the Assessment of Secondary Productivity in Fresh Waters. Int. Biol. Program Handbook 17. Oxford, Blackwell Scientific Publications, pp. 228-255.

Rigler, F. H. 1973. A dynamic view of the phosphorus cycle in lakes. In E. J. Griffith, A. Beeton, J. M. Spencer, and D. T. Mitchell, eds. Environmental Phosphorus Handbook. New York, John Wiley & Sons, pp. 539-572.

Rigler, F. H., M. E. MacCallum, and J. C. Roff, 1974. Production of zooplankton in Char Lake. J. Fish. Res. Bd. Canada 31:637-646.

Riley, G. A. 1939. Limnological studies in Connecticut. Part I. General limnological survey. Part II. The copper cycle. Ecol. Monogr. 9:66-94.

Ringelberg, J. 1964. The positively phototactic reaction of Daphnia magna Straus: A contribution to the understanding of diurnal vertical migration. Netherlands J. Sea Res. 2:319-406.

Ringelberg, J. 1980. Aspects of red pigmentation in zooplankton, especially copepods. In W. C. Kerfoot, Ed. Evolution and Ecology of Zooplankton Communities. Hanover, NH, Univ. Press New England, pp. 91-97.

Ringelberg, J. 1981. On the variation in carotenoid content of copepods. Limnol. Oceanogr. 26:995-997.

Ritchie, J. C., J. R. McHenry, and A. C. Gill. 1973. Dating recent reservoir sediments. Limnol. Oceanogr. 18:254-263.

Rivkin, R. B. and E. Swift. 1980. Characterization of alkaline phosphatase and organic phosphorus utilization in the oceanic dinoflagellate Pyrocystis noctiluca. Mar. Biol. 61:1-8.

Roback, S. S. 1974. Insects (Arthropoda: Insecta). In C. W. Hart, Jr., and S. L. H. Fuller, eds. Pollution Ecology of Freshwater Invertebrates. New York, Academic Press, pp. 313-376.

Robarts, R. D. 1979. Heterotrophic utilization of acetate and glucose in Swartvlei, South Africa. J. Limnol. Soc. South Africa 5:84-88.

Robbins, J. A. and E. Callender. 1975. Diagenesis of manganese in Lake Michigan sediments. Amer. J. Sci. 275:512-533.

Robbins, J. A. and D. N. Edgington. 1975. Determination of recent sedimentation rates in Lake Michigan using Pb-210 and Cs-137. Geochim. Cosmochim. Acta 39:285-304.

Robbins, J. A., J. R. Krezoski, and S. C. Mozley. 1977. Radioactivity in sediments of the Great Lakes: Post-depositional redistribution by deposit-feeding organisms. Earth Planetary Sci. Lett. 36:325-333.

Robbins, J. A., E. Landstrom, and M. Wahlgren. 1972. Tributary inputs of soluble trace metals to Lake Michigan. Proc. Conf. Great Lakes Res., Int. Assoc. Great Lakes Res., 15:270-290.

Robertson, C. K. 1979. Quantitative comparison of the significance of methane in the carbon cycles of two small lakes. Arch. Hydrobiol. Beih. Ergebn. Limnol. 12:123-135.

Robertson, J. D. 1941. The function and metabolism of calcium in the Invertebrata. Biol. Rev. 16:106-133.

Robertson, R. N. 1960. Ion transport and respiration. Biol. Rev. 35:231-264.

Rodda, J. C., R. A. Downing, and F. M. Law. 1976. Systematic Hydrology. London, Newnes-Butterworths, 399 pp.

Rodewald-Rudescu, L. 1974. Das Schilfrohr Phragmites communis Trinius. Die Binnengewässer 27, 302 pp.

Rodgers, G. K. 1966. The thermal bar in Lake Ontario, spring 1965 and winter 1965-66. Publ. Great Lakes Res. Div., Univ. Mich. 15:369-374.

Rodhe, W. 1948. Environmental requirements of freshwater plankton algae. Experimental studies in the ecology of phytoplankton. Symbol. Bot. Upsalien. 10(1), 149 pp.

Rodhe, W. 1949. The ionic composition of lake waters. Verh. Int. Ver. Limnol. 10:377-386.

Rodhe, W. 1951. Minor constituents in lake waters. Verh. Int. Ver. Limnol. 11:317-323.

Rodhe, W. 1955. Can plankton production proceed during winter darkness in subarctic lakes? Verh. Int. Ver. Limnol. 12:117-122.

Rodhe, W. 1958. Primärproduktion und Seetypen. Verh. Int. Ver. Limnol. 13:121-141.

Rodhe, W. 1969. Crystallization of eutrophication concepts in Northern Europe. In Eutrophication: Causes, Consequences, Correctives. Washington, D.C., National Academy of Sciences, pp. 50-64.

Rodhe, W. 1974. Plankton, planktic, planktonic. Limnol. Oceanogr. 19:360.

Rodhe, W., R. A. Vollenweider, and A. Nauwerck. 1958. The primary production and standing crop of phytoplankton. In A. A. Buzzati-Traverso, ed. Perspectives in Marine Biology. Berkeley, University of California Press, pp. 299-322.

Rodhe, W., J. E. Hobbie, and R. T. Wright. 1966. Phototrophy and heterotrophy in high mountain lakes. Verh. Int. Ver. Limnol. 16:302-313.

Rodina, A. G. 1965. Metody Vodnoi Mikrobiologii. Moscow, Izdatel'stvo Nauka, 363 pp.

Rodina, A. G. 1972. Methods in Aquatic Microbiology. Translat. and rev. R. R. Colwell and M. S. Zambruski. Baltimore, University Park Press, 461 pp.

Roemer, S. C. and K. D. Hoagland. 1979. Seasonal attenuation of quantum irradiance (400-700 nm) in three Nebraska reservoirs. Hydrobiologia 63:81-92.

Rogers, K. H. and C. M. Breen. 1982. Decomposition of Potamogeton crispus L.: The effects of drying on the pattern of mass and nutrient loss. Aquatic Bot. 12:1-12.

Rogerson, A. 1981. The ecological energetics of Amoeba proteus (Protozoa). Hydrobiologia 85:117-128.

Rohlich, G. A. 1969. Engineering aspects of nutrient removal. In Eutrophication: Causes, Consequences, Correctives. Washington, D.C., Nat. Acad. Sciences. pp. 371–382.

Romagoux, J.-C. 1979. Caracteristiques du microphytobenthos d'un lac volcanique meromitique (Lac Pavin, France). I. Biomasse chlorophyllienne et déterminisme du cycle annuel. Int. Rev. ges. Hydrobiol. 64:303–343.

Romanenko, V. I. 1966. Microbiological processes in the formation and breakdown of organic matter in the Rybinsk Reservoir. Trudy Inst. Biol. Vnutrennikh Vod 13:137–158. (Translated into English, Israel Program for Scientific Translations, Jerusalem, 1969.)

Romanenko, V. I. 1967. Sootnoshenie mezhdu fotosintezom fitoplanktona i destruktsiei organicheskogo veshchestva v vodokhranilishchakh. Trudy Inst. Biol. Vnutrennikh Vod 15:61–74.

Romanenko, V. I., and E. G. Dobrynin. 1973. Potreblenie kisloroda, temnovaia assimiliatsiia CO_2 i intensivnost fotosinteza v natural'nykh i profil'trovannykh probakh vody. Mikrobiologiya 42:573–575.

Romanenko, V. I., and S. I. Kuznetsov. 1974. Ekologiia Mikroorganizmov Presnykh Vodoemov. Laboratornoe Rukovodstvo. Leningrad, Izdatel'stvo Nauka, 194 pp.

Roos, P. J., A. F. Post, and J. M. Revier. 1981. Dynamics and architecture of reed periphyton. Verh. Int. Ver. Limnol. 21:948–953.

Ross, H. H. 1963. Stream communities and terrestrial biomes. Arch. Hydrobiol. 59:235–242.

Rossolimo, L. 1935. Die Boden-Gasausscheidung und das Sauerstoffregime der Seen. Verh. Int. Ver. Limnol. 7:539–561.

Rosswall, T., ed. 1973. Modern Methods in the Study of Microbial Ecology. Bull. Ecological Research Committee, Swedish Natural Science Research Council 17, 511 pp.

Roth, J. C. 1968. Benthic and limnetic distribution of three Chaoborus species in a southern Michigan lake (Diptera, Chaoboridae). Limnol. Oceanogr. 13:242–249.

Round, F. E. 1957a. Studies on bottom-living algae in some lakes of the English Lake District. Part I. Some chemical features of the sediments related to algal productivities. J. Ecol. 45:133–148.

Round, F. E. 1957b. Studies on bottom-living algae in some lakes of the English Lake District. Part II. The distribution of Bacillariophyceae on the sediments. J. Ecol. 45:343–360.

Round, F. E. 1957c. Studies on bottom-living algae in some lakes of the English Lake District. Part III. The distribution on the sediments of algal groups other than the Bacillariophyceae. J. Ecol. 45:649–664.

Round, F. E. 1960. Studies on bottom-living algae in some lakes of the English Lake District. Part IV. The seasonal cycles of the Bacillariophyceae. J. Ecol. 48:529–547.

Round, F. E. 1961a. Studies on bottom-living algae in some lakes of the English Lake District. Part V. The seasonal cycles of the Cyanophyceae. J. Ecol. 49:31–38.

Round, F. E. 1961b. Studies on bottom-living algae in some lakes of the English Lake District. Part VI. The effect of depth on the epipelic algal community. J. Ecol. 49:245–254.

Round, F. E. 1961c. The diatoms of a core from Esthwaite Water. New Phytol. 60:43–59.

Round, F. E. 1964a. The ecology of benthic algae. In D. F. Jackson, ed. Algae and Man. New York, Plenum Press, pp. 138–184.

Round, F. E. 1964b. The diatom sequence in lake deposits: Some problems of interpretation. Verh. Int. Ver. Limnol. 15:1012–1020.

Round, F. E. 1965. The epipsammon; A relatively unknown freshwater algal association. Brit. Phycol. Bull. 2:456–462.

Round, F. E. 1971. The growth and succession of algal populations in freshwaters. Mitt. Int. Ver. Limnol. 19:70–99.

Round, F. E. 1972. Patterns of seasonal succession of freshwater epipelic algae. Brit. Phycol. J. 7:213–220.

Round, F. E., and J. W. Eaton. 1966. Persistent, vertical-migration rhythms in benthic microflora. III. The rhythm of epipelic algae in a freshwater pond. J. Ecol. 54:609–615.

Round, F. E., and C. M. Happey. 1965. Persistent, vertical-migration rhythms in benthic microflora. IV. A diurnal rhythm of the epipelic diatom association in non-tidal flowing water. Brit. Phycol. Bull. 2:463–471.

Ruck, P. 1965. The components of the visual system of a dragonfly. J. Gen. Physiol. 49:289–307.

Rudd, J. W. M. and R. D. Hamilton. 1975. Factors controlling rates of methane oxidation and the distribution of the methane oxidizers in a small stratified lake. Arch. Hydrobiol. 75:522–538.

Rudd, J. W. M. and R. D. Hamilton. 1978. Methane cycling in a eutrophic shield lake and its effects on whole lake metabolism. Limnol. Oceanogr. 23:337–348.

Rudd, J. W. M., R. D. Hamilton, and N. E. R. Campbell. 1974. Measurement of microbial oxidation of methane in lake water. Limnol. Oceanogr. 19:519–524.

Rudd, J. W. M., A. Furutani, R. J. Flett, and R. D. Hamilton. 1976. Factors controlling methane oxidation in Shield lakes: The role of nitrogen fixation and oxygen. Limnol. Oceanogr. 21:357–364.

Rudd, J. W. M. and C. D. Taylor. 1980. Methane cycling in aquatic environments. Advances Aquatic Microbiol. 2:77–150.

Rudescu, L., C. Niculescu, and I. P. Chivu. 1965. Monografia Stufului din Delta Dunării. Bucharest, Editura Acad. Rep. Soc. Romania, 542 pp.

Ruschke, R. 1968. Die Bedeutung von Wassermyxobakterien für den Abbau organischen Materials. Mitt. Int. Ver. Limnol. 14:164–167.

Rusness, D., and R. H. Burris. 1970. Acetylene reduction (nitrogen fixation) in Wisconsin lakes. Limnol. Oceanogr. 15:808-813.

Ruttner, F. 1931. Hydrographische und hydrochemische Beobachtungen auf Java, Sumatra und Bali. Arch. Hydrobiol. (suppl.) 8:197-454.

Ruttner, F. 1933. Über metalimnische Sauerstoffminimum. Die Naturwissenschaften 1933(21-23):401-404.

Ruttner, F. 1947. Zur Frage der Karbonatassimilation der Wasserpflanzen. I. Die beiden Haupttypen der Kohlenstoffaufnahme. Öst. Bot. Z. 94:265-294.

Ruttner, F. 1948. Zur Frage der Karbonatassimilation der Wasserpflanzen. II. Das Verhalten von Elodea canadensis und Fontinalis antipyretica in Lösungen von Natrium- bzw. Kaliumbikarbonat. Öst. Bot. Z. 95:208-238.

Ruttner, F. 1960. Über die Kohlenstoffaufnahme bei Algen aus der Rhodophyceen-Gattung Batrachospermum. Schweiz. Z. Hydrol. 22:280-291.

Ruttner, F. 1963. Fundamentals of Limnology. (Translat. D. G. Frey and F. E. J. Fry.) Toronto, University of Toronto Press, 295 pp.

Ruttner-Kolisko, A. 1964. Über die labile Periode im Fortpflanzungszyklus der Rädertiere. Int. Rev. ges. Hydrobiol. 49:473-482.

Ruttner-Kolisko, A. 1972. Rotatoria. In Das Zooplankton der Binnengewässer. 1. Teil. Die Binnengewässer 26(Pt. 1):99-234.

Ruttner-Kolisko, A. 1975. The vertical distribution of planktonic rotifers in a small alpine lake with a sharp oxygen depletion (Lunzer Obersee). Verh. Int. Ver. Limnol. 19:1286-1294.

Ruttner-Kolisko, A. 1980. The abundance and distribution of Filinia terminalis in various types of lakes as related to temperature, oxygen, and food. Hydrobiologia 73:169-175.

Ryhänen, R. 1968. Die Bedeutung der Humussubstanzen im Stoffhaushalt der Gewässer Finnlands. Mitt. Int. Ver. Limnol. 14:168-178.

Safferman, R. S., and M.-E. Morris. 1963. Algal virus: Isolation. Science 140:679-680.

Sager, P. E., and A. D. Hasler. 1969. Species diversity in lacustrine phytoplankton. I. The components of the index of diversity from Shannon's formula. Amer. Nat. 102:51-59.

Sakamoto, M. 1966. Primary production of phytoplankton community in some Japanese lakes and its dependence on lake depth. Arch. Hydrobiol. 62:1-28.

Sale, P. J. M. and R. G. Wetzel. 1982. Growth and metabolism of Typha species in relation to cutting treatments. Aquatic Bot. (In press)

Sandercock, G. A. 1967. A study of selected mechanisms for the coexistence of Diaptomus spp. in Clarke Lake, Ontario. Limnol. Oceanogr. 12:97-112.

Sand-Jensen, K. 1977. Effect of epiphytes on eelgrass photosynthesis. Aquatic Bot. 3:55-63.

Sand-Jensen, K. 1978. Metabolic adaptation and vertical zonation of Littorella uniflora (L.) Aschers. and Isoetes lacustris L. Aquatic Bot. 4:1-10.

Sand-Jensen, K. and M. Søndergaard. 1979. Distribution and quantitative development of aquatic macrophytes in relation to sediment characteristics in oligotrophic Lake Kalgaard, Denmark. Freshwat. Biol. 9:1-11.

Sand-Jensen, K. and M. Søndergaard. 1981. Phytoplankton and epiphyte development and their shading effect on submerged macrophytes in lakes of different nutrient status. Int. Rev. ges. Hydrobiol. 66:529-552.

Sandner, H., and J. Wilkialis. 1972. Leech communities (Hirudinea) in the Mazurian and Bialystok regions and the Pomeranian Lake District. Ekol. Polska 20:345-365.

Sanger, J. E. 1971a. Identification and quantitative measurement of plant pigments in soil humus layers. Ecology 52:959-963.

Sanger, J. E. 1971b. Quantitative investigations of leaf pigments from their inception in buds through autumn coloration to decomposition in falling leaves. Ecology 52:1075-1089.

Sanger, J. E., and E. Gorham. 1970. The diversity of pigments in lake sediments and its ecological significance. Limnol. Oceanogr. 15:59-69.

Sanger, J. E., and E. Gorham. 1973. A comparison of the abundance and diversity of fossil pigments in wetland peats and woodland humus layers. Ecology 54:605-611.

Särkkä, J. and J. Aho. 1980. Distribution of aquatic Oligochaeta in the Finnish Lake District. Freshwat. Biol. 10:197-206.

Sarvala, J. 1979. Effect of temperature on the duration of egg, nauplius and copepodite development of some freshwater benthic Copepoda. Freshwat. Biol. 9:515-534.

Sastroutomo, S. S. 1980a. Environmental control of turion formation in curly pondweed (Potamogeton crispus). Physiol. Plant. 49:261-264.

Sastroutomo, S. S. 1980b. Dormancy and germination in axillary turions of Hydrilla verticillata. Bot. Mag. Tokyo 93:265-273.

Satake, K. and Y. Saijo. 1978. Mechanism of lamination in bottom sediment of the strongly acid Lake Katanuma. Arch. Hydrobiol. 83:429-442.

Sauberer, F. 1939. Beiträge zur Kenntnis des Lichtklimas einiger Alpenseen. Int. Rev. ges. Hydrobiol. 39:20-55.

Sauberer, F. 1950. Die spektrale Strahlungsdurchlässigkeit des Eises. Wetter u. Leben 2:193-197.

Sauberer, F. 1962. Empfehlungen für die Durchführung von Strahlungsmessungen an und in Gewässern. Mitt. Int. Ver. Limnol. 11, 77 pp.

Saunders, G. W. 1957. Interrelations of dissolved organic matter and phytoplankton. Bot. Rev. 23:389-410.

Saunders, G. W. 1963. The biological characteristics of fresh water. Publ. Great Lakes Res. Div., Univ. Mich. 10:245-257.

Saunders, G. W. 1969. Some aspects of feeding in

zooplankton. *In* Eutrophication: Causes, Consequences, Correctives. Washington, D.C., National Academy of Sciences, pp. 556–573.

Saunders, G. W. 1971. Carbon flow in the aquatic system. *In* J. Cairns, Jr., ed. The Structure and Function of Fresh-Water Microbial Communities. Res. Div. Monogr. 3, Virginia Polytechnic Inst. pp. 31–45.

Saunders, G. W. 1972a. The transformation of artificial detritus in lake water. Mem. Ist. Ital. Idrobiol. 29 (suppl.):261–288.

Saunders, G. W., 1972b. Summary of the general conclusions of the symposium. Mem. Ist. Ital. Idrobiol. 29(suppl.):533–540.

Saunders, G. W. 1972c. Potential heterotrophy in a natural population of *Oscillatoria agardhii* var. *isothrix* Skuja. Limnol. Oceanogr., 17:704–711.

Saunders, G. W., and T. A. Storch. 1971. Coupled oscillatory control mechanism in a planktonic system. Nature (New Biol.), 230:58–60.

Saunders, G. W., K. W. Cummins, D. Z. Gak, E. Pieczyńska, V. Straškrabová, and R. G. Wetzel. 1980. Organic matter and decomposers. *In* E. D. Le Cren and R. H. Lowe-McConnell, eds. The Functioning of Freshwater Ecosystems. Cambridge, Cambridge Univ. Press, pp. 341–392.

Savostin, P. 1972. Microbial transformation of silicates. Z. Pflanzenernährung Bodenkunde, 132:37–45.

Sawyer, C. N. 1947. Fertilization of lakes by agricultural and urban drainage. J. New England Water Works Assoc. 61:109–127.

Sawyer, R. T. 1974. Leeches (Annelida: Hirudinea). *In* C. W. Hart, Jr., and S. L. H. Fuller, eds. Pollution Ecology of Freshwater Invertebrates. New York, Academic Press, pp. 81–142.

Schelske, C. L. 1962. Iron, organic matter, and other factors limiting primary productivity in a marl lake. Science, 136:45–46.

Schelske, C. L., L. E. Feldt, M. A. Santiago, and E. F. Stoermer. 1972. Nutrient enrichment and its effects on phytoplankton production and species composition in Lake Superior. Proc. Conf. Great Lakes Res., Int. Assoc. Great Lakes Res., 15:149–165.

Schelske, C. L., F. F. Hooper, and E. J. Haertl. 1962. Responses of a marl lake to chelated iron and fertilizer. Ecology, 43:646–653.

Schelske, C. L., and E. F. Stoermer. 1971. Eutrophication, silica depletion, and predicted changes in algal quality in Lake Michigan. Science. 173:423–424.

Schelske, C. L., and E. F. Stoermer. 1972. Phosphorus, silica, and eutrophication of Lake Michigan. *In* G. E. Likens, ed. Nutrients and Eutrophication. Special Symposium, Amer. Soc. Limnol. Oceanogr., 1:157–171.

Schiegl, W. E. 1972. Deuterium content of peat as a paleoclimatic recorder. Science, 175:512–513.

Schindler, D. W. 1968. Feeding, assimilation and respiration rates of *Daphnia magna* under various environmental conditions and their relation to production estimates. J. Anim. Ecol., 37:369–385.

Schindler, D. W. 1970. Production of phytoplankton and zooplankton in Canadian Shield lakes. *In* Z. Kajak and A. Hillbricht-Ilkowska, eds. Productivity Problems of Freshwaters. Warsaw, PWN Polish Scientific Publishers, pp. 311–331.

Schindler, D. W. 1974. Eutrophication and recovery in experimental lakes: Implications for lake management. Science, 184:897–899.

Schindler, D. W. 1978. Predictive eutrophication models. Limnol. Oceanogr. 23:1080–1081.

Schindler, D. W., G. J. Brunskill, S. Emerson, W. S. Broecker, and T.-H. Peng. 1972. Atmospheric carbon dioxide: Its role in maintaining phytoplankton standing crops. Science. 177:1192–1194.

Schindler, D. W., V. E. Frost, and R. V. Schmidt. 1973. Production of epilithiphyton in two lakes of the Experimental Lakes Area, northwestern Ontario. J. Fish. Res. Bd. Canada, 30:1511–1524.

Schindler, D. W., E. J. Fee, and T. Ruszczynski. 1978. Phosphorus input and its consequences for phytoplankton standing crop and production in the Experimental Lakes Area and in similar lakes. J. Fish. Res. Bd. Can. 35:190–196.

Schmidt, W. 1915. Über den Energie-gehalt der Seen. Mit Beispielen vom Lunzer Untersee nach Messungen mit einen enfachen Temperaturlot. Int. Rev. Hydrobiol. Suppl., 6. (Not seen in original.)

Schmidt, W. 1928. Über Temperatur und Stabilitätsverhältnisse von Seen. Geographiska Annaler, 10:145–177.

Schmitt, M. R. and M. S. Adams. 1981. Dependence of rates of apparent photosynthesis on tissue phosphorus concentrations in *Myriophyllum spicatum* L. Aquatic Bot. 11:379–387.

Schnitzer, M. 1971. Metal-organic matter interactions in soils and waters. *In* S. J. Faust and J. V. Hunter, eds. Organic Compounds in Aquatic Environments. New York, Marcel Dekker, Inc., pp. 297–315.

Schnitzer, M., and S. U. Khan. 1972. Humic Substances in the Environment. New York, Marcel Dekker, Inc., 327 pp.

Schönborn, W. 1962. Über Planktismus und Zyklomorphose bei *Difflugia limnetica* (Levander) Penard. Limnologica, 1:21–34.

Schreiner, S. P. 1980. Effect of water hyacinth on the physicochemistry of a south Georgia pond. J. Aquatic Plant Manage. 18:9–12.

Schreiter, T. 1928. Untersuchungen über den Einfluss einer Helodeawucherung auf das Netzplankton des Hirschberger Grossteiches in Böhmen in den Jahren 1921 bis 1925 incl. Sborník Výzk. Úst. Zeměd. RČS 61: 98 pp.

Schröder, R. 1973. Die Freisetzung von Pflanzennährstoffen im Schilfgebiet und ihr Transport in das Freiwasser am Beispiel des Bodensee-Untersees. Arch. Hydrobiol., 71:145–158.

Schröder, R. 1975. Release of plant nutrients from reed borders and their transport into the open waters of the Bodensee-Untersee. Symp. Biol. Hungarica 15:21–27.

Schults, D. W. and K. W. Malueg. 1971. Uptake of radiophosphorus by rooted aquatic plants. Proc. Third Nat. Symp. Radioecology. pp. 417–424.

Schultz, D. M., and J. G. Quinn. 1973. Fatty acid composition of organic detritus from *Spartina alterniflora*. Estuarine and Coastal Mar. Sci. 1:177–190.

Schwoerbel, J., and G. C. Tillmanns. 1964a. Konzentrationsabhängige Aufnahme von wasserlöslichem PO_4-P bei submersen Wasserpflanzen. Naturwissenschaften 51:319–320.

Schwoerbel, J., and G. C. Tillmanns. 1964b. Untersuchungen über die Stoffwechseldynamik in Fliessgewässern. I. Die Rolle höherer Wasserpflanzen: *Callitriche hamulata* Kütz. Arch. Hydrobiol. (suppl.) 28:245–258.

Schwoerbel, J., and G. C. Tillmanns. 1964c. Untersuchungen über die Stoffwechseldynamik in Fliessgewässern. II. Experimentelle Untersuchungen über die Ammoniumaufnahme und pH-Änderung im Wasser durch *Callitriche hamulta* Kütz. und *Fontinalis antipyretica* L. Arch. Hydrobiol. (suppl.) 28:259–267.

Schwoerbel, J., and G. C. Tillmanns. 1972. Ammonium-Adaptation bei submersen Phanerogamen in situ. Arch. Hydrobiol. (suppl.) 42:139–141.

Schwoerbel, J. and G. C. Tillmanns. 1977. Nitrataufnahme aus dem Wasser und Nitratreduktase-Aktivität bei *Fontinalis antipyretica* L. im Hell-Dunkel-Wechsel. Arch. Hydrobiol. (suppl.) 48:412–423.

Scott, J. T., G. E. Myer, R. Stewart, and E. G. Walther. 1969. On the mechanism of Langmuir circulations and their role in epilimnion mixing. Limnol. Oceanogr. 14:493–503.

Scott, W. 1924. The dirunal oxygen pulse in Eagle (Winona) Lake. Proc. Indiana Acad. Sci. 33(1923):311–314.

Sculthorpe, C. D. 1967. The Biology of Aquatic Vascular Plants. New York, St. Martin's Press, 610 pp.

Seadler, A. W. and N. A. Alldridge. 1977. The translocation of radioactive phosphorus by the aquatic vascular plant *Najas minor*. Ohio J. Sci. 77:76–80.

Sebestyén, O. 1949. Studies of detritus drifts in Lake Balaton. Verh. Int. Ver. Limnol. 10:414–419.

Sederholm, H., A. Mauranen, and L. Montonen. 1973. Some observations on the microbial degradation of humous substances in water. Verh. Int. Ver. Limnol., 18:1301–1305.

Seghers, B. H. 1975. Role of gill rakers in size-selective predation by lake whitefish, *Coregonus clupeaformis* (Mitchell). Verh. Int. Ver. Limnol. 19:2401–2405.

Seitz, A. 1980. The coexistence of three species of *Daphnia* in the Klostersee. I. Field studies on the dynamics of reproduction. Oecologia 45:117–130.

Seitz, W. R. 1982. Fluorescence methods for studying speciation of pollutants in water. Trends Anal. Chem. (In press)

Senft, W. H. 1978. Dependence of light-saturated rates of algal photosynthesis on intracellular concentrations of phosphorus. Limnol. Oceanogr. 23:709–718.

Serruya, C., U. Pollingher, and M. Gophen. 1975. N and P distribution in Lake Kinneret (Israel) with emphasis on dissolved organic nitrogen. Oikos 26:1–8.

Šesták, Z., J. Čatský, and P. G. Jarvis, eds. 1971. Plant Photosynthetic Production Manual of Methods. The Hague, W. Junk N. V. Publishers, 818 pp.

Shan, R. K. 1974. Reproduction in laboratory stocks of *Pleuroxus* (Chydoridae, Cladocera) under the influence of photoperiod and light intensity. Int. Rev. ges. Hydrobiol. 59:643–666.

Shannon, C. E. and W. Weaver. 1949. The Mathematical Theory of Communication. Urbana, Univ. Illinois Press, 117 pp.

Shannon, E. L. 1953. The production of root hairs by aquatic plants. Amer. Midland. Nat. 50:474–479.

Shapiro, J. 1957. Chemical and biological studies on the yellow organic acids of lake water. Limnol. Oceanogr. 2:161–179.

Shapiro, J. 1960. The cause of a metalimnetic minimum of dissolved oxygen. Limnol. Oceanogr. 5:216–227.

Shapiro, J. 1964. Effect of yellow organic acids on iron and other metals in water. J. Amer. Water Wks. Assoc. 56:1062–1082.

Shapiro, J. 1966. The relation of humic color to iron in natural waters. Verh. Int. Ver. Limnol. 16:477–484.

Shapiro, J. 1969. Iron in natural waters—its characteristics and biological availability as determined with the ferrigram. Verh. Int. Ver. Limnol. 17:456–466.

Shapiro, J. and G. E. Glass. 1975. Synergistic effects of phosphate and manganese on growth of Lake Superior algae. Verh. Int. Ver. Limnol. 19:395–404.

Sheath, R. G. and J. A. Hellebust. 1978. Comparison of algae in the euplankton, tychoplankton, and periphyton of a tundra pond. Can. J. Bot. 56:1472–1483.

Sheldon, R. B. and C. W. Boylen. 1975. Factors affecting the contribution by epiphytic algae to the primary productivity of an oligotrophic freshwater lake. Appl. Microbiol. 30:657–667.

Sheldon, R. B. and C. W. Boylen. 1977. Maximum depth inhabited by aquatic vascular plants. Amer. Midland Nat. 97:248–254.

Shelford, V. E. 1918. Conditions of existence. In Freshwater Biology. H. B. Ward and G. C. Whipple, eds. New York, John Wiley & Sons, Inc., pp. 21–60.

Shilo, M. 1971. Biological agents which cause lysis of

blue-green algae. Mitt. Int. Ver. Limnol. *19*:206–213.

Shtegman, B. K., ed. 1966. Produtsirovanie i Krugovorot Organicheskogo Veshchestva vo Vnutrennikh Vodoemakh. (Production and circulation of organic matter in inland waters.) (Translated into English, Israel Program for Scientific Translations, Jerusalem, 1969.) Trudy Inst. Biol. Vnutrennikh Vod *13*, 287 pp.

Shukla, S. S., J. K. Syers, J. D. H. Williams, D. E. Armstrong, and R. F. Harris. 1971. Sorption of inorganic phosphate by lake sediments. Proc. Soil Sci. Soc. Amer. *35*:244–249.

Shulman, M. D., and R. A. Bryson. 1961. The vertical variation of wind-driven currents in Lake Mendota. Limnol. Oceanogr. *6*:347–355.

Siebeck, O. 1960. Untersuchungen über die Vertikalwanderung planktischer Crustaceen unter Berücksichtigung der Strahlungsverhältnisse. Int. Rev. ges. Hydrobiol. *45*:381–454.

Siebeck, O. 1968. "Uferflucht" und optische Orientierung pelagischer Crustaceen. Arch. Hydrobiol. (suppl.) *35*:1–118.

Siebeck, O., and J. Ringelberg. 1969. Spatial orientation of planktonic crustaceans. I. The swimming behaviour in a horizontal plane. 2. The swimming behaviour in a vertical plane. Verh. Int. Ver. Limnol. *17*:831–847.

Simpson, J. H. and J. D. Woods. 1970. Temperature microstructure in a freshwater thermocline. Nature *226*:832–834.

Simpson, T. L. and J. J. Gilbert. 1974. Gemmulation, gemmule hatching, and sexual reproduction in fresh-water sponges. II. Life cycle events in young, larva-produced sponges of *Spongilla lacustris* and an unidentified species. Trans. Amer. Microsc. Soc. *93*:39–45.

Simola, H. 1977. Diatom succession in the formation of annually laminated sediments in Lovojärvi, a small eutrophicated lake. Ann. Bot. Fennica *14*:143–148.

Siver, P. A. 1978. Development of diatom communities on *Potamogeton robbinsii* Oakes. Rhodora *80*:417–430.

Siver, P. A. 1980. Microattachment patterns of diatoms on leaves of *Potamogeton robbinski* Oakes. Trans. Amer. Microsc. Soc. *99*:217–220.

Sjörs, H. 1961. Surface patterns in Boreal peatland. Endeavour *20*:217–224.

Skorik, L. V., K. S. Vladimirova, G. A. Yenaki, and I. K. Palamarchuk. 1972. Organic matter in Kiev Reservoir soils and its role in the development of benthic algae. Hydrobiol. J. *8*:63–66.

Sládeček, V. 1973. System of water quality from the biological point of view. Arch. Hydrobiol. Beih. Ergebn. Limnol. *7*, 218 pp.

Sládeček, V., and A. Sládečková. 1963. Limnological study of the Reservoir Sedlice near Želiv. XXIII. Periphyton production. Sbornik Vysoké Školy Chem.-Technol. Praze, Technol. Vody *7*:77–133.

Sládeček, V., and A. Sládečková. 1964. Determination of the periphyton production by means of the glass slide method. Hydrobiologia *23*:125–158.

Sládečková, A. 1960. Limnological study of the Reservoir Sedlice near Želiv. XI. Periphyton stratification during the first year-long period (June 1957–July 1958). Sci. Pap. Inst. Chem. Technol. Prague, Faculty of Technol. Fuel Water *4*:143–261.

Sládečková, A. 1962. Limnological investigation methods for the periphyton ("Aufwuchs") community. Bot. Rev. *28*:286–350.

Slobodkin, L. B. 1954. Population dynamics in *Daphnia obtusa* Kurz. Ecol. Monogr. *24*:69–88.

Slobodkin, L. B. 1960. Ecological energy relationships at the population level. Amer. Nat. *94*:213–236.

Slobodkin, L. B. 1962. Energy in animal ecology. Adv. Ecol. Res. *1*:69–101.

Smayda, T. J., and B. J. Boleyn. 1965. Experimental observations on the flotation of marine diatoms. I. *Thalassiosira* cf. *nana, Thalassiosira rotula* and *Nitzschia seriata.* Limnol. Oceanogr. *10*:499–509.

Smayda, T. J., and B. J. Boleyn. 1966a. Experimental observations on the flotation of marine diatoms. II. *Skeletonema costatum* and *Rhizosolenia setigera.* Limnol. Oceanogr. *11*:18–34.

Smayda, T. J., and B. J. Boleyn. 1966b. Experimental observations on the flotation of marine diatoms. III. *Bacteriastrum hyalinum* and *Chaetoceros lauderi.* Limnol. Oceanogr. *11*:35–43.

Smith, F. A. 1967. Rates of photosynthesis in Characean cells. I. Photosynthetic $^{14}CO_2$ fixation by *Nitella translucens.* J. Exp. Bot. *18*:509–517.

Smith, F. A. and N. A. Walker. 1980. Photosynthesis by aquatic plants: Effects of unstirred layers in relation to the assimilation of CO_2 and HCO_3^- and to carbon isotopic discrimination. New Phytol. *86*:245–259.

Smith, I. R. 1975. Turbulence in lakes and rivers. Sci. Publ. Freshwater Biol. Assoc. U.K. *29*. 79 pp.

Smith, I. R., and I. J. Sinclair. 1972. Deep water waves in lakes. Freshwat. Biol. *2*:387–399.

Smith, M. W. 1969. Changes in environment and biota of a natural lake after fertilization. J. Fish. Res. Bd. Canada *26*:3101–3132.

Smith, R. C., and W. H. Wilson, Jr. 1972. Photon scalar irradiance. Appl. Optics *11*:934–938.

Smith, R. C., J. E. Tyler, and C. R. Goldman. 1973. Optical properties and color of Lake Tahoe and Crater Lake. Limnol. Oceanogr. *18*:189–199.

Smith, R. L., and M. J. Klug. 1981. Electron donors utilized by sulfate-reducing bacteria in eutrophic lake sediments. Appl. Environ. Microbiol. *42*:116–121.

Smith, R. L. and M. J. Klug. 1981. Reduction of sulfur compounds in sediments of a eutrophic lake basin. Appl. Environ. Microbiol. *41*:1230–1237.

Smol, J. P. 1980. Fossil synuracean (Chrysophyceae) scales in lake sediments: A new group of paleoindicators. Can. J. Bot. *58*:458–465.

Smyly, W. J. P. 1973. Bionomics of *Cyclops strenuus abyssorum* Sars (Copepoda: Cyclopoida). Oecologia 11:163–186.

Smyly, W. J. P. 1980. Food and feeding of aquatic larvae of the midge *Chaoborus flavicans* (Meigen) (Diptera; Chaoboridae) in the laboratory. Hydrobiologia 70:179–188.

Snell, T. W. 1980. Blue-green algae and selection in rotifer populations. Oecologia 46:343–346.

Soeder, C. J. 1965. Some aspects of phytoplankton growth and activity. Mem. Ist. Ital. Idrobiol. 18(Suppl.):47–59.

Sokolova, G. A. 1961. Rol' zhalezobakterii v dinamike zheleza v Glubokom Ozere. Trudy Vses. Gidrobiol. Obshch. 11:5–11.

Solski, A. 1962. Mineralizacja roślin wodnych. I. Uwalnianie fosforu i potasu przez wymywanie. Pol. Arch. Hydrobiol. 10:167–196.

Sommers, L. E., R. F. Harris, J. D. H. Williams, D. E. Armstrong, and J. K. Syers. 1970. Determination of total organic phosphorus in lake sediments. Limnol. Oceanogr. 15:301–304.

Søndergaard, M. 1979. Light and dark respiration and the effect of the lacunal system on refixation of CO_2 in submerged aquatic plants. Aquatic Bot. 6:269–283.

Søndergaard, M. 1981. Kinetics of extracellular release of ^{14}C-labelled organic carbon by submerged macrophytes. Oikos 36:331–347.

Søndergaard, M. and K. Sand-Jensen. 1978. Total autotrophic production in oligotrophic Lake Kalgaard, Denmark. Verh. Int. Ver. Limnol. 20:667–673.

Søndergaard, M. and K. Sand-Jensen. 1979. Carbon uptake by leaves and roots of *Littorella uniflora* (L.) Aschers. Aquatic Bot. 6:1–12.

Søndergaard, M. and R. G. Wetzel. 1980. Photorespiration and internal recycling of CO_2 in the submersed angiosperm *Scirpus subterminalis*. Can. J. Bot. 58:591–598.

Sorokin, Ju. I. 1964a. On the primary production and bacterial activities in the Black Sea. J. Cons. Int. Explor. Mer 29:41–60.

Sorokin, Ju. I. 1964b. On the trophic role of chemosynthesis in water bodies. Int. Rev. ges. Hydrobiol. 49:307–324.

Sorokin, Ju. I. 1965. On the trophic role of chemosynthesis and bacterial biosynthesis in water bodies. Mem. Ist. Ital. Idrobiol. 18(suppl.):187–205.

Sorokin, Ju. I. 1966. Vzaimosviaz mikrobiologicheskikh protessov krugovorota sepy i ugleroda v meromikticheskom Ozere Belovod. *In* Plankton i Bentos. Trudy Inst. Biol. Vnutrennikh Vod 12:332–355.

Sorokin, Ju. I. 1968. The use of ^{14}C in the study of nutrition of aquatic animals. Mitt. Int. Ver. Limnol. 16, 41 pp.

Sorokin, Ju. I. 1970. Interrelations between sulphur and carbon turnover in meromictic lakes. Arch. Hydrobiol. 66:391–446.

Sorokin, Y. I., and H. Kadota, eds. 1972. Techniques for the Assessment of Microbial Production and Decomposition in Fresh Waters. Int. Biol. Program Handbook 23. Oxford, Blackwell Scientific Publications, 112 pp.

Sorokin, Ju, I. and E. B. Paveljeva. 1972. On the quantitative characteristics of the pelagic ecosystem of Dalnee Lake (Kamchatka). Hydrobiologia 40:519–552.

Sorsa, K. 1979. Primary production of epipelic algae in Lake Suomunjärvi, Finnish North Karelia. Ann. Bot. Fennici 16:351–366.

Spain, J. D., G. M. Wernert, and D. W. Hubbard, 1976. The structure of the spring thermal bar in Lake Superior, II. J. Great Lakes Res. 2:296–306.

Sparrow, F. K., Jr. 1968. Ecology of freshwater fungi. *In* G. C. Ainsworth and A. S. Sussman, eds. The Fungi. Vol. III. The Fungal Population. New York, Academic Press, pp. 41–93.

Spence, D. H. N. 1976. Light and plant response in fresh water. *In* G. C. Evans, R. Bainbridge, and O. Rackham, Eds. Light as an Ecological Factor: II. Oxford, Blackwell Scientific Publs., pp. 93–133.

Spence, D. H. N. 1982. The zonation of plants in freshwater lakes. Adv. Ecol. Res. 12:37–125.

Spence, D. H. N., and J. Chrystal. 1970a. Photosynthesis and zonation of freshwater macrophytes. I. Depth distribution and shade tolerance. New Phytol. 69:205–215.

Spence, D. H. N., and J. Chrystal. 1970b. Photosynthesis and zonation of freshwater macrophytes. II. Adaptability of species of deep and shallow water. New Phytol. 69:217–227.

Spence, D. H. N., R. M. Campbell, and J. Chrystal. 1971. Productivity of submerged freshwater macrophytes. Hidrobiologia 12:169–176.

Spencer, M. J. 1978. Microbial activity and biomass relationships in 26 oligotrophic to mesotrophic lakes in South Island, New Zealand. Verh. Int. Ver. Limnol. 20:1175–1181.

Sprules, W. G. 1972. Effects of size-selective predation and food competition on high altitude zooplankton communities. Ecology 53:375–386.

Stabel, H.-H., K. Moaledj, and J. Overbeck. 1979. On the degradation of dissolved organic molecules from Plusssee by oligocarbophilic bacteria. Arch. Hydrobiol. Beih. Ergebn. Limnol. 12:95–104.

Stadelmann, P. 1971. Stickstoffkreislauf und Primärproduktion im mesotrophen Vierwaldstättersee (Horwer Bucht) und im eutrophen Rotsee, mit besonderer Berücksichtigung des Nitrats als limitierenden Faktors. Schweiz. Z. Hydrol. 33:1–65.

Stahl, J. B. 1959. The developmental history of the chironomid and *Chaoborus* faunas of Myers Lake. Invest. Indiana Lakes Streams 5:47–102.

Stahl, J. B. 1966a. The ecology of *Chaoborus* in Myers Lake, Indiana. Limnol. Oceanogr. 11:177–183.

Stahl, J. B. 1966b. Coexistence in *Chaoborus* and its

ecological significance. Invest. Indiana Lakes Streams 7:99–113.

Stahl, J. B. 1969. The uses of chironomids and other midges in interpreting lake histories. Mitt. Int. Ver. Limnol. 17:111–125.

Stallard, R. F. 1980. Major element geochemistry of the Amazon River system. Ph.D. Disseration, Massachusetts Institute of Technology.

Stallard, R. F., and J. M. Edmond, 1981. Geochemistry of the Amazon. I. Precipitation chemistry and the marine contribution to the dissolved load at the time of peak discharge. J. Geophys. Res. 86:9844–9858.

Stangenberg, M. 1968. Bacteriostatic effects of some algae- and Lemna minor extracts. Hydrobiologia 32:88–96.

Stangenberg-Oporowska, K. 1967. Sodium contents in carp-pond waters in Poland. Pol. Arch. Hydrobiol. 14:11–17.

Stanier, R. Y. 1973. Autotrophy and heterotrophy in unicellular blue-green algae. In N. G. Carr and B. A. Whitton, eds. The Biology of Blue-Green Algae. Berkeley, University of California Press, pp. 501–518.

Stanier, R. Y. and G. Cohen-Bazire. 1977. Phototrophic prokaryotes: The cyanobacteria. Ann. Rev. Microbiol. 31:225–274.

Stanley, D. W. 1976. Productivity of epipelic algae in tundra ponds and a lake near Barrow, Alaska. Ecology 57:1015–1024.

Stanley, R. A. 1972. Photosynthesis in Eurasian watermilfoil (Myriophyllum spicatum L.) Plant Physiol. 50:149–151.

Starkweather, P. L. 1975. Diel patterns of grazing in Daphnia pulex Leydig. Verh. Int. Ver. Limnol. 19:2851–2857.

Starkweather, P. L. 1980a. Behavioral determinants of diet quantity and diet quality in Brachionus calyciflorus. In W. C. Kerfoot, Editor. Evolution and Ecology of Zooplankton Communities. Hanover, NH, Univ. Press New England, pp. 151–157.

Starkweather, P. L. 1980b. Aspects of the feeding behavior and trophic ecology of suspension-feeding rotifers. Hydrobiologia 73:63–72.

Starkweather, P. L. and K. G. Bogdan. 1980. Detrital feeding in natural zooplankton communities: Discrimination between live and dead algal foods. Hydrobiologia 73:83–85.

Starkweather, P. L. and J. J. Gilbert. 1977. Feeding in the rotifer Brachionus calyciflorus. II. Effect of food density on feeding rates using Euglena gracilis and Rhodotorula glutinis. Oecologia 28:133–139.

Starkweather, P. L. and J. J. Gilbert. 1978a. Feeding in the rotifer Brachionus calyciflorus. IV. Selective feeding on tracer particles as a factor in trophic ecology and in situ technique. Verh. Int. Ver. Limnol. 20:2389–2394.

Starkweather, P. L. and J. J. Gilbert. 1978b. Radiotracer determination of feeding in Brachionus calyciflorus: The importance of gut passage times. Arch. Hydrobiol. Beih. Ergebn. Limnol. 8:261–263.

Steele, J. H., and I. E. Baird. 1968. Production ecology of a sandy beach. Limnol. Oceanogr. 13:14–25.

Steemann Nielsen, E. 1944. Dependence of freshwater plants on quantity of carbon dioxide and hydrogen ion concentration. Dansk Bot. Ark. 11:1–25.

Steemann Nielsen, E. 1947. Photosynthesis of aquatic plants with special reference to the carbon sources. Dansk Bot. Ark. 12:5–71.

Steemann Nielsen, E. 1962a. Inactivation of the photochemical mechanism in photosynthesis as a means to protect the cells against too high light intensities. Physiol. Plant 15:161–171.

Steemann Nielsen, E. 1962b. On the maximum quantity of plankton chlorophyll per surface unit of a lake or the sea. Int. Rev. ges. Hydrobiol. 47:333–338.

Steemann Nielsen, E., and E. G. Jørgensen. 1962. The physiological background for using chlorophyll measurements in hydrobiology and a theory explaining daily variations in chlorophyll concentration. Arch. Hydrobiol. 58:349–357.

Steemann Nielsen, E., and E. G. Jørgensen. 1968a. The adaptation of plankton algae. I. General part. Physiol. Plant 21:401–413.

Steemann Nielsen, E., and E. G. Jørgensen. 1968b. The adaptation of plankton algae. III. With special consideration of the importance in nature. Physiol. Plant. 21:647–654.

Steemann Nielsen, E., and M. Willemoës. 1971. How to measure the illumination rate when investigating the rate of photosynthesis of unicellular algae under various light conditions. Int. Rev. ges. Hydrobiol. 56:541–556.

Stein, J. R., ed. 1973. Handbook of Phycological Methods. Culture Methods and Growth Measurements. Cambridge, Cambridge University Press, 448 pp.

Steinberg, C. 1978a. Bakterien und ihre Aktivität in der Oberfläche von Profundalsedimenten des Walchensees (Oberbayern). Arch. Hydrobiol. 84:29–41.

Steinberg, C. 1978b. Bacteria, bacterial activities, and uptake kinetics in surficial sediments of a small eutrophicated mountain lake. Verh. Int. Ver. Limnol. 20:2260–2263.

Steinberg, C. 1980. Species of dissolved metals derived from oligotrophic hard water. Water Res. 14:1239–1250.

Stemler, A., and Govindjee. 1973. Bicarbonate ion as a critical factor in photosynthetic oxygen evolution. Plant Physiol. 51:119–123.

Stenson, J. A. E. 1978a. Differential predation by fish on two species of Chaoborus (Diptera, Chaoboridae). Oikos 31:98–101.

Stenson, J. A. E. 1978b. Relations between vertebrate and invertebrate zooplankton predators in some arctic lakes. Astarte 11:21-26.

Štěpánek, M. 1959. Limnological study of the reservoir Sedlice near Želiv. IX. Transmission and transparency of water. Sci. Pap. Inst. Chem. Technol., Prague, Fac. Technol. Fuel Water 3(Pt. 2):363-430.

Sterzyński, W. 1979. Fecundity and body size of planktic rotifers in 30 Polish lakes of various trophic state. Ekol. Polska 27:307-321.

Stevens, R. J. and M. P. Parr. 1977. The significance of alkaline phosphatase activity in Lough Neagh. Freshwat. Biol. 7:351-355.

Stevenson, F. J. 1972. Role and function of humus in soil with emphasis on adsorption of herbicides and chelation of micronutrients. BioScience 22:643-650.

Stevenson, L. H. 1978. A case for bacterial dormancy in aquatic systems. Microbial Ecol. 4:127-133.

Stewart, A. J. and R. G. Wetzel. 1981. Dissolved humic materials: Photodegradation, sediment effects, and reactivity with phosphate and calcium carbonate precipitation. Arch. Hydrobiol. 92:265-286.

Stewart, A. J. and R. G. Wetzel. 1982a. Phytoplankton contribution to alkaline phosphatase activity. Arch. Hydrobiol. 93:265-271.

Stewart, A. J. and R. G. Wetzel. 1982b. Influence of dissolved humic materials on carbon assimilation and alkaline phosphatase activity in natural algal-bacterial assemblages. Freshwat. Biol. 12:369-380.

Stewart, K. M. 1973. Detailed time variations in mean temperature and heat content of some Madison lakes. Limnol. Oceanogr. 18:218-226.

Stewart, K. M. 1976. Oxygen deficits, clarity, and eutrophication in some Madison lakes. Int. Rev. ges. Hydrobiol. 61:563-579.

Stewart, K. M. and E. Hollan. 1975. Meromixis in Ulmener Maar (Germany). Verh. Int. Ver. Limnol. 19:1211-1219.

Stewart, R., and R. K. Schmitt. 1968. Wave interaction and Langmuir circulations. Proc. Conf. Great Lakes Res., Int. Assoc. Great Lakes Res. 11:496-499.

Stewart, W. D. P. 1968. Nitrogen input into aquatic ecosystems. In D. F. Jackson, ed. Algae, Man, and the Environment. Syracuse, N.Y., Syracuse University Press, pp. 53-72.

Stewart, W. D. P. 1969. Biological and ecological aspects of nitrogen fixation by free-living microorganisms. Proc. Roy. Soc. London (ser. B) 172:367-388.

Stewart, W. D. P. 1973. Nitrogen fixation. In N. G. Carr and B. A. Whitton, eds. The Biology of the Blue-Green Algae. Berkeley, University of California Press, pp. 260-278.

Stöber, W. 1967. Formation of silicic acid in aqueous suspensions of different silica modifications. In Equilibrium concepts in natural water systems. Adv. Chem. Ser. 67:161-182.

Stocking, C. R. 1956. Vascular conduction in submerged plants. In W. Ruhland, ed. Handbuch der Pflanzenphysiologie. Band 3. Pflanze und Wasser. Berlin, Springer-Verlag, pp. 587-595.

Stockner, J. G. 1971. Preliminary characterization of lakes in the Experimental Lakes Area, northwestern Ontario using diatom occurrence in sediments. J. Fish. Res. Bd. Canada 28:265-275.

Stockner, J. G. 1972. Paleolimnology as a means of assessing eutrophication. Verh. Int. Ver. Limnol. 18:1018-1030.

Stockner, J. G. 1975. Phytoplankton heterogeneity and paleolimnology of Babine Lake, British Columbia, Canada. Verh. Int. Ver. Limnol. 19:2236-2250.

Stockner, J. G., and W. W. Benson. 1967. The succession of diatom assemblages in the recent sediments of Lake Washington. Limnol. Oceanogr. 12:513-532.

Stoermer, E. F., B. G. Ladewski, and C. L. Schelske. 1978. Population responses of Lake Michigan phytoplankton to nitrogen and phosphorus enrichment. Hydrobiologia 57:249-265.

Stommel, H. 1949. Horizontal diffusion due to oceanic turbulence. J. Mar. Res. 8:199-225.

Stout, G. E., ed. 1967. Isotope Techniques in the Hydrologic Cycle. Geophys. Monogr. Ser. 11. Washington, D.C., American Geophysical Union, 199 pp.

Strain, H. 1951. The pigments of algae. In G. M. Smith, Editor. Manual of Phycology. Waltham, MA, Chronica Botanica Co., pp. 243-262.

Straškraba, M. 1963. Share of the littoral region in the productivity of two fishponds in southern Bohemia. Rozpravy Českosl. Akad. Věd, Řada Matem. Přír. Věd, 73(13), 64 pp.

Straškraba, M. 1965. The effect of fish on the number of invertebrates in ponds and streams. Mitt. Int. Ver. Limnol. 13:106-127.

Straškraba, M. 1967. Quantitative study on the littoral zooplankton of the Poltruba Backwater with an attempt to disclose the effect on fish. Rozpravy Českosl. Akad. Věd, Řada Matem. Přír. Věd 77(11):7-34.

Straškraba, M. 1968. Der Anteil der höheren Pflanzen an der Produktion der stehenden Gewässer. Mitt. Int. Ver. Limnol. 14:212-230.

Straškraba, M., J. Hrbáček, and P. Javornický. 1973. Effect of an upstream reservoir on the stratification conditions in Slapy Reservoir. Hydrobiol. Studies 2:7-82.

Straškraba, M., and E. Piezyńska. 1970. Field experiments on shading effect by emergents on littoral phytoplankton and periphyton production. Rozpravy Českosl. Akad. Věd, Řada Matem. Přír. Věd 80(6):7-32.

Straškrabová, V. and J. Komárková. 1979. Seasonal changes of bacterioplankton in a reservoir related to algae. I. Numbers and biomass. Int. Rev. ges. Hydrobiol. 64:285-302.

Strayer, R. F. and J. M. Tiedje. 1978a. Kinetic parameters of the conversion of methane precursors to methane in a hypereutrophic lake sediment. Appl. Environ. Microbiol. 36:330-340.

Strayer, R. F. and J. M. Tiedje. 1978b. In situ methane production in a small, hypereutrophic, hard-water lake: Loss of methane from sediments by vertical diffusion and ebullition. Limnol. Oceanogr. 23:1201-1206.

Streit, B. 1975a. Experimentelle Untersuchungen zum Stoffhaushalt von Ancylus fluviatilis (Gastropoda—Basommatophora). 1. Ingestion, Assimilation, Wachstum und Eiablage. Arch. Hydrobiol. (suppl.) 47:458-514.

Streit, B. 1975b. Experimentelle Untersuchungen zum Stoffhaushalt von Ancylus fluviatilis (Gastropoda—Basommatophora). 2. Untersuchungen über Einbau und Umsatz des Kohlenstoffs. Arch. Hydrobiol. (suppl.) 48:1-46.

Streit, B. 1977. Morphometric relationships and feeding habits of two species of Chaetogaster, Ch. limnaei and Ch. diastrophus (Oligochaeta). Arch. Hydrobiol. (suppl.) 48:424-437.

Streit, B. and P. Schröder. 1978. Dominierende Benthosinvertebraten in der Geröllbrandungszone des Bodensees: Phänologie, Nahrungsökologie und Biomasse. Arch. Hydrobiol. (suppl.) 55:211-234.

Strekal, T. A., and W. F. McDiffett. 1974. Factors affecting germination, growth, and distribution of the freshwater sponge, Spongilla fragilis Leidy (Porifera). Biol. Bull. 146:267-278.

Strickland, J. D. H. 1958. Solar radiation penetrating the ocean. A review of requirements, data and methods of measurement, with particular reference to photosynthetic productivity. J. Fish. Res. Bd. Can. 15:453-493.

Strickland, J. D. H. 1960. Measuring the production of marine phytoplankton. Bull. Fish. Res. Bd. Can. 122, 172 pp.

Strickland, J. D. H., and T. R. Parsons. 1972. A practical handbook of seawater analysis. 2nd ed. Bull. Fish. Res. Bd. Can. 167, 310 pp.

Strickler, J. R. 1975. Swimming of planktonic Cyclops species (Copepoda, Crustacea): Pattern, movements and their control. In T. Y. -T. Wu, C. J. Brokaw, and C. Brennen, eds. Swimming and Flying in Nature, vol. 2. New York, Plenum Press, pp. 599-613.

Strickler, J. R. 1977. Observation of swimming performances of planktonic copepods. Limnol. Oceanogr. 22:165-170.

Strickler, J. R. and S. Twombly. 1975. Reynolds number, diapause, and predatory copepods. Verh. Int. Ver. Limnol. 19:2943-2950.

Strøm, K. M. 1933. Nutrition of algae. Experiments upon: The feasibility of the Schreiber Method in fresh waters; the relative importance of iron and manganese in the nutritive medium; the nutri-

tive substance given off by lake bottom muds. Arch. Hydrobiol. 25:38-47.

Strøm, K. M. 1945. The temperature of maximum density in fresh waters. An attempt to determine its lowering with increased pressure from observations in deep lakes. Geofysiske Publikasjoner 16(8), 14 pp.

Strøm, K. M. 1947. Correlation between χ^{18} and pH in lakes. Nature 159:782-783.

Strøm, K. M. 1955. Waters and sediments in the deep of lakes. Mem. Ist. Ital. Idrobiol. 8(suppl.):345-356.

Strome, D. J. and M. C. Miller. 1978. Photolytic changes in dissolved humic substances. Verh. Int. Ver. Limnol. 20:1248-1254.

Strong, D. R., Jr. 1972. Life history variation among populations of an amphipod (Hyalella azteca). Ecology 53:1103-1111.

Stross, R. G. 1966. Light and temperature requirements for diapause development and release in Daphnia. Ecology 47:368-374.

Stross, R. G. 1981. Photomorphogenesis in underwater environments. In H. Smith, ed. Plants and the Daylight Spectrum. London, Academic Press, pp. 377-397.

Stross, R. G., and J. C. Hill. 1965. Diapause induction in Daphnia requires two stimuli. Science 150:1462-1464.

Stuckey, R. L., J. R. Wehrmeister, and R. J. Bartolotta. 1978. Submersed aquatic vascular plants in ice-covered ponds of central Ohio. Rhodora 80:575-580.

Stuiver, M. 1967. The sulfur cycle in lake waters during thermal stratification. Geochim. Cosmochim. Acta 31:2151-2167.

Stuiver, M. 1968. Oxygen-18 content of atmospheric precipitation during last 11,000 years in the Great Lakes region. Science 162:994-997.

Stuiver, M. 1978a. Radiocarbon timescale tested against magnetic and other dating methods. Nature 273:271-274.

Stuiver, M. 1978b. Carbon-14 dating: A comparison of beta and ion counting. Science 202:881-883.

Stuiver, M. and P. D. Quay. 1980. Changes in atmospheric carbon-14 attributed to a variable sun. Science 207:11-19.

Stull, E. A., E. deAmezaga, and C. R. Goldman. 1973. The contribution of individual species of algae to primary productivity of Castle Lake, California. Verh. Int. Ver. Limnol. 18:1776-1783.

Stumm, W. 1966. Redox potential as an environmental parameter; conceptual significance and operational limitation. 3rd Int. Conf. on Water Poll. Res., Water Poll. Control Federation. Sec. 1, Paper 13, 16 pp.

Stumm, W., and J. O. Leckie. 1971. Phosphate exchange with sediments; Its role in the productivity of surface waters. Proc. Water Poll. Res. Conf. III, Art. 16, 16 pp.

Stumm, W., and G. F. Lee. 1960. The chemistry of aqueous iron. Schweiz. Z. Hydrol. 22:295-319.

Stumm, W., and J. J. Morgan. 1981. Aquatic Chemistry. An Introduction Emphasizing Chemical Equilibria in Natural Waters. 2nd Edition. J. Wiley & Sons, New York. 780 pp.

Sturm, M. 1979. Origin and composition of clastic varves. In C. Schlüchter, ed. Moraines and Varves: Origin, Genesis, Classification. Rotterdam, A. A. Balkema, pp. 281-285.

Sturm, M. and A. Matter. 1978. Turbidites and varves in Lake Brienz (Switzerland): Deposition of clastic detritus by density currents. In A. Matter and M. E. Tucker, eds. Modern and Ancient Lake Sediments. Spec. Publs. Int. Assoc. Sedimentol. 2:147-168.

Suberkropp, K. and M. J. Klug. 1976. Fungi and bacteria associated with leaves during processing in a woodland stream. Ecology 57:707-719.

Suberkropp, K. and M. J. Klug. 1980. The maceration of deciduous leaf litter by aquatic hyphomycetes. Can. J. Bot. 58:1025-1031.

Suberkropp, K., M. J. Klug, and K. W. Cummins. 1975. Community processing of leaf litter in woodland streams. Verh. Int. Ver. Limnol. 19:1653-1658.

Suberkropp, K., G. L. Godshalk, and M. J. Klug. 1976. Changes in the chemical composition of leaves during processing in a woodland stream. Ecology 57:720-727.

Suess, E. 1968. Calcium carbonate interaction with organic compounds. Ph.D. Diss., Lehigh University, Bethlehem, Pa.

Suess, E. 1970. Interaction of organic compounds with calcium carbonate. I. Association phenomena and geochemical implications. Geochim. Cosmochim. Acta 34:157-168.

Sugawara, K. 1961. Na, Cl and Na/Cl in inland waters. Japan. J. Limnol. 22:49-65.

Sugihara, G. 1980. Minimal community structure: An explanation of species abundance patterns. Amer. Nat. 116:770-787.

Sutton, D. L. 1977. Grass carp (Ctenopharyngodon idella Val.) in North America. Aquatic Bot. 3:157-164.

Suzuki, Y., Y. Sugimura, and Y. Miyake. 1981. Selenium content and its chemical form in river waters of Japan. Japan. J. Limnol. 42:89-91.

Swain, F. M. 1963. Geochemistry of humus. In I. A. Breger, ed. Organic Geochemistry. New York, Macmillan Company, pp. 87-147.

Swain, F. M. 1965. Geochemistry of some Quaternary lake sediments of North America. In H. E. Wright, Jr., and D. G. Frey, eds. The Quaternary of the United States. Princeton, N.J., Princeton University Press, pp. 765-781.

Swanepoel, J. H. and J. F. Vermaak. 1977. Preliminary results on the uptake and release of ^{32}P by Potamogeton pectinatus. J. Limnol. Soc. South Africa 3:63-65.

Swift, M. C. 1976. Energetics of vertical migration in Chaoborus trivittatus larvae. Ecology 57:900-914.

Swift, M. C. and U. T. Hammer. 1979. Zooplankton population dynamics and Diaptomus production in Waldsea Lake, a saline meromictic lake in Saskatchewan. J. Fish. Res. Bd. Canada 36:1431-1438.

Swüste, H. F. J., R. Cremer, and S. Parma. 1973. Selective predation by larvae of Chaoborus flavicans (Diptera, Chaoboridae). Verh. Int. Ver. Limnol. 18:1559-1563.

Symons, F. 1972. On the changes in the structure of two algal populations: Species diversity and stability. Hydrobiologia 40:499-502.

Symposium on the Classification of Brackish Waters. 1959. Venice, 1958. Societas Internationalis Limnologiae. Arch. Oceanogr. Limnol. (suppl.) 11, 248 pp.

Szczepańska, W. 1973. Production of helophytes in different types of lakes. Pol. Arch. Hydrobiol. 20:51-57.

Szczepańska, W. 1976. Development of the underground parts of Phragmites communis Trin. and Typha latifolia L. Polskie Arch. Hydrobiol. 23:227-232.

Szczepańska, W. and A. Szczepański. 1976. Growth of Phragmites communis Trin., Typha latifolia L., and Typha angustifolia L. in relation to the fertility of soils. Polskie Arch. Hydrobiol. 23:233-248.

Szczepański, A. 1965. Deciduous leaves as a source of organic matter in lakes. Bull. Acad. Polon. Sci., Cl. II, 13:215-217.

Szczepański, A. 1968. Scattering of light and visibility in water of different types of lakes. Pol. Arch. Hydrobiol. 15:51-77.

Szczepański, A. 1969. Biomass of underground parts of the reed Phragmites communis Trin. Bull. Acad. Polon. Sci., Cl. II, 17:245-246.

Szczepański, A., and W. Szczepańska. 1966. Primary production and its dependence on the quantity of periphyton. Bull. Acad. Polon. Sci., Cl. II, 14:45-50.

Szilágyi, M. 1973. The redox properties and the determination of the normal potential of the peatwater system. Soil Sci. 115:434-437.

Szlauer, L. 1963. Diurnal migrations of minute invertebrates inhabiting the zone of submerged hydrophytes in a lake. Schweiz. Z. Hydrol. 25:56-64.

Szumiec, M. 1961. Eingangsmessungen der Strahlungsintensität in Teichen. Acta Hydrobiol. 3:133-142.

Takahashi, M., and S. Ichimura. 1968. Vertical distribution and organic matter production of photosynthetic sulfur bacteria in Japanese lakes. Limnol. Oceanogr. 13:644-655.

Tal, E. 1962. Preliminary study of fluctuations of vitamin B_{12} content in fish ponds in the Beith-Shean Valley. Bamidgeh (Israel) 14:19-26.

Talling, J. F. 1960. Self-shading effects in natural populations of a planctonic diatom. Wetter u. Leben 12:235-242.

Talling, J. F. 1965. The photosynthetic activity of

phytoplankton in East African lakes. Int. Rev. ges. Hydrobiol. 50:1-32.

Talling, J. F. 1969. The incidence of vertical mixing, and some biological and chemical consequences, in tropical African lakes. Verh. Int. Ver. Limnol. 17:998-1012.

Talling, J. F. 1971. The underwater light climate as a controlling factor in the production ecology of freshwater phytoplankton. Mitt. Int. Ver. Limnol. 19:214-243.

Talling, J. F. 1973. The application of some electrochemical methods to the measurement of photosynthesis and respiration in fresh waters. Freshwat. Biol. 3:335-362.

Talling, J. F. 1976. The depletion of carbon dioxide from lake water by phytoplankton. J. Ecol. 64:79-121.

Talling, J. F. 1979. Factor interactions and implications for the prediction of lake metabolism. Arch. Hydrobiol. Beih. Ergebn. Limnol. 13:96-109.

Talling, J. F., and I. B. Talling. 1965. The chemical composition of African lake waters. Int. Rev. ges. Hydrobiol. 50:421-463.

Talling, J. F., R. B. Wood, M. V. Prosser, and R. M. Baxter. 1973. The upper limit of photosynthetic productivity by phytoplankton: Evidence from Ethiopian soda lakes. Freshwat. Biol. 3:53-76.

Tam, A. C. and C. K. N. Patel. 1979. Optical absorptions of light and heavy water by laser optoacoustic spectroscopy. Appl. Optics 18:3348-3358.

Tan, T. L. 1973. Physiologie der Nitratreduktion bei *Pseudomonas aeruginosa*. Z. Allg. Mikrobiol. 13:83-94.

Tan, T. L., and J. Overbeck. 1973. Ökologische Untersuchungen über nitratreduzierende Bakterien im Wasser des Pluss-sees (Schleswig-Holstein). Z. Allg. Mikrobiol. 13:71-82.

Tanaka, N., Y. Ueda, M. Onizawa, and H. Kadota. 1977. Bacterial populations in water masses of different organic matter concentrations in Lake Biwa. Japan. J. Limnol. 38:41-47.

Tappa, D. W. 1965. The dynamics of the association of six limnetic species of *Daphnia* in Aziscoos Lake, Maine. Ecol. Monogr. 35:395-423.

Tarapchak, S. J., S. M. Bigelow, and C. Rubitschun. 1982. Overestimation of orthophosphorus concentrations in surface waters of southern Lake Michigan: Effects of acid and ammonium molybdate. Can. J. Fish. Aquat. Sci. 39:296-304.

Taub, F. B., and A. M. Dollar. 1968. The nutritional inadequacy of *Chlorella* and *Chlamydomonas* as food for *Daphnia pulex*. Limnol. Oceanogr. 13:607-617.

Taylor, B. E. and M. Slatkin. 1981. Estimating birth and death rates of zooplankton. Limnol. Oceanogr. 26:143-158.

Taylor, W. D. and J. Berger. 1980. Microspatial heterogeneity in the distribution of ciliates in a small pond. Microbial Ecol. 6:27-34.

Taylor, W. D., and D. R. S. Lean. 1981. Radiotracer experiments on phosphorus uptake and release

by limnetic microzooplankton. Can. J. Fish. Aquat. Sci. 38:1316-1321.

Teal, J. M. 1957. Community metabolism in a temperate cold spring. Ecol. Monogr. 27:283-302.

Tessenow, U. 1966. Untersuchungen über den Kieselsäurehaushalt der Binnengewässer. Arch. Hydrobiol. (suppl.) 32:1-136.

Tessenow, U. and Y. Baynes. 1978. Redoxchemische Einflüsse von *Isoëtes lacustris* L. im Litoralsediment des Feldsees (Hochschwarzwald). Arch. Hydrobiol. 82:20-48.

Thienemann, A. 1925. Die Binnengewässer Mitteleuropas. Eine limnologische Einführung. Die Binnengewässer, 1, 255 pp.

Thienemann, A. 1926. Der Nahrungskreislauf im Wasser. Verh. Deutsch. Zool. Gesell. 31:29-79.

Thienemann, A. 1927. Der Bau des Seebeckens in seiner Bedeutung für den Ablauf des Lebens im See. Verh. Zool.-Bot. Ges. Wien 77:87-91.

Thienemann, A. 1928. Der Sauerstoff im eutrophen und oligotrophen See. Ein Beitrag zur Seetypenlehre. Die Binnengewässer 4, 175 pp.

Thienemann, A. 1931. Der Productionsbegriff in der Biologie. Arch. Hydrobiol. 22:616-622.

Thienemann, A. 1954. *Chironomus*. Leben, Verbreitung und wirtschaftliche Bedeutung der Chironomiden. Die Binnengewässer 20, 834 pp.

Thomas, E. A. 1951. Sturmeinfluss auf das Tiefenwasser des Zürichsees im Winter. Schweiz. Z. Hydrol. 13:5-23.

Thomas, E. A. 1960. Sauerstoffminima und Stoffkreisläufe im ufernahen Oberflächenwasser des Zürichsees (Cladophora- und Phragmites-Gürtel). Monatsbull. Schweiz. Ver. Gas- Wasserfachmännern, 1960(6):1-8.

Thompson, R. 1978. European paleomagnetic secular variation 13,000-0 B.P. Polskie Arch. Hydrobiol. 25:413-418.

Thompson, R., R. W. Battarbee, P. E. O'Sullivan, and F. Oldfield. 1975. Magnetic susceptibility of lake sediments. Limnol. Oceanogr. 20:687-698.

Thompson, R., J. Bloemendal, J. A. Dearing, F. Oldfield, T. A. Rummery, J. C. Stober, and G. M. Turner. 1980. Environmental applications of magnetic measurements. Science 207:481-486.

Thorp, J. H. and G. T. Barthalmus. 1975. Effects of crowding on growth rate and symbiosis in green hydra. Ecology 56:206-212.

Threlkeld, S. T. 1976. Starvation and the size structure of zooplankton communities. Freshwat. Biol. 6:489-496.

Threlkeld, S. T. 1979. Estimating cladoceran birth rates: The importance of egg mortality and the egg age distribution. Limnol. Oceanogr. 24:601-612.

Thunmark, S. 1945. Zur Soziologie des Süsswasserplanktons. Eine methodologisch-ökologische Studie. Folia Limnol. Scandin. 3, 66 pp.

Thunmark, S. 1948. Sjöar och myrar i Lenhovda socken. Lenhovda. En värendssocken berättar. Moheda. 48 pp. (Unable to obtain original; discussed in Lillieroth, 1950).

Thut, R. N. 1969. A study of the profundal bottom fauna of Lake Washington. Ecol. Monogr. *39*:79–100.

Tietema, T. 1980. Ecophysiology of the sand sedge *Carex arenaria* L. II. The distribution of ^{14}C assimilates. Acta Bot. Neerl. *29*:165–178.

Tillman, D. L., and J. R. Barnes. 1973. The reproductive biology of the leech *Helobdella stagnalis* (L.) in Utah Lake, Utah. Freshwat. Biol. *3*:137–145.

Tilman, D. 1977. Resource competition between phytoplanktonic algae: An experimental and theoretical approach. Ecology *58*:338–348.

Tilman, D. 1978. The role of nutrient competition in a predictive theory of phytoplankton population dynamics. Mitt. Int. Ver. Limnol. *21*:585–592.

Tilman, D. 1980. Resources: A graphical-mechanistic approach to competition and predation. Amer. Nat. *116*:362–393.

Tilman, D. and S. S. Kilham. 1976. Phosphate and silicate growth and uptake kinetics of the diatoms *Asterionella formosa* and *Cyclotella meneghiniana* in batch and semicontinuous culture. J. Phycol. *12*:375–383.

Tilzer, M. 1972. Dynamik und Produktivität von Phytoplankton und pelagischen Bakterien in einem Hochgebirgssee (Vorderer Finstertaler See, Österreich). Arch. Hydrobiol. (suppl.) *40*:201–273.

Tilzer, M. M. 1973. Diurnal periodicity in the phytoplankton assemblage of a high mountain lake. Limnol. Oceanogr. *18*:15–30.

Tilzer, M. M., C. R. Goldman, and E. de Amezaga. 1975. The efficiency of photosynthetic light energy utilization by lake phytoplankton. Verh. Int. Ver. Limnol. *19*:800–807.

Tilzer, M. M., H. W. Paerl, and C. R. Goldman. 1977. Sustained viability of aphotic phytoplankton in Lake Tahoe (California-Nevada). Limnol. Oceanogr. *22*:84–91.

Timm, T. 1962a. Maloshchetnikovie chervi Chudsko Pskovskogo Ozera. Gidrobiol. Issledovaniia *3*:106–108.

Timm, T. 1962b. O rasprostranenii maloshchetnikovikh chervei (Oligochaeta) v ozerakh Estonii. Gidrobiol. Issledovaniia *3*:162–168.

Tippett, R. 1970. Artificial surfaces as a method of studying populations of benthic micro-algae in fresh water. Brit. Phycol. J. *5*:187–199.

Tippett, R. 1964. An investigation into the nature of the layering of deep-water sediments in two eastern Ontario lakes. Can. J. Bot. *42*:1693–1709.

Tirén, T., J. Thorin, and H. Nômmik. 1976. Denitrification measurements in lakes. Acta Agriculturae Scand. *26*:175–184.

Tison, D. L., F. E. Palmer, and J. T. Staley. 1977. Nitrogen fixation in lakes of the Lake Washington drainage basin. Water Res. *11*:843–847.

Tison, D. L. and D. H. Pope. 1980. Effect of temperature on mineralization by heterotrophic bacteria. Appl. Environ. Microbiol. *39*:584–587.

Tison, D. L., D. H. Pope, and C. W. Boylen. 1980. Influence of seasonal temperature on the temperature optima of bacteria in sediments of Lake George, New York. Appl. Environ. Microbiol. *39*:675–677.

Titman, D. 1975. A fluorometric technique for measuring sinking rates of freshwater phytoplankton. Limnol. Oceanogr. *20*:869–875.

Titman, D. 1976. Ecological competition between algae: Experimental confirmation of resource-based theory. Science *192*:463–465.

Titman, D. and P. Kilham. 1976. Sinking in freshwater phytoplankton: Some ecological implications of cell nutrient status and physical mixing processes. Limnol. Oceanogr. *21*:409–417.

Titus, J. E. and M. S. Adams. 1979. Coexistence and the comparative light relations of the submersed macrophytes *Myriophyllum spicatum* L. and *Vallisneria americana* Michx. Oecologia *40*:273–286.

Tjepkema, J. D. and H. J. Evans. 1976. Nitrogen fixation associated with *Juncus balticus* and other plants of Oregon wetlands. Soil Biol. Biochem. *8*:505–509.

Toerien, D. F., and B. Cavari. 1982. Effect of temperature on heterotrophic glucose uptake, mineralization, and turnover rates in lake sediments. Appl. Environ. Microbiol. *43*:1–5.

Toetz, D. W. 1973a. The limnology of nitrogen in an Oklahoma reservoir: nitrogenase activity and related limnological factors. Amer. Midland Nat. *89*:369–380.

Toetz, D. W. 1973b. The kinetics of NH_4 uptake by *Ceratophyllum*. Hydrobiologia *41*:275–290.

Toetz, D. W. 1974. Uptake and translocation of ammonia by freshwater hydrophytes. Ecology *55*:199–201.

Toetz, D., L. Varga, and B. Huss. 1977. Observations on uptake of nitrate and ammonia by reservoir phytoplankton. Arch. Hydrobiol. *79*:182–192.

Toetz, D. and B. Cole. 1980. Ammonia mineralization and cycling in Shagawa Lake, Minnesota. Arch. Hydrobiol. *88*:9–23.

Tolonen, K., A. Siiriänen, and R. Thompson. 1975. Prehistoric field erosion sediment in Lake Lojärvi, S. Finland and its palaeomagnetic dating. Ann. Bot. Fenn. *12*:161–164.

Tressler, W. L. 1957. The Ostracoda of Great Slave Lake. J. Wash. Acad. Sci. *47*:415–423.

Trifonova, N. A., et al. 1969. Materialy k Soveshchaniiu po Prognozirovaniiu Soderzhaniia Biolennykh Elementov i Organicheskogo Veshchestva v Vodokhranilishchakh. Rybinsk., Inst. Biol. Vnutrennikh Vod. 175 pp.

Triska, F. J. and J. R. Sedell. 1976. Decomposition of four species of leaf litter in response to nitrate manipulation. Ecology *57*:783–792.

Triska, F. J., J. R. Sedell, and B. Buckley. 1975. The processing of conifer and hardwood leaves in two coniferous forest streams. II. Biochemical and nutrient changes. Verh. Int. Ver. Limnol. *19*:1628–1639.

Trojanowski, J. 1969. Biological degradation of lignin. Int. Biodetn. Bull. *5*:119–124.

Truesdale, G. A., A. L. Downing, and G. F. Lowden. 1955. The solubility of oxygen in pure water and sea-water. J. Appl. Chem. 5:53–62.

Trussell, R. P. 1972. The percent un-ionized ammonia in aqueous ammonia solutions at different pH levels and temperatures. J. Fish. Res. Bd. Can. 29:1505–1507.

Tsuchiya, T. and H. Iwaki. 1979. Impact of nutrient enrichment in a waterchestnut ecosystem at Takahama-iri Bay of Lake Kasumigaura. II. Role of waterchestnut in primary productivity and nutrient uptake. Water, Air, Soil Poll. 12:503–510.

Tucker, A. 1957. The relation of phytoplankton periodicity to the nature of the physico-chemical environment with special reference to phosphorus. I. Morphometrical, physical and chemical conditions. II. Seasonal and vertical distribution of the phytoplankton in relation to the environment. Amer. Midland Nat. 57:300–370.

Tudorancea, C., R. H. Green, and J. Huebner. 1979. Structure, dynamics and production of the benthic fauna in Lake Manitoba. Hydrobiologia 64:59–95.

Tuschall, J. R., Jr. and P. L. Brezonik. 1980. Characterization of organic nitrogen in natural waters: Its molecular size, protein content, and interactions with heavy metals. Limnol. Oceanogr. 25:495–504.

Tutin, W. 1969. The usefulness of pollen analysis in interpretation of stratigraphic horizons, both Late-glacial and Post-glacial. Mitt. Int. Ver. Limnol. 17:154–164.

Twilley, R. R., M. M. Brinson, and G. J. Davis. 1977. Phosphorus absorption, translocation, and secretion in Nuphar luteum. Limnol. Oceanogr. 22:1022–1032.

Tyler, J. E. 1961a. Sun-altitude effect on the distribution of underwater light. Limnol. Oceanogr. 6:24–25.

Tyler, J. E. 1961b. Scattering properties of distilled and natural waters. Limnol. Oceanogr., 6:451–456.

Tyler, J. E. 1968. The Secchi disc. Limnol. Oceanogr. 13:1–6.

Tyler, J. E., and R. C. Smith. 1970. Measurements of Spectral Irradiance Underwater. New York, Gordan and Breach Science Publ., 103 pp.

Úlehlová, B. 1970. An ecological study of aquatic habitats in northwest Overijssel, the Netherlands. Acta Bot. Neerl. 19:830–858.

Úlehlová, B. 1971. Decomposition and humification of plant material in the vegetation of Stratiotes aloides in NW Overijssel, Holland. Hidrobiologia 12:279–285.

Úlehlová, B. 1976. Microbial decomposers and decomposition processes in wetlands. Československá Akad. Věd Studie ČSAV 17. 112 pp.

Ultsch, G. R. 1973. The effects of water hyacinths (Eichhornia crassipes) on the microenvironment of aquatic communities. Arch. Hydrobiol. 72:460–473.

Ultsch, G. R. and D. S. Anthony. 1973. The role of the aquatic exchange of carbon dioxide in the ecology of the water hyacinth (Eichhorina crassipes). Florida Sci. 36:16–22.

Valanne, N., E.-M. Aro, and E. Rintamäki, 1982. Leaf and chloroplast structure of two aquatic Ranunculus species. Aquatic Bot. 12:13–22.

van der Valk, A. G. and L. C. Bliss. 1971. Hydrarch succession and net primary production of Oxbow lakes in central Alberta. Can. J. Bot. 49:1177–1199.

Vallentyne, J. R. 1956. Epiphasic carotenoids in postglacial lake sediments. Limnol. Oceanogr. 1:252–262.

Vallentyne, J. R. 1957a. Principles of modern limnology. Amer. Sci. 45:218–244.

Vallentyne, J. R. 1957b. The molecular nature of organic matter in lakes and oceans, with lesser reference to sewage and terrestrial soils. J. Fish. Res. Bd. Can. 14:33–82.

Vallentyne, J. R. 1960. Fossil pigments. In M. B. Allen, ed. Comparative Biochemistry of Photoreactive Systems. New York, Academic Press, pp. 83–105.

Vallentyne, J. R. 1962. Solubility and the decomposition of organic matter in nature. Arch. Hydrobiol. 58:423–434.

Vallentyne, J. R. 1967. A simplified model of a lake for instructional use. J. Fish. Res. Bd. Can. 24:2473–2479.

Vallentyne, J. R. 1969. Sedimentary organic matter and paleolimnology. Mitt. Int. Ver. Limnol. 17:104–110.

Vallentyne, J. R. 1970. Phosphorus and the control of eutrophication. Can. Res. Development (May-June, 1970):36–43, 49.

Vallentyne, J. R. 1972. Freshwater supplies and pollution: Effects of the demophoric explosion on water and man. In N. Polunin, ed. The Environmental Future. London, Macmillan Press Ltd., pp. 181–211.

Vallentyne, J. R. 1974. The algal bowl—lakes and man. Ottawa, Misc. Special Publ. 22, Dept. of the Environment, 185 pp.

Van, T. K., W. T. Haller, and G. Bowes. 1976. Comparison of the photosynthetic characteristics of three submersed aquatic plants. Plant Physiol. 58:761–768.

Van, T. K., W. T. Haller, and L. A. Garrard, 1978. The effect of daylength and temperature on Hydrilla growth and tuber production. J. Aquat. Plant Manage. 16:57–59.

Van Baalen, C. 1968. The effects of ultraviolet irradiation on a coccoid blue-green alga: Survival, photosynthesis, and photoreactivation. Plant Physiol. 43:1689–1695.

Verber, J. L. 1964. Initial current studies in Lake Michigan. Limnol. Oceanogr. 9:426–430.

Verber, J. L. 1966. Inertial currents in the Great Lakes. Publ. Great Lakes Res. Div., Univ. Mich. 15:375–379.

Verdouw, H. and E. M. J. Dekkers. 1980. Iron and

manganese in Lake Vechten (The Netherlands); Dynamics and role in the cycle of reducing power. Arch. Hydrobiol. 89:509–532.

Verduin, J. 1959. Photosynthesis by aquatic communities in northwestern Ohio. Ecology 40:377–383.

Verduin, J. 1961. Separation rate and neighbor diffusivity. Science 134:837–838.

Vernon, L. P. 1964. Bacterial photosynthesis. Ann. Rev. Plant Physiol. 15:73–100.

Verstreate, D. R., T. A. Storch, and V. L. Dunham. 1980. A comparison of the influence of iron on the growth and nitrate metabolism of *Anabaena* and *Scenedesmus*. Physiol. Plant. 50:47–51.

Vijverberg, J. 1977. Population structure, life histories and abundance of copepods in Tjeukemeer, the Netherlands. Freshwat. Biol. 7:579–597.

Vinberg, G. G. 1934. K voprosu o metalimnianom minimume kisloroda. Trudy Limnol. Sta. Kosine 18:137–142.

Vinberg, G. G. 1960. Pervichnaya Produktsiia Voedoemov. Minsk, Izdatel'stvo Akademii Nauk, 329 pp. (English translat. 1963. The Primary Production of Bodies of Water. U.S. Atomic Energy Comm., Div. Tech. Info. AEC-tr-5692, 601 pp.)

Vinberg, G. G., and V. P. Liakhnovich. 1965. Udobrenie prudov. (Fertilization of fish ponds.) Moscow, Izdatel'stvo "Pishchevaia Promyshlennost," 271 pp. (English translat. 1969. Fish. Res. Bd. Canada Translation Ser. No. 1339, 482 pp.)

Vinberg, G. G., et al. 1972. Biological productivity of different types of lakes. *In* Z. Kajak and A. Hillbricht-Ilkowska, eds. Productivity Problems of Freshwaters. Warsaw, PWN Polish Scientific Publishers, pp. 382–404.

Vincent, W. F. 1980. The physiological ecology of a *Scenedesmus* population in the hypolimnion of a hypertrophic pond. II. Heterotrophy. Br. Phycol. J. 15:35–41.

Vincent, W. F. and C. R. Goldman. 1980. Evidence for algal heterotrophy in Lake Tahoe, California-Nevada. Limnol. Oceanogr. 25:89–99.

Viner, A. B. 1975. The supply of minerals to tropical rivers and lakes (Uganda). *In* A. D. Hasler, ed. Coupling of land and water systems. New York, Springer-Verlag, pp. 227–261.

Viner, A. B. 1975. The sediments of Lake George (Uganda). I. Redox potentials, oxygen consumption and carbon dioxide output. Arch. Hydrobiol. 76:181–197.

Vinyard, G. L. and R. A. Menger. 1980. *Chaoborus americanus* predation on various zooplankters; functional response and behavioral observations. Oecologia 45:90–93.

Visser, S. A. 1964a. Origin of nitrates in tropical rainwater. Nature 201:35–36.

Visser, S. A. 1964b. Oxidation-reduction potentials and capillary activities of humic acids. Nature 204:581.

Vivekanandan, E., M. A. Haniffa, T. J. Pandian, and R. Raghuraman. 1974. Studies on energy transformation in the freshwater snail *Pila globosa*. 1.

Influence of feeding rate. Freshwat. Biol. 4:275–280.

Vlymen, W. J. 1970. Energy expenditure of swimming copepods. Limnol. Oceanogr. 15:348–356.

Vollenweider, R. A. 1950. Ökologische Untersuchungen von planktischen Algen auf experimenteller Grundlage. Schweiz. Z. Hydrol. 12:193–262.

Vollenweider, R. A. 1955. Ein Nomogramm zur Bestimmung des Transmissionskoeffizienten sowie einige Bemerkungen zur Methode seiner Berechnung in der Limnologie. Schweiz. Z. Hydrol. 17:205–216.

Vollenweider, R. A. 1964. Über oligomiktische Verhältnisse des Lago Maggiore und einiger anderer insubrischer Seen. Mem. Ist. Ital. Idrobiol. 17:191–206.

Vollenweider, R. A. 1965. Calculation models of photosynthesis-depth curves and some implications regarding day rate estimates in primary production measurements. Mem. Ist. Ital. Idrobiol. 18(suppl.):425–457.

Vollenweider, R. A. 1966. Advances in defining critical loading levels for phosphorus in lake eutrophication. Mem. Ist. Ital. Idrobiol. 33:53–83.

Vollenweider, R. A. 1968. Scientific Fundamentals of the Eutrophication of Lakes and Flowing Waters, with Particular Reference to Nitrogen and Phosphorus as Factors in Eutrophication. Paris, Rep. Organisation for Economic Cooperation and Development, DAS/CSI/68.27, 192 pp.; Annex, 21 pp.; Bibliography, 61 pp.

Vollenweider, R. A. 1969a. Möglichkeiten und Grenzen elementarer Modelle der Stoffbilanz von Seen. Arch. Hydrobiol 66:1–36.

Vollenweider, R. A., ed. 1969b. A Manual on Methods for Measuring Primary Production in Aquatic Environments. Int. Biol. Program Handbook 12. Oxford, Blackwell Scientific Publications, 213 pp.

Vollenweider, R. A. 1972. Input-Output Models. Mimeographed report, Burlington, Ontario, Can. Cent. Inland Waters, 40 pp.

Vollenweider, R. A. 1975. Input-output models, with special reference to the phosphorus loading concept in limnology. Schweiz. Z. Hydrol. 37:53–84.

Vollenweider, R. A. 1976. Advances in defining critical loading levels for phosphorus in lake eutrophication. Mem. Ist. Ital. Idrobiol. 33:53–83.

Vollenweider, R. A. 1979. Das Nährstoffbelastungskonzept als Grundlage für den externen Eingriff in den Eutrophierungsprozess stehender Gewässer und Talsperren. Z. Wasser- u. Abwasser-Forschung 12:46–56.

Vollenweider, R. A., M. Munawar, and P. Stadelmann. 1974. A comparative review of phytoplankton and primary production in the Laurentian Great Lakes. J. Fish. Res. Bd. Can. 31:739–762.

Vollenweider, R. A. and J. Kerekes. 1980. The loading concept as a basis for controlling eutrophication philosophy and preliminary results of the OECD Programme on eutrophication. Prog. Water Technol. 12:5–18.

Vollenweider, R. A., W. Rast, and J. Kerekes. 1980. The phosphorus loading concept and Great Lakes eutrophication. In R. C. Loehr, C. S. Martin, and W. Rast, Editors. Phosphorus Management Strategies for Lakes. Ann Arbor, MI, Ann Arbor Science Publs., pp. 207–234.

Von Damm, K. L., L. K. Benninger, and K. K. Turekian. 1979. The ^{210}Pb chronology of a core from Mirror Lake, New Hampshire. Limnol. Oceanogr. 24:434–439.

Wachs, B. 1967. Die Oligochaeten-Fauna der Fliessgewässer unter besonderer Berücksichtigung der Beziehungen zwischen der Tubificiden-Besiedlung und dem Substrat. Arch. Hydrobiol. 63:310–386.

Waite, T. D. and C. Kurucz. 1977. The kinetics of bacterial degradation of the aquatic weed 'Hydrilla' sp. Water, Air, Soil Poll. 7:33–43.

Walker, K. F. and G. E. Likens. 1975. Meromixis and a reconsidered typology of lake circulation patterns. Verh. Int. Ver. Limnol. 19:442–458.

Walker, K. F., W. D. Williams, and U. T. Hammer. 1970. The Miller method for oxygen determination applied to saline waters. Limnol. Oceanogr. 15:814–815.

Walker, T. A. 1980. A correction to the Poole and Atkins Secchi disc/light-attenuation formula. J. Mar. Biol. Ass. U.K. 60:769–771.

Wallace, R. L. 1978. Substrate selection by larvae of the sessile rotifer Ptygura beauchampi. Ecology 59:221–227.

Wallace, J. B. and R. W. Merritt. 1980. Filter-feeding ecology of aquatic insects. Ann. Rev. Entomol. 25:103–132.

Walsby, A. E. 1972. Structure and function of gas vacuoles. Bacteriol. Rev. 36:1–32.

Walsby, A. E. 1975. Gas vesicles. Ann. Rev. Plant Physiol. 26:427–439.

Walsby, A. E. and M. J. Booker. 1980. Changes in buoyancy of a planktonic blue-green alga in response to light intensity. Br. Phycol. J. 15:311–319.

Walsh, G. E. 1965a. Studies on dissolved carbohydrate in Cape Cod waters. I. General survey. Limnol. Oceanogr. 10:570–576.

Walsh, G. E. 1965b. Studies on dissolved carbohydrate in Cape Cod waters. II. Diurnal fluctuation in Oyster Pond. Limnol. Oceanogr. 10:577–582.

Walsh, G. E. 1966. Studies on dissolved carbohydrate in Cape Cod waters. III. Seasonal variation in Oyster Pond and Wequaquet Lake, Massachusetts. Limnol. Oceanogr. 11:249–256.

Walz, N. 1978. The energy balance of the freshwater mussel Dreissena polymorpha Pallas in laboratory experiments and in Lake Constance. I. Pattern of activity, feeding and assimilation efficiency. Arch. Hydrobiol. (suppl.) 55:83–105.

Wang, T. S., S. W. Li, and Y. L. Ferng. 1978. Catalytic polymerization of phenolic compounds by clay minerals. Soil Sci. 126:15–21.

Wang, T. S. C., M.-M. Kao, and P. M. Huang. 1980. The effect of pH on the catalytic synthesis of humic substances by illite. Soil Sci. 129:333–338.

Wangersky, P. J. 1963. Manganese in ecology. In V. Schultz and A. W. Klement, Jr., eds. Radioecology. First National Symposium on Radioecology. New York, Reinhold Publishing Corp., pp. 499–508.

Ward, A. K., and R. G. Wetzel. 1975. Sodium: Some effects on blue-green algal growth. J. Phycol. 11:357–363.

Ward, A. K. and R. G. Wetzel. 1980a. Interactions of light and nitrogen source among planktonic blue-green algae. Arch. Hydrobiol. 90:1–25.

Ward, A. K. and R. G. Wetzel. 1980b. Photosynthetic responses of blue-green algal populations to variable light intensities. Arch. Hydrobiol. 90:129–138.

Warren, C. E., and G. E. Davis. 1971. Laboratory stream research: Objectives, possibilities, and constraints. Ann. Rev. Ecol. Systematics 2:111–114.

Watanabe, I., W. L. Barraquio, M. R. de Guzman, and D. A. Cabrera. 1979. Nitrogen-fixing (acetylene reduction) activity and population of aerobic heterotrophic nitrogen-fixing bacteria associated with wetland rice. Appl. Environ. Microbiol. 37:813–819.

Waters, T. F. 1977. Secondary production in inland waters. Adv. Ecol. Res. 10:91–164.

Waters, T. F. and J. C. Hokenstrom. 1980. Annual production and drift of the stream amphipod Gammarus pseudolimnaeus in Valley Creek, Minnesota. Limnol. Oceanogr. 25:700–710.

Watson, N. H. F. and B. N. Smallman. 1971. The role of photoperiod and temperature in the induction and termination of an arrested development in two species of freshwater cyclopid copepods. Can. J. Zool. 49:855–862.

Watts, W. A. and T. C. Winter. 1966. Plant macrofossils from Kirchner Marsh, Minnesota—A paleoecology study. Geol. Soc. Amer. Bull. 77:1339–1360.

Waughman, G. J. and D. J. Bellamy. 1980. Nitrogen fixation and the nitrogen balance in peatland ecosystems. Ecology 61:1185–1198.

Wavre, M., and R. O. Brinkhurst. 1971. Interactions between some tubificid oligochaetes and bacteria found in the sediments of Toronto Harbour, Ontario. J. Fish. Res. Bd. Can. 28:335–341.

Weast, R. C., ed. 1970. Handbook of Chemistry and Physics. 51st ed. Cleveland, Ohio, Chemical Rubber Co., 2367 pp.

Weaver, C. I. and R. G. Wetzel. 1980. Carbonic anhydrase levels and internal lacunar CO_2 concentrations in aquatic macrophytes. Aquatic Bot. 8:173–186.

Webb, M. G. 1961. The effects of thermal stratification on the distribution of benthic protozoa in a freshwater pond. J. Anim. Ecol. *30*:137–151.

Weber, J. A. 1972. The importance of turions in the propagation of *Myriophyllum exalbescens* (Haloragidaceae) in Douglas Lake, Michigan. Mich. Bot. *11*:115–121.

Weber, J. A. and L. D. Noodén. 1974. Turion formation and germination in *Myriophyllum verticillatum*: Phenology and its interpretation. Mich. Bot. *13*:151–158.

Weber, J. A. and L. D. Noodén. 1976a. Environmental and hormonal control of turion formation in *Myriophyllum verticillatum*. Plant Cell Physiol. *17*:721–731.

Weber, J. A. and L. D. Noodén. 1976b. Environmental and hormonal control of turion germination in *Myriophyllum verticillatum*. Am. J. Bot. *63*:936–944.

Weert, R. van der, and G. E. Kamerling. 1974. Evapotranspiration of water hyacinth *(Eichhornia crassipes)*. J. Hydrol. *22*:201–212.

Weibel, S. R. 1969. Urban drainage as a factor in eutrophication. *In* Eutrophication: Causes, Consequences, Correctives. Washington, D.C., National Academy of Sciences, pp. 383–403.

Weiler, R. R. 1974. Exchange of carbon dioxide between the atmosphere and Lake Ontario. J. Fish. Res. Bd. Can. *31*:329–332.

Weinmann, G. 1970. Gelöste Kohlenhydrate und andere organische Stoffe in natürlichen Gewässern und in Kulturen von *Scenedesmus quadricauda*. Arch. Hydrobiol. (suppl.) *37*:164–242.

Weiss, R. F. 1970. The solubility of nitrogen, oxygen and argon in water and seawater. Deep-Sea Res. *17*:721–735.

Welch, H. E. 1968. Relationships between assimilation efficiencies and growth efficiencies for aquatic consumers. Ecology *49*:755–759.

Welch, P. S. 1936. Limnological investigation of a strongly basic bog lake surrounded by an extensive acid-forming bog mat. Pap. Mich. Acad. Sci. Arts Lett. *21*:727–751.

Welch, P. S. 1948. Limnological Methods. Philadelphia, Blakiston Co., 381 pp.

Welch, P. S., and F. E. Eggleton. 1932. Limnological investigations on northern Michigan lakes. II. A further study of depression individuality in Douglas Lake. Pap. Mich. Acad. Sci. Arts Lett. *15*(1931):491–508.

Wells, L. 1960. Seasonal abundance and vertical movements of planktonic Crustacea in Lake Michigan. Fish. Bull. U.S. Fish. Wildl. Serv. *60*(172):343–369.

Wells, L. 1968. Daytime distribution of *Pontoporeia affinis* off bottom in Lake Michigan. Limnol. Oceanogr. *13*:703–705.

Welsh, R. P. H. and P. Denny. 1979. The translocation of ^{32}P in two submerged aquatic angiosperm species. New Phytol. *82*:645–656.

Welton, J. S. 1979. Life-history and production of the amphipod *Gammarus pulex* in a Dorset chalk stream. Freshwat. Biol. *9*:263–275.

Wentz, D. A., and G. F. Lee. 1969. Sedimentary phosphorus in lake cores—analytical procedure. Environ. Sci. Technol. *3*:750–754.

Werner, D., ed. 1977. The Biology of Diatoms. Berkeley, Univ. California Press, 498 pp.

Werner, E. E. 1974. The fish size, prey size, handling time relation in several sunfishes and some implications. J. Fish. Res. Bd. Can. *31*:1531–1536.

Werner, E. E., and D. J. Hall. 1974. Optimal foraging and the size selection of prey by the bluegill sunfish *(Lepomis macrochirus)*. Ecology 55:1042–1052.

Werner, E. E. and G. G. Mittelbach. 1981. Optimal foraging: Field tests of diet choice and habitat switching. Amer. Zool. *21*:813–829.

Werner, E. E., G. G. Mittelbach, and D. J. Hall. 1981. The role of foraging profitability and experience in habitat use by the bluegill sunfish. Ecology 62:116–125.

Westlake, D. F. 1963. Comparisons of plant productivity. Biol. Rev. *38*:385–425.

Westlake, D. F. 1965a. Theoretical aspects of comparability of productivity data. Mem. Ist. Ital. Idrobiol. *18*(suppl.):313–322.

Westlake, D. F. 1965b. Some basic data for investigations of the productivity of aquatic macrophytes. Mem. Ist. Ital. Idrobiol. *18*:(suppl.):229–248.

Westlake, D. F. 1965c. Some problems in the measurement of radiation under water: A review. Photochem. Photobiol. *4*:849–868.

Westlake, D. F. 1966. The biomass and productivity of *Glyceria maxima*. I. Seasonal changes in biomass. J. Ecol. *54*:745–753.

Westlake, D. F. 1967. Some effects of low-velocity currents on the metabolism of aquatic macrophytes. J. Exp. Bot. *18*:187–205.

Westlake, D. F. 1968. Methods used to determine the annual production of reedswamp plants with extensive rhizomes. *In* Methods of Productivity Studies in Root Systems and Rhizosphere Organisms. Leningrad, Publishing House Nauka, pp. 226–234.

Westlake, D. F., et al. 1980. Primary production. *In* E. D. Le Cren and R. H. Lowe-McConnell, eds. The Functioning of Freshwater Ecosystems. Cambridge, Cambridge Univ. Press, pp. 141–246.

Wetzel, R. G. 1960. Marl encrustation on hydrophytes in several Michigan lakes. Oikos *11*:223–236.

Wetzel, R. G. 1964. A comparative study of the primary productivity of higher aquatic plants, periphyton, and phytoplankton in a large, shallow lake. Int. Rev. ges. Hydrobiol. *49*:1–64.

Wetzel, R. G. 1965a. Nutritional aspects of algal productivity in marl lakes with particular reference to enrichment bioassays and their interpretation. Mem. Ist. Ital. Idrobiol. *18*(suppl.):137–157.

Wetzel, R. G. 1965b. Techniques and problems of primary productivity measurements in higher aquatic plants and periphyton. Mem. Ist. Ital, Idrobiol. 18(suppl.):249–267.

Wetzel, R. G. 1966a. Variations in productivity of Goose and hypereutrophic Sylvan lakes, Indiana. Invest. Indiana Lakes Streams 7:147–184.

Wetzel, R. G. 1966b. Productivity and nutrient relationships in marl lakes of northern Indiana. Verh. Int. Ver. Limnol. 16:321–332.

Wetzel, R. G. 1967. Dissolved organic compounds and their utilization in two marl lakes. In Problems of organic matter determination in freshwater. Hidrológiai Közlöny 47:298–303.

Wetzel, R. G. 1968. Dissolved organic matter and phytoplanktonic productivity in marl lakes. Mitt. Int. Ver. Limnol. 14:261–270.

Wetzel, R. G. 1969. Factors influencing photosynthesis and excretion of dissolved organic matter by aquatic macrophytes in hard-water lakes. Verh. Int. Ver. Limnol. 17:72–85.

Wetzel, R. G. 1970. Recent and postglacial production rates of a marl lake. Limnol. Oceanogr. 15:491–503.

Wetzel, R. G. 1972. The role of carbon in hard-water marl lakes. In G. E. Likens, ed. Nutrients and Eutrophication: The Limiting-Nutrient Controversy. Special Symposium, Amer. Soc. Limnol. Oceanogr. 1:84–97.

Wetzel, R. G. 1973. Productivity investigations of interconnected lakes. I. The eight lakes of the Oliver and Walters chains, northeastern Indiana. Hydrobiol. Stud. 3:91–143.

Wetzel, R. G. 1975a. General Secretary's Report—19th Congress of the Societas Internationalis Limnologiae. Verh. Int. Ver. Limnol. 19:3232–3292.

Wetzel, R. G. 1975b. Primary production. In B. A. Whitton, ed. River Ecology. Oxford, Blackwell Scientific Publs., pp. 230–247.

Wetzel, R. G. 1978. Foreword and introduction. In Good, R. E., D. F. Whigham, and R. L. Simpson, eds. Freshwater Wetlands: Ecological Processes and Management Potential. New York, Academic Press. pp. xiii–xvii.

Wetzel, R. G. 1979. The role of the littoral zone and detritus in lake metabolism. Arch. Hydrobiol. Beih. Ergebn. Limnol. 13:145–161.

Wetzel, R. G. 1981. Longterm dissolved and particulate alkaline phosphatase activity in a hardwater lake in relation to lake stability and phosphorus enrichments. Verh. Int. Ver. Limnol. 21:337–349.

Wetzel, R. G., and H. L. Allen. 1970. Functions and interactions of dissolved organic matter and the littoral zone in lake metabolism and eutrophication. In Z. Kajak and A. Hillbricht-Ilkowska, eds. Productivity Problems of Freshwaters. Warsaw, PWN Polish Scientific Publishers, pp. 333–347.

Wetzel, R. G., and R. A. Hough. 1973. Productivity and role of aquatic macrophytes in lakes: An assessment. Pol. Arch. Hydrobiol. 20:9–19.

Wetzel, R. G. and G. E. Likens. 1979. Limnological Analyses. Philadelphia, W. B. Saunders Co., 357 pp.

Wetzel, R. G., and B. A. Manny. 1972a. Secretion of dissolved organic carbon and nitrogen by aquatic macrophytes. Verh. Int. Ver. Limnol. 18:162–170.

Wetzel, R. G., and B. A. Manny. 1972b. Decomposition of dissolved organic carbon and nitrogen compounds from leaves in an experimental hard-water stream. Limnol. Oceanogr. 17:927–931.

Wetzel, R. G. and B. A. Manny. 1978. Postglacial rates of sedimentation, nutrient and fossil pigment deposition in a hardwater marl lake of Michigan. Polskie Arch. Hydrobiol. 25:453–469.

Wetzel, R. G., and D. G. McGregor. 1968. Axenic culture and nutritional studies of aquatic macrophytes. Amer. Midland Nat. 80:52–64.

Wetzel, R. G., and A. Otsuki. 1974. Allochthonous organic carbon of a marl lake. Arch. Hydrobiol. 73:31–56.

Wetzel, R. G., and P. H. Rich. 1973. Carbon in freshwater systems. In G. M. Woodwell and E. V. Pecan, eds. Carbon and the Biosphere. Proc. Brookhaven Symp. in Biol. 24. Brookhaven, N.Y., Tech. Information Center, U.S. Atomic Energy Commission CONF-720510, pp. 241–263.

Wetzel, R. G., and D. F. Westlake. 1969. Periphyton. In R. A. Vollenweider, ed. A. Manual on Methods for Measuring Primary Production in Aquatic Environments. Int. Biol. Program Handbook 12. Oxford, Blackwell Scientific Publications, pp. 33–40.

Wetzel, R. G., P. H. Rich, M. C. Miller, and H. L. Allen. 1972. Metabolism of dissolved and particulate detrital carbon in a temperate hard-water lake. Mem. Ist. Ital. Idrobiol. 29(suppl.):185–243.

Wetzel, R. G., B. A. Manny, W. S. White, R. A. Hough, and K. R. McKinley. 1982. Wintergreen Lake: A study in hypereutrophy. (In preparation.)

Whigham, D. and R. Simpson. 1977. Growth, mortality, and biomass partitioning in freshwater tidal wetland populations of wild rice (Zizania aquatica var. aquatica). Bull. Torrey Bot. Club 104:347–351.

Whipple, G. C. 1898. Classifications of lakes according to temperature. Amer. Nat. 32:25–33.

Whipple, G. C. 1927. The Microscopy of Drinking Water. 4th ed. New York, John Wiley and Sons, Inc., 586 pp.

White, W. S. 1974. Role of calcium carbonate precipitation in lake metabolism. Ph.D. Diss., Michigan State University, East Lansing, 141 pp.

White, W. S., and R. G. Wetzel. 1975. Nitrogen, phosphorus, particulate and colloidal carbon content of sedimenting seston of a hard-water lake. Verh. Int. Ver. Limnol. 19:330–339.

Whitehead, H. C., and J. H. Feth. 1961. Recent chemical analyses of waters from several closed-basin lakes and their tributaries in the western

United States. Bull. Geol. Soc. Amer. 72:1421–1426.

Whiteside, M. C. 1969. Chydorid (Cladocera) remains in surficial sediments of Danish lakes and their significance to paleolimnological interpretations. Mitt. Int. Ver. Limnol. 17:193–201.

Whiteside, M. C. 1974. Chydorid (Cladocera) ecology: Seasonal patterns and abundance of populations in Elk Lake, Minnesota. Ecology 55:538–550.

Whiteside, M. C., J. B. Williams, and C. P. White. 1978. Seasonal abundance and pattern of chydorid Cladocera in mud and vegetative habitats. Ecology 59:1177–1188.

Whitney, L. V. 1937. Microstratification in the waters of inland lakes in summer. Science 85:224–225.

Whitton, B. A. 1975. Algae. In B. A. Whitton, Ed. River Ecology. Oxford, Blackwell Scientific Publications, pp. 81–105.

Wiederholm, T. 1980. Use of benthos in lake monitoring. J. Water Poll. Control Fed. 52:537–547.

Wiederholm, T. and L. Eriksson. 1979. Subfossil chironomids as evidence of eutrophication in Ekoln Bay, central Sweden. Hydrobiologia 62:195–208.

Wierzbicka, M. 1966. Les résultats des recherches concernant l'état de repos (resting stage) des Cyclopoida. Verh. Int. Ver. Limnol. 16:592–599.

Wiessner, W. 1962. Inorganic micronutrients. In R. A. Lewin, ed. Physiology and Biochemistry of Algae. New York, Academic Press, pp. 267–286.

Wiggins, G. B. 1977. Larvae of the North American caddisfly genera. Toronto, Univ. Toronto Press,

Wilkialis, J. and R. W. Davies. 1980. The population ecology of the leech (Hirudinoidea: Glossiphoniidae) Theromyzon tessulatum. Can. J. Zool. 58:906–912.

Wilkins, A. S. 1972. Physiological factors in the regulation of alkaline phosphatase synthesis in Escherichia coli. J. Bacteriol. 110:616–623.

Wilkinson, R. E. 1963. Effects of light intensity and temperature on the growth of waterstargrass, coontail, and duckweed. Weeds 11:287–290.

Williams, J. B. 1982. Temporal and spatial patterns of abundance of the Chydoridae (Cladocera) in Lake Itasca, Minnesota. Ecology 63:345–353.

Williams, J. D. H., and T. Mayer. 1972. Effects of sediment diagenesis and regeneration of phosphorus with special reference to lakes Erie and Ontario. In H. E. Allen and J. R. Kramer, eds. Nutrients in Natural Waters. New York, John Wiley and Sons, Inc., pp. 281–315.

Williams, J. D. H., J. K. Syers, R. F. Harris, and D. E. Armstrong. 1970. Adsorption and desorption of inorganic phosphorus by lake sediments in a 0.1M NaCl system. Environ. Sci. Technol. 4:517–519.

Williams, J. D. H., J. K. Syers, S. S. Shukla, R. F. Harris, and D. E. Armstrong. 1971a. Levels of inorganic and total phosphorus in lake sediments as related to other sediment parameters. Environ. Sci. Technol. 5:1113–1120.

Williams, J. D. H., J. K. Syers, R. F. Harris, and D. E. Armstrong. 1971b. Fractionation of inorganic phosphate in calcareous lake sediments. Proc. Soil Sci. Soc. Amer. 35:250–255.

Williams, J. D. H., J. K. Syers, D. E. Armstrong, and R. F. Harris. 1971c. Characterization of inorganic phosphate in noncalcareous lake sediments. Proc. Soil Sci. Soc. Amer. 35:556–561.

Williams, W. D., and H. F. Wan. 1972. Some distinctive features of Australian inland waters. Water Res. 6:829–836.

Williamson, C. E. 1980. The predatory behavior of Mesocyclops edax: Predator preferences, prey defenses, and starvation-induced changes. Limnol. Oceanogr. 25:903–909.

Willoughby, L. G. 1965. Some observations on the location of sites of fungal activity at Blelham Tarn. Hydrobiologia 25:352–356.

Willoughby, L. G. 1974. Decomposition of litter in fresh water. In C. H. Dickinson and G. J. F. Pugh, eds. Biology of Plant Litter Decomposition. Vol. 2. New York, Academic Press, pp. 659–681.

Willoughby, L. G. and J. H. Marcus. 1979. Feeding and growth of the isopod Asellus aquaticus on actinomycetes, considered as model filamentous bacteria. Freshwat. Biol. 9:441–449.

Wilson, E. O. and W. H. Bossert. 1971. A primer of population biology. Stamford, CT, Sinauer Associates, Inc. Publs., 192 pp.

Wilson, M. S., and H. C. Yeatman. 1959. Free-living Copepoda. In W. T. Edmondson, ed. Fresh-Water Biology. 2nd ed. New York, John Wiley and Sons, Inc., pp. 735–861.

Wilson, L. R. 1937. A quantitative and ecological study of the larger aquatic plants of Sweeney Lake, Oneida County, Wisconsin. Bull. Torr. Bot. Club 64:199–208.

Wilson, L. R. 1941. The larger aquatic vegetation of Trout Lake, Vilas County, Wisconsin. Trans. Wis. Acad. Sci. Arts Lett. 33:135–146.

Winberg, G. G. 1970. Some interim results of Soviet IBP investigations on lakes. In Z. Kajak and A. Hillbricht-Ilkowska, eds. Productivity Problems of Freshwaters. Warsaw, PWN Polish Scientific Publishers, pp. 363–381.

Winberg, G. G., ed. 1971. Methods for the Estimation of Production of Aquatic Animals. New York, Academic Press, 175 pp.

Winberg, G. G., et al. 1970. Biological productivity of different types of lakes. In Z. Kajak and A. Hillbricht-Ilkowska, eds. Productivity Problems of Freshwaters. Warsaw, PWN Polish Scientific Publishers, pp. 383–404.

Winberg, G. G., et al. 1971. Symbols, Units and Conversion Factors in Studies of Freshwater Productivity. Int. Biol. Program, Sec. PF—Productivity of Freshwaters, 23 pp.

Winberg, G. G., et al. 1973a. The progress and state of research on the metabolism, growth, nutrition, and production of fresh-water invertebrate animals. Hydrobiol. J. 9:77–84.

Winberg, G. G., et al. 1973b. Biological productivity of two subarctic lakes. Freshwat. Biol. 3:177–197.

Winfrey, M. R. and J. G. Zeikus. 1977. Effect of sulfate on carbon and electron flow during microbial methanogenesis in freshwater sediments. Appl. Environ. Microbiol. 33:275–281.

Winfrey, M. R. and J. G. Zeikus. 1979. Microbial methanogenesis and acetate metabolism in a meromictic lake. Appl. Environ. Microbiol. 37:213–221.

Winfrey, M. R., D. R. Nelson, S. C. Klevickis, and J. G. Zeikus. 1977. Association of hydrogen metabolism with methanogenesis in Lake Mendota sediments. Appl. Environ. Microbiol. 33:312–318.

Winner, R. W. and J. S. Greber. 1980. Prey selection by *Chaoborus punctipennis* under laboratory conditions. Hydrobiologia 68:231–233.

Winston, R. D. and P. R. Gorham. 1979. Turions and dormancy states in *Utricularia vulgaris*. Can. J. Bot. 57:2740–2749.

Winter, K. 1978. Short-term fixation of ^{14}carbon by the submerged aquatic angiosperm *Potamogeton pectinatus*. J. Exp. Bot. 29:1169–1172.

Wohler, J. R. 1966. Productivity of the duckweeds. M.Sc. Thesis, University of Pittsburgh, 69 pp.

Wohlschlag, D. E. 1950. Vegetation and invertebrate life in a marl lake. Invest. Indiana Lakes Streams 3:321–372.

Wolfe, J. M. and E. L. Rice. 1979. Allelopathic interactions among algae. J. Chem. Ecol. 5:533–542.

Wolfe, R. S. 1971. Microbial formation of methane. Adv. Microbiol. Physiol. 6:107–145.

Wolk, C. P. 1968. Movement of carbon from vegetative cells to heterocysts in *Anabaena cylindrica*. J. Bacteriol. 96:2138–2143.

Wolk, C. P. 1973. Physiology and cytological chemistry of blue-green algae. Bacteriol. Rev. 37:32–101.

Wolverton, B. C. and R. C. McDonald. 1979. Upgrading facultative wastewater lagoons with vascular aquatic plants. J. Water Poll. Control Fed. 51:305–313.

Wolverton, B. C. and R. C. McDonald. 1978. Bioaccumulation and detection of trace levels of cadmium in aquatic systems by *Eichhornia crassipes*. Environ. Health Perspectives 27:161–164.

Wood, K. G. 1974. Carbon dixode diffusivity across the air-water interface. Arch. Hydrobiol. 73:57–69.

Wood, K. G. 1977. Chemical enhancement of CO_2 flux across the air-water interface. Arch. Hydrobiol. 79:103–110.

Woodwell, G. M., and R. H. Whittaker. 1968. Primary production in terrestrial ecosystems. Amer. Zool. 8:19–30.

Woodwell, G. M., P. H. Rich, and C. A. S. Hall. 1973. The carbon cycle of estuaries. *In* G. M. Woodwell and E. V. Pecan, eds. Carbon and the Biosphere. Proc. 24th Brookhaven Symposium in Biology. Brookhaven, N.Y., U.S. Atomic Energy Commission. Symp. Ser. CONF-720510, pp. 221–240.

Wong, B., and F. J. Ward. 1972. Size selection of *Daphnia pulicaria* by yellow perch *(Perca flavescens)* fry in West Blue Lake, Manitoba. J. Fish. Res. Bd. Canada, 29:1761–1764.

Worthington, E. B. 1931. Vertical movements of fresh-water macroplankton. Int. Rev. ges. Hydrobiol. 25:394–436.

Wright, J., W. J. O'Brien, and G. L. Vinyard. 1980. Adaptive value of vertical migration: A simulation model argument for predation hypothesis. *In* W. C. Kerfoot, ed. Evolution and Ecology of Zooplankton Communities. Hanover, NH, Univ. Press New England, pp. 138–147.

Wright, J. C. 1965. The population dynamics and production of *Daphnia* in Canyon Ferry Reservoir, Montana. Limnol. Oceanogr. 10:583–590.

Wright, R. T. 1964. Dynamics of a phytoplankton community in an ice-covered lake. Limnol. Oceanogr. 9:163–178.

Wright, R. T. 1975. Studies on glycolic acid metabolism by freshwater bacteria. Limnol. Oceanogr. 20:626–633.

Wright, R. T. 1978. Measurement and significance of specific activity in the heterotrophic bacteria of natural waters. Appl. Environ. Microbiol. 36:297–305.

Wright, R. T., and J. E. Hobbie. 1966. Use of glucose and acetate by bacteria and algae in aquatic ecosystems. Ecology 47:447–464.

Wright, S. 1955. Limnological Survey of Western Lake Erie. Spec. Sci. Rept. Fish. U.S. Fish Wildl. Serv., *139*, 341 pp.

Wuim-Andersen, S. 1971. Photosynthetic uptake of free CO_2 by the roots of *Lobelia dortmanna*. Physiol. Plant 25:245–248.

Wunderlich, W. O. 1971. The dynamics of density-stratified reservoirs. *In* G. E. Hall, ed. Reservoir Fisheries and Limnology. Washington, D.C., Spec. Publ. 8, Amer. Fish. Soc. pp. 219–231.

Yaskawa, K., T. Nakajima, N. Kawai, M. Torii, N. Natsuhara, and S. Horie. 1973. Palaeomagnetism of a core from Lake Biwa (I). J. Geomag. Geoelectr. 25:447–474.

Yentsch, C. S. 1960. The influence of phytoplankton pigments on the color of sea water. Deep-Sea Res. 7:1–9.

Yoshida, T., M. Aizaki, T. Asami, and N. Makishima. 1979. Biological nitrogen fixation and denitrification in Lake Kasumiga-ura. Japan. J. Limnol. 40:1–9. (In Japanese)

Yoshimura, S. 1935. Relation between depth for maximum amount of excess oxygen during the summer stagnation period and the transparency of freshwater lakes of Japan. Proc. Imperial Acad. Japan 11:356–358.

Yoshimura, S., 1936a. A contribution to the knowledge of deep water temperatures of Japanese lakes. I. Summer temperatures. Jap. J. Astronomy Geophys. 13:61–120.

Yoshimura, S. 1936b. Contributions to the knowledge of iron dissolved in the lake waters of Japan. Second report. Japan. J. Geol. Geogr. 13:39–56.

Young, J. O. 1973. The prey and predators of *Phaenocora typhlops* (Vejdovsky) (Turbellaria: Neorhabdocoela) living in a small pond. J. Anim. Ecol. 42:637–643.

Young, J. O. 1981. A comparative study of the food niches of lake-dwelling triclads and leeches. Hydrobiologia 84:91–102.

Zafar, A. R. 1959. Taxonomy of lakes. Hydrobiologia 13:287–299.

Zago, M. S. A. 1976. A preliminary investigation on the cyclomorphosis of *Daphnia gessneri* Herbst, 1967, in a Brazilian reservoir. Bolm. Zool, Univ. S. Paulo 1:147–160.

Zaitsev, Yu. P. 1970. Marine neustonology. Kiev. Akad. Nauk Ukrainskoi SSR, Naukova Dumka, 207 pp. (Translat. into English, Programs for Scientific Translation, Jerusalem, 1971.)

Zaret, T. M. 1972a. Predator-prey interaction in a tropical lacustrine ecosystem. Ecology 53:248–257.

Zaret, T. M. 1972b. Predators, invisible prey, and the nature of polymorphism in the Cladocera (Class Crustacea). Limnol. Oceanogr. 17:171–184.

Zaret, T. M. 1975. Strategies for existence of zooplankton prey in homogeneous environments. Verh. Int. Ver. Limnol. 19:1484–1489.

Zaret, T. M. 1980. Predation and Freshwater Communities. New Haven, Yale Univ. Press, 187 pp.

Zaret, T. M. and W. C. Kerfoot. 1975. Fish predation on *Bosmina longirostris*: Body-size selection versus visibility selection. Ecology 56:232–237.

Zaret, T. M. and J. S. Suffern. 1976. Vertical migration in zooplankton as a predator avoidance mechanism. Limnol. Oceanogr. 21:804–813.

Zehnder, A. J. B. 1978. Ecology of methane formation. *In* R. Mitchell, ed. Water Pollution Microbiology, vol. 2. New York, John Wiley & Sons, pp. 349–376.

Zeikus, J. G. 1977. Biology of methanogenic bacteria. Bacteriol. Rev. 41:514–541.

Zeikus, J. G. 1981. Lignin metabolism and the carbon cycle. Polymer biosynthesis, biodegradation, and environmental recalcitrance. Adv. Microbial Ecol. 5:211–243.

Zettler, F. W. and T. E. Freeman. 1972. Plant pathogens as biocontrols of aquatic weeds. Ann. Rev. Phytopath. 10:455–470.

Zicker, E. L., K. C. Berger, and A. D. Hasler. 1956. Phosphorus release from bog lake muds. Limnol. Oceanogr. 1:296–303.

Zimmerman, A. P. 1981. Electron intensity, the role of humic acids in extracellular electron transport and chemical determination of pE in natural waters. Hydrobiologia 78:259–265.

Zinder, S. H. and T. D. Brock. 1978. Microbial transformations of sulfur in the environment. *In* J. O. Nriagu, Editor. Sulfur in the Environment. II. Ecological Impacts. New York, John Wiley & Sons, pp. 445–466.

Zmyslowska, I. and M. Sobierajska. 1977. Microbiological studies of the Kortowskie Lake. Polskie Arch. Hydrobiol. 24:61–71.

ZoBell, C. E. 1964. Geochemical aspects of the microbial modification of carbon compounds. *In* U. Colombo and G. D. Hobson, eds. Advances in Organic Geochemistry. New York, Macmillan Co., pp. 339–356.

ZoBell, C. E. 1973. Microbial biogeochemistry of oxygen. *In* A. A. Imshenetskii, ed. Geokhimicheskaia Deiatel'nost' Mikroorganizmov v Vodoemakh i Mestorozhdeniiach Poleznykh Iskopaemykh. Moscow. Tipografiia Izdatel'stva Sovetskoe Radio, pp. 3–76.

Züllig, H. 1961. Die Bestimmung von Myxoxanthophyll in Bohrprofilen zum Nachweis vergangener Blaualgenentfaltungen. Verh. Int. Ver. Limnol. 14:263–270.

Züllig, H. 1981. On the use of carotenoid stratigraphy in lake sediments for detecting past developments of phytoplankton. Limnol. Oceanogr. 26:970–976.

Zumberge, J. H. 1952. The Lakes of Minnesota. Their Origin and Classification. Minneapolis, University of Minnesota Press, 99 pp.

Zygmuntowa, J. 1972. Occurrence of free amino acids in pond water. Acta Hydrobiol. 14:317–325.

INDEX